BUSINESS STATISTICS OF THE UNITED STATES

BUSINESS STATISTICS OF THE UNITED STATES

Patterns of Economic Change

24th Edition
2019
Edited by Susan Ockert

Bernan Press

Lanham, MD

Published by Bernan Press
An imprint of The Rowman & Littlefield Publishing Group, Inc.
4501 Forbes Boulevard, Suite 200, Lanham, Maryland 20706
www.rowman.com
800-462-6420

6 Tinworth Street, London SE11 5AL, United Kingdom

ISBN: 978-1-64143-338-9
E-ISBN: 978-1-64143-339-6

∞™ The paper used in this publication meets the minimum requirements
of American National Standard for Information Sciences—Permanence of
Paper for Printed Library Materials, ANSI/NISO Z39.48-1992.

Printed in the United States of America

CONTENTS

PREFACE

Business Statistics of the United States: Patterns of Economic Change, 24th Edition, 2019 is a basic desk reference for anyone requiring statistics on the U.S. economy. It contains over 3,000 economic time series portraying the period from World War II to December 2018 in industry, product, and demographic detail. In the case of about 200 key series, the period from 1929 through 1948 is depicted as well.

Additionally, for three important statistical series that have been compiled monthly on a continuous basis, monthly and annual data are included going back to 1919 for industrial production and 1913 for consumer and producer prices. Annual data for money and credit have also been taken back before 1929, including money supply figures back to 1892 and some interest rates back to 1919. Annual estimates of the unemployment rate from 1890 to 1929, not official but using a method similar to that used by BLS for its 1929-48 unemployment estimates, are also included.

The National Income and Product Accounts (NIPAs) were comprehensively revised in July 2013, changing some measurement concepts, changing base periods for price indexes and real output, and as a result altering all the data back to 1929. *Business Statistics* incorporates fully revised data for over 700 series from this data set and productivity and costs series that are based on the revised NIPAs, all extended and updated through 2016. The important new concepts introduced in the NIPA revision are explained in the Notes and Definitions to Chapters 1, 4, 5, and 6. In all, this edition provides, in one volume, unparalleled, up-to-date background information on the course of the U.S. economy.

The data are predominantly from federal government sources. Of equal importance are the extensive background notes for each chapter, which help users to understand the data, use them appropriately, and, if desired, seek additional information from the source agencies.

Chapters are organized according to subject matter. For most time series, users will find an initial page displaying the latest two years of monthly data, or three years of quarterly data, for the most recent completed years, along with annual averages for the last completed 40, 50, or even 60 years.

For the most important series for which we have long-term historical data, this initial page will be labeled with "A" after the table number, and identified in the title by the additional words "Recent Data". It will be immediately followed by pages labeled with "B" after the table number and identified by the words "Historical Data", which will show all available historical data annually back as far as possible, and monthly or quarterly back to the earliest available postwar year. In some cases, additional tables of historical data will be included.

The leading article "Cycle and Growth Perspectives" provides other information and techniques to assist in using and interpreting the data.

THE PLAN OF THE BOOK

The history of the U.S. economy is told in major U.S. government sets of statistical data: the national income and product accounts compiled by the Bureau of Economic Analysis (BEA); the data on labor force, employment, hours, earnings, and productivity compiled by the Bureau of Labor Statistics (BLS); the price indexes collected by BLS; and the financial market and industrial production data compiled primarily by the Board of Governors of the Federal Reserve System (FRB). All of these sets exist in annual and either monthly or quarterly form beginning in 1946, 1947, or 1948; many are available, at least annually, as far back as 1929, and a few even farther back; and three are available monthly back to the teens of the 20th century.

In Part A, *Business Statistics* presents the aggregate United States economy in a number of important dimensions. The presentations begin with the national income and product accounts, or NIPAs. The NIPAs comprise a comprehensive, thorough, and internally consistent body of data. They measure the value of the total output of the U.S. economy (the gross domestic product, or GDP) and they allocate that value between its quantity, or "real," and price components. They show how the value of aggregate demand is generated by consumers, business investors, government, and foreign customers; how much of aggregate demand is supplied by imports and how much by domestic production; and how the income generated in domestic production is distributed between labor and capital.

Production estimates covering only the sectors of the economy traditionally labeled "industrial"—manufacturing, mining, and utilities—are shown in Chapter 2, after the presentation of the overall NIPAs in Chapter 1.

Median income—the best measure of the economic well-being of the typical American—income distribution, and poverty statistics from a Census household survey are presented in Chapter 3.

Then, more detail from the NIPAs is presented for the demand components of economic activity. GDP, by its product-side definition, consists of the sum of consumption expenditures, business investment in fixed capital and inventories, government purchases of goods and services, and exports minus imports—as in the elementary economics blackboard identity "GDP = C + I + G + X – M." Chapters on each of these components—consumption, investment, government, and foreign trade—are presented in Part A.

Following these chapters, there is a chapter on prices, two chapters on the compensation of labor and capital inputs and the amount and productivity of labor input, one chapter on energy inputs into production and consumption, and one chapter on money, interest, assets, and debt.

At the end of Part A, comparisons of output, prices, and labor markets among major industrial countries are presented, along with statistics on the value of the dollar against other currencies.

While GDP is initially defined and measured by adding up its demand categories and subtracting imports, this output is produced in industries—some in the old-line heavy industries such as manufacturing, mining, and utilities, but an increasing share in the huge and heterogeneous group known as "service-providing" industries. Part A gives a number of measures of activity classified by industry or industrial sector: industrial production, profits, and employment-related data. Further industry information is provided in Part B, including new quarterly indicators of GDP by industry.

Industry data collection is important because demands for goods and services are channeled into demands for labor and capital through the industries responsible for producing the requested goods and services. These data are reported using the North American Industry Classification System (NAICS). This system, introduced in 1997 to replace the older Standard Industrial Classification System (SIC), delineates industries that are better defined in relation to today's demands and more closely related to each other by technology. Notable examples include more detailed data available on service industries, a more rational grouping of the Computer and electronic product manufacturing subsector, and the creation of the Information sector. See the References at the end of this Preface.

NAICS industries are groupings of producing units—not of products as such—and are grouped according to similarity of production processes. This is done in order to collect consistent data on inputs and outputs, which are then used to measure important concepts, such as productivity and input-output parameters. Emphasis on the production process helps to explain a number of ways in which the NAICS differs from the SIC.

- Manufacturing activities at retail locations, such as bakeries, have been classified separately from retail activity and put into the Food manufacturing industry.

- Central administrative offices of companies have a new sector of their own, Management of companies and enterprises (sector 55). For example, the headquarters office of a food-producing corporation is considered part of the new sector instead of part of the Food manufacturing industry.

- Reproduction of packaged software, which was classified as a business service in the SIC, is now classified in sector 334, Computer and electronic product manufacturing, as a manufacturing process.

- Electronic markets and agents and brokers, formerly undifferentiated components of wholesale trade industries, have a sector of their own (425).

- Retail trade in NAICS (sectors 44 and 45) now includes establishments such as office supply stores, computer and software stores, building materials dealers, plumbing supply stores, and electrical supply stores, that display merchandise and use mass-media advertising to sell to individuals as well as to businesses, which were formerly classified in wholesale trade.

In Part B, *Business Statistics* shows GDP, income, employment, hours, and earnings by industry, followed by statistics for key sectors such as petroleum, housing, manufacturing, retail trade, and services.

Notes and definitions. Productive use of economic data requires accurate knowledge about the sources and meaning of the data. The notes and definitions for each chapter, shown immediately after that chapter's tables, contain definitions, descriptions of recent data revisions, and references to sources of additional technical information. They also include information about data availability and revision and release schedules, which helps users to readily access the latest current values if they need to keep up with the data month by month or quarter by quarter.

A NOTE ON THE IMPORTANCE OF ECONOMIC STATISTICS

To retrieve money and credit data for earlier years, the editor of *Business Statistics* had to consult old printed volumes, where she encountered some inspiring prefatory words in the Federal Reserve Board volume *Banking and Monetary Statistics* (1943). These words were written by the Fed's longtime statistics chief E. A. Goldenweiser, in the stately cadences of an earlier era, about the financial statistics collected in that volume. But they well express the hope and expectation of statisticians and economists that their work can lead to better economic decisions:

"These serried ranks of organized statistics on banking and finance, even though they may inspire awe, should also inspire confidence. They are an augury that credit policy can be based in the future, as in the past, on fact rather than on fancy."

THE HISTORY OF *BUSINESS STATISTICS*

The history of *Business Statistics* began with the publication, many years ago, of the first edition of a volume with the same name by the U.S. Department of Commerce's Bureau of Economic Analysis (BEA). After 27 periodic editions, the last of which appeared in 1992, BEA found it necessary, for budgetary and other reasons, to discontinue both the publication and the maintenance of the database from which the publication was derived.

The individual statistical series gathered together here are all publicly available. However, the task of gathering them from the numerous different sources within the government and assembling them into one coherent database is impractical for most data users. Even when current data are readily available, obtaining the full historical time series is often time-consuming and difficult. Definitions and other documentation can also be inconvenient to find. Believing that a *Business Statistics* compilation was too valuable to be lost to the public, Bernan Press published the first edition of the present publication, edited by Dr. Courtenay M. Slater, in 1995. The first edition received a warm welcome from users of economic data. Dr. Slater, formerly chief economist of the Department of Commerce, continued to develop *Business Statistics* through four subsequent annual editions. A previous editor worked with Dr. Slater on the fourth and fifth editions. In subsequent editions, the current editor has continued in the tradition established by Dr. Slater of ensuring high-quality data, while revising and expanding the book's scope to include significant new aspects of the U.S. economy and longer historical background.

Nearly all of the statistical data in this book are from federal government sources and all are available in the public domain. Sources are given in the applicable notes and definitions.

The data in this volume meet the publication standards of the federal statistical agencies from which they were obtained. Every effort has been made to select data that are accurate, meaningful, and useful. All statistical data are subject to error arising from sampling variability, reporting errors, incomplete coverage, imputation, and other causes. The responsibility of the editor and publisher of this volume is limited to reasonable care in the reproduction and presentation of data obtained from established sources.

The 2019 edition has been edited by Susan Ockert.

Susan Ockert has worked as an economist in the military, for the federal government, and at regional and state levels. She earned her Masters of Economics at George Mason University in Fairfax, Virginia as well as her Masters of International Management at the Thunderbird University in Glendale, Arizona. She currently teaches economics at the collegiate level.

REFERENCES

The NAICS is explained and laid out in *North American Industry Classification System: United States, 2007*, from the Executive Office of the President, Office of Management and Budget. This presents the second five-year updating of the system, which was first introduced in 1997. Changes introduced in recent updatings have not significantly affected the definitions of the industry divisions presented in *Business Statistics*.

Information on differences between NAICS and SIC can be found in *North American Industry Classification System: United States, 1997*, from the Executive Office of the President, Office of Management and Budget (which contains matches between the 1997 NAICS and the 1987 SIC); and *North American Industry Classification System: United States, 2002* (which contains matches that show the relatively few changes from the 1997 NAICS to the 2002 NAICS).

All three of these volumes are available from Bernan Press. These volumes fully describe the development and application of the new classification system and are the sources for the material presented in this volume. Information is also available on the NAICS Web site at <http://www.census.gov/naics>.

CYCLE AND GROWTH PERSPECTIVES

This 24th edition of *Business Statistics of the United States* presents comprehensive and detailed data on U.S. economic performance through December 2018, going back to the end of World War II. These numbers show changes in many dimensions—prices and quantities, to take the two most obvious. There are patterns of fluctuation and, simultaneously, patterns of enduring change. This article will present some ways of looking at these movements and trying to make sense of them.

We begin with a general discussion of business cycles in the United States economy, including a chronology of the cycles occurring in the years covered by this volume.

- Following this is an example of the use of monthly and quarterly data, such as those provided in this book, to track recession and recovery.

- Then, other examples are provided of important analytical techniques for extracting key information from statistical series, particularly for focusing on growth issues.

- The last section of this article concerns measuring the economic well-being of the typical American.

BUSINESS CYCLES IN THE U.S. ECONOMY: RECESSIONS AND EXPANSIONS

The study of economic fluctuations in the United States was pioneered by Wesley C. Mitchell and Arthur F. Burns early in the twentieth century, and was carried on subsequently by other researchers affiliated with the National Bureau of Economic Research (NBER), an independent, nonpartisan research organization. These analysts observed that indicators of the general state of business activity tended to move up and down over periods that were longer than a year and were therefore not accounted for by seasonal variation. Although these periods of expansion and contraction were not uniform in length, and thus not "cycles" in any strict mathematical sense, their recurrent nature and certain generic similarities caused them to be called "business cycles." NBER has identified 32 complete peak-to-peak business cycles over the period beginning with December 1854.

The first NBER-established business cycle dates, identifying the months in which peaks and troughs in general economic activity occurred, were published in 1929. The dates of current

cycles are established, typically within a year or so of their occurrence, by the NBER Business Cycle Dating Committee. This group, first formed in 1978, consists of eight economists who are university professors, associated with research organizations, or both.

Business cycle dates are based on monthly data, and have been identified for periods long before the availability of quarterly data on real gross national product (GDP). It is important to understand that even in the period since 1947 for which quarterly real GDP exists, the NBER identification of a recession does not always coincide with the frequently cited definition, "two consecutive quarters of decline in real GDP."

The NBER monthly and quarterly dates of the cycles from 1891 to the latest announced turning point—the June 2009 trough ending the recession that began in December 2007—are shown in Table A-1 below. The quarterly turning points are identified by Roman numerals. NBER considers that the trough month is both the end of the decline and the beginning of the expansion, based on the concept that the actual turning point was some particular day within that month. Thus, the latest recession ended in June 2009, and the current expansion also began in June 2009.

TABLE A-1. BUSINESS CYCLE REFERENCE DATES 1891–2009

May 1891 (II)	January 1893 (I)
June 1894 (II)	December 1895 (IV)
June 1897 (II)	June 1899 (III)
December 1900 (IV)	September 1902 (IV)
August 1904 (III)	May 1907 (II)
June 1908 (II)	January 1910 (I)
January 1912 (IV)	January 1913 (I)
December 1914 (IV)	August 1918 (III)
March 1919 (I)	January 1920 (I)
July 1921 (III)	May 1923 (II)
July 1924 (III)	October 1926 (III)

TROUGH	PEAK
November 1927 (IV)	August 1929 (III)
March 1933 (I)	May 1937 (II)
June 1938 (II)	February 1945 (I)
October 1945 (IV)	November 1948 (IV)
October 1949 (IV)	July 1953 (II)
May 1954 (II)	August 1957 (III)
April 1958 (II)	April 1960 (II)
February 1961 (I)	December 1969 (IV)
November 1970 (IV)	November 1973 (IV)
March 1975 (I)	January 1980 (I)
July 1980 (III)	July 1981 (III)
November 1982 (IV)	July 1990 (III)
March 1991 (I)	March 2001 (I)
November 2001 (IV)	December 2007 (IV)
June 2009 (II)	

SOURCE: National Bureau of Economic Research, http://www.nber.org/cycles.
For additional information on NBER and its business cycle studies, see the NBER Web site at <http://www.nber.org/cycles>.

THE U.S. ECONOMY 1929–1948

While there has been much discussion in recent years of the "Great Depression," there is surprisingly little familiarity with the basic statistical record of the period between 1929 and the end of World War II. Reproducing material from Chapter 18 of the 15th edition of *Business Statistics,* we present here an explanation of that period from the NBER business cycle perspective, along with a graphic overview and narration of its economic developments as recorded in the statistics now presented in Part A of this volume.

The NBER chronology, presented in Table A-1 in the preceding article "Cycle and Growth Perspectives," may surprise readers who are looking for "The Great Depression" and are not familiar with the NBER approach to business cycles. As NBER perceives it, a downtrend in economic activity began in August 1929 (before the stock market crash) and lasted until March 1933. This 43-month period has been called the "Great Contraction": it was the longest period of economic decline since the 1870s, and more than double the length of the longest recession identified and completed since that time (the one from December 2007 to June 2009, which lasted 18 months). For the rest of the 1930s—excepting a 13-month recession in 1937–1938—the economy is viewed by the NBER as having been in an expansion phase. The NBER chronology does not use the term "depression."

However, the term "Great Depression" is often colloquially used for the entire 1929–1939 period, even though the economy was expanding for most of the period following March 1933. This is because economic activity during that time, though increasing, remained below the likely capacity of the economy, as is indicated in Figures A-2 and A-3 below.

It should also be noted that NBER construes the entire period from June 1938 through February 1945 as a business cycle expansion. The recovery from the 1937–1938 recession merged into a further, continued rise in activity that reflected the outbreak of war in Europe in September 1939, a consequent preparedness effort in the United States, and an increase in demand from abroad for U.S. output. The United States entered the war after being attacked by Japan in December 1941, launching an all-out war production effort at that time.

The February 1945 end of the "wartime expansion" (as NBER terms the period June 1938–February 1945) preceded the end of the war (the European war ended in May 1945 and the Pacific war concluded in August 1945). A brief recession associated with demobilization occurred from February to October 1945, followed by the first postwar expansion, which lasted from October 1945 to November 1948.

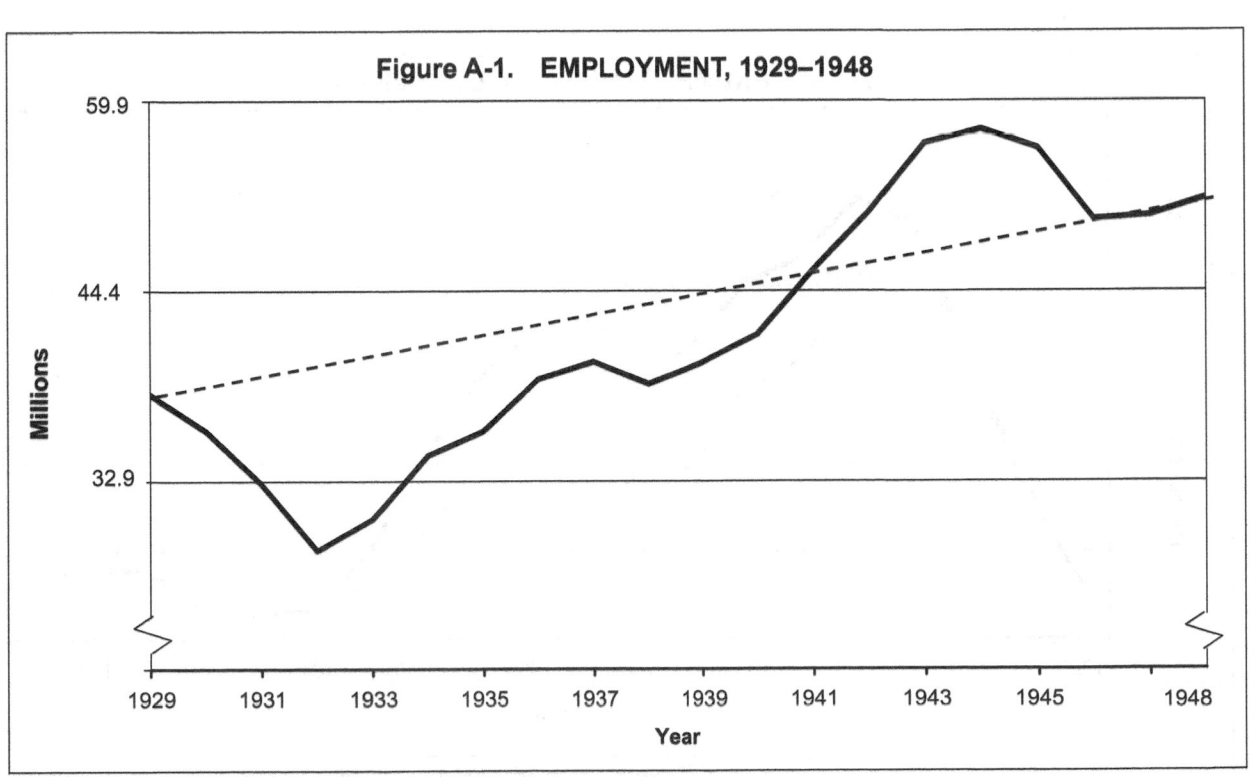

Figure A-1. EMPLOYMENT, 1929–1948

In the graphs that follow, summary annual statistics for the years 1929 through 1948 are presented, depicting the economy during two of its most tumultuous decades. This period encompasses the "great contraction," the economic collapse that began in August 1929 and lasted until March 1933; a subsequent period of recovery, which failed to return many important economic indicators to their expected trend levels and was interrupted by a new, though less severe, recession; apprehension of coming war, then the outbreak of war in Europe in September 1939, leading to increased war-related production; United States participation in all-out war, beginning with the Japanese attack on Pearl Harbor on December 7, 1941, and ending in 1945, with supercharged production and employment rates; then, rapid demobilization, and return to a high peacetime rate of economic activity by 1948.

For a first overall view of the economy during these decades, Figure A-1 shows total U.S. employment as calculated by the Bureau of Economic Analysis, including all private and government jobs, both civilian and military. In this graph a trend line is shown connecting the two peacetime high employment levels of 1929 and 1948.

- More than one-fifth of all the jobs held in the U.S. economy in 1929 were gone by 1932. In the subsequent recovery, total employment was back at the 1929 level by 1936, but only because of government employment, including over 3 ½ million work relief jobs; private industry employment would not recover to its 1929 level until 1941.

- Because the recovery was incomplete and economic activity did not recover to a trend level until after the decade's end, the entire decade of the 1930s is often described as "The Great Depression."

- During the war, men were drafted into the armed forces, practically all of the unemployed were put back to work, and women were drawn into the labor force, resulting in a period of what might be called "super-employment." After the war, employment fell back to a more normal trend level.

- The unemployment rates for the prewar period calculated by the Bureau of Labor Statistics, unlike the employment data used in Figure A-1, count people on government work relief programs as unemployed. By this reckoning, unemployment rose from 3.2 percent of the civilian labor force in 1929 to a peak of 24.9 percent in 1933, and got no lower than 14.3 percent for the rest of the 1930s, as shown in Figure A-2.

- State and local governments started hiring people for work relief in 1930. The Federal government's programs began in 1933, and in 1936 and 1938 agencies such as the WPA and the CCC employed over 3 ½ million people. When workers in these programs are counted as employed rather than unemployed, the high point for unemployment was 22.5 percent in 1932, and it was reduced to 9.1 percent in 1937, as shown by the dashed line in Figure

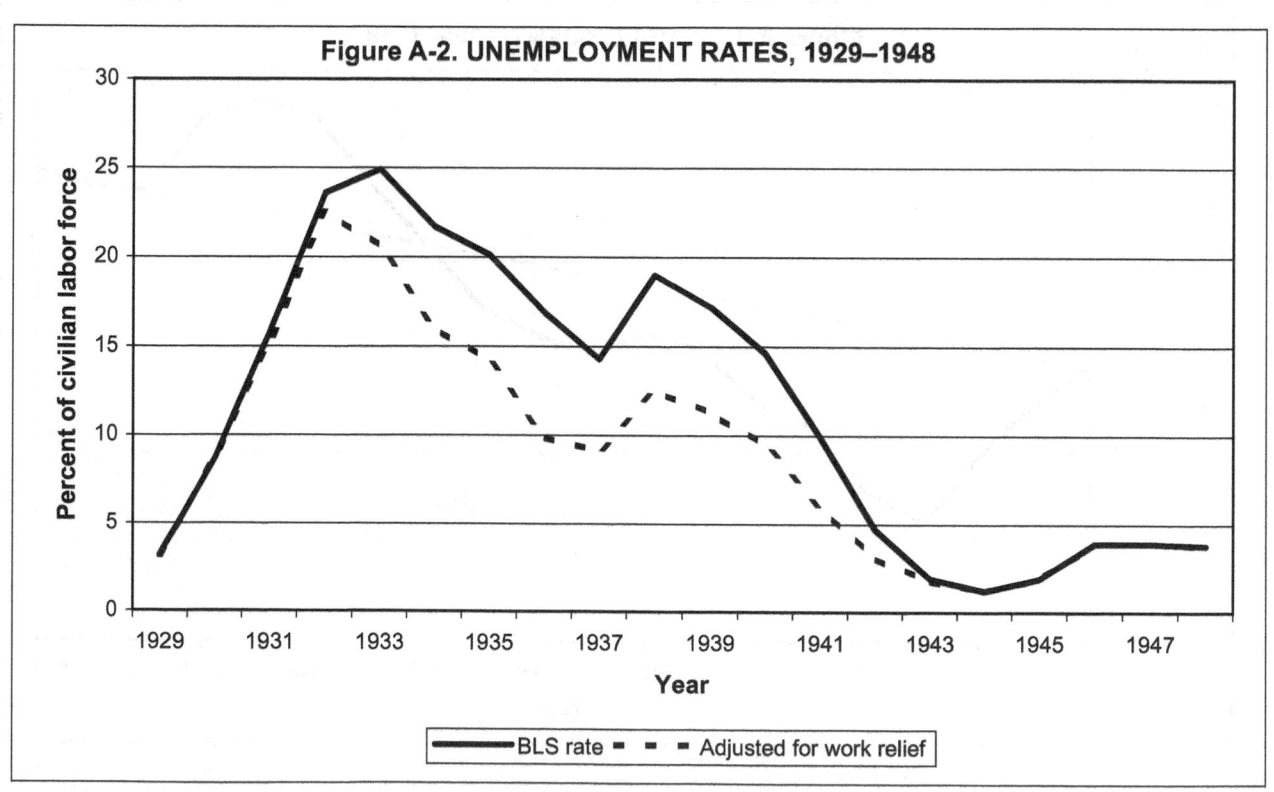

Figure A-2. UNEMPLOYMENT RATES, 1929–1948

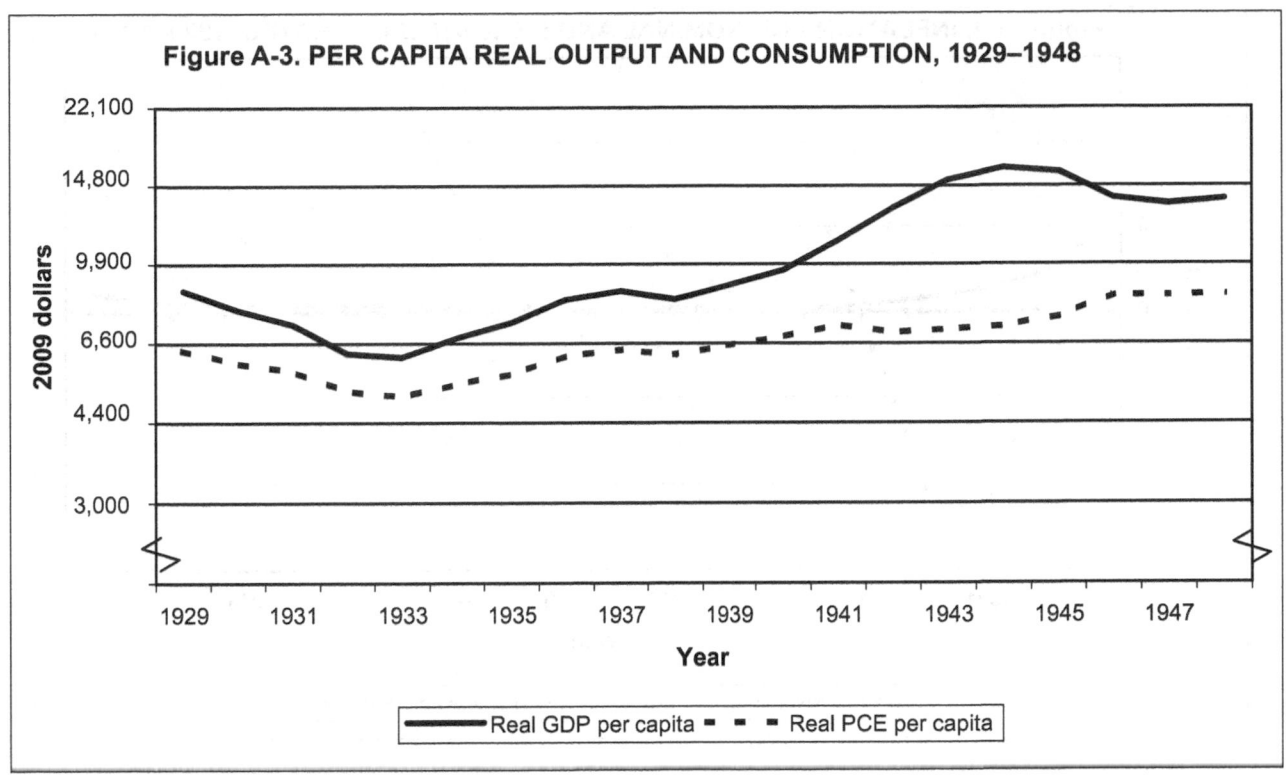

Figure A-3. PER CAPITA REAL OUTPUT AND CONSUMPTION, 1929–1948

A-2; see Notes and Definitions to Chapter 10 for further explanation.)

- By 1943, the economy reached a state of over-full employment, which would have been associated with runaway inflation if not for comprehensive price, wage, and production controls; the years of full wartime production, 1943 through 1945, all saw unemployment rates below 2 percent. After demobilization, the unemployment rate returned to just under 4 percent.

- Real GDP declined 26.7 percent from 1929 to 1933—28.9 percent on a per capita basis. Per capita output recovered almost to the 1929 level in 1937 but fell back 4.1 percent in the 1938 recession. Output then nearly doubled from 1939 to the peak war production year, 1944; the per capita annual growth rate was 12.4 percent. Output fell back during the demobilization, but in 1948 was still at a per capita level representing a 5.0 percent per year growth rate since 1939.

- Per capita personal consumption expenditures fell 20.9 percent from 1929 to 1933. The decline would have been greater if consumers had not dipped into their assets to keep their living standards from declining as steeply as their incomes; the personal saving rate was negative in 1932 and 1933. Real

per capita consumption recovered to the 1929 level by 1937. Despite rationing and shortages, real per capita consumption spending declined little during the war years. In 1948, it was 29.3 percent above the 1939 level.

- Why was the 1929-1933 contraction so deep and long-lasting? By some, blame is placed on U.S. government tax increases and imposition of new trade barriers (the Smoot-Hawley Tariff), which undoubtedly made their contribution. A substantial number of well-regarded economists, however, point primarily to deflation and its interaction with debt. The price index for personal consumption expenditures, declined 27.2 percent from 1929 to 1933, for an annual average deflation rate of 7.6 percent. Current and prospective price declines make debt more burdensome and debtors more likely to default, as interest payments remain fixed while incomes and asset values decline.

- Current and prospective price declines also make borrowing prohibitively expensive; a low nominal interest rate becomes high in real terms (since the real rate is the nominal rate minus the inflation rate, and subtraction means changing the sign and adding). A dollar in the hands of a prospective lender will be worth more in terms of purchasing power if he simply holds on to it than if he invests it in some real economic asset or activity whose price will be lower at the

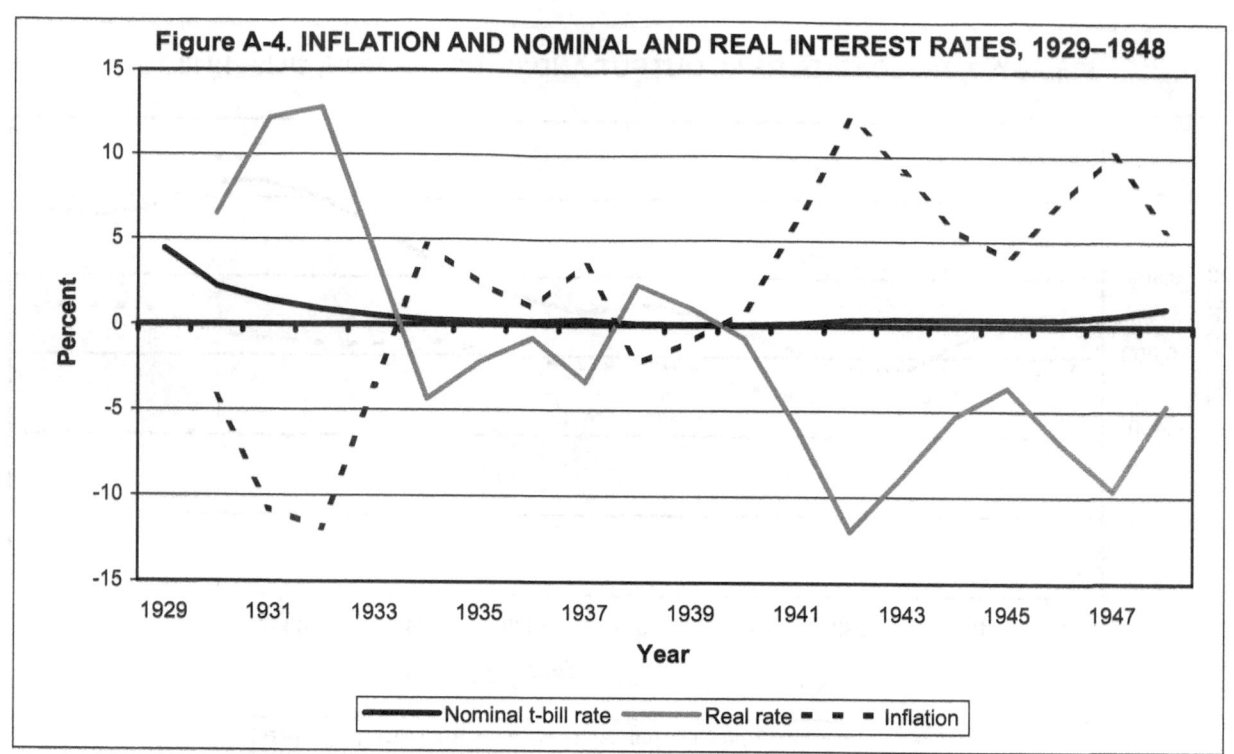

Figure A-4. INFLATION AND NOMINAL AND REAL INTEREST RATES, 1929–1948

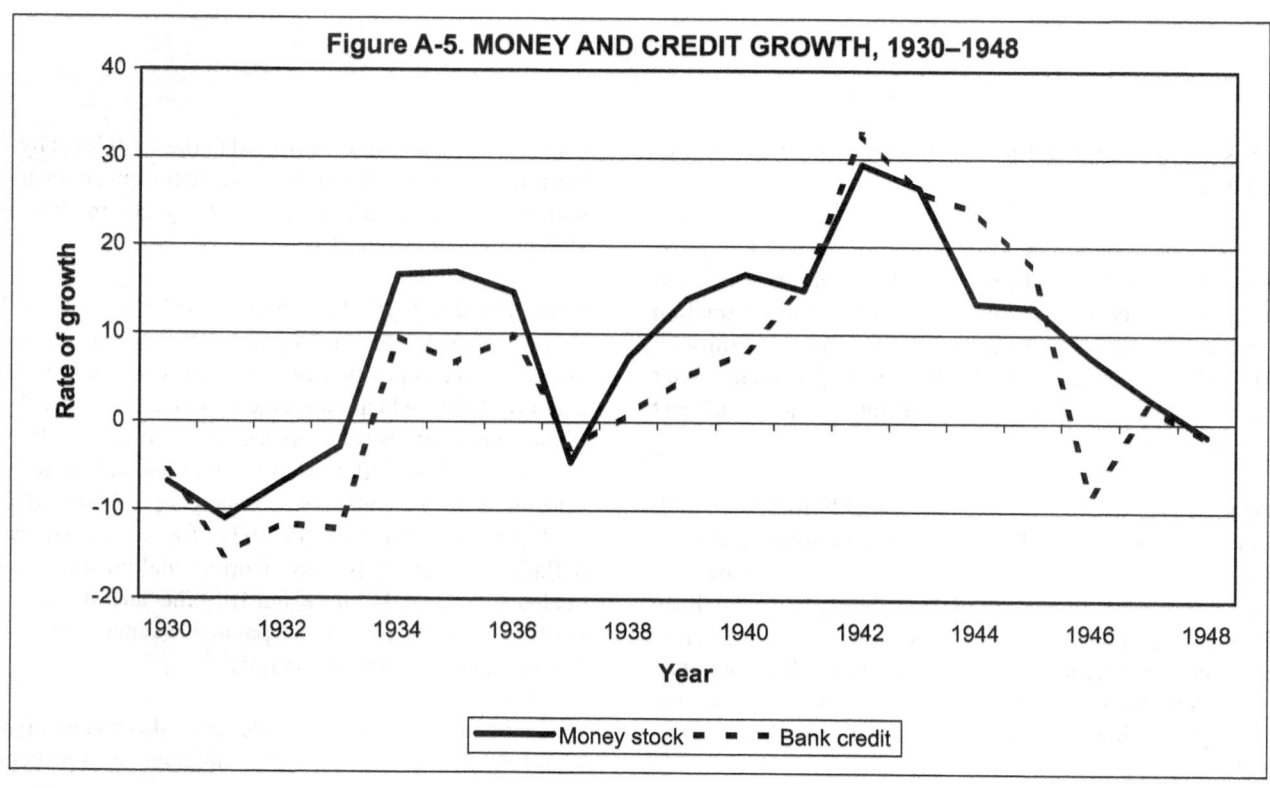

Figure A-5. MONEY AND CREDIT GROWTH, 1930–1948

end of the year. There is no feasible way for a central bank to lower interest rates below zero in order to reduce real rates in the presence of deflation. As Figure A-4 shows, real rates were high from 1930 through 1933 despite near-zero market nominal rates on Treasury bills.

- Deflation was rooted in deep and sustained declines in the dollar volumes of money and credit, as shown in Tables 12-1B and 12-4B and illustrated in Figure A-5. The money stock (currency and demand deposits) declined 27 percent from June 30, 1929, to June 30, 1933. Loans and investments at all commercial banks declined 29 percent over the same period. Commercial paper outstanding plunged 76 percent between December 1929 and December 1932, and bankers' acceptances were down 59 percent.

- Positive money and credit growth after 1933 was reflected in moderate rates of price increase and moderately negative—which means stimulative—real interest rates. As can be seen in Figure A-5, there was a monetary component to the 1937-38 recession. There was also a premature federal budget retrenchment at that time.

- During the war, monetary policy was pre-empted by wartime needs. Interest rates were held low to facilitate the financing of huge federal deficits.

- Worker incomes increased during the wartime boom as employment, average hours, and hourly wages all rose. But inflation and consumer spending were held down by price and production controls and rationing. Personal saving rates soared when there was little to buy and the government emphasized the sale of "war bonds"; the saving rate reached 26 percent in 1943 and 1944, then fell back to 4.2 percent in 1947, almost equal to its 1929 level.

THE GREAT RECESSION 2007–2009 AND RECOVERY 2010–2018

The Great Recession began in December 2007 and ended in June 2009, which makes it the longest recession since World War II. A recession, as defined by the National Bureau of Economic Research (NBER), is a "significant decline in economic activity spread across the economy, lasting more than a few months, normally visible in read GDP, real income, industrial production and wholesale-retail sales."

During the most current recession, housing starts fell, millions of jobs vanished, real GDP shrank, and the major stock indexes declined, thus reducing the net worth of households and nonprofits trillions of dollars in just two years. In October 2009, the unemployment rate reached double digits (10.0 percent) compared to an unemployment rate of 4.7 percent just two years earlier in October 2007. The last time unemployment was in double digits was June 1983.

Even though the Great Recession officially ended in June 2009, the return to a 'normal' U.S. economy has been sluggish. Employment numbers have gradually increased and unemployment decreased. By August 2019, the unemployment rate was down to 3.7 percent. Chapter 10 provides detailed information on employment.

Employment declined by nearly 6.3 million between December 2007 and June 2009. Employment in 2018 surpassed 2007 employment by 6.7 percent. However different industries, like construction, are still struggling to recover its workforce. In 2018, employment in construction was still more than 5 percent below its high in 2006, just before the recession. Chapter 15 describes employment by sector and industry.

Housing data such as new construction, building permits and new house prices is presented in Chapter 16. New housing starts declined 73 percent between 2005 and 2009. By 2018, housing starts were still 40 percent below the high in of 2.1 million in 2005.

In 2017, median household income increased for the fifth consecutive year to $61,372. As income has risen, poverty has declined. In 2017, the poverty rate declined to 12.3 percent— the lowest rate since 2006. The number of uninsured people remained steady at 8.8 percent. Chapter 3 discusses income distribution and poverty.

International trade, imports and exports, and current account balance information is depicted in Chapter 7 while foreign exchange data is found in Chapter 13. The United States has been a debtor nation since the 1980s meaning it must borrow from the world to pay its current account deficit. During the Great Recession, this deficit ballooned but declined significantly by 2015. However, it has increased again in the past few years.

GENERAL NOTES

These notes provide general information about the data in Tables 1-1 through 16-14. Specific notes with information about data sources, definitions, methodology, revisions, and sources of additional information follow the tables in each chapter.

MAIN DIVISIONS OF THE BOOK

The tables are presented in two parts:

Part A (Tables 1-1 through 13-2) pertains principally to the U.S. economy as a whole. Generally, each table presents, on its initial page, annual averages as far back as data availability and space permit, and quarterly or monthly values for the most recent year or years. For many important series, this initial page is followed by full annual and quarterly or monthly histories as far back as they are available on a continuous, consistent basis. Some chapters present data for the United States only in aggregate, while others—such as the chapters concerning industrial production and capacity utilization (chapter 2), capital expenditures (chapter 5), profits (chapter 9), and employment, hours, and earnings (chapter 10)—also have detail for industry groups.

Data by industry are classified using the North American Industry Classification System (NAICS), as far back as such data are made available by the source agencies.

Part B focuses on the individual industries that together produce the gross domestic product (GDP).

- Chapter 14 contains data on the value of GDP, quantity production trends, and factor income by NAICS industry group.

- Chapter 15 provides further detail on payroll employment, hours, and earnings classified according to NAICS.

- Chapter 16 presents various data sets for key economic sectors. Some of the tables are based on definitions of products, rather than of producing establishments, and are valid for either classification system.

Characteristics of the Tables and the Data

The subtitles or column headings for the data tables normally indicate whether the data are *seasonally adjusted, not seasonally adjusted,* or *at a seasonally adjusted annual rate.* These descriptions refer to the monthly or quarterly data, rather than the annual data; annual data by definition require no seasonal adjustment. Annual values are normally calculated as totals or averages, as appropriate, of unadjusted data. Such annual values are shown in either or both adjusted or unadjusted data columns.

Seasonal adjustment removes from a monthly or quarterly time series the average impact of variations that normally occur at about the same time each year, due to occurrences such as weather, holidays, and tax payment dates.

A simplified example of the process of seasonal adjustment, or deseasonalizing, can indicate its importance in the interpretation of economic time series. Statisticians compare actual monthly data for a number of years with "moving average" trends of the monthly data for the 12 months centered on each month's data. For example, they may find that in November, sales values are usually about 95 percent of the moving average, while in December, usual sales values are 110 percent of the average. Suppose that actual November sales in the current year are $100 and December sales are $105. The seasonally adjusted value for November will be $105 ($100/0.95) while the value for December will be $95 ($105/1.10). Thus, an apparent increase in the unadjusted data turns out to be a decrease when adjusted for the usual seasonal pattern.

The statistical method used to achieve the seasonal adjustment may vary from one data set to another. Many of the data are adjusted by a computer method known as X-12-ARIMA, developed by the Census Bureau. A description of the method is found in "New Capabilities and Methods of the X-12-ARIMA Seasonal Adjustment Program," by David F. Findley, Brian C. Monsell, William R. Bell, Mark C. Otto and Bor-Chung Chen (*Journal of Business and Economic Statistics*, April 1998). This article can be downloaded from the Bureau of the Census Web site at <https://www.census.gov>.

Production and sales data presented at *annual rates*—such as NIPA data in dollars, or motor vehicle data in number of units—show values at their annual equivalents: the values that would be registered if the seasonally adjusted rate of activity measured during a particular month or quarter were maintained for a full year. Specifically, seasonally adjusted monthly values are multiplied by 12 and quarterly values by 4 to yield seasonally adjusted annual rates.

Percent changes at seasonally adjusted annual rates for quarterly time periods are calculated using a compound interest formula, by raising the quarter-to-quarter percent change in a seasonally adjusted series to the fourth power. See the article "Cycle and Growth Perspectives" for an explanation of compound annual growth rates.

Indexes. In many of the most important data sets presented in this volume, aggregate measures of prices and quantities are expressed in the form of indexes. The most basic and familiar form of index, the original Consumer Price Index, begins with a "market basket" of goods and services purchased in a base period, with each product category valued at its dollar prices—the amount spent on that category by the average consumer. The value weight ascribed to each component of the market basket is moved forward by the observed change in the price of the item selected to represent that component. These weighted component prices—which constitute the quantities in the base period repriced in the prices of subsequent periods—are aggregated, divided by the base period aggregate, and multiplied by 100 to provide an index number. An index calculated in this way is known as a *Laspeyres index*. In general, economists believe that Laspeyres price indexes have an upward bias, showing more price increase than they would if account were taken of consumers' ability to change spending patterns and maintain the same level of satisfaction in response to changing relative prices.

A *Paasche index* is one that uses the weights of the current period. Since the weights in the Paasche index change in each period, Paasche indexes only provide acceptable indications of change relative to the base period. Paasche indexes for two periods neither of which is the base period cannot be correctly compared: for example, a Paasche price index for a recent period might increase from the period just preceding even if no prices changed between those two periods, if there was a change in the composition of output toward prices that had previously increased more from the base period. When the national income and product account (NIPA) measures of real output were Laspeyres measures, using the weights of a single base year, the implicit deflators (current-dollar values divided by constant-dollar values) were Paasche indexes. Just as Laspeyres price indexes are upward-biased, Paasche price indexes are downward-biased because they overestimate consumers' ability to maintain the same level of satisfaction by changing spending patterns.

In recent years, government statisticians—with the aid of complex computer programs—have developed measures of real output and prices that minimize bias by using the weights of both periods and updating the weights for each period-to-period comparison. Such measures are described as chained indexes and are used in the NIPAs, the index of industrial production, and an experimental consumer price index. Chained measures are discussed more fully in the notes and definitions for Chapter 1, Chapter 2, and Chapter 8. The "Fisher Ideal" index, the "superlative" index, and the "Tornqvist formula" are all types of chained indexes that use weights for both periods under comparison.

Detail may not sum to totals due to rounding. Since annual data are typically calculated by source agencies as the annual totals or averages of not-seasonally-adjusted data, they therefore will not be precisely equal to the annual totals or averages of monthly seasonally-adjusted data. Also, seasonal adjustment procedures are typically multiplicative rather than additive, which may also prevent seasonally adjusted data from adding or averaging to the annual figure. Percent changes and growth rates may have been calculated using unrounded data and therefore differ from those using the published figures.

The data in this volume are from federal government sources and may be reproduced freely. A list of data sources is shown below.

The tables in this volume incorporate data revisions and corrections released by the source agencies through mid-2019, including the July annual revision of the NIPAs and the resulting August revision of productivity and costs.

DATA SOURCES

The source agencies for the data in this volume are listed below. The specific source or sources for each particular data set are identified at the beginning of the notes and definitions for the relevant data pages.

BOARD OF GOVERNORS OF THE FEDERAL RESERVE SYSTEM

20th Street & Constitution Avenue NW
Washington, DC 20551

Data Inquiries and Publication Sales:
Publications Services
Mail Stop 127
Board of Governors of the Federal Reserve System
Washington, DC 20551
Phone: (202) 452-3245

Quarterly Publication:
As of 2006, the *Federal Reserve Bulletin* is available free of charge and only on the Federal Reserve Web site.

URL:
https://www.federalreserve.gov

Census Bureau
U.S. Department of Commerce
4700 Silver Hill Road
Washington, DC 20233

URL:
https://www.census.gov

Ordering Data Products:
Call Center: (301) 763-INFO (4636)

E-mail Questions:
webmaster@census.gov

Bureau of Economic Analysis
U.S. Department of Commerce
Washington, DC 20230

Data Inquiries:
Public Information Office
Phone: (202) 606-9900

Monthly Publication:
Survey of Current Business
Available online.

URL:
https://www.bea.gov

Bureau of Labor Statistics
U.S. Department of Labor
2 Massachusetts Avenue NE
Washington, DC 20212-0001
(202) 691-5200

URL:
https://www.bls.gov

Data Inquiries:
Blsdata_staff@bls.gov

Monthly Publications available online:
Monthly Labor Review
Employment and Earnings
Compensation and Working Conditions
Producer Price Indexes
CPI Detailed Report

Employment and Training Administration
U.S. Department of Labor
200 Constitution Avenue NW
Washington, DC 20210
(877) US2-JOBS
URL:
https://www.doleta.gov
https://www.itsc.state.md.us

Energy Information Administration
U.S. Department of Energy
1000 Independence Avenue SW
Washington, DC 20585

Data Inquiries and Publications:
National Energy Information Center
Phone: (202) 586-8800
E-mail: infoctr@eia.doe.gov

Monthly Publication:
Monthly Energy Review, as of 2007 available only on the EIA
Web site, free of charge.

URL:
https://www.eia.doe.gov

Federal Housing Finance Agency
FHFAinfo@FHFA.gov
(202) 414-6921,6922
(202) 414-6376

URL:
https://www.fhfa.gov/hpi

U.S. Department of the Treasury
Office of International Affairs
Treasury International Capital System

URL:
https://www.treas.gov/tic

To order government publications
Superintendent of Documents
Government Printing Office
Washington, DC 20402
(202) 512-1800

URL:
https://bookstore.gpo.gov

PART A: THE U.S. ECONOMY

CHAPTER 1: NATIONAL INCOME AND PRODUCT

SECTION 1A: GROSS DOMESTIC PRODUCT: VALUES, QUANTITIES, AND PRICES

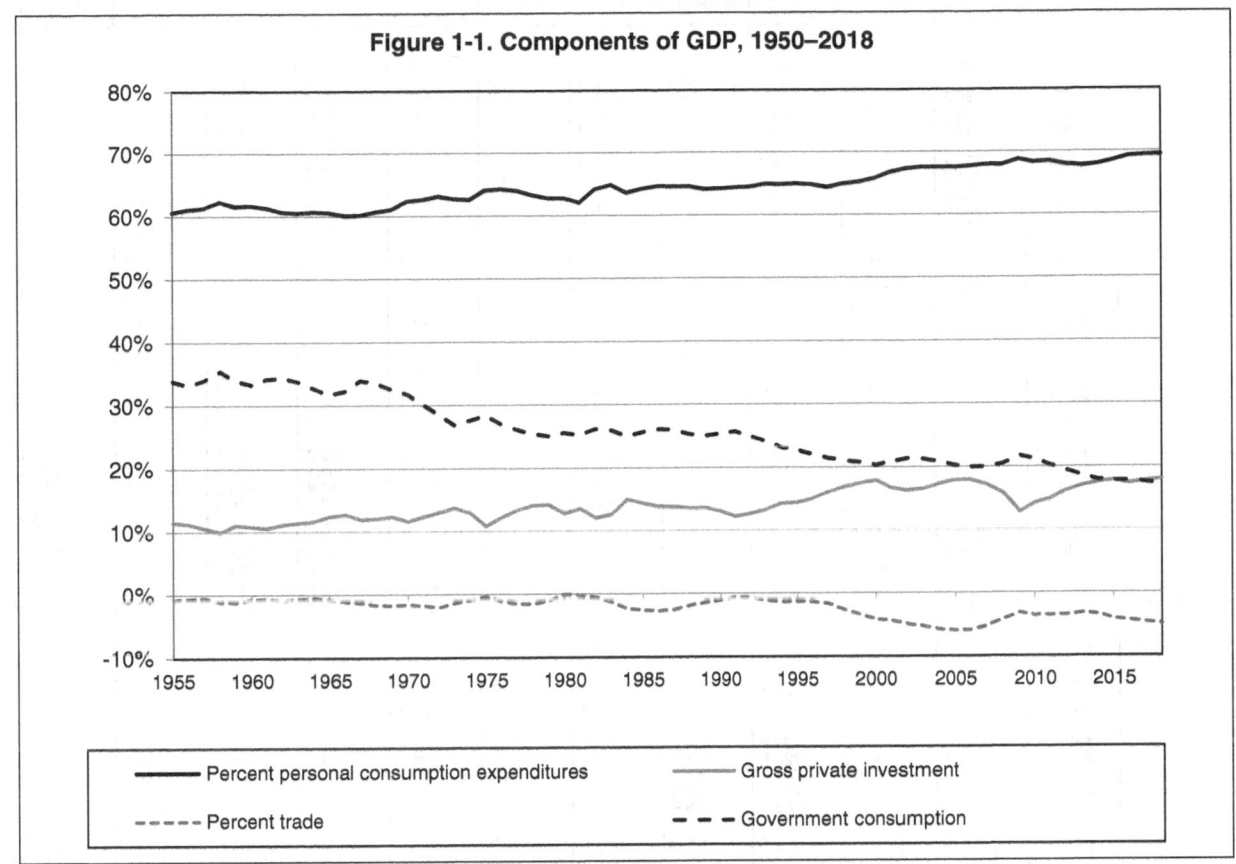

Figure 1-1. Components of GDP, 1950–2018

- As presented in the preface, the output approach to GDP consists of four components: consumption, business investment, government taxes and spending and the foreign trade balance (exports – imports). GDP is one of the most comprehensive and closely watched economic statistics in not only the United States but also globally. Figure 1-1 illustrates the trend of these four components. For example, personal consumption's share of GDP rose from 61 percent in 1955 to 68 percent in 2018. Meanwhile, government's percent share of GDP fell significantly in the same time period. For more details of GDP, see the Notes and Definitions at the end of this chapter (Table 1-1A).

- Real GDP per capita—the constant-dollar average value of production for each man, woman, and child in the population—rose from $17,372 (2012 dollars) in 1955 to $51,794 in 2007. Reflecting the recession that began in December 2007, it declined 4.4 percent from 2007 to 2009. Over the nine years since then, per capita GDP recovered and finally surpassed its 2007 peak in 2013 (Table 1-3A).

- With the impositions of tariffs in January 2018, imports became more expensive. Tariffs have been imposed on solar panels, washing machines, steel and aluminum as well as products from Canada, Mexico, China and the European Union. China then imposed their own tariffs. Americans enjoy a vast selection of goods, many of which are not available in the United States. Tariffs raise prices for those consumers and businesses. The impact of these tariffs on the U.S. trade balance has been negative.

Table 1-1A. Gross Domestic Product: Recent Data

(Billions of dollars, quarterly data are at seasonally adjusted annual rates.)   NIPA Tables 1.1.5, 1.2.5

Year and quarter	Gross domestic product	Personal consump-tion expen-ditures	Gross private domestic investment				Exports and imports of goods and services			Government consumption expenditures and gross investment			Addendum: final sales of domestic product
			Total	Fixed investment		Change in private inventories	Net exports	Exports	Imports	Total	Federal	State and local	
				Nonresi-dential	Residential								
1955	425.5	258.3	73.8	43.4	25.4	5.0	0.5	17.7	17.2	93.0	61.0	31.9	420.5
1956	449.4	271.1	77.7	49.7	24.0	4.0	2.4	21.3	18.9	98.2	63.1	35.1	445.4
1957	474.0	286.3	76.5	53.1	22.6	0.8	4.1	24.0	19.9	107.2	68.5	38.7	473.3
1958	481.2	295.6	70.9	48.5	22.8	-0.4	0.5	20.6	20.0	114.1	71.5	42.6	481.6
1959	521.7	317.1	85.7	53.1	28.6	3.9	0.4	22.7	22.3	118.5	73.6	44.9	517.7
1960	542.4	331.2	86.5	56.4	26.9	3.2	4.2	27.0	22.8	120.5	72.9	47.6	539.1
1961	562.2	341.5	86.6	56.6	27.0	3.0	4.9	27.6	22.7	129.2	77.4	51.9	559.2
1962	603.9	362.6	97.0	61.2	29.6	6.1	4.1	29.1	25.0	140.3	85.5	54.8	597.8
1963	637.5	382.0	103.3	64.8	32.9	5.6	4.9	31.1	26.1	147.2	87.9	59.3	631.8
1964	684.5	410.6	112.2	72.2	35.1	4.8	6.9	35.0	28.1	154.8	90.3	64.5	679.6
1965	742.3	443.0	129.6	85.2	35.2	9.2	5.6	37.1	31.5	164.1	93.2	70.9	733.0
1966	813.4	479.9	144.2	97.2	33.4	13.6	3.9	40.9	37.1	185.4	106.6	78.9	799.8
1967	860.0	506.7	142.7	99.2	33.6	9.9	3.6	43.5	39.9	207.0	120.0	87.0	850.1
1968	940.7	556.9	156.9	107.7	40.2	9.1	1.4	47.9	46.6	225.5	128.0	97.6	931.6
1969	1 017.6	603.6	173.6	120.0	44.4	9.2	1.4	51.9	50.5	239.0	131.2	107.8	1 008.4
1970	1 073.3	646.7	170.0	124.6	43.4	2.0	3.9	59.7	55.8	252.6	132.8	119.8	1 071.3
1971	1 164.9	699.9	196.8	130.4	58.2	8.3	0.6	63.0	62.3	267.5	134.5	133.0	1 156.6
1972	1 279.1	768.2	228.1	146.6	72.4	9.1	-3.4	70.8	74.2	286.2	141.6	144.5	1 270.0
1973	1 425.4	849.6	266.9	172.7	78.3	15.9	4.1	95.3	91.2	304.8	146.2	158.6	1 409.5
1974	1 545.2	930.2	274.5	191.1	69.5	14.0	-0.8	126.7	127.5	341.4	158.8	182.5	1 531.3
1975	1 684.9	1 030.5	257.3	196.8	66.7	-6.3	16.0	138.7	122.7	381.1	173.7	207.4	1 691.2
1976	1 873.4	1 147.7	323.2	219.3	86.8	17.1	-1.6	149.5	151.1	404.2	184.8	219.4	1 856.3
1977	2 081.8	1 274.0	396.6	259.1	115.2	22.3	-23.1	159.3	182.4	434.3	200.3	234.0	2 059.5
1978	2 351.6	1 422.3	478.4	314.6	138.0	25.8	-25.4	186.9	212.3	476.3	218.9	257.4	2 325.8
1979	2 627.3	1 585.4	539.7	373.8	147.8	18.0	-22.5	230.1	252.7	524.8	240.6	284.2	2 609.3
1980	2 857.3	1 750.7	530.1	406.9	129.5	-6.3	-13.1	280.8	293.8	589.6	274.9	314.7	2 863.6
1981	3 207.0	1 934.0	631.2	472.9	128.5	29.8	-12.5	305.2	317.8	654.4	314.0	340.4	3 177.2
1982	3 343.8	2 071.3	581.0	485.1	110.8	-14.9	-20.0	283.2	303.2	711.5	348.3	363.1	3 358.7
1983	3 634.0	2 281.6	637.5	482.2	161.1	-5.8	-51.6	277.0	328.6	766.6	382.4	384.2	3 639.8
1984	4 037.6	2 492.3	820.1	564.3	190.4	65.4	-102.7	302.4	405.1	827.9	411.8	416.1	3 972.2
1985	4 339.0	2 712.8	829.7	607.8	200.1	21.8	-114.0	303.2	417.2	910.5	452.9	457.6	4 317.2
1986	4 579.6	2 886.3	849.1	607.8	234.8	6.6	-131.9	321.0	452.9	976.1	481.7	494.4	4 573.1
1987	4 855.2	3 076.3	892.2	615.2	249.8	27.1	-144.8	363.9	508.7	1 031.5	502.8	528.7	4 828.1
1988	5 236.4	3 330.0	937.0	662.3	256.2	18.5	-109.4	444.6	554.0	1 078.8	511.4	567.4	5 218.0
1989	5 641.6	3 576.8	999.7	716.0	256.0	27.7	-86.7	504.3	591.0	1 151.9	534.1	617.8	5 613.9
1990	5 963.1	3 809.0	993.4	739.2	239.7	14.5	-77.9	551.9	629.7	1 238.6	562.4	676.2	5 948.6
1991	6 158.1	3 943.4	944.3	723.6	221.2	-0.4	-28.6	594.9	623.5	1 299.0	582.9	716.0	6 158.5
1992	6 520.3	4 197.6	1 013.0	741.9	254.7	16.3	-34.7	633.1	667.8	1 344.5	588.5	756.0	6 504.0
1993	6 858.6	4 452.0	1 106.8	799.2	286.8	20.8	-65.2	654.8	720.0	1 364.9	580.2	784.8	6 837.7
1994	7 287.2	4 721.0	1 256.5	868.9	323.8	63.8	-92.5	720.9	813.4	1 402.3	574.7	827.6	7 223.5
1995	7 639.7	4 962.6	1 317.5	962.2	324.1	31.2	-89.8	812.8	902.6	1 449.4	576.7	872.7	7 608.6
1996	8 073.1	5 244.6	1 432.1	1 043.2	358.1	30.8	-96.4	867.6	964.0	1 492.8	579.2	913.7	8 042.3
1997	8 577.6	5 536.8	1 595.6	1 149.1	375.6	70.9	-102.0	953.8	1 055.8	1 547.1	583.3	963.8	8 506.6
1998	9 062.8	5 877.2	1 736.7	1 254.1	418.8	63.7	-162.7	953.0	1 115.7	1 611.6	585.5	1 026.1	8 999.1
1999	9 630.7	6 279.1	1 887.1	1 364.5	461.8	60.8	-255.8	992.8	1 248.6	1 720.4	611.3	1 109.0	9 569.8
2000	10 252.3	6 762.1	2 038.4	1 498.4	485.4	54.5	-375.1	1 096.3	1 471.3	1 826.8	633.7	1 193.1	10 197.8
2001	10 581.8	7 065.6	1 934.8	1 460.1	513.1	-38.3	-367.9	1 024.6	1 392.6	1 949.3	670.1	1 279.2	10 620.1
2002	10 936.4	7 342.7	1 930.4	1 352.8	557.6	20.0	-425.4	998.7	1 424.1	2 088.7	743.0	1 345.7	10 916.4
2003	11 458.2	7 723.1	2 027.1	1 375.9	637.1	14.1	-503.1	1 036.2	1 539.3	2 211.2	826.3	1 384.9	11 444.2
2004	12 213.7	8 212.7	2 281.3	1 467.4	749.8	64.1	-619.1	1 177.6	1 796.7	2 338.9	891.7	1 447.1	12 149.7
2005	13 036.6	8 747.1	2 534.7	1 621.0	856.2	57.5	-721.2	1 305.2	2 026.4	2 476.0	947.5	1 528.5	12 979.1
2006	13 814.6	9 260.3	2 701.0	1 793.8	838.2	69.0	-770.9	1 472.6	2 243.5	2 624.2	1 000.7	1 623.5	13 745.6
2007	14 451.9	9 706.4	2 673.0	1 948.6	690.5	34.0	-718.4	1 660.9	2 379.3	2 790.8	1 050.5	1 740.3	14 417.9
2008	14 712.8	9 976.3	2 477.6	1 990.9	516.0	-29.2	-723.1	1 837.1	2 560.1	2 982.0	1 150.6	1 831.4	14 742.1
2009	14 448.9	9 842.2	1 929.7	1 690.4	390.0	-150.8	-396.5	1 582.0	1 978.4	3 073.5	1 218.2	1 855.3	14 599.7
2010	14 992.1	10 185.8	2 165.5	1 735.0	376.6	53.9	-513.9	1 846.3	2 360.2	3 154.6	1 297.9	1 856.7	14 938.1
2011	15 542.6	10 641.1	2 332.6	1 907.5	378.8	46.3	-579.5	2 103.0	2 682.5	3 148.4	1 298.9	1 849.4	15 496.3
2012	16 197.0	11 006.8	2 621.8	2 118.5	432.0	71.2	-568.6	2 191.3	2 759.9	3 137.0	1 286.5	1 850.5	16 125.8
2013	16 784.9	11 317.2	2 826.0	2 211.5	510.0	104.5	-490.8	2 273.4	2 764.2	3 132.4	1 226.6	1 905.8	16 680.3
2014	17 527.3	11 822.8	3 044.2	2 400.1	560.2	84.0	-507.7	2 371.7	2 879.4	3 168.0	1 215.0	1 953.0	17 443.3
2015	18 224.8	12 284.3	3 223.1	2 457.4	633.8	131.9	-519.8	2 266.8	2 786.6	3 237.3	1 221.5	2 015.7	18 092.9
2016	18 715.0	12 748.5	3 178.7	2 453.1	698.5	27.1	-518.8	2 220.6	2 739.4	3 306.7	1 234.1	2 072.6	18 688.0
2017	19 519.4	13 312.1	3 370.7	2 584.7	755.7	30.2	-575.3	2 356.7	2 932.1	3 412.0	1 269.3	2 142.7	19 489.2
2018	20 580.2	13 998.7	3 628.3	2 786.9	786.7	54.7	-638.2	2 510.3	3 148.5	3 591.5	1 347.3	2 244.2	20 525.5
2016													
1st quarter	18 424.3	12 523.5	3 149.1	2 415.6	686.6	46.9	-522.2	2 164.9	2 687.1	3 273.8	1 227.5	2 046.3	18 377.4
2nd quarter	18 637.3	12 688.3	3 152.9	2 441.8	692.0	19.1	-495.3	2 208.1	2 703.4	3 291.4	1 226.2	2 065.2	18 618.1
3rd quarter	18 806.7	12 822.6	3 166.6	2 471.6	697.7	-2.7	-499.7	2 254.4	2 754.1	3 317.5	1 237.5	2 080.0	18 809.5
4th quarter	18 991.9	12 959.8	3 246.2	2 483.5	717.8	44.9	-558.0	2 255.1	2 813.1	3 343.9	1 245.2	2 098.7	18 946.9
2017													
1st quarter	19 190.4	13 104.4	3 288.2	2 531.1	743.7	13.4	-570.9	2 303.3	2 874.2	3 368.7	1 248.4	2 120.3	19 177.0
2nd quarter	19 356.6	13 212.5	3 335.0	2 567.4	748.8	18.8	-583.7	2 313.2	2 896.9	3 392.9	1 263.6	2 129.3	19 337.8
3rd quarter	19 611.7	13 345.1	3 401.8	2 591.6	753.4	56.8	-550.6	2 360.1	2 910.7	3 415.4	1 270.2	2 145.2	19 554.9
4th quarter	19 918.9	13 586.3	3 457.7	2 648.9	777.1	31.7	-596.1	2 450.3	3 046.5	3 471.0	1 295.1	2 175.9	19 887.2
2018													
1st quarter	20 163.2	13 728.4	3 542.4	2 717.3	783.7	41.5	-629.0	2 476.6	3 105.6	3 521.4	1 318.2	2 203.2	20 121.7
2nd quarter	20 510.2	13 939.8	3 561.6	2 782.0	789.5	-10.0	-568.4	2 543.6	3 112.0	3 577.1	1 340.4	2 236.7	20 520.1
3rd quarter	20 749.8	14 114.6	3 684.0	2 807.7	789.0	87.3	-671.4	2 510.3	3 181.6	3 622.6	1 358.6	2 263.9	20 662.4
4th quarter	20 897.8	14 211.9	3 725.2	2 840.7	784.4	100.1	-684.1	2 510.5	3 194.7	3 644.8	1 371.8	2 273.0	20 797.7

Table 1-1B. Gross Domestic Product: Historical Data

(Billions of dollars, quarterly data are at seasonally adjusted annual rates.) NIPA Tables 1.1.5, 1.2.5

| Year and quarter | Gross domestic product | Personal consumption expenditures | Gross private domestic investment | | | | Exports and imports of goods and services | | | Government consumption expenditures and gross investment | | | Addendum: Final sales of domestic product |
| | | | Total | Fixed investment | | Change in private inventories | Net exports | Exports | Imports | Total | Federal | State and local | |
				Nonresidential	Residential								
1929	104.6	77.4	17.2	11.6	4.1	1.5	0.4	5.9	5.6	9.6	1.9	7.7	103.0
1930	92.2	70.1	11.4	9.2	2.5	-0.2	0.3	4.4	4.1	10.3	2.0	8.3	92.4
1931	77.4	60.7	6.5	5.8	1.9	-1.1	0.0	2.9	2.9	10.2	2.0	8.1	78.5
1932	59.5	48.7	1.8	3.3	0.9	-2.4	0.0	2.0	1.9	9.0	2.0	7.0	61.9
1933	57.2	45.9	2.3	3.0	0.7	-1.4	0.1	2.0	1.9	8.9	2.4	6.5	58.6
1934	66.8	51.5	4.3	3.8	1.0	-0.6	0.3	2.6	2.2	10.7	3.4	7.3	67.4
1935	74.3	55.9	7.4	4.8	1.4	1.1	-0.2	2.8	3.0	11.2	3.6	7.6	73.1
1936	84.9	62.2	9.4	6.4	1.8	1.2	-0.1	3.0	3.2	13.4	5.8	7.7	83.7
1937	93.0	66.8	13.0	8.2	2.2	2.6	0.1	4.0	4.0	13.1	5.3	7.9	90.4
1938	87.4	64.3	7.9	6.3	2.2	-0.6	1.0	3.8	2.8	14.2	5.9	8.3	88.0
1939	93.5	67.2	10.2	6.9	3.2	0.2	0.8	4.0	3.1	15.2	6.2	9.0	93.3
1940	102.9	71.3	14.6	8.5	3.6	2.4	1.5	4.9	3.4	15.6	6.8	8.8	100.6
1941	129.4	81.1	19.4	10.8	4.2	4.3	1.0	5.5	4.4	27.9	19.1	8.8	125.0
1942	166.0	89.0	11.8	7.5	2.4	1.9	-0.3	4.4	4.6	65.5	56.7	8.8	164.1
1943	203.1	99.9	7.4	6.6	1.6	-0.7	-2.2	4.0	6.3	98.1	89.5	8.6	203.9
1944	224.6	108.6	9.2	8.6	1.5	-0.9	-2.0	4.9	6.9	108.7	100.1	8.6	225.5
1945	228.2	120.0	12.4	12.1	1.8	-1.5	-0.8	6.8	7.5	96.6	87.3	9.3	229.6
1946	227.8	144.3	33.1	19.2	8.0	6.0	7.2	14.2	7.0	43.2	32.1	11.1	221.8
1947	249.9	162.0	37.1	25.5	12.2	-0.6	10.8	18.7	7.9	40.0	25.8	14.2	250.5
1948	274.8	175.0	50.3	28.9	15.8	5.7	5.5	15.5	10.1	44.0	27.1	16.9	269.1
1949	272.8	178.5	39.1	26.9	14.8	-2.7	5.2	14.5	9.2	50.0	30.5	19.6	275.5
1947													
1st quarter	243.2	156.2	35.9	24.8	10.5	0.5	10.9	18.4	7.5	40.3	27.0	13.3	242.7
2nd quarter	246.0	160.0	34.5	25.2	10.6	-1.2	11.3	19.5	8.2	40.1	26.4	13.7	247.2
3rd quarter	249.6	163.5	34.9	25.4	12.5	-2.9	11.8	19.4	7.7	39.4	25.0	14.3	252.5
4th quarter	259.7	167.7	43.3	26.5	15.3	1.5	9.3	17.6	8.3	39.5	24.5	15.0	258.3
1948													
1st quarter	265.7	170.4	47.2	28.2	15.3	3.6	7.3	16.9	9.6	40.9	25.4	15.5	262.1
2nd quarter	272.6	174.1	50.3	28.1	16.4	5.9	5.2	15.2	10.0	42.9	26.6	16.3	266.7
3rd quarter	279.2	177.1	52.5	29.1	16.3	7.2	4.9	15.4	10.5	44.7	27.5	17.2	272.0
4th quarter	280.4	177.9	51.3	30.2	15.2	6.0	4.5	14.6	10.1	46.6	28.7	17.9	274.4
1949													
1st quarter	275.0	176.8	43.1	28.6	14.2	0.4	6.5	16.1	9.6	48.6	30.2	18.4	274.6
2nd quarter	271.4	178.4	36.2	27.5	13.9	-5.1	6.3	15.6	9.4	50.4	31.2	19.1	276.5
3rd quarter	272.9	177.8	39.5	26.1	14.7	-1.3	5.2	14.1	8.9	50.4	30.5	19.9	274.2
4th quarter	270.6	180.2	37.5	25.6	16.5	-4.7	3.0	12.1	9.1	49.9	29.8	20.1	275.3
1950													
1st quarter	280.8	182.9	46.7	26.4	18.3	2.0	2.2	11.7	9.5	49.0	28.5	20.4	278.8
2nd quarter	290.4	186.8	52.3	28.9	20.6	2.8	1.6	11.9	10.2	49.6	28.9	20.7	287.6
3rd quarter	308.2	200.5	58.6	31.9	22.6	4.2	-0.7	12.3	13.0	49.7	28.4	21.3	304.0
4th quarter	319.9	197.9	68.4	32.9	21.5	14.0	-0.2	13.5	13.7	53.7	31.8	21.9	305.9
1951													
1st quarter	336.0	209.2	64.6	33.1	21.1	10.4	0.2	15.0	14.9	62.0	39.6	22.5	325.6
2nd quarter	344.1	204.9	67.4	34.1	18.5	14.8	1.9	17.1	15.2	69.8	46.6	23.2	329.3
3rd quarter	351.4	207.6	62.0	34.8	17.4	9.7	3.7	18.1	14.3	78.0	54.3	23.7	341.6
4th quarter	356.2	211.6	57.1	34.6	17.7	4.7	4.2	18.2	14.0	83.3	59.3	24.0	351.5
1952													
1st quarter	359.8	213.0	58.1	35.1	18.3	4.7	3.7	18.7	15.0	85.0	60.9	24.1	355.1
2nd quarter	361.0	217.1	53.0	35.8	18.8	-1.5	2.0	16.6	14.6	89.0	64.1	24.9	362.6
3rd quarter	367.7	219.6	57.2	32.9	18.8	5.6	0.0	15.2	15.3	90.9	66.2	24.8	362.1
4th quarter	380.8	227.7	60.7	35.8	19.6	5.3	-1.0	15.3	16.3	93.4	68.1	25.3	375.5
1953													
1st quarter	388.0	231.2	61.7	37.8	20.0	3.9	-0.7	15.1	15.8	95.8	69.7	26.1	384.1
2nd quarter	391.7	233.0	62.1	38.5	20.1	3.6	-1.3	15.2	16.4	97.9	72.0	26.0	388.2
3rd quarter	391.2	233.7	61.4	39.6	19.5	2.3	-0.6	15.8	16.3	96.6	70.0	26.7	388.8
4th quarter	386.0	233.1	56.4	39.2	19.3	-2.0	-0.3	15.2	15.5	96.7	69.5	27.2	388.0
1954													
1st quarter	385.3	235.2	55.7	38.3	19.4	-2.0	-0.4	14.4	14.8	94.8	66.7	28.1	387.3
2nd quarter	386.1	237.9	55.4	38.2	20.7	-3.4	0.3	16.4	16.2	92.5	63.7	28.8	389.5
3rd quarter	391.0	240.3	59.0	38.9	22.2	-2.1	0.6	15.9	15.3	91.1	61.4	29.8	393.1
4th quarter	399.7	245.1	62.1	38.9	23.6	-0.3	1.1	16.6	15.5	91.4	61.2	30.1	400.1
1955													
1st quarter	413.1	251.4	68.7	39.5	25.4	3.8	1.1	17.3	16.2	91.9	60.7	31.1	409.3
2nd quarter	421.5	256.5	72.7	42.1	26.0	4.6	-0.2	16.9	17.1	92.5	60.8	31.7	416.9
3rd quarter	430.2	260.7	74.7	44.8	25.6	4.3	0.7	18.1	17.4	94.1	61.9	32.2	425.9
4th quarter	437.1	264.6	78.9	47.1	24.6	7.2	0.2	18.3	18.1	93.3	60.6	32.8	429.9
1956													
1st quarter	439.7	266.2	78.3	47.8	24.1	6.4	0.4	19.4	18.9	94.8	61.0	33.9	433.3
2nd quarter	446.0	268.8	77.0	49.1	24.4	3.6	1.9	20.9	19.0	98.2	63.5	34.7	442.4
3rd quarter	451.2	272.1	78.3	50.7	24.0	3.6	2.6	21.8	19.3	98.3	62.8	35.5	447.6
4th quarter	460.5	277.4	77.1	51.4	23.5	2.2	4.5	23.1	18.5	101.3	65.2	36.2	458.2
1957													
1st quarter	469.8	281.9	77.7	52.5	23.1	2.2	4.8	24.9	20.1	105.3	67.9	37.5	467.6
2nd quarter	472.0	284.2	77.9	52.6	22.6	2.7	4.1	24.4	20.3	105.8	67.5	38.3	469.4
3rd quarter	479.5	288.8	79.3	54.1	22.5	2.8	4.0	23.8	19.8	107.4	68.4	39.1	476.7
4th quarter	474.9	290.4	71.0	53.2	22.3	-4.5	3.4	23.0	19.6	110.1	70.0	40.0	479.4
1958													
1st quarter	467.5	289.9	66.7	49.4	21.4	-4.0	1.1	20.5	19.5	109.8	68.7	41.1	471.5
2nd quarter	472.0	292.8	65.1	47.8	21.5	-4.2	0.5	20.5	20.1	113.6	71.5	42.1	476.2
3rd quarter	485.8	297.9	72.0	47.4	23.0	1.5	0.9	20.6	19.7	115.1	71.9	43.2	484.3
4th quarter	499.6	301.8	80.0	49.3	25.4	5.2	-0.3	20.6	20.8	118.0	73.9	44.1	494.3

Table 1-1B. Gross Domestic Product: Historical Data—*Continued*

(Billions of dollars, quarterly data are at seasonally adjusted annual rates.)

NIPA Tables 1.1.5, 1.2.5

| Year and quarter | Gross domestic product | Personal consumption expenditures | Gross private domestic investment | | | | Exports and imports of goods and services | | | Government consumption expenditures and gross investment | | | Addendum: Final sales of domestic product |
| | | | Total | Fixed investment | | Change in private inventories | Net exports | Exports | Imports | Total | Federal | State and local | |
				Nonresidential	Residential								
1959													
1st quarter	510.3	309.4	83.2	50.9	28.3	3.9	0.5	21.9	21.4	117.2	72.4	44.8	506.5
2nd quarter	522.7	315.5	89.4	52.7	29.4	7.3	-0.8	21.8	22.5	118.5	73.6	45.0	515.4
3rd quarter	525.0	320.7	83.6	54.4	28.8	0.4	1.2	24.1	22.9	119.5	74.5	45.0	524.7
4th quarter	528.6	322.8	86.5	54.4	28.0	4.1	0.6	23.1	22.5	118.6	73.8	44.8	524.5
1960													
1st quarter	542.6	326.4	96.5	56.3	29.0	11.2	2.9	26.1	23.3	117.0	71.0	46.0	531.4
2nd quarter	541.1	332.2	87.1	57.2	26.7	3.2	3.4	26.9	23.5	118.4	71.1	47.3	537.9
3rd quarter	545.6	332.1	86.4	56.2	25.9	4.3	4.7	27.6	22.9	122.4	74.2	48.3	541.3
4th quarter	540.2	334.0	76.0	55.9	25.9	-5.8	5.9	27.6	21.7	124.3	75.3	49.0	546.0
1961													
1st quarter	545.0	334.5	78.4	55.0	25.9	-2.5	5.9	27.6	21.7	126.2	75.3	50.9	547.6
2nd quarter	555.5	339.5	84.1	56.2	26.1	1.8	4.7	26.6	21.9	127.3	76.2	51.1	553.8
3rd quarter	567.7	342.3	90.9	56.7	27.6	6.7	4.5	27.8	23.3	130.0	78.1	51.8	561.0
4th quarter	580.6	349.6	92.9	58.5	28.5	6.0	4.6	28.4	23.9	133.5	79.9	53.6	574.6
1962													
1st quarter	594.0	354.8	98.1	59.7	29.0	9.4	4.0	28.3	24.3	137.1	83.3	53.8	584.6
2nd quarter	600.4	360.5	96.7	61.4	29.9	5.4	4.8	29.7	24.9	138.4	84.1	54.3	594.9
3rd quarter	609.0	364.3	98.2	62.1	29.9	6.2	4.5	29.6	25.1	142.1	87.0	55.0	602.8
4th quarter	612.3	370.6	95.0	61.8	29.8	3.4	3.1	28.7	25.6	143.6	87.6	56.0	608.9
1963													
1st quarter	621.7	374.3	99.7	62.0	30.8	6.9	4.0	29.2	25.2	143.7	86.2	57.5	614.8
2nd quarter	629.8	378.4	101.7	63.9	32.9	4.8	5.6	31.4	25.9	144.1	85.8	58.4	624.9
3rd quarter	644.4	385.4	104.6	65.7	33.2	5.7	4.5	31.2	26.7	149.9	89.9	60.1	638.7
4th quarter	653.9	390.0	107.2	67.7	34.5	5.1	5.7	32.5	26.8	151.0	89.7	61.3	648.9
1964													
1st quarter	669.8	399.6	110.5	69.1	36.2	5.1	7.2	34.2	27.0	152.6	90.1	62.5	664.7
2nd quarter	678.7	407.5	110.5	71.1	35.0	4.5	6.3	34.0	27.7	154.3	90.1	64.2	674.2
3rd quarter	692.0	416.4	112.6	73.4	34.5	4.7	6.9	35.4	28.4	156.0	90.8	65.2	687.3
4th quarter	697.3	419.0	115.0	75.3	34.6	5.0	7.2	36.5	29.3	156.2	90.0	66.2	692.3
1965													
1st quarter	717.8	429.7	126.5	80.2	34.8	11.5	4.5	33.0	28.5	157.0	89.5	67.5	706.2
2nd quarter	730.2	436.6	127.1	83.4	35.1	8.6	6.7	38.4	31.7	159.8	90.1	69.7	721.6
3rd quarter	749.3	445.8	131.2	86.7	35.3	9.3	5.5	37.6	32.0	166.8	94.5	72.3	740.0
4th quarter	771.9	459.7	133.8	90.6	35.5	7.6	5.8	39.7	33.9	172.6	98.7	73.9	764.3
1966													
1st quarter	795.7	470.1	144.2	94.4	35.9	13.9	4.4	39.4	35.0	177.0	101.2	75.8	781.9
2nd quarter	805.0	475.2	143.5	96.8	34.4	12.3	4.6	40.9	36.2	181.7	104.0	77.6	792.7
3rd quarter	819.6	484.3	143.2	98.3	33.0	11.9	2.7	40.9	38.2	189.5	109.9	79.5	807.8
4th quarter	833.3	490.1	145.9	99.2	30.2	16.5	3.8	42.6	38.8	193.6	111.2	82.5	816.8
1967													
1st quarter	844.2	494.3	142.8	98.0	29.4	15.4	4.5	43.9	39.4	202.6	117.9	84.7	828.7
2nd quarter	849.0	503.5	137.5	98.3	32.9	6.3	4.2	43.2	39.0	203.8	117.9	85.9	842.6
3rd quarter	865.2	510.7	142.8	98.8	34.8	9.3	3.3	42.8	39.5	208.4	121.0	87.4	856.0
4th quarter	881.4	518.2	147.7	101.7	37.5	8.4	2.2	43.9	41.7	213.3	123.2	90.1	873.0
1968													
1st quarter	909.4	536.3	152.3	105.6	38.3	8.4	1.1	45.5	44.4	219.7	126.4	93.3	900.9
2nd quarter	934.3	550.0	158.9	105.3	39.6	14.1	1.7	47.2	45.4	223.6	127.2	96.5	920.3
3rd quarter	950.8	566.1	155.7	107.6	40.4	7.7	1.7	49.9	48.2	227.4	128.5	98.8	943.1
4th quarter	968.0	575.0	160.8	112.2	42.6	6.0	0.9	49.1	48.2	231.3	129.7	101.6	962.0
1969													
1st quarter	993.3	587.0	172.4	116.0	44.9	11.5	0.2	44.0	43.8	233.7	129.6	104.1	981.8
2nd quarter	1 009.0	598.3	172.7	118.4	45.1	9.2	1.1	53.8	52.7	236.9	129.8	107.1	999.8
3rd quarter	1 030.0	608.6	177.6	122.4	45.0	10.2	1.2	53.6	52.4	242.5	133.4	109.1	1 019.8
4th quarter	1 038.1	620.6	171.6	123.3	42.5	5.8	3.1	56.3	53.1	242.8	132.0	110.8	1 032.4
1970													
1st quarter	1 051.2	631.7	168.1	123.8	42.5	1.8	3.5	57.0	53.5	247.9	133.6	114.3	1 049.4
2nd quarter	1 067.4	641.6	171.5	125.0	41.4	5.1	5.2	60.4	55.2	249.1	131.8	117.4	1 062.3
3rd quarter	1 086.1	653.5	173.9	126.3	42.6	5.1	4.1	60.5	56.4	254.6	132.4	122.2	1 081.0
4th quarter	1 088.6	660.2	166.8	123.5	47.2	-4.0	3.0	60.9	57.9	258.7	133.5	125.2	1 092.6
1971													
1st quarter	1 135.2	679.2	189.5	126.3	51.0	12.3	4.6	63.2	58.7	261.9	133.3	128.6	1 122.9
2nd quarter	1 156.3	693.2	197.3	129.5	57.0	10.9	-0.4	62.9	63.3	266.1	134.3	131.9	1 145.4
3rd quarter	1 177.7	705.6	202.1	131.2	60.7	10.2	0.2	65.7	65.5	269.8	135.6	134.2	1 167.5
4th quarter	1 190.3	721.7	198.4	134.7	64.0	-0.3	-1.9	60.0	61.9	272.1	134.7	137.4	1 190.6
1972													
1st quarter	1 230.6	738.9	213.0	140.6	69.2	3.2	-3.5	68.6	72.2	282.2	141.4	140.8	1 227.4
2nd quarter	1 266.4	757.4	226.8	144.0	70.8	12.0	-4.3	67.2	71.4	286.5	144.2	142.2	1 254.4
3rd quarter	1 290.6	775.8	233.1	147.0	72.4	13.7	-2.6	71.5	74.1	284.3	138.8	145.6	1 276.9
4th quarter	1 328.9	800.5	239.7	155.0	77.3	7.5	-3.1	76.1	79.2	291.7	142.2	149.6	1 321.4
1973													
1st quarter	1 377.5	825.0	254.3	162.8	80.9	10.6	-1.4	84.0	85.4	299.6	146.4	153.2	1 366.9
2nd quarter	1 413.9	840.5	268.2	171.3	78.7	18.2	2.5	91.9	89.5	302.7	146.5	156.2	1 395.7
3rd quarter	1 433.8	858.9	264.3	176.6	77.9	9.8	6.4	97.6	91.1	304.2	144.2	159.9	1 424.1
4th quarter	1 476.3	873.9	280.9	180.1	75.7	25.0	9.0	107.6	98.7	312.6	147.6	165.0	1 451.3
1974													
1st quarter	1 491.2	891.9	268.4	183.4	72.4	12.5	6.4	116.7	110.3	324.6	152.7	171.9	1 478.7
2nd quarter	1 530.1	920.4	277.4	188.8	71.2	17.4	-2.7	126.7	129.4	335.0	154.9	180.1	1 512.6
3rd quarter	1 560.0	949.3	271.0	194.5	70.9	5.6	-7.0	126.6	133.6	346.7	160.4	186.3	1 554.4
4th quarter	1 599.7	959.1	281.3	197.6	63.3	20.4	0.0	136.6	136.6	359.2	167.4	191.9	1 579.2

Table 1-1B. Gross Domestic Product: Historical Data—*Continued*

(Billions of dollars, quarterly data are at seasonally adjusted annual rates.)

NIPA Tables 1.1.5, 1.2.5

| Year and quarter | Gross domestic product | Personal consump-tion expen-ditures | Gross private domestic investment | | | | Exports and imports of goods and services | | | Government consumption expenditures and gross investment | | | Addendum: Final sales of domestic product |
| | | | Total | Fixed investment | | Change in private inventories | Net exports | Exports | Imports | Total | Federal | State and local | |
				Nonresi-dential	Residential								
1975													
1st quarter	1 616.1	985.2	244.3	193.1	61.2	-10.0	16.5	141.4	124.9	370.1	168.6	201.5	1 626.1
2nd quarter	1 651.9	1 013.6	243.3	193.3	64.0	-14.0	21.6	136.8	115.2	373.4	169.4	204.0	1 665.8
3rd quarter	1 709.8	1 047.2	265.2	197.8	68.8	-1.4	12.0	134.1	122.1	385.4	176.1	209.3	1 711.2
4th quarter	1 761.8	1 076.2	276.2	202.9	73.0	0.3	13.8	142.5	128.7	395.6	180.8	214.8	1 761.5
1976													
1st quarter	1 820.5	1 109.9	304.6	209.5	80.4	14.7	4.7	143.6	138.9	401.3	181.6	219.7	1 805.8
2nd quarter	1 852.3	1 129.5	322.3	215.0	84.8	22.4	-0.5	146.6	147.1	401.0	182.5	218.5	1 829.9
3rd quarter	1 886.6	1 158.8	328.3	222.6	84.9	20.8	-4.1	151.8	155.8	403.5	184.9	218.6	1 865.8
4th quarter	1 934.3	1 192.4	337.7	230.2	96.9	10.5	-6.6	156.1	162.7	410.8	190.2	220.6	1 923.8
1977													
1st quarter	1 988.6	1 228.2	360.3	243.3	102.2	14.8	-21.1	155.4	176.4	421.2	194.2	227.0	1 973.8
2nd quarter	2 055.9	1 256.0	389.7	253.7	116.5	19.5	-21.1	161.9	183.0	431.4	198.9	232.4	2 036.4
3rd quarter	2 118.5	1 286.9	414.1	263.3	120.0	30.9	-20.6	162.3	182.9	438.0	201.9	236.1	2 087.6
4th quarter	2 164.3	1 324.8	422.3	275.9	122.2	24.1	-29.6	157.8	187.4	446.7	206.3	240.5	2 140.1
1978													
1st quarter	2 202.8	1 354.1	434.8	282.4	126.9	25.5	-38.7	164.6	203.3	452.6	208.8	243.8	2 177.3
2nd quarter	2 331.6	1 411.4	470.6	309.4	137.0	24.3	-22.6	186.2	208.8	472.3	217.0	255.3	2 307.4
3rd quarter	2 395.1	1 442.2	492.4	325.1	142.3	25.0	-23.8	191.3	215.1	484.2	222.1	262.2	2 370.1
4th quarter	2 476.9	1 481.4	515.8	341.4	145.8	28.5	-16.4	205.4	221.8	496.2	227.8	268.4	2 448.5
1979													
1st quarter	2 526.6	1 517.1	525.8	356.7	145.3	23.9	-18.2	211.7	229.8	501.8	231.7	270.1	2 502.8
2nd quarter	2 591.2	1 557.6	539.3	364.3	147.6	27.4	-22.2	220.9	243.1	516.5	237.6	278.9	2 563.8
3rd quarter	2 667.6	1 611.9	545.6	383.0	150.5	12.1	-23.0	234.3	257.3	533.1	243.7	289.4	2 655.4
4th quarter	2 723.9	1 655.0	547.9	391.3	148.0	8.6	-26.8	253.7	280.5	547.8	249.3	298.4	2 715.3
1980													
1st quarter	2 789.8	1 702.3	554.6	404.5	140.2	9.9	-35.8	268.5	304.3	568.8	261.1	307.7	2 779.9
2nd quarter	2 707.4	1 704.7	510.0	394.7	116.0	7.0	-15.0	277.4	292.6	586.5	276.5	312.0	2 780.6
3rd quarter	2 856.5	1 763.8	495.1	405.7	123.2	-33.9	5.5	284.7	279.2	592.2	276.1	316.1	2 890.3
4th quarter	2 985.6	1 831.9	551.5	422.8	137.8	-9.1	-6.7	292.5	299.2	608.9	285.8	323.1	2 994.7
1981													
1st quarter	3 124.2	1 885.7	619.4	443.0	137.6	38.8	-14.3	305.5	319.7	633.4	297.2	336.1	3 085.5
2nd quarter	3 162.5	1 917.5	609.8	462.9	135.3	11.7	-13.5	308.5	322.0	648.7	311.9	336.8	3 150.8
3rd quarter	3 260.6	1 958.1	652.3	482.1	126.2	44.0	-7.6	302.3	309.9	657.8	317.4	340.3	3 216.7
4th quarter	3 280.8	1 974.4	643.4	503.8	114.8	24.8	-14.8	304.7	319.4	677.7	329.3	348.4	3 256.0
1982													
1st quarter	3 274.3	2 014.2	588.3	500.1	109.7	-21.5	-16.3	293.2	309.5	688.1	334.9	353.2	3 295.8
2nd quarter	3 332.0	2 039.6	593.6	490.1	107.7	-4.2	-4.4	294.7	299.1	703.1	342.9	360.2	3 336.1
3rd quarter	3 366.3	2 085.7	593.0	478.7	108.4	5.8	-29.6	279.6	309.3	717.3	351.5	365.8	3 360.5
4th quarter	3 402.6	2 145.6	549.2	471.5	117.5	-39.8	-29.6	265.3	294.9	737.4	364.1	373.3	3 442.4
1983													
1st quarter	3 473.4	2 184.6	565.5	462.1	138.5	-35.1	-24.5	270.7	295.3	747.9	370.5	377.4	3 508.5
2nd quarter	3 578.8	2 249.4	613.8	466.5	155.0	-7.7	-45.4	272.5	317.9	761.1	380.3	380.7	3 586.5
3rd quarter	3 689.2	2 319.9	652.3	485.4	171.1	-4.2	-65.2	278.2	343.4	782.2	394.4	387.8	3 693.4
4th quarter	3 794.7	2 372.5	718.5	514.7	179.9	23.9	-71.4	286.6	358.0	775.1	384.2	390.9	3 770.8
1984													
1st quarter	3 908.1	2 418.2	790.9	531.5	186.4	73.0	-95.0	293.0	388.0	794.0	392.4	401.6	3 835.1
2nd quarter	4 009.6	2 475.9	818.9	558.3	191.3	69.3	-104.3	302.2	406.5	819.1	408.3	410.8	3 940.3
3rd quarter	4 084.3	2 513.5	838.9	576.6	190.9	71.3	-103.8	305.7	409.6	835.7	414.0	421.7	4 012.9
4th quarter	4 148.6	2 561.8	831.7	590.9	192.9	48.0	-107.8	308.6	416.4	862.8	432.5	430.2	4 100.6
1985													
1st quarter	4 230.2	2 636.0	809.9	599.4	194.2	16.2	-91.3	306.0	397.3	875.6	434.8	440.8	4 213.9
2nd quarter	4 294.9	2 681.8	827.0	609.1	196.3	21.6	-114.4	304.1	418.6	900.5	447.3	453.2	4 273.2
3rd quarter	4 386.8	2 754.1	822.2	604.4	201.4	16.3	-116.9	297.3	414.2	927.4	463.1	464.3	4 370.5
4th quarter	4 444.1	2 779.4	859.5	618.1	208.4	33.1	-133.4	305.4	438.9	938.6	466.4	472.1	4 411.0
1986													
1st quarter	4 507.9	2 823.6	863.5	613.5	219.5	30.4	-126.0	313.4	439.4	946.8	464.0	482.8	4 477.5
2nd quarter	4 545.3	2 851.5	855.2	605.0	234.6	15.7	-128.9	315.1	444.0	967.5	477.8	489.7	4 529.7
3rd quarter	4 607.7	2 917.2	835.8	602.0	240.9	-7.0	-139.0	320.5	459.4	993.6	495.1	498.5	4 614.7
4th quarter	4 657.6	2 952.8	842.1	610.6	244.3	-12.8	-133.6	335.0	468.6	996.4	489.8	506.6	4 670.4
1987													
1st quarter	4 722.2	2 983.5	871.2	596.6	246.7	28.0	-141.2	336.5	477.7	1 008.7	492.1	516.5	4 694.2
2nd quarter	4 806.2	3 053.3	874.6	608.4	249.7	16.5	-147.0	355.4	502.3	1 025.2	501.2	524.0	4 789.6
3rd quarter	4 884.6	3 117.4	876.5	625.5	250.0	1.0	-145.5	371.9	517.3	1 036.2	504.1	532.1	4 883.5
4th quarter	5 008.0	3 150.9	946.5	630.6	252.8	63.1	-145.4	392.1	537.5	1 056.0	513.7	542.3	4 944.9
1988													
1st quarter	5 073.4	3 231.9	908.6	641.5	250.1	17.0	-124.0	418.7	542.7	1 056.9	505.8	551.1	5 056.4
2nd quarter	5 190.0	3 291.7	934.5	659.4	255.5	19.6	-106.6	439.5	546.1	1 070.4	506.9	563.5	5 170.4
3rd quarter	5 282.8	3 361.9	942.0	666.3	257.5	18.2	-99.3	453.6	552.8	1 078.2	507.4	570.8	5 264.7
4th quarter	5 399.5	3 434.5	962.7	681.9	261.7	19.1	-107.7	466.6	574.3	1 109.9	525.6	584.3	5 380.4
1989													
1st quarter	5 511.3	3 490.2	1 005.5	696.5	260.9	48.1	-101.0	485.2	586.2	1 116.6	519.9	596.7	5 463.1
2nd quarter	5 612.5	3 553.8	1 001.0	709.0	255.8	36.3	-88.2	507.2	595.4	1 145.8	534.3	611.5	5 576.2
3rd quarter	5 695.4	3 609.4	996.5	731.1	255.5	9.8	-75.1	509.4	584.4	1 164.6	541.4	623.2	5 685.5
4th quarter	5 747.2	3 653.7	995.8	727.4	251.9	16.6	-82.8	515.4	598.2	1 180.5	540.8	639.7	5 730.7
1990													
1st quarter	5 872.7	3 737.9	1 010.8	740.9	256.0	14.0	-88.5	538.2	626.8	1 212.5	553.7	658.8	5 858.7
2nd quarter	5 960.0	3 783.4	1 014.7	734.1	246.9	33.7	-68.8	545.9	614.8	1 230.7	563.9	666.8	5 926.3
3rd quarter	6 015.1	3 846.7	1 000.8	744.4	234.5	21.9	-75.0	555.1	630.1	1 242.6	562.2	680.3	5 993.3
4th quarter	6 004.7	3 867.9	947.5	737.5	221.3	-11.3	-79.1	568.2	647.3	1 268.5	569.7	698.8	6 016.1

Table 1-1B. Gross Domestic Product: Historical Data—*Continued*

(Billions of dollars, quarterly data are at seasonally adjusted annual rates.)   **NIPA Tables 1.1.5, 1.2.5**

Year and quarter	Gross domestic product	Personal consumption expenditures	Gross private domestic investment				Exports and imports of goods and services			Government consumption expenditures and gross investment			Addendum: Final sales of domestic product
			Total	Fixed investment		Change in private inventories	Net exports	Exports	Imports	Total	Federal	State and local	
				Nonresidential	Residential								
1991													
1st quarter	6 035.2	3 873.6	924.6	729.8	210.3	-15.5	-47.1	573.2	620.3	1 284.2	581.4	702.8	6 050.7
2nd quarter	6 126.9	3 926.9	926.5	726.8	217.8	-18.0	-23.2	590.7	613.9	1 296.6	586.6	709.9	6 144.9
3rd quarter	6 205.9	3 973.3	947.5	720.2	226.5	0.8	-21.1	600.6	621.7	1 306.3	586.3	719.9	6 205.1
4th quarter	6 264.5	4 000.0	978.8	717.6	230.1	31.1	-23.1	615.2	638.3	1 308.8	577.4	731.4	6 233.4
1992													
1st quarter	6 363.1	4 100.4	956.8	714.2	242.4	0.2	-20.5	625.3	645.8	1 326.4	580.3	746.1	6 362.9
2nd quarter	6 470.8	4 155.7	1 013.1	736.7	253.2	23.2	-32.8	626.2	659.0	1 334.8	580.9	753.9	6 447.6
3rd quarter	6 566.6	4 227.0	1 024.2	748.5	255.1	20.5	-38.5	639.4	677.9	1 354.0	594.2	759.8	6 546.1
4th quarter	6 680.8	4 307.2	1 058.0	768.3	268.3	21.3	-47.1	641.4	688.5	1 362.8	598.4	764.4	6 659.5
1993													
1st quarter	6 729.5	4 349.5	1 083.8	776.5	271.4	35.9	-55.7	643.6	699.3	1 351.8	580.3	771.5	6 693.5
2nd quarter	6 808.9	4 418.6	1 094.5	792.4	278.0	24.1	-63.2	653.1	716.3	1 359.1	576.7	782.3	6 784.8
3rd quarter	6 882.1	4 487.2	1 095.9	798.3	290.9	6.6	-68.4	650.9	719.3	1 367.4	578.7	788.7	6 875.5
4th quarter	7 013.7	4 552.7	1 153.1	829.6	306.9	16.6	-73.4	671.6	745.0	1 381.4	584.9	796.5	6 997.1
1994													
1st quarter	7 115.7	4 621.2	1 201.7	840.7	315.6	45.4	-80.6	681.2	761.8	1 373.4	567.0	806.3	7 070.3
2nd quarter	7 246.9	4 683.2	1 264.9	855.6	327.9	81.4	-90.6	707.0	797.6	1 389.4	569.4	820.0	7 165.5
3rd quarter	7 331.1	4 752.8	1 251.7	872.1	326.4	53.2	-96.9	736.9	833.8	1 423.4	586.5	836.9	7 277.9
4th quarter	7 455.3	4 826.7	1 307.6	907.0	325.4	75.1	-101.9	758.6	860.6	1 422.9	575.8	847.1	7 380.1
1995													
1st quarter	7 522.3	4 862.4	1 327.6	944.6	321.8	61.2	-105.3	781.6	886.9	1 437.6	579.1	858.5	7 461.1
2nd quarter	7 581.0	4 933.6	1 304.0	956.8	313.5	33.8	-109.5	798.9	908.3	1 452.9	581.0	871.9	7 547.2
3rd quarter	7 683.1	4 998.7	1 303.2	965.5	326.4	11.3	-74.4	831.4	905.8	1 455.7	579.3	876.3	7 671.8
4th quarter	7 772.6	5 055.7	1 335.1	982.1	334.6	18.4	-69.8	839.4	909.2	1 451.6	567.3	884.3	7 754.2
1996													
1st quarter	7 868.5	5 130.6	1 355.4	1 003.7	344.7	6.9	-88.8	847.9	936.7	1 471.3	579.8	891.5	7 861.6
2nd quarter	8 032.8	5 220.5	1 418.4	1 026.5	361.4	30.5	-93.7	859.0	952.8	1 487.7	582.1	905.5	8 002.3
3rd quarter	8 131.4	5 274.5	1 474.4	1 059.0	364.3	51.1	-114.2	859.6	973.8	1 496.7	577.8	919.0	8 080.3
4th quarter	8 259.8	5 352.8	1 480.1	1 083.6	361.8	34.7	-88.8	903.8	992.6	1 515.7	576.9	938.8	8 225.0
1997													
1st quarter	8 362.7	5 433.1	1 522.4	1 107.3	365.4	49.7	-108.8	918.4	1 027.2	1 516.0	570.7	945.3	8 312.9
2nd quarter	8 518.8	5 471.3	1 590.2	1 129.6	372.3	88.4	-85.2	954.5	1 039.7	1 542.5	587.2	955.4	8 430.5
3rd quarter	8 662.8	5 579.2	1 625.3	1 178.4	379.0	67.9	-96.8	974.1	1 070.9	1 555.2	586.0	969.2	8 594.9
4th quarter	8 765.9	5 663.6	1 644.5	1 181.1	385.8	77.7	-117.0	968.3	1 085.3	1 574.8	589.2	985.6	8 688.3
1998													
1st quarter	8 866.5	5 721.3	1 712.3	1 212.4	394.8	105.1	-135.2	963.0	1 098.2	1 568.0	572.2	995.9	8 761.4
2nd quarter	8 969.7	5 832.6	1 695.8	1 247.1	411.3	37.3	-162.3	947.3	1 109.6	1 603.7	587.1	1 016.6	8 932.4
3rd quarter	9 121.1	5 926.8	1 741.6	1 261.7	427.6	52.4	-174.6	935.3	1 109.9	1 627.3	588.6	1 038.6	9 068.7
4th quarter	9 294.0	6 028.2	1 797.0	1 295.4	441.5	60.0	-178.7	966.3	1 145.0	1 647.5	594.2	1 053.2	9 234.0
1999													
1st quarter	9 417.3	6 102.5	1 853.1	1 322.3	447.4	83.4	-207.7	962.4	1 170.1	1 669.4	595.5	1 073.9	9 333.9
2nd quarter	9 524.2	6 225.3	1 848.3	1 354.0	459.3	35.1	-244.7	973.5	1 218.2	1 695.2	599.8	1 095.4	9 489.0
3rd quarter	9 681.9	6 328.9	1 893.7	1 386.6	466.6	40.5	-275.3	1 003.2	1 278.5	1 734.5	614.9	1 119.6	9 641.3
4th quarter	9 899.4	6 459.6	1 953.1	1 395.0	473.8	84.2	-295.6	1 032.0	1 327.7	1 782.3	635.2	1 147.1	9 815.1
2000													
1st quarter	10 002.9	6 613.6	1 950.7	1 450.3	484.2	16.2	-352.1	1 054.2	1 406.3	1 790.7	620.4	1 170.4	9 986.7
2nd quarter	10 247.7	6 707.5	2 075.8	1 498.7	486.6	90.4	-358.7	1 092.5	1 451.2	1 823.1	642.0	1 181.1	10 157.2
3rd quarter	10 319.8	6 815.4	2 060.0	1 519.7	483.1	57.2	-387.8	1 124.3	1 512.1	1 832.3	634.1	1 198.3	10 262.6
4th quarter	10 439.0	6 912.1	2 067.2	1 525.1	487.8	54.3	-401.5	1 114.1	1 515.6	1 861.2	638.4	1 222.9	10 384.7
2001													
1st quarter	10 472.9	6 986.9	1 971.3	1 505.2	496.7	-30.6	-390.8	1 095.1	1 485.8	1 905.4	653.1	1 252.3	10 503.5
2nd quarter	10 597.8	7 036.3	1 973.0	1 473.6	511.0	-11.6	-358.5	1 052.6	1 411.1	1 947.0	666.1	1 280.9	10 609.4
3rd quarter	10 596.3	7 064.7	1 944.9	1 452.6	522.4	-30.1	-366.0	996.4	1 362.4	1 952.7	674.3	1 278.4	10 626.4
4th quarter	10 660.3	7 174.7	1 850.1	1 408.9	522.1	-80.8	-356.4	954.5	1 310.9	1 992.0	686.8	1 305.2	10 741.1
2002													
1st quarter	10 789.0	7 209.9	1 912.7	1 374.0	538.3	0.3	-372.6	970.7	1 343.3	2 038.9	713.9	1 325.0	10 788.7
2nd quarter	10 893.2	7 302.1	1 933.3	1 357.3	554.8	21.2	-415.7	1 003.9	1 419.6	2 073.5	734.7	1 338.8	10 872.0
3rd quarter	10 992.1	7 390.9	1 933.2	1 348.9	558.9	25.4	-432.5	1 017.2	1 449.7	2 100.4	748.2	1 352.2	10 966.6
4th quarter	11 071.5	7 467.7	1 942.5	1 331.2	578.3	33.0	-480.8	1 003.2	1 484.0	2 142.0	775.1	1 366.9	11 038.5
2003													
1st quarter	11 183.5	7 555.8	1 960.2	1 332.7	601.4	26.1	-504.9	1 008.2	1 513.0	2 172.4	792.3	1 380.0	11 157.4
2nd quarter	11 312.9	7 642.6	1 972.4	1 366.8	612.0	-6.5	-501.5	1 007.9	1 509.5	2 199.4	825.5	1 374.0	11 319.4
3rd quarter	11 567.3	7 802.6	2 044.3	1 392.1	651.7	0.4	-500.8	1 038.1	1 538.9	2 221.2	832.7	1 388.5	11 566.9
4th quarter	11 769.3	7 891.5	2 131.3	1 411.9	683.1	36.3	-505.4	1 090.5	1 595.8	2 251.8	854.6	1 397.3	11 732.9
2004													
1st quarter	11 920.2	8 027.7	2 154.1	1 401.8	706.2	46.0	-549.0	1 137.7	1 686.7	2 287.3	871.3	1 416.0	11 874.1
2nd quarter	12 109.0	8 133.0	2 262.6	1 445.5	743.2	74.0	-608.0	1 167.8	1 775.8	2 321.4	884.2	1 437.2	12 035.0
3rd quarter	12 303.3	8 264.3	2 318.3	1 490.5	763.6	64.2	-636.4	1 182.7	1 819.1	2 357.2	902.2	1 455.0	12 239.2
4th quarter	12 522.4	8 425.6	2 390.1	1 531.7	786.3	72.1	-682.9	1 222.3	1 905.2	2 389.7	909.3	1 480.3	12 450.3
2005													
1st quarter	12 761.3	8 523.0	2 486.1	1 568.3	815.3	102.5	-674.6	1 264.8	1 939.4	2 426.9	931.5	1 495.4	12 658.9
2nd quarter	12 910.0	8 671.4	2 476.5	1 603.8	843.8	28.8	-690.8	1 296.0	1 986.8	2 452.9	939.0	1 513.9	12 881.2
3rd quarter	13 142.9	8 849.2	2 531.1	1 643.1	875.6	12.4	-732.5	1 307.1	2 039.6	2 495.1	956.1	1 539.0	13 130.5
4th quarter	13 332.3	8 944.9	2 645.3	1 668.7	890.2	86.4	-786.9	1 353.0	2 139.9	2 529.1	963.3	1 565.8	13 245.9
2006													
1st quarter	13 603.9	9 090.7	2 709.7	1 735.3	896.2	78.2	-777.2	1 413.0	2 190.2	2 580.7	996.6	1 584.1	13 525.7
2nd quarter	13 749.8	9 210.2	2 709.3	1 774.4	859.3	75.5	-780.6	1 460.0	2 240.6	2 610.9	996.6	1 614.3	13 674.3
3rd quarter	13 867.5	9 333.0	2 709.4	1 815.9	814.7	78.8	-805.6	1 477.8	2 283.5	2 630.7	994.9	1 635.7	13 788.7
4th quarter	14 037.2	9 407.5	2 675.4	1 849.6	782.4	43.5	-720.3	1 539.6	2 259.9	2 674.7	1 014.6	1 660.1	13 993.8

Table 1-1B. Gross Domestic Product: Historical Data—*Continued*

(Billions of dollars, quarterly data are at seasonally adjusted annual rates.)

NIPA Tables 1.1.5, 1.2.5

Year and quarter	Gross domestic product	Personal consump-tion expen-ditures	Gross private domestic investment				Exports and imports of goods and services			Government consumption expenditures and gross investment			Addendum: Final sales of domestic product
			Total	Fixed investment		Change in private inventories	Net exports	Exports	Imports	Total	Federal	State and local	
				Nonresi-dential	Residential								
2007													
1st quarter	14 208.6	9 549.4	2 664.3	1 892.0	750.9	21.4	-724.3	1 577.1	2 301.4	2 719.2	1 017.2	1 702.0	14 187.2
2nd quarter	14 382.4	9 644.7	2 699.2	1 937.7	719.3	42.2	-731.9	1 622.1	2 354.0	2 770.3	1 042.0	1 728.3	14 340.1
3rd quarter	14 535.0	9 753.8	2 686.0	1 967.4	673.7	44.9	-713.8	1 683.7	2 397.5	2 809.0	1 058.3	1 750.7	14 490.1
4th quarter	14 681.5	9 877.8	2 642.6	1 997.1	618.2	27.3	-703.7	1 760.5	2 464.2	2 864.9	1 084.6	1 780.3	14 654.2
2008													
1st quarter	14 651.0	9 934.3	2 563.7	2 013.7	566.7	-16.7	-756.2	1 811.0	2 567.2	2 909.3	1 110.3	1 799.0	14 667.7
2nd quarter	14 805.6	10 052.8	2 540.6	2 024.0	538.8	-22.3	-758.9	1 909.7	2 668.7	2 971.1	1 145.5	1 825.6	14 827.9
3rd quarter	14 835.2	10 081.0	2 498.2	2 007.0	507.1	-15.8	-771.6	1 921.8	2 693.3	3 027.5	1 168.7	1 858.9	14 851.0
4th quarter	14 559.5	9 837.3	2 307.9	1 918.7	451.4	-62.2	-605.7	1 705.7	2 311.4	3 020.0	1 177.9	1 842.2	14 621.8
2009													
1st quarter	14 394.5	9 756.1	2 014.9	1 761.4	404.5	-151.0	-396.1	1 514.3	1 910.4	3 019.7	1 183.0	1 836.7	14 545.5
2nd quarter	14 352.9	9 760.2	1 863.7	1 685.0	374.7	-196.0	-338.6	1 518.3	1 856.9	3 067.6	1 210.8	1 856.7	14 548.9
3rd quarter	14 420.3	9 895.4	1 841.4	1 656.0	389.4	-204.0	-405.5	1 591.1	1 996.6	3 089.0	1 225.5	1 863.5	14 624.3
4th quarter	14 628.0	9 957.1	1 998.7	1 659.3	391.5	-52.1	-445.6	1 704.3	2 149.8	3 117.8	1 253.4	1 864.4	14 680.1
2010													
1st quarter	14 721.4	10 040.5	2 038.2	1 660.0	379.4	-1.2	-489.2	1 746.9	2 236.1	3 131.9	1 275.7	1 856.2	14 722.6
2nd quarter	14 926.1	10 131.8	2 148.8	1 715.7	396.3	36.7	-519.1	1 810.0	2 329.1	3 164.7	1 302.6	1 862.1	14 889.4
3rd quarter	15 079.9	10 220.6	2 236.5	1 762.4	361.2	112.9	-535.1	1 865.6	2 400.7	3 157.9	1 302.3	1 855.6	14 967.0
4th quarter	15 240.8	10 350.5	2 238.4	1 801.9	369.3	67.3	-512.2	1 962.6	2 474.8	3 164.1	1 311.1	1 853.0	15 173.5
2011													
1st quarter	15 285.8	10 485.4	2 206.0	1 805.1	368.8	32.1	-561.5	2 030.3	2 591.7	3 156.0	1 304.7	1 851.2	15 253.8
2nd quarter	15 496.2	10 612.1	2 297.4	1 862.0	374.3	61.0	-581.9	2 105.1	2 686.9	3 168.6	1 311.8	1 856.7	15 435.1
3rd quarter	15 591.9	10 705.4	2 322.8	1 953.8	381.1	-12.0	-573.9	2 138.8	2 712.6	3 137.5	1 288.0	1 849.5	15 603.9
4th quarter	15 796.5	10 761.6	2 504.1	2 009.0	391.2	104.0	-600.7	2 137.8	2 738.5	3 131.4	1 291.2	1 840.3	15 692.5
2012													
1st quarter	16 019.8	10 922.4	2 567.8	2 073.4	414.1	80.3	-615.1	2 164.6	2 779.7	3 144.7	1 295.6	1 849.0	15 939.5
2nd quarter	16 152.3	10 964.9	2 636.9	2 126.2	419.3	91.4	-580.5	2 192.1	2 772.6	3 131.0	1 288.2	1 842.9	16 060.9
3rd quarter	16 257.2	11 014.2	2 644.1	2 126.0	433.7	84.4	-540.8	2 201.8	2 742.6	3 139.6	1 293.3	1 846.3	16 172.7
4th quarter	16 358.9	11 125.7	2 638.3	2 148.6	460.9	28.7	-537.8	2 206.6	2 744.5	3 132.7	1 269.1	1 863.7	16 330.1
2013													
1st quarter	16 569.6	11 223.2	2 738.2	2 170.9	485.1	82.3	-516.9	2 239.2	2 756.0	3 125.0	1 240.0	1 885.0	16 487.3
2nd quarter	16 637.9	11 239.6	2 775.3	2 180.3	507.2	87.8	-508.9	2 250.7	2 759.6	3 132.0	1 232.3	1 899.6	16 550.2
3rd quarter	16 848.7	11 330.9	2 880.0	2 220.7	523.1	136.2	-496.3	2 269.4	2 765.7	3 134.1	1 218.4	1 915.7	16 712.5
4th quarter	17 083.1	11 475.1	2 910.5	2 274.0	524.6	111.9	-441.1	2 334.4	2 775.5	3 138.5	1 215.6	1 923.0	16 971.3
2014													
1st quarter	17 104.6	11 574.2	2 899.2	2 314.9	532.1	52.3	-506.3	2 335.9	2 842.2	3 137.4	1 211.0	1 926.4	17 052.3
2nd quarter	17 432.9	11 756.9	3 030.4	2 383.6	550.4	96.4	-507.6	2 387.1	2 894.8	3 153.3	1 209.0	1 944.2	17 336.5
3rd quarter	17 721.7	11 915.4	3 107.6	2 439.8	568.4	99.4	-492.3	2 391.9	2 884.2	3 190.9	1 228.2	1 962.7	17 622.2
4th quarter	17 849.9	12 044.5	3 139.5	2 462.0	589.7	87.8	-524.4	2 371.9	2 896.3	3 190.3	1 211.7	1 978.6	17 762.1
2015													
1st quarter	17 984.2	12 091.6	3 235.7	2 459.6	605.2	171.0	-532.5	2 287.2	2 819.7	3 189.4	1 214.4	1 975.0	17 813.2
2nd quarter	18 219.4	12 248.0	3 231.8	2 465.7	623.4	142.7	-499.3	2 304.3	2 803.6	3 238.9	1 220.4	2 018.5	18 076.7
3rd quarter	18 344.7	12 376.2	3 241.2	2 471.6	646.1	123.6	-533.0	2 260.1	2 793.1	3 260.3	1 222.4	2 037.9	18 221.1
4th quarter	18 350.8	12 421.3	3 183.6	2 432.8	660.5	90.2	-514.6	2 215.6	2 730.2	3 260.5	1 229.0	2 031.6	18 260.6
2016													
1st quarter	18 424.3	12 523.5	3 149.1	2 415.6	686.6	46.9	-522.2	2 164.9	2 687.1	3 273.8	1 227.5	2 046.3	18 377.4
2nd quarter	18 637.3	12 688.3	3 152.9	2 441.8	692.0	19.1	-495.3	2 208.1	2 703.4	3 291.4	1 226.2	2 065.2	18 618.1
3rd quarter	18 806.7	12 822.4	3 166.6	2 471.6	697.7	-2.7	-499.7	2 254.4	2 754.1	3 317.5	1 237.5	2 080.0	18 809.5
4th quarter	18 991.9	12 959.8	3 246.2	2 483.5	717.8	44.9	-558.0	2 255.1	2 813.1	3 343.9	1 245.2	2 098.7	18 946.9
2017													
1st quarter	19 190.4	13 104.4	3 288.2	2 531.1	743.7	13.4	-570.9	2 303.3	2 874.2	3 368.7	1 248.4	2 120.3	19 177.0
2nd quarter	19 356.6	13 212.5	3 335.0	2 567.4	748.8	18.8	-583.7	2 313.2	2 896.9	3 392.9	1 263.6	2 129.3	19 337.8
3rd quarter	19 611.7	13 345.1	3 401.8	2 591.6	753.4	56.8	-550.6	2 360.1	2 910.7	3 415.4	1 270.2	2 145.2	19 554.9
4th quarter	19 918.9	13 586.3	3 457.7	2 648.9	777.1	31.7	-596.1	2 450.3	3 046.5	3 471.0	1 295.1	2 175.9	19 887.2
2018													
1st quarter	20 163.2	13 728.4	3 542.4	2 717.3	783.7	41.5	-629.0	2 476.6	3 105.6	3 521.4	1 318.2	2 203.2	20 121.7
2nd quarter	20 510.2	13 939.8	3 561.6	2 782.0	789.5	-10.0	-568.4	2 543.6	3 112.0	3 577.1	1 340.4	2 236.7	20 520.1
3rd quarter	20 749.8	14 114.6	3 684.0	2 807.7	789.0	87.3	-671.4	2 510.3	3 181.6	3 622.6	1 358.6	2 263.9	20 662.4
4th quarter	20 897.8	14 211.9	3 725.2	2 840.7	784.4	100.1	-684.1	2 510.5	3 194.7	3 644.8	1 371.8	2 273.0	20 797.7

Table 1-2A. Real Gross Domestic Product: Recent Data

(Billions of chained [2012] dollars, quarterly data are at seasonally adjusted annual rates.)

NIPA Tables 1.1.6, 1.2.6

Year and quarter	Gross domestic product	Personal consump-tion expen-ditures	Gross private domestic investment					Exports and imports of goods and services			Government consumption expen-ditures and gross investment			Residual	Adden-dum: final sales of domestic product
			Total	Fixed investment		Change in private inven-tories		Net exports	Exports	Imports	Total	Federal	State and local		
				Nonresi-dential	Resi-dential										
1955	2 871.2	1 740.0	331.1	...	...	...	...	...	79.6	102.4	971.6	...	...	-169.3	2 866.6
1956	2 932.4	1 790.8	330.2	...	...	...	...	...	92.7	110.7	973.9	...	...	-159.9	2 936.4
1957	2 994.1	1 835.2	317.8	...	...	...	...	...	100.8	115.3	1 018.7	...	...	-166.9	3 014.2
1958	2 972.0	1 851.1	294.8	...	...	...	...	...	87.2	120.8	1 052.9	...	...	-192.2	2 997.1
1959	3 178.2	1 956.7	351.4	...	...	...	...	...	96.2	133.5	1 079.6	...	...	-190.2	3 180.0
1960	3 260.0	2 010.4	352.7	...	...	...	...	...	112.9	135.3	1 086.4	...	...	-182.4	3 266.5
1961	3 343.5	2 051.6	353.8	...	...	...	...	...	113.5	134.4	1 145.2	...	...	-199.9	3 352.4
1962	3 548.4	2 153.1	395.9	...	...	...	...	...	119.2	149.7	1 220.1	...	...	-218.6	3 540.2
1963	3 702.9	2 241.9	422.8	...	...	...	...	...	127.7	153.7	1 249.7	...	...	-211.6	3 698.6
1964	3 916.3	2 375.3	456.1	...	...	...	...	...	142.8	161.9	1 280.3	...	...	-198.7	3 918.1
1965	4 170.8	2 526.0	519.2	...	...	...	...	...	146.8	179.1	1 321.3	...	...	-206.0	4 149.2
1966	4 445.9	2 669.3	565.9	...	...	...	...	...	156.9	205.7	1 436.4	...	...	-237.8	4 402.8
1967	4 567.8	2 749.1	546.1	...	...	...	...	...	160.5	220.7	1 550.4	...	...	-261.4	4 546.7
1968	4 792.3	2 907.5	578.7	...	...	...	...	...	173.2	253.6	1 603.2	...	...	-256.4	4 777.1
1969	4 942.1	3 015.9	610.9	...	...	...	...	...	181.6	268.0	1 604.8	...	...	-242.3	4 927.8
1970	4 951.3	3 086.9	573.8	...	...	...	...	...	201.0	279.4	1 571.0	...	...	-209.2	4 971.9
1971	5 114.3	3 204.8	632.9	...	...	...	...	...	204.5	294.4	1 541.5	...	...	-207.2	5 108.2
1972	5 383.3	3 401.0	704.2	...	...	...	...	...	220.4	327.5	1 533.6	...	...	-181.7	5 375.9
1973	5 687.2	3 569.4	781.3	...	...	...	...	...	261.9	342.7	1 528.7	...	...	-161.7	5 655.4
1974	5 656.5	3 539.5	729.5	...	...	...	...	...	282.7	334.9	1 562.3	...	...	-160.2	5 639.3
1975	5 644.8	3 619.7	611.4	...	...	...	...	...	280.8	297.7	1 596.7	...	...	-149.5	5 698.0
1976	5 949.0	3 821.5	728.0	...	...	...	...	...	293.1	355.9	1 605.0	...	...	-187.0	5 925.8
1977	6 224.1	3 983.0	831.9	...	...	...	...	...	300.2	394.8	1 624.3	...	...	-176.2	6 187.7
1978	6 568.6	4 157.3	928.1	...	...	...	...	...	331.8	429.0	1 670.9	...	...	-149.4	6 526.5
1979	6 776.6	4 256.1	960.7	...	...	...	...	...	364.7	436.2	1 701.0	...	...	-105.9	6 761.6
1980	6 759.2	4 242.8	864.0	...	...	...	...	...	404.0	407.1	1 731.8	...	...	-64.1	6 804.7
1981	6 930.7	4 301.6	940.1	...	...	...	...	...	408.9	417.8	1 748.5	...	...	-101.0	6 900.1
1982	6 805.8	4 364.6	822.0	...	...	...	...	...	377.6	412.5	1 780.0	...	...	-100.4	6 865.5
1983	7 117.7	4 611.7	898.7	...	...	...	...	...	367.8	464.5	1 846.4	...	...	-132.3	7 159.9
1984	7 632.8	4 854.3	1 144.0	...	...	...	...	...	397.8	577.6	1 911.3	...	...	-197.5	7 541.5
1985	7 951.1	5 105.6	1 143.2	...	...	...	...	...	411.1	615.1	2 038.8	...	...	-165.0	7 940.2
1986	8 226.4	5 316.4	1 145.0	...	...	...	...	...	442.6	667.5	2 149.5	...	...	-170.3	8 241.0
1987	8 511.0	5 496.9	1 177.6	...	...	...	...	...	491.0	707.2	2 211.9	...	...	-201.8	8 492.5
1988	8 866.5	5 726.5	1 206.6	...	...	...	...	...	570.6	735.0	2 239.4	...	...	-169.6	8 860.6
1989	9 192.1	5 893.5	1 255.4	...	...	...	...	...	636.6	767.3	2 303.0	...	...	-168.1	9 172.3
1990	9 365.5	6 012.2	1 223.0	...	...	...	...	...	692.8	794.8	2 376.7	...	...	-165.5	9 365.1
1991	9 355.4	6 023.0	1 142.1	...	...	...	...	...	738.6	793.6	2 405.2	...	...	-157.8	9 379.0
1992	9 684.9	6 244.7	1 225.3	...	...	...	...	...	789.8	849.2	2 416.4	...	...	-164.9	9 683.1
1993	9 951.5	6 462.2	1 323.3	...	...	...	...	...	815.7	922.6	2 396.6	...	...	-151.9	9 943.6
1994	10 352.4	6 712.6	1 479.8	...	...	...	...	...	887.7	1 032.7	2 398.6	...	...	-182.1	10 284.1
1995	10 630.3	6 910.7	1 527.3	...	...	...	...	...	979.0	1 115.3	2 410.8	...	...	-122.7	10 608.5
1996	11 031.4	7 150.5	1 660.8	...	...	...	...	...	1 059.0	1 212.3	2 433.1	...	...	-99.5	11 008.6
1997	11 521.9	7 419.7	1 850.2	...	...	...	...	...	1 185.2	1 375.6	2 473.0	...	...	-126.0	11 442.2
1998	12 038.3	7 813.8	2 026.3	...	...	...	...	...	1 212.8	1 536.4	2 533.0	...	...	-103.1	11 966.9
1999	12 610.5	8 225.4	2 198.7	...	...	...	...	...	1 273.3	1 710.0	2 615.4	...	...	-79.7	12 543.1
2000	13 131.0	8 643.4	2 346.7	...	...	...	...	...	1 379.5	1 930.3	2 663.0	...	...	-50.0	13 068.0
2001	13 262.1	8 861.1	2 214.6	...	...	...	...	...	1 299.7	1 876.2	2 762.3	...	...	53.8	13 309.5
2002	13 493.1	9 088.7	2 195.5	1 472.7	692.6	24.3	-667.3		1 277.1	1 944.4	2 885.2	969.6	1 927.6	-86.1	13 476.4
2003	13 879.1	9 377.5	2 290.4	1 509.4	755.5	19.9	-735.0		1 305.0	2 040.1	2 947.2	1 032.7	1 922.2	-47.5	13 864.7
2004	14 406.4	9 729.3	2 502.6	1 594.0	830.9	82.6	-841.4		1 431.2	2 272.6	2 992.7	1 077.5	1 920.1	-5.9	14 335.7
2005	14 912.5	10 075.9	2 670.6	1 716.4	885.4	63.7	-887.8		1 533.2	2 421.0	3 015.5	1 099.1	1 920.1	35.1	14 852.3
2006	15 338.3	10 384.5	2 752.4	1 854.2	818.9	87.1	-905.0		1 676.4	2 581.5	3 063.5	1 125.0	1 941.6	31.8	15 263.0
2007	15 626.0	10 615.3	2 684.1	1 982.1	665.8	40.6	-823.6		1 822.3	2 646.0	3 118.6	1 147.0	1 974.7	25.7	15 588.7
2008	15 604.7	10 592.8	2 462.9	1 994.2	504.6	-32.7	-661.6		1 925.4	2 587.1	3 195.6	1 218.8	1 978.7	-2.4	15 639.7
2009	15 208.8	10 460.0	1 942.0	1 704.3	395.3	-177.3	-484.8		1 763.8	2 248.6	3 307.3	1 293.0	2 015.6	-15.5	15 373.0
2010	15 598.8	10 643.0	2 216.5	1 781.0	383.0	57.3	-565.9		1 977.9	2 543.8	3 307.2	1 346.1	1 961.3	-11.3	15 546.6
2011	15 840.7	10 843.8	2 362.1	1 935.4	382.5	46.7	-568.1		2 119.0	2 687.1	3 203.3	1 311.1	1 892.2	-3.7	15 796.5
2012	16 197.0	11 006.8	2 621.8	2 118.5	432.0	71.2	-568.6		2 191.3	2 759.9	3 137.0	1 286.5	1 850.5	0.0	16 125.8
2013	16 495.4	11 166.9	2 801.5	2 206.0	485.5	108.7	-532.8		2 269.6	2 802.4	3 061.0	1 215.3	1 845.3	-0.4	16 386.2
2014	16 912.0	11 497.4	2 959.2	2 365.3	504.1	86.3	-577.2		2 365.3	2 942.5	3 033.4	1 183.8	1 848.6	0.7	16 822.3
2015	17 403.8	11 921.2	3 104.3	2 408.2	555.3	132.4	-721.6		2 376.5	3 098.1	3 091.8	1 182.7	1 907.5	10.7	17 267.1
2016	17 688.9	12 247.5	3 064.0	2 425.3	591.2	23.0	-783.7		2 376.1	3 159.8	3 147.7	1 187.8	1 957.9	21.1	17 647.6
2017	18 108.1	12 566.9	3 198.9	2 531.2	611.9	31.7	-849.8		2 458.8	3 308.5	3 169.6	1 197.0	1 970.6	17.2	18 058.4
2018	18 638.2	12 944.6	3 360.5	2 692.3	602.9	48.1	-920.0		2 532.9	3 453.0	3 223.9	1 232.2	1 990.0	2.1	18 571.3
2016															
1st quarter	17 556.8	12 124.2	3 054.7	2 389.8	593.0	51.1	-777.7		2 345.1	3 122.7	3 143.0	1 190.6	1 950.5	18.4	17 492.6
2nd quarter	17 639.4	12 211.3	3 041.6	2 413.6	590.1	10.8	-760.9		2 367.9	3 128.9	3 137.5	1 182.5	1 953.0	22.1	17 607.5
3rd quarter	17 735.1	12 289.1	3 045.5	2 446.8	586.2	-14.7	-761.4		2 403.4	3 164.9	3 151.0	1 188.2	1 960.8	19.8	17 726.7
4th quarter	17 824.2	12 365.3	3 114.0	2 451.2	595.5	44.8	-834.6		2 388.1	3 222.7	3 159.3	1 189.9	1 967.4	24.2	17 763.5
2017															
1st quarter	17 925.3	12 438.9	3 140.3	2 490.5	612.4	8.7	-831.5		2 423.5	3 255.0	3 157.3	1 186.4	1 968.9	29.5	17 895.1
2nd quarter	18 021.0	12 512.9	3 167.9	2 517.4	608.9	16.6	-850.0		2 432.9	3 282.9	3 168.0	1 195.9	1 970.1	20.6	17 985.3
3rd quarter	18 163.6	12 586.3	3 225.2	2 532.6	605.9	70.2	-833.7		2 459.5	3 293.2	3 167.1	1 196.1	1 969.0	3.6	18 082.5
4th quarter	18 322.5	12 729.7	3 262.1	2 584.2	620.4	31.1	-883.8		2 519.2	3 403.0	3 186.1	1 209.8	1 974.5	15.3	18 270.7
2018															
1st quarter	18 438.3	12 782.9	3 311.8	2 639.5	612.1	40.5	-884.2		2 524.0	3 408.2	3 201.1	1 218.1	1 981.2	8.2	18 380.4
2nd quarter	18 598.1	12 909.2	3 296.6	2 689.9	606.3	-28.0	-850.5		2 559.9	3 410.4	3 221.4	1 229.9	1 989.9	4.5	18 595.6
3rd quarter	18 732.7	13 019.8	3 404.2	2 703.9	600.1	87.2	-962.4		2 519.3	3 481.8	3 238.0	1 238.7	1 997.7	2.4	18 630.9
4th quarter	18 783.5	13 066.3	3 429.5	2 735.8	593.0	93.0	-983.0		2 528.5	3 511.6	3 234.9	1 242.1	1 991.4	-6.5	18 678.3

Note: Chained (2012) dollar series are calculated as the product of the chain-type quantity index and the 2012 current-dollar value of the corresponding series, divided by 100. Because the formula for the chain-type quantity indexes uses weights from more than one period, the corresponding chained-dollar estimates are usually not additive. The residual column is the difference between the total and the sum of the most detailed components shown in the Bureau of Economic Analysis (BEA) published data.

. . . = Not available.

Table 1-2B. Real Gross Domestic Product: Historical Data

(Billions of chained [2012] dollars, quarterly data are at seasonally adjusted annual rates.)   NIPA Tables 1.1.6, 1.2.6

Year and quarter	Gross domestic product	Personal consumption expenditures	Gross private domestic investment				Exports and imports of goods and services			Government consumption expenditures and gross investment			Residual	Addendum: Final sales of domestic product
			Total	Nonresidential	Residential	Change in private inventories	Net exports	Exports	Imports	Total	Federal	State and local		
1929	1 109.4	830.8	120.4	...	...	...	...	43.5	58.0	180.4	...	...	-20.9	1 121.3
1930	1 015.1	786.3	82.0	...	...	...	...	36.0	50.5	198.7	...	...	-31.5	1 045.6
1931	950.0	761.8	53.3	...	...	...	...	29.9	44.0	206.8	...	...	-44.0	984.8
1932	827.5	693.6	19.7	...	...	...	...	23.4	36.5	200.2	...	...	-42.0	874.0
1933	817.3	678.3	26.6	...	...	...	...	23.6	38.1	193.8	...	...	-51.1	848.1
1934	905.6	726.7	44.0	...	...	...	...	26.2	38.9	217.8	...	...	-62.6	928.4
1935	986.2	770.8	76.5	...	...	...	...	27.6	50.9	224.6	...	...	-72.1	986.5
1936	1 113.3	849.2	96.8	...	...	...	...	29.0	50.4	259.9	...	...	-81.3	1 115.3
1937	1 170.3	880.6	119.2	...	...	...	...	36.5	56.7	249.7	...	...	-75.8	1 162.0
1938	1 131.6	866.6	82.0	...	...	...	...	36.1	44.0	268.7	...	...	-73.8	1 146.6
1939	1 222.4	915.0	103.1	...	...	...	...	38.1	46.3	292.4	...	...	-82.0	1 230.9
1940	1 330.2	962.4	140.4	...	...	...	...	43.4	47.4	302.9	...	...	-92.5	1 315.3
1941	1 565.8	1 030.5	171.9	...	...	...	...	44.5	58.3	509.8	...	...	-166.7	1 535.2
1942	1 861.5	1 006.0	95.7	...	...	...	...	29.4	52.9	1 184.0	...	...	-414.0	1 861.3
1943	2 178.4	1 034.1	59.7	...	...	...	...	24.8	66.7	1 775.0	...	...	-643.7	2 207.2
1944	2 351.6	1 063.5	71.3	...	...	...	...	26.6	69.8	1 992.3	...	...	-726.7	2 383.2
1945	2 328.6	1 129.2	91.7	...	...	...	...	37.3	74.2	1 749.6	...	...	-594.5	2 367.0
1946	2 058.4	1 268.8	220.3	...	...	...	...	80.5	61.5	616.0	...	...	-98.6	2 029.2
1947	2 034.8	1 293.1	212.1	...	...	...	...	91.8	58.4	522.6	...	...	-21.7	2 065.1
1948	2 118.5	1 322.3	267.3	...	...	...	...	72.3	68.1	551.8	...	...	-52.0	2 099.3
1949	2 106.6	1 359.1	206.7	...	...	...	...	71.6	65.7	611.1	...	...	-63.0	2 146.9
1947														
1st quarter	2 033.1	1 275.0	218.1	...	...	...	...	97.8	60.3	524.1	...	...	-28.0	2 049.7
2nd quarter	2 027.6	1 296.2	201.4	...	...	...	...	96.5	61.4	523.6	...	...	-24.2	2 060.3
3rd quarter	2 023.5	1 300.4	195.5	...	...	...	...	91.6	54.5	523.5	...	...	-15.6	2 074.0
4th quarter	2 055.1	1 300.8	233.3	...	...	...	...	81.2	57.4	519.3	...	...	-19.2	2 076.3
1948														
1st quarter	2 086.0	1 307.3	257.2	...	...	...	...	77.5	64.7	527.0	...	...	-32.9	2 084.0
2nd quarter	2 120.5	1 322.5	273.1	...	...	...	...	70.3	67.3	545.6	...	...	-51.5	2 095.8
3rd quarter	2 132.6	1 324.4	275.9	...	...	...	...	71.9	70.8	557.0	...	...	-58.4	2 100.3
4th quarter	2 135.0	1 335.0	262.7	...	...	...	...	69.4	69.7	577.5	...	...	-65.1	2 117.1
1949														
1st quarter	2 105.6	1 337.2	223.5	...	...	...	...	77.7	67.6	592.8	...	...	-55.5	2 125.3
2nd quarter	2 098.4	1 357.7	193.8	...	...	...	...	77.0	66.6	618.6	...	...	-59.5	2 152.6
3rd quarter	2 120.0	1 360.8	209.9	...	...	...	...	70.6	63.9	621.8	...	...	-74.2	2 150.2
4th quarter	2 102.3	1 380.7	199.5	...	...	...	...	61.1	64.9	611.0	...	...	-62.6	2 159.7
1950														
1st quarter	2 184.9	1 403.7	246.9	...	...	...	...	60.1	66.7	599.6	...	...	-68.3	2 193.0
2nd quarter	2 251.5	1 426.8	273.3	...	...	...	...	60.8	70.3	610.6	...	...	-63.6	2 252.2
3rd quarter	2 338.5	1 500.1	296.8	...	...	...	...	62.4	86.8	600.7	...	...	-52.9	2 332.4
4th quarter	2 383.3	1 454.9	334.5	...	...	...	...	67.5	87.0	643.2	...	...	-88.0	2 311.9
1951														
1st quarter	2 415.7	1 490.0	300.4	...	...	...	...	70.9	87.0	711.6	...	...	-106.4	2 381.7
2nd quarter	2 457.5	1 447.9	307.3	...	...	...	...	77.8	84.1	806.5	...	...	-150.5	2 395.4
3rd quarter	2 508.2	1 464.8	284.7	...	...	...	...	79.7	77.0	895.1	...	...	-173.9	2 475.6
4th quarter	2 513.7	1 473.3	262.0	...	...	...	...	79.0	75.0	942.4	...	...	-183.0	2 512.8
1952														
1st quarter	2 540.6	1 476.7	268.9	...	...	...	...	83.2	83.7	970.5	...	...	-193.8	2 535.2
2nd quarter	2 546.0	1 505.7	248.6	...	...	...	...	74.1	83.2	1 003.6	...	...	-198.1	2 578.0
3rd quarter	2 564.4	1 512.9	261.9	...	...	...	...	68.2	88.3	1 013.7	...	...	-224.7	2 555.3
4th quarter	2 648.6	1 566.3	280.8	...	...	...	...	68.7	96.2	1 030.6	...	...	-222.2	2 639.5
1953														
1st quarter	2 697.9	1 584.7	286.8	...	...	...	...	67.5	93.9	1 063.3	...	...	-227.3	2 696.0
2nd quarter	2 718.7	1 594.2	288.4	...	...	...	...	67.9	98.7	1 085.1	...	...	-234.7	2 717.7
3rd quarter	2 703.4	1 590.4	282.4	...	...	...	...	70.7	98.4	1 073.6	...	...	-225.2	2 712.1
4th quarter	2 662.5	1 579.7	262.0	...	...	...	...	68.3	93.5	1 070.9	...	...	-216.4	2 697.9
1954														
1st quarter	2 649.8	1 585.6	260.1	...	...	...	...	65.3	88.1	1 043.0	...	...	-208.5	2 682.2
2nd quarter	2 652.6	1 606.1	259.5	...	...	...	...	74.6	96.0	1 007.7	...	...	-187.1	2 692.4
3rd quarter	2 682.6	1 627.9	272.0	...	...	...	...	72.2	90.2	985.9	...	...	-175.5	2 719.3
4th quarter	2 735.1	1 662.0	284.0	...	...	...	...	75.6	91.1	980.0	...	...	-171.8	2 763.6
1955														
1st quarter	2 813.2	1 699.3	312.3	...	...	...	...	78.4	96.5	980.6	...	...	-175.4	2 816.4
2nd quarter	2 859.0	1 731.9	331.3	...	...	...	...	76.5	102.0	970.9	...	...	-171.2	2 852.8
3rd quarter	2 897.6	1 753.3	336.4	...	...	...	...	81.6	103.7	977.2	...	...	-166.7	2 894.7
4th quarter	2 915.0	1 775.3	344.2	...	...	...	...	81.9	107.4	957.8	...	...	-163.8	2 902.4
1956														
1st quarter	2 903.7	1 778.2	334.4	...	...	...	...	85.8	111.9	957.5	...	...	-163.0	2 897.8
2nd quarter	2 927.7	1 784.1	331.8	...	...	...	...	91.6	111.4	976.9	...	...	-161.1	2 930.9
3rd quarter	2 925.0	1 788.1	328.8	...	...	...	...	94.7	112.3	968.3	...	...	-154.7	2 933.0
4th quarter	2 973.2	1 812.8	325.7	...	...	...	...	98.9	107.2	992.9	...	...	-160.8	2 983.9
1957														
1st quarter	2 992.2	1 825.4	320.7	...	...	...	...	105.3	115.7	1 014.2	...	...	-163.6	3 010.6
2nd quarter	2 985.7	1 828.6	320.4	...	...	...	...	102.2	116.6	1 008.9	...	...	-166.3	3 000.3
3rd quarter	3 014.9	1 843.0	327.7	...	...	...	...	99.4	114.3	1 017.5	...	...	-170.7	3 023.5
4th quarter	2 983.7	1 843.8	302.5	...	...	...	...	96.4	114.7	1 034.3	...	...	-167.0	3 022.5

Note: Chained (2012) dollar series are calculated as the product of the chain-type quantity index and the 2012 current-dollar value of the corresponding series, divided by 100. Because the formula for the chain-type quantity indexes uses weights from more than one period, the corresponding chained-dollar estimates are usually not additive. The residual column is the difference between the total and the sum of the most detailed components shown in the Bureau of Economic Analysis (BEA) published data.

. . . = Not available.

Table 1-2B. Real Gross Domestic Product: Historical Data—*Continued*

(Billions of chained [2012] dollars, quarterly data are at seasonally adjusted annual rates.)   **NIPA Tables 1.1.6, 1.2.6**

| Year and quarter | Gross domestic product | Personal consump-tion expen-ditures | Gross private domestic investment | | | | Exports and imports of goods and services | | | Government consumption expenditures and gross investment | | | Residual | Adden-dum: Final sales of domestic product |
| | | | Total | Fixed investment | | Change in private inventories | Net exports | Exports | Imports | Total | Federal | State and local | | |
				Nonresi-dential	Residential									
1958														
1st quarter	2 906.3	1 818.3	282.1	...	...	...	...	86.8	116.5	1 025.3	...	...	-175.5	2 948.4
2nd quarter ...	2 925.4	1 833.3	276.4	...	...	...	...	87.3	121.4	1 050.9	...	...	-188.6	2 965.0
3rd quarter ...	2 993.1	1 863.7	297.6	...	...	...	...	87.5	119.6	1 057.0	...	...	-199.5	3 008.1
4th quarter	3 063.1	1 889.1	323.0	...	...	...	...	87.2	125.8	1 078.3	...	...	-205.3	3 066.8
1959														
1st quarter	3 121.9	1 923.7	340.6	...	...	...	...	93.5	129.1	1 067.3	...	...	-187.6	3 127.0
2nd quarter ...	3 192.4	1 953.4	367.1	...	...	...	...	92.8	135.3	1 080.5	...	...	-200.4	3 173.1
3rd quarter ...	3 194.7	1 973.8	343.4	...	...	...	...	101.7	136.6	1 089.8	...	...	-181.5	3 216.6
4th quarter	3 203.8	1 976.0	354.6	...	...	...	...	96.7	133.1	1 080.8	...	...	-191.3	3 203.2
1960														
1st quarter	3 275.8	1 994.9	390.2	...	...	...	...	109.1	138.2	1 062.4	...	...	-192.9	3 239.4
2nd quarter ...	3 258.1	2 020.1	353.8	...	...	...	...	112.3	139.2	1 074.3	...	...	-177.6	3 266.5
3rd quarter ...	3 274.0	2 012.0	352.9	...	...	...	...	114.7	135.1	1 100.6	...	...	-191.4	3 272.9
4th quarter	3 232.0	2 014.6	313.8	...	...	...	...	115.3	128.7	1 108.2	...	...	-167.7	3 287.1
1961														
1st quarter	3 253.8	2 013.9	322.0	...	...	...	...	114.5	128.1	1 124.7	...	...	-182.4	3 292.8
2nd quarter ...	3 309.1	2 043.8	344.5	...	...	...	...	109.2	129.7	1 127.5	...	...	-195.7	3 323.3
3rd quarter ...	3 372.6	2 053.8	371.3	...	...	...	...	114.1	138.3	1 151.6	...	...	-210.8	3 359.9
4th quarter	3 438.7	2 095.1	377.5	...	...	...	...	116.0	141.6	1 176.8	...	...	-210.7	3 433.5
1962														
1st quarter	3 500.1	2 117.3	397.8	...	...	...	...	115.3	145.9	1 199.6	...	...	-224.5	3 475.6
2nd quarter ...	3 531.7	2 143.3	394.7	...	...	...	...	121.9	149.0	1 205.6	...	...	-210.2	3 527.1
3rd quarter ...	3 575.1	2 160.6	401.4	...	...	...	...	121.7	150.9	1 233.4	...	...	-220.9	3 565.5
4th quarter	3 586.8	2 191.2	389.7	...	...	...	...	117.8	152.8	1 242.0	...	...	-218.6	3 592.7
1963														
1st quarter	3 626.0	2 206.5	410.2	...	...	...	...	119.6	149.2	1 225.1	...	...	-220.9	3 609.6
2nd quarter ...	3 666.7	2 227.3	416.7	...	...	...	...	129.1	152.8	1 226.0	...	...	-202.7	3 665.5
3rd quarter ...	3 747.3	2 257.5	429.5	...	...	...	...	128.5	156.6	1 280.4	...	...	-219.3	3 742.3
4th quarter	3 771.8	2 276.3	434.7	...	...	...	...	133.5	156.1	1 267.2	...	...	-203.7	3 776.9
1964														
1st quarter	3 851.4	2 321.0	451.0	...	...	...	...	140.3	156.0	1 271.1	...	...	-197.3	3 854.4
2nd quarter ...	3 893.3	2 362.1	449.4	...	...	...	...	139.7	159.5	1 282.5	...	...	-201.3	3 897.9
3rd quarter ...	3 954.1	2 405.7	459.9	...	...	...	...	144.1	163.8	1 284.1	...	...	-200.2	3 954.2
4th quarter	3 966.3	2 412.6	464.2	...	...	...	...	147.0	168.2	1 283.5	...	...	-196.1	3 965.8
1965														
1st quarter	4 062.3	2 466.4	508.5	...	...	...	...	129.9	162.1	1 282.0	...	...	-215.8	4 028.1
2nd quarter ...	4 113.6	2 493.6	509.8	...	...	...	...	151.6	181.3	1 297.2	...	...	-195.8	4 096.5
3rd quarter ...	4 205.1	2 536.4	527.6	...	...	...	...	148.2	182.0	1 340.2	...	...	-208.7	4 181.8
4th quarter	4 302.0	2 607.7	531.1	...	...	...	...	157.3	190.9	1 365.6	...	...	-203.7	4 290.4
1966														
1st quarter	4 406.7	2 646.2	573.5	...	...	...	...	153.4	195.8	1 391.1	...	...	-222.8	4 362.4
2nd quarter ...	4 421.7	2 653.0	565.0	...	...	...	...	158.0	200.7	1 417.4	...	...	-229.7	4 382.3
3rd quarter ...	4 459.2	2 683.3	560.8	...	...	...	...	156.1	211.9	1 456.0	...	...	-237.3	4 426.7
4th quarter	4 495.8	2 694.5	564.1	...	...	...	...	160.0	214.3	1 481.1	...	...	-261.5	4 439.8
1967														
1st quarter	4 535.6	2 710.0	550.3	...	...	...	...	162.0	217.7	1 543.7	...	...	-279.9	4 486.0
2nd quarter ...	4 538.4	2 747.1	530.4	...	...	...	...	159.8	215.9	1 537.3	...	...	-249.8	4 533.7
3rd quarter ...	4 581.3	2 761.1	545.9	...	...	...	...	158.6	218.8	1 554.4	...	...	-259.7	4 563.7
4th quarter	4 615.9	2 778.0	557.7	...	...	...	...	161.5	230.3	1 566.2	...	...	-256.0	4 603.4
1968														
1st quarter	4 710.0	2 844.6	568.9	...	...	...	...	166.1	243.9	1 594.2	...	...	-255.3	4 699.0
2nd quarter ...	4 788.7	2 887.9	590.6	...	...	...	...	168.9	247.6	1 602.5	...	...	-275.5	4 746.1
3rd quarter ...	4 825.8	2 942.0	574.5	...	...	...	...	180.9	262.5	1 607.3	...	...	-250.2	4 817.4
4th quarter	4 844.8	2 955.3	580.8	...	...	...	...	176.9	260.3	1 608.7	...	...	-244.7	4 845.8
1969														
1st quarter	4 920.6	2 988.1	615.0	...	...	...	...	156.3	235.7	1 612.4	...	...	-262.5	4 896.1
2nd quarter ...	4 935.6	3 007.4	611.4	...	...	...	...	190.5	282.2	1 607.3	...	...	-238.2	4 920.5
3rd quarter ...	4 968.2	3 022.0	623.8	...	...	...	...	187.4	278.6	1 610.6	...	...	-242.5	4 947.6
4th quarter	4 943.9	3 046.2	593.7	...	...	...	...	192.3	275.5	1 588.9	...	...	-226.2	4 946.8
1970														
1st quarter	4 936.6	3 065.1	576.0	...	...	...	...	194.8	274.6	1 581.3	...	...	-210.4	4 960.0
2nd quarter ...	4 943.6	3 079.0	577.2	...	...	...	...	202.3	280.0	1 562.9	...	...	-218.7	4 949.1
3rd quarter ...	4 989.2	3 106.0	586.6	...	...	...	...	203.2	279.4	1 569.5	...	...	-217.6	4 995.0
4th quarter	4 935.7	3 097.5	555.5	...	...	...	...	203.6	283.7	1 570.3	...	...	-189.9	4 983.8
1971														
1st quarter	5 069.7	3 157.0	620.2	...	...	...	...	205.4	280.4	1 547.3	...	...	-229.2	5 045.8
2nd quarter ...	5 097.2	3 186.0	637.8	...	...	...	...	203.9	301.5	1 543.7	...	...	-214.8	5 079.2
3rd quarter ...	5 139.1	3 211.4	645.5	...	...	...	...	214.1	308.2	1 542.9	...	...	-206.6	5 125.0
4th quarter	5 151.2	3 264.7	628.2	...	...	...	...	194.3	287.3	1 532.1	...	...	-178.4	5 183.0
1972														
1st quarter	5 246.0	3 307.8	669.6	...	...	...	...	216.4	329.3	1 540.5	...	...	-173.5	5 258.3
2nd quarter ...	5 365.0	3 370.7	707.6	...	...	...	...	210.1	317.6	1 546.7	...	...	-199.8	5 341.4
3rd quarter ...	5 415.7	3 422.7	717.6	...	...	...	...	222.8	324.2	1 517.3	...	...	-191.1	5 389.0
4th quarter	5 506.4	3 503.0	722.1	...	...	...	...	232.2	338.8	1 529.8	...	...	-162.3	5 514.7

Note: Chained (2012) dollar series are calculated as the product of the chain-type quantity index and the 2012 current-dollar value of the corresponding series, divided by 100. Because the formula for the chain-type quantity indexes uses weights from more than one period, the corresponding chained-dollar estimates are usually not additive. The residual column is the difference between the total and the sum of the most detailed components shown in the Bureau of Economic Analysis (BEA) published data.

. . . = Not available.

Table 1-2B. Real Gross Domestic Product: Historical Data—*Continued*

(Billions of chained [2012] dollars, quarterly data are at seasonally adjusted annual rates.)   NIPA Tables 1.1.6, 1.2.6

Year and quarter	Gross domestic product	Personal consumption expenditures	Gross private domestic investment Total	Fixed investment Nonresidential	Fixed investment Residential	Change in private inventories	Net exports	Exports	Imports	Government Total	Government Federal	Government State and local	Residual	Addendum: Final sales of domestic product
1973														
1st quarter	5 642.7	3 567.0	764.5	...	...	...	...	249.2	354.5	1 543.0	...	...	-159.8	5 631.7
2nd quarter ...	5 704.1	3 565.3	797.0	...	...	...	...	261.2	344.3	1 532.1	...	...	-172.5	5 656.4
3rd quarter	5 674.1	3 577.9	768.1	...	...	...	...	262.1	334.8	1 514.1	...	...	-144.4	5 662.9
4th quarter	5 728.0	3 567.2	795.5	...	...	...	...	275.0	337.3	1 525.5	...	...	-170.0	5 670.5
1974														
1st quarter	5 678.7	3 535.3	750.1	...	...	...	...	278.5	325.8	1 553.0	...	...	-147.6	5 663.3
2nd quarter ...	5 692.2	3 548.0	747.1	...	...	...	...	292.1	343.1	1 560.8	...	...	-158.9	5 668.7
3rd quarter	5 638.4	3 563.3	708.4	...	...	...	...	276.4	337.1	1 563.9	...	...	-151.2	5 652.6
4th quarter	5 616.5	3 511.2	712.7	...	...	...	...	283.6	333.7	1 571.4	...	...	-183.0	5 572.7
1975														
1st quarter	5 548.2	3 540.6	597.9	...	...	...	...	285.6	300.2	1 588.5	...	...	-142.3	5 612.8
2nd quarter ...	5 587.8	3 598.9	580.1	...	...	...	...	277.3	276.2	1 575.1	...	...	-128.9	5 667.2
3rd quarter	5 683.4	3 650.0	625.0	...	...	...	...	272.6	299.2	1 604.5	...	...	-163.4	5 720.9
4th quarter	5 760.0	3 689.3	642.5	...	...	...	...	288.0	315.2	1 618.6	...	...	-163.6	5 791.0
1976														
1st quarter	5 889.5	3 763.0	704.5	...	...	...	...	286.0	334.9	1 621.1	...	...	-190.9	5 871.0
2nd quarter ...	5 932.7	3 797.7	732.4	...	...	...	...	288.9	349.4	1 603.1	...	...	-201.0	5 890.3
3rd quarter	5 965.3	3 837.7	734.9	...	...	...	...	297.3	363.7	1 598.4	...	...	-194.5	5 930.0
4th quarter	6 008.5	3 887.4	740.4	...	...	...	...	300.3	375.7	1 597.4	...	...	-161.5	6 011.8
1977														
1st quarter	6 079.5	3 933.3	774.7	...	...	...	...	295.5	393.5	1 611.0	...	...	-175.8	6 069.8
2nd quarter ...	6 197.7	3 954.6	830.3	...	...	...	...	303.4	397.1	1 625.7	...	...	-170.5	6 165.3
3rd quarter	6 309.5	3 992.0	872.3	...	...	...	...	305.8	391.6	1 632.3	...	...	-191.2	6 232.3
4th quarter	6 309.7	4 052.0	850.4	...	...	...	...	296.0	397.1	1 628.4	...	...	-167.3	6 283.4
1978														
1st quarter	6 329.8	4 074.8	867.0	...	...	...	...	303.0	423.8	1 628.8	...	...	-176.3	6 290.2
2nd quarter ...	6 574.4	4 161.9	923.3	...	...	...	...	334.5	425.1	1 669.6	...	...	-145.3	6 535.3
3rd quarter	6 640.5	4 179.4	950.2	...	...	...	...	338.2	430.7	1 685.1	...	...	-138.1	6 597.0
4th quarter	6 729.8	4 213.1	972.1	...	...	...	...	351.6	436.6	1 700.2	...	...	-137.9	6 683.3
1979														
1st quarter	6 741.9	4 234.9	973.8	...	...	...	...	351.9	435.4	1 683.4	...	...	-117.6	6 705.0
2nd quarter ...	6 749.1	4 232.2	972.9	...	...	...	...	352.8	437.8	1 699.8	...	...	-128.4	6 705.7
3rd quarter	6 799.2	4 273.3	956.5	...	...	...	...	365.5	431.1	1 704.8	...	...	-91.1	6 803.3
4th quarter	6 816.2	4 284.0	939.5	...	...	...	...	388.5	440.5	1 716.0	...	...	-86.6	6 832.4
1980														
1st quarter	6 837.6	4 277.9	933.1	...	...	...	...	399.4	440.8	1 741.2	...	...	-90.0	6 845.3
2nd quarter ...	6 696.8	4 181.5	853.8	...	...	...	...	406.9	408.8	1 744.6	...	...	-95.7	6 710.3
3rd quarter	6 688.8	4 227.4	797.4	...	...	...	...	406.0	379.3	1 720.9	...	...	-21.5	6 799.9
4th quarter	6 813.5	4 284.5	871.7	...	...	...	...	403.6	399.7	1 720.5	...	...	-49.5	6 863.2
1981														
1st quarter	6 947.0	4 298.8	952.4	...	...	...	...	411.2	416.7	1 743.5	...	...	-113.9	6 891.4
2nd quarter ...	6 895.6	4 299.2	912.9	...	...	...	...	413.4	417.4	1 746.9	...	...	-77.4	6 907.5
3rd quarter	6 978.1	4 319.0	964.4	...	...	...	...	404.6	412.6	1 740.8	...	...	-112.2	6 916.2
4th quarter	6 902.1	4 289.5	930.4	...	...	...	...	406.4	424.4	1 763.0	...	...	-100.5	6 885.3
1982														
1st quarter	6 794.9	4 321.1	842.4	...	...	...	...	388.8	412.3	1 761.0	...	...	-70.8	6 869.5
2nd quarter ...	6 825.9	4 334.3	841.8	...	...	...	...	391.1	405.5	1 767.7	...	...	-96.5	6 861.7
3rd quarter	6 799.8	4 363.3	834.1	...	...	...	...	373.8	424.3	1 781.9	...	...	-136.6	6 817.5
4th quarter	6 802.5	4 439.7	769.6	...	...	...	...	356.6	407.7	1 809.3	...	...	-97.7	6 913.3
1983														
1st quarter	6 892.1	4 483.6	796.3	...	...	...	...	362.4	417.0	1 826.7	...	...	-97.0	6 991.7
2nd quarter ...	7 049.0	4 574.9	866.8	...	...	...	...	363.5	449.1	1 841.8	...	...	-139.0	7 094.1
3rd quarter	7 189.9	4 657.0	921.1	...	...	...	...	368.9	484.1	1 873.6	...	...	-143.2	7 228.0
4th quarter	7 339.9	4 731.2	1 010.7	...	...	...	...	376.4	507.8	1 843.6	...	...	-149.8	7 325.7
1984														
1st quarter	7 483.4	4 770.5	1 108.3	...	...	...	...	384.2	548.6	1 863.1	...	...	-207.8	7 378.3
2nd quarter ...	7 612.7	4 837.3	1 144.4	...	...	...	...	393.8	571.6	1 903.6	...	...	-201.6	7 513.5
3rd quarter	7 686.1	4 873.2	1 169.1	...	...	...	...	402.5	586.4	1 919.0	...	...	-202.9	7 582.3
4th quarter	7 749.2	4 936.3	1 154.0	...	...	...	...	410.7	603.6	1 959.3	...	...	-177.7	7 692.0
1985														
1st quarter	7 824.2	5 020.2	1 122.3	...	...	...	...	411.8	590.3	1 980.6	...	...	-142.1	7 822.8
2nd quarter ...	7 893.1	5 066.3	1 141.4	...	...	...	...	410.9	619.3	2 024.6	...	...	-162.6	7 885.2
3rd quarter	8 013.7	5 162.5	1 133.7	...	...	...	...	404.9	613.3	2 070.7	...	...	-171.5	8 010.3
4th quarter	8 073.2	5 173.6	1 175.5	...	...	...	...	417.0	637.3	2 079.3	...	...	-184.0	8 042.6
1986														
1st quarter	8 148.6	5 218.9	1 175.1	...	...	...	...	430.0	636.7	2 095.7	...	...	-178.1	8 122.9
2nd quarter ...	8 185.3	5 275.7	1 154.6	...	...	...	...	434.6	663.8	2 139.4	...	...	-176.5	8 187.5
3rd quarter	8 263.6	5 369.0	1 123.5	...	...	...	...	444.6	682.2	2 186.0	...	...	-165.2	8 300.9
4th quarter	8 308.0	5 402.0	1 126.9	...	...	...	...	461.2	687.5	2 177.0	...	...	-161.2	8 352.7
1987														
1st quarter	8 369.9	5 407.4	1 157.3	...	...	...	...	461.6	683.4	2 189.7	...	...	-209.7	8 347.8
2nd quarter ...	8 460.2	5 481.2	1 157.9	...	...	...	...	480.6	700.5	2 206.7	...	...	-192.1	8 460.5
3rd quarter	8 533.6	5 543.7	1 158.1	...	...	...	...	501.5	714.2	2 209.0	...	...	-162.2	8 560.1
4th quarter	8 680.2	5 555.5	1 237.0	...	...	...	...	520.2	730.6	2 242.1	...	...	-243.2	8 601.6

Note: Chained (2012) dollar series are calculated as the product of the chain-type quantity index and the 2012 current-dollar value of the corresponding series, divided by 100. Because the formula for the chain-type quantity indexes uses weights from more than one period, the corresponding chained-dollar estimates are usually not additive. The residual column is the difference between the total and the sum of the most detailed components shown in the Bureau of Economic Analysis (BEA) published data.

. . . = Not available.

Table 1-2B. Real Gross Domestic Product: Historical Data—*Continued*

(Billions of chained [2012] dollars, quarterly data are at seasonally adjusted annual rates.)   **NIPA Tables 1.1.6, 1.2.6**

Year and quarter	Gross domestic product	Personal consump-tion expen-ditures	Gross private domestic investment				Exports and imports of goods and services			Government consumption expenditures and gross investment			Residual	Adden-dum: Final sales of domestic product
			Total	Fixed investment		Change in private inventories	Net exports	Exports	Imports	Total	Federal	State and local		
				Nonresi-dential	Residential									
1988														
1st quarter	8 725.0	5 653.6	1 177.6	...	...	...	...	549.1	727.2	2 223.4	...	...	-173.4	8 726.3
2nd quarter ...	8 839.6	5 695.3	1 205.7	...	...	...	...	564.5	718.8	2 230.1	...	...	-167.4	8 834.1
3rd quarter	8 891.4	5 745.9	1 212.4	...	...	...	...	575.6	735.7	2 230.3	...	...	-162.2	8 884.5
4th quarter	9 009.9	5 811.3	1 230.9	...	...	...	...	593.2	758.2	2 273.6	...	...	-175.4	8 997.5
1989														
1st quarter	9 101.5	5 838.2	1 272.6	...	...	...	...	610.9	761.7	2 263.6	...	...	-190.5	9 048.8
2nd quarter ...	9 171.0	5 865.5	1 260.3	...	...	...	...	637.6	765.3	2 298.4	...	...	-177.9	9 136.6
3rd quarter	9 238.9	5 922.3	1 249.0	...	...	...	...	644.2	764.3	2 319.1	...	...	-145.8	9 247.1
4th quarter	9 257.1	5 948.0	1 239.7	...	...	...	...	653.9	778.1	2 330.8	...	...	-158.2	9 256.6
1990														
1st quarter	9 358.3	5 998.1	1 252.1	...	...	...	...	682.3	802.8	2 368.0	...	...	-158.6	9 360.8
2nd quarter ...	9 392.3	6 016.3	1 252.4	...	...	...	...	690.8	801.4	2 371.8	...	...	-185.9	9 362.9
3rd quarter	9 398.5	6 040.2	1 228.2	...	...	...	...	696.0	798.2	2 375.5	...	...	-174.8	9 387.3
4th quarter	9 312.9	5 994.2	1 159.5	...	...	...	...	702.1	776.8	2 391.4	...	...	-142.7	9 349.5
1991														
1st quarter	9 269.4	5 971.7	1 120.9	...	...	...	...	706.4	767.9	2 405.7	...	...	-143.9	9 313.8
2nd quarter ...	9 341.6	6 021.2	1 120.9	...	...	...	...	732.2	781.6	2 415.5	...	...	-138.7	9 390.6
3rd quarter	9 388.8	6 051.2	1 143.1	...	...	...	...	749.5	803.6	2 408.7	...	...	-161.2	9 411.6
4th quarter	9 421.6	6 048.2	1 183.6	...	...	...	...	766.4	821.3	2 391.1	...	...	-187.2	9 399.9
1992														
1st quarter	9 534.3	6 161.4	1 161.7	...	...	...	...	780.2	827.3	2 412.3	...	...	-157.1	9 555.9
2nd quarter ...	9 637.7	6 203.2	1 227.4	...	...	...	...	780.6	840.9	2 407.6	...	...	-170.0	9 625.5
3rd quarter	9 733.0	6 269.7	1 237.2	...	...	...	...	797.2	854.1	2 422.9	...	...	-167.2	9 726.0
4th quarter	9 834.5	6 344.4	1 274.9	...	...	...	...	801.2	874.6	2 422.7	...	...	-165.4	9 825.0
1993														
1st quarter	9 851.0	6 368.8	1 305.0	...	...	...	...	803.0	893.5	2 392.6	...	...	-178.9	9 816.2
2nd quarter ...	9 908.3	6 426.7	1 312.5	...	...	...	...	812.6	912.0	2 392.8	...	...	-160.7	9 892.2
3rd quarter	9 955.6	6 498.2	1 303.6	...	...	...	...	810.5	924.0	2 396.2	...	...	-131.3	9 974.9
4th quarter	10 091.0	6 555.3	1 372.1	...	...	...	...	836.6	961.1	2 404.7	...	...	-136.7	10 091.0
1994														
1st quarter	10 189.0	6 630.3	1 424.4	...	...	...	...	844.8	983.5	2 374.9	...	...	-166.0	10 145.1
2nd quarter ...	10 327.0	6 681.8	1 494.4	...	...	...	...	873.3	1 020.1	2 386.7	...	...	-204.9	10 232.0
3rd quarter	10 387.4	6 732.8	1 470.0	...	...	...	...	906.0	1 048.8	2 426.9	...	...	-171.8	10 336.0
4th quarter	10 506.4	6 805.6	1 530.5	...	...	...	...	926.9	1 078.3	2 405.8	...	...	-185.7	10 423.3
1995														
1st quarter	10 543.6	6 822.5	1 546.5	...	...	...	...	944.8	1 102.0	2 414.2	...	...	-167.4	10 478.2
2nd quarter ...	10 575.1	6 882.3	1 514.4	...	...	...	...	958.0	1 112.5	2 422.3	...	...	-138.0	10 547.3
3rd quarter	10 665.1	6 944.7	1 505.4	...	...	...	...	999.5	1 116.3	2 416.0	...	...	-95.2	10 673.0
4th quarter	10 737.5	6 993.1	1 542.8	...	...	...	...	1 013.7	1 130.6	2 390.7	...	...	-90.1	10 735.4
1996														
1st quarter	10 817.9	7 057.6	1 567.6	...	...	...	...	1 025.9	1 167.3	2 406.5	...	...	-74.8	10 832.8
2nd quarter ...	10 998.3	7 133.6	1 642.9	...	...	...	...	1 042.6	1 193.4	2 436.4	...	...	-98.5	10 981.6
3rd quarter	11 097.0	7 176.8	1 717.5	...	...	...	...	1 051.0	1 232.1	2 436.4	...	...	-129.5	11 035.5
4th quarter	11 212.2	7 233.9	1 715.2	...	...	...	...	1 116.6	1 256.4	2 453.2	...	...	-95.1	11 184.4
1997														
1st quarter	11 284.6	7 310.2	1 749.9	...	...	...	...	1 137.8	1 310.1	2 440.5	...	...	-80.9	11 253.0
2nd quarter ...	11 472.1	7 343.1	1 857.0	...	...	...	...	1 183.3	1 354.1	2 472.1	...	...	-173.8	11 354.1
3rd quarter	11 615.6	7 468.2	1 883.0	...	...	...	...	1 210.7	1 404.0	2 483.5	...	...	-121.8	11 539.4
4th quarter	11 715.4	7 557.4	1 911.0	...	...	...	...	1 209.0	1 434.2	2 496.0	...	...	-127.6	11 622.4
1998														
1st quarter	11 832.5	7 633.9	1 994.4	...	...	...	...	1 214.6	1 488.0	2 487.1	...	...	-165.7	11 705.7
2nd quarter ...	11 942.0	7 768.3	1 982.1	...	...	...	...	1 201.2	1 522.1	2 530.6	...	...	-74.5	11 902.4
3rd quarter	12 091.6	7 869.6	2 033.0	...	...	...	...	1 195.6	1 542.7	2 549.7	...	...	-84.9	12 036.4
4th quarter	12 287.0	7 983.3	2 095.8	...	...	...	...	1 239.9	1 592.7	2 564.7	...	...	-87.5	12 222.9
1999														
1st quarter	12 403.3	8 060.8	2 153.3	1 427.0	632.0	118.0	...	1 238.3	1 633.9	2 582.2	...	...	-115.4	12 314.3
2nd quarter ...	12 498.7	8 178.3	2 154.7	1 464.9	641.9	50.0	...	1 250.1	1 680.0	2 591.9	...	...	-46.2	12 463.7
3rd quarter	12 662.4	8 270.6	2 213.8	1 506.9	646.7	62.4	...	1 286.5	1 742.0	2 623.3	...	...	-52.2	12 616.4
4th quarter	12 877.6	8 391.8	2 273.1	1 513.6	651.4	119.2	...	1 318.2	1 784.0	2 664.2	...	...	-105.0	12 777.9
2000														
1st quarter	12 924.2	8 520.7	2 257.1	1 569.0	654.6	21.4	...	1 337.2	1 853.9	2 645.4	...	...	-3.7	12 912.0
2nd quarter ...	13 160.8	8 603.0	2 390.7	1 618.0	651.3	123.6	...	1 375.2	1 911.4	2 671.3	...	...	-91.5	13 055.1
3rd quarter	13 178.4	8 687.5	2 367.3	1 635.6	641.1	87.3	...	1 408.8	1 978.1	2 659.7	...	...	-54.0	13 109.9
4th quarter	13 260.5	8 762.2	2 371.8	1 641.8	641.8	82.3	...	1 396.8	1 977.8	2 675.7	...	...	-50.6	13 194.8
2001														
1st quarter	13 222.7	8 797.3	2 264.2	1 627.1	644.6	-45.6	...	1 376.2	1 946.1	2 716.8	...	...	60.0	13 261.1
2nd quarter ...	13 300.0	8 818.1	2 267.4	1 593.4	654.6	-2.7	...	1 330.8	1 886.6	2 764.1	...	...	8.9	13 305.3
3rd quarter	13 244.8	8 848.3	2 218.4	1 574.6	658.6	-46.7	...	1 266.0	1 849.0	2 761.4	...	...	46.5	13 285.9
4th quarter	13 280.9	8 980.6	2 108.5	1 529.3	654.4	-117.9	...	1 225.9	1 823.1	2 807.0	...	...	99.8	13 386.0
2002														
1st quarter	13 397.0	9 008.1	2 170.7	1 492.6	675.9	-7.8	-623.9	1 252.8	1 876.7	2 855.5	946.9	1 921.6	-102.7	13 413.4
2nd quarter ...	13 478.2	9 054.3	2 206.4	1 475.8	692.6	33.1	-652.1	1 287.8	1 939.9	2 877.5	965.3	1 924.2	-91.6	13 453.2
3rd quarter	13 538.1	9 119.9	2 203.1	1 471.0	694.7	28.7	-669.8	1 294.5	1 964.3	2 892.8	974.8	1 929.8	-67.9	13 513.1
4th quarter	13 559.0	9 172.4	2 201.7	1 451.2	707.2	43.4	-723.4	1 273.4	1 996.8	2 915.1	991.3	1 934.7	-82.1	13 525.7

Note: Chained (2012) dollar series are calculated as the product of the chain-type quantity index and the 2012 current-dollar value of the corresponding series, divided by 100. Because the formula for the chain-type quantity indexes uses weights from more than one period, the corresponding chained-dollar estimates are usually not additive. The residual column is the difference between the total and the sum of the most detailed components shown in the Bureau of Economic Analysis (BEA) published data.

. . . = Not available.

Table 1-2B. Real Gross Domestic Product: Historical Data—*Continued*

(Billions of chained [2012] dollars, quarterly data are at seasonally adjusted annual rates.)   NIPA Tables 1.1.6, 1.2.6

Year and quarter	Gross domestic product	Personal consumption expenditures	Gross private domestic investment Total	Fixed investment Nonresidential	Fixed investment Residential	Change in private inventories	Net exports	Exports	Imports	Government Total	Federal	State and local	Residual	Addendum: Final sales of domestic product
2003														
1st quarter	13 634.3	9 215.5	2 218.7	1 458.4	720.2	39.9	-712.0	1 274.7	1 986.6	2 918.5	1 002.2	1 926.2	-79.9	13 604.0
2nd quarter	13 751.5	9 319.0	2 234.0	1 498.7	731.7	-9.2	-741.9	1 272.8	2 014.7	2 946.1	1 036.7	1 916.7	-49.1	13 761.4
3rd quarter	13 985.1	9 455.7	2 315.4	1 530.5	773.6	2.8	-739.9	1 308.2	2 048.0	2 953.1	1 036.4	1 924.3	-33.2	13 985.7
4th quarter	14 145.6	9 519.8	2 393.4	1 549.8	796.6	46.2	-746.4	1 364.4	2 110.8	2 971.0	1 055.7	1 921.6	-27.7	14 107.8
2004														
1st quarter	14 221.1	9 604.5	2 398.2	1 534.4	805.9	58.7	-776.9	1 401.6	2 178.5	2 984.4	1 067.2	1 922.7	-21.3	14 171.3
2nd quarter	14 329.5	9 664.3	2 493.9	1 572.8	831.4	97.7	-844.0	1 423.2	2 267.1	2 992.4	1 073.6	1 924.0	-14.1	14 245.7
3rd quarter	14 465.0	9 771.1	2 532.6	1 617.8	837.4	83.1	-863.2	1 434.0	2 297.2	2 997.9	1 085.5	1 916.6	0.1	14 394.7
4th quarter	14 609.9	9 877.4	2 585.5	1 651.0	848.9	91.1	-881.6	1 465.9	2 347.5	2 996.2	1 083.6	1 917.0	11.9	14 531.1
2005														
1st quarter	14 771.6	9 935.0	2 658.2	1 675.0	868.9	124.9	-869.8	1 500.2	2 370.1	3 011.2	1 095.7	1 919.3	11.2	14 661.9
2nd quarter	14 839.8	10 047.8	2 622.8	1 702.0	885.0	24.7	-874.7	1 527.0	2 401.7	3 009.5	1 094.5	1 918.8	38.2	14 815.2
3rd quarter	14 972.1	10 145.3	2 657.5	1 737.5	894.7	7.9	-886.8	1 530.7	2 417.4	3 019.4	1 102.9	1 920.0	54.0	14 959.2
4th quarter	15 066.6	10 175.4	2 743.8	1 750.9	892.9	97.4	-919.8	1 574.9	2 494.7	3 021.8	1 103.3	1 922.1	36.8	14 972.8
2006														
1st quarter	15 267.0	10 288.9	2 784.6	1 809.7	884.6	102.7	-913.0	1 634.7	2 547.7	3 060.1	1 131.9	1 930.8	32.1	15 183.2
2nd quarter	15 302.7	10 341.0	2 766.6	1 841.3	840.3	92.7	-908.8	1 666.7	2 575.5	3 059.2	1 124.1	1 938.2	32.0	15 222.6
3rd quarter	15 326.4	10 403.8	2 756.0	1 872.9	793.8	97.3	-932.5	1 667.2	2 599.7	3 054.8	1 113.9	1 944.5	32.6	15 239.1
4th quarter	15 456.9	10 504.5	2 702.5	1 892.7	756.9	55.7	-866.0	1 737.1	2 603.1	3 079.9	1 130.2	1 952.9	30.5	15 407.3
2007														
1st quarter	15 493.3	10 563.3	2 683.9	1 926.8	722.7	27.5	-872.9	1 761.4	2 634.2	3 084.5	1 123.5	1 964.6	39.8	15 461.4
2nd quarter	15 582.1	10 582.8	2 713.5	1 969.6	694.3	56.4	-862.3	1 788.4	2 650.8	3 112.5	1 141.9	1 973.8	27.6	15 531.5
3rd quarter	15 666.7	10 642.5	2 686.1	2 000.5	650.1	43.4	-819.4	1 842.1	2 661.5	3 126.5	1 151.7	1 977.8	21.2	15 629.2
4th quarter	15 762.0	10 672.8	2 653.1	2 031.3	596.1	35.0	-740.0	1 897.4	2 637.4	3 150.8	1 170.8	1 982.5	14.2	15 732.7
2008														
1st quarter	15 671.4	10 644.4	2 583.3	2 039.4	548.7	1.6	-732.1	1 913.5	2 645.6	3 157.7	1 188.4	1 971.4	3.9	15 673.6
2nd quarter	15 752.3	10 661.7	2 536.4	2 043.5	523.8	-25.4	-647.1	1 974.7	2 621.8	3 184.5	1 213.6	1 972.8	1.0	15 782.2
3rd quarter	15 667.0	10 581.9	2 485.5	2 005.4	497.2	-19.3	-626.8	1 961.2	2 588.0	3 209.5	1 228.8	1 982.5	5.0	15 682.0
4th quarter	15 328.0	10 483.4	2 246.4	1 888.6	448.8	-87.6	-640.4	1 852.4	2 492.8	3 230.5	1 244.3	1 987.8	-19.3	15 421.2
2009														
1st quarter	15 155.9	10 459.7	1 986.8	1 746.3	405.3	-179.0	-543.8	1 702.7	2 246.5	3 266.2	1 260.1	2 007.7	-23.7	15 328.3
2nd quarter	15 134.1	10 417.3	1 872.3	1 693.2	380.4	-228.3	-445.2	1 707.9	2 153.1	3 313.2	1 289.7	2 024.9	-20.9	15 343.9
3rd quarter	15 189.2	10 489.2	1 868.0	1 683.3	398.1	-245.1	-473.0	1 770.0	2 242.9	3 321.9	1 301.3	2 021.8	-1.0	15 410.3
4th quarter	15 356.1	10 473.6	2 040.7	1 694.5	397.2	-56.8	-477.3	1 874.5	2 351.8	3 328.0	1 321.0	2 007.9	-16.3	15 409.4
2010														
1st quarter	15 415.1	10 525.4	2 087.2	1 706.4	384.4	-5.0	-505.9	1 902.6	2 408.6	3 315.2	1 336.1	1 979.5	-13.8	15 420.0
2nd quarter	15 557.3	10 609.1	2 196.7	1 762.3	404.4	33.3	-572.6	1 947.6	2 520.2	3 326.5	1 353.9	1 972.8	-11.1	15 528.1
3rd quarter	15 672.0	10 683.3	2 294.7	1 810.1	368.6	128.8	-611.0	2 001.6	2 612.6	3 303.9	1 348.1	1 955.8	-14.4	15 555.1
4th quarter	15 750.6	10 754.0	2 287.4	1 845.2	374.7	72.1	-574.1	2 059.6	2 633.7	3 283.3	1 346.2	1 937.0	-6.1	15 683.0
2011														
1st quarter	15 712.8	10 799.7	2 244.2	1 842.7	373.4	32.8	-573.7	2 077.1	2 650.8	3 243.2	1 327.7	1 915.5	-5.0	15 685.0
2nd quarter	15 825.1	10 823.7	2 336.1	1 890.4	377.6	71.3	-555.8	2 110.8	2 666.6	3 221.4	1 322.9	1 898.4	-4.6	15 757.0
3rd quarter	15 820.7	10 866.0	2 343.8	1 978.9	384.4	-17.7	-563.8	2 133.2	2 697.0	3 175.5	1 294.4	1 881.1	-4.2	15 840.6
4th quarter	16 004.1	10 885.9	2 524.4	2 029.4	394.4	100.4	-579.0	2 155.2	2 734.1	3 173.2	1 299.4	1 873.8	-0.9	15 903.4
2012														
1st quarter	16 129.4	10 973.3	2 577.2	2 081.3	418.3	72.8	-580.5	2 168.5	2 749.0	3 159.6	1 299.4	1 860.1	4.6	16 051.7
2nd quarter	16 198.8	10 989.6	2 636.5	2 128.0	421.9	87.1	-570.4	2 192.2	2 762.5	3 143.0	1 289.1	1 854.0	-0.3	16 112.2
3rd quarter	16 220.7	11 007.5	2 648.5	2 120.9	432.7	98.0	-573.7	2 203.6	2 777.3	3 138.2	1 291.7	1 846.5	-2.8	16 125.8
4th quarter	16 239.1	11 056.9	2 624.9	2 144.0	455.2	27.0	-549.8	2 200.8	2 750.6	3 107.2	1 265.9	1 841.4	-1.5	16 213.4
2013														
1st quarter	16 383.0	11 114.2	2 722.8	2 171.6	471.8	81.2	-532.7	2 225.9	2 758.6	3 079.4	1 236.9	1 842.3	-3.2	16 304.1
2nd quarter	16 403.2	11 122.2	2 753.2	2 177.5	486.8	85.8	-545.8	2 252.7	2 798.5	3 074.0	1 226.8	1 846.8	2.1	16 315.2
3rd quarter	16 531.7	11 167.4	2 859.8	2 214.7	495.5	147.6	-551.9	2 266.8	2 818.6	3 057.4	1 209.1	1 847.8	0.6	16 383.6
4th quarter	16 663.6	11 263.6	2 870.1	2 260.0	487.7	120.0	-500.9	2 333.2	2 834.1	3 033.4	1 188.2	1 844.4	-1.1	16 541.7
2014														
1st quarter	16 616.5	11 308.0	2 838.3	2 291.7	484.3	58.3	-549.5	2 316.7	2 866.2	3 020.8	1 187.0	1 833.1	2.8	16 553.6
2nd quarter	16 841.5	11 431.8	2 958.1	2 353.3	499.8	98.6	-572.2	2 366.8	2 939.0	3 024.6	1 179.9	1 843.8	3.8	16 736.5
3rd quarter	17 047.1	11 554.8	3 018.6	2 400.8	507.1	106.3	-569.2	2 377.3	2 946.6	3 044.9	1 193.0	1 851.0	-1.3	16 936.3
4th quarter	17 143.0	11 695.0	3 021.9	2 415.5	525.2	82.0	-617.9	2 400.3	3 018.2	3 043.4	1 175.5	1 866.6	-2.7	17 062.8
2015														
1st quarter	17 277.6	11 792.1	3 116.5	2 406.1	535.5	173.2	-696.1	2 373.6	3 069.7	3 059.0	1 179.3	1 878.3	6.0	17 105.1
2nd quarter	17 405.7	11 886.0	3 118.9	2 413.2	548.9	149.1	-694.3	2 396.4	3 090.7	3 089.3	1 181.8	1 906.0	9.2	17 252.3
3rd quarter	17 463.2	11 976.6	3 114.8	2 420.4	563.9	121.4	-743.4	2 372.6	3 116.0	3 105.2	1 181.2	1 922.2	11.0	17 337.3
4th quarter	17 468.9	12 030.2	3 067.1	2 393.2	573.0	86.0	-752.8	2 363.2	3 116.0	3 113.6	1 188.6	1 923.3	16.6	17 373.8
2016														
1st quarter	17 556.8	12 124.2	3 054.7	2 389.8	593.0	51.1	-777.7	2 345.1	3 122.7	3 143.0	1 190.6	1 950.5	18.4	17 492.6
2nd quarter	17 639.4	12 211.3	3 041.6	2 413.6	590.1	10.8	-760.9	2 367.9	3 128.9	3 137.5	1 182.5	1 953.0	22.1	17 607.5
3rd quarter	17 735.1	12 289.1	3 045.5	2 446.8	586.2	-14.7	-761.4	2 403.4	3 164.9	3 151.0	1 188.2	1 960.8	19.8	17 726.7
4th quarter	17 824.2	12 365.3	3 114.0	2 451.2	595.5	44.8	-834.6	2 388.1	3 222.7	3 159.3	1 189.9	1 967.4	24.2	17 763.5
2017														
1st quarter	17 925.3	12 438.9	3 140.3	2 490.5	612.4	8.7	-831.5	2 423.5	3 255.0	3 157.3	1 186.4	1 968.9	29.5	17 895.1
2nd quarter	18 021.0	12 512.9	3 167.9	2 517.4	608.9	16.6	-850.0	2 432.9	3 282.9	3 168.0	1 195.9	1 970.1	20.6	17 985.3
3rd quarter	18 163.6	12 586.3	3 225.2	2 532.6	605.9	70.2	-833.7	2 459.5	3 293.2	3 167.1	1 196.1	1 969.0	3.6	18 082.5
4th quarter	18 322.5	12 729.7	3 262.1	2 584.2	620.4	31.1	-883.8	2 519.2	3 403.0	3 186.1	1 209.8	1 974.5	15.3	18 270.7

Note: Chained (2012) dollar series are calculated as the product of the chain-type quantity index and the 2012 current-dollar value of the corresponding series, divided by 100. Because the formula for the chain-type quantity indexes uses weights from more than one period, the corresponding chained-dollar estimates are usually not additive. The residual column is the difference between the total and the sum of the most detailed components shown in the Bureau of Economic Analysis (BEA) published data.

Table 1-3A. U.S. Population and Per Capita Product and Income: Recent Data

(Dollars, except as noted; quarterly data are at seasonally adjusted annual rates.)

NIPA Table 7.1

Year and quarter	Population (mid-period, thousands)	Current dollars — Gross domestic product	Personal income	Disposable personal income	PCE Total	Durable goods	Nondurable goods	Services	Chained (2012) dollars — Gross domestic product	Disposable personal income	PCE Total	Durable goods	Nondurable goods	Services
1955	165 275	2 574	1 961	1 763	1 563	247	645	671	17 372	11 873	10 528	474	3 872	6 457
1956	168 221	2 671	2 068	1 851	1 612	239	666	707	17 432	12 223	10 646	448	3 938	6 649
1957	171 274	2 768	2 149	1 921	1 672	245	686	740	17 482	12 317	10 715	444	3 945	6 756
1958	174 141	2 763	2 176	1 955	1 698	227	701	770	17 066	12 241	10 630	404	3 932	6 860
1959	177 130	2 945	2 274	2 036	1 790	253	721	816	17 943	12 560	11 047	444	4 039	7 082
1960	180 760	3 001	2 335	2 080	1 832	252	727	853	18 035	12 629	11 122	444	4 024	7 212
1961	183 742	3 060	2 398	2 141	1 858	241	733	885	18 197	12 861	11 166	422	4 039	7 355
1962	186 590	3 237	2 513	2 236	1 943	265	748	930	19 017	13 281	11 539	463	4 099	7 585
1963	189 300	3 367	2 603	2 315	2 018	286	760	971	19 561	13 585	11 843	498	4 128	7 795
1964	191 927	3 566	2 752	2 481	2 140	310	796	1 034	20 405	14 350	12 376	537	4 262	8 146
1965	194 347	3 819	2 936	2 640	2 279	342	840	1 098	21 460	15 052	12 998	596	4 421	8 491
1966	196 599	4 137	3 155	2 818	2 441	365	905	1 171	22 614	15 672	13 577	638	4 614	8 809
1967	198 752	4 327	3 350	2 982	2 549	372	931	1 246	22 982	16 181	13 832	641	4 662	9 071
1968	200 745	4 686	3 641	3 208	2 774	423	995	1 356	23 873	16 748	14 483	705	4 811	9 458
1969	202 736	5 019	3 948	3 432	2 977	446	1 057	1 475	24 377	17 148	14 876	724	4 895	9 780
1970	205 089	5 233	4 218	3 715	3 153	439	1 116	1 599	24 142	17 734	15 051	697	4 947	10 048
1971	207 692	5 609	4 491	4 002	3 370	493	1 154	1 723	24 625	18 321	15 430	757	4 976	10 267
1972	209 924	6 093	4 880	4 291	3 659	555	1 226	1 878	25 644	18 999	16 201	842	5 123	10 740
1973	211 939	6 725	5 383	4 758	4 009	616	1 350	2 043	26 834	19 989	16 841	921	5 220	11 140
1974	213 898	7 224	5 852	5 146	4 349	609	1 502	2 238	26 445	19 583	16 547	854	5 046	11 250
1975	215 981	7 801	6 340	5 657	4 771	658	1 617	2 497	26 136	19 869	16 759	848	5 040	11 561
1976	218 086	8 590	6 890	6 098	5 262	773	1 732	2 757	27 278	20 306	17 523	945	5 229	11 941
1977	220 289	9 450	7 532	6 634	5 783	871	1 854	3 058	28 254	20 740	18 081	1 019	5 295	12 310
1978	222 629	10 563	8 371	7 340	6 388	958	2 022	3 408	29 505	21 455	18 674	1 061	5 428	12 739
1979	225 106	11 672	9 252	8 057	7 043	1 005	2 273	3 765	30 104	21 630	18 907	1 044	5 507	12 985
1980	227 726	12 547	10 204	8 888	7 688	994	2 518	4 176	29 681	21 542	18 631	950	5 434	13 045
1981	230 008	13 943	11 326	9 823	8 408	1 061	2 719	4 628	30 132	21 849	18 702	950	5 448	13 116
1982	232 218	14 399	12 021	10 494	8 919	1 090	2 783	5 047	29 308	22 113	18 795	939	5 452	13 261
1983	234 333	15 508	12 721	11 216	9 737	1 259	2 897	5 581	30 374	22 669	19 680	1 064	5 579	13 816
1984	236 394	17 080	13 929	12 330	10 543	1 447	3 052	6 043	32 289	24 016	20 535	1 205	5 759	14 221
1985	238 506	18 192	14 779	13 027	11 374	1 595	3 175	6 605	33 337	24 518	21 407	1 314	5 880	14 816
1986	240 683	19 028	15 510	13 691	11 992	1 751	3 217	7 024	34 179	25 219	22 089	1 427	6 035	15 136
1987	242 843	19 993	16 313	14 297	12 668	1 820	3 353	7 494	35 047	25 548	22 636	1 442	6 084	15 680
1988	245 061	21 368	17 479	15 414	13 589	1 939	3 519	8 131	36 181	26 508	23 368	1 511	6 187	16 236
1989	247 387	22 805	18 698	16 403	14 458	1 998	3 757	8 703	37 157	27 027	23 823	1 530	6 294	16 595
1990	250 181	23 835	19 641	17 264	15 225	1 987	3 974	9 264	37 435	27 250	24 031	1 506	6 297	16 891
1991	253 530	24 290	20 056	17 734	15 554	1 882	4 024	9 648	36 900	27 086	23 757	1 406	6 195	16 928
1992	256 922	25 379	21 099	18 714	16 338	1 978	4 107	10 253	37 696	27 841	24 306	1 467	6 232	17 373
1993	260 282	26 350	21 738	19 245	17 104	2 119	4 191	10 795	38 234	27 935	24 828	1 557	6 308	17 679
1994	263 455	27 660	22 574	19 943	17 919	2 305	4 325	11 290	39 295	28 356	25 479	1 661	6 475	18 001
1995	266 588	28 658	23 600	20 792	18 615	2 385	4 426	11 805	39 875	28 954	25 923	1 706	6 558	18 310
1996	269 714	29 932	24 762	21 658	19 445	2 507	4 603	12 335	40 900	29 528	26 511	1 812	6 669	18 622
1997	272 958	31 424	25 984	22 570	20 284	2 621	4 730	12 933	42 211	30 246	27 183	1 937	6 782	18 990
1998	276 154	32 818	27 545	23 806	21 283	2 822	4 813	13 647	43 593	31 651	28 295	2 145	6 953	19 623
1999	279 328	34 478	28 647	24 666	22 479	3 063	5 125	14 292	45 146	32 312	29 447	2 393	7 224	20 137
2000	282 398	36 305	30 640	26 262	23 945	3 232	5 455	15 259	46 498	33 568	30 607	2 570	7 373	20 920
2001	285 225	37 100	31 574	27 230	24 772	3 301	5 554	15 917	46 497	34 149	31 067	2 678	7 427	21 175
2002	287 955	37 980	31 807	28 153	25 499	3 422	5 603	16 474	46 858	34 848	31 563	2 848	7 493	21 361
2003	290 626	39 426	32 645	29 192	26 574	3 502	5 866	17 206	47 756	35 446	32 267	3 026	7 685	21 641
2004	293 262	41 648	34 219	30 643	28 004	3 685	6 211	18 109	49 125	36 302	33 176	3 247	7 865	22 094
2005	295 993	44 044	35 806	31 710	29 552	3 813	6 603	19 136	50 381	36 527	34 041	3 395	8 052	22 600
2006	298 818	46 231	38 089	33 549	30 990	3 876	6 965	20 148	51 330	37 621	34 752	3 512	8 238	22 996
2007	301 696	47 902	39 801	34 855	32 173	3 938	7 222	21 013	51 794	38 119	35 186	3 645	8 298	23 214
2008	304 543	48 311	40 855	35 906	32 758	3 608	7 436	21 715	51 240	38 125	34 783	3 403	8 091	23 291
2009	307 240	47 028	39 250	35 500	32 034	3 294	7 056	21 684	49 501	37 728	34 045	3 167	7 887	23 012
2010	309 780	48 396	40 518	36 524	32 881	3 386	7 324	22 171	50 354	38 163	34 357	3 316	7 945	23 105
2011	312 033	49 811	42 709	38 052	34 102	3 504	7 770	22 828	50 766	38 777	34 752	3 460	7 957	23 338
2012	314 255	51 541	44 582	39 780	35 025	3 641	7 935	23 449	51 541	39 780	35 025	3 641	7 935	23 449
2013	316 421	53 046	44 817	39 521	35 766	3 759	8 029	23 978	52 131	38 996	35 291	3 837	8 022	23 436
2014	318 717	54 993	47 038	41 440	37 095	3 897	8 223	24 975	53 063	40 300	36 074	4 084	8 174	23 830
2015	321 026	56 770	48 961	42 925	38 266	4 068	8 144	26 054	54 213	41 656	37 135	4 357	8 389	24 417
2016	323 317	57 885	49 862	43 812	39 430	4 184	8 175	27 071	54 711	42 090	37 881	4 591	8 529	24 809
2017	325 410	59 984	51 869	45 583	40 909	4 341	8 458	28 109	55 647	43 031	38 619	4 875	8 682	25 144
2018	327 436	62 853	54 420	48 075	42 752	4 506	8 824	29 422	56 922	44 455	39 533	5 148	8 886	25 618
2016														
1st quarter	322 476	57 134	49 423	43 463	38 836	4 124	8 073	26 639	54 444	42 077	37 597	4 469	8 493	24 671
2nd quarter	322 998	57 701	49 626	43 603	39 283	4 159	8 190	26 934	54 612	41 964	37 806	4 539	8 547	24 765
3rd quarter	323 606	58 116	49 985	43 899	39 623	4 218	8 198	27 208	54 805	42 073	37 975	4 648	8 545	24 837
4th quarter	324 187	58 583	50 411	44 281	39 976	4 233	8 241	27 502	54 981	42 249	38 143	4 705	8 530	24 962
2017														
1st quarter	324 648	59 111	51 145	44 980	40 365	4 267	8 356	27 742	55 214	42 695	38 315	4 738	8 583	25 052
2nd quarter	325 107	59 539	51 520	45 319	40 641	4 302	8 386	27 953	55 431	42 919	38 489	4 820	8 662	25 081
3rd quarter	325 667	60 220	51 987	45 693	40 978	4 348	8 467	28 163	55 773	43 095	38 648	4 903	8 694	25 138
4th quarter	326 218	61 060	52 821	46 335	41 648	4 447	8 624	28 577	56 166	43 414	39 022	5 038	8 788	25 305
2018														
1st quarter	326 670	61 723	53 694	47 343	42 025	4 453	8 705	28 867	56 443	44 082	39 131	5 060	8 792	25 388
2nd quarter	327 138	62 696	54 182	47 849	42 611	4 514	8 824	29 274	56 851	44 311	39 461	5 151	8 867	25 564
3rd quarter	327 697	63 320	54 711	48 343	43 072	4 532	8 889	29 651	57 165	44 594	39 731	5 187	8 931	25 736
4th quarter	328 237	63 667	55 091	48 762	43 298	4 526	8 877	29 895	57 226	44 831	39 807	5 195	8 954	25 782

Table 1-3B. U.S. Population and Per Capita Product and Income: Historical Data

(Dollars, except as noted; quarterly data are at seasonally adjusted annual rates.)

NIPA Table 7.1

Year and quarter	Population (mid-period, thousands)	Current dollars							Chained (2012) dollars					
		Gross domestic product	Personal income	Disposable personal income	Personal consumption expenditures				Gross domestic product	Disposable personal income	Personal consumption expenditures			
					Total	Durable goods	Nondurable goods	Services			Total	Durable goods	Nondurable goods	Services
1929	121 878	858	700	686	635	81	278	276	9 103	7 361	6 817	251	2 769	4 063
1930	123 188	748	621	609	569	62	248	260	8 240	6 822	6 383	206	2 597	3 936
1931	124 149	623	529	521	489	48	208	233	7 652	6 540	6 136	177	2 549	3 806
1932	124 949	476	403	397	390	32	161	197	6 623	5 648	5 551	133	2 310	3 553
1933	125 690	455	376	369	366	30	159	177	6 502	5 452	5 396	129	2 288	3 401
1934	126 485	528	428	421	407	36	189	182	7 160	5 938	5 745	147	2 459	3 546
1935	127 362	583	478	469	439	43	205	191	7 744	6 468	6 052	177	2 582	3 644
1936	128 181	662	540	530	485	53	228	205	8 685	7 237	6 625	214	2 858	3 844
1937	128 961	721	579	564	518	57	240	221	9 075	7 437	6 829	224	2 911	4 000
1938	129 969	672	532	517	495	47	230	217	8 706	6 974	6 668	184	2 930	3 937
1939	131 028	713	562	551	513	55	235	224	9 329	7 496	6 984	217	3 039	4 057
1940	132 122	779	601	588	540	63	245	232	10 068	7 942	7 284	246	3 143	4 179
1941	133 402	969	734	716	608	77	279	252	11 737	9 106	7 724	281	3 299	4 384
1942	134 860	1 231	940	903	660	57	322	282	13 803	10 212	7 460	176	3 272	4 575
1943	136 739	1 485	1 142	1 020	731	55	357	319	15 931	10 556	7 562	156	3 248	4 873
1944	138 397	1 622	1 226	1 099	785	56	381	348	16 992	10 754	7 684	142	3 276	5 083
1945	139 928	1 629	1 256	1 117	857	65	418	374	16 642	10 516	8 070	156	3 441	5 297
1946	141 389	1 609	1 291	1 169	1 020	121	489	410	14 558	10 289	8 974	279	3 713	5 471
1947	144 126	1 732	1 349	1 212	1 123	151	538	433	14 118	9 684	8 972	322	3 621	5 389
1948	146 631	1 872	1 456	1 325	1 193	167	566	460	14 448	10 019	9 018	337	3 586	5 437
1949	149 188	1 826	1 415	1 303	1 195	178	546	471	14 120	9 927	9 110	358	3 577	5 472
1947														
1st quarter	143 143	1 699	1 325	1 191	1 091	145	523	423	14 203	9 727	8 907	311	3 603	5 380
2nd quarter	143 790	1 711	1 319	1 184	1 113	148	535	430	14 101	9 590	9 014	317	3 648	5 434
3rd quarter	144 449	1 728	1 369	1 233	1 132	151	544	437	14 008	9 801	9 002	319	3 648	5 397
4th quarter	145 122	1 790	1 383	1 240	1 155	162	551	443	14 161	9 618	8 964	339	3 584	5 344
1948														
1st quarter	145 709	1 824	1 417	1 272	1 169	162	560	448	14 316	9 758	8 972	336	3 581	5 380
2nd quarter	146 289	1 863	1 449	1 319	1 190	164	569	458	14 495	10 021	9 040	337	3 602	5 442
3rd quarter	146 921	1 900	1 480	1 356	1 205	172	568	465	14 515	10 140	9 015	341	3 565	5 448
4th quarter	147 607	1 899	1 478	1 353	1 205	169	567	469	14 464	10 150	9 044	334	3 595	5 477
1949														
1st quarter	148 254	1 855	1 430	1 310	1 193	165	558	470	14 202	9 906	9 020	327	3 593	5 488
2nd quarter	148 847	1 823	1 417	1 303	1 199	177	550	471	14 098	9 916	9 121	354	3 588	5 493
3rd quarter	149 485	1 826	1 408	1 299	1 189	183	537	470	14 182	9 938	9 103	369	3 542	5 458
4th quarter	150 167	1 802	1 404	1 299	1 200	189	539	472	13 999	9 949	9 194	382	3 583	5 450
1950														
1st quarter	150 786	1 862	1 501	1 391	1 213	195	540	478	14 490	10 672	9 309	395	3 614	5 495
2nd quarter	151 319	1 919	1 505	1 388	1 235	197	547	490	14 879	10 603	9 429	397	3 644	5 614
3rd quarter	151 973	2 028	1 553	1 429	1 319	246	570	503	15 388	10 690	9 871	489	3 691	5 677
4th quarter	152 658	2 096	1 604	1 457	1 297	218	568	511	15 612	10 707	9 530	426	3 592	5 687
1951														
1st quarter	153 291	2 192	1 658	1 499	1 365	232	600	532	15 759	10 673	9 720	438	3 643	5 819
2nd quarter	153 902	2 236	1 710	1 538	1 332	200	594	538	15 968	10 868	9 408	374	3 576	5 835
3rd quarter	154 610	2 273	1 730	1 551	1 343	195	603	545	16 223	10 941	9 474	365	3 634	5 881
4th quarter	155 344	2 293	1 752	1 561	1 362	196	614	552	16 181	10 870	9 484	362	3 659	5 868
1952														
1st quarter	155 976	2 307	1 758	1 561	1 365	197	608	561	16 288	10 821	9 468	363	3 614	5 918
2nd quarter	156 587	2 306	1 777	1 574	1 386	198	617	571	16 259	10 916	9 616	369	3 684	5 986
3rd quarter	157 267	2 338	1 814	1 609	1 396	187	626	583	16 306	11 088	9 620	345	3 733	6 038
4th quarter	157 986	2 410	1 848	1 639	1 441	213	635	593	16 765	11 274	9 914	403	3 770	6 096
1953														
1st quarter	158 574	2 447	1 869	1 658	1 458	222	634	602	17 013	11 365	9 993	415	3 779	6 133
2nd quarter	159 160	2 461	1 886	1 676	1 464	220	632	612	17 082	11 471	10 016	411	3 783	6 183
3rd quarter	159 889	2 447	1 877	1 670	1 461	216	624	621	16 908	11 364	9 947	407	3 731	6 191
4th quarter	160 639	2 403	1 868	1 663	1 451	210	622	619	16 574	11 273	9 834	393	3 719	6 116
1954														
1st quarter	161 328	2 389	1 860	1 673	1 458	204	628	626	16 425	11 282	9 828	379	3 739	6 147
2nd quarter	161 983	2 384	1 851	1 666	1 469	208	626	635	16 376	11 247	9 916	396	3 714	6 227
3rd quarter	162 731	2 403	1 854	1 670	1 477	204	628	644	16 485	11 314	10 004	396	3 744	6 306
4th quarter	163 524	2 445	1 878	1 692	1 499	214	633	652	16 726	11 475	10 164	418	3 788	6 351
1955														
1st quarter	164 204	2 516	1 909	1 718	1 531	234	636	661	17 132	11 613	10 349	454	3 812	6 399
2nd quarter	164 864	2 557	1 947	1 751	1 556	247	644	665	17 341	11 822	10 505	477	3 866	6 419
3rd quarter	165 612	2 598	1 983	1 782	1 574	256	646	672	17 496	11 986	10 587	490	3 875	6 456
4th quarter	166 419	2 626	2 004	1 799	1 590	249	655	686	17 516	12 067	10 668	475	3 934	6 553
1956														
1st quarter	167 109	2 632	2 029	1 817	1 593	237	661	694	17 376	12 139	10 641	451	3 962	6 588
2nd quarter	167 788	2 658	2 055	1 839	1 602	237	663	702	17 449	12 204	10 633	450	3 936	6 625
3rd quarter	168 573	2 677	2 077	1 858	1 614	235	667	712	17 352	12 210	10 608	440	3 921	6 663
4th quarter	169 416	2 718	2 112	1 888	1 638	245	671	722	17 550	12 337	10 700	452	3 932	6 717
1957														
1st quarter	170 172	2 761	2 129	1 903	1 656	250	678	728	17 583	12 321	10 727	458	3 939	6 713
2nd quarter	170 869	2 762	2 148	1 920	1 663	247	682	735	17 473	12 357	10 702	446	3 937	6 743
3rd quarter	171 638	2 794	2 165	1 937	1 682	244	695	743	17 566	12 364	10 738	441	3 974	6 762
4th quarter	172 417	2 754	2 150	1 925	1 684	241	690	753	17 305	12 226	10 694	433	3 930	6 804

Table 1-3B. U.S. Population and Per Capita Product and Income: Historical Data—*Continued*

(Dollars, except as noted; quarterly data are at seasonally adjusted annual rates.)   NIPA Table 7.1

| Year and quarter | Population (mid-period, thousands) | Current dollars | | | | | | | Chained (2012) dollars | | | | | |
| | | Gross domestic product | Personal income | Disposable personal income | Personal consumption expenditures | | | | Gross domestic product | Disposable personal income | Personal consumption expenditures | | | |
					Total	Durable goods	Nondurable goods	Services			Total	Durable goods	Nondurable goods	Services
1958														
1st quarter	173 064	2 702	2 145	1 925	1 675	228	693	755	16 793	12 073	10 506	403	3 880	6 769
2nd quarter	173 729	2 717	2 151	1 934	1 685	222	698	766	16 839	12 110	10 553	395	3 901	6 842
3rd quarter	174 483	2 784	2 193	1 970	1 707	226	705	776	17 154	12 323	10 681	402	3 955	6 907
4th quarter	175 288	2 850	2 215	1 990	1 722	232	708	782	17 475	12 455	10 777	415	3 991	6 922
1959														
1st quarter	176 045	2 899	2 238	2 006	1 758	248	716	794	17 734	12 473	10 927	435	4 025	6 978
2nd quarter	176 727	2 957	2 277	2 039	1 785	257	719	809	18 064	12 625	11 053	450	4 043	7 055
3rd quarter	177 481	2 958	2 280	2 040	1 807	261	722	824	18 000	12 554	11 121	457	4 041	7 120
4th quarter	178 268	2 965	2 302	2 057	1 811	247	726	837	17 972	12 588	11 085	433	4 048	7 174
1960														
1st quarter	179 319	3 026	2 324	2 071	1 820	253	723	844	18 268	12 659	11 125	444	4 035	7 197
2nd quarter	180 401	2 999	2 338	2 083	1 842	257	731	853	18 060	12 664	11 198	452	4 054	7 233
3rd quarter	181 301	3 009	2 341	2 084	1 832	253	726	853	18 059	12 625	11 098	446	4 010	7 188
4th quarter	182 019	2 968	2 339	2 084	1 835	246	727	862	17 756	12 570	11 068	433	3 997	7 229
1961														
1st quarter	182 634	2 984	2 355	2 101	1 832	231	730	870	17 816	12 646	11 027	408	4 008	7 276
2nd quarter	183 337	3 030	2 380	2 124	1 852	236	732	883	18 049	12 789	11 148	415	4 046	7 353
3rd quarter	184 103	3 083	2 409	2 151	1 859	242	731	886	18 319	12 905	11 156	423	4 032	7 345
4th quarter	184 894	3 140	2 446	2 186	1 891	252	736	902	18 598	13 101	11 331	442	4 069	7 444
1962														
1st quarter	185 553	3 201	2 475	2 209	1 912	257	743	912	18 863	13 183	11 411	449	4 088	7 487
2nd quarter	186 203	3 224	2 507	2 233	1 936	263	745	927	18 967	13 279	11 511	460	4 090	7 572
3rd quarter	186 926	3 258	2 522	2 242	1 949	265	749	935	19 126	13 298	11 558	461	4 106	7 607
4th quarter	187 680	3 262	2 546	2 260	1 975	275	754	946	19 111	13 363	11 675	480	4 113	7 672
1963														
1st quarter	188 299	3 302	2 566	2 279	1 988	279	757	952	19 257	13 434	11 718	488	4 122	7 681
2nd quarter	188 906	3 334	2 584	2 296	2 003	285	756	962	19 410	13 514	11 791	497	4 122	7 738
3rd quarter	189 631	3 398	2 611	2 323	2 032	288	766	978	19 761	13 610	11 905	501	4 144	7 843
4th quarter	190 362	3 435	2 651	2 361	2 049	293	763	993	19 814	13 779	11 958	507	4 123	7 919
1964														
1st quarter	190 954	3 508	2 693	2 411	2 093	304	778	1 010	20 169	14 005	12 155	525	4 181	8 020
2nd quarter	191 560	3 543	2 733	2 474	2 127	310	791	1 026	20 324	14 337	12 331	537	4 247	8 107
3rd quarter	192 256	3 600	2 773	2 505	2 166	319	806	1 041	20 567	14 471	12 513	553	4 314	8 185
4th quarter	192 938	3 614	2 809	2 533	2 172	308	808	1 056	20 558	14 584	12 504	534	4 307	8 269
1965														
1st quarter	193 467	3 710	2 858	2 564	2 221	336	815	1 070	20 997	14 716	12 748	582	4 336	8 341
2nd quarter	193 994	3 764	2 900	2 599	2 251	334	829	1 088	21 205	14 841	12 854	581	4 366	8 444
3rd quarter	194 647	3 850	2 964	2 671	2 290	343	843	1 105	21 604	15 196	13 031	599	4 417	8 531
4th quarter	195 279	3 953	3 022	2 724	2 354	354	873	1 127	22 030	15 452	13 354	622	4 566	8 647
1966														
1st quarter	195 763	4 065	3 080	2 765	2 402	369	889	1 143	22 510	15 566	13 518	650	4 587	8 715
2nd quarter	196 277	4 101	3 122	2 788	2 421	355	903	1 163	22 528	15 566	13 517	623	4 619	8 787
3rd quarter	196 877	4 163	3 180	2 835	2 460	367	913	1 179	22 650	15 708	13 629	641	4 638	8 832
4th quarter	197 481	4 220	3 239	2 882	2 482	368	913	1 200	22 766	15 844	13 644	639	4 611	8 901
1967														
1st quarter	197 967	4 264	3 286	2 926	2 497	359	921	1 217	22 911	16 043	13 689	625	4 651	8 961
2nd quarter	198 455	4 278	3 314	2 956	2 537	375	927	1 234	22 869	16 131	13 842	651	4 674	9 025
3rd quarter	199 012	4 348	3 375	3 004	2 566	375	933	1 257	23 020	16 239	13 874	644	4 657	9 123
4th quarter	199 572	4 417	3 423	3 043	2 597	379	941	1 276	23 129	16 311	13 920	643	4 665	9 174
1968														
1st quarter	199 995	4 547	3 513	3 120	2 682	404	968	1 310	23 551	16 548	14 223	682	4 746	9 299
2nd quarter	200 452	4 661	3 604	3 196	2 744	414	987	1 343	23 889	16 782	14 407	695	4 794	9 422
3rd quarter	200 997	4 731	3 688	3 231	2 817	436	1 009	1 372	24 009	16 790	14 637	724	4 857	9 516
4th quarter	201 538	4 803	3 759	3 283	2 853	436	1 017	1 400	24 039	16 875	14 664	718	4 845	9 595
1969														
1st quarter	201 955	4 919	3 824	3 316	2 907	446	1 034	1 427	24 365	16 882	14 796	730	4 889	9 658
2nd quarter	202 419	4 985	3 906	3 384	2 956	447	1 048	1 461	24 383	17 008	14 857	726	4 888	9 756
3rd quarter	202 986	5 074	3 998	3 485	2 998	446	1 064	1 488	24 475	17 306	14 888	722	4 893	9 807
4th quarter	203 584	5 099	4 061	3 542	3 048	446	1 079	1 523	24 284	17 387	14 963	717	4 911	9 900
1970														
1st quarter	204 086	5 151	4 124	3 612	3 095	439	1 100	1 556	24 189	17 526	15 019	705	4 941	9 982
2nd quarter	204 721	5 214	4 201	3 686	3 134	445	1 107	1 583	24 148	17 688	15 040	711	4 923	10 012
3rd quarter	205 419	5 287	4 258	3 768	3 181	448	1 118	1 615	24 288	17 909	15 120	711	4 941	10 091
4th quarter	206 130	5 281	4 287	3 795	3 203	424	1 138	1 641	23 945	17 805	15 027	661	4 982	10 108
1971														
1st quarter	206 763	5 490	4 371	3 895	3 285	474	1 141	1 670	24 520	18 104	15 269	730	4 987	10 147
2nd quarter	207 362	5 576	4 471	3 985	3 343	487	1 152	1 704	24 581	18 315	15 364	744	4 984	10 220
3rd quarter	208 000	5 662	4 525	4 033	3 392	497	1 156	1 739	24 707	18 354	15 439	762	4 958	10 285
4th quarter	208 642	5 705	4 598	4 092	3 459	514	1 168	1 777	24 689	18 511	15 647	792	4 977	10 413
1972														
1st quarter	209 142	5 884	4 717	4 144	3 533	529	1 181	1 823	25 083	18 551	15 816	808	4 981	10 563
2nd quarter	209 637	6 041	4 790	4 202	3 613	544	1 213	1 856	25 592	18 700	16 079	827	5 104	10 666
3rd quarter	210 181	6 140	4 901	4 310	3 691	559	1 237	1 894	25 767	19 014	16 284	846	5 161	10 782
4th quarter	210 737	6 306	5 111	4 507	3 799	586	1 272	1 941	26 129	19 724	16 623	888	5 244	10 946

Table 1-3B. U.S. Population and Per Capita Product and Income: Historical Data—*Continued*

(Dollars, except as noted; quarterly data are at seasonally adjusted annual rates.)   NIPA Table 7.1

Year and quarter	Population (mid-period, thousands)	Current dollars							Chained (2012) dollars					
		Gross domestic product	Personal income	Disposable personal income	Personal consumption expenditures				Gross domestic product	Disposable personal income	Personal consumption expenditures			
					Total	Durable goods	Nondurable goods	Services			Total	Durable goods	Nondurable goods	Services
1973														
1st quarter	211 192	6 522	5 185	4 587	3 906	624	1 304	1 978	26 718	19 831	16 890	943	5 264	11 062
2nd quarter	211 663	6 680	5 314	4 704	3 971	620	1 329	2 021	26 949	19 954	16 844	929	5 208	11 131
3rd quarter	212 191	6 757	5 432	4 800	4 048	616	1 366	2 065	26 741	19 994	16 862	919	5 220	11 177
4th quarter	212 708	6 940	5 598	4 940	4 108	603	1 399	2 106	26 929	20 164	16 771	895	5 186	11 191
1974														
1st quarter	213 144	6 996	5 662	4 992	4 184	594	1 449	2 141	26 643	19 788	16 586	872	5 106	11 147
2nd quarter	213 602	7 163	5 772	5 075	4 309	611	1 489	2 209	26 649	19 562	16 611	875	5 062	11 235
3rd quarter	214 147	7 285	5 926	5 203	4 433	637	1 530	2 266	26 330	19 530	16 639	878	5 060	11 266
4th quarter	214 700	7 451	6 048	5 314	4 467	592	1 541	2 334	26 160	19 454	16 354	792	4 957	11 352
1975														
1st quarter	215 135	7 512	6 128	5 394	4 579	613	1 563	2 404	25 789	19 385	16 458	808	4 956	11 435
2nd quarter	215 652	7 660	6 260	5 698	4 700	634	1 599	2 467	25 911	20 233	16 688	821	5 054	11 544
3rd quarter	216 289	7 905	6 410	5 704	4 842	679	1 646	2 517	26 277	19 880	16 876	870	5 088	11 564
4th quarter	216 848	8 125	6 561	5 830	4 963	707	1 659	2 597	26 562	19 984	17 013	893	5 063	11 700
1976														
1st quarter	217 314	8 377	6 701	5 953	5 107	751	1 691	2 665	27 101	20 182	17 316	936	5 154	11 811
2nd quarter	217 776	8 506	6 802	6 024	5 187	762	1 713	2 712	27 242	20 254	17 438	938	5 222	11 864
3rd quarter	218 338	8 641	6 954	6 148	5 307	778	1 743	2 786	27 321	20 359	17 577	946	5 249	11 977
4th quarter	218 917	8 836	7 102	6 268	5 447	801	1 780	2 866	27 446	20 434	17 758	959	5 290	12 109
1977														
1st quarter	219 427	9 063	7 241	6 380	5 597	838	1 807	2 952	27 706	20 433	17 925	993	5 274	12 229
2nd quarter	219 956	9 347	7 424	6 535	5 710	860	1 835	3 014	28 177	20 575	17 979	1 014	5 261	12 245
3rd quarter	220 573	9 604	7 606	6 706	5 834	879	1 858	3 096	28 605	20 803	18 098	1 026	5 272	12 341
4th quarter	221 201	9 784	7 854	6 912	5 989	907	1 915	3 167	28 525	21 140	18 318	1 044	5 370	12 423
1978														
1st quarter	221 719	9 935	8 020	7 063	6 107	894	1 944	3 269	28 549	21 256	18 378	1 014	5 381	12 587
2nd quarter	222 281	10 400	8 270	7 266	6 350	973	2 001	3 376	29 577	21 427	18 724	1 087	5 408	12 744
3rd quarter	222 933	10 743	8 490	7 430	6 469	972	2 044	3 453	29 787	21 530	18 747	1 069	5 434	12 798
4th quarter	223 583	11 078	8 704	7 598	6 626	994	2 099	3 532	30 100	21 609	18 844	1 075	5 488	12 828
1979														
1st quarter	224 152	11 272	8 947	7 815	6 768	996	2 162	3 610	30 077	21 815	18 893	1 059	5 499	12 929
2nd quarter	224 737	11 530	9 101	7 936	6 931	988	2 225	3 718	30 031	21 561	18 832	1 033	5 474	12 975
3rd quarter	225 418	11 834	9 364	8 145	7 151	1 024	2 317	3 809	30 163	21 594	18 957	1 058	5 521	12 984
4th quarter	226 117	12 046	9 593	8 331	7 319	1 013	2 385	3 921	30 145	21 565	18 946	1 028	5 533	13 052
1980														
1st quarter	226 754	12 303	9 843	8 587	7 507	1 023	2 469	4 014	30 154	21 579	18 866	1 009	5 515	13 037
2nd quarter	227 389	12 302	9 980	8 696	7 497	932	2 489	4 076	29 451	21 329	18 389	897	5 425	12 919
3rd quarter	228 070	12 525	10 297	8 972	7 733	988	2 528	4 218	29 328	21 505	18 535	934	5 396	13 029
4th quarter	228 689	13 055	10 690	9 296	8 010	1 033	2 585	4 392	29 794	21 742	18 735	960	5 402	13 192
1981														
1st quarter	229 155	13 634	10 948	9 504	8 229	1 074	2 680	4 476	30 316	21 667	18 760	986	5 439	13 090
2nd quarter	229 674	13 770	11 145	9 653	8 349	1 049	2 712	4 588	30 023	21 643	18 719	945	5 450	13 150
3rd quarter	230 301	14 158	11 546	9 997	8 502	1 094	2 732	4 677	30 300	22 050	18 754	971	5 448	13 116
4th quarter	230 903	14 209	11 661	10 134	8 551	1 026	2 753	4 772	29 892	22 016	18 577	899	5 454	13 109
1982														
1st quarter	231 395	14 150	11 789	10 265	8 704	1 065	2 765	4 874	29 365	22 023	18 674	926	5 453	13 150
2nd quarter	231 906	14 368	11 954	10 403	8 795	1 075	2 754	4 966	29 434	22 107	18 690	927	5 431	13 195
3rd quarter	232 498	14 479	12 097	10 591	8 971	1 084	2 795	5 092	29 247	22 156	18 767	932	5 441	13 263
4th quarter	233 074	14 599	12 244	10 714	9 205	1 134	2 818	5 254	29 186	22 171	19 049	972	5 483	13 435
1983														
1st quarter	233 546	14 872	12 385	10 882	9 354	1 151	2 813	5 389	29 511	22 335	19 198	980	5 490	13 586
2nd quarter	234 028	15 292	12 571	11 035	9 612	1 237	2 876	5 499	30 120	22 443	19 549	1 050	5 543	13 743
3rd quarter	234 603	15 725	12 793	11 320	9 889	1 289	2 934	5 665	30 647	22 724	19 850	1 087	5 613	13 911
4th quarter	235 153	16 137	13 134	11 622	10 089	1 358	2 962	5 769	31 213	23 176	20 120	1 138	5 668	14 022
1984														
1st quarter	235 605	16 587	13 493	11 960	10 264	1 407	2 998	5 858	31 762	23 594	20 248	1 179	5 672	14 057
2nd quarter	236 082	16 984	13 821	12 252	10 487	1 446	3 060	5 982	32 246	23 937	20 490	1 204	5 784	14 139
3rd quarter	236 657	17 258	14 102	12 479	10 621	1 445	3 062	6 114	32 478	24 194	20 592	1 201	5 774	14 282
4th quarter	237 232	17 487	14 297	12 628	10 799	1 492	3 089	6 218	32 665	24 333	20 808	1 237	5 805	14 404
1985														
1st quarter	237 673	17 798	14 538	12 720	11 091	1 548	3 124	6 419	32 920	24 224	21 122	1 278	5 829	14 638
2nd quarter	238 176	18 032	14 676	13 045	11 260	1 567	3 161	6 532	33 140	24 644	21 271	1 290	5 862	14 741
3rd quarter	238 789	18 371	14 838	13 072	11 534	1 660	3 185	6 689	33 560	24 503	21 619	1 369	5 895	14 901
4th quarter	239 387	18 564	15 062	13 270	11 610	1 603	3 231	6 777	33 725	24 701	21 612	1 319	5 933	14 983
1986														
1st quarter	239 861	18 794	15 291	13 513	11 772	1 633	3 249	6 891	33 972	24 976	21 758	1 343	5 997	15 015
2nd quarter	240 368	18 910	15 430	13 643	11 863	1 694	3 193	6 976	34 053	25 242	21 949	1 388	6 037	15 074
3rd quarter	240 962	19 122	15 592	13 768	12 106	1 850	3 200	7 057	34 294	25 341	22 282	1 502	6 035	15 162
4th quarter	241 539	19 283	15 728	13 839	12 225	1 826	3 225	7 173	34 396	25 318	22 365	1 475	6 069	15 292
1987														
1st quarter	242 009	19 512	15 933	14 071	12 328	1 729	3 295	7 304	34 585	25 502	22 344	1 385	6 065	15 511
2nd quarter	242 520	19 818	16 153	14 044	12 590	1 811	3 349	7 430	34 885	25 210	22 601	1 439	6 099	15 631
3rd quarter	243 120	20 091	16 406	14 394	12 822	1 894	3 376	7 552	35 101	25 598	22 802	1 494	6 089	15 732
4th quarter	243 721	20 548	16 758	14 678	12 928	1 846	3 392	7 690	35 615	25 879	22 794	1 450	6 081	15 847

Table 1-3B. U.S. Population and Per Capita Product and Income: Historical Data—*Continued*

(Dollars, except as noted; quarterly data are at seasonally adjusted annual rates.)   NIPA Table 7.1

Year and quarter	Population (mid-period, thou-sands)	Current dollars							Chained (2012) dollars					
		Gross domestic product	Personal income	Dispos-able personal income	Personal consumption expenditures				Gross domestic product	Dispos-able personal income	Personal consumption expenditures			
					Total	Durable goods	Nondur-able goods	Services			Total	Durable goods	Nondur-able goods	Services
1988														
1st quarter	244 208	20 775	17 037	14 981	13 234	1 926	3 433	7 875	35 728	26 207	23 151	1 516	6 135	16 031
2nd quarter	244 716	21 208	17 305	15 271	13 451	1 934	3 488	8 030	36 122	26 421	23 273	1 513	6 170	16 139
3rd quarter	245 354	21 531	17 633	15 568	13 702	1 917	3 549	8 236	36 239	26 607	23 419	1 490	6 197	16 339
4th quarter	245 966	21 952	17 937	15 834	13 963	1 977	3 603	8 384	36 631	26 792	23 626	1 525	6 246	16 435
1989														
1st quarter	246 460	22 362	18 410	16 166	14 161	1 974	3 662	8 526	36 929	27 042	23 688	1 515	6 262	16 514
2nd quarter	247 017	22 721	18 593	16 299	14 387	1 997	3 756	8 634	37 127	26 902	23 745	1 531	6 257	16 543
3rd quarter	247 698	22 993	18 767	16 459	14 572	2 041	3 780	8 750	37 299	27 006	23 909	1 562	6 302	16 605
4th quarter	248 374	23 139	19 018	16 684	14 710	1 980	3 830	8 900	37 271	27 161	23 948	1 510	6 356	16 717
1990														
1st quarter	248 936	23 591	19 358	17 018	15 016	2 070	3 913	9 032	37 593	27 307	24 095	1 570	6 334	16 765
2nd quarter	249 711	23 868	19 620	17 238	15 151	1 996	3 928	9 227	37 612	27 412	24 093	1 514	6 321	16 916
3rd quarter	250 595	24 003	19 787	17 390	15 350	1 970	4 003	9 378	37 505	27 306	24 103	1 493	6 309	16 991
4th quarter	251 482	23 877	19 797	17 408	15 380	1 912	4 051	9 417	37 032	26 977	23 836	1 447	6 224	16 891
1991														
1st quarter	252 258	23 925	19 791	17 489	15 356	1 870	4 020	9 465	36 746	26 962	23 673	1 403	6 207	16 826
2nd quarter	253 063	24 211	19 972	17 656	15 518	1 878	4 038	9 602	36 914	27 072	23 793	1 405	6 224	16 939
3rd quarter	253 965	24 436	20 109	17 785	15 645	1 907	4 034	9 704	36 969	27 086	23 827	1 422	6 212	16 952
4th quarter	254 835	24 583	20 351	18 001	15 697	1 874	4 005	9 818	36 971	27 218	23 734	1 394	6 138	16 995
1992														
1st quarter	255 585	24 896	20 728	18 424	16 043	1 941	4 060	10 042	37 304	27 685	24 107	1 444	6 211	17 223
2nd quarter	256 439	25 233	21 036	18 668	16 205	1 954	4 084	10 168	37 583	27 866	24 190	1 450	6 216	17 299
3rd quarter	257 386	25 513	21 199	18 805	16 423	1 990	4 122	10 311	37 815	27 892	24 359	1 475	6 230	17 418
4th quarter	258 277	25 867	21 430	18 957	16 677	2 025	4 161	10 490	38 077	27 923	24 565	1 499	6 271	17 551
1993														
1st quarter	259 039	25 979	21 467	19 085	16 791	2 038	4 166	10 587	38 029	27 945	24 586	1 509	6 259	17 566
2nd quarter	259 826	26 206	21 691	19 214	17 006	2 108	4 181	10 716	38 135	27 946	24 735	1 554	6 290	17 601
3rd quarter	260 714	26 397	21 777	19 249	17 211	2 135	4 191	10 886	38 186	27 875	24 925	1 566	6 330	17 745
4th quarter	261 547	26 816	22 014	19 432	17 407	2 194	4 226	10 987	38 582	27 980	25 063	1 598	6 354	17 803
1994														
1st quarter	262 250	27 133	22 147	19 578	17 621	2 245	4 259	11 117	38 852	28 089	25 282	1 632	6 422	17 901
2nd quarter	263 020	27 553	22 480	19 827	17 805	2 276	4 289	11 240	39 263	28 289	25 404	1 644	6 454	17 976
3rd quarter	263 870	27 783	22 657	20 021	18 012	2 309	4 357	11 346	39 366	28 362	25 516	1 656	6 488	18 037
4th quarter	264 678	28 167	23 007	20 342	18 236	2 387	4 394	11 455	39 695	28 681	25 713	1 709	6 537	18 089
1995														
1st quarter	265 388	28 344	23 291	20 561	18 322	2 341	4 397	11 584	39 729	28 849	25 708	1 670	6 543	18 165
2nd quarter	266 142	28 485	23 486	20 680	18 538	2 356	4 422	11 760	39 735	28 849	25 860	1 682	6 557	18 294
3rd quarter	267 000	28 776	23 705	20 888	18 722	2 407	4 433	11 882	39 944	29 020	26 010	1 724	6 559	18 368
4th quarter	267 820	29 022	23 913	21 038	18 877	2 435	4 449	11 992	40 092	29 100	26 111	1 748	6 572	18 414
1996														
1st quarter	268 487	29 307	24 288	21 302	19 109	2 458	4 511	12 141	40 292	29 302	26 287	1 763	6 599	18 550
2nd quarter	269 251	29 834	24 698	21 579	19 389	2 512	4 604	12 273	40 848	29 487	26 494	1 814	6 664	18 604
3rd quarter	270 128	30 102	24 899	21 776	19 526	2 515	4 614	12 396	41 080	29 630	26 568	1 821	6 690	18 644
4th quarter	270 991	30 480	25 158	21 971	19 753	2 545	4 680	12 527	41 375	29 692	26 694	1 850	6 723	18 689
1997														
1st quarter	271 709	30 778	25 540	22 220	19 996	2 597	4 715	12 685	41 532	29 896	26 904	1 895	6 745	18 805
2nd quarter	272 487	31 263	25 771	22 409	20 079	2 556	4 689	12 833	42 102	30 075	26 948	1 884	6 729	18 893
3rd quarter	273 391	31 687	26 099	22 657	20 407	2 644	4 745	13 018	42 487	30 328	27 317	1 961	6 814	19 055
4th quarter	274 246	31 964	26 520	22 991	20 652	2 688	4 769	13 195	42 719	30 679	27 557	2 005	6 840	19 204
1998														
1st quarter	274 950	32 248	27 040	23 417	20 809	2 683	4 752	13 373	43 035	31 245	27 765	2 016	6 872	19 376
2nd quarter	275 703	32 534	27 425	23 716	21 155	2 790	4 787	13 578	43 315	31 588	28 176	2 112	6 936	19 553
3rd quarter	276 564	32 980	27 726	23 954	21 430	2 839	4 826	13 765	43 721	31 807	28 455	2 165	6 965	19 745
4th quarter	277 400	33 504	27 984	24 134	21 731	2 975	4 885	13 871	44 293	31 962	28 779	2 286	7 039	19 795
1999														
1st quarter	278 103	33 863	28 230	24 354	21 943	2 948	4 980	14 015	44 600	32 169	28 985	2 286	7 151	19 907
2nd quarter	278 864	34 153	28 402	24 475	22 324	3 065	5 086	14 173	44 820	32 154	29 327	2 388	7 199	20 041
3rd quarter	279 751	34 609	28 700	24 694	22 623	3 114	5 148	14 362	45 263	32 270	29 564	2 438	7 209	20 197
4th quarter	280 592	35 280	29 250	25 138	23 021	3 123	5 284	14 614	45 894	32 657	29 907	2 460	7 334	20 403
2000														
1st quarter	281 304	35 559	30 064	25 767	23 510	3 274	5 305	14 932	45 944	33 198	30 290	2 590	7 252	20 668
2nd quarter	282 002	36 339	30 467	26 105	23 785	3 198	5 444	15 143	46 669	33 482	30 507	2 536	7 377	20 869
3rd quarter	282 769	36 496	30 912	26 500	24 102	3 224	5 508	15 370	46 605	33 779	30 723	2 572	7 406	21 009
4th quarter	283 518	36 820	31 111	26 671	24 380	3 230	5 564	15 585	46 771	33 810	30 905	2 583	7 457	21 132
2001														
1st quarter	284 169	36 854	31 615	27 034	24 587	3 261	5 533	15 792	46 531	34 038	30 958	2 620	7 396	21 186
2nd quarter	284 838	37 207	31 645	27 050	24 703	3 228	5 584	15 891	46 693	33 899	30 958	2 611	7 413	21 182
3rd quarter	285 584	37 104	31 517	27 618	24 738	3 235	5 572	15 930	46 378	34 591	30 983	2 634	7 428	21 156
4th quarter	286 311	37 233	31 518	27 216	25 059	3 478	5 525	16 056	46 386	34 067	31 367	2 847	7 472	21 175
2002														
1st quarter	286 935	37 601	31 567	27 820	25 127	3 402	5 526	16 199	46 690	34 759	31 394	2 809	7 487	21 251
2nd quarter	287 574	37 880	31 814	28 159	25 392	3 399	5 597	16 396	46 868	34 916	31 485	2 821	7 475	21 340
3rd quarter	288 303	38 127	31 822	28 201	25 636	3 475	5 608	16 554	46 958	34 798	31 633	2 898	7 469	21 381
4th quarter	289 007	38 309	32 023	28 430	25 839	3 412	5 681	16 746	46 916	34 919	31 738	2 865	7 540	21 472

Table 1-3B. U.S. Population and Per Capita Product and Income: Historical Data—*Continued*

(Dollars, except as noted; quarterly data are at seasonally adjusted annual rates.)   **NIPA Table 7.1**

Year and quarter	Population (mid-period, thousands)	Current dollars							Chained (2012) dollars					
		Gross domestic product	Personal income	Disposable personal income	Personal consumption expenditures				Gross domestic product	Disposable personal income	Personal consumption expenditures			
					Total	Durable goods	Nondurable goods	Services			Total	Durable goods	Nondurable goods	Services
2003														
1st quarter	289 609	38 616	32 164	28 638	26 090	3 365	5 811	16 914	47 078	34 929	31 821	2 861	7 585	21 517
2nd quarter	290 253	38 976	32 447	28 930	26 331	3 472	5 754	17 105	47 378	35 276	32 106	2 985	7 629	21 588
3rd quarter	290 974	39 754	32 794	29 527	26 815	3 580	5 927	17 309	48 063	35 783	32 497	3 109	7 750	21 694
4th quarter	291 669	40 351	33 171	29 670	27 056	3 590	5 972	17 494	48 499	35 792	32 639	3 146	7 776	21 763
2004														
1st quarter	292 237	40 789	33 459	29 995	27 470	3 642	6 088	17 741	48 663	35 887	32 866	3 195	7 821	21 886
2nd quarter	292 875	41 345	33 986	30 480	27 770	3 640	6 156	17 974	48 927	36 219	32 998	3 196	7 835	22 008
3rd quarter	293 603	41 905	34 382	30 757	28 148	3 696	6 220	18 232	49 267	36 365	33 280	3 271	7 872	22 161
4th quarter	294 334	42 545	35 042	31 333	28 626	3 761	6 379	18 486	49 637	36 733	33 559	3 324	7 932	22 316
2005														
1st quarter	294 957	43 265	35 051	31 077	28 896	3 763	6 414	18 718	50 081	36 225	33 683	3 333	7 990	22 377
2nd quarter	295 588	43 676	35 540	31 493	29 336	3 850	6 494	18 993	50 204	36 491	33 992	3 414	8 036	22 541
3rd quarter	296 340	44 351	35 978	31 843	29 862	3 887	6 702	19 273	50 523	36 506	34 235	3 472	8 045	22 704
4th quarter	297 086	44 877	36 648	32 421	30 109	3 752	6 799	19 558	50 715	36 881	34 251	3 360	8 137	22 777
2006														
1st quarter	297 736	45 691	37 582	33 147	30 533	3 876	6 861	19 795	51 277	37 517	34 557	3 486	8 203	22 864
2nd quarter	298 408	46 077	37 948	33 421	30 865	3 851	6 960	20 054	51 281	37 524	34 654	3 479	8 201	22 973
3rd quarter	299 180	46 352	38 202	33 662	31 195	3 879	7 062	20 254	51 228	37 524	34 774	3 520	8 227	23 019
4th quarter	299 946	46 799	38 620	33 961	31 364	3 899	6 977	20 488	51 532	37 922	35 021	3 561	8 320	23 129
2007														
1st quarter	300 609	47 266	39 376	34 498	31 767	3 922	7 083	20 762	51 540	38 161	35 140	3 601	8 330	23 189
2nd quarter	301 284	47 737	39 753	34 789	32 012	3 935	7 190	20 887	51 719	38 173	35 126	3 630	8 293	23 174
3rd quarter	302 062	48 119	39 887	34 926	32 291	3 946	7 244	21 101	51 866	38 108	35 233	3 665	8 298	23 237
4th quarter	302 829	48 481	40 185	35 204	32 618	3 948	7 372	21 298	52 049	38 037	35 244	3 683	8 270	23 255
2008														
1st quarter	303 494	48 275	40 577	35 520	32 733	3 801	7 423	21 509	51 637	38 060	35 073	3 556	8 188	23 313
2nd quarter	304 160	48 677	41 566	36 463	33 051	3 741	7 581	21 730	51 790	38 671	35 053	3 523	8 183	23 337
3rd quarter	304 902	48 656	40 889	35 979	33 063	3 594	7 649	21 820	51 384	37 766	34 706	3 396	8 044	23 269
4th quarter	305 616	47 640	40 391	35 664	32 188	3 298	7 091	21 799	50 154	38 006	34 302	3 139	7 948	23 244
2009														
1st quarter	306 237	47 005	39 151	35 226	31 858	3 280	6 908	21 670	49 491	37 767	34 156	3 138	7 929	23 115
2nd quarter	306 866	46 772	39 376	35 691	31 806	3 241	6 973	21 592	49 318	38 094	33 948	3 109	7 866	22 999
3rd quarter	307 573	46 884	39 140	35 450	32 173	3 365	7 131	21 676	49 384	37 577	34 103	3 252	7 873	22 989
4th quarter	308 285	47 450	39 332	35 632	32 298	3 290	7 211	21 798	49 811	37 481	33 974	3 168	7 880	22 944
2010														
1st quarter	308 901	47 657	39 740	35 883	32 504	3 306	7 268	21 930	49 903	37 616	34 074	3 201	7 906	22 984
2nd quarter	309 468	48 231	40 368	36 449	32 739	3 373	7 262	22 104	50 271	38 166	34 282	3 295	7 934	23 062
3rd quarter	310 088	48 631	40 750	36 699	32 960	3 394	7 297	22 269	50 540	38 361	34 453	3 341	7 948	23 171
4th quarter	310 718	49 050	41 201	37 053	33 312	3 471	7 468	22 373	50 691	38 498	34 610	3 427	7 991	23 196
2011														
1st quarter	311 237	49 113	42 277	37 695	33 689	3 495	7 650	22 543	50 485	38 825	34 699	3 451	8 004	23 247
2nd quarter	311 763	49 705	42 556	37 920	34 039	3 473	7 799	22 766	50 760	38 675	34 718	3 421	7 974	23 326
3rd quarter	312 389	49 912	42 902	38 193	34 269	3 492	7 801	22 976	50 644	38 766	34 784	3 445	7 931	23 412
4th quarter	313 002	50 468	43 065	38 368	34 382	3 554	7 824	23 004	51 131	38 811	34 779	3 520	7 913	23 346
2012														
1st quarter	313 520	51 097	43 971	39 289	34 838	3 630	7 944	23 264	51 446	39 472	35 000	3 600	7 958	23 442
2nd quarter	314 041	51 434	44 411	39 675	34 915	3 610	7 906	23 400	51 582	39 765	34 994	3 601	7 936	23 457
3rd quarter	314 660	51 666	44 224	39 427	35 004	3 628	7 914	23 462	51 550	39 402	34 982	3 641	7 924	23 417
4th quarter	315 277	51 887	45 651	40 667	35 289	3 691	7 963	23 635	51 508	40 415	35 070	3 716	7 909	23 446
2013														
1st quarter	315 750	52 477	44 343	39 119	35 544	3 765	8 053	23 726	51 886	38 739	35 199	3 806	8 002	23 394
2nd quarter	316 251	52 610	44 696	39 377	35 540	3 748	7 949	23 843	51 868	38 966	35 169	3 817	7 981	23 374
3rd quarter	316 877	53 171	44 913	39 629	35 758	3 752	8 020	23 986	52 171	39 057	35 242	3 842	8 017	23 388
4th quarter	317 515	53 803	45 214	39 867	36 140	3 762	8 076	24 303	52 481	39 133	35 474	3 875	8 072	23 533
2014														
1st quarter	317 838	53 815	46 059	40 571	36 416	3 787	8 136	24 493	52 280	39 638	35 578	3 932	8 082	23 570
2nd quarter	318 370	54 757	46 772	41 251	36 928	3 893	8 240	24 795	52 899	40 110	35 907	4 067	8 146	23 708
3rd quarter	319 009	55 552	47 397	41 761	37 351	3 936	8 283	25 132	53 438	40 497	36 221	4 133	8 198	23 906
4th quarter	319 651	55 842	47 917	42 174	37 680	3 972	8 233	25 475	53 631	40 950	36 587	4 202	8 270	24 132
2015														
1st quarter	320 157	56 173	48 347	42 400	37 768	4 005	8 062	25 701	53 966	41 350	36 832	4 266	8 327	24 262
2nd quarter	320 683	56 814	48 911	42 859	38 194	4 078	8 160	25 956	54 277	41 592	37 065	4 351	8 361	24 381
3rd quarter	321 315	57 093	49 270	43 211	38 517	4 098	8 227	26 193	54 349	41 816	37 274	4 398	8 432	24 477
4th quarter	321 947	56 999	49 314	43 227	38 582	4 091	8 127	26 363	54 260	41 866	37 367	4 414	8 437	24 548
2016														
1st quarter	322 476	57 134	49 423	43 463	38 836	4 124	8 073	26 639	54 444	42 077	37 597	4 469	8 493	24 671
2nd quarter	322 998	57 701	49 626	43 603	39 283	4 159	8 190	26 934	54 612	41 964	37 806	4 539	8 547	24 765
3rd quarter	323 606	58 116	49 985	43 899	39 623	4 218	8 198	27 208	54 805	42 073	37 975	4 648	8 545	24 837
4th quarter	324 187	58 583	50 411	44 281	39 976	4 233	8 241	27 502	54 981	42 249	38 143	4 705	8 530	24 962
2017														
1st quarter	324 648	59 111	51 145	44 980	40 365	4 267	8 356	27 742	55 214	42 695	38 315	4 738	8 583	25 052
2nd quarter	325 107	59 539	51 520	45 319	40 641	4 302	8 386	27 953	55 431	42 919	38 489	4 820	8 662	25 081
3rd quarter	325 667	60 220	51 987	45 693	40 978	4 348	8 467	28 163	55 773	43 095	38 648	4 903	8 694	25 138
4th quarter	326 218	61 060	52 821	46 335	41 648	4 447	8 624	28 577	56 166	43 414	39 022	5 038	8 788	25 305

Table 1-4. Contributions to Percent Change in Real Gross Domestic Product: Recent Data

(Percent, percentage points.) NIPA Table 1.1.2

Year and quarter	Percent change at seasonally adjusted annual rate, real GDP	Personal consumption expenditures	Gross private domestic investment				Exports and imports of goods and services			Government consumption expenditures and gross investment		
			Total	Fixed investment		Change in private inventories	Net exports	Exports	Imports	Total	Federal	State and local
				Nonresidential	Residential							
1955	7.1	4.50	3.45	1.07	0.88	1.50	-0.04	0.43	-0.47	-0.78	-1.31	0.54
1956	2.1	1.77	-0.05	0.70	-0.47	-0.28	0.36	0.69	-0.33	0.05	-0.19	0.25
1957	2.1	1.50	-0.64	0.20	-0.31	-0.53	0.24	0.41	-0.18	1.01	0.55	0.46
1958	-0.7	0.52	-1.16	-1.04	0.05	-0.17	-0.87	-0.67	-0.19	0.76	0.08	0.68
1959	6.9	3.51	2.83	0.81	1.18	0.83	0.00	0.44	-0.44	0.60	0.30	0.30
1960	2.6	1.67	0.06	0.56	-0.37	-0.13	0.70	0.76	-0.06	0.14	-0.23	0.37
1961	2.6	1.25	0.05	0.08	0.02	-0.05	0.05	0.03	0.03	1.21	0.66	0.54
1962	6.1	3.00	1.82	0.81	0.46	0.55	-0.21	0.25	-0.45	1.51	1.24	0.27
1963	4.4	2.48	1.08	0.59	0.57	-0.07	0.23	0.34	-0.11	0.57	0.03	0.53
1964	5.8	3.57	1.27	1.09	0.31	-0.12	0.35	0.57	-0.22	0.57	-0.05	0.62
1965	6.5	3.80	2.26	1.76	-0.13	0.64	-0.29	0.14	-0.44	0.73	0.10	0.63
1966	6.6	3.38	1.56	1.40	-0.40	0.56	-0.28	0.35	-0.63	1.94	1.34	0.60
1967	2.7	1.76	-0.62	-0.04	-0.11	-0.47	-0.21	0.12	-0.33	1.81	1.32	0.49
1968	4.9	3.39	0.99	0.55	0.53	-0.09	-0.29	0.40	-0.68	0.82	0.21	0.61
1969	3.1	2.20	0.93	0.79	0.14	0.00	-0.03	0.25	-0.28	0.02	-0.34	0.36
1970	0.2	1.39	-1.03	-0.10	-0.23	-0.70	0.33	0.54	-0.21	-0.50	-0.80	0.30
1971	3.3	2.29	1.63	-0.01	1.08	0.56	-0.18	0.10	-0.28	-0.45	-0.80	0.35
1972	5.3	3.66	1.90	0.97	0.87	0.06	-0.19	0.42	-0.61	-0.12	-0.37	0.25
1973	5.6	2.97	1.95	1.51	-0.04	0.48	0.80	1.08	-0.28	-0.07	-0.39	0.32
1974	-0.5	-0.50	-1.24	0.10	-1.08	-0.26	0.73	0.56	0.17	0.47	0.06	0.41
1975	-0.2	1.36	-2.91	-1.13	-0.54	-1.24	0.86	-0.05	0.91	0.49	0.05	0.43
1976	5.4	3.41	2.91	0.66	0.88	1.37	-1.05	0.36	-1.41	0.12	0.01	0.10
1977	4.6	2.59	2.47	1.26	0.97	0.24	-0.70	0.19	-0.89	0.26	0.21	0.05
1978	5.5	2.68	2.22	1.72	0.38	0.12	0.05	0.80	-0.76	0.60	0.23	0.37
1979	3.2	1.44	0.72	1.34	-0.22	-0.40	0.64	0.80	-0.16	0.36	0.20	0.16
1980	-0.3	-0.19	-2.07	0.00	-1.19	-0.89	1.64	0.95	0.69	0.36	0.38	-0.02
1981	2.5	0.85	1.64	0.87	-0.37	1.13	-0.15	0.12	-0.26	0.20	0.43	-0.23
1982	-1.8	0.88	-2.46	-0.43	-0.72	-1.31	-0.59	-0.71	0.12	0.37	0.35	0.01
1983	4.6	3.51	1.60	-0.06	1.38	0.28	-1.32	-0.22	-1.10	0.79	0.65	0.14
1984	7.2	3.30	4.73	2.18	0.65	1.90	-1.54	0.61	-2.16	0.74	0.33	0.41
1985	4.2	3.20	-0.01	0.91	0.11	-1.03	-0.39	0.24	-0.63	1.37	0.78	0.59
1986	3.5	2.58	0.03	-0.24	0.58	-0.31	-0.29	0.53	-0.82	1.14	0.61	0.53
1987	3.5	2.15	0.53	0.01	0.10	0.41	0.17	0.77	-0.60	0.62	0.38	0.24
1988	4.2	2.65	0.45	0.63	-0.05	-0.13	0.81	1.23	-0.41	0.26	-0.15	0.42
1989	3.7	1.86	0.72	0.71	-0.16	0.17	0.51	0.97	-0.46	0.58	0.15	0.43
1990	1.9	1.28	-0.45	0.14	-0.38	-0.21	0.40	0.78	-0.37	0.65	0.20	0.45
1991	-0.1	0.12	-1.09	-0.48	-0.35	-0.26	0.62	0.61	0.01	0.25	0.01	0.24
1992	3.5	2.36	1.11	0.33	0.49	0.28	-0.04	0.66	-0.70	0.10	-0.15	0.25
1993	2.8	2.24	1.24	0.84	0.32	0.07	-0.56	0.31	-0.87	-0.17	-0.32	0.15
1994	4.0	2.51	1.90	0.91	0.38	0.61	-0.41	0.84	-1.25	0.02	-0.31	0.32
1995	2.7	1.91	0.55	1.15	-0.15	-0.44	0.12	1.02	-0.90	0.10	-0.21	0.31
1996	3.8	2.26	1.49	1.13	0.35	0.02	-0.15	0.86	-1.01	0.18	-0.09	0.27
1997	4.4	2.45	2.01	1.38	0.11	0.52	-0.31	1.26	-1.57	0.30	-0.06	0.36
1998	4.5	3.42	1.76	1.44	0.38	-0.07	-1.14	0.26	-1.39	0.44	-0.06	0.50
1999	4.8	3.42	1.62	1.36	0.29	-0.03	-0.87	0.52	-1.39	0.58	0.13	0.46
2000	4.1	3.32	1.31	1.31	0.03	-0.03	-0.83	0.86	-1.69	0.33	0.02	0.31
2001	1.0	1.66	-1.11	-0.31	0.04	-0.84	-0.22	-0.61	0.39	0.67	0.24	0.43
2002	1.7	1.71	-0.16	-0.94	0.29	0.48	-0.64	-0.17	-0.47	0.82	0.47	0.35
2003	2.9	2.13	0.76	0.30	0.47	-0.02	-0.45	0.20	-0.64	0.41	0.45	-0.03
2004	3.8	2.53	1.64	0.67	0.57	0.41	-0.67	0.88	-1.55	0.30	0.31	-0.01
2005	3.5	2.39	1.26	0.92	0.41	-0.07	-0.29	0.69	-0.97	0.15	0.15	0.00
2006	2.9	2.05	0.60	1.00	-0.50	0.10	-0.10	0.94	-1.04	0.30	0.17	0.13
2007	1.9	1.49	-0.48	0.89	-1.13	-0.25	0.53	0.93	-0.41	0.34	0.14	0.20
2008	-0.1	-0.14	-1.52	0.08	-1.14	-0.46	1.04	0.66	0.38	0.48	0.46	0.02
2009	-2.5	-0.85	-3.52	-1.95	-0.74	-0.83	1.13	-1.01	2.14	0.70	0.47	0.23
2010	2.6	1.20	1.86	0.52	-0.08	1.42	-0.49	1.35	-1.84	0.00	0.35	-0.35
2011	1.6	1.29	0.94	1.00	0.00	-0.05	-0.01	0.90	-0.91	-0.66	-0.23	-0.44
2012	2.2	1.03	1.64	1.16	0.31	0.17	0.00	0.46	-0.46	-0.42	-0.16	-0.26
2013	1.8	0.99	1.11	0.54	0.34	0.23	0.22	0.48	-0.26	-0.47	-0.44	-0.03
2014	2.5	1.99	0.95	0.95	0.12	-0.12	-0.25	0.57	-0.81	-0.17	-0.19	0.02
2015	2.9	2.48	0.85	0.25	0.33	0.28	-0.77	0.06	-0.83	0.35	-0.01	0.35
2016	1.6	1.85	-0.23	0.09	0.23	-0.55	-0.30	0.00	-0.30	0.32	0.03	0.29
2017	2.4	1.78	0.75	0.57	0.13	0.04	-0.28	0.41	-0.69	0.12	0.05	0.07
2018	2.9	2.05	0.87	0.84	-0.06	0.09	-0.29	0.37	-0.66	0.30	0.19	0.11
2016												
1st quarter	2.0	2.11	-0.26	-0.08	0.50	-0.68	-0.50	-0.38	-0.11	0.67	0.05	0.63
2nd quarter	1.9	1.95	-0.28	0.52	-0.07	-0.72	0.35	0.45	-0.10	-0.12	-0.18	0.06
3rd quarter	2.2	1.74	0.09	0.72	-0.10	-0.53	0.05	0.71	-0.66	0.31	0.13	0.18
4th quarter	2.0	1.70	1.50	0.09	0.24	1.18	-1.36	-0.30	-1.06	0.19	0.04	0.15
2017												
1st quarter	2.3	1.63	0.57	0.84	0.43	-0.70	0.13	0.72	-0.58	-0.04	-0.08	0.03
2nd quarter	2.2	1.63	0.59	0.57	-0.09	0.11	-0.31	0.20	-0.51	0.24	0.21	0.03
3rd quarter	3.2	1.61	1.25	0.32	-0.08	1.00	0.35	0.54	-0.18	-0.02	0.01	-0.02
4th quarter	3.5	3.12	0.80	1.08	0.37	-0.64	-0.80	1.19	-1.99	0.42	0.30	0.12
2018												
1st quarter	2.5	1.15	1.07	1.15	-0.21	0.13	0.00	0.10	-0.10	0.33	0.18	0.15
2nd quarter	3.5	2.70	-0.30	1.04	-0.15	-1.20	0.67	0.71	-0.04	0.44	0.25	0.19
3rd quarter	2.9	2.34	2.27	0.29	-0.16	2.14	-2.05	-0.78	-1.27	0.36	0.19	0.17
4th quarter	1.1	0.97	0.53	0.64	-0.18	0.07	-0.35	0.18	-0.53	-0.07	0.07	-0.14

Table 1-5A. Chain-Type Quantity Indexes for Gross Domestic Product and Domestic Purchases: Recent Data

(Index numbers, 2012 = 100.)

NIPA Tables 1.1.3, 1.4.3, 2.3.3

Year and quarter	Gross domestic product												Gross domestic purchases
	Gross domestic product, total	Personal consumption expenditures		Private fixed investment			Exports and imports of goods and services		Government consumption expenditures and gross investment			
		Total	Excluding food and energy	Total	Nonresidential	Residential	Exports	Imports	Total	Federal	State and local	
1955	17.7	15.8	12.9	12.8	7.8	48.9	3.6	3.7	31.0	40.1	21.7	17.5
1956	18.1	16.3	13.2	13.0	8.3	45.0	4.2	4.0	31.0	39.6	22.4	17.8
1957	18.5	16.7	13.5	12.9	8.5	42.4	4.6	4.2	32.5	41.1	23.7	18.1
1958	18.3	16.8	13.6	12.1	7.7	42.8	4.0	4.4	33.6	41.3	25.7	18.2
1959	19.6	17.8	14.4	13.7	8.3	53.6	4.4	4.8	34.4	42.2	26.6	19.4
1960	20.1	18.3	14.9	13.9	8.8	49.9	5.2	4.9	34.6	41.5	27.7	19.8
1961	20.6	18.6	15.3	13.9	8.8	50.1	5.2	4.9	36.5	43.5	29.4	20.3
1962	21.9	19.6	16.2	15.1	9.5	54.9	5.4	5.4	38.9	47.4	30.3	21.6
1963	22.9	20.4	17.0	16.3	10.1	61.3	5.8	5.6	39.8	47.6	32.1	22.5
1964	24.2	21.6	18.1	17.8	11.2	65.0	6.5	5.9	40.8	47.4	34.2	23.7
1965	25.8	23.0	19.3	19.7	13.0	63.4	6.7	6.5	42.1	47.7	36.5	25.3
1966	27.4	24.3	20.5	20.9	14.6	58.0	7.2	7.5	45.8	52.8	38.7	27.1
1967	28.2	25.0	21.1	20.7	14.6	56.5	7.3	8.0	49.4	58.1	40.7	27.9
1968	29.6	26.4	22.4	22.1	15.3	64.2	7.9	9.2	51.1	59.0	43.1	29.4
1969	30.5	27.4	23.3	23.4	16.4	66.2	8.3	9.7	51.2	57.6	44.6	30.3
1970	30.6	28.0	23.8	23.0	16.2	62.7	9.2	10.1	50.1	54.0	45.8	30.2
1971	31.6	29.1	24.8	24.5	16.2	79.4	9.3	10.7	49.1	50.6	47.3	31.3
1972	33.2	30.9	26.6	27.3	17.6	93.2	10.1	11.9	48.9	49.0	48.3	33.0
1973	35.1	32.4	28.3	29.7	19.9	92.6	12.0	12.4	48.7	47.2	49.7	34.6
1974	34.9	32.2	28.2	28.0	20.1	74.5	12.9	12.1	49.8	47.5	51.5	34.2
1975	34.9	32.9	28.8	25.3	18.3	65.5	12.8	10.8	50.9	47.8	53.4	33.8
1976	36.7	34.7	30.5	27.8	19.3	80.0	13.4	12.9	51.2	47.9	53.8	36.0
1977	38.4	36.2	32.0	31.5	21.4	96.4	13.7	14.3	51.8	48.9	54.1	37.9
1978	40.6	37.8	33.7	35.2	24.4	102.8	15.1	15.5	53.3	50.0	55.8	40.0
1979	41.8	38.7	34.7	37.2	26.8	98.9	16.6	15.8	54.2	51.1	56.6	41.0
1980	41.7	38.5	34.7	35.0	26.8	78.2	18.4	14.8	55.2	53.3	56.5	40.2
1981	42.8	39.1	35.4	35.9	28.5	71.8	18.7	15.1	55.7	55.7	55.4	41.3
1982	42.0	39.7	35.9	33.7	27.6	58.8	17.2	14.9	56.7	57.7	55.4	40.8
1983	43.9	41.9	38.3	36.3	27.5	83.5	16.8	16.8	58.9	61.3	56.2	43.2
1984	47.1	44.1	40.6	42.1	32.1	95.8	18.2	20.9	60.9	63.2	58.3	46.9
1985	49.1	46.4	42.9	44.4	34.2	98.0	18.8	22.3	65.0	68.1	61.6	49.0
1986	50.8	48.3	44.9	45.2	33.6	110.1	20.2	24.2	68.5	72.1	64.7	50.8
1987	52.5	49.9	46.7	45.5	33.6	112.3	22.4	25.6	70.5	74.7	66.1	52.4
1988	54.7	52.0	48.7	47.0	35.3	111.3	26.0	26.6	71.4	73.6	68.7	54.1
1989	56.8	53.5	50.3	48.5	37.3	107.7	29.1	27.8	73.4	74.8	71.4	55.8
1990	57.8	54.6	51.4	47.8	37.7	98.6	31.6	28.8	75.8	76.4	74.3	56.6
1991	57.8	54.7	51.5	45.3	36.2	89.8	33.7	28.8	76.7	76.4	75.9	56.2
1992	59.8	56.7	53.7	47.8	37.3	102.2	36.0	30.8	77.0	75.2	77.5	58.2
1993	61.4	58.7	55.8	51.5	40.1	110.5	37.2	33.4	76.4	72.6	78.5	60.1
1994	63.9	61.0	58.1	55.7	43.2	120.4	40.5	37.4	76.5	69.9	80.7	62.7
1995	65.6	62.8	60.0	59.1	47.4	116.3	44.7	40.4	76.9	68.1	82.9	64.3
1996	68.1	65.0	62.3	64.3	51.7	125.9	48.3	43.9	77.6	67.3	84.9	66.8
1997	71.1	67.4	64.9	69.8	57.3	128.9	54.1	49.8	78.8	66.7	87.6	70.0
1998	74.3	71.0	68.7	77.0	63.5	140.0	55.3	55.7	80.7	66.1	91.5	73.9
1999	77.9	74.7	72.5	84.0	69.8	148.8	58.1	62.0	83.4	67.4	95.1	77.9
2000	81.1	78.5	76.6	89.9	76.3	149.8	63.0	69.9	84.9	67.6	97.6	81.7
2001	81.9	80.5	78.8	88.7	74.6	151.2	59.3	68.0	88.1	70.2	101.2	82.7
2002	83.3	82.6	81.0	85.6	69.5	160.3	58.3	70.5	92.0	75.4	104.2	84.6
2003	85.7	85.2	83.7	89.4	71.2	174.9	59.6	73.9	93.9	80.3	103.9	87.3
2004	88.9	88.4	87.1	95.7	75.2	192.3	65.3	82.3	95.4	83.8	103.8	91.0
2005	92.1	91.5	90.3	102.7	81.0	204.9	70.0	87.7	96.1	85.4	103.8	94.3
2006	94.7	94.3	93.3	105.3	87.5	189.5	76.5	93.5	97.7	87.4	104.9	96.9
2007	96.5	96.4	95.6	104.0	93.6	154.1	83.2	95.9	99.4	89.2	106.7	98.1
2008	96.3	96.2	95.7	98.0	94.1	116.8	87.9	93.7	101.9	94.7	106.9	97.0
2009	93.9	95.0	94.3	82.3	80.5	91.5	80.5	81.5	105.4	100.5	108.9	93.6
2010	96.3	96.7	96.0	84.9	84.1	88.7	90.3	92.2	105.4	104.6	106.0	96.4
2011	97.8	98.5	98.2	90.9	91.4	88.5	96.7	97.4	102.1	101.9	102.3	97.9
2012	100.0	100.0	100.0	100.0	100.0	100.0	100.0	100.0	100.0	100.0	100.0	100.0
2013	101.8	101.5	101.4	105.6	104.1	112.4	103.6	101.5	97.6	94.5	99.7	101.6
2014	104.4	104.5	104.7	112.5	111.6	116.7	107.9	106.6	96.7	92.0	99.9	104.3
2015	107.5	108.3	108.8	116.3	113.7	128.5	108.5	112.3	98.6	91.9	103.1	108.0
2016	109.2	111.3	111.9	118.5	114.5	136.8	108.4	114.5	100.3	92.3	105.8	110.1
2017	111.8	114.2	115.0	123.5	119.5	141.6	112.2	119.9	101.0	93.0	106.5	112.9
2018	115.1	117.6	118.4	129.1	127.1	139.5	115.6	125.1	102.8	95.8	107.5	116.4
2016												
1st quarter	108.4	110.2	110.8	117.3	112.8	137.3	107.0	113.1	100.2	92.5	105.4	109.3
2nd quarter	108.9	110.9	111.5	118.1	113.9	136.6	108.1	113.4	100.0	91.9	105.5	109.7
3rd quarter	109.5	111.7	112.2	119.1	115.5	135.7	109.7	114.7	100.4	92.4	106.0	110.2
4th quarter	110.0	112.3	113.1	119.7	115.7	137.8	109.0	116.8	100.7	92.5	106.3	111.2
2017												
1st quarter	110.7	113.0	113.9	122.0	117.6	141.8	110.6	117.9	100.6	92.2	106.4	111.7
2nd quarter	111.3	113.7	114.4	122.8	118.8	141.0	111.0	119.0	101.0	93.0	106.5	112.4
3rd quarter	112.1	114.4	115.1	123.3	119.5	140.2	112.2	119.3	101.0	93.0	106.4	113.2
4th quarter	113.1	115.7	116.4	125.9	122.0	143.6	115.0	123.3	101.6	94.0	106.7	114.4
2018												
1st quarter	113.8	116.1	116.9	127.6	124.6	141.7	115.2	123.5	102.0	94.7	107.1	115.1
2nd quarter	114.8	117.3	118.1	129.2	127.0	140.3	116.8	123.6	102.7	95.6	107.5	115.8
3rd quarter	115.7	118.3	119.2	129.4	127.6	138.9	115.0	126.2	103.2	96.3	108.0	117.2
4th quarter	116.0	118.7	119.6	130.3	129.1	137.3	115.4	127.2	103.1	96.5	107.6	117.6

Table 1-5B. Chain-Type Quantity Indexes for Gross Domestic Product and Domestic Purchases: Historical Data

(Index numbers, 2012 = 100.)

NIPA Tables 1.1.3, 1.4.3, 2.3.3

Year and quarter	Gross domestic product, total	Personal consumption expenditures		Private fixed investment			Exports and imports of goods and services		Government consumption expenditures and gross investment			Gross domestic purchases
		Total	Excluding food and energy	Total	Nonresidential	Residential	Exports	Imports	Total	Federal	State and local	
1929	6.9	7.5	6.2	5.9	4.1	18.6	2.0	2.1	5.8	2.3	11.3	6.8
1930	6.3	7.1	5.8	4.6	3.4	11.5	1.6	1.8	6.3	2.5	12.4	6.2
1931	5.9	6.9	5.5	3.3	2.3	9.8	1.4	1.6	6.6	2.6	13.0	5.8
1932	5.1	6.3	5.0	2.0	1.4	5.5	1.1	1.3	6.4	2.6	12.4	5.1
1933	5.0	6.2	4.8	1.8	1.3	4.6	1.1	1.4	6.2	3.2	11.1	5.0
1934	5.6	6.6	5.1	2.3	1.7	6.0	1.2	1.4	6.9	4.2	11.6	5.6
1935	6.1	7.0	5.4	2.9	2.1	8.2	1.3	1.8	7.2	4.3	12.0	6.1
1936	6.9	7.7	5.9	3.8	2.7	10.2	1.3	1.8	8.3	6.4	12.0	6.9
1937	7.2	8.0	6.2	4.4	3.2	11.0	1.7	2.1	8.0	5.8	12.0	7.2
1938	7.0	7.9	6.0	3.6	2.5	11.0	1.6	1.6	8.6	6.4	12.7	6.9
1939	7.5	8.3	6.4	4.2	2.7	15.4	1.7	1.7	9.3	6.9	14.0	7.4
1940	8.2	8.7	6.7	5.0	3.3	17.3	2.0	1.7	9.7	7.8	13.5	8.1
1941	9.7	9.4	7.3	5.8	4.0	18.4	2.0	2.1	16.3	20.6	12.7	9.6
1942	11.5	9.1	7.1	3.4	2.5	9.5	1.3	1.9	37.7	61.6	11.6	11.5
1943	13.4	9.4	7.5	2.7	2.1	5.9	1.1	2.4	56.6	98.1	10.6	13.6
1944	14.5	9.7	7.7	3.2	2.6	5.1	1.2	2.5	63.5	111.7	10.2	14.7
1945	14.4	10.3	8.2	4.3	3.6	5.8	1.7	2.7	55.8	96.5	10.6	14.4
1946	12.7	11.5	9.1	7.6	5.2	23.5	3.7	2.2	19.6	27.0	11.6	12.2
1947	12.6	11.7	9.3	9.1	6.0	30.1	4.2	2.1	16.7	20.2	13.2	12.0
1948	13.1	12.0	9.6	9.9	6.3	35.9	3.3	2.5	17.6	21.3	14.0	12.7
1949	13.0	12.3	9.9	9.1	5.7	33.3	3.3	2.4	19.5	23.0	16.0	12.6
1947												
1st quarter	12.6	11.6	. . .	8.9	6.1	27.6	4.5	2.2	16.7	20.6	12.8	11.9
2nd quarter	12.5	11.8	. . .	8.7	6.0	26.2	4.4	2.2	16.7	20.4	13.0	11.9
3rd quarter	12.5	11.8	. . .	9.0	5.9	30.3	4.2	2.0	16.7	20.2	13.3	11.9
4th quarter	12.7	11.8	. . .	9.7	6.0	36.1	3.7	2.1	16.6	19.7	13.5	12.2
1948												
1st quarter	12.9	11.9	. . .	10.0	6.4	35.6	3.5	2.3	16.8	20.2	13.5	12.5
2nd quarter	13.1	12.0	. . .	10.0	6.2	37.7	3.2	2.4	17.4	21.0	13.9	12.8
3rd quarter	13.2	12.0	. . .	9.9	6.2	36.5	3.3	2.6	17.8	21.5	14.1	12.8
4th quarter	13.2	12.1	. . .	9.8	6.3	33.8	3.2	2.5	18.4	22.5	14.4	12.9
1949												
1st quarter	13.0	12.1	. . .	9.3	6.0	31.3	3.5	2.4	18.9	22.9	15.0	12.6
2nd quarter	13.0	12.3	. . .	9.0	5.8	30.8	3.5	2.4	19.7	23.7	15.9	12.6
3rd quarter	13.1	12.4	. . .	9.0	5.5	33.4	3.2	2.3	19.8	23.3	16.5	12.7
4th quarter	13.0	12.5	. . .	9.3	5.5	37.5	2.8	2.4	19.5	22.3	16.8	12.7
1950												
1st quarter	13.5	12.8	. . .	9.9	5.6	41.7	2.7	2.4	19.1	21.3	17.2	13.2
2nd quarter	13.9	13.0	. . .	10.7	6.1	45.7	2.8	2.5	19.5	21.8	17.4	13.7
3rd quarter	14.4	13.6	. . .	11.5	6.6	48.2	2.8	3.1	19.1	21.2	17.4	14.3
4th quarter	14.7	13.2	. . .	11.3	6.6	45.9	3.1	3.2	20.5	23.8	17.4	14.5
1951												
1st quarter	14.9	13.5	. . .	10.8	6.4	43.6	3.2	3.2	22.7	28.2	17.2	14.7
2nd quarter	15.2	13.2	. . .	10.4	6.5	37.7	3.6	3.0	25.7	33.8	17.5	14.9
3rd quarter	15.5	13.3	. . .	10.2	6.6	35.4	3.6	2.8	28.5	39.3	17.6	15.1
4th quarter	15.5	13.4	. . .	10.1	6.5	35.7	3.6	2.7	30.0	42.3	17.6	15.1
1952												
1st quarter	15.7	13.4	. . .	10.3	6.5	36.6	3.8	3.0	30.9	44.1	17.6	15.3
2nd quarter	15.7	13.7	. . .	10.5	6.6	37.3	3.4	3.0	32.0	45.8	18.0	15.4
3rd quarter	15.8	13.7	. . .	9.9	6.1	37.0	3.1	3.2	32.3	46.9	17.5	15.6
4th quarter	16.4	14.2	. . .	10.6	6.7	38.9	3.1	3.5	32.9	47.6	17.9	16.2
1953												
1st quarter	16.7	14.4	. . .	11.1	7.0	39.6	3.1	3.4	33.9	49.3	18.3	16.5
2nd quarter	16.8	14.5	. . .	11.2	7.1	39.7	3.1	3.6	34.6	50.7	18.3	16.7
3rd quarter	16.7	14.4	. . .	11.2	7.2	38.2	3.2	3.6	34.2	49.5	18.8	16.5
4th quarter	16.4	14.4	. . .	11.1	7.2	37.9	3.1	3.4	34.1	48.9	19.2	16.3
1954												
1st quarter	16.4	14.4	. . .	10.9	7.0	38.3	3.0	3.2	33.2	46.4	19.9	16.2
2nd quarter	16.4	14.6	. . .	11.1	6.9	40.8	3.4	3.5	32.1	44.1	20.0	16.2
3rd quarter	16.6	14.8	. . .	11.5	7.1	43.3	3.3	3.3	31.4	42.2	20.5	16.3
4th quarter	16.9	15.1	. . .	11.8	7.1	46.0	3.4	3.3	31.2	41.7	20.6	16.6
1955												
1st quarter	17.4	15.4	. . .	12.2	7.2	49.4	3.6	3.5	31.3	40.9	21.5	17.1
2nd quarter	17.7	15.7	. . .	12.7	7.7	50.1	3.5	3.7	31.0	40.0	21.7	17.5
3rd quarter	17.9	15.9	. . .	13.0	8.0	49.0	3.7	3.8	31.1	40.4	21.7	17.7
4th quarter	18.0	16.1	. . .	13.1	8.3	47.0	3.7	3.9	30.5	39.1	21.8	17.8

. . . = Not available.

Table 1-5B. Chain-Type Quantity Indexes for Gross Domestic Product and Domestic Purchases: Historical Data—*Continued*

(Index numbers, 2012 = 100.)

NIPA Tables 1.1.3, 1.4.3, 2.3.3

Year and quarter	Gross domestic product, total	Personal consumption expenditures		Private fixed investment			Exports and imports of goods and services		Government consumption expenditures and gross investment			Gross domestic purchases
		Total	Excluding food and energy	Total	Nonresidential	Residential	Exports	Imports	Total	Federal	State and local	
1956												
1st quarter	17.9	16.2	. . .	12.9	8.2	45.8	3.9	4.1	30.5	38.9	22.1	17.7
2nd quarter	18.1	16.2	. . .	13.0	8.3	45.6	4.2	4.0	31.1	39.8	22.4	17.8
3rd quarter	18.1	16.2	. . .	13.0	8.4	44.7	4.3	4.1	30.9	39.2	22.5	17.8
4th quarter	18.4	16.5	. . .	12.9	8.4	44.0	4.5	3.9	31.7	40.5	22.7	18.0
1957												
1st quarter	18.5	16.6	. . .	12.9	8.4	43.4	4.8	4.2	32.3	41.3	23.2	18.1
2nd quarter	18.4	16.6	. . .	12.8	8.4	42.4	4.7	4.2	32.2	40.8	23.4	18.1
3rd quarter	18.6	16.7	. . .	13.0	8.6	41.8	4.5	4.1	32.4	41.0	23.8	18.3
4th quarter	18.4	16.8	. . .	12.7	8.4	41.8	4.4	4.2	33.0	41.5	24.4	18.1
1958												
1st quarter	17.9	16.5	. . .	12.0	7.9	40.1	4.0	4.2	32.7	40.2	25.0	17.7
2nd quarter	18.1	16.7	. . .	11.7	7.6	40.3	4.0	4.4	33.5	41.5	25.4	17.9
3rd quarter	18.5	16.9	. . .	11.9	7.5	43.2	4.0	4.3	33.7	41.4	25.9	18.3
4th quarter	18.9	17.2	. . .	12.6	7.8	47.7	4.0	4.6	34.4	42.3	26.4	18.8
1959												
1st quarter	19.3	17.5	14.1	13.4	8.0	53.1	4.3	4.7	34.0	41.4	26.5	19.1
2nd quarter	19.7	17.7	14.4	13.8	8.2	54.9	4.2	4.9	34.4	42.2	26.6	19.6
3rd quarter	19.7	17.9	14.6	13.9	8.5	53.9	4.6	5.0	34.7	42.8	26.6	19.5
4th quarter	19.8	18.0	14.6	13.7	8.4	52.4	4.4	4.8	34.5	42.3	26.5	19.6
1960												
1st quarter	20.2	18.1	14.8	14.2	8.7	54.0	5.0	5.0	33.9	40.7	26.9	19.9
2nd quarter	20.1	18.4	15.0	13.9	8.9	49.5	5.1	5.0	34.2	40.8	27.6	19.8
3rd quarter	20.2	18.3	14.9	13.6	8.7	48.1	5.2	4.9	35.1	42.1	28.0	19.9
4th quarter	20.0	18.3	15.0	13.6	8.7	48.0	5.3	4.7	35.3	42.3	28.3	19.6
1961												
1st quarter	20.1	18.3	14.9	13.5	8.6	48.2	5.2	4.6	35.9	42.4	29.3	19.7
2nd quarter	20.4	18.6	15.2	13.7	8.8	48.4	5.0	4.7	35.9	42.8	29.1	20.1
3rd quarter	20.8	18.7	15.3	14.1	8.8	51.0	5.2	5.0	36.7	44.1	29.3	20.5
4th quarter	21.2	19.0	15.7	14.5	9.1	52.8	5.3	5.1	37.5	44.9	30.1	20.9
1962												
1st quarter	21.6	19.2	15.8	14.8	9.3	53.7	5.3	5.3	38.2	46.5	29.9	21.3
2nd quarter	21.8	19.5	16.1	15.2	9.6	55.4	5.6	5.4	38.4	46.7	30.1	21.5
3rd quarter	22.1	19.6	16.2	15.3	9.7	55.3	5.6	5.5	39.3	48.1	30.4	21.7
4th quarter	22.1	19.9	16.5	15.3	9.6	55.2	5.4	5.5	39.6	48.4	30.8	21.9
1963												
1st quarter	22.4	20.0	16.6	15.5	9.7	57.2	5.5	5.4	39.1	46.7	31.3	22.1
2nd quarter	22.6	20.2	16.9	16.2	10.0	61.3	5.9	5.5	39.1	46.5	31.6	22.2
3rd quarter	23.1	20.5	17.1	16.6	10.2	62.5	5.9	5.7	40.8	49.1	32.4	22.8
4th quarter	23.3	20.7	17.3	17.1	10.5	64.4	6.1	5.7	40.4	47.9	32.9	22.9
1964												
1st quarter	23.8	21.1	17.7	17.6	10.7	68.4	6.4	5.7	40.5	47.6	33.4	23.3
2nd quarter	24.0	21.5	18.0	17.6	11.0	65.0	6.4	5.8	40.9	47.6	34.1	23.6
3rd quarter	24.4	21.9	18.3	17.9	11.4	63.9	6.6	5.9	40.9	47.3	34.5	23.9
4th quarter	24.5	21.9	18.4	18.1	11.6	62.8	6.7	6.1	40.9	46.9	34.8	24.0
1965												
1st quarter	25.1	22.4	18.9	18.9	12.3	63.1	5.9	5.9	40.9	46.5	35.2	24.7
2nd quarter	25.4	22.7	19.1	19.4	12.8	63.7	6.9	6.6	41.4	46.6	36.0	25.0
3rd quarter	26.0	23.0	19.4	19.9	13.3	64.0	6.8	6.6	42.7	48.3	37.1	25.6
4th quarter	26.6	23.7	19.9	20.4	13.8	62.7	7.2	6.9	43.5	49.4	37.6	26.1
1966												
1st quarter	27.2	24.0	20.3	21.2	14.4	64.3	7.0	7.1	44.3	50.6	38.0	26.8
2nd quarter	27.3	24.1	20.3	20.9	14.6	59.3	7.2	7.3	45.2	52.0	38.3	26.9
3rd quarter	27.5	24.4	20.6	21.0	14.8	57.3	7.1	7.7	46.4	54.0	38.8	27.2
4th quarter	27.8	24.5	20.7	20.4	14.8	51.3	7.3	7.8	47.2	54.6	39.8	27.4
1967												
1st quarter	28.0	24.6	20.8	20.0	14.6	49.9	7.4	7.9	49.2	58.1	40.3	27.7
2nd quarter	28.0	25.0	21.1	20.5	14.5	55.6	7.3	7.8	49.0	57.5	40.5	27.7
3rd quarter	28.3	25.1	21.3	20.8	14.5	58.6	7.2	7.9	49.6	58.5	40.6	28.0
4th quarter	28.5	25.2	21.4	21.4	14.8	62.0	7.4	8.3	49.9	58.5	41.3	28.2
1968												
1st quarter	29.1	25.8	21.9	21.9	15.2	62.2	7.6	8.8	50.8	59.5	42.1	28.9
2nd quarter	29.6	26.2	22.3	21.8	15.0	63.7	7.7	9.0	51.1	59.2	42.9	29.3
3rd quarter	29.8	26.7	22.7	22.2	15.2	65.0	8.3	9.5	51.2	58.9	43.5	29.6
4th quarter	29.9	26.9	22.8	22.7	15.7	65.9	8.1	9.4	51.3	58.6	43.9	29.7

. . . = Not available.

Table 1-5B. Chain-Type Quantity Indexes for Gross Domestic Product and Domestic Purchases: Historical Data—Continued

(Index numbers, 2012 = 100.)

NIPA Tables 1.1.3, 1.4.3, 2.3.3

Year and quarter	Gross domestic product, total	Personal consumption expenditures		Private fixed investment			Exports and imports of goods and services		Government consumption expenditures and gross investment			Gross domestic purchases
		Total	Excluding food and energy	Total	Nonresidential	Residential	Exports	Imports	Total	Federal	State and local	
1969												
1st quarter	30.4	27.1	23.0	23.3	16.1	68.1	7.1	8.5	51.4	58.4	44.3	30.2
2nd quarter	30.5	27.3	23.2	23.4	16.2	67.5	8.7	10.2	51.2	57.7	44.7	30.3
3rd quarter	30.7	27.5	23.4	23.8	16.6	66.9	8.6	10.1	51.3	57.8	44.8	30.5
4th quarter	30.5	27.7	23.5	23.2	16.5	62.3	8.8	10.0	50.7	56.5	44.7	30.3
1970												
1st quarter	30.5	27.8	23.6	23.1	16.4	62.4	8.9	10.0	50.4	55.5	45.1	30.2
2nd quarter	30.5	28.0	23.8	22.6	16.3	58.5	9.2	10.1	49.8	54.0	45.3	30.2
3rd quarter	30.8	28.2	24.0	23.0	16.4	61.9	9.3	10.1	50.0	53.4	46.4	30.5
4th quarter	30.5	28.1	23.8	23.0	15.8	68.2	9.3	10.3	50.1	53.2	46.6	30.2
1971												
1st quarter	31.3	28.7	24.4	23.5	15.9	71.7	9.4	10.2	49.3	51.5	46.8	30.9
2nd quarter	31.5	28.9	24.6	24.4	16.1	78.6	9.3	10.9	49.2	50.9	47.1	31.2
3rd quarter	31.7	29.2	24.9	24.8	16.2	82.2	9.8	11.2	49.2	50.6	47.3	31.5
4th quarter	31.8	29.7	25.4	25.5	16.5	85.3	8.9	10.4	48.8	49.3	47.9	31.6
1972												
1st quarter	32.4	30.1	25.9	26.5	17.1	90.9	9.9	11.9	49.1	49.7	48.0	32.2
2nd quarter	33.1	30.6	26.3	27.0	17.3	92.6	9.6	11.5	49.3	50.2	47.9	32.9
3rd quarter	33.4	31.1	26.8	27.3	17.6	93.0	10.2	11.7	48.4	47.9	48.3	33.2
4th quarter	34.0	31.8	27.4	28.5	18.4	96.3	10.6	12.3	48.8	48.0	49.0	33.7
1973												
1st quarter	34.8	32.4	28.1	29.6	19.2	99.6	11.4	12.8	49.2	48.7	49.2	34.5
2nd quarter	35.2	32.4	28.2	29.8	19.9	94.4	11.9	12.5	48.8	47.9	49.3	34.7
3rd quarter	35.0	32.5	28.3	29.8	20.2	90.3	12.0	12.1	48.3	46.2	49.8	34.5
4th quarter	35.4	32.4	28.3	29.5	20.4	86.2	12.5	12.2	48.6	46.3	50.4	34.7
1974												
1st quarter	35.1	32.1	28.3	28.9	20.4	80.6	12.7	11.8	49.5	47.4	51.1	34.3
2nd quarter	35.1	32.2	28.3	28.5	20.3	77.4	13.3	12.4	49.8	47.2	51.7	34.3
3rd quarter	34.8	32.4	28.3	28.1	20.1	74.8	12.6	12.2	49.9	47.5	51.6	34.1
4th quarter	34.7	31.9	27.8	26.6	19.6	65.1	12.9	12.1	50.1	48.0	51.6	33.9
1975												
1st quarter	34.3	32.2	28.2	25.1	18.5	61.4	13.0	10.9	50.6	47.5	53.1	33.2
2nd quarter	34.5	32.7	28.5	24.8	18.0	63.1	12.7	10.0	50.2	47.0	52.8	33.3
3rd quarter	35.1	33.2	29.0	25.4	18.2	67.3	12.4	10.8	51.1	48.2	53.4	34.1
4th quarter	35.6	33.5	29.6	25.9	18.4	70.2	13.1	11.4	51.6	48.4	54.2	34.6
1976												
1st quarter	36.4	34.2	30.1	26.9	18.8	76.8	13.1	12.1	51.7	47.9	54.8	35.5
2nd quarter	36.6	34.5	30.3	27.4	19.1	78.7	13.2	12.7	51.1	47.8	53.8	35.9
3rd quarter	36.8	34.9	30.6	27.7	19.5	77.5	13.6	13.2	51.0	47.8	53.4	36.1
4th quarter	37.1	35.3	30.9	29.0	19.9	86.8	13.7	13.6	50.9	47.9	53.3	36.4
1977												
1st quarter	37.5	35.7	31.4	30.0	20.6	89.4	13.5	14.3	51.4	48.3	53.8	37.1
2nd quarter	38.3	35.9	31.8	31.6	21.2	99.2	13.8	14.4	51.8	48.9	54.1	37.7
3rd quarter	39.0	36.3	32.1	32.0	21.6	98.9	14.0	14.2	52.0	49.4	54.1	38.3
4th quarter	39.0	36.8	32.6	32.5	22.3	97.8	13.5	14.4	51.9	49.0	54.2	38.5
1978												
1st quarter	39.1	37.0	32.8	32.8	22.5	98.7	13.8	15.4	51.9	49.1	54.2	38.8
2nd quarter	40.6	37.8	33.8	35.1	24.2	103.4	15.3	15.4	53.2	50.0	55.8	39.9
3rd quarter	41.0	38.0	34.0	36.0	25.0	104.6	15.4	15.6	53.7	50.3	56.4	40.3
4th quarter	41.5	38.3	34.3	36.8	25.8	104.4	16.0	15.8	54.2	50.8	57.0	40.8
1979												
1st quarter	41.6	38.5	34.4	37.1	26.4	101.9	16.1	15.8	53.7	50.7	56.0	40.9
2nd quarter	41.7	38.5	34.5	36.9	26.4	100.1	16.1	15.9	54.2	51.2	56.5	40.9
3rd quarter	42.0	38.8	34.9	37.6	27.2	98.9	16.7	15.6	54.3	51.3	56.7	41.0
4th quarter	42.1	38.9	35.0	37.2	27.2	94.9	17.7	16.0	54.7	51.4	57.4	41.0
1980												
1st quarter	42.2	38.9	34.9	36.7	27.5	87.6	18.2	16.0	55.5	52.7	57.6	41.0
2nd quarter	41.3	38.0	34.0	33.7	26.3	71.4	18.6	14.8	55.6	53.8	56.9	39.8
3rd quarter	41.3	38.4	34.6	34.1	26.4	73.5	18.5	13.7	54.9	53.2	56.0	39.5
4th quarter	42.1	38.9	35.2	35.4	27.0	80.4	18.4	14.5	54.8	53.4	55.7	40.4
1981												
1st quarter	42.9	39.1	35.5	35.7	27.5	78.4	18.8	15.1	55.6	54.5	56.1	41.3
2nd quarter	42.6	39.1	35.3	36.0	28.1	75.9	18.9	15.1	55.7	55.9	55.1	41.0
3rd quarter	43.1	39.2	35.6	36.0	28.8	70.0	18.5	14.9	55.5	55.7	54.9	41.5
4th quarter	42.6	39.0	35.2	35.9	29.4	62.7	18.5	15.4	56.2	56.6	55.3	41.2

Table 1-5B. Chain-Type Quantity Indexes for Gross Domestic Product and Domestic Purchases: Historical Data—*Continued*

(Index numbers, 2012 = 100.)   NIPA Tables 1.1.3, 1.4.3, 2.3.3

Year and quarter	Gross domestic product, total	Personal consumption expenditures		Private fixed investment			Exports and imports of goods and services		Government consumption expenditures and gross investment			Gross domestic purchases
		Total	Excluding food and energy	Total	Nonresidential	Residential	Exports	Imports	Total	Federal	State and local	
1982												
1st quarter	42.0	39.3	35.4	34.9	28.8	59.1	17.7	14.9	56.1	56.7	55.2	40.6
2nd quarter	42.1	39.4	35.6	33.8	27.9	57.2	17.8	14.7	56.4	56.9	55.4	40.7
3rd quarter	42.0	39.6	35.9	33.0	27.1	57.2	17.1	15.4	56.8	57.9	55.4	40.9
4th quarter	42.0	40.3	36.7	33.1	26.6	61.7	16.3	14.8	57.7	59.2	55.8	40.9
1983												
1st quarter	42.6	40.7	37.1	33.8	26.2	72.3	16.5	15.1	58.2	60.2	56.0	41.5
2nd quarter	43.5	41.6	37.9	35.1	26.6	80.7	16.6	16.3	58.7	61.3	55.9	42.7
3rd quarter	44.4	42.3	38.6	37.0	27.7	88.6	16.8	17.5	59.7	62.9	56.4	43.7
4th quarter	45.3	43.0	39.3	39.1	29.4	92.3	17.2	18.4	58.8	60.8	56.4	44.8
1984												
1st quarter	46.2	43.3	39.9	40.3	30.3	95.1	17.5	19.9	59.4	61.3	57.1	45.9
2nd quarter	47.0	43.9	40.4	41.9	31.8	96.7	18.0	20.7	60.7	63.2	57.9	46.8
3rd quarter	47.5	44.3	40.8	42.7	32.7	95.6	18.4	21.2	61.2	63.1	58.9	47.3
4th quarter	47.8	44.8	41.4	43.5	33.5	95.7	18.7	21.9	62.5	65.2	59.5	47.7
1985												
1st quarter	48.3	45.6	42.1	43.9	33.9	95.9	18.8	21.4	63.1	65.8	60.2	48.0
2nd quarter	48.7	46.0	42.6	44.5	34.4	96.7	18.8	22.4	64.5	67.5	61.3	48.7
3rd quarter	49.5	46.9	43.5	44.3	34.0	98.5	18.5	22.2	66.0	69.5	62.3	49.4
4th quarter	49.8	47.0	43.5	45.1	34.6	100.7	19.0	23.1	66.3	69.5	62.8	49.8
1986												
1st quarter	50.3	47.4	40.9	45.2	34.2	104.8	19.6	23.1	66.8	69.3	63.9	50.2
2nd quarter	50.5	47.9	44.5	45.3	33.5	110.8	19.8	24.1	68.2	71.6	64.5	50.6
3rd quarter	51.0	48.8	45.5	45.1	33.2	112.3	20.3	24.7	69.7	74.1	65.1	51.1
4th quarter	51.3	49.1	45.7	45.4	33.5	112.6	21.0	24.9	69.4	73.2	65.4	51.3
1987												
1st quarter	51.7	49.1	45.8	44.6	32.7	112.4	21.1	24.8	69.8	73.7	65.7	51.6
2nd quarter	52.2	49.8	46.5	45.3	33.3	112.8	21.9	25.4	70.3	74.7	65.9	52.1
3rd quarter	52.7	50.4	47.2	46.1	34.3	111.9	22.9	25.9	70.4	74.6	66.1	52.5
4th quarter	53.6	50.5	47.3	46.1	34.2	112.1	23.7	26.5	71.5	75.8	67.0	53.4
1988												
1st quarter	53.9	51.4	48.1	46.1	34.5	110.1	25.1	26.3	70.9	73.7	67.7	53.4
2nd quarter	54.6	51.7	48.4	47.0	35.3	111.4	25.8	26.0	71.1	73.1	68.5	53.9
3rd quarter	54.9	52.2	48.9	47.2	35.5	111.5	26.3	26.7	71.1	72.8	68.9	54.2
4th quarter	55.6	52.8	49.5	47.8	36.0	112.2	27.1	27.5	72.5	74.8	69.7	55.0
1989												
1st quarter	56.2	53.0	49.7	48.2	36.6	111.1	27.9	27.6	72.2	73.4	70.2	55.4
2nd quarter	56.6	53.3	50.1	48.3	37.1	107.7	29.1	27.7	73.3	74.9	71.0	55.6
3rd quarter	57.0	53.8	50.6	49.1	38.0	107.2	29.4	27.7	73.9	75.5	71.7	56.0
4th quarter	57.2	54.0	50.7	48.5	37.6	104.9	29.8	28.2	74.3	75.2	72.7	56.1
1990												
1st quarter	57.8	54.5	51.4	49.0	38.1	105.8	31.1	29.1	75.5	76.4	73.8	56.7
2nd quarter	58.0	54.7	51.4	48.1	37.6	101.6	31.5	29.0	75.6	76.5	73.9	56.8
3rd quarter	58.0	54.9	51.6	47.7	37.9	96.2	31.8	28.9	75.7	76.2	74.4	56.8
4th quarter	57.5	54.5	51.3	46.4	37.3	90.6	32.0	28.1	76.2	76.3	75.2	56.1
1991												
1st quarter	57.2	54.3	51.1	45.2	36.5	85.8	32.2	27.8	76.7	77.2	75.4	55.7
2nd quarter	57.7	54.7	51.4	45.3	36.3	88.5	33.4	28.3	77.0	77.5	75.7	56.1
3rd quarter	58.0	55.0	51.7	45.4	36.0	91.5	34.2	29.1	76.8	76.4	76.1	56.4
4th quarter	58.2	54.9	51.8	45.6	36.0	93.3	35.0	29.8	76.2	74.5	76.6	56.6
1992												
1st quarter	58.9	56.0	53.0	46.1	35.9	98.6	35.6	30.0	76.9	74.8	77.6	57.2
2nd quarter	59.5	56.4	53.3	47.6	37.0	102.0	35.6	30.5	76.7	74.6	77.5	57.9
3rd quarter	60.1	57.0	54.0	48.1	37.6	102.0	36.4	30.9	77.2	75.7	77.5	58.5
4th quarter	60.7	57.6	54.6	49.5	38.6	105.9	36.6	31.7	77.2	75.7	77.5	59.2
1993												
1st quarter	60.8	57.9	54.9	49.9	39.0	106.0	36.6	32.4	76.3	73.2	77.8	59.4
2nd quarter	61.2	58.4	55.5	50.8	39.7	107.4	37.1	33.0	76.3	72.5	78.4	59.8
3rd quarter	61.5	59.0	56.1	51.6	40.1	111.4	37.0	33.5	76.4	72.2	78.8	60.2
4th quarter	62.3	59.6	56.6	53.7	41.5	117.2	38.2	34.8	76.7	72.3	79.2	61.1
1994												
1st quarter	62.9	60.2	57.3	54.3	41.9	118.9	38.6	35.6	75.7	69.8	79.5	61.7
2nd quarter	63.8	60.7	57.7	55.4	42.6	122.8	39.9	37.0	76.1	69.5	80.4	62.6
3rd quarter	64.1	61.2	58.3	55.8	43.3	121.0	41.3	38.0	77.4	71.2	81.4	62.9
4th quarter	64.9	61.8	59.0	57.2	45.0	119.1	42.3	39.1	76.7	69.3	81.7	63.7
1995												
1st quarter	65.1	62.0	59.2	58.4	46.7	116.4	43.1	39.9	77.0	69.1	82.3	64.0
2nd quarter	65.3	62.5	59.7	58.4	47.1	112.8	43.7	40.3	77.2	68.9	82.9	64.1
3rd quarter	65.8	63.1	60.3	59.2	47.4	117.0	45.6	40.4	77.0	68.3	83.0	64.4
4th quarter	66.3	63.5	60.8	60.3	48.3	119.2	46.3	41.0	76.2	66.0	83.5	64.8

Table 1-5B. Chain-Type Quantity Indexes for Gross Domestic Product and Domestic Purchases: Historical Data—*Continued*

(Index numbers, 2012 = 100.)   NIPA Tables 1.1.3, 1.4.3, 2.3.3

Year and quarter	Gross domestic product, total	Personal consumption expenditures		Private fixed investment			Exports and imports of goods and services		Government consumption expenditures and gross investment			Gross domestic purchases
		Total	Excluding food and energy	Total	Nonresidential	Residential	Exports	Imports	Total	Federal	State and local	
1996												
1st quarter	66.8	64.1	61.3	61.9	49.6	122.3	46.8	42.3	76.7	67.3	83.3	65.5
2nd quarter	67.9	64.8	62.1	63.8	50.9	127.7	47.6	43.2	77.7	67.9	84.5	66.6
3rd quarter	68.5	65.2	62.6	65.2	52.5	127.4	48.0	44.6	77.7	67.0	85.2	67.4
4th quarter	69.2	65.7	63.1	66.3	53.8	126.0	51.0	45.5	78.2	66.7	86.4	67.8
1997												
1st quarter	69.7	66.4	63.9	67.5	55.1	126.6	51.9	47.5	77.8	65.8	86.5	68.5
2nd quarter	70.8	66.7	64.2	68.8	56.2	128.4	54.0	49.1	78.8	67.2	87.1	69.6
3rd quarter	71.7	67.9	65.4	71.2	58.7	129.6	55.2	50.9	79.2	66.9	88.0	70.5
4th quarter	72.3	68.7	66.2	71.8	59.1	131.0	55.2	52.0	79.6	66.8	88.8	71.3
1998												
1st quarter	73.1	69.4	67.0	73.9	61.0	133.7	55.4	53.9	79.3	65.2	89.6	72.3
2nd quarter	73.7	70.6	68.2	76.4	63.0	138.3	54.8	55.2	80.7	66.5	91.1	73.3
3rd quarter	74.7	71.5	69.2	77.9	64.0	142.4	54.6	55.9	81.3	66.2	92.3	74.3
4th quarter	75.9	72.5	70.4	80.0	65.9	145.7	56.6	57.7	81.8	66.6	92.9	75.5
1999												
1st quarter	76.6	73.2	71.0	81.4	67.4	146.3	56.5	59.2	82.3	66.6	93.9	76.5
2nd quarter	77.2	74.3	72.1	83.4	69.1	148.6	57.0	60.9	82.6	66.5	94.5	77.2
3rd quarter	78.2	75.1	72.9	85.3	71.1	149.7	58.7	63.1	83.6	67.6	95.4	78.4
4th quarter	79.5	76.2	74.1	85.8	71.4	150.8	60.2	64.6	84.9	68.9	96.7	79.7
2000												
1st quarter	79.8	77.4	75.6	88.2	74.1	151.5	61.0	67.2	84.3	66.7	97.4	80.3
2nd quarter	81.3	78.2	76.2	90.2	76.4	150.8	62.8	69.3	85.2	68.7	97.2	81.8
3rd quarter	81.4	78.9	77.0	90.6	77.2	148.4	64.3	71.7	84.8	67.4	97.6	82.1
4th quarter	81.9	79.6	77.6	90.8	77.5	148.6	63.7	71.7	85.3	67.5	98.4	82.7
2001												
1st quarter	81.6	79.9	78.0	90.3	76.8	149.2	62.8	70.5	86.6	69.0	99.6	82.4
2nd quarter	82.1	80.1	78.4	89.2	75.2	151.5	60.7	68.4	88.1	70.0	101.5	82.8
3rd quarter	81.8	80.4	78.6	88.6	74.3	152.5	57.8	67.0	88.0	70.4	101.0	82.6
4th quarter	82.0	81.6	80.0	86.6	72.2	151.5	55.9	66.1	89.5	71.3	102.9	82.9
2002												
1st quarter	82.7	81.8	80.3	85.8	70.5	156.4	57.2	68.0	91.0	73.6	103.8	83.8
2nd quarter	83.2	82.3	80.6	85.7	69.7	160.3	58.8	70.3	91.7	75.0	104.0	84.4
3rd quarter	83.6	82.9	81.2	85.6	69.4	160.8	59.1	71.2	92.2	75.8	104.3	84.8
4th quarter	83.7	83.3	81.7	85.2	68.5	163.7	58.1	72.4	92.9	77.1	104.6	85.3
2003												
1st quarter	84.2	83.7	82.1	86.0	68.8	166.7	58.2	72.0	93.0	77.9	104.1	85.7
2nd quarter	84.9	84.7	83.2	88.1	70.7	169.4	58.1	73.0	93.9	80.6	103.6	86.5
3rd quarter	86.3	85.9	84.5	91.0	72.2	179.1	59.7	74.2	94.1	80.6	104.0	87.9
4th quarter	87.3	86.5	85.0	92.6	73.2	184.4	62.3	76.5	94.7	82.1	103.8	88.9
2004												
1st quarter	87.8	87.3	85.8	92.4	72.4	186.5	64.0	78.9	95.1	83.0	103.9	89.5
2nd quarter	88.5	87.8	86.4	94.9	74.2	192.5	64.9	82.1	95.4	83.5	104.0	90.5
3rd quarter	89.3	88.8	87.5	96.9	76.4	193.8	65.4	83.2	95.6	84.4	103.6	91.5
4th quarter	90.2	89.7	88.5	98.7	77.9	196.5	66.9	85.1	95.5	84.2	103.6	92.4
2005												
1st quarter	91.2	90.3	88.9	100.4	79.1	201.1	68.5	85.9	96.0	85.2	103.7	93.3
2nd quarter	91.6	91.3	90.0	102.1	80.3	204.9	69.7	87.0	95.9	85.1	103.7	93.8
3rd quarter	92.4	92.2	91.0	103.9	82.0	207.1	69.9	87.6	96.3	85.7	103.8	94.6
4th quarter	93.0	92.4	91.3	104.3	82.6	206.7	71.9	90.4	96.3	85.8	103.9	95.4
2006												
1st quarter	94.3	93.5	92.6	106.3	85.4	204.8	74.6	92.3	97.5	88.0	104.3	96.5
2nd quarter	94.5	94.0	92.9	105.7	86.9	194.5	76.1	93.3	97.5	87.4	104.7	96.7
3rd quarter	94.6	94.5	93.4	105.0	88.4	183.7	76.1	94.2	97.4	86.6	105.1	97.0
4th quarter	95.4	95.4	94.4	104.3	89.3	175.2	79.3	94.3	98.2	87.8	105.5	97.4
2007												
1st quarter	95.7	96.0	94.9	104.2	91.0	167.3	80.4	95.4	98.3	87.3	106.2	97.6
2nd quarter	96.2	96.1	95.2	104.7	93.0	160.7	81.6	96.0	99.2	88.8	106.7	98.1
3rd quarter	96.7	96.7	95.8	104.1	94.4	150.5	84.1	96.4	99.7	89.5	106.9	98.4
4th quarter	97.3	97.0	96.2	103.1	95.9	138.0	86.6	95.6	100.4	91.0	107.1	98.5
2008												
1st quarter	96.8	96.7	96.0	101.5	96.3	127.0	87.3	95.9	100.7	92.4	106.5	97.9
2nd quarter	97.3	96.9	96.3	100.7	96.5	121.2	90.1	95.0	101.5	94.3	106.6	97.9
3rd quarter	96.7	96.1	95.8	98.1	94.7	115.1	89.5	93.8	102.3	95.5	107.1	97.2
4th quarter	94.6	95.2	94.6	91.7	89.1	103.9	84.5	90.3	103.0	96.7	107.4	95.2
2009												
1st quarter	93.6	95.0	94.4	84.4	82.4	93.8	77.7	81.4	104.1	97.9	108.5	93.7
2nd quarter	93.4	94.6	93.9	81.3	79.9	88.1	77.9	78.0	105.6	100.2	109.4	93.0
3rd quarter	93.8	95.3	94.6	81.6	79.5	92.1	80.8	81.3	105.9	101.1	109.3	93.4
4th quarter	94.8	95.2	94.3	82.0	80.0	92.0	85.5	85.2	106.1	102.7	108.5	94.5

Table 1-5B. Chain-Type Quantity Indexes for Gross Domestic Product and Domestic Purchases: Historical Data—Continued

(Index numbers, 2012 = 100.)

NIPA Tables 1.1.3, 1.4.3, 2.3.3

Year and quarter	Gross domestic product, total	Personal consumption expenditures		Private fixed investment			Exports and imports of goods and services		Government consumption expenditures and gross investment			Gross domestic purchases
		Total	Excluding food and energy	Total	Nonresidential	Residential	Exports	Imports	Total	Federal	State and local	
2010												
1st quarter	95.2	95.6	94.8	82.0	80.5	89.0	86.8	87.3	105.7	103.9	107.0	95.0
2nd quarter	96.1	96.4	95.7	85.0	83.2	93.6	88.9	91.3	106.0	105.2	106.6	96.2
3rd quarter	96.8	97.1	96.4	85.4	85.4	85.3	91.3	94.7	105.3	104.8	105.7	97.1
4th quarter	97.2	97.7	97.1	87.0	87.1	86.7	94.0	95.4	104.7	104.6	104.7	97.4
2011												
1st quarter	97.0	98.1	97.7	86.9	87.0	86.4	94.8	96.1	103.4	103.2	103.5	97.1
2nd quarter	97.7	98.3	98.1	88.9	89.2	87.4	96.3	96.6	102.7	102.8	102.6	97.7
3rd quarter	97.7	98.7	98.4	92.7	93.4	80.0	97.3	97.7	101.2	100.6	101.7	97.7
4th quarter	98.8	98.9	98.8	95.0	95.8	91.3	98.4	99.1	101.2	101.0	101.3	98.9
2012												
1st quarter	99.6	99.7	99.8	98.0	98.2	96.8	99.0	99.6	100.7	101.0	100.5	99.7
2nd quarter	100.0	99.8	99.7	100.0	100.4	97.6	100.0	100.1	100.2	100.2	100.2	100.0
3rd quarter	100.1	100.0	99.9	100.1	100.1	100.2	100.6	100.6	100.0	100.4	99.8	100.2
4th quarter	100.3	100.5	100.5	101.9	101.2	105.4	100.4	99.7	99.1	98.4	99.5	100.1
2013												
1st quarter	101.1	101.0	100.9	103.7	102.5	109.2	101.6	100.0	98.2	96.1	99.6	100.9
2nd quarter	101.3	101.0	101.0	104.5	102.8	112.7	102.8	101.4	98.0	95.4	99.8	101.1
3rd quarter	102.1	101.5	101.5	106.3	104.5	114.7	103.4	102.1	97.5	94.0	99.9	101.9
4th quarter	102.9	102.3	102.3	107.7	106.7	112.9	106.5	102.7	96.7	92.4	99.7	102.4
2014												
1st quarter	102.6	102.7	102.0	108.8	108.2	112.1	105.7	103.9	96.3	92.3	99.1	102.4
2nd quarter	104.0	103.9	104.1	111.8	111.1	115.7	108.0	106.5	96.4	91.7	99.6	103.9
3rd quarter	105.2	105.0	105.4	114.0	113.3	117.4	108.5	106.8	97.1	92.7	100.0	105.1
4th quarter	105.8	106.3	106.6	115.3	114.0	121.6	109.5	109.4	97.0	91.4	100.9	105.9
2015												
1st quarter	106.7	107.1	107.4	115.4	113.6	123.9	108.3	111.2	97.5	91.7	101.5	107.2
2nd quarter	107.5	108.0	108.5	116.2	113.9	127.1	109.4	112.0	98.5	91.9	103.0	107.9
3rd quarter	107.8	108.8	109.4	117.2	114.3	130.5	108.3	112.9	99.0	91.8	103.9	108.5
4th quarter	107.9	109.3	110.0	116.5	113.0	132.6	107.8	112.9	99.3	92.4	103.9	108.6
2016												
1st quarter	108.4	110.2	110.8	117.3	112.8	137.3	107.0	113.1	100.2	92.5	105.4	109.3
2nd quarter	108.9	110.9	111.5	118.1	113.9	136.6	108.1	113.4	100.0	91.9	105.5	109.7
3rd quarter	109.5	111.7	112.2	119.1	115.5	135.7	109.7	114.7	100.4	92.4	106.0	110.2
4th quarter	110.0	112.3	113.1	119.7	115.7	137.8	109.0	116.8	100.7	92.5	106.3	111.2
2017												
1st quarter	110.7	113.0	113.9	122.0	117.6	141.8	110.6	117.9	100.6	92.2	106.4	111.7
2nd quarter	111.3	113.7	114.4	122.8	118.8	141.0	111.0	119.0	101.0	93.0	106.5	112.4
3rd quarter	112.1	114.4	115.1	123.3	119.5	140.2	112.2	119.3	101.0	93.0	106.4	113.2
4th quarter	113.1	115.7	116.4	125.9	122.0	143.6	115.0	123.3	101.6	94.0	106.7	114.4
2018												
1st quarter	113.8	116.1	116.9	127.6	124.6	141.7	115.2	123.5	102.0	94.7	107.1	115.1
2nd quarter	114.8	117.3	118.1	129.2	127.0	140.3	116.8	123.6	102.7	95.6	107.5	115.8
3rd quarter	115.7	118.3	119.2	129.4	127.6	138.9	115.0	126.2	103.2	96.3	108.0	117.2
4th quarter	116.0	118.7	119.6	130.3	129.1	137.3	115.4	127.2	103.1	96.5	107.6	117.6

Table 1-6A. Chain-Type Price Indexes for Gross Domestic Product and Domestic Purchases: Recent Data

(Index numbers, 2012 = 100.)

NIPA Tables 1.1.4, 1.6.4, 2.3.4

Year and quarter	Gross domestic product, total	Personal consumption expenditures		Private fixed investment			Exports and imports of goods and services		Government consumption expenditures and gross investment			Gross domestic purchases
		Total	Excluding food and energy	Total	Nonresidential	Residential	Exports	Imports	Total	Federal	State and local	
1955	14.8	14.8	15.3	22.0	26.3	12.0	22.2	16.8	9.6	11.8	8.0	14.5
1956	15.3	15.1	15.7	23.3	28.2	12.3	22.9	17.1	10.1	12.4	8.5	14.9
1957	15.8	15.6	16.2	24.1	29.6	12.4	23.8	17.3	10.5	12.9	8.8	15.4
1958	16.2	16.0	16.5	24.1	29.8	12.3	23.6	16.6	10.8	13.4	9.0	15.8
1959	16.4	16.2	16.9	24.4	30.2	12.4	23.6	16.7	11.0	13.6	9.1	16.0
1960	16.6	16.5	17.2	24.5	30.4	12.5	24.0	16.9	11.1	13.7	9.3	16.2
1961	16.8	16.6	17.4	24.5	30.3	12.5	24.3	16.9	11.3	13.8	9.5	16.4
1962	17.0	16.8	17.6	24.5	30.3	12.5	24.4	16.7	11.5	14.0	9.8	16.6
1963	17.2	17.0	17.9	24.4	30.3	12.4	24.3	17.0	11.8	14.4	10.0	16.8
1964	17.5	17.3	18.1	24.6	30.5	12.5	24.5	17.4	12.1	14.8	10.2	17.0
1965	17.8	17.5	18.4	25.0	30.9	12.8	25.3	17.6	12.4	15.2	10.5	17.3
1966	18.3	18.0	18.8	25.5	31.3	13.3	26.1	18.0	12.9	15.7	11.0	17.8
1967	18.8	18.4	19.4	26.1	32.1	13.8	27.1	18.1	13.4	16.0	11.6	18.3
1968	19.6	19.2	20.2	27.1	33.2	14.5	27.7	18.4	14.1	16.8	12.2	19.1
1969	20.6	20.0	21.1	28.4	34.6	15.5	28.6	18.8	14.9	17.7	13.1	20.0
1970	21.7	21.0	22.1	29.6	36.3	16.0	29.7	20.0	16.1	19.1	14.1	21.1
1971	22.8	21.8	23.2	31.1	38.0	16.9	30.8	21.2	17.4	20.7	15.2	22.2
1972	23.8	22.6	23.9	32.4	39.3	18.0	32.1	22.7	18.7	22.5	16.2	23.2
1973	25.1	23.8	24.8	34.2	40.9	19.6	36.4	26.6	19.9	24.1	17.2	24.5
1974	27.3	26.3	26.8	37.6	44.9	21.6	44.8	38.1	21.9	26.0	19.2	27.0
1975	29.8	28.5	29.0	42.1	50.8	23.6	49.4	41.2	23.9	28.3	21.0	29.5
1976	31.5	30.0	30.8	44.4	53.6	25.1	51.0	42.5	25.2	30.0	22.0	31.1
1977	33.4	32.0	32.8	47.7	57.1	27.7	53.1	46.2	26.7	31.9	23.4	33.1
1978	35.8	34.2	34.9	51.5	60.9	31.1	56.3	49.5	28.5	34.0	24.9	35.5
1979	38.8	37.3	37.5	56.1	65.8	34.6	63.1	57.9	30.9	36.6	27.1	38.6
1980	42.3	41.3	40.9	61.4	71.6	38.3	69.5	72.2	34.0	40.1	30.1	42.6
1981	46.3	45.0	44.5	67.1	78.5	41.4	74.7	76.1	37.4	43.8	33.2	46.5
1982	49.1	47.5	47.4	70.7	82.9	43.6	75.0	73.5	40.0	46.9	35.4	49.2
1983	51.1	49.5	49.8	70.9	82.8	44.7	75.3	70.8	41.5	48.5	37.0	50.9
1984	52.9	51.3	51.9	71.7	83.0	46.0	76.0	70.1	43.3	50.6	38.5	52.6
1985	54.6	53.1	54.0	72.5	83.9	47.3	73.8	67.8	44.7	51.7	40.1	54.2
1986	55.7	54.3	55.9	74.2	85.4	49.4	72.5	67.8	45.4	52.0	41.3	55.3
1987	57.0	56.0	57.7	75.7	86.3	51.5	74.1	71.9	46.6	52.3	43.2	56.9
1988	59.1	58.2	60.1	77.6	88.5	53.3	77.9	75.4	48.2	54.0	44.6	58.9
1989	61.4	60.7	62.6	79.6	90.6	55.0	79.2	77.0	50.0	55.5	46.8	61.2
1990	63.7	63.4	65.2	81.3	92.5	56.3	79.7	79.2	52.1	57.3	49.2	63.7
1991	65.8	65.5	67.5	82.6	94.3	57.0	80.5	78.6	54.0	59.3	51.0	65.7
1992	67.3	67.2	69.5	82.6	94.0	57.7	80.2	78.6	55.6	60.8	52.7	67.2
1993	68.9	68.9	71.4	83.6	94.2	60.1	80.3	78.0	57.0	62.2	54.0	68.7
1994	70.4	70.3	73.0	84.9	94.9	62.2	81.2	78.8	58.5	63.9	55.4	70.1
1995	71.9	71.8	74.6	86.2	95.8	64.5	83.0	80.9	60.1	65.8	56.9	71.7
1996	73.2	73.3	76.0	86.2	95.3	65.9	81.9	79.5	61.4	66.9	58.2	72.9
1997	74.4	74.6	77.4	86.2	94.7	67.4	80.5	76.8	62.6	68.0	59.5	74.0
1998	75.3	75.2	78.4	85.6	93.2	69.2	78.6	72.6	63.6	68.8	60.6	74.5
1999	76.3	76.3	79.4	85.7	92.3	71.8	78.0	73.0	65.8	70.5	63.0	75.6
2000	78.1	78.2	80.8	86.8	92.7	75.0	79.5	76.2	68.6	72.9	66.0	77.6
2001	79.8	79.7	82.3	87.6	92.3	78.6	78.8	74.2	70.6	74.2	68.3	79.0
2002	81.0	80.8	83.6	87.8	91.9	80.5	78.2	73.2	72.4	76.6	69.8	80.1
2003	82.6	82.4	84.8	88.6	91.2	84.3	79.4	75.5	75.0	80.0	72.1	81.8
2004	84.8	84.4	86.5	91.1	92.1	90.2	82.3	79.1	78.2	82.8	75.4	84.1
2005	87.4	86.8	88.4	94.8	94.4	96.7	85.1	83.7	82.1	86.2	79.6	87.0
2006	90.1	89.2	90.4	98.2	96.7	102.4	87.8	86.9	85.7	88.9	83.6	89.8
2007	92.5	91.4	92.4	99.7	98.3	103.7	91.1	89.9	89.5	91.6	88.1	92.2
2008	94.3	94.2	94.2	100.5	99.8	102.2	95.4	99.0	93.3	94.4	92.6	94.8
2009	95.0	94.1	95.3	99.3	99.2	98.7	89.7	88.0	92.9	94.2	92.0	94.6
2010	96.1	95.7	96.6	97.7	97.4	98.3	93.3	92.8	95.4	96.4	94.7	95.9
2011	98.1	98.1	98.1	98.7	98.6	99.0	99.2	99.8	98.3	99.1	97.7	98.2
2012	100.0	100.0	100.0	100.0	100.0	100.0	100.0	100.0	100.0	100.0	100.0	100.0
2013	101.8	101.3	101.5	101.0	100.3	105.1	100.2	98.6	102.3	100.9	103.3	101.5
2014	103.6	102.8	103.1	102.9	101.5	111.1	100.3	97.9	104.4	102.6	105.6	103.1
2015	104.7	103.0	104.4	103.7	102.0	114.1	95.4	89.9	104.7	103.3	105.7	103.5
2016	105.8	104.1	106.1	103.6	101.1	118.1	93.5	86.7	105.1	103.9	105.9	104.2
2017	107.8	105.9	107.8	105.4	102.1	123.5	95.9	88.6	107.6	106.0	108.7	106.1
2018	110.4	108.1	109.9	107.8	103.5	130.5	99.1	91.2	111.4	109.3	112.8	108.6
2016												
1st quarter	104.9	103.3	105.3	103.0	101.1	115.8	92.3	86.1	104.2	103.1	104.9	103.4
2nd quarter	105.6	103.9	105.8	103.4	101.2	117.3	93.3	86.4	104.9	103.7	105.7	104.0
3rd quarter	106.0	104.3	106.4	103.6	101.0	119.0	93.8	87.0	105.3	104.1	106.1	104.4
4th quarter	106.5	104.8	106.7	104.2	101.3	120.5	94.4	87.3	105.8	104.7	106.7	104.9
2017												
1st quarter	107.0	105.4	107.2	104.6	101.6	121.5	95.1	88.3	106.7	105.2	107.7	105.5
2nd quarter	107.4	105.6	107.5	105.2	102.0	123.0	95.1	88.3	107.1	105.7	108.1	105.8
3rd quarter	108.0	106.0	107.9	105.8	102.3	124.3	96.0	88.4	107.8	106.2	108.9	106.3
4th quarter	108.7	106.7	108.5	106.0	102.5	125.3	97.3	89.5	108.9	107.1	110.2	107.0
2018												
1st quarter	109.3	107.4	109.1	106.9	103.0	128.0	98.1	91.1	110.0	108.2	111.2	107.8
2nd quarter	110.2	108.0	109.7	107.6	103.4	130.2	99.4	91.3	111.0	109.0	112.4	108.5
3rd quarter	110.8	108.4	110.1	108.2	103.8	131.5	99.6	91.4	111.9	109.7	113.3	109.0
4th quarter	111.2	108.8	110.6	108.4	103.8	132.3	99.3	91.0	112.7	110.5	114.1	109.4

Table 1-6B. Chain-Type Price Indexes for Gross Domestic Product and Domestic Purchases: Historical Data

(Index numbers, 2012 = 100.)   NIPA Tables 1.1.4, 1.6.4, 2.3.4

Year and quarter	Gross domestic product											Gross domestic purchases
	Gross domestic product, total	Personal consumption expenditures		Private fixed investment			Exports and imports of goods and services		Government consumption expenditures and gross investment			
		Total	Excluding food and energy	Total	Nonresi-dential	Residential	Exports	Imports	Total	Federal	State and local	
1929	9.4	9.3	9.6	14.0	13.5	5.1	13.7	9.6	5.3	6.6	3.7	9.1
1930	9.0	8.9	9.2	13.4	12.8	5.0	12.4	8.2	5.2	6.3	3.6	8.8
1931	8.1	8.0	8.5	12.3	11.9	4.5	9.7	6.6	4.9	6.2	3.4	7.9
1932	7.2	7.0	7.5	10.4	10.8	3.7	8.4	5.3	4.5	5.9	3.0	7.0
1933	7.0	6.8	7.2	9.5	10.6	3.6	8.4	5.1	4.6	6.0	3.1	6.8
1934	7.4	7.1	7.3	9.5	10.9	4.0	9.8	5.8	4.9	6.4	3.4	7.1
1935	7.5	7.3	7.4	9.4	11.0	4.0	10.0	5.9	5.0	6.4	3.4	7.3
1936	7.6	7.3	7.5	9.5	11.0	4.1	10.4	6.3	5.2	7.0	3.4	7.4
1937	7.9	7.6	7.8	10.2	11.9	4.5	11.1	7.0	5.3	7.1	3.5	7.6
1938	7.7	7.4	7.8	9.9	12.0	4.7	10.5	6.5	5.3	7.1	3.5	7.5
1939	7.6	7.3	7.7	9.9	11.9	4.7	10.4	6.8	5.2	7.0	3.5	7.4
1940	7.7	7.4	7.8	10.1	12.1	4.9	11.3	7.2	5.1	6.8	3.5	7.5
1941	8.2	7.9	8.2	10.8	12.9	5.3	12.3	7.6	5.5	7.2	3.7	7.9
1942	8.9	8.8	9.1	12.0	14.3	5.7	14.9	8.7	5.5	7.2	4.1	8.6
1943	9.3	9.7	9.8	12.6	14.9	6.2	16.3	9.4	5.5	7.1	4.4	9.0
1944	9.5	10.2	10.5	13.1	15.4	6.8	18.4	9.9	5.5	7.0	4.5	9.2
1945	9.8	10.6	11.0	13.4	15.8	7.2	18.2	10.2	5.5	7.0	4.7	9.4
1946	11.0	11.4	11.7	14.7	17.4	7.8	17.6	11.3	7.0	9.2	5.1	10.7
1947	12.3	12.5	12.7	17.4	20.1	9.4	20.4	13.6	7.6	9.9	5.8	11.9
1948	12.9	13.2	13.4	18.8	21.8	10.2	21.5	14.8	7.9	9.9	6.5	12.6
1949	12.9	13.1	13.4	19.0	22.3	10.3	20.2	14.1	8.2	10.3	6.5	12.6
1947												
1st quarter	12.0	12.3	...	16.5	19.3	8.9	18.7	12.5	7.7	10.2	5.6	11.7
2nd quarter	12.1	12.3	...	17.2	19.9	9.4	20.1	13.3	7.7	10.1	5.7	11.8
3rd quarter	12.3	12.0	...	17.7	20.4	9.6	21.1	14.0	7.5	9.7	5.8	12.0
4th quarter	12.6	12.9	...	18.0	20.8	9.8	21.7	14.5	7.6	9.7	6.0	12.2
1948												
1st quarter	12.7	13.0	...	18.2	20.9	10.0	21.8	14.9	7.8	9.8	6.2	12.4
2nd quarter	12.9	13.2	...	18.6	21.5	10.1	21.7	14.9	7.9	9.8	6.4	12.5
3rd quarter	13.1	13.4	...	19.1	22.2	10.3	21.4	14.8	8.0	9.9	6.6	12.7
4th quarter	13.1	13.3	...	19.2	22.5	10.4	21.1	14.5	8.1	10.0	6.7	12.7
1949												
1st quarter	13.1	13.2	...	19.1	22.4	10.5	20.7	14.2	8.2	10.3	6.6	12.7
2nd quarter	13.0	13.1	...	19.1	22.3	10.4	20.3	14.1	8.1	10.3	6.5	12.6
3rd quarter	12.9	13.1	...	18.9	22.3	10.2	20.0	14.0	8.1	10.2	6.5	12.5
4th quarter	12.9	13.1	...	18.8	22.1	10.2	19.8	14.0	8.2	10.4	6.5	12.6
1950												
1st quarter	12.8	13.0	...	18.8	22.2	10.2	19.5	14.3	8.2	10.4	6.4	12.5
2nd quarter	12.9	13.1	...	19.1	22.4	10.5	19.5	14.6	8.1	10.3	6.4	12.6
3rd quarter	13.2	13.4	...	19.6	22.8	10.9	19.7	15.1	8.3	10.4	6.6	12.9
4th quarter	13.4	13.6	...	20.0	23.6	10.8	20.1	15.8	8.4	10.4	6.8	13.1
1951												
1st quarter	13.8	14.0	...	20.8	24.4	11.2	21.3	17.1	8.7	10.9	7.0	13.5
2nd quarter	13.9	14.2	...	21.1	24.8	11.3	22.0	18.0	8.7	10.7	7.2	13.6
3rd quarter	13.9	14.2	...	21.2	25.0	11.4	22.7	18.5	8.7	10.7	7.3	13.7
4th quarter	14.1	14.4	...	21.4	25.3	11.5	23.0	18.6	8.8	10.9	7.4	13.8
1952												
1st quarter	14.1	14.4	...	21.5	25.4	11.5	22.5	17.9	8.8	10.7	7.4	13.9
2nd quarter	14.2	14.4	...	21.5	25.4	11.6	22.4	17.6	8.9	10.9	7.5	13.9
3rd quarter	14.3	14.5	...	21.5	25.4	11.8	22.4	17.3	9.0	11.0	7.6	14.0
4th quarter	14.4	14.5	...	21.4	25.4	11.7	22.3	17.0	9.1	11.1	7.6	14.0
1953												
1st quarter	14.4	14.6	...	21.4	25.4	11.7	22.4	16.8	9.0	11.0	7.7	14.0
2nd quarter	14.4	14.6	...	21.5	25.6	11.7	22.4	16.7	9.0	11.0	7.7	14.1
3rd quarter	14.5	14.7	...	21.7	25.8	11.8	22.3	16.6	9.0	11.0	7.7	14.1
4th quarter	14.5	14.8	...	21.6	25.8	11.8	22.2	16.6	9.0	11.1	7.7	14.2
1954												
1st quarter	14.6	14.8	...	21.7	25.9	11.7	22.1	16.8	9.1	11.2	7.6	14.2
2nd quarter	14.6	14.8	...	21.7	26.0	11.7	22.0	16.9	9.2	11.2	7.8	14.3
3rd quarter	14.6	14.8	...	21.7	25.8	11.9	22.0	17.0	9.2	11.3	7.9	14.3
4th quarter	14.6	14.7	...	21.7	25.9	11.9	22.0	17.0	9.3	11.4	7.9	14.3
1955												
1st quarter	14.7	14.8	...	21.7	25.8	11.9	22.0	16.8	9.4	11.5	7.8	14.3
2nd quarter	14.7	14.8	...	21.9	26.0	12.0	22.1	16.8	9.5	11.8	7.9	14.4
3rd quarter	14.8	14.9	...	22.1	26.4	12.1	22.2	16.8	9.6	11.9	8.0	14.5
4th quarter	14.9	14.9	...	22.4	26.9	12.1	22.4	16.9	9.7	12.0	8.1	14.6
1956												
1st quarter	15.1	15.0	...	22.9	27.6	12.2	22.6	16.9	9.9	12.2	8.3	14.7
2nd quarter	15.2	15.1	...	23.1	27.9	12.4	22.8	17.0	10.1	12.4	8.4	14.9
3rd quarter	15.4	15.2	...	23.5	28.5	12.4	23.1	17.1	10.2	12.5	8.5	15.0
4th quarter	15.5	15.3	...	23.7	28.9	12.4	23.3	17.3	10.2	12.5	8.6	15.1
1957												
1st quarter	15.7	15.4	...	23.9	29.3	12.3	23.7	17.4	10.4	12.8	8.7	15.3
2nd quarter	15.8	15.5	...	24.0	29.5	12.3	23.8	17.4	10.5	12.9	8.8	15.4
3rd quarter	15.9	15.7	...	24.1	29.7	12.4	23.9	17.3	10.6	13.0	8.9	15.5
4th quarter	16.0	15.7	...	24.2	29.9	12.4	23.9	17.1	10.6	13.1	8.9	15.6

. . . = Not available.

Table 1-6B. Chain-Type Price Indexes for Gross Domestic Product and Domestic Purchases: Historical Data—Continued

(Index numbers, 2012 = 100.)

NIPA Tables 1.1.4, 1.6.4, 2.3.4

Year and quarter	Gross domestic product, total	Personal consumption expenditures		Private fixed investment			Exports and imports of goods and services		Government consumption expenditures and gross investment			Gross domestic purchases
		Total	Excluding food and energy	Total	Nonresidential	Residential	Exports	Imports	Total	Federal	State and local	
1958												
1st quarter	16.1	15.9	. . .	24.0	29.6	12.3	23.6	16.7	10.7	13.3	8.9	15.7
2nd quarter	16.2	16.0	. . .	24.1	29.8	12.3	23.5	16.5	10.8	13.4	9.0	15.8
3rd quarter	16.2	16.0	. . .	24.2	29.9	12.3	23.5	16.5	10.9	13.5	9.0	15.8
4th quarter	16.3	16.0	. . .	24.2	30.0	12.3	23.6	16.6	10.9	13.6	9.0	15.8
1959												
1st quarter	16.3	16.1	16.7	24.2	30.0	12.4	23.4	16.6	11.0	13.6	9.1	15.9
2nd quarter	16.4	16.2	16.8	24.3	30.2	12.4	23.5	16.7	11.0	13.6	9.1	16.0
3rd quarter	16.4	16.3	16.9	24.4	30.3	12.4	23.7	16.7	11.0	13.6	9.1	16.0
4th quarter	16.5	16.3	17.0	24.5	30.4	12.4	23.9	16.9	11.0	13.6	9.1	16.1
1960												
1st quarter	16.5	16.4	17.1	24.5	30.4	12.4	24.0	16.9	11.0	13.5	9.2	16.1
2nd quarter	16.6	16.4	17.2	24.5	30.4	12.5	23.9	16.8	11.0	13.5	9.3	16.2
3rd quarter	16.7	16.5	17.2	24.5	30.4	12.5	24.0	16.9	11.1	13.7	9.3	16.2
4th quarter	16.7	16.6	17.3	24.5	30.3	12.5	24.0	16.9	11.2	13.9	9.4	16.3
1961												
1st quarter	16.8	16.6	17.3	24.5	30.3	12.4	24.1	16.9	11.2	13.8	9.4	16.3
2nd quarter	16.8	16.6	17.4	24.5	30.3	12.5	24.4	16.9	11.3	13.8	9.5	16.4
3rd quarter	16.8	16.7	17.4	24.5	30.3	12.5	24.3	16.9	11.3	13.8	9.6	16.4
4th quarter	16.9	16.7	17.5	24.5	30.3	12.5	24.5	16.9	11.3	13.8	9.6	16.4
1962												
1st quarter	17.0	16.8	17.6	24.5	30.3	12.5	24.6	16.7	11.4	13.9	9.7	16.5
2nd quarter	17.0	16.8	17.6	24.5	30.3	12.5	24.3	16.7	11.5	14.0	9.8	16.6
3rd quarter	17.0	16.9	17.7	24.5	30.3	12.5	24.3	16.6	11.5	14.1	9.8	16.6
4th quarter	17.1	16.9	17.7	24.5	30.3	12.5	24.3	16.7	11.6	14.1	9.8	16.6
1963												
1st quarter	17.2	17.0	17.8	24.5	30.3	12.5	24.4	16.9	11.7	14.3	9.9	16.7
2nd quarter	17.2	17.0	17.8	24.4	30.3	12.4	24.3	17.0	11.8	14.3	10.0	16.7
3rd quarter	17.2	17.1	17.9	24.4	30.3	12.3	24.3	17.1	11.7	14.2	10.0	16.8
4th quarter	17.3	17.1	18.0	24.4	30.3	12.4	24.3	17.1	11.9	14.6	10.1	16.9
1964												
1st quarter	17.4	17.2	18.1	24.4	30.4	12.3	24.4	17.3	12.0	14.7	10.1	16.9
2nd quarter	17.4	17.3	18.1	24.6	30.5	12.5	24.4	17.4	12.0	14.7	10.2	17.0
3rd quarter	17.5	17.3	18.2	24.6	30.5	12.5	24.6	17.4	12.2	14.9	10.2	17.1
4th quarter	17.6	17.4	18.2	24.8	30.7	12.8	24.8	17.4	12.2	14.9	10.3	17.1
1965												
1st quarter	17.7	17.4	18.3	24.8	30.7	12.8	25.4	17.6	12.3	15.0	10.4	17.2
2nd quarter	17.7	17.5	18.3	24.9	30.8	12.8	25.3	17.5	12.3	15.0	10.5	17.3
3rd quarter	17.8	17.6	18.4	24.9	30.9	12.8	25.3	17.6	12.4	15.2	10.5	17.4
4th quarter	17.9	17.6	18.5	25.2	31.0	13.1	25.2	17.8	12.6	15.5	10.6	17.5
1966												
1st quarter	18.1	17.8	18.6	25.1	31.0	12.9	25.7	17.9	12.7	15.6	10.8	17.6
2nd quarter	18.2	17.9	18.7	25.5	31.3	13.4	25.9	18.1	12.8	15.6	11.0	17.8
3rd quarter	18.4	18.0	18.9	25.5	31.4	13.3	26.2	18.0	13.0	15.8	11.1	17.9
4th quarter	18.5	18.2	19.0	25.8	31.6	13.6	26.6	18.1	13.1	15.8	11.2	18.0
1967												
1st quarter	18.6	18.2	19.1	25.9	31.8	13.6	27.1	18.1	13.1	15.8	11.4	18.1
2nd quarter	18.7	18.3	19.3	26.0	32.0	13.7	27.1	18.1	13.3	15.9	11.5	18.2
3rd quarter	18.9	18.5	19.4	26.1	32.2	13.8	27.0	18.1	13.4	16.1	11.6	18.4
4th quarter	19.1	18.7	19.6	26.4	32.5	14.0	27.2	18.1	13.6	16.4	11.8	18.6
1968												
1st quarter	19.3	18.9	19.9	26.7	32.7	14.3	27.4	18.2	13.8	16.5	12.0	18.8
2nd quarter	19.5	19.0	20.1	26.9	33.1	14.4	27.9	18.3	14.0	16.7	12.2	19.0
3rd quarter	19.7	19.2	20.3	27.1	33.3	14.4	27.6	18.4	14.1	17.0	12.3	19.2
4th quarter	20.0	19.5	20.5	27.7	33.8	15.0	27.8	18.5	14.4	17.2	12.5	19.4
1969												
1st quarter	20.2	19.6	20.8	28.0	34.1	15.3	28.2	18.6	14.5	17.3	12.7	19.6
2nd quarter	20.4	19.9	21.0	28.3	34.4	15.5	28.2	18.7	14.7	17.5	13.0	19.9
3rd quarter	20.7	20.1	21.3	28.5	34.8	15.5	28.6	18.8	15.1	17.9	13.2	20.2
4th quarter	21.0	20.4	21.5	28.9	35.2	15.8	29.3	19.3	15.3	18.2	13.4	20.4
1970												
1st quarter	21.3	20.6	21.7	29.1	35.6	15.8	29.3	19.5	15.7	18.7	13.7	20.7
2nd quarter	21.6	20.8	22.0	29.7	36.2	16.3	29.9	19.7	15.9	18.9	14.0	21.0
3rd quarter	21.8	21.0	22.2	29.7	36.4	15.9	29.8	20.2	16.2	19.3	14.2	21.2
4th quarter	22.1	21.3	22.5	30.0	36.9	16.0	29.9	20.4	16.5	19.5	14.5	21.5
1971												
1st quarter	22.4	21.5	22.8	30.5	37.4	16.5	30.8	20.9	16.9	20.1	14.8	21.8
2nd quarter	22.7	21.8	23.1	31.0	37.9	16.8	30.8	21.0	17.2	20.5	15.1	22.1
3rd quarter	22.9	22.0	23.3	31.3	38.2	17.1	30.7	21.2	17.5	20.8	15.3	22.3
4th quarter	23.1	22.1	23.4	31.6	38.5	17.4	30.9	21.6	17.8	21.2	15.5	22.5
1972												
1st quarter	23.5	22.3	23.7	31.9	38.9	17.6	31.7	21.9	18.3	22.1	15.8	22.9
2nd quarter	23.6	22.5	23.8	32.2	39.2	17.7	32.0	22.5	18.5	22.3	16.0	23.0
3rd quarter	23.8	22.7	24.0	32.5	39.4	18.0	32.1	22.9	18.7	22.5	16.3	23.3
4th quarter	24.1	22.9	24.2	33.0	39.7	18.6	32.8	23.4	19.1	23.0	16.5	23.5

. . . = Not available.

Table 1-6B. Chain-Type Price Indexes for Gross Domestic Product and Domestic Purchases: Historical Data—Continued

(Index numbers, 2012 = 100.)

NIPA Tables 1.1.4, 1.6.4, 2.3.4

Year and quarter	Gross domestic product, total	Personal consumption expenditures		Private fixed investment			Exports and imports of goods and services		Government consumption expenditures and gross investment			Gross domestic purchases
		Total	Excluding food and energy	Total	Nonresidential	Residential	Exports	Imports	Total	Federal	State and local	
1973												
1st quarter	24.4	23.1	24.3	33.3	40.0	18.8	33.8	24.1	19.4	23.4	16.8	23.8
2nd quarter	24.8	23.6	24.7	33.8	40.6	19.3	35.3	25.9	19.8	23.8	17.1	24.3
3rd quarter	25.3	24.0	25.0	34.6	41.2	19.9	37.3	27.2	20.1	24.3	17.4	24.7
4th quarter	25.7	24.5	25.3	35.0	41.7	20.3	39.2	29.2	20.5	24.8	17.7	25.2
1974												
1st quarter	26.2	25.2	25.8	35.7	42.5	20.7	41.9	33.9	20.9	25.0	18.2	25.8
2nd quarter	26.8	25.9	26.4	36.8	43.8	21.2	43.4	37.7	21.5	25.5	18.8	26.6
3rd quarter	27.7	26.6	27.2	38.2	45.6	21.9	45.8	39.7	22.2	26.2	19.5	27.4
4th quarter	28.5	27.3	27.8	39.6	47.6	22.5	48.1	40.9	22.9	27.1	20.1	28.2
1975												
1st quarter	29.1	27.8	28.4	41.0	49.3	23.1	49.5	41.6	23.3	27.6	20.5	28.8
2nd quarter	29.6	28.2	28.8	41.9	50.6	23.5	49.4	41.7	23.7	28.0	20.9	29.2
3rd quarter	30.1	28.7	29.2	42.4	51.2	23.7	49.2	40.8	24.0	28.4	21.2	29.7
4th quarter	30.6	29.2	29.7	43.0	51.9	24.1	49.5	40.8	24.4	29.0	21.4	30.2
1976												
1st quarter	30.9	29.5	30.2	43.4	52.5	24.3	50.2	41.5	24.8	29.5	21.7	30.5
2nd quarter	31.2	29.7	30.5	44.1	53.2	25.0	50.8	42.1	25.0	29.7	21.9	30.8
3rd quarter	31.6	30.2	31.0	44.7	53.9	25.4	51.1	42.9	25.2	30.0	22.1	31.2
4th quarter	32.2	30.7	31.5	45.4	54.7	25.8	52.0	43.3	25.7	30.8	22.4	31.7
1977												
1st quarter	32.7	31.2	32.0	46.3	55.7	26.5	52.6	44.9	26.1	31.3	22.8	32.3
2nd quarter	33.2	31.8	32.5	47.2	56.6	27.2	53.4	46.1	26.5	31.6	23.2	32.9
3rd quarter	33.7	32.2	33.0	48.1	57.6	28.1	53.1	46.7	26.8	31.8	23.6	33.4
4th quarter	34.2	32.7	33.5	49.1	58.6	29.0	53.3	47.2	27.4	32.7	24.0	33.9
1978												
1st quarter	34.8	33.2	34.0	50.0	59.4	29.8	54.4	48.0	27.8	33.1	24.3	34.5
2nd quarter	35.5	33.9	34.6	51.0	60.4	30.7	55.7	49.1	28.3	33.7	24.7	35.2
3rd quarter	36.1	34.5	35.2	52.0	61.4	31.5	56.6	49.9	28.7	34.3	25.1	35.8
4th quarter	36.8	35.2	35.9	53.1	62.5	32.3	58.5	50.8	29.2	34.9	25.5	36.5
1979												
1st quarter	37.5	35.8	36.4	54.2	63.8	33.0	60.2	52.8	29.8	35.5	26.1	37.2
2nd quarter	38.4	36.8	37.2	55.5	65.2	34.1	62.7	55.6	30.4	36.1	26.7	38.1
3rd quarter	39.2	37.7	37.8	56.9	66.6	35.2	64.2	59.7	31.3	36.9	27.6	39.1
4th quarter	39.9	38.6	38.6	58.0	67.8	36.1	65.3	63.7	31.9	37.7	28.1	40.0
1980												
1st quarter	40.8	39.8	39.6	59.4	69.3	37.0	67.3	69.0	32.7	38.5	28.9	41.1
2nd quarter	41.8	40.8	40.5	60.8	70.9	37.9	68.2	71.5	33.7	40.0	29.6	42.1
3rd quarter	42.7	41.7	41.4	62.0	72.4	38.8	70.1	73.5	34.4	40.4	30.5	43.1
4th quarter	43.8	42.8	42.4	63.4	73.9	39.6	72.5	74.8	35.4	41.6	31.3	44.1
1981												
1st quarter	45.0	43.9	43.3	65.2	76.0	40.6	74.3	76.7	36.3	42.4	32.3	45.3
2nd quarter	45.8	44.6	44.1	66.6	77.8	41.2	74.6	77.1	37.1	43.4	33.0	46.2
3rd quarter	46.7	45.3	44.9	67.7	79.2	41.7	74.7	75.1	37.8	44.4	33.5	46.9
4th quarter	47.5	46.0	45.7	69.0	80.9	42.3	75.0	75.3	38.4	45.2	34.0	47.7
1982												
1st quarter	48.2	46.6	46.4	69.8	81.9	42.9	75.4	75.1	39.1	45.9	34.6	48.3
2nd quarter	48.8	47.1	47.0	70.6	82.9	43.6	75.4	73.8	39.8	46.8	35.2	48.9
3rd quarter	49.5	47.8	47.8	71.1	83.3	43.9	74.8	72.9	40.3	47.2	35.7	49.6
4th quarter	50.0	48.3	48.5	71.2	83.5	44.1	74.4	72.3	40.8	47.8	36.2	50.0
1983												
1st quarter	50.4	48.7	49.1	71.0	83.1	44.4	74.7	70.8	40.9	47.9	36.4	50.3
2nd quarter	50.8	49.2	49.5	70.8	82.7	44.5	75.0	70.8	41.3	48.2	36.8	50.7
3rd quarter	51.3	49.8	50.2	70.8	82.6	44.7	75.4	70.9	41.8	48.8	37.2	51.2
4th quarter	51.7	50.2	50.6	71.0	82.6	45.1	76.1	70.5	42.0	49.1	37.5	51.5
1984												
1st quarter	52.2	50.7	51.1	71.2	82.7	45.4	76.2	70.7	42.6	49.7	38.0	52.0
2nd quarter	52.7	51.2	51.7	71.6	83.0	45.8	76.7	71.1	43.0	50.2	38.4	52.5
3rd quarter	53.2	51.6	52.2	71.9	83.2	46.2	76.0	69.8	43.6	51.0	38.7	52.9
4th quarter	53.5	51.9	52.6	72.0	83.3	46.6	75.1	69.0	44.0	51.6	39.1	53.2
1985												
1st quarter	54.1	52.5	53.3	72.2	83.5	46.9	74.3	67.3	44.2	51.4	39.6	53.7
2nd quarter	54.4	52.9	53.8	72.3	83.7	47.0	74.0	67.6	44.5	51.5	40.0	54.0
3rd quarter	54.8	53.4	54.3	72.6	84.0	47.3	73.4	67.5	44.8	51.8	40.3	54.4
4th quarter	55.0	53.7	54.7	73.1	84.4	47.9	73.3	68.9	45.1	52.1	40.7	54.8
1986												
1st quarter	55.3	54.1	55.3	73.4	84.6	48.5	72.9	69.0	45.2	52.0	40.8	55.1
2nd quarter	55.5	54.1	55.7	73.9	85.1	49.0	72.5	66.9	45.2	51.9	41.0	55.1
3rd quarter	55.8	54.3	56.1	74.5	85.7	49.7	72.1	67.3	45.5	51.9	41.4	55.4
4th quarter	56.1	54.7	56.5	75.0	86.1	50.2	72.6	68.2	45.8	52.0	41.9	55.8
1987												
1st quarter	56.4	55.2	56.9	75.3	86.1	50.8	72.9	69.9	46.1	51.9	42.5	56.2
2nd quarter	56.8	55.7	57.4	75.5	86.1	51.2	74.0	71.7	46.5	52.2	43.0	56.7
3rd quarter	57.2	56.2	57.9	75.7	86.1	51.7	74.2	72.5	46.9	52.5	43.5	57.1
4th quarter	57.7	56.7	58.5	76.4	87.0	52.2	75.4	73.6	47.1	52.7	43.8	57.6

Table 1-6B. Chain-Type Price Indexes for Gross Domestic Product and Domestic Purchases: Historical Data—Continued

(Index numbers, 2012 = 100.)

NIPA Tables 1.1.4, 1.6.4, 2.3.4

Year and quarter	Gross domestic product, total	Personal consumption expenditures		Private fixed investment			Exports and imports of goods and services		Government consumption expenditures and gross investment			Gross domestic purchases
		Total	Excluding food and energy	Total	Nonresidential	Residential	Exports	Imports	Total	Federal	State and local	
1988												
1st quarter	58.1	57.2	59.1	76.9	87.8	52.6	76.3	74.6	47.5	53.4	44.0	58.1
2nd quarter	58.7	57.8	59.8	77.4	88.2	53.1	77.9	76.0	48.0	53.9	44.4	58.6
3rd quarter	59.4	58.5	60.5	77.7	88.6	53.5	78.8	75.1	48.3	54.2	44.8	59.2
4th quarter	60.0	59.1	61.2	78.5	89.4	54.0	78.7	75.7	48.8	54.6	45.3	59.8
1989												
1st quarter	60.5	59.8	61.8	78.9	89.9	54.3	79.4	77.0	49.3	55.0	45.9	60.4
2nd quarter	61.2	60.6	62.4	79.4	90.3	55.0	79.5	77.8	49.9	55.4	46.5	61.1
3rd quarter	61.7	61.0	62.9	79.8	90.8	55.2	79.1	76.5	50.2	55.7	47.0	61.5
4th quarter	62.1	61.4	63.4	80.2	91.3	55.6	78.8	76.9	50.7	55.9	47.6	62.0
1990												
1st quarter	62.8	62.3	64.2	80.7	91.8	56.0	78.9	78.1	51.2	56.3	48.3	62.7
2nd quarter	63.5	62.9	64.9	81.0	92.1	56.2	79.0	76.7	51.9	57.3	48.8	63.3
3rd quarter	64.0	63.7	65.5	81.5	92.8	56.4	79.8	78.9	52.3	57.4	49.4	64.0
4th quarter	64.5	64.5	66.1	81.8	93.4	56.5	80.9	83.3	53.0	58.0	50.2	64.7
1991												
1st quarter	65.1	64.9	66.7	82.6	94.3	56.7	81.1	80.7	53.4	58.6	50.4	65.1
2nd quarter	65.6	65.2	67.2	82.7	94.4	57.0	80.7	78.5	53.7	58.9	50.7	65.4
3rd quarter	66.1	65.7	67.8	82.8	94.3	57.3	80.1	77.3	54.2	59.6	51.1	65.9
4th quarter	66.5	66.1	68.3	82.5	94.1	57.1	80.3	77.7	54.7	60.2	51.6	66.3
1992												
1st quarter	66.7	66.6	68.8	82.4	94.0	56.9	80.1	78.1	55.0	60.3	51.9	66.6
2nd quarter	67.1	67.0	69.3	82.5	93.9	57.5	80.2	78.4	55.4	60.5	52.6	67.0
3rd quarter	67.5	67.4	69.7	82.7	94.0	57.9	80.2	79.4	55.9	61.0	52.9	67.4
4th quarter	67.9	67.9	70.3	83.0	94.0	58.6	80.1	78.7	56.3	61.4	53.3	67.8
1993												
1st quarter	68.3	68.3	70.7	83.2	94.1	59.3	80.1	78.3	56.5	61.6	53.6	68.2
2nd quarter	68.7	68.8	71.3	83.6	94.1	59.9	80.4	78.5	56.8	61.9	53.9	68.6
3rd quarter	69.1	69.1	71.7	83.8	94.1	60.4	80.3	77.8	57.1	62.3	54.1	68.9
4th quarter	69.5	69.5	72.0	84.0	94.4	60.6	80.3	77.5	57.4	62.9	54.4	69.2
1994												
1st quarter	69.9	69.7	72.4	84.5	94.6	61.4	80.7	77.5	57.8	63.2	54.8	69.5
2nd quarter	70.2	70.1	72.9	84.7	94.8	61.8	81.0	78.2	58.2	63.7	55.1	69.9
3rd quarter	70.6	70.6	73.3	85.0	95.0	62.5	81.4	79.5	58.7	64.0	55.6	70.4
4th quarter	71.0	70.9	73.6	85.4	95.1	63.3	81.9	79.8	59.1	64.6	56.1	70.8
1995												
1st quarter	71.4	71.3	74.0	85.9	95.5	64.0	82.7	80.5	59.5	65.1	56.4	71.1
2nd quarter	71.7	71.7	74.5	86.2	95.8	64.3	83.4	81.7	60.0	65.5	56.8	71.5
3rd quarter	72.0	72.0	74.8	86.5	96.1	64.6	83.2	81.1	60.2	65.9	57.0	71.8
4th quarter	72.4	72.3	75.2	86.4	95.9	65.0	82.8	80.4	60.7	66.8	57.3	72.1
1996												
1st quarter	72.7	72.7	75.5	86.2	95.6	65.2	82.6	80.2	61.1	67.0	57.8	72.5
2nd quarter	73.0	73.2	75.8	86.0	95.2	65.5	82.4	79.8	61.1	66.6	57.9	72.7
3rd quarter	73.4	73.5	76.2	86.3	95.2	66.2	81.8	79.0	61.4	67.0	58.3	73.0
4th quarter	73.7	74.0	76.6	86.2	95.0	66.5	80.9	79.0	61.8	67.2	58.7	73.4
1997												
1st quarter	74.0	74.3	76.9	86.2	94.9	66.8	80.7	78.4	62.1	67.4	59.1	73.7
2nd quarter	74.4	74.5	77.3	86.3	94.9	67.1	80.7	76.7	62.4	67.9	59.3	73.9
3rd quarter	74.6	74.7	77.5	86.3	94.8	67.7	80.5	76.2	62.6	68.0	59.5	74.1
4th quarter	74.8	74.9	77.8	86.2	94.4	68.2	80.1	75.6	63.1	68.5	60.0	74.3
1998												
1st quarter	74.9	74.9	78.0	85.8	93.8	68.4	79.3	73.8	63.1	68.2	60.1	74.2
2nd quarter	75.1	75.1	78.2	85.6	93.4	68.8	78.9	72.9	63.4	68.7	60.3	74.3
3rd quarter	75.4	75.3	78.5	85.5	93.0	69.5	78.2	71.9	63.8	69.1	60.8	74.6
4th quarter	75.6	75.5	78.8	85.6	92.8	70.1	77.9	71.9	64.2	69.3	61.3	74.8
1999												
1st quarter	75.9	75.7	79.0	85.6	92.7	70.8	77.7	71.6	64.7	69.5	61.8	75.0
2nd quarter	76.2	76.1	79.3	85.7	92.4	71.6	77.9	72.5	65.4	70.1	62.7	75.4
3rd quarter	76.5	76.5	79.5	85.6	92.0	72.2	78.0	73.4	66.1	70.8	63.4	75.8
4th quarter	76.9	77.0	79.9	85.8	92.2	72.7	78.3	74.5	66.9	71.6	64.1	76.3
2000												
1st quarter	77.4	77.6	80.3	86.4	92.4	74.0	78.8	75.9	67.7	72.3	65.0	76.9
2nd quarter	77.8	78.0	80.6	86.7	92.6	74.7	79.4	75.9	68.3	72.6	65.6	77.3
3rd quarter	78.3	78.5	81.0	87.0	92.9	75.3	79.8	76.4	68.9	73.1	66.3	77.8
4th quarter	78.7	78.9	81.3	87.2	92.9	76.0	79.8	76.6	69.6	73.5	67.2	78.2
2001												
1st quarter	79.2	79.4	81.8	87.2	92.5	77.1	79.6	76.4	70.1	73.6	68.0	78.7
2nd quarter	79.8	79.8	82.1	87.5	92.5	78.1	79.1	74.8	70.4	74.0	68.2	79.0
3rd quarter	80.0	79.8	82.3	87.7	92.3	79.3	78.7	73.7	70.7	74.4	68.4	79.2
4th quarter	80.3	79.9	82.8	87.8	92.1	79.8	77.9	72.0	71.0	74.9	68.5	79.3
2002												
1st quarter	80.5	80.0	83.0	87.9	92.1	79.7	77.5	71.6	71.4	75.4	69.0	79.5
2nd quarter	80.8	80.7	83.5	87.7	92.0	80.1	78.0	73.2	72.1	76.1	69.6	79.9
3rd quarter	81.2	81.0	83.9	87.6	91.7	80.5	78.6	73.8	72.6	76.8	70.1	80.3
4th quarter	81.6	81.4	84.2	88.1	91.7	81.8	78.8	74.3	73.5	78.2	70.7	80.8

Table 1-6B. Chain-Type Price Indexes for Gross Domestic Product and Domestic Purchases: Historical Data—Continued

(Index numbers, 2012 = 100.)   NIPA Tables 1.1.4, 1.6.4, 2.3.4

Year and quarter	Gross domestic product, total	Personal consumption expenditures		Private fixed investment			Exports and imports of goods and services		Government consumption expenditures and gross investment			Gross domestic purchases
		Total	Excluding food and energy	Total	Nonresidential	Residential	Exports	Imports	Total	Federal	State and local	
2003												
1st quarter	82.0	82.0	84.4	88.5	91.4	83.5	79.1	76.2	74.4	79.1	71.6	81.4
2nd quarter	82.3	82.0	84.6	88.4	91.2	83.7	79.2	74.9	74.7	79.6	71.7	81.5
3rd quarter	82.7	82.5	85.0	88.4	91.0	84.3	79.4	75.1	75.2	80.4	72.2	81.9
4th quarter	83.2	82.9	85.3	89.0	91.1	85.8	79.9	75.6	75.8	81.0	72.7	82.3
2004												
1st quarter	83.8	83.6	85.9	89.8	91.4	87.7	81.2	77.4	76.6	81.7	73.6	83.1
2nd quarter	84.5	84.2	86.4	90.8	91.9	89.4	82.1	78.4	77.6	82.4	74.7	83.8
3rd quarter	85.1	84.6	86.7	91.6	92.1	91.2	82.5	79.2	78.6	83.1	75.9	84.4
4th quarter	85.7	85.3	87.2	92.4	92.8	92.7	83.4	81.2	79.8	83.9	77.2	85.2
2005												
1st quarter	86.4	85.8	87.8	93.4	93.6	93.9	84.3	81.9	80.6	85.0	77.9	85.9
2nd quarter	87.0	86.3	88.2	94.3	94.2	95.4	84.9	82.8	81.5	85.8	78.9	86.5
3rd quarter	87.8	87.2	88.5	95.3	94.6	97.9	85.4	84.4	82.6	86.7	80.2	87.5
4th quarter	88.5	87.9	89.1	96.4	95.3	99.7	85.9	85.8	83.7	87.3	81.5	88.3
2006												
1st quarter	89.1	88.4	89.6	97.3	95.9	101.3	86.4	86.0	84.3	88.0	82.0	88.9
2nd quarter	89.8	89.1	90.2	97.9	96.4	102.2	87.6	87.0	85.3	88.7	83.3	89.6
3rd quarter	90.5	89.7	90.7	98.4	97.0	102.6	88.7	87.8	86.1	89.3	84.1	90.3
4th quarter	90.8	89.6	91.1	99.2	97.7	103.3	88.7	86.8	86.8	89.8	85.0	90.4
2007												
1st quarter	91.8	90.4	91.7	99.7	98.2	103.9	89.6	87.4	88.2	90.5	86.6	91.3
2nd quarter	92.3	91.1	92.1	99.7	98.4	103.6	90.7	88.8	89.0	91.3	87.6	91.9
3rd quarter	92.7	91.7	92.5	99.7	98.3	103.6	91.4	90.1	89.9	91.9	88.5	92.4
4th quarter	93.2	92.0	93.1	100.6	98.3	103.7	92.8	93.4	90.9	92.6	89.8	93.2
2008												
1st quarter	93.6	93.3	93.7	99.7	98.7	103.4	94.7	97.1	92.1	93.4	91.3	94.0
2nd quarter	93.9	94.3	94.1	99.8	99.0	102.9	96.8	101.8	93.3	94.4	92.5	94.8
3rd quarter	94.7	95.3	94.5	100.3	100.0	102.1	98.0	104.1	94.3	95.1	93.8	95.7
4th quarter	94.9	93.8	94.6	102.1	101.6	100.6	92.1	92.8	93.5	94.6	92.7	94.9
2009												
1st quarter	95.0	93.3	94.7	101.3	100.9	99.8	89.0	85.1	92.5	93.9	91.5	94.2
2nd quarter	94.9	93.7	95.1	99.7	99.5	99.2	88.9	86.3	92.6	93.9	91.7	94.3
3rd quarter	94.9	94.3	95.4	98.3	98.4	97.8	89.9	89.1	93.0	94.2	92.2	94.6
4th quarter	95.3	95.1	96.0	98.0	97.9	98.5	91.0	91.5	93.7	94.9	92.9	95.2
2010												
1st quarter	95.5	95.4	96.3	97.6	97.3	98.7	91.9	92.9	94.5	95.5	93.8	95.5
2nd quarter	95.9	95.5	96.5	97.6	97.4	98.0	93.0	92.4	95.1	96.2	94.4	95.7
3rd quarter	96.3	95.7	96.7	97.6	97.4	98.0	93.2	91.9	95.6	96.6	94.9	95.9
4th quarter	96.8	96.3	96.9	98.0	97.7	98.6	95.3	94.0	96.4	97.4	95.7	96.5
2011												
1st quarter	97.3	97.1	97.4	98.2	98.0	98.8	97.8	97.8	97.3	98.3	96.6	97.3
2nd quarter	98.0	98.0	98.0	98.7	98.5	99.1	99.7	100.8	98.4	99.2	97.8	98.2
3rd quarter	98.5	98.5	98.4	98.9	98.7	99.1	100.3	100.6	98.8	99.5	98.3	98.6
4th quarter	98.7	98.9	98.8	99.0	99.0	99.2	99.2	100.2	98.7	99.4	98.2	98.9
2012												
1st quarter	99.3	99.5	99.4	99.4	99.6	99.0	99.8	101.1	99.5	99.7	99.4	99.5
2nd quarter	99.7	99.8	99.9	99.9	99.9	99.4	100.0	100.4	99.6	99.9	99.4	99.8
3rd quarter	100.3	100.1	100.1	100.3	100.2	100.3	99.9	98.7	100.0	100.1	100.0	100.1
4th quarter	100.7	100.6	100.6	100.4	100.2	101.3	100.3	99.8	100.8	100.2	101.2	100.6
2013												
1st quarter	101.1	101.0	101.0	100.5	100.0	102.8	100.6	99.9	101.5	100.2	102.3	101.0
2nd quarter	101.4	101.1	101.3	100.8	100.1	104.2	99.9	98.6	101.9	100.4	102.9	101.2
3rd quarter	102.0	101.5	101.7	101.1	100.3	105.6	100.1	98.1	102.5	100.8	103.7	101.6
4th quarter	102.6	101.9	102.2	101.6	100.6	107.6	100.1	97.9	103.5	102.3	104.3	102.1
2014												
1st quarter	103.0	102.4	102.5	102.3	101.0	109.9	100.8	99.1	103.9	102.0	105.1	102.6
2nd quarter	103.5	102.8	103.0	102.6	101.3	110.2	100.9	98.5	104.3	102.5	105.4	103.1
3rd quarter	104.0	103.1	103.4	103.2	101.6	112.1	100.6	97.9	104.8	102.9	106.0	103.4
4th quarter	104.1	103.0	103.6	103.6	101.9	112.3	98.8	95.9	104.8	103.1	106.0	103.4
2015												
1st quarter	104.1	102.5	103.8	103.7	102.2	113.0	96.4	91.8	104.3	103.0	105.1	103.1
2nd quarter	104.7	103.0	104.3	103.7	102.2	113.6	96.2	90.7	104.8	103.3	105.9	103.5
3rd quarter	105.0	103.3	104.6	103.8	102.1	114.6	95.3	89.6	105.0	103.5	106.0	103.7
4th quarter	105.0	103.3	104.9	103.5	101.7	115.3	93.8	87.6	104.7	103.4	105.6	103.6
2016												
1st quarter	104.9	103.3	105.3	103.0	101.1	115.8	92.3	86.1	104.2	103.1	104.9	103.4
2nd quarter	105.6	103.9	105.8	103.4	101.2	117.3	93.3	86.4	104.9	103.7	105.7	104.0
3rd quarter	106.0	104.3	106.4	103.6	101.0	119.0	93.8	87.0	105.3	104.1	106.1	104.4
4th quarter	106.5	104.8	106.7	104.2	101.3	120.5	94.4	87.3	105.8	104.7	106.7	104.9
2017												
1st quarter	107.0	105.4	107.2	104.6	101.6	121.5	95.1	88.3	106.7	105.2	107.7	105.5
2nd quarter	107.4	105.6	107.5	105.2	102.0	123.0	95.1	88.3	107.1	105.7	108.1	105.8
3rd quarter	108.0	106.0	107.9	105.8	102.3	124.3	96.0	88.4	107.8	106.2	108.9	106.3
4th quarter	108.7	106.7	108.5	106.0	102.5	125.3	97.3	89.5	108.9	107.1	110.2	107.0

Table 1-7. Final Sales

(Quarterly dollar data are at seasonally adjusted annual rates.)

NIPA Tables 1.4.4, 1.4.5, 1.4.6

Year and quarter	Final sales of domestic product			Final sales to domestic purchasers		
	Billions of dollars	Billions of chained (2012) dollars	Chain-type price index, 2012 = 100	Billions of dollars	Billions of chained (2012) dollars	Chain-type price index, 2012 = 100
1955	420.5	2 866.6	14.7	420.0	2 929.8	14.3
1956	445.4	2 936.4	15.2	443.0	2 990.5	14.8
1957	473.3	3 014.2	15.7	469.2	3 063.0	15.3
1958	481.6	2 997.1	16.1	481.1	3 072.3	15.7
1959	517.7	3 180.0	16.3	517.3	3 259.9	15.9
1960	539.1	3 266.5	16.5	534.9	3 325.7	16.1
1961	559.2	3 352.4	16.7	554.3	3 412.0	16.2
1962	597.8	3 540.2	16.9	593.7	3 612.0	16.4
1963	631.8	3 698.6	17.1	626.9	3 766.2	16.6
1964	679.6	3 918.1	17.3	672.7	3 977.9	16.9
1965	733.0	4 149.2	17.7	727.4	4 226.9	17.2
1966	799.8	4 402.8	18.2	795.9	4 499.5	17.7
1967	850.1	4 546.7	18.7	846.5	4 657.1	18.2
1968	931.6	4 777.1	19.5	930.2	4 907.7	19.0
1969	1 008.4	4 927.8	20.5	1 007.0	5 064.4	19.9
1970	1 071.3	4 971.9	21.5	1 067.4	5 093.0	21.0
1971	1 156.6	5 108.2	22.6	1 156.0	5 242.4	22.1
1972	1 270.0	5 375.9	23.6	1 273.4	5 527.0	23.0
1973	1 409.5	5 655.4	24.9	1 405.4	5 769.1	24.4
1974	1 531.3	5 639.3	27.2	1 532.1	5 710.1	26.8
1975	1 691.2	5 698.0	29.7	1 675.2	5 719.9	29.3
1976	1 856.3	5 925.8	31.3	1 857.9	6 011.3	30.9
1977	2 059.5	6 187.7	33.3	2 082.6	6 318.8	33.0
1978	2 325.8	6 526.5	35.6	2 351.2	6 657.9	35.3
1979	2 609.3	6 761.6	38.6	2 631.9	6 851.8	38.4
1980	2 863.6	6 804.7	42.1	2 876.7	6 783.2	42.4
1981	3 177.2	6 900.1	46.0	3 189.8	6 887.9	46.3
1982	3 358.7	6 865.5	48.9	3 378.7	6 894.2	49.0
1983	3 639.8	7 159.9	50.8	3 691.4	7 278.9	50.7
1984	3 972.2	7 541.5	52.7	4 074.9	7 772.3	52.4
1985	4 317.2	7 940.2	54.4	4 431.2	8 203.0	54.0
1986	4 573.1	8 241.0	55.5	4 704.9	8 528.9	55.2
1987	4 828.1	8 492.5	56.9	4 972.8	8 767.5	56.7
1988	5 218.0	8 860.6	58.9	5 327.3	9 067.0	58.8
1989	5 613.9	9 172.3	61.2	5 700.6	9 333.9	61.1
1990	5 948.6	9 365.1	63.5	6 026.5	9 489.7	63.5
1991	6 158.5	9 379.0	65.7	6 187.1	9 445.2	65.5
1992	6 504.0	9 683.1	67.2	6 538.7	9 753.9	67.0
1993	6 837.7	9 943.6	68.8	6 902.9	10 069.3	68.6
1994	7 223.5	10 284.1	70.2	7 315.9	10 451.4	70.0
1995	7 608.6	10 608.5	71.7	7 698.3	10 764.1	71.5
1996	8 042.3	11 008.6	73.1	8 138.7	11 182.0	72.8
1997	8 506.6	11 442.2	74.3	8 608.6	11 652.3	73.9
1998	8 999.1	11 966.9	75.2	9 161.8	12 313.5	74.4
1999	9 569.8	12 543.1	76.3	9 825.7	13 001.0	75.6
2000	10 197.8	13 068.0	78.0	10 572.9	13 635.7	77.5
2001	10 620.1	13 309.5	79.8	10 988.0	13 908.0	79.0
2002	10 916.4	13 476.4	81.0	11 341.8	14 162.1	80.1
2003	11 444.2	13 864.7	82.5	11 947.3	14 615.3	81.7
2004	12 149.7	14 335.7	84.8	12 768.7	15 184.1	84.1
2005	12 979.1	14 852.3	87.4	13 700.3	15 745.1	87.0
2006	13 745.6	15 263.0	90.1	14 516.5	16 172.2	89.8
2007	14 417.9	15 588.7	92.5	15 136.3	16 418.1	92.2
2008	14 742.1	15 639.7	94.3	15 465.2	16 306.5	94.8
2009	14 599.7	15 373.0	95.0	14 996.2	15 863.7	94.5
2010	14 938.1	15 546.6	96.1	15 452.0	16 112.5	95.9
2011	15 496.3	15 796.5	98.1	16 075.8	16 364.6	98.2
2012	16 125.8	16 125.8	100.0	16 694.4	16 694.4	100.0
2013	16 680.3	16 386.2	101.8	17 171.1	16 919.3	101.5
2014	17 443.3	16 822.3	103.7	17 951.0	17 398.0	103.2
2015	18 092.9	17 267.1	104.8	18 612.8	17 977.1	103.5
2016	18 688.0	17 647.6	105.9	19 206.8	18 415.5	104.3
2017	19 489.2	18 058.4	107.9	20 064.6	18 882.3	106.3
2018	20 525.5	18 571.3	110.5	21 163.7	19 456.8	108.8
2016						
1st quarter	18 377.4	17 492.6	105.1	18 899.6	18 255.2	103.5
2nd quarter	18 618.1	17 607.5	105.7	19 113.5	18 356.5	104.1
3rd quarter	18 809.5	17 726.7	106.1	19 309.2	18 475.6	104.5
4th quarter	18 946.9	17 763.5	106.7	19 504.9	18 574.9	105.0
2017						
1st quarter	19 177.0	17 895.1	107.2	19 748.0	18 703.6	105.6
2nd quarter	19 337.8	17 985.3	107.5	19 921.6	18 810.1	105.9
3rd quarter	19 554.9	18 082.5	108.1	20 105.4	18 893.4	106.4
4th quarter	19 887.2	18 270.7	108.9	20 483.3	19 122.1	107.1
2018						
1st quarter	20 121.7	18 380.4	109.5	20 750.7	19 233.5	107.9
2nd quarter	20 520.1	18 595.6	110.4	21 088.5	19 421.1	108.6
3rd quarter	20 662.4	18 630.9	110.9	21 333.8	19 554.0	109.1
4th quarter	20 797.7	18 678.3	111.4	21 481.9	19 618.8	109.5

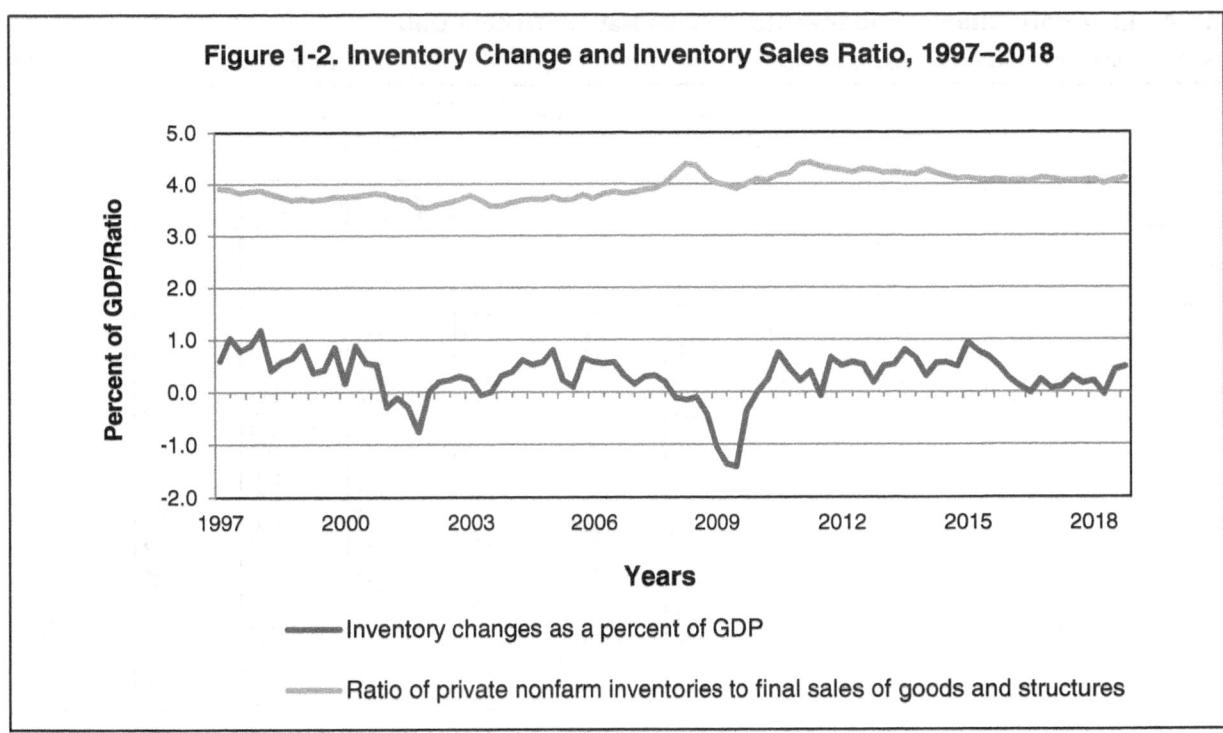

Figure 1-2. Inventory Change and Inventory Sales Ratio, 1997–2018

- The change in inventories is part of business investment component that, although small, can be volatile and even negative, reflecting the fact that production is frequently more variable than sales. When inventory change is positive, the economy is producing more than is sold. This is the normal condition, as inventories grow to accommodate a growing level of sales. When inventory change is negative, production is below the rate of sales—which could be because sales rose faster than anticipated, causing involuntary inventory reduction, or on the other hand because production was deliberately cut back in order to get rid of unwanted stocks. Furthermore, it is the <u>change</u> in inventory —the change in an already volatile number—that drives the GDP's <u>growth rate</u>.

- In 2009, following a rapid and unwanted rise in inventories, stocks were reduced at a record rate of $150 billion (2012 dollars), contributing to the 2.7 percent decline in real GDP. In 2010, the resumption of inventory-building contributed to the 2.6 percent GDP increase. (Tables 1-2A, 1-8A and 1-4A).

- Inventory investment has been moderate and variable during the recovery. (Tables 1-8A and B)

Table 1-8A. Inventory Change and Inventory-Sales Ratios: Recent Data

(Quarterly dollar data are at seasonally adjusted annual rates.)   **NIPA Tables 5.7.5B, 5.7.6B, 5.8.5B, 5.8.6B**

Year and quarter	Change in private inventories				Ratio, inventories at end of quarter to monthly rate of sales during the quarter					
	Billions of current dollars		Billions of chained dollars [1]		Total private inventories to final sales of domestic business		Nonfarm inventories to final sales of domestic business		Nonfarm inventories to final sales of goods and structures	
	Farm	Nonfarm	Farm	Nonfarm	Current dollars	Chained dollars[1]	Current dollars	Chained dollars[1]	Current dollars	Chained dollars[1]
OLD BASIS										
1960	0.6	2.7	1.7	9.9	4.22	3.37	2.89	2.34	3.93	4.10
1961	0.9	2.1	2.5	7.8	4.12	3.30	2.81	2.29	3.84	4.03
1962	0.6	5.5	1.7	20.4	4.14	3.32	2.82	2.34	3.87	4.11
1963	0.5	5.1	1.4	19.0	3.95	3.25	2.78	2.32	3.81	4.07
1964	-1.2	6.0	-3.8	21.8	3.79	3.17	2.75	2.31	3.79	4.07
1965	0.8	8.4	2.5	30.5	3.77	3.09	2.72	2.29	3.71	3.99
1966	-0.5	14.1	-1.4	50.2	3.92	3.26	2.92	2.50	4.01	4.38
1967	0.9	9.0	2.5	31.5	3.90	3.35	2.99	2.59	4.14	4.57
1968	1.4	7.7	4.1	26.5	3.79	3.33	2.90	2.58	4.03	4.56
1969	0.0	9.2	0.0	30.9	3.88	3.40	2.98	2.68	4.18	4.76
1970	-0.8	2.8	-2.4	8.6	3.81	3.39	2.96	2.68	4.23	4.83
1971	1.7	6.6	4.1	20.6	3.76	3.33	2.88	2.64	4.12	4.75
1972	0.3	8.8	0.4	26.7	3.74	3.18	2.77	2.55	3.94	4.53
1973	1.5	14.4	1.8	40.5	4.20	3.22	2.98	2.62	4.22	4.67
1974	-2.8	16.8	-4.4	39.0	4.52	3.46	3.54	2.88	5.11	5.26
1975	3.4	-9.6	5.8	-21.4	4.04	3.26	3.16	2.66	4.59	4.88
1976	-0.8	18.0	-1.6	37.6	3.95	3.23	3.19	2.68	4.68	4.92
1977	4.5	17.8	7.2	35.7	3.89	3.22	3.16	2.67	4.63	4.87
1978	1.4	24.4	2.1	44.9	3.98	3.14	3.15	2.63	4.54	4.73
1979	3.6	14.4	4.2	23.8	4.19	3.16	3.34	2.64	4.85	4.75
1980	-6.1	-0.2	-7.2	-0.4	4.24	3.13	3.44	2.65	5.08	4.82
1981	8.8	21.0	10.1	28.2	4.17	3.28	3.49	2.75	5.20	5.03
1982	5.8	-20.7	7.5	-27.4	3.97	3.23	3.30	2.68	5.10	4.99
1983	-15.4	9.6	-17.4	10.7	3.69	2.98	3.08	2.52	4.77	4.64
1984	5.7	59.7	6.3	72.3	3.72	3.05	3.16	2.60	4.89	4.72
1985	5.8	16.1	7.0	18.4	3.51	2.98	3.01	2.53	4.74	4.65
1986	-1.5	8.0	-2.3	10.4	3.25	2.90	2.82	2.47	4.49	4.53
1987	-6.4	33.6	-9.1	41.1	3.33	2.90	2.90	2.50	4.65	4.62
1988	-11.9	30.4	-13.1	35.4	3.29	2.81	2.86	2.47	4.63	4.55
1989	0.0	27.7	0.0	30.8	3.23	2.80	2.83	2.47	4.63	4.56
1990	2.4	12.2	2.5	14.0	3.22	2.83	2.82	2.49	4.71	4.66
1991	-1.3	0.9	-1.8	0.4	3.06	2.83	2.70	2.50	4.62	4.74
1992	6.3	10.1	7.1	11.0	2.92	2.73	2.56	2.39	4.41	4.53
1993	-6.2	27.0	-7.5	29.8	2.85	2.68	2.51	2.38	4.33	4.47
1994	12.0	51.8	13.7	55.9	2.89	2.72	2.56	2.40	4.39	4.47
1995	-11.1	42.2	-13.0	44.3	2.88	2.69	2.58	2.40	4.45	4.46
1996	8.6	22.1	8.3	23.2	2.76	2.62	2.47	2.33	4.24	4.28
1997	3.3	67.7	3.7	73.6	2.70	2.65	2.42	2.37	4.17	4.32
NEW BASIS										
1997	3.3	67.7	6.2	88.0	2.60	2.97	2.34	2.57	3.86	4.64
1998	1.3	62.5	2.8	86.5	2.47	2.93	2.25	2.55	3.69	4.56
1999	-2.7	63.5	-6.3	87.7	2.49	2.91	2.27	2.55	3.75	4.55
2000	-1.4	55.9	-3.0	77.6	2.51	2.90	2.29	2.56	3.82	4.60
2001	0.0	-38.3	0.2	-51.4	2.33	2.81	2.13	2.47	3.55	4.43
2002	-2.5	22.5	-5.2	27.1	2.38	2.83	2.16	2.50	3.71	4.55
2003	0.1	14.0	0.2	19.2	2.33	2.71	2.10	2.39	3.57	4.28
2004	8.8	55.3	13.5	70.3	2.39	2.73	2.17	2.40	3.70	4.30
2005	0.2	57.3	0.4	61.5	2.42	2.71	2.21	2.39	3.79	4.27
2006	-3.6	72.6	-6.3	89.1	2.42	2.71	2.21	2.41	3.82	4.31
2007	0.6	33.3	-1.4	40.6	2.51	2.69	2.28	2.40	4.01	4.31
2008	-0.5	-28.7	-0.6	-31.5	2.50	2.76	2.28	2.45	4.14	4.56
2009	-6.7	-144.1	-10.0	-165.2	2.37	2.58	2.17	2.29	3.99	4.28
2010	-12.2	66.1	-17.8	70.2	2.49	2.58	2.27	2.31	4.21	4.31
2011	-2.1	48.3	-2.2	48.4	2.58	2.57	2.32	2.30	4.30	4.27
2012	-18.7	89.9	-18.7	89.9	2.56	2.57	2.33	2.33	4.27	4.28
2013	11.7	92.8	10.4	98.2	2.52	2.60	2.30	2.36	4.19	4.27
2014	-3.2	87.2	-3.5	90.1	2.46	2.58	2.23	2.36	4.10	4.26
2015	1.6	130.3	1.1	131.3	2.39	2.67	2.20	2.45	4.09	4.46
2016	-5.8	32.9	-6.6	28.5	2.35	2.63	2.19	2.42	4.11	4.38
2017	-3.7	33.9	-4.8	35.3	2.33	2.57	2.18	2.36	4.06	4.24
2018	-7.8	62.5	-9.0	55.2	2.34	2.55	2.19	2.36	4.11	4.23
2016										
1st quarter	-13.6	60.5	-13.1	62.9	2.37	2.67	2.18	2.45	4.06	4.44
2nd quarter	-3.5	22.7	-4.7	14.9	2.36	2.65	2.19	2.43	4.07	4.40
3rd quarter	-0.1	-2.7	-1.1	-13.7	2.33	2.62	2.17	2.41	4.05	4.35
4th quarter	-6.1	51.1	-7.4	50.1	2.35	2.63	2.19	2.42	4.11	4.38
2017										
1st quarter	-3.7	17.1	-4.6	12.5	2.36	2.60	2.19	2.39	4.10	4.33
2nd quarter	-3.6	22.4	-4.0	20.0	2.35	2.59	2.18	2.39	4.06	4.29
3rd quarter	-4.0	60.8	-5.3	73.8	2.33	2.59	2.18	2.39	4.06	4.29
4th quarter	-3.5	35.2	-5.2	34.9	2.33	2.57	2.18	2.36	4.06	4.24
2018										
1st quarter	-5.0	46.4	-5.1	44.8	2.35	2.56	2.19	2.36	4.09	4.23
2nd quarter	-6.0	-3.9	-6.6	-21.7	2.30	2.52	2.16	2.32	4.00	4.14
3rd quarter	-8.7	96.1	-10.5	94.6	2.32	2.53	2.18	2.34	4.07	4.20
4th quarter	-11.3	111.4	-13.7	103.0	2.34	2.55	2.19	2.36	4.11	4.23

[1]Before 1997, 2005 dollars; 1997 forward, 2012 dollars.

Table 1-8B. Inventory Change and Inventory-Sales Ratios: Historical Data

(Quarterly dollar data are at seasonally adjusted annual rates.)   NIPA Tables 5.7.5B, 5.7.6B, 5.8.5B, 5.8.6B

| Year and quarter | Change in private inventories | | | | Ratio, inventories at end of quarter to monthly rate of sales during the quarter | | | | | |
| | Billions of current dollars | | Billions of chained dollars [1] | | Total private inventories to final sales of domestic business | | Nonfarm inventories to final sales of domestic business | | Nonfarm inventories to final sales of goods and structures | |
	Farm	Nonfarm	Farm	Nonfarm	Current dollars	Chained dollars[1]	Current dollars	Chained dollars[1]	Current dollars	Chained dollars[1]
OLD BASIS										
1929	-0.1	1.7	-0.6	10.6	...	...	...	...	...	...
1930	-0.3	0.0	-1.7	-2.4	...	...	...	...	...	...
1931	0.5	-1.6	3.1	-13.4	...	...	...	...	...	...
1932	0.1	-2.5	0.8	-21.2	...	...	...	...	...	...
1933	-0.1	-1.3	-1.1	-9.4	...	...	...	...	...	...
1934	-0.8	0.2	-5.7	1.0	...	...	...	...	...	...
1935	0.7	0.4	4.0	2.5	...	...	...	...	...	...
1936	-0.9	2.0	-5.4	15.5	...	...	...	...	...	...
1937	0.9	1.7	4.1	7.9	...	...	...	...	...	...
1938	0.4	-1.0	2.8	-6.0	...	...	...	...	...	...
1939	-0.1	0.3	-1.1	2.7	...	...	...	...	...	...
1940	0.5	1.9	2.8	12.4	...	...	...	...	...	...
1941	0.4	3.9	1.9	24.3	...	...	...	...	...	...
1942	1.3	0.6	5.6	3.0	...	...	...	...	...	...
1943	-0.2	-0.5	-1.1	-2.3	...	...	...	...	...	...
1944	-0.3	-0.6	-1.5	-2.5	...	...	...	...	...	...
1945	-0.9	-0.6	-3.3	-3.8	...	...	...	...	...	...
1946	-0.2	6.2	-1.0	31.8	...	...	...	...	...	...
1947	-1.8	1.2	-4.5	4.4	...	...	...	...	...	...
1948	2.7	3.0	6.1	12.4	...	...	...	...	...	...
1949	-0.6	-2.1	-1.9	-8.6	...	...	...	...	...	...
1950	-0.1	5.9	-0.3	22.8	5.81	3.67	3.04	2.30	3.94	3.97
1951	1.0	8.9	2.1	28.4	5.85	3.76	3.15	2.46	4.07	4.24
1952	1.4	2.1	3.1	7.4	5.24	3.74	3.06	2.45	3.98	4.23
1953	0.7	1.2	2.0	4.5	5.04	3.74	3.08	2.45	4.04	4.22
1954	0.2	-2.1	0.5	-7.5	4.74	3.52	2.88	2.28	3.81	3.94
1947										
1st quarter	-1.1	1.6	-0.4	6.6	5.93	3.88	2.82	2.27	3.71	3.99
2nd quarter	-2.7	1.5	-4.5	4.2	5.89	3.84	2.82	2.27	3.69	4.00
3rd quarter	-2.5	-0.4	-6.8	-3.7	6.02	3.76	2.76	2.23	3.61	3.93
4th quarter	-0.8	2.3	-6.3	10.5	6.30	3.74	2.81	2.25	3.65	3.93
1948										
1st quarter	1.3	2.4	1.6	11.7	6.00	3.76	2.87	2.27	3.71	3.96
2nd quarter	2.8	3.0	7.7	11.9	6.02	3.81	2.89	2.30	3.76	4.03
3rd quarter	3.4	3.7	8.3	15.0	5.88	3.88	2.97	2.34	3.85	4.10
4th quarter	3.1	2.9	6.6	11.1	5.74	3.91	3.00	2.36	3.90	4.13
1949										
1st quarter	-0.2	0.6	-1.2	-0.4	5.66	3.90	2.97	2.36	3.88	4.14
2nd quarter	-1.2	-4.0	-3.0	-15.1	5.38	3.82	2.83	2.30	3.70	4.02
3rd quarter	-0.9	-0.4	-2.3	-0.6	5.39	3.81	2.82	2.29	3.69	4.00
4th quarter	-0.2	-4.5	-0.9	-18.4	5.23	3.73	2.77	2.23	3.61	3.87
1950										
1st quarter	-0.1	2.2	-1.0	9.7	5.26	3.69	2.76	2.22	3.61	3.84
2nd quarter	-1.3	4.2	-2.5	16.2	5.27	3.62	2.75	2.20	3.57	3.79
3rd quarter	0.5	3.7	0.6	15.2	5.21	3.47	2.72	2.12	3.50	3.63
4th quarter	0.7	13.4	1.5	50.2	5.81	3.67	3.04	2.30	3.94	3.97
1951										
1st quarter	1.2	9.2	2.6	29.0	5.94	3.68	3.12	2.33	4.01	4.01
2nd quarter	0.9	13.8	2.5	44.3	5.99	3.81	3.25	2.45	4.23	4.28
3rd quarter	0.8	9.0	1.6	29.3	5.89	3.79	3.20	2.47	4.16	4.29
4th quarter	1.1	3.6	1.7	11.0	5.85	3.76	3.15	2.46	4.07	4.24
1952										
1st quarter	1.0	3.8	2.7	12.8	5.77	3.79	3.16	2.49	4.11	4.31
2nd quarter	1.9	-3.4	3.6	-10.5	5.63	3.75	3.08	2.44	4.01	4.24
3rd quarter	2.2	3.4	4.5	11.2	5.63	3.85	3.15	2.51	4.13	4.38
4th quarter	0.5	4.8	1.5	16.2	5.24	3.74	3.06	2.45	3.98	4.23
1953										
1st quarter	0.8	3.1	2.5	10.6	5.06	3.69	3.02	2.42	3.93	4.17
2nd quarter	-0.7	4.3	-0.2	14.1	5.00	3.69	3.06	2.43	3.99	4.20
3rd quarter	0.7	1.6	1.8	5.8	4.99	3.72	3.09	2.45	4.05	4.23
4th quarter	2.1	-4.1	3.8	-12.3	5.04	3.74	3.08	2.45	4.04	4.22
1954										
1st quarter	0.8	-2.8	2.1	-9.2	5.02	3.72	3.04	2.42	4.01	4.20
2nd quarter	-0.2	-3.2	0.5	-10.8	4.91	3.66	2.98	2.38	3.93	4.11
3rd quarter	0.7	-2.8	0.8	-9.1	4.85	3.59	2.93	2.32	3.89	4.04
4th quarter	-0.5	0.2	-1.5	-0.9	4.74	3.52	2.88	2.28	3.81	3.94
1955										
1st quarter	-0.1	3.9	-1.7	14.0	4.68	3.46	2.86	2.25	3.77	3.88
2nd quarter	-1.1	5.7	-2.1	20.0	4.57	3.44	2.87	2.26	3.73	3.83
3rd quarter	-1.3	5.6	-2.0	17.9	4.47	3.42	2.89	2.26	3.77	3.85
4th quarter	0.3	6.9	-1.6	22.7	4.43	3.44	2.95	2.29	3.87	3.94
1956										
1st quarter	0.0	6.4	-2.9	20.8	4.50	3.48	3.00	2.33	3.96	4.02
2nd quarter	-1.5	5.0	-3.4	16.2	4.57	3.48	3.03	2.35	3.99	4.04
3rd quarter	-0.8	4.3	-3.2	13.2	4.50	3.48	3.02	2.36	3.98	4.07
4th quarter	-1.6	3.9	-3.3	12.1	4.47	3.45	3.03	2.35	4.00	4.07

[1]Before 1997, 2005 dollars; 1997 forward, 2009 dollars.
. . . = Not available.

Table 1-8B. Inventory Change and Inventory-Sales Ratios: Historical Data—*Continued*

(Quarterly dollar data are at seasonally adjusted annual rates.)

NIPA Tables 5.7.5B, 5.7.6B, 5.8.5B, 5.8.6B

Year and quarter	Change in private inventories				Ratio, inventories at end of quarter to monthly rate of sales during the quarter					
	Billions of current dollars		Billions of chained dollars [1]		Total private inventories to final sales of domestic business		Nonfarm inventories to final sales of domestic business		Nonfarm inventories to final sales of goods and structures	
	Farm	Nonfarm	Farm	Nonfarm	Current dollars	Chained dollars[1]	Current dollars	Chained dollars[1]	Current dollars	Chained dollars[1]
1957										
1st quarter	0.3	1.9	-0.9	5.7	4.44	3.44	3.02	2.35	3.97	4.05
2nd quarter	0.7	2.0	-0.2	6.8	4.48	3.48	3.04	2.38	4.00	4.11
3rd quarter	0.5	2.3	1.8	7.5	4.46	3.47	3.03	2.38	3.99	4.10
4th quarter	-1.0	-3.5	1.1	-10.2	4.46	3.48	3.01	2.37	4.01	4.14
1958										
1st quarter	2.2	-6.3	7.6	-20.4	4.67	3.56	3.04	2.41	4.06	4.20
2nd quarter	1.6	-5.8	6.9	-18.6	4.65	3.56	3.00	2.38	4.00	4.17
3rd quarter	2.2	-0.7	6.2	-2.5	4.58	3.48	2.92	2.32	3.92	4.07
4th quarter	1.8	3.4	2.5	10.6	4.50	3.44	2.90	2.30	3.88	4.00
1959										
1st quarter	-0.3	4.2	-3.5	14.9	4.37	3.35	2.83	2.25	3.79	3.93
2nd quarter	-1.9	9.2	-5.3	33.8	4.33	3.34	2.86	2.27	3.85	3.98
3rd quarter	-2.3	2.6	-5.4	9.6	4.25	3.31	2.84	2.27	3.82	3.96
4th quarter	-1.8	5.9	-4.6	21.1	4.26	3.37	2.90	2.32	3.94	4.08
1960										
1st quarter	0.6	10.6	0.4	39.0	4.29	3.37	2.93	2.35	3.96	4.10
2nd quarter	0.9	2.4	1.9	9.0	4.21	3.36	2.91	2.34	3.95	4.11
3rd quarter	1.0	3.3	3.1	12.1	4.27	3.41	2.95	2.38	3.99	4.15
4th quarter	-0.2	-5.6	1.5	-20.2	4.22	3.37	2.89	2.34	3.93	4.10
1961										
1st quarter	0.6	-3.1	2.5	-11.3	4.19	3.35	2.86	2.32	3.90	4.07
2nd quarter	0.6	1.2	2.2	4.7	4.12	3.33	2.83	2.30	3.89	4.07
3rd quarter	0.9	5.7	2.6	21.0	4.17	3.35	2.85	2.32	3.89	4.08
4th quarter	1.4	4.6	2.8	16.8	4.12	3.30	2.81	2.29	3.84	4.03
1962										
1st quarter	1.5	7.9	2.5	29.0	4.15	3.32	2.82	2.32	3.85	4.07
2nd quarter	0.2	5.2	0.8	19.2	4.09	3.29	2.80	2.31	3.84	4.06
3rd quarter	0.2	5.9	1.4	22.0	4.17	3.32	2.83	2.33	3.87	4.09
4th quarter	0.4	3.0	2.2	11.3	4.14	3.32	2.82	2.34	3.87	4.11
1963										
1st quarter	1.9	4.9	7.5	18.2	4.12	3.34	2.83	2.35	3.87	4.12
2nd quarter	0.1	4.8	0.7	17.4	4.04	3.29	2.79	2.32	3.83	4.08
3rd quarter	-0.6	6.4	-1.7	23.6	4.00	3.27	2.79	2.33	3.81	4.07
4th quarter	0.6	4.5	-0.8	16.7	3.95	3.25	2.78	2.32	3.81	4.07
1964										
1st quarter	-0.6	5.8	-3.9	21.1	3.85	3.19	2.74	2.29	3.75	4.01
2nd quarter	-1.2	5.7	-4.3	20.7	3.79	3.17	2.73	2.29	3.75	4.02
3rd quarter	-1.8	6.5	-4.2	23.8	3.78	3.15	2.73	2.29	3.74	4.00
4th quarter	-1.1	6.1	-3.1	21.7	3.79	3.17	2.75	2.31	3.79	4.07
1965										
1st quarter	0.3	11.2	0.9	40.8	3.81	3.17	2.76	2.33	3.79	4.08
2nd quarter	1.1	7.4	2.9	26.9	3.82	3.15	2.75	2.32	3.79	4.08
3rd quarter	0.8	8.5	3.1	30.5	3.78	3.14	2.76	2.32	3.78	4.06
4th quarter	1.0	6.6	3.2	23.7	3.77	3.09	2.72	2.29	3.71	3.99
1966										
1st quarter	0.6	13.2	1.1	46.8	3.80	3.09	2.73	2.31	3.71	4.00
2nd quarter	-1.7	14.0	-2.7	50.2	3.86	3.16	2.80	2.38	3.83	4.16
3rd quarter	-0.6	12.4	-1.7	43.7	3.91	3.19	2.84	2.42	3.88	4.22
4th quarter	-0.4	16.9	-2.3	59.9	3.92	3.26	2.92	2.50	4.01	4.38
1967										
1st quarter	1.7	13.7	3.4	48.6	3.94	3.32	2.98	2.55	4.11	4.50
2nd quarter	2.1	4.2	7.6	14.7	3.92	3.31	2.95	2.54	4.07	4.46
3rd quarter	0.5	8.7	1.4	29.7	3.92	3.34	2.97	2.57	4.10	4.52
4th quarter	-0.9	9.3	-2.2	33.1	3.90	3.35	2.99	2.59	4.14	4.57
1968										
1st quarter	3.0	5.4	7.3	19.1	3.87	3.32	2.94	2.57	4.08	4.51
2nd quarter	4.3	9.8	13.4	33.9	3.87	3.34	2.93	2.58	4.07	4.55
3rd quarter	0.2	7.5	1.2	25.1	3.81	3.32	2.90	2.56	4.03	4.51
4th quarter	-2.0	8.0	-5.5	27.8	3.79	3.33	2.90	2.58	4.03	4.56
1969										
1st quarter	1.8	9.7	3.9	33.0	3.79	3.32	2.90	2.59	4.02	4.54
2nd quarter	1.4	7.8	4.4	26.6	3.83	3.35	2.91	2.61	4.06	4.61
3rd quarter	-1.2	11.4	-2.5	38.6	3.83	3.38	2.94	2.64	4.10	4.67
4th quarter	-1.9	7.7	-5.8	25.5	3.88	3.40	2.98	2.68	4.18	4.76
1970										
1st quarter	1.6	0.2	3.4	0.0	3.86	3.38	2.96	2.66	4.18	4.74
2nd quarter	0.4	4.7	1.9	14.9	3.85	3.40	2.96	2.68	4.20	4.80
3rd quarter	-1.8	6.9	-4.4	21.9	3.84	3.39	2.97	2.67	4.22	4.79
4th quarter	-3.5	-0.5	-10.3	-2.2	3.81	3.39	2.96	2.68	4.23	4.83
1971										
1st quarter	2.5	9.7	6.6	31.3	3.83	3.38	2.95	2.68	4.21	4.81
2nd quarter	4.2	6.7	10.7	20.6	3.81	3.39	2.94	2.68	4.20	4.81
3rd quarter	2.3	7.9	5.8	25.2	3.80	3.38	2.93	2.68	4.18	4.80
4th quarter	-2.3	2.0	-6.8	5.4	3.76	3.33	2.88	2.64	4.12	4.75

[1] Before 1997, 2005 dollars; 1997 forward, 2009 dollars.

Table 1-8B. Inventory Change and Inventory-Sales Ratios: Historical Data—*Continued*

(Quarterly dollar data are at seasonally adjusted annual rates.)   **NIPA Tables 5.7.5B, 5.7.6B, 5.8.5B, 5.8.6B**

Year and quarter	Change in private inventories				Ratio, inventories at end of quarter to monthly rate of sales during the quarter					
	Billions of current dollars		Billions of chained dollars[1]		Total private inventories to final sales of domestic business		Nonfarm inventories to final sales of domestic business		Nonfarm inventories to final sales of goods and structures	
	Farm	Nonfarm	Farm	Nonfarm	Current dollars	Chained dollars[1]	Current dollars	Chained dollars[1]	Current dollars	Chained dollars[1]
1972										
1st quarter	-0.5	3.7	0.1	11.1	3.72	3.30	2.84	2.62	4.06	4.69
2nd quarter	2.0	10.0	6.8	30.6	3.75	3.27	2.83	2.60	4.03	4.64
3rd quarter	1.0	12.7	1.9	38.5	3.77	3.26	2.83	2.60	4.05	4.66
4th quarter	-1.4	8.9	-7.1	26.5	3.74	3.18	2.77	2.55	3.94	4.53
1973										
1st quarter	-4.2	14.8	-10.7	42.9	3.86	3.11	2.79	2.52	3.92	4.45
2nd quarter	5.0	13.2	12.0	38.3	4.03	3.15	2.86	2.54	4.02	4.51
3rd quarter	2.6	7.2	3.5	20.0	4.11	3.16	2.87	2.56	4.05	4.55
4th quarter	2.8	22.2	2.3	60.9	4.20	3.22	2.98	2.62	4.22	4.67
1974										
1st quarter	-3.3	15.9	-5.2	39.7	4.26	3.27	3.13	2.68	4.42	4.76
2nd quarter	1.0	16.4	0.9	39.6	4.29	3.31	3.28	2.72	4.67	4.86
3rd quarter	-0.1	5.7	1.7	11.7	4.45	3.34	3.40	2.75	4.84	4.93
4th quarter	-8.7	29.2	-15.0	64.9	4.52	3.46	3.54	2.88	5.11	5.26
1975										
1st quarter	7.1	-17.1	16.4	-38.3	4.32	3.42	3.41	2.82	4.94	5.16
2nd quarter	3.1	-17.1	5.4	-38.6	4.26	3.35	3.30	2.75	4.81	5.06
3rd quarter	0.8	-2.2	-1.0	-5.0	4.18	3.31	3.24	2.71	4.69	4.95
4th quarter	2.5	-2.1	2.3	-3.7	4.04	3.26	3.16	2.66	4.59	4.88
1976										
1st quarter	-0.5	15.2	0.1	33.2	4.00	3.22	3.14	2.64	4.56	4.83
2nd quarter	-1.8	24.3	-1.9	51.8	4.07	3.26	3.21	2.69	4.67	4.91
3rd quarter	2.0	18.8	5.4	38.9	4.04	3.28	3.23	2.70	4.74	4.96
4th quarter	-3.0	13.6	-9.8	26.5	3.95	3.23	3.19	2.68	4.68	4.92
1977										
1st quarter	-1.3	16.1	-7.0	33.0	3.96	3.22	3.21	2.68	4.72	4.92
2nd quarter	5.3	14.2	11.3	27.8	3.88	3.19	3.16	2.65	4.62	4.84
3rd quarter	4.4	26.5	16.3	53.8	3.86	3.22	3.16	2.67	4.63	4.88
4th quarter	9.8	14.4	8.4	28.1	3.89	3.22	3.16	2.67	4.63	4.87
1978										
1st quarter	-0.6	26.1	-3.6	48.9	4.02	3.26	3.21	2.72	4.78	5.00
2nd quarter	0.4	23.9	0.2	44.2	3.92	3.14	3.12	2.62	4.55	4.75
3rd quarter	7.2	17.8	11.1	32.3	3.95	3.15	3.13	2.62	4.54	4.74
4th quarter	-1.5	30.0	0.5	54.1	3.98	3.14	3.15	2.63	4.54	4.73
1979										
1st quarter	4.2	19.6	6.6	33.0	4.15	3.17	3.22	2.65	4.66	4.78
2nd quarter	3.1	24.3	5.3	40.4	4.19	3.21	3.29	2.69	4.78	4.87
3rd quarter	6.6	5.5	7.0	8.2	4.16	3.16	3.28	2.64	4.74	4.73
4th quarter	0.3	8.3	-1.9	13.4	4.19	3.16	3.34	2.64	4.85	4.75
1980										
1st quarter	-0.6	10.5	-2.6	17.4	4.25	3.17	3.45	2.66	5.02	4.78
2nd quarter	-5.2	13.0	-5.9	19.3	4.40	3.28	3.57	2.76	5.26	5.02
3rd quarter	-12.5	-21.4	-14.2	-32.6	4.34	3.18	3.50	2.68	5.15	4.88
4th quarter	-6.1	-3.0	-6.0	-5.8	4.24	3.13	3.44	2.65	5.08	4.82
1981										
1st quarter	6.8	31.9	10.3	44.6	4.26	3.17	3.48	2.67	5.13	4.84
2nd quarter	9.9	1.8	10.9	1.7	4.24	3.18	3.47	2.67	5.15	4.87
3rd quarter	11.8	32.1	14.5	42.6	4.20	3.22	3.48	2.71	5.16	4.93
4th quarter	6.5	18.3	4.5	23.8	4.17	3.28	3.49	2.75	5.20	5.03
1982										
1st quarter	5.1	-26.5	6.7	-34.3	4.19	3.28	3.46	2.74	5.20	5.03
2nd quarter	4.3	-8.5	6.4	-12.2	4.15	3.28	3.42	2.74	5.17	5.04
3rd quarter	9.0	-3.2	12.0	-5.5	4.12	3.33	3.43	2.77	5.26	5.15
4th quarter	4.6	-44.4	4.9	-57.5	3.97	3.23	3.30	2.68	5.10	4.99
1983										
1st quarter	-7.3	-27.8	-10.4	-37.3	3.89	3.15	3.19	2.62	4.98	4.89
2nd quarter	-13.0	5.3	-12.6	6.5	3.81	3.09	3.14	2.58	4.89	4.80
3rd quarter	-32.4	28.2	-37.7	34.3	3.73	3.03	3.12	2.55	4.86	4.73
4th quarter	-8.8	32.7	-9.0	39.4	3.69	2.98	3.08	2.52	4.77	4.64
1984										
1st quarter	5.5	67.5	7.4	81.9	3.78	3.02	3.15	2.56	4.87	4.71
2nd quarter	5.6	63.7	6.4	77.2	3.76	3.03	3.16	2.57	4.88	4.69
3rd quarter	6.8	64.5	8.1	78.9	3.75	3.06	3.18	2.60	4.93	4.75
4th quarter	5.0	43.0	3.3	51.0	3.72	3.05	3.16	2.60	4.89	4.72
1985										
1st quarter	7.9	8.3	8.4	8.8	3.60	3.00	3.06	2.55	4.77	4.65
2nd quarter	4.2	17.4	4.8	20.3	3.55	2.99	3.04	2.55	4.76	4.66
3rd quarter	6.6	9.7	9.1	12.0	3.47	2.96	2.98	2.51	4.67	4.60
4th quarter	4.3	28.8	5.9	32.7	3.51	2.98	3.01	2.53	4.74	4.65
1986										
1st quarter	2.5	27.8	1.7	33.3	3.42	2.97	2.94	2.52	4.65	4.63
2nd quarter	-3.5	19.1	-5.9	22.3	3.37	2.97	2.92	2.53	4.63	4.65
3rd quarter	-3.1	-3.9	-4.3	-5.5	3.29	2.92	2.85	2.49	4.52	4.55
4th quarter	-1.8	-11.0	-0.6	-8.5	3.25	2.90	2.82	2.47	4.49	4.53

[1] Before 1997, 2005 dollars; 1997 forward, 2009 dollars.

Table 1-8B. Inventory Change and Inventory-Sales Ratios: Historical Data—*Continued*

(Quarterly dollar data are at seasonally adjusted annual rates.)

NIPA Tables 5.7.5B, 5.7.6B, 5.8.5B, 5.8.6B

Year and quarter	Change in private inventories				Ratio, inventories at end of quarter to monthly rate of sales during the quarter					
	Billions of current dollars		Billions of chained dollars [1]		Total private inventories to final sales of domestic business		Nonfarm inventories to final sales of domestic business		Nonfarm inventories to final sales of goods and structures	
	Farm	Nonfarm	Farm	Nonfarm	Current dollars	Chained dollars[1]	Current dollars	Chained dollars[1]	Current dollars	Chained dollars[1]
1987										
1st quarter	-7.7	35.6	-9.7	44.5	3.31	2.94	2.86	2.51	4.62	4.66
2nd quarter	-10.7	27.2	-14.5	33.1	3.30	2.90	2.86	2.49	4.60	4.60
3rd quarter	-4.0	5.0	-5.2	4.1	3.25	2.85	2.82	2.45	4.52	4.51
4th quarter	-3.3	66.4	-7.0	82.5	3.33	2.90	2.90	2.50	4.65	4.62
1988										
1st quarter	-4.4	21.4	-7.6	25.1	3.30	2.86	2.86	2.47	4.61	4.56
2nd quarter	-12.2	31.9	-15.2	38.0	3.29	2.83	2.85	2.46	4.60	4.52
3rd quarter	-11.2	29.4	-11.4	32.6	3.30	2.83	2.85	2.47	4.63	4.57
4th quarter	-19.7	38.8	-18.3	45.9	3.29	2.81	2.86	2.47	4.63	4.55
1989										
1st quarter	7.2	41.0	7.8	46.6	3.32	2.83	2.89	2.48	4.70	4.58
2nd quarter	2.9	33.2	3.6	37.6	3.28	2.83	2.87	2.48	4.65	4.56
3rd quarter	-5.5	15.5	-5.8	16.9	3.21	2.79	2.82	2.45	4.55	4.49
4th quarter	-4.6	21.2	-5.6	22.3	3.23	2.80	2.83	2.47	4.63	4.56
1990										
1st quarter	1.8	12.1	1.7	13.3	3.18	2.78	2.78	2.45	4.53	4.50
2nd quarter	-1.5	35.3	-2.6	39.9	3.17	2.81	2.78	2.47	4.58	4.60
3rd quarter	4.0	17.9	4.4	20.3	3.20	2.82	2.82	2.48	4.67	4.63
4th quarter	5.2	-16.6	6.7	-17.7	3.22	2.83	2.82	2.49	4.71	4.66
1991										
1st quarter	0.0	-15.6	0.7	-18.1	3.18	2.85	2.78	2.51	4.65	4.71
2nd quarter	-0.7	-17.3	-0.5	-20.6	3.08	2.81	2.70	2.47	4.55	4.66
3rd quarter	-10.9	11.7	-13.8	13.1	3.04	2.81	2.69	2.48	4.54	4.67
4th quarter	6.4	24.7	6.1	27.1	3.06	2.83	2.70	2.50	4.62	4.74
1992										
1st quarter	8.0	-7.8	10.5	-7.5	3.00	2.78	2.63	2.44	4.52	4.64
2nd quarter	9.4	13.8	10.9	13.2	2.99	2.77	2.62	2.43	4.50	4.61
3rd quarter	5.1	15.4	4.7	16.6	2.96	2.75	2.60	2.41	4.47	4.57
4th quarter	2.5	18.9	2.4	21.6	2.92	2.73	2.56	2.39	4.41	4.53
1993										
1st quarter	-5.6	41.6	-6.2	46.5	2.95	2.75	2.58	2.42	4.47	4.60
2nd quarter	-4.9	29.0	-5.3	32.3	2.92	2.74	2.57	2.41	4.44	4.56
3rd quarter	-12.5	19.1	-14.2	20.6	2.89	2.72	2.54	2.40	4.42	4.55
4th quarter	-1.7	18.3	-4.1	19.9	2.85	2.68	2.51	2.38	4.33	4.47
1994										
1st quarter	16.5	28.8	19.4	31.1	2.87	2.69	2.51	2.38	4.33	4.48
2nd quarter	17.1	64.3	21.0	70.0	2.86	2.71	2.53	2.39	4.37	4.49
3rd quarter	10.7	42.5	10.6	46.1	2.86	2.71	2.53	2.39	4.36	4.48
4th quarter	3.7	71.5	3.7	76.3	2.89	2.72	2.56	2.40	4.39	4.47
1995										
1st quarter	-5.8	67.0	-5.9	71.2	2.93	2.75	2.62	2.43	4.49	4.52
2nd quarter	-13.7	47.4	-13.6	50.1	2.94	2.75	2.64	2.44	4.55	4.55
3rd quarter	-20.4	31.7	-26.4	32.8	2.90	2.71	2.60	2.41	4.49	4.50
4th quarter	-4.4	22.8	-6.2	22.9	2.88	2.69	2.58	2.40	4.45	4.46
1996										
1st quarter	1.1	5.8	-2.9	6.2	2.84	2.66	2.55	2.38	4.39	4.42
2nd quarter	11.8	18.6	7.4	19.8	2.81	2.63	2.51	2.35	4.30	4.35
3rd quarter	17.1	34.0	27.4	35.0	2.81	2.65	2.50	2.35	4.29	4.34
4th quarter	4.5	30.2	1.3	31.7	2.76	2.62	2.47	2.33	4.24	4.28
1997										
1st quarter	-0.5	50.3	-10.8	53.3	2.74	2.62	2.45	2.33	4.21	4.28
2nd quarter	0.8	87.6	8.2	95.7	2.73	2.65	2.44	2.36	4.21	4.33
3rd quarter	8.3	59.6	10.1	65.5	2.69	2.63	2.41	2.34	4.14	4.27
4th quarter	4.7	73.1	7.2	79.8	2.70	2.65	2.42	2.37	4.17	4.32
NEW BASIS										
1997										
1st quarter	-0.6	50.3	-13.5	45.4	2.66	2.93	2.37	2.54	3.91	4.61
2nd quarter	0.8	87.6	13.1	130.9	2.64	2.97	2.36	2.57	3.90	4.67
3rd quarter	8.3	59.6	15.1	82.4	2.60	2.94	2.33	2.55	3.83	4.60
4th quarter	4.6	73.0	10.0	93.4	2.60	2.97	2.34	2.57	3.86	4.64
1998										
1st quarter	5.6	99.5	9.4	144.4	2.60	3.00	2.34	2.61	3.87	4.71
2nd quarter	-4.9	42.2	-4.0	57.0	2.55	2.97	2.30	2.58	3.81	4.65
3rd quarter	0.5	51.9	4.5	65.5	2.51	2.96	2.28	2.57	3.75	4.62
4th quarter	3.8	56.2	1.3	79.1	2.47	2.93	2.25	2.55	3.69	4.56
1999										
1st quarter	4.1	79.3	-2.9	114.9	2.48	2.95	2.25	2.57	3.71	4.61
2nd quarter	-0.2	35.3	1.6	46.9	2.46	2.92	2.24	2.55	3.68	4.56
3rd quarter	-9.1	49.6	-18.1	71.1	2.47	2.91	2.25	2.54	3.71	4.54
4th quarter	-5.7	90.0	-5.9	118.0	2.49	2.91	2.27	2.55	3.75	4.55
2000										
1st quarter	-17.7	33.9	-30.9	40.7	2.48	2.87	2.26	2.52	3.75	4.51
2nd quarter	5.9	84.5	10.4	112.1	2.48	2.88	2.27	2.53	3.76	4.54
3rd quarter	-1.1	58.4	-3.5	86.1	2.48	2.89	2.28	2.55	3.79	4.58
4th quarter	7.3	47.0	11.8	71.5	2.51	2.90	2.29	2.56	3.82	4.60

[1] Before 1997, 2005 dollars; 1997 forward, 2009 dollars.

Table 1-8B. Inventory Change and Inventory-Sales Ratios: Historical Data—*Continued*

(Quarterly dollar data are at seasonally adjusted annual rates.)

NIPA Tables 5.7.5B, 5.7.6B, 5.8.5B, 5.8.6B

| Year and quarter | Change in private inventories | | | | Ratio, inventories at end of quarter to monthly rate of sales during the quarter | | | | | |
| | Billions of current dollars | | Billions of chained dollars [1] | | Total private inventories to final sales of domestic business | | Nonfarm inventories to final sales of domestic business | | Nonfarm inventories to final sales of goods and structures | |
	Farm	Nonfarm	Farm	Nonfarm	Current dollars	Chained dollars[1]	Current dollars	Chained dollars[1]	Current dollars	Chained dollars[1]
2001										
1st quarter	5.6	-36.2	10.0	-50.7	2.48	2.88	2.26	2.53	3.79	4.56
2nd quarter	-3.0	-8.6	-5.0	0.8	2.44	2.88	2.23	2.53	3.72	4.55
3rd quarter	1.8	-31.9	2.6	-46.8	2.41	2.88	2.20	2.53	3.68	4.55
4th quarter	-4.3	-76.6	-6.9	-108.8	2.33	2.81	2.13	2.47	3.55	4.43
2002										
1st quarter	4.1	-3.8	4.1	-10.2	2.31	2.81	2.10	2.47	3.54	4.45
2nd quarter	-10.3	31.5	-16.4	43.0	2.32	2.82	2.12	2.48	3.60	4.50
3rd quarter	-2.9	28.3	-5.6	31.5	2.34	2.82	2.14	2.48	3.64	4.50
4th quarter	-0.8	33.8	-3.0	44.1	2.38	2.83	2.16	2.50	3.71	4.55
2003										
1st quarter	2.8	23.2	4.9	35.1	2.42	2.83	2.20	2.49	3.78	4.53
2nd quarter	-1.4	-5.1	-1.5	-7.9	2.37	2.79	2.16	2.45	3.69	4.45
3rd quarter	0.9	-0.4	0.3	2.5	2.33	2.72	2.10	2.40	3.57	4.30
4th quarter	-2.1	38.4	-2.9	47.3	2.33	2.71	2.10	2.39	3.57	4.28
2004										
1st quarter	4.4	41.6	8.6	50.7	2.36	2.73	2.13	2.40	3.64	4.31
2nd quarter	18.4	55.5	23.7	77.1	2.39	2.74	2.16	2.41	3.68	4.33
3rd quarter	7.7	56.5	12.6	71.5	2.40	2.73	2.17	2.40	3.71	4.31
4th quarter	4.7	67.4	9.0	82.0	2.39	2.73	2.17	2.40	3.70	4.30
2005										
1st quarter	-6.7	109.2	-7.6	126.7	2.42	2.74	2.20	2.42	3.75	4.32
2nd quarter	0.9	27.9	1.2	23.0	2.38	2.71	2.17	2.39	3.69	4.25
3rd quarter	4.3	8.0	5.0	4.1	2.39	2.68	2.17	2.37	3.70	4.20
4th quarter	2.3	84.1	2.9	92.4	2.42	2.71	2.21	2.39	3.79	4.27
2006										
1st quarter	3.7	74.5	2.6	98.0	2.39	2.70	2.18	2.38	3.72	4.22
2nd quarter	-8.1	83.6	-11.0	97.5	2.42	2.71	2.22	2.40	3.81	4.28
3rd quarter	-6.3	85.1	-9.4	101.2	2.45	2.74	2.24	2.43	3.85	4.34
4th quarter	-3.6	47.1	-7.1	59.7	2.42	2.71	2.21	2.41	3.82	4.31
2007										
1st quarter	7.9	13.4	5.3	22.6	2.44	2.71	2.22	2.40	3.85	4.32
2nd quarter	-4.1	46.4	5.9	50.4	2.46	2.71	2.24	2.41	3.89	4.32
3rd quarter	-1.7	46.6	-15.5	54.3	2.46	2.70	2.24	2.40	3.91	4.32
4th quarter	0.4	26.9	-1.4	35.2	2.51	2.69	2.28	2.40	4.01	4.31
2008										
1st quarter	-8.6	-8.1	-6.1	6.5	2.61	2.71	2.37	2.42	4.21	4.37
2nd quarter	3.8	-26.1	2.1	-26.6	2.71	2.69	2.46	2.39	4.39	4.32
3rd quarter	4.1	-20.0	2.3	-20.7	2.68	2.71	2.44	2.41	4.36	4.38
4th quarter	-1.5	-60.8	-0.8	-85.0	2.50	2.76	2.28	2.45	4.14	4.56
2009										
1st quarter	-4.8	-146.2	-8.6	-168.1	2.43	2.75	2.21	2.44	4.02	4.55
2nd quarter	-6.3	-189.7	-9.5	-215.3	2.40	2.68	2.19	2.38	3.98	4.44
3rd quarter	-10.9	-193.1	-13.5	-228.3	2.35	2.60	2.14	2.30	3.91	4.28
4th quarter	-4.6	-47.4	-8.4	-49.0	2.37	2.58	2.17	2.29	3.99	4.28
2010										
1st quarter	-6.6	5.3	-14.0	6.1	2.42	2.58	2.20	2.29	4.10	4.28
2nd quarter	-11.7	48.4	-16.5	45.2	2.40	2.57	2.19	2.28	4.07	4.26
3rd quarter	-16.5	129.4	-20.6	141.9	2.45	2.59	2.23	2.31	4.17	4.34
4th quarter	-13.9	81.2	-20.0	87.7	2.49	2.58	2.27	2.31	4.21	4.31
2011										
1st quarter	-2.1	34.2	-2.1	34.3	2.60	2.58	2.34	2.31	4.38	4.34
2nd quarter	-5.7	66.8	-4.1	74.4	2.62	2.59	2.37	2.32	4.42	4.34
3rd quarter	-1.2	-10.8	-2.2	-15.4	2.58	2.55	2.33	2.29	4.34	4.27
4th quarter	0.7	103.2	-0.4	100.2	2.58	2.57	2.32	2.30	4.30	4.27
2012										
1st quarter	-6.4	86.7	-10.8	83.7	2.57	2.55	2.32	2.29	4.28	4.23
2nd quarter	-15.4	106.8	-16.4	103.3	2.54	2.56	2.30	2.31	4.23	4.25
3rd quarter	-29.9	114.3	-25.1	122.8	2.58	2.59	2.34	2.34	4.29	4.30
4th quarter	-23.1	51.8	-22.5	49.8	2.56	2.57	2.33	2.33	4.27	4.28
2013										
1st quarter	2.4	79.9	0.1	80.7	2.52	2.57	2.31	2.33	4.22	4.25
2nd quarter	15.0	72.7	12.4	73.7	2.52	2.59	2.30	2.35	4.22	4.28
3rd quarter	17.4	118.8	15.0	132.5	2.52	2.61	2.30	2.37	4.21	4.30
4th quarter	11.9	100.0	14.1	105.9	2.52	2.60	2.30	2.36	4.19	4.27
2014										
1st quarter	-3.5	55.7	-5.1	63.3	2.57	2.62	2.33	2.38	4.28	4.32
2nd quarter	0.0	96.3	-3.1	102.0	2.54	2.60	2.30	2.37	4.21	4.27
3rd quarter	-2.0	101.5	-1.5	108.1	2.51	2.59	2.27	2.36	4.15	4.25
4th quarter	-7.4	95.2	-4.4	87.0	2.46	2.58	2.23	2.36	4.10	4.26
2015										
1st quarter	4.5	166.5	3.5	170.3	2.44	2.63	2.22	2.40	4.11	4.37
2nd quarter	1.7	141.0	1.7	147.7	2.43	2.64	2.22	2.41	4.09	4.38
3rd quarter	-1.2	124.8	-0.9	122.1	2.40	2.66	2.21	2.43	4.07	4.41
4th quarter	1.3	88.9	0.2	85.1	2.39	2.67	2.20	2.45	4.09	4.46
2016										
1st quarter	-13.6	60.5	-13.1	62.9	2.37	2.67	2.18	2.45	4.06	4.44
2nd quarter	-3.5	22.7	-4.7	14.9	2.36	2.65	2.19	2.43	4.07	4.40
3rd quarter	-0.1	-2.7	-1.1	-13.7	2.33	2.62	2.17	2.41	4.05	4.35
4th quarter	-6.1	51.1	-7.4	50.1	2.35	2.63	2.19	2.42	4.11	4.38

[1] Before 1997, 2005 dollars; 1997 forward, 2009 dollars.

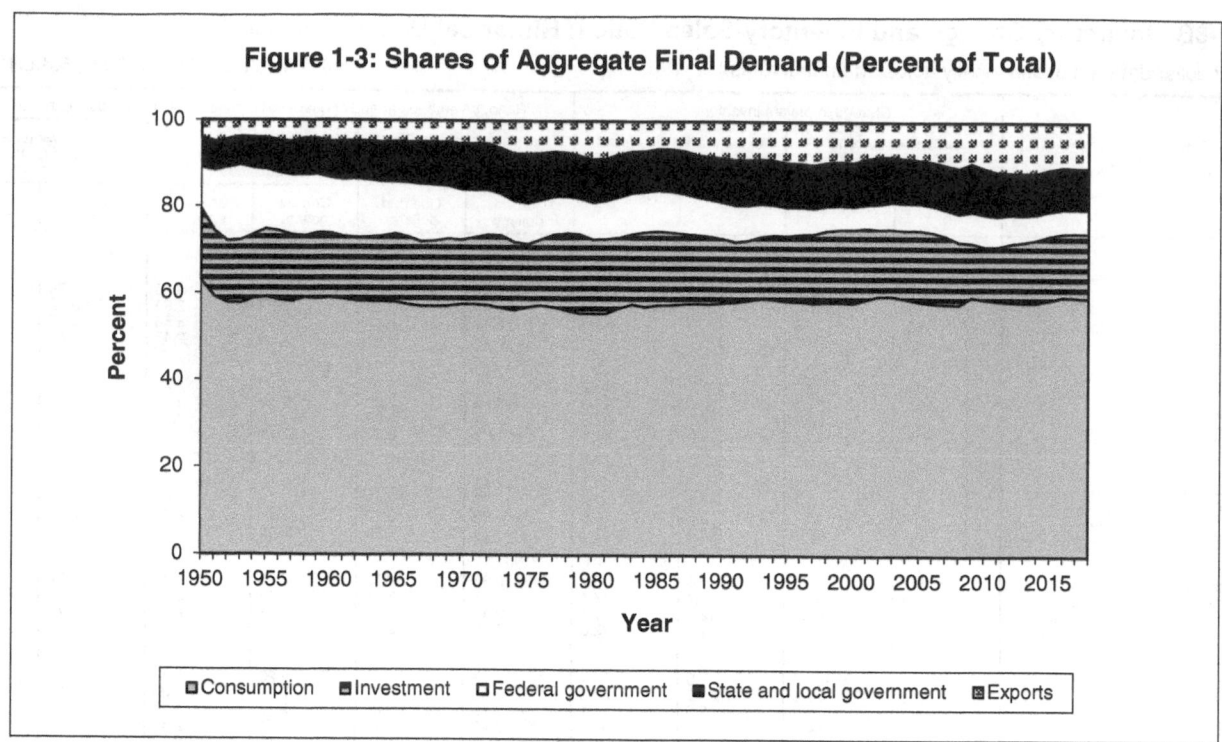

Figure 1-3: Shares of Aggregate Final Demand (Percent of Total)

Legend: □ Consumption   ■ Investment   □ Federal government   ■ State and local government   ▧ Exports

- People often want to know how various sources of demand, especially consumption spending, affect GDP. Often, it is said that consumption "accounts for" about 70 percent, or sometimes two-thirds, of GDP. That is based on a NIPA table showing the components of GDP as a percent of total GDP. In such a calculation, personal consumption expenditures (PCE) amounted to 68.0 percent of total GDP in 2018, up from a range of 60 to 65 percent earlier in the postwar period. (Table 1-1A)

- Is "amounted to" the same as "accounted for"? No, because the value of PCE includes imported goods and services—coffee, cocoa, and bananas; crude oil from the Middle East, transformed into gasoline; clothing, cars, and toys manufactured overseas—even the money that U.S. consumers spend when they travel. These imports included in PCE don't account for demand for U.S. GDP, which excludes imports; instead, they account for demand for GDP in China and other countries that export to us.

- It is not possible with NIPA data to precisely estimate and subtract out the import content of PCE, and so there is no way to precisely estimate PCE's contribution to GDP. What we can do is calculate PCE and the other sources of aggregate final demand for U.S. output as shares of total final demand; this is shown in Table 1-10 and Figure 1-3. These shares also approximate their contributions to GDP (excluding inventory change) if each final demand source has, in aggregate, about the same percentage import content as total GDP.

- The share of imports in the total supply of goods and services in the U.S. market rose from around 4 percent in the 1950s to a high of 14.8 percent in 2008. By 2018, the import share had decreased to 13.3 percent. Only in 2009, the year of maximum inventory liquidation, did the import share temporarily dip back to 12.0 percent. (Table 1-9)

Table 1-9. Shares of Aggregate Supply

(Billions of dollars, quarterly dollar data are at seasonally adjusted rates, percents.)   **NIPA Table 1.1.5**

Year and quarter	Aggregate supply (billions of dollars)			Shares of aggregate supply (percent)	
	Total	GDP	Imports	Domestic production (GDP)	Imports
1950	311.4	299.8	11.6	96.3	3.7
1951	361.5	346.9	14.6	96.0	4.0
1952	382.6	367.3	15.3	96.0	4.0
1953	405.2	389.2	16.0	96.1	3.9
1954	405.9	390.5	15.4	96.2	3.8
1955	442.7	425.5	17.2	96.1	3.9
1956	468.3	449.4	18.9	96.0	4.0
1957	493.9	474.0	19.9	96.0	4.0
1958	501.2	481.2	20.0	96.0	4.0
1959	544.0	521.7	22.3	95.9	4.1
1960	565.2	542.4	22.8	96.0	4.0
1961	584.9	562.2	22.7	96.1	3.9
1962	628.9	603.9	25.0	96.0	4.0
1963	663.6	637.5	26.1	96.1	3.9
1964	712.6	684.5	28.1	96.1	3.9
1965	773.8	742.3	31.5	95.9	4.1
1966	850.5	813.4	37.1	95.6	4.4
1967	899.9	860.0	39.9	95.6	4.4
1968	987.3	940.7	46.6	95.3	4.7
1969	1 068.1	1 017.6	50.5	95.3	4.7
1970	1 129.1	1 073.3	55.8	95.1	4.9
1971	1 227.2	1 164.9	62.3	94.9	5.1
1972	1 353.3	1 279.1	74.2	94.5	5.5
1973	1 516.6	1 425.4	91.2	94.0	6.0
1974	1 672.7	1 545.2	127.5	92.4	7.6
1975	1 807.6	1 684.9	122.7	93.2	6.8
1976	2 024.5	1 873.4	151.1	92.5	7.5
1977	2 264.2	2 081.8	182.4	91.9	8.1
1978	2 563.9	2 351.6	212.3	91.7	8.3
1979	2 880.0	2 627.3	252.7	91.2	8.8
1980	3 151.1	2 857.3	293.8	90.7	9.3
1981	3 524.8	3 207.0	317.8	91.0	9.0
1982	3 647.0	3 343.8	303.2	91.7	8.3
1983	3 962.6	3 634.0	328.6	91.7	8.3
1984	4 442.7	4 037.6	405.1	90.9	9.1
1985	4 756.2	4 339.0	417.2	91.2	8.8
1986	5 032.5	4 579.6	452.9	91.0	9.0
1987	5 363.9	4 855.2	508.7	90.5	9.5
1988	5 790.4	5 236.4	554.0	90.4	9.6
1989	6 232.6	5 641.6	591.0	90.5	9.5
1990	6 592.8	5 963.1	629.7	90.4	9.6
1991	6 781.6	6 158.1	623.5	90.8	9.2
1992	7 188.1	6 520.3	667.8	90.7	9.3
1993	7 578.6	6 858.6	720.0	90.5	9.5
1994	8 100.6	7 287.2	813.4	90.0	10.0
1995	8 542.3	7 639.7	902.6	89.4	10.6
1996	9 037.1	8 073.1	964.0	89.3	10.7
1997	9 633.4	8 577.6	1 055.8	89.0	11.0
1998	10 178.5	9 062.8	1 115.7	89.0	11.0
1999	10 879.3	9 630.7	1 248.6	88.5	11.5
2000	11 723.6	10 252.3	1 471.3	87.5	12.5
2001	11 974.4	10 581.8	1 392.6	88.4	11.6
2002	12 360.5	10 936.4	1 424.1	88.5	11.5
2003	12 997.5	11 458.2	1 539.3	88.2	11.8
2004	14 010.4	12 213.7	1 796.7	87.2	12.8
2005	15 063.0	13 036.6	2 026.4	86.5	13.5
2006	16 058.1	13 814.6	2 243.5	86.0	14.0
2007	16 831.2	14 451.9	2 379.3	85.9	14.1
2008	17 272.9	14 712.8	2 560.1	85.2	14.8
2009	16 427.3	14 448.9	1 978.4	88.0	12.0
2010	17 352.3	14 992.1	2 360.2	86.4	13.6
2011	18 225.1	15 542.6	2 682.5	85.3	14.7
2012	18 956.9	16 197.0	2 759.9	85.4	14.6
2013	19 549.1	16 784.9	2 764.2	85.9	14.1
2014	20 406.7	17 527.3	2 879.4	85.9	14.1
2015	21 011.4	18 224.8	2 786.6	86.7	13.3
2016	21 454.4	18 715.0	2 739.4	87.2	12.8
2017	22 451.5	19 519.4	2 932.1	86.9	13.1
2018	23 728.7	20 580.2	3 148.5	86.7	13.3
2016					
1st quarter	21 111.4	18 424.3	2 687.1	87.3	12.7
2nd quarter	21 340.7	18 637.3	2 703.4	87.3	12.7
3rd quarter	21 560.8	18 806.7	2 754.1	87.2	12.8
4th quarter	21 805.0	18 991.9	2 813.1	87.1	12.9
2017					
1st quarter	22 064.6	19 190.4	2 874.2	87.0	13.0
2nd quarter	22 253.5	19 356.6	2 896.9	87.0	13.0
3rd quarter	22 522.4	19 611.7	2 910.7	87.1	12.9
4th quarter	22 965.4	19 918.9	3 046.5	86.7	13.3
2018					
1st quarter	23 268.8	20 163.2	3 105.6	86.7	13.3
2nd quarter	23 622.2	20 510.2	3 112.0	86.8	13.2
3rd quarter	23 931.4	20 749.8	3 181.6	86.7	13.3
4th quarter	24 092.5	20 897.8	3 194.7	86.7	13.3

Table 1-10. Shares of Aggregate Final Demand

(Billions of dollars, quarterly dollar data are at seasonally adjusted rates, percents.)   NIPA Table 1.1.5

Year and quarter	Aggregate final demand (billions of dollars)							Shares of aggregate final demand (percent)					
	Total	Consumption (PCE)	Nonresidential fixed investment	Residential investment	Exports	Government consumption and gross investment — Federal	Government consumption and gross investment — State and local	PCE	Nonresidential investment	Residential investment	Exports	Federal government	State and local government
1955	437.7	258.3	43.4	25.4	17.7	61.0	31.9	59.0	9.9	5.8	4.0	13.9	7.3
1956	464.3	271.1	49.7	24.0	21.3	63.1	35.1	58.4	10.7	5.2	4.6	13.6	7.6
1957	493.2	286.3	53.1	22.6	24.0	68.5	38.7	58.0	10.8	4.6	4.9	13.9	7.8
1958	501.6	295.6	48.5	22.8	20.6	71.5	42.6	58.9	9.7	4.5	4.1	14.3	8.5
1959	540.0	317.1	53.1	28.6	22.7	73.6	44.9	58.7	9.8	5.3	4.2	13.6	8.3
1960	562.0	331.2	56.4	26.9	27.0	72.9	47.6	58.9	10.0	4.8	4.8	13.0	8.5
1961	582.0	341.5	56.6	27.0	27.6	77.4	51.9	58.7	9.7	4.6	4.7	13.3	8.9
1962	622.8	362.6	61.2	29.6	29.1	85.5	54.8	58.2	9.8	4.8	4.7	13.7	8.8
1963	658.0	382.0	64.8	32.9	31.1	87.9	59.3	58.1	9.8	5.0	4.7	13.4	9.0
1964	707.7	410.6	72.2	35.1	35.0	90.3	64.5	58.0	10.2	5.0	4.9	12.8	9.1
1965	764.6	443.0	85.2	35.2	37.1	93.2	70.9	57.9	11.1	4.6	4.9	12.2	9.3
1966	836.9	479.9	97.2	33.4	40.9	106.6	78.9	57.3	11.6	4.0	4.9	12.7	9.4
1967	890.0	506.7	99.2	33.6	43.5	120.0	87.0	56.9	11.1	3.8	4.9	13.5	9.8
1968	978.3	556.9	107.7	40.2	47.9	128.0	97.6	56.9	11.0	4.1	4.9	13.1	10.0
1969	1 058.9	603.6	120.0	44.4	51.9	131.2	107.8	57.0	11.3	4.2	4.9	12.4	10.2
1970	1 127.0	646.7	124.6	43.4	59.7	132.8	119.8	57.4	11.1	3.9	5.3	11.8	10.6
1971	1 219.0	699.9	130.4	58.2	63.0	134.5	133.0	57.4	10.7	4.8	5.2	11.0	10.9
1972	1 344.1	768.2	146.6	72.4	70.8	141.6	144.5	57.2	10.9	5.4	5.3	10.5	10.8
1973	1 500.7	849.6	172.7	78.3	95.3	146.2	158.6	56.6	11.5	5.2	6.4	9.7	10.6
1974	1 658.6	930.2	191.1	69.5	126.7	158.8	182.5	56.1	11.5	4.2	7.6	9.6	11.0
1975	1 813.8	1 030.5	196.8	66.7	138.7	173.7	207.4	56.8	10.9	3.7	7.6	9.6	11.4
1976	2 007.5	1 147.7	219.3	86.8	149.5	184.8	219.4	57.2	10.9	4.3	7.4	9.2	10.9
1977	2 241.9	1 274.0	259.1	115.2	159.3	200.3	234.0	56.8	11.6	5.1	7.1	8.9	10.4
1978	2 538.1	1 422.3	314.6	138.0	186.9	218.9	257.4	56.0	12.4	5.4	7.4	8.6	10.1
1979	2 861.9	1 585.4	373.8	147.8	230.1	240.6	284.2	55.4	13.1	5.2	8.0	8.4	9.9
1980	3 157.5	1 750.7	406.9	129.5	280.8	274.9	314.7	55.4	12.9	4.1	8.9	8.7	10.0
1981	3 495.0	1 934.0	472.9	128.5	305.2	314.0	340.4	55.3	13.5	3.7	8.7	9.0	9.7
1982	3 661.8	2 071.3	485.1	110.8	283.2	348.3	363.1	56.6	13.2	3.0	7.7	9.5	9.9
1983	3 968.5	2 281.6	482.2	161.1	277.0	382.4	384.2	57.5	12.2	4.1	7.0	9.6	9.7
1984	4 377.3	2 492.3	564.3	190.4	302.4	411.8	416.1	56.9	12.9	4.3	6.9	9.4	9.5
1985	4 734.4	2 712.8	607.8	200.1	303.2	452.9	457.6	57.3	12.8	4.2	6.4	9.6	9.7
1986	5 026.0	2 886.3	607.8	234.8	321.0	481.7	494.4	57.4	12.1	4.7	6.4	9.6	9.8
1987	5 336.7	3 076.3	615.2	249.8	363.9	502.8	528.7	57.6	11.5	4.7	6.8	9.4	9.9
1988	5 771.9	3 330.0	662.3	256.2	444.6	511.4	567.4	57.7	11.5	4.4	7.7	8.9	9.8
1989	6 205.0	3 576.8	716.0	256.0	504.3	534.1	617.8	57.6	11.5	4.1	8.1	8.6	10.0
1990	6 578.4	3 809.0	739.2	239.7	551.9	562.4	676.2	57.9	11.2	3.6	8.4	8.5	10.3
1991	6 782.0	3 943.4	723.6	221.2	594.9	582.9	716.0	58.1	10.7	3.3	8.8	8.6	10.6
1992	7 171.8	4 197.6	741.9	254.7	633.1	588.5	756.0	58.5	10.3	3.6	8.8	8.2	10.5
1993	7 557.8	4 452.0	799.2	286.8	654.8	580.2	784.8	58.9	10.6	3.8	8.7	7.7	10.4
1994	8 036.9	4 721.0	868.9	323.8	720.9	574.7	827.6	58.7	10.8	4.0	9.0	7.2	10.3
1995	8 511.1	4 962.6	962.2	324.1	812.8	576.7	872.7	58.3	11.3	3.8	9.5	6.8	10.3
1996	9 006.4	5 244.6	1 043.2	358.1	867.6	579.2	913.7	58.2	11.6	4.0	9.6	6.4	10.1
1997	9 562.4	5 536.8	1 149.1	375.6	953.8	583.3	963.8	57.9	12.0	3.9	10.0	6.1	10.1
1998	10 114.7	5 877.2	1 254.1	418.8	953.0	585.5	1 026.1	58.1	12.4	4.1	9.4	5.8	10.1
1999	10 818.5	6 279.1	1 364.5	461.8	992.8	611.3	1 109.0	58.0	12.6	4.3	9.2	5.7	10.3
2000	11 669.0	6 762.1	1 498.4	485.4	1 096.3	633.7	1 193.1	57.9	12.8	4.2	9.4	5.4	10.2
2001	12 012.7	7 065.6	1 460.1	513.1	1 024.6	670.1	1 279.2	58.8	12.2	4.3	8.5	5.6	10.6
2002	12 340.5	7 342.7	1 352.8	557.6	998.7	743.0	1 345.7	59.5	11.0	4.5	8.1	6.0	10.9
2003	12 983.5	7 723.1	1 375.9	637.1	1 036.2	826.3	1 384.9	59.5	10.6	4.9	8.0	6.4	10.7
2004	13 946.3	8 212.7	1 467.4	749.8	1 177.6	891.7	1 447.1	58.9	10.5	5.4	8.4	6.4	10.4
2005	15 005.5	8 747.1	1 621.0	856.2	1 305.2	947.5	1 528.5	58.3	10.8	5.7	8.7	6.3	10.2
2006	15 989.1	9 260.3	1 793.8	838.2	1 472.6	1 000.7	1 623.5	57.9	11.2	5.2	9.2	6.3	10.2
2007	16 797.2	9 706.4	1 948.6	690.5	1 660.9	1 050.5	1 740.3	57.8	11.6	4.1	9.9	6.3	10.4
2008	17 302.3	9 976.3	1 990.9	516.0	1 837.1	1 150.6	1 831.4	57.7	11.5	3.0	10.6	6.6	10.6
2009	16 578.1	9 842.2	1 690.4	390.0	1 582.0	1 218.2	1 855.3	59.4	10.2	2.4	9.5	7.3	11.2
2010	17 298.3	10 185.8	1 735.0	376.6	1 846.3	1 297.9	1 856.7	58.9	10.0	2.2	10.7	7.5	10.7
2011	18 178.7	10 641.1	1 907.5	378.8	2 103.0	1 298.9	1 849.4	58.5	10.5	2.1	11.6	7.1	10.2
2012	18 885.6	11 006.8	2 118.5	432.0	2 191.3	1 286.5	1 850.5	58.3	11.2	2.3	11.6	6.8	9.8
2013	19 444.5	11 317.2	2 211.5	510.0	2 273.4	1 226.6	1 905.8	58.2	11.4	2.6	11.7	6.3	9.8
2014	20 322.8	11 822.8	2 400.1	560.2	2 371.7	1 215.0	1 953.0	58.2	11.8	2.8	11.7	6.0	9.6
2015	20 879.5	12 284.3	2 457.4	633.8	2 266.8	1 221.5	2 015.7	58.8	11.8	3.0	10.9	5.9	9.7
2016	21 427.4	12 748.5	2 453.1	698.5	2 220.6	1 234.1	2 072.6	59.5	11.4	3.3	10.4	5.8	9.7
2017	22 421.2	13 312.1	2 584.7	755.7	2 356.7	1 269.3	2 142.7	59.4	11.5	3.4	10.5	5.7	9.6
2018	23 674.1	13 998.7	2 786.9	786.7	2 510.3	1 347.3	2 244.2	59.1	11.8	3.3	10.6	5.7	9.5
2016													
1st quarter	21 064.4	12 523.5	2 415.6	686.6	2 164.9	1 227.5	2 046.3	59.5	11.5	3.3	10.3	5.8	9.7
2nd quarter	21 321.6	12 688.3	2 441.8	692.0	2 208.1	1 226.2	2 065.2	59.5	11.5	3.2	10.4	5.8	9.7
3rd quarter	21 563.6	12 822.4	2 471.6	697.7	2 254.4	1 237.5	2 080.0	59.5	11.5	3.2	10.5	5.7	9.6
4th quarter	21 760.1	12 959.8	2 483.5	717.8	2 255.1	1 245.2	2 098.7	59.6	11.4	3.3	10.4	5.7	9.6
2017													
1st quarter	22 051.2	13 104.4	2 531.1	743.7	2 303.3	1 248.4	2 120.3	59.4	11.5	3.4	10.4	5.7	9.6
2nd quarter	22 234.8	13 212.5	2 567.4	748.8	2 313.2	1 263.6	2 129.3	59.4	11.5	3.4	10.4	5.7	9.6
3rd quarter	22 465.6	13 345.1	2 591.6	753.4	2 360.1	1 270.2	2 145.2	59.4	11.5	3.4	10.5	5.7	9.5
4th quarter	22 933.6	13 586.3	2 648.9	777.1	2 450.3	1 295.1	2 175.9	59.2	11.6	3.4	10.7	5.6	9.5
2018													
1st quarter	23 227.4	13 728.4	2 717.3	783.7	2 476.6	1 318.2	2 203.2	59.1	11.7	3.4	10.7	5.7	9.5
2nd quarter	23 632.0	13 939.8	2 782.0	789.5	2 543.6	1 340.4	2 236.7	59.0	11.8	3.3	10.8	5.7	9.5
3rd quarter	23 844.1	14 114.6	2 807.7	789.0	2 510.3	1 358.6	2 263.9	59.2	11.8	3.3	10.5	5.7	9.5
4th quarter	23 992.3	14 211.9	2 840.7	784.4	2 510.5	1 371.8	2 273.0	59.2	11.8	3.3	10.5	5.7	9.5

SECTION 1B: INCOME AND VALUE ADDED

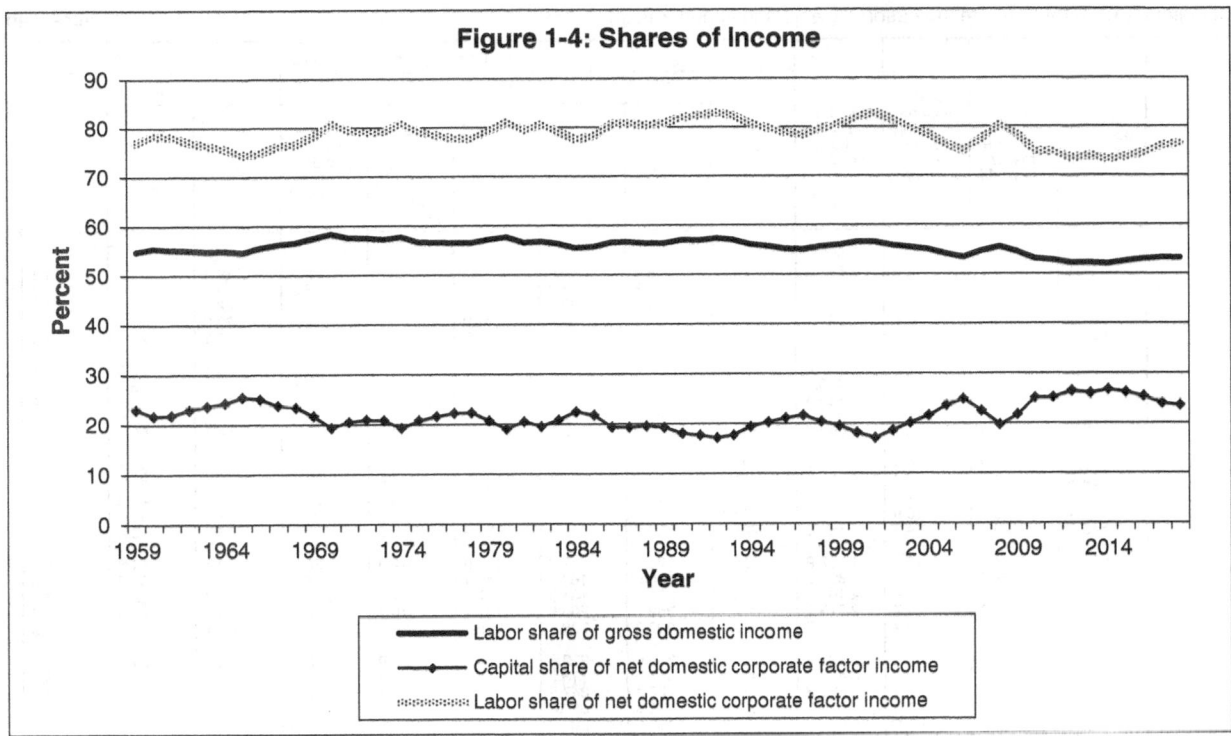

Figure 1-4: Shares of Income

Legend:
- Labor share of gross domestic income
- Capital share of net domestic corporate factor income
- Labor share of net domestic corporate factor income

- Recent years have seen a decline in the share of national income paid to workers—not just the share paid out as worker wages, but in total compensation, including all health and other benefits. One indicator of this can be seen in a later chapter, in Tables 9-3A and B, where Bureau of Labor Statistics calculations (based on BEA data) document a declining labor share in the value of output of private nonfarm business. In the tables in Section 1b immediately following this page, BEA compilations show the composition of domestic income, national income, and corporate value added. Table 1-16 shows income share percentages calculated by the editors based on these data. Labor shares rise temporarily in recession years (when the "pie" is smaller) but aside from that effect they have declined over much of the period, and the decline has accelerated in recent years.

- Employee compensation has fallen since the early 1970s to a record low of 52.0 percent of gross domestic income in 2014, as shown in the middle line in Figure 1-4. In 2018, the share grew to 53.2 percent. One factor in this decline has been a rising share of output required to replace used-up capital ("consumption of fixed capital"), reflecting increased use of capital per unit of production. In addition, the "net operating surplus," which is largely profits, interest, and proprietor's income, reached a new peak in 2014 but fell again by 2018. (Tables 1-12 and 1-16)

- Narrowing the focus to the corporate business sector, net income shares become clearer. (Proprietor's, i.e. unincorporated business; income includes a labor compensation component of uncertain size.) Total net factor income (that is, net of capital consumption) can be completely attributed to either labor or capital. These shares are also charted in Figure 1-4. Between 1959 and 2004, the shares vary cyclically but seem stable over time, averaging 79 percent for labor and 21 percent for capital. Since then, they have broken out of that range. (Tables 1-14 and 1-16)

Table 1-11. Relation of Gross Domestic Product, Gross and Net National Product, National Income, and Personal Income

(Billions of dollars, quarterly data are at seasonally adjusted annual rates.)

NIPA Table 1.7.5

Year and quarter	Gross domestic product	Plus: Income receipts from the rest of the world	Less: Income payments to the rest of the world	Equals: Gross national product	Less: Consumption of fixed capital									Equals: Net national product
					Total	Private					Government			
						Total	Domestic business			House-holds and institutions	Total	General government	Govern-ment enter-prises	
							Total	Capital consump-tion allow-ances	Less: Capital consump-tion ad-justment					
1960	542.4	4.9	1.8	545.5	67.9	48.2	40.0	39.7	-0.3	8.2	19.7	18.6	1.1	477.6
1961	562.2	5.3	1.8	565.7	70.6	49.8	41.3	41.7	0.4	8.5	20.8	19.7	1.1	495.1
1962	603.9	5.9	1.8	608.0	74.1	51.8	42.9	46.6	3.6	8.8	22.3	21.1	1.2	533.9
1963	637.5	6.5	2.1	641.9	78.0	54.2	44.9	49.9	5.0	9.3	23.9	22.6	1.3	563.9
1964	684.5	7.2	2.3	689.4	82.4	57.3	47.4	53.2	5.8	9.9	25.1	23.7	1.4	607.0
1965	742.3	7.9	2.6	747.6	88.0	61.6	50.9	57.0	6.1	10.7	26.4	25.0	1.5	659.6
1966	813.4	8.1	3.0	818.5	95.3	67.2	55.6	61.9	6.2	11.5	28.1	26.5	1.6	723.2
1967	860.0	8.7	3.3	865.4	103.6	73.3	60.9	66.9	6.0	12.4	30.2	28.5	1.8	761.8
1968	940.7	10.1	4.0	946.7	113.4	80.6	67.0	72.5	5.6	13.6	32.8	30.8	2.0	833.4
1969	1 017.6	11.8	5.7	1 023.7	124.9	89.4	74.2	79.7	5.5	15.2	35.5	33.3	2.2	898.8
1970	1 073.3	12.8	6.4	1 079.7	136.8	98.3	81.8	85.9	4.1	16.5	38.6	36.1	2.5	942.9
1971	1 164.9	14.0	6.4	1 172.4	148.9	107.6	89.4	92.3	2.9	18.2	41.3	38.5	2.8	1 023.5
1972	1 279.1	16.3	7.7	1 287.7	161.0	117.5	97.2	102.3	5.2	20.3	43.5	40.4	3.1	1 126.7
1973	1 425.4	23.5	10.9	1 438.0	178.7	131.5	108.2	111.9	3.7	23.3	47.2	43.7	3.5	1 259.3
1974	1 545.2	29.8	14.3	1 560.8	206.9	153.2	126.1	123.9	-2.2	27.0	53.7	49.4	4.3	1 353.9
1975	1 684.9	28.0	15.0	1 697.9	238.5	178.8	147.9	135.9	-12.0	30.9	59.7	54.6	5.1	1 459.4
1976	1 873.4	32.4	15.5	1 890.3	260.2	196.5	162.2	147.4	-14.8	34.3	63.7	58.2	5.5	1 630.0
1977	2 081.8	37.2	16.9	2 102.1	289.8	221.1	181.5	167.0	-14.5	39.6	68.7	62.7	6.0	1 812.3
1978	2 351.6	46.3	24.7	2 373.2	327.2	252.1	205.5	189.4	-16.2	46.6	75.1	68.5	6.6	2 046.0
1979	2 627.3	68.3	36.4	2 659.3	373.9	290.7	236.4	216.1	-20.2	54.4	83.1	75.6	7.5	2 285.4
1980	2 857.3	79.1	44.9	2 891.5	428.4	335.0	272.8	245.8	-26.9	62.2	93.5	84.7	8.7	2 463.1
1981	3 207.0	92.0	59.1	3 240.0	487.2	381.9	313.5	301.2	-12.3	68.5	105.3	95.4	9.9	2 752.7
1982	3 343.8	101.0	64.5	3 380.3	537.0	420.4	347.5	343.1	-4.4	72.9	116.6	105.7	10.8	2 843.3
1983	3 634.0	101.9	64.8	3 671.1	562.6	438.8	362.8	385.6	22.8	76.0	123.8	112.4	11.4	3 108.5
1984	4 037.6	121.9	85.6	4 073.9	598.4	463.5	382.7	435.6	52.9	80.9	134.9	122.8	12.1	3 475.5
1985	4 339.0	112.7	87.3	4 364.3	640.1	496.4	410.3	493.5	83.2	86.2	143.7	130.9	12.8	3 724.2
1986	4 579.6	111.3	94.4	4 596.6	685.3	531.6	438.1	512.0	73.9	93.4	153.7	140.1	13.6	3 911.3
1987	4 855.2	123.3	105.8	4 872.7	730.4	566.3	464.5	533.1	68.6	101.8	164.1	149.6	14.5	4 142.3
1988	5 236.4	152.1	129.5	5 259.1	784.5	607.9	499.2	563.7	64.5	108.8	176.6	161.1	15.5	4 474.6
1989	5 641.6	177.7	152.9	5 666.4	838.3	649.6	532.6	589.5	56.9	117.0	188.6	172.0	16.6	4 828.1
1990	5 963.1	188.8	154.2	5 997.8	888.5	688.4	564.6	596.0	31.4	123.8	200.1	182.3	17.8	5 109.3
1991	6 158.1	168.4	136.8	6 189.7	932.4	721.5	592.3	611.6	19.3	129.2	210.9	192.1	18.8	5 257.3
1992	6 520.3	152.1	121.0	6 551.4	960.2	742.9	608.2	628.4	20.2	134.7	217.4	197.6	19.8	5 591.2
1993	6 858.6	155.6	124.4	6 889.7	1 003.5	778.2	634.2	658.9	24.7	144.0	225.3	204.2	21.1	5 886.2
1994	7 287.2	184.5	161.6	7 310.2	1 055.6	822.5	669.0	701.6	32.6	153.5	233.1	210.9	22.2	6 254.6
1995	7 639.7	229.8	201.9	7 667.7	1 122.4	880.7	717.8	753.9	36.1	162.9	241.7	218.2	23.5	6 545.3
1996	8 073.1	246.4	215.5	8 104.0	1 175.3	929.1	758.5	808.3	49.8	170.6	246.2	221.6	24.6	6 928.7
1997	8 577.6	280.1	256.8	8 600.9	1 239.3	987.8	807.9	877.0	69.2	179.9	251.6	225.9	25.7	7 361.5
1998	9 062.8	286.8	269.4	9 080.2	1 309.7	1 052.2	860.1	941.4	81.3	192.1	257.6	230.8	26.8	7 770.5
1999	9 630.7	320.2	294.7	9 656.2	1 398.9	1 132.2	924.0	1 023.8	99.8	208.2	266.7	238.3	28.4	8 257.3
2000	10 252.3	380.6	345.6	10 287.4	1 511.2	1 231.5	1 004.6	1 096.1	91.5	227.0	279.7	249.4	30.3	8 776.1
2001	10 581.8	324.1	275.3	10 630.6	1 599.5	1 311.7	1 064.9	1 180.1	115.2	246.8	287.8	256.0	31.8	9 031.1
2002	10 936.4	314.8	269.6	10 981.7	1 658.0	1 361.8	1 098.8	1 283.9	185.1	263.0	296.2	262.9	33.2	9 323.7
2003	11 458.2	353.8	295.4	11 516.6	1 719.1	1 411.9	1 125.4	1 303.5	178.1	286.6	307.1	272.2	34.9	9 797.5
2004	12 213.7	446.9	368.8	12 291.9	1 821.8	1 497.1	1 178.8	1 332.0	153.2	318.3	324.7	286.9	37.9	10 470.1
2005	13 036.6	566.0	488.1	13 114.6	1 971.0	1 622.6	1 266.4	1 169.5	-97.0	356.2	348.4	307.0	41.4	11 143.6
2006	13 814.6	712.0	661.5	13 865.1	2 124.1	1 751.8	1 362.6	1 245.1	-117.4	389.2	372.3	327.6	44.7	11 740.9
2007	14 451.9	866.6	757.6	14 560.9	2 252.8	1 852.5	1 446.7	1 326.8	-119.8	405.8	400.3	352.1	48.2	12 308.1
2008	14 712.8	848.8	694.2	14 867.5	2 358.8	1 931.8	1 524.9	1 584.1	59.3	427.0	427.0	375.7	51.3	12 508.6
2009	14 448.9	647.8	505.8	14 590.9	2 371.5	1 928.7	1 531.2	1 562.1	30.9	397.5	442.8	389.8	53.0	12 219.4
2010	14 992.1	715.2	519.5	15 187.8	2 390.9	1 933.8	1 537.1	1 605.9	68.8	396.7	457.2	402.2	55.0	12 796.8
2011	15 542.6	789.2	552.8	15 779.0	2 474.5	1 997.3	1 595.1	1 831.3	236.2	402.2	477.2	419.3	57.8	13 304.5
2012	16 197.0	799.7	567.4	16 429.3	2 576.0	2 082.4	1 670.4	1 665.9	-4.5	412.0	493.6	433.0	60.7	13 853.3
2013	16 784.9	823.4	592.7	17 015.6	2 681.2	2 176.6	1 738.8	1 753.6	14.8	437.8	504.6	441.8	62.9	14 334.4
2014	17 527.3	853.5	612.5	17 768.3	2 815.0	2 298.5	1 831.2	1 866.4	35.2	467.3	516.6	451.4	65.2	14 953.3
2015	18 224.8	837.7	613.1	18 449.4	2 916.5	2 393.7	1 908.0	1 981.9	73.9	485.7	522.8	455.6	67.1	15 532.9
2016	18 715.0	861.7	643.5	18 933.2	2 991.6	2 463.2	1 954.4	2 074.4	120.1	508.9	528.4	459.3	69.1	15 941.6
2017	19 519.4	957.9	714.6	19 762.7	3 121.4	2 578.2	2 041.0	2 246.5	205.5	537.2	543.2	471.3	71.9	16 641.3
2018	20 580.2	1 106.2	838.3	20 848.1	3 291.4	2 725.8	2 151.4	2 560.8	409.4	574.4	565.7	490.3	75.4	17 556.7
2016														
1st quarter	18 424.3	826.4	624.9	18 625.8	2 949.8	2 426.6	1 929.1	2 048.7	119.6	497.5	523.2	455.2	68.0	15 676.0
2nd quarter	18 637.3	861.8	648.0	18 851.0	2 979.8	2 452.4	1 947.3	2 067.6	120.3	505.2	527.4	458.6	68.8	15 871.2
3rd quarter	18 806.7	860.1	655.3	19 011.5	3 002.3	2 472.7	1 959.9	2 083.9	124.0	512.8	529.5	460.1	69.4	16 009.2
4th quarter	18 991.9	898.4	645.7	19 244.6	3 034.4	2 501.1	1 981.1	2 097.4	116.3	519.9	533.3	463.2	70.1	16 210.2
2017														
1st quarter	19 190.4	898.4	665.2	19 423.6	3 064.9	2 527.9	2 003.2	2 122.8	119.7	524.8	537.0	466.1	70.9	16 358.7
2nd quarter	19 356.6	924.9	708.4	19 573.1	3 101.7	2 561.0	2 027.8	2 147.8	120.0	533.1	540.7	469.2	71.5	16 471.4
3rd quarter	19 611.7	982.1	725.9	19 867.9	3 141.4	2 596.1	2 054.2	2 171.0	116.8	541.9	545.4	473.2	72.2	16 726.5
4th quarter	19 918.9	1 026.2	758.9	20 186.2	3 177.7	2 628.0	2 078.8	2 544.5	465.6	549.1	549.8	476.8	72.9	17 008.5
2018														
1st quarter	20 163.2	1 070.5	789.5	20 444.1	3 220.2	2 664.1	2 103.8	2 524.9	421.1	560.4	556.1	482.2	73.8	17 223.9
2nd quarter	20 510.2	1 111.4	845.8	20 775.8	3 271.6	2 708.0	2 136.4	2 547.2	410.8	571.6	563.7	488.7	75.0	17 504.2
3rd quarter	20 749.8	1 116.0	843.6	21 022.1	3 315.8	2 746.8	2 167.4	2 572.1	404.7	579.3	569.1	493.1	75.9	17 706.3
4th quarter	20 897.8	1 127.0	874.4	21 150.4	3 358.1	2 784.2	2 198.0	2 598.9	400.9	586.3	573.9	497.1	76.8	17 792.3

Table 1-12. Gross Domestic Income by Type of Income

(Billions of dollars, quarterly data are at seasonally adjusted annual rates.)

NIPA Tables 1.1.5, 1.10

Year and quarter	Gross domestic income	Compensation of employees			Taxes on production and imports	Less: Subsidies	Net operating surplus	Private enterprises				
		Total	Wages and salaries	Supplements to wages and salaries			Total	Total	Net interest and miscellaneous payments, domestic industries	Business current transfer payments, net	Proprietors' income with IVA and CCAdj	Rental income of persons with CCAdj
1960	543.7	301.4	273.0	28.4	44.5	1.1	130.9	130.5	10.1	1.7	50.6	16.5
1961	563.1	310.6	280.7	29.9	47.0	2.0	137.0	136.7	11.7	1.9	53.2	17.2
1962	603.9	332.3	299.5	32.8	50.4	2.3	149.4	149.0	13.4	2.1	55.2	18.0
1963	638.9	350.5	314.9	35.6	53.4	2.2	159.3	158.5	14.4	2.5	56.4	18.7
1964	684.5	376.0	337.8	38.2	57.3	2.7	171.6	170.7	16.5	3.0	59.1	18.8
1965	741.5	405.4	363.8	41.7	60.7	3.0	190.4	189.6	18.6	3.5	63.7	19.3
1966	808.3	449.2	400.3	48.9	63.2	3.9	204.5	204.2	21.3	3.4	67.9	19.9
1967	856.6	481.8	429.0	52.8	67.9	3.8	207.2	206.9	24.3	3.6	69.5	20.3
1968	937.5	530.7	472.0	58.8	76.4	4.2	221.2	220.7	26.5	4.2	73.8	20.1
1969	1 016.0	584.4	518.3	66.1	83.9	4.5	227.4	227.2	33.2	4.8	77.0	20.3
1970	1 068.0	623.3	551.5	71.8	91.4	4.8	221.2	222.2	40.2	4.4	77.8	20.7
1971	1 155.3	665.0	584.5	80.4	100.5	4.7	245.6	247.1	44.4	4.2	83.9	21.8
1972	1 271.9	731.3	638.8	92.5	107.9	6.6	278.3	279.2	49.0	4.8	95.1	22.7
1973	1 419.2	812.7	708.8	103.9	117.2	5.2	315.9	317.8	58.0	5.7	112.5	23.1
1974	1 537.8	887.7	772.3	115.4	124.9	3.3	321.6	324.0	73.6	6.8	112.2	23.2
1975	1 671.6	947.3	814.9	132.4	135.3	4.5	355.0	359.0	85.3	9.0	118.2	22.3
1976	1 852.8	1 048.4	899.8	148.6	146.4	5.1	402.9	405.3	87.1	9.1	131.0	20.3
1977	2 062.4	1 165.9	994.2	171.7	159.7	7.1	454.1	457.1	101.9	8.1	144.5	15.9
1978	2 328.3	1 316.9	1 120.7	196.2	170.9	8.9	522.3	524.6	116.0	10.4	166.0	16.5
1979	2 582.3	1 477.3	1 253.4	223.9	180.1	8.5	559.5	562.2	139.6	12.8	179.4	16.1
1980	2 812.9	1 622.3	1 373.5	248.8	200.3	9.8	571.6	576.0	183.4	14.0	171.6	19.0
1981	3 169.0	1 792.7	1 511.5	281.2	235.6	11.5	664.9	669.6	231.4	16.9	179.7	23.8
1982	3 335.0	1 893.2	1 587.7	305.5	240.9	15.0	678.9	682.3	270.7	19.3	171.2	23.8
1983	3 577.1	2 012.7	1 677.7	335.0	263.3	21.3	759.8	761.7	284.6	21.7	186.3	24.4
1984	3 996.0	2 216.1	1 845.1	371.0	289.8	21.1	912.8	913.7	330.3	29.2	228.2	24.7
1985	4 284.7	2 387.5	1 982.8	404.0	308.1	21.4	970.3	968.4	350.7	34.1	241.1	26.2
1986	4 499.6	2 543.8	2 104.1	439.7	323.4	24.9	972.0	969.4	373.8	36.0	256.5	18.3
1987	4 811.4	2 723.8	2 257.6	466.1	347.5	30.3	1 040.0	1 037.2	382.9	33.3	286.5	16.6
1988	5 233.4	2 948.9	2 440.6	508.2	374.5	29.5	1 155.1	1 149.6	411.3	32.8	325.5	22.5
1989	5 573.6	3 140.9	2 584.3	556.6	398.9	27.4	1 223.0	1 215.8	467.7	38.3	341.1	21.5
1990	5 867.6	3 342.7	2 743.5	599.2	425.0	27.0	1 238.4	1 234.7	472.5	39.2	353.2	28.2
1991	6 065.2	3 453.3	2 817.2	636.0	457.1	27.5	1 249.9	1 241.7	433.9	38.9	354.2	38.6
1992	6 404.4	3 671.2	2 968.5	702.7	483.4	30.1	1 319.6	1 309.2	404.7	39.7	400.2	60.6
1993	6 702.6	3 820.6	3 082.7	737.9	503.1	36.7	1 412.1	1 400.6	395.5	39.4	428.0	90.1
1994	7 147.3	4 010.2	3 240.6	769.6	545.2	32.5	1 568.7	1 556.1	398.3	40.7	456.6	113.7
1995	7 546.7	4 202.2	3 422.1	780.1	557.9	34.8	1 699.1	1 682.6	418.3	45.0	481.2	124.9
1996	8 015.0	4 421.1	3 620.6	800.5	580.8	35.2	1 873.1	1 855.4	428.9	52.6	543.8	142.5
1997	8 566.0	4 713.2	3 881.2	832.0	611.6	33.8	2 035.6	2 017.5	474.3	50.1	584.0	147.1
1998	9 118.1	5 075.7	4 186.2	889.5	639.5	36.4	2 129.5	2 112.6	537.5	64.1	640.2	165.2
1999	9 663.8	5 409.9	4 465.2	944.8	673.6	45.2	2 226.6	2 209.7	553.8	67.8	696.4	178.5
2000	10 348.8	5 854.6	4 832.4	1 022.2	708.6	45.8	2 320.3	2 308.8	645.2	85.4	753.9	183.5
2001	10 695.0	6 046.3	4 961.6	1 084.7	727.7	58.7	2 380.1	2 374.5	651.9	99.2	831.0	202.4
2002	11 009.1	6 143.4	5 004.1	1 139.3	760.0	41.4	2 489.1	2 481.5	565.1	80.7	869.8	211.1
2003	11 471.9	6 362.3	5 147.0	1 215.3	805.6	49.1	2 634.0	2 628.9	526.9	76.3	896.9	231.5
2004	12 235.8	6 729.3	5 430.8	1 298.5	868.1	46.4	2 863.0	2 862.9	475.8	82.0	962.0	248.9
2005	13 091.7	7 077.1	5 703.0	1 374.7	942.4	60.9	3 161.5	3 166.0	598.9	94.1	978.0	232.0
2006	14 022.5	7 491.3	6 068.3	1 422.9	997.0	51.5	3 461.6	3 468.6	728.5	81.6	1 049.6	202.3
2007	14 434.2	7 889.4	6 407.3	1 482.1	1 036.8	54.6	3 309.8	3 323.9	851.9	98.3	994.0	184.4
2008	14 530.0	8 068.7	6 545.9	1 522.7	1 049.7	52.6	3 105.3	3 123.5	896.3	114.0	960.9	256.7
2009	14 256.8	7 767.2	6 257.3	1 509.9	1 026.8	58.3	3 149.6	3 165.7	737.4	124.4	938.5	327.3
2010	14 931.0	7 933.0	6 380.1	1 552.9	1 063.1	55.8	3 599.9	3 619.9	647.2	126.8	1 108.7	394.2
2011	15 595.8	8 234.0	6 634.0	1 600.0	1 103.7	60.0	3 843.6	3 863.0	629.8	128.1	1 229.3	478.6
2012	16 438.4	8 575.4	6 936.1	1 639.2	1 136.1	58.0	4 208.9	4 224.3	668.1	98.8	1 347.3	518.0
2013	16 945.2	8 843.6	7 122.6	1 721.0	1 188.7	59.7	4 291.4	4 307.3	624.6	110.3	1 403.6	557.0
2014	17 816.4	9 259.7	7 485.8	1 773.9	1 240.8	58.1	4 559.0	4 570.0	669.3	132.9	1 447.7	604.6
2015	18 479.7	9 709.2	7 867.8	1 841.5	1 277.1	57.3	4 634.1	4 639.7	753.7	156.7	1 422.2	648.1
2016	18 827.0	9 972.7	8 095.9	1 876.8	1 312.8	61.8	4 611.7	4 614.3	741.3	168.2	1 423.7	681.4
2017	19 587.0	10 424.5	8 474.9	1 949.5	1 364.5	61.1	4 737.7	4 740.2	805.9	145.4	1 518.2	718.8
2018	20 569.4	10 941.4	8 901.4	2 040.0	1 441.8	64.4	4 959.2	4 965.7	893.5	153.7	1 588.8	756.8
2016												
1st quarter	18 673.5	9 855.4	7 994.7	1 860.7	1 297.1	60.7	4 631.9	4 633.9	749.3	166.2	1 415.2	669.9
2nd quarter	18 718.3	9 912.4	8 044.4	1 868.0	1 301.2	62.4	4 587.3	4 590.6	734.1	172.0	1 410.2	680.2
3rd quarter	18 880.6	10 005.8	8 124.7	1 881.1	1 322.3	63.1	4 613.4	4 616.1	735.2	154.7	1 429.5	683.6
4th quarter	19 035.5	10 117.3	8 219.8	1 897.5	1 330.5	61.0	4 614.3	4 616.6	746.7	179.9	1 440.0	692.1
2017												
1st quarter	19 307.0	10 240.3	8 323.3	1 917.0	1 340.8	59.9	4 720.9	4 722.9	798.4	163.6	1 494.8	707.4
2nd quarter	19 496.9	10 347.1	8 410.6	1 936.5	1 355.1	58.7	4 751.7	4 753.9	797.5	149.5	1 512.2	709.9
3rd quarter	19 638.4	10 469.7	8 510.9	1 958.8	1 371.4	63.2	4 719.0	4 721.6	794.7	130.5	1 523.1	722.0
4th quarter	19 905.6	10 640.8	8 654.9	1 985.9	1 390.6	62.8	4 759.3	4 762.5	833.0	138.0	1 542.9	736.0
2018												
1st quarter	20 252.2	10 798.9	8 789.5	2 009.4	1 415.8	59.3	4 876.7	4 880.2	880.4	144.0	1 567.5	743.8
2nd quarter	20 460.1	10 888.8	8 857.7	2 031.1	1 433.9	58.6	4 924.2	4 929.4	895.3	145.4	1 573.3	754.0
3rd quarter	20 716.9	11 007.3	8 955.3	2 052.0	1 442.3	58.4	5 009.8	5 017.1	903.2	166.1	1 590.0	765.2
4th quarter	20 848.6	11 070.4	9 003.1	2 067.4	1 475.2	81.3	5 026.2	5 036.1	895.0	159.2	1 624.4	764.1

Table 1-13A. National Income by Type of Income: Recent Data

(Billions of dollars, quarterly data are at seasonally adjusted annual rates.)

NIPA Tables 1.7.5, 1.12

Year and quarter	National income, total	Compensation of employees Total	Wages and salaries Total	Wages and salaries Government	Wages and salaries Other	Supplements to wages and salaries Total	Employer contributions for: Employee pension and insurance funds	Employer contributions for: Government social insurance	Proprietors' income with IVA and CCAdj Total	Proprietors' income with IVA and CCAdj Farm	Proprietors' income with IVA and CCAdj Nonfarm	Rental income of persons with CCAdj
1955	376.9	230.6	212.2	36.6	175.6	18.4	13.2	5.2	44.3	10.7	33.6	13.4
1956	400.1	249.3	229.0	38.8	190.2	20.2	14.6	5.7	45.8	10.6	35.2	13.7
1957	418.5	262.6	240.0	41.0	198.9	22.6	16.3	6.4	47.8	10.6	37.2	14.1
1958	420.8	264.7	241.3	44.1	197.2	23.4	17.0	6.3	50.2	12.4	37.7	14.8
1959	458.8	285.8	259.8	46.1	213.8	26.0	18.1	7.9	50.3	10.0	40.3	15.6
1960	478.9	301.3	272.9	49.2	223.7	28.4	19.2	9.3	50.6	10.6	39.9	16.5
1961	496.0	310.4	280.5	52.5	228.0	29.9	20.3	9.6	53.2	11.2	42.0	17.2
1962	533.9	332.2	299.4	56.3	243.0	32.8	21.7	11.2	55.2	11.2	44.0	18.0
1963	565.4	350.4	314.9	60.0	254.8	35.6	23.2	12.4	56.4	11.0	45.4	18.7
1964	607.0	376.0	337.8	64.9	272.9	38.2	25.6	12.6	59.1	9.8	49.4	18.8
1965	658.8	405.4	363.8	69.9	293.8	41.7	28.6	13.1	63.7	12.0	51.6	19.3
1966	718.1	449.2	400.3	78.4	321.9	48.9	32.1	16.8	67.9	13.0	54.9	19.9
1967	758.4	481.8	429.0	86.5	342.5	52.8	34.8	18.0	69.5	11.6	57.8	20.3
1968	830.2	530.8	472.0	96.7	375.3	58.8	38.8	20.0	73.8	11.7	62.2	20.1
1969	897.2	584.5	518.3	105.6	412.7	66.1	43.4	22.8	77.0	12.8	64.2	20.3
1970	937.5	623.3	551.6	117.2	434.3	71.8	47.9	23.8	77.8	12.9	64.9	20.7
1971	1 014.0	665.0	584.5	126.8	457.8	80.4	54.0	26.4	83.9	13.4	70.5	21.8
1972	1 119.5	731.3	638.8	137.9	500.9	92.5	61.4	31.2	95.1	17.0	78.1	22.7
1973	1 253.2	812.7	708.8	148.8	560.0	103.9	64.1	39.8	112.5	29.1	83.4	23.1
1974	1 346.4	887.7	772.3	160.5	611.8	115.4	70.7	44.7	112.2	23.5	88.7	23.2
1975	1 446.0	947.2	814.8	176.2	638.6	132.4	85.7	46.7	118.2	22.0	96.2	22.3
1976	1 609.4	1 048.3	899.7	188.9	710.8	148.6	94.2	54.4	131.0	17.2	113.8	20.3
1977	1 792.8	1 165.8	994.2	202.6	791.6	171.7	110.6	61.1	144.5	16.0	128.5	15.9
1978	2 022.7	1 316.8	1 120.6	220.0	900.6	196.2	124.7	71.5	166.0	19.9	146.1	16.5
1979	2 240.3	1 477.2	1 253.3	237.1	1 016.2	223.9	141.3	82.6	179.4	22.2	157.3	16.1
1980	2 418.6	1 622.2	1 373.4	261.5	1 112.0	248.8	159.9	88.9	171.6	11.7	159.9	19.0
1981	2 714.7	1 792.5	1 511.4	285.8	1 225.5	281.2	177.5	103.6	179.7	19.0	160.7	23.8
1982	2 834.5	1 893.0	1 587.5	307.5	1 280.0	305.5	195.7	109.8	171.2	13.3	157.9	23.8
1983	3 051.5	2 012.5	1 677.5	324.8	1 352.7	335.0	215.1	119.9	186.3	6.2	180.1	24.4
1984	3 433.9	2 215.9	1 844.9	348.1	1 496.8	371.0	231.9	139.0	228.2	20.9	207.3	24.7
1985	3 669.9	2 387.3	1 982.6	373.9	1 608.7	404.8	257.0	147.7	241.1	21.0	220.1	26.2
1986	3 831.2	2 542.1	2 102.3	397.2	1 705.1	439.7	281.9	157.9	256.5	22.8	233.7	18.3
1987	4 098.5	2 722.4	2 256.3	423.1	1 833.2	466.1	299.9	166.3	286.5	28.9	257.6	16.6
1988	4 471.6	2 948.0	2 439.8	452.0	1 987.7	508.2	323.6	184.6	325.5	26.8	298.7	22.5
1989	4 760.1	3 139.6	2 583.1	481.1	2 101.9	556.6	362.9	193.7	341.1	33.0	308.1	21.5
1990	5 013.8	3 340.4	2 741.2	519.0	2 222.2	599.2	392.7	206.5	353.2	32.2	321.0	28.2
1991	5 164.4	3 450.5	2 814.5	548.8	2 265.7	636.0	420.9	215.1	354.2	26.8	327.4	38.6
1992	5 475.2	3 668.2	2 965.5	572.0	2 393.5	702.7	474.3	228.4	400.2	34.8	365.4	60.6
1993	5 730.3	3 817.3	3 079.3	589.0	2 490.3	737.9	498.3	239.7	428.0	31.4	396.6	90.1
1994	6 114.6	4 006.2	3 236.6	609.5	2 627.1	769.6	515.5	254.1	456.6	34.7	422.0	113.7
1995	6 452.3	4 198.1	3 418.0	629.0	2 789.0	780.1	515.9	264.1	481.2	22.0	459.2	124.9
1996	6 870.6	4 416.9	3 616.5	648.1	2 968.4	800.5	525.7	274.8	543.8	37.3	506.4	142.5
1997	7 349.9	4 708.8	3 876.8	671.9	3 205.0	832.0	542.4	289.6	584.0	32.4	551.6	147.1
1998	7 825.7	5 071.1	4 181.6	701.3	3 480.3	889.5	582.3	307.2	640.2	28.5	611.7	165.2
1999	8 290.4	5 402.8	4 458.0	733.8	3 724.2	944.8	621.4	323.3	696.4	28.1	668.3	178.5
2000	8 872.6	5 848.1	4 825.9	779.8	4 046.1	1 022.2	677.0	345.2	753.9	31.5	722.4	183.5
2001	9 144.2	6 039.1	4 954.4	822.0	4 132.4	1 084.7	726.7	358.0	831.0	32.1	798.9	202.4
2002	9 396.4	6 135.6	4 996.3	872.9	4 123.4	1 139.3	773.2	366.0	869.8	19.9	849.8	211.1
2003	9 811.2	6 354.1	5 138.7	914.0	4 224.8	1 215.3	832.8	382.5	896.9	36.5	860.4	231.5
2004	10 492.2	6 720.1	5 421.6	952.3	4 469.2	1 298.5	889.7	408.8	962.0	51.5	910.5	248.9
2005	11 198.7	7 066.6	5 691.9	991.3	4 700.6	1 374.7	946.7	428.1	978.0	46.8	931.2	232.0
2006	11 948.8	7 479.9	6 057.0	1 034.5	5 022.4	1 422.9	975.6	447.3	1 049.6	33.1	1 016.6	202.3
2007	12 290.4	7 878.9	6 396.8	1 088.5	5 308.2	1 482.1	1 020.4	461.7	994.0	40.3	953.8	184.4
2008	12 325.8	8 057.0	6 534.2	1 143.9	5 390.4	1 522.7	1 051.3	471.4	960.9	40.2	920.7	256.7
2009	12 027.2	7 758.5	6 248.6	1 175.2	5 073.4	1 509.9	1 051.8	458.1	938.5	28.1	910.5	327.3
2010	12 735.8	7 924.9	6 372.1	1 191.2	5 180.9	1 552.9	1 083.9	469.0	1 108.7	39.0	1 069.7	394.2
2011	13 357.7	8 225.9	6 625.9	1 194.9	5 431.1	1 600.0	1 107.3	492.7	1 229.3	64.9	1 164.4	478.6
2012	14 094.7	8 566.7	6 927.5	1 198.3	5 729.2	1 639.2	1 125.9	513.3	1 347.3	60.9	1 286.4	518.0
2013	14 494.7	8 834.2	7 113.2	1 208.0	5 905.2	1 721.0	1 194.7	526.3	1 403.6	88.3	1 315.3	557.0
2014	15 242.5	9 249.1	7 475.2	1 236.9	6 238.3	1 773.9	1 227.5	546.4	1 447.7	69.8	1 377.9	604.6
2015	15 787.9	9 698.2	7 856.7	1 275.6	6 581.0	1 841.5	1 272.3	569.2	1 422.2	56.0	1 366.2	648.1
2016	16 053.6	9 960.3	8 083.5	1 308.0	6 775.5	1 876.8	1 295.6	581.2	1 423.7	35.6	1 388.1	681.4
2017	16 708.8	10 411.0	8 462.1	1 348.0	7 114.1	1 949.5	1 343.9	605.7	1 518.2	38.1	1 480.1	718.8
2018	17 545.9	10 928.0	8 888.5	1 402.6	7 485.9	2 040.0	1 417.2	622.8	1 588.8	27.2	1 561.6	756.8
2016												
1st quarter	15 925.2	9 843.5	7 982.8	1 294.2	6 688.5	1 860.7	1 286.5	574.2	1 415.2	36.5	1 378.7	669.9
2nd quarter	15 952.3	9 900.1	8 032.1	1 302.5	6 729.6	1 868.0	1 290.5	577.5	1 410.2	38.3	1 371.9	680.2
3rd quarter	16 083.1	9 993.2	8 112.2	1 314.1	6 798.1	1 881.1	1 298.0	583.1	1 429.5	36.5	1 393.0	683.6
4th quarter	16 253.9	10 104.0	8 206.9	1 321.2	6 885.7	1 897.5	1 307.5	590.0	1 440.0	31.2	1 408.9	692.1
2017												
1st quarter	16 475.3	10 227.0	8 310.6	1 331.4	6 979.2	1 917.0	1 320.4	596.6	1 494.8	44.5	1 450.3	707.4
2nd quarter	16 611.7	10 334.0	8 397.7	1 340.3	7 057.4	1 936.5	1 334.3	602.2	1 512.2	42.1	1 470.1	709.9
3rd quarter	16 753.1	10 456.0	8 497.9	1 353.0	7 144.9	1 958.8	1 350.8	607.9	1 523.1	34.1	1 489.0	722.0
4th quarter	16 995.2	10 628.0	8 642.0	1 367.2	7 274.9	1 985.9	1 370.0	615.9	1 542.9	31.8	1 511.1	736.0
2018												
1st quarter	17 313.0	10 786.0	8 776.7	1 380.4	7 396.3	2 009.4	1 391.8	617.6	1 567.5	28.1	1 539.4	743.8
2nd quarter	17 454.0	10 876.0	8 845.0	1 394.1	7 450.9	2 031.1	1 410.9	620.2	1 573.3	27.5	1 545.8	754.0
3rd quarter	17 673.5	10 994.0	8 942.2	1 412.6	7 529.6	2 052.0	1 426.6	625.4	1 590.0	17.4	1 572.6	765.2
4th quarter	17 743.1	11 057.4	8 990.0	1 423.3	7 566.8	2 067.4	1 439.3	628.1	1 624.4	35.9	1 588.4	764.1

Table 1-13A. National Income by Type of Income: Recent Data—*Continued*

(Billions of dollars, quarterly data are at seasonally adjusted annual rates.)

NIPA Tables 1.7.5, 1.12

| Year and quarter | Corporate profits with IVA and CCAdj | | | | | Net interest and miscellaneous payments | Taxes on production and imports | Less: Subsidies | Business current transfer payments, net | | | Current surplus of government enterprises | Addendum: Net national factor income |
| | Total | Taxes on corporate income | Profits after tax | | | | | | Total [1] | To persons | To government | | |
			Total	Net dividends	Undistributed corporate profits								
1955	50.2	21.8	28.4	10.7	17.6	5.9	31.5	0.2	1.3	0.9	0.4	...	344.3
1956	49.6	21.6	28.1	11.7	16.4	6.5	34.2	0.7	1.7	1.2	0.4	...	364.9
1957	49.1	20.9	28.3	12.3	16.0	7.7	36.6	1.1	1.8	1.4	0.4	...	381.2
1958	43.9	18.4	25.4	12.1	13.3	9.2	37.7	1.4	1.7	1.2	0.5	...	382.7
1959	55.5	22.8	32.7	13.5	19.2	9.3	41.1	1.1	1.7	1.3	0.3	...	416.6
1960	54.7	21.9	32.8	14.3	18.5	10.3	44.5	1.1	1.7	1.3	0.4	...	433.3
1961	55.9	22.2	33.7	14.6	19.2	12.1	47.0	2.0	1.9	1.4	0.5	...	448.8
1962	64.0	23.3	40.7	15.8	24.9	13.8	50.4	2.3	2.1	1.5	0.6	...	483.3
1963	70.5	25.5	45.0	17.1	27.9	14.8	53.4	2.2	2.5	1.9	0.7	...	510.9
1964	77.7	26.6	51.1	19.8	31.4	17.0	57.3	2.7	3.0	2.2	0.8	...	548.6
1965	89.3	29.8	59.5	21.5	38.0	19.1	60.7	3.0	3.5	2.3	1.2	...	596.8
1966	96.1	32.2	63.9	22.3	41.6	21.8	63.2	3.9	3.4	2.1	1.3	...	655.1
1967	93.9	31.0	62.9	23.4	39.5	24.9	67.9	3.8	3.6	2.3	1.4	...	690.4
1968	101.7	37.2	64.6	26.0	38.6	27.0	76.4	4.2	4.2	2.8	1.4	...	753.4
1969	98.4	37.0	61.5	27.3	34.2	32.7	83.9	4.5	4.8	3.3	1.5	...	813.0
1970	86.2	31.3	55.0	27.8	27.2	39.5	91.4	4.8	4.4	2.9	1.4	...	847.6
1971	100.6	34.8	65.8	28.4	37.5	44.2	100.5	4.7	4.2	2.7	1.5	...	915.5
1972	117.2	39.1	78.1	30.1	48.0	48.0	107.9	6.6	4.8	3.1	1.7	...	1 014.4
1973	133.4	45.6	87.8	34.2	53.5	55.7	117.2	5.2	5.7	3.9	1.8	...	1 137.4
1974	125.7	47.2	78.5	38.8	39.7	71.7	124.9	3.3	6.8	4.7	2.1	...	1 220.5
1975	138.9	46.3	92.6	38.3	54.3	83.7	135.3	4.5	9.0	6.8	2.2	...	1 310.3
1976	174.3	59.4	114.9	44.9	70.0	87.4	146.4	5.1	9.1	6.7	2.4	...	1 461.4
1977	205.9	68.6	137.3	50.7	00.0	100.2	159.7	7.1	8.1	5.1	3.0	...	1 635.2
1978	238.6	77.9	160.7	57.8	102.9	114.8	170.9	8.9	10.4	6.5	3.9	...	1 852.7
1979	249.0	80.7	168.2	66.8	101.4	137.0	180.1	8.5	12.8	8.2	4.5	...	2 058.7
1980	223.6	75.5	148.1	75.8	72.3	182.2	200.3	9.8	14.0	8.6	5.4	...	2 218.6
1981	247.5	70.3	177.2	87.8	89.4	234.8	235.6	11.5	16.9	11.2	5.7	...	2 478.3
1982	229.9	51.3	178.6	92.9	85.6	274.8	240.9	15.0	19.3	12.4	6.9	...	2 592.7
1983	279.8	66.4	213.3	97.7	115.7	286.8	263.3	21.3	21.7	13.8	7.9	...	2 789.8
1984	337.9	81.5	256.4	106.9	149.5	330.2	289.8	21.1	29.2	19.7	9.4	...	3 136.9
1985	354.5	81.6	272.9	115.3	157.5	338.2	308.1	21.4	34.1	22.3	11.8	...	3 347.2
1986	324.4	91.9	232.5	124.0	108.5	353.1	323.4	24.9	36.0	22.9	13.0	...	3 494.3
1987	366.0	112.7	253.3	130.1	123.2	353.7	347.5	30.3	33.3	20.2	13.1	...	3 745.2
1988	414.5	124.3	290.2	147.3	142.9	377.9	374.5	29.5	32.8	20.6	12.2	...	4 088.3
1989	414.3	124.4	289.9	179.6	110.3	426.6	398.9	27.4	38.3	23.2	15.1	...	4 343.2
1990	417.7	121.8	295.9	192.7	103.2	433.4	425.0	27.0	39.2	22.2	17.0	...	4 572.9
1991	452.6	117.8	334.8	201.3	133.5	391.8	457.1	27.5	38.9	17.6	21.3	...	4 687.6
1992	477.2	131.9	345.3	206.3	139.0	365.6	483.4	30.1	39.7	16.3	23.6	-0.1	4 971.8
1993	524.6	155.0	369.5	221.3	148.2	353.1	503.1	36.7	39.4	14.1	25.6	-0.3	5 213.0
1994	624.8	172.7	452.1	256.4	195.7	347.3	545.2	32.5	40.7	13.3	27.9	-0.4	5 548.6
1995	706.2	194.4	511.8	282.3	229.4	357.4	557.9	34.8	45.0	18.7	24.9	1.4	5 867.7
1996	789.5	211.4	578.1	323.6	254.5	361.9	580.8	35.2	52.6	22.9	30.7	-1.0	6 254.7
1997	869.7	224.8	645.0	360.1	284.9	394.4	611.6	33.8	50.1	19.4	29.8	0.8	6 704.0
1998	808.5	221.8	586.6	383.6	203.0	456.7	639.5	36.4	64.1	26.0	34.3	3.8	7 141.7
1999	834.9	227.4	607.5	373.5	234.1	464.8	673.6	45.2	67.8	34.0	36.5	-2.6	7 577.4
2000	786.6	233.4	553.2	410.2	142.9	541.1	708.6	45.8	85.4	42.4	41.9	1.0	8 113.1
2001	758.7	170.1	588.6	397.9	190.8	539.1	727.7	58.7	99.2	46.8	43.9	8.6	8 370.3
2002	911.7	160.6	751.1	424.9	326.2	461.4	760.0	41.4	80.7	34.2	46.2	0.3	8 589.4
2003	1 056.3	213.7	842.5	456.0	386.5	434.6	805.6	49.1	76.3	26.3	48.3	1.7	8 973.2
2004	1 289.3	278.5	1 010.8	582.2	428.6	368.1	868.1	46.4	82.0	16.8	52.4	12.8	9 588.4
2005	1 488.6	379.8	1 108.8	602.0	506.8	462.3	942.4	60.9	94.1	25.8	53.4	14.9	10 227.6
2006	1 646.3	430.4	1 215.8	755.1	460.8	550.6	997.0	51.5	81.6	20.8	57.5	3.3	10 928.7
2007	1 533.2	392.1	1 141.1	853.5	287.6	633.6	1 036.8	54.6	98.3	30.8	60.9	6.6	11 224.1
2008	1 285.8	256.1	1 029.7	840.3	189.4	672.4	1 049.7	52.6	114.0	35.8	69.5	8.7	11 232.8
2009	1 386.8	204.2	1 182.6	622.1	560.6	539.3	1 026.8	58.3	124.4	39.0	86.1	-0.7	10 950.4
2010	1 728.7	272.5	1 456.2	643.2	813.0	465.2	1 063.1	55.8	126.8	43.7	84.1	-1.0	11 621.8
2011	1 809.8	281.1	1 528.7	779.1	749.6	461.7	1 103.7	60.0	128.1	48.5	86.8	-7.3	12 205.3
2012	1 997.4	334.9	1 662.5	948.7	713.9	503.7	1 136.1	58.0	98.8	40.4	69.0	-10.6	12 933.2
2013	2 010.7	362.8	1 647.9	1 009.0	638.9	465.9	1 188.7	59.7	110.3	38.4	84.7	-12.8	13 271.4
2014	2 120.2	407.3	1 712.9	1 096.1	616.8	516.1	1 240.8	58.1	132.9	42.9	98.1	-8.1	13 937.8
2015	2 061.5	396.6	1 664.9	1 164.9	500.0	586.8	1 277.1	57.3	156.7	50.3	109.5	-3.0	14 416.8
2016	2 011.5	377.6	1 633.9	1 175.9	458.0	560.0	1 312.8	61.8	168.2	59.7	103.8	4.8	14 637.0
2017	2 005.9	319.4	1 686.5	1 239.6	446.9	608.0	1 364.5	61.1	145.4	48.1	97.1	0.2	15 262.6
2018	2 074.6	219.8	1 854.9	1 312.6	542.3	672.6	1 441.8	64.4	153.7	53.2	101.4	-0.9	16 021.3
2016													
1st quarter	2 022.2	373.3	1 649.0	1 168.9	480.1	573.9	1 297.1	60.7	166.2	59.5	103.9	2.8	14 524.7
2nd quarter	1 998.1	373.8	1 624.3	1 166.7	457.6	556.0	1 301.2	62.4	172.0	61.3	108.7	2.0	14 544.7
3rd quarter	2 013.0	391.7	1 621.3	1 183.3	438.0	552.6	1 322.3	63.1	154.7	60.6	90.7	3.4	14 671.9
4th quarter	2 012.6	371.5	1 641.0	1 184.8	456.2	557.6	1 330.5	61.0	179.9	57.2	111.8	10.9	14 806.7
2017													
1st quarter	1 995.4	322.8	1 672.5	1 219.5	453.1	607.7	1 340.8	59.9	163.6	51.2	113.5	-1.1	15 032.8
2nd quarter	2 008.0	314.1	1 693.9	1 246.8	447.1	603.6	1 355.1	58.7	149.5	47.6	90.7	11.2	15 168.0
3rd quarter	2 019.0	335.3	1 683.7	1 242.7	441.0	596.3	1 371.4	63.2	130.5	46.2	91.4	-7.1	15 317.0
4th quarter	2 001.4	305.4	1 696.0	1 249.5	446.5	624.5	1 390.6	62.8	138.0	47.3	92.8	-2.1	15 532.6
2018													
1st quarter	2 052.3	207.6	1 844.7	1 266.3	578.4	666.4	1 415.8	59.3	144.0	50.6	103.6	-10.2	15 816.0
2nd quarter	2 056.4	222.6	1 833.8	1 291.9	541.9	678.6	1 433.9	58.6	145.4	53.0	96.7	-4.4	15 938.5
3rd quarter	2 104.2	230.3	1 873.9	1 329.7	544.2	677.0	1 442.3	58.4	166.1	54.4	115.0	-3.2	16 130.7
4th quarter	2 085.6	218.5	1 867.1	1 362.5	504.6	668.4	1 475.2	81.3	159.2	54.7	90.4	14.1	16 199.9

[1] Includes net transfer payments to the rest of the world, not shown separately.

. . . = Not available.

Table 1-13B. National Income by Type of Income: Historical Data

(Billions of dollars, quarterly data are at seasonally adjusted annual rates.)

NIPA Tables 1.7.5, 1.12

Year and quarter	National income, total	Compensation of employees							Proprietors' income with IVA and CCAdj			Rental income of persons with CCAdj
		Total	Wages and salaries			Supplements to wages and salaries			Total	Farm	Nonfarm	
			Total	Government	Other	Total	Employer contributions for:					
							Employee pension and insurance funds	Government social insurance				
1929	94.2	51.4	50.5	5.0	45.5	0.9	0.9	0.0	14.0	5.7	8.4	6.1
1930	83.1	47.2	46.2	5.2	41.0	1.0	0.9	0.0	10.9	3.9	7.0	5.4
1931	67.7	40.1	39.2	5.3	33.9	0.9	0.9	0.0	8.3	3.0	5.3	4.4
1932	51.3	31.3	30.5	5.0	25.5	0.8	0.8	0.0	5.0	1.8	3.3	3.6
1933	48.9	29.8	29.0	5.2	23.9	0.8	0.8	0.0	5.3	2.2	3.0	2.9
1934	58.3	34.6	33.7	6.1	27.6	0.8	0.8	0.0	7.0	2.6	4.4	2.5
1935	66.3	37.7	36.7	6.5	30.2	0.9	0.9	0.0	10.1	4.9	5.2	2.6
1936	75.1	43.3	42.0	7.9	34.1	1.3	1.1	0.2	10.4	3.9	6.4	2.7
1937	83.7	48.3	46.1	7.5	38.6	2.2	1.1	1.0	12.5	5.6	6.9	3.0
1938	77.0	45.4	43.0	8.3	34.8	2.4	1.2	1.2	10.6	4.0	6.6	3.5
1939	82.4	48.6	46.0	8.2	37.7	2.6	1.3	1.3	11.1	4.0	7.1	3.7
1940	91.5	52.7	49.9	8.5	41.4	2.9	1.5	1.4	12.2	4.1	8.2	3.8
1941	117.3	66.2	62.1	10.2	51.9	4.1	2.4	1.7	16.7	6.0	10.6	4.4
1942	152.4	88.0	82.1	16.0	66.1	5.9	3.9	2.0	23.3	9.7	13.7	5.5
1943	187.2	112.7	105.8	26.6	79.2	6.9	4.6	2.3	28.2	11.5	16.7	6.0
1944	200.9	124.3	116.7	33.0	83.8	7.6	5.0	2.5	29.3	11.4	17.9	6.3
1945	201.3	126.3	117.5	34.9	82.6	8.8	5.3	3.5	30.8	11.8	19.0	6.6
1946	201.3	122.5	112.0	20.7	91.3	10.5	5.4	5.1	35.7	14.2	21.5	6.9
1947	218.7	132.4	123.1	17.5	105.6	9.3	5.3	3.9	34.6	14.3	20.2	6.9
1948	244.8	144.3	135.6	19.0	116.5	8.8	5.8	3.0	39.3	16.7	22.6	7.5
1949	239.7	144.3	134.7	20.8	113.9	9.6	6.3	3.3	34.7	12.0	22.7	7.8
1947												
1st quarter	213.0	129.0	118.9	17.1	101.7	10.1	5.5	4.7	36.2	16.0	20.2	6.7
2nd quarter	215.6	130.8	121.3	17.3	103.9	9.5	5.2	4.3	33.3	12.4	20.9	6.8
3rd quarter	219.2	132.8	124.0	17.6	106.4	8.8	5.2	3.5	33.9	14.1	19.8	7.0
4th quarter	227.2	136.9	128.3	18.0	110.3	8.6	5.4	3.2	35.0	14.9	20.0	7.2
1948												
1st quarter	235.8	140.4	131.7	18.3	113.4	8.7	5.5	3.2	36.4	14.5	21.9	7.3
2nd quarter	244.0	142.1	133.4	18.6	114.7	8.7	5.7	3.0	40.2	18.0	22.3	7.5
3rd quarter	248.8	146.8	138.1	19.3	118.8	8.8	5.8	2.9	40.6	17.9	22.7	7.5
4th quarter	250.7	147.9	139.1	19.9	119.2	8.8	6.0	2.9	39.9	16.4	23.5	7.6
1949												
1st quarter	243.3	146.3	136.7	20.4	116.4	9.5	6.0	3.5	35.4	12.7	22.7	7.5
2nd quarter	239.7	144.7	135.0	20.7	114.3	9.7	6.1	3.5	34.9	12.1	22.8	7.7
3rd quarter	239.9	143.6	134.0	21.1	112.9	9.6	6.3	3.3	34.2	11.6	22.6	7.9
4th quarter	236.0	142.8	133.2	21.2	112.0	9.6	6.5	3.0	34.2	11.5	22.7	8.1
1950												
1st quarter	246.3	147.0	136.9	21.2	115.6	10.1	6.9	3.3	36.2	12.2	24.0	8.5
2nd quarter	258.5	153.6	143.0	21.8	121.2	10.6	7.3	3.3	36.8	12.2	24.6	8.7
3rd quarter	275.6	162.3	151.0	23.0	128.1	11.3	7.9	3.4	38.5	13.1	25.4	8.8
4th quarter	286.1	170.2	158.1	24.5	133.5	12.1	8.5	3.6	38.5	13.9	24.6	9.1
1951												
1st quarter	297.8	178.6	165.3	26.7	138.6	13.2	9.2	4.0	40.9	15.0	25.9	9.3
2nd quarter	305.2	184.9	171.0	28.5	142.5	13.9	9.8	4.1	42.5	15.4	27.1	9.6
3rd quarter	310.6	187.9	173.5	30.2	143.3	14.4	10.3	4.1	43.2	15.1	28.1	9.9
4th quarter	317.0	191.5	176.6	31.5	145.1	14.9	10.7	4.1	43.7	15.7	28.0	10.1
1952												
1st quarter	318.5	196.1	181.0	32.4	148.6	15.1	11.0	4.1	41.7	13.7	28.0	10.4
2nd quarter	320.4	197.4	182.1	33.1	149.0	15.3	11.2	4.1	43.0	14.5	28.4	10.7
3rd quarter	327.0	201.5	186.0	33.8	152.2	15.6	11.4	4.1	44.7	16.0	28.8	10.9
4th quarter	338.5	209.3	193.5	34.2	159.3	15.9	11.6	4.2	42.8	13.0	29.7	11.3
1953												
1st quarter	344.5	213.1	197.1	34.3	162.7	16.1	11.9	4.2	42.9	13.0	29.9	11.6
2nd quarter	347.3	216.2	200.0	34.4	165.7	16.2	12.0	4.2	42.3	12.4	29.9	11.9
3rd quarter	345.6	216.5	200.1	34.3	165.9	16.4	12.1	4.3	41.2	11.7	29.5	12.2
4th quarter	338.0	215.1	198.7	34.4	164.3	16.4	12.2	4.2	41.6	11.5	30.0	12.5
1954												
1st quarter	339.6	213.1	196.4	34.4	162.0	16.7	12.1	4.6	42.7	12.9	29.8	12.8
2nd quarter	339.9	212.7	196.0	34.7	161.2	16.7	12.1	4.6	42.1	11.7	30.4	13.0
3rd quarter	343.3	213.2	196.3	35.1	161.2	16.9	12.3	4.6	42.2	11.8	30.5	13.1
4th quarter	352.2	217.5	200.4	35.4	165.0	17.2	12.5	4.6	42.0	10.7	31.3	13.3
1955												
1st quarter	364.6	221.9	204.1	35.5	168.6	17.8	12.8	5.0	43.7	10.9	32.7	13.3
2nd quarter	374.0	228.1	209.9	36.2	173.6	18.2	13.1	5.1	44.3	11.3	33.0	13.4
3rd quarter	381.4	233.9	215.2	37.4	177.8	18.7	13.4	5.2	44.6	10.7	33.9	13.4
4th quarter	387.7	238.4	219.5	37.3	182.2	18.9	13.7	5.3	44.5	10.0	34.5	13.5
1956												
1st quarter	391.2	242.8	223.3	37.8	185.4	19.5	14.0	5.6	44.9	10.2	34.7	13.6
2nd quarter	397.0	247.5	227.5	38.5	189.0	20.0	14.3	5.7	45.3	10.4	34.9	13.6
3rd quarter	401.9	250.4	230.0	39.2	190.8	20.5	14.8	5.7	46.4	10.9	35.5	13.7
4th quarter	410.1	256.4	235.4	39.7	195.7	21.0	15.2	5.8	46.8	10.9	35.8	13.8
1957												
1st quarter	416.8	260.2	238.3	40.2	198.0	22.0	15.6	6.4	46.9	9.9	37.0	13.9
2nd quarter	418.9	262.1	239.6	40.7	198.9	22.4	16.0	6.4	47.7	10.5	37.2	14.0
3rd quarter	422.6	264.7	241.8	41.5	200.3	22.9	16.5	6.4	48.8	11.0	37.8	14.1
4th quarter	415.8	263.3	240.1	41.6	198.5	23.2	16.9	6.3	47.7	10.7	37.0	14.3

Table 1-13B. National Income by Type of Income: Historical Data—*Continued*

(Billions of dollars, quarterly data are at seasonally adjusted annual rates.)　　　　NIPA Tables 1.7.5, 1.12

Year and quarter	Total	Taxes on corporate income	Total	Net dividends	Undistributed corporate profits	Net interest and miscellaneous payments	Taxes on production and imports	Less: Subsidies	Total 1	To persons	To government	Current surplus of government enterprises	Addendum: Net national factor income
1929	10.8	1.4	9.5	5.8	3.7	4.6	6.8	0.0	0.5	0.4	0.1	...	86.9
1930	7.5	0.8	6.7	5.5	1.2	4.8	7.0	0.1	0.5	0.4	0.1	...	75.7
1931	3.0	0.5	2.5	4.1	-1.6	4.8	6.7	0.1	0.5	0.4	0.1	...	60.6
1932	-0.2	0.4	-0.6	2.5	-3.1	4.5	6.6	0.1	0.6	0.5	0.1	...	44.2
1933	-0.2	0.5	-0.7	2.0	-2.7	4.0	6.9	0.2	0.5	0.4	0.1	...	41.7
1934	2.5	0.7	1.8	2.6	-0.8	4.0	7.6	0.5	0.5	0.4	0.1	...	50.6
1935	4.0	1.0	3.1	2.8	0.2	4.1	8.0	0.6	0.5	0.4	0.1	...	58.4
1936	6.2	1.4	4.8	4.5	0.3	3.8	8.5	0.3	0.5	0.4	0.1	...	66.4
1937	7.1	1.5	5.6	4.7	0.9	3.7	8.9	0.3	0.5	0.4	0.1	...	74.5
1938	5.0	1.0	4.0	3.2	0.8	3.6	8.9	0.5	0.4	0.3	0.2	...	68.1
1939	6.6	1.4	5.2	3.8	1.4	3.6	9.1	0.8	0.4	0.3	0.2	...	73.6
1940	9.9	2.8	7.0	4.0	3.0	3.3	9.8	0.7	0.5	0.3	0.2	...	82.0
1941	15.7	7.6	8.1	4.4	3.7	3.3	11.1	0.5	0.5	0.4	0.2	...	106.2
1942	20.8	11.4	9.4	4.3	5.1	3.2	11.5	0.5	0.5	0.3	0.2	...	140.8
1943	24.9	14.1	10.8	4.4	6.4	2.9	12.4	0.6	0.6	0.4	0.3	...	174.7
1944	25.0	12.9	12.0	4.6	7.4	2.4	13.7	1.0	0.8	0.4	0.4	...	187.4
1945	20.5	10.7	9.8	4.6	5.2	2.3	15.1	1.1	0.9	0.5	0.4	...	186.5
1946	18.2	9.1	9.1	5.6	3.6	1.8	16.8	1.4	0.7	0.4	0.3	...	185.1
1947	24.2	11.2	13.0	6.4	6.6	2.3	18.1	0.4	0.7	0.4	0.3	...	200.3
1948	31.4	12.3	19.1	7.2	11.9	2.5	19.7	0.5	0.7	0.4	0.3	...	225.0
1949	29.1	10.0	19.0	7.4	11.6	2.8	20.9	0.5	0.7	0.4	0.3	...	218.6
1947													
1st quarter	21.5	11.5	10.0	6.1	3.9	2.0	17.7	0.5	0.5	0.2	0.3	...	195.3
2nd quarter	24.5	10.9	13.6	6.4	7.2	2.3	17.7	0.4	0.7	0.4	0.3	...	197.6
3rd quarter	24.6	10.7	13.9	6.6	7.3	2.5	18.0	0.4	0.8	0.5	0.3	...	200.7
4th quarter	26.1	11.7	14.5	6.6	7.0	2.5	19.0	0.3	0.9	0.5	0.3	...	207.0
1948													
1st quarter	30.0	12.0	18.0	7.2	10.8	2.4	19.0	0.4	0.7	0.4	0.4	...	216.5
2nd quarter	31.8	12.6	19.1	6.8	12.3	2.4	19.7	0.3	0.7	0.4	0.3	...	224.0
3rd quarter	31.2	12.4	18.8	7.3	11.5	2.5	20.0	0.6	0.7	0.4	0.3	...	228.7
4th quarter	32.7	12.0	20.7	7.6	13.1	2.5	20.3	0.9	0.7	0.4	0.3	...	230.6
1949													
1st quarter	31.0	10.8	20.2	7.4	12.7	2.7	20.4	0.6	0.7	0.3	0.3	...	222.9
2nd quarter	28.7	9.5	19.1	7.4	11.7	2.7	20.8	0.4	0.7	0.3	0.3	...	218.7
3rd quarter	29.9	9.9	20.0	7.3	12.7	2.8	21.3	0.5	0.7	0.4	0.3	...	218.5
4th quarter	26.6	9.7	16.9	7.6	9.3	2.8	21.2	0.5	0.7	0.4	0.3	...	214.5
1950													
1st quarter	30.1	13.4	16.6	8.5	8.1	3.0	21.6	0.7	0.7	0.4	0.3	...	224.8
2nd quarter	33.9	16.1	17.8	8.6	9.2	3.1	22.5	0.8	0.8	0.5	0.3	...	236.0
3rd quarter	38.4	19.7	18.7	9.4	9.4	3.1	24.4	0.8	0.9	0.6	0.3	...	251.1
4th quarter	41.8	21.6	20.2	9.7	10.5	3.1	23.4	1.1	0.9	0.7	0.3	...	262.8
1951													
1st quarter	40.7	26.0	14.7	8.6	6.1	3.3	25.1	1.2	1.1	0.8	0.3	...	272.8
2nd quarter	40.6	22.0	18.6	8.9	9.7	3.5	24.1	1.1	1.2	0.9	0.3	...	281.1
3rd quarter	41.1	19.9	21.2	8.8	12.4	3.7	24.5	0.9	1.2	0.9	0.3	...	285.8
4th quarter	42.4	21.3	21.1	9.0	12.2	3.8	25.3	1.0	1.2	0.9	0.3	...	291.5
1952													
1st quarter	40.0	19.5	20.5	8.4	12.1	3.8	26.1	0.8	1.2	0.9	0.3	...	292.0
2nd quarter	38.1	18.4	19.8	9.0	10.8	3.9	26.9	0.8	1.2	0.9	0.3	...	293.1
3rd quarter	38.1	18.3	19.8	8.9	10.9	4.0	27.3	0.8	1.2	0.9	0.3	...	299.3
4th quarter	42.4	20.2	22.2	9.1	13.1	4.1	28.1	0.7	1.2	0.9	0.3	...	309.9
1953													
1st quarter	43.3	21.3	22.1	8.8	13.3	4.3	28.7	0.7	1.2	0.9	0.3	...	315.3
2nd quarter	42.5	21.4	21.1	9.5	11.6	4.4	29.2	0.5	1.2	0.9	0.3	...	317.4
3rd quarter	41.3	20.9	20.4	9.4	11.0	4.6	29.3	0.7	1.2	0.8	0.3	...	315.8
4th quarter	34.0	16.2	17.8	9.2	8.5	4.8	29.2	0.2	1.1	0.8	0.3	...	307.9
1954													
1st quarter	36.5	16.1	20.4	9.7	10.7	5.0	28.7	0.1	1.0	0.6	0.3	...	310.1
2nd quarter	37.8	16.6	21.2	9.2	12.1	5.3	28.8	0.7	0.9	0.5	0.3	...	310.9
3rd quarter	39.9	17.7	22.2	9.6	12.7	5.5	28.7	0.3	0.9	0.5	0.4	...	314.0
4th quarter	43.6	19.0	24.6	9.7	14.8	5.7	29.3	0.2	0.9	0.6	0.4	...	322.1
1955													
1st quarter	48.8	21.1	27.7	10.3	17.4	5.8	30.2	0.3	1.1	0.7	0.4	...	333.5
2nd quarter	50.1	21.4	28.7	10.4	18.3	6.0	31.2	0.3	1.2	0.9	0.4	...	341.8
3rd quarter	50.2	21.9	28.3	11.0	17.2	5.9	31.9	0.0	1.4	1.0	0.4	...	348.1
4th quarter	51.6	22.8	28.8	11.2	17.6	5.9	32.5	0.2	1.5	1.1	0.4	...	354.0
1956													
1st quarter	49.4	21.7	27.7	11.5	16.2	6.2	33.0	0.3	1.6	1.2	0.4	...	356.9
2nd quarter	49.6	22.1	27.5	11.6	16.0	6.4	33.6	0.6	1.6	1.2	0.4	...	362.3
3rd quarter	49.3	20.7	28.6	11.6	17.0	6.7	34.6	0.9	1.7	1.3	0.4	...	366.5
4th quarter	50.2	21.9	28.4	12.0	16.3	6.7	35.7	1.2	1.7	1.3	0.4	...	373.8
1957													
1st quarter	51.7	22.3	29.3	12.1	17.2	7.5	36.1	1.3	1.8	1.4	0.4	...	380.2
2nd quarter	50.3	21.4	28.9	12.3	16.6	7.7	36.6	1.2	1.8	1.4	0.4	...	381.7
3rd quarter	49.3	20.9	28.5	12.5	16.0	7.8	37.0	1.1	1.8	1.4	0.4	...	384.8
4th quarter	45.3	19.0	26.3	12.3	14.0	7.6	36.7	1.0	1.8	1.4	0.4	...	378.2

1Includes net transfer payments to the rest of the world, not shown separately.
. . . = Not available.

Table 1-13B. National Income by Type of Income: Historical Data—*Continued*

(Billions of dollars, quarterly data are at seasonally adjusted annual rates.)

NIPA Tables 1.7.5, 1.12

Year and quarter	National income, total	Compensation of employees Total	Wages and salaries Total	Government	Other	Supplements to wages and salaries Total	Employer contributions for: Employee pension and insurance funds	Government social insurance	Proprietors' income with IVA and CCAdj Total	Farm	Nonfarm	Rental income of persons with CCAdj
1958												
1st quarter	409.7	259.8	236.7	42.0	194.7	23.0	16.7	6.3	50.3	13.3	36.9	14.6
2nd quarter	411.6	259.4	236.3	43.1	193.2	23.2	16.9	6.3	50.3	13.0	37.3	14.8
3rd quarter	424.5	267.2	243.8	46.1	197.7	23.4	17.0	6.4	50.1	12.1	38.0	14.9
4th quarter	437.2	272.3	248.4	45.3	203.2	23.8	17.4	6.4	49.9	11.2	38.7	15.0
1959												
1st quarter	448.3	279.7	254.0	45.5	208.6	25.7	18.0	7.8	50.1	10.9	39.2	15.0
2nd quarter	462.4	286.5	260.6	45.8	214.7	25.9	18.0	7.9	50.3	9.9	40.5	15.4
3rd quarter	459.6	286.9	260.9	46.2	214.7	26.0	18.2	7.9	50.4	9.4	41.0	15.8
4th quarter	464.8	290.2	263.9	46.7	217.2	26.3	18.4	7.9	50.6	10.1	40.5	16.0
1960												
1st quarter	478.1	298.7	270.7	47.7	223.0	28.0	18.7	9.3	49.7	9.6	40.1	16.2
2nd quarter	478.8	301.7	273.4	48.6	224.8	28.3	19.1	9.3	50.7	10.6	40.1	16.4
3rd quarter	480.2	302.6	274.0	49.9	224.1	28.6	19.3	9.3	50.9	11.0	39.9	16.5
4th quarter	478.5	302.1	273.3	50.5	222.8	28.8	19.6	9.2	51.0	11.3	39.7	16.7
1961												
1st quarter	480.3	303.1	273.8	51.1	222.7	29.3	19.9	9.5	52.3	11.5	40.8	16.9
2nd quarter	489.9	307.3	277.6	51.8	225.8	29.7	20.1	9.6	52.6	10.9	41.8	17.1
3rd quarter	499.7	312.3	282.3	52.8	229.5	30.1	20.4	9.6	53.3	11.0	42.3	17.3
4th quarter	514.2	318.9	288.4	54.2	234.2	30.5	20.7	9.8	54.5	11.5	43.0	17.5
1962												
1st quarter	523.4	325.5	293.3	55.3	238.0	32.2	21.2	11.0	55.3	11.9	43.5	17.7
2nd quarter	530.5	331.4	298.7	56.0	242.8	32.7	21.5	11.2	55.0	11.1	43.9	18.0
3rd quarter	536.9	334.2	301.2	56.5	244.7	33.1	21.8	11.2	54.9	10.7	44.2	18.2
4th quarter	544.9	337.7	304.2	57.6	246.6	33.4	22.1	11.3	55.8	11.3	44.5	18.3
1963												
1st quarter	551.8	342.7	308.0	58.6	249.3	34.7	22.5	12.2	56.0	11.4	44.6	18.5
2nd quarter	560.9	347.6	312.4	59.4	253.0	35.2	22.9	12.3	55.8	10.8	45.0	18.7
3rd quarter	569.4	352.6	316.8	60.2	256.7	35.8	23.3	12.5	56.2	10.7	45.6	18.8
4th quarter	579.4	358.8	322.2	61.9	260.3	36.5	23.9	12.6	57.6	11.1	46.5	18.8
1964												
1st quarter	591.7	365.2	328.2	63.1	265.1	37.0	24.5	12.4	57.9	9.9	48.0	18.8
2nd quarter	601.8	372.6	334.8	64.2	270.7	37.8	25.2	12.6	58.8	9.5	49.3	18.8
3rd quarter	613.2	380.0	341.4	65.7	275.7	38.6	25.9	12.7	59.3	9.3	50.0	18.9
4th quarter	621.4	386.1	346.7	66.8	279.9	39.4	26.6	12.8	60.5	10.4	50.1	18.8
1965												
1st quarter	640.0	392.9	352.8	67.5	285.4	40.1	27.3	12.8	61.9	11.4	50.5	19.1
2nd quarter	651.2	399.9	358.9	68.5	290.3	41.0	28.1	13.0	63.2	12.0	51.2	19.3
3rd quarter	662.4	408.3	366.2	70.5	295.7	42.1	29.0	13.2	64.0	12.2	51.8	19.5
4th quarter	681.8	420.5	377.1	73.2	304.0	43.3	29.9	13.4	65.6	12.5	53.0	19.5
1966												
1st quarter	702.0	433.1	385.8	74.9	310.8	47.3	30.9	16.5	69.2	14.9	54.4	19.8
2nd quarter	711.8	444.4	395.9	76.9	319.0	48.5	31.8	16.7	67.2	12.7	54.5	19.8
3rd quarter	723.1	455.7	406.1	79.8	326.3	49.5	32.5	17.0	67.2	12.3	54.9	20.0
4th quarter	735.5	463.8	413.5	82.0	331.5	50.3	33.1	17.2	68.1	12.2	55.9	20.0
1967												
1st quarter	741.8	470.0	418.8	83.5	335.4	51.2	33.6	17.6	68.9	11.9	57.0	20.3
2nd quarter	748.4	475.6	423.6	85.0	338.6	52.0	34.3	17.8	68.5	11.1	57.4	20.4
3rd quarter	763.8	485.3	432.0	87.1	344.9	53.3	35.1	18.2	70.5	12.0	58.6	20.4
4th quarter	779.7	496.2	441.6	90.4	351.2	54.7	36.1	18.6	70.0	11.5	58.4	20.3
1968												
1st quarter	800.5	510.9	454.2	92.8	361.5	56.7	37.2	19.4	71.7	11.6	60.1	20.1
2nd quarter	821.5	524.0	465.9	95.2	370.7	58.1	38.3	19.8	73.2	11.2	61.9	20.1
3rd quarter	841.0	537.8	478.3	98.7	379.6	59.5	39.4	20.1	75.0	11.8	63.2	20.2
4th quarter	858.0	550.3	489.4	100.1	389.4	60.9	40.4	20.5	75.5	12.0	63.5	20.0
1969												
1st quarter	874.6	562.8	499.1	101.3	397.8	63.7	41.6	22.1	75.6	11.6	64.1	20.2
2nd quarter	890.1	576.7	511.4	103.1	408.3	65.3	42.8	22.5	77.1	12.5	64.6	20.3
3rd quarter	908.1	593.4	526.4	108.2	418.3	67.0	44.0	23.0	78.0	13.1	64.8	20.4
4th quarter	916.3	604.9	536.5	109.8	426.6	68.5	45.1	23.4	77.3	14.1	63.2	20.4
1970												
1st quarter	921.1	615.0	545.1	114.4	430.7	69.9	46.2	23.7	77.1	13.5	63.5	20.4
2nd quarter	933.2	620.2	549.1	116.5	432.6	71.2	47.3	23.9	76.7	12.4	64.3	20.2
3rd quarter	947.7	628.2	555.7	118.3	437.4	72.5	48.5	23.9	78.5	13.2	65.3	20.9
4th quarter	948.0	630.0	556.4	119.7	436.6	73.7	49.8	23.9	78.9	12.5	66.4	21.2
1971												
1st quarter	984.6	647.9	570.5	123.9	446.6	77.4	51.3	26.1	80.3	13.1	67.2	21.1
2nd quarter	1 004.4	659.8	580.4	125.6	454.8	79.3	52.9	26.4	82.8	13.3	69.6	21.7
3rd quarter	1 022.4	670.1	588.8	128.0	460.8	81.3	54.8	26.5	84.6	13.0	71.6	22.0
4th quarter	1 044.7	682.2	598.4	129.6	468.8	83.7	56.9	26.8	87.9	14.2	73.6	22.5
1972												
1st quarter	1 077.9	708.4	618.5	134.2	484.3	89.9	59.3	30.6	87.9	13.1	74.8	23.1
2nd quarter	1 098.2	722.5	630.4	135.7	494.7	92.1	61.1	31.0	91.3	15.4	75.9	20.2
3rd quarter	1 127.9	735.7	642.3	138.4	503.9	93.4	62.3	31.2	95.5	17.2	78.4	23.9
4th quarter	1 174.0	758.8	664.0	143.3	520.7	94.8	63.0	31.8	105.6	22.4	83.3	23.8

Table 1-13B. National Income by Type of Income: Historical Data—*Continued*

(Billions of dollars, quarterly data are at seasonally adjusted annual rates.)

NIPA Tables 1.7.5, 1.12

| Year and quarter | Corporate profits with IVA and CCAdj | | | | | Net interest and miscel- laneous payments | Taxes on production and imports | Less: Subsidies | Business current transfer payments, net | | | Current surplus of govern- ment enterprises | Addendum: Net national factor income |
| | Total | Taxes on corporate income | Profits after tax | | | | | | Total 1 | To persons | To govern- ment | | |
			Total	Net dividends	Undis- tributed corporate profits								
1958													
1st quarter	39.5	16.3	23.1	12.1	11.0	8.1	36.8	1.1	1.8	1.3	0.5	...	372.2
2nd quarter	40.6	16.7	23.9	12.2	11.7	8.7	37.3	1.3	1.7	1.3	0.5	...	373.8
3rd quarter	44.8	18.9	25.9	12.1	13.7	9.5	37.8	1.5	1.7	1.2	0.5	...	386.5
4th quarter	50.6	21.8	28.7	11.8	16.9	10.6	38.9	1.6	1.6	1.2	0.4	...	398.3
1959													
1st quarter	54.3	22.8	31.5	12.8	18.7	8.5	39.8	1.1	1.6	1.3	0.3	...	407.5
2nd quarter	59.1	25.0	34.1	13.2	20.9	9.5	40.3	0.9	1.7	1.3	0.3	...	420.8
3rd quarter	54.2	22.1	32.1	13.8	18.3	9.5	41.7	1.1	1.7	1.3	0.3	...	416.8
4th quarter	54.5	21.4	33.1	14.1	19.0	9.8	42.4	1.1	1.7	1.4	0.3	...	421.2
1960													
1st quarter	58.6	24.2	34.4	14.1	20.3	9.8	43.7	1.0	1.7	1.3	0.4	...	433.1
2nd quarter	54.8	22.1	32.7	14.2	18.5	9.9	44.3	1.3	1.7	1.3	0.4	...	433.5
3rd quarter	53.8	21.1	32.7	14.4	18.2	10.5	44.9	1.0	1.7	1.3	0.4	...	434.3
4th quarter	51.5	20.1	31.4	14.3	17.1	10.9	45.3	1.2	1.8	1.3	0.5	...	432.3
1961													
1st quarter	50.2	20.0	30.3	14.3	16.0	11.3	45.8	1.6	1.9	1.3	0.5	...	433.9
2nd quarter	54.5	21.4	33.1	14.3	18.7	11.9	46.5	2.0	1.9	1.3	0.5	...	443.4
3rd quarter	57.2	22.7	34.5	14.6	19.9	12.2	47.3	2.2	1.9	1.4	0.5	...	452.4
4th quarter	61.8	24.7	37.1	15.2	21.9	12.9	48.4	2.3	2.0	1.4	0.6	...	465.7
1962													
1st quarter	62.6	23.1	39.4	15.3	24.1	12.7	49.4	2.3	2.0	1.5	0.6	...	473.8
2nd quarter	62.5	22.9	39.7	15.8	23.9	13.7	49.9	2.4	2.1	1.5	0.6	...	480.6
3rd quarter	64.2	23.6	40.7	16.0	24.7	14.2	50.9	2.2	2.1	1.6	0.6	...	485.7
4th quarter	66.9	23.6	43.2	16.2	27.0	14.6	51.3	2.2	2.2	1.6	0.6	...	493.2
1963													
1st quarter	67.2	23.7	43.5	16.6	26.9	14.3	52.0	2.0	2.4	1.8	0.6	...	498.7
2nd quarter	70.2	25.4	44.8	16.9	27.9	14.5	52.9	2.2	2.5	1.8	0.7	...	506.8
3rd quarter	71.7	26.1	45.5	17.2	28.3	15.0	53.9	2.3	2.5	1.9	0.7	...	514.4
4th quarter	73.0	26.7	46.2	17.7	28.5	15.5	54.7	2.4	2.6	2.0	0.7	...	523.6
1964													
1st quarter	77.1	26.4	50.7	19.0	31.7	16.2	55.7	2.7	2.7	2.1	0.6	...	535.1
2nd quarter	77.3	26.4	50.9	19.6	31.3	16.7	56.7	2.9	2.8	2.2	0.6	...	544.3
3rd quarter	78.4	27.0	51.5	20.0	31.4	17.5	57.9	2.6	3.2	2.2	1.0	...	554.1
4th quarter	78.0	26.5	51.5	20.5	31.0	17.6	58.8	2.7	3.3	2.3	1.0	...	561.1
1965													
1st quarter	85.8	28.3	57.5	20.4	37.1	18.6	60.2	2.9	3.5	2.3	1.2	...	578.4
2nd quarter	88.0	29.2	58.8	21.2	37.7	19.0	60.3	3.0	3.5	2.3	1.2	...	589.4
3rd quarter	89.3	29.8	59.5	21.9	37.6	19.5	60.5	3.0	3.5	2.2	1.2	...	600.6
4th quarter	94.0	31.7	62.3	22.6	39.7	19.5	61.8	3.1	3.4	2.2	1.3	...	619.0
1966													
1st quarter	97.4	32.7	64.7	22.7	41.9	20.7	61.4	3.6	3.4	2.2	1.3	...	640.2
2nd quarter	96.4	32.6	63.7	22.5	41.3	21.3	62.8	3.9	3.4	2.1	1.3	...	649.1
3rd quarter	94.8	32.2	62.7	22.2	40.4	22.1	63.7	4.1	3.4	2.1	1.3	...	659.8
4th quarter	95.9	31.4	64.5	21.8	42.6	23.2	64.9	4.2	3.4	2.1	1.3	...	671.1
1967													
1st quarter	93.3	30.6	62.7	22.9	39.8	24.0	65.6	4.0	3.5	2.2	1.3	...	676.4
2nd quarter	92.2	30.3	61.8	23.5	38.3	24.8	66.8	3.9	3.6	2.2	1.4	...	681.5
3rd quarter	93.2	30.6	62.6	23.9	38.7	25.2	68.7	3.7	3.7	2.3	1.4	...	694.7
4th quarter	96.8	32.4	64.4	23.2	41.1	25.7	70.6	3.7	3.8	2.4	1.4	...	709.0
1968													
1st quarter	98.1	36.5	61.7	24.9	36.8	26.0	73.4	4.0	3.9	2.5	1.4	...	726.7
2nd quarter	102.1	37.1	65.1	25.8	39.2	26.3	75.4	4.2	4.1	2.7	1.4	...	745.7
3rd quarter	102.6	37.1	65.5	26.5	39.0	27.1	77.7	4.2	4.3	2.9	1.4	...	762.8
4th quarter	104.1	38.0	66.1	26.8	39.3	28.4	79.1	4.2	4.4	3.0	1.4	...	778.3
1969													
1st quarter	103.5	38.7	64.8	26.7	38.1	31.4	80.6	4.3	4.7	3.3	1.5	...	793.5
2nd quarter	100.3	37.5	62.9	27.1	35.8	32.2	82.9	4.5	4.8	3.4	1.5	...	806.6
3rd quarter	97.4	36.1	61.4	27.3	34.0	33.3	85.2	4.7	4.8	3.4	1.5	...	822.6
4th quarter	92.4	35.6	56.8	27.9	29.0	34.1	86.7	4.7	4.8	3.3	1.5	...	829.2
1970													
1st quarter	84.7	31.1	53.6	27.9	25.7	36.2	88.5	4.7	4.6	3.1	1.5	...	833.3
2nd quarter	88.5	31.2	57.3	27.8	29.6	38.5	90.5	4.8	4.5	3.0	1.5	...	844.2
3rd quarter	88.4	32.1	56.2	27.8	28.4	40.9	92.5	4.7	4.2	2.8	1.4	...	856.8
4th quarter	83.5	30.7	52.8	27.7	25.1	42.4	94.1	4.8	4.1	2.8	1.4	...	855.9
1971													
1st quarter	96.6	34.4	62.2	28.5	33.8	43.4	97.7	4.8	4.1	2.7	1.5	...	889.4
2nd quarter	98.7	35.3	63.4	28.2	35.2	44.3	98.9	4.8	4.1	2.7	1.4	...	907.4
3rd quarter	101.4	34.6	66.9	28.5	38.4	44.4	101.7	4.5	4.2	2.7	1.5	...	922.6
4th quarter	105.8	35.0	70.8	28.3	42.5	44.5	103.7	4.6	4.2	2.8	1.5	...	942.8
1972													
1st quarter	111.5	37.1	74.4	29.3	45.0	45.4	104.6	6.1	4.5	2.9	1.6	...	976.3
2nd quarter	113.4	37.5	76.0	29.6	46.3	46.6	106.8	6.2	4.7	3.0	1.7	...	994.0
3rd quarter	118.2	38.8	79.4	30.3	49.1	48.9	108.9	7.2	4.9	3.2	1.7	...	1 022.2
4th quarter	125.7	43.1	82.7	31.0	51.7	51.1	111.5	7.1	5.0	3.3	1.7	...	1 065.0

1Includes net transfer payments to the rest of the world, not shown separately.
. . . = Not available.

Table 1-13B. National Income by Type of Income: Historical Data—Continued

(Billions of dollars, quarterly data are at seasonally adjusted annual rates.)

NIPA Tables 1.7.5, 1.12

Year and quarter	National income, total	Compensation of employees							Proprietors' income with IVA and CCAdj			Rental income of persons with CCAdj
		Total	Wages and salaries			Supplements to wages and salaries			Total	Farm	Nonfarm	
			Total	Government	Other	Total	Employer contributions for:					
							Employee pension and insurance funds	Government social insurance				
1973												
1st quarter	1 211.0	785.4	683.4	145.0	538.3	102.1	63.3	38.7	104.1	21.8	82.4	23.5
2nd quarter	1 235.7	803.2	700.2	147.2	553.0	103.0	63.7	39.3	109.9	27.6	82.4	23.3
3rd quarter	1 264.0	820.4	716.2	149.6	566.6	104.2	64.3	39.9	113.8	29.5	84.3	22.3
4th quarter	1 301.9	841.7	735.3	153.4	582.0	106.4	65.3	41.1	122.2	37.5	84.7	23.3
1974												
1st quarter	1 313.9	858.3	748.2	155.6	592.5	110.2	66.6	43.6	115.7	28.6	87.1	23.5
2nd quarter	1 334.6	878.5	765.3	158.2	607.1	113.2	68.7	44.5	108.6	20.1	88.5	22.9
3rd quarter	1 360.6	900.2	783.1	161.3	621.8	117.1	71.8	45.3	111.0	21.5	89.5	23.3
4th quarter	1 376.5	913.8	792.5	166.9	625.6	121.3	75.9	45.4	113.5	23.7	89.8	23.0
1975												
1st quarter	1 383.4	918.4	791.9	170.5	621.4	126.5	80.7	45.7	112.7	19.7	93.0	22.7
2nd quarter	1 413.8	931.2	800.4	174.5	625.9	130.8	84.8	46.0	114.3	20.1	94.2	22.4
3rd quarter	1 471.7	956.0	821.3	177.8	643.5	134.7	87.7	47.0	120.7	23.8	96.9	22.2
4th quarter	1 515.3	983.4	845.8	182.2	663.6	137.6	89.4	48.2	125.2	24.3	100.8	21.9
1976												
1st quarter	1 565.6	1 014.7	871.2	184.8	686.4	143.5	90.4	53.1	126.4	19.4	107.0	21.6
2nd quarter	1 591.3	1 035.5	889.4	187.3	702.1	146.1	92.2	53.9	128.6	17.0	111.6	20.5
3rd quarter	1 623.6	1 058.4	908.5	189.5	719.0	149.9	95.1	54.9	132.7	16.1	116.6	20.0
4th quarter	1 656.9	1 084.8	929.9	194.1	735.8	154.9	99.2	55.7	136.2	16.3	119.9	19.1
1977												
1st quarter	1 704.1	1 113.3	950.0	196.7	753.3	163.2	104.2	59.1	138.6	16.0	122.6	16.4
2nd quarter	1 770.3	1 150.3	980.9	199.7	781.2	169.4	108.8	60.6	140.5	14.1	126.4	16.3
3rd quarter	1 825.0	1 182.1	1 007.5	203.7	803.8	174.6	112.9	61.7	142.4	11.8	130.6	15.6
4th quarter	1 871.9	1 217.6	1 038.2	210.3	827.9	179.4	116.4	63.0	156.5	22.0	134.5	15.1
1978												
1st quarter	1 910.5	1 250.9	1 063.8	213.6	850.3	187.1	119.4	67.6	158.5	18.5	140.0	15.9
2nd quarter	2 006.3	1 299.1	1 106.0	217.3	888.8	193.1	122.6	70.4	166.1	20.9	145.2	15.6
3rd quarter	2 054.8	1 336.9	1 138.0	222.4	915.6	198.9	126.3	72.6	168.3	20.8	147.6	17.1
4th quarter	2 119.1	1 380.2	1 174.4	226.8	947.6	205.8	130.3	75.4	170.9	19.5	151.4	17.5
1979												
1st quarter	2 172.6	1 422.5	1 208.6	231.3	977.2	214.0	134.7	79.2	180.1	23.8	156.3	17.3
2nd quarter	2 212.5	1 454.7	1 234.4	233.5	1 000.9	220.4	139.0	81.4	179.2	21.8	157.3	15.6
3rd quarter	2 264.9	1 496.7	1 269.4	239.4	1 030.0	227.3	143.5	83.9	180.7	22.3	158.5	15.4
4th quarter	2 311.3	1 534.9	1 300.9	244.3	1 056.6	234.0	148.1	85.9	177.8	20.8	157.0	16.1
1980												
1st quarter	2 353.0	1 573.6	1 333.8	250.2	1 083.6	239.8	152.9	86.9	167.6	13.3	154.3	15.8
2nd quarter	2 356.2	1 599.2	1 353.9	260.6	1 093.3	245.3	157.7	87.6	160.3	3.1	157.2	17.1
3rd quarter	2 422.8	1 628.6	1 377.3	264.2	1 113.1	251.3	162.2	89.1	173.4	11.7	161.7	19.9
4th quarter	2 542.6	1 687.6	1 428.7	270.8	1 157.9	258.9	166.7	92.2	185.0	18.7	166.3	23.2
1981												
1st quarter	2 630.5	1 739.6	1 467.7	277.3	1 190.3	271.9	171.0	100.9	188.2	17.6	170.6	24.0
2nd quarter	2 676.3	1 774.8	1 496.7	282.6	1 214.1	278.1	175.3	102.7	176.4	18.2	158.2	23.3
3rd quarter	2 769.4	1 815.1	1 530.6	288.6	1 241.9	284.5	179.7	104.8	182.3	23.3	158.9	23.1
4th quarter	2 782.5	1 840.7	1 550.6	294.8	1 255.8	290.1	184.1	106.0	171.8	17.0	154.9	24.7
1982												
1st quarter	2 786.1	1 864.7	1 567.5	300.8	1 266.8	297.1	188.5	108.6	165.2	14.4	150.8	23.5
2nd quarter	2 833.6	1 884.4	1 581.7	305.4	1 276.3	302.6	193.2	109.5	169.4	12.8	156.6	21.1
3rd quarter	2 852.9	1 904.8	1 596.4	309.5	1 286.9	308.3	198.0	110.3	170.7	11.9	158.8	24.4
4th quarter	2 865.4	1 918.1	1 604.4	314.5	1 289.9	313.8	203.1	110.7	179.5	14.2	165.3	26.3
1983												
1st quarter	2 924.4	1 947.2	1 622.4	318.7	1 303.7	324.8	208.4	116.4	183.3	12.9	170.4	25.8
2nd quarter	3 004.2	1 986.3	1 654.7	322.6	1 332.1	331.6	213.2	118.4	182.6	7.9	174.8	24.7
3rd quarter	3 085.1	2 029.6	1 691.4	326.7	1 364.7	338.2	217.5	120.7	183.8	0.2	183.6	23.7
4th quarter	3 192.4	2 086.8	1 741.5	331.2	1 410.4	345.3	221.3	124.0	195.3	3.8	191.5	23.5
1984												
1st quarter	3 320.8	2 143.8	1 784.0	337.8	1 446.2	359.8	224.7	135.1	221.9	19.5	202.4	23.2
2nd quarter	3 408.8	2 195.2	1 828.1	344.8	1 483.4	367.0	228.9	138.1	231.8	21.3	210.5	21.7
3rd quarter	3 474.3	2 241.6	1 867.1	351.9	1 515.2	374.6	234.0	140.5	232.4	20.4	211.9	24.4
4th quarter	3 531.6	2 282.9	1 900.5	358.1	1 542.4	382.4	240.0	142.4	226.9	22.5	204.4	29.3
1985												
1st quarter	3 593.1	2 323.5	1 932.1	363.9	1 568.2	391.4	247.0	144.4	240.7	22.8	217.8	26.9
2nd quarter	3 641.2	2 363.6	1 963.6	370.4	1 593.2	400.0	253.8	146.2	238.5	20.4	218.0	25.3
3rd quarter	3 702.9	2 406.4	1 997.4	378.1	1 619.3	409.0	260.4	148.6	240.7	19.3	221.5	27.1
4th quarter	3 742.5	2 455.7	2 037.0	383.0	1 654.0	418.7	266.9	151.8	244.3	21.4	223.0	25.6
1986												
1st quarter	3 793.2	2 491.1	2 063.5	387.9	1 675.6	427.6	273.1	154.5	245.6	18.3	227.4	22.0
2nd quarter	3 804.1	2 517.2	2 081.8	393.6	1 688.3	435.4	279.1	156.3	251.6	19.6	232.0	21.1
3rd quarter	3 842.3	2 555.2	2 111.5	399.4	1 712.0	443.8	284.9	158.9	264.4	26.6	237.8	16.6
4th quarter	3 885.3	2 604.7	2 152.5	407.9	1 744.6	452.2	290.4	161.8	264.2	26.6	237.6	13.4
1987												
1st quarter	3 953.9	2 647.8	2 192.7	413.5	1 779.1	455.1	292.4	162.7	275.4	27.3	248.2	14.5
2nd quarter	4 053.5	2 692.0	2 229.9	419.9	1 810.0	462.1	297.5	164.6	282.3	29.0	253.3	14.1
3rd quarter	4 146.6	2 740.3	2 270.8	425.1	1 845.7	469.5	302.4	167.1	289.5	28.8	260.7	17.5
4th quarter	4 239.9	2 809.6	2 331.7	433.9	1 897.8	477.8	307.2	170.6	298.6	30.4	268.2	20.3

Table 1-13B. National Income by Type of Income: Historical Data—*Continued*

(Billions of dollars, quarterly data are at seasonally adjusted annual rates.)

NIPA Tables 1.7.5, 1.12

Year and quarter	Corporate profits with IVA and CCAdj					Net interest and miscellaneous payments	Taxes on production and imports	Less: Subsidies	Business current transfer payments, net			Current surplus of government enterprises	Addendum: Net national factor income
	Total	Taxes on corporate income	Profits after tax						Total [1]	To persons	To government		
			Total	Net dividends	Undistributed corporate profits								
1973													
1st quarter	133.5	46.0	87.5	32.1	55.4	51.6	114.6	5.9	5.6	3.6	1.9	. . .	1 098.1
2nd quarter	131.2	46.0	85.2	33.5	51.7	53.4	116.2	5.7	5.9	3.9	2.0	. . .	1 121.1
3rd quarter	133.0	44.0	89.0	35.0	54.0	57.3	118.4	4.7	5.6	4.1	1.6	. . .	1 146.8
4th quarter	135.8	46.5	89.3	36.4	52.9	60.5	119.7	4.6	5.8	4.2	1.6	. . .	1 183.5
1974													
1st quarter	130.3	44.6	85.7	37.7	48.0	64.9	120.8	3.6	6.3	4.3	2.0	. . .	1 192.8
2nd quarter	128.9	46.7	82.3	38.8	43.5	69.8	124.1	2.9	6.6	4.5	2.1	. . .	1 208.8
3rd quarter	124.3	51.5	72.8	39.4	33.4	73.4	127.1	3.2	7.0	4.9	2.1	. . .	1 232.2
4th quarter	119.3	46.2	73.2	39.2	33.9	78.5	127.7	3.6	7.3	5.2	2.1	. . .	1 248.1
1975													
1st quarter	117.8	38.3	79.4	38.4	41.0	82.6	128.8	4.2	8.4	6.1	2.2	. . .	1 254.1
2nd quarter	128.9	41.4	87.5	38.0	49.5	83.1	133.0	4.3	9.1	6.9	2.2	. . .	1 279.9
3rd quarter	149.8	52.0	97.7	38.0	59.7	84.3	138.2	4.6	9.3	7.1	2.2	. . .	1 333.0
4th quarter	159.0	53.3	105.7	38.9	66.8	84.7	141.1	4.9	9.4	7.2	2.2	. . .	1 374.1
1976													
1st quarter	175.1	60.8	114.3	41.9	72.3	84.1	141.7	5.1	9.5	7.0	2.5	. . .	1 422.0
2nd quarter	173.2	59.4	113.8	43.9	69.9	86.6	144.9	4.8	9.4	7.0	2.4	. . .	1 444.4
3rd quarter	174.7	59.0	115.8	45.8	69.9	88.4	147.7	5.1	9.1	6.7	2.4	. . .	1 474.3
4th quarter	174.3	58.5	115.9	47.9	68.0	90.4	151.3	5.5	8.6	6.1	2.5	. . .	1 504.9
1977													
1st quarter	184.1	62.9	121.2	48.6	72.6	97.1	154.8	5.8	8.2	5.5	2.7	. . .	1 549.6
2nd quarter	204.9	68.4	130.5	49.9	80.6	101.0	150.0	5.0	7.8	5.0	2.8	. . .	1 613.0
3rd quarter	219.6	70.8	148.8	51.5	97.3	105.3	161.5	6.4	8.0	4.9	3.2	. . .	1 665.1
4th quarter	214.5	71.8	142.6	52.8	89.8	109.3	164.3	10.3	8.3	5.0	3.3	. . .	1 713.0
1978													
1st quarter	209.9	66.2	143.7	54.6	89.2	110.8	166.9	8.7	9.4	5.6	3.7	. . .	1 746.1
2nd quarter	239.7	80.0	159.7	56.1	103.7	113.7	173.1	8.4	10.0	6.2	3.9	. . .	1 834.3
3rd quarter	246.2	80.2	166.0	59.0	107.0	115.4	169.7	8.3	10.7	6.7	4.0	. . .	1 884.0
4th quarter	258.4	85.0	173.4	61.6	111.8	119.4	173.9	10.4	11.4	7.3	4.1	. . .	1 946.4
1979													
1st quarter	249.7	81.9	167.8	63.6	104.2	125.0	176.4	8.4	12.3	7.9	4.4	. . .	1 994.5
2nd quarter	252.1	82.1	170.0	65.8	104.3	130.8	178.5	8.8	12.7	8.2	4.5	. . .	2 032.5
3rd quarter	250.1	81.0	169.0	67.6	101.4	138.9	180.9	8.1	13.0	8.4	4.6	. . .	2 081.8
4th quarter	244.0	77.9	166.1	70.2	95.9	153.1	184.6	8.9	13.1	8.5	4.6	. . .	2 126.0
1980													
1st quarter	237.5	85.4	152.1	73.4	78.7	168.6	189.5	9.2	13.2	8.2	5.0	. . .	2 163.1
2nd quarter	207.0	64.9	142.1	76.6	65.5	176.1	196.9	9.6	13.4	8.3	5.1	. . .	2 159.8
3rd quarter	214.6	72.1	142.5	75.7	66.8	182.9	204.3	10.1	13.7	8.7	5.0	. . .	2 219.4
4th quarter	235.1	79.5	155.6	77.4	78.2	201.3	210.6	10.3	15.7	9.3	6.4	. . .	2 332.2
1981													
1st quarter	244.4	78.1	166.2	81.7	84.5	203.8	230.8	10.6	16.4	10.4	6.1	. . .	2 399.9
2nd quarter	240.8	69.1	171.8	86.3	85.5	223.7	235.5	10.7	16.6	11.0	5.6	. . .	2 439.0
3rd quarter	257.8	71.6	186.2	90.4	95.8	252.3	237.5	11.1	17.1	11.5	5.6	. . .	2 530.6
4th quarter	246.9	62.4	184.5	92.8	91.8	259.5	238.8	13.5	17.5	11.9	5.7	. . .	2 543.7
1982													
1st quarter	221.8	50.6	171.2	92.8	78.4	272.2	237.4	14.0	18.6	12.0	6.6	. . .	2 547.5
2nd quarter	236.2	52.7	183.5	92.2	91.2	282.1	238.3	13.6	19.2	12.3	6.9	. . .	2 593.1
3rd quarter	233.7	53.2	180.5	92.5	87.9	274.6	241.8	13.0	19.5	12.5	7.0	. . .	2 608.1
4th quarter	227.7	48.6	179.1	94.1	85.0	270.3	246.3	19.4	20.0	12.7	7.2	. . .	2 621.9
1983													
1st quarter	244.6	50.2	194.3	95.0	99.4	275.1	250.7	19.9	20.3	12.8	7.5	. . .	2 676.0
2nd quarter	274.7	65.3	209.4	96.0	113.4	277.4	261.2	21.6	20.9	13.2	7.7	. . .	2 745.7
3rd quarter	291.2	74.0	217.3	98.6	118.6	291.2	267.5	22.2	21.9	14.0	7.9	. . .	2 819.7
4th quarter	308.6	76.1	232.4	101.2	131.3	303.6	273.7	21.5	23.6	15.3	8.4	. . .	2 917.8
1984													
1st quarter	336.6	88.4	248.2	104.0	144.1	309.8	281.6	21.2	26.7	17.6	9.1	. . .	3 035.4
2nd quarter	338.2	87.1	251.1	106.6	144.5	327.6	287.7	21.0	28.6	19.2	9.4	. . .	3 114.5
3rd quarter	333.6	74.7	258.8	107.4	151.5	341.7	292.2	20.9	30.1	20.6	9.6	. . .	3 173.7
4th quarter	343.1	75.6	267.5	109.4	158.1	341.9	297.5	21.2	31.2	21.5	9.7	. . .	3 224.1
1985													
1st quarter	347.9	80.5	267.4	113.6	153.8	341.3	301.0	21.1	32.2	21.9	10.3	. . .	3 280.2
2nd quarter	351.5	78.8	272.7	115.2	157.6	337.3	305.7	21.0	38.1	22.1	16.0	. . .	3 316.3
3rd quarter	368.9	84.7	284.1	115.8	168.3	334.1	311.9	21.3	32.8	22.3	10.5	. . .	3 377.2
4th quarter	349.7	82.4	267.3	116.8	150.5	340.0	313.9	22.0	33.1	22.6	10.5	. . .	3 415.3
1986													
1st quarter	340.7	87.8	252.9	122.0	130.9	354.9	317.5	23.1	42.1	23.3	18.7	. . .	3 454.3
2nd quarter	326.6	88.4	238.2	124.3	113.9	355.3	319.5	24.2	34.5	23.5	11.1	. . .	3 471.8
3rd quarter	315.6	90.0	225.6	124.8	100.8	354.0	326.2	25.5	33.2	23.0	10.2	. . .	3 505.8
4th quarter	314.7	101.2	213.4	124.9	88.6	348.1	330.4	26.8	34.0	22.0	12.0	. . .	3 545.1
1987													
1st quarter	325.9	101.5	224.3	126.0	98.3	345.6	336.0	28.3	34.0	21.3	12.7	. . .	3 609.1
2nd quarter	362.9	115.1	247.8	127.5	120.2	351.0	344.4	30.4	34.4	20.6	13.8	. . .	3 702.4
3rd quarter	386.4	119.9	266.5	131.3	135.2	356.9	352.4	31.3	32.5	19.7	12.8	. . .	3 790.5
4th quarter	388.8	114.3	274.5	135.3	139.2	361.4	357.4	31.1	32.3	19.3	13.0	. . .	3 878.7

[1] Includes net transfer payments to the rest of the world, not shown separately.
. . . = Not available.

Table 1-13B. National Income by Type of Income: Historical Data—*Continued*

(Billions of dollars, quarterly data are at seasonally adjusted annual rates.)

Year and quarter	National income, total	Compensation of employees							Proprietors' income with IVA and CCAdj			Rental income of persons with CCAdj
		Total	Wages and salaries			Supplements to wages and salaries			Total	Farm	Nonfarm	
			Total	Government	Other	Total	Employer contributions for:					
							Employee pension and insurance funds	Government social insurance				
1988												
1st quarter	4 329.2	2 857.6	2 366.2	441.7	1 924.4	491.4	312.1	179.2	319.2	33.3	285.9	21.1
2nd quarter	4 421.4	2 923.1	2 421.5	448.5	1 973.0	501.6	318.6	183.0	322.5	27.4	295.1	20.3
3rd quarter	4 514.3	2 975.8	2 462.9	454.5	2 008.3	513.0	326.8	186.2	333.6	28.8	304.8	21.2
4th quarter	4 621.4	3 035.4	2 508.6	463.4	2 045.2	526.9	336.8	190.0	326.8	17.7	309.2	27.4
1989												
1st quarter	4 697.6	3 079.1	2 540.8	469.5	2 071.3	538.3	348.1	190.1	348.1	36.7	311.4	23.1
2nd quarter	4 732.4	3 114.6	2 563.5	476.9	2 086.6	551.1	358.8	192.3	339.5	32.6	306.9	21.2
3rd quarter	4 785.4	3 155.0	2 592.3	485.1	2 107.1	562.8	368.2	194.6	337.2	30.3	306.9	20.0
4th quarter	4 825.0	3 209.7	2 635.7	493.0	2 142.7	574.1	376.4	197.7	339.7	32.5	307.2	21.8
1990												
1st quarter	4 927.3	3 272.8	2 684.4	503.5	2 180.9	588.3	386.3	202.0	345.6	34.5	311.1	23.1
2nd quarter	5 014.2	3 334.7	2 738.8	518.7	2 220.1	595.9	390.3	205.6	351.1	32.7	318.4	26.7
3rd quarter	5 042.9	3 372.5	2 769.3	523.9	2 245.5	603.2	394.7	208.5	358.9	31.8	327.2	31.6
4th quarter	5 070.6	3 381.5	2 772.2	530.1	2 242.1	609.3	399.6	209.6	357.0	29.7	327.3	31.3
1991												
1st quarter	5 093.1	3 387.5	2 771.7	539.6	2 232.1	615.8	405.2	210.6	347.5	26.2	321.3	33.0
2nd quarter	5 133.7	3 427.6	2 800.1	546.0	2 254.1	627.5	413.6	213.9	353.0	27.9	325.2	35.5
3rd quarter	5 185.5	3 469.9	2 828.1	551.6	2 276.5	641.8	425.1	216.8	353.6	24.5	329.1	39.4
4th quarter	5 245.1	3 517.0	2 858.0	557.8	2 300.2	659.0	439.7	219.3	362.6	28.6	333.9	46.4
1992												
1st quarter	5 369.1	3 595.7	2 912.9	564.2	2 348.7	682.8	456.8	226.0	378.7	32.9	345.8	50.0
2nd quarter	5 451.9	3 651.9	2 952.0	569.8	2 382.2	699.9	471.0	229.0	395.2	35.4	359.8	57.3
3rd quarter	5 492.3	3 688.5	2 976.6	575.2	2 401.4	712.0	481.4	230.6	407.9	36.7	371.1	64.3
4th quarter	5 587.6	3 736.9	3 020.8	578.8	2 442.0	716.1	488.1	228.0	418.8	34.1	384.8	70.7
1993												
1st quarter	5 596.5	3 745.6	3 015.9	582.8	2 433.1	729.6	491.7	238.0	418.9	28.6	390.3	80.7
2nd quarter	5 696.8	3 800.2	3 066.5	586.6	2 479.8	733.8	495.6	238.2	429.6	34.9	394.6	86.9
3rd quarter	5 744.2	3 834.1	3 093.6	591.8	2 501.9	740.5	500.2	240.3	423.4	26.1	397.4	92.5
4th quarter	5 883.6	3 889.3	3 141.4	594.8	2 546.5	747.9	505.7	242.2	440.0	35.8	404.2	100.2
1994												
1st quarter	5 949.8	3 913.4	3 152.7	601.1	2 551.6	760.7	511.4	249.3	450.2	41.8	408.4	109.0
2nd quarter	6 073.8	3 991.5	3 223.3	607.2	2 616.2	768.2	515.4	252.8	454.1	36.7	417.5	113.5
3rd quarter	6 163.9	4 031.8	3 258.7	612.5	2 646.2	773.0	517.6	255.5	456.8	32.1	424.6	116.3
4th quarter	6 271.1	4 088.1	3 311.6	617.2	2 694.4	776.5	517.4	259.0	465.5	28.1	437.4	115.9
1995												
1st quarter	6 331.9	4 134.4	3 358.1	622.4	2 735.7	776.3	515.6	260.8	467.4	20.3	447.1	118.3
2nd quarter	6 397.0	4 172.9	3 395.2	627.5	2 767.7	777.7	514.9	262.8	472.0	18.4	453.6	121.6
3rd quarter	6 497.6	4 219.7	3 438.9	630.9	2 808.0	780.8	515.6	265.3	484.5	21.4	463.1	125.7
4th quarter	6 582.7	4 265.3	3 479.9	635.3	2 844.6	785.4	517.8	267.7	501.0	28.1	472.9	134.0
1996												
1st quarter	6 701.0	4 314.5	3 524.4	640.0	2 884.4	790.1	521.0	269.1	523.3	36.9	486.4	139.5
2nd quarter	6 827.4	4 387.5	3 590.0	645.6	2 944.3	797.6	524.4	273.2	547.5	44.2	503.3	141.6
3rd quarter	6 909.7	4 451.2	3 647.2	650.7	2 996.5	804.0	527.3	276.7	547.5	33.5	514.0	143.7
4th quarter	7 044.3	4 514.5	3 704.3	656.0	3 048.2	810.2	530.0	280.3	556.8	34.7	522.1	145.3
1997												
1st quarter	7 159.1	4 592.4	3 776.4	662.3	3 114.1	816.0	532.7	283.3	575.6	37.8	537.8	144.2
2nd quarter	7 273.8	4 660.0	3 835.6	668.0	3 167.7	824.3	537.4	286.9	575.1	28.1	547.0	144.6
3rd quarter	7 422.5	4 738.9	3 903.0	674.1	3 228.9	835.9	544.7	291.2	588.5	32.4	556.1	147.4
4th quarter	7 544.4	4 844.0	3 992.4	683.2	3 309.2	851.6	554.7	297.0	596.6	31.2	565.4	152.1
1998												
1st quarter	7 641.9	4 942.0	4 074.9	689.8	3 385.1	867.1	566.7	300.4	617.4	27.9	589.5	156.8
2nd quarter	7 767.1	5 028.3	4 145.5	697.4	3 448.1	882.8	578.0	304.9	629.6	26.4	603.2	162.6
3rd quarter	7 898.7	5 114.9	4 217.4	705.6	3 511.8	897.4	587.9	309.5	645.0	26.6	618.4	168.0
4th quarter	7 995.3	5 199.4	4 288.7	712.4	3 576.2	910.7	596.6	314.1	668.7	33.0	635.7	173.4
1999												
1st quarter	8 142.8	5 288.0	4 365.7	718.1	3 647.6	922.3	604.1	318.2	681.9	33.6	648.3	176.1
2nd quarter	8 215.0	5 344.3	4 409.7	727.0	3 682.7	934.6	613.9	320.7	690.1	28.1	661.9	179.6
3rd quarter	8 309.7	5 423.3	4 473.0	738.6	3 734.5	950.3	626.3	324.0	699.8	26.2	673.6	179.4
4th quarter	8 494.2	5 555.5	4 583.6	751.4	3 832.3	971.8	641.4	330.5	713.9	24.5	689.4	178.9
2000												
1st quarter	8 730.1	5 750.0	4 752.3	765.2	3 987.1	997.7	657.1	340.6	720.9	27.1	693.9	179.8
2nd quarter	8 824.0	5 792.7	4 780.0	778.6	4 001.4	1 012.7	671.3	341.4	751.3	33.4	717.9	180.2
3rd quarter	8 942.1	5 904.9	4 872.3	783.8	4 088.4	1 032.7	684.6	348.1	760.8	31.9	728.9	182.9
4th quarter	8 994.3	5 944.7	4 898.9	791.5	4 107.4	1 045.8	694.9	350.8	782.4	33.7	748.7	191.0
2001												
1st quarter	9 137.5	6 050.8	4 980.1	801.8	4 178.4	1 070.7	712.9	357.8	809.8	34.0	775.7	197.1
2nd quarter	9 168.7	6 041.7	4 962.2	814.0	4 148.2	1 079.5	721.6	358.0	827.7	32.5	795.2	202.6
3rd quarter	9 121.3	6 025.4	4 936.8	828.9	4 107.9	1 088.6	731.2	357.4	845.6	32.6	813.0	204.8
4th quarter	9 149.4	6 038.6	4 938.6	843.3	4 095.3	1 100.0	741.3	358.7	840.9	29.2	811.7	205.1
2002												
1st quarter	9 257.0	6 061.1	4 946.2	855.1	4 091.1	1 114.9	753.2	361.6	859.7	17.9	841.8	209.1
2nd quarter	9 356.4	6 129.3	4 998.5	868.2	4 130.3	1 130.8	764.8	366.0	863.7	12.8	850.9	211.0
3rd quarter	9 417.0	6 157.8	5 013.0	879.8	4 133.2	1 144.8	777.3	367.5	872.4	21.3	851.2	210.4
4th quarter	9 554.9	6 194.2	5 027.5	888.4	4 139.2	1 166.6	797.6	369.0	883.2	27.7	855.5	213.6

Table 1-13B. National Income by Type of Income: Historical Data—*Continued*

(Billions of dollars, quarterly data are at seasonally adjusted annual rates.)

NIPA Tables 1.7.5, 1.12

| Year and quarter | Corporate profits with IVA and CCAdj | | Profits after tax | | | Net interest and miscellaneous payments | Taxes on production and imports | Less: Subsidies | Business current transfer payments, net | | | Current surplus of government enterprises | Addendum: Net national factor income |
	Total	Taxes on corporate income	Total	Net dividends	Undistributed corporate profits				Total [1]	To persons	To government		
1988													
1st quarter	394.2	112.6	281.6	138.5	143.1	369.1	365.2	30.4	31.1	19.6	11.5	. . .	3 961.1
2nd quarter	408.6	120.0	288.6	143.1	145.5	366.4	372.5	29.8	31.7	20.1	11.6	. . .	4 040.9
3rd quarter	415.2	129.1	286.1	150.3	135.8	380.4	377.5	29.2	33.2	20.8	12.4	. . .	4 126.3
4th quarter	439.8	135.4	304.5	157.3	147.2	395.5	382.6	28.6	35.1	21.9	13.2	. . .	4 225.0
1989													
1st quarter	423.7	137.2	286.5	169.2	117.3	416.3	391.0	28.0	37.2	23.3	14.0	. . .	4 290.2
2nd quarter	415.8	123.0	292.9	177.8	115.1	426.5	397.5	27.4	37.7	23.1	14.5	. . .	4 317.6
3rd quarter	416.6	118.9	297.7	182.5	115.2	433.4	403.9	27.1	39.1	23.2	15.9	. . .	4 362.2
4th quarter	401.0	118.6	282.4	189.0	93.4	430.3	403.0	27.3	39.0	23.1	16.0	. . .	4 402.6
1990													
1st quarter	413.9	116.8	297.1	192.7	104.4	435.0	419.5	27.1	39.4	23.3	16.1	. . .	4 490.3
2nd quarter	433.9	121.7	312.2	193.1	119.1	432.8	419.5	27.0	39.0	22.9	16.0	. . .	4 579.2
3rd quarter	407.7	125.1	282.6	194.7	87.9	430.0	426.8	26.9	38.9	21.9	17.0	. . .	4 600.8
4th quarter	415.4	123.7	291.7	190.1	101.6	435.9	434.2	27.0	39.5	20.7	18.8	. . .	4 621.1
1991													
1st quarter	455.0	120.2	334.8	196.7	138.1	407.7	444.0	27.1	39.8	19.2	20.6	. . .	4 630.6
2nd quarter	452.7	116.0	336.7	200.9	135.7	393.8	451.6	27.2	38.7	17.9	20.7	. . .	4 662.6
3rd quarter	450.7	116.9	333.8	203.6	130.2	390.9	461.3	27.5	38.5	17.0	21.5	. . .	4 704.6
4th quarter	452.0	118.2	333.8	203.9	129.9	374.8	471.5	28.1	38.7	16.4	22.2	. . .	4 752.7
1992													
1st quarter	478.2	130.4	347.8	202.9	144.8	370.3	476.4	28.6	38.6	16.5	22.4	-0.2	4 872.9
2nd quarter	479.6	132.4	347.3	205.5	141.7	367.0	481.2	29.2	38.4	16.5	22.0	-0.1	4 951.0
3rd quarter	465.2	127.8	337.4	207.3	130.1	360.6	486.0	30.4	38.9	16.3	22.7	-0.1	4 986.5
4th quarter	485.7	137.1	348.6	209.4	139.2	364.6	489.9	32.2	43.0	15.9	27.1	0.0	5 076.7
1993													
1st quarter	483.6	141.7	341.9	213.0	128.9	363.5	489.7	35.5	40.2	15.1	25.1	-0.1	5 092.4
2nd quarter	511.9	154.1	357.8	217.1	140.7	357.3	497.6	37.6	39.3	14.4	25.1	-0.2	5 185.8
3rd quarter	526.3	146.4	379.9	223.3	156.5	350.1	504.9	37.7	38.7	13.7	25.4	-0.4	5 226.5
4th quarter	576.5	178.0	398.5	231.8	166.7	341.5	520.3	36.0	39.3	13.1	26.9	-0.6	5 347.4
1994													
1st quarter	584.3	155.9	428.4	244.2	184.2	341.5	531.5	33.6	40.7	12.8	28.7	-0.8	5 398.4
2nd quarter	607.3	164.1	443.2	253.3	189.9	342.9	544.4	32.4	39.7	12.8	27.7	-0.8	5 509.3
3rd quarter	641.4	180.2	461.2	260.9	200.3	345.1	550.5	31.9	41.3	13.2	28.4	-0.4	5 591.3
4th quarter	666.0	190.4	475.6	267.4	208.3	359.6	554.6	32.2	41.2	14.2	26.8	0.2	5 695.2
1995													
1st quarter	667.8	194.7	473.0	270.6	202.4	362.6	555.3	34.0	43.6	16.0	25.9	1.7	5 750.5
2nd quarter	692.4	191.0	501.4	277.0	224.4	358.1	553.6	34.6	44.9	17.8	25.1	1.9	5 816.9
3rd quarter	726.4	198.0	528.4	285.4	243.0	355.1	558.9	35.1	45.8	19.6	24.6	1.6	5 911.4
4th quarter	738.3	194.0	544.3	296.4	247.9	353.7	563.8	35.5	45.5	21.4	23.9	0.2	5 992.2
1996													
1st quarter	772.3	201.7	570.6	306.2	264.4	352.5	570.4	35.5	46.8	22.8	24.7	-0.7	6 102.1
2nd quarter	784.8	213.4	571.4	318.7	252.6	358.2	577.7	35.4	48.2	23.5	25.8	-1.1	6 219.6
3rd quarter	790.0	213.7	576.3	329.9	246.4	363.6	581.8	35.2	48.9	23.2	26.8	-1.2	6 296.0
4th quarter	810.9	216.8	594.1	339.6	254.5	373.4	593.2	34.8	66.7	21.9	45.5	-0.8	6 401.0
1997													
1st quarter	839.0	218.2	620.8	347.7	273.2	381.8	595.7	34.4	46.9	19.7	27.8	-0.6	6 532.9
2nd quarter	861.6	222.5	639.2	356.2	282.9	389.8	610.4	33.6	47.4	18.7	28.5	0.1	6 631.1
3rd quarter	896.6	234.2	662.4	364.4	298.0	396.4	616.6	33.4	53.5	19.0	33.4	1.2	6 767.8
4th quarter	881.6	224.2	657.4	372.2	285.3	409.8	623.8	33.8	52.5	20.4	29.6	2.5	6 884.2
1998													
1st quarter	812.7	222.1	590.5	383.5	207.0	442.0	629.1	33.8	58.9	22.8	31.1	5.0	6 970.9
2nd quarter	809.4	218.9	590.5	386.4	204.1	459.0	635.5	35.0	60.4	25.1	30.4	4.9	7 088.9
3rd quarter	817.6	225.5	592.2	385.2	207.0	467.0	643.0	36.8	62.6	27.1	31.8	3.7	7 212.6
4th quarter	794.1	220.7	573.4	379.4	194.0	458.8	650.3	39.9	74.3	28.9	43.9	1.6	7 294.3
1999													
1st quarter	847.1	226.4	620.6	368.7	251.9	451.4	657.5	42.4	65.4	31.2	35.6	-1.4	7 444.5
2nd quarter	840.2	223.5	616.7	366.6	250.1	454.5	667.1	45.0	67.0	32.8	36.2	-2.0	7 508.6
3rd quarter	823.7	227.6	596.1	372.1	224.1	466.3	679.0	46.4	67.8	34.8	36.2	-3.2	7 592.5
4th quarter	828.7	231.9	596.8	386.6	210.2	487.0	690.7	46.9	71.1	37.1	37.8	-3.7	7 763.9
2000													
1st quarter	808.5	243.3	565.2	400.6	164.7	521.3	698.6	45.1	81.2	39.8	41.0	0.4	7 980.6
2nd quarter	798.1	241.7	556.5	412.0	144.5	542.0	707.3	45.5	84.2	41.5	41.8	0.9	8 064.3
3rd quarter	781.2	222.8	558.5	416.0	142.5	550.4	711.3	45.8	86.5	43.3	42.0	1.1	8 180.1
4th quarter	758.4	225.9	532.5	412.3	120.2	550.9	717.1	47.0	89.6	45.2	42.8	1.6	8 227.4
2001													
1st quarter	748.0	188.6	559.4	404.2	155.2	558.1	724.2	55.2	98.5	47.9	42.8	7.8	8 363.8
2nd quarter	778.0	182.6	595.3	396.5	198.9	551.5	724.1	62.0	100.5	48.3	43.2	8.9	8 401.5
3rd quarter	746.7	162.8	583.9	393.7	190.3	535.1	725.3	71.2	100.3	47.0	44.2	9.1	8 357.6
4th quarter	762.2	146.4	615.8	397.1	218.6	511.6	737.1	46.4	97.6	43.9	45.2	8.5	8 358.4
2002													
1st quarter	852.5	147.3	705.2	412.3	292.9	480.9	744.0	42.6	87.7	39.1	46.5	2.1	8 463.3
2nd quarter	883.2	153.6	729.6	424.3	305.3	470.8	751.3	39.8	81.4	35.2	45.9	0.4	8 558.0
3rd quarter	914.6	161.8	752.8	428.2	324.7	446.8	768.5	41.3	77.4	32.3	45.8	-0.6	8 602.0
4th quarter	996.5	179.8	816.7	434.7	382.1	446.9	776.3	41.9	76.2	30.2	46.8	-0.8	8 734.4

[1]Includes net transfer payments to the rest of the world, not shown separately.
. . . = Not available.

Table 1-13B. National Income by Type of Income: Historical Data—*Continued*

(Billions of dollars, quarterly data are at seasonally adjusted annual rates.)

NIPA Tables 1.7.5, 1.12

Year and quarter	National income, total	Compensation of employees Total	Wages and salaries Total	Wages and salaries Government	Wages and salaries Other	Supplements to wages and salaries Total	Employer contributions for: Employee pension and insurance funds	Employer contributions for: Government social insurance	Proprietors' income with IVA and CCAdj Total	Farm	Nonfarm	Rental income of persons with CCAdj
2003												
1st quarter	9 593.5	6 213.6	5 027.4	901.2	4 126.3	1 186.2	812.2	374.0	874.9	28.4	846.4	223.0
2nd quarter	9 723.3	6 307.7	5 101.8	911.9	4 189.8	1 205.9	826.2	379.8	891.5	36.4	855.1	227.7
3rd quarter	9 881.9	6 394.1	5 169.1	917.8	4 251.4	1 225.0	839.9	385.0	905.0	39.4	865.5	232.0
4th quarter	10 046.0	6 500.8	5 256.5	924.9	4 331.6	1 244.3	852.8	391.4	916.2	41.6	874.5	243.3
2004												
1st quarter	10 225.4	6 542.0	5 275.2	936.6	4 338.6	1 266.8	868.0	398.8	945.7	55.4	890.3	247.5
2nd quarter	10 416.9	6 669.1	5 381.4	948.3	4 433.1	1 287.7	881.6	406.1	962.1	54.3	907.8	252.1
3rd quarter	10 614.6	6 804.3	5 493.4	956.6	4 536.7	1 310.9	897.1	413.9	963.2	46.1	917.1	248.5
4th quarter	10 711.8	6 864.8	5 536.4	967.9	4 568.5	1 328.4	912.2	416.2	977.1	50.2	926.8	247.4
2005												
1st quarter	10 937.4	6 926.9	5 575.3	977.8	4 597.5	1 351.6	930.6	421.1	956.6	47.1	909.5	240.1
2nd quarter	11 083.5	7 006.0	5 638.3	986.1	4 652.2	1 367.7	943.1	424.6	963.4	50.4	913.0	237.3
3rd quarter	11 265.5	7 122.9	5 738.6	995.5	4 743.0	1 384.4	953.4	430.9	986.8	48.2	938.5	225.6
4th quarter	11 508.4	7 210.6	5 815.4	1 005.8	4 809.6	1 395.2	959.5	435.6	1 005.4	41.4	963.9	225.0
2006												
1st quarter	11 799.7	7 380.2	5 971.9	1 014.6	4 957.3	1 408.3	964.7	443.6	1 048.1	33.8	1 014.2	216.7
2nd quarter	11 898.5	7 428.1	6 012.7	1 025.8	4 986.9	1 415.4	970.1	445.3	1 050.9	32.7	1 018.2	206.3
3rd quarter	12 011.3	7 487.2	6 061.8	1 041.9	5 019.8	1 425.4	978.5	447.0	1 047.7	32.4	1 015.3	199.5
4th quarter	12 085.8	7 624.0	6 181.4	1 055.8	5 125.7	1 442.6	989.2	453.3	1 051.9	33.3	1 018.5	186.8
2007												
1st quarter	12 183.1	7 806.8	6 344.5	1 071.6	5 272.9	1 462.2	1 002.5	459.7	1 016.2	41.8	974.4	174.2
2nd quarter	12 309.2	7 845.4	6 370.5	1 080.5	5 290.0	1 474.8	1 015.0	459.8	995.9	36.6	959.3	183.8
3rd quarter	12 304.0	7 885.1	6 397.4	1 093.6	5 303.9	1 487.6	1 026.8	460.9	985.1	37.2	947.9	187.5
4th quarter	12 365.3	7 978.2	6 474.6	1 108.5	5 366.1	1 503.6	1 037.3	466.3	978.9	45.4	933.5	191.9
2008												
1st quarter	12 377.2	8 055.3	6 539.8	1 124.8	5 415.0	1 515.5	1 045.8	469.7	964.7	48.6	916.1	220.5
2nd quarter	12 401.1	8 054.8	6 532.6	1 136.0	5 396.6	1 522.2	1 051.9	470.4	968.6	45.5	923.1	245.0
3rd quarter	12 440.9	8 074.0	6 546.7	1 151.2	5 395.6	1 527.2	1 054.4	472.8	958.2	35.5	922.8	268.1
4th quarter	12 083.9	8 043.8	6 517.9	1 163.4	5 354.4	1 526.0	1 053.3	472.7	952.2	31.2	921.0	293.0
2009												
1st quarter	11 863.4	7 729.8	6 225.0	1 162.8	5 062.2	1 504.8	1 048.8	456.0	914.3	24.0	890.3	303.4
2nd quarter	11 918.8	7 761.4	6 253.9	1 176.0	5 077.8	1 507.5	1 048.1	459.4	902.3	24.4	878.0	318.4
3rd quarter	12 056.5	7 746.2	6 236.9	1 179.0	5 057.9	1 509.3	1 051.5	457.8	935.3	27.4	907.9	338.5
4th quarter	12 270.2	7 796.7	6 278.6	1 183.1	5 095.6	1 518.1	1 058.9	459.2	1 002.2	36.5	965.7	349.0
2010												
1st quarter	12 402.8	7 767.2	6 232.8	1 185.8	5 047.0	1 534.4	1 070.3	464.1	1 056.9	31.5	1 025.4	371.8
2nd quarter	12 633.0	7 910.1	6 360.2	1 196.6	5 163.6	1 549.9	1 080.4	469.5	1 104.0	35.3	1 068.7	387.9
3rd quarter	12 894.2	7 981.8	6 422.1	1 191.3	5 230.8	1 559.7	1 088.9	470.8	1 129.4	41.3	1 088.2	399.5
4th quarter	13 013.3	8 040.7	6 473.3	1 191.1	5 282.2	1 567.4	1 095.9	471.5	1 144.4	47.8	1 096.7	417.6
2011												
1st quarter	13 092.4	8 170.0	6 577.2	1 193.1	5 384.2	1 592.8	1 102.0	490.8	1 181.0	66.8	1 114.2	453.4
2nd quarter	13 286.7	8 206.1	6 607.9	1 199.2	5 408.7	1 598.2	1 106.4	491.8	1 212.2	61.2	1 151.0	471.6
3rd quarter	13 446.7	8 287.1	6 681.5	1 197.6	5 483.8	1 605.7	1 109.5	496.2	1 253.5	66.4	1 187.1	485.0
4th quarter	13 605.2	8 240.4	6 637.1	1 189.5	5 447.6	1 603.3	1 111.1	492.2	1 270.4	65.1	1 205.3	504.6
2012												
1st quarter	13 999.6	8 455.9	6 835.8	1 198.7	5 637.2	1 620.1	1 112.1	508.0	1 322.8	60.9	1 261.9	509.9
2nd quarter	14 076.7	8 503.0	6 875.2	1 195.8	5 679.5	1 627.8	1 117.8	510.0	1 356.5	59.0	1 297.5	517.5
3rd quarter	14 043.4	8 545.3	6 905.3	1 196.8	5 708.5	1 640.0	1 128.8	511.2	1 345.8	60.3	1 285.5	520.5
4th quarter	14 258.9	8 762.6	7 093.6	1 202.0	5 891.6	1 669.0	1 145.1	523.9	1 364.2	63.4	1 300.8	524.0
2013												
1st quarter	14 310.7	8 731.7	7 043.0	1 206.4	5 836.6	1 688.7	1 168.4	520.4	1 388.3	94.1	1 294.2	537.9
2nd quarter	14 444.4	8 818.7	7 102.1	1 206.8	5 895.3	1 716.6	1 190.7	525.9	1 411.8	95.3	1 316.5	550.2
3rd quarter	14 521.8	8 844.7	7 112.0	1 204.8	5 907.2	1 732.7	1 205.9	526.8	1 412.5	94.3	1 318.2	564.6
4th quarter	14 701.9	8 941.7	7 195.6	1 214.0	5 981.6	1 746.2	1 214.0	532.1	1 401.7	69.6	1 332.0	575.2
2014												
1st quarter	14 826.7	9 113.0	7 357.4	1 221.9	6 135.6	1 755.6	1 216.1	539.5	1 413.4	68.3	1 345.1	588.4
2nd quarter	15 148.7	9 171.3	7 408.2	1 230.6	6 177.6	1 763.1	1 221.3	541.8	1 452.7	79.0	1 373.6	599.2
3rd quarter	15 436.0	9 279.6	7 502.0	1 241.5	6 260.5	1 777.6	1 229.9	547.7	1 468.3	66.1	1 402.2	609.9
4th quarter	15 558.5	9 432.5	7 633.3	1 253.7	6 379.6	1 799.2	1 242.5	556.7	1 456.6	65.9	1 390.7	621.0
2015												
1st quarter	15 641.7	9 563.3	7 745.0	1 263.2	6 481.8	1 818.4	1 257.0	561.4	1 432.9	54.0	1 378.9	623.1
2nd quarter	15 770.0	9 667.2	7 830.2	1 272.8	6 557.4	1 837.0	1 269.5	567.5	1 406.3	55.7	1 350.5	647.7
3rd quarter	15 851.5	9 748.7	7 898.2	1 280.1	6 618.1	1 850.5	1 278.3	572.2	1 425.9	58.9	1 367.0	658.1
4th quarter	15 888.3	9 813.4	7 953.3	1 286.4	6 666.9	1 860.0	1 284.2	575.8	1 423.7	55.5	1 368.2	663.5
2016												
1st quarter	15 925.2	9 843.5	7 982.8	1 294.2	6 688.5	1 860.7	1 286.5	574.2	1 415.2	36.5	1 378.7	669.9
2nd quarter	15 952.3	9 900.1	8 032.1	1 302.5	6 729.6	1 868.0	1 290.5	577.5	1 410.2	38.3	1 371.9	680.2
3rd quarter	16 083.1	9 993.2	8 112.2	1 314.1	6 798.1	1 881.1	1 298.0	583.1	1 429.5	36.5	1 393.0	683.6
4th quarter	16 253.9	10 104.5	8 206.9	1 321.2	6 885.7	1 897.5	1 307.5	590.0	1 440.0	31.2	1 408.9	692.1
2017												
1st quarter	16 475.3	10 227.6	8 310.6	1 331.4	6 979.2	1 917.0	1 320.4	596.6	1 494.8	44.5	1 450.3	707.4
2nd quarter	16 611.7	10 334.2	8 397.7	1 340.3	7 057.4	1 936.5	1 334.3	602.2	1 512.2	42.1	1 470.1	709.9
3rd quarter	16 753.1	10 456.7	8 497.9	1 353.0	7 144.9	1 958.8	1 350.8	607.9	1 523.1	34.1	1 489.0	722.0
4th quarter	16 995.2	10 628.0	8 642.0	1 367.2	7 274.9	1 985.9	1 370.0	615.9	1 542.9	31.8	1 511.1	736.0

Table 1-13B. National Income by Type of Income: Historical Data—*Continued*

(Billions of dollars, quarterly data are at seasonally adjusted annual rates.)

NIPA Tables 1.7.5, 1.12

| Year and quarter | Corporate profits with IVA and CCAdj | | | | | Net interest and miscellaneous payments | Taxes on production and imports | Less: Subsidies | Business current transfer payments, net | | | Current surplus of government enterprises | Addendum: Net national factor income |
| | Total | Taxes on corporate income | Profits after tax | | | | | | Total¹ | To persons | To government | | |
			Total	Net dividends	Undistributed corporate profits								
2003													
1st quarter	988.6	199.9	788.7	435.3	353.4	467.2	788.6	47.1	77.7	29.6	47.1	0.9	8 767.2
2nd quarter	1 024.8	196.4	828.4	435.6	392.8	446.7	800.0	57.1	76.4	28.0	47.7	0.7	8 898.4
3rd quarter	1 075.7	217.6	858.1	457.8	400.3	428.0	813.0	45.9	75.9	25.5	48.6	1.8	9 034.8
4th quarter	1 135.9	241.0	894.9	495.4	399.5	396.5	820.9	46.0	75.1	22.0	49.8	3.3	9 192.6
2004													
1st quarter	1 242.4	251.4	991.0	532.7	458.3	362.7	847.3	44.2	80.1	17.5	51.3	11.3	9 340.2
2nd quarter	1 271.7	271.6	1 000.1	560.3	439.7	365.2	859.9	43.7	79.6	15.4	52.5	11.6	9 520.1
3rd quarter	1 330.4	292.7	1 037.7	567.2	470.5	364.5	871.3	45.4	78.5	15.7	53.0	9.8	9 711.0
4th quarter	1 312.9	298.4	1 014.5	668.5	346.0	380.2	893.8	52.3	89.7	18.4	52.6	18.6	9 782.3
2005													
1st quarter	1 442.1	375.8	1 066.2	575.2	491.1	421.7	915.1	56.7	94.6	23.7	52.0	18.9	9 987.4
2nd quarter	1 457.0	364.0	1 092.9	592.1	500.9	449.4	937.3	60.7	97.8	26.7	53.2	17.9	10 113.1
3rd quarter	1 483.0	370.3	1 112.7	598.4	514.3	469.5	952.1	62.0	93.2	27.3	53.5	12.4	10 287.8
4th quarter	1 572.5	409.1	1 163.3	642.3	521.0	508.5	965.3	64.2	91.1	25.6	55.0	10.5	10 521.9
2006													
1st quarter	1 629.6	421.7	1 208.0	714.1	493.9	523.0	981.8	55.7	79.5	21.6	56.0	2.0	10 797.6
2nd quarter	1 643.1	432.9	1 210.2	747.2	463.0	558.0	991.7	51.5	77.8	19.7	57.1	1.0	10 886.4
3rd quarter	1 688.7	451.5	1 237.1	765.6	471.5	558.4	1 004.1	49.9	83.9	19.9	58.0	6.0	10 981.5
4th quarter	1 623.6	415.6	1 208.1	793.4	414.7	562.9	1 010.5	48.7	85.2	22.1	58.9	4.2	11 049.2
2007													
1st quarter	1 528.8	418.9	1 109.9	820.3	289.6	592.6	1 025.9	49.5	101.1	26.5	59.2	15.3	11 118.7
2nd quarter	1 593.0	413.6	1 179.4	858.8	320.6	631.7	1 033.1	58.2	98.1	30.0	59.9	8.2	11 249.8
3rd quarter	1 525.9	376.8	1 149.1	868.1	281.0	655.0	1 035.8	55.9	99.1	32.5	60.8	5.8	11 238.6
4th quarter	1 485.0	359.0	1 126.0	866.8	259.2	655.2	1 052.6	54.7	94.8	34.1	63.7	-2.9	11 289.2
2008													
1st quarter	1 388.4	298.2	1 090.2	903.0	187.2	657.0	1 045.7	51.9	115.0	34.7	65.2	15.1	11 285.9
2nd quarter	1 364.6	285.5	1 079.2	866.5	212.7	675.7	1 054.7	51.7	107.9	35.4	66.1	6.3	11 308.8
3rd quarter	1 372.6	270.9	1 101.7	820.8	280.9	673.6	1 058.5	52.0	106.5	36.1	65.3	5.0	11 346.5
4th quarter	1 017.8	170.0	847.8	770.9	76.9	683.2	1 040.0	54.6	126.5	36.9	81.2	8.4	10 990.1
2009													
1st quarter	1 255.1	172.2	1 082.8	702.8	380.0	595.7	1 015.9	55.4	121.8	37.8	82.3	1.7	10 798.3
2nd quarter	1 283.8	195.6	1 088.2	620.7	467.4	566.7	1 017.3	55.5	140.1	38.6	99.1	2.4	10 832.6
3rd quarter	1 456.5	206.6	1 249.9	584.1	665.8	516.7	1 028.8	67.1	116.7	39.4	81.2	-4.0	10 993.2
4th quarter	1 551.9	242.3	1 309.6	580.5	729.1	477.9	1 045.3	55.5	119.0	40.2	81.8	-3.1	11 177.7
2010													
1st quarter	1 642.6	256.6	1 386.0	593.9	792.1	464.4	1 044.6	54.8	127.8	41.0	82.4	4.3	11 302.9
2nd quarter	1 640.8	262.5	1 378.3	613.0	765.3	477.8	1 062.1	55.5	125.6	42.4	81.4	1.8	11 520.7
3rd quarter	1 802.8	279.4	1 523.4	661.7	861.7	459.8	1 069.1	56.0	128.9	44.1	86.1	-1.3	11 773.4
4th quarter	1 828.8	291.6	1 537.2	704.3	832.9	458.8	1 076.4	56.9	125.0	47.4	86.5	-8.9	11 890.3
2011													
1st quarter	1 670.4	285.4	1 384.9	748.0	636.9	466.9	1 091.5	58.9	139.1	48.8	88.9	1.5	11 941.7
2nd quarter	1 791.8	285.4	1 506.4	761.0	745.3	454.0	1 105.5	59.9	125.2	49.5	87.2	-11.5	12 135.7
3rd quarter	1 818.5	256.7	1 561.7	787.5	774.2	451.3	1 103.9	60.2	126.7	48.9	87.9	-10.1	12 295.3
4th quarter	1 958.5	296.8	1 661.7	819.8	841.9	474.6	1 114.0	61.1	121.4	47.0	83.3	-8.9	12 448.5
2012													
1st quarter	2 025.6	320.1	1 705.5	879.7	825.8	516.8	1 130.9	58.4	112.2	43.5	77.0	-8.4	12 831.1
2nd quarter	2 006.9	334.5	1 672.4	902.1	770.3	530.3	1 133.9	58.1	102.1	41.0	69.9	-8.8	12 914.2
3rd quarter	1 985.4	342.0	1 643.4	906.9	736.6	495.0	1 131.3	56.3	91.3	39.1	63.0	-10.8	12 892.1
4th quarter	1 971.7	342.8	1 628.9	1 106.1	522.8	472.8	1 148.4	59.4	89.9	38.1	66.0	-14.2	13 095.3
2013													
1st quarter	1 983.5	360.8	1 622.7	870.8	751.9	467.2	1 174.6	59.4	102.1	37.8	78.8	-14.5	13 108.6
2nd quarter	2 000.2	357.3	1 642.9	1 127.8	515.2	453.8	1 180.8	60.1	105.3	37.9	80.6	-13.2	13 234.7
3rd quarter	2 011.1	364.9	1 646.2	958.4	687.9	465.7	1 195.0	60.0	104.4	38.5	77.3	-11.4	13 298.6
4th quarter	2 047.9	368.1	1 679.8	1 079.0	600.8	477.0	1 204.1	59.4	129.3	39.5	102.0	-12.2	13 443.6
2014													
1st quarter	1 967.4	403.7	1 563.8	1 052.4	511.3	485.4	1 220.5	58.7	109.8	40.9	81.8	-12.9	13 567.7
2nd quarter	2 138.3	425.8	1 712.4	1 093.2	619.3	504.4	1 237.5	58.5	115.3	42.2	82.3	-9.3	13 865.9
3rd quarter	2 191.0	398.3	1 792.7	1 105.6	687.2	532.8	1 248.4	58.2	174.6	43.6	137.9	-6.9	14 081.6
4th quarter	2 184.2	401.5	1 782.7	1 133.1	649.6	541.7	1 257.0	57.0	132.0	45.0	90.2	-3.2	14 236.0
2015													
1st quarter	2 132.0	418.8	1 713.1	1 150.9	562.2	564.0	1 263.8	56.0	126.0	46.3	85.3	-5.6	14 315.2
2nd quarter	2 103.7	420.1	1 683.7	1 131.4	552.3	596.5	1 275.0	56.4	136.5	48.4	92.4	-4.4	14 421.5
3rd quarter	2 062.3	389.1	1 673.2	1 157.2	516.1	602.9	1 277.8	57.8	138.5	51.3	87.7	-0.6	14 497.9
4th quarter	1 948.0	358.4	1 589.7	1 220.1	369.6	583.9	1 291.9	58.8	226.1	55.0	172.5	-1.5	14 432.6
2016													
1st quarter	2 022.2	373.3	1 649.0	1 168.9	480.1	573.9	1 297.1	60.7	166.2	59.5	103.9	2.8	14 524.7
2nd quarter	1 998.1	373.8	1 624.3	1 166.7	457.6	556.0	1 301.2	62.4	172.0	61.3	108.7	2.0	14 544.7
3rd quarter	2 013.0	391.7	1 621.3	1 183.3	438.0	552.6	1 322.3	63.1	154.7	60.6	90.7	3.4	14 671.9
4th quarter	2 012.6	371.5	1 641.0	1 184.8	456.2	557.6	1 330.5	61.0	179.9	57.2	111.8	10.9	14 806.7
2017													
1st quarter	1 995.4	322.8	1 672.5	1 219.5	453.1	607.7	1 340.8	59.9	163.6	51.2	113.5	-1.1	15 032.8
2nd quarter	2 008.0	314.1	1 693.9	1 246.8	447.1	603.6	1 355.1	58.7	149.5	47.6	90.7	11.2	15 168.0
3rd quarter	2 019.0	335.3	1 683.7	1 242.7	441.0	596.3	1 371.4	63.2	130.5	46.2	91.4	-7.1	15 317.0
4th quarter	2 001.4	305.4	1 696.0	1 249.5	446.5	624.5	1 390.6	62.8	138.0	47.3	92.8	-2.1	15 532.6

¹Includes net transfer payments to the rest of the world, not shown separately.

Table 1-14. Gross and Net Value Added of Domestic Corporate Business

(Billions of dollars, quarterly data are at seasonally adjusted annual rates.)   NIPA Table 1.14

Year and quarter	Gross value added of corporate business, total	Consumption of fixed capital	Net value added			Net operating surplus			Corporate profits with IVA and CCAdj		Profits after tax			Gross value added of financial corporate business
			Total	Compensation of employees	Taxes on production and imports less subsidies	Total	Net interest and miscellaneous payments	Business current transfer payments	Total	Taxes on corporate income	Total	Net dividends	Undistributed	
1955	233.1	19.6	213.5	144.6	19.8	49.0	0.1	1.1	47.8	21.8	26.0	9.2	16.9	11.7
1956	250.1	22.1	228.0	158.2	21.5	48.3	0.0	1.4	46.8	21.6	25.2	9.9	15.4	12.8
1957	261.7	24.6	237.1	166.5	22.8	47.8	0.2	1.6	46.0	20.9	25.2	10.4	14.8	13.7
1958	256.8	26.4	230.4	164.0	23.1	43.3	0.6	1.4	41.3	18.4	22.9	10.4	12.5	14.6
1959	287.4	27.6	259.8	180.3	25.6	54.0	-0.2	1.4	52.8	22.8	30.0	11.6	18.4	15.8
1960	299.9	28.5	271.4	190.7	27.8	52.9	-0.2	1.5	51.5	21.9	29.7	12.3	17.4	17.4
1961	308.7	29.5	279.2	195.6	28.9	54.7	0.4	1.7	52.6	22.2	30.4	12.2	18.2	18.3
1962	336.0	30.8	305.1	211.0	31.2	62.9	0.8	1.9	60.3	23.3	37.0	13.2	23.8	19.1
1963	357.4	32.4	324.9	222.7	33.2	69.1	0.4	2.3	66.4	25.5	40.9	14.4	26.5	19.5
1964	386.0	34.4	351.6	239.2	35.6	76.8	0.8	2.7	73.2	26.6	46.7	16.6	30.1	21.5
1965	423.8	37.1	386.6	259.9	37.8	89.0	1.2	3.1	84.6	29.8	54.8	18.2	36.6	23.0
1966	465.1	40.9	424.2	288.5	38.9	96.8	2.3	2.9	91.6	32.2	59.4	19.5	39.9	25.0
1967	491.2	45.1	446.1	308.4	41.4	96.3	4.0	3.1	89.1	31.0	58.1	20.2	37.9	28.0
1968	542.1	50.0	492.1	340.2	47.8	104.1	4.3	3.7	96.1	37.2	58.9	22.6	36.3	31.2
1969	590.7	55.7	534.9	377.5	52.9	104.6	8.5	4.3	91.8	37.0	54.9	23.4	31.4	36.0
1970	612.1	61.8	550.3	398.0	57.0	95.3	12.5	3.7	79.1	31.3	47.9	23.9	24.0	39.2
1971	660.9	67.6	593.3	421.7	62.8	108.8	12.6	3.4	92.8	34.8	58.0	23.7	34.3	42.8
1972	732.9	73.4	659.5	468.2	67.3	124.0	12.4	3.9	107.7	39.1	68.6	25.2	43.4	47.0
1973	820.2	82.0	738.2	526.1	74.1	138.0	14.7	4.8	118.5	45.6	72.9	27.4	45.4	51.4
1974	889.6	96.0	793.5	577.3	78.6	137.6	23.1	6.3	108.2	47.2	61.0	29.0	32.0	59.5
1975	965.5	113.5	852.0	607.8	84.5	159.7	27.2	8.2	124.2	46.3	78.0	31.8	46.2	67.3
1976	1 087.3	125.0	962.3	682.8	91.4	188.1	22.7	7.6	157.8	59.4	98.4	35.9	62.4	72.6
1977	1 232.5	140.5	1 092.1	771.7	100.0	220.3	27.7	6.0	186.7	68.5	118.2	39.6	78.5	85.1
1978	1 406.9	159.3	1 247.6	884.7	108.7	254.2	30.6	7.9	215.7	77.9	137.8	46.6	91.2	103.0
1979	1 564.3	183.4	1 380.9	1 004.4	115.0	261.5	36.3	10.8	214.4	80.7	133.6	50.8	82.9	114.0
1980	1 702.2	212.1	1 490.1	1 102.0	128.6	259.6	59.4	12.0	188.1	75.5	112.6	59.0	53.6	127.9
1981	1 934.8	244.9	1 689.9	1 220.6	154.4	314.9	82.8	14.3	217.8	70.3	147.5	72.3	75.2	146.1
1982	2 018.4	272.5	1 745.9	1 275.1	161.3	309.4	95.5	16.6	197.3	51.3	146.0	76.5	69.4	163.7
1983	2 172.4	285.9	1 886.5	1 353.0	177.4	356.1	91.9	19.5	244.7	66.4	178.3	85.5	92.7	186.6
1984	2 434.6	303.2	2 131.5	1 501.1	195.6	434.8	106.6	26.9	301.3	81.5	219.8	94.6	125.3	208.8
1985	2 601.3	326.9	2 274.4	1 615.9	209.0	449.5	102.2	30.9	316.4	81.6	234.8	103.5	131.3	232.3
1986	2 706.6	349.9	2 356.7	1 723.4	219.6	413.7	99.5	29.3	284.9	91.9	193.0	106.1	86.9	241.9
1987	2 894.9	371.3	2 523.6	1 847.6	233.4	442.6	99.7	24.9	318.0	112.7	205.3	113.3	91.9	253.2
1988	3 142.4	399.9	2 742.5	2 002.3	252.0	488.2	104.4	26.4	357.5	124.3	233.2	115.4	117.8	268.6
1989	3 320.0	427.1	2 892.8	2 119.3	267.5	506.0	125.2	33.6	347.2	124.4	222.7	148.0	74.7	299.9
1990	3 467.1	454.7	3 012.5	2 234.9	284.5	493.1	117.4	34.0	341.6	121.8	219.8	167.7	52.1	306.6
1991	3 554.4	479.6	3 074.8	2 277.8	307.9	489.1	79.3	33.7	376.1	117.8	258.3	177.0	81.3	323.5
1992	3 741.9	494.3	3 247.6	2 420.5	325.9	501.2	63.2	34.0	404.1	131.9	272.2	178.3	93.9	365.9
1993	3 918.3	517.2	3 401.2	2 515.7	343.8	541.7	62.3	31.8	447.6	155.0	292.6	200.7	91.9	383.9
1994	4 209.4	547.6	3 661.8	2 649.0	376.1	636.7	57.7	32.2	546.8	172.7	374.1	218.3	155.9	393.5
1995	4 473.4	590.4	3 882.9	2 787.9	384.8	710.3	60.2	36.9	613.3	194.4	418.8	249.6	169.2	432.7
1996	4 769.1	626.3	4 142.8	2 953.7	398.4	790.7	59.4	43.7	687.5	211.4	476.1	283.2	192.9	473.8
1997	5 141.8	670.1	4 471.8	3 176.6	416.9	878.3	81.7	34.4	762.2	224.8	537.4	312.9	224.5	533.8
1998	5 473.4	715.0	4 758.4	3 447.5	430.8	880.2	121.6	52.8	705.7	221.8	483.9	341.4	142.5	600.1
1999	5 802.2	769.6	5 032.5	3 684.3	454.0	894.2	127.2	53.9	713.2	227.4	485.8	331.8	154.0	632.3
2000	6 214.6	838.6	5 376.0	4 008.9	479.8	887.3	174.0	72.4	640.9	233.4	407.5	380.8	26.7	702.0
2001	6 202.0	888.6	5 313.4	4 013.8	469.4	830.2	155.4	84.8	589.9	170.1	419.8	357.0	62.8	736.0
2002	6 293.6	914.6	5 379.0	3 972.7	497.5	908.7	92.3	61.5	754.9	160.6	594.3	376.8	217.5	754.4
2003	6 515.5	932.5	5 583.0	4 040.9	524.3	1 017.8	70.0	50.5	897.3	213.7	683.6	423.9	259.7	791.6
2004	6 948.3	973.0	5 975.3	4 240.2	569.4	1 165.8	16.9	54.6	1 094.2	278.5	815.7	519.8	295.8	825.5
2005	7 473.4	1 042.0	6 431.4	4 443.0	616.3	1 372.2	50.3	58.9	1 262.9	379.8	883.1	341.1	542.0	913.6
2006	8 008.7	1 119.3	6 889.5	4 681.2	654.5	1 553.7	98.9	48.3	1 406.5	430.4	976.1	677.2	299.0	1 006.5
2007	8 180.7	1 188.9	6 991.8	4 894.2	676.3	1 421.2	156.5	69.4	1 195.4	392.1	803.3	713.1	90.2	961.0
2008	8 094.3	1 256.6	6 837.7	4 940.3	685.0	1 212.4	220.8	96.0	895.7	256.1	639.5	659.8	-20.3	807.6
2009	7 803.9	1 263.8	6 540.1	4 607.5	655.5	1 277.2	141.4	97.8	1 038.0	204.2	833.9	503.9	329.9	939.4
2010	8 232.1	1 271.2	6 960.9	4 700.8	684.9	1 575.2	132.0	100.2	1 343.0	272.5	1 070.4	521.8	548.6	988.9
2011	8 627.5	1 324.6	7 302.9	4 928.0	718.7	1 656.2	151.0	108.1	1 397.2	281.1	1 116.1	624.2	492.0	1 012.2
2012	9 182.7	1 392.5	7 790.3	5 182.7	743.2	1 864.4	210.2	62.2	1 592.1	334.9	1 257.2	768.8	488.4	1 123.6
2013	9 478.1	1 450.2	8 027.9	5 352.4	793.1	1 882.4	202.4	68.1	1 611.9	362.8	1 249.1	870.6	378.5	1 104.0
2014	10 056.4	1 527.5	8 528.9	5 645.2	828.3	2 055.3	245.1	94.9	1 715.3	407.3	1 308.0	934.4	373.6	1 269.5
2015	10 495.4	1 592.6	8 902.8	5 941.8	850.6	2 110.3	326.8	124.6	1 659.0	396.6	1 262.4	1 002.1	260.3	1 356.4
2016	10 662.4	1 630.6	9 031.9	6 095.5	869.4	2 067.0	324.4	143.0	1 599.6	377.6	1 222.0	1 003.7	218.3	1 424.5
2017	11 030.5	1 704.0	9 326.5	6 412.9	897.0	2 016.6	352.8	111.9	1 551.9	319.4	1 232.4	1 038.5	194.0	1 423.2
2018	11 576.3	1 796.3	9 780.0	6 750.3	953.6	2 076.1	381.3	121.8	1 573.0	219.8	1 353.2	466.3	886.9	1 513.2
2016														
1st quarter	10 586.1	1 609.6	8 976.6	6 017.0	859.2	2 100.4	326.9	140.2	1 633.3	373.3	1 260.1	1 018.7	241.4	1 352.9
2nd quarter	10 604.5	1 624.7	8 979.8	6 054.2	861.7	2 063.9	321.7	148.2	1 594.0	373.8	1 220.2	997.9	222.2	1 423.1
3rd quarter	10 693.4	1 635.0	9 058.4	6 115.9	875.4	2 067.2	321.5	132.5	1 613.1	391.7	1 221.4	999.4	221.9	1 445.4
4th quarter	10 765.7	1 653.0	9 112.6	6 194.9	881.3	2 036.4	327.6	150.9	1 557.9	371.5	1 186.4	998.9	187.5	1 476.6
2017														
1st quarter	10 873.4	1 672.1	9 201.3	6 291.4	881.8	2 028.1	338.5	130.9	1 558.7	322.8	1 235.9	1 022.6	213.3	1 374.6
2nd quarter	10 996.0	1 692.8	9 303.3	6 362.0	891.4	2 049.9	348.7	116.5	1 584.7	314.1	1 270.6	1 077.7	192.8	1 399.8
3rd quarter	11 060.1	1 715.0	9 345.0	6 440.8	901.6	2 002.6	357.6	93.7	1 551.3	335.3	1 216.0	956.9	259.2	1 442.3
4th quarter	11 192.5	1 736.1	9 456.4	6 557.6	913.2	1 985.7	366.4	106.6	1 512.7	305.4	1 207.3	1 096.7	110.6	1 476.0
2018														
1st quarter	11 392.7	1 756.5	9 636.2	6 669.5	937.8	2 028.9	374.1	110.2	1 544.5	207.6	1 336.9	52.1	1 284.8	1 468.4
2nd quarter	11 506.4	1 783.5	9 722.9	6 718.6	950.4	2 054.0	379.8	112.8	1 561.4	222.6	1 338.8	331.8	1 006.9	1 504.1
3rd quarter	11 667.2	1 809.6	9 857.5	6 789.7	956.0	2 111.8	383.4	135.8	1 592.6	230.3	1 362.3	779.9	582.4	1 540.0
4th quarter	11 738.8	1 835.6	9 903.1	6 823.2	970.3	2 109.7	387.9	128.5	1 593.3	218.5	1 374.9	701.5	673.4	1 540.1

Table 1-15. Gross Value Added of Nonfinancial Domestic Corporate Business in Current and Chained Dollars

(Billions of dollars, quarterly data are at seasonally adjusted annual rates.)   NIPA Table 1.14

Year and quarter	Current-dollar gross value added													Gross value added in billions of chained (2009) dollars
	Total	Consumption of fixed capital	Net value added											
			Total	Compensation of employees	Taxes on production and imports less subsidies	Net operating surplus								
						Total	Net interest and miscellaneous payments	Business current transfer payments	Corporate profits with IVA and CCAdj					
									Total	Taxes on corporate income	Profits after tax			
											Total	Net dividends	Undistributed	
1960	282.5	27.6	254.9	180.4	26.6	47.9	3.3	1.3	43.3	19.1	24.2	10.5	13.7	1 288.2
1961	290.4	28.6	261.9	184.5	27.6	49.7	3.8	1.4	44.5	19.4	25.1	10.6	14.5	1 319.1
1962	316.8	29.8	287.0	199.3	29.9	57.9	4.5	1.6	51.8	20.6	31.2	11.6	19.6	1 430.8
1963	337.8	31.3	306.5	210.1	31.7	64.7	4.8	1.6	58.3	22.8	35.5	12.4	23.1	1 519.0
1964	364.5	33.3	331.3	225.7	33.9	71.7	5.3	1.9	64.6	23.9	40.7	14.0	26.7	1 625.5
1965	400.7	35.8	364.9	245.4	36.0	83.5	6.0	2.1	75.4	27.1	48.3	16.2	32.1	1 762.0
1966	440.1	39.4	400.7	272.9	37.0	90.9	7.2	2.6	81.1	29.5	51.6	16.8	34.8	1 892.2
1967	463.2	43.5	419.7	291.1	39.3	89.4	8.6	2.7	78.1	27.8	50.2	17.3	33.0	1 946.4
1968	510.9	48.1	462.9	320.9	45.5	96.4	10.1	2.9	83.4	33.5	49.8	19.0	30.8	2 072.7
1969	554.6	53.5	501.1	356.1	50.2	94.9	13.5	3.0	78.3	33.3	45.0	19.0	25.9	2 155.6
1970	572.9	59.3	513.6	374.5	54.2	84.9	17.9	3.2	63.9	27.3	36.6	18.3	18.3	2 136.8
1971	618.1	64.7	553.4	396.2	59.5	97.7	18.9	3.6	75.2	30.0	45.2	18.1	27.0	2 221.0
1972	685.9	70.2	615.8	439.9	63.7	112.2	20.0	3.9	88.3	33.8	54.5	19.7	34.8	2 389.1
1973	768.8	78.2	690.7	495.1	70.1	125.5	23.7	4.5	97.2	40.4	56.8	20.8	36.1	2 534.8
1974	830.0	91.4	738.7	542.9	74.4	121.3	30.1	3.7	87.4	42.8	44.6	21.5	23.0	2 496.5
1975	898.3	107.7	790.6	569.0	80.2	141.3	32.4	4.7	104.3	41.9	62.4	24.6	37.8	2 461.9
1976	1 014.7	118.3	896.4	640.0	86.7	169.7	30.0	6.7	133.0	53.5	79.6	27.8	51.8	2 663.6
1977	1 147.4	132.6	1 014.8	723.3	94.6	196.9	33.0	8.7	155.1	60.6	94.6	30.9	63.6	2 860.6
1978	1 303.9	150.2	1 153.8	829.5	102.7	221.6	36.6	9.2	175.8	67.6	108.2	35.9	72.3	3 046.3
1979	1 450.3	172.4	1 277.8	942.4	108.8	226.7	43.6	9.0	174.0	70.6	103.4	37.6	65.8	3 144.1
1980	1 574.4	100.9	1 075.5	1 030.7	121.5	223.3	58.0	9.8	155.7	68.2	87.5	44.7	42.8	3 113.9
1981	1 788.7	229.1	1 559.6	1 139.9	146.7	273.1	72.5	10.7	189.9	66.0	123.9	52.5	71.4	3 240.6
1982	1 854.7	253.9	1 600.8	1 183.3	152.9	264.6	83.2	8.1	173.3	48.8	124.5	54.1	70.5	3 170.0
1983	1 985.8	264.6	1 721.2	1 250.1	168.0	303.1	82.6	10.0	210.5	61.7	148.8	63.2	85.6	3 324.9
1984	2 225.9	279.0	1 946.9	1 388.2	185.0	373.7	93.8	10.9	268.9	75.9	193.0	67.2	125.8	3 621.0
1985	2 369.0	299.0	2 070.0	1 490.1	196.6	383.3	96.9	15.4	271.1	71.1	200.0	72.0	128.0	3 787.4
1986	2 464.7	317.9	2 146.8	1 578.2	204.6	364.1	107.0	26.9	230.2	76.2	154.0	72.9	81.1	3 886.1
1987	2 641.7	334.7	2 307.0	1 685.5	216.8	404.7	115.7	29.7	259.4	94.2	165.1	76.3	88.9	4 092.1
1988	2 873.7	358.2	2 515.5	1 825.3	233.8	456.4	135.4	27.0	294.0	104.0	190.0	82.2	107.8	4 343.6
1989	3 020.0	380.2	2 639.9	1 934.8	248.2	456.9	159.4	23.5	274.0	101.2	172.8	105.4	67.5	4 426.2
1990	3 160.5	402.6	2 757.9	2 037.5	263.5	456.8	169.5	24.9	262.5	98.5	164.0	118.3	45.7	4 490.3
1991	3 231.0	423.6	2 807.4	2 071.1	285.7	450.6	156.4	26.1	268.1	88.6	179.5	125.5	54.0	4 467.0
1992	3 376.1	436.3	2 939.8	2 188.7	302.5	448.6	130.9	30.7	287.1	94.4	192.6	134.3	58.4	4 602.9
1993	3 534.4	455.8	3 078.6	2 271.0	319.3	488.2	117.0	29.5	341.7	108.0	233.7	149.2	84.5	4 716.7
1994	3 815.9	482.3	3 333.5	2 398.7	350.7	584.1	116.3	34.7	433.2	132.4	300.8	158.0	142.9	5 007.3
1995	4 040.6	520.8	3 519.8	2 524.6	358.7	636.6	125.6	30.2	480.8	140.3	340.5	178.0	162.5	5 249.1
1996	4 295.3	554.2	3 741.1	2 667.7	371.7	701.7	118.9	37.3	545.5	152.9	392.6	197.6	195.1	5 556.6
1997	4 608.0	594.6	4 013.4	2 862.6	388.9	762.0	126.6	38.5	596.9	161.4	435.5	215.9	219.6	5 928.1
1998	4 873.3	634.1	4 239.2	3 093.8	402.9	742.5	146.4	34.1	562.0	158.7	403.3	241.0	162.3	6 250.7
1999	5 169.8	679.1	4 490.7	3 310.0	424.6	756.1	159.8	45.6	550.7	166.4	384.3	224.7	159.6	6 570.2
2000	5 512.6	737.5	4 775.2	3 597.3	449.9	728.0	198.0	45.5	484.5	165.1	319.4	251.3	68.1	6 886.2
2001	5 466.0	781.5	4 684.5	3 582.3	441.5	660.7	219.8	55.1	385.8	106.2	279.7	245.4	34.2	6 713.7
2002	5 539.2	802.9	4 736.3	3 540.5	467.0	728.8	199.8	51.9	477.1	91.0	386.1	254.8	131.3	6 782.6
2003	5 723.9	816.1	4 907.8	3 594.4	491.1	822.3	171.2	61.5	589.6	126.5	463.2	293.4	169.8	6 930.4
2004	6 122.8	849.7	5 273.1	3 761.3	531.7	980.0	162.5	62.2	755.3	179.2	576.1	364.5	211.6	7 272.9
2005	6 559.8	911.0	5 648.8	3 928.7	575.7	1 144.4	176.9	78.9	888.7	262.6	626.1	170.8	455.3	7 537.1
2006	7 002.2	981.5	6 020.7	4 127.5	611.8	1 281.4	186.9	68.1	1 026.4	296.3	730.1	471.1	259.1	7 825.4
2007	7 219.7	1 044.0	6 175.7	4 307.1	631.3	1 237.4	250.3	58.6	928.5	277.7	650.8	484.6	166.2	7 896.1
2008	7 286.7	1 105.0	6 181.7	4 364.3	637.4	1 180.0	310.2	45.8	824.0	208.0	616.0	474.2	141.8	7 809.2
2009	6 864.6	1 110.7	5 753.8	4 094.9	608.5	1 050.4	284.2	60.0	706.2	162.3	543.9	351.4	192.5	7 255.1
2010	7 243.2	1 119.8	6 123.4	4 166.6	638.2	1 318.6	283.3	71.3	964.0	204.0	760.0	375.5	384.5	7 568.0
2011	7 615.3	1 169.3	6 446.0	4 372.7	670.8	1 402.4	283.2	79.3	1 039.9	209.3	830.6	441.0	389.6	7 774.1
2012	8 059.1	1 230.6	6 828.5	4 608.3	695.1	1 525.1	291.2	81.8	1 152.1	245.7	906.4	517.9	388.5	8 059.1
2013	8 374.1	1 279.7	7 094.4	4 768.1	742.0	1 584.3	280.1	83.9	1 220.2	263.6	956.6	531.9	424.7	8 261.5
2014	8 786.9	1 349.2	7 437.6	5 026.2	767.6	1 643.8	294.0	80.0	1 269.8	290.9	978.9	597.5	381.3	8 523.6
2015	9 139.0	1 405.9	7 733.2	5 290.1	784.1	1 658.9	310.9	97.3	1 250.8	283.4	967.3	641.1	326.2	8 812.5
2016	9 238.0	1 433.3	7 804.6	5 426.6	799.8	1 578.2	333.7	69.2	1 175.3	262.9	912.3	690.7	221.7	8 841.0
2017	9 607.3	1 495.6	8 111.7	5 698.5	835.0	1 578.2	319.5	95.5	1 163.2	232.5	930.7	681.2	249.5	9 014.2
2018	10 063.1	1 575.2	8 487.9	6 007.6	876.4	1 603.9	347.4	83.3	1 173.2	155.6	1 017.6	196.8	820.8	9 188.5
2016														
1st quarter	9 233.3	1 416.4	7 816.8	5 375.7	790.3	1 650.8	331.5	66.0	1 253.3	262.6	990.7	687.4	303.3	8 922.0
2nd quarter	9 181.4	1 428.8	7 752.6	5 399.8	792.7	1 560.1	336.4	64.1	1 159.6	267.3	892.3	691.3	200.9	8 787.6
3rd quarter	9 248.0	1 436.4	7 811.6	5 439.9	805.4	1 566.3	335.8	62.1	1 168.4	264.2	904.2	689.0	215.2	8 825.3
4th quarter	9 289.1	1 451.6	7 837.5	5 491.2	810.8	1 535.5	331.2	84.7	1 119.7	257.5	862.2	695.0	167.2	8 829.0
2017														
1st quarter	9 498.7	1 467.7	8 031.0	5 620.0	821.0	1 590.0	320.9	87.2	1 182.0	241.4	940.6	691.4	249.2	8 973.7
2nd quarter	9 596.2	1 485.6	8 110.7	5 665.5	829.8	1 615.3	316.3	97.7	1 201.3	228.4	972.9	724.7	248.2	9 049.2
3rd quarter	9 617.8	1 505.6	8 112.2	5 716.5	839.3	1 556.5	316.9	95.9	1 143.6	241.5	902.1	602.3	299.8	8 997.6
4th quarter	9 716.5	1 523.7	8 192.8	5 791.8	849.9	1 551.1	323.9	101.2	1 125.9	218.6	907.3	706.5	200.7	9 036.2
2018														
1st quarter	9 924.3	1 540.9	8 383.4	5 955.6	862.0	1 565.8	336.2	93.9	1 135.7	140.9	994.8	-214.2	1 208.9	9 110.8
2nd quarter	10 002.3	1 564.1	8 438.2	5 987.2	873.6	1 577.4	345.3	84.2	1 147.9	157.3	990.6	82.6	908.1	9 101.9
3rd quarter	10 127.2	1 586.4	8 540.8	6 033.1	878.8	1 629.0	351.2	76.9	1 200.9	166.8	1 034.1	514.6	519.5	9 242.8
4th quarter	10 198.7	1 609.4	8 589.3	6 054.6	891.2	1 643.4	356.8	78.3	1 208.3	157.5	1 050.8	404.0	646.8	9 298.6

Table 1-16. Shares of Income

(Billions of dollars, percent, quarterly data are at seasonally adjusted annual rates.)

NIPA Tables 1.10, 1.14

| Year and quarter | Gross domestic income, total economy | | | | | | | | | Net factor income, domestic corporate business | | | | |
| | Billions of dollars | | | | | Percent of total | | | | Billions of dollars | | | Percent of total | |
	Total	Compensation of employees	Taxes on production and imports less subsidies	Net operating surplus	Consumption of fixed capital	Compensation of employees	Taxes on production and imports less subsidies	Net operating surplus	Consumption of fixed capital	Total	Compensation of employees	Net operating surplus	Compensation	Net operating surplus
1959	521.5	285.9	40.0	130.1	65.4	54.8	7.7	24.9	12.5	234.3	180.3	54.0	77.0	23.0
1960	543.7	301.4	43.4	130.9	67.9	55.4	8.0	24.1	12.5	243.6	190.7	52.9	78.3	21.7
1961	563.1	310.6	45.0	137.0	70.6	55.2	8.0	24.3	12.5	250.3	195.6	54.7	78.1	21.9
1962	603.9	332.3	48.1	149.4	74.1	55.0	8.0	24.7	12.3	273.9	211.0	62.9	77.0	23.0
1963	638.9	350.5	51.2	159.3	78.0	54.9	8.0	24.9	12.2	291.8	222.7	69.1	76.3	23.7
1964	684.5	376.0	54.6	171.6	82.4	54.9	8.0	25.1	12.0	316.0	239.2	76.8	75.7	24.3
1965	741.5	405.4	57.7	190.4	88.0	54.7	7.8	25.7	11.9	348.9	259.9	89.0	74.5	25.5
1966	808.3	449.2	59.3	204.5	95.3	55.6	7.3	25.3	11.8	385.3	288.5	96.8	74.9	25.1
1967	856.6	481.8	64.1	207.2	103.6	56.2	7.5	24.2	12.1	404.7	308.4	96.3	76.2	23.8
1968	937.5	530.7	72.2	221.2	113.4	56.6	7.7	23.6	12.1	444.3	340.2	104.1	76.6	23.4
1969	1 016.0	584.4	79.4	227.4	124.9	57.5	7.8	22.4	12.3	482.1	377.5	104.6	78.3	21.7
1970	1 068.0	623.3	86.6	221.2	136.8	58.4	8.1	20.7	12.8	493.3	398.0	95.3	80.7	19.3
1971	1 155.3	665.0	95.8	245.6	148.9	57.6	8.3	21.3	12.9	530.5	421.7	108.8	79.5	20.5
1972	1 271.9	731.3	101.3	278.3	161.0	57.5	8.0	21.9	12.7	592.2	468.2	124.0	79.1	20.9
1973	1 419.2	812.7	112.0	315.9	178.7	57.3	7.9	22.3	12.6	664.1	526.1	138.0	79.2	20.8
1974	1 537.8	887.7	121.6	321.6	206.9	57.7	7.9	20.9	13.5	714.9	577.3	137.6	80.8	19.2
1975	1 671.6	947.3	130.8	355.0	238.5	56.7	7.8	21.2	14.3	767.5	607.8	159.7	79.2	20.8
1976	1 852.8	1 048.4	141.3	402.9	260.2	56.6	7.6	21.7	14.0	870.9	682.8	188.1	78.4	21.6
1977	2 062.4	1 165.9	152.6	454.1	289.8	56.5	7.4	22.0	14.1	992.0	771.7	220.3	77.8	22.2
1978	2 328.3	1 316.9	162.0	522.3	327.2	56.6	7.0	22.4	14.1	1 138.9	884.7	254.2	77.7	22.3
1979	2 582.3	1 477.3	171.6	559.5	373.9	57.2	6.6	21.7	14.5	1 265.9	1 004.4	261.5	79.3	20.7
1980	2 812.9	1 622.3	190.5	571.6	428.4	57.7	6.8	20.3	15.2	1 361.6	1 102.0	259.6	80.9	19.1
1981	3 169.0	1 792.7	224.1	664.9	487.2	56.6	7.1	21.0	15.4	1 535.5	1 220.6	314.9	79.5	20.5
1982	3 335.0	1 893.2	225.9	678.9	537.0	56.8	6.8	20.4	16.1	1 584.5	1 275.1	309.4	80.5	19.5
1983	3 577.1	2 012.7	242.0	759.8	562.6	56.3	6.8	21.2	15.7	1 709.1	1 353.0	356.1	79.2	20.8
1984	3 996.0	2 216.1	268.7	912.8	598.4	55.5	6.7	22.8	15.0	1 935.9	1 501.1	434.8	77.5	22.5
1985	4 284.7	2 387.5	286.7	970.3	640.1	55.7	6.7	22.6	14.9	2 065.4	1 615.9	449.5	78.2	21.8
1986	4 499.6	2 543.8	298.5	972.0	685.3	56.5	6.6	21.6	15.2	2 137.1	1 723.4	413.7	80.6	19.4
1987	4 811.4	2 723.8	317.2	1 040.0	730.4	56.6	6.6	21.6	15.2	2 290.2	1 847.6	442.6	80.7	19.3
1988	5 233.4	2 948.9	345.0	1 155.1	784.5	56.3	6.6	22.1	15.0	2 490.5	2 002.3	488.2	80.4	19.6
1989	5 573.6	3 140.9	371.5	1 223.0	838.3	56.4	6.7	21.9	15.0	2 625.3	2 119.3	506.0	80.7	19.3
1990	5 867.6	3 342.7	398.0	1 238.4	888.5	57.0	6.8	21.1	15.1	2 728.0	2 234.9	493.1	81.9	18.1
1991	6 065.2	3 453.3	429.6	1 249.9	932.4	56.9	7.1	20.6	15.4	2 766.9	2 277.8	489.1	82.3	17.7
1992	6 404.4	3 671.2	453.3	1 319.6	960.2	57.3	7.1	20.6	15.0	2 921.7	2 420.5	501.2	82.8	17.2
1993	6 702.6	3 820.6	466.4	1 412.1	1 003.5	57.0	7.0	21.1	15.0	3 057.4	2 515.7	541.7	82.3	17.7
1994	7 147.3	4 010.2	512.7	1 568.7	1 055.6	56.1	7.2	21.9	14.8	3 285.7	2 649.0	636.7	80.6	19.4
1995	7 546.7	4 202.2	523.1	1 699.1	1 122.4	55.7	6.9	22.5	14.9	3 498.2	2 787.9	710.3	79.7	20.3
1996	8 015.0	4 421.1	545.6	1 873.1	1 175.3	55.2	6.8	23.4	14.7	3 744.4	2 953.7	790.7	78.9	21.1
1997	8 566.0	4 713.2	577.8	2 035.6	1 239.3	55.0	6.7	23.8	14.5	4 054.9	3 176.6	878.3	78.3	21.7
1998	9 118.1	5 075.7	603.1	2 129.5	1 309.7	55.7	6.6	23.4	14.4	4 327.7	3 447.5	880.2	79.7	20.3
1999	9 663.8	5 409.9	628.4	2 226.6	1 398.9	56.0	6.5	23.0	14.5	4 578.5	3 684.3	894.2	80.5	19.5
2000	10 348.8	5 854.6	662.8	2 320.3	1 511.2	56.6	6.4	22.4	14.6	4 896.2	4 008.9	887.3	81.9	18.1
2001	10 695.0	6 046.3	669.0	2 380.1	1 599.5	56.5	6.3	22.3	15.0	4 844.0	4 013.8	830.2	82.9	17.1
2002	11 009.1	6 143.4	718.6	2 489.1	1 658.0	55.8	6.5	22.6	15.1	4 881.4	3 972.7	908.7	81.4	18.6
2003	11 471.9	6 362.3	756.5	2 634.0	1 719.1	55.5	6.6	23.0	15.0	5 058.7	4 040.9	1 017.8	79.9	20.1
2004	12 235.8	6 729.3	821.7	2 863.0	1 821.8	55.0	6.7	23.4	14.9	5 406.0	4 240.2	1 165.8	78.4	21.6
2005	13 091.7	7 077.7	881.5	3 161.5	1 971.0	54.1	6.7	24.1	15.1	5 815.2	4 443.0	1 372.2	76.4	23.6
2006	14 022.5	7 491.3	945.5	3 461.6	2 124.1	53.4	6.7	24.7	15.1	6 234.9	4 681.2	1 553.7	75.1	24.9
2007	14 434.2	7 889.4	982.2	3 309.8	2 252.8	54.7	6.8	22.9	15.6	6 315.4	4 894.2	1 421.2	77.5	22.5
2008	14 530.0	8 068.7	997.1	3 105.3	2 358.8	55.5	6.9	21.4	16.2	6 152.7	4 940.3	1 212.4	80.3	19.7
2009	14 256.8	7 767.2	968.5	3 149.6	2 371.5	54.5	6.8	22.1	16.6	5 884.7	4 607.5	1 277.2	78.3	21.7
2010	14 931.0	7 933.0	1 007.3	3 599.9	2 390.9	53.1	6.7	24.1	16.0	6 276.0	4 700.8	1 575.2	74.9	25.1
2011	15 595.8	8 234.0	1 043.7	3 843.6	2 474.5	52.8	6.7	24.6	15.9	6 584.2	4 928.0	1 656.2	74.8	25.2
2012	16 438.4	8 575.4	1 078.1	4 208.9	2 576.0	52.2	6.6	25.6	15.7	7 047.1	5 182.7	1 864.4	73.5	26.5
2013	16 945.2	8 843.6	1 129.0	4 291.4	2 681.2	52.2	6.7	25.3	15.8	7 234.8	5 352.4	1 882.4	74.0	26.0
2014	17 816.4	9 259.7	1 182.7	4 559.0	2 815.0	52.0	6.6	25.6	15.8	7 700.5	5 645.2	2 055.3	73.3	26.7
2015	18 479.7	9 709.2	1 219.8	4 634.1	2 916.5	52.5	6.6	25.1	15.8	8 052.1	5 941.8	2 110.3	73.8	26.2
2016	18 827.0	9 972.7	1 251.0	4 611.7	2 991.6	53.0	6.6	24.5	15.9	8 162.5	6 095.5	2 067.0	74.7	25.3
2017	19 587.0	10 424.5	1 303.4	4 737.7	3 121.4	53.2	6.7	24.2	15.9	8 429.5	6 412.9	2 016.6	76.1	23.9
2018	20 569.4	10 941.4	1 377.4	4 959.2	3 291.4	53.2	6.7	24.1	16.0	8 826.4	6 750.3	2 076.1	76.5	23.5
2016														
1st quarter	18 673.5	9 855.4	1 236.4	4 631.9	2 949.8	52.8	6.6	24.8	15.8	8 117.4	6 017.0	2 100.4	74.1	25.9
2nd quarter	18 718.3	9 912.4	1 238.8	4 587.3	2 979.8	53.0	6.6	24.5	15.9	8 118.1	6 054.2	2 063.9	74.6	25.4
3rd quarter	18 880.6	10 005.8	1 259.2	4 613.4	3 002.3	53.0	6.7	24.4	15.9	8 183.1	6 115.9	2 067.2	74.7	25.3
4th quarter	19 035.5	10 117.3	1 269.5	4 614.3	3 034.4	53.1	6.7	24.2	15.9	8 231.3	6 194.9	2 036.4	75.3	24.7
2017														
1st quarter	19 307.0	10 240.3	1 280.9	4 720.9	3 064.9	53.0	6.6	24.5	15.9	8 319.5	6 291.4	2 028.1	75.6	24.4
2nd quarter	19 496.9	10 347.1	1 296.4	4 751.7	3 101.7	53.1	6.6	24.4	15.9	8 411.9	6 362.0	2 049.9	75.6	24.4
3rd quarter	19 638.4	10 469.7	1 308.2	4 719.0	3 141.4	53.3	6.7	24.0	16.0	8 443.4	6 440.8	2 002.6	76.3	23.7
4th quarter	19 905.6	10 640.8	1 327.8	4 759.3	3 177.7	53.5	6.7	23.9	16.0	8 543.3	6 557.6	1 985.7	76.8	23.2
2018														
1st quarter	20 252.2	10 798.2	1 356.5	4 876.7	3 220.2	53.3	6.7	24.1	15.9	8 698.4	6 669.5	2 028.9	76.7	23.3
2nd quarter	20 460.1	10 888.8	1 375.3	4 924.2	3 271.6	53.2	6.7	24.1	16.0	8 772.6	6 718.6	2 054.0	76.6	23.4
3rd quarter	20 716.9	11 007.3	1 383.9	5 009.8	3 315.8	53.1	6.7	24.2	16.0	8 901.5	6 789.7	2 111.8	76.3	23.7
4th quarter	20 848.6	11 070.4	1 393.9	5 026.2	3 358.1	53.1	6.7	24.1	16.1	8 932.9	6 823.2	2 109.7	76.4	23.6

NOTES AND DEFINITIONS, CHAPTER 1

TABLES 1-1 THROUGH 1-16

National Income and Product

SOURCE: U.S. DEPARTMENT OF COMMERCE, BUREAU OF ECONOMIC ANALYSIS (BEA)

The mission of the Bureau of Economic Analysis (BEA) of the Department of Commerce is to promote a better understanding of the U.S. economy by providing timely, relevant, and accurate economic accounts. BEA quarterly and annually collects and produces data on how the U.S. economy is functioning. Featured in the NIPAs is gross domestic product (GDP), which measures the value of the goods and services produced by the U.S. economy in a given time period. GDP is one of the most comprehensive and closely watched economic statistics.

The National Income and Product Accounts (NIPA) are a set of economic accounts that provide information on the value and composition of output produced in the United States during a given period and on the types and uses of the income generated by that production. NIPA was created during the Great Depression to try to understand why the U.S. economy was in freefall and how to stop it.

U.S. input – output (I-O) accounts are a primary component of the U.S. economic accounts. I-O analysis is an economic tool that measures the relationships between various industries in the economy. These tables provides a detailed snapshot of the economy. More specifically, I-O tables show the commodity inputs that are used by each industry to produce its outputs; the commodities produced by each industry and the use of commodities by consumers.

NIPA is a snapshot of the economy at a certain time period. Over the past eight decades, new products have been created that use different inputs. To incorporate these changes, BEA conducts comprehensive revisions, called benchmarks, every five years. These comprehensive updates seek to introduce major improvements to maintain and improve the NIPA data.

The 2019 annual update of the National Income and Product Accounts was released on July 26, 2019. Data for 2014 through 2018 were revised. The reference year remained 2012. The updated statistics largely reflect the incorporation of newly available and revised source data and improvements to existing methodologies.

The 2018 comprehensive benchmark revision in July 2018 incorporated three major types of revisions: 1) statistical changes to introduce new and improved technologies and newly available source data; 2) changes in definitions to more accurately portray the evolving U.S. economy to more accurately portray the evolving U.S. economy and to provide consistent comparisons with data for other national economies; and 3) changes in presentations to reflect the definitional and statistical changes, where necessary, or to provide additional data or perspective for users. Also output and price measures will be expressed with 2012 equal to 100. The estimates for most tables showing real, or chained-dollar, estimates will begin with 2002.

Major statistical changes to the NIPA data include the 2012 benchmark input-output (I-O) accounts which incorporated the quinquennial economic census. The I-O accounts tracks the flows of detailed inputs and outputs throughout the economy. Another change is the incorporation of the improved methodology for seasonal adjustments over historical time spans and extending back improvements in previous annual updates. Other changes incorporated the handling of issues such as

- fixed investment in software, medical equipment, and communications equipment

- savings institutions and credit unions

- state and local defined benefit pension plans

Major changes in definitions in this comprehensive update are as follows:

- reclassifications of research and development (R & D) for software originals

- recognition of capital services

- reclassification of payments by Federal Reserve banks to the U.S. government

- reclassification of 'other' state and local personal taxes

The above revisions will be incorporated into the presentations produced by BEA.

Currently, BEA has adopted an expansion of the basic concept of production embodied in the accounts and changed the method of accounting for an important category of pension plans. Also, the constant-dollar or "real" estimates were restated to 2012 dollars, so that the levels shown are different even if the indicated rates of change are the same.

Fixed investment, one of the basic building blocks of the estimation of gross domestic product, was expanded in this revision by recognizing expenditures for research and development by business, government, and nonprofit institutions serving households (NPISHs) as investment.

The workers and capital that produce this R&D have always been reported in the labor and capital tabulations, but were treated as if they were costs of producing this year's output only. In the national income accounting as in conventional business accounting, the full costs of the R&D were subtracted from current-year business receipts in calculating business profit.

Now, in the NIPAs, the estimated value of the research over the whole life of the product—for example, from the estimated future sales of a new drug—is added to the value produced this year. However, this R&D will lose value over time and, as with a piece of machinery, its use and obsolescence, called consumption of fixed capital (CFC), is estimated and charged against GDP in each year that this capital is in use. For that reason, this change adds more to gross national product than to net national product and income.

The basic concept of the NIPAs requires equivalence (except for measurement error) between the value of output and the income and other charges against that value. The higher values of production are balanced on the income side of the accounts by the higher capital consumption allowances and by addition, to profits and other operating surplus (nonlabor income) components, of the net R&D investment—the difference between gross investment value and current year's CFC. (This does not change the tax accounting treatment. The firm still gets to charge off the entire expenditure in the year it is made, which is the most favorable treatment possible.)

Similarly, BEA is now capitalizing private spending for certain entertainment, literary, and other artistic originals as fixed investment. This includes movies, books, music, and long-lived TV programs such as situation comedies and drama (but not news, sports, games, or reality shows).

In a third expansion of the GDP investment definition, all ownership transfer costs for residential and nonresidential structures are capitalized (instead of only broker's commissions as in the previous treatment).

The other important conceptual change introduced in this revision was the adoption of accrual instead of cash accounting for defined benefit pension plans. This does not increase income and saving (or dis-saving) overall, but relocates it. Pension benefits are credited to the labor income and saving of employees when earned instead of when paid—matched by an equal and opposite change in the surplus or deficit of the employing government or business.

All the quarterly NIPA data published here, including indexes of quantity and price, are seasonally adjusted; this is not specifically noted on each table. All of the quarterly level values in current and constant dollars are also expressed at annual rates, which means that seasonally adjusted quarterly levels have been multiplied by 4 so that their scale will be comparable with the annual values shown, and this is noted on the pertinent *Business Statistics* tables. Where quarterly percent changes are shown, they are expressed as seasonally adjusted annual rates, which means that the quarter-to-quarter change in the level of GDP is calculated and then raised to the 4th power in order to express what that rate of change would be if continued over an entire year; this is noted in the column headings. For further information on these concepts see "General Notes" at the front of this volume.

Each table is notated, just above the table at the right-hand side, with the numbers of the NIPA tables from which the data are drawn.

All data and references are available on the BEA website, <http://www.bea.gov>.

DEFINITIONS AND NOTES ON THE DATA:

Basic concepts of total output and income (Tables 1-1, 1-11, 1-12, and 1-13)

The NIPAs depict the U.S. economy in several different dimensions. The basic concept, and the measure that is now most frequently cited, is gross domestic product (GDP), which is the market value of all goods and services produced by labor and property located in the United States.

In principle, GDP can be measured by summing the values created by each industry in the economy. However, it can also, and more readily, be measured by summing all the final demands for the economy's output. This final-demand approach also has the advantage of depicting the origins of demand for economic production, whether from consumers, businesses, or government.

Since production for the market necessarily generates incomes equal to its value, there is also an income total that corresponds to the production value total. This income can be measured and its distribution among labor, capital, and other income recipients can be depicted.

The relationships of several of these major concepts are illustrated in Table 1-11. The definitions of these concepts are as follows:

Gross domestic product (GDP), the featured measure of the value of U.S. output, is the market value of the goods and services produced by labor and property located in the United States. Market values represent output valued at the prices paid by the final customer, and therefore include taxes on production and imports, such as sales taxes, customs duties, and taxes on property.

The term "gross" in gross domestic product and gross national product is used to indicate that capital consumption allowances (economic depreciation) have not been deducted.

GDP is primarily measured by summing the values of all of the final demands in the economy, net of the demands met by imports; this is shown in Table 1-1. Specifically, GDP is the sum of personal consumption expenditures (PCE), gross private domestic investment (including change in private inventories and before deduction of charges for consumption of fixed capital), net exports of goods and services, and government consumption expenditures and gross investment. GDP measured in this way excludes duplication involving "intermediate" purchases of goods and services (goods and services purchased by industries and used in production), the value of which is already included in the value of the final products. Production of any intermediate goods that are not used in further production in the current period is captured in the measurement of inventory change.

In concept, GDP is equal to the sum of the economic value added by (also referred to as "gross product originating in") all industries in the United States. This, in turn, also makes it the conceptual equivalent of *gross domestic income (GDI)*, a new concept introduced in the 2003 revision. GDI is the sum of the incomes earned in each domestic industry, plus the taxes on production and imports and less the subsidies that account for the difference between output value and factor input value. This derivation is shown in Table 1-12. Since the incomes and taxes can be measured directly, they can be summed to a total that is equivalent to GDP in concept but differs due to imperfections in measurement. The difference between the two is known as the *statistical discrepancy*. It is defined as GDP minus GDI, and is shown in Tables 1-11 and 1-12.

Gross national product (GNP) refers to all goods and services produced by labor and property supplied by U.S. residents—whether located in the United States or abroad—expressed at market prices. It is equal to GDP, plus *income receipts from the rest of the world*, less *income payments to the rest of the world*. *Domestic* production and income refer to the location of the factors of production, with only factors located in the United States included; *national* production and income refer to the ownership of the factors of production, with only factors owned by United States residents included.

Before the comprehensive NIPA revisions that were made in 1991, GNP was the commonly used measure of U.S. production. However, GDP is clearly preferable to GNP when used in conjunction with indicators such as employment, hours worked, and capital utilized—for example, in the calculation of labor and capital productivity—because it is confined to production taking place within the borders of the United States.

The income-side aggregate corresponding to GNP is *gross national income (GNI)*, shown as an addendum to Table 1-11.

It consists of gross domestic income plus income receipts from the rest of the world, less income payments to the rest of the world. It is used as the denominator for a national saving-income ratio, presented in Chapter 5. National income is the preferred measure for calculating and comparing saving, since it is the income aggregate from which that saving arises. As with GDP and gross domestic income, the statistical discrepancy indicates the difference between the product-side and income-side measurement of the same concept.

Net national product is the market value, net of depreciation, of goods and services attributable to the labor and property supplied by U.S. residents. It is equal to GNP minus the *consumption of fixed capital (CFC)*. CFC relates only to fixed capital located in the United States. (Investment in that capital is measured by private fixed investment and government gross investment.) As of the 2009 comprehensive revision, CFC represents only the normal using-up of capital in the process of production, and no longer includes extraordinary disaster losses such as those caused by Hurricane Katrina and the 9/11 attacks. These losses are still estimated and used to write down the estimates of the capital stock, but they no longer have negative effects on our calculation of current income from production.

National income has been redefined and now includes all net incomes (net of the consumption of fixed capital) earned by U.S. residents, and also includes not only "factor incomes"—net incomes received by labor and capital as a result of their participation in the production process, but also "nonfactor charges"—taxes on production and imports, business transfer payments, and the current surplus of government enterprises, less subsidies. This change has been made to conform with the international guidelines for national accounts, *System of National Accounts (SNA) 1993*. According to *SNA 1993*, these charges cannot be eliminated from the input and output prices.

Since national income now includes the nonfactor charges, it is conceptually equivalent to *net national product* and differs only by the amount of the statistical discrepancy.

The concept formerly known as "national income," which excludes the nonfactor charges, is still included in the accounts as an addendum item, now called "net national factor income." It is shown in Table 1-13. *Net national factor income* consists of compensation of employees, proprietors' income with inventory valuation and capital consumption adjustments (IVA and CCadj, respectively), rental income of persons with capital consumption adjustment, corporate profits with inventory valuation and capital consumption adjustments, and net interest.

By definition, national income and its components exclude all income from capital gains (increases in the value of owned assets). Such increases have no counterpart on the production side of the accounts. This exclusion is partly accomplished by means of the inventory valuation and capital consumption adjustments, which will be described in the definitions of the components of product and income.

DEFINITIONS AND NOTES ON THE DATA:

Imputation

The term *imputation* will appear from time to time in the definitions of product and income components. Imputed values are values estimated by BEA statisticians for certain important product and income components that are not explicitly valued in the source data, usually because a market transaction in money terms is not involved. Imputed values appear on both the product and income side of the accounts; they add equal amounts to income and spending, so that no imputed saving is created.

One important example is the imputed rent on owner-occupied housing. The building of such housing is counted as investment, yet in the monetary accounts of the household sector, there is no income from that investment nor any rental paid for it. In the NIPAs, the rent that each such dwelling would earn if rented is estimated and added to both national and personal income (as part of rental income receipts) and to personal consumption expenditures (as part of expenditures on housing services).

Another important example is imputed interest. For example, many individuals keep monetary balances in a bank or other financial institution, receiving either no interest or below-market interest, but receiving the institution's services, such as clearing checks and otherwise facilitating payments, with little or no charge. In this case, where is the product generated by the institution's workers and capital? In the NIPAs, the depositor is imputed a market-rate-based interest return on his or her balance, which is then imputed as a service charge received by the institution, and therefore included in the value of the institution's output.

DEFINITIONS AND NOTES ON THE DATA:

Components of product (Tables 1-1 through 1-6)

Personal consumption expenditures (PCE) are goods and services purchased by persons residing in the United States. PCE consists mainly of purchases of new goods and services by individuals from businesses. It includes purchases that are financed by insurance, such as government-provided and private medical insurance. In addition, PCE includes purchases of new goods and services by nonprofit institutions, net purchases of used goods ("net" here indicates purchases of used goods from business less sales of used goods to business) by individuals and nonprofit institutions, and purchases abroad of goods and services by U.S. residents traveling or working in foreign countries. PCE also includes purchases for certain goods and services provided by

government agencies. (See the notes and definitions for Chapter 4 for additional information.) In the 2009 revision, new detail was provided on the allocation of PCE between the household and nonprofit sectors.

Gross private domestic investment consists of gross private fixed investment and change in private inventories.

Private fixed investment consists of both nonresidential and residential fixed investment. The term "residential" refers to the construction and equipping of living quarters for permanent occupancy. Hotels and motels are included in *nonresidential fixed investment*, as described subsequently in this section.

Private fixed investment consists of purchases of fixed assets, which are commodities that will be used in a production process for more than one year, including replacements and additions to the capital stock, and now also including research and development and other intellectual property. It is measured "gross," before a deduction for consumption of existing fixed capital. It covers all investment by private businesses and nonprofit institutions in the United States, regardless of whether the investment is owned by U.S. residents. The residential component includes investment in owner-occupied housing; the homeowner is treated equivalently to a business in these investment accounts. (However, when GDP by sector is calculated, owner-occupied housing is no longer included in the business sector. It is allocated to the households and institutions sector.) Private fixed investment does not include purchases of the same types of equipment and structures by government agencies, which are included in government gross investment, nor does it include investment by U.S. residents in other countries.

Nonresidential fixed investment is the total of nonresidential structures, nonresidential equipment, and intellectual property.

Nonresidential structures consists of new construction, brokers' commissions and other ownership transfer costs on sales of structures, and net purchases of used structures by private business and by nonprofit institutions from government agencies (that is, purchases of used structures from government minus sales of used structures to government). New construction also includes hotels and motels and mining exploration, shafts, and wells.

Nonresidential equipment consists of private business purchases on capital account of new machinery, equipment, and vehicles; dealers' margins on sales of used equipment; and net purchases of used equipment from government agencies, persons, and the rest of the world (that is, purchases of such equipment minus sales of such equipment). It does not include the estimated personal-use portion of equipment purchased for both business and personal use, which is allocated to PCE.

Intellectual property consists of purchases and in-house production of software; research and development; and private expenditures for specified entertainment, literary, and artistic originals. (See above for further description.)

Residential private fixed investment consists of both residential structures and residential producers' durable equipment (including such equipment as appliances owned by landlords and rented to tenants). Investment in structures consists of new units, improvements to existing units, purchases of manufactured homes, brokers' commissions and other ownership transfer costs on the sale of residential property, and net purchases of used residential structures from government agencies (that is, purchases of such structures from government minus sales of such structures to government). As noted above, it includes investment in owner-occupied housing.

Change in private inventories is the change in the physical volume of inventories held by businesses, with that change being valued at the average price of the period. It differs from the change in the book value of inventories reported by most businesses; an *inventory valuation adjustment (IVA)* converts book value change using historical cost valuations to the change in physical volume, valued at average replacement cost.

Net exports of goods and services is *exports of goods and services* less *imports of goods and services*. It does not include income payments or receipts or transfer payments to and from the rest of the world.

Government consumption expenditures is the estimated value of the services produced by governments (federal, state, and local) for current consumption. Since these are generally not sold, there is no market valuation and they are priced at the cost of inputs. The input costs consist of the compensation of general government employees; the estimated consumption of general government fixed capital including software and R&D (CFC, or economic depreciation); and the cost of goods and services purchased by government less the value of sales to other sectors. The value of investment in equipment and structures produced by government workers and capital is also subtracted, and is instead included in government investment. Government sales to other sectors consist primarily of receipts of tuition payments for higher education and receipts of charges for medical care.

This definition of government consumption expenditures differs in concept—but not in the amount contributed to GDP—from the treatment in existence before the 2003 revision of the NIPAs. In the current definition, goods and services purchased by government are considered to be intermediate output. In the previous definition, they were considered to be final sales. Since their value is added to the other components to yield total government consumption expenditures, the dollar total contributed to GDP is the same. The only practical difference is that the goods purchased disappear from the goods account and appear in the services account instead. In the industry sector accounts, the value added by government is also unchanged. It

continues to be measured as the sum of compensation and CFC, or equivalently as gross government output less the value of goods and services purchased.

Gross government investment consists of general government and government enterprise expenditures for fixed assets (structures, equipment, software, and R&D). Government inventory investment is included in government consumption expenditures.

DEFINITIONS AND NOTES ON THE DATA:

Real values, quantity and price indexes (Tables 1-2 through 1-7)

Real, or chained (2012) dollar, estimates are estimates from which the effect of price change has been removed. Prior to the 1996 comprehensive revision, constant-dollar measures were obtained by combining real output measures for different goods and services using the relative prices of a single year as weights for the entire time span of the series. In the recent environment of rapid technological change, which has caused the prices of computers and electronic components to decline rapidly relative to other prices, this method distorts the measurement of economic growth and causes excessive revisions of growth rates at each benchmark revision. The current, chained-dollar measure changes the relative price weights each year, as relative prices shift over time. As a result, recent changes in relative prices do not change historical growth rates.

Chained-dollar estimates, although expressed for continuity's sake as if they had occurred according to the prices of a single year (currently 2009), are usually not additive. This means that because of the changes in price weights each year, the chained (2009) dollar components in any given table for any year other than 2009 usually do not add to the chained (2009) dollar total. The amount of the difference for the major components of GDP is called the *residual* and is shown in Table 1-2. It is specific to each individual BEA tabulation, corresponding to the sum of the lowest level of aggregation on that tabulation. In time periods close to the base year, residuals are usually quite small; over longer periods, the differences become much larger. For this reason, BEA no longer publishes chained-dollar estimates prior to 1999, except for selected aggregate series. For the more detailed components of GDP, historical trends and fluctuations in real volumes are represented by *chain-type quantity indexes,* which are presented in Tables 1-5, 4-3, 5-4, 5-6, 6-6, 6-7, 6-11, 6-16, 7-2, 7-5, and 14-2.

Chain-weighting leads to complexity in estimating the contribution of economic sectors to an overall change in output: it becomes difficult, for someone without access to the complicated statistical methods that BEA uses, to find the correct answers to questions such as "How much are government spending cuts contributing to the drag on GDP growth?" Because of this, BEA is now calculating and publishing estimates of the arithmetic contribution of each major component to the total change in real GDP. *Business Statistics* reproduces these calculations in Table 1-4. (As will be explained later, users of these calculations should, however, bear in mind that imports are treated as a negative contribution to GDP instead of being subtracted from the demand components that give rise to them.) For further information, see J. Steven Landefeld, Brent R. Moulton, and Cindy M. Vojtech, "Chained-Dollar Indexes: Issues, Tips on Their Use, and Upcoming Changes," *Survey of Current Business* (November 2003); and J. Steven Landefeld and Robert P. Parker, "BEA's Chain Indexes, Time Series, and Measures of Long-Term Economic Growth," *Survey of Current Business* (May 1997).

GDP price indexes measure price changes between any two adjacent years (or quarters) for a fixed "market basket" of goods and services consisting of the average quantities purchased in those two years (or quarters). The annual measures are chained together to form an index with prices in 2009 set to equal 100. Using average quantities as weights while changing weights each period eliminates the substitution bias that arises in more conventional indexes, in which weights are taken from a single base period that usually takes place early in the period under measurement. Generally, using a single, early base period leads to an overstatement of price increase. The CPI-U and the CPI-W are examples of such conventional indexes, technically known as "Laspeyres" indexes. See the "General Notes" at the beginning of this volume and the notes and definitions for Chapter 8 for further explanation.

The chain-type formula guarantees that a GDP price index change will differ only trivially from the change in the implicit deflator (ratio of current-dollar to real value, expressed as a percent). Therefore, *Business Statistics* is no longer publishing a separate table of implicit deflators.

DEFINITIONS AND NOTES ON THE DATA:

BEA Aggregates of sales and purchases (Tables 1-1 and 1-5 through 1-7)

Final sales of domestic product is GDP minus change in private inventories. It is the sum of personal consumption expenditures, gross private domestic fixed investment, government consumption expenditures and gross investment, and net exports of goods and services.

Gross domestic purchases is the market value of goods and services purchased by U.S. residents, regardless of where those goods and services were produced. It is GDP minus net exports (that is, minus exports plus imports) of goods and services; equivalently, it is the sum of personal consumption expenditures, gross private domestic investment, and government consumption expenditures and gross investment. The price index for gross

domestic purchases is therefore a measure of price change for goods and services purchased by (rather than produced by) U.S. residents.

Final sales to domestic purchasers is gross domestic purchases minus change in private inventories.

DEFINITIONS AND NOTES ON THE DATA:

U.S. Population and per capita product and income estimates (Table 1-3)

In Table 1-3, annual and quarterly measures of product, income, and consumption spending are expressed in per capita terms— the aggregate dollar amount divided by the U.S. population. Population data from 1991 forward reflect the results of the 2000 and 2010 Censuses.

DEFINITIONS AND NOTES ON THE DATA:

Inventory-sales ratios (Table 1-8)

Inventories to sales ratios. The ratios shown in Table 1-8 are based on the inventory estimates underlying the measurement of inventory change in the NIPAs. They include data and estimates for not only the inventories held in manufacturing and trade (which are shown in Chapter 16), but also stocks held by all other businesses in the U.S. economy.

For the current-dollar ratios, inventories at the end of each quarter are valued in the prices that prevailed at the end of that quarter. For the constant-dollar ratios, they are valued in chained (2009) dollars. In both cases, the inventory-sales ratio is the value of the inventories at the end of the quarter divided by quarterly total sales at <u>monthly</u> rates (quarterly totals divided by 3). In other words, they represent how many months' supply businesses had on hand at the end of the period. This makes them comparable in concept and order of magnitude to the ratios shown in Chapter 16. Annual ratios are those for the fourth quarter.

Inventory data consistent with the 2013 revision are only available from 1997 to date. Tables 1-8A and B show the old estimates for earlier years with an overlap period.

DEFINITIONS AND NOTES ON THE DATA:

Shares of aggregate supply and demand (Tables 1-9 and 1-10)

These tables, developed by the editor of *Business Statistics*, are not official NIPA calculations. They are components of current-dollar GDP rearranged in order to highlight relationships that are not always apparent in the official presentation of the NIPAs.

Aggregate supply combines domestic production (GDP) and imports, the two sources that, between them, supply the goods and services demanded by consumers, businesses, and governments in the U.S. economy. In this table the user can observe the growing share of imports that satisfy demands in the U.S. marketplace.

Aggregate final demand is the sum of all final (that is, excluding inventory change) demands for goods and services in the U.S. market—consumption spending, fixed investment, exports, and government consumption and investment. It is different from *final sales of domestic product* (Tables 1-1 and 1-7) because imports are not subtracted; in this table, imports are considered a source of supply, not a negative element of demand. It is different from *gross domestic purchases* (Tables 1-5 and 1-6) because it includes exports, since they are a source of demand for U.S. output, but excludes inventory change. It is like *final sales to domestic purchasers* (Table 1-7) in excluding inventory change, but different because it also includes exports.

Table 1-10 provides alternative data on the question of the relative importance of consumption spending and other final demands to the U.S. economy. NIPA statistics on PCE as a percent of GDP are frequently cited, but there is a problem with this, since PCE includes the value of imports while GDP does not.

DEFINITIONS AND NOTES ON THE DATA:

Components of income (Tables 1-11, 1-12 and 1-13)

There are now two different presentations of aggregate income for the United States: *gross domestic income* (Table 1-12) and *national income* (Table 1-13). As noted above, domestic income refers to income generated from production within the United States, while national income refers to income received by residents of the United States. This means that some of the income components differ between the two tables. Domestic income payments include payments to the rest of the world from domestic industries. National income payments exclude payments to the rest of the world but include payments received by U.S. residents from the rest of the world. These differences are seen in employee compensation, interest, and corporate profits. Taxes on production and imports, taxes on corporate income, business transfer payments, subsidies, proprietors' income, rental income, and the current surplus of government enterprises are the same in both accounts.

All income entries are now calculated on an accrual basis, associating the income with the period in which it was earned rather than when it was paid.

A third important income aggregate is *personal income.* The derivation of this well-known statistic from national income is shown in Table 1-11. See Chapter 4 and its notes and definitions for data and more information.

Compensation of employees is the income accruing to employees as remuneration for their work. It is the sum of wage and salary accruals and supplements to wages and salaries. In the domestic income account, it refers to all payments generated by domestic production, including those to workers residing in the "rest of the world." In the national and personal income accounts, there is a slightly different "compensation of employees," including that received by U.S. residents from the rest of the world but excluding that paid from the domestic production account to workers residing elsewhere.

Wages and salaries consists of the monetary remuneration of employees, including the compensation of corporate officers; corporate directors' fees paid to directors who are also employees of the corporation; commissions, tips, and bonuses; voluntary employee contributions to certain deferred compensation plans, such as 401(k) plans; and receipts-in-kind that represent income. As of the 2003 revision, it also includes judicial fees to jurors and witnesses, compensation of prison inmates, and marriage fees to justices of the peace, all of which were formerly included in "other labor income."

In concept, wages and salaries include the value of the exercise by employees of "nonqualified stock options," in which an employee is allowed to buy stock for less than its current market price. (Actual measurement of these values involves a number of problems, particularly in the short run. Such stock options are not included in the monthly wage data from the Bureau of Labor Statistics, which are the main source for current extrapolations of wages and salaries, and are not consistently reported in corporate financial statements. They are, however, generally included in the unemployment insurance wage data that are used to correct the preliminary wage and salary estimates.) Another form of stock option, the "incentive stock option," leads to a capital gain only and is thus not included in the definition of wages and salaries.

Supplements to wages and salaries consists of *employer contributions for employee pension and insurance funds* and *employer contributions for government social insurance.*

Employer contributions for employee pension and insurance funds consists of employer payments (including payments-in-kind) to private pension and profit-sharing plans, private group health and life insurance plans, privately administered workers' compensation plans, government employee retirement plans, and supplemental unemployment benefit plans. They are now measured in the period in which the employee earns the obligation rather than when the employer makes the cash payment into the fund. This includes the major part of what was once called "other labor income." The remainder of "other labor income" has been reclassified as wages and salaries, as noted above.

Employer contributions for government social insurance consists of employer payments under the following federal, state, and local government programs: old-age, survivors, and disability insurance (Social Security); hospital insurance (Medicare); unemployment insurance; railroad retirement; pension benefit guaranty; veterans' life insurance; publicly administered workers' compensation; military medical insurance; and temporary disability insurance.

Taxes on production and imports is included in the gross domestic income account to make it comparable in concept to gross domestic product. It consists of federal excise taxes and customs duties and of state and local sales taxes, property taxes (including residential real estate taxes), motor vehicle license taxes, severance taxes, special assessments, and other taxes. It is equal to the former "indirect business taxes and nontax liabilities" less most of the nontax liabilities, which have now been reclassified as "business transfer payments."

Subsidies (payments by government to business other than purchases of goods and services) are now presented separately from the current surplus of government enterprises, which is presented as a component of net operating surplus. However, for data representing the years before 1959, subsidies continue to be presented as net of the current surplus of government enterprises, since detailed data to separate the series for this period are not available.

Net operating surplus is a new aggregate introduced in the 2003 NIPA revision—a grouping of the business income components of the gross domestic income account. It represents the net income accruing to business capital. It is equal to gross domestic income minus compensation of employees, taxes on production and imports less subsidies (that is, the taxes are taken out of income and the subsidies are put in), and consumption of fixed capital (CFC). Net operating surplus consists of the surplus for private enterprises and the current surplus of government enterprises. The net operating surplus of private enterprises comprises net interest and miscellaneous payments, business current transfer payments, proprietors' income, rental income of persons, and corporate profits.

Net interest and miscellaneous payments, domestic industries consists of interest paid by domestic private enterprises and of rents and royalties paid by private enterprises to government, less interest received by domestic private enterprises. Interest received does not include interest received by noninsured pension plans, which are recorded as being directly received by persons in personal income. Both interest categories include monetary and imputed interest. In the *national* account, interest paid to the rest of the world is subtracted from the interest paid by domestic industries and interest received from the rest of the world is

added. Interest payments on mortgage and home improvement loans and on home equity loans are included as net interest in the private enterprises account.

It should be noted that net interest does not include interest paid by federal, state, or local governments. In fact, government interest does not enter into the national and domestic income accounts, though it does appear as a component of personal income. The NIPAs draw a distinction between interest paid by government and that paid by business.

The reasoning behind this distinction is that interest paid by <u>business</u> is one of the income counterparts of the production side of the account. The value of business production (as measured by its output of goods and services) includes the value added by business capital, and interest paid by business to its lenders is part of the total return to business capital.

However, there is no product flow in the accounts that is a counterpart to the payment of interest by <u>government</u>. The output of government does not have a market value. For purposes of GDP measurement, BEA estimates the government contribution to GDP as the sum of government's compensation of employees, purchases of goods and services, and consumption of government fixed capital. (See above, and also the notes and definitions to Chapter 6.) This implies an estimate (described as "conservative" by BEA) that the <u>net</u> return to government capital is zero—that is, that the gross return is just sufficient to pay down the depreciation. Consequently, this assumption generates no income, imputed or actual, that might correspond to the government's interest payment.

Supporting the distinction between business and government interest payments, it may be noted that most federal government debt was not incurred to finance investment, but rather to finance wars, to avoid tax increases and spending cuts during recessions, or to stimulate the economy. Furthermore, some of the largest and most productive government investments—highways—are typically financed by taxes on a pay-as-you-go basis and not by borrowing.

Business current transfer payments, net consists of payments to persons, government, and the rest of the world by private business for which no current services are performed. Net insurance settlements—actual insured losses (or claims payable) less a normal level of losses—are treated as transfer payments. Payments to government consist of federal deposit insurance premiums, fines, regulatory and inspection fees, tobacco settlements, and other miscellaneous payments previously classified as "nontaxes." Taxes paid by domestic corporations to foreign governments, formerly classified as transfer payments, are now counted as taxes on corporate income.

In the NIPAs, capital income other than interest—corporate profits, proprietors' income, and rental income—is converted from the basis usually shown in the books of business, and reported to the Internal Revenue Service, to a basis that more closely represents income from current production. In the business accounts that provide the source data, depreciation of structures and equipment typically reflects a historical cost basis and a possibly arbitrary service life allowed by law to be used for tax purposes. BEA adjusts these values to reflect the average actual life of the capital goods and the cost of replacing them in the current period's prices. This conversion is done for all three forms of capital income. In addition, corporate and proprietors' incomes also require an adjustment for inventory valuation to exclude any profits or losses that might appear in the books, should the cost of inventory acquisition not be valued in the current period's prices. These two adjustments are called the *capital consumption adjustment (CCAdj)* and the *inventory valuation adjustment (IVA)*. They are described in more detail below.

Proprietors' income with inventory valuation and capital consumption adjustments is the current-production income (including income-in-kind) of sole proprietorships and partnerships and of tax-exempt cooperatives. The imputed net rental income of owner-occupants of farm dwellings is included, but the imputed net rental income of owner-occupants of nonfarm dwellings is included in rental income of persons. Fees paid to outside directors of corporations are included. Proprietors' income excludes dividends and monetary interest received by nonfinancial business and rental incomes received by persons not primarily engaged in the real estate business; these incomes are included in dividends, net interest, and rental income of persons, respectively.

Rental income of persons with capital consumption adjustment is the net current-production income of persons from the rental of real property (except for the income of persons primarily engaged in the real estate business), the imputed net rental income of owner-occupants of nonfarm dwellings, and the royalties received by persons from patents, copyrights, and rights to natural resources. Consistent with the classification of investment in owner-occupied housing as business investment, the homeowner is considered to be paying himself or herself the rental value of the house (classified as PCE for services) and receiving as net income the amount of the rental that remains after paying interest and other costs.

Corporate profits with inventory valuation and capital consumption adjustments, often referred to as "economic profits," is the current-production income, net of economic depreciation, of organizations treated as corporations in the NIPAs. These organizations consist of all entities required to file federal corporate tax returns, including mutual financial institutions and cooperatives subject to federal income tax; private noninsured pension funds; nonprofit institutions that primarily serve business; Federal Reserve Banks, which accrue income stemming from the conduct of monetary policy; and federally sponsored credit

agencies. This income is measured as receipts less expenses as defined in federal tax law, except for the following differences: receipts exclude capital gains and dividends received; expenses exclude depletion and capital losses and losses resulting from bad debts; inventory withdrawals are valued at replacement cost; and depreciation is on a consistent accounting basis and is valued at replacement cost.

Since *national* income is defined as the income of U.S. residents, its profits component includes income earned abroad by U.S. corporations and excludes income earned by the rest of the world within the United States.

Taxes on corporate income consists of taxes on corporate income paid to government and to the rest of the world.

Taxes on corporate income paid to government is the sum of federal, state, and local income taxes on all income subject to taxes. This income includes capital gains and other income excluded from profits before tax. These taxes are measured on an accrual basis, net of applicable tax credits.

Taxes on corporate income paid to the rest of the world consists of nonresident taxes, which are those paid by domestic corporations to foreign governments. These taxes were formerly classified as "business transfer payments to the rest of the world."

Profits after tax is total corporate profits with IVA and CCAdj less taxes on corporate income. It consists of dividends and undistributed corporate profits.

Dividends is payments in cash or other assets, excluding those made using corporations' own stock, that are made by corporations to stockholders. In the domestic account, these are payments by domestic industries to stockholders in the United States and abroad; in the national account, these are dividends received by U.S. residents from domestic and foreign industries. The payments are measured net of dividends received by U.S. corporations. Dividends paid to state and local government social insurance funds and general government are included.

Undistributed profits is corporate profits after tax with IVA and CCAdj less dividends.

The *inventory valuation adjustment (IVA)* is the difference between the cost of inventory withdrawals valued at replacement cost and the cost as valued in the source data used to determine profits before tax, which in many cases charge inventories at acquisition cost. It is calculated separately for corporate profits and for nonfarm proprietors' income. Its behavior is determined by price changes, especially for materials. When prices are rising, which has been typical of much of the postwar period, the business-reported value of inventory change will include a capital gains component, which needs to be removed from reported inventory change on the product side in order to correctly measure the change in the volume of inventories, and from reported profits on the income side of the accounts in order to remove the capital gains element. At such times, the IVA will be a negative figure, which is added to reported profits to yield economic profits. Occasionally, falling prices—especially for petroleum and products—will result in a positive IVA. No adjustment is needed for farm proprietors' income, as farm inventories are measured on a current-market-cost basis.

Consumption of fixed capital (CFC) is a charge for the using-up of private and government fixed capital located in the United States. It is not based on the depreciation schedules allowed in tax law, but instead on studies of prices of used equipment and structures in resale markets and other service life information.

For general government and for nonprofit institutions that primarily serve individuals, CFC on their capital assets is recorded in government consumption expenditures and in personal consumption expenditures, respectively. It is considered to be the value of the current services of the fixed capital assets owned and used by these entities.

Private capital consumption allowances consists of tax-return-based depreciation charges for corporations and nonfarm proprietorships; BEA estimates for R&D and other intellectual property; and historical cost depreciation (calculated by BEA using a geometric pattern of price declines) for farm proprietorships, rental income of persons, and nonprofit institutions.

The *private capital consumption adjustment (CCAdj)* is the difference between private capital consumption allowances and private consumption of fixed capital. The CCAdj has two parts:

- The first component of CCAdj converts tax-return-based depreciation to consistent historical cost accounting based on actual service lives of capital. In the postwar period, this has usually been a large positive number, that is, a net addition to profits and subtraction from reported depreciation. This is the case because U.S. tax law typically allows depreciation periods shorter than actual service lives. Tax depreciation was accelerated even further for 2001 through 2004. This component is a reallocation of gross business saving from depreciation to profits; gross saving is unchanged, with exactly offsetting changes in capital consumption and net saving.

- The second component is analogous to the IVA: it converts reported business capital consumption allowances from the historical cost basis to a replacement cost basis. It is determined by the price behavior of capital goods. These prices have had an upward drift in the postwar period, which has been much more stable than the changes in materials prices. Hence, this component is consistently negative (serving to reduce economic profits relative to the reported data) but less volatile than the IVA.

In 1982 through 2004, positive values for the first component outweighed negative values for the second, resulting in a net positive CCAdj, an addition to profits. However, when the accelerated depreciation expired in 2005, there was a sharp decline in the consistent-accounting adjustment and it was outweighed by the price adjustment. This led to negative CCAdjs in 2005 through 2007, reducing profits from the reported numbers.

DEFINITIONS AND NOTES ON THE DATA:

Gross value added of domestic corporate business (Tables 1-14 and 1-15)

Gross value added is the term now used for what was formerly called "gross domestic product originating." It represents the share of the GDP that is produced in the specified sector or industry. Tables 1-14 and 1-15 show the current-dollar value of gross value added for all domestic corporate business and its financial and nonfinancial components. For the total and for nonfinancial corporations, consumption of fixed capital and net value added are shown, as is the allocation of net value added among employee compensation, taxes and transfer payments, and capital income. Constant-dollar values are also shown for nonfinancial corporations.

The data for nonfinancial corporations are often considered to be somewhat sturdier than data for the other sectors of the economy, since they exclude sectors whose outputs are difficult to evaluate—households, institutions, general government, and financial business—as well as excluding all noncorporate business, in which the separate contributions of labor and capital are not readily measured.

DATA AVAILABILITY AND REVISIONS

Annual data are available beginning with 1929. Quarterly data begin with 1946 for current-dollar values and 1947 for quantity and price measures such as real GDP and the GDP price index. Not all data are available for all time periods.

New data are normally released toward the end of each month. The "advance" estimate of GDP for each calendar quarter is released at the end of the month after the quarter's end. The "second" estimate, including more complete product data and the first estimates of corporate profits, is released at the end of the second month after the quarter's end, and a "third" estimate including still more complete data at the end of the third month. Wage and salary and related income-side components may be revised for previous quarters as well.

At the end of each July, there is an "annual" revision, incorporating more complete data and other improvements, affecting at least the previous 3 years. Every five years, there is a "comprehensive" revision, such as the 2013 revision incorporated in this volume, corresponding with updated statistics from the quinquennial benchmark input-output accounts and incorporating a revision of the base year for constant-price and index numbers.

The most recent data are published each month in the *Survey of Current Business*. Current and historical data may be obtained from the BEA Web site at <http://www.bea.gov> and the STAT-USA subscription Web site at <http://www.stat-usa.gov>.

REFERENCES

The 2019 annual revision of the NIPAs is presented and described in a July 26, 2019 release on the BEA Web site.

Other documentation available on the BEA Web site at <http://www.bea.gov> includes the following: "NIPA Handbook: Concepts and Methods of the U.S. National Income and Product Accounts, October 2009"; separate chapters from the Handbook on Personal Consumption Expenditures, Private Fixed Investment, and Change in Private Inventories; "Measuring the Economy: A Primer on GDP and the National Income and Product Accounts"; "An Introduction to the National Income and Product Accounts"; and "Taking the Pulse of the Economy: Measuring GDP," *Journal of Economic Perspectives*, Spring 2008.

The treatment of employee stock options is discussed in Carol Moylan, "Treatment of Employee Stock Options in the U.S. National Economic Accounts," available on the BEA Web site at <http://www.bea.gov>.

The data for 1929 through 1946 published here have been calculated after the fact and differ from the national income data that were currently available during the 1930s and 1940s. For an article on what was available at that time and the history of the NIPAs during that period, see Rosemary D. Marcuss and Richard E. Kane, "U.S. National Income and Product Statistics: Born of the Great Depression and World War II," *Survey of Current Business,* February 2007, pp. 32-46.

CHAPTER 2: INDUSTRIAL PRODUCTION AND CAPACITY UTILIZATION

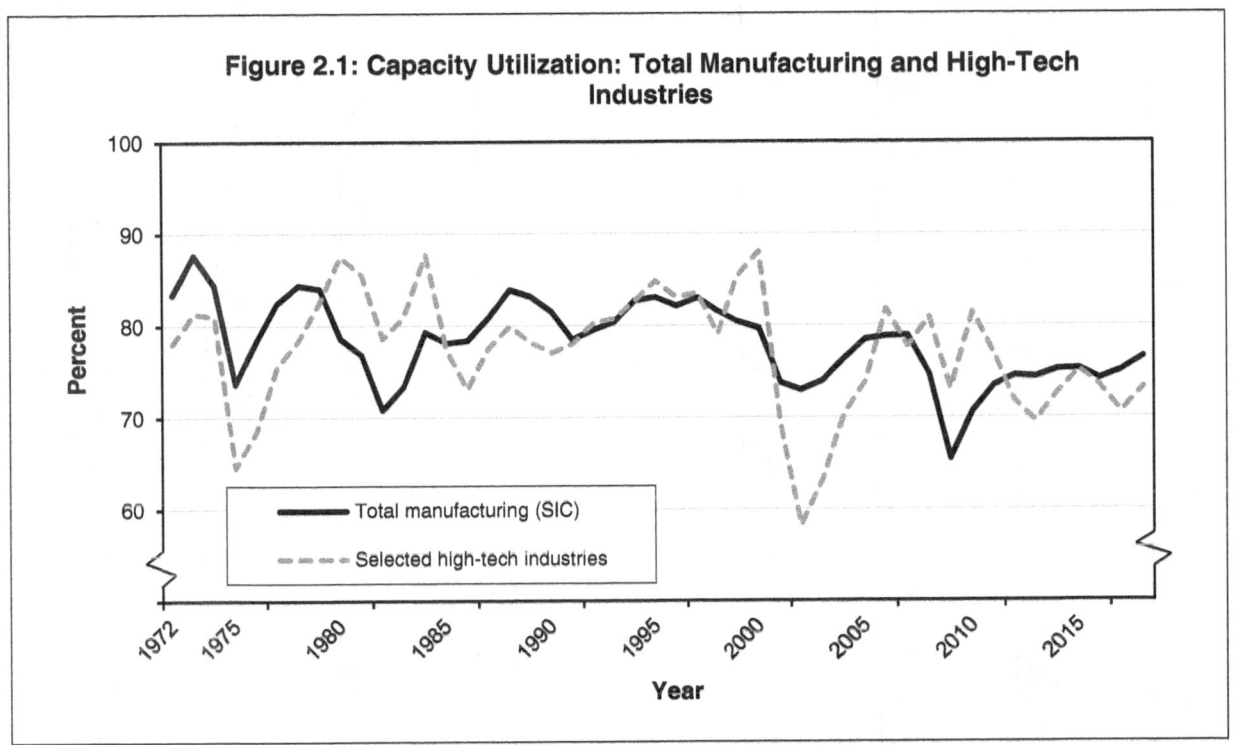

Figure 2.1: Capacity Utilization: Total Manufacturing and High-Tech Industries

- Manufacturing capacity utilization is a key statistic for the U.S. economy, despite being limited to a sector that by some measures has diminished in importance. The Federal Reserve also provides measures of capacity utilization for "total industry"—manufacturing, mining, and utilities. However, mining and utilities are less significant in the context of business cycle analysis, and much of the variation in capacity use by utilities is a result of transitory weather variations, not economic factors. Manufacturing utilization is an important indicator of inflationary pressure and also measures an important element in the demand for new capital goods. (Tables 2-3 and 2-4A)

- In 2018, industrial production reached an all-time high of 108.6 while increasing for most categories listed on Table 2-2. However, it decreased for printing and support, petroleum and coal products, and miscellaneous manufacturing. (Table 2-2)

- With higher interest rates spurred by Federal Reserve policy, business investment on equipment, construction product, land, raw materials, etc. will become more expensive. Consequently, industrial production may decline. Also, with tariffs on such raw materials as steel and aluminum, manufacturing may decline because of increased prices.

- Manufacturing capacity utilization plunged in 2008 and 2009 following the great recession. In November 2007, capacity utilization reached 81.1 but by May 2009, it collapsed to 67.0. During the 18 months of the Great Recession, the rate declined significantly. Manufacturing capacity in both 2017 and 2018 fluctuated monthly, never able to build up steam. (Table 2-4A)

- Capacity utilization in the high-tech industries (computers and office equipment, communications equipment, and semiconductors and related electronic components) peaked in 2000, before the Tech Bubble, at 88.0 but fell nearly 34 percent by 2002. Capacity utilization has recovered but not to its 2000 level as seen in Figure 2-1. (Tables 2-3 and 2-4A)

Table 2-1. Industrial Production Indexes by Market Groups

(Seasonally adjusted, 2012 = 100.)

Year and month	Total industrial production	Final products and nonindustrial supplies Total	Consumer goods Total	Durable consumer goods Total	Automotive products	Home electronics	Appliances, furniture, and carpeting	Miscellaneous durable goods	Nondurable consumer goods Total
1970	38.6	39.7	52.8	39.6	35.3	1.0	76.1	61.0	59.4
1971	39.1	40.2	55.9	44.9	45.0	1.1	80.3	64.3	61.1
1972	42.9	44.0	60.4	50.3	48.5	1.2	94.3	72.2	65.0
1973	46.4	47.3	63.1	54.1	52.7	1.5	101.9	75.5	66.9
1974	46.3	47.1	61.3	49.2	45.6	1.3	92.9	72.8	67.0
1975	42.2	43.8	58.8	44.7	44.0	1.2	79.9	64.2	65.7
1976	45.5	46.9	63.6	50.5	50.0	1.4	90.1	71.6	69.9
1977	48.9	50.7	67.5	56.7	56.6	1.6	101.2	79.4	72.4
1978	51.6	53.8	69.7	58.1	56.2	1.8	106.6	82.2	74.9
1979	53.2	55.6	68.6	56.0	50.6	1.8	107.2	82.9	74.4
1980	51.8	54.8	65.9	48.7	39.0	1.8	99.8	76.1	74.5
1981	52.5	55.9	66.4	49.4	40.2	1.9	98.5	77.1	74.8
1982	49.8	54.4	66.2	46.6	39.0	1.7	89.0	72.7	76.1
1983	51.1	56.0	68.6	51.7	45.3	2.5	98.3	74.8	77.0
1984	55.7	60.7	71.8	57.9	50.6	3.1	109.3	82.9	78.5
1985	56.4	62.2	72.5	57.9	50.5	3.2	108.8	83.1	79.5
1986	56.9	63.4	74.9	61.6	54.4	4.0	114.1	85.8	81.3
1987	59.9	66.6	78.1	65.3	58.2	4.0	120.3	91.1	84.2
1988	63.0	69.8	81.1	68.5	61.0	5.1	122.3	95.2	87.0
1989	63.6	70.6	81.4	70.1	63.0	5.7	123.7	95.7	86.7
1990	64.2	71.4	81.8	68.0	59.3	6.2	120.7	94.7	88.2
1991	63.2	70.3	81.7	65.0	55.7	7.2	112.4	91.3	89.5
1992	65.1	72.2	84.2	71.3	64.6	8.3	118.8	94.4	90.3
1993	67.2	74.5	86.8	77.1	71.0	10.6	125.6	98.5	91.5
1994	70.8	77.8	90.4	84.0	78.0	13.9	134.5	104.2	93.7
1995	74.0	81.0	92.9	87.2	79.5	20.3	133.7	107.4	96.0
1996	77.4	84.2	94.6	89.9	80.8	24.8	135.0	111.1	97.2
1997	83.0	89.8	98.1	96.1	87.2	35.0	141.3	113.2	99.6
1998	87.8	95.0	101.7	103.1	93.6	45.9	150.5	117.1	101.9
1999	91.7	97.9	103.9	111.3	104.1	58.6	155.4	120.1	101.8
2000	95.2	100.9	105.9	114.5	105.7	67.7	159.7	123.7	103.3
2001	92.3	98.4	104.8	109.2	101.6	69.5	152.8	114.7	103.8
2002	92.6	98.1	107.0	116.2	112.4	71.3	156.2	118.2	104.2
2003	93.8	99.4	108.4	119.8	117.9	79.4	156.4	118.3	104.9
2004	96.4	101.5	109.6	121.5	117.1	92.4	161.3	120.6	105.9
2005	99.6	105.7	112.5	122.0	113.6	105.4	164.4	124.2	109.5
2006	101.8	108.3	113.1	121.4	109.5	127.8	159.4	126.3	110.5
2007	104.4	110.3	113.1	120.9	112.6	137.1	149.6	122.2	110.6
2008	100.8	105.6	107.2	106.6	93.3	153.0	130.6	113.1	107.2
2009	89.2	93.2	99.2	86.0	74.9	132.6	98.9	93.1	102.9
2010	94.1	96.2	100.3	94.2	90.2	87.1	97.7	99.6	101.9
2011	97.1	98.2	101.4	97.7	96.4	91.4	97.1	100.4	102.3
2012	100.0	100.0	100.0	100.0	100.0	100.0	100.0	100.0	100.0
2013	102.0	100.8	100.7	105.5	108.8	104.9	102.3	102.5	99.5
2014	105.2	102.0	101.5	110.7	118.5	101.8	103.9	103.8	99.2
2015	104.1	101.4	102.9	115.0	126.2	108.3	108.5	103.5	99.9
2016	102.1	100.3	103.5	117.5	130.1	107.4	110.0	104.8	100.0
2017	104.4	102.3	104.1	119.3	131.8	107.6	110.2	107.5	100.3
2018	108.6	104.7	106.3	122.0	135.6	109.9	108.9	110.5	102.3
2017									
January	103.0	101.2	103.4	120.1	133.1	106.7	113.3	106.8	99.2
February	102.6	100.6	102.3	120.0	133.5	105.4	110.4	107.3	97.8
March	103.3	101.2	103.0	117.7	128.9	106.2	109.4	107.2	99.3
April	104.3	102.3	103.7	120.0	133.1	107.6	110.1	107.8	99.6
May	104.4	102.6	104.6	119.5	132.9	107.2	109.5	106.9	100.8
June	104.6	102.4	104.1	119.1	131.0	105.4	111.2	107.7	100.3
July	104.5	102.4	104.1	118.0	128.6	105.0	109.7	108.4	100.6
August	104.0	102.1	103.9	118.1	129.1	106.9	109.3	108.0	100.3
September	104.1	102.4	104.0	119.3	131.4	110.2	110.0	108.0	100.2
October	105.6	103.3	105.2	120.3	133.9	110.8	110.4	107.2	101.4
November	106.2	103.4	105.2	119.8	132.5	112.5	110.1	107.7	101.5
December	106.5	103.7	105.8	119.8	133.0	107.8	109.6	107.4	102.2
2018									
January	106.3	103.7	105.9	120.0	133.5	106.0	109.2	107.7	102.3
February	106.6	103.9	105.8	122.2	136.0	109.1	110.5	109.9	101.6
March	107.3	104.1	106.1	122.7	137.7	106.2	108.6	109.8	102.0
April	108.2	105.1	107.5	121.9	135.9	105.5	109.4	109.8	103.8
May	107.4	103.5	105.0	117.9	127.9	105.4	108.0	109.8	101.7
June	108.2	104.3	105.8	121.1	134.5	106.8	108.4	109.8	101.9
July	108.7	104.4	106.0	120.9	132.6	109.4	110.0	110.8	102.2
August	109.5	105.0	106.4	122.8	136.7	110.3	108.7	111.2	102.3
September	109.7	105.0	106.5	123.9	138.4	110.5	109.0	111.8	102.2
October	109.9	105.4	106.7	122.7	136.0	113.3	108.5	111.9	102.6
November	110.5	105.8	107.4	123.0	136.8	115.3	108.5	111.5	103.4
December	110.6	105.7	106.8	125.0	140.6	121.5	107.8	111.8	102.3

Table 2-1. Industrial Production Indexes by Market Groups—*Continued*

(Seasonally adjusted, 2012 = 100.)

	Final products and nonindustrial supplies—*Continued*												
Year and month	Consumer goods—*Continued*						Business equipment				Defense and space equipment	Construc-tion supplies	Business supplies
	Nondurable consumer goods—*Continued*					Consumer energy products	Total	Transit	Information processing	Industrial and other			
	Nondurable non-energy consumer goods												
	Total	Foods and tobacco	Clothing	Chemical products	Paper products								
1970	62.1	60.8	498.0	31.8	78.9	48.9	15.8	50.2	1.0	47.9	47.4	62.7	38.7
1971	63.7	62.6	495.1	33.6	80.5	51.1	15.0	48.6	1.0	45.8	42.6	64.7	39.9
1972	67.9	66.1	538.8	36.7	81.2	54.0	17.1	53.0	1.1	52.1	41.4	73.5	43.9
1973	70.1	67.8	549.1	39.5	83.9	54.6	19.8	63.3	1.3	59.2	45.4	79.8	46.6
1974	69.8	68.3	514.9	41.7	82.5	56.2	20.9	61.7	1.6	61.7	46.9	77.9	46.4
1975	68.0	67.0	501.2	40.1	77.8	57.5	18.7	54.3	1.5	54.0	47.3	65.9	42.6
1976	72.4	71.4	527.7	43.7	80.9	60.6	19.9	56.2	1.8	56.2	45.9	70.9	45.4
1977	75.0	72.7	553.5	45.5	88.4	62.8	23.0	66.7	2.4	61.2	41.1	77.1	49.3
1978	77.9	75.4	566.8	48.3	92.7	63.8	25.9	77.3	3.0	65.5	41.9	81.5	52.0
1979	76.8	74.8	535.0	48.4	93.8	65.5	29.2	90.2	3.8	69.6	44.8	83.6	53.8
1980	77.7	76.0	544.3	48.3	94.9	62.7	29.9	85.1	4.8	67.3	53.2	77.4	52.5
1981	78.2	76.2	544.4	49.0	97.2	62.2	30.8	79.7	5.7	67.1	57.6	76.1	53.7
1982	79.8	78.8	542.8	49.1	99.8	62.6	28.1	61.1	6.4	58.1	68.9	69.1	53.0
1983	80.8	78.9	558.4	49.7	104.2	62.9	28.3	60.1	7.4	53.8	69.3	73.9	55.6
1984	82.2	79.8	560.4	50.8	109.5	64.9	32.5	62.0	9.1	61.3	79.4	80.3	60.5
1985	83.5	82.2	537.0	51.3	115.1	64.7	33.7	64.1	9.8	62.1	88.9	82.2	61.9
1986	85.6	83.5	537.0	55.3	116.6	65.8	33.2	58.6	9.9	61.8	94.4	84.9	63.9
1987	88.6	85.4	542.4	59.6	123.3	68.2	35.5	59.5	11.3	64.0	96.4	90.4	67.8
1988	91.2	87.9	534.5	63.2	126.6	71.8	39.2	66.4	12.7	69.5	97.2	92.4	70.3
1989	90.7	87.3	510.3	64.5	127.1	72.2	40.5	69.8	13.1	71.8	97.3	92.1	71.3
1990	92.8	89.8	501.4	66.9	129.6	71.5	42.0	74.7	14.2	71.3	93.9	91.4	72.9
1991	93.9	90.3	499.7	69.3	130.2	73.5	41.4	77.6	14.4	67.5	87.0	86.4	71.8
1992	95.2	91.7	511.5	69.1	131.8	72.7	43.2	75.8	16.3	68.7	88.8	90.1	73.6
1993	95.9	91.0	521.0	71.0	134.9	75.8	45.3	70.9	17.9	73.3	76.4	94.1	75.9
1994	98.4	94.5	530.5	72.5	133.4	76.7	48.5	68.6	20.4	78.2	71.8	101.0	78.8
1995	100.7	96.0	527.9	75.8	134.3	78.7	53.1	68.0	24.6	83.1	69.3	103.3	82.0
1996	101.4	96.6	514.0	79.3	133.7	82.1	58.3	71.9	30.1	86.1	67.6	108.0	85.2
1997	104.6	98.2	512.7	83.8	145.5	81.5	67.1	84.4	38.4	91.4	66.7	113.2	91.2
1998	107.5	100.9	487.1	88.0	152.4	81.2	74.8	99.1	46.0	94.3	69.7	119.2	96.4
1999	106.7	98.9	466.9	89.5	155.4	84.2	78.9	97.0	55.8	92.1	67.6	122.2	100.5
2000	108.2	100.3	451.6	92.9	154.8	85.8	85.0	86.1	68.0	97.6	60.1	124.8	104.4
2001	108.5	100.4	403.1	98.0	150.4	86.9	80.1	83.0	66.9	88.3	66.0	119.1	100.6
2002	108.2	98.5	308.9	106.2	148.2	90.6	75.0	75.1	59.2	86.8	66.7	119.0	100.7
2003	109.0	99.8	291.0	108.7	142.8	91.1	75.2	70.8	62.7	86.4	71.0	118.7	102.6
2004	109.7	100.9	246.2	111.5	143.6	92.7	79.4	74.8	69.3	88.7	69.7	121.5	105.1
2005	113.4	104.3	236.4	117.3	142.1	96.1	85.4	81.9	76.0	93.6	76.4	127.4	109.0
2006	114.7	104.6	229.3	121.5	141.6	96.5	93.4	95.1	87.0	97.3	76.1	130.6	110.8
2007	113.5	104.4	184.2	122.4	135.6	100.5	99.0	98.3	97.3	101.2	90.7	129.4	112.1
2008	109.1	100.9	148.8	118.1	126.9	100.0	96.7	88.3	100.6	99.1	98.0	117.5	107.2
2009	103.5	99.5	108.8	108.9	112.3	100.4	80.0	71.1	90.4	79.1	93.8	90.4	95.0
2010	101.8	99.8	107.6	103.9	106.8	102.3	86.1	80.2	92.6	85.6	100.9	93.6	98.0
2011	102.0	99.7	105.2	105.7	103.9	103.1	91.1	82.9	92.9	93.8	98.0	95.9	99.0
2012	100.0	100.0	100.0	100.0	100.0	100.0	100.0	100.0	100.0	100.0	100.0	100.0	100.0
2013	97.6	101.5	92.2	90.9	95.6	105.3	99.9	105.9	100.2	97.3	97.2	103.1	101.8
2014	97.3	101.2	88.2	91.5	94.2	104.9	101.7	115.5	97.8	97.6	93.9	106.4	102.8
2015	98.8	103.1	83.8	93.6	91.7	102.8	99.6	120.7	98.9	91.3	91.7	107.1	101.4
2016	98.4	103.9	80.0	91.8	89.2	105.1	94.4	110.5	98.7	86.1	89.1	108.1	102.0
2017	98.9	106.0	73.4	91.2	86.1	104.2	97.8	109.7	104.2	90.4	90.9	111.8	104.3
2018	100.0	107.4	71.4	93.6	80.5	109.7	101.0	111.6	108.2	93.8	93.1	114.9	105.2
2017													
January	98.2	105.1	78.3	89.9	87.8	101.8	95.7	109.1	102.6	87.6	91.8	110.9	103.0
February	98.3	105.6	76.8	89.4	88.7	94.6	95.4	109.3	101.6	87.2	90.5	111.4	103.0
March	97.9	105.0	75.2	89.6	87.8	103.3	96.2	106.9	103.2	89.0	90.9	110.8	103.8
April	98.4	105.9	73.3	89.6	87.3	102.9	98.7	110.8	104.6	91.5	91.5	111.5	104.3
May	98.7	105.9	73.2	90.8	87.0	107.4	98.3	110.9	103.9	90.9	91.5	111.2	104.5
June	98.7	105.8	72.2	91.0	86.4	105.2	98.1	110.9	104.4	90.5	91.2	111.5	104.4
July	99.1	106.3	72.5	91.6	85.2	105.2	97.8	109.4	104.5	90.5	91.1	111.3	104.8
August	99.1	106.4	72.4	91.4	84.9	103.7	97.7	110.7	104.8	89.6	91.0	110.7	104.3
September	99.5	106.7	71.7	92.2	85.1	101.8	98.5	110.3	104.6	91.2	90.5	112.5	104.0
October	99.9	106.8	71.9	93.1	85.9	105.7	99.2	110.3	105.4	92.2	90.4	112.9	104.8
November	99.4	106.4	71.1	92.6	84.9	108.1	99.4	110.1	105.4	92.7	90.5	113.1	105.4
December	99.7	106.7	72.2	93.5	82.0	110.4	98.8	108.3	105.0	92.5	89.9	113.8	105.6
2018													
January	99.2	106.0	73.2	93.2	82.3	112.3	99.4	109.6	106.9	92.4	90.2	111.9	105.1
February	100.9	108.4	73.2	93.8	84.5	103.4	99.4	110.5	106.7	92.0	90.2	115.1	105.1
March	99.9	107.4	73.0	92.7	83.7	108.3	99.6	111.3	107.3	91.7	90.6	114.4	105.3
April	100.4	108.3	73.6	93.1	81.6	114.6	100.2	111.3	107.9	92.6	92.1	114.7	106.0
May	99.8	107.6	73.1	93.1	79.7	107.3	98.5	105.6	107.5	92.0	91.5	115.1	105.2
June	99.9	108.0	71.1	93.2	77.6	108.1	100.3	112.0	108.1	92.5	92.3	114.7	105.2
July	100.6	108.6	73.7	94.0	78.3	107.0	100.7	111.5	109.3	92.9	92.9	114.5	105.1
August	100.3	108.1	71.5	93.8	79.8	108.4	102.0	112.2	108.7	95.1	93.9	115.3	105.1
September	100.1	107.9	69.8	93.5	80.3	108.5	102.5	114.1	108.5	95.3	94.5	114.8	104.8
October	99.4	106.2	69.1	93.8	80.9	113.0	102.8	112.3	108.5	96.6	95.3	115.3	105.6
November	99.4	105.9	68.3	94.6	80.4	116.3	103.0	112.2	108.4	97.1	95.4	115.5	105.4
December	99.7	107.0	67.0	94.2	77.7	110.1	103.2	116.0	110.2	95.3	98.1	117.5	104.8

Table 2-1. Industrial Production Indexes by Market Groups—*Continued*

(Seasonally adjusted, 2012 = 100.)

Year and month		Materials										
	Total	Non-energy materials										Energy materials
		Total	Durable				Nondurable					
			Total	Consumer parts	Equipment parts	Other	Total	Textile	Paper	Chemicals		
1970	36.6	30.8	20.5	45.7	3.6	64.2	61.6	151.6	66.7	44.4		65.5
1971	37.2	31.3	20.6	50.6	3.6	61.8	64.2	158.8	69.7	47.0		66.1
1972	40.9	35.1	23.3	56.3	4.1	69.8	70.8	167.2	74.3	54.7		68.6
1973	44.6	38.9	26.6	65.3	4.9	77.5	74.3	163.0	80.1	60.3		70.5
1974	44.5	38.8	26.3	57.8	5.1	77.2	75.4	152.1	84.3	62.0		70.3
1975	39.7	33.3	22.0	46.6	4.4	64.5	67.5	148.9	73.3	51.9		69.9
1976	43.1	37.0	24.5	59.5	4.8	69.2	74.8	166.3	80.6	59.5		71.6
1977	46.2	40.1	26.7	65.2	5.4	73.4	80.2	177.2	84.2	66.2		74.0
1978	48.5	42.7	28.9	69.0	6.0	78.3	83.2	175.2	88.3	70.2		74.9
1979	49.8	43.9	29.8	65.0	6.6	80.2	84.7	173.8	91.9	73.1		76.9
1980	47.9	41.3	27.6	50.1	6.8	74.1	81.9	169.7	92.6	67.6		77.6
1981	48.2	41.4	27.6	47.5	7.0	74.3	82.6	165.9	94.4	68.1		78.4
1982	44.5	37.2	24.0	40.4	6.4	62.5	78.3	151.6	95.1	61.1		75.1
1983	45.7	39.8	25.7	49.1	6.5	65.5	83.8	169.9	101.1	67.6		72.8
1984	50.0	44.4	29.8	58.4	7.8	72.6	87.4	169.4	107.3	71.7		77.4
1985	49.9	44.4	30.0	60.8	7.8	72.0	86.7	160.3	106.6	70.5		77.2
1986	49.9	45.2	30.3	60.2	8.0	73.4	89.4	166.7	111.1	74.0		74.4
1987	52.6	48.2	32.5	61.9	8.7	78.6	94.7	186.0	116.4	80.4		76.3
1988	55.5	51.3	35.1	66.9	9.5	84.2	98.2	184.2	120.3	85.2		79.2
1989	55.9	51.6	35.2	63.5	9.9	84.5	99.0	188.0	120.2	86.4		80.1
1990	56.3	51.7	35.3	59.1	10.2	85.0	99.3	178.7	120.6	87.3		81.7
1991	55.5	50.6	34.3	55.8	10.2	81.3	98.2	178.7	118.1	86.0		82.0
1992	57.3	53.1	36.6	62.8	10.7	85.7	100.7	189.3	120.9	88.2		81.2
1993	59.3	55.5	39.0	71.0	11.4	89.1	101.8	197.2	120.9	88.6		81.5
1994	63.0	59.9	43.3	82.0	13.0	95.6	104.9	209.0	125.5	91.4		83.0
1995	66.4	63.8	47.6	85.2	15.7	99.2	105.9	205.3	128.7	92.1		84.4
1996	69.9	67.9	52.8	87.5	19.4	102.2	104.9	199.4	124.6	92.6		85.7
1997	75.4	74.9	60.0	94.3	24.5	107.3	109.9	208.6	126.0	99.3		85.5
1998	79.8	80.4	66.4	97.7	30.0	110.2	111.3	206.9	127.2	99.5		86.1
1999	84.8	86.9	74.3	107.3	37.2	111.9	112.7	203.3	128.9	102.4		86.0
2000	89.0	91.9	81.1	107.7	45.7	112.9	112.0	195.2	126.9	102.7		87.5
2001	85.5	87.3	77.7	95.9	46.6	105.0	104.8	172.6	120.3	95.1		86.6
2002	86.6	89.1	79.3	100.9	47.8	105.1	106.7	175.3	120.5	98.4		86.4
2003	87.7	90.7	82.0	100.5	53.1	104.3	105.8	166.3	116.6	99.0		86.4
2004	90.7	95.0	87.5	101.6	60.3	108.3	107.8	158.8	117.9	104.2		86.5
2005	92.8	98.9	92.5	104.1	67.9	110.7	109.1	162.5	117.4	105.3		85.8
2006	94.7	101.3	96.1	103.9	73.8	113.0	109.3	147.3	116.4	107.3		86.8
2007	98.0	105.6	100.2	104.3	83.4	112.9	114.1	134.8	116.4	117.4		88.6
2008	95.6	100.7	97.3	90.7	87.3	107.7	105.7	116.8	111.3	104.1		89.3
2009	84.9	83.8	76.5	64.9	71.9	84.5	95.4	95.8	97.8	92.0		86.7
2010	91.9	93.0	87.9	82.2	82.4	94.0	101.1	106.4	99.7	102.1		90.6
2011	95.9	97.3	95.4	89.5	93.7	98.6	100.3	102.2	99.3	101.0		94.1
2012	100.0	100.0	100.0	100.0	100.0	100.0	100.0	100.0	100.0	100.0		100.0
2013	103.3	101.9	102.4	103.8	101.6	102.5	101.1	104.5	100.1	100.9		105.1
2014	108.5	103.9	106.6	109.1	107.7	105.1	99.7	103.3	98.5	98.4		114.0
2015	106.9	102.1	104.4	110.9	103.9	102.5	98.5	96.8	97.2	96.0		112.6
2016	103.7	100.9	102.1	108.5	101.1	100.5	99.0	94.2	96.0	96.8		105.9
2017	106.5	102.9	104.0	109.1	103.2	102.6	101.1	94.5	94.7	99.8		110.1
2018	112.8	106.1	107.6	112.2	108.2	105.7	103.7	97.7	93.0	104.6		121.3
2017												
January	104.8	102.5	103.7	111.0	103.4	101.4	100.7	95.9	95.7	98.7		105.9
February	104.6	102.3	103.6	110.0	102.6	102.0	100.3	94.7	97.3	97.7		105.9
March	105.4	102.1	103.3	109.0	102.5	101.7	100.4	93.8	95.1	99.0		108.3
April	106.2	103.2	104.2	109.8	103.5	102.7	101.7	94.6	96.5	100.7		108.5
May	106.0	102.8	103.3	109.5	102.8	101.4	101.9	94.5	94.1	101.6		109.0
June	106.7	103.3	103.7	109.4	103.3	102.1	102.4	94.2	94.7	102.6		109.8
July	106.6	102.8	103.0	106.0	102.5	102.2	102.3	93.6	93.8	102.7		110.5
August	105.9	102.1	103.2	108.1	102.3	102.1	100.3	94.7	95.9	97.4		109.8
September	105.5	101.0	104.2	108.4	103.0	103.4	96.4	95.3	93.8	91.0		110.6
October	108.0	103.5	104.6	108.8	103.5	103.8	101.8	95.0	92.6	101.5		113.0
November	109.0	104.4	105.6	109.5	104.4	104.9	102.7	92.9	92.9	102.9		114.2
December	109.5	104.4	105.3	110.4	104.5	104.0	102.9	95.0	93.9	102.4		115.4
2018												
January	108.9	103.5	105.1	109.3	105.1	103.7	101.1	96.3	93.9	99.7		115.4
February	109.5	104.5	106.6	113.1	105.5	105.0	101.4	98.1	92.2	99.8		115.2
March	110.5	105.1	106.8	114.1	105.7	105.0	102.5	95.9	93.6	102.4		117.1
April	111.5	105.8	107.4	112.9	107.3	105.5	103.5	96.5	93.0	103.6		118.3
May	111.6	105.4	106.5	108.8	107.0	105.3	103.7	96.2	93.0	104.5		119.3
June	112.4	106.0	107.2	110.9	107.9	105.5	104.1	96.9	92.4	105.7		120.4
July	113.3	106.5	107.5	110.7	108.8	105.7	104.8	96.6	93.9	106.1		122.0
August	114.5	106.9	108.0	112.0	110.0	105.6	105.1	99.6	92.6	107.1		124.4
September	114.8	106.9	108.5	112.6	110.2	106.1	104.4	98.2	92.6	105.8		125.1
October	114.9	106.9	108.6	112.2	110.3	106.4	104.3	101.4	92.8	105.7		125.3
November	115.7	107.3	109.1	113.6	110.0	107.0	104.6	100.0	92.8	106.6		126.9
December	115.9	108.1	110.0	115.6	110.3	107.9	105.3	96.8	93.1	107.8		125.9

Table 2-1. Industrial Production Indexes by Market Groups—Continued

(Seasonally adjusted, 2012 = 100.)

	Special aggregates											
	Energy						Non-energy					Total non-energy, excluding high-tech
Year and month								Selected high-tech				
	Total	Consumer energy products	Commercial energy products	Oil and gas well drilling	Converted fuels	Primary energy	Total	Total	Computers and peripheral equipment	Communications equipment	Semiconductors and related components	
1970	57.5	48.9	34.1	. . .	61.1	71.6	36.0	0.1	. . .	. . .	. . .	57.4
1971	58.6	51.1	35.8	. . .	62.7	71.0	36.5	0.1	. . .	2.1	. . .	58.5
1972	61.3	54.0	37.8	62.0	66.6	72.2	40.3	0.1	0.0	2.1	0.1	64.3
1973	62.9	54.6	39.7	58.4	69.3	73.5	44.0	0.1	0.0	2.3	0.1	69.8
1974	63.3	56.2	39.8	67.8	68.1	74.1	43.8	0.1	0.1	2.5	0.1	69.0
1975	63.6	57.5	41.1	76.5	65.8	75.3	39.1	0.1	0.1	2.4	0.1	61.5
1976	65.7	60.6	43.2	86.2	69.5	75.4	42.6	0.2	0.1	2.6	0.1	66.7
1977	68.3	62.8	44.6	109.4	72.2	77.8	46.2	0.2	0.2	3.1	0.1	71.7
1978	69.4	63.8	45.9	121.9	71.6	79.7	49.2	0.2	0.2	3.4	0.1	75.7
1979	71.5	65.5	48.2	130.0	74.2	81.4	50.6	0.3	0.3	4.1	0.1	77.1
1980	71.7	62.7	47.1	153.7	73.0	83.1	48.9	0.4	0.5	4.9	0.1	73.4
1981	72.8	62.2	48.0	184.7	72.1	84.7	49.5	0.5	0.7	5.3	0.2	73.6
1982	70.2	62.6	48.6	162.5	66.5	82.4	46.7	0.5	0.8	5.5	0.2	68.5
1983	68.1	62.9	49.5	125.9	66.3	78.9	48.9	0.6	1.1	5.9	0.2	71.1
1984	72.1	64.9	52.2	137.1	70.4	84.0	53.8	0.8	1.6	6.3	0.3	77.2
1985	71.9	64.7	53.9	125.6	70.2	83.7	54.7	0.9	1.9	6.1	0.3	78.3
1986	69.4	65.8	55.3	61.2	67.9	80.6	56.0	0.9	2.0	5.8	0.3	80.1
1987	71.5	68.2	58.2	59.0	71.4	81.4	59.2	1.1	2.6	6.4	0.4	83.8
1988	74.5	71.8	60.3	70.8	74.6	84.1	62.4	1.3	3.2	7.5	0.4	87.6
1989	75.3	72.2	62.4	61.7	77.4	83.8	62.9	1.4	3.4	7.9	0.4	88.0
1990	76.5	71.5	64.1	65.3	77.8	86.2	63.4	1.5	3.6	9.4	0.5	88.1
1991	76.9	73.5	65.0	50.8	77.9	86.6	62.2	1.7	3.8	10.1	0.6	85.8
1992	76.1	72.7	64.6	40.5	79.2	84.5	64.6	2.0	4.7	12.0	0.7	89.1
1993	77.3	75.8	66.4	51.0	81.0	83.9	66.9	2.4	5.8	14.2	0.8	90.4
1994	70.7	76.7	69.1	51.2	82.3	85.5	70.9	3.0	7.3	17.8	1.0	94.4
1995	80.3	78.7	71.5	49.8	83.4	87.1	74.5	4.2	10.2	22.0	1.4	96.8
1996	82.3	82.1	73.7	56.7	84.9	88.3	78.2	6.0	14.5	28.0	2.2	98.3
1997	82.7	81.5	76.7	66.9	86.4	87.2	84.8	8.8	20.8	39.1	3.4	103.2
1998	83.0	81.2	77.8	61.5	87.8	87.3	90.6	12.2	29.0	47.4	4.9	107.0
1999	83.7	84.2	80.9	50.0	89.4	86.1	95.1	17.7	38.4	60.0	8.0	108.3
2000	85.6	85.8	84.1	60.2	91.6	87.2	99.1	24.5	46.1	81.1	11.8	109.1
2001	85.6	86.9	85.1	71.6	89.5	87.0	95.4	25.9	47.7	76.7	13.3	103.9
2002	85.9	90.6	86.6	52.8	91.3	85.7	95.8	26.4	48.8	58.4	15.8	104.2
2003	86.7	91.1	91.5	61.7	91.6	85.5	97.1	31.9	54.0	63.3	20.9	104.2
2004	87.5	92.7	94.4	68.3	93.8	84.7	100.1	38.4	56.3	74.1	27.0	106.2
2005	88.1	96.1	97.1	75.8	95.4	83.1	104.2	45.6	68.6	76.9	33.4	109.5
2006	89.2	96.5	97.7	87.1	95.1	84.6	106.9	54.8	87.9	96.1	38.3	111.2
2007	91.6	100.5	100.2	91.1	99.6	85.5	109.6	65.3	104.5	99.5	48.8	112.8
2008	92.2	100.0	100.1	98.2	97.5	87.1	104.2	75.4	127.2	100.1	58.4	106.0
2009	89.2	100.4	98.2	63.3	92.0	85.3	89.4	66.6	113.7	87.5	51.5	90.7
2010	92.7	102.3	99.4	71.5	97.5	88.7	94.7	80.9	93.6	92.4	75.4	95.5
2011	95.9	103.1	100.4	88.4	96.3	93.5	97.5	91.1	80.8	100.7	90.6	97.9
2012	100.0	100.0	100.0	100.0	100.0	100.0	100.0	100.0	100.0	100.0	100.0	100.0
2013	104.8	105.3	102.8	101.9	99.3	106.5	100.8	110.7	100.3	116.8	111.0	100.4
2014	111.8	104.9	104.0	110.4	100.5	117.4	102.3	122.7	102.6	109.4	134.3	101.5
2015	109.3	102.8	104.5	77.3	101.0	115.4	101.8	126.7	111.8	117.5	134.7	100.9
2016	104.1	105.1	106.7	45.2	99.5	106.4	100.7	133.9	111.7	121.4	145.7	99.6
2017	107.5	104.2	107.3	65.2	97.9	113.1	102.7	137.4	127.8	117.1	149.2	101.6
2018	117.0	109.7	111.3	75.8	102.6	126.8	105.1	146.1	134.1	120.4	161.4	103.7
2017												
January	103.7	101.8	105.3	55.0	96.0	108.0	102.1	136.8	121.6	120.2	148.5	100.9
February	102.3	94.6	103.8	58.2	92.2	109.8	102.0	135.5	123.7	116.6	147.4	100.9
March	106.1	103.3	107.5	61.9	98.4	110.4	101.8	135.4	125.1	114.7	147.6	100.6
April	106.2	102.9	106.8	65.8	98.1	110.8	102.9	137.3	126.6	114.5	150.8	101.8
May	107.6	107.4	108.5	67.6	98.1	111.5	102.7	137.2	128.6	115.2	149.5	101.5
June	107.8	105.2	107.4	70.6	97.6	113.0	102.9	137.1	131.9	116.1	147.9	101.7
July	108.2	105.2	107.3	70.2	99.1	113.2	102.7	135.4	127.0	117.0	146.0	101.5
August	107.3	103.7	106.4	68.3	97.5	112.9	102.3	135.5	126.8	117.8	145.7	101.2
September	107.4	101.8	106.4	67.1	96.0	114.7	102.3	137.0	130.0	118.3	147.1	101.1
October	109.9	105.7	107.7	66.1	99.1	116.7	103.6	138.9	131.6	118.5	150.0	102.4
November	111.3	108.1	109.0	65.2	100.5	117.8	103.8	140.2	128.0	118.4	153.7	102.6
December	112.8	110.4	111.4	66.2	102.4	118.8	103.8	142.2	132.9	117.8	155.9	102.5
2018												
January	113.3	112.3	112.0	68.0	103.7	118.3	103.3	142.4	135.7	116.9	155.9	102.0
February	111.1	103.4	108.5	70.5	96.7	120.7	104.5	142.3	134.3	115.9	156.6	103.3
March	113.7	108.3	111.1	72.8	100.4	121.9	104.4	142.7	134.9	114.8	157.6	103.2
April	116.0	114.6	113.1	74.9	102.1	122.8	104.9	143.3	135.8	114.4	158.8	103.7
May	115.1	107.3	110.5	77.4	103.2	123.8	104.1	143.4	133.3	115.2	159.4	102.8
June	116.1	108.1	111.2	79.2	100.6	126.4	104.8	146.5	135.9	117.7	162.9	103.5
July	116.9	107.0	110.7	77.0	103.3	127.4	105.2	148.3	135.1	121.0	164.7	103.9
August	119.0	108.4	112.1	77.3	104.3	130.4	105.7	150.5	134.5	124.0	167.5	104.3
September	119.3	108.5	110.7	77.0	105.3	131.0	105.7	150.1	131.1	125.6	167.1	104.3
October	120.6	113.0	113.4	77.7	105.6	131.2	105.6	149.2	131.9	126.0	165.0	104.2
November	122.4	116.3	113.2	78.9	107.2	132.7	105.8	147.5	133.7	125.7	161.4	104.5
December	120.1	110.1	109.9	79.0	98.9	134.7	106.6	147.4	132.8	127.7	160.5	105.3

. . . = Not available.

Table 2-2. Industrial Production Indexes by NAICS Industry Groups

(Seasonally adjusted, 2012 = 100.)

Year and month	Total industrial production	Manu-facturing (SIC)	Manufacturing (NAICS) Total	Durable goods manufacturing Total	Wood products	Nonmetallic mineral products	Primary metals	Fabricated metal products	Machinery	Computer and electronic products	Electrical equipment, appliances, and components	Motor vehicles and parts	Aerospace and miscellaneous transport equipment
1970	38.6	35.9	...	...	...	...	...	...	...	...	...	...	...
1971	39.1	36.5	...	...	...	...	...	...	...	...	...	...	...
1972	42.9	40.3	38.1	25.5	100.0	94.8	115.4	69.1	54.9	0.5	80.7	45.9	53.5
1973	46.4	43.9	41.7	28.7	96.8	101.8	134.3	76.3	63.4	0.6	90.9	52.5	60.9
1974	46.3	43.8	41.6	28.5	87.9	100.7	137.6	75.0	66.5	0.7	88.7	45.1	62.0
1975	42.2	39.2	37.1	24.8	81.0	90.1	106.7	64.8	57.9	0.6	71.2	39.3	58.8
1976	45.5	42.7	40.5	27.1	91.3	95.1	113.2	69.4	60.4	0.7	80.4	50.2	55.1
1977	48.9	46.4	44.0	29.8	98.4	101.3	114.6	75.3	65.9	0.9	88.6	57.1	55.5
1978	51.6	49.2	46.7	32.1	99.7	108.0	122.0	79.0	70.9	1.2	94.0	59.5	61.2
1979	53.2	50.7	48.2	33.7	96.1	107.8	124.7	82.5	74.9	1.4	97.9	54.4	71.3
1980	51.8	48.9	46.3	32.2	89.0	97.2	109.3	77.8	71.2	1.7	92.1	40.1	76.7
1981	52.5	49.4	46.8	32.5	87.0	93.1	109.4	77.3	70.5	2.0	90.9	39.0	73.0
1982	49.8	46.7	44.1	29.7	78.2	82.4	76.8	69.2	58.9	2.2	81.9	35.2	68.3
1983	51.1	49.0	46.2	31.2	90.5	88.6	79.0	69.7	53.2	2.6	84.7	44.9	65.3
1984	55.7	53.7	50.9	35.6	97.0	95.6	86.9	75.9	62.0	3.2	95.3	53.8	69.7
1985	56.4	54.6	51.6	36.4	98.0	97.3	80.0	76.9	62.2	3.4	93.8	55.9	74.4
1986	56.9	55.8	52.8	37.0	106.5	101.3	78.3	76.4	61.2	3.5	95.6	55.9	78.3
1987	59.9	59.0	55.8	39.2	115.9	106.9	84.5	77.9	62.5	4.0	96.8	57.9	81.2
1988	63.0	62.1	58.9	42.1	115.7	109.1	94.5	81.8	68.8	4.4	101.5	61.9	85.4
1989	63.6	62.6	59.5	42.6	113.8	108.2	92.3	81.2	71.4	4.6	100.0	61.2	90.6
1990	64.2	63.1	60.0	42.7	112.5	106.6	91.3	80.2	69.6	5.0	97.4	57.6	90.9
1991	63.2	61.9	58.9	41.4	105.2	98.1	85.7	76.6	65.3	5.2	92.4	55.1	87.7
1992	65.1	64.2	61.3	43.6	111.1	102.4	88.2	79.0	65.2	5.9	97.9	62.7	81.1
1993	67.2	66.5	63.6	46.1	112.4	104.6	92.4	82.0	70.0	6.5	104.0	69.3	75.6
1994	70.8	70.4	67.6	50.0	119.1	110.5	99.4	89.1	76.7	7.7	111.6	79.6	67.9
1995	74.0	74.0	71.2	54.1	122.1	113.7	100.5	94.6	82.1	9.8	114.2	81.9	64.5
1996	77.4	77.6	74.9	59.1	126.0	121.1	102.8	98.0	84.9	12.8	117.6	82.6	67.0
1997	83.0	84.2	81.2	66.1	129.4	125.2	107.1	102.4	89.6	17.1	122.0	89.1	74.8
1998	87.8	89.8	86.6	73.0	135.1	131.5	108.9	105.7	92.0	21.9	126.4	93.7	86.7
1999	91.7	94.3	91.1	79.3	140.1	132.7	108.8	106.5	90.1	29.0	128.4	103.8	83.5
2000	95.2	98.2	95.1	85.0	138.3	132.8	105.0	110.7	94.8	37.6	134.9	103.3	73.2
2001	92.3	94.6	91.8	81.6	129.6	127.8	95.0	102.7	83.9	39.1	121.5	94.4	77.9
2002	92.6	95.1	92.4	82.0	135.2	127.9	95.5	100.5	80.8	38.8	111.9	103.9	73.7
2003	93.8	96.4	93.9	84.2	134.9	129.1	93.8	99.4	80.4	44.6	109.1	107.6	70.4
2004	96.4	99.4	97.0	88.2	138.4	133.0	101.8	99.8	83.7	52.5	111.0	108.1	69.4
2005	99.6	103.4	101.2	93.4	147.7	138.6	99.1	104.4	88.9	59.7	112.9	108.8	77.3
2006	101.8	106.1	104.0	97.8	148.9	141.2	101.9	110.2	93.1	68.0	113.6	107.1	82.0
2007	104.4	109.0	107.3	102.7	139.2	139.7	104.0	114.9	97.1	78.4	118.4	106.3	96.3
2008	100.8	103.8	102.4	99.2	119.0	123.4	104.2	110.7	94.5	84.8	113.9	85.1	98.2
2009	89.2	89.5	88.4	80.6	90.9	93.0	77.5	85.2	73.5	75.6	89.5	62.3	89.4
2010	94.1	94.7	94.2	89.2	94.1	95.9	95.1	90.7	82.1	85.6	93.1	82.7	90.6
2011	97.1	97.5	97.2	94.7	94.3	97.4	102.0	97.1	92.5	92.6	97.6	90.4	90.5
2012	100.0	100.0	100.0	100.0	100.0	100.0	100.0	100.0	100.0	100.0	100.0	100.0	100.0
2013	102.0	100.9	101.1	102.1	105.8	105.2	103.3	101.8	95.4	103.2	100.0	100.0	100.0
2014	105.2	102.0	102.3	105.1	108.4	109.1	104.0	103.6	96.7	107.4	101.8	107.2	102.9
2015	104.1	101.5	101.9	103.9	112.7	109.8	96.8	100.2	89.0	108.1	101.3	123.2	106.8
2016	102.1	100.7	101.1	101.7	116.9	111.3	92.5	96.5	82.2	110.4	101.0	124.8	99.5
2017	104.4	102.7	103.2	104.0	124.1	115.3	93.7	97.9	87.9	115.2	101.8	124.7	99.9
2018	108.6	105.0	106.0	107.5	127.1	119.6	97.6	102.5	92.6	120.9	103.7	129.9	100.1
2017													
January	103.0	102.0	102.5	103.6	122.1	113.3	93.9	97.0	85.1	113.9	103.2	126.2	100.3
February	102.6	102.0	102.4	103.3	123.3	115.4	94.0	97.4	84.0	112.8	102.3	126.2	100.1
March	103.3	101.7	102.2	103.1	123.0	115.7	93.1	97.0	85.8	114.1	101.6	122.1	99.9
April	104.3	102.8	103.3	104.5	122.6	114.4	93.5	97.3	88.1	115.9	102.7	126.9	100.3
May	104.4	102.6	103.2	103.8	122.5	113.6	91.5	97.6	88.6	115.1	101.9	126.0	100.3
June	104.6	102.8	103.3	103.8	121.8	114.4	92.8	97.1	88.7	114.9	101.5	124.7	100.3
July	104.5	102.6	103.1	103.3	123.5	114.7	92.8	97.5	87.7	114.6	102.2	121.0	100.1
August	104.0	102.3	102.8	103.4	124.0	113.0	93.8	97.6	86.8	115.0	100.5	122.9	100.4
September	104.1	102.1	102.7	104.3	125.8	116.8	95.2	98.4	89.1	115.3	102.3	124.1	99.6
October	105.6	103.5	104.1	104.9	125.8	116.7	94.8	99.1	89.6	116.6	101.8	125.8	99.5
November	106.2	103.8	104.4	105.2	127.4	117.7	95.5	99.5	90.4	117.4	101.2	125.5	99.5
December	106.5	103.7	104.4	105.0	127.0	118.6	94.3	99.6	90.8	117.2	100.0	125.6	98.8
2018													
January	106.3	103.3	104.0	105.1	126.2	116.1	95.0	99.5	90.5	118.4	101.8	126.9	98.4
February	106.6	104.4	105.2	106.3	128.7	121.3	95.7	101.2	90.2	118.6	102.8	130.4	98.1
March	107.3	104.5	105.2	106.6	128.5	120.0	96.6	101.4	90.2	119.2	101.6	132.8	97.9
April	108.2	104.9	105.7	107.1	127.7	120.8	96.5	101.6	91.7	119.9	103.0	130.8	99.3
May	107.4	104.1	105.0	105.7	127.1	120.7	96.2	101.4	90.6	119.7	103.0	121.5	99.0
June	108.2	104.8	105.8	107.1	127.4	119.6	96.1	102.5	91.1	120.9	104.0	129.1	99.8
July	108.7	105.2	106.2	107.2	126.7	119.9	96.1	102.6	92.2	122.4	105.1	126.7	100.2
August	109.5	105.7	106.7	108.4	128.2	119.6	97.7	103.3	94.2	122.4	104.9	130.4	101.2
September	109.7	105.7	106.7	108.7	128.2	117.6	98.8	103.3	94.4	122.3	104.8	132.8	101.3
October	109.9	105.6	106.6	108.9	126.4	119.6	99.5	103.9	95.2	122.1	104.5	130.3	101.7
November	110.5	105.8	106.8	109.2	125.3	118.2	101.5	104.4	96.5	121.5	104.4	130.0	101.4
December	110.6	106.4	107.5	110.0	125.1	121.8	101.1	104.5	94.6	123.2	104.8	136.3	102.6

... = Not available.

Table 2-2. Industrial Production Indexes by NAICS Industry Groups—*Continued*

(Seasonally adjusted, 2012 = 100.)

Year and month	Durable goods manufacturing—*Continued*		Manufacturing (NAICS)—*Continued* Nondurable goods manufacturing									Other manu-facturing (non-NAICS)
	Furniture and related products	Miscel-laneous manu-facturing	Total	Food, beverage, and tobacco products	Textile and product mills	Apparel and leather	Paper	Printing and support	Petroleum and coal products	Chemicals	Plastics and rubber products	
1970	. . .	. . .	. . .	. . .	. . .	. . .	. . .	. . .	. . .	. . .	. . .	. . .
1971	. . .	. . .	. . .	. . .	. . .	. . .	. . .	. . .	. . .	. . .	. . .	119.0
1972	77.6	39.6	64.2	64.0	158.4	499.6	76.4	66.3	68.7	48.3	39.6	122.8
1973	81.3	40.8	67.2	64.6	157.1	507.7	82.6	69.7	67.5	52.9	44.7	123.4
1974	75.1	39.9	67.5	65.8	144.4	477.7	86.2	67.6	72.8	54.9	43.1	123.4
1975	64.5	37.1	62.6	64.4	139.6	466.3	74.7	63.1	71.8	48.3	36.4	117.4
1976	71.9	40.5	68.3	68.8	155.7	490.3	82.4	67.7	79.5	54.0	40.2	121.1
1977	82.4	44.1	73.0	70.3	170.0	514.0	86.0	73.3	85.1	58.8	47.5	132.7
1978	88.8	45.1	75.6	72.8	169.6	525.3	89.9	77.6	86.0	61.8	49.6	137.3
1979	88.6	45.3	76.1	72.3	169.2	492.9	91.2	80.0	91.8	63.2	48.9	140.2
1980	85.5	43.0	73.7	73.5	162.3	502.3	91.0	80.6	81.4	59.7	43.8	145.0
1981	84.8	44.6	74.4	74.2	158.8	502.2	92.2	82.6	77.6	60.7	46.3	148.4
1982	79.9	45.1	73.3	76.4	147.5	499.2	90.7	88.9	74.2	56.8	45.6	150.2
1983	87.6	45.1	76.7	76.5	165.8	513.3	96.5	95.5	75.4	60.7	49.6	154.5
1984	98.1	48.8	80.2	77.6	169.8	512.9	101.4	104.0	77.0	64.3	57.3	161.6
1985	99.2	49.5	80.7	80.1	163.9	490.9	99.2	108.2	75.9	63.8	59.5	168.0
1986	103.6	50.7	83.0	81.1	170.1	489.4	103.1	113.6	75.5	66.7	62.0	171.4
1987	111.1	54.5	87.4	82.7	186.7	495.3	106.4	122.0	79.1	71.8	68.7	181.2
1988	109.9	59.6	90.4	85.0	185.0	487.0	110.6	125.9	81.5	76.0	71.7	180.4
1989	109.5	60.4	90.9	84.6	187.8	465.9	111.8	126.4	80.7	77.4	74.2	177.9
1990	107.3	63.3	92.4	87.0	180.2	456.4	111.7	131.2	80.8	79.2	76.3	175.8
1991	99.2	64.5	92.1	87.7	177.8	453.5	111.9	127.1	79.6	79.0	75.4	168.6
1992	107.0	67.3	94.5	89.0	187.6	466.6	114.7	134.0	79.2	80.1	81.2	165.1
1993	111.5	71.2	95.9	88.8	194.8	475.9	116.0	134.4	79.8	81.1	86.0	100.0
1994	115.2	71.7	99.2	91.6	205.2	481.7	121.0	135.9	82.0	83.2	94.2	164.9
1995	117.1	74.4	100.9	94.2	203.0	478.2	122.7	137.9	83.5	84.6	96.6	164.8
1996	118.1	78.0	101.2	93.5	198.1	467.2	118.9	138.9	85.5	86.3	99.8	163.3
1997	130.8	80.2	105.0	95.6	209.1	466.7	121.4	141.7	88.3	91.4	105.9	177.1
1998	140.1	84.9	106.7	98.6	207.5	444.3	122.4	143.5	86.6	92.9	109.7	187.6
1999	144.7	86.8	107.3	96.8	208.3	423.2	123.3	142.9	90.0	94.8	115.5	193.0
2000	147.2	91.6	107.8	98.1	204.1	408.0	120.1	144.2	89.2	96.2	116.7	192.5
2001	138.0	90.7	104.7	98.0	182.7	358.6	113.3	139.4	88.7	94.5	110.0	180.0
2002	142.7	95.9	106.0	97.0	183.6	282.4	114.6	136.2	93.5	99.9	113.9	173.9
2003	140.5	98.8	106.2	98.9	176.3	262.9	111.9	130.8	92.1	101.6	114.0	169.0
2004	144.5	98.8	107.8	99.1	174.1	230.8	112.9	131.2	97.5	105.6	115.4	169.7
2005	149.8	105.3	110.5	102.7	176.6	225.7	112.4	131.0	100.1	109.3	116.6	169.2
2006	147.7	108.7	111.2	102.9	160.2	220.6	111.2	129.5	101.9	111.9	117.4	167.2
2007	142.2	105.9	112.5	103.1	141.7	179.8	111.7	132.2	104.7	117.5	114.1	157.7
2008	128.3	107.7	105.8	100.2	124.5	144.8	106.9	123.7	100.1	108.6	103.4	143.9
2009	93.1	99.8	97.7	99.5	98.7	106.8	95.4	103.6	98.8	98.1	86.4	120.4
2010	91.9	103.4	99.8	99.8	105.4	107.8	97.4	103.5	97.9	101.3	94.2	111.3
2011	95.0	102.9	99.9	99.6	103.1	106.0	97.3	101.8	100.2	101.4	95.5	106.1
2012	100.0	100.0	100.0	100.0	100.0	100.0	100.0	100.0	100.0	100.0	100.0	100.0
2013	100.8	103.2	100.0	101.7	103.7	92.7	100.2	100.3	104.0	96.6	101.2	95.0
2014	101.3	100.0	99.3	101.6	105.8	88.8	99.3	98.5	100.2	95.6	103.9	93.8
2015	106.3	100.4	99.6	103.8	100.9	84.7	98.6	97.5	98.1	95.2	105.8	90.4
2016	106.6	100.7	100.4	105.0	99.0	80.9	97.7	99.2	104.4	94.7	107.1	88.0
2017	106.3	99.5	102.3	107.6	99.7	74.6	97.0	99.8	107.1	96.6	109.3	87.5
2018	106.3	99.4	104.3	109.7	100.9	72.5	96.0	97.6	106.9	100.4	110.2	78.9
2017												
January	109.5	100.2	101.3	106.5	100.5	79.3	97.8	100.5	105.9	95.2	107.4	89.0
February	107.4	100.3	101.4	107.0	99.4	77.8	99.1	101.2	105.5	94.4	109.5	89.6
March	107.3	100.1	101.2	106.3	99.1	76.4	97.1	100.4	106.0	95.2	107.6	88.6
April	107.1	101.5	102.1	107.3	100.0	74.6	98.2	101.0	108.5	95.9	108.5	88.1
May	105.4	98.2	102.4	107.4	99.7	74.4	96.5	100.1	109.5	97.0	107.9	88.2
June	106.4	99.3	102.7	107.4	99.7	73.4	96.9	99.9	109.4	97.4	109.3	88.8
July	105.2	99.6	102.9	107.9	99.6	73.7	95.7	99.6	107.1	98.2	109.7	88.0
August	105.6	99.9	102.2	108.3	99.4	73.6	98.0	100.2	106.4	95.6	109.3	87.3
September	104.5	99.2	100.9	108.3	99.9	72.9	96.4	98.1	102.8	93.4	110.3	86.3
October	105.8	99.1	103.2	108.4	99.9	73.3	95.2	98.8	107.9	98.6	109.7	86.9
November	105.8	98.4	103.6	108.3	98.9	72.3	95.8	98.9	107.7	99.1	111.7	85.3
December	105.6	97.8	103.8	108.8	100.2	73.4	97.0	99.4	108.2	99.1	110.7	83.4
2018												
January	104.9	96.7	102.9	108.1	101.1	74.5	96.8	98.7	107.2	97.8	109.8	83.1
February	105.6	98.6	103.9	110.5	102.3	74.5	95.2	99.5	106.3	98.4	111.3	84.0
March	105.2	98.6	103.7	109.4	100.8	74.2	95.9	97.3	106.6	99.0	111.0	82.9
April	105.6	99.3	104.3	110.5	100.9	74.8	96.6	98.2	106.5	99.8	110.2	81.3
May	106.0	100.0	104.1	109.8	100.0	74.2	95.9	98.4	106.8	100.3	109.0	79.3
June	106.2	99.3	104.4	110.2	100.8	72.2	95.2	98.3	107.2	100.7	109.1	76.7
July	106.8	99.3	105.1	110.9	100.7	74.7	96.6	98.6	107.1	101.5	110.5	76.5
August	106.4	99.6	104.9	110.3	102.3	72.5	95.8	96.8	107.9	101.5	109.7	76.7
September	107.4	99.1	104.5	110.2	100.8	70.9	95.7	96.7	107.5	100.8	109.9	77.0
October	107.8	100.7	104.2	108.6	102.2	70.2	96.0	97.0	106.9	101.2	109.9	77.5
November	107.6	100.3	104.3	108.2	100.7	69.3	95.8	96.2	105.9	102.2	110.6	77.0
December	106.7	100.8	104.8	109.3	98.8	68.1	96.3	95.8	107.1	102.1	111.5	75.6

. . . = Not available.

Table 2-3. Capacity Utilization by NAICS Industry Groups

(Output as a percent of capacity, seasonally adjusted.)

Year and month	Total industry	Total manufacturing (SIC)	Manufacturing (NAICS) Total	Durable goods manufacturing Total	Wood products	Nonmetallic mineral products	Primary metals	Fabricated metal products	Machinery	Computer and electronic products	Electrical equipment, appliances, and components	Motor vehicles and parts	Aerospace and miscellaneous transportation equipment
1970	81.2	79.4	...	77.7	...	73.5	79.4	78.4	79.7	...	...	66.3	75.4
1971	79.6	77.9	...	75.4	...	74.9	73.0	78.4	73.8	...	...	79.1	67.0
1972	84.7	83.4	83.3	82.0	92.2	79.0	82.9	85.0	83.1	80.7	89.8	84.3	65.6
1973	88.3	87.7	87.8	88.6	87.9	83.9	94.7	90.9	92.4	85.2	97.7	91.8	73.5
1974	85.1	84.4	84.5	84.7	77.8	81.4	96.5	86.2	91.9	83.1	91.5	76.7	74.0
1975	75.8	73.7	73.5	71.8	70.0	72.6	75.2	72.2	77.7	68.7	70.7	66.1	70.7
1976	79.8	78.4	78.4	76.5	79.5	77.4	78.6	75.9	79.9	72.1	78.8	81.8	66.2
1977	83.4	82.5	82.5	81.1	86.0	82.4	79.3	79.9	85.4	77.8	85.5	89.9	66.2
1978	85.1	84.4	84.4	83.8	84.8	86.4	84.2	80.7	88.6	80.4	87.7	90.7	72.5
1979	85.0	84.0	84.0	84.0	79.8	84.4	86.1	81.7	90.1	83.5	88.8	80.9	81.4
1980	80.8	78.7	78.4	77.5	72.4	75.1	76.0	75.2	83.5	86.8	81.7	59.7	84.1
1981	79.5	76.9	76.5	75.1	70.6	72.4	77.0	72.8	80.5	84.7	78.8	57.3	76.7
1982	73.6	70.9	70.3	66.4	63.8	64.8	56.5	64.7	66.6	81.1	69.6	50.7	69.8
1983	74.9	73.5	72.9	68.8	75.2	70.2	59.8	66.7	60.5	82.0	72.8	67.3	66.5
1984	80.4	79.4	79.0	76.9	81.1	75.4	69.5	73.6	70.9	87.6	82.3	81.7	69.8
1985	79.2	78.1	77.6	75.8	79.7	75.7	67.6	74.1	70.5	80.4	79.0	83.5	72.4
1986	78.6	78.4	78.0	75.4	83.7	78.1	69.6	73.8	70.0	77.5	80.2	79.1	74.0
1987	81.1	80.9	80.5	77.6	85.7	80.6	78.3	75.4	71.8	79.9	82.4	78.0	75.4
1988	84.2	83.9	83.7	81.9	84.3	81.5	87.9	80.1	80.1	81.3	87.2	82.6	79.7
1989	83.7	83.2	83.1	81.7	82.1	80.6	85.1	79.7	83.8	78.8	86.3	81.0	85.3
1990	82.4	81.5	81.4	79.3	80.2	79.2	85.6	77.6	81.4	78.6	84.0	71.9	84.7
1991	79.9	78.6	78.4	77.4	76.0	73.1	80.8	73.9	76.3	78.3	79.2	64.0	83.8
1992	80.6	79.6	79.6	77.1	78.7	76.8	81.9	76.0	75.1	78.9	82.3	72.4	78.9
1993	81.5	80.5	80.4	78.6	78.8	78.5	85.0	76.9	78.6	78.3	86.7	78.7	74.3
1994	83.5	82.8	82.8	81.5	82.8	81.4	90.6	81.2	83.2	79.9	91.9	86.1	68.1
1995	83.9	83.1	83.2	82.1	80.8	81.7	88.9	83.1	85.3	82.9	91.3	83.4	65.5
1996	83.4	82.1	82.2	81.6	81.2	85.0	88.7	82.6	84.4	81.4	89.9	80.5	68.1
1997	84.1	83.0	82.9	82.3	80.9	84.0	90.0	81.7	84.8	81.5	88.4	81.8	74.2
1998	82.8	81.6	81.3	80.7	81.0	84.4	86.4	80.0	81.8	77.6	85.7	79.2	82.6
1999	81.8	80.5	80.2	80.2	81.4	82.2	84.4	78.1	76.3	81.2	83.3	83.5	76.9
2000	81.5	79.7	79.3	79.7	78.2	79.3	82.4	79.2	77.7	83.7	86.0	81.7	66.6
2001	76.2	73.8	73.3	71.6	72.4	74.4	73.8	73.2	68.4	69.5	77.0	72.8	70.8
2002	74.9	73.0	72.6	70.1	75.1	73.8	75.8	72.2	67.1	60.3	73.4	78.7	67.1
2003	76.0	74.0	73.6	71.1	76.1	74.2	74.1	73.6	68.8	63.9	74.4	78.5	64.2
2004	78.2	76.5	76.2	74.2	79.2	75.1	81.9	75.8	72.9	71.1	77.0	77.0	63.6
2005	80.1	78.5	78.3	76.7	80.8	76.3	79.5	79.5	77.0	73.1	80.7	77.3	70.3
2006	80.6	78.8	78.8	77.9	78.2	75.2	78.8	84.4	80.3	77.6	82.5	72.3	73.6
2007	80.8	78.9	79.1	78.8	73.0	70.9	79.8	86.8	83.4	76.8	87.5	71.9	85.3
2008	77.8	74.7	74.6	74.9	63.1	61.9	77.6	81.9	81.0	77.7	85.8	58.0	84.4
2009	68.5	65.5	65.3	61.4	50.3	47.9	54.1	64.6	63.1	70.8	69.1	43.8	74.0
2010	73.5	70.7	70.9	68.8	55.8	51.8	68.7	71.5	70.4	77.0	74.4	59.1	73.4
2011	76.1	73.5	73.8	72.6	60.6	54.9	76.9	77.3	77.1	76.3	79.5	64.1	71.1
2012	76.9	74.5	75.0	75.1	67.9	57.9	76.1	77.7	81.8	75.8	80.9	69.6	76.7
2013	77.2	74.4	74.9	74.9	72.8	61.8	75.8	78.2	77.1	73.9	79.9	71.2	77.0
2014	78.6	75.2	75.7	76.2	72.4	64.8	73.0	80.0	77.0	73.9	80.5	75.8	81.4
2015	76.9	75.3	75.8	75.3	74.4	66.4	66.8	78.2	71.1	72.3	77.4	79.4	82.0
2016	75.0	74.2	74.6	73.1	76.4	66.2	63.9	75.9	66.5	69.9	75.7	79.5	75.7
2017	76.5	75.1	75.4	74.2	79.3	66.8	67.1	77.2	72.3	69.9	75.9	77.4	75.9
2018	78.7	76.6	77.0	76.1	79.5	68.2	72.0	80.5	77.3	71.7	76.1	78.5	75.6
2017													
January	75.5	74.8	75.1	74.0	78.4	66.2	65.7	76.3	69.5	69.9	77.4	79.5	76.2
February	75.2	74.7	75.0	73.8	79.1	67.2	66.1	76.6	68.7	69.0	76.7	79.3	76.0
March	75.7	74.5	74.7	73.6	78.8	67.3	65.7	76.4	70.3	69.6	76.1	76.5	75.9
April	76.4	75.3	75.6	74.5	78.5	66.5	66.2	76.6	72.2	70.5	76.8	79.3	76.2
May	76.5	75.1	75.4	74.1	78.3	65.9	65.1	76.9	72.8	69.9	76.1	78.5	76.2
June	76.6	75.2	75.5	74.0	77.9	66.3	66.3	76.5	72.9	69.6	75.8	77.4	76.2
July	76.5	75.1	75.3	73.6	78.8	66.4	66.5	76.9	72.2	69.4	76.2	74.9	76.1
August	76.2	74.8	75.1	73.7	79.1	65.4	67.5	77.0	71.5	69.5	74.9	75.8	76.3
September	76.1	74.7	75.0	74.3	80.2	67.4	68.8	77.7	73.5	69.5	76.1	76.4	75.7
October	77.3	75.7	76.0	74.7	80.1	67.3	68.8	78.2	74.0	70.2	75.6	77.2	75.6
November	77.6	75.9	76.2	74.9	81.1	67.8	69.5	78.5	74.7	70.6	75.1	76.8	75.6
December	77.9	75.8	76.2	74.7	80.7	68.3	68.8	78.5	75.2	70.4	74.1	76.7	75.1
2018													
January	77.6	75.5	75.9	74.7	80.1	66.7	69.5	78.4	75.1	71.1	75.3	77.2	74.7
February	77.8	76.3	76.7	75.6	81.5	69.7	70.2	79.8	74.9	71.0	76.0	79.2	74.5
March	78.2	76.3	76.7	75.7	81.2	68.8	71.0	79.9	75.0	71.3	75.0	80.5	74.3
April	78.8	76.6	77.0	76.0	80.5	69.1	71.1	80.0	76.3	71.6	75.9	79.2	75.3
May	78.1	76.0	76.4	75.0	80.0	69.0	71.0	79.7	75.5	71.4	75.8	73.5	75.0
June	78.6	76.5	77.0	75.9	79.9	68.3	71.0	80.6	76.0	71.9	76.4	78.0	75.5
July	78.8	76.7	77.2	75.9	79.2	68.4	71.1	80.5	77.0	72.6	77.1	76.5	75.7
August	79.3	77.0	77.5	76.7	79.9	68.1	72.3	81.0	78.7	72.5	76.8	78.6	76.3
September	79.3	76.9	77.4	76.8	79.7	66.9	73.2	81.0	79.0	72.2	76.6	80.1	76.2
October	79.3	76.8	77.3	76.9	78.3	67.9	73.7	81.4	79.7	71.8	76.2	78.5	76.4
November	79.6	76.9	77.4	77.0	77.3	67.0	75.2	81.7	80.9	71.3	76.1	78.9	76.1
December	79.5	77.3	77.8	77.5	77.0	68.9	75.0	81.8	79.3	72.0	76.2	82.1	76.9

. . . = Not available.

Table 2-3. Capacity Utilization by NAICS Industry Groups—Continued

(Output as a percent of capacity, seasonally adjusted.)

| Year and month | Durable goods manufacturing—Continued | | Manufacturing (NAICS)—Continued / Nondurable goods manufacturing | | | | | | | | | Other manufacturing (non-NAICS) |
	Furniture and related products	Miscellaneous manufacturing	Total	Food, beverage, and tobacco products	Textile and product mills	Apparel and leather	Paper	Printing and support	Petroleum and coal products	Chemicals	Plastics and rubber products	
1970	83.6	. . .	82.2	84.1	. . .	. . .	86.2	. . .	96.0	76.3	79.3	. . .
1971	84.3	. . .	81.8	84.1	. . .	. . .	86.9	. . .	94.6	75.6	79.8	. . .
1972	93.6	80.2	85.3	85.2	88.8	81.6	91.4	92.3	93.1	80.3	88.5	85.6
1973	94.6	79.5	86.6	84.7	86.3	82.2	94.9	94.0	90.3	83.8	92.5	84.7
1974	83.5	74.6	84.2	83.9	76.4	76.3	95.0	88.0	92.5	83.9	84.2	82.7
1975	69.7	67.6	76.1	80.1	72.7	74.6	80.7	79.6	83.7	71.7	70.1	77.3
1976	76.3	72.4	81.2	83.2	81.6	78.2	87.7	82.3	86.1	77.9	77.3	77.6
1977	84.6	77.6	84.4	82.9	88.9	81.7	90.2	86.0	87.7	81.4	88.4	83.2
1978	86.1	78.8	85.3	83.6	88.1	84.2	92.2	87.4	86.4	82.6	88.7	85.1
1979	81.0	78.4	83.9	81.2	87.6	79.2	90.9	86.0	88.5	83.0	83.7	85.6
1980	75.0	73.4	79.7	80.9	83.6	80.0	88.3	83.7	76.2	76.7	74.1	86.8
1981	71.8	75.7	78.8	80.3	80.7	78.8	86.8	81.2	72.8	76.1	77.5	87.5
1982	66.3	73.7	76.4	81.3	74.4	77.9	83.9	82.4	71.2	69.3	74.5	87.4
1983	72.2	70.8	79.4	80.8	84.1	81.1	88.9	84.4	74.4	72.9	81.5	88.0
1984	79.0	75.9	82.1	81.3	85.8	81.2	91.1	87.0	78.4	76.0	91.2	89.5
1985	77.2	73.9	80.5	82.6	81.6	77.9	87.8	84.8	80.0	73.4	87.0	90.4
1986	78.9	73.9	81.8	82.8	84.2	79.7	90.1	85.5	81.8	76.0	85.3	88.8
1987	82.7	77.1	84.7	83.7	91.3	82.1	90.2	89.1	82.3	81.0	89.5	90.5
1988	80.4	81.7	86.2	85.3	88.7	82.1	91.3	90.3	83.2	84.4	89.0	88.6
1989	79.0	79.7	84.9	83.7	88.3	80.2	90.4	88.9	84.2	83.3	87.1	85.4
1990	76.1	79.5	84.2	83.9	83.4	79.0	89.2	88.9	84.7	82.9	83.8	83.7
1991	70.6	78.2	82.3	83.0	81.6	80.3	87.5	84.2	82.7	81.1	78.8	80.8
1992	76.5	77.1	82.7	82.6	85.6	82.8	88.3	86.4	85.4	79.7	82.1	80.1
1993	79.1	77.4	82.7	81.1	88.0	83.8	88.9	84.9	89.1	79.1	86.6	81.4
1994	80.5	77.2	84.6	83.3	90.2	84.9	90.7	84.1	89.0	80.5	91.3	81.5
1995	79.9	79.6	84.5	84.3	86.2	84.5	89.6	83.3	89.6	80.8	89.6	82.2
1996	78.7	81.1	83.1	82.4	82.3	83.2	85.5	83.1	91.8	80.4	88.3	80.6
1997	83.3	79.4	83.8	82.4	84.5	83.0	87.4	81.4	95.3	81.3	89.0	85.6
1998	83.0	79.8	82.2	82.8	81.8	77.4	87.3	79.8	92.9	78.6	87.6	86.8
1999	80.6	77.5	80.1	78.4	81.7	76.0	86.5	77.7	90.7	77.2	86.3	87.2
2000	77.5	77.1	78.9	77.3	79.4	78.8	84.3	77.0	89.5	76.4	82.3	87.5
2001	70.8	73.8	75.7	76.2	71.5	75.1	80.0	74.8	88.5	72.1	76.6	82.9
2002	72.6	74.6	75.9	75.4	73.8	66.8	82.0	74.9	88.9	73.8	78.7	81.6
2003	71.5	75.4	76.8	77.1	73.0	69.5	81.8	75.4	89.8	73.9	79.2	81.5
2004	76.9	76.0	78.7	77.2	75.0	70.2	83.2	77.5	92.8	76.5	83.1	82.4
2005	80.2	78.7	80.3	79.2	78.4	75.4	83.7	77.5	92.0	77.1	83.9	81.9
2006	79.1	77.9	79.8	78.9	75.4	75.3	83.6	77.5	89.4	77.4	82.0	79.8
2007	75.3	73.8	79.3	78.4	72.4	75.9	83.7	77.0	87.6	78.0	78.6	76.3
2008	71.1	74.0	74.1	76.3	67.9	75.2	82.0	69.6	81.7	71.3	70.0	77.3
2009	57.6	69.7	69.8	76.2	56.6	59.9	77.1	60.0	78.5	66.3	60.2	69.6
2010	61.2	75.3	73.3	77.1	63.8	66.8	81.5	63.0	80.5	70.3	68.7	66.2
2011	66.9	76.8	75.2	78.7	65.8	70.4	81.2	63.4	84.5	72.1	69.9	65.4
2012	71.7	74.8	75.0	78.5	66.4	69.2	83.1	63.6	82.9	70.5	75.3	63.1
2013	73.4	77.0	74.9	78.4	70.5	65.6	83.0	66.0	84.1	68.4	79.3	62.2
2014	75.9	75.6	75.1	76.6	72.7	65.8	83.8	67.5	81.8	70.5	81.8	63.7
2015	80.3	78.4	76.3	76.8	69.6	66.6	85.7	69.5	81.4	72.3	84.4	63.8
2016	79.5	80.9	76.2	76.5	69.0	66.8	87.0	73.2	77.8	72.6	84.2	64.2
2017	78.6	80.5	76.8	76.8	70.7	65.0	87.4	75.6	79.0	73.6	83.2	66.3
2018	77.7	79.2	78.0	77.0	72.3	66.2	86.9	73.9	80.6	76.7	81.7	62.3
2017												
January	81.1	81.2	76.2	76.6	70.7	67.5	88.0	75.4	76.6	72.8	82.6	66.2
February	79.5	81.3	76.2	76.8	70.0	66.5	89.2	76.1	76.4	72.0	84.0	66.9
March	79.4	81.1	76.0	76.2	69.9	65.6	87.4	75.6	77.1	72.6	82.4	66.4
April	79.2	82.2	76.7	76.8	70.7	64.3	88.4	76.3	79.2	73.1	82.9	66.2
May	78.0	79.6	76.9	76.8	70.6	64.4	86.9	75.7	80.2	73.9	82.3	66.5
June	78.7	80.4	77.1	76.6	70.7	63.8	87.3	75.7	80.6	74.2	83.3	67.2
July	77.8	80.6	77.2	76.9	70.8	64.4	86.2	75.6	79.2	74.7	83.5	66.9
August	78.1	80.8	76.7	77.1	70.7	64.6	88.3	76.0	79.0	72.8	83.0	66.6
September	77.3	80.2	75.8	77.0	71.2	64.2	86.8	74.5	76.6	71.1	83.7	66.1
October	78.2	80.1	77.5	77.0	71.3	64.8	85.8	75.1	80.7	75.1	83.1	66.7
November	78.1	79.4	77.7	76.9	70.7	64.2	86.4	75.2	80.7	75.5	84.5	65.8
December	77.9	78.8	77.9	77.1	71.6	65.4	87.5	75.6	81.3	75.5	83.5	64.5
2018												
January	77.3	77.9	77.2	76.5	72.3	66.6	87.3	75.0	80.7	74.5	82.7	64.5
February	77.7	79.3	77.9	78.1	73.3	66.9	85.9	75.6	80.1	75.0	83.7	65.4
March	77.4	79.2	77.8	77.3	72.2	66.9	86.6	73.9	80.4	75.5	83.2	64.8
April	77.5	79.6	78.2	77.9	72.3	67.7	87.3	74.5	80.4	76.1	82.4	63.8
May	77.7	80.0	78.0	77.3	71.7	67.4	86.7	74.6	80.5	76.6	81.2	62.3
June	77.7	79.3	78.2	77.5	72.2	65.8	86.2	74.5	80.9	76.9	81.1	60.5
July	78.0	79.2	78.7	77.8	72.2	68.3	87.5	74.6	80.7	77.6	81.8	60.5
August	77.6	79.2	78.4	77.3	73.3	66.6	86.8	73.2	81.3	77.6	81.0	60.8
September	78.2	78.7	78.1	77.1	72.2	65.3	86.8	73.1	81.0	77.1	80.8	61.2
October	78.3	79.7	77.8	75.8	73.1	64.9	87.1	73.3	80.5	77.4	80.5	61.8
November	78.0	79.2	77.8	75.4	72.1	64.3	87.0	72.6	79.7	78.1	80.8	61.6
December	77.2	79.4	78.1	76.0	70.7	63.4	87.6	72.3	80.6	78.1	81.1	60.6

. . . = Not available.

Table 2-3. Capacity Utilization by NAICS Industry Groups—Continued

(Output as a percent of capacity, seasonally adjusted.)

Year and month	Mining	Utilities	Selected high-tech industries				Measures excluding selected high-tech industries		Stage-of-process groups		
			Total	Computers and peripheral equipment	Communications equipment	Semiconductors and related electronic components	Total industry	Manufacturing	Crude	Primary and semi-finished	Finished
1970	89.5	96.4	82.9	. . .	. . .	. . .	81.0	79.3	84.6	81.4	78.1
1971	88.1	95.0	73.6	. . .	. . .	. . .	80.0	78.3	83.6	81.6	75.7
1972	90.8	95.4	78.0	81.4	73.3	81.4	84.9	83.6	88.4	88.1	79.6
1973	91.6	93.1	81.4	81.9	75.1	88.5	88.6	87.9	90.0	92.1	83.2
1974	91.1	86.7	81.0	88.3	73.3	83.5	85.2	84.6	91.0	87.3	80.3
1975	89.5	85.2	64.5	69.0	62.1	63.3	76.2	74.1	84.0	75.2	73.7
1976	89.6	85.7	68.5	76.9	62.0	68.6	80.3	78.8	87.0	80.2	76.9
1977	89.5	86.9	75.6	77.8	72.5	76.7	83.8	82.8	89.1	84.6	79.9
1978	89.7	87.2	78.4	77.7	77.2	80.3	85.4	84.7	88.7	86.3	82.3
1979	91.2	87.2	82.6	78.1	85.3	85.1	85.1	84.1	90.0	85.9	81.7
1980	91.3	85.5	87.5	87.0	91.6	84.6	80.5	78.2	89.4	78.8	79.4
1981	90.9	84.4	85.5	85.3	89.3	83.1	79.3	76.4	89.3	77.1	77.5
1982	84.1	80.0	78.6	70.4	87.6	82.0	73.3	70.4	82.3	70.4	73.1
1983	79.8	79.3	81.0	75.9	86.0	82.7	74.6	73.0	79.9	74.5	73.0
1984	85.8	81.9	87.8	85.4	85.9	91.3	80.0	78.8	85.8	81.2	77.2
1985	84.4	81.7	77.5	76.2	79.6	77.1	79.3	78.2	83.8	79.8	76.6
1986	77.6	80.9	73.1	72.6	76.5	71.3	78.9	78.8	79.2	79.7	77.1
1987	80.3	83.5	77.6	73.7	80.3	80.0	81.4	81.2	82.8	82.8	78.7
1988	84.1	86.8	79.9	76.2	83.6	81.5	84.5	84.2	86.3	85.8	81.6
1989	85.1	86.8	78.2	74.8	80.3	79.8	84.0	83.6	86.8	84.6	81.6
1990	86.9	86.6	77.1	71.4	81.8	79.5	82.7	81.9	87.9	82.6	80.5
1991	85.4	87.8	78.0	73.7	78.3	80.9	80.0	78.6	85.5	80.0	78.2
1992	85.2	86.4	80.3	78.2	79.4	82.0	80.6	79.6	85.9	81.5	78.2
1993	85.8	88.2	80.8	79.9	81.9	80.5	81.6	80.5	85.8	83.3	78.4
1994	86.8	88.3	82.7	74.3	84.8	85.8	83.6	82.8	87.8	86.3	79.2
1995	87.6	89.3	84.9	78.2	79.7	90.4	83.8	83.0	89.0	86.4	79.7
1996	90.5	90.7	83.1	87.0	76.4	84.5	83.4	82.1	89.1	85.6	79.3
1997	91.8	90.1	83.5	83.4	79.3	85.6	84.1	83.0	90.4	86.0	80.3
1998	89.3	92.6	79.1	79.3	84.6	76.3	83.2	81.9	87.1	84.2	80.3
1999	86.2	94.2	85.3	84.5	87.5	84.5	81.5	80.0	86.1	84.3	78.0
2000	90.5	94.3	88.0	81.3	90.2	89.5	80.9	78.9	88.5	84.0	76.9
2001	89.8	90.1	69.3	70.8	70.3	68.1	76.7	74.3	85.5	77.4	72.6
2002	86.0	87.6	58.2	68.7	43.1	63.7	76.3	74.4	83.2	77.4	70.5
2003	87.8	85.7	63.1	72.5	44.0	72.2	76.9	74.9	85.0	78.2	71.3
2004	88.2	84.5	70.5	77.3	53.6	77.9	78.6	76.9	86.5	80.2	73.4
2005	88.5	85.1	73.9	75.8	58.9	80.8	80.5	78.8	86.7	81.9	75.7
2006	90.1	83.7	81.8	78.2	77.8	85.6	80.5	78.7	88.1	81.5	76.4
2007	89.4	85.9	77.8	76.4	76.5	78.0	80.9	79.0	88.7	81.2	77.1
2008	90.0	84.2	80.9	79.1	81.6	81.6	77.6	74.3	87.5	77.0	73.9
2009	80.3	80.6	73.1	89.9	80.3	64.7	68.3	65.1	77.9	65.8	68.1
2010	83.9	83.0	81.4	99.4	81.5	77.8	73.3	70.2	83.2	71.8	71.2
2011	85.9	81.4	76.9	75.6	81.9	75.9	76.1	73.3	84.5	74.4	73.7
2012	87.3	78.4	71.8	68.2	77.3	71.2	77.1	74.7	85.5	74.7	74.8
2013	87.2	79.9	69.6	63.4	82.6	66.8	77.4	74.6	86.0	75.5	73.8
2014	90.5	80.8	72.7	70.5	70.3	74.2	78.7	75.3	88.4	76.7	74.6
2015	84.2	79.9	75.2	80.3	72.2	75.3	77.0	75.3	82.7	76.3	75.1
2016	77.6	78.8	73.5	78.1	71.7	73.1	75.1	74.2	78.4	75.2	73.6
2017	84.3	77.0	70.6	76.7	65.2	71.4	76.6	75.3	83.7	75.7	74.2
2018	90.2	79.3	73.2	73.9	62.1	78.3	78.8	76.7	88.8	77.5	75.4
2017											
January	80.4	75.6	71.5	79.3	68.9	70.7	75.6	74.9	80.8	75.0	73.9
February	81.9	71.5	70.5	79.4	66.5	70.0	75.3	74.8	81.6	74.3	73.7
March	82.5	77.1	70.1	79.1	65.1	70.1	75.8	74.6	82.3	75.3	73.5
April	83.4	76.6	70.8	78.7	64.6	71.6	76.5	75.4	83.5	75.5	74.5
May	84.0	77.8	70.6	78.6	64.6	71.1	76.6	75.3	83.9	75.7	74.3
June	84.8	77.2	70.4	79.3	64.8	70.5	76.7	75.4	84.6	75.8	74.2
July	84.7	77.8	69.4	75.2	65.0	69.8	76.7	75.2	84.6	75.6	74.2
August	84.1	76.6	69.3	74.0	65.0	69.9	76.3	75.0	83.3	75.3	74.1
September	84.9	76.3	69.9	74.9	65.0	70.7	76.3	74.9	82.2	75.4	74.4
October	86.1	78.2	70.8	75.0	64.7	72.4	77.4	75.9	85.1	76.5	74.8
November	87.3	78.9	71.4	72.2	64.2	74.4	77.8	76.0	86.4	76.9	74.6
December	87.7	80.6	72.3	74.3	63.6	75.6	78.0	76.0	86.6	77.3	74.6
2018											
January	86.6	81.7	72.3	75.4	62.7	75.8	77.7	75.6	85.2	77.2	74.6
February	88.1	75.8	72.1	74.2	61.8	76.2	78.0	76.5	86.3	76.6	75.3
March	88.6	78.6	72.2	74.3	60.8	76.8	78.3	76.5	87.2	77.3	75.0
April	89.0	81.7	72.4	74.7	60.2	77.4	79.0	76.8	87.5	78.2	75.4
May	89.0	79.2	72.3	73.2	60.2	77.7	78.2	76.1	87.9	77.2	74.5
June	90.2	78.5	73.7	74.6	61.1	79.3	78.7	76.5	89.1	77.1	75.2
July	90.4	78.3	74.4	74.3	62.4	80.1	78.9	76.8	89.5	77.2	75.5
August	91.8	79.2	75.2	74.0	63.4	81.3	79.4	77.0	90.6	77.6	75.8
September	92.1	78.7	74.8	72.2	63.7	80.9	79.3	77.0	90.7	77.4	75.8
October	91.6	80.6	74.1	72.6	63.4	79.6	79.4	76.9	90.2	77.9	75.6
November	91.8	82.6	72.9	73.7	62.7	77.7	79.7	77.0	90.4	78.5	75.5
December	93.3	76.8	72.5	73.2	63.1	76.9	79.6	77.4	91.5	77.4	76.0

. . . = Not available.

Table 2-4A. Industrial Production and Capacity Utilization, Historical Data, 1948–2016

(Seasonally adjusted.)

Year and month	Production indexes, 2007 = 100										Capacity utilization (output as percent of capacity)	
			Market groups									
	Total industry	Manufac-turing (SIC)	Consumer goods			Business equipment	Defense and space equipment	Construction supplies	Business supplies	Materials	Total industry	Manufac-turing (SIC)
			Total	Durable	Nondurable							
1949	14.0	13.2	21.2	14.8	24.5	5.1	7.4	27.1	13.3	12.7	. . .	74.2
1950	16.2	15.4	24.2	19.8	26.5	5.5	8.7	32.6	14.8	15.2	. . .	82.8
1951	17.6	16.6	23.9	17.2	27.4	6.7	21.4	34.0	15.7	16.8	. . .	85.8
1952	18.3	17.3	24.5	16.6	28.5	7.6	30.0	33.7	15.7	17.0	. . .	85.4
1953	19.8	18.9	25.9	19.5	29.3	7.9	36.0	36.2	16.7	18.9	. . .	89.3
1954	18.8	17.6	25.8	18.1	29.7	6.9	31.7	35.6	16.9	17.5	. . .	80.1
1955	21.2	19.9	28.8	22.2	32.0	7.5	29.0	41.0	18.9	20.6	. . .	87.0
1956	22.1	20.7	29.8	21.5	34.0	8.7	28.4	42.2	20.0	21.2	. . .	86.1
1957	22.4	20.9	30.5	21.5	35.1	9.0	29.6	41.6	20.3	21.2	. . .	83.6
1958	21.0	19.5	30.3	19.1	36.1	7.6	29.7	40.2	20.1	19.1	. . .	75.0
1959	23.5	21.9	33.2	22.6	38.6	8.6	31.3	45.0	21.9	22.0	. . .	81.6
1960	24.0	22.4	34.5	23.9	39.9	8.8	32.2	43.9	22.6	22.3	. . .	80.1
1961	24.1	22.4	35.2	23.5	41.2	8.5	32.7	44.3	23.3	22.3	. . .	77.3
1962	26.1	24.4	37.5	26.6	43.1	9.3	37.9	47.0	24.8	24.3	. . .	81.4
1963	27.7	25.9	39.6	28.9	45.1	9.7	40.9	49.2	26.4	25.8	. . .	83.5
1964	29.6	27.7	41.8	31.0	47.3	10.9	39.6	52.2	28.3	27.9	. . .	85.6
1965	32.5	30.7	45.1	36.4	49.3	12.5	43.8	55.4	30.1	31.1	. . .	89.5
1966	35.4	33.5	47.4	38.5	51.7	14.4	51.4	57.7	32.4	33.9	. . .	91.1
1967	36.1	34.1	48.6	37.0	54.3	14.7	58.7	59.3	34.2	33.6	87.0	87.2
1968	38.1	36.0	51.5	41.2	56.5	15.4	58.8	62.3	36.2	35.8	87.3	87.1
1969	39.9	37.6	53.4	43.0	58.4	16.4	56.0	65.0	38.5	38.0	87.4	86.6
1970	38.6	35.9	52.8	39.6	59.4	15.8	47.4	62.7	38.7	36.6	81.2	79.4
1971	39.1	36.5	55.9	44.9	61.1	15.0	42.6	64.7	39.9	37.2	79.6	77.9
1972	42.9	40.3	60.4	50.3	65.0	17.1	41.4	73.5	43.9	40.9	84.7	83.4
1973	46.4	43.9	63.1	54.1	66.9	19.8	45.4	79.8	46.6	44.6	88.3	87.7
1974	46.3	43.8	61.3	49.2	67.0	20.9	46.9	77.9	46.4	44.5	85.1	84.4
1975	42.2	39.2	58.8	44.7	65.7	18.7	47.3	65.9	42.6	39.7	75.8	73.7
1976	45.5	42.7	63.6	50.5	69.9	19.9	45.9	70.9	45.4	43.1	79.8	78.4
1977	48.9	46.4	67.5	56.7	72.4	23.0	41.1	77.1	49.3	46.2	83.4	82.5
1978	51.6	49.2	69.7	58.1	74.9	25.9	41.9	81.5	52.0	48.5	85.1	84.4
1979	53.2	50.7	68.6	56.0	74.4	29.2	44.8	83.6	53.8	49.8	85.0	84.0
1980	51.8	48.9	65.9	48.7	74.5	29.9	53.2	77.4	52.5	47.9	80.8	78.7
1981	52.5	49.4	66.4	49.4	74.8	30.8	57.6	76.1	53.7	48.2	79.5	76.9
1982	49.8	46.7	66.2	46.6	76.1	28.1	68.9	69.1	53.0	44.5	73.6	70.9
1983	51.1	49.0	68.6	51.7	77.0	28.3	69.3	73.9	55.6	45.7	74.9	73.5
1984	55.7	53.7	71.8	57.9	78.5	32.5	79.4	80.3	60.5	50.0	80.4	79.4
1985	56.4	54.6	72.5	57.9	79.5	33.7	88.9	82.2	61.9	49.9	79.2	78.1
1986	56.9	55.8	74.9	61.6	81.3	33.2	94.4	84.9	63.9	49.9	78.6	78.4
1987	59.9	59.0	78.1	65.3	84.2	35.5	96.4	90.4	67.8	52.6	81.1	80.9
1988	63.0	62.1	81.1	68.5	87.0	39.2	97.2	92.4	70.3	55.5	84.2	83.9
1989	63.6	62.6	81.4	70.1	86.7	40.5	97.3	92.1	71.3	55.9	83.7	83.2
1990	64.2	63.1	81.8	68.0	88.2	42.0	93.9	91.4	72.9	56.3	82.4	81.5
1991	63.2	61.9	81.7	65.0	89.5	41.4	87.0	86.4	71.8	55.5	79.9	78.6
1992	65.1	64.2	84.2	71.3	90.3	43.2	80.8	90.1	73.6	57.3	80.6	79.6
1993	67.2	66.5	86.8	77.1	91.5	45.3	76.4	94.1	75.9	59.3	81.5	80.5
1994	70.8	70.4	90.4	84.0	93.7	48.5	71.8	101.0	78.8	63.0	83.5	82.8
1995	74.0	74.0	92.9	87.2	96.0	53.1	69.3	103.3	82.0	66.4	83.9	83.1
1996	77.4	77.6	94.6	89.9	97.2	58.3	67.6	108.0	85.2	69.9	83.4	82.1
1997	83.0	84.2	98.1	96.1	99.6	67.1	66.7	113.2	91.2	75.4	84.1	83.0
1998	87.8	89.8	101.7	103.1	101.9	74.8	69.7	119.2	96.4	79.8	82.8	81.6
1999	91.7	94.3	103.9	111.3	101.8	78.9	67.6	122.2	100.5	84.8	81.8	80.5
2000	95.2	98.2	105.9	114.5	103.3	85.0	60.1	124.8	104.4	89.0	81.5	79.7
2001	92.3	94.6	104.8	109.2	103.8	80.1	66.0	119.1	100.6	85.5	76.2	73.8
2002	92.6	95.1	107.0	116.2	104.2	75.0	66.7	119.0	100.7	86.6	74.9	73.0
2003	93.8	96.4	108.4	119.8	104.9	75.2	71.0	118.7	102.6	87.7	76.0	74.0
2004	96.4	99.4	109.6	121.5	105.9	79.4	69.7	121.5	105.1	90.7	78.2	76.5
2005	99.6	103.4	112.5	122.0	109.5	85.4	76.4	127.4	109.0	92.8	80.1	78.5
2006	101.8	106.1	113.1	121.4	110.5	93.4	76.1	130.6	110.8	94.7	80.6	78.8
2007	104.4	109.0	113.1	120.9	110.6	99.0	90.7	129.4	112.1	98.0	80.8	78.9
2008	100.8	103.8	107.2	106.6	107.2	96.7	98.0	117.5	107.2	95.6	77.8	74.7
2009	89.2	89.5	99.2	86.0	102.9	80.0	93.8	90.4	95.0	84.9	68.5	65.5
2010	94.1	94.7	100.3	94.2	101.9	86.1	100.9	93.6	98.0	91.9	73.5	70.7
2011	97.1	97.5	101.4	97.7	102.3	91.1	98.0	95.9	99.0	95.9	76.1	73.5
2012	100.0	100.0	100.0	100.0	100.0	100.0	100.0	100.0	100.0	100.0	76.9	74.5
2013	102.0	100.9	100.7	105.5	99.5	99.9	97.2	103.1	101.8	103.3	77.2	74.4
2014	105.2	102.0	101.5	110.7	99.2	101.7	93.9	106.4	102.8	108.5	78.6	75.2
2015	104.1	101.5	102.9	115.0	99.9	99.6	91.7	107.1	101.4	106.9	76.9	75.3
2016	102.1	100.7	103.5	117.5	100.0	94.4	89.1	108.1	102.0	103.7	75.0	74.2
1950												
January	14.4	13.7	22.1	16.6	25.1	4.7	6.9	28.2	13.8	13.1	. . .	74.9
February	14.5	13.8	22.2	16.6	25.1	4.9	6.9	29.3	14.2	12.9	. . .	75.4
March	15.0	14.0	22.6	17.3	25.4	4.9	7.0	30.0	14.2	13.8	. . .	76.4
April	15.5	14.6	23.3	18.5	25.8	5.0	7.2	31.6	14.5	14.4	. . .	79.2
May	15.8	15.0	23.8	19.5	26.0	5.3	7.4	31.9	14.6	14.8	. . .	81.0
June	16.3	15.4	24.4	21.1	26.2	5.5	7.7	33.0	14.7	15.3	. . .	83.1
July	16.8	15.9	25.2	22.0	26.8	5.7	8.1	33.9	15.1	15.8	. . .	85.5
August	17.3	16.5	26.0	22.6	27.8	6.0	8.9	34.5	15.4	16.2	. . .	88.4
September	17.2	16.4	25.3	21.6	27.4	5.8	9.8	34.6	15.3	16.4	. . .	87.2
October	17.3	16.5	25.1	21.2	27.2	5.9	10.6	34.9	15.5	16.6	. . .	87.5
November	17.3	16.4	25.0	20.7	27.2	6.0	11.3	34.9	15.6	16.4	. . .	87.0
December	17.6	16.7	25.4	20.5	28.0	6.1	12.3	35.0	15.8	16.7	. . .	88.1

. . . = Not available.

Table 2-4A. Industrial Production and Capacity Utilization, Historical Data, 1948–2016—*Continued*

(Seasonally adjusted.)

Year and month	Production indexes, 2007 = 100										Capacity utilization (output as percent of capacity)	
			Market groups									
	Total industry	Manufac-turing (SIC)	Consumer goods			Business equipment	Defense and space equipment	Construction supplies	Business supplies	Materials	Total industry	Manufac-turing (SIC)
			Total	Durable	Nondurable							
1951												
January	17.7	16.8	25.6	20.1	28.4	6.2	13.6	35.3	16.0	16.5	. . .	88.3
February	17.8	16.8	25.5	20.2	28.4	6.2	15.7	34.9	15.8	16.6	. . .	88.3
March	17.9	16.9	25.2	20.1	27.8	6.3	17.9	35.1	16.1	17.0	. . .	88.4
April	17.9	17.0	24.7	19.2	27.7	6.5	19.6	34.9	16.4	17.1	. . .	88.2
May	17.8	16.8	24.3	18.2	27.4	6.5	20.2	34.7	16.3	17.2	. . .	87.4
June	17.8	16.8	24.0	17.3	27.4	6.6	21.3	34.5	16.0	17.3	. . .	86.6
July	17.5	16.5	23.2	15.5	27.1	6.7	22.5	33.6	15.8	17.0	. . .	84.9
August	17.3	16.3	22.6	14.5	26.9	6.8	23.2	33.5	15.6	16.7	. . .	83.6
September	17.4	16.4	22.9	15.1	26.9	6.9	24.0	33.4	15.5	16.8	. . .	83.7
October	17.4	16.3	22.7	15.0	26.8	7.0	25.0	33.1	15.2	16.6	. . .	83.1
November	17.5	16.5	23.1	15.3	27.2	7.2	26.4	32.8	15.3	16.6	. . .	83.6
December	17.6	16.6	23.3	15.4	27.4	7.2	27.0	32.9	15.3	16.6	. . .	83.9
1952												
January	17.8	16.7	23.5	15.4	27.6	7.4	27.5	33.4	15.4	17.0	. . .	84.4
February	17.9	16.8	23.6	15.3	28.0	7.5	27.7	33.6	15.4	16.9	. . .	84.6
March	18.0	16.9	23.8	15.7	28.0	7.6	27.8	33.4	15.4	16.9	. . .	84.7
April	17.8	16.8	23.8	15.5	28.0	7.6	28.1	32.9	15.3	16.6	. . .	83.6
May	17.7	16.7	23.7	16.0	27.7	7.6	29.0	32.4	15.2	16.3	. . .	83.1
June	17.5	16.6	24.4	16.1	28.7	7.6	29.9	32.2	15.5	15.4	. . .	81.9
July	17.2	16.2	24.0	14.6	28.8	7.3	30.1	32.1	15.7	15.1	. . .	79.8
August	18.3	17.3	24.4	16.0	28.9	7.4	30.6	34.3	15.8	17.0	. . .	85.1
September	19.0	17.9	25.1	17.7	28.9	7.6	31.1	34.6	16.0	18.1	. . .	87.7
October	19.2	18.2	25.5	18.3	29.1	7.7	32.1	35.1	16.3	18.0	. . .	88.8
November	19.6	18.6	26.0	19.4	29.4	7.8	32.8	35.7	16.4	18.5	. . .	90.2
December	19.7	18.7	26.0	19.5	29.4	7.8	33.8	35.9	16.3	18.6	. . .	90.5
1953												
January	19.7	18.8	26.2	20.3	29.4	7.9	34.3	36.5	16.1	18.5	. . .	90.5
February	19.9	19.0	26.4	20.5	29.5	7.9	35.0	37.0	16.4	18.9	. . .	91.1
March	20.0	19.1	26.5	20.7	29.4	8.0	35.7	37.1	16.8	19.2	. . .	91.4
April	20.1	19.2	26.4	20.5	29.5	8.0	36.2	37.3	16.8	19.3	. . .	91.5
May	20.2	19.3	26.5	20.6	29.6	7.9	36.8	36.5	16.9	19.7	. . .	91.7
June	20.1	19.1	26.1	19.8	29.5	7.9	37.0	36.2	16.9	19.7	. . .	90.7
July	20.4	19.3	26.1	19.7	29.4	8.0	37.3	36.7	17.0	20.0	. . .	91.0
August	20.3	19.2	26.0	19.5	29.4	8.0	37.1	36.6	17.0	19.4	. . .	90.6
September	19.9	18.8	25.6	18.7	29.2	7.9	37.0	35.9	16.9	18.9	. . .	88.3
October	19.7	18.6	25.6	18.6	29.3	7.9	36.6	36.0	16.7	18.4	. . .	87.2
November	19.2	18.2	25.2	17.9	29.1	7.6	34.3	35.4	16.7	17.9	. . .	84.7
December	18.7	17.7	24.8	17.2	28.8	7.5	34.6	34.5	16.4	17.5	. . .	82.3
1954												
January	18.6	17.5	24.9	17.0	29.1	7.3	33.9	35.2	16.4	17.3	. . .	81.3
February	18.7	17.5	25.2	17.4	29.3	7.2	33.6	35.3	16.6	17.2	. . .	80.8
March	18.6	17.4	25.3	17.4	29.4	7.1	33.2	35.0	16.6	17.1	. . .	80.2
April	18.4	17.3	25.3	17.6	29.2	6.9	32.6	35.0	16.7	17.0	. . .	79.4
May	18.6	17.4	25.4	17.9	29.2	6.9	32.1	35.5	16.6	17.3	. . .	79.8
June	18.6	17.5	25.6	18.1	29.4	6.8	31.6	34.6	16.8	17.5	. . .	79.9
July	18.6	17.4	25.7	17.9	29.6	6.8	31.5	34.6	16.6	17.5	. . .	79.4
August	18.6	17.4	25.7	18.0	29.6	6.8	30.8	34.4	16.7	17.4	. . .	78.8
September	18.6	17.5	25.9	18.1	29.9	6.7	30.5	36.1	17.2	17.3	. . .	79.2
October	18.9	17.7	26.0	18.4	30.0	6.7	30.2	37.3	17.4	17.7	. . .	79.7
November	19.2	18.0	26.6	18.9	30.5	6.8	30.1	37.7	17.6	18.0	. . .	80.9
December	19.4	18.2	27.0	19.6	30.8	6.8	29.6	38.0	17.8	18.3	. . .	81.8
1955												
January	19.9	18.7	27.6	21.0	30.9	6.9	29.4	38.5	18.0	19.0	. . .	83.5
February	20.1	18.9	27.8	21.2	30.9	7.0	29.4	39.1	18.2	19.4	. . .	84.1
March	20.6	19.3	28.3	21.8	31.4	7.1	29.3	40.4	18.7	20.0	. . .	85.8
April	20.8	19.6	28.5	22.2	31.6	7.3	29.2	40.7	18.6	20.3	. . .	86.7
May	21.2	20.0	28.9	22.7	31.9	7.4	29.2	40.8	18.9	20.7	. . .	87.9
June	21.2	20.0	28.6	22.3	31.8	7.5	28.9	41.6	19.1	20.8	. . .	87.6
July	21.4	20.1	28.7	22.6	31.7	7.5	28.9	41.6	19.0	21.0	. . .	87.7
August	21.3	20.1	28.8	22.6	31.8	7.6	28.7	41.6	18.9	21.0	. . .	87.3
September	21.5	20.2	29.0	22.7	32.1	7.6	28.8	41.9	19.3	21.3	. . .	87.5
October	21.8	20.4	29.5	22.8	32.9	8.0	28.6	41.9	19.5	21.5	. . .	88.4
November	21.9	20.5	29.6	22.6	33.1	8.0	28.6	42.3	19.8	21.4	. . .	88.3
December	22.0	20.8	29.7	22.5	33.4	8.1	28.7	42.6	19.6	21.5	. . .	89.0
1956												
January	22.1	20.7	29.8	22.3	33.6	8.2	28.4	43.3	19.8	21.6	. . .	88.2
February	21.9	20.6	29.7	21.9	33.7	8.3	28.1	43.1	19.8	21.2	. . .	87.4
March	21.9	20.5	29.7	21.9	33.7	8.4	27.5	43.0	20.0	21.1	. . .	87.0
April	22.1	20.8	29.8	22.2	33.7	8.7	27.7	42.8	20.2	21.3	. . .	87.8
May	21.9	20.6	29.6	21.7	33.8	8.6	27.7	42.2	20.0	20.9	. . .	86.3
June	21.7	20.4	29.5	21.2	33.9	8.7	27.7	41.8	20.0	20.6	. . .	85.3
July	21.0	19.6	29.6	21.2	34.0	8.7	27.7	39.6	20.1	19.0	. . .	81.5
August	21.9	20.5	29.8	21.2	34.2	8.8	28.1	41.6	20.1	20.7	. . .	84.9
September	22.4	20.8	29.7	20.9	34.3	8.8	28.4	42.7	20.1	21.8	. . .	86.0
October	22.6	21.0	30.0	21.2	34.5	8.9	29.1	42.4	20.3	22.2	. . .	86.5
November	22.4	20.9	29.8	20.9	34.4	9.0	29.5	42.1	20.3	21.7	. . .	85.8
December	22.7	21.3	30.1	21.7	34.3	9.1	30.2	42.8	20.4	22.1	. . .	86.8

. . . = Not available.

Table 2-4A. Industrial Production and Capacity Utilization, Historical Data, 1948–2016—*Continued*

(Seasonally adjusted.)

Year and month	Total industry	Manufacturing (SIC)	Consumer goods Total	Durable	Nondurable	Business equipment	Defense and space equipment	Construction supplies	Business supplies	Materials	Total industry	Manufacturing (SIC)
1957												
January	22.6	21.2	30.3	21.8	34.5	9.3	30.3	42.2	20.4	21.6	. . .	86.2
February	22.8	21.5	30.6	22.2	34.9	9.5	30.5	43.5	20.5	21.8	. . .	87.0
March	22.8	21.4	30.7	22.1	35.1	9.4	30.4	42.8	20.4	21.7	. . .	86.4
April	22.5	21.1	30.4	21.4	34.9	9.2	30.5	41.9	20.4	21.4	. . .	85.0
May	22.4	21.0	30.5	21.3	35.0	9.1	30.1	41.7	20.6	21.3	. . .	84.2
June	22.5	21.1	30.6	21.7	35.1	9.1	30.2	42.0	20.4	21.4	. . .	84.6
July	22.6	21.1	30.7	21.5	35.4	9.1	29.9	42.1	20.5	21.5	. . .	84.3
August	22.6	21.1	30.9	22.1	35.4	9.1	29.9	41.7	20.5	21.6	. . .	84.2
September	22.4	21.0	30.9	22.0	35.4	9.0	29.3	41.5	20.5	21.3	. . .	83.2
October	22.1	20.6	30.4	21.2	35.1	8.8	28.6	41.0	20.2	21.0	. . .	81.4
November	21.6	20.1	30.3	21.1	34.9	8.5	27.7	40.5	20.0	20.2	. . .	79.4
December	21.2	19.7	30.1	20.2	35.3	8.3	27.5	39.8	20.0	19.5	. . .	77.5
1958												
January	20.8	19.3	29.7	19.4	35.2	8.1	27.8	39.3	19.9	19.0	. . .	75.7
February	20.3	18.9	29.5	18.7	35.2	7.8	28.0	38.1	19.8	18.4	. . .	73.8
March	20.1	18.7	29.3	18.1	35.2	7.6	28.6	37.9	19.9	17.9	. . .	72.7
April	19.7	18.4	29.0	17.4	35.2	7.5	29.0	37.4	19.7	17.4	. . .	71.3
May	19.9	18.6	29.4	18.0	35.5	7.3	29.3	38.6	19.7	17.7	. . .	71.9
June	20.5	19.1	30.0	18.5	36.1	7.3	30.2	40.2	19.9	18.4	. . .	73.9
July	20.8	19.3	30.4	18.8	36.5	7.4	30.2	39.9	19.9	18.9	. . .	74.3
August	21.2	19.7	30.5	19.1	36.6	7.5	30.5	41.7	20.2	19.5	. . .	75.7
September	21.4	19.9	30.2	17.8	36.8	7.5	30.7	41.8	20.4	19.9	. . .	76.2
October	21.6	20.0	30.5	18.7	36.8	7.6	30.6	41.8	20.8	20.3	. . .	76.4
November	22.3	20.7	31.9	21.9	37.2	7.8	30.9	43.4	20.9	20.8	. . .	79.1
December	22.3	20.8	32.1	22.0	37.2	7.8	30.9	42.8	20.8	20.9	. . .	79.0
1959												
January	22.6	21.1	32.4	22.1	37.7	8.0	31.0	43.7	21.3	21.3	. . .	80.2
February	23.1	21.5	32.7	22.2	38.1	8.1	30.8	44.9	21.5	21.9	. . .	81.4
March	23.4	21.9	32.7	22.6	37.9	8.2	30.9	46.0	21.7	22.5	. . .	82.5
April	23.9	22.3	33.2	22.7	38.6	8.4	31.1	47.3	21.7	23.1	. . .	84.0
May	24.3	22.7	33.4	23.2	38.6	8.7	31.3	47.7	21.7	23.7	. . .	84.9
June	24.3	22.7	33.2	23.4	38.3	8.9	31.4	47.5	22.0	23.6	. . .	84.8
July	23.7	22.3	33.6	23.8	38.6	8.9	31.5	45.7	22.2	22.2	. . .	83.0
August	22.9	21.4	33.7	23.2	39.0	8.8	31.4	42.8	22.1	20.5	. . .	79.5
September	22.9	21.3	33.6	22.5	39.3	8.8	31.5	42.3	22.3	20.4	. . .	79.0
October	22.7	21.1	33.4	23.0	38.8	8.7	31.5	42.4	22.2	20.2	. . .	78.2
November	22.8	21.3	32.7	20.2	39.3	8.5	31.5	43.9	22.2	21.0	. . .	78.5
December	24.3	22.8	33.7	22.7	39.5	8.7	31.7	47.3	22.4	23.3	. . .	83.6
1960												
January	24.9	23.4	34.8	25.2	39.6	9.0	31.9	47.0	22.7	24.0	. . .	85.6
February	24.7	23.2	34.4	24.9	39.2	9.1	32.1	46.5	22.7	23.7	. . .	84.6
March	24.5	22.9	34.5	24.4	39.6	9.1	32.3	45.1	22.6	23.3	. . .	83.2
April	24.3	22.7	34.7	24.4	40.0	9.0	32.1	45.1	23.0	22.7	. . .	82.3
May	24.2	22.6	34.9	24.6	40.1	9.0	32.4	44.6	23.1	22.5	. . .	81.5
June	23.9	22.3	34.7	24.3	39.9	8.9	31.6	43.8	22.9	22.2	. . .	80.2
July	23.8	22.3	34.3	23.4	40.0	8.8	32.4	44.2	22.9	22.2	. . .	79.7
August	23.8	22.2	34.4	23.7	40.0	8.7	32.6	43.1	22.7	22.1	. . .	79.1
September	23.6	22.0	34.3	23.4	39.9	8.6	32.5	42.7	22.6	21.7	. . .	77.9
October	23.5	21.9	34.6	23.6	40.3	8.5	32.2	42.8	22.7	21.6	. . .	77.5
November	23.2	21.5	34.0	22.9	39.9	8.5	32.3	42.3	22.7	21.1	. . .	75.8
December	22.8	21.1	33.7	22.1	39.8	8.3	31.8	41.8	22.3	20.5	. . .	74.3
1961												
January	22.8	21.2	33.5	21.3	40.0	8.4	32.1	41.4	22.6	20.7	. . .	74.1
February	22.8	21.1	33.7	21.2	40.2	8.3	31.9	41.4	22.7	20.5	. . .	73.5
March	22.9	21.2	33.7	21.2	40.2	8.3	31.8	42.3	22.9	20.7	. . .	73.9
April	23.4	21.7	34.5	22.7	40.6	8.4	31.9	43.4	23.0	21.3	. . .	75.4
May	23.7	22.1	34.9	23.4	40.9	8.4	31.9	43.5	23.1	22.0	. . .	76.4
June	24.1	22.4	35.3	24.2	41.0	8.4	31.9	44.5	23.3	22.3	. . .	77.3
July	24.3	22.7	35.6	24.6	41.2	8.5	32.1	45.2	23.5	22.6	. . .	78.1
August	24.6	23.0	35.8	24.7	41.6	8.5	32.3	45.7	23.7	23.1	. . .	79.0
September	24.5	22.8	35.2	23.3	41.3	8.7	33.0	46.1	23.6	23.1	. . .	78.2
October	25.0	23.3	36.2	24.6	42.1	8.6	33.7	46.4	23.9	23.5	. . .	79.6
November	25.4	23.7	36.8	25.7	42.5	8.9	34.5	46.1	24.1	23.8	. . .	80.8
December	25.6	24.0	36.9	26.2	42.5	8.9	35.1	46.4	24.4	24.2	. . .	81.6
1962												
January	25.4	23.7	36.6	25.6	42.3	8.9	35.5	44.3	24.4	24.0	. . .	80.2
February	25.8	24.1	36.8	25.7	42.6	9.0	36.2	46.8	24.6	24.4	. . .	81.4
March	25.9	24.3	37.1	26.1	42.8	9.1	36.6	47.4	24.5	24.4	. . .	81.9
April	26.0	24.3	37.4	26.7	42.9	9.2	37.0	46.8	24.5	24.4	. . .	81.7
May	26.0	24.3	37.6	26.9	43.2	9.2	37.2	46.7	24.9	24.0	. . .	81.3
June	25.9	24.2	37.4	26.5	43.0	9.3	37.6	47.1	24.9	24.0	. . .	80.9
July	26.2	24.4	38.0	27.0	43.6	9.3	38.3	46.9	24.8	24.2	. . .	81.5
August	26.2	24.5	37.6	26.7	43.2	9.4	38.9	47.7	25.0	24.2	. . .	81.4
September	26.4	24.7	37.8	27.0	43.4	9.4	38.9	48.3	25.2	24.4	. . .	81.8
October	26.4	24.6	37.8	27.1	43.2	9.5	39.1	47.5	25.2	24.4	. . .	81.4
November	26.5	24.8	38.0	27.1	43.4	9.5	39.5	47.7	25.3	24.6	. . .	81.8
December	26.5	24.9	38.2	27.4	43.6	9.4	39.6	48.1	25.2	24.5	. . .	81.7

. . . = Not available.

Table 2-4A. Industrial Production and Capacity Utilization, Historical Data, 1948–2016—Continued

(Seasonally adjusted.)

Year and month	Production indexes, 2007 = 100										Capacity utilization (output as percent of capacity)	
	Total industry	Manufac-turing (SIC)	Market groups								Total industry	Manufac-turing (SIC)
			Consumer goods			Business equipment	Defense and space equipment	Construction supplies	Business supplies	Materials		
			Total	Durable	Nondurable							
1963												
January	26.7	25.0	38.6	27.6	44.2	9.4	41.5	46.9	25.4	24.6	. . .	81.9
February	27.0	25.2	39.1	27.9	44.7	9.5	41.2	47.0	25.6	25.0	. . .	82.4
March	27.2	25.4	39.2	28.0	45.0	9.5	40.9	47.5	25.4	25.3	. . .	82.6
April	27.4	25.7	39.4	28.2	45.1	9.5	40.9	49.2	26.2	25.6	. . .	83.5
May	27.7	25.9	39.5	28.7	45.0	9.5	40.9	50.2	26.5	26.2	. . .	84.0
June	27.8	26.0	39.7	29.2	45.0	9.6	40.9	50.0	26.4	26.3	. . .	83.9
July	27.7	25.9	39.6	29.1	44.8	9.7	40.4	49.8	26.5	26.0	. . .	83.3
August	27.8	26.0	39.9	29.2	45.3	9.9	40.6	50.0	26.7	25.8	. . .	83.5
September	28.1	26.2	40.0	29.7	45.2	9.9	40.8	49.7	27.0	26.3	. . .	83.8
October	28.2	26.5	40.3	29.8	45.5	10.1	40.8	50.4	27.2	26.5	. . .	84.3
November	28.4	26.5	40.3	30.0	45.5	10.1	40.6	51.0	27.5	26.6	. . .	84.3
December	28.3	26.5	40.6	30.1	45.9	10.1	40.7	50.2	27.3	26.5	. . .	84.0
1964												
January	28.6	26.8	41.0	30.2	46.4	10.4	40.3	50.5	27.6	26.7	. . .	84.5
February	28.8	26.9	40.9	30.4	46.2	10.3	40.0	51.9	27.7	27.1	. . .	84.7
March	28.8	26.9	40.7	30.1	46.0	10.5	39.9	52.0	27.9	27.1	. . .	84.4
April	29.2	27.4	41.7	30.9	47.1	10.7	39.8	52.2	28.3	27.4	. . .	85.6
May	29.4	27.5	42.0	31.1	47.6	10.8	39.1	52.5	28.5	27.6	. . .	85.6
June	29.5	27.6	42.0	31.4	47.3	10.9	38.9	52.1	28.6	27.8	. . .	85.4
July	29.7	27.8	42.6	32.0	47.9	11.0	38.8	53.1	28.6	27.9	. . .	85.9
August	29.9	28.0	42.5	32.2	47.6	11.0	39.0	52.6	28.4	28.4	. . .	86.1
September	30.0	28.1	42.0	31.4	47.3	11.1	39.2	52.2	28.5	28.9	. . .	86.2
October	29.6	27.7	41.2	28.2	47.9	11.0	39.5	52.4	28.5	28.3	. . .	84.6
November	30.5	28.5	42.7	32.0	48.1	11.4	39.9	53.6	28.7	29.2	. . .	86.8
December	30.8	29.1	43.6	33.9	48.3	11.6	40.2	52.9	29.0	29.6	. . .	88.0
1965												
January	31.2	29.4	44.2	34.5	49.0	11.6	40.7	53.3	29.3	29.9	. . .	88.6
February	31.4	29.6	44.3	34.9	48.9	11.8	41.1	54.7	29.4	30.0	. . .	88.7
March	31.8	29.9	44.7	35.9	49.0	11.9	41.8	55.1	29.7	30.5	. . .	89.3
April	31.9	30.1	44.6	35.8	48.8	12.0	42.4	54.3	29.7	30.8	. . .	89.3
May	32.2	30.4	44.9	36.1	49.1	12.2	43.3	54.9	30.0	30.9	. . .	89.4
June	32.4	30.5	45.0	36.4	49.2	12.4	43.8	54.9	30.2	31.3	. . .	89.5
July	32.7	31.0	45.0	36.7	49.0	12.6	44.5	56.5	30.2	31.6	. . .	90.3
August	32.9	31.0	45.0	36.3	49.2	12.6	44.9	55.9	30.4	31.9	. . .	89.9
September	33.0	31.1	45.6	37.0	49.8	12.8	44.9	55.4	30.5	31.6	. . .	89.6
October	33.3	31.4	45.9	37.3	49.9	13.0	45.5	56.3	30.8	31.9	. . .	89.8
November	33.4	31.5	46.1	37.6	50.2	13.3	45.9	57.1	31.0	31.8	. . .	89.6
December	33.8	32.0	46.4	38.3	50.2	13.5	46.4	58.3	31.5	32.2	. . .	90.5
1966												
January	34.2	32.3	46.6	38.5	50.5	13.8	47.4	58.2	31.4	32.6	. . .	90.9
February	34.4	32.5	46.8	38.4	50.7	13.8	48.1	57.6	31.8	33.0	. . .	90.9
March	34.9	32.9	47.1	38.7	51.1	14.0	48.5	58.8	32.1	33.6	. . .	91.6
April	34.9	33.1	47.3	39.4	51.0	14.2	49.5	58.8	31.8	33.5	. . .	91.5
May	35.3	33.4	47.3	38.8	51.3	14.3	50.4	59.1	32.3	33.9	. . .	91.6
June	35.4	33.5	47.5	38.8	51.6	14.5	51.2	58.2	32.7	34.1	. . .	91.5
July	35.6	33.7	47.4	38.1	51.9	14.7	51.8	58.8	33.1	34.2	. . .	91.4
August	35.6	33.7	47.2	37.5	52.0	14.8	52.5	57.1	32.9	34.5	. . .	91.1
September	36.0	34.0	47.4	37.7	52.1	14.9	53.2	57.1	33.1	34.8	. . .	91.2
October	36.2	34.3	48.4	39.6	52.5	14.9	54.1	57.1	33.1	34.9	. . .	91.6
November	36.0	34.0	48.1	38.4	52.8	14.7	55.0	57.2	33.2	34.4	. . .	90.1
December	36.1	34.1	48.0	38.0	52.9	14.9	55.6	57.1	33.3	34.4	. . .	90.0
1967												
January	36.2	34.2	48.5	37.1	54.1	14.8	56.6	58.6	34.0	34.3	89.4	89.8
February	35.8	33.9	47.8	36.1	53.7	14.8	57.2	58.1	33.7	33.6	88.0	88.4
March	35.6	33.7	48.0	36.4	53.7	14.8	57.8	58.2	33.8	33.0	87.1	87.5
April	35.9	33.9	48.8	36.7	54.8	14.7	58.4	58.0	34.1	33.3	87.5	87.7
May	35.6	33.7	47.8	36.2	53.6	14.8	58.8	58.9	33.5	33.0	86.4	86.6
June	35.6	33.6	48.0	35.8	54.1	14.7	58.7	59.2	33.7	32.9	86.0	86.1
July	35.6	33.5	47.9	36.2	53.7	14.4	59.0	59.5	33.8	32.9	85.4	85.3
August	36.2	34.1	48.4	36.7	54.2	14.7	59.1	60.1	34.7	33.9	86.6	86.5
September	36.2	34.1	48.5	36.5	54.4	14.6	59.3	60.6	34.8	33.6	86.1	86.1
October	36.5	34.4	49.1	37.2	55.1	14.5	59.7	60.3	35.1	34.1	86.4	86.4
November	37.0	35.0	50.2	39.3	55.5	14.9	59.8	60.7	35.1	34.4	87.3	87.5
December	37.4	35.4	51.0	40.9	55.9	15.0	59.8	60.7	35.1	34.9	87.8	88.0
1968												
January	37.3	35.3	50.3	39.7	55.5	15.1	59.6	61.1	35.2	35.0	87.4	87.4
February	37.5	35.4	50.6	40.3	55.6	15.1	60.3	61.7	35.5	35.0	87.3	87.4
March	37.6	35.5	50.9	40.2	56.1	15.2	59.1	61.8	35.5	35.2	87.3	87.2
April	37.7	35.5	50.8	40.2	56.0	15.2	57.9	62.2	35.8	35.4	87.1	86.9
May	38.1	36.0	51.1	40.8	56.1	15.4	58.7	62.4	36.2	36.0	87.7	87.6
June	38.2	36.1	51.4	41.1	56.4	15.4	59.0	62.4	36.3	36.1	87.7	87.4
July	38.2	35.9	51.2	40.8	56.4	15.2	59.0	62.5	36.3	36.2	87.2	86.7
August	38.3	36.1	51.9	41.3	57.0	15.3	59.2	62.8	36.7	35.9	87.1	86.8
September	38.4	36.1	52.1	41.8	57.1	15.5	59.1	62.3	36.8	36.0	87.1	86.5
October	38.5	36.4	52.4	42.3	57.3	15.6	57.6	62.2	37.0	36.1	86.9	86.6
November	39.0	36.9	53.1	43.3	57.7	15.7	58.2	63.7	37.4	36.7	87.7	87.5
December	39.1	36.9	52.8	43.7	57.0	15.8	57.9	64.9	37.7	36.9	87.6	87.1

. . . = Not available.

Table 2-4A. Industrial Production and Capacity Utilization, Historical Data, 1948–2016—*Continued*

(Seasonally adjusted.)

Year and month	Total industry	Manufac- turing (SIC)	Production indexes, 2007 = 100								Capacity utilization (output as percent of capacity)	
			Market groups								Total industry	Manufac- turing (SIC)
			Consumer goods			Business equipment	Defense and space equipment	Construction supplies	Business supplies	Materials		
			Total	Durable	Nondurable							
1969												
January	39.3	37.1	53.1	43.6	57.5	16.0	57.9	65.3	37.8	37.0	87.8	87.3
February	39.6	37.4	53.5	43.5	58.2	16.0	57.5	65.9	37.7	37.4	88.1	87.7
March	39.9	37.7	53.9	43.8	58.7	16.2	57.8	66.2	38.9	37.6	88.5	88.0
April	39.8	37.5	53.1	42.6	58.1	16.4	57.3	65.5	38.4	37.7	87.8	87.3
May	39.6	37.4	52.7	42.1	57.8	16.2	57.2	65.1	38.7	37.6	87.2	86.6
June	40.0	37.6	53.3	43.5	57.9	16.4	56.4	65.5	39.0	38.1	87.7	86.8
July	40.2	37.9	54.2	43.5	59.3	16.6	56.2	64.9	38.8	38.2	87.9	87.1
August	40.3	37.9	54.1	43.9	58.9	16.5	55.4	64.8	39.0	38.5	87.8	86.9
September	40.3	37.9	53.7	43.5	58.5	16.7	55.0	64.8	38.9	38.6	87.4	86.4
October	40.3	37.9	53.7	43.7	58.4	16.7	54.5	65.1	39.0	38.6	87.1	86.2
November	39.9	37.5	53.3	41.9	58.8	16.3	53.5	64.7	38.8	38.4	86.0	85.0
December	39.8	37.3	53.3	41.7	59.0	16.3	52.9	64.3	39.3	38.2	85.5	84.2
1970												
January	39.1	36.5	52.4	39.5	58.9	16.1	52.1	62.3	39.2	37.3	83.6	82.1
February	39.0	36.5	53.0	40.1	59.4	16.1	51.3	62.1	38.9	37.1	83.3	81.9
March	39.0	36.4	52.9	40.5	59.0	16.2	50.1	62.6	39.1	37.0	82.9	81.4
April	38.9	36.3	53.1	40.4	59.4	16.1	49.0	63.3	38.9	36.8	82.5	80.9
May	38.9	36.2	53.4	40.5	59.8	16.1	47.9	63.3	38.8	36.7	82.1	80.4
June	38.7	36.1	53.5	41.3	59.5	16.0	47.1	63.0	38.8	36.5	81.6	79.9
July	38.8	36.2	53.7	41.3	59.7	16.0	46.3	63.8	38.9	36.7	81.5	79.9
August	38.8	36.0	52.8	40.1	59.0	15.9	46.0	63.5	38.6	37.1	81.1	79.1
September	38.5	35.7	52.6	39.0	59.3	15.6	45.5	63.4	38.9	36.9	80.3	78.2
October	37.7	34.9	51.9	36.7	59.7	15.1	44.9	62.7	38.7	35.9	78.5	76.2
November	37.5	34.7	51.6	36.8	59.1	15.0	44.5	61.9	38.8	35.7	77.8	75.5
December	38.3	35.6	53.8	41.1	60.1	15.2	44.0	62.6	38.8	36.5	79.3	77.3
1971												
January	38.6	35.9	54.6	42.7	60.4	14.9	44.4	62.7	39.0	37.0	79.7	77.7
February	38.6	35.9	54.6	43.7	59.8	15.0	43.3	63.1	39.4	36.8	79.4	77.6
March	38.5	35.8	54.7	43.7	60.0	14.8	42.9	62.8	39.1	36.8	79.1	77.2
April	38.7	36.0	55.2	44.0	60.5	14.7	43.0	63.4	39.5	37.1	79.3	77.4
May	38.9	36.3	55.2	44.7	60.2	14.6	43.5	63.7	39.5	37.5	79.5	77.7
June	39.1	36.4	55.6	45.1	60.6	14.7	42.8	64.4	39.4	37.7	79.6	77.7
July	39.0	36.4	56.5	45.9	61.5	14.7	42.5	65.0	40.4	36.8	79.2	77.7
August	38.8	36.0	55.8	45.4	60.7	14.9	42.5	63.9	39.9	36.6	78.5	76.5
September	39.4	36.7	56.3	45.4	61.6	15.3	42.1	66.3	40.6	37.3	79.6	77.9
October	39.7	37.3	57.1	46.2	62.3	15.4	41.8	67.2	40.8	37.4	80.0	78.9
November	39.9	37.4	57.6	46.7	62.7	15.5	41.5	67.4	41.3	37.4	80.2	79.0
December	40.3	37.7	57.9	46.9	63.2	15.6	40.8	68.5	41.4	38.2	80.9	79.5
1972												
January	41.3	38.7	58.8	48.3	63.8	16.1	40.9	70.4	42.2	39.4	82.6	81.3
February	41.7	39.0	59.1	48.5	64.2	16.3	41.0	70.7	42.9	39.8	83.2	81.7
March	42.0	39.3	59.1	48.1	64.4	16.5	41.3	71.2	43.3	40.2	83.6	82.1
April	42.4	39.7	59.9	49.5	64.7	16.8	41.5	72.0	43.3	40.6	84.3	82.8
May	42.4	39.8	59.6	49.1	64.5	16.8	41.1	72.4	43.5	40.7	84.0	82.7
June	42.5	39.9	59.6	49.1	64.5	16.9	41.1	73.0	44.0	40.7	84.1	82.8
July	42.5	39.9	60.1	50.2	64.6	16.9	41.1	74.1	44.0	40.5	83.8	82.6
August	43.1	40.5	60.7	50.6	65.3	17.2	41.0	74.6	44.6	41.1	84.7	83.4
September	43.4	40.8	61.0	51.1	65.5	17.4	41.3	75.2	44.5	41.5	85.1	83.9
October	44.0	41.4	61.9	52.3	66.2	17.7	41.3	76.4	45.2	41.9	86.0	84.9
November	44.5	41.9	62.4	53.5	66.2	18.1	42.5	76.9	45.4	42.5	86.7	85.7
December	45.0	42.5	62.9	54.6	66.3	18.3	43.2	76.7	45.5	43.2	87.5	86.6
1973												
January	45.3	42.7	62.7	54.4	66.2	18.6	43.8	77.5	45.8	43.6	87.8	86.9
February	46.0	43.4	63.5	55.4	66.9	19.0	44.7	79.0	46.3	44.2	88.8	88.0
March	46.0	43.5	63.7	55.5	67.0	19.1	44.5	79.5	46.3	44.1	88.5	87.9
April	45.9	43.5	63.1	54.7	66.6	19.3	44.3	79.3	46.3	44.2	88.1	87.5
May	46.2	43.7	63.4	54.5	67.2	19.5	44.6	79.7	46.6	44.4	88.3	87.7
June	46.3	43.7	62.9	54.4	66.6	19.6	45.0	79.8	46.7	44.5	88.1	87.4
July	46.4	43.9	62.8	54.4	66.4	19.9	46.0	80.5	46.9	44.7	88.2	87.5
August	46.4	43.8	62.2	52.7	66.4	19.9	46.0	80.6	46.9	44.8	87.7	87.0
September	46.8	44.2	63.2	54.5	67.0	20.3	46.0	80.5	47.0	45.0	88.2	87.5
October	47.1	44.5	63.4	54.1	67.5	20.6	47.0	80.4	47.4	45.3	88.5	87.9
November	47.3	44.9	63.7	54.1	67.9	20.7	46.6	80.9	47.4	45.6	88.7	88.3
December	47.2	44.9	62.6	52.9	66.9	20.8	46.3	81.6	47.1	45.8	88.3	88.1
1974												
January	46.9	44.5	61.7	50.2	67.2	20.8	46.1	81.8	47.2	45.4	87.4	87.1
February	46.8	44.4	61.5	50.1	66.9	20.7	46.5	80.8	46.9	45.4	86.9	86.5
March	46.8	44.3	61.6	50.1	67.0	20.9	46.5	81.0	47.0	45.2	86.7	86.2
April	46.6	44.1	61.4	49.8	66.9	20.7	46.4	80.3	47.0	45.1	86.2	85.6
May	47.0	44.5	62.0	50.0	67.7	21.0	46.6	80.6	47.3	45.4	86.6	86.0
June	46.9	44.5	62.2	50.6	67.7	21.0	45.9	80.2	47.4	45.2	86.4	85.8
July	46.9	44.4	62.1	50.5	67.6	21.0	46.5	78.6	47.1	45.4	86.2	85.5
August	46.5	44.1	62.1	50.5	67.6	21.0	47.4	77.7	46.9	44.6	85.2	84.6
September	46.5	44.1	61.6	50.4	66.9	21.4	47.5	77.1	46.6	44.8	85.1	84.4
October	46.3	43.8	61.6	49.7	67.2	21.4	48.2	75.7	46.4	44.5	84.6	83.6
November	44.8	42.5	59.8	47.3	65.9	21.1	47.9	73.3	45.3	42.6	81.7	80.9
December	43.2	40.6	57.9	43.4	65.2	20.2	47.4	69.9	44.3	41.0	78.6	77.2

Table 2-4A. Industrial Production and Capacity Utilization, Historical Data, 1948–2016—*Continued*

(Seasonally adjusted.)

Year and month	Production indexes, 2007 = 100											Capacity utilization (output as percent of capacity)	
	Total industry	Manufac-turing (SIC)	Market groups									Total industry	Manufac-turing (SIC)
			Consumer goods			Business equipment	Defense and space equipment	Construction supplies	Business supplies	Materials		
			Total	Durable	Nondurable							
1975													
January	42.6	39.7	56.6	41.6	64.1	19.8	47.8	69.6	43.6	40.6	77.4	75.5
February	41.7	38.6	55.8	40.4	63.6	19.1	45.3	67.4	42.7	39.7	75.5	73.2
March	41.2	38.1	56.0	40.9	63.6	18.8	45.7	64.9	42.1	39.1	74.6	72.1
April	41.2	38.1	57.1	42.5	64.5	18.6	45.5	64.3	42.2	38.9	74.5	71.9
May	41.2	38.0	57.4	43.6	64.3	18.4	47.7	64.4	42.0	38.6	74.2	71.8
June	41.4	38.4	58.2	44.0	65.3	18.3	48.4	64.0	42.1	38.9	74.6	72.2
July	41.8	38.9	59.7	46.2	66.3	18.4	47.5	64.9	42.5	39.0	75.2	73.1
August	42.3	39.3	60.1	46.9	66.5	18.3	47.4	65.7	42.9	39.7	75.8	73.7
September	42.8	39.9	60.9	47.8	67.3	18.5	48.8	66.4	43.0	40.2	76.6	74.8
October	43.0	40.1	61.0	47.6	67.6	18.6	48.7	66.8	43.3	40.4	76.8	75.0
November	43.1	40.2	61.2	47.7	67.9	18.6	46.8	67.1	43.4	40.6	76.8	75.1
December	43.6	40.8	61.8	48.5	68.2	18.9	48.1	67.1	43.8	41.2	77.6	75.9
1976													
January	44.2	41.3	62.6	49.3	69.0	19.1	48.1	68.8	44.3	41.8	78.6	76.7
February	44.7	41.9	62.8	50.0	69.1	19.2	48.1	69.9	44.4	42.4	79.2	77.7
March	44.7	42.0	62.7	49.9	68.9	19.3	48.1	68.8	44.6	42.6	79.1	77.6
April	45.0	42.3	62.9	49.9	69.1	19.5	47.2	69.7	44.8	42.9	79.4	78.0
May	45.2	42.5	63.4	50.0	69.9	19.7	46.7	70.5	45.1	42.9	79.6	78.2
June	45.2	42.5	63.1	49.7	69.7	19.7	46.1	71.0	44.8	43.1	79.4	78.0
July	45.5	42.8	63.6	50.0	70.2	19.9	44.9	72.6	45.5	43.2	79.7	78.5
August	45.8	43.1	63.6	50.7	69.9	20.2	44.9	71.8	45.6	43.7	80.1	78.8
September	45.9	43.2	63.7	50.4	70.1	20.2	44.5	72.3	46.6	43.7	80.1	78.8
October	45.9	43.2	64.2	51.0	70.5	20.2	44.5	72.2	46.9	43.5	80.0	78.6
November	46.6	43.7	65.3	52.7	71.2	20.8	44.2	72.5	47.1	44.1	80.9	79.3
December	47.1	44.2	65.9	54.0	71.6	21.2	43.4	73.0	47.5	44.6	81.6	80.0
1977													
January	46.8	44.1	65.8	53.8	71.4	21.3	42.8	71.8	47.3	44.2	81.0	79.6
February	47.5	44.9	66.6	54.4	72.3	21.8	42.7	73.4	47.9	44.8	82.0	80.8
March	48.1	45.5	66.7	55.9	71.6	22.1	41.8	75.0	48.2	45.7	82.8	81.7
April	48.6	46.0	67.1	56.3	72.0	22.4	42.0	76.7	48.8	46.1	83.4	82.4
May	49.0	46.4	67.2	56.6	71.9	22.9	41.9	77.8	49.3	46.5	83.9	82.9
June	49.3	46.8	67.7	57.7	72.0	23.4	41.9	78.4	49.8	46.6	84.3	83.3
July	49.4	46.8	67.8	57.7	72.2	23.5	41.8	78.5	49.9	46.7	84.2	83.1
August	49.4	47.0	67.9	57.8	72.4	23.6	41.5	79.1	50.2	46.6	84.0	83.3
September	49.7	47.1	68.0	58.0	72.4	23.7	41.6	78.8	50.4	46.9	84.1	83.1
October	49.8	47.2	68.6	58.0	73.3	23.6	37.9	78.8	50.4	47.1	84.1	83.1
November	49.8	47.3	68.6	58.0	73.4	23.6	37.5	79.2	50.5	47.1	83.9	83.0
December	49.9	47.8	69.3	58.4	74.1	23.9	40.1	80.0	50.9	46.7	83.8	83.6
1978													
January	49.2	47.1	67.5	55.3	73.2	23.5	40.6	78.4	50.7	46.3	82.4	82.2
February	49.5	47.3	68.6	56.7	74.1	24.0	38.5	78.2	50.9	46.2	82.6	82.2
March	50.4	48.1	69.9	58.4	75.2	24.7	42.0	79.6	51.6	46.9	83.9	83.4
April	51.4	48.9	70.4	59.5	75.3	25.3	41.8	81.5	51.8	48.3	85.4	84.5
May	51.6	49.0	69.9	58.6	75.1	25.4	42.0	81.3	52.1	48.7	85.4	84.5
June	52.0	49.4	70.4	58.9	75.5	25.8	42.4	82.1	52.5	49.0	85.8	84.9
July	52.0	49.4	70.0	59.0	74.9	26.0	42.4	82.0	52.4	49.0	85.5	84.6
August	52.2	49.6	69.9	58.7	74.9	26.5	42.8	82.1	52.5	49.1	85.6	84.7
September	52.3	49.8	70.0	58.3	75.3	26.8	42.8	82.4	52.6	49.2	85.6	84.8
October	52.7	50.2	69.9	58.6	75.1	27.3	42.4	83.2	52.9	49.7	86.1	85.2
November	53.1	50.7	70.1	58.7	75.3	27.8	42.1	83.7	53.2	50.1	86.5	85.7
December	53.4	51.1	70.3	58.9	75.5	28.1	42.6	85.1	53.5	50.3	86.7	86.2
1979													
January	53.0	50.6	70.1	59.3	74.9	28.4	42.9	83.0	53.5	49.7	85.9	85.2
February	53.3	50.8	69.6	58.7	74.6	28.7	43.8	83.5	54.0	50.1	86.2	85.3
March	53.5	51.1	69.9	58.5	75.1	28.9	43.4	84.6	54.3	50.1	86.2	85.5
April	52.9	50.3	68.6	55.6	74.8	28.3	42.4	83.3	54.0	49.8	85.1	83.9
May	53.3	50.9	69.1	57.3	74.5	29.1	43.1	83.8	54.2	50.1	85.5	84.7
June	53.3	51.0	68.8	56.6	74.5	29.2	43.5	84.1	53.9	50.1	85.3	84.6
July	53.2	51.0	68.3	55.9	74.1	29.5	44.3	84.1	54.0	50.0	85.0	84.5
August	52.9	50.4	67.5	53.5	74.2	29.2	45.0	83.4	54.2	49.7	84.3	83.2
September	52.9	50.5	67.9	55.4	73.8	30.0	45.4	83.5	53.5	49.4	84.2	83.2
October	53.2	50.7	68.1	55.1	74.2	29.6	46.9	84.1	54.0	49.8	84.5	83.2
November	53.2	50.6	67.9	54.2	74.4	29.6	48.1	83.9	54.3	49.7	84.2	82.9
December	53.3	50.7	67.9	53.9	74.6	29.8	49.1	84.3	54.3	49.6	84.1	82.9
1980													
January	53.5	51.0	67.7	52.9	74.8	30.3	49.7	84.0	53.9	50.0	84.3	83.1
February	53.5	50.9	67.9	52.6	75.3	30.4	51.7	82.7	54.0	49.9	84.2	82.8
March	53.3	50.5	67.4	51.7	75.1	30.2	52.2	81.7	53.8	49.9	83.8	82.0
April	52.2	49.4	66.3	49.5	74.6	29.9	52.7	78.0	52.9	48.7	81.9	80.1
May	51.0	48.0	64.9	46.4	74.2	29.4	52.9	74.8	51.9	47.3	79.8	77.6
June	50.3	47.3	64.6	45.6	74.2	29.1	53.4	73.3	51.2	46.5	78.6	76.2
July	49.9	46.8	64.6	45.5	74.1	29.1	53.6	72.7	51.4	45.7	77.8	75.2
August	50.1	47.1	64.8	45.9	74.3	29.1	53.8	73.7	51.6	45.9	78.0	75.6
September	50.9	47.9	65.4	48.0	74.0	29.6	53.8	75.6	52.3	46.8	79.1	76.6
October	51.6	48.7	65.9	48.9	74.4	30.2	54.5	77.2	52.5	47.4	79.9	77.7
November	52.5	49.6	66.2	50.1	74.2	30.7	55.1	78.8	53.1	48.5	81.1	79.0
December	52.8	49.8	66.2	49.4	74.5	30.7	55.3	78.7	53.7	49.0	81.3	79.0

Table 2-4A. Industrial Production and Capacity Utilization, Historical Data, 1948–2016—*Continued*

(Seasonally adjusted.)

Year and month	Production indexes, 2007 = 100										Capacity utilization (output as percent of capacity)	
	Total industry	Manufac-turing (SIC)	Market groups								Total industry	Manufac-turing (SIC)
			Consumer goods			Business equipment	Defense and space equipment	Construction supplies	Business supplies	Materials		
			Total	Durable	Nondurable							
1981												
January	52.5	49.6	66.2	49.2	74.7	30.8	55.0	78.5	53.7	48.4	80.7	78.5
February	52.2	49.3	66.1	49.0	74.5	30.4	54.9	77.7	53.2	48.3	80.1	77.8
March	52.5	49.5	66.1	49.7	74.1	30.7	55.0	78.0	53.2	48.6	80.3	77.8
April	52.3	49.7	66.2	50.3	74.0	31.1	55.0	78.0	53.5	47.9	79.8	78.0
May	52.6	50.0	66.9	51.1	74.6	31.1	55.6	78.0	54.1	48.2	80.0	78.2
June	52.8	49.7	66.4	50.6	74.1	30.9	56.0	76.8	54.3	48.8	80.2	77.5
July	53.2	49.8	66.9	50.9	74.7	31.0	56.9	77.0	54.6	49.2	80.5	77.5
August	53.2	49.9	66.9	50.6	74.9	31.0	57.8	76.8	54.3	49.1	80.3	77.3
September	52.9	49.7	66.3	49.4	74.7	31.0	59.1	76.3	54.4	48.7	79.6	76.8
October	52.5	49.2	66.7	49.2	75.4	30.9	60.5	74.0	54.0	48.1	78.9	75.9
November	51.9	48.6	66.6	48.4	75.8	30.4	62.1	72.8	53.7	47.2	77.8	74.8
December	51.3	47.8	65.9	46.3	75.8	30.0	63.7	71.5	53.6	46.6	76.7	73.3
1982												
January	50.3	46.6	64.7	45.0	74.7	28.8	63.2	69.0	52.8	45.8	75.0	71.4
February	51.3	47.9	66.5	46.7	76.6	29.8	66.6	71.5	53.9	46.3	76.4	73.2
March	50.9	47.5	66.1	46.5	76.0	29.4	67.5	70.0	53.6	46.0	75.7	72.5
April	50.5	47.2	66.1	47.4	75.5	29.0	68.2	69.5	53.4	45.4	74.8	71.8
May	50.1	47.1	66.2	47.4	75.6	28.8	69.3	69.9	53.1	44.9	74.2	71.6
June	50.0	47.0	66.6	47.7	76.1	28.3	69.4	69.3	53.1	44.8	73.9	71.4
July	49.8	47.0	66.8	48.0	76.2	28.2	70.3	69.2	53.1	44.5	73.6	71.2
August	49.4	46.6	66.6	47.3	76.4	27.5	70.0	69.2	53.1	44.0	72.8	70.5
September	49.2	46.5	66.6	46.6	76.7	27.4	70.8	69.4	53.3	43.8	72.5	70.3
October	48.8	45.9	66.6	46.0	77.1	26.8	70.4	68.4	53.0	43.3	71.8	69.4
November	48.6	45.6	66.5	45.9	76.9	26.7	70.7	68.0	53.0	43.1	71.5	68.9
December	48.2	45.5	65.6	45.9	75.6	26.9	70.2	67.4	52.7	42.7	70.9	68.6
1983												
January	49.2	46.6	67.1	47.8	76.7	27.0	69.8	69.9	53.5	43.7	72.2	70.2
February	48.9	46.6	66.2	47.7	75.4	26.9	68.7	69.5	53.4	43.5	71.7	70.1
March	49.3	47.0	66.5	48.4	75.5	27.1	68.8	70.3	54.2	43.9	72.3	70.7
April	49.9	47.5	67.7	49.4	76.8	27.2	68.2	71.2	54.8	44.4	73.1	71.4
May	50.2	48.1	68.0	50.4	76.8	27.5	68.1	72.6	54.9	44.8	73.6	72.3
June	50.5	48.5	68.3	51.3	76.7	27.8	67.7	73.9	55.2	45.1	74.0	72.8
July	51.3	49.2	69.2	52.4	77.4	28.3	68.5	75.4	55.9	45.8	75.1	73.8
August	51.8	49.6	69.7	53.4	77.7	28.6	69.0	75.7	56.5	46.5	75.9	74.4
September	52.6	50.5	70.6	54.4	78.5	29.4	69.9	76.7	57.4	47.1	77.0	75.7
October	53.1	51.1	70.3	55.1	77.6	29.8	70.6	77.9	57.7	47.7	77.6	76.5
November	53.3	51.3	70.3	55.1	77.6	30.0	71.0	77.7	57.9	48.0	77.8	76.7
December	53.5	51.4	70.4	56.4	77.1	30.3	71.8	77.9	58.0	48.2	78.1	76.9
1984												
January	54.6	52.4	71.9	57.8	78.6	31.0	73.8	78.4	59.1	49.2	79.6	78.2
February	54.8	52.9	71.7	57.9	78.2	31.2	75.3	80.2	59.3	49.4	79.8	78.9
March	55.1	53.2	72.0	58.1	78.6	31.5	75.8	79.7	59.8	49.7	80.1	79.1
April	55.5	53.5	72.2	58.0	79.0	31.8	77.6	80.2	59.8	50.0	80.5	79.4
May	55.7	53.6	71.9	57.5	78.8	32.0	78.3	80.4	60.6	50.4	80.7	79.4
June	55.9	53.8	71.8	57.7	78.6	32.3	79.3	81.0	61.0	50.5	80.9	79.6
July	56.1	54.1	71.9	58.3	78.3	32.7	78.5	80.7	61.2	50.6	81.0	79.8
August	56.1	54.2	71.5	58.7	77.6	33.1	81.0	81.0	61.3	50.6	80.9	79.7
September	56.0	54.0	71.3	57.9	77.7	33.2	82.8	81.3	61.2	50.3	80.5	79.3
October	55.9	54.2	71.9	57.4	78.8	33.4	83.4	81.0	61.5	49.9	80.2	79.4
November	56.2	54.4	72.1	58.3	78.7	33.7	82.9	80.8	61.8	50.1	80.3	79.5
December	56.2	54.6	72.5	59.0	78.9	34.0	84.4	81.6	61.4	49.9	80.2	79.5
1985												
January	56.1	54.4	72.0	58.2	78.6	33.8	84.7	80.2	61.4	50.0	79.9	79.0
February	56.3	54.2	72.5	57.7	79.5	33.6	85.5	80.3	62.0	50.2	80.0	78.5
March	56.4	54.7	72.4	58.4	79.1	34.0	87.0	82.5	61.9	50.1	79.9	78.9
April	56.3	54.4	72.0	57.5	78.9	33.6	87.2	82.4	62.2	50.1	79.5	78.3
May	56.3	54.6	72.1	57.6	79.1	33.7	87.7	82.7	62.4	50.1	79.4	78.2
June	56.4	54.7	72.4	57.5	79.6	33.8	89.0	83.2	62.0	50.0	79.3	78.2
July	56.0	54.3	72.1	57.5	79.1	33.6	88.3	82.8	61.6	49.7	78.6	77.5
August	56.3	54.6	72.5	58.1	79.4	33.7	89.7	83.0	62.2	49.8	78.7	77.8
September	56.5	54.7	72.9	58.1	79.9	33.6	90.3	82.9	62.6	50.0	78.9	77.7
October	56.3	54.6	72.7	57.8	79.9	33.5	91.3	83.0	62.1	49.7	78.5	77.4
November	56.5	54.9	73.1	59.2	79.8	33.8	92.4	83.0	62.3	49.8	78.6	77.8
December	57.0	55.1	73.9	59.3	80.9	33.8	93.3	82.7	63.2	50.5	79.3	78.0
1986												
January	57.3	55.7	74.7	60.7	81.4	33.8	93.9	84.5	63.6	50.6	79.6	78.8
February	56.9	55.4	74.1	60.3	80.6	33.5	92.6	83.7	63.1	50.4	78.9	78.2
March	56.5	55.3	73.8	60.3	80.2	33.5	93.4	83.9	62.8	49.8	78.3	77.9
April	56.6	55.5	74.3	60.2	81.0	33.2	93.5	84.7	63.4	49.7	78.3	78.2
May	56.7	55.6	74.7	60.4	81.5	33.1	93.9	85.2	63.9	49.7	78.4	78.2
June	56.5	55.4	74.7	61.1	81.2	32.8	94.3	84.4	64.3	49.5	78.1	77.9
July	56.8	55.6	75.1	61.8	81.5	33.0	94.8	84.8	64.3	49.8	78.4	78.2
August	56.7	55.8	75.1	62.1	81.3	33.1	95.0	85.7	64.4	49.6	78.2	78.3
September	56.9	55.9	75.2	62.7	81.1	33.1	94.8	85.8	64.5	49.8	78.3	78.4
October	57.1	56.2	75.5	62.7	81.6	33.1	95.1	85.7	64.9	50.1	78.5	78.6
November	57.4	56.4	76.1	63.6	82.0	33.1	95.5	86.0	65.0	50.3	78.8	78.8
December	57.9	56.9	76.8	64.7	82.4	33.4	95.7	86.5	65.9	50.7	79.3	79.3

Table 2-4A. Industrial Production and Capacity Utilization, Historical Data, 1948–2016—Continued

(Seasonally adjusted.)

Year and month	Production indexes, 2007 = 100										Capacity utilization (output as percent of capacity)	
	Total industry	Manufacturing (SIC)	Market groups								Total industry	Manufacturing (SIC)
			Consumer goods			Business equipment	Defense and space equipment	Construction supplies	Business supplies	Materials		
			Total	Durable	Nondurable							
1987												
January	57.7	56.7	76.0	64.3	81.5	33.5	96.0	87.5	65.4	50.6	79.0	78.9
February	58.4	57.6	77.0	65.4	82.4	34.2	96.6	89.1	65.9	51.2	79.8	79.9
March	58.5	57.6	77.2	65.1	82.9	34.2	96.3	88.7	66.3	51.3	79.8	79.7
April	58.9	57.9	77.1	64.7	83.0	34.5	96.4	89.2	67.0	51.8	80.2	80.0
May	59.3	58.3	77.7	65.0	83.6	34.8	96.1	89.9	67.8	52.0	80.5	80.4
June	59.5	58.5	77.8	64.4	84.2	34.9	95.8	90.1	68.1	52.4	80.7	80.4
July	60.0	59.0	78.4	64.3	85.0	35.2	95.5	90.5	68.6	52.7	81.1	80.9
August	60.5	59.4	78.8	64.9	85.4	35.8	96.4	91.2	68.9	53.2	81.7	81.2
September	60.6	59.7	78.5	65.4	84.6	36.3	96.8	91.6	69.1	53.4	81.7	81.5
October	61.5	60.6	79.9	67.6	85.8	37.1	96.5	92.9	69.6	54.1	82.8	82.6
November	61.8	61.0	80.0	67.5	85.9	37.4	96.9	92.9	69.6	54.6	83.1	82.9
December	62.1	61.4	80.1	66.9	86.3	37.8	97.7	93.7	69.8	54.9	83.4	83.3
1988												
January	62.1	61.3	80.4	66.6	87.0	37.8	99.8	92.1	70.1	54.7	83.4	83.1
February	62.4	61.4	80.8	66.6	87.5	38.1	98.3	92.9	70.6	54.9	83.7	83.2
March	62.5	61.5	80.7	67.2	87.1	38.5	97.6	93.3	70.5	55.1	83.8	83.4
April	62.9	62.1	81.3	68.7	87.2	38.9	96.7	92.9	70.5	55.5	84.3	84.1
May	62.8	62.0	80.9	68.7	86.7	39.1	96.6	93.1	70.0	55.5	84.1	83.9
June	63.0	62.1	80.9	69.0	86.6	39.4	95.8	92.5	70.2	55.7	84.3	84.0
July	63.0	62.0	80.9	67.4	87.2	39.1	96.9	92.5	70.5	55.8	84.3	84.0
August	63.3	62.1	81.4	68.0	87.7	39.2	96.6	91.8	71.0	56.1	84.7	84.0
September	63.1	62.3	80.9	69.3	86.4	39.6	97.0	92.2	70.6	55.8	84.4	84.2
October	63.4	62.6	81.6	70.0	87.1	40.0	97.0	92.5	70.9	55.9	84.7	84.6
November	63.5	62.8	81.7	70.8	86.8	40.0	96.8	92.9	70.9	56.1	84.8	84.7
December	63.8	63.1	82.1	71.6	87.0	40.1	97.4	93.2	71.1	56.4	85.1	85.0
1989												
January	64.0	63.6	82.2	73.3	86.4	40.5	97.4	94.7	71.1	56.5	85.2	85.6
February	63.7	63.0	82.2	72.5	86.7	40.3	97.3	92.4	71.4	56.1	84.7	84.6
March	63.9	62.9	82.3	71.4	87.4	40.1	96.7	92.6	72.0	56.3	84.8	84.4
April	63.9	63.1	82.2	72.0	87.0	40.8	98.0	92.7	71.5	56.3	84.7	84.4
May	63.5	62.5	81.3	70.4	86.5	40.2	98.5	91.8	71.2	56.1	84.0	83.5
June	63.5	62.6	81.4	69.6	87.0	40.8	98.4	92.1	71.5	55.8	83.8	83.5
July	62.9	61.9	79.7	67.4	85.5	40.3	98.8	92.1	70.9	55.6	82.9	82.3
August	63.5	62.5	81.0	69.9	86.2	40.8	99.3	92.1	71.3	56.0	83.5	82.8
September	63.3	62.3	80.7	69.8	85.8	40.6	98.8	91.8	71.5	55.7	83.0	82.4
October	63.3	62.2	80.9	68.5	86.8	40.1	95.1	92.2	71.5	55.8	82.7	82.1
November	63.5	62.3	81.3	69.0	87.0	40.4	93.8	92.2	71.9	56.0	82.8	82.0
December	63.8	62.4	82.5	69.7	88.6	41.0	95.9	91.2	72.2	56.0	83.1	81.9
1990												
January	63.4	62.2	80.7	65.8	87.8	40.5	96.3	93.0	72.8	55.8	82.4	81.5
February	64.0	63.1	81.8	69.9	87.4	41.3	96.0	93.5	72.7	56.3	83.0	82.5
March	64.4	63.5	82.3	71.2	87.6	42.0	95.5	93.3	73.1	56.5	83.2	82.7
April	64.3	63.3	82.1	69.7	87.9	41.8	94.8	92.5	73.1	56.5	83.0	82.3
May	64.4	63.4	81.9	70.2	87.5	42.3	94.0	91.9	73.4	56.6	82.9	82.2
June	64.6	63.6	82.8	71.2	88.2	42.4	93.7	92.2	73.3	56.7	83.0	82.3
July	64.5	63.5	82.1	69.2	88.2	42.6	94.3	91.4	73.6	56.7	82.8	82.0
August	64.7	63.7	82.3	69.0	88.6	42.8	92.8	91.4	73.5	57.0	82.9	82.0
September	64.8	63.6	83.0	69.4	89.4	42.9	92.5	91.1	73.5	56.9	82.9	81.9
October	64.3	63.1	81.7	67.0	88.6	42.6	92.8	90.0	73.4	56.6	82.1	81.1
November	63.6	62.4	80.9	63.5	89.1	41.7	91.3	90.0	73.1	56.0	81.1	80.0
December	63.2	62.0	80.3	62.4	88.7	41.4	92.4	89.4	72.6	55.6	80.4	79.3
1991												
January	62.9	61.5	80.7	62.5	89.2	41.1	91.3	86.1	72.4	55.2	80.0	78.6
February	62.4	61.1	79.9	61.1	88.8	40.9	90.3	85.8	71.8	55.0	79.3	78.0
March	62.1	60.7	80.0	61.4	88.7	40.9	89.9	84.8	70.8	54.5	78.8	77.3
April	62.2	60.9	80.0	62.8	88.1	40.8	87.4	85.4	71.3	54.8	78.9	77.5
May	62.9	61.3	81.4	63.9	89.6	41.1	85.8	85.5	71.9	55.3	79.6	78.0
June	63.4	61.9	82.5	65.6	90.4	41.6	86.2	87.1	72.3	55.6	80.2	78.7
July	63.5	62.1	82.1	66.6	89.4	41.7	85.5	86.7	71.9	56.0	80.2	78.8
August	63.6	62.3	82.2	65.7	89.9	41.5	86.1	87.7	72.3	56.0	80.2	78.9
September	64.1	63.0	83.4	68.5	90.4	42.2	85.7	88.0	72.7	56.3	80.8	79.6
October	64.0	62.8	83.2	68.1	90.2	41.8	85.9	87.1	72.5	56.4	80.5	79.3
November	63.9	62.7	83.3	68.3	90.3	41.8	85.4	87.9	72.6	56.2	80.3	79.1
December	63.7	62.6	82.3	67.7	89.1	41.8	84.7	87.8	72.5	56.2	79.8	78.8
1992												
January	63.3	62.2	81.4	64.8	89.2	41.1	83.6	88.0	72.4	56.1	79.2	78.1
February	63.8	62.8	82.3	67.6	89.2	41.9	83.1	88.7	72.5	56.3	79.6	78.6
March	64.3	63.4	83.1	69.0	89.7	42.4	82.6	89.3	73.0	56.8	80.2	79.3
April	64.8	63.8	83.8	70.3	90.2	42.8	81.2	90.0	73.5	57.2	80.6	79.5
May	65.0	64.1	84.4	72.4	90.1	43.2	80.9	90.8	73.7	57.3	80.7	79.7
June	65.0	64.3	83.9	71.4	89.9	43.2	80.7	90.4	73.7	57.5	80.5	79.8
July	65.6	64.9	85.0	73.3	90.6	43.7	79.9	91.1	74.2	57.9	81.0	80.3
August	65.3	64.6	85.1	72.7	91.0	43.5	79.7	91.3	74.1	57.3	80.5	79.7
September	65.4	64.7	84.6	72.4	90.4	43.7	79.6	91.2	74.3	57.8	80.5	79.6
October	65.9	65.1	85.9	74.0	91.5	43.8	79.4	91.3	74.5	58.1	80.9	79.9
November	66.2	65.4	86.1	74.7	91.5	44.2	79.4	91.0	74.8	58.4	81.1	80.1
December	66.3	65.3	86.0	75.5	91.1	44.6	79.3	91.3	75.1	58.3	81.1	79.8

Table 2-4A. Industrial Production and Capacity Utilization, Historical Data, 1948–2016—_Continued_

(Seasonally adjusted.)

Year and month	Production indexes, 2007 = 100										Capacity utilization (output as percent of capacity)	
			Market groups									
	Total industry	Manufacturing (SIC)	Consumer goods			Business equipment	Defense and space equipment	Construction supplies	Business supplies	Materials	Total industry	Manufacturing (SIC)
			Total	Durable	Nondurable							
1993												
January	66.6	65.9	86.5	76.6	91.3	44.8	78.7	91.6	75.2	58.6	81.3	80.4
February	66.9	66.1	86.6	76.3	91.6	44.8	78.2	93.2	75.6	59.0	81.5	80.5
March	66.8	65.9	86.6	76.7	91.4	44.8	77.3	92.4	76.1	58.8	81.3	80.2
April	67.0	66.3	86.7	77.1	91.4	45.2	77.4	92.7	76.1	59.1	81.4	80.5
May	66.8	66.2	86.1	77.2	90.5	45.2	76.5	93.8	75.8	58.9	81.1	80.3
June	66.9	66.1	86.2	76.4	91.0	44.9	76.0	93.7	75.9	59.2	81.1	80.0
July	67.1	66.3	86.9	76.3	92.1	44.9	76.9	94.3	76.0	59.2	81.2	80.1
August	67.0	66.2	86.8	75.5	92.2	44.5	75.4	94.5	76.1	59.2	81.0	79.8
September	67.3	66.6	87.1	77.0	92.0	45.3	75.7	95.0	76.4	59.4	81.3	80.2
October	67.9	67.2	87.6	79.0	91.8	46.3	75.1	95.9	76.5	59.9	81.8	80.8
November	68.1	67.4	87.6	79.7	91.6	46.5	75.0	96.7	76.6	60.3	82.0	80.9
December	68.5	67.8	87.8	80.0	91.7	46.8	74.4	97.8	77.1	60.7	82.3	81.2
1994												
January	68.8	67.9	88.5	81.3	92.1	47.1	73.6	97.5	77.5	60.8	82.4	81.2
February	68.8	68.0	88.6	81.2	92.3	46.6	72.4	96.8	77.7	61.0	82.2	81.1
March	69.5	68.9	89.4	81.9	93.1	47.1	73.2	98.6	78.2	61.7	82.9	81.9
April	69.9	69.5	89.5	82.9	92.9	47.5	73.4	100.1	78.5	62.1	83.1	82.4
May	70.2	69.9	90.1	83.2	93.7	47.6	72.3	100.9	78.6	62.5	83.3	82.6
June	70.7	70.1	90.8	83.9	94.3	47.9	71.4	101.0	79.3	63.0	83.6	82.6
July	70.8	70.4	90.3	84.0	93.6	48.5	71.0	102.0	79.0	63.2	83.4	82.7
August	71.2	71.0	91.4	85.5	94.5	48.6	70.0	101.9	79.2	63.6	83.7	83.0
September	71.5	71.3	90.9	85.8	93.7	49.1	70.7	102.9	79.6	64.0	83.7	83.1
October	72.1	72.0	91.8	86.9	94.6	50.0	70.5	103.3	80.2	64.4	84.2	83.7
November	72.5	72.6	91.9	86.7	94.7	50.6	71.4	103.5	80.4	65.0	84.4	84.0
December	73.3	73.4	92.6	87.6	95.3	51.1	71.6	104.6	81.0	65.9	85.0	84.6
1995												
January	73.4	73.5	92.5	88.0	95.0	51.4	71.5	104.6	81.2	66.0	84.9	84.4
February	73.3	73.3	02.7	87.5	05.5	51.4	70.3	103.3	81.3	65.9	84.5	83.9
March	73.4	73.5	92.6	87.3	95.4	51.9	70.4	103.3	81.5	65.9	84.3	83.8
April	73.4	73.3	92.3	87.1	95.1	51.9	69.9	102.7	81.4	66.1	84.0	83.3
May	73.6	73.5	92.4	86.3	95.6	52.4	69.9	102.0	81.7	66.2	83.9	83.1
June	73.9	73.8	92.9	86.9	96.0	53.1	70.3	102.2	82.1	66.2	83.9	83.2
July	73.6	73.4	92.4	85.0	96.1	52.9	69.6	102.0	82.1	66.0	83.3	82.3
August	74.5	74.2	93.8	87.8	96.9	53.8	69.3	102.9	83.0	66.7	84.0	82.8
September	74.8	74.8	93.9	88.7	96.7	54.3	68.9	104.5	83.0	67.0	84.0	83.2
October	74.7	74.8	93.1	87.7	96.1	54.3	67.9	104.4	83.2	67.2	83.6	82.7
November	74.9	74.8	93.5	87.8	96.5	54.5	66.6	104.5	83.6	67.3	83.4	82.4
December	75.2	75.1	93.8	88.4	96.7	54.9	66.5	105.0	83.5	67.6	83.4	82.3
1996												
January	74.7	74.5	92.6	86.0	96.0	54.2	65.7	103.4	83.2	67.5	82.5	81.2
February	75.8	75.7	94.2	88.4	97.2	55.6	67.5	104.6	84.1	68.4	83.4	82.1
March	75.8	75.6	93.6	85.6	97.5	55.5	67.8	105.8	84.3	68.5	82.9	81.5
April	76.5	76.4	94.4	89.6	97.1	56.8	67.7	106.4	84.2	69.0	83.3	82.0
May	77.0	77.0	94.6	90.1	97.1	57.6	67.8	107.5	85.0	69.6	83.5	82.2
June	77.7	77.8	95.3	91.9	97.4	58.5	67.4	109.3	85.3	70.2	83.8	82.6
July	77.6	78.0	94.5	92.1	96.2	59.2	68.0	108.6	85.3	70.1	83.3	82.3
August	78.0	78.5	94.4	91.2	96.4	59.5	68.0	109.7	86.2	70.7	83.4	82.3
September	78.6	79.1	95.4	92.2	97.5	60.2	68.1	110.0	86.6	71.0	83.6	82.5
October	78.5	79.0	94.8	90.1	97.5	60.0	68.0	110.2	86.8	71.2	83.1	82.0
November	79.2	79.7	95.9	91.4	98.5	60.9	67.6	111.0	87.6	71.7	83.5	82.2
December	79.7	80.4	96.2	92.8	98.3	62.0	67.6	110.4	88.1	72.2	83.6	82.5
1997												
January	79.8	80.5	95.9	92.2	98.1	62.3	66.7	109.4	88.7	72.4	83.3	82.1
February	80.8	81.6	96.4	93.7	98.2	63.4	67.1	111.6	89.6	73.4	83.9	82.8
March	81.3	82.4	97.2	94.9	98.8	64.4	66.7	112.6	89.8	73.8	84.0	83.2
April	81.4	82.3	96.2	92.5	98.4	64.8	66.7	112.3	90.1	74.1	83.6	82.5
May	81.8	82.9	96.9	93.5	99.0	65.6	66.5	113.0	90.7	74.4	83.7	82.7
June	82.2	83.5	96.9	95.3	98.3	66.7	66.3	112.8	91.0	74.8	83.6	82.7
July	82.9	84.0	97.7	95.3	99.4	67.1	66.4	112.9	91.6	75.5	83.8	82.7
August	83.7	85.1	98.6	97.4	99.8	68.6	66.7	113.7	91.9	76.3	84.2	83.2
September	84.5	85.8	99.3	98.2	100.5	69.1	66.5	114.2	93.0	77.1	84.4	83.4
October	85.2	86.6	100.9	99.4	102.2	70.1	66.8	114.9	94.0	77.3	84.6	83.5
November	85.9	87.5	101.2	101.9	101.7	71.5	66.4	115.4	94.4	78.2	84.8	83.8
December	86.2	87.9	100.7	101.6	101.1	71.8	67.4	116.7	94.6	78.7	84.5	83.5
1998												
January	86.6	88.6	101.2	102.1	101.6	72.8	68.0	117.5	94.6	78.9	84.4	83.5
February	86.8	88.7	101.1	101.9	101.5	73.1	68.4	118.1	94.9	79.0	83.9	83.0
March	86.8	88.6	101.2	102.2	101.6	73.5	68.1	117.3	95.4	78.9	83.5	82.3
April	87.1	89.0	101.9	102.6	102.3	73.7	68.2	117.8	95.7	79.1	83.2	82.1
May	87.7	89.5	102.3	103.3	102.6	74.1	69.0	119.2	96.4	79.7	83.2	81.9
June	87.1	88.8	101.1	98.4	102.8	74.1	69.3	118.7	96.5	79.1	82.2	80.7
July	86.8	88.5	100.2	94.4	103.1	73.7	70.4	119.4	97.1	78.8	81.4	79.9
August	88.6	90.6	103.1	105.6	102.9	76.0	70.7	119.8	97.8	80.3	82.7	81.4
September	88.5	90.4	102.2	105.6	101.6	76.0	70.2	119.4	97.8	80.4	82.1	80.7
October	89.2	91.3	102.7	107.6	101.6	76.9	71.6	120.7	98.1	81.1	82.3	81.0
November	89.1	91.4	102.1	107.3	100.8	76.9	71.4	120.9	98.4	81.3	81.9	80.7
December	89.4	91.9	102.1	108.2	100.4	77.1	70.8	122.3	98.4	81.9	81.8	80.8

Table 2-4A. Industrial Production and Capacity Utilization, Historical Data, 1948–2016—*Continued*

(Seasonally adjusted.)

Year and month	Production indexes, 2007 = 100										Capacity utilization (output as percent of capacity)	
	Total industry	Manufac-turing (SIC)	Market groups									
			Consumer goods			Business equipment	Defense and space equipment	Construction supplies	Business supplies	Materials	Total industry	Manufac-turing (SIC)
			Total	Durable	Nondurable							
1999												
January	89.9	92.2	103.1	108.2	101.9	77.1	70.5	121.6	99.2	82.1	81.8	80.6
February	90.3	93.0	103.5	109.1	102.1	77.6	71.1	121.7	99.4	82.7	81.9	80.9
March	90.5	92.9	103.3	108.8	101.8	77.5	70.7	120.4	99.8	83.3	81.7	80.4
April	90.7	93.3	103.2	110.1	101.3	78.0	69.9	120.8	100.0	83.6	81.6	80.4
May	91.4	94.1	104.2	111.1	102.3	78.9	69.1	121.2	100.6	84.1	81.8	80.7
June	91.2	93.8	103.1	110.5	101.0	78.5	68.1	121.3	100.5	84.5	81.4	80.1
July	91.8	94.2	102.7	110.8	100.3	79.1	67.8	122.2	101.2	85.6	81.6	80.1
August	92.2	94.8	104.1	113.4	101.3	79.7	67.5	122.0	101.2	85.6	81.7	80.3
September	91.8	94.4	103.3	111.7	100.9	79.3	65.5	122.1	101.2	85.4	81.0	79.6
October	93.0	95.8	105.4	115.3	102.3	80.0	65.1	123.6	102.1	86.4	81.8	80.5
November	93.4	96.4	105.2	114.5	102.5	80.0	63.7	124.3	102.6	87.3	81.9	80.7
December	94.2	97.1	106.4	114.5	104.0	80.6	62.8	125.6	103.3	87.9	82.2	80.9
2000												
January	94.2	97.2	105.0	116.4	101.5	82.0	62.8	126.0	103.2	88.0	82.1	80.7
February	94.5	97.4	105.7	116.1	102.5	82.5	61.3	126.2	103.2	88.2	82.0	80.5
March	94.8	98.0	105.3	115.6	102.2	83.6	60.7	126.3	103.9	88.7	82.0	80.7
April	95.5	98.7	106.3	117.2	102.9	84.7	59.5	126.9	105.1	89.1	82.3	80.9
May	95.6	98.6	106.3	116.6	103.2	85.3	58.8	124.9	105.1	89.4	82.2	80.5
June	95.7	98.8	106.4	116.2	103.5	85.4	59.4	124.5	104.9	89.5	82.0	80.3
July	95.6	98.9	106.0	114.4	103.5	86.0	60.5	125.0	105.0	89.2	81.6	80.1
August	95.3	98.2	105.5	114.2	102.9	85.6	59.0	124.2	104.7	89.2	81.1	79.3
September	95.7	98.7	106.4	114.5	104.1	86.4	56.9	124.3	104.6	89.4	81.2	79.3
October	95.4	98.4	105.4	113.2	103.2	86.5	59.1	123.8	104.3	89.3	80.7	78.8
November	95.4	98.1	105.8	110.9	104.5	86.3	61.0	123.5	104.5	89.1	80.4	78.3
December	95.2	97.5	106.4	109.2	105.8	85.7	61.6	121.9	104.1	88.6	80.0	77.6
2001												
January	94.5	97.0	105.4	107.2	105.2	85.6	63.3	122.2	104.0	87.7	79.2	76.8
February	93.9	96.3	104.7	106.9	104.4	85.2	63.1	120.9	102.5	87.3	78.4	76.1
March	93.7	96.1	104.7	109.4	103.5	84.7	64.9	121.1	101.9	86.9	78.0	75.7
April	93.4	95.8	105.1	109.6	104.0	82.8	65.5	120.8	101.2	86.7	77.6	75.2
May	92.9	95.2	105.1	110.5	103.6	81.4	65.8	120.1	100.7	86.1	76.9	74.5
June	92.3	94.6	105.0	109.8	103.8	80.5	66.9	119.2	100.2	85.3	76.2	73.8
July	91.8	94.2	104.5	110.1	103.0	79.7	67.8	119.2	100.1	84.6	75.6	73.3
August	91.7	93.7	104.7	109.4	103.5	78.2	66.8	117.9	99.7	84.9	75.3	72.8
September	91.3	93.5	104.3	108.5	103.2	77.2	67.3	117.8	99.8	84.7	74.8	72.5
October	90.9	93.0	104.7	107.4	104.1	76.0	67.2	116.4	99.2	84.2	74.3	71.9
November	90.5	92.7	104.7	109.3	103.5	75.4	66.5	116.1	98.5	83.6	73.8	71.6
December	90.5	93.0	105.3	111.9	103.5	74.8	66.5	117.1	98.7	83.4	73.7	71.7
2002												
January	91.1	93.5	106.4	112.2	104.8	74.5	65.9	116.7	98.4	84.3	74.0	72.1
February	91.1	93.5	105.6	112.5	103.6	74.4	65.5	117.1	98.5	84.8	73.9	71.9
March	91.8	94.2	106.5	113.6	104.4	74.8	65.4	118.9	99.5	85.4	74.4	72.4
April	92.2	94.4	106.3	115.3	103.6	74.4	65.3	118.8	100.1	86.4	74.6	72.5
May	92.6	94.9	106.6	115.8	103.9	74.8	65.4	119.4	100.6	86.8	74.9	72.9
June	93.4	96.0	107.9	117.2	105.1	75.4	66.1	120.3	101.2	87.6	75.5	73.6
July	93.2	95.6	107.7	118.5	104.4	75.1	66.3	118.6	101.2	87.4	75.3	73.4
August	93.2	95.8	107.2	117.6	104.0	75.5	66.5	119.2	101.2	87.6	75.3	73.5
September	93.4	96.0	107.4	117.6	104.3	75.5	67.7	120.2	101.8	87.5	75.4	73.6
October	93.1	95.6	107.2	116.9	104.2	75.2	68.2	119.8	102.3	86.9	75.2	73.3
November	93.6	96.0	108.0	119.7	104.3	75.5	67.8	119.7	102.0	87.6	75.6	73.6
December	93.1	95.6	107.1	117.9	103.7	74.6	70.5	118.8	101.9	87.3	75.2	73.3
2003												
January	93.8	96.2	108.4	119.7	104.9	74.8	70.5	119.3	102.9	87.8	75.8	73.8
February	94.0	96.2	108.8	117.9	106.1	74.8	71.4	118.3	103.1	87.8	75.9	73.7
March	93.7	96.3	108.8	118.3	106.0	75.0	71.4	117.9	103.0	87.3	75.8	73.8
April	93.1	95.5	107.9	117.1	105.1	74.2	70.8	116.6	101.9	87.0	75.3	73.2
May	93.1	95.6	107.6	117.2	104.7	74.2	71.5	118.2	102.3	86.9	75.3	73.3
June	93.2	96.0	107.9	118.3	104.6	74.5	71.7	118.6	101.8	87.1	75.5	73.7
July	93.7	96.3	108.8	121.3	104.9	74.8	71.0	118.0	102.4	87.3	75.8	73.9
August	93.5	95.9	108.0	119.0	104.6	75.0	71.4	118.8	102.2	87.4	75.8	73.6
September	94.1	96.6	108.9	122.0	104.7	75.8	71.5	118.6	102.2	87.9	76.2	74.2
October	94.2	96.8	108.2	121.2	104.1	75.7	71.4	119.2	102.7	88.5	76.3	74.3
November	94.9	97.7	109.0	122.7	104.7	77.1	70.8	120.7	103.3	89.0	76.9	75.1
December	94.9	97.5	109.0	122.6	104.7	76.8	69.1	120.5	103.0	89.1	76.9	75.0
2004												
January	95.1	97.5	109.5	124.0	104.9	77.1	67.2	120.5	103.5	89.2	77.1	75.0
February	95.7	98.3	110.0	123.7	105.7	78.2	68.5	120.4	104.4	89.7	77.6	75.6
March	95.2	98.1	108.8	122.6	104.5	78.0	68.6	120.4	103.5	89.5	77.2	75.5
April	95.6	98.5	109.6	122.7	105.5	78.1	68.6	120.2	104.2	89.8	77.6	75.8
May	96.4	99.3	110.0	121.7	106.4	79.0	69.2	121.6	105.0	90.7	78.2	76.4
June	95.6	98.5	108.4	118.9	105.2	78.8	68.3	120.7	104.9	90.0	77.6	75.9
July	96.3	99.4	108.8	119.5	105.4	80.5	69.5	122.2	105.5	90.7	78.2	76.6
August	96.4	99.9	109.3	120.4	105.8	80.2	70.1	122.3	105.5	90.6	78.2	76.9
September	96.5	99.9	109.2	119.7	106.0	80.4	71.0	121.3	105.3	90.8	78.3	76.9
October	97.4	100.9	110.3	122.3	106.6	81.0	71.2	123.3	105.9	91.7	79.0	77.6
November	97.6	100.8	110.1	120.6	106.9	80.6	71.9	122.9	106.3	92.3	79.1	77.5
December	98.3	101.5	110.9	121.7	107.5	81.4	72.7	122.9	107.1	93.0	79.7	77.9

Table 2-4A. Industrial Production and Capacity Utilization, Historical Data, 1948–2016—*Continued*

(Seasonally adjusted.)

Year and month	Production indexes, 2007 = 100										Capacity utilization (output as percent of capacity)	
	Total industry	Manufac-turing (SIC)	Market groups								Total industry	Manufac-turing (SIC)
			Consumer goods			Business equipment	Defense and space equipment	Construction supplies	Business supplies	Materials		
			Total	Durable	Nondurable							
2005												
January	98.8	102.3	111.3	120.8	108.4	82.9	72.6	124.4	108.1	93.0	80.0	78.4
February	99.5	103.1	111.9	123.6	108.3	83.7	75.2	125.2	108.1	93.7	80.5	79.0
March	99.3	102.6	111.4	121.2	108.3	83.7	76.6	123.9	108.2	93.8	80.3	78.4
April	99.5	103.0	111.3	120.1	108.6	84.6	77.7	126.1	108.8	93.6	80.3	78.6
May	99.6	103.4	112.0	120.2	109.3	85.4	77.4	126.3	108.8	93.3	80.3	78.7
June	100.0	103.5	112.9	120.8	110.4	85.2	77.9	125.3	109.2	93.6	80.5	78.6
July	99.7	103.1	112.5	118.6	110.5	84.7	77.4	126.2	108.7	93.3	80.2	78.2
August	99.9	103.6	113.0	121.9	110.1	85.8	78.2	126.7	109.0	93.3	80.3	78.4
September	98.1	102.6	113.2	124.5	109.7	83.3	75.3	128.5	109.2	89.7	78.7	77.5
October	99.3	104.1	113.2	125.8	109.3	87.8	75.9	131.1	109.6	90.8	79.6	78.5
November	100.3	104.9	113.0	124.0	109.6	88.9	76.4	131.8	110.0	92.6	80.3	78.9
December	100.9	105.1	113.8	122.8	111.0	88.2	76.7	132.9	110.5	93.5	80.7	78.9
2006												
January	101.1	105.9	113.0	124.3	109.5	89.9	75.1	133.7	110.6	93.8	80.7	79.4
February	101.1	105.6	112.4	122.1	109.3	90.2	75.8	132.6	110.6	94.1	80.6	79.1
March	101.3	105.6	112.9	123.1	109.8	91.0	74.0	132.4	111.0	94.1	80.6	78.9
April	101.7	106.1	113.3	123.0	110.3	92.6	74.5	131.8	111.0	94.3	80.8	79.2
May	101.6	105.6	113.0	122.1	110.2	92.5	73.9	130.6	110.8	94.5	80.6	78.7
June	102.0	106.0	113.6	122.9	110.7	93.3	74.6	130.1	110.9	94.8	80.8	78.9
July	101.9	105.7	112.6	118.7	110.7	94.0	75.9	130.4	110.9	95.0	80.6	78.5
August	102.3	106.4	113.6	121.9	111.0	94.8	75.7	129.6	110.7	95.2	80.8	78.9
September	102.1	106.4	113.1	119.8	110.9	94.8	76.4	129.2	110.6	95.1	80.5	78.8
October	102.1	106.0	113.0	118.2	111.3	95.1	77.6	127.8	110.7	95.0	80.2	78.3
November	102.0	106.1	113.1	118.8	111.3	95.5	78.9	127.5	110.3	94.7	80.0	78.2
December	103.0	107.6	113.7	121.5	111.2	97.0	80.4	131.2	111.1	95.7	80.6	79.1
2007												
January	102.5	107.0	112.9	118.4	111.1	95.1	82.1	128.6	111.0	95.6	80.0	78.5
February	103.5	107.5	114.2	119.9	112.4	96.0	82.8	129.1	111.0	96.6	80.6	78.6
March	103.8	108.4	113.0	120.5	110.0	97.8	82.1	130.7	112.3	96.9	80.6	79.1
April	104.5	109.1	113.9	122.6	111.2	98.9	84.6	130.1	113.3	97.6	81.0	79.4
May	104.5	109.0	113.5	121.9	110.8	99.0	86.7	129.9	112.8	98.0	80.9	79.1
June	104.6	109.4	113.6	123.3	110.6	99.0	90.2	130.5	112.1	97.9	80.8	79.2
July	104.5	109.5	113.4	122.1	110.7	99.0	92.0	130.0	111.4	98.0	80.6	79.1
August	104.8	109.2	113.2	121.6	110.5	99.1	94.1	129.4	111.6	98.5	80.7	78.7
September	105.2	109.7	113.3	120.1	111.0	100.7	96.8	129.6	112.3	98.7	81.0	78.9
October	104.7	109.3	112.1	119.7	109.7	100.1	97.2	128.4	112.0	98.8	80.6	78.6
November	105.3	110.0	112.1	120.2	109.5	101.0	99.5	128.3	112.5	99.6	81.1	78.9
December	105.3	110.1	111.8	120.2	109.1	102.1	100.2	128.5	112.0	99.6	81.1	78.9
2008												
January	105.1	109.6	111.4	117.3	109.5	102.4	100.8	128.4	111.7	99.2	80.9	78.6
February	104.7	108.9	111.2	116.5	109.4	102.4	100.5	125.9	111.5	98.9	80.7	78.1
March	104.5	108.6	109.7	113.5	108.3	103.2	100.9	124.4	111.1	99.1	80.6	77.8
April	103.7	107.4	108.9	111.0	108.1	101.1	100.4	122.1	110.0	98.6	80.0	77.0
May	103.1	106.8	108.3	109.8	107.6	101.5	99.6	121.0	109.3	97.9	79.6	76.7
June	102.8	106.1	108.1	110.4	107.2	100.9	100.3	119.6	108.5	97.9	79.5	76.2
July	102.3	104.8	107.3	107.9	107.0	99.0	98.1	119.3	107.6	97.8	79.0	75.4
August	100.7	103.6	105.5	103.3	105.9	97.0	97.4	116.9	106.9	96.2	77.8	74.6
September	96.4	100.0	104.8	102.4	105.2	89.3	94.8	113.7	105.4	90.0	74.4	72.1
October	97.3	99.3	105.1	99.9	106.3	86.7	94.8	111.9	104.4	92.5	75.0	71.8
November	96.1	97.0	104.2	96.1	106.4	88.0	94.5	106.4	102.0	91.2	74.0	70.2
December	93.3	93.6	102.0	91.7	104.8	88.6	94.1	100.2	98.4	87.7	71.7	67.8
2009												
January	91.0	90.8	99.8	81.9	104.8	83.4	94.1	96.0	96.9	86.2	70.0	65.9
February	90.5	90.6	99.8	83.6	104.4	83.1	93.8	94.3	95.7	85.6	69.4	65.9
March	89.0	89.0	99.3	83.3	103.8	81.5	92.3	91.1	94.8	84.0	68.3	64.7
April	88.3	88.3	98.9	83.4	103.3	79.9	91.7	89.4	94.4	83.4	67.7	64.4
May	87.4	87.4	97.7	80.0	102.7	77.6	91.8	89.0	93.9	82.9	67.0	63.8
June	87.1	87.1	97.2	79.3	102.3	76.9	92.1	89.0	93.9	82.6	66.7	63.7
July	88.0	88.4	98.4	87.0	101.5	78.0	93.4	89.4	93.8	83.6	67.4	64.7
August	89.0	89.4	99.3	88.0	102.4	79.0	94.3	90.1	94.3	84.8	68.2	65.6
September	89.7	90.2	100.0	91.6	102.2	79.5	95.4	89.7	94.6	85.7	68.8	66.2
October	90.0	90.4	100.4	90.3	103.1	79.9	95.0	88.5	95.4	85.9	69.1	66.5
November	90.3	91.2	99.8	91.8	101.9	79.8	95.7	90.1	95.6	86.8	69.5	67.2
December	90.6	91.1	99.6	91.1	101.9	81.2	95.8	88.0	96.7	87.1	69.9	67.2
2010												
January	91.7	92.1	100.4	92.3	102.5	82.4	98.2	89.8	97.0	88.3	70.8	68.0
February	92.0	92.0	99.5	90.7	101.9	82.0	98.8	89.9	96.9	89.5	71.2	68.1
March	92.6	93.1	99.7	91.8	101.9	83.4	101.2	91.4	96.8	90.0	71.8	69.0
April	92.9	93.9	99.0	92.3	100.8	84.5	102.2	94.2	97.5	90.4	72.3	69.7
May	94.3	95.2	100.9	95.5	102.2	86.2	102.1	94.6	98.4	91.7	73.5	70.8
June	94.4	95.1	100.5	94.5	102.1	86.6	101.8	94.6	98.4	92.2	73.8	70.9
July	94.9	95.7	101.2	98.3	101.8	87.6	103.2	94.3	98.4	92.4	74.2	71.4
August	95.1	95.7	100.9	95.4	102.3	87.5	103.4	94.8	98.4	93.2	74.6	71.6
September	95.4	95.7	100.5	95.5	101.8	87.8	101.3	94.8	98.4	93.8	74.9	71.7
October	95.1	95.8	100.5	96.0	101.7	88.1	100.1	95.0	97.7	93.4	74.8	71.9
November	95.1	95.9	100.0	94.6	101.3	87.9	99.7	95.7	98.8	93.5	74.9	72.0
December	96.1	96.3	101.0	94.2	102.8	88.6	99.1	94.5	99.3	94.8	75.6	72.5

Table 2-4A. Industrial Production and Capacity Utilization, Historical Data, 1948–2016—*Continued*

(Seasonally adjusted.)

Year and month	Production indexes, 2007 = 100									Capacity utilization (output as percent of capacity)	
	Total industry	Manufac-turing (SIC)	Market groups								
			Consumer goods			Business equipment	Defense and space equipment	Construction supplies	Business supplies	Materials	Total industry	Manufac-turing (SIC)
			Total	Durable	Nondurable							
2011												
January	95.9	96.5	101.1	95.3	102.7	89.4	98.7	94.1	98.9	94.4	75.6	72.7
February	95.5	96.6	100.9	97.5	101.8	89.7	98.0	93.7	98.9	93.7	75.3	72.8
March	96.5	97.1	101.1	99.1	101.6	89.4	97.1	94.5	99.1	95.4	76.0	73.3
April	96.1	96.6	101.0	95.4	102.4	88.7	96.7	94.4	98.8	95.0	75.7	72.9
May	96.3	96.7	101.3	96.7	102.4	89.5	96.8	95.6	98.9	95.0	75.8	73.0
June	96.6	96.8	101.1	96.1	102.4	89.5	96.2	95.9	98.8	95.6	75.9	73.1
July	97.1	97.4	101.9	97.6	102.9	90.7	97.1	96.8	99.1	95.8	76.2	73.5
August	97.7	97.8	102.3	98.5	103.3	91.6	98.0	96.6	99.3	96.4	76.6	73.8
September	97.6	98.1	101.9	98.7	102.7	92.1	98.0	96.8	99.6	96.3	76.4	74.0
October	98.3	98.7	102.1	100.3	102.6	93.4	99.1	96.9	99.2	97.3	76.8	74.3
November	98.2	98.4	101.0	98.6	101.6	94.1	100.6	96.9	98.3	97.7	76.6	74.0
December	98.8	99.1	100.9	99.2	101.4	95.6	99.9	98.4	98.9	98.3	76.9	74.5
2012												
January	99.4	99.9	100.9	101.8	100.7	97.2	100.3	99.1	99.2	99.1	77.2	75.0
February	99.6	100.2	100.7	101.1	100.7	97.8	101.0	100.2	99.7	99.3	77.2	75.1
March	99.2	99.8	99.2	100.0	98.9	98.6	101.0	100.1	99.1	99.1	76.7	74.7
April	99.9	100.3	100.0	100.5	99.8	99.6	100.0	101.3	100.2	99.7	77.2	75.0
May	100.1	99.9	100.4	99.8	100.6	100.1	98.9	100.3	100.5	99.8	77.2	74.6
June	100.1	100.1	100.1	99.8	100.2	101.5	98.1	99.6	100.5	99.8	77.0	74.7
July	100.3	100.0	100.3	99.9	100.3	100.4	100.4	99.2	100.4	100.4	77.1	74.5
August	99.9	99.8	99.7	98.8	99.9	101.0	100.4	99.7	100.0	99.7	76.6	74.2
September	99.9	99.7	99.5	97.7	100.0	100.9	99.9	99.6	100.0	99.9	76.5	74.1
October	100.1	99.4	99.4	98.6	99.6	99.8	99.8	99.2	99.9	100.7	76.6	73.8
November	100.6	100.1	99.9	100.0	99.9	101.1	100.1	100.6	100.1	101.0	76.8	74.2
December	101.0	100.9	100.0	102.1	99.4	102.1	99.9	101.2	100.4	101.4	77.0	74.7
2013												
January	100.8	100.6	100.0	101.4	99.6	99.9	99.2	101.5	100.6	101.6	76.7	74.4
February	101.4	101.0	100.5	103.6	99.7	100.8	99.1	103.2	100.6	102.1	77.1	74.7
March	101.8	100.9	101.0	104.5	100.1	101.3	98.8	102.8	101.0	102.6	77.3	74.5
April	101.6	100.5	100.7	104.2	99.8	100.5	98.4	102.2	101.2	102.5	77.1	74.2
May	101.7	100.8	100.5	104.7	99.5	100.0	97.5	102.0	101.7	102.9	77.0	74.4
June	102.0	101.1	100.8	106.1	99.5	100.0	97.5	102.9	101.9	103.0	77.1	74.5
July	101.5	100.1	99.8	103.8	98.8	97.9	96.2	102.6	101.7	103.2	76.7	73.8
August	102.2	101.0	100.3	106.5	98.8	99.6	96.6	103.2	102.2	103.8	77.2	74.4
September	102.7	101.2	100.7	107.2	99.1	100.4	96.5	104.3	102.4	104.4	77.5	74.5
October	102.5	101.2	101.0	107.3	99.4	99.9	96.5	104.2	102.6	104.0	77.3	74.5
November	102.8	101.2	101.0	108.5	99.1	99.7	95.1	104.9	102.5	104.8	77.4	74.5
December	103.2	101.2	101.9	108.5	100.2	98.9	94.5	103.4	103.1	105.1	77.6	74.5
2014												
January	102.7	100.0	100.1	104.9	98.9	98.9	93.7	103.1	102.5	105.3	77.2	73.6
February	103.6	101.1	101.4	108.6	99.5	100.4	93.8	103.8	103.1	106.0	77.8	74.4
March	104.6	101.9	101.9	109.4	100.0	101.6	94.7	105.2	103.3	107.3	78.5	75.0
April	104.6	101.8	101.4	108.6	99.6	101.2	94.1	104.4	102.9	107.8	78.4	74.9
May	105.0	102.0	101.3	110.7	99.0	101.9	94.1	106.1	103.0	108.3	78.6	75.2
June	105.4	102.4	101.4	111.3	98.9	101.5	94.1	106.8	103.0	109.1	78.8	75.5
July	105.6	102.9	101.5	114.3	98.2	102.8	94.5	107.7	102.8	109.1	78.9	75.8
August	105.5	102.3	100.9	111.1	98.3	101.9	93.4	107.7	102.5	109.4	78.7	75.5
September	105.8	102.3	101.2	111.0	98.8	101.3	93.9	108.1	102.6	109.8	78.8	75.5
October	105.8	102.2	101.0	110.7	98.5	102.4	93.9	107.7	102.4	109.8	78.7	75.5
November	106.7	103.0	103.0	114.1	100.2	103.9	93.4	107.7	103.1	110.0	79.2	76.1
December	106.5	102.7	102.7	113.4	99.9	102.6	93.1	108.5	102.1	110.3	79.0	75.9
2015												
January	106.0	102.2	102.8	113.4	100.1	102.2	91.9	107.8	102.2	109.4	78.6	75.7
February	105.4	101.5	102.5	110.7	100.4	101.5	93.4	106.5	102.1	109.1	78.1	75.2
March	105.1	101.9	103.5	113.2	101.1	101.6	93.1	105.9	101.4	108.3	77.8	75.5
April	104.5	101.7	102.6	114.0	99.7	100.7	92.5	106.4	101.8	107.9	77.3	75.5
May	104.1	101.7	102.4	116.5	98.9	101.0	92.1	106.8	101.4	107.0	76.9	75.5
June	103.7	101.3	102.3	113.8	99.4	100.3	92.0	107.2	101.1	106.5	76.6	75.2
July	104.3	102.0	104.0	119.8	100.0	100.3	91.7	107.0	100.8	106.9	77.0	75.8
August	104.2	101.7	103.8	116.6	100.6	99.9	91.6	107.6	101.1	106.6	76.9	75.5
September	103.8	101.2	103.3	116.0	100.1	98.6	90.7	106.4	101.2	106.5	76.6	75.2
October	103.4	101.2	103.0	116.0	99.7	97.8	90.4	107.8	101.5	105.9	76.3	75.1
November	102.7	100.9	102.8	115.3	99.6	96.5	90.1	107.6	101.6	104.8	75.7	74.9
December	102.1	100.6	102.2	115.1	98.9	95.3	90.4	108.2	101.1	104.3	75.3	74.6
2016												
January	103.0	101.3	103.7	116.8	100.3	95.9	89.6	108.9	101.9	105.0	75.9	75.1
February	102.2	100.7	103.4	116.3	100.1	95.0	88.3	108.2	101.6	104.1	75.3	74.6
March	101.4	100.5	102.4	115.0	99.2	94.4	87.7	108.0	100.7	103.3	74.7	74.3
April	101.5	100.1	103.2	116.1	100.0	94.0	87.1	108.0	101.1	103.1	74.7	74.0
May	101.4	100.1	103.1	115.1	100.1	94.0	88.3	107.3	101.3	103.0	74.6	73.9
June	101.9	100.4	103.9	117.5	100.5	94.1	87.8	107.1	101.6	103.5	74.9	74.0
July	102.1	100.7	104.0	119.2	100.1	94.3	88.3	108.0	102.2	103.7	75.1	74.1
August	102.0	100.3	103.9	118.1	100.4	93.8	89.1	107.3	102.3	103.6	75.0	73.8
September	102.0	100.7	103.9	118.7	100.1	93.7	89.6	107.8	102.8	103.5	74.9	74.0
October	102.2	101.0	103.6	119.8	99.5	94.2	90.5	108.4	102.5	103.9	75.0	74.1
November	102.1	101.1	102.9	118.7	98.8	94.2	91.3	109.0	102.6	103.7	74.9	74.1
December	102.9	101.4	104.6	118.9	100.9	95.2	91.2	108.9	103.4	104.1	75.5	74.4

Table 2-4B. Industrial Production: Historical Data, 1919–1947

(Seasonally adjusted, 2012 = 100.)

Year and month	January	February	March	April	May	June	July	August	September	October	November	December	Annual averages
1919													
Industrial production, total ...	5.01	4.79	4.65	4.74	4.76	5.07	5.37	5.46	5.34	5.29	5.21	5.29	5.08
Manufacturing	4.95	4.85	4.67	4.77	4.74	5.08	5.37	5.49	5.29	5.18	5.37	5.26	5.08
1920													
Industrial production, total ..	5.79	5.79	5.68	5.37	5.51	5.57	5.43	5.46	5.26	5.04	4.62	4.35	5.32
Manufacturing	5.81	5.81	5.68	5.34	5.52	5.52	5.31	5.34	5.21	4.92	4.41	4.10	5.25
1921													
Industrial production, total ..	4.10	4.02	3.90	3.90	4.02	3.99	3.96	4.10	4.13	4.38	4.32	4.29	4.09
Manufacturing	3.89	3.89	3.78	3.78	3.91	3.89	3.89	4.04	4.10	4.28	4.33	4.28	4.00
1922													
Industrial production, total ..	4.46	4.65	4.90	4.74	4.98	5.23	5.23	5.12	5.40	5.70	5.95	6.12	5.21
Manufacturing	4.38	4.48	4.69	4.90	5.16	5.39	5.44	5.24	5.39	5.62	5.83	6.01	5.21
1923													
Industrial production, total ..	5.98	6.06	6.26	6.40	6.48	6.42	6.37	6.26	6.12	6.09	6.09	5.95	6.21
Manufacturing	5.91	5.96	6.22	6.30	6.43	6.35	6.25	6.09	6.06	5.96	5.96	5.88	6.11
1924													
Industrial production, total ...	6.09	6.20	6.09	5.90	5.65	5.40	5.32	5.51	5.70	5.84	5.95	6.12	5.82
Manufacturing	5.99	6.12	6.01	5.86	5.57	5.29	5.21	5.39	5.60	5.75	5.88	6.09	5.73
1925													
Industrial production, total ..	6.31	6.31	6.31	6.37	6.34	6.29	6.45	6.34	6.26	6.51	6.65	6.73	6.41
Manufacturing	6.27	6.30	6.32	6.32	6.27	6.25	6.38	6.27	6.30	6.56	6.74	6.89	6.41
1926													
Industrial production, total ..	6.62	6.62	6.70	6.70	6.65	6.73	6.76	6.84	6.95	6.95	6.92	6.90	6.78
Manufacturing	6.74	6.69	6.66	6.66	6.61	6.69	6.69	6.76	6.89	6.84	6.79	6.76	6.73
1927													
Industrial production, total ..	6.87	6.92	7.01	6.84	6.90	6.87	6.78	6.78	6.67	6.54	6.54	6.56	6.77
Manufacturing	6.71	6.76	6.82	6.79	6.82	6.82	6.76	6.71	6.61	6.51	6.48	6.56	6.70
1928													
Industrial production, total ..	6.70	6.76	6.81	6.78	6.87	6.92	7.01	7.14	7.20	7.34	7.48	7.62	7.05
Manufacturing	6.69	6.76	6.79	6.76	6.87	6.95	7.05	7.21	7.23	7.36	7.52	7.65	7.07
1929													
Industrial production, total ..	7.73	7.70	7.73	7.86	8.00	8.06	8.17	8.09	8.03	7.89	7.50	7.17	7.83
Manufacturing	7.70	7.67	7.75	7.85	8.01	8.11	8.19	8.14	7.98	7.91	7.46	7.02	7.82
1930													
Industrial production, total ..	7.17	7.14	7.03	6.98	6.87	6.67	6.37	6.23	6.12	5.95	5.82	5.68	6.50
Manufacturing	7.10	7.05	6.82	6.92	6.79	6.61	6.30	6.06	6.01	5.81	5.65	5.52	6.39
1931													
Industrial production, total ..	5.65	5.68	5.79	5.82	5.73	5.59	5.51	5.32	5.07	4.87	4.82	4.79	5.39
Manufacturing	5.52	5.57	5.68	5.68	5.62	5.42	5.34	5.16	4.92	4.69	4.61	4.59	5.23
1932													
Industrial production, total ..	4.65	4.54	4.49	4.18	4.04	3.90	3.79	3.90	4.15	4.29	4.29	4.21	4.20
Manufacturing	4.54	4.41	4.25	3.99	3.89	3.76	3.63	3.73	3.99	4.10	4.12	4.04	4.04
1933													
Industrial production, total ..	4.13	4.15	3.90	4.18	4.87	5.62	6.15	5.90	5.57	5.29	4.98	5.01	4.98
Manufacturing	3.99	3.89	3.65	4.04	4.74	5.52	6.09	5.73	5.44	5.13	4.79	4.85	4.82
1934													
Industrial production, total ..	5.18	5.43	5.68	5.68	5.79	5.68	5.29	5.23	4.93	5.15	5.21	5.54	5.40
Manufacturing	5.00	5.29	5.49	5.60	5.68	5.55	5.11	5.08	4.79	4.98	5.05	5.42	5.25
1935													
Industrial production, total ..	5.98	6.09	6.06	5.95	5.95	6.04	6.04	6.26	6.42	6.62	6.76	6.84	6.25
Manufacturing	5.91	6.01	5.94	5.88	5.86	5.83	6.01	6.25	6.40	6.58	6.74	6.82	6.19
1936													
Industrial production, total ..	6.73	6.56	6.65	7.06	7.20	7.34	7.48	7.59	7.73	7.84	8.06	8.31	7.38
Manufacturing	6.69	6.48	6.61	7.02	7.21	7.36	7.49	7.62	7.75	7.88	8.09	8.35	7.38
1937													
Industrial production, total ..	8.28	8.39	8.58	8.58	8.61	8.50	8.56	8.50	8.22	7.62	6.87	6.26	8.08
Manufacturing	8.35	8.48	8.53	8.66	8.71	8.55	8.60	8.50	8.19	7.52	6.64	5.94	8.05
1938													
Industrial production, total ..	6.12	6.06	6.06	5.95	5.82	5.87	6.20	6.54	6.73	6.90	7.17	7.26	6.39
Manufacturing	5.83	5.81	5.83	5.65	5.62	5.62	5.96	6.35	6.56	6.76	7.08	7.13	6.18

Table 2-4B. Industrial Production: Historical Data, 1919–1947—*Continued*

(Seasonally adjusted, 2012 = 100.)

Year and month	January	February	March	April	May	June	July	August	September	October	November	December	Annual averages
1939													
Industrial production, total	7.26	7.31	7.34	7.31	7.28	7.45	7.67	7.78	8.25	8.67	8.89	8.89	7.84
Products	7.71	7.74	7.83	7.83	7.89	7.97	8.12	8.20	8.46	8.69	8.87	8.92	8.19
Consumer goods	12.08	12.08	12.28	12.28	12.32	12.44	12.60	12.64	12.84	12.96	13.07	13.11	12.56
Materials	6.51	6.62	6.62	6.41	6.38	6.69	6.98	7.05	7.86	8.60	8.76	8.76	7.27
Manufacturing	6.79	6.84	6.84	6.76	6.76	6.92	7.13	7.33	7.75	8.19	8.35	8.45	7.34
1940													
Industrial production, total	8.78	8.50	8.31	8.47	8.72	9.00	9.11	9.17	9.36	9.50	9.72	10.05	9.06
Products	8.84	8.75	8.66	8.75	8.87	9.07	9.07	9.18	9.44	9.61	9.79	10.13	9.18
Consumer goods	13.07	13.03	12.88	12.96	13.03	13.15	13.11	13.11	13.47	13.71	13.99	14.42	13.33
Materials	8.53	8.06	7.75	7.98	8.35	8.79	9.02	9.10	9.20	9.30	9.48	9.82	8.78
Manufacturing	8.32	8.06	7.88	7.96	8.29	8.66	8.76	8.89	9.07	9.20	9.36	9.69	8.68
1941													
Industrial production, total	10.30	10.61	10.94	10.97	11.46	11.55	11.69	11.82	11.82	11.94	11.99	12.18	11.44
Products	10.45	10.76	11.05	11.28	11.66	11.77	11.86	12.00	12.03	12.15	12.20	12.38	11.63
Consumer goods	14.70	15.13	15.53	15.81	16.28	16.32	16.32	16.36	16.28	16.28	16.32	16.32	15.97
Materials	9.95	10.23	10.59	10.36	11.01	11.16	11.32	11.40	11.40	11.45	11.52	11.76	11.01
Manufacturing	9.87	10.24	10.50	10.78	11.14	11.27	11.40	11.48	11.46	11.56	11.59	11.74	11.09
1942													
Industrial production, total	12.43	12.66	12.79	12.43	12.46	12.49	12.79	13.18	13.49	13.93	14.26	14.59	13.13
Products	12.75	13.07	13.21	12.38	12.41	12.46	12.81	13.18	13.64	14.13	14.59	15.05	13.31
Consumer goods	16.64	16.44	16.44	14.18	14.02	13.79	13.99	14.10	14.14	14.38	14.46	14.78	14.78
Materials	11.78	11.96	12.07	12.38	12.35	12.38	12.61	12.97	13.13	13.49	13.70	13.85	12.72
Manufacturing	12.00	12.29	12.47	12.10	12.16	12.23	12.62	13.04	13.43	13.92	14.31	14.70	12.94
1943													
Industrial production, total	14.73	15.12	15.23	15.42	15.54	15.45	15.92	16.26	16.67	16.92	17.14	16.92	15.94
Products	15.11	15.46	15.63	15.83	15.97	16.15	16.58	16.87	17.33	17.59	17.85	17.47	16.49
Consumer goods	14.34	14.46	14.50	14.66	14.90	15.13	15.37	15.45	15.57	15.45	15.41	15.09	15.03
Materials	14.06	14.42	14.57	14.70	14.68	14.32	14.86	15.14	15.53	15.79	15.92	15.92	14.99
Manufacturing	14.90	15.21	15.34	15.52	15.63	15.68	16.04	16.38	16.82	17.18	17.42	17.13	16.11
1944													
Industrial production, total	17.11	17.25	17.22	17.22	17.11	17.06	17.03	17.25	17.14	17.20	17.06	17.00	17.14
Products	17.70	17.87	17.87	17.93	17.90	17.96	18.10	18.31	18.16	18.19	17.96	17.82	17.98
Consumer goods	15.25	15.33	15.53	15.61	15.73	15.81	15.89	16.24	15.89	15.93	15.89	15.89	15.75
Materials	16.07	16.20	16.12	16.07	15.84	15.61	15.37	15.56	15.53	15.58	15.61	15.68	15.77
Manufacturing	17.39	17.52	17.49	17.47	17.34	17.31	17.31	17.55	17.44	17.49	17.34	17.34	17.42
1945													
Industrial production, total	16.84	16.78	16.67	16.37	15.92	15.56	15.20	13.62	12.41	11.91	12.35	12.41	14.67
Products	17.67	17.47	17.21	16.90	16.44	16.12	15.86	14.13	12.66	12.35	12.52	12.55	15.16
Consumer goods	16.12	16.01	15.97	16.09	16.12	16.24	16.20	15.49	16.24	16.48	16.96	17.15	16.26
Materials	15.50	15.66	15.68	15.45	15.04	14.65	14.11	12.79	11.83	11.19	11.96	12.09	13.83
Manufacturing	17.11	17.03	16.85	16.48	15.97	15.52	15.16	13.32	11.84	11.38	11.69	11.82	14.51
1946													
Industrial production, total	11.71	11.13	12.30	12.07	11.63	12.35	12.77	13.24	13.49	13.74	13.82	13.90	12.68
Products	12.41	12.29	12.58	12.78	12.64	12.78	13.07	13.50	13.87	14.10	14.25	14.39	13.22
Consumer goods	17.91	18.62	18.50	18.70	18.78	18.82	19.25	19.89	20.28	20.52	20.80	20.92	19.42
Materials	10.59	9.33	11.73	10.96	10.13	11.55	12.22	12.69	12.76	13.05	13.02	13.10	11.76
Manufacturing	11.02	10.32	11.61	11.64	11.09	11.72	12.10	12.65	12.91	13.19	13.35	13.40	12.08
1947													
Industrial production, total	14.07	14.15	14.23	14.12	14.18	14.18	14.10	14.18	14.29	14.43	14.62	14.68	14.27
Products	14.54	14.59	14.65	14.65	14.68	14.65	14.62	14.71	14.85	15.05	15.26	15.34	14.80
Consumer goods	20.48	20.40	20.48	20.40	20.25	20.28	20.36	20.52	20.72	21.04	21.35	21.39	20.64
Materials	13.10	13.20	13.64	13.31	13.36	13.23	13.10	13.08	13.26	13.36	13.67	13.51	13.32
Manufacturing	13.35	13.40	13.45	13.48	13.37	13.37	13.30	13.37	13.43	13.61	13.81	13.84	13.48

NOTES AND DEFINITIONS, CHAPTER 2

TABLES 2-1 THROUGH 2-4

Industrial Production and Capacity Utilization

SOURCE: BOARD OF GOVERNORS OF THE FEDERAL RESERVE SYSTEM

Ever since the Federal Reserve was founded in 1913, understanding current business conditions has been a central focus in pursuing its mission of a stable and secure financial system. The index of industrial production is one of the oldest continuous statistical series maintained by the federal government, and one of the few economic indicators for which monthly data are available before the post-World-War-II period.

The Federal Reserve's monthly index of industrial production and the related capacity indexes and capacity utilization rates cover manufacturing, mining, and electric and gas utilities. The industrial sector, together with construction, accounts for the bulk of the variation in national output over the course of the business cycle. The industrial detail provided by these measures helps illuminate structural developments in the economy.

Around the 15th day of each month, the Federal Reserve issues estimates of industrial production and capacity utilization for the previous month. The production estimates are in the form of index numbers (2012 = 100) that reflect the monthly levels of total output of the nation's factories, mines, and gas and electric utilities expressed as a percent of the monthly average in the 2012 base year. Capacity estimates are expressed as index numbers, 2012 output (not 2012 capacity) = 100, and capacity utilization is measured by the production index as a percent of the capacity index. Since, for each component industry, the bases of those two indexes are the same, this procedure yields production as a percent of capacity. Monthly estimates are subject to revision in subsequent months, as well as to annual and comprehensive revisions in subsequent years.

Definitions and notes on the data

The *industrial production index* measures changes in the physical volume or quantity of output of manufacturing, mining, and electric and gas utilities. *Capacity utilization* is calculated by dividing a seasonally adjusted industrial production index for an industry or group of industries by a related index of productive capacity.

The index of industrial production measures a large portion of the goods output of the national economy on a monthly basis. This portion, together with construction, has also accounted for the bulk of the variation in output over the course of many historical business cycles. The substantial industrial detail included in the index illuminates structural developments in the economy.

The total industrial production index and the indexes for its major components are constructed from individual industry series (312 series for data from 1997 forward) based on the 2012 North American Industry Classification System (NAICS). See the Preface to this volume for information on NAICS.

The Federal Reserve has been able to provide a longer continuous historical series on the NAICS basis than some other government agencies. In a major research effort, the Fed and the Census Bureau's Center for Economic Studies re-coded data from seven Censuses of Manufactures, beginning in 1963, to establish benchmark NAICS data for output, value added, and capacity utilization. The resulting indexes are shown annually for the last 48 years (52 years for aggregate levels) in Tables 2-1 through 2-4.

The Fed's featured indexes for total industry and total manufacturing are on the Standard Industrial Classification (SIC) basis and do not observe the reclassifications under NAICS of the logging industry to the Agriculture sector and the publishing industry to the Information sector. (The reason cited by the Fed was to avoid "changing the scope or historical continuity of these statistics.") One advantage of the SIC index for capacity utilization is that it is a continuous series back to 1949 (shown in Table 2-4A). On the new NAICS basis, production and capacity utilization are shown back to 1970 in Tables 2-2 and 2-3.

The individual series components of the indexes are grouped in two ways: market groups and industry groups.

Market groups. For analyzing market trends and product flows, the individual series are grouped into two major divisions: *final products and nonindustrial supplies* and *materials*. *Final products* consists of products purchased by consumers, businesses, or government for final use. *Nonindustrial supplies* are expected to become inputs in nonindustrial sectors: the two major subgroups are *construction supplies* and *business supplies*. *Materials* comprises industrial output that requires further processing within the industrial sector. This twofold division distinguishes between products that are ready to ship outside the industrial sector and those that will stay within the sector for further processing.

Final products are divided into *consumer goods* and *equipment*, and *equipment* is divided into *business equipment* and *defense and space equipment*. Further subdivisions of each market group are based on type of product and the market destination for the product.

Industry groups are typically groupings by 3-digit NAICS industries and major aggregates of these industries—for example, *durable goods* and *nondurable goods manufacturing*, *mining*, and *utilities*. Indexes are also calculated for *stage-of-process* industry groups—*crude, primary and semifinished*, and *finished* processing. The stage-of-process grouping was a new feature in the 2002 revision, replacing the two narrower and less well-defined "primary processing manufacturing" and "advanced processing manufacturing" groups that were previously published. *Crude processing* consists of logging, much of mining, and certain basic manufacturing activities in the chemical, paper, and metals industries. *Primary and semifinished processing* represents industries that produce materials and parts used as inputs by other industries. *Finished processing* includes industries that produce goods in their finished form for use by consumers, business investment, or government.

The indexes of industrial production are constructed using data from a variety of sources. Current monthly estimates of production are based on measures of physical output where possible and appropriate. For a few high-tech industries, the estimated value of nominal output is deflated by a corresponding price index. For industries in which such direct measurement is not possible on a monthly basis, output is inferred from production-worker hours, adjusted for trends in worker productivity derived from annual and benchmark revisions. (Between the 1960s and 1997, electric power consumption was used as a monthly output indicator for some industries instead of hours. However, the coverage of the electric power consumption survey deteriorated, and in the 2005 revision, the decision was made to resume the use of hours in those industries, beginning with the data for 1997.)

In annual and benchmark revisions, the individual indexes are revised using data from the quinquennial Economic Census which includes Manufactures and Mineral Industries, the Annual Survey of Manufactures, and the quarterly Survey of Plant Capacity, prepared by the Census Bureau; deflators from the Producer Price Indexes and other sources; the *Minerals Yearbook*, prepared by the Department of the Interior; publications from the Department of Energy; and other sources.

The weights used in compiling the indexes are based on Census value added—the difference between the value of production and the cost of materials and supplies consumed. Census value added differs in some respects from the economic concept of industry value added used in the national income and product accounts (NIPAs). Industry value added as defined in the NIPAs is not available in sufficient detail for the industrial production indexes. See Chapter 14 for data and a description of NIPA value added (equivalently, gross domestic product) by major industry group.

Before 1972, a linked-Laspeyres formula (base period prices) is used to compute the weighted individual indexes. Beginning with 1972, the index uses a version of the Fisher-ideal index formula—a chain-weighting (continually updated average price) system similar to that in the NIPAs. See the "General Notes" article at the front of this book and the notes and definitions for Chapter 1 for more information. Chain-weighting keeps the index from being distorted by the use of obsolete relative prices.

For the purpose of these value-added weights, value added per unit of output is based on data from the Censuses of Manufacturing and Mineral Industries, the Census Bureau's Annual Survey of Manufactures, and revenue and expense data reported by the Department of Energy and the American Gas Association, which are projected into recent years by using changes in relevant Producer Price Indexes.

To separate seasonal movements from cyclical patterns and underlying trends, each component of the index is seasonally adjusted by the Census X-12-ARIMA method.

The index does not cover production on farms, in the construction industry, in transportation, or in various trade and service industries. A number of groups and subgroups include data for individual series not published separately.

Capacity utilization is calculated for the manufacturing, mining, and electric and gas utilities industries. Output is measured by seasonally adjusted indexes of industrial production. The capacity indexes attempt to capture the concept of sustainable maximum output, which is defined as the greatest level of output that a plant can maintain within the framework of a realistic work schedule, taking account of normal downtime and assuming sufficient availability of inputs to operate the machinery and equipment in place. The 89 individual industry capacity indexes are based on a variety of data, including capacity data measured in physical units compiled by government agencies and trade associations, Census Bureau surveys of utilization rates and investment, and estimates of growth of the capital stock.

In the "Explanatory Note" to its monthly release, the statistics cover output, capacity, and capacity utilization in the U.S. industrial sector, which is defined by the Federal Reserve to comprise manufacturing, mining, and electric and gas utilities. Mining is defined as all industries in sector 21 of the North American Industry Classification System (NAICS); electric and gas utilities are those in NAICS sectors 2211 and 2212. Manufacturing comprises NAICS manufacturing industries (sector 31-33) plus the logging industry and the newspaper, periodical, book, and directory publishing industries. Logging and publishing are classified elsewhere in NAICS (under agriculture and information respectively), but historically they were considered to be manufacturing and were included in the industrial sector under the Standard Industrial Classification (SIC) system. In December 2002 the Federal Reserve reclassified all its industrial output data from the SIC system to NAICS.

Revisions

Revisions normally occur annually. New annual data have been incorporated in addition to the data from the Annual Survey of Manufacturing. The Census Bureau benchmarked the Service Annual Survey to the 2012 Economic Census, which resulted in updated estimates for 2008 through 2012. For logging, the IP indexes were updated with 2015 data from the U.S. Forest Service. In addition, the indexes for metallic and nonmetallic minerals were updated with revised annual data for 2015 from the Department of the Interior's U.S. Geological Survey (USGS). Data on prices from the Bureau of Labor Statistics (BLS) were also incorporated into most of the manufacturing indexes.

Data availability

Data are available monthly in Federal Reserve release G.17. Current and historical data and background information are available on the Federal Reserve Web site at <http://www.federalreserve.gov>.

Chain-weighting makes it difficult for the user to analyze in detail the sources of aggregate output change. An "Explanatory Note," included in each month's index release, provides some assistance for the user, including a reference to an Internet location with the exact contribution of a monthly change in a component index to the monthly change in the total index.

REFERENCES

The G.17 release each month contains extensive explanatory material, as well as references for further detail. An earlier detailed description of the industrial production index, together with a history of the index, a glossary of terms, and a bibliography is presented in *Industrial Production—1986 Edition*, available from Publication Services, Mail Stop 127, Board of Governors of the Federal Reserve System, Washington, DC 20551.

CHAPTER 3: INCOME DISTRIBUTION AND POVERTY

SECTION 3A: HOUSEHOLD AND FAMILY INCOME

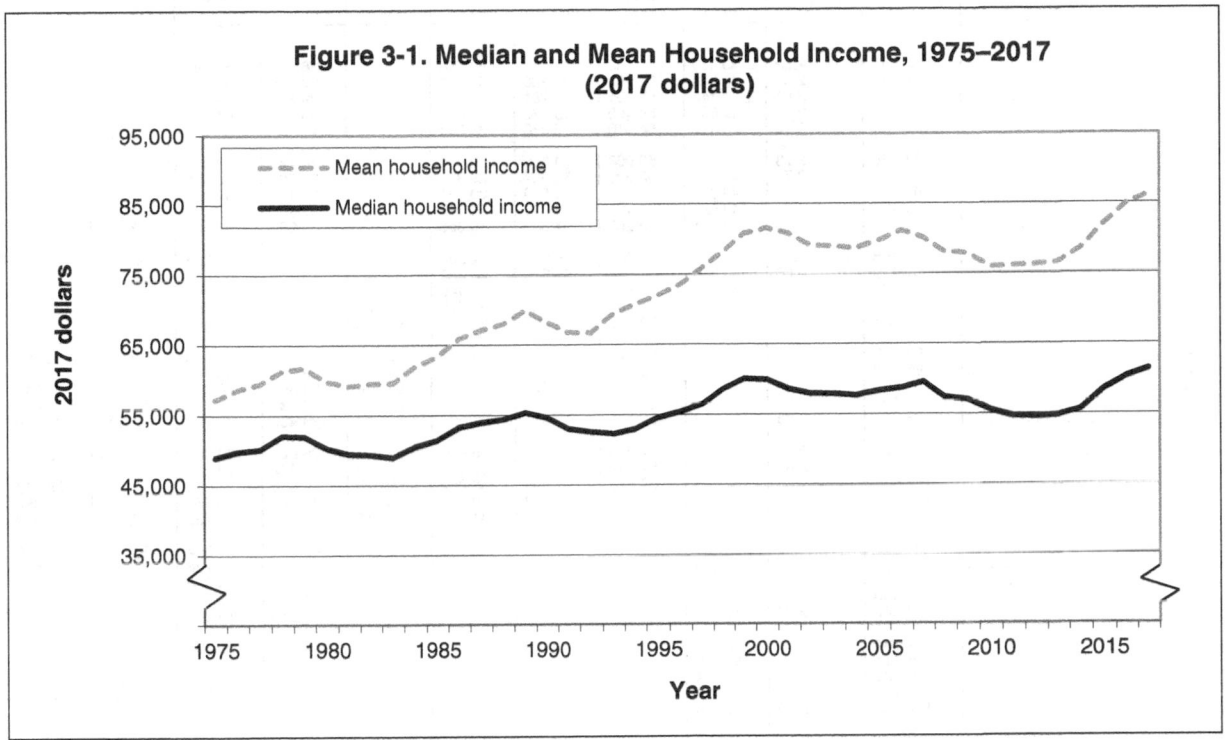

Figure 3-1. Median and Mean Household Income, 1975–2017 (2017 dollars)

- Measured as cash income before taxes, median household income was $61,372 in 2017, an increase of 1.8 percent from the 2016 median of $60,309. Income in 2016 finally surpassed the previous all-time high in 1999 at $60,309, an increase of 0.41 percent. From 1999 to 2007, median household income declined 0.9 percent. Median income is the income of the household at the middle of the income distribution. (Table 3-1)

- The median income of non-Hispanic White, Black, and Hispanic-origin households increased 2.6 percent, -0.2 percent, and 3.7 percent, respectively, between 2016 and 2017. While median household income increased for Hispanic households for the fifth consecutive year, median household income declined for Black households in 2017 after increasing the previous five years. (Table 3-1)

- The 2017 female-to-male earnings ratio remained steady at 0.805. This is the highest that the male to female ratio has ever been. (Table 3-1)

- The Gini index is a statistical measure of income inequality ranging from 0 to 1, with a measure of 1 indicating perfect inequality (one household having all the income and the rest having none) and a measure of 0 indicating perfect equality (all households having an equal share of income). Household income inequality, as measured by the "Gini coefficient," decreased slightly from .464 in 2016 to .463 in 2017, indicating a slight decrease in income inequality. (Table 3-3)

- Income inequality is also measured by determining the percentage share of aggregate income. For the lowest fifth, households received 3.8 percent of this income in 2017. The highest fifth received 48.8 percent in 2017. (Table 3-4)

Table 3-1. Median and Mean Household Income and Median Earnings

(2017 dollars.)

Year	Number of households (thousands)	Average size of household (number of people)	Median household income All races	White Total	White Not Hispanic	Black	Asian¹	Hispanic (any race)	Mean household income, all races	Median earnings of year-round, full-time workers (2018 dollars) Male workers	Female workers	Ratio, female to male
1975	72 867	2.89	49 020	51 263	51 649	30 774	. . .	36 827	57 241	52 999	31 173	0.588
1976	74 142	2.86	49 833	52 202	53 266	31 041	. . .	37 589	58 616	52 854	31 814	0.602
1977	76 030	2.81	50 148	52 735	53 781	31 119	. . .	39 341	59 489	54 043	31 843	0.589
1978	77 330	2.78	52 089	54 150	55 170	32 542	. . .	40 813	61 308	54 392	32 331	0.594
1979	80 776	2.76	51 990	54 510	55 278	32 004	. . .	41 191	61 758	53 736	32 060	0.597
1980	82 368	2.73	50 301	53 068	54 008	30 573	. . .	38 773	59 825	52 863	31 803	0.602
1981	83 527	2.72	49 502	52 302	53 057	29 350	. . .	39 707	59 138	52 580	31 146	0.592
1982	83 918	2.73	49 368	51 683	52 549	29 291	. . .	37 148	59 495	51 585	31 851	0.617
1983	85 407	2.71	49 021	51 408	. . .	29 173	. . .	37 335	59 621	51 359	32 661	0.636
1984	86 789	2.69	50 511	53 287	54 393	30 356	. . .	38 290	61 888	52 320	33 306	0.637
1985	88 458	2.67	51 455	54 265	55 486	32 285	. . .	38 050	63 324	52 712	34 039	0.646
1986	89 479	2.66	53 309	56 045	57 319	32 289	. . .	39 295	65 860	54 077	34 755	0.643
1987	91 124	2.64	53 945	56 837	58 400	32 440	66 706	40 025	67 087	53 707	35 005	0.652
1988	92 830	2.62	54 390	57 498	59 083	32 778	64 463	40 673	67 959	53 253	35 173	0.660
1989	93 347	2.63	55 329	58 200	59 452	34 613	69 103	41 959	69 903	52 314	35 926	0.687
1990	94 312	2.63	54 621	56 970	58 273	34 068	70 139	40 733	68 229	50 489	36 158	0.716
1991	95 669	2.62	53 025	55 565	56 892	33 103	64 155	39 939	66 747	51 784	36 176	0.699
1992	96 426	2.66	52 615	55 316	57 173	32 210	64 920	38 808	66 704	51 861	36 710	0.708
1993	97 107	2.67	52 334	55 214	57 246	32 721	64 238	38 338	69 399	50 937	36 430	0.715
1994	98 990	2.65	52 942	55 837	57 639	34 503	66 427	38 432	70 777	50 629	36 436	0.720
1995	99 627	2.65	54 600	57 308	59 571	35 880	65 076	36 629	72 005	50 466	36 047	0.714
1996	101 018	2.64	55 394	57 999	60 537	36 649	67 543	38 872	73 547	50 169	37 005	0.738
1997	102 528	2.62	56 533	59 538	61 990	38 269	69 128	40 680	75 915	51 444	38 152	0.742
1998	103 874	2.61	58 612	61 667	63 969	38 212	70 296	42 702	78 161	53 276	38 982	0.732
1999	106 434	2.60	60 062	62 467	65 170	41 192	75 211	45 377	80 785	53 709	38 839	0.723
2000	108 209	2.58	59 938	62 688	65 124	42 348	79 590	47 345	81 557	53 175	39 200	0.737
2001	109 297	2.58	58 609	61 786	64 268	40 902	74 442	46 586	80 788	53 123	40 548	0.763
2002	111 278	2.57	57 947	. . .	. . .	. . .	. . .	45 232	79 048	53 875	41 269	0.766
2003	112 000	2.57	57 875	. . .	. . .	. . .	. . .	44 086	78 916	54 334	41 049	0.755
2004	113 343	2.57	57 674	. . .	. . .	. . .	. . .	44 583	78 660	53 072	40 640	0.766
2005	114 384	2.57	58 291	. . .	. . .	. . .	. . .	45 256	79 704	52 075	40 086	0.770
2006	116 011	2.56	58 746	. . .	. . .	. . .	. . .	46 046	81 134	51 506	39 628	0.769
2007	116 783	2.56	59 534	. . .	. . .	. . .	. . .	45 841	80 128	53 466	41 602	0.778
2008	117 181	2.57	57 412	. . .	. . .	. . .	. . .	43 271	78 094	52 920	40 797	0.771
2009	117 538	2.59	57 010	. . .	. . .	. . .	. . .	43 566	77 853	53 975	41 549	0.770
2010	119 927	2.56	55 520	. . .	. . .	. . .	. . .	42 399	75 932	54 027	41 562	0.769
2011	121 084	2.55	54 673	. . .	. . .	. . .	. . .	42 188	76 107	52 650	40 543	0.770
2012	122 459	2.54	54 569	. . .	. . .	. . .	. . .	41 721	76 237	52 838	40 422	0.765
2013	123 931	2.53	54 744	. . .	. . .	. . .	. . .	43 176	76 565	52 717	40 888	0.776
2014	124 587	2.54	55 613	. . .	. . .	. . .	. . .	44 040	78 500	52 220	41 066	0.786
2015	125 819	2.53	58 476	. . .	. . .	. . .	. . .	46 714	82 012	52 988	42 155	0.796
2016	128 224	2.54	60 309	. . .	. . .	. . .	. . .	48 700	84 931	52 751	42 448	0.805
2017	127 586	2.53	61 372	. . .	. . .	. . .	. . .	50 486	86 220	52 146	41 977	0.805
By race												
Race alone												
2002	. . .	. . .	. . .	61 605	64 084	39 661	71 908	. . .	. . .	. . .	. . .	. . .
2003	. . .	. . .	. . .	60 965	63 832	39 607	74 417	. . .	. . .	. . .	. . .	. . .
2004	. . .	. . .	. . .	60 697	63 627	39 151	74 807	. . .	. . .	. . .	. . .	. . .
2005	. . .	. . .	. . .	61 094	63 900	38 828	76 873	. . .	. . .	. . .	. . .	. . .
2006	. . .	. . .	. . .	61 759	63 892	38 963	78 291	. . .	. . .	. . .	. . .	. . .
2007	. . .	. . .	. . .	61 765	65 089	40 196	78 343	. . .	. . .	. . .	. . .	. . .
2008	. . .	. . .	. . .	59 705	63 378	39 054	74 913	. . .	. . .	. . .	. . .	. . .
2009	. . .	. . .	. . .	59 397	62 374	37 319	74 982	. . .	. . .	. . .	. . .	. . .
2010	. . .	. . .	. . .	58 261	61 361	36 195	72 402	. . .	. . .	. . .	. . .	. . .
2011	. . .	. . .	. . .	57 033	60 526	35 203	71 139	. . .	. . .	. . .	. . .	. . .
2012	. . .	. . .	. . .	57 446	60 979	35 641	73 415	. . .	. . .	. . .	. . .	. . .
2013	. . .	. . .	. . .	58 242	61 417	36 467	70 687	. . .	. . .	. . .	. . .	. . .
2014	. . .	. . .	. . .	58 939	62 453	36 689	77 006	. . .	. . .	. . .	. . .	. . .
2015	. . .	. . .	. . .	62 194	65 133	38 178	79 842	. . .	. . .	. . .	. . .	. . .
2016	. . .	. . .	. . .	63 188	66 440	40 339	83 182	. . .	. . .	. . .	. . .	. . .
2017	. . .	. . .	. . .	65 273	68 145	40 258	81 331	. . .	. . .	. . .	. . .	. . .
Race alone or in combination												
2002	. . .	. . .	. . .	. . .	. . .	39 867	71 442	. . .	. . .	. . .	. . .	. . .
2003	. . .	. . .	. . .	. . .	. . .	39 666	73 833	. . .	. . .	. . .	. . .	. . .
2004	. . .	. . .	. . .	. . .	. . .	39 333	74 735	. . .	. . .	. . .	. . .	. . .
2005	. . .	. . .	. . .	. . .	. . .	38 949	76 815	. . .	. . .	. . .	. . .	. . .
2006	. . .	. . .	. . .	. . .	. . .	39 162	77 879	. . .	. . .	. . .	. . .	. . .
2007	. . .	. . .	. . .	. . .	. . .	40 403	78 074	. . .	. . .	. . .	. . .	. . .
2008	. . .	. . .	. . .	. . .	. . .	39 199	74 833	. . .	. . .	. . .	. . .	. . .
2009	. . .	. . .	. . .	. . .	. . .	37 509	74 528	. . .	. . .	. . .	. . .	. . .
2010	. . .	. . .	. . .	. . .	. . .	36 231	71 577	. . .	. . .	. . .	. . .	. . .
2011	. . .	. . .	. . .	. . .	. . .	35 353	70 993	. . .	. . .	. . .	. . .	. . .
2012	. . .	. . .	. . .	. . .	. . .	36 066	72 929	. . .	. . .	. . .	. . .	. . .
2013	. . .	. . .	. . .	. . .	. . .	36 653	71 005	. . .	. . .	. . .	. . .	. . .
2014	. . .	. . .	. . .	. . .	. . .	36 953	77 557	. . .	. . .	. . .	. . .	. . .
2015	. . .	. . .	. . .	. . .	. . .	38 501	79 423	. . .	. . .	. . .	. . .	. . .
2016	. . .	. . .	. . .	. . .	. . .	40 927	82 560	. . .	. . .	. . .	. . .	. . .
2017	. . .	. . .	. . .	. . .	. . .	40 594	80 961	. . .	. . .	. . .	. . .	. . .

¹For 1987 through 2001, Asian and Pacific Islander.
. . . = Not available.

Table 3-2. Median Family Income by Type of Family

(2017 dollars, except as noted.)

Year	All families	Married couples			Male householder [1]	Female householder [1]	4-person families	Average size of family (number of people)
		Total	Wife in paid labor force	Wife not in paid labor force				
1950	29 584	30 716	35 681	29 549	27 766	17 132	32 757	3.54
1951	30 640	31 697	38 256	30 020	28 517	18 339	34 051	3.54
1952	31 557	32 944	39 751	30 924	29 326	18 131	35 475	3.53
1953	34 182	35 222	43 554	33 175	33 143	19 782	. . .	3.59
1954	33 281	34 606	42 617	32 354	32 059	18 322	. . .	3.59
1955	35 442	36 894	45 101	34 704	33 613	19 823	39 461	3.58
1956	37 759	39 283	47 056	36 692	32 917	21 755	42 017	3.60
1957	37 981	39 442	46 968	36 964	35 037	21 132	41 974	3.64
1958	37 864	39 561	46 253	37 090	31 708	20 402	42 315	3.65
1959	39 991	41 799	49 499	39 252	34 055	20 405	44 811	3.67
1960	40 821	42 659	50 119	40 095	35 301	21 558	45 724	3.70
1961	41 242	43 413	51 691	40 213	36 452	21 523	46 290	3.67
1962	42 409	44 595	53 125	41 041	40 664	22 294	48 105	3.68
1963	43 889	46 305	54 705	42 414	40 103	22 552	50 133	3.70
1964	45 516	48 032	56 610	43 916	40 133	23 960	51 884	3.70
1965	47 476	49 578	58 668	44 985	41 955	24 103	53 229	3.69
1966	49 983	52 013	61 357	47 302	42 683	26 610	55 351	3.67
1967	51 048	54 317	64 066	48 976	43 848	27 632	57 876	3.63
1968	53 450	56 621	66 169	50 868	45 332	27 722	60 893	3.60
1969	55 916	59 283	68 934	52 632	49 437	28 584	62 970	3.58
1970	55 743	59 410	69 353	52 563	50 913	28 773	63 087	3.57
1971	55 665	59 481	69 564	52 737	47 206	27 678	62 923	3.53
1972	58 412	62 547	73 025	55 469	54 150	28 071	67 303	3.48
1973	59 595	64 426	75 350	56 464	53 121	28 667	67 799	3.44
1974	58 003	62 593	72 924	54 986	52 410	29 168	67 295	3.42
1975	56 991	61 760	71 606	52 974	53 984	28 431	65 836	3.39
1976	58 758	63 648	73 579	54 724	50 516	28 326	68 016	3.37
1977	59 153	65 091	74 890	55 658	53 644	28 692	69 181	3.33
1978	60 997	66 875	76 450	55 865	55 208	29 520	70 637	3.31
1979	61 863	67 680	78 520	55 922	53 086	31 205	71 101	3.29
1980	59 711	65 727	76 344	53 886	49 759	29 562	69 110	3.27
1981	58 103	65 050	75 903	52 749	51 617	28 444	68 188	3.25
1982	57 351	63 680	74 261	52 128	49 292	28 107	67 596	3.26
1983	57 694	64 046	75 362	51 380	51 275	27 671	68 501	3.24
1984	59 565	66 729	78 122	53 140	52 561	28 851	70 075	3.23
1985	60 424	67 756	79 370	53 499	49 285	29 760	71 409	3.21
1986	63 074	70 241	82 105	55 248	53 448	29 220	74 333	3.19
1987	64 106	72 198	84 353	55 144	52 179	30 393	76 766	3.17
1988	64 311	72 697	85 323	54 380	53 595	30 658	78 016	3.16
1989	65 487	73 783	86 644	55 025	53 302	31 472	78 025	3.17
1990	64 489	72 775	85 328	55 208	52 984	30 887	75 613	3.18
1991	63 257	72 156	84 783	52 936	49 901	29 380	75 784	3.17
1992	62 811	71 942	85 484	51 821	47 359	29 239	75 997	3.19
1993	61 913	72 041	85 776	50 620	44 337	29 220	75 653	3.20
1994	63 638	73 774	87 475	51 157	45 537	29 924	77 142	3.19
1995	65 071	75 408	89 446	51 875	48 643	31 551	79 614	3.20
1996	66 019	77 580	91 118	52 672	49 319	31 076	80 406	3.19
1997	68 087	78 817	92 685	55 039	50 354	32 117	81 504	3.18
1998	70 447	81 666	96 092	56 013	53 782	33 406	84 501	3.18
1999	72 069	83 389	98 113	56 792	55 108	35 070	88 191	3.15
2000	72 417	84 360	98 829	57 072	53 853	36 708	89 458	3.14
2001	71 349	83 741	98 312	56 602	50 784	35 732	87 825	3.15
2002	70 615	83 527	99 481	54 795	51 566	36 104	85 716	3.13
2003	70 383	83 210	100 431	54 941	50 813	35 472	86 967	3.13
2004	70 328	82 771	99 979	54 918	52 506	35 084	85 889	3.13
2005	70 708	82 928	99 096	55 939	51 729	34 281	88 472	3.13
2006	71 185	84 588	100 900	55 767	50 998	35 136	89 476	3.13
2007	72 716	86 030	102 439	56 092	52 571	35 906	89 687	3.15
2008	70 215	83 023	98 862	55 356	49 729	34 387	87 277	3.15
2009	68 819	82 035	98 437	54 573	47 531	34 096	85 218	3.16
2010	67 869	81 395	98 472	54 908	48 681	32 849	84 222	3.14
2011	66 601	80 600	97 232	55 066	47 044	33 051	82 544	3.13
2012	66 575	80 794	98 170	54 424	45 307	32 823	85 247	3.12
2013	69 007	80 462	99 393	54 639	46 877	33 104	84 696	3.14
2014	69 062	83 761	103 629	56 776	49 333	32 928	86 454	3.14
2015	73 149	87 248	107 295	57 952	51 498	35 310	91 199	3.14
2016	74 271	88 678	108 363	59 956	52 677	37 446	92 698	3.14
2017	75 938	90 148	110 893	61 901	52 950	37 098	94 799	3.14

[1]No spouse present.
. . . = Not available.

Table 3-3. Shares of Aggregate Income Received by Each Fifth and Top 5 Percent of Households

Year	Money income							Equivalence-adjusted income						
	Share of aggregate income (percent)						Gini coefficient	Share of aggregate income (percent)						Gini coefficient
	Lowest fifth	Second fifth	Third fifth	Fourth fifth	Highest fifth	Top 5 percent		Lowest fifth	Second fifth	Third fifth	Fourth fifth	Highest fifth	Top 5 percent	
1967	4.0	10.8	17.3	24.2	43.6	17.2	0.397	5.6	12.0	17.1	23.2	42.1	. . .	0.362
1968	4.2	11.1	17.6	24.5	42.6	16.3	0.386	5.8	12.3	17.4	23.4	41.1	. . .	0.351
1969	4.1	10.9	17.5	24.5	43.0	16.6	0.391	5.8	12.2	17.3	23.4	41.3	. . .	0.353
1970	4.1	10.8	17.4	24.5	43.3	16.6	0.394	5.7	12.1	17.3	23.4	41.5	. . .	0.357
1971	4.1	10.6	17.3	24.5	43.5	16.7	0.396	5.7	12.0	17.2	23.4	41.7	. . .	0.359
1972	4.1	10.4	17.0	24.5	43.9	17.0	0.401	5.6	11.9	17.2	23.4	41.9	. . .	0.362
1973	4.2	10.4	17.0	24.5	43.9	16.9	0.400	5.6	12.0	17.2	23.5	41.7	. . .	0.360
1974	4.3	10.6	17.0	24.6	43.5	16.5	0.395	5.8	12.1	17.3	23.6	41.2	. . .	0.354
1975	4.3	10.4	17.0	24.7	43.6	16.5	0.397	5.6	11.9	17.3	23.6	41.6	. . .	0.359
1976	4.3	10.3	17.0	24.7	43.7	16.6	0.398	5.6	11.8	17.4	23.8	41.5	. . .	0.359
1977	4.2	10.2	16.9	24.7	44.0	16.8	0.402	5.5	11.7	17.3	23.7	41.7	. . .	0.362
1978	4.2	10.2	16.8	24.7	44.1	16.8	0.402	5.4	11.8	17.3	23.7	41.8	. . .	0.363
1979	4.1	10.2	16.8	24.6	44.2	16.9	0.404	5.3	11.7	17.2	23.8	41.9	. . .	0.366
1980	4.2	10.2	16.8	24.7	44.1	16.5	0.403	5.2	11.6	17.3	24.0	41.9	. . .	0.367
1981	4.1	10.1	16.7	24.8	44.3	16.5	0.406	5.0	11.4	17.2	24.0	42.4	. . .	0.373
1982	4.0	10.0	16.5	24.5	45.0	17.0	0.412	4.7	11.1	17.0	23.9	43.2	. . .	0.384
1983	4.0	9.9	16.4	24.6	45.1	17.0	0.414	4.6	11.0	16.9	24.0	43.5	. . .	0.389
1984	4.0	9.9	16.3	24.6	45.2	17.1	0.415	4.6	11.0	16.8	24.0	43.6	. . .	0.389
1985	3.9	9.8	16.2	24.4	45.6	17.6	0.419	4.6	10.9	16.7	23.7	44.1	. . .	0.394
1986	3.8	9.7	16.2	24.3	46.1	18.0	0.425	4.5	10.8	16.6	23.8	44.3	. . .	0.397
1987	3.8	9.6	16.1	24.3	46.2	18.2	0.426	4.4	10.8	16.7	23.8	44.4	. . .	0.399
1988	3.8	9.6	16.0	24.2	46.3	18.3	0.426	4.4	10.7	16.5	23.7	44.7	. . .	0.402
1989	3.8	9.5	15.8	24.0	46.8	18.9	0.431	4.4	10.5	16.3	23.4	45.4	. . .	0.408
1990	3.8	9.6	15.9	24.0	46.6	18.5	0.428	4.4	10.6	16.3	23.5	45.1	. . .	0.406
1991	3.8	9.6	15.9	24.2	46.5	18.1	0.428	4.3	10.6	16.5	23.7	45.0	. . .	0.406
1992	3.8	9.4	15.8	24.2	46.9	18.6	0.433	4.1	10.3	16.3	23.7	45.5	. . .	0.413
1993	3.6	9.0	15.1	23.5	48.9	21.0	0.454	3.9	9.8	15.6	23.0	47.7	. . .	0.436
1994	3.6	8.9	15.0	23.4	49.1	21.2	0.456	4.0	9.8	15.6	22.8	47.8	. . .	0.436
1995	3.7	9.1	15.2	23.3	48.7	21.0	0.450	4.1	9.9	15.6	22.8	47.6	. . .	0.433
1996	3.6	9.0	15.1	23.3	49.0	21.4	0.455	4.0	9.8	15.5	22.7	47.9	. . .	0.437
1997	3.6	8.9	15.0	23.2	49.4	21.7	0.459	4.0	9.8	15.4	22.6	48.3	. . .	0.440
1998	3.6	9.0	15.0	23.2	49.2	21.4	0.456	4.0	9.8	15.4	22.7	48.1	. . .	0.439
1999	3.6	8.9	14.9	23.2	49.4	21.5	0.458	4.0	9.7	15.3	22.6	48.4	. . .	0.441
2000	3.6	8.9	14.8	23.0	49.8	22.1	0.462	4.1	9.8	15.2	22.3	48.6	. . .	0.442
2001	3.5	8.7	14.6	23.0	50.1	22.4	0.466	4.0	9.6	15.2	22.4	48.8	. . .	0.446
2002	3.5	8.8	14.8	23.3	49.7	21.7	0.462	4.0	9.6	15.2	22.7	48.4	. . .	0.443
2003	3.4	8.7	14.8	23.4	49.8	21.4	0.464	3.9	9.5	15.2	22.8	48.6	. . .	0.445
2004	3.4	8.7	14.7	23.2	50.1	21.8	0.466	3.8	9.6	15.2	22.7	48.7	. . .	0.447
2005	3.4	8.6	14.6	23.0	50.4	22.2	0.469	3.8	9.5	15.1	22.6	49.1	. . .	0.450
2006	3.4	8.6	14.5	22.9	50.5	22.3	0.470	3.8	9.4	14.9	22.5	49.3	. . .	0.452
2007	3.4	8.7	14.8	23.4	49.7	21.2	0.463	3.8	9.5	15.3	22.9	48.5	. . .	0.444
2008	3.4	8.6	14.7	23.3	50.0	21.5	0.466	3.7	9.4	15.1	22.8	48.9	21.4	0.450
2009	3.4	8.6	14.6	23.2	50.3	21.7	0.468	3.6	9.3	15.0	22.9	49.4	21.7	0.456
2010	3.3	8.5	14.6	23.4	50.3	21.3	0.470	3.4	9.2	15.0	23.1	49.2	21.0	0.456
2011	3.2	8.4	14.3	23.0	51.1	22.3	0.477	3.4	9.0	14.8	22.8	50.0	22.1	0.463
2012	3.2	8.3	14.4	23.0	51.0	22.3	0.477	3.4	9.0	14.8	22.9	49.9	22.1	0.463
2013	3.1	8.2	14.3	23.0	51.4	22.2	0.482	3.5	8.8	14.7	22.8	50.3	22.1	0.467
2014	3.1	8.2	14.3	23.2	51.2	21.9	0.480	3.3	9.0	14.8	22.9	50.0	21.8	0.464
2015	3.1	8.2	14.3	23.2	51.1	22.1	0.479	3.4	9.0	14.8	22.9	49.8	21.8	0.462
2016	3.1	8.3	14.2	22.9	51.5	22.5	0.481	3.5	9.1	14.7	22.5	50.2	22.4	0.464
2017	3.1	8.2	14.3	23.0	51.5	22.3	0.482	3.5	9.0	14.7	22.7	50.1	21.8	0.463

. . . = Not available.

Table 3-4. Shares of Aggregate Income Received by Each Fifth and Top 5 Percent of Families

Year	Number of families (thousands)	Share of aggregate income (percent)						Mean family income (2017 dollars)						Gini coefficient
		Lowest fifth	Second fifth	Third fifth	Fourth fifth	Highest fifth	Top 5 percent	Lowest fifth	Second fifth	Third fifth	Fourth fifth	Highest fifth	Top 5 percent	
1950	39 929	4.5	12.0	17.4	23.4	42.7	17.3	...	...	...	...	...	...	0.379
1951	40 578	5.0	12.4	17.6	23.4	41.6	16.8	...	...	...	...	...	...	0.363
1952	40 832	4.9	12.3	17.4	23.4	41.9	17.4	...	...	...	...	...	...	0.368
1953	41 202	4.7	12.5	18.0	23.9	40.9	15.7	...	...	...	...	...	...	0.359
1954	41 951	4.5	12.1	17.7	23.9	41.8	16.3	...	...	...	...	...	...	0.371
1955	42 889	4.8	12.3	17.8	23.7	41.3	16.4	...	...	...	...	...	...	0.363
1956	43 497	5.0	12.5	17.9	23.7	41.0	16.1	...	...	...	...	...	...	0.358
1957	43 696	5.1	12.7	18.1	23.8	40.4	15.6	...	...	...	...	...	...	0.351
1958	44 232	5.0	12.5	18.0	23.9	40.6	15.4	...	...	...	...	...	...	0.354
1959	45 111	4.9	12.3	17.9	23.8	41.1	15.9	...	...	...	...	...	...	0.361
1960	45 539	4.8	12.2	17.8	24.0	41.3	15.9	...	...	...	...	...	...	0.364
1961	46 418	4.7	11.9	17.5	23.8	42.2	16.6	...	...	...	...	...	...	0.374
1962	47 059	5.0	12.1	17.6	24.0	41.3	15.7	...	...	...	...	...	...	0.362
1963	47 540	5.0	12.1	17.7	24.0	41.2	15.8	...	...	...	...	...	...	0.362
1964	47 956	5.1	12.0	17.7	24.0	41.2	15.9	...	...	...	...	...	...	0.361
1965	48 509	5.2	12.2	17.8	23.9	40.9	15.5	...	...	...	...	...	...	0.356
1966	49 214	5.6	12.4	17.8	23.8	40.5	15.6	15 562	34 547	49 432	66 175	112 939	173 419	0.349
1967	50 111	5.4	12.2	17.5	23.5	41.4	16.4	15 804	35 341	50 797	68 127	120 295	190 326	0.358
1968	50 823	5.6	12.4	17.7	23.7	40.5	15.6	17 035	37 153	53 079	70 986	121 297	186 983	0.348
1969	51 586	5.6	12.4	17.7	23.7	40.6	15.6	17 576	38 815	55 543	74 405	127 429	195 723	0.349
1970	52 227	5.4	12.2	17.6	23.8	40.9	15.6	17 310	38 275	55 342	74 652	128 373	195 477	0.353
1971	53 296	5.5	12.0	17.6	23.8	41.1	15.7	17 325	37 713	55 178	74 728	128 710	196 250	0.355
1972	54 373	5.5	11.9	17.5	23.9	41.4	15.9	18 082	39 453	57 949	79 136	137 248	210 410	0.359
1973	55 053	5.5	11.9	17.5	24.0	41.1	15.5	18 554	40 190	59 051	80 701	138 461	208 960	0.356
1974	55 698	5.7	12.0	17.6	24.1	40.6	14.8	18 882	39 858	58 192	79 726	134 339	195 947	0.355
1975	56 245	5.6	11.9	17.7	24.2	40.7	14.9	18 141	38 352	57 025	78 078	131 576	193 013	0.357
1976	56 710	5.6	11.9	17.7	24.2	40.7	14.9	18 580	39 321	58 656	80 143	134 886	197 419	0.358
1977	57 215	5.5	11.7	17.6	24.3	40.9	14.9	18 501	39 547	59 478	82 047	138 222	201 506	0.363
1978	57 804	5.4	11.7	17.6	24.2	41.1	15.1	18 762	40 689	61 097	84 178	142 841	209 318	0.363
1979	59 550	5.4	11.6	17.5	24.1	41.4	15.3	18 944	40 958	61 768	85 039	145 859	215 905	0.365
1980	60 309	5.3	11.6	17.6	24.4	41.1	14.6	18 209	39 639	59 882	83 078	139 878	198 848	0.365
1981	61 019	5.3	11.4	17.5	24.6	41.2	14.4	17 656	38 394	58 723	82 446	138 335	193 300	0.369
1982	61 393	5.0	11.3	17.2	24.4	42.2	15.3	16 655	37 789	57 701	81 855	141 441	202 366	0.380
1983	62 015	4.9	11.2	17.2	24.5	42.4	15.3	16 400	37 734	58 056	82 741	143 466	206 850	0.382
1984	62 706	4.8	11.1	17.1	24.5	42.5	15.4	16 948	38 804	59 869	85 612	148 792	215 126	0.383
1985	63 558	4.8	11.0	16.9	24.3	43.1	16.1	17 144	39 340	60 738	87 063	154 725	231 221	0.389
1986	64 491	4.7	10.9	16.9	24.1	43.4	16.5	17 639	40 727	63 096	90 154	162 382	246 587	0.392
1987	65 204	4.6	10.7	16.8	24.0	43.8	17.2	17 609	41 215	63 999	91 596	167 393	262 044	0.393
1988	65 837	4.6	10.7	16.7	24.0	44.0	17.2	17 770	41 382	64 397	92 484	169 692	265 138	0.395
1989	66 090	4.6	10.6	16.5	23.7	44.6	17.9	18 115	42 145	65 474	94 199	177 367	284 126	0.401
1990	66 322	4.6	10.8	16.6	23.8	44.3	17.4	17 937	41 837	64 433	92 662	172 207	270 201	0.396
1991	67 173	4.5	10.7	16.6	24.1	44.2	17.1	17 133	40 668	63 102	91 521	168 144	260 175	0.397
1992	68 216	4.3	10.5	16.5	24.0	44.7	17.6	16 463	39 708	62 732	91 184	169 684	267 156	0.404
1993	68 506	4.1	9.9	15.7	23.3	47.0	20.3	16 315	39 182	62 092	92 044	185 973	320 983	0.429
1994	69 313	4.2	10.0	15.7	23.3	46.9	20.1	17 044	40 325	63 680	94 132	189 702	325 451	0.426
1995	69 597	4.4	10.1	15.8	23.2	46.5	20.0	18 050	41 588	65 113	95 268	191 401	328 254	0.421
1996	70 241	4.2	10.0	15.8	23.1	46.8	20.3	17 774	41 901	66 280	96 847	196 072	339 235	0.425
1997	70 884	4.2	9.9	15.7	23.0	47.2	20.7	18 420	43 161	68 098	99 856	205 150	359 046	0.429
1998	71 551	4.2	9.9	15.7	23.0	47.3	20.7	18 881	44 438	70 334	103 145	212 298	371 581	0.430
1999	73 206	4.3	9.9	15.6	23.0	47.2	20.3	19 641	45 655	72 063	106 362	217 990	375 855	0.429
2000	73 778	4.3	9.8	15.4	22.7	47.7	21.1	20 158	46 091	72 438	106 760	223 993	396 919	0.433
2001	74 340	4.2	9.7	15.4	22.9	47.7	21.0	19 460	45 061	71 531	106 379	221 574	389 053	0.435
2002	75 616	4.2	9.7	15.5	23.0	47.6	20.8	19 153	44 436	70 873	105 410	217 663	380 936	0.434
2003	76 232	4.1	9.6	15.5	23.2	47.6	20.5	18 532	43 951	70 832	106 500	218 206	376 053	0.436
2004	76 866	4.0	9.6	15.4	23.0	47.9	20.9	18 472	43 967	70 565	105 426	219 414	382 253	0.438
2005	77 418	4.0	9.6	15.3	22.9	48.1	21.1	18 581	44 212	70 749	105 815	221 824	388 350	0.440
2006	78 454	4.0	9.5	15.1	22.9	48.5	21.5	18 939	44 813	71 280	107 732	228 384	404 269	0.444
2007	77 908	4.1	9.7	15.6	23.3	47.3	20.1	19 043	45 396	72 821	108 894	221 067	375 243	0.432
2008	78 874	4.0	9.6	15.5	23.1	47.8	20.5	18 154	43 513	70 285	105 184	217 308	373 130	0.438
2009	78 867	3.9	9.4	15.3	23.2	48.2	20.7	17 511	42 428	68 612	104 179	217 019	372 250	0.443
2010	79 559	3.8	9.4	15.4	23.5	47.9	20.0	16 814	41 588	67 784	103 417	210 831	352 523	0.440
2011	80 529	3.8	9.3	15.1	23.0	48.9	21.3	16 692	41 022	66 664	101 836	216 198	376 326	0.450
2012	80 944	3.8	9.2	15.1	23.0	48.9	21.3	16 616	40 843	66 813	102 122	216 663	376 871	0.451
2013	82 316	3.6	9.1	15.0	22.9	49.4	21.2	16 979	41 648	67 368	102 458	217 851	378 098	0.455
2014	81 730	3.6	9.2	15.1	23.2	48.9	20.8	16 697	42 164	69 338	106 875	224 934	383 579	0.452
2015	82 199	3.7	9.2	15.2	23.2	48.6	20.9	17 969	44 181	72 945	111 246	233 092	400 244	0.448
2016	82 854	3.7	9.2	15.0	22.9	49.2	21.4	18 593	45 910	74 472	113 642	244 636	426 060	0.452
2017	83 103	3.8	9.2	15.1	23.1	48.8	20.7	18 944	46 346	75 840	115 834	245 039	416 639	0.449

... = Not available.

SECTION 3B: POVERTY

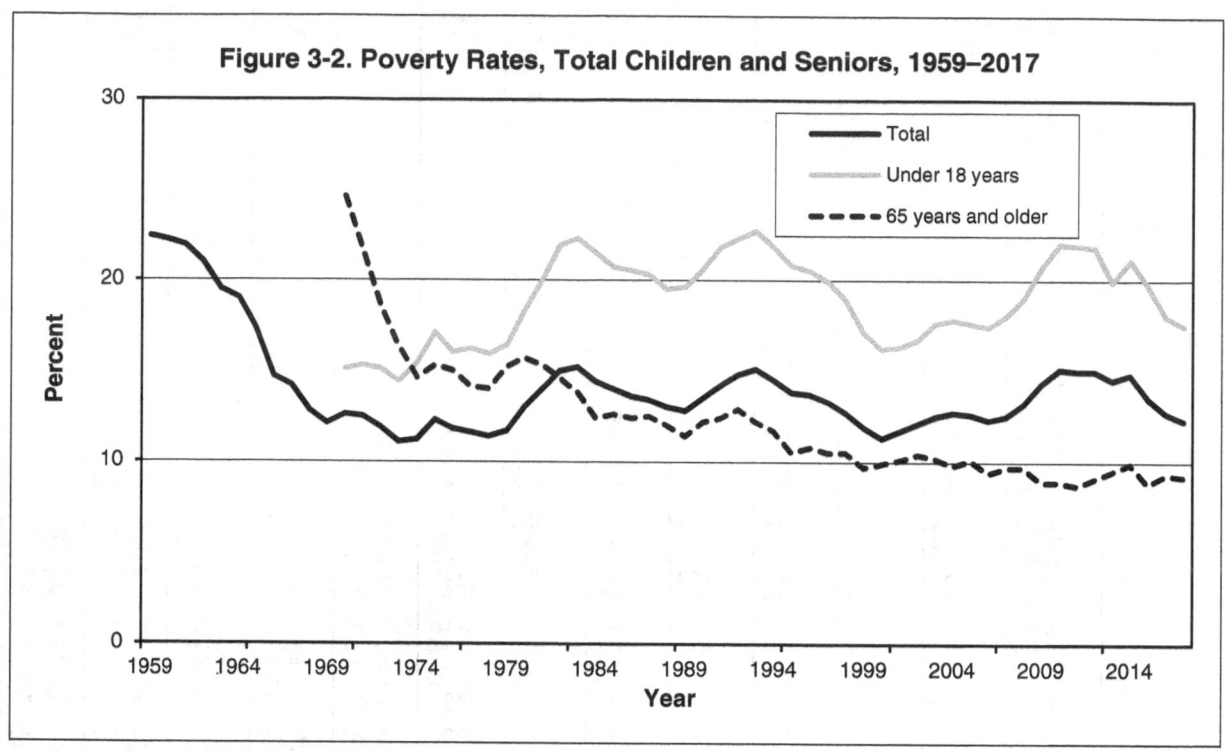

Figure 3-2. Poverty Rates, Total Children and Seniors, 1959–2017

- In 2017, the official poverty rate was 12.3 percent, down 0.4 percentage points from 12.7 percent in 2016. This is the second consecutive annual decline in poverty. Since 2010, the poverty rate has fallen from 15.1 percent to 12.3 percent. There were nearly 39.7 million people in poverty in 2017 down nearly 46.7 million in 2014. (Table 3-6)

- Poverty thresholds vary by family size and age. For example, for a family of three, the poverty threshold was $19,515 in 2017 while for a family of six, the poverty level was $33,618. These levels are adjusted yearly using the Consumer Price Index for all Urban Consumers. (Table 3-5)

- According to the official calculation, the poverty rate for married couples with children under 18 was 6.4 percent. The number of persons aged 65 and over in poverty declined slightly in 2017 after increasing 0.5 percent in 2016. In 1959, the first poverty rate calculations showed more than one-third of all seniors living in poverty, compared with 17.0 percent of working-age adults and 27.3 percent of children. Between 1959 and 1974, ad hoc legislative changes raised Social Security benefits by a cumulative 104 percent—exceeding the 69 percent increase in consumer prices—and since then, each year's benefits have been indexed to the rate of change in the Consumer Price Index, Urban Wage Earners and Clerical Workers (CPI-W). As a result, since the early 1980s, officially-measured senior poverty has been consistently lower than the population average. (Table 3-7 and Table 3-8)

- In 2016, the official poverty rate for Hispanics (who may be of any race) was 18.3 percent compared with 8.7 percent for non-Hispanic Whites, 21.2 percent for Blacks, and 10.0 percent for Asians. (Table 3-6)

- Workers who worked year-round and had full-time jobs made up 8.8 percent of the total poor in 2017, an increase of 8.5 percent in 2016. (Table 3-9)

Table 3-5. Weighted Average Poverty Thresholds by Family Size

(Dollars.)

Year	Unrelated individuals			Families of 2 people			Families, all ages							CPI-U, all items (1982–1984 = 100)
	All ages	Under 65 years	65 years and older	All ages	House-holder under 65 years	House-holder 65 years and older	3 people	4 people	5 people	6 people	7 people	8 people	9 people or more	
1959	1 467	1 503	1 397	1 894	1 952	1 761	2 324	2 973	3 506	3 944	...	...	...	29.2
1960	1 490	1 526	1 418	1 924	1 982	1 788	2 359	3 022	3 560	4 002	...	...	...	29.6
1961	1 506	1 545	1 433	1 942	2 005	1 808	2 383	3 054	3 597	4 041	...	...	...	29.9
1962	1 519	1 562	1 451	1 962	2 027	1 828	2 412	3 089	3 639	4 088	...	...	...	30.3
1963	1 539	1 581	1 470	1 988	2 052	1 850	2 442	3 128	3 685	4 135	...	...	...	30.6
1964	1 558	1 601	1 488	2 015	2 079	1 875	2 473	3 169	3 732	4 193	...	...	...	31.0
1965	1 582	1 626	1 512	2 048	2 114	1 906	2 514	3 223	3 797	4 264	...	...	...	31.5
1966	1 628	1 674	1 556	2 107	2 175	1 961	2 588	3 317	3 908	4 388	...	...	...	32.5
1967	1 675	1 722	1 600	2 168	2 238	2 017	2 661	3 410	4 019	4 516	...	...	...	33.4
1968	1 748	1 797	1 667	2 262	2 333	2 102	2 774	3 553	4 188	4 706	...	...	...	34.8
1969	1 840	1 893	1 757	2 383	2 458	2 215	2 924	3 743	4 415	4 958	...	...	...	36.7
1970	1 954	2 010	1 861	2 525	2 604	2 348	3 099	3 968	4 680	5 260	...	...	...	38.8
1971	2 040	2 098	1 940	2 633	2 716	2 448	3 229	4 137	4 880	5 489	...	...	...	40.5
1972	2 109	2 168	2 005	2 724	2 808	2 530	3 339	4 275	5 044	5 673	...	...	...	41.8
1973	2 247	2 307	2 130	2 895	2 984	2 688	3 548	4 540	5 358	6 028	...	...	...	44.4
1974	2 495	2 562	2 364	3 211	3 312	2 982	3 936	5 038	5 950	6 699	...	...	...	49.3
1975	2 724	2 797	2 581	3 506	3 617	3 257	4 293	5 500	6 499	7 316	...	...	...	53.8
1976	2 884	2 959	2 730	3 711	3 826	3 445	4 540	5 815	6 876	7 760	...	...	...	56.9
1977	3 075	3 152	2 906	3 951	4 072	3 666	4 833	6 191	7 320	8 261	...	...	...	60.6
1978	3 311	3 392	3 127	4 249	4 383	3 944	5 201	6 662	7 880	8 891	...	...	...	65.2
1979	3 689	3 778	3 479	4 725	4 878	4 390	5 784	7 412	8 775	9 914	...	...	...	72.6
1980	4 190	4 290	3 949	5 363	5 537	4 983	6 565	8 414	9 966	11 269	12 761	14 199	16 896	82.4
1981	4 620	4 729	4 359	5 917	6 111	5 498	7 250	9 287	11 007	12 449	14 110	15 655	18 572	90.9
1982	4 901	5 019	4 626	6 281	6 487	5 836	7 693	9 862	11 684	13 207	15 036	16 719	19 698	96.5
1983	5 061	5 180	4 775	6 483	6 697	6 023	7 938	10 178	12 049	13 630	15 500	17 170	20 310	99.6
1984	5 278	5 400	4 979	6 762	6 983	6 282	8 277	10 609	12 566	14 207	16 096	17 961	21 247	103.9
1985	5 469	5 593	5 156	6 998	7 231	6 503	8 573	10 989	13 007	14 696	16 656	18 512	22 083	107.6
1986	5 572	5 701	5 255	7 138	7 372	6 630	8 737	11 203	13 259	14 986	17 049	18 791	22 497	109.6
1987	5 778	5 909	5 447	7 397	7 641	6 872	9 056	11 611	13 737	15 509	17 649	19 515	23 105	113.6
1988	6 022	6 155	5 674	7 704	7 958	7 157	9 435	12 092	14 304	16 146	18 232	20 253	24 129	118.3
1989	6 310	6 451	5 947	8 076	8 343	7 501	9 885	12 674	14 990	16 921	19 162	21 328	25 480	124.0
1990	6 652	6 800	6 268	8 509	8 794	7 905	10 419	13 359	15 792	17 839	20 241	22 582	26 848	130.7
1991	6 932	7 086	6 532	8 865	9 165	8 241	10 860	13 924	16 456	18 587	21 058	23 582	27 942	136.2
1992	7 143	7 299	6 729	9 137	9 443	8 487	11 186	14 335	16 952	19 137	21 594	24 053	28 745	140.3
1993	7 363	7 518	6 930	9 414	9 728	8 740	11 522	14 763	17 449	19 718	22 383	24 838	29 529	144.5
1994	7 547	7 710	7 108	9 661	9 976	8 967	11 821	15 141	17 900	20 235	22 923	25 427	30 300	148.2
1995	7 763	7 929	7 309	9 933	10 259	9 219	12 158	15 569	18 408	20 804	23 552	26 237	31 280	152.4
1996	7 995	8 163	7 525	10 233	10 564	9 491	12 516	16 036	18 952	21 389	24 268	27 091	31 971	156.9
1997	8 183	8 350	7 698	10 473	10 805	9 712	12 802	16 400	19 380	21 886	24 802	27 593	32 566	160.5
1998	8 316	8 480	7 818	10 634	10 972	9 862	13 003	16 660	19 680	22 228	25 257	28 166	33 339	163.0
1999	8 499	8 667	7 990	10 864	11 213	10 075	13 289	17 030	20 128	22 730	25 918	28 970	34 436	166.6
2000	8 791	8 959	8 259	11 235	11 589	10 418	13 740	17 604	20 815	23 533	26 750	29 701	35 150	172.2
2001	9 039	9 214	8 494	11 569	11 920	10 715	14 128	18 104	21 405	24 195	27 517	30 627	36 286	177.1
2002	9 183	9 359	8 628	11 756	12 110	10 885	14 348	18 392	21 744	24 576	28 001	30 907	37 062	179.9
2003	9 393	9 573	8 825	12 015	12 384	11 133	14 680	18 810	22 245	25 122	28 544	31 589	37 656	184.0
2004	9 646	9 827	9 060	12 335	12 714	11 430	15 066	19 307	22 830	25 787	29 233	32 641	39 062	188.9
2005	9 973	10 160	9 367	12 755	13 145	11 815	15 577	19 971	23 613	26 683	30 249	33 610	40 288	195.3
2006	10 294	10 488	9 669	13 167	13 569	12 201	16 079	20 614	24 382	27 560	31 205	34 774	41 499	201.6
2007	10 590	10 787	9 944	13 540	13 954	12 550	16 530	21 203	25 080	28 323	32 233	35 816	42 739	207.3
2008	10 991	11 201	10 326	14 051	14 489	13 030	17 163	22 025	26 049	29 456	33 529	37 220	44 346	215.3
2009	10 956	11 161	10 289	13 991	14 439	12 982	17 098	21 954	25 991	29 405	33 372	37 252	44 366	214.5
2010	11 137	11 344	10 458	14 216	14 676	13 194	17 373	22 315	26 442	29 904	34 019	37 953	45 224	218.1
2011	11 484	11 702	10 788	14 657	15 139	13 609	17 916	23 021	27 251	30 847	35 085	39 064	46 572	224.9
2012	11 720	11 945	11 011	14 937	15 450	13 892	18 284	23 492	27 827	31 471	35 743	39 688	47 297	229.6
2013	11 888	12 119	11 173	15 142	15 679	14 095	18 552	23 834	28 265	31 925	36 384	40 484	48 065	233.0
2014	12 071	12 316	11 354	15 379	15 934	14 326	18 850	24 230	28 695	32 473	36 927	40 968	49 021	236.7
2015	12 082	12 331	11 367	15 391	15 952	14 342	18 871	24 257	28 741	32 542	36 998	41 029	49 177	237.0
2016	12 228	12 486	11 511	15 569	16 151	14 522	19 105	24 563	29 111	32 928	37 458	41 781	49 721	240.0
2017	12 488	12 752	11 756	15 877	16 493	14 828	19 515	25 094	29 714	33 618	38 173	42 684	50 681	245.1

. . . = Not available.

Table 3-6. Poverty Status of People by Race and Hispanic Origin

(Thousands of people, percent of population.)

Year	Number of people, all races	Below poverty level											
		All races		White		White, not Hispanic		Black		Asian [1]		Hispanic (any race)	
		Number	Poverty rate (percent)	Number	Poverty rate (percent)	Number	Poverty rate (percent)	Number	Poverty rate (percent)	Number	Poverty rate (percent)	Number	Poverty rate (percent)
1970	202 183	25 420	12.6	17 484	9.9	...	...	7 548	33.5	...	...	...	...
1971	204 554	25 559	12.5	17 780	9.9	...	...	7 396	32.5	...	...	...	...
1972	206 004	24 460	11.9	16 203	9.0	...	...	7 710	33.3	...	...	2 414	22.8
1973	207 621	22 973	11.1	15 142	8.4	12 864	7.5	7 388	31.4	...	...	2 366	21.9
1974	209 362	23 370	11.2	15 736	8.6	13 217	7.7	7 182	30.3	...	...	2 575	23.0
1975	210 864	25 877	12.3	17 770	9.7	14 883	8.6	7 545	31.3	...	...	2 991	26.9
1976	212 303	24 975	11.8	16 713	9.1	14 025	8.1	7 595	31.1	...	...	2 783	24.7
1977	213 867	24 720	11.6	16 416	8.9	13 802	8.0	7 726	31.3	...	...	2 700	22.4
1978	215 656	24 497	11.4	16 259	8.7	13 755	7.9	7 625	30.6	...	...	2 607	21.6
1979	222 903	26 072	11.7	17 214	9.0	14 419	8.1	8 050	31.0	...	...	2 921	21.8
1980	225 027	29 272	13.0	19 699	10.2	16 365	9.1	8 579	32.5	...	...	3 491	25.7
1981	227 157	31 822	14.0	21 553	11.1	17 987	9.9	9 173	34.2	...	...	3 713	26.5
1982	229 412	34 398	15.0	23 517	12.0	19 362	10.6	9 697	35.6	...	...	4 301	29.9
1983	231 700	35 303	15.2	23 984	12.1	19 538	10.8	9 882	35.7	...	...	4 633	28.0
1984	233 816	33 700	14.4	22 955	11.5	18 300	10.0	9 490	33.8	...	...	4 806	28.4
1985	236 594	33 064	14.0	22 860	11.4	17 839	9.7	8 926	31.3	...	...	5 236	29.0
1986	238 554	32 370	13.6	22 183	11.0	17 244	9.4	8 983	31.1	...	...	5 117	27.3
1987	240 982	32 221	13.4	21 195	10.4	16 029	8.7	9 520	32.4	1 021	16.1	5 422	28.0
1988	243 530	31 745	13.0	20 715	10.1	15 565	8.4	9 356	31.3	1 117	17.3	5 357	26.7
1989	245 992	31 528	12.8	20 785	10.0	15 599	8.3	9 302	30.7	939	14.1	5 430	26.2
1990	248 644	33 585	13.5	22 326	10.7	16 622	8.8	9 837	31.9	858	12.2	6 006	28.1
1991	251 192	35 708	14.2	23 747	11.3	17 741	9.4	10 242	32.7	996	13.8	6 339	28.7
1992	256 549	38 014	14.8	25 259	11.9	18 202	9.6	10 827	33.4	985	12.7	7 592	29.6
1993	259 278	39 265	15.1	26 226	12.2	18 882	9.9	10 877	33.1	1 134	15.3	8 126	30.6
1994	261 616	38 059	14.5	25 379	11.7	18 110	9.4	10 196	30.6	974	14.6	8 416	30.7
1995	263 733	36 425	13.8	24 423	11.2	16 267	8.5	9 872	29.3	1 411	14.6	8 574	30.3
1996	266 218	36 529	13.7	24 650	11.2	16 462	8.6	9 694	28.4	1 454	14.5	8 697	29.4
1997	268 480	35 574	13.3	24 396	11.0	16 491	8.6	9 116	26.5	1 468	14.0	8 308	27.1
1998	271 059	34 476	12.7	23 454	10.5	15 799	8.2	9 091	26.1	1 360	12.5	8 070	25.6
1999	276 208	32 791	11.9	22 169	9.8	14 735	7.7	8 441	23.6	1 285	10.7	7 876	22.7
2000	278 944	31 581	11.3	21 645	9.5	14 366	7.4	7 982	22.5	1 258	9.9	7 747	21.5
2001	281 475	32 907	11.7	22 739	9.9	15 271	7.8	8 136	22.7	1 275	10.2	7 997	21.4
2002	285 317	34 570	12.1	...	...	...	...	...	...	...	...	8 555	21.8
2003	287 699	35 861	12.5	...	...	...	...	...	...	...	...	9 051	22.5
2004	290 617	37 040	12.7	...	...	...	...	...	...	...	...	9 122	21.9
2005	293 135	36 950	12.6	...	...	...	...	...	...	...	...	9 368	21.8
2006	296 450	36 460	12.3	...	...	...	...	...	...	...	...	9 243	20.6
2007	298 699	37 276	12.5	...	...	...	...	...	...	...	...	9 890	21.5
2008	301 041	39 829	13.2	...	...	...	...	...	...	...	...	10 987	23.2
2009	303 820	43 569	14.3	...	...	...	...	...	...	...	...	12 350	25.3
2010	306 130	46 343	15.1	...	...	...	...	...	...	...	...	13 522	26.5
2011	308 456	46 247	15.0	...	...	...	...	...	...	...	...	13 244	25.3
2012	310 648	46 496	15.0	...	...	...	...	...	...	...	...	13 616	25.6
2013	312 965	45 318	14.5	...	...	...	...	...	...	...	...	12 744	23.5
2014	315 804	46 657	14.8	...	...	...	...	...	...	...	...	13 104	23.6
2015	318 454	43 123	13.5	...	...	...	...	...	...	...	...	12 133	21.4
2016	319 911	40 616	12.7	...	...	...	...	...	...	...	...	11 137	19.4
2017	322 549	39 698	12.3	...	...	...	...	...	...	...	...	10 790	18.3
By race													
Race alone													
2002	...	...	...	23 466	10.2	15 567	8.0	8 602	24.1	1 161	10.1	...	...
2003	...	...	...	24 272	10.5	15 902	8.2	8 781	24.4	1 401	11.8	...	...
2004	...	...	...	25 327	10.8	16 908	8.7	9 014	24.7	1 201	9.8	...	...
2005	...	...	...	24 872	10.6	16 227	8.3	9 168	24.9	1 402	11.1	...	...
2006	...	...	...	24 416	10.3	16 013	8.2	9 048	24.3	1 353	10.3	...	...
2007	...	...	...	25 120	10.5	16 032	8.2	9 237	24.5	1 349	10.2	...	...
2008	...	...	...	26 990	11.2	17 024	8.6	9 379	24.7	1 576	11.8	...	...
2009	...	...	...	29 830	12.3	18 530	9.4	9 944	25.8	1 746	12.5	...	...
2010	...	...	...	31 083	13.0	19 251	9.9	10 746	27.4	1 899	12.2	...	...
2011	...	...	...	30 849	12.8	19 171	9.8	10 929	27.6	1 973	12.3	...	...
2012	...	...	...	30 816	12.7	18 940	9.7	10 911	27.2	1 921	11.7	...	...
2013	...	...	...	29 936	12.3	18 796	9.6	11 041	27.2	1 785	10.5	...	...
2014	...	...	...	31 088	12.7	19 653	10.1	10 755	26.2	2 137	12.0	...	...
2015	...	...	...	28 566	11.6	17 786	9.1	10 020	24.1	2 078	11.4	...	...
2016	...	...	...	27 113	11.0	17 263	8.8	9 234	22.0	1 908	10.1	...	...
2017	...	...	...	26 436	10.7	16 993	8.7	8 993	21.2	1 953	10.0	...	...
Race alone or in combination													
2010	...	...	...	...	...	...	...	11 597	27.4	2 064	12.0	...	...
2011	...	...	...	...	...	...	...	11 730	27.5	2 189	12.3	...	...
2012	...	...	...	...	...	...	...	11 809	27.1	2 072	11.4	...	...
2013	...	...	...	...	...	...	...	11 162	27.1	1 974	10.4	...	...
2014	...	...	...	...	...	...	...	11 581	26.0	2 268	11.5	...	...
2015	...	...	...	...	...	...	...	10 797	23.9	2 234	11.1	...	...
2016	...	...	...	...	...	...	...	9 965	21.8	2 062	9.9	...	...
2017	...	...	...	...	...	...	...	9 820	21.2	2 104	9.8	...	...

[1] For 1987 through 2001, Asian and Pacific Islander.
... = Not available.

Table 3-7. Poverty Status of Families by Type of Family

(Thousands of families, percent.)

Year	Married couple families				Families with no spouse present						Unrelated individuals	
	Number of families		Poverty rate (percent)		Male householder			Female householder				
						Poverty rate (percent)			Poverty rate (percent)		Below poverty level	Poverty rate
	Total	Total below poverty level	Total	With children under 18 years	Familes below poverty level	Total	With children under 18 years	Familes below poverty level	Total	With children under 18 years		
1959	39 335	...	...	...	...	...	...	1 916	42.6	59.9	4 928	46.1
1960	39 624	...	...	...	...	...	...	1 955	42.4	56.3	4 926	45.2
1961	40 405	...	...	...	...	...	...	1 954	42.1	56.0	5 119	45.9
1962	40 923	...	...	...	...	...	...	2 034	42.9	59.7	5 002	45.4
1963	41 311	...	...	...	...	...	...	1 972	40.4	55.7	4 938	44.2
1964	41 648	...	...	...	...	...	...	1 822	36.4	49.7	5 143	42.7
1965	42 107	...	...	...	...	...	...	1 916	38.4	52.2	4 827	39.8
1966	42 553	...	...	...	...	...	...	1 721	33.1	47.1	4 701	38.3
1967	43 292	...	...	...	...	...	...	1 774	33.3	44.5	4 998	38.1
1968	43 842	...	...	...	...	...	...	1 755	32.3	44.6	4 694	34.0
1969	44 436	...	...	...	...	...	...	1 827	32.7	44.9	4 972	34.0
1970	44 739	...	...	...	...	...	...	1 952	32.5	43.8	5 090	32.9
1971	45 752	...	...	...	...	...	...	2 100	33.9	44.9	5 154	31.6
1972	46 314	...	...	...	...	...	...	2 158	32.7	44.5	4 883	29.0
1973	46 812	2 482	5.3	...	154	10.7	...	2 193	32.2	43.2	4 674	25.6
1974	47 069	2 474	5.3	6.0	125	8.9	15.4	2 324	32.1	43.7	4 553	24.1
1975	47 318	2 904	6.1	7.2	116	8.0	11.7	2 430	32.5	44.0	5 088	25.1
1976	47 497	2 606	5.5	6.4	162	10.8	15.4	2 543	33.0	44.1	5 344	24.9
1977	47 385	2 524	5.3	6.3	177	11.1	14.8	2 610	31.7	41.8	5 216	22.6
1978	47 692	2 474	5.2	5.9	152	9.2	14.7	2 654	31.4	42.2	5 435	22.1
1979	49 112	2 640	5.4	6.1	176	10.2	15.5	2 645	30.4	39.6	5 743	21.9
1980	49 294	3 032	6.2	7.7	213	11.0	18.0	2 972	32.7	42.9	6 227	22.9
1981	49 630	3 394	6.8	8.7	205	10.3	14.0	3 252	34.6	44.3	6 490	23.4
1982	49 908	3 789	7.6	9.8	290	14.4	20.6	3 434	36.3	47.8	6 458	23.1
1983	50 081	3 815	7.6	10.1	268	13.2	20.2	3 564	36.0	47.1	6 740	23.1
1984	50 350	3 488	6.9	9.4	292	13.1	18.1	3 498	34.5	45.7	6 609	21.8
1985	50 933	3 438	6.7	8.9	311	12.9	17.1	3 474	34.0	45.4	6 725	21.5
1986	51 537	3 123	6.1	8.0	287	11.4	17.8	3 613	34.6	46.0	6 846	21.6
1987	51 675	3 011	5.8	7.7	340	12.0	16.8	3 654	34.2	45.5	6 857	20.8
1988	52 100	2 897	5.6	7.2	336	11.8	18.0	3 642	33.4	44.7	7 070	20.6
1989	52 317	2 931	5.6	7.3	348	12.1	18.1	3 504	32.2	42.8	6 760	19.2
1990	52 147	2 981	5.7	7.8	349	12.0	18.8	3 768	33.4	44.5	7 446	20.7
1991	52 457	3 158	6.0	8.3	392	13.0	19.6	4 161	35.6	47.1	7 773	21.1
1992	53 090	3 385	6.4	8.6	484	15.8	22.5	4 275	35.4	46.2	8 075	21.9
1993	53 181	3 481	6.5	9.0	488	16.8	22.5	4 424	35.6	46.1	8 388	22.1
1994	53 865	3 272	6.1	8.3	549	17.0	22.6	4 232	34.6	44.0	8 287	21.5
1995	53 570	2 982	5.6	7.5	493	14.0	19.7	4 057	32.4	41.5	8 247	20.9
1996	53 604	3 010	5.6	7.5	531	13.8	20.0	4 167	32.6	41.9	8 452	20.8
1997	54 321	2 821	5.2	7.1	507	13.0	18.7	3 995	31.6	41.0	8 687	20.8
1998	54 778	2 879	5.3	6.9	476	12.0	16.6	3 831	29.9	38.7	8 478	19.9
1999	56 290	2 748	4.9	6.4	485	11.8	16.3	3 559	27.8	35.7	8 400	19.1
2000	56 598	2 637	4.7	6.0	485	11.3	15.3	3 278	25.4	33.0	8 653	19.0
2001	56 755	2 760	4.9	6.1	583	13.1	17.7	3 470	26.4	33.6	9 226	19.9
2002	57 327	3 052	5.3	6.8	564	12.1	16.6	3 613	26.5	33.7	9 618	20.4
2003	57 725	3 115	5.4	7.0	636	13.5	19.1	3 856	28.0	35.5	9 713	20.4
2004	57 983	3 216	5.5	7.0	657	13.4	17.1	3 962	28.3	35.9	9 926	20.4
2005	58 189	2 944	5.1	6.5	669	13.0	17.6	4 044	28.7	36.2	10 425	21.1
2006	58 964	2 910	4.9	6.4	671	13.2	17.9	4 087	28.3	36.5	9 977	20.0
2007	58 395	2 849	4.9	6.7	696	13.6	17.5	4 078	28.3	37.0	10 189	19.7
2008	59 137	3 261	5.5	7.5	723	13.8	17.6	4 163	28.7	37.2	10 710	20.8
2009	58 428	3 409	5.8	8.3	942	16.9	23.7	4 441	29.9	38.5	11 678	22.0
2010	58 667	3 681	6.3	9.0	892	15.8	24.1	4 827	31.7	40.9	12 449	22.9
2011	58 963	3 652	6.2	8.8	950	16.1	21.9	4 894	31.2	40.9	12 416	22.8
2012	59 224	3 705	6.3	8.9	1 023	16.4	22.6	4 793	30.9	40.9	12 558	22.4
2013	59 692	3 476	5.8	7.6	1 008	15.9	19.7	4 646	30.6	39.6	13 181	23.3
2014	60 015	3 735	6.2	8.2	969	15.7	22.0	4 764	30.6	39.8	13 374	23.1
2015	60 258	3 245	5.4	7.5	939	14.9	22.1	4 404	28.2	36.5	12 671	21.5
2016	60 821	3 096	5.1	6.6	847	13.1	17.3	4 138	26.6	35.6	12 336	21.0
2017	61 254	3 005	4.9	6.4	793	12.4	16.2	3 959	25.7	34.4	12 593	20.7

. . . = Not available.

Table 3-8. Poverty Status of People by Sex and Age

(Thousands of people, percent of population.)

Year	Poverty status of people by sex				Poverty status of people by age					
	Males below poverty level		Females below poverty level		Children under 18 years below poverty level		People 18 to 64 years below poverty level		People 65 years and older below poverty level	
	Number (thousands)	Poverty rate (percent)	Number (thousands)	Poverty rate (percent)	Number (thousands)	Poverty rate (percent)	Number (thousands)	Poverty rate (percent)	Number (thousands)	Poverty rate (percent)
1959	...	...	...	...	17 552	27.3	16 457	17.0	5 481	35.2
1966	12 225	13.0	16 265	16.3	12 389	17.6	11 007	10.5	5 114	28.5
1967	11 813	12.5	15 951	15.8	11 656	16.6	10 725	10.0	5 388	29.5
1968	10 793	11.3	14 578	14.3	10 954	15.6	9 803	9.0	4 632	25.0
1969	10 292	10.6	13 978	13.6	9 691	14.0	9 669	8.7	4 787	25.3
1970	10 879	11.1	14 632	14.0	10 440	15.1	10 187	9.0	4 793	24.6
1971	10 708	10.8	14 841	14.1	10 551	15.3	10 735	9.3	4 273	21.6
1972	10 190	10.2	14 258	13.4	10 284	15.1	10 438	8.8	3 738	18.6
1973	9 642	9.6	13 316	12.5	9 642	14.4	9 977	8.3	3 354	16.3
1974	9 945	9.8	13 429	12.5	10 156	15.4	10 132	8.3	3 085	14.6
1975	10 908	10.7	14 970	13.8	11 104	17.1	11 456	9.2	3 317	15.3
1976	10 373	10.1	14 603	13.4	10 273	16.0	11 389	9.0	3 313	15.0
1977	10 340	10.0	14 381	13.0	10 288	16.2	11 316	8.8	3 177	14.1
1978	10 017	9.6	14 480	13.0	9 931	15.9	11 332	8.7	3 233	14.0
1979	10 861	10.1	15 211	13.2	10 377	16.4	12 014	8.9	3 682	15.2
1980	12 207	11.2	17 065	14.7	11 543	18.3	13 858	10.1	3 871	15.7
1981	13 360	12.1	18 462	15.8	12 505	20.0	15 464	11.1	3 853	15.3
1982	14 842	13.4	19 556	16.5	13 647	21.9	17 000	12.0	3 751	14.6
1983	15 296	13.6	20 006	16.8	13 911	22.3	17 767	12.4	3 625	13.8
1984	14 537	12.8	19 163	15.9	13 420	21.5	16 952	11.7	3 330	12.4
1985	14 140	12.3	18 923	15.6	13 010	20.7	16 598	11.3	3 456	12.6
1986	13 721	11.8	18 649	15.2	12 876	20.5	16 017	10.8	3 477	12.4
1987	13 781	11.8	18 439	14.9	12 843	20.3	15 815	10.6	3 563	12.5
1988	13 599	11.5	18 146	14.5	12 455	19.5	15 809	10.5	3 481	12.0
1989	13 366	11.2	18 162	14.4	12 590	19.6	15 575	10.2	3 363	11.4
1990	14 211	11.7	19 373	15.2	13 431	20.6	16 496	10.7	3 658	12.2
1991	15 082	12.3	20 626	16.0	14 341	21.8	17 586	11.4	3 781	12.4
1992	16 222	12.9	21 792	16.6	15 294	22.3	18 793	11.9	3 928	12.9
1993	16 900	13.3	22 365	16.9	15 727	22.7	19 781	12.4	3 755	12.2
1994	16 316	12.8	21 744	16.3	15 289	21.8	19 107	11.9	3 663	11.7
1995	15 683	12.2	20 742	15.4	14 665	20.8	18 442	11.4	3 318	10.5
1996	15 611	12.0	20 918	15.4	14 463	20.5	18 638	11.4	3 428	10.8
1997	15 187	11.6	20 387	14.9	14 113	19.9	18 085	10.9	3 376	10.5
1998	14 712	11.1	19 764	14.3	13 467	18.9	17 623	10.5	3 386	10.5
1999	14 079	10.4	18 712	13.2	12 280	17.1	17 289	10.1	3 222	9.7
2000	13 536	9.9	18 045	12.6	11 587	16.2	16 671	9.6	3 323	9.9
2001	14 327	10.4	18 580	12.9	11 733	16.3	17 760	10.1	3 414	10.1
2002	15 162	10.9	19 408	13.3	12 133	16.7	18 861	10.6	3 576	10.4
2003	15 783	11.2	20 078	13.7	12 866	17.6	19 443	10.8	3 552	10.2
2004	16 399	11.5	20 641	13.9	13 041	17.8	20 545	11.3	3 453	9.8
2005	15 950	11.1	21 000	14.1	12 896	17.6	20 450	11.1	3 603	10.1
2006	16 000	11.0	20 460	13.6	12 827	17.4	20 239	10.8	3 394	9.4
2007	16 302	11.1	20 973	13.8	13 324	18.0	20 396	10.9	3 556	9.7
2008	17 698	12.0	22 131	14.4	14 068	19.0	22 105	11.7	3 656	9.7
2009	19 475	13.0	24 094	15.6	15 451	20.7	24 684	12.9	3 433	8.9
2010	20 893	14.0	25 451	16.3	16 286	22.0	26 499	13.8	3 558	8.9
2011	20 501	13.6	25 746	16.3	16 134	21.9	26 492	13.7	3 620	8.7
2012	20 656	13.6	25 840	16.3	16 073	21.8	26 497	13.7	3 926	9.1
2013	20 119	13.1	25 199	15.8	14 659	19.9	26 429	13.6	4 231	9.5
2014	20 708	13.4	25 949	16.1	15 540	21.1	26 527	13.5	4 590	9.9
2015	19 037	12.2	24 086	14.8	14 509	19.7	24 414	12.4	4 201	8.8
2016	17 685	11.3	22 931	14.0	13 253	18.0	22 795	11.6	4 568	9.3
2017	17 365	11.0	22 333	13.6	12 808	17.5	22 209	11.2	4 681	9.2

. . . = Not available.

Table 3-9. Working-Age Poor People by Work Experience

(Thousands of people, percent of population [poverty rate], percent of total poor people.)

Year	Total number of working-age poor people	Worked Number	Worked Percent of total poor	Worked year-round, full-time Number	Worked year-round, full-time Poverty rate (percent)	Worked year-round, full-time Percent of total poor	Worked less than year-round or full-time Number	Worked less than year-round or full-time Poverty rate (percent)	Worked less than year-round or full-time Percent of total poor	Did not work Number	Did not work Poverty rate (percent)	Did not work Percent of total poor
16 Years and Over												
1978	16 914	6 599	39.0	1 309	. . .	7.7	5 290	. . .	31.3	10 315	. . .	61.0
1979	16 803	6 601	39.3	1 394	. . .	8.3	5 207	. . .	31.0	10 202	. . .	60.7
1980	18 892	7 674	40.6	1 644	. . .	8.7	6 030	. . .	31.9	11 218	. . .	59.4
1981	20 571	8 524	41.4	1 881	. . .	9.1	6 643	. . .	32.3	12 047	. . .	58.6
1982	22 100	9 013	40.8	1 999	. . .	9.0	7 014	. . .	31.7	13 087	. . .	59.2
1983	22 741	9 329	41.0	2 064	. . .	9.1	7 265	. . .	31.9	13 412	. . .	59.0
1984	21 541	8 999	41.8	2 076	. . .	9.6	6 923	. . .	32.1	12 542	. . .	58.2
1985	21 243	9 008	42.4	1 972	. . .	9.3	7 036	. . .	33.1	12 235	. . .	57.6
1986	20 688	8 743	42.3	2 007	. . .	9.7	6 736	. . .	32.6	11 945	. . .	57.7
1987	20 546	8 258	40.2	1 821	2.4	8.9	6 436	12.5	31.3	12 288	21.6	59.8
1988	20 323	8 363	41.2	1 929	2.4	9.5	6 434	12.7	31.7	11 960	21.2	58.8
1989	19 952	8 376	42.0	1 908	2.4	9.6	6 468	12.5	32.4	11 576	20.8	58.0
1990	21 242	8 716	41.0	2 076	2.6	9.8	6 639	12.6	31.3	12 526	22.1	59.0
1991	22 530	9 208	40.9	2 103	2.6	9.3	7 105	13.4	31.5	13 322	22.8	59.1
1992	23 951	9 739	40.6	2 211	2.7	9.2	7 529	14.1	31.4	14 212	23.7	59.3
1993	24 832	10 144	40.8	2 408	2.9	9.7	7 737	14.6	31.2	14 688	24.2	59.1
1994	24 108	9 829	40.8	2 520	2.9	10.5	7 309	13.9	30.3	14 279	23.6	59.2
1995	23 077	9 484	41.1	2 418	2.7	10.5	7 066	13.7	30.6	13 593	22.3	58.9
1996	23 472	9 586	40.8	2 263	2.5	9.6	7 322	14.1	31.2	13 886	22.7	59.2
1997	22 753	9 444	41.5	2 345	2.5	10.3	7 098	13.8	31.2	13 309	21.7	58.5
1998	22 256	9 133	41.0	2 804	2.9	12.6	6 330	12.7	28.4	13 123	21.1	59.0
1999	21 762	9 251	42.5	2 559	2.6	11.8	6 692	13.2	30.8	12 511	19.9	57.5
2000	21 080	8 511	40.4	2 439	2.4	11.6	6 072	12.1	28.8	12 569	19.8	59.6
2001	22 245	8 530	38.3	2 567	2.6	11.5	5 964	11.8	26.8	13 715	20.6	61.7
2002	23 601	8 954	37.9	2 635	2.6	11.2	6 318	12.4	26.8	14 647	21.0	62.1
2003	24 266	8 820	36.3	2 636	2.6	10.9	6 183	12.2	25.5	15 446	21.5	63.7
2004	25 256	9 384	37.2	2 891	2.8	11.4	6 493	12.8	25.7	15 872	21.7	62.8
2005	25 381	9 340	36.8	2 894	2.8	11.4	6 446	12.8	25.4	16 041	21.8	63.2
2006	24 896	9 181	36.9	2 906	2.7	11.7	6 275	12.6	25.2	15 715	21.1	63.1
2007	25 297	9 089	35.9	2 768	2.5	10.9	6 320	12.7	25.0	16 208	21.5	64.1
2008	27 216	10 085	37.1	2 754	2.6	10.1	7 331	13.5	26.9	17 131	22.0	62.9
2009	29 625	10 680	36.1	2 641	2.7	8.9	8 039	14.5	27.1	18 945	22.7	63.9
2010	31 731	10 742	33.9	2 640	2.7	8.3	8 102	15.0	25.5	20 989	23.9	66.1
2011	31 630	10 588	33.5	2 770	2.7	8.8	7 818	14.9	24.7	21 042	23.6	66.5
2012	31 933	10 977	34.4	2 904	2.8	9.1	8 073	15.0	25.3	20 956	23.6	65.6
2013	32 011	11 025	34.4	2 812	2.7	8.8	8 213	15.8	25.7	20 986	23.2	65.5
2014	32 508	10 469	32.2	3 142	2.9	9.7	7 327	14.3	22.5	22 039	24.2	67.8
2015	30 123	9 795	32.5	2 576	2.3	8.8	7 220	13.9	24.0	20 328	22.5	67.5
2016	28 696	9 109	31.7	2 448	2.2	8.5	6 661	13.0	21.6	19 587	21.6	68.3
2017	28 167	8 503	30.2	2 468	2.1	8.8	6 035	12.0	21.4	19 644	21.4	69.2

. . . = Not available.

SECTION 3C: ALTERNATIVE MEASURES OF INCOME AND POVERTY

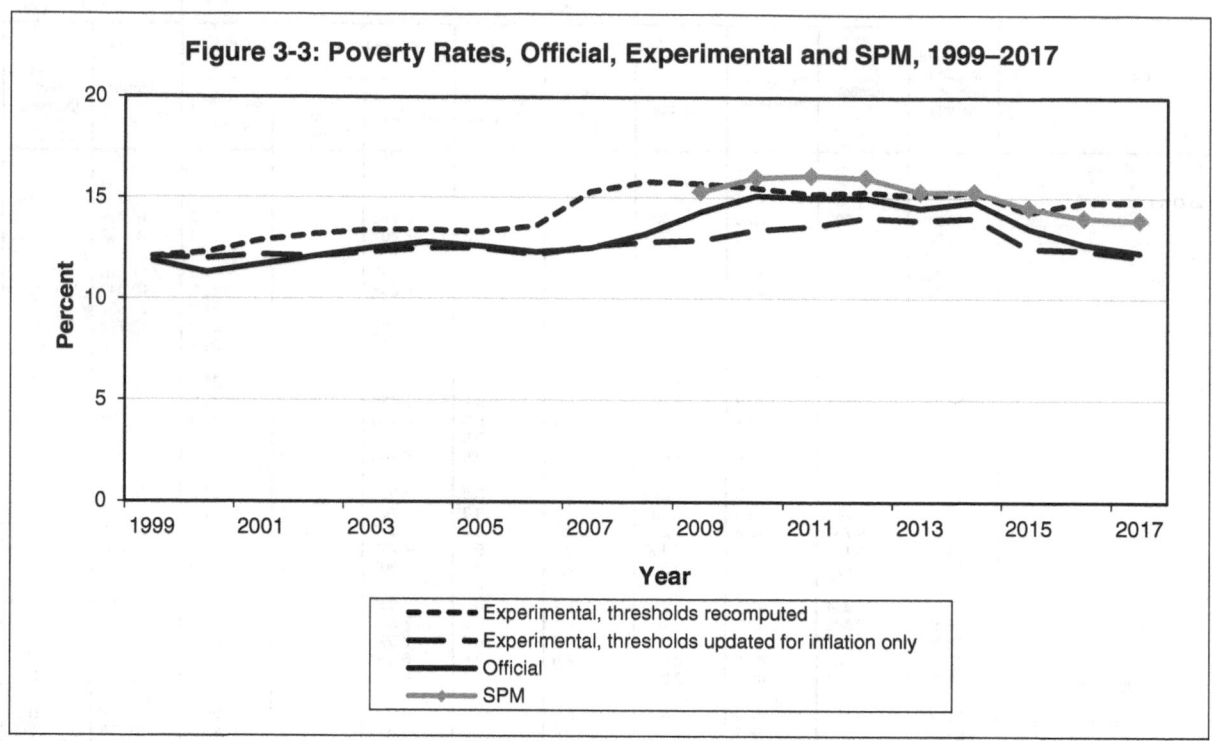

Figure 3-3: Poverty Rates, Official, Experimental and SPM, 1999–2017

- For the years 1999 forward, the Census Bureau has calculated several experimental poverty measures based on recommendations by a special National Academy of Sciences (NAS) panel. And for 2009 through 2017 the Bureau has released a "Supplemental Poverty Measure" (SPM), based on further refinement of these recommendations. In Figure 3-3, two of the experimental measures are shown along with the official rate and the SPM. The SPM and all of the experimental measures use broader and more realistic definitions of both needs and "resources" (income) than the official rate. Also, the SPM and the two experimental measures shown in the graph include adjustment for geographic differences in the cost of living, and adjust for differing health care needs by excluding medical out-of-pocket expenses from the resources of the measurement unit. The thresholds for the experimental measures were initially based on needs estimated for the year 1999. In 2017, the supplemental poverty rate was 13.9 percent while the official rate was 12.3 percent. (Table 3-10)

- The SPM defines poverty as the lack of economic resources for consumption of basic needs such as food, housing, clothing, and utilities (FCSU). To determine family resources, gross money income from private and public sources is supplemented with benefits such as food stamps, housing subsidies, and tax credits. Deducted from family income are medical out-of-pocket expenses including health insurance premiums, income and Social Security payroll taxes, child support payments, work-related expenses and child care costs.

- As seen in Table 3-11, numerous experimental poverty measures have evolved since 1999. These measures address medical out-of-pocket expenses, geographic adjustment, and the measurement of inflation by the Consumer Price Index or the Personal Consumption Expenditures. (Table 3-11)

Table 3-10. Number and Percent of People in Poverty Using the Supplemental Poverty Measure, 2015–2017, and Comparison With Official Measures in 2017

(Numbers in thousands; percent of population.)

Characteristic	SPM 2015		SPM 2016		SPM 2017		Official 2017[1]	
	Number	Percent	Number	Percent	Number	Percent	Number	Percent
All People	46 250	14.5	44 752	14.0	44 972	13.9	39 804	12.3
Sex								
Male	21 678	13.9	20 693	13.2	20 717	13.1	17 427	11.0
Female	24 572	15.1	24 059	14.7	24 255	14.7	22 377	13.6
Age								
Under 18 years	12 026	16.2	11 281	15.2	11 521	15.6	12 914	17.5
18 to 64 years	27 719	14.1	26 303	13.3	26 244	13.2	22 209	11.2
65 years and older	6 506	13.7	7 168	14.5	7 207	14.1	4 681	9.2
Type of Unit								
In married-couple unit	17 341	9.1	16 516	8.6	16 879	8.7	11 020	5.7
In female householder unit	11 623	27.0	11 655	27.3	11 408	26.9	11 111	26.2
In male householder unit	2 683	18.8	2 635	17.5	2 382	16.3	1 641	11.2
Cohabitating	3 970	15.4	3 261	13.0	3 558	13.3	6 729	25.1
Unrelated individuals	10 632	23.7	10 685	23.6	10 745	23.5	9 303	20.4
Race and Hispanic Origin								
White	31 493	12.8	30 717	12.5	30 433	12.3	26 522	10.7
White, not Hispanic	20 082	10.3	19 446	9.9	19 249	9.8	17 037	8.7
Black	9 527	22.8	9 086	21.6	9 394	22.1	9 007	21.2
Asian	2 929	16.1	2 774	14.7	2 948	15.1	1 953	10.0
Hispanic, any race	12 862	22.6	12 670	22.0	12 654	21.4	10 835	18.3
Nativity								
Native born	36 789	13.3	35 515	12.8	35 538	12.8	33 198	12.0
Foreign born	9 461	22.0	9 237	21.1	9 435	20.8	6 607	14.5
Naturalized citizen	3 355	16.7	3 205	15.7	3 513	16.1	2 213	10.1
Not a citizen	6 106	26.6	6 032	25.7	5 921	25.1	4 394	18.7
Educational Attainment								
Total, age 25 and over	27 951	13.0	27 929	12.9	27 801	12.6	22 163	10.1
No high school diploma	6 916	29.5	6 356	28.2	6 429	28.7	5 485	24.5
High school, no college	9 647	15.6	10 139	16.2	10 038	16.0	7 942	12.7
Some college	6 723	11.7	6 615	11.5	6 263	10.8	5 075	8.8
Bachelor's degree or higher	4 665	6.5	4 819	6.5	5 072	6.6	3 661	4.8
Tenure								
Owner	19 460	9.3	19 149	9.1	19 764	9.2	15 185	7.1
Owner/ mortgage	10 323	7.7	10 122	7.4	10 492	7.6	7 152	5.1
Owner/no-mortgage/rent-free	9 985	12.8	9 825	12.7	9 886	12.5	8 718	11.0
Renter	25 942	24.3	24 806	23.3	24 594	23.5	23 934	22.8
Residence								
Inside metropolitan statistical areas	40 298	14.7	39 120	14.1	39 472	14.1	33 408	11.9
Inside principal cities	18 714	18.0	17 971	17.4	18 216	17.5	16 241	15.6
Outside principal cities	21 585	12.6	21 148	12.2	21 257	12.1	17 167	9.8
Outside metropolitan statistical areas	5 951	13.4	5 633	12.9	5 500	12.8	6 396	14.8
Region								
Northeast	8 033	14.4	6 874	12.4	7 976	14.2	6 381	11.4
Midwest	7 401	11.0	7 424	11.1	7 198	10.7	7 661	11.4
South	18 816	15.7	17 966	14.8	18 147	14.8	16 662	13.6
West	12 000	15.8	12 489	16.3	11 652	15.1	9 100	11.8
Health Insurance Coverage								
With private insurance	18 814	8.8	17 898	8.3	17 872	8.2	11 219	5.2
With public, no private insurance	19 658	26.0	19 646	25.8	19 851	25.6	21 838	28.1
Not insured	7 777	26.8	7 208	25.7	7 249	25.4	6 748	23.6
Work Experience								
Total, 18 to 64 years	27 719	14.1	26 303	13.3	26 244	13.2	22 209	11.2
All workers	12 949	8.6	12 111	8.0	12 172	8.0	8 135	5.3
Full-time, year-round	5 251	5.0	5 099	4.7	5 368	4.9	2 422	2.2
Less than full-time, year-round	7 699	17.3	7 012	16.3	6 804	16.0	5 714	13.4
Did not work at least 1 week	14 770	31.4	14 193	30.8	14 072	30.6	14 073	30.7
Disability Status								
Total, 18 to 64 years	27 719	14.1	26 303	13.3	26 244	13.2	22 209	11.2
With a disability	4 054	26.5	3 905	25.4	3 550	23.5	3 764	24.9
With no disability	23 589	13.0	22 350	12.4	22 656	12.4	18 412	10.1

[1]Differs from published official rates because these figures include unrelated individuals under 15 years of age as does the SPM.

Table 3-11. Official and Experimental Poverty Rates and Supplemental Poverty Measure

(Percent of population.)

Measurement method	Capital gains included in income								
	1999	2000	2001	2002 (new tax model)	2003	2004	2005	2006	2007
Official measure	11.9	11.3	11.7	12.1	12.5	12.8	12.6	12.3	12.5
Experimental									
MSI-GA-CPI	12.1	12.0	12.2	12.1	12.3	12.5	12.5	12.2	12.6
MIT-GA-CPI	12.7	12.5	12.5	12.6	12.7	13.0	13.0	12.6	13.0
MSI-NGA-CPI	12.2	12.1	12.3	12.3	12.4	12.7	12.6	12.4	12.6
MIT-NGA-CPI	12.8	12.7	12.7	12.8	12.7	13.1	13.0	12.8	12.9
MSI-GA-CE	12.1	12.3	12.9	13.2	13.4	13.4	13.3	13.6	15.3
MIT-GA-CE	12.7	12.8	13.2	13.7	13.9	14.0	14.1	14.1	15.9
MSI-NGA-CE	12.2	12.5	13.0	13.4	13.5	13.4	13.5	13.7	15.1
MIT-NGA-CE	12.8	13.0	13.4	13.9	14.1	14.1	14.2	14.2	16.0
Supplemental poverty measure	. . .	. . .	. . .	. . .	. . .	. . .	. . .	. . .	. . .

Measurement method	Capital gains and losses excluded from income										
	2007	2008	2009	2010	2011	2012	2013	2014	2015	2016	2017
Official measure	12.5	13.2	14.3	15.1	15.0	15.0	14.5	14.8	13.5	12.7	12.3
Experimental											
MSI-GA-CPI	12.6	12.8	12.9	13.4	13.6	14.0	13.9	14.0	12.5	12.4	12.1
MIT-GA-CPI	13.0	13.2	13.2	13.8	13.9	14.5	14.3	14.4	12.9	12.6	12.4
MSI-NGA-CPI	12.6	12.8	12.8	13.5	13.5	14.0	13.8	14.0	12.7	12.3	12.0
MIT-NGA-CPI	13.0	13.1	13.3	14.0	13.8	14.4	14.4	14.6	13.1	12.5	12.2
MSI-GA-CE	15.3	15.8	15.7	15.5	15.2	15.3	15.1	15.3	14.3	14.8	14.8
MIT-GA-CE	15.9	16.9	17.3	17.2	16.9	16.8	16.6	16.9	16.1	16.6	16.8
MSI-NGA-CE	15.2	15.7	15.7	15.6	15.0	15.3	15.1	15.3	14.4	14.7	14.7
MIT-NGA-CE	16.1	17.1	17.3	17.3	16.9	16.9	16.6	16.9	15.9	16.7	17.0
Supplemental poverty measure	. . .	. . .	15.3	16.0	16.1	16.0	15.3	15.3	14.3	14.0	13.9

Note: The Census Bureau changed the way it modeled taxes in the experimental measures, effective with the revised 2002 estimates. Consequently, comparisons of 2002 and later data with earlier years may be affected.

MSI means "Medical out-of-pocket expenses subtracted from income."
MIT means "Medical out-of-pocket expenses in the thresholds."
GA means "Geographic adjustment (of poverty thresholds)."
NGA means "No geographic adjustment (of poverty thresholds)."
CPI means "Thresholds were adjusted since 1999 using the Consumer Price Index for All Urban Consumers."
CE means "Thresholds were recomputed since 1999 using data from the Consumer Expenditure Survey."

See notes and definitions for further explanation.
. . . = Not available.

SECTION 3D: HEALTH INSURANCE COVERAGE

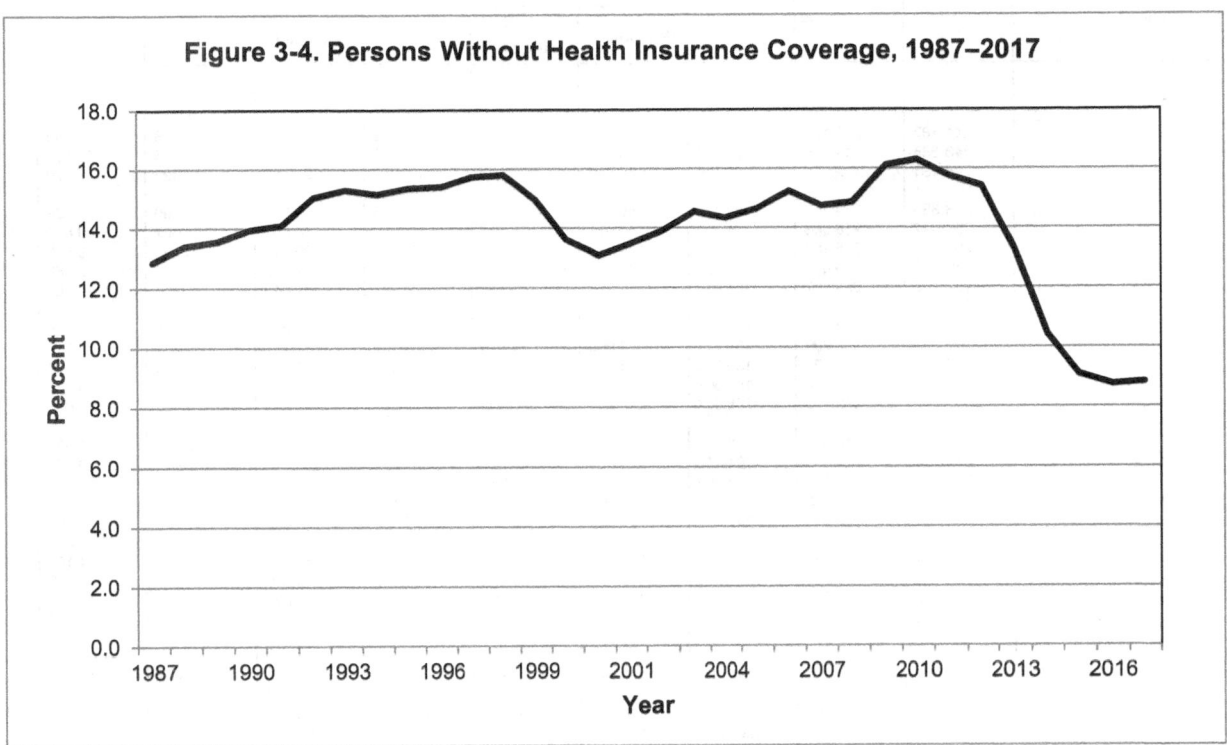

Figure 3-4. Persons Without Health Insurance Coverage, 1987–2017

- Health insurance is a means for financing a person's health care expenses. Over time, changes in the rate of health insurance coverage and the distribution of coverage types may reflect economic trends, shifts in the demographic composition of the populace and policy changes that impact access to care. Such a policy change was the Patient Protection and Affordable Care Act (ACA) in 2010.

- The uninsured rate remained steady at 8.8 percent between 2016 and 2017 as measured by the CPS ASEC. In 2017, 28.5 million people did not have health insurance. Although this was an increase from 2016, it was still much lower than the nearly 50 million people who did not have health insurance in 2010. (Tables 3-12 and 3-13)

- Private health insurance coverage continued to be more prevalent than government coverage, at 67.2 percent and 37.7 percent, respectively in 2017. Of the subtypes of health insurance coverage, employer-based insurance covered 56.0 percent of the population for some or all of the calendar year, followed by Medicaid (19.3 percent), Medicare (17.2 percent), direct-purchase (16.0 percent), and military coverage (4.8 percent). (Table 3-12)

- In 2017, the percentage of uninsured children under age of 19 remained steady at 5.4 percent. The uninsured rate for children in poverty (7.8 percent) was significantly higher than the rate for children not in poverty (4.9 percent). Non-Hispanic Whites had the lowest uninsured rate among race and Hispanic origin groups at 6.3 percent followed by 7.3 percent for Asians and 10.6 percent for Blacks. (Table 3-13)

Table 3-12. Health Insurance Coverage

(Number in thousands. People as of March of the following year.)

Year	Total	Total covered	Private health insurance			Government health insurance				Not covered
			Total	Employment	Direct purchase	Total	Medicaid	Medicare	Military	
NUMBER										
1987	241 187	210 161	182 160	149 739	. . .	56 282	20 211	30 458	10 542	31 026
1988	243 685	211 005	182 019	150 940	. . .	56 850	20 728	30 925	10 105	32 680
1989	246 191	212 807	183 610	151 644	. . .	57 382	21 185	31 495	9 870	33 384
1990	248 886	214 167	182 135	150 215	. . .	60 965	24 261	32 260	9 922	34 719
1991	251 447	216 003	181 375	150 077	. . .	63 882	26 880	32 907	9 820	35 444
1992	256 830	218 189	181 466	148 796	. . .	66 244	29 416	33 230	9 510	38 641
1993	259 753	220 040	182 351	148 318	. . .	68 554	31 749	33 097	9 560	39 713
1994	262 105	222 387	184 318	159 634	31 349	70 163	31 645	33 901	11 165	39 718
1995	264 314	223 733	185 881	161 453	30 188	69 776	31 877	34 655	9 375	40 581
1996	266 792	225 699	188 224	164 096	28 419	69 000	31 451	35 227	8 712	41 093
1997	269 094	226 735	189 955	166 419	27 431	66 685	28 956	35 590	8 527	42 359
1998	271 743	228 800	192 507	170 105	26 165	66 087	27 854	35 887	8 747	42 943
1999	550 891	472 175	398 557	349 227	56 608	133 279	55 243	73 056	17 056	78 716
2000	279 517	242 932	205 575	181 862	28 432	68 183	28 062	37 787	8 937	36 585
2001	282 082	244 059	204 142	179 984	28 398	70 330	30 166	37 870	9 580	38 023
2002	285 933	246 157	204 163	179 563	29 287	72 825	31 934	38 359	9 892	39 776
2003	288 280	246 332	201 989	177 362	28 826	76 119	34 326	39 284	10 124	41 948
2004	291 166	249 414	203 014	177 924	29 161	79 480	38 055	39 757	10 584	41 752
2005	293 834	250 799	203 205	178 391	28 980	80 283	38 191	40 167	11 164	43 035
2006	296 824	251 610	203 942	178 880	29 033	80 343	38 370	40 336	10 543	45 214
2007	299 106	255 018	203 903	178 971	28 500	83 147	39 685	41 387	10 955	44 088
2008	301 483	256 702	202 626	177 543	28 513	87 586	42 831	43 031	11 562	44 781
2009	304 280	255 295	196 245	170 762	29 098	93 245	47 847	43 434	12 414	48 985
2010	306 553	256 603	196 147	169 372	30 347	95 525	48 533	44 906	12 927	49 950
2011	308 827	260 214	197 323	170 102	30 244	99 497	50 835	46 922	13 712	48 613
2012	311 116	263 165	198 812	170 877	30 622	101 493	50 903	48 884	13 702	47 951
2013	313 401	271 606	201 038	174 418	35 755	108 287	54 919	49 020	14 016	41 795
2014	316 168	283 200	208 600	175 027	46 165	115 470	61 650	50 546	14 143	32 968
2015	318 868	289 903	214 238	177 540	52 057	118 395	62 384	51 865	14 849	28 966
2016	320 372	292 320	216 203	178 455	51 961	119 361	62 303	53 372	14 638	28 052
2017	323 156	294 613	217 007	181 036	51 821	121 965	62 492	55 623	15 532	28 543

. . . = Not available.

Table 3-13. Percentage of People by Type of Health Insurance by Selected Demographics, 2016–2017

(Number in thousands. Percent. People as of March of the following year.)

Characteristic	Total 2016	Total 2017	Any health insurance 2016	Any health insurance 2017	Private health insurance 2016	Private health insurance 2017	Government health insurance 2016	Government health insurance 2017	Uninsured 2016	Uninsured 2017
TOTAL	320 372	323 156	91.2	91.2	67.5	67.2	37.3	37.7	8.8	8.8
Age										
Under age 19	78 150	78 106	94.6	94.6	62.9	63.3	41.5	41.9	5.4	5.4
Age 19 to 64	192 948	193 971	87.9	87.8	73.1	72.9	21.1	21.3	12.1	12.2
Age 65 and over	49 274	51 080	98.8	98.7	52.8	51.1	93.6	93.7	1.2	1.3
Family Status										
In families	259 863	260 709	91.8	91.7	68.7	68.3	36.4	36.9	8.2	8.3
Householder	82 854	83 103	91.6	91.2	71.2	70.0	36.3	37.0	8.4	8.8
Related children under 18	72 674	72 532	94.8	94.7	63.0	63.4	41.5	41.8	5.2	5.3
Related children under 6	23 531	23 574	94.2	94.0	58.9	59.2	45.1	44.9	5.8	6.0
In unrelated subfamilies	1 208	1 054	86.5	87.7	48.5	52.5	48.6	45.4	13.5	12.3
Unrelated individuals	59 301	61 393	88.7	88.8	62.8	62.5	40.6	41.2	11.3	11.2
Residence										
Inside metropolitan statistical areas	276 682	280 048	91.3	91.2	68.5	68.0	35.9	36.6	8.7	8.8
Inside principal cities	103 365	104 068	90.2	89.6	64.0	63.1	37.9	38.2	9.8	10.4
Outside principal cities	173 317	175 980	92.0	92.2	71.2	70.8	34.8	35.6	8.0	7.8
Outside metropolitan statistical areas	43 689	43 108	90.6	90.8	61.1	61.9	45.6	45.5	9.4	9.2
Race and Hispanic Origin										
White	246 310	247 695	91.6	91.5	69.4	69.0	36.6	37.1	8.4	8.5
White, not Hispanic	195 453	195 530	93.7	93.7	73.9	73.2	35.9	36.6	6.3	6.3
Black	42 040	42 564	89.5	89.4	56.5	56.5	43.7	44.1	10.5	10.6
Asian	10 007	19 484	92.4	92.7	74.2	72.2	27.1	29.6	7.6	7.3
Hispanic (any race)	57 670	59 227	84.0	83.9	52.4	53.5	40.1	39.5	16.0	16.1
Nativity										
Native born	276 518	277 748	92.7	92.5	68.7	68.2	38.1	38.7	7.3	7.5
Foreign born	43 854	45 408	82.0	83.2	59.9	60.6	31.7	32.0	18.0	16.8
Naturalized citizen	20 409	21 854	91.5	91.1	67.3	65.6	37.2	37.5	8.5	8.9
Not a citizen	23 445	23 554	73.8	75.9	53.5	55.9	27.0	27.0	26.2	24.1
WORKING AGE ADULTS 19 TO 64										
Marital Status										
Married	101 822	101 580	91.2	90.9	80.1	79.7	17.9	18.3	8.8	9.1
Widowed	3 633	3 586	86.1	86.6	58.7	57.2	33.5	36.0	13.9	13.4
Divorced	19 460	19 510	86.1	86.4	64.3	65.4	26.8	26.3	13.9	13.6
Separated	4 495	4 372	80.8	79.7	55.9	55.4	31.0	29.9	19.2	20.3
Never married	63 537	64 923	84.0	84.0	66.5	66.6	23.2	23.1	16.0	16.0
Disability Status										
With a disability	15 248	14 957	91.2	91.2	43.5	44.8	58.6	57.8	8.8	8.8
Without a disability	176 842	178 063	87.6	87.5	75.9	75.5	17.5	17.8	12.4	12.5
Work Experience										
All workers	149 105	150 487	88.8	88.7	80.1	80.2	13.9	14.0	11.2	11.3
Worked full-time, year round	107 577	109 511	90.2	90.2	84.5	84.4	10.4	10.9	9.8	9.8
Less than full-time, year round	41 528	40 976	85.2	84.6	69.0	68.9	23.1	22.4	14.8	15.4
Did not work at least one week	43 843	43 484	85.0	84.9	49.1	47.9	45.6	46.5	15.0	15.1
Educational Attainment										
Total, 26 to 64 years old	163 133	164 049	88.1	88.1	73.4	73.4	20.8	20.9	11.9	11.9
No high school diploma	15 389	15 150	72.7	73.7	40.9	42.4	37.7	37.5	27.3	26.3
High school graduate (includes equivalency)	45 401	44 772	84.8	84.5	65.0	65.4	26.3	26.3	15.2	15.5
Some college, no degree	26 594	26 109	88.4	88.0	71.8	70.6	23.8	24.7	11.6	12.0
Associate's degree	17 739	17 659	90.7	90.5	77.9	77.2	19.5	19.5	9.3	9.5
Bachelor's degree	36 528	38 465	93.2	92.8	86.8	85.5	11.6	12.4	6.8	7.2
Graduate or professional degree	21 482	21 894	95.2	95.8	90.0	90.4	9.8	10.3	4.8	4.2

. . . = Not available.

NOTES AND DEFINITIONS, CHAPTER 3

TABLES 3-1 THROUGH 3-13

Income Distribution and Poverty

SOURCE: U.S. DEPARTMENT OF COMMERCE, BUREAU OF THE CENSUS

All data in this chapter are derived from the Current Population Survey (CPS), which is also the source of the data on labor force, employment, and unemployment used in Chapter 10. (See the notes and definitions for Tables 10-1 through 10-5.) In March of each year (with some data also collected in February and April), the households in this monthly survey are asked additional questions concerning earnings and other income in the previous year. This additional information, informally known as the "March Supplement," is now formally known as the Current Population Survey Annual Social and Economic Supplement (CPS-ASEC). It was previously called the Annual Demographic Supplement.

The population represented by the income and poverty survey is the civilian noninstitutional population of the United States and members of the armed forces in the United States living off post or with their families on post, but excluding all other members of the armed forces. This is slightly different from the population base for the civilian employment and unemployment data, which excludes those armed forces households. As it is a survey of households, homeless persons are not included.

TABLES 3-1 THROUGH 3-10 AND 3-13

Definitions: Racial classification and Hispanic origin

In 2002 and all earlier years, the CPS required respondents to report identification with only one race group. Since 2003, the CPS has allowed respondents to choose more than one race group. Income data for 2002 were collected in early 2003; thus, in the data for 2002 and all subsequent years, an individual could report identification with more than one race group. In the 2000 census, about 2.6 percent of people reported identification with more than one race.

Therefore, data from 2002 onward that are classified by race are not strictly comparable with race-classified data for 2001 and earlier years. As alternative approaches to dealing with this problem, the Census Bureau has tabulated two different race concepts for each racial category in a number of cases. In the case of Blacks, for example, this means there is one income measure for "Black alone," consisting of persons who report Black and no other race, and one for "Black alone or in combination," which includes all the "Black alone" reporters plus those who report Black in combination with any other race. The tables in this volume show both the "alone" and the "alone or in combination" values where available.

The race classifications now used in the CPS are *White, Black, Asian, American Indian and Alaska Native,* and *Native Hawaiian and Other Pacific Islander.* (Native Hawaiians and other Pacific Islanders were included in the "Asian" category in the data for 1987 through 2001.) The last two of these five racial groups are too small to provide reliable data for a single year, but in new Census Bureau tables available on the website, household income and poverty data for all five groups are presented in 2- and 3-year averages. Table 3-1 displays some of these data.

Hispanic origin is a separate question in the survey—not a racial classification—and Hispanics may be of any race. A subgroup of *White non-Hispanic* is shown in some tables. According to the Census 2010 population results, "Being Hispanic was reported by 11.6 percent of White householders who reported only one race, 4.5 percent of Black householders who reported only one race, and 3.5 percent of Asian householders who reported only one race. Data users should exercise caution when interpreting aggregate results for the Hispanic population or for race groups because these populations consist of many distinct groups that differ in socioeconomic characteristics, culture, and recent immigration status." ("Income, and Poverty in the United States: 2017", footnote 2, p. 4 and "Health Insurance Coverage in the United States: 2017" footnote 2, p. 4.)

Definitions: General

A *household* consists of all persons who occupy a housing unit. A household includes the related family members and all the unrelated persons, if any (such as lodgers, foster children, wards, or employees), who share the housing unit. A person living alone in a housing unit or a group of unrelated persons sharing a housing unit as partners is also counted as a household. The count of households excludes group quarters.

Earnings includes all income from work, including wages, salaries, armed forces pay, commissions, tips, piece-rate payments, and cash bonuses, before deductions such as taxes, bonds, pensions, and union dues. This category also includes net income from nonfarm self-employment and farm self-employment. Wage and salary supplements that are paid directly by the employer, such as the employer share of Social Security taxes and the cost of employer-provided health insurance, are not included.

Income, in the official definition used in the survey, is money income, including *earnings* from work as defined above; unemployment compensation; workers' compensation; Social Security; Supplemental Security Income; cash public assistance (welfare payments); veterans' payments; survivor benefits; disability benefits; pension or retirement income; interest income; dividends (but not capital gains); rents, royalties, and payments from estates or trusts; educational assistance, such as scholarships or grants; child support; alimony; financial assistance from outside of the household; and other cash income regularly received, such as foster child payments, military family allotments, and foreign government pensions. Receipts not counted as income include capital gains or losses, withdrawals of bank deposits, money borrowed, tax refunds, gifts, and lump-sum inheritances or insurance payments.

A *year-round, full-time worker* is a person who worked 35 or more hours per week and 50 or more weeks during the previous calendar year.

A *family* is a group of two or more persons related by birth, marriage, or adoption who reside together.

Unrelated individuals are persons 15 years old and over who are not living with any relatives. In the official poverty measure, the poverty status of unrelated individuals is determined independently of and is not affected by the incomes of other persons with whom they may share a household.

Median income is the amount of income that divides a ranked income distribution into two equal groups, with half having incomes above the median, and half having incomes below the median. The median income for persons is based on persons 15 years old and over with income. Since median income is updated annually to inflation adjusted current dollars, comparison between editions is not possible.

Mean income is the amount obtained by dividing the total aggregate income of a group by the number of units in that group. In this survey, as in most surveys of incomes, means are higher than medians because of the skewed nature of the income distribution.

Historical income figures are shown in constant *2017 dollars*. All constant-dollar figures are converted from current-dollar values using the *CPI-U-RS* (the Consumer Price Index, All Urban, Research Series), which measures changes in prices for past periods using the methodologies of the current CPI, and is similar in concept and behavior to the deflators used in the NIPAs for consumer income and spending. See Chapter 8 for CPI-U-RS data and the corresponding notes and definitions.

Definitions: Income distribution

Income distribution is portrayed by dividing the total ranked distribution of families or households into *fifths*, also known as *quintiles*, and also by separately tabulating the top 5 percent (which is included in the highest fifth). The households or families are arrayed from those with the lowest income to those with the highest income, then divided into five groups, with each group containing one-fifth of the total number of households. Within each quintile, incomes are summed and calculated as a share of total income for all quintiles, and are averaged to show the average (mean) income within that quintile.

A statistical measure that summarizes the dispersion of income across the entire income distribution is the *Gini coefficient* (also known as Gini ratio, Gini index, or index of income concentration), which can take values ranging from 0 to 1. A Gini value of 1 indicates "perfect" inequality; that is, one household has all the income and the rest have none. A value of 0 indicates "perfect" equality, a situation in which each household has the same income.

A new "equivalence-adjusted" measure of household income inequality was introduced recently and is displayed in Table 3-3. For a Census-defined household, a given level of money income can have different implications for that household's well-being, depending on the size of the household and how many children, if any, are in the household. Since there have been substantial changes over past decades in the average size and composition of households, some have questioned the pertinence of standard income distribution tables. In response, the Census Bureau now also reports measures of income inequality for households using "equivalence-adjusted" income.

The equivalence adjustment is based on a three-parameter scale reflecting the size of the household and the facts that children consume less than adults; that as family size increases, expenses do not increase at the same rate; and that the increase in expenses is larger for the first child of a single-parent family than the first child of a two-adult family. The same equivalence concept is used in the poverty thresholds for the NAS-based alternative poverty estimates shown in Tables 3-11 and described later in these Notes.

As can be seen in Table 3-3, the equivalence-adjusted measures generally show somewhat less inequality than the raw money income measures, but they show a greater rise in inequality over the period 1967–2017.

Definitions: Poverty

The *number of people below poverty level*, or the number of poor people, is the number of people with family or individual incomes below a specified level that is intended to measure the cost of a minimum standard of living. These minimum levels vary by size and composition of family and are known as *poverty thresholds*.

The official poverty thresholds are based on a definition developed in 1964 by Mollie Orshansky of the Social Security Administration. She calculated food budgets for families of various sizes and compositions, using an "economy food plan" developed by the U.S. Department of Agriculture (the cheapest of four plans developed). Reflecting a 1955 Department of Agriculture survey that found that families of three or more persons spent about one-third of their after-tax incomes on food, Orshansky multiplied the costs of the food plan by 3 to arrive at a set of thresholds for poverty income for families of three or larger. For 2-person families, the multiplier was 3.7; for 1-person families, the threshold was 80 percent of the two-person threshold.

These poverty thresholds have been adjusted each year for price increases, using the percent change in the Consumer Price Index for All Urban Consumers (CPI-U). See Chapter 8 for additional information on the Consumer Price Index.

For more information on the Orshansky thresholds (the description of which has been simplified here), see Gordon Fisher, "The Development of the Orshansky Thresholds and Their Subsequent History as the Official U.S. Poverty Measure" (May 1992), available on the Census Bureau Web site at http://www.census.gov/hhes/poverty/povmeas/papers/orshansky.html.

The *poverty rate* for a demographic group is the number of poor people or poor families in that group expressed as a percentage of the total number of people or families in the group.

Average poverty thresholds. The thresholds used to calculate the official poverty rates vary not only with the size of the family but with the number of children in the family. For example, the threshold for a three-person family in 2018 was $19,985 if there were no children in the family but $20,231 if the family consisted of 1 adult and 2 children. There are 48 different threshold values depending on size of household, number of children, and whether the householder is 65 years old or over (with <u>lower</u> thresholds for the older householders). To give a general sense of the "poverty line," the Census Bureau also publishes the <u>average</u> threshold for each size family, based on the actual mix of family types in that year. These are the values shown in Table 3-5 to represent the history of poverty thresholds. The preliminary <u>average</u> value for 3-person families in 2017, as shown in Table 3-5, was $19,515, a weighted average of the values actually used for the 3 different possible family compositions.

A person with *work experience* (Table 3-9 is one who, during the preceding calendar year and on a part-time or full-time basis, did any work for pay or profit or worked without pay on a family-operated farm or business at any time during the year). A *year-round* worker is one who worked for 50 weeks or more during the preceding calendar year. A person is classified as having worked *full time* if he or she worked 35 hours or more per week during a majority of the weeks worked. A *year-round, full-time worker* is a person who worked 35 or more hours per week and 50 or more weeks during the previous calendar year.

Toward better measures of income and poverty

The definition of the official poverty rate is established by the Office of Management of Budget in the Executive Office of the President and has not been substantially changed since 1969. Criticisms of the current definition are legion. In response to these criticisms, the Census Bureau has published extensive research work illustrating the effects of various ways of modifying income definitions and poverty thresholds. Some of the results of this work are published here in Tables 3-10 and 3-11 and explained in the notes and definitions below.

Improving the income concept

One type of criticism accepted the general concept of the Orshansky threshold but recommended making the income definition more realistic by including some or all of the following: capital gains; taxes and tax credits; noncash food, housing, and health benefits provided by government and employers; and the value of homeownership. There is debate about whether it is appropriate to use income data augmented in this way in conjunction with the official thresholds. The original 1964 thresholds made no allowance for health insurance or other health expenses—in effect, they assumed that the poor would get free medical care, or at least that the poverty calculation was not required to allow for medical needs—and because of the imprecision of Orshansky's multiplier it is not clear to what extent they include housing expenses in a way that is comparable with the inclusion of a homeownership component in income. Nevertheless, the Census Bureau has calculated and published income and poverty figures based on broadened income definitions and either the official thresholds or thresholds that are closely related to the official ones. Some of these calculations are presented in Table 3-10 and 3-11 and described below.

Still accepting the validity of the basic Orshansky threshold concept, some critics have also argued that use of the CPI-U in the official measure to update the thresholds each year has overstated the price increase, and that an inflator such as the CPI-U-RS should be used instead. (See the notes and definitions for Chapter 8.) Use of the CPI-U-RS leads to lower poverty thresholds beginning in the late 1970s, when the CPI began to be distorted by housing and other biases subsequently corrected by new methods; these newer methods were not carried backward to revise the <u>official</u> CPI-U. Use of the CPI-U-RS, which does carry current methods back and thereby revises the CPI time series, eliminates a presumed upward bias in the poverty rate <u>relative to the poverty rates estimated before the bias emerged</u>. This is a bias in the behavior of the time series <u>given the concept of the Orshansky threshold</u>, not necessarily a bias in the current <u>level</u> of poverty, since all the other criticisms of the Orshansky thresholds need to be considered when assessing the general adequacy of today's poverty measurements.

Improving the concepts of income and poverty together

Another type of criticism argues that the official thresholds are also no longer relevant to today's needs, and that the concepts of income (or "resources") and of the threshold level that depicts a minimum adequate standard of living need to be rethought together. These critics cite the availability of more up-to-date and detailed information about consumer spending at various income levels. The Consumer Expenditure Survey (CEX), originally designed to provide the weights for the Consumer Price Index, is now conducted quarterly and provides extensive data on consumer spending patterns. To give just one example of the information available now that was not available to Mollie Orshansky, the CEX indicates that food now accounts for one-sixth, not one-third, of total family expenditures, even among low-income families. (For data from, and notes on, the Consumer Expenditure Survey, see Bernan Press, *Handbook of U.S. Labor Statistics*.)

A special panel of the National Academy of Sciences (NAS) undertook a study that reconsidered both resources and thresholds. The Census Bureau has calculated and published experimental poverty rates for 1999 through 2015, developed following NAS recommendations, which are presented in Table 3-10. Beginning in November 2011 the Bureau has issued "The Research Supplemental Poverty Measure" or SPM for the most recent two years, a further refinement of the NAS recommendations; SPM measures are presented in Tables 3-10.

Supplemental Poverty Measure—Tables 3-10

In March of 2010, an Interagency Technical Working Group produced suggestions for a Supplemental Poverty Measure (SPM), which were first implemented in the "Research Supplemental Poverty Measure: 2010" issued by the Census Bureau in November 2011. The SPM goes beyond the experimental measures just described, and takes advantage of new data from questions introduced in the CPS-ASEC in recent years. In the new SPM:

- The measurement unit now covers not just the family but "all related individuals who live at the same address, including any co-resident unrelated children who are cared for by the family (such as foster children) and any co-habitors and their children." This redefinition adds unrelated individuals under the age of 15 to the universe measured by the official rate. The "official" measures shown in Table 3-10 have been adjusted to include these individuals, and are therefore slightly higher than the regular published official rates shown in Tables 3-6 through 3-9.

- The SPM thresholds are calculated separately for three housing status groups: owners with mortgages, owners without mortgages, and renters. For each of these three groups the basic threshold represents the 33rd percentile of expenditures on food, clothing, shelter, and utilities (FCSU), averaged over a five-year period, by consumer units with two children, multiplied by 1.2 to allow for other needs. (Before averaging, the expenditures are converted to their dollar values in the prices of the threshold year using the CPI-U.) The consumer units selected for calculating the threshold include not only two-adult two-child families but also other types with two children, such as single-parent families. But they are adjusted to a four-person basis, using the equivalence scales, before averaging them to determine the basic threshold appropriate to a two-adult two-child household. Then that threshold is used as the basis for calculating thresholds for other size measurement units, again using the three-parameter equivalence scales for family size and composition, and for geographic differences.

- The thresholds are updated each year, using an updated five-year moving average of FCSU at the 33rd percentile, expressed in the prices of the new threshold year. Thus the thresholds are adjusted for inflation, but in addition, the "real" (constant-dollar) purchasing power of the poverty threshold will change (gradually) over time as the real standard of FCSU spending in the 33rd percentile changes. The real value of the threshold will likely change more slowly and gradually than in the "experimental" measures described above, because of using five instead of three years in the average.

- Family resources include cash income, plus in-kind benefits that families can use to meet their FCSU needs (for example, the Supplemental Nutrition Assistance Program [SNAP] formerly known as food stamps), minus income and payroll taxes, plus tax credits, minus childcare and other work-related expenses, minus child support payments to another household, and minus out-of-pocket medical expenses.

The SPM is not intended to replace the official poverty measure and is not to be used in calculations affecting program eligibility and funding distribution. According to the report referenced above, it is "designed to provide information on aggregate levels of economic need at a national level or within large subpopulations or areas...providing further understanding of economic conditions and trends." Census Bureau presentation of the SPM has focused on the different distribution of the poverty population indicated by the new measures, as shown in Table 3-10.

Experimental measures based on NAS recommendations—Table 3-11

Several alternative poverty rates that redefined both resources and needs were presented in "Alternative Poverty Estimates in the United States: 2003" (see below for complete reference) and have been updated through 2017 on the Census Bureau Web site at www.census.gov, under the general heading "NAS-based Experimental Poverty Estimates."

To derive these estimates, a baseline set of poverty thresholds for the year 1999, based on data from the Consumer Expenditure Survey (CEX) for the years 1997–1999, was developed as follows:

- A reference family type was selected: a 2-adult, 2-child family.

- The definition of necessities for the purpose of the poverty threshold was expenditures on food, clothing, shelter, and utilities (FCSU), augmented by a multiplier of between 1.15 and 1.25 to include other needs such as household and personal supplies. The definition of food includes food away from home. Expenses on shelter include interest (but not principal repayment) for homeowners and rent for renters.

- The numerical estimates were derived as follows: FCSU expenditures for reference families falling between the

30th and 35th percentile of the distribution of expenditures in the CEX, based on 1997–1999 data but expressed in 1999 dollars, were calculated and expanded by the multiplier percentages. The midpoint of these estimates was then selected as the poverty threshold; it turned out to be 96.725 percent of FCSU expenditures by the median household.

- For health care, three different treatments were developed; these treatments are described below.

- Three-parameter equivalence scale adjustments were used to convert the threshold for the reference family to thresholds for other family sizes and compositions, accounting for the differing needs of adults and children and the economies of scale of living in larger families.

- For some of the experimental measures, thresholds were adjusted geographically to reflect differences in the cost of living (in practice, difference in housing costs) in different areas.

The family incomes to be compared with these poverty thresholds were defined and measured to include the effects of all taxes, tax credits, and in-kind benefits such as food stamps, but not the value of homeownership, and to allow for expenses such as child care that are necessary to hold a job.

The eight experimental measures shown in Table 3-11 are as originally presented in Table B-3 in "Alternative Poverty Estimates in the United States: 2003," and most recently updated in the table "Official and National Academy of Sciences (NAS) Based Poverty Rates: 1999 to 2015" on the Census Bureau Web site. The following abbreviations are used to define the alternative definitions of poverty that are presented in this table.

- *MSI* indicates that in calculating the poverty rate, medical out-of-pocket expenses are subtracted from family income before comparing that income to the family's threshold.

- *MIT* indicates that poverty thresholds were increased to take the family's potential medical out-of-pocket expenses into account, using the CEX and the 1996 Medical Expenditures Panel Survey, with the amounts depending on family size, age, and health insurance coverage.

- *GA* indicates that the thresholds were adjusted for geographic differences in the cost of living. Measures labeled *NGA* were not.

- *CPI* indicates that the thresholds established for 1999 were updated to succeeding years using the percent change in the CPI-U. In effect, the thresholds for these measures have been held constant in real (inflation-adjusted) terms since 1999, just as the official thresholds have been held constant in real

terms—in intention, and in fact except for bias in the price indexes—since 1964.

- *CE* indicates that the thresholds were updated using the percent change in median FCSU expenditures from the latest available 12 quarters of CEX data. This means that if the actual real FCSU spending of middle-income families rises (or falls), the real standard of living represented by the poverty thresholds will rise (or fall) commensurately.

Notes on the data

The following are the principal changes that may affect year-to-year comparability of all income and poverty data from the CPS.

- Beginning in 1952, the estimates are based on 1950 census population controls. Earlier figures were based on 1940 census population controls.

- Beginning in 1962, 1960 census–based sample design and population controls are fully implemented.

- With 1971 and 1972 data, 1970 census–based sample design and population controls were introduced.

- With 1983–1985 data, 1980 census–based sample design was introduced; 1980 population controls were introduced; and these were extended back to 1979 data.

- With 1993 data, there was a major redesign of the CPS, including the introduction of computer-assisted interviewing. The limits used to "code" reported income amounts were changed, resulting in reporting of higher income values for the highest-income families and, consequently, an exaggerated year-to-year increase in income inequality. (It is possible that this jump actually reflects in one year an increase that had emerged more gradually, so that the distribution measures for 1993 and later years may be properly comparable with data for decades earlier even if they should not be directly compared with 1992.) In addition, 1990 census–based population controls were introduced, and these were extended back to the 1992 data.

- With 1995 data, the 1990 census–based sample design was implemented and the sample was reduced by 7,000 households.

- Data for 2001 implemented population controls based on the 2000 census, which were carried back to 2000 and 1999 data as well. Data from 2000 forward also incorporate results from a 28,000-household sample expansion.

- Beginning with the data for 2010 presented here, population controls based on the 2010 census were implemented. (This resulted in some revision of CPS data initially published for 2010.)

For more information on these and other changes that could affect comparability, see "Current Population Survey Technical Paper 63RV: Design and Methodology" (March 2002) and footnotes to CPS historical income tables, both available on the Census Bureau Web site at www.census.gov/hhes/income.

Data availability

Data embodying the official definitions of income and poverty are published annually in late summer or early fall by the Census Bureau, as part of a series with the general title *Current Population Reports: Consumer Income, P60*. Most of the data up to 2012 in this chapter were derived from report P60-245, "Income, Poverty, and Health Insurance Coverage in the United States: 2012" (September 2013), and from the "Historical Income Tables" and "Historical Poverty Tables" on the Census Web site (see below for the address). In 2013, the Census Bureau split the previous report in two. Health insurance coverage is now published separately from income and poverty. The historical tables can be found in the "CPS-ASEC" section under the "Data" category.

The data in Table 3-11 were first published in two reports, both issued in June 2005: P60-228, "Alternative Income Estimates in the United States: 2003," and P60-227, "Alternative Poverty Estimates in the United States: 2003." Updates available on the website are cited in the table description above. Corrections to the 2006 data on the website were obtained from Census Bureau staff.

The NAS-based experimental data used in Table 3-11 are from the Census Bureau Web site, as specified in the description above.

The SPM data in Tables 3-10 are from reports entitled "The Research Supplemental Poverty Measure." Reports for 2012, 2011, and 2010 respectively are numbered P60-247, P60-244, and P60-241, and are available on the Census Web site.

All these reports and related data, including historical tabulations, used in *Business Statistics* are available on the Census Bureau Web site at <http://www.census.gov>, under the general headings of "Income" and "Poverty."

References

Definitions and descriptions of the concepts and data of all series are provided in the source documents listed above and in the references contained therein.

TABLES 3-12 THROUGH 3-13

Health Insurance Coverage

SOURCE: U.S. DEPARTMENT OF COMMERCE, BUREAU OF THE CENSUS

The Current Population Survey Annual Social and Economic Supplement (CPS ASEC) and the American Community Survey (ACS) is used to produce official estimates of income and poverty, and it serves as the most widely-cited source of estimates on health insurance coverage.

Due to questions of the validity of the health insurance data in previous reports, the Census Bureau implemented changes in 2014 to CPS ACS, including a complete redesign of the health insurance questions that replaced the existing questions in the CPS ASEC. Due the differences in measurement, health insurance estimates from calendar year 2013 are different from estimates in previous years.

Health insurance coverage in the CPS ASEC refers to comprehensive coverage during the calendar year. The American Community Survey (ACS) health insurance coverage status is at the time the individual is interviewed. Therefore, two uninsured measures are reported.

Since the passage of the Patient Protection and Affordable Care Act (ACA) in 2010, several provisions of the ACA have gone into effect at different times. For example, in 2010, the Young Adult Provision enabled adults under age 26 to remain as dependents on their parents' health insurance plans. In 2014 policy changes associated with the ACA provide the option for states to expand Medicaid to people whose income-to-poverty ratio full under a particular threshold. The decreases in the uninsured rates in 2013 and 2014 are consistent with what some of the provisions of the ACA intended.

Definitions

Health insurance is a means for financing a person's health care expense.

Private insurance is a plan provided through an employer or a union and coverage purchased directly by an individual from an insurance company or though an exchange.

Government health insurance includes federal programs such as Medicare, Medicaid, the Children's Health Insurance Program (CHIP), individual state health programs, TRICARE, CHAMPVA (Civilian Health and Medical Program of the Department of Veterans Affairs), as well care care provided by the Department of Veterans Affairs and the military.

Uninsured are people if, for an entire year, they were not covered by any type of health insurance. Additionally, people were considered uninsured if they only had coverage through the Indian Health Service (IHS).

Data availability

Detailed health insurance questions have been asked in the CPS since 1988 as part of a mandate to collect data on non-cash basis. However, as noted, comparing older results with newer ones is uncertain due to the changes in questions in 2013.

CHAPTER 4: CONSUMER INCOME AND SPENDING

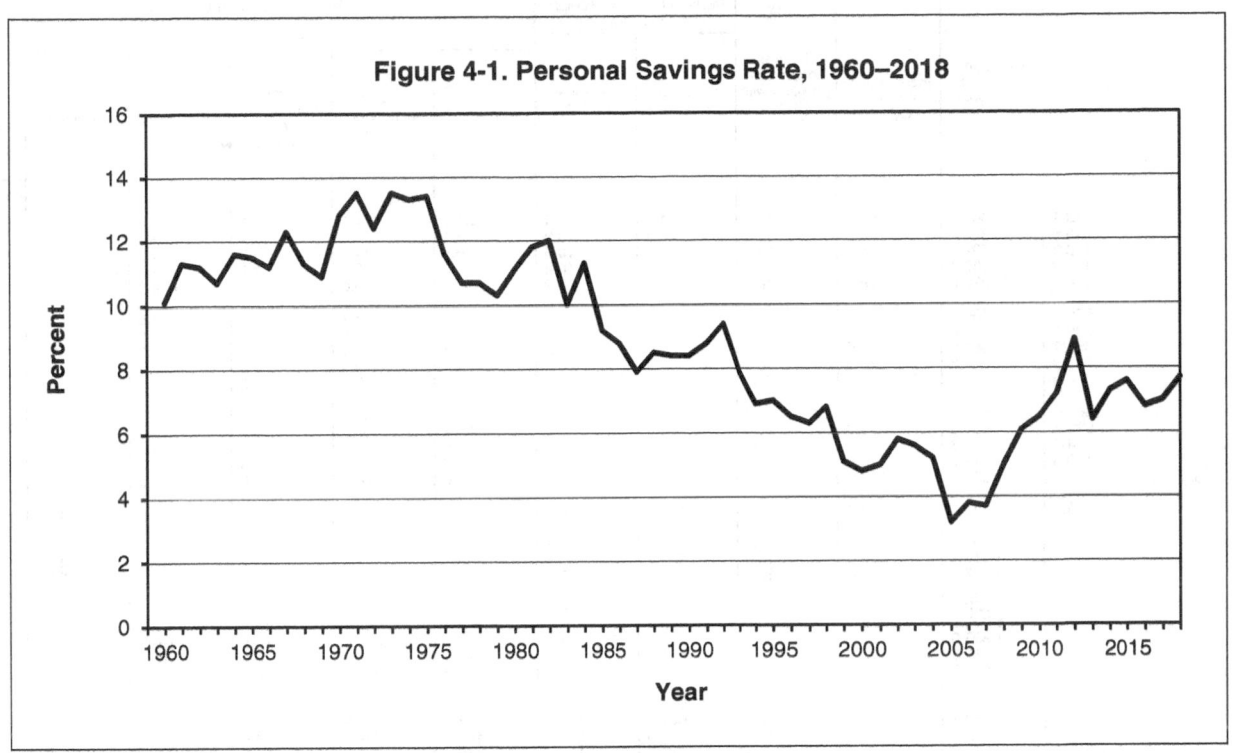

Figure 4-1. Personal Savings Rate, 1960–2018

- The personal saving rate—now defined to include saving in individual pension plans as the earned benefits rather than the employer contributions—got as high as 13.5 percent in the early 1970s. (Benefits from social insurance systems such as Social Security, unlike those in private and government employee retirement plans, do not figure in personal saving; any excess of receipts over payments is counted as government saving.) Since then the personal saving rate entered a downtrend, and hit the lowest rates since the Great Depression in the real estate bubble years 2005 through 2007, saving as a percent of disposable personal income has increased somewhat since then and reached a post-recession high of 8.9 percent in 2012. Although it has declined steadily since then, personal savings as a percent of disposable personal income increased to 7.7 percent in 2018 from 7.0 percent in 2017 and 6.8 percent in 2016. (Tables 4-1A and B)

- It should be noted that personal income does not, by definition, include any capital gains. Despite that, taxes on realized capital gains are deducted (along with all other income taxes) from personal income to get after-tax income. Capital gains, both realized and unrealized, can be a source of spending power in addition to current disposable income; during the housing price bubble they were converted into cash by asset sales, refinancing, and home equity loans, all of which were reflected in low saving rates. (Table 12-7)

- Government social benefits were 5.8 percent of personal income in 1960, rising to 14.1 percent in the high-employment year 2007. In the recession they rose further, to 18.2 percent in 2010, falling back somewhat to 16.4 percent by 2018. It should be understood that this category counts as income all spending financed by government health insurance programs such as Medicare and Medicaid, simultaneously counting that spending as personal consumption expenditures rather than direct government spending in the NIPAs. (Table 4-1)

Table 4-1A. Personal Income and Its Disposition: Recent Data

(Billions of current dollars, except as noted; quarterly data are at seasonally adjusted annual rates.)   NIPA Table 2.1

	Personal income											
					Personal income receipts on assets		Personal current transfer receipts					
								Government social benefits to persons				
Year and quarter	Total	Compensation of employees	Proprietors' income with IVA and CCAdj	Rental income of persons with CCAdj	Personal interest income	Personal dividend income	Total	Social Security and Medicare	Government unemployment insurance	Veterans	Medicaid	Other government benefits to persons
1960	422.1	301.3	50.6	16.5	31.1	13.4	24.4	11.1	3.1	4.5	...	5.7
1961	440.6	310.4	53.2	17.2	33.4	13.9	28.1	12.6	4.4	4.9	...	6.2
1962	468.8	332.2	55.2	18.0	37.1	15.0	28.8	14.3	3.2	4.6	...	6.8
1963	492.8	350.4	56.4	18.7	40.5	16.2	30.3	15.2	3.1	4.8	...	7.3
1964	528.2	376.0	59.1	18.8	44.9	18.2	31.3	16.0	2.8	4.6	...	7.9
1965	570.7	405.4	63.7	19.3	49.3	20.2	33.9	18.1	2.4	4.8	...	8.6
1966	620.3	449.2	67.9	19.9	54.3	20.7	37.5	20.8	1.9	4.8	1.9	8.1
1967	665.7	481.8	69.5	20.3	59.5	21.5	45.8	25.8	2.2	5.5	2.7	9.4
1968	730.9	530.8	73.8	20.1	65.3	23.5	53.3	30.5	2.2	5.9	4.0	10.8
1969	800.3	584.5	77.0	20.3	76.1	24.2	59.0	33.1	2.3	6.6	4.6	12.4
1970	865.0	623.3	77.8	20.7	90.6	24.3	71.7	38.7	4.2	7.5	5.5	16.0
1971	932.8	665.0	83.9	21.8	100.1	25.0	85.4	44.6	6.2	8.4	6.7	19.4
1972	1 024.5	731.3	95.1	22.7	109.8	26.8	94.8	49.7	6.0	9.4	8.2	21.4
1973	1 140.8	812.7	112.5	23.1	125.5	29.9	108.6	60.9	4.6	10.2	9.6	23.3
1974	1 251.8	887.7	112.2	23.2	147.4	33.2	128.6	70.3	7.0	11.6	11.2	28.4
1975	1 369.4	947.2	118.2	22.3	168.0	32.9	163.1	81.5	18.1	14.0	13.9	35.7
1976	1 502.6	1 048.3	131.0	20.3	181.0	39.0	177.6	93.3	16.4	13.8	15.5	38.7
1977	1 659.2	1 165.8	144.5	15.9	206.9	44.7	189.5	105.3	13.1	13.3	16.7	40.9
1978	1 863.7	1 316.8	166.0	16.5	235.1	50.7	203.4	116.9	9.4	13.6	18.6	44.9
1979	2 082.7	1 477.2	179.4	16.1	269.7	57.4	227.3	132.5	9.7	14.2	21.1	49.9
1980	2 323.6	1 622.2	171.6	19.0	332.9	64.0	271.5	154.8	16.1	14.7	23.9	62.1
1981	2 605.1	1 792.5	179.7	23.8	412.2	73.6	307.8	182.1	15.9	15.8	27.7	66.3
1982	2 791.6	1 893.0	171.2	23.8	479.5	77.6	343.1	204.6	25.2	16.3	30.2	66.8
1983	2 981.1	2 012.5	186.3	24.4	516.3	83.3	370.5	222.2	26.4	16.4	33.9	71.5
1984	3 292.7	2 215.9	228.2	24.7	590.1	90.6	380.9	237.7	16.0	16.3	36.6	74.3
1985	3 524.9	2 387.3	241.1	26.2	628.9	97.4	403.1	253.0	15.9	16.6	39.7	78.0
1986	3 733.1	2 542.1	256.5	18.3	662.1	106.0	428.6	268.9	16.5	16.6	43.6	83.0
1987	3 961.6	2 722.4	286.5	16.6	679.0	112.2	447.9	282.6	14.6	16.5	47.8	86.4
1988	4 283.4	2 948.0	325.5	22.5	721.7	129.7	476.9	300.2	13.3	16.7	53.0	93.6
1989	4 625.6	3 139.6	341.1	21.5	806.5	157.8	521.1	325.6	14.4	17.2	60.8	103.1
1990	4 913.8	3 340.4	353.2	28.2	836.5	168.8	574.7	351.7	18.2	17.7	73.1	113.9
1991	5 084.9	3 450.5	354.2	38.6	823.5	180.2	650.5	381.7	26.8	18.1	96.9	127.0
1992	5 420.9	3 668.2	400.2	60.6	809.8	189.1	731.8	414.4	39.6	18.6	116.2	142.9
1993	5 657.9	3 817.3	428.0	90.1	802.3	204.7	778.9	444.7	34.8	19.3	130.1	150.0
1994	5 947.1	4 006.2	456.6	113.7	814.6	235.2	815.7	476.6	23.9	19.7	139.4	156.1
1995	6 291.4	4 198.1	481.2	124.9	878.6	258.0	864.7	508.9	21.7	20.5	149.6	164.0
1996	6 678.5	4 416.9	543.8	142.5	899.0	302.2	906.3	536.9	22.3	21.4	158.2	167.6
1997	7 092.5	4 708.8	584.0	147.1	947.1	337.9	935.4	563.5	20.1	22.3	163.1	166.4
1998	7 606.7	5 071.1	640.2	165.2	1 015.5	355.4	957.9	574.8	19.7	23.3	170.2	170.0
1999	8 001.9	5 402.8	696.4	178.5	1 012.7	346.6	992.2	588.6	20.5	24.1	184.6	174.4
2000	8 652.6	5 848.1	753.9	183.5	1 102.2	383.5	1 044.9	620.5	20.7	25.0	199.5	179.1
2001	9 005.6	6 039.1	831.0	202.4	1 104.3	369.3	1 145.8	667.7	31.9	26.6	227.3	192.4
2002	9 159.0	6 135.6	869.8	211.1	1 010.1	398.8	1 251.0	706.6	53.5	29.5	250.0	211.3
2003	9 487.5	6 354.1	896.9	231.5	1 005.0	432.1	1 321.0	740.2	53.2	31.8	264.5	231.2
2004	10 035.1	6 720.1	962.0	248.9	950.4	561.7	1 404.5	789.9	36.4	34.1	289.8	254.3
2005	10 598.2	7 066.6	978.0	232.0	1 100.4	577.8	1 490.9	844.8	31.8	36.4	304.4	273.5
2006	11 381.7	7 479.9	1 049.6	202.3	1 235.8	722.8	1 593.0	943.2	30.4	38.9	299.1	281.5
2007	12 007.8	7 878.9	994.0	184.4	1 368.6	815.3	1 697.3	1 003.9	32.7	41.7	324.2	294.9
2008	12 442.2	8 057.0	960.9	256.7	1 396.3	804.6	1 919.3	1 067.1	51.1	45.0	338.3	417.7
2009	12 059.1	7 758.5	938.5	327.3	1 299.3	553.0	2 107.7	1 157.5	131.2	51.5	369.6	398.0
2010	12 551.6	7 924.9	1 108.7	394.2	1 238.5	543.9	2 281.4	1 203.6	138.9	58.0	396.9	484.2
2011	13 326.8	8 225.9	1 229.3	478.6	1 269.4	681.5	2 310.1	1 248.9	107.2	63.3	406.0	484.8
2012	14 010.1	8 566.7	1 347.3	518.0	1 330.5	835.1	2 322.6	1 316.8	83.6	70.1	417.5	434.4
2013	14 181.1	8 834.2	1 403.6	557.0	1 273.0	793.3	2 385.9	1 371.8	62.5	79.1	440.0	432.5
2014	14 991.7	9 249.1	1 447.7	604.6	1 349.0	953.2	2 498.6	1 434.6	35.5	84.2	490.9	453.5
2015	15 717.8	9 698.2	1 422.2	648.1	1 437.9	1 032.9	2 633.0	1 505.3	32.1	92.6	535.9	467.1
2016	16 121.2	9 960.3	1 423.7	681.4	1 457.4	1 064.0	2 714.6	1 556.7	31.7	95.9	562.7	467.6
2017	16 878.8	10 411.6	1 518.2	718.8	1 551.6	1 130.0	2 800.1	1 615.4	29.8	104.0	577.4	473.5
2018	17 819.2	10 928.5	1 588.8	756.8	1 702.7	1 227.5	2 918.3	1 703.3	27.1	109.9	597.7	480.3
2016												
1st quarter	15 937.6	9 843.5	1 415.2	669.9	1 447.1	1 043.5	2 684.2	1 536.3	32.3	94.2	550.2	471.3
2nd quarter	16 029.0	9 900.1	1 410.2	680.2	1 449.1	1 056.2	2 704.2	1 549.6	31.9	94.8	558.6	469.3
3rd quarter	16 175.5	9 993.2	1 429.5	683.6	1 457.9	1 071.5	2 723.2	1 562.4	31.6	96.3	566.5	466.4
4th quarter	16 342.6	10 104.5	1 440.0	692.1	1 475.6	1 084.7	2 746.6	1 578.5	30.8	98.3	575.6	463.4
2017												
1st quarter	16 604.1	10 227.6	1 494.8	707.4	1 545.4	1 085.3	2 772.4	1 593.8	30.7	100.7	573.2	474.1
2nd quarter	16 749.6	10 334.2	1 512.2	709.9	1 523.5	1 133.5	2 780.6	1 606.9	29.6	103.6	569.0	471.6
3rd quarter	16 930.4	10 456.7	1 523.1	722.0	1 528.9	1 142.4	2 815.7	1 621.8	29.6	105.5	583.7	475.1
4th quarter	17 231.2	10 628.0	1 542.9	736.0	1 608.6	1 158.8	2 831.5	1 639.0	29.2	106.3	584.0	473.1
2018												
1st quarter	17 540.3	10 786.0	1 567.5	743.8	1 669.6	1 182.0	2 884.8	1 670.7	28.8	108.8	589.8	486.7
2nd quarter	17 725.0	10 876.1	1 573.3	754.0	1 694.6	1 214.7	2 910.1	1 689.9	27.2	109.0	600.4	483.5
3rd quarter	17 928.5	10 994.3	1 590.0	765.2	1 719.3	1 238.4	2 929.4	1 712.7	26.6	109.9	602.9	477.4
4th quarter	18 082.8	11 057.4	1 624.4	764.1	1 727.2	1 274.8	2 949.0	1 739.7	26.0	111.8	597.6	473.8

. . . = Not available.

Table 4-1A. Personal Income and Its Disposition: Recent Data—*Continued*

(Billions of current dollars, except as noted; quarterly data are at seasonally adjusted annual rates.)   NIPA Table 2.1

Year and quarter	Personal income—Continued: Personal current transfer receipts—Continued From business, net	Less: Contributions for government social insurance, domestic	Less: Personal current taxes	Equals: Disposable personal income	Less: Personal outlays Total	Personal consumption expenditures	Personal interest payments	Personal current transfer payments Total	To government	To the rest of the world, net	Equals: Personal saving Billions of dollars	Percent of disposable personal income	Billions of chained (2009) dollars: Percent income excluding current transfers	Disposable personal income
1960	1.3	16.4	46.1	376.1	338.2	331.2	6.2	0.8	0.3	0.5	37.9	10.1	2 406.3	2 282.9
1961	1.4	17.0	47.3	393.3	348.9	341.5	6.5	1.0	0.5	0.5	44.4	11.3	2 470.1	2 363.2
1962	1.5	19.1	51.6	417.3	370.6	362.6	6.9	1.1	0.5	0.6	46.7	11.2	2 603.7	2 478.1
1963	1.9	21.7	54.6	438.2	391.1	382.0	7.9	1.2	0.5	0.7	47.1	10.7	2 702.8	2 571.7
1964	2.2	22.4	52.1	476.1	420.8	410.6	8.9	1.3	0.6	0.7	55.3	11.6	2 861.6	2 754.2
1965	2.3	23.4	57.7	513.0	454.2	443.0	9.9	1.4	0.6	0.8	58.8	11.5	3 047.8	2 925.3
1966	2.1	31.3	66.4	554.0	492.1	479.9	10.6	1.6	0.8	0.8	61.9	11.2	3 229.9	3 081.1
1967	2.3	34.9	73.0	592.8	519.8	506.7	11.1	2.0	1.0	1.0	73.0	12.3	3 351.4	3 216.1
1968	2.8	38.7	87.0	643.9	571.0	556.9	12.1	2.0	1.0	1.0	72.9	11.3	3 523.4	3 362.1
1969	3.3	44.1	104.5	695.8	619.8	603.6	13.9	2.2	1.1	1.1	76.1	10.9	3 687.4	3 476.5
1970	2.9	46.4	103.1	762.0	664.4	646.7	15.1	2.6	1.3	1.3	97.6	12.8	3 772.6	3 637.0
1971	2.7	51.2	101.7	831.1	719.2	699.9	16.4	2.8	1.5	1.4	111.9	13.5	3 867.5	3 805.2
1972	3.1	59.2	123.6	900.8	789.3	768.2	18.0	3.2	1.8	1.4	111.5	12.4	4 102.3	3 988.4
1973	3.9	75.5	132.4	1 008.4	872.6	849.6	19.6	3.4	1.8	1.6	135.8	13.5	4 319.9	4 236.5
1974	4.7	85.2	151.0	1 100.8	954.5	930.2	20.9	3.4	2.1	1.4	146.3	13.3	4 256.2	4 188.7
1975	6.8	89.3	147.6	1 221.8	1 057.8	1 030.5	23.4	3.8	2.5	1.3	164.0	13.4	4 212.8	4 291.4
1976	6.7	101.3	172.7	1 330.0	1 175.6	1 147.7	23.5	4.4	3.0	1.4	154.4	11.6	4 389.7	4 428.5
1977	5.1	113.1	197.9	1 461.4	1 305.4	1 274.0	26.6	4.8	3.5	1.4	155.9	10.7	4 579.1	4 568.8
1978	6.5	131.3	229.6	1 634.1	1 459.0	1 422.3	31.3	5.4	3.9	1.6	175.1	10.7	4 834.2	4 776.4
1979	8.2	152.7	280.9	1 810.0	1 627.0	1 585.4	35.5	6.0	4.3	1.7	186.5	10.3	4 958.6	4 869.1
1980	8.6	166.2	299.5	2 024.1	1 800.1	1 750.7	42.5	6.9	5.0	2.0	224.1	11.1	4 952.5	4 905.6
1981	11.2	195.7	345.8	2 259.3	1 993.9	1 934.0	48.4	11.5	6.0	5.6	265.5	11.8	5 085.0	5 025.4
1982	12.4	208.9	354.7	2 436.9	2 143.5	2 071.3	58.5	13.8	7.1	6.7	293.3	12.0	5 133.5	5 135.0
1983	13.8	226.0	352.9	2 628.2	2 364.2	2 281.6	67.4	15.1	8.1	7.0	264.0	10.0	5 248.7	5 312.2
1984	19.7	257.5	377.9	2 914.8	2 584.5	2 492.3	75.0	17.1	9.2	7.9	330.3	11.3	5 632.9	5 677.1
1985	22.3	281.4	417.8	3 107.1	2 822.1	2 712.8	90.6	18.8	10.4	8.3	284.9	9.2	5 833.4	5 847.6
1986	22.9	303.4	437.8	3 295.3	3 004.7	2 886.3	97.3	21.1	12.0	9.1	290.6	8.8	6 044.4	6 069.8
1987	20.2	323.1	489.6	3 472.0	3 196.6	3 076.3	97.1	23.2	13.2	10.0	275.4	7.9	6 242.4	6 204.1
1988	20.6	361.5	505.9	3 777.5	3 457.0	3 330.0	101.3	25.6	14.8	10.8	320.5	8.5	6 510.6	6 496.0
1989	23.2	385.2	567.7	4 057.8	3 717.9	3 576.8	113.1	28.0	16.5	11.6	340.0	8.4	6 724.9	6 686.2
1990	22.2	410.1	594.7	4 319.1	3 958.0	3 809.0	118.4	30.6	18.4	12.2	361.1	8.4	6 813.9	6 817.4
1991	17.6	430.2	588.9	4 496.0	4 100.0	3 943.4	119.9	36.7	22.6	14.1	396.0	8.8	6 746.0	6 867.0
1992	16.3	455.0	612.8	4 808.1	4 354.2	4 197.6	116.1	40.5	26.0	14.5	453.9	9.4	6 951.8	7 152.9
1993	14.1	477.4	648.8	5 009.2	4 611.5	4 452.0	113.9	45.6	28.6	17.1	397.7	7.9	7 061.7	7 271.1
1994	13.3	508.2	693.1	5 254.0	4 890.6	4 721.0	119.9	49.8	30.9	18.9	363.4	6.9	7 277.4	7 470.6
1995	18.7	532.8	748.4	5 543.0	5 155.9	4 962.6	140.4	52.9	32.6	20.3	387.1	7.0	7 530.8	7 718.9
1996	22.9	555.1	837.1	5 841.4	5 459.2	5 244.6	157.0	57.6	34.9	22.6	382.3	6.5	7 838.6	7 964.2
1997	19.4	587.2	937.1	6 160.7	5 770.4	5 536.8	169.7	63.9	38.2	25.7	390.3	6.3	8 224.9	8 255.8
1998	26.0	624.7	1 032.4	6 574.1	6 127.7	5 877.2	180.9	69.5	39.9	29.7	446.5	6.8	8 804.9	8 740.4
1999	34.0	661.3	1 111.9	6 890.0	6 540.6	6 279.1	187.5	74.1	44.1	30.0	349.4	5.1	9 137.9	9 025.6
2000	42.4	705.8	1 236.3	7 416.3	7 058.0	6 762.1	214.8	81.0	48.8	32.3	358.3	4.8	9 669.9	9 479.5
2001	46.8	733.2	1 239.0	7 766.6	7 374.9	7 065.6	220.0	89.3	52.9	36.3	391.6	5.0	9 798.3	9 740.1
2002	34.2	751.5	1 052.2	8 106.8	7 633.1	7 342.7	195.7	94.7	56.4	38.4	473.7	5.8	9 746.0	10 034.5
2003	26.3	779.3	1 003.5	8 484.0	8 012.5	7 723.1	190.9	98.5	60.5	38.0	471.5	5.6	9 884.0	10 301.4
2004	16.8	829.2	1 048.7	8 986.4	8 522.6	8 212.7	202.2	107.7	66.9	40.8	463.8	5.2	10 204.6	10 645.9
2005	25.8	873.3	1 212.4	9 385.8	9 089.1	8 747.1	230.5	111.5	71.6	39.9	296.7	3.2	10 461.1	10 811.6
2006	20.8	922.5	1 356.8	10 024.9	9 639.3	9 260.3	258.4	120.5	75.3	45.2	385.6	3.8	10 953.7	11 241.9
2007	30.8	961.4	1 492.2	10 515.6	10 123.9	9 706.4	284.6	132.9	79.2	53.7	391.6	3.7	11 242.3	11 500.3
2008	35.8	988.4	1 507.2	10 935.0	10 390.1	9 976.3	268.8	144.9	81.0	64.0	544.9	5.0	11 135.2	11 610.8
2009	39.0	964.3	1 152.0	10 907.1	10 240.6	9 842.2	254.0	144.3	80.1	64.3	666.5	6.1	10 534.5	11 591.7
2010	43.7	983.7	1 237.3	11 314.3	10 573.5	10 185.8	242.8	144.8	79.4	65.4	740.9	6.5	10 685.4	11 822.1
2011	48.5	916.7	1 453.2	11 873.6	11 023.7	10 641.1	232.1	150.6	82.2	68.4	849.8	7.2	11 177.0	12 099.8
2012	40.4	950.5	1 508.9	12 501.2	11 393.6	11 006.8	232.4	154.4	86.3	68.1	1 107.6	8.9	11 647.1	12 501.2
2013	38.4	1 104.3	1 675.8	12 505.3	11 703.9	11 317.2	229.5	157.2	87.5	69.7	801.4	6.4	11 600.6	12 339.1
2014	42.9	1 153.6	1 784.0	13 207.7	12 237.0	11 822.8	243.8	170.4	93.4	77.1	970.8	7.3	12 107.6	12 844.3
2015	50.3	1 204.7	1 937.8	13 780.0	12 731.2	12 284.3	264.1	182.8	100.0	82.8	1 048.8	7.6	12 649.3	13 372.7
2016	59.7	1 239.9	1 956.1	14 165.1	13 206.3	12 748.5	273.7	184.1	102.0	82.1	958.8	6.8	12 822.4	13 608.4
2017	48.1	1 299.6	2 045.8	14 833.0	13 802.1	13 312.1	299.3	190.7	103.6	87.1	1 030.9	7.0	13 245.3	14 002.8
2018	53.2	1 356.5	2 077.6	15 741.5	14 531.1	13 998.7	336.7	195.8	106.9	88.9	1 210.4	7.7	13 729.6	14 556.2
2016														
1st quarter	59.5	1 225.3	1 922.0	14 015.6	12 977.5	12 523.5	267.4	186.6	101.9	84.7	1 038.1	7.4	12 773.2	13 568.7
2nd quarter	61.3	1 232.4	1 945.3	14 083.7	13 138.6	12 688.3	271.1	179.3	101.9	77.4	945.1	6.7	12 764.9	13 554.3
3rd quarter	60.6	1 244.0	1 969.6	14 205.9	13 280.4	12 822.4	275.5	182.5	102.0	80.5	925.5	6.5	12 834.8	13 615.0
4th quarter	57.2	1 258.0	1 987.4	14 355.2	13 428.6	12 959.8	281.0	187.9	102.1	85.8	926.5	6.5	12 917.8	13 696.7
2017														
1st quarter	51.2	1 280.0	2 001.5	14 602.6	13 576.8	13 104.4	286.5	185.9	102.8	83.1	1 025.8	7.0	13 080.6	13 860.9
2nd quarter	47.6	1 292.0	2 016.0	14 733.5	13 699.7	13 212.5	294.8	192.4	103.2	89.1	1 033.9	7.0	13 184.2	13 953.4
3rd quarter	46.2	1 304.6	2 049.8	14 880.6	13 841.8	13 345.1	305.8	191.0	103.8	87.2	1 038.8	7.0	13 268.6	14 034.5
4th quarter	47.3	1 321.8	2 115.8	15 115.4	14 090.2	13 586.3	310.3	193.7	104.6	89.2	1 025.2	6.8	13 447.6	14 162.4
2018														
1st quarter	50.6	1 344.0	2 074.9	15 465.4	14 245.2	13 728.4	322.3	194.5	105.7	88.8	1 220.2	7.9	13 599.1	14 400.3
2nd quarter	53.0	1 350.9	2 071.7	15 653.3	14 465.9	13 939.8	329.6	196.4	106.5	89.9	1 187.4	7.6	13 670.5	14 495.9
3rd quarter	54.4	1 362.4	2 086.5	15 842.0	14 655.6	14 114.6	341.5	199.6	107.3	92.3	1 186.4	7.5	13 785.6	14 613.3
4th quarter	54.7	1 368.7	2 077.4	16 005.4	14 757.8	14 211.9	353.5	192.5	107.9	84.5	1 247.6	7.8	13 863.6	14 715.2

Table 4-1B. Personal Income and Its Disposition: Historical Data

(Billions of current dollars, except as noted; quarterly data are at seasonally adjusted annual rates.)   NIPA Table 2.1

| Year and quarter | Personal income | | | | | | | Less: Personal current taxes | Equals: Disposable personal income | Less: Personal outlays | Equals: Personal saving | | Billions of chained (2009) dollars | |
	Total	Compensation of employees	Proprietors' income with IVA and CCAdj	Rental income of persons with CCAdj	Personal income receipts on assets	Personal current transfer receipts	Less: Contributions for government social insurance, domestic				Billions of dollars	Percent of disposable personal income	Personal income excluding current transfers	Disposable personal income
1929	85.3	51.4	14.0	6.1	12.7	1.2	0.1	1.7	83.6	79.6	4.0	4.7	902.9	897.1
1930	76.5	47.2	10.9	5.4	12.0	1.2	0.1	1.6	75.0	71.6	3.3	4.5	844.3	840.4
1931	65.7	40.1	8.3	4.4	10.6	2.3	0.1	1.0	64.7	61.8	2.8	4.4	795.0	811.9
1932	50.3	31.3	5.0	3.6	8.7	1.7	0.1	0.7	49.6	49.7	-0.1	-0.2	691.2	705.7
1933	47.2	29.8	5.3	2.9	7.7	1.7	0.1	0.8	46.4	46.8	-0.3	-0.7	672.3	685.3
1934	54.1	34.6	7.0	2.5	8.3	1.8	0.1	0.9	53.2	52.3	0.9	1.7	739.1	751.1
1935	60.8	37.7	10.1	2.6	8.6	2.0	0.1	1.1	59.8	56.8	3.0	5.1	811.3	823.8
1936	69.2	43.3	10.4	2.7	10.1	3.1	0.3	1.3	67.9	63.1	4.8	7.1	903.2	927.6
1937	74.7	48.3	12.5	3.0	10.4	2.0	1.5	1.9	72.8	67.9	4.9	6.7	958.3	959.0
1938	69.1	45.4	10.6	3.5	8.8	2.4	1.6	1.9	67.2	65.3	2.0	2.9	899.1	906.4
1939	73.6	48.6	11.1	3.7	9.5	2.5	1.8	1.5	72.1	68.2	3.9	5.4	968.2	982.2
1940	79.4	52.7	12.2	3.8	9.8	2.7	1.9	1.7	77.7	72.4	5.3	6.8	1 035.9	1 049.3
1941	97.9	66.2	16.7	4.4	10.3	2.7	2.3	2.3	95.6	82.3	13.3	13.9	1 210.3	1 214.7
1942	126.7	88.0	23.3	5.5	10.1	2.7	2.9	4.9	121.8	90.0	31.9	26.2	1 402.3	1 377.1
1943	156.2	112.7	28.2	6.0	10.5	2.5	3.8	16.7	139.5	100.8	38.6	27.7	1 590.9	1 443.4
1944	169.7	124.3	29.3	6.3	11.0	3.1	4.3	17.7	152.1	109.7	42.4	27.9	1 630.8	1 488.3
1945	175.8	126.3	30.8	6.6	11.8	5.6	5.3	19.4	156.3	121.2	35.2	22.5	1 601.4	1 471.5
1946	182.5	122.5	35.7	6.9	13.5	10.5	6.6	17.2	165.3	145.8	19.6	11.8	1 513.4	1 454.7
1947	194.5	132.4	34.6	6.9	15.4	10.8	5.6	19.8	174.7	163.7	11.0	6.3	1 467.3	1 395.7
1948	213.5	144.3	39.3	7.5	16.8	10.3	4.6	19.2	194.3	177.1	17.2	8.8	1 536.8	1 469.1
1949	211.1	144.3	34.7	7.8	17.9	11.2	4.9	16.7	194.3	180.7	13.6	7.0	1 523.1	1 481.0
1950	233.7	158.3	37.5	8.8	20.7	14.0	5.5	18.9	214.8	194.8	20.0	9.3	1 655.1	1 617.9
1951	264.2	185.7	42.6	9.7	21.4	11.4	6.6	27.1	237.2	211.3	25.9	10.9	1 782.7	1 672.4
1952	282.5	201.1	43.0	10.8	22.6	11.9	6.9	32.0	250.5	222.6	27.8	11.1	1 869.6	1 730.5
1953	299.2	215.2	42.0	12.0	24.5	12.5	7.1	33.2	266.0	236.8	29.2	11.0	1 955.3	1 814.0
1954	302.2	214.1	42.3	13.1	26.5	14.3	8.1	30.2	272.1	243.9	28.1	10.3	1 947.0	1 839.9
1955														
1st quarter	313.5	221.9	43.7	13.3	28.2	15.3	8.9	31.4	282.1	255.9	26.2	9.3	2 015.9	1 906.9
2nd quarter	321.1	228.1	44.3	13.4	28.7	15.6	9.0	32.4	288.6	261.2	27.5	9.5	2 062.7	1 949.1
3rd quarter	328.5	233.9	44.6	13.4	29.8	15.9	9.2	33.4	295.1	265.6	29.5	10.0	2 102.7	1 985.1
4th quarter	333.6	238.4	44.5	13.5	30.4	16.0	9.3	34.2	299.4	269.8	29.6	9.9	2 130.7	2 008.2
1956														
1st quarter	339.0	242.8	44.9	13.6	31.2	16.3	9.9	35.4	303.6	271.6	32.1	10.6	2 155.6	2 028.5
2nd quarter	344.8	247.5	45.3	13.6	31.8	16.6	10.0	36.2	308.6	274.3	34.2	11.1	2 178.2	2 047.7
3rd quarter	350.1	250.4	46.4	13.7	32.5	17.1	10.0	36.9	313.2	277.6	35.5	11.3	2 188.6	2 058.3
4th quarter	357.8	256.4	46.8	13.8	33.6	17.3	10.1	37.9	319.9	283.1	36.8	11.5	2 224.2	2 090.1
1957														
1st quarter	362.4	260.2	46.9	13.9	34.5	18.2	11.4	38.6	323.8	287.7	36.1	11.1	2 228.6	2 096.7
2nd quarter	367.1	262.1	47.7	14.0	35.4	19.4	11.4	39.0	328.1	290.1	38.1	11.6	2 237.2	2 111.5
3rd quarter	371.7	264.7	48.8	14.1	35.9	19.6	11.5	39.2	332.5	294.7	37.8	11.4	2 247.0	2 122.2
4th quarter	370.8	263.3	47.7	14.3	36.0	20.8	11.3	38.8	332.0	296.4	35.6	10.7	2 222.2	2 108.0
1958														
1st quarter	371.3	259.8	50.3	14.6	35.9	22.1	11.3	38.2	333.1	295.9	37.2	11.2	2 190.1	2 089.4
2nd quarter	373.7	259.4	50.3	14.8	36.6	23.9	11.3	37.7	336.0	298.8	37.2	11.1	2 190.2	2 103.9
3rd quarter	382.6	267.2	50.1	14.9	37.6	24.2	11.4	38.9	343.7	303.8	39.9	11.6	2 242.4	2 150.1
4th quarter	388.2	272.3	49.9	15.0	38.9	23.7	11.5	39.4	348.8	307.8	41.0	11.8	2 281.6	2 183.1
1959														
1st quarter	394.0	279.7	50.1	15.0	38.8	24.0	13.7	40.8	353.2	315.4	37.9	10.7	2 299.9	2 195.8
2nd quarter	402.4	286.5	50.3	15.4	40.1	23.9	13.9	42.0	360.4	321.6	38.8	10.8	2 343.0	2 231.2
3rd quarter	404.7	286.9	50.4	15.8	41.4	24.2	13.9	42.7	362.1	327.1	35.0	9.7	2 342.0	2 228.2
4th quarter	410.4	290.2	50.6	16.0	42.6	24.8	13.9	43.7	366.6	329.4	37.2	10.2	2 359.7	2 244.1
1960														
1st quarter	416.7	298.7	49.7	16.2	43.7	24.6	16.4	45.3	371.4	333.1	38.3	10.3	2 396.2	2 270.0
2nd quarter	421.7	301.7	50.7	16.4	44.1	25.3	16.5	46.0	375.7	339.2	36.5	9.7	2 410.8	2 284.6
3rd quarter	424.4	302.6	50.9	16.5	44.9	26.0	16.5	46.5	377.8	339.3	38.6	10.2	2 413.1	2 289.0
4th quarter	425.8	302.1	51.0	16.7	45.3	27.0	16.4	46.4	379.4	341.3	38.0	10.0	2 405.1	2 288.0
1961														
1st quarter	430.2	303.1	52.3	16.9	45.7	28.8	16.7	46.5	383.7	341.9	41.7	10.9	2 416.1	2 309.7
2nd quarter	436.4	307.3	52.6	17.1	46.6	29.7	16.9	46.9	389.4	346.8	42.6	10.9	2 448.6	2 344.7
3rd quarter	443.4	312.3	53.3	17.3	47.7	29.9	17.1	47.4	396.0	349.8	46.2	11.7	2 481.1	2 375.9
4th quarter	452.3	318.9	54.5	17.5	49.3	29.4	17.3	48.1	404.2	357.2	47.0	11.6	2 534.5	2 422.3
1962														
1st quarter	459.3	325.5	55.3	17.7	49.7	30.0	18.9	49.4	409.9	362.5	47.4	11.6	2 561.9	2 446.1
2nd quarter	466.8	331.4	55.0	18.0	51.5	30.0	19.1	50.9	415.8	368.4	47.5	11.4	2 597.0	2 472.5
3rd quarter	471.5	334.2	54.9	18.2	53.0	30.4	19.2	52.3	419.2	372.5	46.7	11.1	2 615.7	2 485.8
4th quarter	477.8	337.7	55.8	18.3	54.1	31.2	19.3	53.6	424.2	379.0	45.2	10.7	2 640.5	2 507.9
1963														
1st quarter	483.2	342.7	56.0	18.5	54.7	32.6	21.3	54.1	429.1	382.9	46.2	10.8	2 656.2	2 529.7
2nd quarter	488.1	347.6	55.8	18.7	55.9	31.7	21.5	54.3	433.7	387.3	46.4	10.7	2 686.1	2 553.0
3rd quarter	495.2	352.6	56.2	18.8	57.3	32.0	21.8	54.6	440.6	394.7	45.9	10.4	2 713.5	2 580.9
4th quarter	504.6	358.8	57.6	18.8	58.9	32.5	22.0	55.2	449.4	399.6	49.8	11.1	2 755.2	2 622.9

Table 4-1B. Personal Income and Its Disposition: Historical Data—*Continued*

(Billions of current dollars, except as noted; quarterly data are at seasonally adjusted annual rates.)   NIPA Table 2.1

| Year and quarter | Personal income | | | | | | | Less: Personal current taxes | Equals: Disposable personal income | Less: Personal outlays | Equals: Personal saving | | Billions of chained (2009) dollars | |
	Total	Compensation of employees	Proprietors' income with IVA and CCAdj	Rental income of persons with CCAdj	Personal income receipts on assets	Personal current transfer receipts	Less: Contributions for government social insurance, domestic				Billions of dollars	Percent of disposable personal income	Personal income excluding current transfers	Disposable personal income
1964														
1st quarter	514.1	365.2	57.9	18.8	60.7	33.6	22.0	53.8	460.4	409.3	51.1	11.1	2 791.5	2 674.3
2nd quarter	523.6	372.6	58.8	18.8	62.4	33.2	22.3	49.7	473.8	417.6	56.3	11.9	2 842.1	2 746.4
3rd quarter	533.2	380.0	59.3	18.9	64.0	33.5	22.5	51.6	481.6	426.8	54.7	11.4	2 886.6	2 782.2
4th quarter	541.9	386.1	60.5	18.8	65.4	33.7	22.7	53.2	488.7	429.6	59.0	12.1	2 926.1	2 813.8
1965														
1st quarter	553.0	392.9	61.9	19.1	66.9	35.0	22.9	57.0	496.0	440.5	55.5	11.2	2 972.9	2 847.0
2nd quarter	562.6	399.9	63.2	19.3	68.7	34.6	23.2	58.4	504.1	447.9	56.3	11.2	3 014.9	2 879.1
3rd quarter	576.9	408.3	64.0	19.5	70.5	38.1	23.6	57.0	519.9	457.3	62.6	12.0	3 065.1	2 957.8
4th quarter	590.2	420.5	65.6	19.5	71.8	36.9	24.0	58.3	532.0	471.4	60.6	11.4	3 138.6	3 017.4
1966														
1st quarter	602.9	433.1	69.2	19.8	73.3	37.8	30.4	61.5	541.4	481.9	59.5	11.0	3 180.5	3 047.2
2nd quarter	612.8	444.4	67.2	19.8	74.4	37.9	30.8	65.6	547.2	487.3	59.9	11.0	3 209.9	3 055.3
3rd quarter	626.0	455.7	67.2	20.0	75.5	39.6	31.9	67.9	558.2	496.6	61.6	11.0	3 249.2	3 092.6
4th quarter	639.6	463.8	68.1	20.0	76.7	43.1	32.2	70.6	569.1	502.6	66.5	11.7	3 279.6	3 128.9
1967														
1st quarter	650.5	470.0	68.9	20.3	78.9	46.1	33.6	71.2	579.3	507.1	72.2	12.5	3 313.6	3 176.0
2nd quarter	657.6	475.6	68.5	20.4	80.5	47.2	34.6	70.9	586.7	516.8	69.9	11.9	3 330.7	3 201.2
3rd quarter	671.6	485.3	70.5	20.4	81.9	48.7	35.2	73.8	597.8	523.8	74.0	12.4	3 367.5	3 231.7
4th quarter	683.2	496.2	70.0	20.3	82.6	50.0	36.0	75.9	607.3	531.4	75.9	12.5	3 393.8	3 255.3
1968														
1st quarter	702.5	510.9	71.7	20.1	85.1	52.4	37.6	78.6	624.0	549.8	74.2	11.9	3 448.2	3 309.6
2nd quarter	722.4	524.0	73.2	20.1	87.6	55.9	38.4	81.7	640.7	563.9	76.8	12.0	3 499.2	3 364.0
3rd quarter	741.3	537.8	75.0	20.2	90.0	57.3	39.1	91.9	649.4	580.6	68.8	10.6	3 554.2	3 374.8
4th quarter	757.5	550.3	75.5	20.0	92.7	58.7	39.7	95.9	661.7	589.8	71.9	10.9	3 591.9	3 400.9
1969														
1st quarter	772.4	562.8	75.6	20.2	96.2	60.4	42.9	102.6	669.8	602.3	67.4	10.1	3 623.9	3 409.3
2nd quarter	790.6	576.7	77.1	20.3	98.7	61.5	43.7	105.7	685.0	614.3	70.6	10.3	3 664.5	3 442.7
3rd quarter	811.6	593.4	78.0	20.4	101.5	62.9	44.6	104.1	707.5	625.0	82.5	11.7	3 717.6	3 512.9
4th quarter	826.8	604.9	77.3	20.4	105.0	64.3	45.3	105.6	721.1	637.4	83.7	11.6	3 742.4	3 539.7
1970														
1st quarter	841.7	615.0	77.1	20.4	109.2	66.1	46.0	104.6	737.2	648.9	88.3	12.0	3 763.6	3 576.9
2nd quarter	860.0	620.2	76.7	20.2	113.1	76.1	46.3	105.5	754.5	659.1	95.4	12.6	3 762.3	3 621.0
3rd quarter	874.7	628.2	78.5	20.9	117.5	76.3	46.7	100.7	774.0	671.3	102.7	13.3	3 794.4	3 678.9
4th quarter	883.7	630.0	78.9	21.2	120.0	80.1	46.5	101.5	782.2	678.2	103.9	13.3	3 770.6	3 670.1
1971														
1st quarter	903.7	647.9	80.3	21.1	122.8	82.0	50.5	98.3	805.3	697.6	107.7	13.4	3 819.1	3 743.3
2nd quarter	927.1	659.8	82.8	21.7	124.2	89.6	51.0	100.7	826.4	712.1	114.2	13.8	3 849.0	3 797.8
3rd quarter	941.1	670.1	84.6	22.0	126.1	89.6	51.3	102.3	838.8	725.1	113.7	13.6	3 875.6	3 817.7
4th quarter	959.3	682.2	87.9	22.5	127.4	91.3	51.9	105.5	853.8	741.8	112.1	13.1	3 926.5	3 862.2
1972														
1st quarter	986.5	708.4	87.9	23.1	131.0	94.3	58.1	119.8	866.8	759.4	107.4	12.4	3 993.7	3 879.9
2nd quarter	1 004.2	722.5	91.3	20.2	134.1	94.9	58.8	123.4	880.8	778.4	102.4	11.6	4 046.9	3 920.2
3rd quarter	1 030.1	735.7	95.5	23.9	138.4	96.0	59.5	124.3	905.8	797.2	108.7	12.0	4 121.0	3 996.4
4th quarter	1 077.0	758.8	105.6	23.8	142.8	106.4	60.4	127.1	949.9	822.2	127.6	13.4	4 247.3	4 156.6
1973														
1st quarter	1 095.1	785.4	104.1	23.5	146.6	109.1	73.6	126.4	968.7	847.0	121.7	12.6	4 263.1	4 188.1
2nd quarter	1 124.9	803.2	109.9	23.3	151.6	111.5	74.7	129.2	995.7	863.0	132.7	13.3	4 298.7	4 223.5
3rd quarter	1 152.5	820.4	113.8	22.3	158.8	113.3	76.1	134.1	1 018.4	881.8	136.6	13.4	4 329.2	4 242.6
4th quarter	1 190.7	841.7	122.2	23.3	164.6	116.5	77.6	140.0	1 050.7	898.4	152.3	14.5	4 384.9	4 289.0
1974														
1st quarter	1 206.8	858.3	115.7	23.5	170.5	121.9	83.1	142.8	1 064.0	915.3	148.7	14.0	4 300.7	4 217.8
2nd quarter	1 232.9	878.5	108.6	22.9	177.6	129.9	84.7	148.9	1 084.0	944.4	139.6	12.9	4 251.8	4 178.5
3rd quarter	1 269.1	900.2	111.0	23.3	183.9	137.1	86.4	154.9	1 114.2	973.9	140.2	12.6	4 249.1	4 182.3
4th quarter	1 298.5	913.8	113.5	23.0	190.4	144.3	86.6	157.6	1 140.9	984.3	156.5	13.7	4 225.4	4 176.7
1975														
1st quarter	1 318.4	918.4	112.7	22.7	196.4	155.9	87.6	158.0	1 160.4	1 011.8	148.6	12.8	4 177.9	4 170.5
2nd quarter	1 350.0	931.2	114.3	22.4	198.5	171.5	88.0	121.1	1 228.9	1 040.4	188.5	15.3	4 184.1	4 363.4
3rd quarter	1 386.5	956.0	120.7	22.2	202.5	174.8	89.8	152.8	1 233.6	1 074.8	158.9	12.9	4 223.3	4 299.9
4th quarter	1 422.7	983.4	125.2	21.9	206.5	177.6	91.8	158.5	1 264.2	1 104.0	160.1	12.7	4 268.1	4 333.5
1976														
1st quarter	1 456.1	1 014.7	126.4	21.6	210.5	181.8	98.9	162.5	1 293.6	1 137.3	156.3	12.1	4 320.6	4 385.9
2nd quarter	1 481.2	1 035.5	128.6	20.5	217.1	180.0	100.4	169.3	1 311.9	1 157.1	154.8	11.8	4 374.8	4 410.8
3rd quarter	1 518.4	1 058.4	132.7	20.0	222.8	186.7	102.2	176.1	1 342.3	1 186.9	155.4	11.6	4 410.1	4 445.2
4th quarter	1 554.9	1 084.8	136.2	19.1	229.7	188.8	103.8	182.7	1 372.1	1 221.1	151.0	11.0	4 453.5	4 473.4
1977														
1st quarter	1 588.8	1 113.3	138.6	16.4	238.8	191.0	109.3	188.8	1 400.0	1 258.2	141.9	10.1	4 476.5	4 483.5
2nd quarter	1 633.0	1 150.3	140.5	16.3	246.6	191.4	112.1	195.7	1 437.3	1 287.0	150.4	10.5	4 539.0	4 525.6
3rd quarter	1 677.7	1 182.1	142.4	15.6	255.4	196.6	114.3	198.6	1 479.2	1 318.8	160.4	10.8	4 594.6	4 588.5
4th quarter	1 737.4	1 217.6	156.5	15.1	265.5	199.3	116.7	208.5	1 528.9	1 357.8	171.1	11.2	4 704.1	4 676.2

Table 4-1B. Personal Income and Its Disposition: Historical Data—*Continued*

(Billions of current dollars, except as noted; quarterly data are at seasonally adjusted annual rates.) NIPA Table 2.1

Year and quarter	Personal income Total	Compensation of employees	Proprietors' income with IVA and CCAdj	Rental income of persons with CCAdj	Personal income receipts on assets	Personal current transfer receipts	Less: Contributions for government social insurance, domestic	Less: Personal current taxes	Equals: Disposable personal income	Less: Personal outlays	Equals: Personal saving Billions of dollars	Percent of disposable personal income	Billions of chained (2009) dollars Personal income excluding current transfers	Disposable personal income
1978														
1st quarter	1 778.1	1 250.9	158.5	15.9	273.3	203.3	123.9	212.0	1 566.1	1 388.5	177.6	11.3	4 738.9	4 712.9
2nd quarter ...	1 838.2	1 299.1	166.1	15.6	281.3	205.1	129.0	223.1	1 615.1	1 447.5	167.6	10.4	4 815.7	4 762.7
3rd quarter	1 892.6	1 336.9	168.3	17.1	289.8	213.9	133.4	236.3	1 656.3	1 479.8	176.5	10.7	4 864.7	4 799.8
4th quarter	1 946.0	1 380.2	170.9	17.5	298.9	217.2	138.8	247.2	1 698.7	1 520.2	178.5	10.5	4 916.7	4 831.4
1979														
1st quarter	2 005.4	1 422.5	180.1	17.3	308.8	222.7	146.0	253.6	1 751.8	1 556.7	195.0	11.1	4 976.1	4 889.8
2nd quarter ...	2 045.4	1 454.7	179.2	15.6	318.8	227.4	150.3	262.0	1 783.4	1 598.3	185.1	10.4	4 939.6	4 845.6
3rd quarter	2 110.9	1 496.7	180.7	15.4	330.4	243.0	155.4	274.8	1 836.0	1 653.9	182.1	9.9	4 952.0	4 867.6
4th quarter	2 169.0	1 534.9	177.8	16.1	350.4	249.1	159.4	285.2	1 883.8	1 698.9	184.9	9.8	4 969.6	4 876.2
1980														
1st quarter	2 231.9	1 573.6	167.6	15.8	377.5	259.4	161.9	284.8	1 947.1	1 751.3	195.8	10.1	4 956.9	4 893.1
2nd quarter ...	2 269.4	1 599.2	160.3	17.1	391.2	264.6	162.9	292.2	1 977.3	1 753.6	223.6	11.3	4 917.8	4 850.1
3rd quarter	2 348.5	1 628.6	173.4	19.9	397.0	296.6	167.0	302.2	2 046.3	1 812.2	234.1	11.4	4 918.0	4 904.5
4th quarter	2 444.7	1 687.6	185.0	23.2	421.9	300.0	173.0	318.9	2 125.8	1 883.0	242.9	11.4	5 016.2	4 972.1
1981														
1st quarter	2 508.9	1 739.6	188.2	24.0	440.5	306.5	189.9	330.9	2 178.0	1 941.4	236.6	10.9	5 020.8	4 965.1
2nd quarter ...	2 559.8	1 774.8	176.4	23.3	469.0	309.9	193.6	342.7	2 217.1	1 975.7	241.4	10.9	5 044.3	4 970.9
3rd quarter	2 659.1	1 815.1	182.3	23.1	508.5	328.5	198.4	356.9	2 302.2	2 019.6	282.6	12.3	5 140.8	5 078.1
4th quarter	2 692.7	1 840.7	171.8	24.7	525.4	331.1	201.0	352.7	2 340.0	2 038.8	301.2	12.9	5 130.6	5 083.7
1982														
1st quarter	2 727.8	1 864.7	165.2	23.5	544.1	336.3	206.0	352.5	2 375.3	2 082.1	293.2	12.3	5 130.7	5 095.9
2nd quarter ...	2 772.3	1 884.4	169.4	21.1	559.9	345.5	208.0	359.7	2 412.5	2 110.9	301.6	12.5	5 156.9	5 126.7
3rd quarter	2 812.5	1 904.8	170.7	24.4	560.8	362.1	210.3	350.1	2 462.3	2 159.3	303.0	12.3	5 126.2	5 151.3
4th quarter	2 853.8	1 918.1	179.5	26.3	563.2	377.9	211.2	356.6	2 497.2	2 221.7	275.5	11.0	5 123.4	5 167.5
1983														
1st quarter	2 892.4	1 947.2	183.3	25.8	574.3	380.7	218.9	350.9	2 541.5	2 263.2	278.4	11.0	5 154.9	5 216.2
2nd quarter ...	2 942.1	1 986.3	182.6	24.7	584.4	386.9	222.9	359.6	2 582.5	2 330.6	251.9	9.8	5 196.7	5 252.4
3rd quarter	3 001.2	2 029.6	183.8	23.7	609.2	382.5	227.7	345.4	2 655.8	2 403.9	251.8	9.5	5 256.8	5 331.2
4th quarter	3 088.6	2 086.8	195.3	23.5	630.2	387.0	234.3	355.7	2 732.9	2 458.9	274.0	10.0	5 387.3	5 449.9
1984														
1st quarter	3 178.9	2 143.8	221.9	23.2	645.6	393.9	249.5	361.2	2 817.8	2 506.0	311.8	11.1	5 494.2	5 558.8
2nd quarter ...	3 262.9	2 195.2	231.8	21.7	671.4	398.3	255.5	370.4	2 892.4	2 566.5	325.9	11.3	5 596.8	5 651.1
3rd quarter	3 337.4	2 241.6	232.4	24.4	698.4	401.0	260.5	384.1	2 953.3	2 607.2	346.1	11.7	5 693.0	5 725.7
4th quarter	3 391.7	2 282.9	226.9	29.3	707.7	409.4	264.5	395.9	2 995.7	2 658.1	337.6	11.3	5 746.5	5 772.5
1985														
1st quarter	3 455.4	2 323.5	240.7	26.9	718.6	420.0	274.3	432.3	3 023.1	2 740.5	282.7	9.3	5 780.7	5 757.4
2nd quarter ...	3 495.5	2 363.6	238.5	25.3	723.8	422.6	278.3	388.5	3 107.0	2 789.6	317.3	10.2	5 805.1	5 869.5
3rd quarter	3 543.1	2 406.4	240.7	27.1	724.1	427.9	283.2	421.5	3 121.5	2 865.2	256.4	8.2	5 839.1	5 851.1
4th quarter	3 605.5	2 455.7	244.3	25.6	738.7	430.9	289.6	428.9	3 176.6	2 893.3	283.3	8.9	5 909.4	5 913.0
1986														
1st quarter	3 667.6	2 491.1	245.6	22.0	762.3	443.2	296.7	426.3	3 241.3	2 940.1	301.2	9.3	5 959.5	5 990.8
2nd quarter ...	3 708.8	2 517.2	251.6	21.1	770.2	449.1	300.4	429.4	3 279.4	2 969.3	310.0	9.5	6 031.1	6 067.4
3rd quarter	3 757.1	2 555.2	264.4	16.6	770.3	456.1	305.5	439.5	3 317.7	3 036.0	281.7	8.5	6 075.5	6 106.1
4th quarter	3 798.8	2 604.7	264.2	13.4	769.8	457.8	311.1	456.0	3 342.8	3 073.2	269.5	8.1	6 112.1	6 115.4
1987														
1st quarter	3 856.0	2 647.8	275.4	14.5	771.1	463.1	315.9	450.7	3 405.3	3 102.2	303.0	8.9	6 149.3	6 171.7
2nd quarter ...	3 917.5	2 692.0	282.3	14.1	781.0	468.1	320.0	511.7	3 405.8	3 172.7	233.2	6.8	6 192.2	6 114.0
3rd quarter	3 988.5	2 740.3	289.5	17.5	796.9	469.2	324.8	489.0	3 499.6	3 238.6	260.9	7.5	6 258.6	6 223.4
4th quarter	4 084.3	2 809.6	298.6	20.3	815.4	472.2	331.7	507.0	3 577.4	3 273.0	304.4	8.5	6 368.7	6 307.3
1988														
1st quarter	4 160.6	2 857.6	319.2	21.1	824.0	489.9	351.1	502.1	3 658.5	3 355.5	303.0	8.3	6 421.1	6 399.9
2nd quarter ...	4 234.8	2 923.1	322.5	20.3	833.7	493.5	358.3	497.8	3 737.0	3 417.4	319.6	8.6	6 473.2	6 465.8
3rd quarter	4 326.3	2 975.8	333.6	21.2	860.5	499.6	364.5	506.7	3 819.6	3 489.6	330.0	8.6	6 540.3	6 528.3
4th quarter	4 411.9	3 035.4	326.8	27.4	887.5	506.8	372.0	517.2	3 894.7	3 565.3	329.4	8.5	6 607.5	6 589.8
1989														
1st quarter	4 537.2	3 079.1	348.1	23.1	933.6	531.4	378.0	552.9	3 984.4	3 626.3	358.0	9.0	6 700.7	6 664.9
2nd quarter ...	4 592.9	3 114.6	339.5	21.2	961.5	538.8	382.6	566.7	4 026.3	3 693.9	332.3	8.3	6 691.3	6 645.2
3rd quarter	4 648.5	3 155.0	337.2	20.0	975.6	547.9	387.2	571.6	4 076.9	3 751.7	325.2	8.0	6 728.3	6 689.4
4th quarter	4 723.7	3 209.7	339.7	21.8	986.6	558.8	393.1	579.8	4 143.9	3 799.5	344.4	8.3	6 780.1	6 746.0
1990														
1st quarter	4 818.9	3 272.8	345.6	23.1	999.5	579.5	401.6	582.5	4 236.3	3 884.5	351.9	8.3	6 802.7	6 797.8
2nd quarter ...	4 899.2	3 334.7	351.1	26.7	1 003.1	590.5	406.9	594.6	4 304.6	3 930.6	374.0	8.7	6 851.7	6 845.1
3rd quarter	4 958.5	3 372.5	358.9	31.6	1 010.0	600.0	414.6	600.7	4 357.8	3 997.1	360.7	8.3	6 843.8	6 842.7
4th quarter	4 978.6	3 381.5	357.0	31.3	1 008.7	617.5	417.4	600.8	4 377.8	4 019.8	357.9	8.2	6 758.5	6 784.3
1991														
1st quarter	4 992.5	3 387.5	347.5	33.0	1 003.9	641.6	421.0	580.8	4 411.7	4 028.3	383.4	8.7	6 707.5	6 801.3
2nd quarter ...	5 054.1	3 427.6	353.0	35.5	1 004.2	661.4	427.7	585.9	4 468.2	4 083.2	384.9	8.6	6 735.3	6 851.0
3rd quarter	5 107.1	3 469.9	353.6	39.4	1 006.8	670.8	433.5	590.2	4 516.8	4 130.5	386.3	8.6	6 756.4	6 879.0
4th quarter	5 186.0	3 517.0	362.6	46.4	999.9	698.8	438.6	598.7	4 587.4	4 158.2	429.1	9.4	6 784.9	6 936.2

Table 4-1B. Personal Income and Its Disposition: Historical Data—*Continued*

(Billions of current dollars, except as noted; quarterly data are at seasonally adjusted annual rates.)　　　　NIPA Table 2.1

| Year and quarter | Personal income | | | | | | | Less: Personal current taxes | Equals: Disposable personal income | Less: Personal outlays | Equals: Personal saving | | Billions of chained (2009) dollars | |
	Total	Compensation of employees	Proprietors' income with IVA and CCAdj	Rental income of persons with CCAdj	Personal income receipts on assets	Personal current transfer receipts	Less: Contributions for government social insurance, domestic				Billions of dollars	Percent of disposable personal income	Personal income excluding current transfers	Disposable personal income
1992														
1st quarter	5 297.8	3 595.7	378.7	50.0	996.8	727.0	450.4	588.9	4 708.9	4 255.1	453.8	9.6	6 868.3	7 075.8
2nd quarter ...	5 394.4	3 651.9	395.2	57.3	1 000.4	745.6	456.0	607.2	4 787.2	4 311.2	476.0	9.9	6 939.3	7 146.0
3rd quarter	5 456.3	3 688.5	407.9	64.3	997.1	757.6	459.1	616.2	4 840.1	4 388.0	452.1	9.3	6 969.4	7 179.1
4th quarter	5 535.0	3 736.9	418.8	70.7	1 000.9	762.0	454.4	638.9	4 896.1	4 462.5	433.5	8.9	7 030.6	7 211.8
1993														
1st quarter	5 560.7	3 745.6	418.9	80.7	1 007.7	781.6	473.8	617.0	4 943.7	4 508.3	435.4	8.8	6 997.7	7 238.8
2nd quarter ...	5 635.8	3 800.2	429.6	86.9	1 007.4	786.0	474.2	643.5	4 992.3	4 578.0	414.3	8.3	7 053.9	7 261.2
3rd quarter	5 677.6	3 834.1	423.4	92.5	1 006.4	799.9	478.8	659.2	5 018.4	4 646.1	372.3	7.4	7 063.7	7 267.5
4th quarter	5 757.7	3 889.3	440.0	100.2	1 006.7	804.3	482.9	675.3	5 082.4	4 713.7	368.7	7.3	7 132.2	7 318.0
1994														
1st quarter	5 807.9	3 913.4	450.2	109.0	1 015.2	818.1	498.0	673.7	5 134.3	4 783.9	350.4	6.8	7 159.1	7 366.3
2nd quarter ...	5 912.7	3 991.5	454.1	113.5	1 035.9	822.9	505.1	697.8	5 214.9	4 850.1	364.8	7.0	7 262.2	7 440.5
3rd quarter	5 978.4	4 031.8	456.8	116.3	1 057.7	826.9	511.0	695.4	5 283.0	4 925.8	357.2	6.8	7 297.6	7 483.9
4th quarter	6 089.4	4 088.1	465.5	115.9	1 090.4	848.0	518.5	705.4	5 384.0	5 002.8	381.2	7.1	7 390.3	7 591.3
1995														
1st quarter	6 181.2	4 134.4	467.4	118.3	1 112.7	874.0	525.5	724.6	5 456.6	5 042.6	414.0	7.6	7 446.7	7 656.2
2nd quarter ...	6 250.7	4 172.9	472.0	121.6	1 130.9	883.5	530.0	746.8	5 503.9	5 124.1	379.9	6.9	7 487.3	7 677.9
3rd quarter	6 329.3	4 219.7	484.5	125.7	1 144.3	890.4	535.4	752.2	5 577.0	5 195.4	381.6	6.8	7 556.3	7 748.3
4th quarter	6 404.3	4 265.3	501.0	134.0	1 158.4	886.0	540.3	770.0	5 634.3	5 261.6	372.7	6.6	7 633.1	7 793.5
1996														
1st quarter	6 521.0	4 314.5	523.3	139.5	1 170.8	916.0	543.2	801.7	5 719.2	5 337.7	381.5	6.7	7 710.1	7 867.3
2nd quarter ,,,	6 649.8	4 387.5	547.5	141.6	1 188.5	936.4	551.6	839.6	5 810.3	5 432.5	377.8	6.5	7 807.2	7 939.5
3rd quarter	6 725.9	4 451.2	547.5	143.7	1 210.7	931.8	559.0	843.5	5 882.3	5 492.0	390.4	6.6	7 883.7	8 003.8
4th quarter	6 817.5	4 514.5	556.8	145.3	1 234.7	932.6	566.5	863.5	5 954.0	5 574.5	379.4	6.4	7 953.0	8 046.4
1997														
1st quarter	6 939.4	4 592.4	575.6	144.2	1 251.4	950.2	574.4	902.1	6 037.2	5 658.1	379.1	6.3	8 058.3	8 123.0
2nd quarter ...	7 022.3	4 660.0	575.1	144.6	1 274.5	950.0	581.9	916.2	6 106.0	5 702.5	403.6	6.6	8 149.7	8 195.0
3rd quarter	7 135.3	4 738.9	588.5	147.4	1 294.9	956.2	590.5	941.1	6 194.2	5 815.8	378.4	6.1	8 271.3	8 291.5
4th quarter	7 273.0	4 844.0	596.6	152.1	1 319.4	963.1	602.2	967.8	6 305.2	5 905.2	400.0	6.3	8 419.9	8 413.6
1998														
1st quarter	7 434.6	4 942.0	617.4	156.8	1 353.9	974.7	610.3	996.1	6 438.4	5 961.5	476.9	7.4	8 619.3	8 590.7
2nd quarter ...	7 561.1	5 028.3	629.6	162.6	1 379.1	981.1	619.7	1 022.4	6 538.7	6 082.6	456.1	7.0	8 763.8	8 708.8
3rd quarter	7 668.1	5 114.9	645.0	168.0	1 385.2	984.5	629.5	1 043.2	6 624.9	6 181.1	443.8	6.7	8 874.5	8 796.5
4th quarter	7 762.9	5 199.4	668.7	173.4	1 365.3	995.4	639.2	1 068.0	6 694.9	6 285.6	409.3	6.1	8 962.3	8 866.2
1999														
1st quarter	7 850.9	5 288.0	681.9	176.1	1 340.5	1 014.6	650.2	1 077.9	6 773.0	6 354.8	418.1	6.2	9 030.1	8 946.3
2nd quarter ...	7 920.4	5 344.3	690.1	179.6	1 344.4	1 017.8	655.7	1 095.2	6 825.3	6 484.9	340.4	5.0	9 068.1	8 966.5
3rd quarter	8 028.8	5 423.3	699.8	179.4	1 358.4	1 030.9	663.0	1 120.6	6 908.2	6 593.6	314.6	4.6	9 144.8	9 027.7
4th quarter	8 207.3	5 555.5	713.9	178.9	1 394.0	1 041.4	676.2	1 154.0	7 053.4	6 729.1	324.3	4.6	9 309.5	9 163.2
2000														
1st quarter	8 457.2	5 750.0	720.9	179.8	1 445.8	1 056.7	696.0	1 208.8	7 248.5	6 892.1	356.3	4.9	9 534.5	9 338.7
2nd quarter ...	8 591.8	5 792.7	751.3	180.2	1 482.3	1 083.7	698.4	1 230.2	7 361.6	6 995.9	365.7	5.0	9 629.8	9 442.0
3rd quarter	8 741.0	5 904.9	760.8	182.9	1 506.0	1 098.0	711.6	1 247.7	7 493.3	7 119.9	373.3	5.0	9 742.4	9 551.6
4th quarter	8 820.4	5 944.7	782.4	191.0	1 508.9	1 110.8	717.3	1 258.7	7 561.7	7 224.0	337.7	4.5	9 773.2	9 585.7
2001														
1st quarter	8 984.0	6 050.8	809.8	197.1	1 503.0	1 155.6	732.3	1 301.9	7 682.1	7 298.0	384.1	5.0	9 856.7	9 672.6
2nd quarter ...	9 013.6	6 041.7	827.7	202.6	1 487.3	1 187.4	733.1	1 308.9	7 704.7	7 347.0	357.7	4.6	9 807.9	9 655.7
3rd quarter	9 000.8	6 025.4	845.6	204.8	1 465.2	1 192.0	732.4	1 113.6	7 887.2	7 376.2	511.0	6.5	9 780.1	9 878.5
4th quarter	9 024.0	6 038.6	840.9	205.1	1 439.3	1 235.2	735.0	1 231.8	7 792.3	7 478.6	313.7	4.0	9 749.4	9 753.7
2002														
1st quarter	9 057.8	6 061.1	859.7	209.1	1 411.0	1 260.0	743.1	1 075.1	7 982.7	7 507.2	475.5	6.0	9 742.5	9 973.5
2nd quarter ...	9 148.8	6 129.3	863.7	211.0	1 414.1	1 282.3	751.5	1 051.0	8 097.9	7 593.0	504.8	6.2	9 754.3	10 041.1
3rd quarter	9 174.4	6 157.8	872.4	210.4	1 396.0	1 292.1	754.3	1 044.1	8 130.3	7 678.3	452.0	5.6	9 726.2	10 032.3
4th quarter	9 254.8	6 194.2	883.2	213.6	1 414.4	1 306.4	757.0	1 038.4	8 216.4	7 753.9	462.5	5.6	9 762.7	10 091.9
2003														
1st quarter	9 315.1	6 213.6	874.9	223.0	1 442.7	1 324.2	763.3	1 021.3	8 293.8	7 841.5	452.2	5.5	9 746.1	10 115.6
2nd quarter ...	9 417.8	6 307.7	891.5	227.7	1 424.2	1 340.6	773.9	1 020.8	8 397.0	7 929.4	467.6	5.6	9 848.9	10 238.9
3rd quarter	9 542.3	6 394.1	905.0	232.0	1 434.0	1 361.0	783.8	950.6	8 591.6	8 090.9	500.8	5.8	9 914.6	10 411.9
4th quarter	9 675.1	6 500.8	916.2	243.3	1 447.7	1 363.2	796.1	1 021.3	8 653.7	8 188.3	465.4	5.4	10 026.9	10 439.3
2004														
1st quarter	9 777.9	6 542.0	945.7	247.5	1 454.8	1 397.1	809.2	1 012.2	8 765.7	8 328.7	437.0	5.0	10 026.9	10 487.4
2nd quarter ...	9 953.6	6 669.1	962.1	252.1	1 478.0	1 416.0	823.6	1 026.7	8 926.9	8 439.3	487.6	5.5	10 145.0	10 607.6
3rd quarter	10 094.7	6 804.3	963.2	248.5	1 492.3	1 425.5	839.2	1 064.3	9 030.4	8 576.7	453.7	5.0	10 249.8	10 676.9
4th quarter	10 314.1	6 864.8	977.1	247.4	1 623.4	1 446.3	844.9	1 091.5	9 222.5	8 745.7	476.8	5.2	10 395.8	10 811.7
2005														
1st quarter	10 338.5	6 926.9	956.6	240.1	1 584.8	1 488.1	858.1	1 172.2	9 166.3	8 839.9	326.3	3.6	10 316.7	10 684.9
2nd quarter ...	10 505.2	7 006.0	963.4	237.3	1 652.1	1 512.6	866.3	1 196.3	9 308.9	9 015.1	293.8	3.2	10 419.8	10 786.5
3rd quarter	10 661.7	7 122.9	986.8	225.6	1 677.6	1 528.3	879.5	1 225.4	9 436.3	9 199.8	236.4	2.5	10 471.1	10 818.3
4th quarter	10 887.6	7 210.6	1 005.4	225.0	1 798.2	1 537.9	889.5	1 255.7	9 631.9	9 301.6	330.2	3.4	10 635.9	10 956.9

Table 4-1B. Personal Income and Its Disposition: Historical Data—*Continued*

(Billions of current dollars, except as noted; quarterly data are at seasonally adjusted annual rates.) **NIPA Table 2.1**

Year and quarter	Personal income							Less: Personal current taxes	Equals: Disposable personal income	Less: Personal outlays	Equals: Personal saving		Billions of chained (2009) dollars	
	Total	Compensation of employees	Proprietors' income with IVA and CCAdj	Rental income of persons with CCAdj	Personal income receipts on assets	Personal current transfer receipts	Less: Contributions for government social insurance, domestic				Billions of dollars	Percent of disposable personal income	Personal income excluding current transfers	Disposable personal income
2006														
1st quarter	11 189.5	7 380.2	1 048.1	216.7	1 869.5	1 588.3	913.2	1 320.3	9 869.2	9 451.7	417.5	4.2	10 866.8	11 170.1
2nd quarter ...	11 324.0	7 428.1	1 050.9	206.3	1 953.9	1 602.8	918.1	1 351.0	9 973.0	9 582.5	390.4	3.9	10 914.7	11 197.4
3rd quarter ...	11 429.4	7 487.2	1 047.7	199.5	1 989.3	1 628.3	922.6	1 358.5	10 070.9	9 718.5	352.4	3.5	10 925.5	11 226.4
4th quarter ...	11 583.9	7 624.0	1 051.9	186.8	2 021.6	1 635.9	936.2	1 397.3	10 186.6	9 804.5	382.1	3.8	11 108.1	11 374.5
2007														
1st quarter	11 836.7	7 806.8	1 016.2	174.2	2 088.5	1 706.7	955.7	1 466.3	10 370.4	9 951.9	418.6	4.0	11 205.5	11 471.4
2nd quarter ...	11 977.0	7 845.4	995.9	183.8	2 198.8	1 710.4	957.3	1 495.6	10 481.3	10 065.1	416.2	4.0	11 265.2	11 500.8
3rd quarter ...	12 048.4	7 885.1	985.1	187.5	2 218.6	1 732.7	960.6	1 498.6	10 549.8	10 177.4	372.4	3.5	11 255.6	11 511.0
4th quarter ...	12 169.1	7 978.2	978.9	191.9	2 229.5	1 762.7	972.1	1 508.3	10 660.8	10 301.3	359.4	3.4	11 244.0	11 518.8
2008														
1st quarter	12 315.0	8 055.3	964.7	220.5	2 255.6	1 802.9	984.0	1 534.8	10 780.2	10 357.7	422.4	3.9	11 263.6	11 550.8
2nd quarter ...	12 642.6	8 054.8	968.6	245.0	2 212.0	2 148.4	986.2	1 552.1	11 090.5	10 469.9	620.6	5.6	11 129.8	11 762.2
3rd quarter ...	12 467.2	8 074.0	958.2	268.1	2 217.0	1 941.4	991.5	1 497.2	10 970.0	10 497.9	472.0	4.3	11 048.8	11 515.0
4th quarter ...	12 344.0	8 043.8	952.2	293.0	2 118.9	1 927.8	991.7	1 444.6	10 899.4	10 234.8	664.7	6.1	11 100.4	11 615.3
2009														
1st quarter	11 989.6	7 729.8	914.3	303.4	1 962.2	2 039.7	959.8	1 202.1	10 787.5	10 155.4	632.1	5.9	10 667.5	11 565.5
2nd quarter ...	12 083.2	7 761.4	902.3	318.4	1 888.7	2 178.6	966.3	1 130.8	10 952.4	10 158.6	793.8	7.2	10 571.4	11 689.8
3rd quarter ...	12 038.3	7 746.2	935.3	338.5	1 805.8	2 176.3	963.8	1 135.0	10 903.4	10 293.5	609.8	5.6	10 453.8	11 557.6
4th quarter ...	12 125.3	7 796.7	1 002.2	349.0	1 752.3	2 192.3	967.2	1 140.4	10 984.9	10 354.6	630.3	5.7	10 448.3	11 554.8
2010														
1st quarter	12 275.9	7 767.2	1 056.9	371.8	1 750.3	2 303.2	973.6	1 191.5	11 084.4	10 434.4	650.0	5.9	10 454.3	11 619.8
2nd quarter ...	12 492.6	7 910.1	1 104.0	387.9	1 764.0	2 311.1	984.5	1 212.9	11 279.7	10 518.3	761.4	6.8	10 661.3	11 811.2
3rd quarter ...	12 636.0	7 981.8	1 129.4	399.5	1 776.5	2 336.1	987.4	1 255.9	11 380.1	10 605.8	774.2	6.8	10 766.2	11 895.3
4th quarter ...	12 801.9	8 040.7	1 144.4	417.6	1 838.5	2 350.1	989.5	1 288.8	11 513.1	10 735.3	777.8	6.8	10 859.2	11 962.0
2011														
1st quarter	13 158.2	8 170.0	1 181.0	453.4	1 903.8	2 361.8	911.8	1 426.1	11 732.1	10 866.5	865.7	7.4	11 120.1	12 083.9
2nd quarter ...	13 267.3	8 206.1	1 212.2	471.6	1 930.2	2 361.6	914.5	1 445.4	11 821.9	10 994.3	827.7	7.0	11 123.1	12 057.6
3rd quarter ...	13 402.0	8 287.1	1 253.5	485.0	1 947.3	2 352.1	922.9	1 470.9	11 931.1	11 087.8	843.3	7.1	11 215.8	12 110.2
4th quarter ...	13 479.5	8 240.4	1 270.4	504.6	2 022.3	2 359.3	917.4	1 470.4	12 009.2	11 146.4	862.7	7.2	11 248.7	12 147.9
2012														
1st quarter	13 785.7	8 455.9	1 322.8	509.9	2 097.0	2 340.4	940.3	1 467.8	12 317.9	11 309.1	1 008.9	8.2	11 498.6	12 375.3
2nd quarter ...	13 946.8	8 503.0	1 356.5	517.5	2 151.7	2 362.8	944.7	1 487.1	12 459.7	11 353.5	1 106.1	8.9	11 610.1	12 487.8
3rd quarter ...	13 915.4	8 545.3	1 345.8	520.5	2 086.7	2 364.7	947.6	1 509.5	12 406.0	11 398.0	1 008.0	8.1	11 543.7	12 398.4
4th quarter ...	14 392.6	8 762.6	1 364.2	524.0	2 327.1	2 384.2	969.4	1 571.4	12 821.2	11 513.9	1 307.3	10.2	11 934.1	12 741.9
2013														
1st quarter	14 001.2	8 731.7	1 388.3	537.9	2 030.5	2 403.5	1 090.6	1 649.3	12 351.9	11 608.6	743.2	6.0	11 485.1	12 231.9
2nd quarter ...	14 135.0	8 818.7	1 411.8	550.2	2 041.1	2 416.2	1 103.1	1 681.9	12 453.1	11 625.5	827.7	6.6	11 596.4	12 323.0
3rd quarter ...	14 232.0	8 844.7	1 412.5	564.6	2 082.0	2 434.5	1 106.3	1 674.5	12 557.5	11 716.7	840.9	6.7	11 627.3	12 376.3
4th quarter ...	14 356.2	8 941.7	1 401.7	575.2	2 111.6	2 443.2	1 117.2	1 697.7	12 658.5	11 864.9	793.6	6.3	11 693.5	12 425.2
2014														
1st quarter	14 639.3	9 113.0	1 413.4	588.4	2 190.1	2 474.1	1 139.8	1 744.4	12 894.9	11 976.2	918.6	7.1	11 885.4	12 598.3
2nd quarter ...	14 890.8	9 171.3	1 452.7	599.2	2 285.3	2 526.8	1 144.5	1 757.8	13 133.0	12 164.8	968.2	7.4	12 022.2	12 769.9
3rd quarter ...	15 120.0	9 279.6	1 468.3	609.9	2 350.1	2 568.2	1 156.1	1 797.9	13 322.0	12 332.5	989.5	7.4	12 172.0	12 919.0
4th quarter ...	15 316.8	9 432.5	1 456.6	621.0	2 383.4	2 597.1	1 173.8	1 835.8	13 481.0	12 474.3	1 006.7	7.5	12 350.6	13 089.7
2015														
1st quarter	15 478.5	9 563.3	1 432.9	623.1	2 405.4	2 642.8	1 188.9	1 904.0	13 574.6	12 526.6	1 048.0	7.7	12 517.9	13 238.4
2nd quarter ...	15 685.0	9 667.2	1 406.3	647.7	2 485.8	2 679.2	1 201.1	1 940.8	13 744.2	12 692.8	1 051.5	7.7	12 621.4	13 338.0
3rd quarter ...	15 831.1	9 748.7	1 425.9	658.1	2 512.7	2 696.4	1 210.6	1 946.7	13 884.3	12 827.6	1 056.7	7.6	12 710.6	13 436.0
4th quarter ...	15 876.4	9 813.4	1 423.7	663.5	2 479.3	2 714.7	1 218.2	1 959.6	13 916.8	12 877.9	1 038.9	7.5	12 747.3	13 478.6
2016														
1st quarter	15 937.6	9 843.5	1 415.2	669.9	2 490.6	2 743.7	1 225.3	1 922.0	14 015.6	12 977.5	1 038.1	7.4	12 773.2	13 568.7
2nd quarter ...	16 029.0	9 900.1	1 410.2	680.2	2 505.3	2 765.5	1 232.4	1 945.3	14 083.7	13 138.6	945.1	6.7	12 764.9	13 554.3
3rd quarter ...	16 175.5	9 993.2	1 429.5	683.6	2 529.4	2 783.7	1 244.0	1 969.6	14 205.9	13 280.4	925.5	6.5	12 834.8	13 615.0
4th quarter ...	16 342.6	10 104.5	1 440.0	692.1	2 560.2	2 803.8	1 258.0	1 987.4	14 355.2	13 428.6	926.5	6.5	12 917.8	13 696.7
2017														
1st quarter	16 604.1	10 227.6	1 494.8	707.4	2 630.7	2 823.6	1 280.0	2 001.5	14 602.6	13 576.8	1 025.8	7.0	13 080.6	13 860.9
2nd quarter ...	16 749.6	10 334.2	1 512.2	709.9	2 657.1	2 828.2	1 292.0	2 016.0	14 733.5	13 699.7	1 033.9	7.0	13 184.2	13 953.4
3rd quarter ...	16 930.4	10 456.7	1 523.1	722.0	2 671.3	2 861.9	1 304.6	2 049.8	14 880.6	13 841.8	1 038.8	7.0	13 268.6	14 034.5
4th quarter ...	17 231.2	10 628.0	1 542.9	736.0	2 767.4	2 878.8	1 321.8	2 115.8	15 115.4	14 090.2	1 025.2	6.8	13 447.6	14 162.4
2018														
1st quarter	17 540.3	10 786.0	1 567.5	743.8	2 851.6	2 935.4	1 344.0	2 074.9	15 465.4	14 245.2	1 220.2	7.9	13 599.1	14 400.3
2nd quarter ...	17 725.0	10 876.1	1 573.3	754.0	2 909.3	2 963.1	1 350.9	2 071.7	15 653.3	14 465.9	1 187.4	7.6	13 670.5	14 495.9
3rd quarter ...	17 928.5	10 994.3	1 590.0	765.2	2 957.7	2 983.8	1 362.4	2 086.5	15 842.0	14 655.6	1 186.4	7.5	13 785.6	14 613.3
4th quarter ...	18 082.8	11 057.4	1 624.4	764.1	3 002.0	3 003.7	1 368.7	2 077.4	16 005.4	14 757.8	1 247.6	7.8	13 863.6	14 715.2

Table 4-2. Personal Consumption Expenditures by Major Type of Product

(Billions of dollars, quarterly data are at seasonally adjusted rates.)

NIPA Table 2.3.5

Year and quarter	Personal consumption expenditures, total	Total goods	Durable goods					Nondurable goods				
			Durable goods, total	Motor vehicles and parts	Furnishings and household equipment	Recreational goods and vehicles	Other durable goods	Nondurable goods, total	Food and beverages off-premises	Clothing and footwear	Gasoline and other energy goods	Other nondurable goods
1960	331.2	177.0	45.6	19.6	15.4	6.4	4.3	131.4	62.6	25.9	15.8	27.1
1961	341.5	178.8	44.2	17.7	15.6	6.6	4.3	134.6	63.7	26.6	15.7	28.6
1962	362.6	189.0	49.5	21.4	16.4	6.9	4.7	139.5	64.7	27.9	16.3	30.7
1963	382.0	198.2	54.2	24.2	17.5	7.6	4.9	143.9	65.9	28.6	16.9	32.5
1964	410.6	212.3	59.6	25.8	19.5	8.8	5.5	152.7	69.5	31.1	17.7	34.5
1965	443.0	229.7	66.4	29.6	20.7	10.1	6.0	163.3	74.4	32.7	19.1	37.1
1966	479.9	249.6	71.7	29.9	22.6	12.3	6.9	177.9	80.6	35.8	20.7	40.8
1967	506.7	259.0	74.0	29.6	23.7	13.6	7.1	185.0	82.6	37.5	21.9	43.0
1968	556.9	284.6	84.8	35.4	26.1	15.3	8.0	199.8	88.8	41.3	23.2	46.5
1969	603.6	304.7	90.5	37.4	27.6	16.7	8.7	214.2	95.4	44.3	25.0	49.5
1970	646.7	318.8	90.0	34.5	28.2	17.9	9.4	228.8	103.5	45.5	26.3	53.6
1971	699.9	342.1	102.4	43.2	29.9	19.2	10.1	239.7	107.1	49.0	27.6	55.9
1972	768.2	373.8	116.4	49.4	33.5	22.6	11.0	257.4	114.5	53.5	29.4	60.0
1973	849.6	416.6	130.5	54.4	38.0	25.3	12.9	286.1	126.7	59.2	34.3	65.8
1974	930.2	451.5	130.2	48.2	40.9	26.6	14.5	321.4	143.0	62.4	43.8	72.1
1975	1 030.5	491.3	142.2	52.6	42.6	30.4	16.5	349.2	156.6	66.9	48.0	77.7
1976	1 147.7	546.3	168.6	68.2	47.2	34.2	19.1	377.7	167.3	72.2	53.0	85.2
1977	1 274.0	600.4	192.0	79.8	53.4	37.7	21.1	408.4	179.8	79.3	57.8	91.5
1978	1 422.3	663.6	213.3	89.2	59.0	41.7	23.5	450.2	196.1	89.3	61.5	103.3
1979	1 585.4	737.9	226.3	90.2	65.3	45.8	25.1	511.6	218.4	96.4	80.4	116.5
1980	1 750.7	799.8	226.4	84.4	67.8	46.5	27.6	573.4	239.2	103.0	101.9	129.3
1981	1 934.0	869.4	243.9	93.0	71.5	49.9	29.6	625.4	255.3	113.2	113.4	143.5
1982	2 071.3	899.3	253.0	100.0	71.8	51.3	29.9	646.3	267.1	116.7	108.4	154.0
1983	2 281.6	973.8	295.0	122.9	79.8	58.7	33.7	678.8	277.0	126.4	106.5	168.8
1984	2 492.3	1 063.7	342.2	147.2	88.8	67.8	38.3	721.5	291.1	137.6	108.2	184.6
1985	2 712.8	1 137.6	380.4	170.1	94.6	74.1	41.6	757.2	303.0	146.8	110.5	196.9
1986	2 886.3	1 195.6	421.4	187.5	103.5	83.0	47.5	774.2	316.4	157.2	91.2	209.4
1987	3 076.3	1 256.3	442.0	188.2	109.5	91.8	52.5	814.3	324.3	167.7	96.4	225.9
1988	3 330.0	1 337.3	475.1	202.2	115.2	99.9	57.8	862.3	342.8	178.2	99.9	241.4
1989	3 576.8	1 423.8	494.3	207.8	121.4	103.9	61.3	929.5	365.4	190.4	110.4	263.3
1990	3 809.0	1 491.3	497.1	205.1	120.9	105.6	65.5	994.2	391.2	195.2	124.2	283.6
1991	3 943.4	1 497.4	477.2	185.7	118.8	107.7	64.9	1 020.3	403.0	199.1	121.1	297.1
1992	4 197.6	1 563.3	508.1	204.8	124.3	111.0	68.0	1 055.2	404.5	211.2	125.0	314.5
1993	4 452.0	1 642.3	551.5	224.7	131.4	123.5	72.0	1 090.8	413.5	219.1	126.9	331.4
1994	4 721.0	1 746.6	607.2	249.8	140.5	140.3	76.5	1 139.4	432.1	227.4	129.2	350.6
1995	4 962.6	1 815.5	635.7	255.7	146.7	153.7	79.6	1 179.8	443.7	231.2	133.4	371.4
1996	5 244.6	1 917.7	676.3	273.5	153.5	164.9	84.3	1 241.4	461.9	239.5	144.7	395.2
1997	5 536.8	2 006.5	715.5	293.1	160.5	174.6	87.3	1 291.0	474.8	247.5	147.7	421.0
1998	5 877.2	2 108.4	779.3	320.2	173.6	191.4	94.2	1 329.1	487.4	257.8	132.4	451.5
1999	6 279.1	2 287.1	855.6	350.7	191.2	210.9	102.7	1 431.5	515.5	271.1	146.5	498.3
2000	6 762.1	2 453.2	912.6	363.2	208.1	230.9	110.4	1 540.6	540.6	280.8	184.5	534.7
2001	7 065.6	2 525.6	941.5	383.3	214.9	234.9	108.4	1 584.1	564.0	277.9	178.0	564.2
2002	7 342.7	2 598.8	985.4	401.3	225.9	244.8	113.4	1 613.4	575.1	278.8	167.9	591.7
2003	7 723.1	2 722.6	1 017.8	401.5	235.2	259.5	121.7	1 704.8	599.6	285.3	196.4	623.5
2004	8 212.7	2 902.0	1 080.6	409.3	254.3	284.8	132.1	1 821.4	632.6	297.4	232.7	658.7
2005	8 747.1	3 082.9	1 128.6	410.0	271.3	306.4	141.0	1 954.3	668.2	310.5	283.8	691.8
2006	9 260.3	3 239.7	1 158.3	394.9	283.6	326.3	153.5	2 081.3	700.3	320.0	319.7	741.4
2007	9 706.4	3 367.0	1 188.0	400.6	283.5	339.2	164.8	2 179.0	737.3	323.5	345.5	772.6
2008	9 976.3	3 363.2	1 098.8	343.3	264.3	328.1	163.0	2 264.5	769.1	317.4	391.1	786.9
2009	9 842.2	3 180.0	1 012.1	318.6	238.3	297.5	157.7	2 167.9	772.9	304.0	287.0	803.9
2010	10 185.8	3 317.8	1 049.0	344.5	240.9	298.6	165.0	2 268.9	786.9	316.6	336.7	828.7
2011	10 641.1	3 518.1	1 093.5	365.2	246.9	305.4	176.1	2 424.6	819.5	332.6	413.8	858.7
2012	11 006.8	3 637.7	1 144.2	396.6	253.9	311.8	181.9	2 493.5	846.2	345.2	421.9	880.2
2013	11 317.2	3 730.0	1 189.4	417.5	263.6	321.6	186.7	2 540.6	864.0	350.5	418.2	907.8
2014	11 822.8	3 863.0	1 242.1	442.0	276.2	329.9	194.0	2 620.9	896.9	360.8	403.3	959.9
2015	12 284.3	3 920.3	1 305.9	474.2	294.1	336.0	201.5	2 614.4	920.1	368.8	309.4	1 016.1
2016	12 748.5	3 995.9	1 352.6	483.6	309.0	356.7	203.3	2 643.3	937.8	374.7	275.0	1 055.7
2017	13 312.1	4 165.0	1 412.6	502.2	324.7	378.8	206.9	2 752.5	967.5	376.4	308.0	1 100.6
2018	13 998.7	4 364.8	1 475.6	521.5	341.2	394.6	218.3	2 889.2	1 003.4	391.5	349.6	1 144.6
2016												
1st quarter	12 523.5	3 933.2	1 330.0	472.1	305.3	349.8	202.8	2 603.2	929.5	373.7	256.4	1 043.5
2nd quarter	12 688.3	3 988.6	1 343.3	476.0	307.8	356.1	203.4	2 645.4	938.8	374.6	277.1	1 054.9
3rd quarter	12 822.4	4 017.8	1 364.9	489.6	310.8	360.8	203.7	2 652.9	939.0	376.8	275.6	1 061.4
4th quarter	12 959.8	4 044.0	1 372.4	496.8	311.9	360.3	203.4	2 671.6	943.9	373.6	291.0	1 063.1
2017												
1st quarter	13 104.4	4 097.9	1 385.1	492.4	319.4	369.7	203.6	2 712.8	952.3	374.1	305.8	1 080.5
2nd quarter	13 212.5	4 124.9	1 398.7	493.9	321.6	378.5	204.7	2 726.2	960.8	375.2	295.0	1 095.3
3rd quarter	13 345.1	4 173.3	1 415.9	501.6	325.4	380.3	208.6	2 757.4	970.7	376.2	304.1	1 106.4
4th quarter	13 586.3	4 264.0	1 450.5	521.1	332.4	386.5	210.6	2 813.4	986.2	379.9	327.1	1 120.3
2018												
1st quarter	13 728.4	4 298.5	1 454.8	512.8	336.4	390.4	215.2	2 843.7	993.0	383.9	340.3	1 126.6
2nd quarter	13 939.8	4 363.2	1 476.7	520.7	342.1	394.0	220.0	2 886.5	1 000.5	392.6	352.2	1 141.3
3rd quarter	14 114.6	4 398.0	1 485.2	524.0	344.5	397.5	219.3	2 912.8	1 008.0	394.9	357.9	1 152.0
4th quarter	14 211.9	4 399.4	1 485.6	528.5	341.9	396.6	218.7	2 913.8	1 012.1	394.8	348.2	1 158.6

Table 4-2. Personal Consumption Expenditures by Major Type of Product—*Continued*

(Billions of dollars, quarterly data are at seasonally adjusted rates.)

NIPA Table 2.3.5

Year and quarter	Services	Services — Household consumption expenditures for services — Household services, total	Housing and utilities	Health care	Transportation services	Recreation services	Food services and accommodations	Financial services and insurance	Other services	NPISHs — Final consumption expenditures	Gross output	Less: receipts from sales of goods and services
1960	154.2	149.1	56.7	16.0	9.2	6.5	20.5	13.2	27.1	5.1	14.8	9.8
1961	162.7	157.4	60.3	17.1	9.6	6.9	21.0	14.3	28.3	5.2	15.8	10.5
1962	173.6	168.1	64.5	19.1	10.1	7.5	22.3	14.8	29.9	5.5	16.9	11.5
1963	183.9	178.1	68.2	21.0	10.6	7.9	23.3	15.5	31.6	5.8	18.3	12.6
1964	198.4	191.9	72.1	24.1	11.4	8.5	24.9	17.2	33.8	6.4	20.3	13.8
1965	213.3	206.3	76.6	26.0	12.1	9.0	27.2	18.8	36.6	7.0	22.1	15.1
1966	230.3	222.8	81.2	28.7	13.1	9.8	29.6	20.6	39.9	7.5	24.3	16.9
1967	247.7	239.6	86.3	31.9	14.3	10.5	31.0	22.1	43.6	8.1	27.0	18.9
1968	272.2	263.4	92.7	36.6	15.9	11.7	34.6	25.2	46.8	8.8	30.3	21.6
1969	299.0	289.5	101.0	42.1	18.0	12.9	37.5	27.7	50.5	9.4	34.0	24.6
1970	327.9	317.5	109.4	47.7	20.0	14.0	41.6	30.1	54.6	10.5	38.1	27.6
1971	357.8	346.1	120.0	53.7	22.4	15.1	43.8	33.1	58.1	11.7	42.8	31.1
1972	394.3	381.5	131.2	59.8	24.5	16.3	48.9	37.1	63.7	12.8	47.4	34.5
1973	432.9	419.2	143.5	67.2	26.1	18.3	54.8	39.9	69.3	13.8	51.8	38.1
1974	478.6	463.1	158.6	76.1	28.5	20.9	60.6	44.1	74.4	15.5	58.8	43.3
1975	539.2	522.2	176.5	89.0	32.0	23.7	68.8	51.8	80.3	17.0	67.2	50.2
1976	601.4	582.4	194.7	101.8	36.2	26.5	77.8	56.8	88.6	19.0	76.3	57.3
1977	673.6	653.0	217.8	115.7	41.4	29.6	85.7	65.1	97.7	20.6	85.2	64.5
1978	758.7	735.7	244.3	131.2	45.2	32.9	97.1	76.7	108.3	23.0	96.7	73.6
1979	847.5	821.4	273.4	148.8	50.7	36.7	110.9	83.6	117.2	26.1	109.4	83.2
1980	950.9	920.8	312.5	171.7	55.4	40.8	121.7	91.7	127.1	30.0	126.1	96.1
1981	1 064.6	1 030.4	352.1	201.9	59.8	47.1	133.9	98.5	137.2	34.1	145.7	111.5
1982	1 172.0	1 134.0	387.5	225.2	61.7	52.5	142.5	113.7	150.9	38.0	163.8	125.8
1983	1 307.8	1 267.1	421.2	253.1	68.9	59.4	153.6	141.0	169.9	40.8	179.2	138.5
1984	1 428.6	1 383.3	457.5	276.5	80.0	66.1	164.9	150.8	187.5	45.4	194.4	149.0
1985	1 575.2	1 527.3	500.6	302.2	90.1	74.0	174.3	178.2	207.9	47.9	209.3	161.4
1986	1 690.7	1 638.0	537.0	330.2	95.0	80.4	186.7	187.7	221.2	52.6	227.3	174.7
1987	1 820.0	1 764.3	571.6	366.0	103.1	87.3	204.4	189.5	242.4	55.7	246.9	191.2
1988	1 992.7	1 929.4	614.4	410.1	114.2	99.2	225.8	202.9	262.8	63.3	275.5	212.2
1989	2 153.0	2 084.9	655.2	451.2	121.8	110.7	242.6	222.3	281.1	68.0	301.2	233.2
1990	2 317.7	2 241.8	696.5	506.2	126.4	121.8	262.7	230.8	297.5	75.9	334.4	258.6
1991	2 446.0	2 365.9	735.2	555.8	123.7	127.2	273.4	250.1	300.5	80.1	362.9	282.8
1992	2 634.3	2 546.4	771.1	612.8	133.6	139.8	286.3	277.0	325.7	87.9	396.4	308.5
1993	2 809.6	2 719.6	814.9	648.8	145.7	153.4	298.4	314.0	344.3	90.1	418.7	328.7
1994	2 974.4	2 876.6	863.3	680.5	161.0	164.7	308.3	308.3	371.0	97.8	440.1	342.3
1995	3 147.1	3 044.7	913.7	719.9	176.4	181.1	316.1	347.0	390.6	102.3	458.5	356.2
1996	3 326.9	3 216.9	962.4	752.1	192.7	195.6	326.6	372.1	415.5	110.0	483.1	373.1
1997	3 530.3	3 424.7	1 009.8	790.9	211.8	208.3	343.4	408.9	451.5	105.6	502.6	397.0
1998	3 768.8	3 645.0	1 065.5	832.0	225.2	220.2	361.8	446.1	494.2	123.8	542.5	418.7
1999	3 992.0	3 853.8	1 123.1	863.6	241.3	238.1	380.3	486.4	520.9	138.2	576.3	438.1
2000	4 309.0	4 150.9	1 198.6	918.4	261.3	254.4	408.8	543.0	566.5	158.0	621.6	463.6
2001	4 540.0	4 361.0	1 287.5	996.6	259.8	262.3	419.7	525.7	609.5	179.1	676.4	497.4
2002	4 743.9	4 545.5	1 333.6	1 082.9	251.9	271.4	436.3	534.7	634.8	198.3	737.1	538.7
2003	5 000.5	4 795.0	1 394.1	1 154.0	259.6	288.9	462.7	560.3	675.4	205.5	774.6	569.1
2004	5 310.6	5 104.3	1 469.1	1 238.9	271.2	311.5	498.2	605.5	709.9	206.4	819.0	612.6
2005	5 664.2	5 453.9	1 583.6	1 320.5	283.9	328.1	533.6	659.0	745.1	210.3	868.5	658.2
2006	6 020.7	5 781.5	1 682.4	1 391.9	297.1	351.3	570.6	695.0	793.3	239.2	932.2	693.0
2007	6 339.4	6 090.6	1 758.2	1 478.2	307.6	375.6	601.5	737.2	832.4	248.8	983.1	734.4
2008	6 613.1	6 325.8	1 835.4	1 555.3	312.7	389.1	620.2	756.6	856.6	287.3	1 047.4	760.1
2009	6 662.2	6 373.0	1 877.7	1 632.7	297.4	388.4	612.7	711.3	852.9	289.2	1 088.3	799.1
2010	6 868.0	6 573.6	1 903.9	1 699.6	305.2	403.7	635.7	754.4	871.1	294.4	1 128.6	834.2
2011	7 123.0	6 811.1	1 955.9	1 757.1	328.4	409.0	669.5	797.9	893.3	311.9	1 173.5	861.7
2012	7 369.1	7 027.5	1 996.3	1 821.3	341.1	430.8	704.9	820.1	913.0	341.5	1 236.5	895.0
2013	7 587.2	7 234.6	2 055.1	1 858.2	359.9	447.1	732.3	858.4	923.5	352.6	1 271.7	919.1
2014	7 959.8	7 594.2	2 149.9	1 940.5	383.0	466.6	776.9	908.1	969.1	365.6	1 322.4	956.8
2015	8 363.9	7 992.5	2 255.7	2 057.2	398.6	492.1	832.8	956.9	999.2	371.4	1 383.2	1 011.8
2016	8 752.6	8 355.0	2 355.3	2 160.1	418.9	519.5	872.4	977.5	1 051.3	397.5	1 463.0	1 065.5
2017	9 147.0	8 733.3	2 455.0	2 243.4	439.4	539.9	913.8	1 040.4	1 101.3	413.8	1 521.6	1 107.8
2018	9 633.9	9 190.9	2 567.2	2 352.6	462.2	563.2	973.3	1 111.0	1 161.3	443.0	1 597.9	1 154.9
2016												
1st quarter	8 590.3	8 203.6	2 307.2	2 117.2	412.2	512.2	861.5	960.1	1 033.2	386.7	1 424.2	1 037.6
2nd quarter	8 699.6	8 312.6	2 342.5	2 163.9	414.4	514.1	867.8	967.9	1 042.0	387.0	1 453.6	1 066.7
3rd quarter	8 804.6	8 398.6	2 377.7	2 156.5	422.0	522.1	875.9	987.1	1 057.5	406.0	1 474.3	1 068.3
4th quarter	8 915.8	8 505.3	2 393.7	2 202.8	427.0	529.9	884.5	995.1	1 072.4	410.5	1 499.8	1 089.3
2017												
1st quarter	9 006.5	8 590.9	2 407.5	2 211.8	427.6	539.4	904.6	1 012.3	1 087.6	415.6	1 509.8	1 094.3
2nd quarter	9 087.6	8 674.1	2 444.8	2 218.9	435.6	539.0	905.8	1 030.4	1 099.7	413.5	1 507.8	1 094.3
3rd quarter	9 171.8	8 759.2	2 465.8	2 253.2	441.1	541.0	913.9	1 044.8	1 099.2	412.6	1 525.8	1 113.3
4th quarter	9 322.3	8 908.7	2 501.8	2 289.5	453.5	540.4	930.8	1 074.1	1 118.6	413.6	1 542.8	1 129.2
2018												
1st quarter	9 429.8	9 008.0	2 524.3	2 307.7	459.9	551.4	948.6	1 091.2	1 125.0	421.8	1 551.0	1 129.2
2nd quarter	9 576.6	9 140.7	2 558.3	2 341.4	459.4	561.4	968.1	1 102.7	1 149.5	435.9	1 588.2	1 152.3
3rd quarter	9 716.6	9 271.7	2 579.0	2 380.3	462.5	566.4	989.5	1 118.4	1 175.6	444.9	1 619.3	1 174.4
4th quarter	9 812.5	9 343.3	2 607.2	2 381.1	467.1	573.7	987.1	1 131.7	1 195.4	469.2	1 633.0	1 163.7

Table 4-3. Chain-Type Quantity Indexes for Personal Consumption Expenditures by Major Type of Product

(Index numbers, 2012 = 100.)   NIPA Table 2.3.3

Year and quarter	Personal consumption expenditures, total	Total goods	Goods									
			Durable goods					Nondurable goods				
			Durable goods, total	Motor vehicles and parts	Furnishings and household equipment	Recreational goods and vehicles	Other durable goods	Nondurable goods, total	Food and beverages off-premises	Clothing and footwear	Gasoline and other energy goods	Other nondurable goods
1960	18.3	18.3	7.0	16.7	12.2	0.8	7.9	29.2	44.7	14.5	53.2	18.5
1961	18.6	18.4	6.8	15.0	12.2	0.8	8.0	29.8	45.4	14.8	52.9	19.5
1962	19.6	19.3	7.5	17.8	12.9	0.9	8.8	30.7	45.7	15.5	54.6	20.9
1963	20.4	20.1	8.2	20.1	13.8	0.9	9.1	31.3	46.0	15.7	56.5	21.8
1964	21.6	21.3	9.0	21.2	15.3	1.1	9.9	32.8	47.6	16.9	59.5	22.8
1965	23.0	22.8	10.1	24.5	16.3	1.3	10.9	34.5	50.0	17.6	62.3	24.2
1966	24.3	24.3	11.0	24.9	17.7	1.6	12.6	36.4	51.9	18.8	65.7	26.1
1967	25.0	24.7	11.1	24.3	18.1	1.7	12.7	37.2	53.1	19.0	67.4	26.8
1968	26.4	26.3	12.4	28.1	19.2	1.9	13.9	38.7	55.5	19.7	70.2	27.9
1969	27.4	27.1	12.8	29.2	19.6	2.0	14.4	39.8	57.0	20.1	73.3	28.7
1970	28.0	27.3	12.5	26.1	19.6	2.2	15.1	40.7	58.6	19.9	76.1	29.7
1971	29.1	28.4	13.7	31.2	20.5	2.2	15.8	41.4	59.3	20.7	78.7	29.8
1972	30.9	30.3	15.5	35.5	22.6	2.6	16.7	43.1	60.5	22.2	82.7	31.2
1973	32.4	31.9	17.1	38.9	25.0	2.9	19.0	44.4	59.4	23.8	87.5	33.3
1974	32.2	30.7	16.0	32.5	24.9	2.9	19.9	43.3	58.2	23.3	80.4	33.2
1975	32.9	30.9	16.0	32.3	23.6	3.1	21.1	43.7	59.3	23.9	82.2	32.1
1976	34.7	33.1	18.0	38.9	25.1	3.4	23.2	45.7	62.2	25.0	86.6	33.5
1977	36.2	34.5	19.6	43.0	27.4	3.6	24.5	46.8	63.1	26.4	88.2	34.0
1978	37.8	35.9	20.7	44.9	28.8	3.9	25.8	48.5	62.8	29.1	89.6	36.4
1979	38.7	36.5	20.5	42.2	30.0	4.1	25.5	49.7	63.7	30.8	87.0	38.5
1980	38.5	35.6	18.9	36.8	28.8	3.9	23.3	49.6	64.4	31.8	79.3	39.2
1981	39.1	36.0	19.1	37.7	28.3	4.0	23.7	50.3	64.1	33.8	78.3	40.1
1982	39.7	36.2	19.1	38.7	27.0	4.0	23.7	50.8	65.4	34.4	78.4	40.0
1983	41.0	38.6	21.8	46.1	29.3	4.7	25.9	52.4	67.0	36.7	80.2	41.0
1984	44.1	41.3	24.9	53.6	32.3	5.5	28.8	54.6	68.4	39.8	82.3	43.2
1985	46.4	43.5	27.4	60.6	34.0	6.1	31.0	56.2	70.4	41.5	83.9	44.4
1986	48.3	45.9	30.0	65.0	36.8	7.0	34.5	58.2	71.8	44.6	88.2	45.4
1987	49.9	46.8	30.6	62.4	38.3	7.9	35.7	59.2	71.4	46.3	89.9	47.1
1988	52.0	48.5	32.4	66.0	39.7	8.6	36.9	60.8	73.3	47.7	92.4	48.1
1989	53.5	49.7	33.1	65.9	41.8	9.0	37.4	62.4	74.2	50.2	93.9	49.7
1990	54.6	50.0	32.9	64.8	41.2	9.2	37.6	63.2	75.7	49.8	92.2	51.0
1991	54.7	49.0	31.2	57.1	40.2	9.5	36.0	63.0	75.6	49.9	91.5	50.7
1992	56.7	50.6	32.9	61.8	41.6	10.1	36.7	64.2	75.3	52.5	95.1	51.8
1993	58.7	52.7	35.4	65.2	43.7	11.7	38.5	65.8	75.9	54.8	97.5	53.6
1994	61.0	55.5	38.2	69.3	46.1	13.6	40.3	68.4	78.1	57.8	99.1	56.3
1995	62.8	57.1	39.7	68.2	48.2	15.5	41.2	70.1	78.5	60.2	100.9	58.7
1996	65.0	59.7	42.7	71.5	50.3	17.9	44.0	72.1	79.2	63.1	102.5	61.3
1997	67.4	62.5	46.2	76.3	52.6	20.7	46.2	74.2	79.9	65.2	104.6	64.6
1998	71.0	66.7	51.8	84.3	57.0	24.8	50.7	77.0	81.0	69.2	107.5	68.0
1999	74.7	72.0	58.4	92.1	63.6	29.9	56.8	80.9	84.3	73.7	110.0	72.2
2000	78.5	75.7	63.4	95.0	69.7	34.9	61.8	83.5	86.4	77.3	107.1	75.8
2001	80.5	78.0	66.8	99.9	73.3	38.2	60.5	85.0	87.6	78.0	107.0	78.2
2002	82.6	81.0	71.7	105.1	78.6	42.5	64.4	86.5	88.0	80.4	107.9	80.5
2003	85.2	85.0	76.8	108.2	84.2	48.1	70.7	89.6	90.0	84.4	108.0	84.6
2004	88.4	89.3	83.2	111.2	92.2	55.5	77.1	92.5	92.1	88.2	108.9	88.3
2005	91.5	93.0	87.8	109.7	98.3	63.2	83.1	95.6	95.6	92.9	108.4	91.4
2006	94.3	96.5	91.7	105.6	103.3	71.6	89.7	98.7	98.6	96.2	108.1	95.9
2007	96.4	99.2	96.1	107.7	104.1	79.9	94.5	100.4	99.9	98.1	107.9	98.5
2008	96.2	96.2	90.6	94.1	97.8	81.3	91.2	98.8	98.2	97.0	103.7	98.0
2009	95.0	93.2	85.0	87.4	88.5	78.3	88.0	97.2	97.5	92.2	104.3	95.8
2010	96.7	95.8	89.8	90.8	93.4	84.5	92.4	98.7	99.0	96.6	103.8	97.0
2011	98.5	97.9	94.4	93.3	97.2	92.4	96.1	99.6	99.1	99.8	101.4	99.0
2012	100.0	100.0	100.0	100.0	100.0	100.0	100.0	100.0	100.0	100.0	100.0	100.0
2013	101.5	103.1	106.1	104.7	105.8	108.9	104.9	101.8	101.1	100.5	101.9	103.0
2014	104.5	107.4	113.8	110.8	114.8	116.9	113.5	104.5	103.0	103.2	101.9	107.6
2015	108.3	112.4	122.2	118.9	125.3	124.0	122.3	108.0	104.5	106.8	106.7	112.6
2016	111.3	116.5	129.7	122.6	135.2	138.4	124.4	110.6	107.6	108.8	107.2	115.6
2017	114.2	121.0	138.6	128.9	146.1	152.9	128.2	113.3	111.1	109.9	106.2	119.1
2018	117.6	126.0	147.3	134.4	155.2	165.5	137.9	116.7	114.7	114.2	106.0	123.3
2016												
1st quarter	110.2	114.8	126.0	118.8	131.8	132.3	123.9	109.8	105.9	108.5	108.4	115.0
2nd quarter	110.9	116.1	128.1	120.5	134.0	136.9	123.5	110.7	107.4	108.8	107.4	115.9
3rd quarter	111.7	117.2	131.5	124.5	136.8	140.8	124.9	110.9	108.1	109.5	106.9	115.7
4th quarter	112.3	117.8	133.3	126.7	138.1	143.6	125.3	110.9	109.0	108.2	105.9	115.6
2017												
1st quarter	113.0	118.7	134.4	125.3	141.2	147.6	125.2	111.7	109.8	108.2	105.3	117.3
2nd quarter	113.7	120.3	137.0	126.6	144.1	152.5	126.5	112.9	110.3	109.9	107.0	118.8
3rd quarter	114.4	121.5	139.6	129.4	147.4	153.5	130.0	113.5	111.4	109.9	106.4	119.5
4th quarter	115.7	123.7	143.6	134.3	151.7	157.9	131.3	115.0	113.1	111.7	106.2	121.0
2018												
1st quarter	116.1	124.1	144.5	132.3	153.4	161.3	133.7	115.2	113.8	111.2	105.6	121.3
2nd quarter	117.3	125.7	147.3	134.7	155.1	164.9	137.9	116.3	114.3	113.8	106.5	122.7
3rd quarter	118.3	126.8	148.6	134.8	157.2	167.5	138.7	117.4	115.1	116.0	105.8	124.1
4th quarter	118.7	127.3	149.0	135.8	155.2	168.3	141.1	117.9	115.5	116.0	106.3	124.9

Table 4-3. Chain-Type Quantity Indexes for Personal Consumption Expenditures by Major Type of Product—Continued

(Index numbers, 2012 = 100.)

NIPA Table 2.3.3

| Year and quarter | Services | Household consumption expenditures for services | | | | | | | | Nonprofit institutions serving households (NPISHs) | | |
		Household services, total	Housing and utilities	Health care	Transportation services	Recreation services	Food services and accommodations	Financial services and insurance	Other services	Final consumption expenditures	Gross output	Less: receipts from sales of goods and services
1960	17.7	18.9	20.9	14.0	20.6	10.3	26.4	12.1	25.2	2.9	12.6	19.3
1961	18.3	19.6	22.0	14.7	20.9	10.7	26.5	12.7	26.1	2.9	13.1	20.4
1962	19.2	20.5	23.2	16.0	21.7	11.3	27.5	12.7	27.1	3.0	13.8	21.7
1963	20.0	21.4	24.3	17.0	22.6	11.7	28.2	13.2	28.1	3.1	14.6	23.1
1964	21.2	22.6	25.5	19.0	23.8	12.2	29.6	14.1	29.3	3.5	15.7	24.6
1965	22.4	23.9	26.9	19.8	25.1	12.7	31.7	15.0	31.0	3.7	16.7	26.1
1966	23.5	25.1	28.1	20.9	26.5	13.4	33.3	15.6	32.8	3.8	17.8	28.2
1967	24.5	26.1	29.4	21.9	27.9	13.9	33.0	16.3	34.8	4.0	18.8	30.1
1968	25.8	27.5	30.8	23.6	29.7	14.7	35.0	17.3	35.7	4.2	20.1	32.4
1969	26.9	28.8	32.5	25.3	31.7	15.6	35.8	17.5	36.8	4.2	21.2	34.6
1970	28.0	29.9	33.8	26.7	32.8	16.2	37.0	18.1	38.0	4.5	22.2	36.2
1971	28.9	30.9	35.2	28.6	34.2	16.7	37.0	18.8	38.1	4.8	23.8	39.0
1972	30.6	32.7	37.1	30.6	36.1	17.6	39.8	19.8	39.8	5.0	25.2	41.3
1973	32.0	34.2	38.8	32.8	36.8	19.1	41.8	20.9	40.9	5.0	25.9	42.8
1974	32.7	34.9	40.7	33.9	37.0	20.4	41.4	21.4	39.6	5.0	26.2	43.8
1975	33.9	36.3	42.0	35.7	37.5	21.5	43.1	23.5	39.9	5.0	27.2	45.9
1976	35.3	37.8	43.2	37.2	39.3	22.9	45.7	24.8	41.4	5.3	28.7	48.3
1977	36.8	39.4	44.4	39.0	42.1	24.4	46.9	26.3	43.5	5.4	29.8	50.5
1978	38.5	41.2	46.5	40.8	43.1	25.6	48.9	27.6	45.8	5.6	31.4	53.4
1979	39.7	42.5	48.1	42.2	44.4	26.8	50.4	28.4	46.2	5.8	32.3	54.8
1980	40.3	43.1	49.8	43.4	41.8	27.7	50.5	28.8	45.9	6.1	33.3	56.1
1981	40.9	43.7	50.5	45.5	40.3	30.1	51.0	28.7	44.9	6.6	34.7	57.7
1982	41.8	44.4	50.9	45.6	39.4	31.8	51.3	31.1	45.9	7.8	36.2	58.1
1983	43.9	46.5	52.0	47.0	42.3	34.5	53.0	35.4	48.8	9.0	37.6	58.4
1984	45.6	48.1	53.7	47.7	47.3	36.8	54.4	35.6	51.3	10.5	38.9	58.3
1985	48.0	50.6	55.9	49.2	52.3	39.4	55.2	39.1	54.9	11.4	40.2	59.6
1986	49.4	52.0	57.2	50.7	54.5	41.2	56.8	40.5	55.5	12.9	42.2	61.1
1987	51.7	54.3	58.8	52.8	56.8	43.1	59.7	42.9	59.4	13.9	43.9	62.8
1988	54.0	56.5	60.8	54.9	60.1	47.2	63.3	43.7	61.8	16.0	46.3	64.6
1989	55.7	58.1	62.3	55.5	61.4	50.2	65.0	46.2	63.7	18.4	47.9	64.9
1990	57.3	59.5	63.5	57.4	61.3	52.3	67.4	46.9	64.8	22.4	50.8	66.1
1991	58.2	60.2	64.8	58.5	58.4	52.0	67.4	50.8	62.4	25.6	52.7	66.6
1992	60.6	62.3	66.1	60.3	61.4	55.4	69.1	54.2	64.9	30.9	55.5	67.6
1993	62.4	64.1	67.9	60.6	64.7	58.9	70.6	58.3	66.3	33.4	56.9	68.2
1994	64.4	66.0	70.2	61.0	70.7	61.9	71.7	59.7	68.9	36.4	58.1	68.5
1995	66.2	67.8	72.3	62.3	75.8	66.6	71.9	61.6	70.2	38.1	58.9	68.8
1996	68.2	69.7	74.0	63.6	82.0	69.6	72.3	63.6	72.6	40.3	60.4	69.9
1997	70.3	72.2	75.6	65.6	88.0	71.8	73.8	67.1	76.7	37.1	61.3	72.9
1998	73.5	75.4	77.9	67.7	92.0	74.1	75.7	72.9	82.2	40.7	63.1	75.4
1999	76.3	78.2	80.2	68.8	97.3	77.6	77.6	79.9	84.6	43.9	66.4	76.9
2000	80.2	82.0	82.8	71.1	102.2	79.8	81.1	88.3	90.4	47.5	68.9	78.7
2001	82.0	83.6	85.0	74.7	101.1	79.7	81.0	85.9	94.2	51.5	72.2	81.6
2002	83.5	84.9	85.5	79.1	98.1	80.2	82.2	85.4	94.3	56.7	76.5	85.2
2003	85.3	86.6	86.7	81.2	99.0	82.8	85.4	85.9	96.5	60.9	78.3	85.6
2004	87.9	89.3	88.9	84.1	102.1	87.0	89.0	88.8	98.0	61.9	80.3	88.1
2005	90.8	92.3	92.5	86.9	103.1	89.1	92.3	93.6	99.4	61.7	82.3	91.0
2006	93.3	94.5	94.3	88.9	103.5	92.3	95.5	95.8	102.2	69.3	85.4	92.0
2007	95.0	96.3	95.2	91.0	104.9	96.0	96.9	98.6	103.8	71.3	87.4	94.0
2008	96.3	97.0	96.2	93.2	101.0	96.5	96.2	100.6	102.5	81.7	90.7	94.3
2009	95.9	96.5	97.3	95.3	93.8	95.2	92.9	98.7	99.5	84.7	93.0	96.2
2010	97.1	97.6	98.5	96.7	94.1	98.0	95.1	98.8	99.3	87.5	94.9	97.7
2011	98.8	99.2	99.8	98.2	98.2	97.6	97.7	101.4	100.0	91.7	96.7	98.6
2012	100.0	100.0	100.0	100.0	100.0	100.0	100.0	100.0	100.0	100.0	100.0	100.0
2013	100.6	100.6	100.5	100.6	104.5	102.0	101.7	99.4	98.9	101.2	100.7	100.5
2014	103.1	103.2	102.2	103.9	109.8	104.6	105.2	99.7	102.3	101.1	102.4	102.9
2015	106.4	106.7	104.6	109.5	113.9	108.6	109.7	102.0	104.7	99.0	105.0	107.4
2016	108.8	109.2	106.1	113.7	118.8	111.9	112.0	99.7	109.1	102.6	108.8	111.3
2017	111.0	111.4	106.9	116.4	123.0	113.2	114.9	101.6	113.8	102.9	110.4	113.4
2018	113.8	114.1	108.4	119.8	126.7	115.6	119.6	102.6	117.7	107.9	113.3	115.4
2016												
1st quarter	108.0	108.3	105.3	112.1	117.2	111.6	111.5	100.3	107.7	101.2	107.0	109.3
2nd quarter	108.5	108.9	106.1	114.2	117.4	110.8	111.8	99.2	108.2	100.5	108.6	111.8
3rd quarter	109.1	109.3	106.7	113.2	119.5	112.0	112.1	99.7	109.5	104.4	109.3	111.2
4th quarter	109.8	110.1	106.4	115.2	121.0	113.3	112.6	99.6	111.1	104.2	110.4	112.9
2017												
1st quarter	110.4	110.7	106.1	115.4	120.4	114.0	114.5	101.2	112.5	104.2	110.5	113.0
2nd quarter	110.7	111.0	106.9	115.4	122.0	113.4	114.2	101.1	113.9	102.8	109.7	112.4
3rd quarter	111.1	111.5	107.0	116.8	123.7	112.9	114.9	101.6	113.4	102.1	110.5	113.9
4th quarter	112.0	112.5	107.7	117.9	126.1	112.3	116.0	102.3	115.2	102.3	111.0	114.4
2018												
1st quarter	112.5	113.0	107.8	118.4	127.9	113.9	117.6	102.5	114.8	103.6	110.9	113.8
2nd quarter	113.5	113.8	108.4	119.4	126.5	115.7	119.1	102.3	116.8	106.6	113.0	115.4
3rd quarter	114.4	114.8	108.6	121.0	126.3	115.9	121.2	102.6	118.8	108.2	114.6	117.1
4th quarter	114.8	114.9	108.8	120.4	126.1	116.8	120.5	103.0	120.3	113.3	114.8	115.3

Table 4-4. Chain-Type Price Indexes for Personal Consumption Expenditures by Major Type of Product

(Index numbers, 2012 =1000.)   NIPA Table 2.3.4

Year and quarter	Personal consumption expenditures, total	Total goods	Goods									
			Durable goods					Nondurable goods				
			Durable goods, total	Motor vehicles and parts	Furnishings and household equipment	Recreational goods and vehicles	Other durable goods	Nondurable goods, total	Food and beverages off-premises	Clothing and footwear	Gasoline and other energy goods	Other nondurable goods
1960	16.5	26.6	56.9	29.6	50.0	266.2	29.7	18.1	16.5	51.7	7.0	16.6
1961	16.6	26.7	57.0	29.8	50.2	264.8	29.6	18.1	16.6	52.0	7.1	16.6
1962	16.8	26.9	57.3	30.2	50.1	261.9	29.5	18.2	16.7	52.2	7.1	16.7
1963	17.0	27.1	57.5	30.5	50.1	261.3	29.6	18.4	16.9	52.6	7.1	16.9
1964	17.3	27.4	57.8	30.7	50.2	258.2	30.4	18.7	17.3	53.1	7.0	17.2
1965	17.5	27.7	57.3	30.5	49.8	254.6	30.2	19.0	17.6	53.6	7.3	17.4
1966	18.0	28.3	57.2	30.3	50.3	251.8	30.2	19.6	18.3	55.1	7.5	17.8
1967	18.4	28.8	58.1	30.7	51.4	253.3	30.9	20.0	18.4	57.4	7.7	18.2
1968	19.2	29.8	59.9	31.7	53.4	258.3	31.6	20.7	18.9	60.6	7.8	18.9
1969	20.0	30.9	61.6	32.4	55.3	264.2	33.2	21.6	19.8	63.9	8.1	19.6
1970	21.0	32.1	63.0	33.4	56.5	264.6	34.0	22.6	20.9	66.3	8.2	20.5
1971	21.8	33.1	65.1	35.0	57.5	274.0	35.0	23.2	21.4	68.6	8.3	21.3
1972	22.6	33.9	65.9	35.1	58.4	278.0	36.1	23.9	22.4	69.9	8.4	21.8
1973	23.8	35.9	66.8	35.2	59.7	282.2	37.3	25.9	25.2	72.2	9.3	22.4
1974	26.3	40.4	71.2	37.4	64.6	294.7	40.2	29.8	29.0	77.5	12.9	24.6
1975	28.5	43.7	77.6	41.1	71.0	314.2	43.2	32.1	31.2	81.0	13.8	27.5
1976	30.0	45.4	81.8	44.2	74.1	325.1	45.3	33.1	31.8	83.6	14.5	28.9
1977	32.0	47.8	85.5	46.8	76.8	334.4	47.3	35.0	33.7	87.2	15.5	30.6
1978	34.2	50.8	90.3	50.1	80.7	346.4	49.9	37.3	36.9	88.8	16.3	32.2
1979	37.3	55.6	96.3	53.8	85.7	361.8	54.1	41.3	40.5	90.6	21.9	34.3
1980	41.3	61.8	104.6	57.8	92.7	381.9	65.1	46.0	40.9	93.9	30.4	37.5
1981	45.0	66.4	111.6	62.3	99.6	399.0	68.7	49.9	47.1	97.0	34.3	40.6
1982	47.5	68.2	116.0	65.2	104.7	408.2	69.4	51.0	48.3	98.4	32.7	43.8
1983	49.5	69.4	118.3	67.3	107.1	403.3	71.6	51.9	48.8	99.7	31.5	46.8
1984	51.3	70.7	120.1	69.2	108.4	397.6	73.1	53.0	50.3	100.3	31.1	48.5
1985	53.1	71.9	121.4	70.8	109.6	390.6	73.9	54.0	50.9	102.4	31.2	50.4
1986	54.3	71.5	122.7	72.7	110.8	378.3	75.7	53.3	52.1	102.0	24.5	52.4
1987	56.0	73.8	126.2	76.1	112.5	373.5	80.8	55.1	53.7	104.9	25.4	54.5
1988	58.2	75.8	128.3	77.3	114.1	371.1	86.1	56.9	55.3	108.3	25.6	57.0
1989	60.7	78.7	130.6	79.6	114.5	370.1	90.1	59.7	58.2	109.9	27.9	60.1
1990	63.4	81.9	131.9	79.8	115.6	367.1	95.8	63.1	61.1	113.5	31.9	63.2
1991	65.5	83.9	133.9	82.1	116.3	362.4	99.3	65.0	63.0	115.6	31.4	66.5
1992	67.2	84.9	134.8	83.6	117.7	351.3	101.9	65.9	63.5	116.5	31.1	69.0
1993	68.9	85.7	136.1	86.9	118.3	338.7	102.7	66.4	64.3	115.9	30.8	70.2
1994	70.3	86.6	138.8	90.9	120.1	331.6	104.4	66.8	65.4	113.9	30.9	70.7
1995	71.8	87.4	139.8	94.5	119.9	317.7	106.1	67.5	66.8	111.3	31.3	71.9
1996	73.3	88.3	138.4	96.4	120.3	295.2	105.5	69.0	68.9	109.9	33.5	73.3
1997	74.6	88.2	135.4	96.8	120.1	270.5	104.0	69.7	70.2	110.0	33.5	74.1
1998	75.2	86.9	131.5	95.7	120.0	247.6	102.1	69.2	71.1	108.0	29.2	75.5
1999	76.3	87.3	128.0	96.0	118.4	226.1	99.5	70.9	72.2	106.5	31.6	78.4
2000	78.2	89.1	125.7	96.4	117.6	211.9	98.2	74.0	73.9	105.3	40.8	80.1
2001	79.7	89.0	123.3	96.8	115.5	197.5	98.5	74.8	76.1	103.2	39.4	82.0
2002	80.8	88.2	120.1	96.3	113.2	184.7	96.9	74.8	77.2	100.5	36.9	83.5
2003	82.4	88.1	115.8	93.5	110.0	173.0	94.7	76.3	78.7	98.0	43.1	83.7
2004	84.4	89.3	113.5	92.8	108.6	164.5	94.2	79.0	81.2	97.6	50.7	84.7
2005	86.8	91.1	112.3	94.2	108.6	155.6	93.3	82.0	82.6	96.8	62.0	86.0
2006	89.2	92.3	110.4	94.3	108.1	146.2	94.1	84.6	84.0	96.4	70.1	87.8
2007	91.4	93.3	108.0	93.8	107.2	136.2	95.9	87.0	87.2	95.5	75.9	89.1
2008	94.2	96.1	106.0	92.0	106.5	129.4	98.3	91.9	92.6	94.7	89.4	91.2
2009	94.1	93.8	104.0	91.9	106.1	121.8	98.6	89.5	93.7	95.6	65.2	95.3
2010	95.7	95.2	102.1	95.7	101.5	113.4	98.2	92.2	93.9	94.9	76.9	97.0
2011	98.1	98.8	101.3	98.7	100.0	106.0	100.8	97.7	97.7	96.5	96.7	98.5
2012	100.0	100.0	100.0	100.0	100.0	100.0	100.0	100.0	100.0	100.0	100.0	100.0
2013	101.3	99.4	98.0	100.5	98.1	94.7	97.9	100.1	101.0	101.0	97.3	100.2
2014	102.8	98.9	95.4	100.6	94.7	90.5	94.0	100.6	102.9	101.3	93.8	101.3
2015	103.0	95.9	93.4	100.5	92.4	86.9	90.6	97.1	104.1	100.1	68.7	102.5
2016	104.1	94.3	91.1	99.4	90.0	82.7	89.9	95.9	103.0	99.8	60.8	103.8
2017	105.9	94.6	89.0	98.3	87.5	79.4	88.7	97.4	102.9	99.2	68.7	105.0
2018	108.1	95.2	87.5	97.8	86.6	76.5	87.1	99.3	103.4	99.3	78.2	105.5
2016												
1st quarter	103.3	94.2	92.3	100.2	91.2	84.7	90.0	95.0	103.8	99.8	56.0	103.1
2nd quarter	103.9	94.5	91.6	99.6	90.4	83.4	90.6	95.8	103.3	99.7	61.1	103.4
3rd quarter	104.3	94.2	90.7	99.1	89.4	82.1	89.7	95.9	102.7	99.7	61.1	104.2
4th quarter	104.8	94.4	89.9	98.9	88.9	80.4	89.3	96.6	102.3	100.0	65.1	104.5
2017												
1st quarter	105.4	94.9	90.0	99.1	89.0	80.3	89.4	97.4	102.5	100.2	68.8	104.7
2nd quarter	105.6	94.3	89.2	98.4	87.9	79.6	89.0	96.8	102.9	98.9	65.3	104.8
3rd quarter	106.0	94.4	88.7	97.7	86.9	79.5	88.2	97.4	103.0	99.1	67.7	105.2
4th quarter	106.7	94.8	88.2	97.8	86.3	78.5	88.2	98.1	103.0	98.6	73.0	105.2
2018												
1st quarter	107.4	95.2	88.0	97.7	86.4	77.6	88.5	99.0	103.1	100.0	76.4	105.5
2nd quarter	108.0	95.4	87.6	97.5	86.8	76.6	87.7	99.5	103.4	100.0	78.4	105.7
3rd quarter	108.4	95.3	87.4	98.0	86.3	76.1	86.9	99.5	103.5	98.6	80.2	105.4
4th quarter	108.8	95.0	87.1	98.1	86.8	75.5	85.2	99.1	103.6	98.6	77.6	105.4

Table 4-4. Chain-Type Price Indexes for Personal Consumption Expenditures by Major Type of Product—Continued

(Index numbers, 2012 =1000.)

NIPA Table 2.3.4

Year and quarter	Services	Household consumption expenditures for services								Nonprofit institutions serving households (NPISHs)		
		Household services, total	Housing and utilities	Health care	Transportation services	Recreation services	Food services and accommodations	Financial services and insurance	Other services	Final consumption expenditures	Gross output	Less: receipts from sales of goods and services
1960	11.8	11.2	13.6	6.3	13.0	14.6	11.0	13.3	11.7	51.5	9.6	5.7
1961	12.0	11.4	13.7	6.4	13.4	15.0	11.2	13.7	11.9	52.1	9.7	5.8
1962	12.3	11.7	13.9	6.6	13.7	15.4	11.5	14.2	12.1	52.8	9.9	5.9
1963	12.5	11.8	14.0	6.8	13.8	15.7	11.7	14.3	12.3	53.7	10.2	6.1
1964	12.7	12.1	14.2	7.0	14.0	16.1	11.9	14.9	12.6	53.9	10.4	6.3
1965	12.9	12.3	14.3	7.2	14.2	16.5	12.2	15.3	12.9	55.4	10.7	6.5
1966	13.3	12.6	14.5	7.5	14.5	16.9	12.6	16.1	13.3	57.4	11.1	6.7
1967	13.7	13.1	14.7	8.0	15.0	17.5	13.3	16.5	13.7	59.3	11.6	7.0
1968	14.3	13.6	15.1	8.5	15.7	18.4	14.0	17.8	14.3	61.5	12.2	7.4
1969	15.1	14.3	15.6	9.1	16.6	19.2	14.9	19.3	15.0	65.4	13.0	7.9
1970	15.9	15.1	16.2	9.8	17.9	20.1	16.0	20.3	15.8	68.3	13.9	8.5
1971	16.8	16.0	17.1	10.3	19.2	21.0	16.8	21.5	16.7	72.1	14.5	8.9
1972	17.5	16.6	17.7	10.7	19.9	21.5	17.4	22.8	17.5	75.1	15.2	9.4
1973	18.3	17.4	18.5	11.3	20.8	22.3	18.6	23.3	18.6	79.9	16.2	10.0
1974	19.9	18.9	19.5	12.3	22.6	23.8	20.8	25.1	20.6	91.5	18.1	11.1
1975	21.6	20.5	21.1	13.7	25.1	25.6	22.6	26.9	22.1	99.9	19.9	12.2
1976	23.1	21.9	22.6	15.0	26.9	26.9	24.1	27.9	23.4	105.1	21.5	13.3
1977	24.8	23.6	24.6	16.3	28.8	28.1	25.9	30.2	24.6	112.2	23.1	14.3
1978	26.8	25.4	26.3	17.7	30.7	29.8	28.2	33.9	25.9	119.6	24.9	15.4
1979	29.0	27.5	28.5	19.4	33.5	31.8	31.2	35.9	27.8	131.3	27.4	17.0
1980	32.0	30.4	31.4	21.7	38.8	34.2	34.2	38.8	30.3	143.5	30.6	19.1
1981	35.3	33.5	34.9	24.4	43.5	36.4	37.3	41.9	33.5	150.6	34.0	21.6
1982	38.1	36.3	38.1	27.1	46.0	38.4	39.4	44.5	36.1	142.8	36.6	24.2
1983	40.4	38.7	40.6	29.6	47.7	40.0	41.1	48.5	38.1	132.1	38.6	26.5
1984	42.5	40.9	42.7	31.8	49.6	41.8	43.0	51.6	40.0	126.3	40.5	28.6
1985	44.6	43.0	44.9	33.7	50.6	43.6	44.8	55.6	41.5	123.5	42.1	30.3
1986	46.4	44.8	47.0	35.7	51.1	45.2	46.7	56.5	43.7	119.5	43.6	32.0
1987	47.8	46.3	48.7	38.0	53.2	47.0	48.5	53.8	44.7	117.2	45.5	34.0
1988	50.1	48.6	50.7	41.0	55.7	48.8	50.6	56.6	46.6	115.8	48.2	36.7
1989	52.4	51.0	52.7	44.6	58.2	51.2	52.9	58.7	48.3	108.3	50.8	40.2
1990	54.8	53.6	55.0	48.4	60.4	54.0	55.3	60.0	50.3	99.2	53.2	43.7
1991	57.0	55.9	56.8	52.2	62.1	56.8	57.5	60.1	52.7	91.7	55.7	47.4
1992	59.0	58.2	58.4	55.8	63.8	58.6	58.8	62.3	55.0	83.3	57.7	51.0
1993	61.1	60.4	60.1	58.8	66.1	60.5	59.9	65.7	56.9	78.9	59.5	53.8
1994	62.7	62.1	61.6	61.2	66.8	61.8	61.0	67.0	59.0	78.7	61.2	55.8
1995	64.5	63.9	63.3	63.4	68.2	63.1	62.4	68.7	60.9	78.7	62.9	57.9
1996	66.2	65.6	65.2	64.9	68.9	65.2	64.1	71.3	62.7	80.0	64.7	59.6
1997	68.1	67.5	66.9	66.2	70.5	67.3	66.0	74.3	64.5	83.4	66.3	60.9
1998	69.5	68.8	68.5	67.5	71.8	69.0	67.8	74.6	65.8	89.1	68.3	62.1
1999	71.0	70.2	70.2	69.0	72.7	71.2	69.5	74.2	67.4	92.3	70.2	63.6
2000	72.9	72.0	72.5	70.9	75.0	74.0	71.5	75.0	68.6	97.4	72.9	65.8
2001	75.2	74.2	75.9	73.3	75.4	76.4	73.5	74.6	70.9	101.9	75.7	68.1
2002	77.1	76.2	78.1	75.2	75.3	78.6	75.3	76.4	73.7	102.3	77.9	70.7
2003	79.5	78.8	80.6	78.0	76.9	81.0	76.9	79.6	76.6	98.7	80.0	74.3
2004	82.0	81.3	82.8	80.9	77.8	83.1	79.4	83.1	79.3	97.6	82.5	77.7
2005	84.7	84.1	85.8	83.5	80.7	85.5	82.0	85.8	82.1	99.7	85.4	80.8
2006	87.6	87.1	89.4	86.0	84.1	88.4	84.8	88.4	85.0	101.0	88.3	84.1
2007	90.5	90.0	92.5	89.2	86.0	90.8	88.0	91.2	87.8	102.2	91.0	87.2
2008	93.2	92.8	95.5	91.6	90.8	93.6	91.5	91.7	91.6	102.9	93.3	90.1
2009	94.2	94.0	96.6	94.1	93.0	94.7	93.5	87.9	93.9	99.9	94.6	92.8
2010	96.0	95.8	96.8	96.5	95.1	95.7	94.8	93.1	96.1	98.5	96.2	95.4
2011	97.8	97.7	98.1	98.2	98.1	97.3	97.2	96.0	97.9	99.6	98.1	97.6
2012	100.0	100.0	100.0	100.0	100.0	100.0	100.0	100.0	100.0	100.0	100.0	100.0
2013	102.3	102.3	102.4	101.4	101.0	101.7	102.1	105.3	102.3	102.0	102.1	102.2
2014	104.8	104.8	105.4	102.5	102.3	103.6	104.8	111.0	103.7	105.9	104.4	103.9
2015	106.7	106.6	108.1	103.1	102.6	105.2	107.7	114.4	104.5	109.9	106.5	105.3
2016	109.1	108.9	111.2	104.3	103.4	107.8	110.5	119.5	105.5	113.5	108.7	107.0
2017	111.8	111.5	115.0	105.8	104.7	110.8	112.8	124.9	106.0	117.8	111.4	109.1
2018	114.9	114.6	118.6	107.8	107.0	113.1	115.4	132.0	108.1	120.2	114.0	111.8
2016												
1st quarter	108.0	107.8	109.7	103.7	103.1	106.5	109.6	116.8	105.1	111.9	107.7	106.1
2nd quarter	108.8	108.6	110.6	104.0	103.5	107.7	110.1	118.9	105.5	112.8	108.3	106.7
3rd quarter	109.6	109.4	111.7	104.6	103.6	108.2	110.8	120.7	105.8	113.9	109.1	107.3
4th quarter	110.2	109.9	112.7	105.0	103.5	108.6	111.5	121.8	105.7	115.4	109.9	107.9
2017												
1st quarter	110.7	110.5	113.6	105.2	104.2	109.8	112.0	122.0	105.9	116.7	110.5	108.2
2nd quarter	111.5	111.2	114.5	105.6	104.7	110.3	112.5	124.2	105.8	117.8	111.2	108.8
3rd quarter	112.0	111.8	115.4	105.9	104.5	111.2	112.9	125.4	106.1	118.3	111.6	109.2
4th quarter	112.9	112.7	116.4	106.6	105.4	111.7	113.8	128.1	106.4	118.3	112.4	110.3
2018												
1st quarter	113.7	113.5	117.3	107.0	105.4	112.3	114.4	129.8	107.4	119.2	113.1	110.9
2nd quarter	114.5	114.3	118.2	107.7	106.4	112.7	115.4	131.4	107.8	119.8	113.7	111.5
3rd quarter	115.2	115.0	119.0	108.0	107.3	113.4	115.8	133.0	108.4	120.4	114.3	112.1
4th quarter	116.0	115.7	120.0	108.6	108.6	114.0	116.2	133.9	108.8	121.3	115.0	112.7

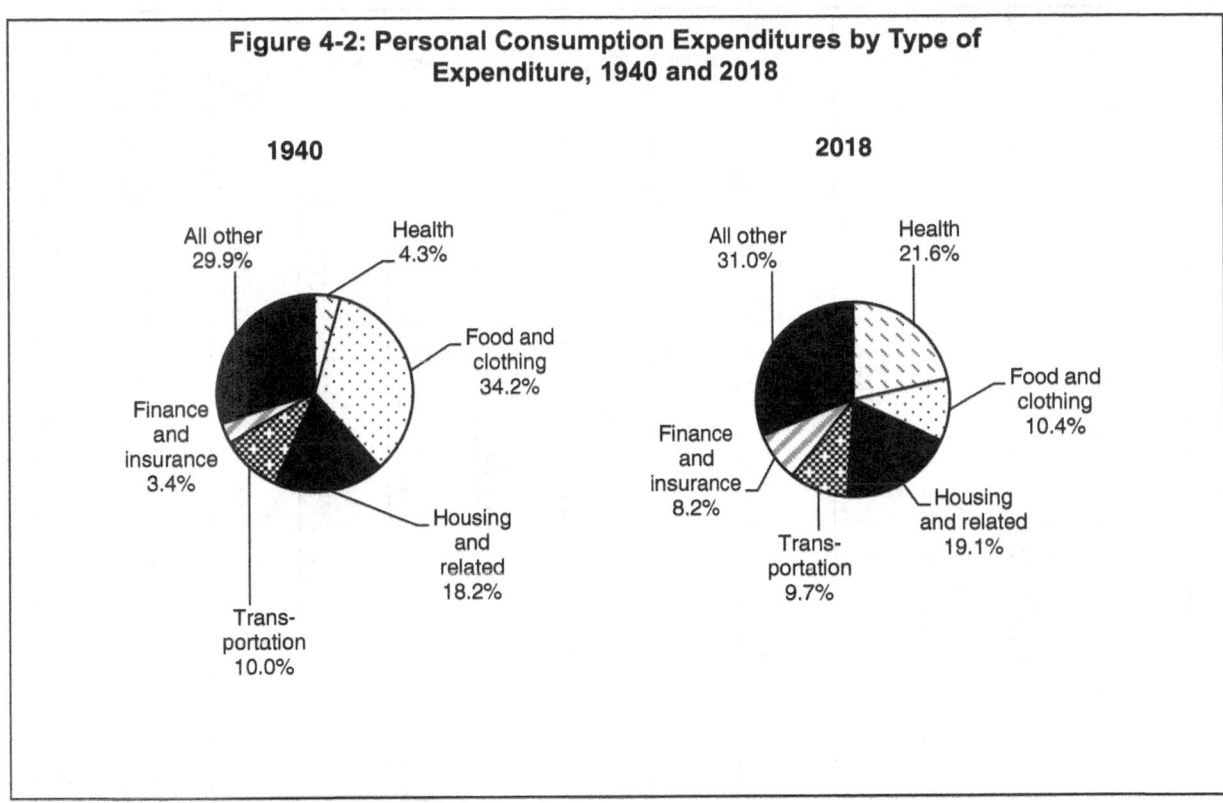

Figure 4-2: Personal Consumption Expenditures by Type of Expenditure, 1940 and 2018

- Figure 4-2 compares the composition of household consumer spending (excluding the net spending of nonprofit institutions) in 1940 and 2018, the latest available data at the time this edition of *Business Statistics* was compiled. (Table 4-5)

- Household spending on health in 2018 made up 21.6 percent of all household consumption spending—more than five times the percentage in 1940. It is important to recognize that household consumption spending, as presented in these data, includes (in health spending and in the total) medical care payments made by government and private insurance on behalf of individuals as well as out-of-pocket consumer payments. (Table 4-5)

- In 2018, a much smaller share was required for food and clothing than in 1940, (10.4 percent vs. 34.2 percent respectively) as seen in Figure 4-2. The shares of housing-related spending and transportation were similar in the two years. The share of finance and insurance spending nearly tripled, and households also spent a larger share on the remaining category, here labeled "all other," which includes communication, recreation, education, food services, accommodations, and other miscellaneous goods and services. (Table 4-5)

Table 4-5. Household Consumption Expenditures by Function

(Billions of dollars.)   NIPA Table 2.5.5

Year	Household consumption expenditures, total	Food and beverages off-premises	Clothing, footwear, and related services	Housing, utilities, and fuels	Furnishings, household equipment, and maintenance	Health	Transportation	Communication	Recreation	Education	Food services and accommodations	Financial services and insurance	Other goods and services	Foreign travel and expenditures, net
1940	70.2	15.9	8.1	12.8	6.0	3.0	7.0	0.7	3.9	0.6	4.4	2.4	5.3	0.1
1941	80.0	18.2	9.5	13.7	7.1	3.2	8.3	0.8	4.5	0.6	5.3	2.5	6.0	0.1
1942	87.7	21.6	11.8	14.8	7.4	3.6	5.4	1.0	4.9	0.7	6.9	2.6	6.8	0.2
1943	98.4	24.3	14.4	15.5	7.4	4.0	5.4	1.2	5.3	0.8	9.1	2.8	8.0	0.3
1944	107.0	26.0	15.7	16.2	8.0	4.5	5.7	1.3	5.8	0.9	10.9	3.0	8.5	0.6
1945	118.2	28.2	17.6	16.9	8.9	4.8	6.6	1.4	6.5	0.9	12.7	3.2	9.3	1.2
1946	142.2	34.9	19.7	18.8	12.7	5.9	12.1	1.6	9.0	1.0	12.7	3.7	10.3	-0.1
1947	159.7	40.2	20.4	21.2	15.5	6.6	15.4	1.7	9.8	1.2	12.5	4.1	11.1	0.0
1948	172.6	41.8	21.8	23.9	16.8	7.6	18.0	1.9	10.3	1.3	12.6	4.6	11.7	0.3
1949	176.0	40.4	21.1	25.6	16.3	7.9	21.2	2.1	10.6	1.4	12.4	4.8	11.8	0.6
1950	189.6	41.4	21.4	28.4	18.7	8.5	24.6	2.3	11.7	1.5	12.7	5.4	12.2	0.7
1951	205.7	46.0	23.1	31.6	19.7	9.3	24.2	2.6	12.3	1.6	15.0	6.2	13.3	0.9
1952	216.4	48.4	23.9	34.6	19.3	10.1	24.4	2.9	13.0	1.7	16.0	6.6	14.3	1.1
1953	229.7	49.3	24.2	37.9	20.0	11.0	28.0	3.2	13.8	1.8	16.5	7.5	15.0	1.5
1954	236.4	50.6	24.3	40.9	19.9	11.9	27.4	3.3	14.3	1.9	16.6	8.3	15.4	1.5
1955	254.9	52.0	25.4	43.8	22.3	12.8	33.4	3.6	15.4	2.1	17.0	9.2	16.3	1.6
1956	267.3	54.2	26.5	46.8	23.6	14.1	32.7	3.9	16.4	2.4	17.7	10.0	17.5	1.7
1957	282.3	57.1	26.8	50.0	23.9	15.5	35.7	4.3	16.8	2.6	18.5	10.6	18.8	1.7
1958	291.3	59.8	27.1	53.3	24.1	17.1	33.7	4.5	17.3	2.9	18.7	11.2	19.9	1.9
1959	312.3	61.6	28.6	56.9	25.6	18.8	38.8	4.8	18.8	3.1	19.7	12.1	21.4	2.0
1960	326.1	62.6	29.3	60.5	26.2	20.4	40.8	5.2	19.7	3.4	20.5	13.2	22.4	2.1
1961	336.2	63.7	29.9	64.0	26.8	21.8	39.3	5.6	20.5	3.6	21.0	14.3	23.7	2.0
1962	357.1	64.7	31.3	68.2	28.3	24.3	44.1	5.9	22.1	4.0	22.3	14.8	24.9	2.3
1963	376.3	65.9	32.1	72.2	29.8	26.4	47.8	6.4	23.9	4.3	23.3	15.5	26.2	2.5
1964	404.2	69.5	34.7	76.2	32.7	29.8	50.7	6.9	26.3	4.8	24.9	17.2	27.8	2.6
1965	436.0	74.4	36.5	81.0	34.7	32.0	56.5	7.5	28.9	5.5	27.2	18.8	30.2	2.9
1966	472.5	80.6	39.8	85.8	37.8	35.0	59.0	8.1	33.4	6.2	29.6	20.6	33.3	3.1
1967	498.6	82.6	41.7	91.1	39.7	38.4	61.0	8.9	35.8	6.9	31.0	22.1	35.8	3.8
1968	548.1	88.8	45.6	97.4	43.2	43.8	69.8	9.7	39.9	7.7	34.6	25.2	38.6	3.7
1969	594.2	95.4	48.8	105.5	45.4	50.0	75.9	10.7	43.7	8.7	37.5	27.7	41.0	4.0
1970	636.3	103.5	49.9	113.8	46.6	56.8	76.5	11.6	47.0	9.9	41.6	30.1	44.3	4.5
1971	688.2	107.1	53.4	124.5	48.8	63.4	88.7	12.8	50.1	10.9	43.8	33.1	47.0	4.8
1972	755.3	114.5	58.0	136.2	53.5	70.4	98.3	14.4	56.2	11.7	48.9	37.1	51.0	5.2
1973	835.8	126.7	63.7	149.7	59.9	78.7	108.6	16.1	63.0	13.0	54.8	39.9	56.9	4.7
1974	914.7	143.0	66.9	166.3	64.6	88.7	112.8	17.7	68.9	14.2	60.6	44.1	62.2	4.7
1975	1 013.5	156.6	71.4	184.8	67.8	102.7	124.4	20.0	77.1	15.9	68.8	51.8	67.8	4.4
1976	1 128.7	167.3	76.9	204.6	75.0	116.8	147.4	22.4	86.2	17.4	77.8	56.8	76.3	3.8
1977	1 253.3	179.8	84.3	228.7	84.0	131.6	168.1	24.3	94.9	18.8	85.7	65.1	83.8	4.3
1978	1 399.2	196.1	94.6	255.6	93.6	149.4	184.5	27.0	106.3	20.9	97.1	76.7	93.2	4.3
1979	1 559.3	218.4	101.9	287.5	104.3	169.9	207.1	29.4	118.9	22.9	110.9	83.6	100.4	4.1
1980	1 720.6	239.2	108.8	327.7	110.7	195.5	226.5	31.8	127.4	25.4	121.7	91.7	110.7	3.5
1981	1 899.8	255.3	119.1	367.6	118.3	228.9	250.7	36.0	141.6	28.3	133.9	98.5	121.2	0.4
1982	2 033.3	267.1	122.7	401.7	121.0	254.9	255.9	41.1	151.4	31.0	142.5	113.7	127.8	2.5
1983	2 240.8	277.0	132.9	434.6	132.3	287.1	284.9	44.8	169.3	34.3	153.6	141.0	143.4	5.4
1984	2 447.0	291.1	144.7	471.1	147.1	315.2	321.8	48.4	190.0	37.7	164.9	150.8	157.7	6.6
1985	2 664.9	303.0	154.3	513.9	156.0	345.3	357.4	53.1	207.2	41.2	174.3	178.2	173.3	7.7
1986	2 833.7	316.4	165.1	548.1	168.7	378.4	362.6	56.9	225.4	44.5	186.7	187.7	190.0	3.3
1987	3 020.6	324.3	176.4	582.6	176.9	419.9	376.6	60.0	247.0	48.8	204.4	189.5	208.2	6.1
1988	3 266.7	342.8	188.1	625.9	186.6	470.7	404.8	63.1	273.4	54.4	225.8	202.9	225.4	2.8
1989	3 508.7	365.4	201.2	666.9	197.5	519.0	428.3	67.3	295.7	60.6	242.6	222.3	245.7	-3.7
1990	3 733.1	391.2	206.5	709.3	200.6	583.7	442.9	70.1	314.7	66.0	262.7	230.8	262.3	-7.7
1991	3 863.3	403.0	210.1	747.5	199.1	638.4	418.3	73.9	326.3	70.6	273.4	250.1	268.0	-15.2
1992	4 109.6	404.5	223.0	783.3	209.4	700.4	451.3	81.1	346.8	76.4	286.3	277.0	290.1	-20.0
1993	4 361.9	413.5	231.1	827.3	221.9	741.7	485.0	85.8	378.4	81.1	298.4	314.0	304.5	-20.7
1994	4 623.2	432.1	240.1	876.1	238.6	779.9	527.3	93.3	414.0	86.4	308.3	327.9	316.9	-17.6
1995	4 860.3	443.7	244.7	926.7	251.7	826.0	552.5	98.9	449.8	92.3	316.1	347.0	332.7	-21.9
1996	5 134.6	461.9	253.5	976.7	263.7	868.3	596.7	108.3	481.5	99.6	326.6	372.1	350.8	-25.2
1997	5 431.2	474.8	262.0	1 023.1	277.3	919.9	639.3	120.1	509.5	107.1	343.4	408.9	368.4	-22.5
1998	5 753.4	487.4	273.1	1 077.0	297.3	979.7	666.2	128.8	544.3	115.2	361.8	446.1	393.7	-17.3
1999	6 140.8	515.5	287.2	1 135.5	320.6	1 033.3	726.3	140.2	589.9	123.9	380.3	486.4	429.4	-27.7
2000	6 604.1	540.6	297.5	1 214.5	344.0	1 109.6	793.1	153.8	633.7	134.3	408.8	543.0	458.5	-27.2
2001	6 886.6	564.0	294.6	1 303.0	353.2	1 209.4	805.6	159.1	647.0	143.6	419.7	525.7	477.9	-16.2
2002	7 144.3	575.1	295.2	1 347.9	366.3	1 317.1	806.8	161.1	669.3	149.5	436.3	534.7	496.8	-11.8
2003	7 517.6	599.6	301.4	1 411.6	381.1	1 410.7	840.1	165.1	704.3	159.5	462.7	560.3	526.5	-5.3
2004	8 006.3	632.6	313.4	1 488.4	406.6	1 514.9	893.8	171.2	758.8	169.0	498.2	605.5	559.1	-5.3
2005	8 536.9	668.2	326.4	1 606.0	430.1	1 612.3	955.3	177.3	805.2	180.5	533.6	659.0	589.5	-6.6
2006	9 021.2	700.3	336.2	1 706.1	448.7	1 715.0	987.9	190.5	855.7	193.1	570.6	695.0	624.5	-2.3
2007	9 457.7	737.3	339.5	1 783.8	455.0	1 823.2	1 028.0	203.0	897.0	206.0	601.5	737.2	657.6	-11.6
2008	9 689.0	769.1	333.1	1 865.5	438.7	1 909.3	1 017.0	214.0	900.0	216.4	620.2	756.6	667.9	-18.8
2009	9 553.0	772.9	319.1	1 900.9	405.9	1 999.3	879.7	212.8	863.4	226.1	612.7	711.3	666.4	-17.4
2010	9 891.4	786.9	331.7	1 928.5	410.7	2 078.1	961.8	220.7	884.4	240.3	635.7	754.4	687.2	-28.9
2011	10 329.3	819.5	347.1	1 982.9	423.7	2 153.2	1 080.1	228.1	901.6	248.2	669.5	797.9	712.1	-34.9
2012	10 665.3	846.2	359.5	2 020.5	438.6	2 230.7	1 135.5	232.5	934.4	250.6	704.9	820.1	725.4	-33.6
2013	10 964.6	864.0	365.3	2 080.5	454.3	2 288.3	1 170.5	237.5	967.4	256.7	732.3	858.4	740.1	-50.5
2014	11 457.2	896.9	376.0	2 176.2	476.6	2 409.1	1 202.0	254.0	1 004.5	263.3	776.9	908.1	766.2	-52.8
2015	11 912.9	920.1	384.2	2 275.4	499.8	2 564.3	1 162.5	259.8	1 046.4	270.8	832.8	956.9	797.8	-58.0
2016	12 351.0	937.8	390.2	2 371.2	520.1	2 691.3	1 161.7	268.5	1 103.6	281.2	872.4	977.5	821.9	-46.4
2017	12 898.3	967.5	392.4	2 472.7	542.4	2 800.7	1 232.0	266.5	1 157.0	293.5	913.8	1 040.4	856.6	-37.1
2018	13 555.7	1 003.4	408.3	2 588.8	572.6	2 933.1	1 311.8	273.5	1 203.5	304.0	973.3	1 111.0	893.4	-20.9

NOTES AND DEFINITIONS, CHAPTER 4

TABLES 4-1 THROUGH Table 4-5

SOURCE: U.S. DEPARTMENT OF COMMERCE, BUREAU OF ECONOMIC ANALYSIS (BEA)

All personal income and personal consumption expenditure series are from the national income and product accounts (NIPAs). All quarterly series are shown at seasonally adjusted annual rates. Current and constant dollar values are in billions of dollars. Indexes of price and quantity are based on the average for the year 2012, which equals 100.

Tables 4-1 through 4-4 cover all income and spending by the personal sector, which includes nonprofit institutions serving households (NPISHs). In a new feature introduced in the 2009 comprehensive revision of the NIPAs, Tables 4-2, 4-3, and 4-4 show the services component of personal consumption expenditures broken down into separate aggregates for "household consumption expenditures for services" and "final consumption expenditures" by NPISHs.

The last table gives further details of the separate accounts for households and NPISHs that are only available annually. Table 4-5 shows a more detailed functional breakdown of household consumption expenditures, not including NPISHs.

In several cases, the notes and definitions below will refer to *imputations* or *imputed values*. See the notes and definitions to Chapter 1 for an explanation of imputation and the role it plays in national and personal income measurement.

TABLES 4-1 THROUGH 4-4

Sources and Disposition of Personal Income; Personal Consumption Expenditures by Major Type of Product

Definitions

Personal income is the income received by persons residing in the United States from participation in production, from government and business transfer payments, and from government interest, which is treated similarly to a transfer payment rather than as income from participation in production. *Persons* denotes the total for individuals, *nonprofit institutions that primarily serve households (NPISHs)*, private noninsured welfare funds, and private trust funds. Personal income, outlays, and saving excluding NPISHs are referred to as *household* income, outlays, and saving. All proprietors' income is treated as received by individuals. Life insurance carriers and private noninsured pension funds are not counted as persons, but their saving is credited to persons.

Income from the sale of illegal goods and services is excluded by definition from national and personal income, and the value of purchases of illegal goods and services is not included in personal consumption expenditures.

Personal income is the sum of compensation received by employees, proprietors' income with inventory valuation and capital consumption adjustments (IVA and CCAdj), rental income of persons with capital consumption adjustment, personal receipts on assets, and personal current transfer receipts, less contributions for social insurance.

Personal income differs from national income in that it includes current transfer payments and interest received by persons, regardless of source, while it excludes the following national income components: employee and employer contributions for social insurance; business transfer payments, interest payments, and other payments on assets other than to persons; taxes on production and imports less subsidies; the current surplus of government enterprises; and undistributed corporate profits with IVA and CCAdj. The relationships of GDP, gross and net national product, national income, and personal income are displayed in Table 1-11.

Compensation of employees is the sum of wages and salaries and supplements to wages and salaries, as defined in the *national* income account (see Table 1-13 and the notes and definitions to Chapter 1).

As in *national* income, the *compensation of employees* component of personal income refers to compensation received by residents of the United States, including compensation from the rest of the world, but excludes compensation from domestic industries to workers residing in the rest of the world.

Wages and salaries consists of the monetary remuneration of employees, including wages and salaries as conventionally defined; the compensation of corporate officers; corporate directors' fees paid to directors who are also employees of the corporation; the value of employee exercise of "nonqualified stock options"; commissions, tips, and bonuses; voluntary employee contributions to certain deferred-compensation plans, such as 401(k) plans; and receipts in kind that represent income. This category also now includes judicial fees to jurors and witnesses, compensation of prison inmates, and marriage fees to justices of the peace, which earlier were classified as "other labor income". As of the 2013 revision, wages and salaries are now measured on an accrual basis, that is to say when earned rather than when paid, consistent with the treatment in the gross domestic income and national income tables.

Supplements to wages and salaries consists of employer contributions to employee pension and insurance funds and to government social insurance funds. In a substantial change introduced in the 2013 revision, defined benefit pension plan transactions are now recorded on an accrual basis instead of a cash transactions basis: employees are now credited with defined pension benefits, based on actuarial estimates of pension costs, at the time they earn them. (This was already the case with defined contribution plans.)

The following two categories, *proprietors' income* and *rental income,* are both measured net of depreciation of the capital (structures, equipment, and intellectual property products) involved. BEA calculates normal depreciation, based on the estimated life of the capital, and subtracts it from the estimated value of receipts to yield net income.

Proprietors' income with inventory valuation and capital consumption adjustments is the currentproduction income (including income-in-kind) of sole proprietors and partnerships and of taxexempt cooperatives. The imputed net rental income of owneroccupants of farm dwellings is included. Dividends and monetary interest received by proprietors of nonfinancial business and rental incomes received by persons not primarily engaged in the real estate business are excluded. These incomes are included in personal income receipts on assets and rental income of persons, respectively. Fees paid to outside directors of corporations are included. The two valuation adjustments are designed to obtain income measures that exclude any element of capital gains: inventory withdrawals are valued at replacement cost, rather than historical cost, and charges for depreciation are on an economically consistent accounting basis and are valued at replacement cost.

Rental income of persons with capital consumption adjustment consists of the net currentproduction income of persons from the rental of real property (other than the incomes of persons primarily engaged in the real estate business), the imputed net rental income of owneroccupants of nonfarm dwellings, and the royalties received by persons from patents, copyrights, and rights to natural resources. The capital consumption adjustment converts charges for depreciation to an economically consistent accounting basis valued at replacement cost. Rental income is net of interest and other expenses, and hence is affected by changing indebtedness and interest payments on owner-occupied and other housing.

Personal income receipts on assets consists of personal interest income and personal dividend income.

Personal interest income is the interest income (monetary and imputed) of persons from all sources, including interest paid by government to government employee retirement plans as well as government interest paid directly to persons.

Personal dividend income is the dividend income of persons from all sources, excluding capital gains distributions. It equals net dividends paid by corporations (dividends paid by corporations minus dividends received by corporations) less a small amount of corporate dividends received by general government. Dividends received by government employee retirement systems are included in personal dividend income.

Personal current transfer receipts is income payments to persons for which no current services are performed. It consists of government social benefits to persons and net receipts from business.

Government social benefits to persons consists of benefits from the following categories of programs:

* *Social Security and Medicare,* consisting of federal oldage, survivors, disability, and health insurance benefits distributed from the Social Security and Medicare trust funds;

* *Medicaid,* the federal-state means-tested program covering medical expenses for lower-income children and adults as well as nursing care expenses;

* *Unemployment insurance;*

* *Veterans' benefits;*

* *Other government benefits to persons,* which includes pension benefit guaranty; workers' compensation; military medical insurance; temporary disability insurance; food stamps; Black Lung benefits; supplemental security income; family assistance, which consists of aid to families with dependent children and (beginning in 1996) assistance programs operating under the Personal Responsibility and Work Opportunity Reconciliation Act of 1996; educational assistance; and the earned income credit. Government payments to nonprofit institutions, other than for work under research and development contracts, also are included. Payments from government employee retirement plans are not included.

Note that the value of Medicare and Medicaid spending, though in practice it is usually paid directly from the government to the health care provider, is treated as if it were cash income to the consumer which is then expended in personal consumption expenditures; this value is not treated in the national accounts as a government purchase of medical services but as a government benefit paid to persons, which then finances personal consumption spending.

Contributions for government social insurance, domestic, which is subtracted to arrive at personal income, includes payments by U.S. employers, employees, selfemployed, and other individuals who participate in the following programs: oldage, survivors, and disability insurance (Social Security); hospital insurance and supplementary medical insurance (Medicare); unemployment insurance; railroad retirement; veterans' life insurance; and temporary disability insurance. Contributions to government employee retirement plans are not included in this item.

In the 2009 comprehensive revision, most transactions between the U.S. government and economic agents in Guam, the U.S. Virgin Islands, American Samoa, Puerto Rico, and the Northern Mariana Islands are treated as government transactions with the rest of the world. Since the NIPAs only cover the 50 states and the District of Columbia, the *domestic* contributions to government social insurance funds are the only ones that need to be subtracted from NIPA payroll data to calculate personal income. The social insurance receipts of governments, shown in Chapter 6, will be somewhat larger than this personal income entry because they will include contributions from residents of those territories and commonwealths.

Personal current taxes is tax payments (net of refunds) by persons residing in the United States that are not chargeable to business expenses, including taxes on income, on realized net capital gains, and on personal property. As of the 1999 revisions, estate and gift taxes are classified as capital transfers and are not included in personal current taxes.

Disposable personal income is personal income minus personal current taxes. It is the income from current production that is available to persons for spending or saving. However, it is not the cash flow available, since it excludes realized capital gains. Disposable personal income in chained (2009) dollars represents the inflation-adjusted value of disposable personal income, using the implicit price deflator for personal consumption expenditures.

Personal income excluding current transfer receipts, also shown in chained (2009) dollars using the implicit price deflator for personal consumption expenditures, is an important business cycle indicator, to which particular attention is paid because it is calculated monthly as well as quarterly and annually, and thus can help establish monthly cycle turning point dates. As a pre-income-tax measure which excludes transfer payments, it is a better measure of income generated by the economy than alternative monthly income indicators such as total or disposable personal income, which are more oriented toward purchasing power.

Personal outlays is the sum of *personal consumption expenditures* (defined below), *personal interest payments,* and *personal current transfer payments.*

Personal interest payments is nonmortgage interest paid by households. As noted above in the definition of rental income, mortgage interest has been subtracted from gross rental or imputed rental receipts of persons to yield a net rental income estimate; hence, it is not included as an interest outlay in this category.

Personal current transfer payments to government includes donations, fees, and fines paid to federal, state, and local governments.

Personal current transfer payments to the rest of the world (net) is personal remittances in cash and in kind to the rest of the world less such remittances from the rest of the world.

Personal saving is derived by subtracting personal outlays from disposable personal income. It is the current net saving of individuals (including proprietors), nonprofit institutions that primarily serve individuals, life insurance carriers, retirement funds (including those of government employees), private noninsured welfare funds, and private trust funds. Conceptually, personal saving may also be viewed as the sum of the net acquisition of financial assets and the change in physical assets less the sum of net borrowing and consumption of fixed capital. In either case, it is defined to exclude both realized and unrealized capital gains.

Note that in the context of national income accounting, the term just defined is *saving,* not "savings." *Saving* refers to a <u>flow</u> of income during a particular time span (such as a year or a quarter) that is not consumed. It is therefore available to finance a commensurate <u>flow</u> of investment during that time span. Strictly defined, "savings" denotes an accumulated <u>stock</u> of monetary funds—possibly the cumulative effects of successive periods of *saving*—available to the owner in asset form, such as in a bank savings account.

Personal consumption expenditures (PCE) is goods and services purchased by persons residing in the United States. Persons are defined as individuals and nonprofit institutions that primarily serve individuals. PCE mostly consists of purchases of new goods and services by individuals from business, including purchases financed by insurance (such as both private and government medical insurance). In addition, PCE includes purchases of new goods and services by nonprofit institutions, net purchases of used goods by individuals and nonprofit institutions, and purchases abroad of goods and services by U.S. residents traveling or working in foreign countries. PCE also includes purchases for certain goods and services provided by the government, primarily tuition payments for higher education, charges for medical care, and charges for water and sanitary services. Finally, PCE includes imputed purchases that keep PCE invariant to changes in the way that certain activities are carried out. For example, to take account of the value of the services provided by owner-occupied housing, PCE includes an imputation equal to what (estimated) rent homeowners would pay if they rented their houses from themselves. (See the discussion of imputation in the notes and definitions to Chapter 1.) Actual purchases of residential structures by individuals are classified as gross private domestic investment.

In the 2009 comprehensive revision, the classification system used for breakdowns of PCE was revised, and new calculations were introduced separating the consumption spending of the household sector proper from that of nonprofit institutions serving households.

Goods is the sum of *Durable* and *Nondurable goods.*

The PCE category *Durable goods* is subdivided into *Motor vehicles and parts, Furnishings and household equipment* (which includes appliances), *Recreational goods and vehicles* (which includes video, audio, photographic, and information processing equipment and media), and *Other durable goods.*

Nondurable goods encompasses *Food and beverages off-premises, Clothing and footwear, Gasoline and other energy goods,* and *Other nondurable goods.* This food and beverages category no longer includes meals and beverages purchased for consumption on the premises, which are now included in services, since they have a high service component and since their prices are much more stable than those of off-premises food and beverages.

Services, total is subdivided into *Household services* and *Final consumption expenditures of nonprofit institutions serving households (NPISHs).* The latter is the difference between the *gross output* of NPISHs and the amounts that they receive from sales of goods and services to households—for example, payments for services of a nonprofit hospital. Such sales of goods and services appear in the appropriate category of household expenditures—for example, *Health* in the case of the hospital services.

The components of *Household services* are *Housing and utilities, Health care, Transportation services, Recreation services, Food services and accommodations* (which includes meals and beverages purchased for consumption on premises, as well as payments for hotels and similar accommodations), *Financial services and insurance,* and *Other services.*

These are the categories used in the quarterly data presented in Tables 4-2, 4-3, and 4-4. This classification system is not particularly helpful with respect to the objective of spending. For example, the *Health care* component of services does not include drugs and medicines, which are included instead in nondurable goods. For a more precise classification of consumption spending by objective, see Table 4-5, Household Consumption Expenditures by Function, described in more detail below. This classification by type of expenditure is only available on an annual basis, and later than the principal quarterly NIPA data.

Data availability

Monthly data on personal income and spending are made available in a BEA press release, usually distributed the first business day following the monthly release of the latest quarterly national income and product account (NIPA) estimates. Monthly and quarterly data are subsequently published each month in the BEA's *Survey of Current Business.* Current and historical data are available on the BEA Web site at <http://www.bea.gov>, and may also be obtained from the STAT-USA subscription Web site at <http://www.stat-usa.gov>.

References

References can be found in the notes and definitions to Chapter 1. A discussion of monthly estimates of personal income and its disposition appears in the November 1979 edition of the *Survey of Current Business.* Additional and more recent information can be found in the articles listed in the notes and definitions for Chapter 1.

TABLE 4-5

Household Consumption Expenditures by Function

SOURCE: BUREAU OF ECONOMIC ANALYSIS (BEA)

In this table, also derived from the NIPAs, annual estimates of the current-dollar value of PCE by households—excluding the "final" consumption expenditures of nonprofit institutions serving households (NPISHs); see definition above—are presented by function.

Definitions

Food and beverages includes food and beverages (including alcoholic beverages) purchased for home consumption and food produced and consumed on farms.

Clothing, footwear, and related services includes purchases, rental, cleaning, and repair of clothing and footwear.

Housing, utilities, and fuels includes rents paid for rental housing, imputed rent of owner-occupied dwellings, and purchase of fuels and utility services.

Furnishings, household equipment, and routine household maintenance includes furniture, floor coverings, household textiles, appliances, tableware etc., tools, and other supplies and services.

Health includes drugs, other medical products and equipment, and outpatient, hospital, and nursing home services.

Transportation includes the purchase and operation of motor vehicles and public transportation services.

Communication includes telephone equipment, postal and delivery services, telecommunication services, and Internet access.

Recreation includes video and audio equipment, computers, and related services; sports vehicles and other sports goods and services; memberships and admissions; magazines, newspapers, books, and stationery; pets and related goods and services; photo goods and services; tour services; and legal gambling. (As noted earlier, purchases of goods and services that are illegal and the incomes from such purchases are outside the scope of the national income and product accounts.)

Education includes educational services and books.

Food services and accommodations includes meals and beverages purchased for consumption on the premises, food furnished to employees, and hotel and other accommodations including housing at schools.

Financial services and insurance consists of financial services and life, household, medical care, motor vehicle, and other insurance.

Other goods and services includes personal goods (such as cosmetics, jewelry, and luggage) and services, social services and religious activities, professional and other services, and tobacco.

Foreign travel and expenditures, net consists of foreign travel spending and other expenditures abroad by U.S. residents minus expenditures in the United States by nonresidents. A negative

figure indicates that foreigners spent more here than U.S. residents spent abroad. Positive values for this foreign travel category, indicating that U.S. residents spent more abroad than foreigners spent here, appear in the 1980s when the dollar was strong against other major currencies. (The international value of the dollar is shown in Table 13-2.) Negative values in subsequent years have resulted in part from the weakening of the dollar, which discouraged U.S. residents' travel abroad and encouraged tourism by foreigners in the United States. The negative sign does not indicate a drain on GDP—these effects of a weaker dollar are in fact positive for real GDP—but rather reflects the fact that the goods and services purchased by foreigners in the United States must be subtracted from the total consumer purchases recorded in the other columns of this table in order to be added to other exports and classified in the category of exports rather than in the consumption spending of U.S. residents.

Data availability and revisions

Data are updated once a year, after the general midyear revision of the NIPAs, and are available on the BEA Web site at <http://www.bea.gov>.

CHAPTER 5: SAVING AND INVESTMENT

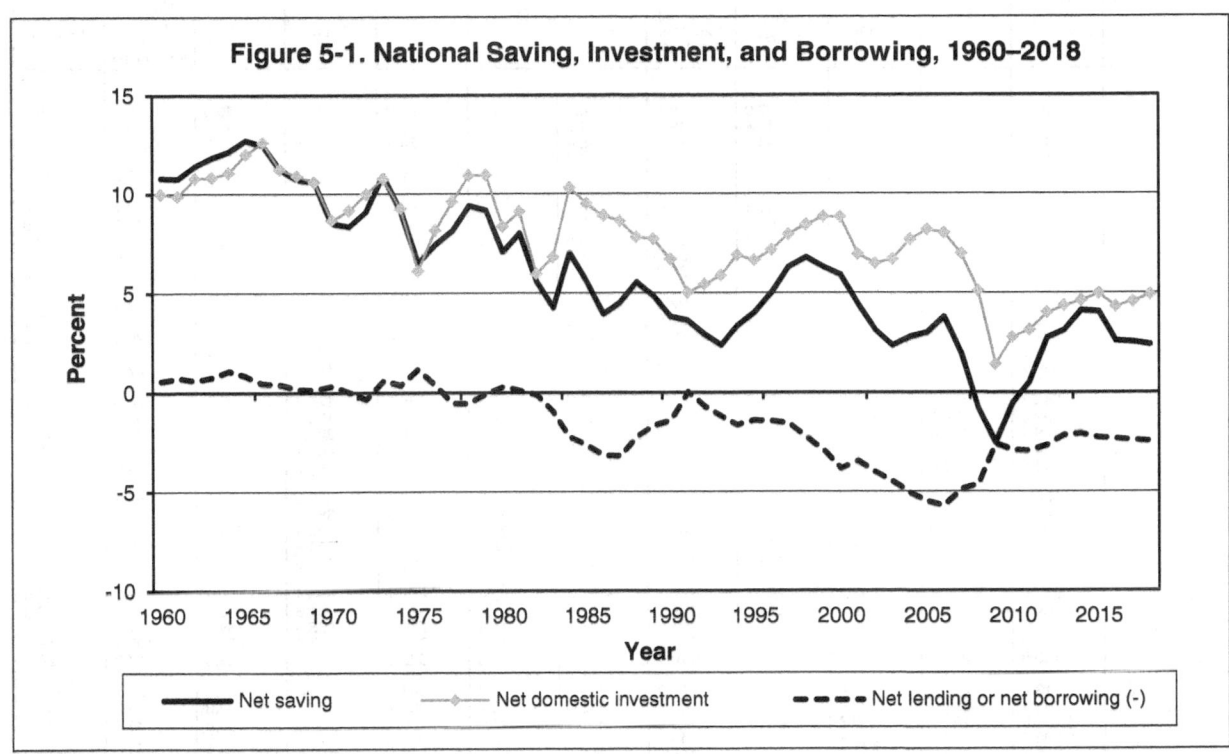

Figure 5-1. National Saving, Investment, and Borrowing, 1960–2018

- Net national saving—the middle line in Figure 5-1—averaged a much lower rate of gross national income (GNI) in the years 1984 through 2000, than in the rates seen in the 1950s and 1960s. Net saving is gross saving by U.S. persons, businesses, and governments, <u>less</u> the consumption of fixed capital. Thus, it represents saving available for investment over and above the replacement of the existing capital stock. (Tables 5-1A and B)

- In the recession that began at the end of 2007, net national saving fell into negative territory, for the first time since the depression years 1931 through 1934. Government deficits soared as the Great Recession cut deeply into tax revenues and called forth both automatic and discretionary anti-recession spending such as unemployment compensation, food stamps (Supplemental Nutrition Assistance Program – SNAP, and Temporary Assistance for Needy Families – TANF. In one year, 2008 to 2009, the budget deficit doubled. (Tables 5-1A and B and Table 6-14B)

- Net saving as a percent of gross national income is currently much lower than it was in the 1960s and 1970s. In 1965, net saving got as high as 12.7 percent. In 2018, the national saving rate declined for the third consecutive year to 2.4 percent. (Table 5-1A)

Table 5-1A. Saving and Investment: Recent Data

(Billions of dollars, except as noted; quarterly data are at seasonally adjusted annual rates.)   NIPA Tables 1.7.5, 5.1

			Gross saving										
		Net saving						Consumption of fixed capital					
			Private			Government			Private			Government	
Year and quarter	Total	Total	Total	Personal saving	Domestic corporate business	Federal	State and local	Total	Total	Domestic business	House-holds and institutions	Federal	State and local
1960	127.0	59.1	56.4	37.9	18.5	0.2	2.5	67.9	48.2	40.0	8.2	15.2	4.5
1961	131.6	61.0	63.5	44.4	19.2	-4.7	2.2	70.6	49.8	41.3	8.5	16.0	4.8
1962	143.4	69.3	71.6	46.7	24.9	-5.4	3.1	74.1	51.8	42.9	8.8	17.2	5.1
1963	154.1	76.1	75.0	47.1	27.9	-2.2	3.3	78.0	54.2	44.9	9.3	18.4	5.5
1964	166.1	83.7	86.6	55.3	31.4	-6.9	3.9	82.4	57.3	47.4	9.9	19.2	5.9
1965	182.9	94.9	96.8	58.8	38.0	-5.5	3.7	88.0	61.6	50.9	10.7	20.0	6.4
1966	196.5	101.2	103.4	61.9	41.6	-7.0	4.7	95.3	67.2	55.6	11.5	21.0	7.1
1967	200.5	96.9	112.5	73.0	39.5	-19.5	4.0	103.6	73.3	60.9	12.4	22.5	7.8
1968	214.6	101.2	111.5	72.9	38.6	-13.8	3.5	113.4	80.6	67.0	13.6	24.2	8.6
1969	233.1	108.2	110.3	76.1	34.2	-5.1	3.1	124.9	89.4	74.2	15.2	25.8	9.6
1970	228.2	91.4	124.8	97.6	27.2	-34.8	1.4	136.8	98.3	81.8	16.5	27.6	10.9
1971	246.1	97.2	149.4	111.9	37.5	-50.9	-1.3	148.9	107.6	89.4	18.2	29.1	12.2
1972	277.6	116.6	159.6	111.5	48.0	-49.0	6.1	161.0	117.5	97.2	20.3	30.2	13.3
1973	335.3	156.6	189.3	135.8	53.5	-38.3	5.6	178.7	131.5	108.2	23.3	32.4	14.8
1974	349.2	142.3	186.0	146.3	39.7	-41.3	-2.3	206.9	153.2	126.1	27.0	35.4	18.4
1975	348.1	109.6	218.3	164.0	54.3	-97.9	-10.7	238.5	178.8	147.9	30.9	38.7	21.0
1976	399.3	139.1	224.4	154.4	70.0	-80.9	-4.4	260.2	196.5	162.2	34.3	41.7	22.0
1977	459.4	169.6	242.5	155.9	86.6	-73.4	0.5	289.8	221.1	181.5	39.6	45.3	23.4
1978	548.0	220.8	278.0	175.1	102.9	-62.0	4.9	327.2	252.1	205.5	46.6	49.7	25.4
1979	613.5	239.6	288.2	186.8	101.4	-47.4	-1.2	373.9	290.7	236.4	54.4	54.6	28.6
1980	630.1	201.7	296.4	224.1	72.3	-88.8	-5.9	428.4	335.0	272.8	62.2	60.4	33.1
1981	743.9	256.6	354.9	265.5	89.4	-88.1	-10.2	487.2	381.9	313.5	68.5	67.5	37.8
1982	725.8	188.9	379.0	293.3	85.6	-167.4	-22.8	537.0	420.4	347.5	72.9	75.4	41.2
1983	716.7	154.1	379.7	264.0	115.7	-207.2	-18.4	562.6	438.8	362.8	76.0	81.1	42.7
1984	881.6	283.2	479.9	330.3	149.5	-196.5	-0.2	598.4	463.5	382.7	80.9	90.6	44.3
1985	881.0	240.8	442.5	284.9	157.5	-199.2	-2.4	640.1	496.4	410.3	86.2	97.0	46.8
1986	864.5	179.2	399.1	290.6	108.5	-215.9	-4.0	685.3	531.6	438.1	93.4	103.6	50.1
1987	948.9	218.5	398.6	275.4	123.2	-165.7	-14.4	730.4	566.3	464.5	101.8	110.1	54.0
1988	1 076.6	292.1	463.4	320.5	142.9	-160.0	-11.3	784.5	607.9	499.2	108.8	118.9	57.7
1989	1 109.8	271.5	450.2	340.0	110.3	-159.4	-19.3	838.3	649.6	532.6	117.0	126.9	61.8
1990	1 113.4	224.8	464.4	361.1	103.2	-203.3	-36.2	888.5	688.4	564.6	123.8	133.5	66.6
1991	1 153.4	221.0	529.5	396.0	133.5	-248.4	-60.1	932.4	721.5	592.3	129.2	140.1	70.8
1992	1 147.6	187.4	592.8	453.9	139.0	-334.5	-71.0	960.2	742.9	608.2	134.7	143.3	74.1
1993	1 163.4	159.9	545.9	397.7	148.2	-313.5	-72.5	1 003.5	778.2	634.2	144.0	147.0	78.3
1994	1 295.1	239.5	559.0	363.4	195.7	-255.6	-63.9	1 055.6	822.5	669.0	153.5	150.3	82.8
1995	1 426.3	303.9	616.5	387.1	229.4	-242.1	-70.4	1 122.4	880.7	717.8	162.9	153.5	88.1
1996	1 578.9	403.6	636.8	382.3	254.5	-179.4	-53.8	1 175.3	929.1	758.5	170.6	153.7	92.5
1997	1 780.5	541.2	675.1	390.3	284.9	-92.0	-42.0	1 239.3	987.8	807.9	179.9	154.3	97.3
1998	1 930.6	620.8	649.5	446.5	203.0	1.4	-30.1	1 309.7	1 052.2	860.1	192.1	155.4	102.2
1999	2 010.3	611.4	583.4	349.4	234.1	66.9	-38.9	1 398.9	1 132.2	924.0	208.2	158.2	108.6
2000	2 127.3	616.1	501.2	358.3	142.9	155.5	-40.6	1 511.2	1 231.5	1 004.6	227.0	163.1	116.6
2001	2 076.9	477.4	582.4	391.6	190.8	14.0	-119.0	1 599.5	1 311.7	1 064.9	246.8	164.4	123.4
2002	2 003.6	345.6	799.9	473.7	326.2	-271.5	-182.9	1 658.0	1 361.8	1 098.8	263.0	166.8	129.3
2003	1 991.7	272.6	858.0	471.5	386.5	-404.1	-181.3	1 719.1	1 411.9	1 125.4	286.6	172.2	134.9
2004	2 164.3	342.5	892.4	463.8	428.6	-400.9	-149.0	1 821.8	1 497.1	1 178.8	318.3	180.5	144.2
2005	2 365.8	394.8	803.5	296.7	506.8	-305.9	-102.8	1 971.0	1 622.6	1 266.4	356.2	190.5	157.9
2006	2 657.9	533.8	846.4	385.6	460.8	-227.6	-85.0	2 124.1	1 751.8	1 362.6	389.2	201.3	171.0
2007	2 536.6	283.8	679.2	391.6	287.6	-266.1	-129.3	2 252.8	1 852.5	1 446.7	405.8	212.8	187.5
2008	2 241.2	-117.7	734.3	544.9	189.4	-631.1	-220.9	2 358.8	1 931.8	1 524.9	407.0	225.9	201.2
2009	2 008.3	-363.2	1 227.1	666.5	560.6	-1 248.9	-341.3	2 371.5	1 928.7	1 531.2	397.5	233.6	209.1
2010	2 312.2	-78.7	1 553.9	740.9	813.0	-1 325.1	-307.5	2 390.9	1 933.8	1 537.1	396.7	243.7	213.4
2011	2 556.9	82.4	1 599.4	849.8	749.6	-1 242.0	-275.1	2 474.5	1 997.3	1 595.1	402.2	254.9	222.2
2012	3 036.0	460.0	1 821.5	1 107.6	713.9	-1 078.6	-282.8	2 576.0	2 082.4	1 670.4	412.0	261.6	232.0
2013	3 218.2	537.0	1 440.3	801.4	638.9	-637.9	-265.4	2 681.2	2 176.6	1 738.8	437.8	265.9	238.8
2014	3 560.3	745.3	1 587.6	970.8	616.8	-604.3	-238.0	2 815.0	2 298.5	1 831.2	467.3	270.2	246.3
2015	3 674.9	758.4	1 548.8	1 048.8	500.0	-570.1	-220.3	2 916.5	2 393.7	1 908.0	485.7	271.4	251.4
2016	3 484.5	493.0	1 416.8	958.8	458.0	-677.0	-246.8	2 991.6	2 463.2	1 954.4	508.9	271.8	256.6
2017	3 626.5	505.0	1 477.8	1 030.9	446.9	-724.7	-248.1	3 121.4	2 578.2	2 041.0	537.2	277.1	261.1
2018	3 795.2	503.8	1 752.7	1 210.4	542.3	-1 009.8	-239.2	3 291.4	2 725.8	2 151.4	574.4	286.1	279.6
2016													
1st quarter	3 570.1	620.3	1 518.2	1 038.1	480.1	-644.5	-253.4	2 949.8	2 426.6	1 929.1	497.5	269.9	253.3
2nd quarter	3 454.7	474.9	1 402.7	945.1	457.6	-674.8	-253.0	2 979.8	2 452.4	1 947.3	505.2	271.1	256.3
3rd quarter	3 435.5	433.2	1 363.5	925.5	438.0	-687.2	-243.0	3 002.3	2 472.7	1 959.9	512.8	272.2	257.4
4th quarter	3 477.9	443.5	1 382.7	926.5	456.2	-701.6	-237.6	3 034.4	2 501.1	1 981.1	519.9	273.9	259.4
2017													
1st quarter	3 600.7	535.8	1 478.9	1 025.8	453.1	-685.0	-258.1	3 064.9	2 527.9	2 003.2	524.8	274.9	262.1
2nd quarter	3 613.8	512.1	1 481.0	1 033.9	447.1	-699.2	-269.7	3 101.7	2 561.0	2 027.8	533.1	276.1	264.6
3rd quarter	3 658.6	517.1	1 479.8	1 038.8	441.0	-707.1	-255.6	3 141.4	2 596.1	2 054.2	541.9	277.7	267.7
4th quarter	3 632.8	455.1	1 471.6	1 025.2	446.5	-807.6	-208.9	3 177.7	2 628.0	2 078.8	549.1	279.8	270.0
2018													
1st quarter	3 826.2	605.9	1 798.6	1 220.2	578.4	-976.3	-216.3	3 220.2	2 664.1	2 103.8	560.4	282.7	273.4
2nd quarter	3 753.8	482.2	1 729.3	1 187.4	541.9	-1 013.8	-233.3	3 271.6	2 708.0	2 136.4	571.6	285.2	278.5
3rd quarter	3 814.9	499.1	1 730.6	1 186.4	544.2	-981.3	-250.1	3 315.8	2 746.8	2 167.4	579.3	287.6	281.5
4th quarter	3 785.9	427.8	1 752.2	1 247.6	504.6	-1 067.6	-256.8	3 358.1	2 784.2	2 198.0	586.3	289.0	284.9

Table 5-1A. Saving and Investment: Recent Data—*Continued*

(Billions of dollars, except as noted; quarterly data are at seasonally adjusted annual rates.)   NIPA Tables 1.7.5, 5.1

Year and quarter	Gross domestic investment, capital account transactions, and net lending, NIPAs Total	Gross domestic investment Total	Private	Government	Capital account transactions, net	Net lending or net borrowing (-), NIPAs	Statistical discrepancy	Net domestic investment	Gross national income	Gross saving as a percent of gross national income	Net saving as a percent of gross national income
1960	125.7	122.5	86.5	36.0	...	3.2	-1.3	54.6	546.8	23.2	10.8
1961	130.7	126.5	86.6	39.9	...	4.2	-0.9	55.9	566.6	23.2	10.8
1962	143.4	139.6	97.0	42.6	...	3.8	0.0	65.5	608.0	23.6	11.4
1963	152.7	147.7	103.3	44.4	...	4.9	-1.5	69.7	643.4	24.0	11.8
1964	166.0	158.5	112.2	46.4	...	7.5	0.0	76.2	689.4	24.1	12.1
1965	183.7	177.5	129.6	47.9	...	6.2	0.7	89.5	746.8	24.5	12.7
1966	201.6	197.8	144.2	53.6	...	3.8	5.1	102.4	813.4	24.2	12.4
1967	203.9	200.4	142.7	57.7	...	3.5	3.4	96.8	862.0	23.3	11.2
1968	217.7	216.2	156.9	59.2	...	1.5	3.1	102.8	943.6	22.7	10.7
1969	234.7	233.1	173.6	59.5	0.0	1.6	1.6	108.2	1 022.1	22.8	10.6
1970	233.6	229.8	170.0	59.8	0.0	3.7	5.3	93.0	1 074.4	21.2	8.5
1971	255.6	255.3	196.8	58.5	0.0	0.3	9.5	106.4	1 162.9	21.2	8.4
1972	284.8	288.8	228.1	60.7	0.0	-4.1	7.2	127.8	1 280.5	21.7	9.1
1973	341.4	332.6	266.9	65.6	0.0	8.8	6.1	153.9	1 431.8	23.4	10.9
1974	356.6	350.7	274.5	76.2	0.0	5.9	7.4	143.8	1 553.3	22.5	9.2
1975	361.5	341.7	257.3	84.4	0.1	19.8	13.3	103.1	1 684.6	20.7	6.5
1976	420.0	412.9	323.2	89.6	0.1	7.0	20.7	152.6	1 869.6	21.4	7.4
1977	478.9	489.8	396.6	93.2	0.1	-11.0	19.4	199.9	2 082.7	22.1	8.1
1978	571.3	583.9	478.4	105.6	0.1	-12.7	23.3	256.7	2 349.9	23.3	9.4
1979	658.6	659.8	539.7	120.1	0.1	-1.3	45.1	285.9	2 614.2	23.5	9.2
1980	674.6	666.0	530.1	135.9	0.1	8.4	44.4	237.6	2 847.1	22.1	7.1
1981	781.9	778.6	631.2	147.3	0.1	3.3	38.1	291.3	3 201.9	23.2	8.0
1982	734.7	738.0	581.0	156.9	0.1	-3.4	8.8	201.0	3 371.4	21.5	5.6
1983	773.6	808.7	637.5	171.2	0.1	-35.2	57.0	246.1	3 614.2	19.8	4.3
1984	923.2	1 013.3	820.1	193.2	0.1	-90.2	41.6	414.9	4 032.3	21.9	7.0
1985	935.2	1 049.5	829.7	219.9	0.1	-114.5	54.3	409.4	4 310.1	20.4	5.6
1986	944.6	1 087.2	849.1	238.1	0.1	-142.8	80.1	401.9	4 516.5	19.1	4.0
1987	992.7	1 146.8	892.2	254.6	0.1	-154.2	43.8	416.4	4 828.9	19.7	4.5
1988	1 079.6	1 195.4	937.0	258.4	0.1	-115.9	3.0	410.9	5 256.1	20.5	5.6
1989	1 177.8	1 270.1	999.7	270.4	0.3	-92.7	68.0	431.9	5 598.4	19.8	4.9
1990	1 208.9	1 283.8	993.4	290.4	7.4	-82.3	95.5	395.3	5 902.3	18.9	3.8
1991	1 246.3	1 238.4	944.3	294.1	5.3	2.6	93.0	306.0	6 096.8	18.9	3.6
1992	1 263.6	1 309.1	1 013.0	296.1	-1.3	-44.3	115.9	348.9	6 435.5	17.8	2.9
1993	1 319.3	1 398.7	1 106.8	291.9	0.9	-80.2	156.0	395.2	6 733.8	17.3	2.4
1994	1 435.1	1 550.7	1 256.5	294.2	1.3	-116.9	140.0	495.0	7 170.3	18.1	3.3
1995	1 519.3	1 625.2	1 317.5	307.7	0.4	-106.3	93.0	502.8	7 574.7	18.8	4.0
1996	1 637.0	1 752.0	1 432.1	320.0	0.2	-115.2	58.1	576.7	8 045.9	19.6	5.0
1997	1 792.1	1 922.2	1 595.6	326.6	0.5	-130.6	11.6	682.9	8 589.3	20.7	6.3
1998	1 875.3	2 080.7	1 736.7	344.0	0.2	-205.6	-55.2	770.9	9 135.5	21.1	6.8
1999	1 977.2	2 255.5	1 887.1	368.5	4.5	-282.8	-33.2	856.6	9 689.4	20.7	6.3
2000	2 030.8	2 427.3	2 038.4	388.9	0.3	-396.8	-96.5	916.0	10 383.9	20.5	5.9
2001	1 963.8	2 346.7	1 934.8	411.9	-12.9	-370.0	-113.1	747.2	10 743.7	19.3	4.4
2002	1 930.9	2 374.1	1 930.4	443.7	0.5	-443.7	-72.7	716.1	11 054.3	18.1	3.1
2003	1 978.1	2 491.3	2 027.1	464.2	2.1	-515.3	-13.7	772.2	11 530.3	17.3	2.4
2004	2 142.2	2 767.5	2 281.3	486.2	-2.8	-622.4	-22.1	945.6	12 314.0	17.6	2.8
2005	2 310.7	3 048.0	2 534.7	513.3	-12.9	-724.5	-55.1	1 077.0	13 169.7	18.0	3.0
2006	2 450.0	3 251.8	2 701.0	550.9	2.1	-803.9	-207.9	1 127.7	14 073.0	18.9	3.8
2007	2 554.3	3 265.0	2 673.0	592.0	-0.1	-710.7	17.7	1 012.2	14 543.2	17.4	2.0
2008	2 424.0	3 107.2	2 477.6	629.6	-5.4	-677.8	182.9	748.4	14 684.6	15.3	-0.8
2009	2 200.5	2 572.6	1 929.7	642.9	0.6	-372.7	192.2	201.1	14 398.7	13.9	-2.5
2010	2 373.3	2 810.0	2 165.5	644.5	0.7	-437.4	61.0	419.1	15 126.7	15.3	-0.5
2011	2 503.6	2 969.2	2 332.6	636.6	1.6	-467.2	-53.2	494.7	15 832.2	16.1	0.5
2012	2 794.7	3 242.8	2 621.8	621.0	-6.5	-441.6	-241.3	666.8	16 670.7	18.2	2.8
2013	3 057.9	3 426.4	2 826.0	600.4	0.8	-369.4	-160.3	745.2	17 175.9	18.7	3.1
2014	3 271.1	3 646.7	3 044.2	602.6	0.4	-376.0	-289.2	831.7	18 057.5	19.7	4.1
2015	3 420.0	3 844.1	3 223.1	621.0	0.4	-424.5	-254.9	927.6	18 704.3	19.6	4.1
2016	3 372.6	3 813.9	3 178.7	635.2	0.5	-441.9	-112.0	822.4	19 045.2	18.3	2.6
2017	3 558.9	4 025.5	3 370.7	654.8	9.5	-476.0	-67.6	904.0	19 830.3	18.2	2.5
2018	3 806.0	4 315.5	3 628.3	687.2	-2.8	-506.7	10.8	1 024.0	20 837.3	18.2	2.4
2016											
1st quarter	3 320.9	3 786.6	3 149.1	637.5	0.6	-466.3	-249.2	836.8	18 875.0	18.9	3.3
2nd quarter	3 373.6	3 786.4	3 152.9	633.5	0.4	-413.2	-81.1	806.6	18 932.1	18.2	2.5
3rd quarter	3 361.6	3 798.0	3 166.6	631.4	0.8	-437.1	-73.9	795.7	19 085.3	18.0	2.3
4th quarter	3 434.2	3 884.8	3 246.2	638.6	0.4	-451.0	-43.7	850.4	19 288.2	18.0	2.3
2017											
1st quarter	3 484.1	3 934.5	3 288.2	646.2	0.6	-451.0	-116.6	869.6	19 540.2	18.4	2.7
2nd quarter	3 473.5	3 989.3	3 335.0	654.4	0.8	-516.6	-140.3	887.7	19 713.4	18.3	2.6
3rd quarter	3 631.9	4 055.8	3 401.8	653.9	35.8	-459.7	-26.7	914.3	19 894.6	18.4	2.6
4th quarter	3 646.1	4 122.4	3 457.7	664.6	0.6	-476.8	13.3	944.6	20 172.9	18.0	2.3
2018											
1st quarter	3 737.1	4 215.0	3 542.4	672.6	0.4	-478.4	-89.1	994.8	20 533.2	18.6	3.0
2nd quarter	3 804.0	4 248.3	3 561.6	686.7	0.4	-444.8	50.1	976.7	20 725.7	18.1	2.3
3rd quarter	3 847.8	4 377.9	3 684.0	693.9	-1.7	-528.4	32.9	1 062.1	20 989.3	18.2	2.4
4th quarter	3 835.1	4 420.6	3 725.2	695.4	-10.5	-575.1	49.2	1 062.5	21 101.2	17.9	2.0

. . . = Not available.

Table 5-1B. Saving and Investment: Historical Data

(Billions of dollars, except as noted; quarterly data are at seasonally adjusted annual rates.)   NIPA Tables 1.7.5, 5.1

Year and quarter	Gross saving Total	Net saving Private	Net saving Government Federal	Net saving Government State and local	Consumption of fixed capital Total	Consumption Private	Consumption Government Federal	Consumption Government State and local	Gross domestic investment Total	Gross domestic investment Private	Gross domestic investment Government	Net lending or net borrow-ing (-), NIPAs	Net domestic invest-ment	Gross national income	Net saving as a percent-age of gross national income
1929	20.1	7.6	0.8	1.3	10.4	9.4	0.2	0.8	20.1	17.2	2.9	0.8	9.7	104.6	9.3
1930	15.8	4.5	0.0	1.1	10.2	9.2	0.2	0.8	14.7	11.4	3.3	0.7	4.5	93.3	6.0
1931	9.1	1.2	-2.4	0.7	9.5	8.6	0.2	0.7	9.6	6.5	3.1	0.2	0.1	77.2	-0.5
1932	4.0	-3.2	-1.6	0.4	8.3	7.5	0.2	0.7	4.1	1.8	2.3	0.2	-4.3	59.6	-7.3
1933	3.9	-3.1	-1.2	0.2	8.0	7.0	0.2	0.7	4.3	2.3	2.0	0.2	-3.7	56.9	-7.2
1934	7.0	0.1	-2.6	1.0	8.4	7.3	0.3	0.9	7.0	4.3	2.7	0.4	-1.5	66.7	-2.1
1935	10.4	3.2	-2.3	1.0	8.5	7.3	0.3	0.9	10.2	7.4	2.9	-0.1	1.7	74.8	2.6
1936	12.3	5.1	-3.6	2.1	8.8	7.5	0.4	0.9	13.6	9.4	4.2	-0.1	4.8	83.9	4.2
1937	17.1	5.8	-0.2	1.8	9.8	8.4	0.4	1.0	16.9	13.0	3.9	0.2	7.1	93.4	7.9
1938	12.7	2.8	-1.8	1.7	10.0	8.6	0.4	1.0	12.2	7.9	4.3	1.2	2.2	87.1	3.1
1939	14.6	5.3	-2.6	1.7	10.1	8.6	0.5	1.1	14.8	10.2	4.6	1.0	4.7	92.5	4.8
1940	19.5	8.3	-1.0	1.6	10.6	9.0	0.5	1.1	19.0	14.6	4.4	1.5	8.4	102.1	8.7
1941	31.1	16.9	0.6	1.5	12.1	10.0	0.9	1.2	30.2	19.4	10.8	1.3	18.1	129.4	14.7
1942	41.6	37.0	-11.4	1.1	14.9	11.2	2.2	1.4	40.7	11.8	29.0	-0.1	25.8	167.3	16.0
1943	47.0	45.0	-17.1	1.0	18.0	11.5	5.0	1.5	47.3	7.4	39.9	-2.1	29.3	205.2	14.1
1944	42.8	49.8	-29.4	1.1	21.3	12.0	7.8	1.5	47.3	9.2	38.2	-2.0	26.0	222.2	9.7
1945	32.6	40.4	-32.0	1.1	23.1	12.5	9.2	1.4	37.7	12.4	25.3	-1.3	14.6	224.4	4.2
1946	41.5	23.1	-8.1	0.8	25.7	14.2	9.9	1.6	37.8	33.1	4.7	4.9	12.1	227.0	7.0
1947	49.2	17.6	2.5	0.1	29.1	17.7	9.5	1.9	42.8	37.1	5.7	9.3	13.7	247.9	8.1
1948	61.5	29.1	1.0	0.1	31.3	20.8	8.4	2.1	58.8	50.3	8.5	2.4	27.5	276.2	10.9
1949	49.6	25.2	-8.5	0.5	32.3	22.6	7.5	2.2	50.4	39.1	11.3	0.9	18.1	272.0	6.3
1950	64.9	29.3	1.9	0.3	33.4	24.3	6.9	2.2	68.0	56.5	11.5	-1.8	34.6	300.0	10.5
1951	79.8	35.9	4.5	1.6	37.7	27.7	7.4	2.6	82.3	62.8	19.6	0.9	44.6	345.4	12.2
1952	79.9	39.5	-2.1	1.9	40.6	29.5	8.3	2.8	81.9	57.3	24.6	0.6	41.3	366.7	10.7
1953	82.0	40.3	-4.2	2.3	43.5	31.3	9.3	2.9	87.1	60.4	26.7	-1.3	43.6	387.3	10.7
1954	80.7	40.7	-8.0	2.0	46.0	33.0	10.1	3.0	83.5	58.1	25.4	0.2	37.5	389.7	8.9
1947															
1st quarter	47.1	16.5	2.5	0.1	28.1	16.3	10.0	1.8	40.8	35.9	4.9	9.4	12.7	241.1	7.9
2nd quarter	46.7	15.7	2.1	0.2	28.7	17.3	9.6	1.8	39.6	34.5	5.1	9.9	10.9	244.3	7.4
3rd quarter	49.1	19.9	-0.4	0.0	29.5	18.2	9.4	1.9	41.0	34.9	6.1	10.1	11.5	248.7	7.9
4th quarter	54.0	18.3	5.6	-0.1	30.2	19.1	9.1	2.0	49.8	43.3	6.6	7.8	19.6	257.4	9.3
1948															
1st quarter	59.0	23.6	5.1	-0.3	30.6	19.9	8.7	2.0	54.7	47.2	7.5	4.9	24.1	266.5	10.6
2nd quarter	62.7	29.0	2.6	0.1	31.1	20.6	8.4	2.1	58.6	50.3	8.2	3.0	27.5	275.1	11.5
3rd quarter	62.0	31.4	-1.1	0.1	31.7	21.2	8.3	2.2	61.3	52.5	8.8	0.9	29.6	280.4	10.8
4th quarter	62.4	32.6	-2.7	0.5	32.0	21.7	8.1	2.2	60.7	51.3	9.4	0.8	28.7	282.7	10.8
1949															
1st quarter	54.7	27.9	-6.0	0.6	32.2	22.1	8.0	2.2	53.4	43.1	10.3	2.2	21.1	275.6	8.2
2nd quarter	48.6	24.9	-9.1	0.4	32.3	22.4	7.6	2.2	47.5	36.2	11.3	1.7	15.3	272.0	6.0
3rd quarter	50.1	26.6	-9.4	0.6	32.3	22.8	7.3	2.2	51.5	39.5	11.9	0.6	19.2	272.1	6.6
4th quarter	44.8	21.5	-9.4	0.3	32.4	23.1	7.1	2.2	49.3	37.5	11.8	-1.0	16.9	268.4	4.6
1950															
1st quarter	53.2	32.3	-11.5	0.0	32.4	23.4	6.9	2.1	57.4	46.7	10.7	-1.0	25.0	278.7	7.4
2nd quarter	61.7	29.8	-0.6	-0.3	32.8	23.9	6.8	2.1	63.4	52.3	11.1	-1.2	30.6	291.3	9.9
3rd quarter	67.1	23.1	9.6	0.8	33.6	24.6	6.8	2.2	70.4	58.6	11.7	-2.7	36.8	309.1	10.8
4th quarter	77.6	32.0	10.0	0.8	34.8	25.5	7.0	2.3	80.8	68.4	12.3	-2.5	45.9	320.9	13.3
1951															
1st quarter	74.3	23.6	12.7	1.7	36.3	26.6	7.2	2.5	79.5	64.6	14.9	-1.7	43.2	334.1	11.4
2nd quarter	82.5	38.5	5.1	1.5	37.3	27.5	7.2	2.6	85.6	67.4	18.2	0.4	48.3	342.5	13.2
3rd quarter	80.7	41.6	-0.5	1.4	38.3	28.2	7.4	2.7	83.1	62.0	21.1	2.2	44.9	348.8	12.2
4th quarter	81.5	40.1	0.7	1.7	39.0	28.7	7.6	2.7	81.2	57.1	24.1	2.7	42.1	356.0	11.9
1952															
1st quarter	82.0	39.4	1.2	1.8	39.5	29.0	7.9	2.7	82.2	58.1	24.1	3.5	42.7	358.1	11.9
2nd quarter	76.0	36.9	-2.4	1.3	40.2	29.3	8.1	2.8	77.3	53.0	24.3	1.1	37.1	360.6	9.9
3rd quarter	79.3	40.9	-4.8	2.1	41.0	29.7	8.4	2.9	82.1	57.2	24.8	-1.1	41.1	368.0	10.4
4th quarter	82.3	40.8	-2.5	2.3	41.7	30.2	8.7	2.8	86.0	60.7	25.3	-1.2	44.3	380.2	10.7
1953															
1st quarter	84.0	41.2	-1.5	1.7	42.5	30.7	8.9	2.9	88.2	61.7	26.5	-1.3	45.7	387.0	10.7
2nd quarter	84.2	41.4	-3.3	2.8	43.3	31.1	9.2	2.9	89.3	62.1	27.2	-1.8	46.0	390.6	10.5
3rd quarter	83.7	40.2	-2.6	2.3	43.8	31.6	9.3	2.9	88.5	61.4	27.0	-1.2	44.6	389.4	10.2
4th quarter	76.0	38.4	-9.2	2.4	44.4	32.0	9.5	2.9	82.4	56.4	25.9	-0.9	38.0	382.4	8.3
1954															
1st quarter	78.9	41.2	-9.7	2.4	44.9	32.4	9.7	2.8	82.1	55.7	26.4	-0.5	37.2	384.6	8.8
2nd quarter	79.2	39.8	-8.4	2.1	45.7	32.8	10.0	3.0	81.4	55.4	25.9	0.3	35.7	385.6	8.7
3rd quarter	80.2	39.7	-7.6	1.8	46.3	33.2	10.2	3.0	83.7	59.0	24.7	-0.1	37.4	389.6	8.7
4th quarter	84.7	42.1	-6.3	1.9	47.0	33.5	10.4	3.0	86.6	62.1	24.5	0.9	39.6	399.2	9.4
1955															
1st quarter	89.9	43.7	-2.9	1.7	47.5	34.0	10.5	3.0	92.8	68.7	24.1	0.5	45.4	412.1	10.3
2nd quarter	95.4	45.8	-0.5	2.0	48.2	34.5	10.6	3.1	97.4	72.7	24.8	-0.2	49.3	422.1	11.2
3rd quarter	97.8	46.8	-0.7	2.5	49.3	35.3	10.8	3.2	99.0	74.7	24.3	0.8	49.7	430.7	11.3
4th quarter	102.3	47.2	1.9	2.6	50.6	36.2	11.1	3.3	103.2	78.9	24.3	0.4	52.5	438.3	11.8
1956															
1st quarter	105.6	48.3	2.3	2.9	52.1	37.3	11.4	3.5	103.9	78.3	25.6	0.9	51.7	443.4	12.1
2nd quarter	107.5	50.2	0.7	3.1	53.5	38.4	11.6	3.6	103.6	77.0	26.6	2.2	50.1	450.5	12.0
3rd quarter	111.6	52.5	1.2	3.1	54.8	39.3	11.8	3.7	106.0	78.3	27.8	3.0	51.2	456.7	12.4
4th quarter	113.4	53.1	1.4	2.9	56.0	40.2	12.0	3.8	105.8	77.1	28.7	4.5	49.8	466.1	12.3

Table 5-1B. Saving and Investment: Historical Data—*Continued*

(Billions of dollars, except as noted; quarterly data are at seasonally adjusted annual rates.)   **NIPA Tables 1.7.5, 5.1**

| Year and quarter | Gross saving | Net saving | | | Consumption of fixed capital | | | | Gross domestic investment and net lending, NIPAs | | | | Net domestic invest-ment | Gross national income | Net saving as a percent-age of gross national income |
| | Total | Private | Government | | Total | Private | Government | | Gross domestic investment | | | Net lending or net borrow-ing (-), NIPAs | | | |
			Federal	State and local			Federal	State and local	Total	Private	Govern-ment				
1957															
1st quarter	114.3	53.3	0.4	3.4	57.3	41.1	12.4	3.8	107.5	77.7	29.8	5.6	50.2	474.2	12.0
2nd quarter	114.4	54.7	-1.6	2.8	58.4	41.9	12.6	4.0	107.6	77.9	29.7	4.8	49.2	477.3	11.7
3rd quarter	114.2	53.8	-1.6	2.5	59.5	42.7	12.8	4.0	109.9	79.3	30.6	4.9	50.5	482.1	11.3
4th quarter	104.3	49.6	-7.7	1.9	60.4	43.4	13.1	4.0	101.9	71.0	30.8	3.6	41.4	476.2	9.2
1958															
1st quarter	100.1	48.2	-10.2	1.0	61.1	44.0	13.1	4.0	98.0	66.7	31.3	1.5	36.9	470.8	8.3
2nd quarter	96.8	48.9	-15.0	0.9	62.0	44.7	13.3	4.1	97.0	65.1	32.0	0.8	35.0	473.6	7.3
3rd quarter	105.0	53.6	-12.5	0.9	62.9	45.2	13.5	4.2	104.9	72.0	32.9	1.0	42.0	487.4	8.6
4th quarter	113.1	57.9	-10.6	1.9	63.8	45.7	13.8	4.2	114.3	80.0	34.3	-0.1	50.5	501.0	9.8
1959															
1st quarter	117.8	56.6	-4.6	1.3	64.5	46.2	14.0	4.3	119.4	83.2	36.2	-1.4	54.9	512.7	10.4
2nd quarter	124.9	59.7	-1.8	2.0	65.1	46.6	14.2	4.3	125.5	89.4	36.2	-2.9	60.4	527.5	11.3
3rd quarter	117.3	53.3	-4.3	2.5	65.8	47.0	14.4	4.3	119.7	83.6	36.1	0.0	54.0	525.4	9.8
4th quarter	120.3	56.2	-5.0	2.6	66.4	47.4	14.6	4.4	121.5	86.5	34.9	-0.8	55.1	531.2	10.1
1960															
1st quarter	132.9	58.6	4.7	2.6	67.0	47.7	14.9	4.4	131.6	96.5	35.1	1.9	64.6	545.1	12.1
2nd quarter	126.3	55.0	1.2	2.5	67.6	48.0	15.1	4.5	122.3	87.1	35.2	1.7	54.7	546.4	10.8
3rd quarter	127.1	56.8	-0.4	2.5	68.2	48.4	15.3	4.6	123.2	86.4	36.8	4.2	55.0	548.5	10.7
4th quarter	121.5	55.1	-4.9	2.5	68.8	48.7	15.5	4.6	112.8	76.0	36.9	4.9	44.0	547.3	9.6
1961															
1st quarter	124.6	57.8	-4.7	2.1	69.5	49.1	15.7	4.7	118.0	78.4	39.6	5.4	48.5	549.8	10.0
2nd quarter	127.0	61.3	-0.4	1.9	70.2	49.5	15.9	4.8	122.8	84.1	38.5	3.2	52.4	560.1	10.1
3rd quarter	134.6	66.1	-4.9	2.4	71.0	50.0	16.2	4.8	130.8	90.9	39.9	4.3	59.8	570.7	11.1
4th quarter	140.2	68.9	-2.9	2.5	71.8	50.5	16.4	4.9	134.5	92.9	41.6	3.9	62.8	586.0	11.7
1962															
1st quarter	141.6	71.5	-5.4	2.8	72.7	51.0	16.7	5.0	140.0	98.1	42.0	3.1	67.3	596.1	11.6
2nd quarter	142.0	71.3	-5.7	2.8	73.5	51.5	16.9	5.1	138.6	96.7	41.9	3.7	65.0	604.1	11.3
3rd quarter	144.4	71.4	-4.9	3.4	74.5	52.1	17.3	5.2	141.3	98.2	43.1	4.8	66.7	611.5	11.4
4th quarter	145.5	72.3	-5.5	3.2	75.6	52.6	17.7	5.3	138.4	95.0	43.5	3.6	62.8	620.5	11.3
1963															
1st quarter	149.1	73.1	-3.5	3.0	76.5	53.2	18.0	5.4	142.9	99.7	43.3	4.0	66.4	628.4	11.5
2nd quarter	153.8	74.3	-1.1	3.1	77.5	53.8	18.3	5.5	145.2	101.7	43.6	4.3	67.7	638.4	12.0
3rd quarter	154.7	74.3	-1.7	3.6	78.5	54.5	18.5	5.5	150.4	104.6	45.8	5.2	71.9	647.9	11.8
4th quarter	158.9	78.4	-2.4	3.4	79.5	55.2	18.7	5.7	152.3	107.2	45.1	6.3	72.8	658.9	12.1
1964															
1st quarter	161.9	82.8	-5.5	4.1	80.6	55.9	18.9	5.8	156.3	110.5	45.8	8.2	75.6	672.4	12.1
2nd quarter	163.0	87.6	-10.2	3.9	81.8	56.8	19.1	5.9	157.0	110.5	46.5	6.0	75.3	683.6	11.9
3rd quarter	166.4	86.2	-6.7	4.0	82.9	57.7	19.3	6.0	159.2	112.6	46.6	8.1	76.3	696.1	12.0
4th quarter	172.9	90.0	-5.1	3.7	84.2	58.7	19.5	6.1	161.7	115.0	46.7	7.6	77.5	705.6	12.6
1965															
1st quarter	180.8	92.6	-1.0	3.5	85.6	59.7	19.7	6.2	172.7	126.5	46.2	5.7	87.1	725.6	13.1
2nd quarter	182.7	93.9	-2.0	3.7	87.1	60.9	19.9	6.4	174.0	127.1	46.9	6.3	86.8	738.3	12.9
3rd quarter	183.3	100.2	-9.4	3.7	88.8	62.1	20.1	6.5	180.1	131.2	48.9	6.6	91.3	751.2	12.6
4th quarter	185.0	100.3	-9.6	3.8	90.5	63.5	20.4	6.6	183.2	133.8	49.5	6.1	92.7	772.2	12.2
1966															
1st quarter	194.3	101.4	-4.1	4.7	92.3	65.0	20.6	6.8	196.1	144.2	51.9	4.6	103.8	794.3	12.8
2nd quarter	194.9	101.2	-5.7	5.0	94.3	66.4	20.9	7.0	195.4	143.5	51.9	3.3	101.1	806.1	12.5
3rd quarter	195.3	102.0	-7.9	4.9	96.3	67.9	21.2	7.2	197.4	143.2	54.2	3.2	101.1	819.4	12.1
4th quarter	201.4	109.1	-10.3	4.3	98.3	69.4	21.5	7.4	202.1	145.9	56.2	4.1	103.8	833.8	12.4
1967															
1st quarter	197.6	112.0	-19.3	4.6	100.3	70.9	21.8	7.5	200.7	142.8	57.9	4.3	100.4	842.0	11.6
2nd quarter	194.0	108.2	-20.5	4.0	102.4	72.5	22.2	7.6	194.1	137.5	56.6	3.3	91.7	850.8	10.8
3rd quarter	201.9	112.6	-18.8	3.5	104.6	74.1	22.7	7.9	200.8	142.8	57.9	3.6	96.1	868.5	11.2
4th quarter	208.5	117.0	-19.5	4.0	107.0	75.8	23.1	8.1	206.0	147.7	58.3	2.8	99.0	886.7	11.4
1968															
1st quarter	207.0	111.0	-17.0	3.5	109.5	77.6	23.6	8.3	210.6	152.3	58.4	1.6	101.1	909.9	10.7
2nd quarter	213.6	116.0	-18.7	4.3	111.9	79.5	23.9	8.5	218.5	158.9	59.5	2.0	106.5	933.5	10.9
3rd quarter	215.5	107.8	-10.1	3.2	114.5	81.6	24.3	8.6	215.2	155.7	59.5	1.8	100.7	955.5	10.6
4th quarter	222.1	111.2	-9.3	2.8	117.5	83.7	24.8	8.9	220.3	160.8	59.6	0.7	102.9	975.5	10.7
1969															
1st quarter	231.0	105.5	2.2	2.8	120.5	86.1	25.2	9.2	233.9	172.4	61.5	1.7	113.4	995.1	11.1
2nd quarter	231.3	106.4	-1.6	3.0	123.5	88.3	25.6	9.5	232.4	172.7	59.7	0.5	108.9	1 013.6	10.6
3rd quarter	237.7	116.5	-8.7	3.6	126.3	90.6	26.0	9.8	237.5	177.6	59.9	1.7	111.2	1 034.4	10.8
4th quarter	232.5	112.7	-12.6	3.2	129.3	92.8	26.4	10.1	228.6	171.6	57.0	2.5	99.3	1 045.6	9.9
1970															
1st quarter	226.9	113.9	-22.3	3.0	132.2	94.9	27.0	10.4	227.1	168.1	59.0	3.9	94.9	1 053.4	9.0
2nd quarter	229.8	125.0	-32.7	2.3	135.3	97.1	27.4	10.8	230.3	171.5	58.8	5.0	95.0	1 068.5	8.8
3rd quarter	231.3	131.2	-39.6	1.4	138.3	99.4	27.9	11.1	234.1	173.9	60.2	3.7	95.8	1 086.1	8.6
4th quarter	224.9	129.0	-44.7	-0.9	141.5	101.7	28.3	11.5	227.9	166.8	61.1	2.2	86.3	1 089.5	7.7
1971															
1st quarter	238.7	141.4	-45.6	-1.8	144.6	104.1	28.7	11.8	247.5	189.5	58.1	4.5	102.9	1 129.3	8.3
2nd quarter	243.2	149.4	-52.5	-1.3	147.6	106.5	29.0	12.1	255.7	197.3	58.3	-0.2	108.1	1 152.0	8.3
3rd quarter	249.0	152.1	-51.8	-1.7	150.4	108.8	29.2	12.3	261.2	202.1	59.2	0.0	110.9	1 172.7	8.4
4th quarter	253.6	154.6	-53.7	-0.4	153.1	111.1	29.3	12.7	256.9	198.4	58.5	-3.1	103.8	1 197.8	8.4

Table 5-1B. Saving and Investment: Historical Data—*Continued*

(Billions of dollars, except as noted; quarterly data are at seasonally adjusted annual rates.)   NIPA Tables 1.7.5, 5.1

Year and quarter	Gross saving	Net saving				Consumption of fixed capital				Gross domestic investment and net lending, NIPAs				Net domestic invest-ment	Gross national income	Net saving as a percent-age of gross national income
	Total	Private	Government			Total	Private	Government		Gross domestic investment			Net lending or net borrow-ing (-), NIPAs			
				Federal	State and local			Federal	State and local	Total	Private	Govern-ment				
1972																
1st quarter	263.0	152.4	-45.8	0.4	155.9	113.4	29.6	12.9	273.0	213.0	60.0	-5.1	117.0	1 233.8	8.7
2nd quarter	265.7	148.8	-51.1	9.1	159.0	116.0	29.9	13.1	287.7	226.8	60.9	-4.8	128.8	1 257.2	8.5
3rd quarter	279.8	157.8	-40.9	0.2	162.7	118.8	30.5	13.4	292.3	233.1	59.2	-3.6	129.6	1 290.6	9.1
4th quarter	302.1	179.3	-58.3	14.7	166.4	121.8	30.8	13.8	302.3	239.7	62.6	-2.8	135.9	1 340.5	10.1
1973																
1st quarter	316.0	177.1	-40.5	9.0	170.5	125.2	31.2	14.2	320.5	254.3	66.2	2.2	150.0	1 381.5	10.5
2nd quarter	324.8	184.4	-40.8	5.8	175.4	129.0	31.8	14.6	333.7	268.2	65.5	5.4	158.3	1 411.1	10.6
3rd quarter	339.0	190.7	-37.3	4.4	181.2	133.4	32.8	15.0	329.0	264.3	64.6	12.4	147.7	1 445.2	10.9
4th quarter	361.4	205.2	-34.5	3.1	187.6	138.3	33.7	15.6	347.1	280.9	66.2	15.3	159.5	1 489.5	11.7
1974																
1st quarter	355.8	196.7	-36.1	0.9	194.3	143.8	34.0	16.5	339.3	268.4	71.0	16.5	145.0	1 508.2	10.7
2nd quarter	346.4	183.1	-38.1	-0.7	202.1	149.7	34.6	17.8	353.4	277.4	76.0	2.6	151.3	1 536.8	9.4
3rd quarter	345.9	173.6	-37.0	-1.9	211.2	156.1	36.0	19.1	349.2	271.0	78.2	0.3	138.0	1 571.7	8.6
4th quarter	348.7	190.5	-54.2	-7.6	220.0	163.0	36.9	20.0	360.8	281.3	79.5	4.4	140.9	1 596.5	8.1
1975																
1st quarter	329.8	189.7	-76.2	-12.3	228.7	170.3	37.8	20.6	327.3	244.3	83.0	17.8	98.6	1 612.1	6.3
2nd quarter	330.0	237.9	-132.9	-10.8	235.8	176.6	38.3	20.9	322.4	243.3	79.2	21.4	86.7	1 649.6	5.7
3rd quarter	360.1	218.6	-90.9	-9.7	242.1	182.0	39.1	21.1	350.7	265.2	85.5	18.3	108.6	1 713.8	6.9
4th quarter	372.6	226.9	-91.7	-10.0	247.4	186.3	39.8	21.4	366.1	276.2	89.9	21.6	118.7	1 762.7	7.1
1976																
1st quarter	392.0	228.6	-81.9	-6.6	251.9	189.6	40.7	21.7	397.5	304.6	92.9	13.1	145.6	1 817.5	7.7
2nd quarter	399.3	224.7	-76.8	-5.3	256.7	193.7	41.1	22.0	411.3	322.3	89.0	8.9	154.6	1 848.0	7.7
3rd quarter	402.1	225.3	-80.8	-5.2	262.7	198.6	42.0	22.1	417.0	328.3	88.7	2.4	154.3	1 886.3	7.4
4th quarter	403.8	219.1	-84.2	-0.7	269.6	204.2	42.9	22.4	425.6	337.7	88.0	3.8	156.1	1 926.5	7.0
1977																
1st quarter	416.0	214.5	-73.5	-2.5	277.5	210.7	44.0	22.8	451.7	360.3	91.4	-8.1	174.2	1 981.7	7.0
2nd quarter	454.4	237.0	-67.6	-0.7	285.7	217.5	45.0	23.2	483.9	389.7	94.2	-8.9	198.3	2 056.0	8.2
3rd quarter	479.3	257.7	-75.1	2.9	293.8	224.5	45.6	23.6	507.8	414.1	93.6	-7.9	214.0	2 118.8	8.8
4th quarter	488.0	260.9	-77.7	2.4	302.4	231.8	46.5	24.0	515.7	422.3	93.4	-19.0	213.3	2 174.3	8.5
1978																
1st quarter	500.5	266.8	-81.6	3.7	311.6	239.4	47.7	24.5	528.3	434.8	93.5	-25.1	216.7	2 222.1	8.5
2nd quarter	540.6	271.3	-62.3	9.6	322.0	247.5	49.4	25.1	575.5	470.6	105.0	-12.9	253.5	2 328.3	9.4
3rd quarter	561.5	283.5	-56.0	2.0	332.0	256.2	50.1	25.7	602.0	492.4	109.7	-11.5	270.1	2 386.7	9.6
4th quarter	589.4	290.3	-48.3	4.1	343.2	265.4	51.4	26.3	629.9	515.8	114.1	-1.4	286.7	2 462.3	10.0
1979																
1st quarter	608.8	299.2	-45.1	-0.1	354.8	275.2	52.5	27.1	635.5	525.8	109.7	-1.9	280.8	2 527.4	10.1
2nd quarter	612.5	289.4	-42.6	-1.6	367.2	285.3	53.9	28.0	656.2	539.3	116.9	-2.8	289.0	2 579.7	9.5
3rd quarter	615.0	283.5	-48.5	-0.3	380.3	295.8	55.5	29.0	671.0	545.6	125.4	2.3	290.7	2 645.2	8.9
4th quarter	617.8	280.8	-53.3	-2.9	393.2	306.6	56.6	30.1	676.2	547.9	128.4	-2.7	283.0	2 704.5	8.3
1980																
1st quarter	611.5	274.4	-66.2	-3.5	406.8	317.7	57.9	31.2	690.3	554.6	135.7	-10.8	283.5	2 759.7	7.4
2nd quarter	607.9	289.2	-93.3	-9.0	421.0	329.0	59.6	32.4	654.0	519.3	134.7	9.7	233.0	2 777.2	6.7
3rd quarter	624.8	300.8	-104.1	-7.3	435.4	340.7	61.0	33.7	629.4	495.1	134.4	28.0	194.1	2 858.1	6.6
4th quarter	676.3	321.1	-91.5	-3.9	450.6	352.5	63.1	35.0	690.5	551.5	139.0	6.6	239.9	2 993.2	7.5
1981																
1st quarter	707.9	321.1	-74.3	-4.5	465.5	364.7	64.7	36.2	765.8	619.4	146.4	1.5	300.2	3 096.0	7.8
2nd quarter	719.4	326.9	-78.7	-9.2	480.4	376.4	66.7	37.3	755.1	609.8	145.3	0.1	274.7	3 156.7	7.6
3rd quarter	775.7	378.4	-86.8	-10.5	494.5	387.8	68.4	38.3	798.0	652.3	145.7	7.0	303.5	3 263.9	8.6
4th quarter	772.5	393.0	-112.4	-16.5	508.4	398.8	70.4	39.3	795.4	643.4	152.0	4.5	287.0	3 291.0	8.0
1982																
1st quarter	736.9	371.6	-136.7	-20.0	521.9	409.4	72.5	40.1	738.6	588.3	150.2	0.5	216.6	3 308.0	6.5
2nd quarter	762.0	392.9	-143.3	-21.5	533.9	418.0	74.9	41.0	750.0	593.6	156.4	17.3	216.1	3 367.5	6.8
3rd quarter	730.6	390.9	-179.3	-23.7	542.7	424.7	76.4	41.7	750.4	593.0	157.4	-14.1	207.6	3 395.7	5.5
4th quarter	673.7	360.5	-210.2	-25.9	549.3	429.5	77.8	42.0	713.0	549.2	163.8	-17.4	163.7	3 414.7	3.6
1983																
1st quarter	691.6	377.7	-208.6	-29.8	552.3	432.3	77.7	42.3	730.2	565.5	164.7	-8.0	177.9	3 476.7	4.0
2nd quarter	700.9	365.3	-203.2	-20.2	559.1	436.0	80.5	42.5	780.8	613.8	167.1	-27.7	221.8	3 563.3	4.0
3rd quarter	703.1	370.5	-219.8	-13.7	566.2	440.6	82.7	42.8	826.8	652.3	174.5	-48.0	260.6	3 651.3	3.8
4th quarter	771.1	405.3	-197.1	-10.1	572.9	446.2	83.6	43.1	896.9	718.5	178.4	-56.9	324.0	3 765.3	5.3
1984																
1st quarter	854.4	455.9	-182.3	-1.5	582.2	452.6	86.1	43.5	975.0	790.9	184.1	-78.6	392.7	3 903.1	7.0
2nd quarter	873.6	470.4	-191.0	2.7	591.5	459.5	87.9	44.0	1 008.4	818.9	189.5	-88.0	416.9	4 000.3	7.1
3rd quarter	898.1	497.5	-199.6	-3.0	603.2	467.0	91.6	44.6	1 032.6	838.9	193.8	-90.5	429.4	4 077.5	7.2
4th quarter	900.3	495.7	-213.1	1.0	616.7	474.9	96.7	45.0	1 037.1	831.7	205.4	-103.5	420.5	4 148.3	6.8
1985																
1st quarter	889.6	436.4	-171.0	0.6	623.6	483.4	94.5	45.8	1 018.1	809.9	208.3	-89.5	394.5	4 216.8	6.3
2nd quarter	887.3	474.9	-221.4	-0.7	634.6	492.0	96.2	46.4	1 045.4	827.0	218.4	-111.5	410.9	4 275.8	5.9
3rd quarter	865.9	424.7	-200.6	-3.3	645.2	500.7	97.4	47.1	1 048.4	822.2	226.3	-121.4	403.2	4 348.0	5.1
4th quarter	880.9	433.9	-203.9	-6.2	657.2	509.6	99.7	47.8	1 086.1	859.5	226.6	-135.3	428.9	4 399.7	5.1
1986																
1st quarter	895.5	432.0	-206.5	1.8	668.2	518.6	100.9	48.6	1 092.5	863.5	229.0	-128.4	424.3	4 461.4	5.1
2nd quarter	871.0	423.9	-228.0	-4.8	679.9	527.4	102.9	49.6	1 089.7	855.2	234.4	-142.0	409.8	4 484.0	4.3
3rd quarter	838.6	382.5	-232.3	-2.4	691.0	536.0	104.4	50.6	1 081.5	835.8	245.7	-150.3	390.6	4 533.2	3.3
4th quarter	852.8	358.1	-196.9	-10.4	702.1	544.3	106.1	51.7	1 085.3	842.1	243.2	-150.4	383.2	4 587.5	3.3

Table 5-1B. Saving and Investment: Historical Data—*Continued*

(Billions of dollars, except as noted; quarterly data are at seasonally adjusted annual rates.)

Year and quarter	Gross saving Total	Net saving Private	Government Federal	Government State and local	Consumption of fixed capital Total	Private	Government Federal	Government State and local	Gross domestic investment Total	Private	Government	Net lending or net borrowing (-), NIPAs	Net domestic investment	Gross national income	Net saving as a percentage of gross national income
1987															
1st quarter	897.9	401.3	-197.6	-18.4	712.6	552.3	107.7	52.6	1 118.8	871.2	247.7	-151.1	406.2	4 666.5	4.0
2nd quarter	926.6	353.4	-144.0	-6.6	723.8	561.2	109.1	53.5	1 128.0	874.6	253.5	-154.5	404.3	4 777.3	4.2
3rd quarter	961.2	396.1	-155.6	-15.4	736.1	570.7	110.9	54.5	1 134.8	876.5	258.3	-154.2	398.7	4 882.7	4.6
4th quarter	1 009.8	443.6	-165.8	-17.1	749.1	581.0	112.8	55.3	1 205.6	946.5	259.1	-157.2	456.5	4 989.0	5.2
1988															
1st quarter	1 029.9	446.1	-166.6	-13.4	763.8	591.9	115.6	56.3	1 162.1	908.6	253.5	-127.0	398.3	5 093.1	5.2
2nd quarter	1 069.0	465.1	-158.7	-15.1	777.7	602.6	117.9	57.2	1 192.7	934.5	258.2	-110.1	415.0	5 199.1	5.6
3rd quarter	1 096.4	465.8	-153.3	-7.5	791.4	613.3	119.9	58.1	1 200.1	942.0	258.1	-106.6	408.7	5 305.7	5.7
4th quarter	1 111.3	476.6	-161.3	-9.1	805.1	623.9	122.2	59.1	1 226.5	962.7	263.8	-119.7	421.4	5 426.5	5.6
1989															
1st quarter	1 145.0	475.4	-140.1	-8.7	818.4	634.4	124.0	60.1	1 269.3	1 005.5	263.8	-106.8	450.9	5 516.1	5.9
2nd quarter	1 104.7	447.4	-161.5	-12.9	831.7	644.7	125.8	61.2	1 268.3	1 001.0	267.2	-93.7	436.6	5 564.1	4.9
3rd quarter	1 100.5	440.3	-165.6	-19.1	844.9	654.8	127.8	62.3	1 271.5	996.5	275.0	-81.8	426.6	5 630.3	4.5
4th quarter	1 089.0	437.8	-170.3	-36.6	858.0	664.6	129.9	63.5	1 271.5	995.8	275.6	-88.5	413.4	5 683.0	4.1
1990															
1st quarter	1 102.7	456.3	-198.6	-24.9	870.0	674.3	131.0	64.7	1 297.7	1 010.8	286.8	-90.9	427.7	5 797.3	4.0
2nd quarter	1 134.9	493.1	-208.0	-32.4	882.2	683.8	132.4	66.0	1 303.0	1 014.7	288.3	-73.0	420.8	5 896.4	4.3
3rd quarter	1 105.6	448.6	-199.6	-38.4	895.0	693.1	134.4	67.4	1 292.4	1 000.8	291.6	-82.5	397.3	5 937.9	3.5
4th quarter	1 110.2	459.5	-206.9	-49.3	906.9	702.3	136.1	68.4	1 242.2	947.5	294.7	-82.7	335.3	5 977.5	3.4
1991															
1st quarter	1 199.7	521.6	-185.8	-55.0	919.0	711.4	138.2	69.4	1 215.7	924.6	291.1	-1.1	296.7	6 012.1	4.7
2nd quarter	1 144.5	520.7	-244.0	-50.4	925.9	719.0	139.5	70.5	1 222.6	926.5	296.0	14.2	293.6	6 062.6	3.6
3rd quarter	1 122.7	516.5	-273.3	-58.3	937.8	725.3	141.1	71.4	1 243.5	947.5	296.0	-29.0	305.7	6 123.3	3.0
4th quarter	1 146.4	559.1	-290.0	-66.6	943.9	730.2	141.6	72.1	1 272.0	978.8	293.2	-17.0	328.1	6 189.0	3.3
1992															
1st quarter	1 156.0	598.7	-325.6	-65.2	948.2	733.7	141.8	72.7	1 256.3	956.8	299.5	-22.0	308.2	6 317.3	3.3
2nd quarter	1 175.0	617.7	-328.7	-69.2	955.2	738.8	142.7	73.7	1 310.2	1 013.1	297.1	-39.8	355.0	6 407.0	3.4
3rd quarter	1 125.4	582.3	-344.2	-76.0	963.4	745.4	143.4	74.5	1 318.3	1 024.2	294.1	-46.8	354.9	6 455.7	2.5
4th quarter	1 134.0	572.7	-339.5	-73.4	974.3	753.6	145.1	75.5	1 351.7	1 058.0	293.7	-68.6	377.4	6 561.9	2.4
1993															
1st quarter	1 131.0	564.3	-337.5	-82.0	986.2	763.4	146.1	76.8	1 372.3	1 083.8	288.5	-60.0	386.2	6 582.6	2.2
2nd quarter	1 168.0	555.1	-308.8	-75.8	997.5	773.2	146.5	77.9	1 387.7	1 094.5	293.2	-74.7	390.1	6 694.4	2.5
3rd quarter	1 143.8	528.8	-318.9	-74.8	1 008.7	783.1	146.9	78.7	1 386.5	1 095.9	290.7	-77.9	377.8	6 752.9	2.0
4th quarter	1 210.9	535.4	-289.0	-57.1	1 021.6	793.2	148.7	79.7	1 448.3	1 153.1	295.2	-108.5	426.7	6 905.2	2.7
1994															
1st quarter	1 232.5	534.6	-269.1	-66.3	1 033.3	803.3	149.0	81.0	1 485.4	1 201.7	283.7	-93.1	452.1	6 983.1	2.9
2nd quarter	1 297.0	554.7	-238.6	-66.1	1 047.0	815.0	149.9	82.0	1 555.2	1 264.9	290.3	-112.8	508.3	7 120.7	3.5
3rd quarter	1 305.9	557.5	-256.8	-57.1	1 062.3	828.4	150.5	83.4	1 553.8	1 251.7	302.1	-122.8	491.5	7 226.2	3.4
4th quarter	1 345.0	589.4	-258.0	-66.3	1 079.9	843.3	151.9	84.7	1 608.2	1 307.6	300.6	-138.8	528.3	7 351.0	3.6
1995															
1st quarter	1 394.2	616.4	-254.4	-66.9	1 099.2	859.8	153.1	86.2	1 634.6	1 327.6	307.0	-119.5	535.4	7 431.0	4.0
2nd quarter	1 394.1	604.2	-243.0	-82.9	1 115.8	874.8	153.4	87.6	1 614.4	1 304.0	310.4	-119.1	498.6	7 512.7	3.7
3rd quarter	1 437.0	624.6	-247.0	-71.3	1 130.8	888.2	153.8	88.8	1 610.4	1 303.2	307.1	-101.8	479.6	7 628.4	4.0
4th quarter	1 479.9	620.6	-224.0	-60.6	1 143.9	900.1	153.8	90.0	1 641.4	1 335.1	306.3	-84.8	497.5	7 726.6	4.3
1996															
1st quarter	1 520.1	645.9	-224.8	-56.6	1 155.6	910.4	154.0	91.1	1 672.2	1 355.4	316.8	-103.1	516.6	7 856.6	4.6
2nd quarter	1 559.9	630.4	-180.2	-57.9	1 167.6	922.0	153.6	91.9	1 736.1	1 418.4	317.8	-109.5	568.6	7 995.0	4.9
3rd quarter	1 596.2	636.8	-170.6	-51.9	1 181.9	934.9	153.9	93.0	1 795.4	1 474.4	321.0	-133.7	613.5	8 091.5	5.1
4th quarter	1 639.4	633.9	-141.9	-48.8	1 196.2	949.0	153.1	94.1	1 804.4	1 480.1	324.2	-114.5	608.2	8 240.5	5.4
1997															
1st quarter	1 696.4	652.3	-121.5	-47.8	1 213.3	964.4	153.6	95.3	1 841.7	1 522.4	319.3	-132.2	628.4	8 372.4	5.8
2nd quarter	1 767.8	686.5	-104.0	-45.9	1 231.2	979.9	154.6	96.7	1 914.2	1 590.2	324.0	-102.5	683.0	8 505.0	6.3
3rd quarter	1 818.3	676.4	-68.5	-37.4	1 247.8	995.5	154.5	97.8	1 956.7	1 625.3	331.5	-123.5	708.9	8 670.3	6.6
4th quarter	1 839.6	685.3	-73.8	-36.8	1 265.0	1 011.2	154.6	99.2	1 976.2	1 644.5	331.7	-164.1	711.2	8 809.4	6.5
1998															
1st quarter	1 910.1	683.9	-24.5	-30.7	1 281.3	1 027.0	154.2	100.2	2 043.6	1 712.3	331.2	-165.1	762.2	8 923.2	7.0
2nd quarter	1 915.7	660.1	-10.4	-34.1	1 300.0	1 043.3	155.5	101.2	2 036.8	1 695.8	341.1	-196.3	736.8	9 067.1	6.8
3rd quarter	1 958.0	650.8	19.9	-31.5	1 318.8	1 060.4	155.6	102.8	2 095.8	1 741.6	354.2	-224.9	777.0	9 217.4	6.9
4th quarter	1 938.5	603.3	20.6	-24.2	1 338.8	1 078.0	156.4	104.4	2 146.5	1 797.0	349.5	-236.0	807.7	9 334.1	6.4
1999															
1st quarter	2 039.7	670.0	46.3	-35.3	1 358.7	1 096.3	156.6	105.9	2 210.1	1 853.1	357.1	-234.0	851.4	9 501.5	7.2
2nd quarter	1 995.7	590.5	66.2	-44.2	1 383.2	1 117.9	157.5	107.7	2 210.9	1 848.3	362.5	-263.6	827.6	9 598.2	6.4
3rd quarter	1 981.2	538.7	71.5	-39.6	1 410.7	1 143.1	158.3	109.3	2 263.8	1 893.7	370.0	-298.6	853.1	9 720.3	5.9
4th quarter	2 024.6	534.5	83.5	-36.6	1 443.2	1 171.6	160.2	111.4	2 337.4	1 953.1	384.3	-335.2	894.2	9 937.3	5.9
2000															
1st quarter	2 146.1	521.0	173.0	-26.9	1 479.0	1 203.5	162.0	113.4	2 339.8	1 950.7	389.2	-371.3	860.8	10 209.1	6.5
2nd quarter	2 133.1	510.2	144.1	-27.4	1 506.1	1 227.5	162.8	115.8	2 461.6	2 075.8	385.8	-379.5	955.5	10 330.1	6.1
3rd quarter	2 150.5	515.8	154.5	-44.7	1 525.0	1 243.5	163.7	117.7	2 446.2	2 060.0	386.2	-416.1	921.2	10 467.1	6.0
4th quarter	2 079.5	457.9	150.3	-63.6	1 534.8	1 251.5	163.9	119.5	2 461.5	2 067.2	394.2	-420.3	926.6	10 529.1	5.2
2001															
1st quarter	2 153.4	539.3	127.4	-79.8	1 566.4	1 281.2	164.0	121.2	2 372.2	1 971.3	400.9	-412.4	805.8	10 703.9	5.5
2nd quarter	2 137.0	556.5	93.1	-106.7	1 594.0	1 307.1	164.3	122.7	2 392.0	1 973.0	418.9	-373.4	797.9	10 762.7	5.0
3rd quarter	2 053.6	701.3	-130.1	-129.2	1 611.6	1 322.8	164.8	124.0	2 348.4	1 944.9	403.5	-355.2	736.8	10 732.8	4.1
4th quarter	1 963.6	532.3	-34.4	-160.3	1 626.0	1 335.7	164.8	125.5	2 274.3	1 850.1	424.3	-339.1	648.4	10 775.4	3.1

Table 5-1B. Saving and Investment: Historical Data—*Continued*

(Billions of dollars, except as noted; quarterly data are at seasonally adjusted annual rates.)   NIPA Tables 1.7.5, 5.1

Year and quarter	Gross saving Total	Net saving Private	Net saving Government Federal	Net saving Government State and local	Consumption of fixed capital Total	Consumption of fixed capital Private	Consumption of fixed capital Government Federal	Consumption of fixed capital Government State and local	Gross domestic investment Total	Gross domestic investment Private	Gross domestic investment Government	Net lending or net borrowing (-), NIPAs	Net domestic investment	Gross national income	Net saving as a percentage of gross national income
2002															
1st quarter	2 005.8	768.4	-233.4	-165.6	1 636.3	1 343.8	165.5	127.1	2 347.4	1 912.7	434.7	-397.6	711.1	10 893.4	3.4
2nd quarter	2 009.9	810.1	-264.1	-185.1	1 648.9	1 354.0	166.2	128.7	2 371.7	1 933.3	438.4	-444.4	722.9	11 005.3	3.3
3rd quarter	1 980.5	776.7	-278.4	-183.0	1 665.2	1 367.8	167.2	130.1	2 380.3	1 933.2	447.1	-447.1	715.0	11 082.2	2.8
4th quarter	2 018.1	844.5	-310.1	-197.8	1 681.5	1 381.7	168.4	131.4	2 397.0	1 942.5	454.5	-485.7	715.5	11 236.4	3.0
2003															
1st quarter	1 947.2	805.7	-341.1	-212.8	1 695.5	1 392.7	169.7	133.0	2 417.3	1 960.2	457.1	-533.0	721.9	11 288.9	2.2
2nd quarter	1 982.0	860.4	-390.8	-196.1	1 708.5	1 403.1	171.2	134.2	2 430.7	1 972.4	458.3	-517.6	722.2	11 431.8	2.4
3rd quarter	1 985.9	901.0	-474.0	-167.2	1 726.1	1 417.5	173.0	135.5	2 513.4	2 044.3	469.1	-516.0	787.3	11 608.0	2.2
4th quarter	2 051.8	864.9	-410.3	-149.1	1 746.3	1 434.4	175.1	136.8	2 603.7	2 131.3	472.4	-494.8	857.4	11 792.3	2.6
2004															
1st quarter	2 068.0	895.3	-443.6	-157.8	1 774.1	1 458.2	177.3	138.5	2 628.4	2 154.1	474.3	-541.4	854.3	11 999.5	2.4
2nd quarter	2 168.2	927.3	-399.2	-163.4	1 803.5	1 482.3	179.4	141.8	2 745.9	2 262.6	483.3	-616.0	942.4	12 220.4	3.0
3rd quarter	2 239.4	924.2	-369.8	-150.7	1 835.8	1 508.4	181.2	146.2	2 808.4	2 318.3	490.1	-620.1	972.6	12 450.4	3.2
4th quarter	2 181.7	822.8	-391.0	-124.0	1 873.9	1 539.5	184.0	150.4	2 887.2	2 390.1	497.1	-712.3	1 013.3	12 585.6	2.4
2005															
1st quarter	2 310.5	817.4	-315.1	-103.3	1 911.5	1 572.4	186.7	152.5	2 987.6	2 486.1	501.6	-683.8	1 076.1	12 849.0	3.1
2nd quarter	2 324.5	794.7	-312.0	-106.7	1 948.5	1 603.4	189.1	156.0	2 986.7	2 476.5	510.2	-707.1	1 038.2	13 031.9	2.9
3rd quarter	2 335.7	750.7	-297.8	-107.1	1 989.9	1 638.1	191.7	160.1	3 048.4	2 531.1	517.3	-676.7	1 058.6	13 255.3	2.6
4th quarter	2 492.4	851.2	-298.9	-94.1	2 034.2	1 676.6	194.7	163.0	3 169.3	2 645.3	524.0	-830.2	1 135.1	13 542.6	3.4
2006															
1st quarter	2 668.9	911.4	-247.2	-66.0	2 070.7	1 708.4	197.4	164.9	3 249.4	2 709.7	539.7	-793.3	1 178.7	13 870.4	4.3
2nd quarter	2 651.8	853.4	-243.0	-65.9	2 107.2	1 737.9	200.0	169.3	3 261.1	2 709.3	551.8	-814.3	1 153.8	14 005.8	3.9
3rd quarter	2 646.3	824.0	-224.8	-93.8	2 140.9	1 765.7	202.7	172.5	3 259.5	2 709.4	550.1	-858.6	1 118.6	14 152.2	3.6
4th quarter	2 664.7	796.8	-195.4	-114.4	2 177.7	1 795.2	205.1	177.4	3 237.4	2 675.4	562.0	-749.4	1 059.7	14 263.5	3.4
2007															
1st quarter	2 585.0	708.1	-226.8	-111.2	2 214.9	1 823.9	208.4	182.6	3 238.6	2 664.3	574.3	-791.0	1 023.7	14 398.0	2.6
2nd quarter	2 632.0	736.7	-242.4	-103.8	2 241.3	1 844.2	211.2	185.9	3 288.9	2 699.2	589.6	-749.2	1 047.5	14 550.6	2.7
3rd quarter	2 507.4	653.4	-271.7	-139.4	2 265.0	1 861.7	214.3	189.0	3 281.9	2 686.0	596.0	-677.5	1 016.9	14 569.0	1.7
4th quarter	2 422.2	618.6	-323.6	-162.8	2 290.0	1 880.2	217.3	192.6	3 250.8	2 642.6	608.2	-624.9	960.8	14 655.3	0.9
2008															
1st quarter	2 336.4	609.6	-424.9	-168.2	2 319.9	1 903.0	221.0	196.0	3 175.3	2 563.7	611.6	-723.7	855.4	14 697.1	0.1
2nd quarter	2 236.1	833.3	-777.9	-166.5	2 347.3	1 923.9	224.5	198.9	3 168.2	2 540.6	627.6	-709.4	820.9	14 748.4	-0.8
3rd quarter	2 234.6	753.0	-646.9	-247.4	2 376.0	1 945.7	227.7	202.6	3 136.0	2 498.2	637.7	-674.4	759.9	14 816.9	-1.0
4th quarter	2 157.5	741.5	-674.7	-301.4	2 392.1	1 954.7	230.3	207.1	2 949.3	2 307.9	641.4	-603.6	557.2	14 476.0	-1.6
2009															
1st quarter	2 008.1	1 012.1	-1 059.6	-331.1	2 386.7	1 946.0	231.3	209.3	2 654.9	2 014.9	640.0	-385.8	268.3	14 250.1	-2.7
2nd quarter	1 982.2	1 261.2	-1 313.5	-335.0	2 369.5	1 927.9	232.5	209.0	2 510.2	1 863.7	646.5	-348.4	140.7	14 288.2	-2.7
3rd quarter	1 949.0	1 275.6	-1 329.8	-355.8	2 359.1	1 916.4	234.0	208.7	2 486.4	1 841.4	644.9	-361.7	127.3	14 415.6	-2.8
4th quarter	2 093.9	1 359.4	-1 292.8	-343.5	2 370.7	1 924.5	236.7	209.5	2 638.8	1 998.7	640.1	-394.9	268.1	14 640.9	-1.9
2010															
1st quarter	2 109.2	1 442.1	-1 374.8	-333.7	2 375.6	1 925.3	239.3	211.0	2 671.3	2 038.2	633.2	-425.5	295.8	14 778.3	-1.8
2nd quarter	2 255.9	1 526.7	-1 320.1	-333.8	2 383.1	1 927.8	242.5	212.7	2 796.8	2 148.8	648.0	-436.2	413.7	15 016.1	-0.8
3rd quarter	2 438.9	1 635.9	-1 299.4	-290.1	2 392.6	1 933.5	244.9	214.2	2 885.9	2 236.5	649.4	-465.9	493.3	15 286.8	0.3
4th quarter	2 444.9	1 610.7	-1 306.0	-272.3	2 412.5	1 948.5	248.2	215.8	2 885.8	2 238.4	647.4	-422.0	473.4	15 425.8	0.2
2011															
1st quarter	2 397.1	1 502.6	-1 262.7	-279.1	2 436.4	1 966.7	251.8	217.9	2 845.3	2 206.0	639.4	-474.7	408.9	15 528.8	-0.3
2nd quarter	2 480.2	1 573.0	-1 298.1	-256.8	2 462.2	1 987.2	254.2	220.7	2 933.6	2 297.4	636.3	-489.6	471.4	15 748.9	0.1
3rd quarter	2 610.0	1 617.6	-1 207.9	-287.6	2 487.9	2 007.7	256.4	223.8	2 958.0	2 322.8	635.1	-446.9	470.0	15 934.6	0.8
4th quarter	2 740.1	1 704.6	-1 199.1	-276.8	2 511.4	2 027.6	257.3	226.5	3 139.8	2 504.1	635.7	-457.6	628.5	16 116.6	1.4
2012															
1st quarter	2 974.4	1 834.6	-1 117.2	-279.6	2 536.5	2 047.9	259.7	228.9	3 194.6	2 567.8	626.8	-488.9	658.1	16 536.1	2.6
2nd quarter	3 054.7	1 876.4	-1 104.7	-281.5	2 564.4	2 071.8	261.2	231.4	3 264.0	2 636.9	627.1	-471.4	699.5	16 641.1	2.9
3rd quarter	2 990.6	1 744.6	-1 057.5	-287.5	2 591.0	2 095.2	262.6	233.2	3 263.5	2 644.1	619.4	-423.9	672.5	16 634.4	2.4
4th quarter	3 124.4	1 830.2	-1 035.1	-282.7	2 612.0	2 114.6	263.0	234.4	3 249.0	2 638.3	610.8	-382.3	637.0	16 871.0	3.0
2013															
1st quarter	3 091.7	1 495.1	-776.8	-261.7	2 635.0	2 135.5	263.6	235.9	3 336.9	2 738.2	598.7	-407.8	701.9	16 945.7	2.7
2nd quarter	3 222.8	1 342.9	-537.1	-247.2	2 664.2	2 161.3	265.1	237.7	3 373.7	2 775.3	598.5	-394.6	709.5	17 108.5	3.3
3rd quarter	3 239.2	1 528.8	-706.7	-277.0	2 694.2	2 188.4	266.2	239.6	3 483.2	2 880.0	603.2	-370.6	789.0	17 216.0	3.2
4th quarter	3 319.1	1 394.4	-531.0	-275.8	2 731.4	2 221.0	268.5	241.9	3 511.8	2 910.5	601.3	-304.5	780.4	17 433.3	3.4
2014															
1st quarter	3 368.0	1 430.0	-579.1	-253.6	2 770.8	2 257.1	269.8	243.9	3 488.0	2 899.2	588.7	-370.0	717.2	17 597.5	3.4
2nd quarter	3 565.8	1 587.4	-594.4	-224.5	2 797.3	2 281.9	269.9	245.6	3 633.1	3 030.4	602.7	-345.2	835.8	17 946.0	4.3
3rd quarter	3 643.9	1 676.7	-632.8	-232.8	2 832.9	2 315.1	270.4	247.4	3 714.1	3 107.6	606.5	-358.8	881.2	18 268.8	4.4
4th quarter	3 663.4	1 656.3	-610.9	-241.0	2 859.1	2 339.9	270.7	248.5	3 751.8	3 139.5	612.3	-430.1	892.7	18 417.7	4.4
2015															
1st quarter	3 723.5	1 610.2	-544.9	-227.8	2 886.0	2 365.1	271.7	249.1	3 841.5	3 235.7	605.7	-432.8	955.5	18 527.7	4.5
2nd quarter	3 701.8	1 603.8	-574.3	-234.8	2 907.1	2 384.9	271.2	251.0	3 858.0	3 231.8	626.3	-398.2	950.9	18 677.1	4.3
3rd quarter	3 635.0	1 572.8	-615.9	-252.2	2 930.3	2 406.2	271.7	252.4	3 871.4	3 241.2	630.2	-455.6	941.1	18 781.8	3.8
4th quarter	3 639.3	1 408.5	-545.2	-166.4	2 942.5	2 418.5	270.9	253.0	3 805.5	3 183.6	621.9	-411.4	863.0	18 830.7	3.7
2016															
1st quarter	3 570.1	1 518.2	-644.5	-253.4	2 949.8	2 426.6	269.9	253.3	3 786.6	3 149.1	637.5	-466.3	836.8	18 875.0	3.3
2nd quarter	3 454.7	1 402.7	-674.8	-253.0	2 979.8	2 452.4	271.1	256.3	3 786.4	3 152.9	633.5	-413.2	806.6	18 932.1	2.5
3rd quarter	3 435.5	1 363.5	-687.2	-243.0	3 002.3	2 472.7	272.2	257.4	3 798.0	3 166.6	631.4	-437.1	795.7	19 085.3	2.3
4th quarter	3 477.9	1 382.7	-701.6	-237.6	3 034.4	2 501.1	273.9	259.4	3 884.8	3 246.2	638.6	-451.0	850.4	19 288.2	2.3
2017															
1st quarter	3 600.7	1 478.9	-685.0	-258.1	3 064.9	2 527.9	274.9	262.1	3 934.5	3 288.2	646.2	-451.0	869.6	19 540.2	2.7
2nd quarter	3 613.8	1 481.0	-699.2	-269.7	3 101.7	2 561.0	276.1	264.6	3 989.3	3 335.0	654.4	-516.6	887.7	19 713.4	2.6
3rd quarter	3 658.6	1 479.8	-707.1	-255.6	3 141.4	2 596.1	277.7	267.7	4 055.8	3 401.8	653.9	-459.7	914.3	19 894.6	2.6
4th quarter	3 632.8	1 471.6	-807.6	-208.9	3 177.7	2 628.0	279.8	270.0	4 122.4	3 457.7	664.6	-476.8	944.6	20 172.9	2.3

Table 5-2. Gross Private Fixed Investment by Type

(Billions of dollars, quarterly data are at seasonally adjusted annual rates.)

NIPA Table 5.3.5

Year and quarter	Total gross private fixed investment	Nonresidential											
			Structures						Equipment				
										Information processing equipment			
		Total	Total	Comm-ercial and health care	Manufac-turing	Power and communi-cation	Mining explor-ation, shafts, and wells	Other non-residential structures	Total	Total	Computers and peripheral equipment	Other information processing	Industrial equipment
1960	83.2	56.4	19.6	4.8	2.9	4.4	2.3	5.2	29.7	4.7	0.2	4.6	9.4
1961	83.6	56.6	19.7	5.5	2.8	4.1	2.3	5.0	28.9	5.1	0.3	4.8	8.8
1962	90.9	61.2	20.8	6.2	2.8	4.1	2.5	5.2	32.1	5.5	0.3	5.1	9.3
1963	97.7	64.8	21.2	6.1	2.9	4.4	2.3	5.6	34.4	6.1	0.7	5.4	10.0
1964	107.3	72.2	23.7	6.8	3.6	4.8	2.4	6.2	38.7	6.8	0.9	5.9	11.4
1965	120.4	85.2	28.3	8.2	5.1	5.4	2.4	7.2	45.8	7.8	1.2	6.7	13.7
1966	130.6	97.2	31.3	8.3	6.6	6.3	2.5	7.8	53.0	9.7	1.7	8.0	16.2
1967	132.8	99.2	31.5	8.2	6.0	7.1	2.4	7.8	53.7	10.1	1.9	8.2	16.9
1968	147.9	107.7	33.6	9.4	6.0	8.3	2.6	7.3	58.5	10.6	1.9	8.7	17.3
1969	164.4	120.0	37.7	11.7	6.8	8.7	2.8	7.8	65.2	12.8	2.4	10.4	19.1
1970	168.0	124.6	40.3	12.5	7.0	10.2	2.8	7.8	66.4	14.3	2.7	11.6	20.3
1971	188.6	130.4	42.7	14.9	6.3	11.0	2.7	7.9	69.1	14.9	2.8	12.2	19.5
1972	219.0	146.6	47.2	17.6	5.9	12.1	3.1	8.6	78.9	16.7	3.5	13.2	21.4
1973	251.0	172.7	55.0	19.8	7.9	13.8	3.5	9.9	95.1	19.9	3.5	16.3	26.0
1974	260.5	191.1	61.2	20.6	10.0	15.1	5.2	10.3	104.3	23.1	3.9	19.2	30.7
1975	263.5	196.8	61.4	17.7	10.6	15.7	7.4	10.1	107.6	23.8	3.6	20.2	31.3
1976	306.1	219.3	65.9	18.1	10.1	18.2	8.6	11.0	121.2	27.5	4.4	23.1	34.1
1977	374.3	259.1	74.6	20.3	11.1	19.3	11.5	12.5	148.7	33.7	5.7	28.0	39.4
1978	452.6	314.6	93.6	25.3	16.2	21.4	15.4	15.2	180.6	42.3	7.6	34.8	47.7
1979	521.7	373.8	117.7	33.5	22.0	24.6	19.0	18.5	208.1	50.3	10.2	40.2	56.2
1980	536.4	406.9	136.2	41.0	20.5	27.3	27.4	20.0	216.4	58.9	12.5	46.4	60.7
1981	601.4	472.9	167.3	48.3	25.4	30.0	42.5	21.2	240.9	69.6	17.1	52.5	65.5
1982	606.0	405.1	177.6	55.8	26.1	29.6	44.8	21.3	234.9	74.2	18.9	55.3	62.7
1983	643.3	482.2	154.3	55.8	19.5	25.8	30.0	23.3	246.5	83.7	23.9	59.8	58.9
1984	754.7	564.3	177.4	70.6	20.9	26.5	31.3	28.1	291.9	101.2	31.6	69.6	68.1
1985	807.8	607.8	194.5	84.1	24.1	26.5	27.9	31.8	307.9	106.6	33.7	72.9	72.5
1986	842.6	607.8	176.5	80.9	21.0	28.3	15.7	30.7	317.7	111.1	33.4	77.7	75.4
1987	865.0	615.2	174.2	80.8	21.2	25.4	13.1	33.7	320.9	112.2	35.8	76.4	76.7
1988	918.5	662.3	182.8	86.3	23.2	25.0	15.7	32.5	346.8	120.8	38.0	82.8	84.2
1989	972.0	716.0	193.7	88.3	28.8	27.5	14.9	34.3	372.2	130.7	43.1	87.6	93.3
1990	978.9	739.2	202.9	87.5	33.6	26.3	17.9	37.6	371.9	129.6	38.6	90.9	92.1
1991	944.7	723.6	183.6	68.9	31.4	31.6	18.5	33.2	360.8	129.2	37.7	91.5	89.3
1992	996.7	741.9	172.6	64.5	29.0	33.9	14.2	31.0	381.7	142.1	44.0	98.1	93.0
1993	1 086.0	799.2	177.2	69.4	23.6	33.2	16.6	34.5	425.1	153.3	47.9	105.4	102.2
1994	1 192.7	868.9	186.8	75.4	28.9	31.2	16.4	34.9	476.4	167.0	52.4	114.6	113.6
1995	1 286.3	962.2	207.3	83.1	35.5	33.1	15.0	40.6	528.1	188.4	66.1	122.3	129.0
1996	1 401.3	1 043.2	224.6	91.5	38.2	29.2	16.8	48.9	565.3	204.7	72.8	131.9	136.5
1997	1 524.7	1 149.1	250.3	104.3	37.6	28.8	22.4	57.2	610.9	222.8	81.4	141.4	140.4
1998	1 673.0	1 254.1	276.0	116.0	40.5	34.2	22.3	63.1	660.0	240.1	87.9	152.2	147.4
1999	1 826.2	1 364.5	285.7	125.4	35.1	40.4	18.3	66.4	713.6	259.8	97.2	162.5	149.1
2000	1 983.9	1 498.4	321.0	139.3	37.6	48.1	23.7	72.2	766.1	293.8	103.2	190.6	162.9
2001	1 973.1	1 460.1	333.5	137.3	37.8	51.1	34.6	72.7	711.5	265.9	87.6	178.4	151.9
2002	1 910.4	1 352.8	287.0	119.4	22.7	51.0	30.2	63.6	659.6	236.7	79.7	157.0	141.7
2003	2 013.0	1 375.9	286.6	115.0	21.4	48.1	38.5	63.6	670.6	242.7	79.9	162.8	143.4
2004	2 217.2	1 467.4	307.7	125.3	23.2	43.1	47.3	68.8	721.9	255.8	84.2	171.6	144.2
2005	2 477.2	1 621.0	353.0	135.9	28.4	48.1	69.4	71.2	794.9	267.0	84.2	182.8	162.4
2006	2 632.0	1 793.8	425.2	156.3	32.3	55.8	96.0	84.7	862.3	288.5	92.6	195.9	181.6
2007	2 639.1	1 948.6	510.3	181.8	40.2	81.6	102.2	104.5	893.4	310.9	95.4	215.5	194.1
2008	2 506.9	1 990.9	571.1	181.9	53.0	95.6	120.3	120.3	845.4	306.3	93.9	212.4	194.3
2009	2 080.4	1 690.4	455.8	126.8	56.8	95.8	79.2	97.3	670.3	275.6	88.9	186.7	153.7
2010	2 111.6	1 735.0	379.8	92.1	40.3	83.8	93.5	70.1	777.0	307.5	99.6	207.9	155.2
2011	2 286.3	1 907.5	404.5	93.4	39.6	81.8	124.7	65.0	881.3	313.3	95.6	217.7	191.5
2012	2 550.5	2 118.5	479.4	103.8	46.8	102.4	152.9	73.6	983.4	331.2	103.5	227.7	211.2
2013	2 721.5	2 211.5	492.5	109.8	49.9	98.9	155.6	78.3	1 027.0	341.7	102.1	239.6	209.3
2014	2 960.2	2 400.1	577.6	127.4	58.1	115.3	188.3	88.4	1 091.9	346.0	101.9	244.1	218.8
2015	3 091.2	2 457.4	572.6	143.4	79.3	113.0	136.9	100.0	1 121.5	353.8	101.6	252.2	218.5
2016	3 151.6	2 453.1	545.8	171.1	75.7	113.8	74.7	110.6	1 093.6	355.4	99.5	255.8	215.1
2017	3 340.5	2 584.7	586.8	180.6	65.8	115.3	109.0	116.1	1 143.7	381.0	107.8	273.2	230.7
2018	3 573.6	2 786.9	633.2	187.4	64.7	118.6	137.9	124.5	1 222.6	408.6	118.8	289.8	245.9
2016													
1st quarter	3 102.2	2 415.6	520.5	159.7	76.3	104.4	76.7	103.4	1 101.4	352.9	100.8	252.1	213.2
2nd quarter	3 133.8	2 441.8	537.1	167.2	77.6	110.6	69.9	111.8	1 092.7	353.0	99.6	253.4	215.0
3rd quarter	3 169.3	2 471.6	559.6	176.8	76.2	119.0	74.6	113.0	1 091.2	357.5	98.1	259.4	214.7
4th quarter	3 201.3	2 483.5	566.0	180.6	72.8	121.0	77.6	114.0	1 088.9	358.1	99.7	258.4	217.6
2017													
1st quarter	3 274.8	2 531.1	580.2	179.7	68.2	122.5	94.5	115.3	1 108.8	366.1	102.4	263.7	222.3
2nd quarter	3 316.1	2 567.4	589.0	181.1	66.2	116.3	109.9	115.5	1 132.9	377.1	107.7	269.5	229.6
3rd quarter	3 345.0	2 591.6	583.7	179.9	64.5	112.7	111.2	115.4	1 149.5	384.1	111.6	272.5	232.8
4th quarter	3 426.0	2 648.9	594.4	181.6	64.3	109.8	120.4	118.3	1 183.6	396.7	109.7	287.1	238.2
2018													
1st quarter	3 500.9	2 717.3	615.9	189.2	63.9	116.7	125.6	120.5	1 201.8	404.4	117.2	287.2	243.1
2nd quarter	3 571.6	2 782.0	640.0	189.1	62.2	121.4	143.3	123.9	1 214.3	405.8	120.1	285.7	242.1
3rd quarter	3 596.7	2 807.7	641.7	186.6	65.9	120.1	142.9	126.1	1 227.9	414.8	120.2	294.6	246.9
4th quarter	3 625.2	2 840.7	635.2	184.8	66.8	116.3	139.9	127.4	1 246.4	409.5	117.7	291.8	251.6

Table 5-2. Gross Private Fixed Investment by Type—Continued

(Billions of dollars, quarterly data are at seasonally adjusted annual rates.)

NIPA Table 5.3.5

Year and quarter	Equipment—Continued		Intellectual property				Residential						
	Transportation equipment	Other nonresidential equipment	Total	Software [1]	Research and development [2]	Entertainment, literary, and artistic originals	Total	Residential structures					Residential equipment
								Total	Permanent site			Other residential structures	
									Total	Single family	Multifamily		
1960	8.5	7.1	7.1	0.1	4.9	2.1	26.9	26.3	17.5	14.9	2.6	8.8	0.5
1961	8.0	7.0	8.0	0.2	5.2	2.7	27.0	26.5	17.4	14.1	3.3	9.1	0.5
1962	9.8	7.5	8.4	0.2	5.6	2.6	29.6	29.1	19.9	15.1	4.8	9.2	0.5
1963	9.4	8.8	9.2	0.4	6.0	2.8	32.9	32.3	22.4	16.0	6.4	9.8	0.6
1964	10.6	9.9	9.8	0.5	6.5	2.8	35.1	34.5	24.1	17.6	6.4	10.4	0.6
1965	13.2	11.0	11.1	0.7	7.2	3.2	35.2	34.5	23.8	17.8	6.0	10.6	0.7
1966	14.5	12.7	12.8	1.0	8.1	3.7	33.4	32.7	21.8	16.6	5.2	10.9	0.7
1967	14.3	12.4	14.0	1.2	9.0	3.8	33.6	32.9	21.5	16.8	4.7	11.4	0.7
1968	17.6	13.0	15.6	1.3	9.9	4.3	40.2	39.3	26.7	19.5	7.2	12.6	0.9
1969	18.9	14.4	17.2	1.8	11.0	4.4	44.4	43.4	29.2	19.7	9.5	14.1	1.0
1970	16.2	15.6	17.9	2.3	11.5	4.1	43.4	42.3	27.1	17.5	9.5	15.2	1.1
1971	18.4	16.3	18.7	2.4	11.9	4.4	58.2	56.9	38.7	25.8	12.9	18.2	1.3
1972	21.8	19.0	20.6	2.8	12.9	4.9	72.4	70.9	50.1	32.8	17.2	20.8	1.5
1973	26.6	22.6	22.7	3.2	14.6	4.9	78.3	76.6	54.6	35.2	19.4	22.0	1.7
1974	26.3	24.3	25.5	3.9	16.4	5.2	69.5	67.6	43.4	29.7	13.7	24.2	1.9
1975	25.2	27.4	27.8	4.8	17.5	5.5	66.7	64.8	36.3	29.6	6.7	28.5	1.9
1976	30.0	29.6	32.2	5.2	19.6	7.4	86.8	84.6	50.8	43.9	6.9	33.9	2.1
1977	39.3	36.3	35.8	5.5	21.8	8.6	115.2	112.8	72.2	62.2	10.0	40.6	2.4
1978	47.3	43.2	40.4	6.3	24.9	9.1	138.0	135.3	85.6	72.8	12.8	49.7	2.7
1979	53.6	47.9	48.1	8.1	29.1	10.9	147.8	144.7	89.3	72.3	17.0	55.4	3.2
1980	48.4	48.3	54.4	9.8	34.2	10.3	129.5	126.1	69.6	52.9	16.7	56.5	3.4
1981	50.6	55.2	64.8	11.8	39.7	13.2	128.5	124.9	69.4	52.0	17.5	55.4	3.6
1982	46.8	51.2	72.7	14.0	44.8	13.9	110.8	107.2	57.0	41.5	15.5	50.2	3.7
1983	53.5	50.4	81.3	16.4	49.6	15.3	161.1	156.9	95.0	72.5	22.4	61.9	4.2
1984	64.4	58.2	95.0	20.4	56.9	17.8	190.4	185.6	114.6	86.4	28.2	71.0	4.7
1985	69.0	59.9	105.3	23.8	63.0	18.6	200.1	195.0	115.9	87.4	28.5	79.1	5.1
1986	70.5	60.7	113.5	25.6	66.5	21.4	234.8	229.3	135.2	104.1	31.0	94.1	5.5
1987	68.1	64.0	120.1	29.0	69.2	21.9	249.8	244.0	142.7	117.2	25.5	101.3	5.8
1988	72.9	69.0	132.7	33.3	76.4	23.0	256.2	250.1	142.4	120.1	22.3	107.7	6.1
1989	67.9	80.2	150.1	40.6	84.1	25.4	256.0	249.9	143.2	120.9	22.3	106.6	6.1
1990	70.0	80.2	164.4	45.4	91.5	27.5	239.7	233.7	132.1	112.9	19.3	101.5	6.0
1991	71.5	70.8	179.1	48.7	101.0	29.4	221.2	215.4	114.6	99.4	15.1	100.8	5.7
1992	74.7	72.0	187.7	51.1	105.4	31.2	254.7	248.8	135.1	122.0	13.1	113.8	5.9
1993	89.4	80.2	196.9	57.2	106.3	33.4	286.8	280.7	150.9	140.1	10.8	129.8	6.1
1994	107.7	88.1	205.7	60.4	109.2	36.1	323.8	317.6	176.4	162.3	14.1	141.2	6.2
1995	116.1	94.7	226.8	65.5	121.2	40.2	324.1	317.7	171.4	153.5	17.9	146.3	6.3
1996	123.2	101.0	253.3	74.5	134.5	44.3	358.1	351.7	191.1	170.8	20.3	160.6	6.3
1997	135.5	112.1	288.0	93.8	148.1	46.1	375.6	369.3	198.1	175.2	22.9	171.3	6.3
1998	147.1	125.4	318.1	109.2	160.6	48.3	418.8	412.1	224.0	199.4	24.6	188.2	6.7
1999	174.4	130.4	365.1	136.6	177.5	51.0	461.8	454.5	251.3	223.8	27.4	203.2	7.3
2000	170.8	138.6	411.3	156.8	199.0	55.6	485.4	477.7	265.0	236.8	28.3	212.7	7.7
2001	154.2	139.5	415.0	157.7	202.7	54.7	513.1	505.2	279.4	249.1	30.3	225.8	7.8
2002	141.6	139.6	406.2	152.5	196.1	57.6	557.6	549.6	298.8	265.9	33.0	250.7	8.0
2003	134.1	150.5	418.7	155.0	201.0	62.7	637.1	628.8	345.7	310.6	35.1	283.1	8.3
2004	159.2	162.7	437.8	166.3	207.4	64.1	749.8	740.8	417.5	377.6	39.9	323.3	9.0
2005	179.6	186.0	473.1	178.6	224.7	69.8	856.2	846.6	480.8	433.5	47.3	365.8	9.6
2006	194.3	198.0	506.3	189.5	245.6	71.2	838.2	828.1	468.8	416.0	52.8	359.3	10.0
2007	188.8	199.6	544.8	206.4	268.0	70.4	690.5	680.6	354.1	305.2	49.0	326.5	9.9
2008	148.7	196.1	574.4	223.8	284.2	66.4	516.0	506.4	230.1	185.8	44.3	276.3	9.6
2009	74.9	166.1	564.4	226.0	274.6	63.7	390.0	381.2	133.9	105.3	28.5	247.3	8.8
2010	135.8	178.5	578.2	226.4	282.4	69.4	376.6	367.4	127.3	112.6	14.7	240.2	9.2
2011	177.8	198.7	621.7	249.8	303.4	68.6	378.8	369.1	123.2	108.2	15.0	245.9	9.8
2012	215.3	225.7	655.7	272.1	313.4	70.2	432.0	421.5	154.5	132.0	22.5	267.0	10.5
2013	242.5	233.6	691.9	283.7	337.9	70.3	510.0	499.0	202.3	170.8	31.5	296.7	11.0
2014	272.8	254.4	730.5	297.5	359.5	73.4	560.2	548.8	235.2	193.6	41.6	313.7	11.3
2015	306.7	242.5	763.3	307.1	378.9	77.3	633.8	622.1	273.7	221.1	52.5	348.4	11.7
2016	293.0	230.0	813.8	327.6	405.2	81.0	698.5	686.4	303.6	242.5	61.1	382.8	12.1
2017	283.0	248.9	854.2	347.9	422.0	84.3	755.7	743.3	330.1	270.2	59.9	413.2	12.5
2018	301.8	266.3	931.1	380.0	461.7	89.4	786.7	773.7	344.4	284.3	60.1	429.3	12.9
2016													
1st quarter	302.7	232.5	793.8	321.4	392.7	79.6	686.6	674.5	301.5	240.4	61.1	373.0	12.1
2nd quarter	296.3	228.4	812.1	325.2	406.4	80.5	692.0	679.9	302.3	241.2	61.1	377.6	12.1
3rd quarter	289.5	229.6	820.9	329.7	409.7	81.5	697.7	685.5	299.6	238.1	61.5	385.9	12.2
4th quarter	283.6	229.6	828.6	334.2	411.9	82.5	717.8	705.6	311.0	250.2	60.7	394.6	12.2
2017													
1st quarter	283.4	237.0	842.1	341.1	417.9	83.2	743.7	731.3	320.4	259.6	60.8	410.8	12.5
2nd quarter	280.1	246.1	845.5	345.7	416.1	83.8	748.8	736.5	327.5	267.7	59.8	409.0	12.3
3rd quarter	280.0	252.6	858.4	350.6	423.2	84.6	753.4	741.0	332.3	273.6	58.7	408.6	12.4
4th quarter	288.7	260.0	870.9	354.3	430.9	85.8	777.1	764.4	340.2	279.7	60.5	424.3	12.7
2018													
1st quarter	294.9	259.3	899.6	367.9	444.4	87.2	783.7	770.9	345.6	286.8	58.8	425.3	12.8
2nd quarter	301.5	265.0	927.7	377.3	461.6	88.9	789.5	776.6	348.0	288.2	59.8	428.6	13.0
3rd quarter	299.7	266.5	938.1	383.8	464.1	90.2	789.0	775.9	345.2	285.8	59.4	430.7	13.0
4th quarter	311.0	274.2	959.1	391.0	476.8	91.3	784.4	771.6	338.8	276.3	62.5	432.7	12.9

[1] Excludes software "embedded," or bundled, in computers and other equipment. Includes software development expenditures.
[2] Excludes software development.

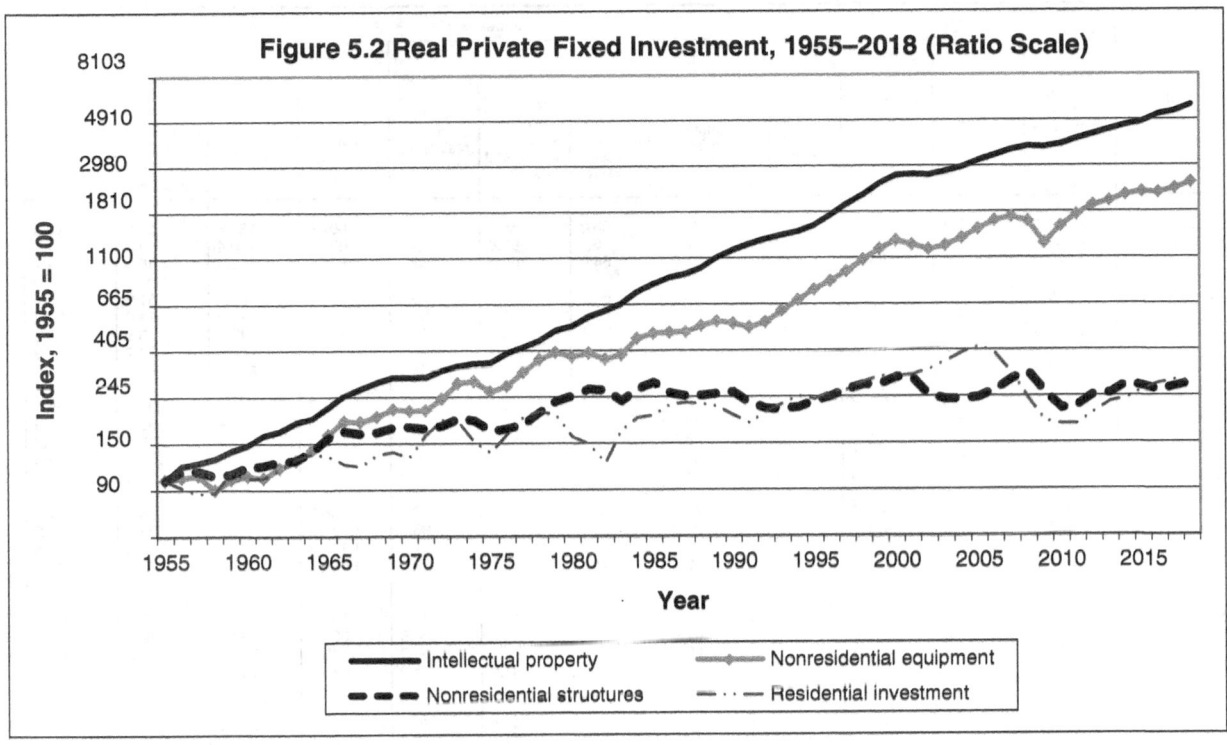

Figure 5.2 Real Private Fixed Investment, 1955–2018 (Ratio Scale)

- Between 1955 and 2006, which was the recent high point for total real (constant-dollar) investment, the quantity of real gross private fixed investment increased eightfold, with an average annual growth rate of 4.2 percent. Every major type of investment grew in real terms between those two years, but as Figure 5-2 indicates, by far the fastest-growing major category has been "intellectual property products." Computer software is the main driver of growth in this category, but it also includes research and development and "entertainment, literary, and artistic originals." (Table 5-3)

- Between 2006 and 2009, real gross investment fell 21.8 percent in real terms—a collapse unprecedented in the postwar period. By 2018, real gross investment surpassed the 2006 level by 22.6 percent, achieving the all-time high. (Table 5-3)

- Intellectual property investment barely hesitated in 2009 and went on to new record peaks in each following year. Equipment purchases continued to increase went on to a new record in 2018. On the other hand, residential investment dropped 40.6 percent between 2007 to 2009. Residential investment increased steadily from 2012 to 2017 but declined slightly in 2018. (Table 5-4)

Table 5-3. Real Gross Private Fixed Investment by Type

(Billions of chained [2012] dollars, quarterly data are at seasonally adjusted annual rates.)   NIPA Table 5.3.6

Year and quarter	Total gross private fixed investment	Nonresidential											
		Total	Structures						Equipment				
			Total	Commer-cial and health care	Manufac-turing	Power and communi-cation	Mining explor-ation, shafts, and wells	Other non-residential structures	Total	Information processing equipment			Industrial equipment
										Total	Computers and peripheral equipment[1]	Other information processing	
2002	2 183.4	1 472.7	473.5	172.6	32.2	83.3	85.5	81.7	607.8	133.3	35.9	98.3	181.4
2003	2 280.6	1 509.4	456.6	161.7	29.9	76.7	97.6	79.7	634.3	150.4	40.2	111.1	182.2
2004	2 440.7	1 594.0	456.3	165.7	30.7	64.0	103.6	82.0	688.6	169.4	45.7	124.7	178.8
2005	2 618.7	1 716.4	466.1	164.3	34.9	66.4	113.4	79.7	760.0	187.6	51.8	136.5	194.2
2006	2 686.8	1 854.2	501.7	174.2	37.2	70.4	122.5	89.1	832.6	217.0	64.7	152.4	210.6
2007	2 653.5	1 982.1	568.6	191.6	44.0	98.1	120.5	105.0	865.8	247.2	73.9	173.3	217.3
2008	2 499.4	1 994.2	605.4	184.6	55.1	107.8	129.4	118.9	824.4	260.6	79.7	180.9	208.3
2009	2 099.8	1 704.3	492.2	128.0	57.9	109.3	92.4	98.1	649.7	247.5	81.1	166.5	162.7
2010	2 164.2	1 781.0	412.8	96.2	42.2	91.7	108.3	72.6	781.2	289.1	94.1	195.1	162.5
2011	2 317.8	1 935.4	424.1	95.8	40.6	84.6	136.7	66.5	886.2	303.2	93.9	209.3	194.9
2012	2 550.5	2 118.5	479.4	103.8	46.8	102.4	152.9	73.6	983.4	331.2	103.5	227.7	211.2
2013	2 692.1	2 206.0	485.5	107.5	48.7	97.8	155.4	76.1	1 029.2	351.8	103.0	248.8	208.4
2014	2 869.2	2 365.3	538.8	121.3	55.0	112.5	167.8	82.4	1 101.1	370.2	102.9	267.7	216.5
2015	2 967.0	2 408.2	522.4	134.2	73.7	108.7	118.8	90.0	1 136.6	394.6	103.7	291.9	217.0
2016	3 023.6	2 425.3	496.4	157.9	69.9	109.0	67.8	96.3	1 122.3	415.5	103.2	314.2	214.6
2017	3 149.7	2 531.2	519.5	162.9	59.3	108.0	95.6	97.9	1 175.6	456.3	112.3	346.5	228.2
2018	3 293.4	2 692.3	540.9	162.8	55.7	107.3	118.6	100.5	1 255.3	498.5	123.5	377.5	238.5
2009													
1st quarter	2 151.9	1 746.3	551.3	147.7	63.2	107.1	119.2	107.0	653.7	238.1	75.6	162.6	173.9
2nd quarter	2 073.9	1 693.2	509.0	134.4	61.8	110.0	90.7	104.5	627.8	238.2	78.5	159.8	163.4
3rd quarter	2 081.6	1 683.3	474.3	121.3	56.8	115.2	78.7	95.9	643.7	249.7	81.3	168.4	157.8
4th quarter	2 092.0	1 694.5	434.1	108.4	50.0	105.1	81.0	85.0	673.5	264.0	89.0	175.1	155.6
2010													
1st quarter	2 091.0	1 706.4	405.1	100.7	47.7	83.0	94.4	76.1	721.1	276.6	93.1	183.6	151.7
2nd quarter	2 167.1	1 762.3	416.8	96.6	45.6	91.7	107.1	73.8	768.1	285.8	96.3	189.6	161.3
3rd quarter	2 178.7	1 810.1	410.7	94.1	40.2	89.4	114.8	71.1	809.5	295.6	95.4	200.2	164.7
4th quarter	2 220.0	1 845.2	418.6	93.5	35.3	102.7	116.8	69.4	825.9	298.5	91.4	207.1	172.2
2011													
1st quarter	2 216.2	1 842.7	387.8	90.5	34.3	77.5	122.8	62.7	846.3	292.9	87.8	205.1	182.8
2nd quarter	2 268.0	1 890.4	414.8	95.0	39.0	83.7	130.5	66.4	858.0	302.1	93.6	208.5	185.4
3rd quarter	2 363.3	1 978.9	439.4	98.2	43.6	88.7	141.1	67.7	909.7	308.2	97.5	210.7	199.1
4th quarter	2 423.7	2 029.4	454.5	99.2	45.5	88.3	152.4	69.3	930.7	309.6	96.8	212.8	212.1
2012													
1st quarter	2 499.4	2 081.3	476.0	101.7	44.5	102.0	156.4	71.4	959.6	327.3	105.3	222.0	210.3
2nd quarter	2 549.8	2 128.0	487.4	104.0	46.6	104.8	158.1	73.9	986.7	333.5	104.6	228.9	215.0
3rd quarter	2 553.6	2 120.9	481.8	105.2	47.4	103.2	151.1	74.8	983.8	327.3	100.6	226.7	207.5
4th quarter	2 599.4	2 144.0	472.5	104.1	48.6	99.4	145.9	74.4	1 003.4	336.8	103.6	233.2	212.1
2013													
1st quarter	2 643.9	2 171.6	462.1	106.3	47.0	81.9	153.3	73.4	1 021.1	349.9	105.9	244.0	210.1
2nd quarter	2 665.3	2 177.5	475.8	103.4	46.2	91.3	159.3	75.4	1 019.0	348.6	100.2	248.5	207.7
3rd quarter	2 711.3	2 214.7	499.8	106.9	51.0	104.2	159.9	77.7	1 019.0	352.1	101.2	251.0	209.2
4th quarter	2 748.0	2 260.0	504.2	113.2	50.7	113.9	148.9	77.7	1 057.9	356.5	104.8	251.8	206.4
2014													
1st quarter	2 775.6	2 291.7	521.9	113.1	49.6	126.1	155.5	78.5	1 066.9	358.9	101.3	257.9	211.4
2nd quarter	2 852.8	2 353.3	540.3	119.2	50.4	119.0	170.9	81.2	1 093.6	372.1	103.5	269.0	217.1
3rd quarter	2 907.3	2 400.8	542.2	124.2	54.6	109.4	171.7	82.4	1 127.0	370.6	102.3	268.7	222.0
4th quarter	2 941.2	2 415.5	551.0	128.8	65.6	95.5	173.1	87.4	1 116.9	379.3	104.7	275.1	215.3
2015													
1st quarter	2 943.1	2 406.1	538.3	128.4	71.5	100.4	153.2	85.6	1 126.2	381.9	102.4	280.2	216.3
2nd quarter	2 964.7	2 413.2	540.2	136.9	75.9	116.1	123.2	91.4	1 130.4	388.2	103.7	285.2	217.8
3rd quarter	2 988.2	2 420.4	520.0	135.3	74.6	113.9	106.7	93.3	1 152.0	402.7	106.5	297.1	216.3
4th quarter	2 971.9	2 393.2	491.1	136.0	72.7	104.2	92.1	89.8	1 137.9	405.7	102.2	305.2	217.6
2016													
1st quarter	2 991.0	2 389.8	476.4	148.7	70.6	100.6	69.3	91.4	1 126.5	406.5	104.1	303.9	212.8
2nd quarter	3 010.9	2 413.6	487.9	153.8	71.3	106.2	63.1	97.8	1 120.0	409.1	102.9	307.9	214.6
3rd quarter	3 038.9	2 446.8	509.0	163.3	70.6	113.7	68.0	98.2	1 120.9	420.2	101.7	321.1	213.9
4th quarter	3 053.7	2 451.2	512.1	165.6	67.1	115.2	71.0	97.9	1 122.0	426.0	104.3	324.1	217.0
2017													
1st quarter	3 111.1	2 490.5	521.1	164.5	62.7	115.6	84.7	98.4	1 139.3	436.4	106.9	331.9	220.8
2nd quarter	3 133.0	2 517.4	523.7	164.2	60.1	109.2	96.5	98.0	1 163.8	451.2	112.3	341.0	227.3
3rd quarter	3 144.1	2 532.6	513.3	161.0	57.4	105.2	96.9	96.9	1 181.4	460.6	116.3	346.1	230.0
4th quarter	3 210.7	2 584.2	519.9	161.8	57.1	102.1	104.2	98.4	1 217.8	477.2	113.8	366.8	234.4
2018													
1st quarter	3 254.0	2 639.5	534.9	167.3	56.2	107.6	108.9	99.0	1 237.5	489.3	121.6	370.0	237.7
2nd quarter	3 295.4	2 689.9	549.1	165.0	53.9	110.5	123.5	100.4	1 247.8	493.9	124.8	371.0	235.2
3rd quarter	3 301.3	2 703.9	546.2	161.5	56.6	107.7	123.0	101.3	1 256.7	506.6	124.9	384.4	238.7
4th quarter	3 323.0	2 735.8	533.4	157.4	55.9	103.2	119.2	101.2	1 279.2	504.2	122.7	384.5	242.5

[1]See notes and definitions.

Table 5-3. Real Gross Private Fixed Investment by Type—Continued

(Billions of chained [2012] dollars, quarterly data are at seasonally adjusted annual rates.)

NIPA Table 5.3.6

Year and quarter	Equipment—Continued		Intellectual property				Total	Residential structures					Residential equipment
	Transportation equipment	Other nonresidential equipment	Total	Software [2]	Research and development [3]	Entertainment, literary, and artistic originals		Total	Permanent site			Other residential structures	
									Total	Single family	Multifamily		
2002	162.4	168.7	421.5	125.5	244.1	60.1	692.6	685.1	379.0	327.1	48.3	303.9	7.7
2003	150.3	180.2	437.7	133.5	246.1	64.6	755.5	747.7	416.1	362.0	49.6	329.2	8.1
2004	171.2	193.3	459.2	149.3	248.1	65.6	830.9	822.1	464.3	405.4	53.4	355.4	9.1
2005	192.1	211.9	493.1	163.4	261.6	71.0	885.4	876.3	497.0	432.8	58.7	376.7	9.5
2006	206.4	220.7	521.5	173.5	279.6	71.8	818.9	809.5	453.6	390.4	59.1	353.5	9.7
2007	197.7	218.0	554.3	191.1	296.1	70.0	665.8	656.6	337.7	283.5	52.3	317.7	9.4
2008	155.0	207.4	575.3	206.7	304.8	66.0	504.6	495.7	224.2	178.1	46.0	270.9	9.1
2009	72.5	165.3	572.4	212.9	297.4	63.4	395.3	386.9	134.5	105.3	29.3	252.3	8.4
2010	141.5	186.2	588.1	220.9	298.5	69.5	383.0	373.8	129.1	114.3	14.7	244.7	9.3
2011	181.8	206.2	624.8	245.2	311.0	68.9	382.5	372.4	124.5	109.1	15.4	247.9	10.0
2012	215.3	225.7	655.7	272.1	313.4	70.2	432.0	421.5	154.5	132.0	22.5	267.0	10.5
2013	238.5	230.6	691.4	287.2	333.8	70.3	485.5	474.1	192.4	161.8	30.6	281.7	11.4
2014	265.0	248.9	724.8	305.3	346.9	72.7	504.1	491.8	210.4	171.8	38.8	281.4	12.4
2015	293.2	232.5	750.7	319.8	355.9	75.4	555.3	541.9	235.2	191.4	44.0	306.7	13.4
2016	277.0	220.2	810.0	346.0	386.9	77.6	591.2	576.7	249.7	201.3	48.7	327.0	14.6
2017	263.3	237.6	839.6	373.8	388.5	79.6	611.9	596.6	259.3	214.7	45.1	337.2	15.6
2018	280.1	250.9	901.6	413.5	409.2	83.3	602.9	587.5	258.2	216.6	42.3	329.2	15.7
2009													
1st quarter	70.8	170.8	554.7	204.8	288.1	63.1	405.3	396.9	147.7	109.4	38.6	249.1	8.4
2nd quarter	63.3	161.9	568.6	209.5	297.1	63.5	380.4	372.2	125.0	93.0	32.2	247.2	8.3
3rd quarter	70.6	163.4	574.5	213.6	299.2	63.0	398.1	389.6	132.7	106.7	25.9	257.0	8.5
4th quarter	85.4	165.3	591.9	223.6	305.1	64.2	397.2	388.7	132.5	111.9	20.5	256.1	8.6
2010													
1st quarter	112.8	177.2	581.9	220.5	295.8	66.4	384.4	375.5	131.9	115.6	16.2	243.6	8.9
2nd quarter	131.2	187.9	579.0	217.0	293.4	69.3	404.4	395.1	136.1	121.6	14.5	258.9	9.4
3rd quarter	159.9	187.8	590.3	220.9	299.6	70.6	368.6	359.3	126.6	112.6	14.0	232.7	9.3
4th quarter	162.2	191.7	601.3	225.3	305.1	71.7	374.7	365.1	121.7	107.6	14.1	243.4	9.5
2011													
1st quarter	172.0	198.5	607.4	232.1	306.5	69.3	373.4	363.6	122.1	108.2	14.0	241.5	9.8
2nd quarter	170.2	199.8	617.4	240.2	309.2	68.3	377.6	367.7	121.7	107.2	14.5	246.0	10.0
3rd quarter	188.0	214.3	629.9	249.8	311.7	68.5	384.4	374.4	125.5	109.4	16.0	248.9	10.1
4th quarter	197.1	212.0	644.3	258.5	316.5	69.5	394.4	384.1	128.7	111.7	17.0	255.4	10.3
2012													
1st quarter	205.6	216.5	645.7	263.1	312.3	70.3	418.3	407.8	139.3	120.5	18.8	268.5	10.4
2nd quarter	215.6	222.7	653.9	272.1	311.5	70.2	421.9	411.5	147.6	126.2	21.4	263.9	10.4
3rd quarter	216.1	232.9	655.2	272.8	312.4	70.1	432.7	422.1	158.1	134.5	23.6	264.1	10.6
4th quarter	223.9	230.6	668.0	280.2	317.6	70.2	455.2	444.5	173.1	146.9	26.2	271.4	10.7
2013													
1st quarter	232.2	229.0	688.6	288.3	330.1	70.2	471.8	460.7	184.3	156.0	28.2	276.4	11.1
2nd quarter	236.6	226.2	682.9	280.2	331.9	70.8	486.8	475.5	192.6	163.0	29.5	282.9	11.3
3rd quarter	237.2	220.7	695.8	288.3	337.1	70.3	495.5	483.9	196.5	165.7	30.8	287.4	11.6
4th quarter	248.2	246.4	698.1	292.1	336.1	70.0	487.7	476.2	196.3	162.6	33.8	279.9	11.5
2014													
1st quarter	252.4	243.8	702.5	296.2	335.3	71.2	484.3	472.6	199.9	165.0	35.0	272.8	11.7
2nd quarter	260.5	243.8	718.5	303.3	343.3	72.1	499.8	487.5	207.6	169.5	38.3	279.9	12.4
3rd quarter	275.6	257.5	731.4	309.8	348.7	73.2	507.1	494.7	209.9	169.9	40.3	284.8	12.6
4th quarter	271.6	250.5	746.6	311.9	360.2	74.4	525.2	512.4	224.1	182.8	41.6	288.3	12.8
2015													
1st quarter	286.0	241.6	741.5	315.1	352.1	74.7	535.5	522.5	227.3	185.5	42.1	295.1	13.1
2nd quarter	291.7	232.8	742.5	318.1	350.0	75.0	548.9	535.6	229.9	187.4	42.7	305.7	13.4
3rd quarter	302.8	231.1	750.5	319.7	355.4	75.8	563.9	550.4	239.3	194.4	45.2	311.0	13.5
4th quarter	292.1	224.7	768.3	326.2	366.0	76.3	573.0	559.3	244.2	198.3	46.1	315.0	13.7
2016													
1st quarter	287.8	222.3	792.0	336.9	378.7	76.7	593.0	578.8	253.7	204.7	49.2	325.0	14.3
2nd quarter	281.0	219.1	809.8	342.7	390.7	76.5	590.1	575.7	251.1	202.3	49.0	324.6	14.4
3rd quarter	272.8	219.8	819.2	349.5	391.9	78.2	586.2	571.7	244.5	195.8	48.9	327.2	14.7
4th quarter	266.3	219.7	819.2	354.9	386.2	79.0	595.5	580.7	249.7	202.2	47.7	331.1	14.9
2017													
1st quarter	263.2	227.0	831.8	364.5	389.6	79.2	612.4	597.2	255.3	208.7	47.0	342.0	15.3
2nd quarter	259.7	235.2	832.3	369.3	385.9	79.2	608.9	593.8	258.1	213.1	45.4	335.6	15.4
3rd quarter	260.4	240.7	842.3	378.1	387.5	79.4	605.9	590.6	259.6	216.5	43.7	330.7	15.7
4th quarter	269.9	247.4	852.0	383.5	390.9	80.6	620.4	604.7	264.3	220.6	44.3	340.2	16.1
2018													
1st quarter	275.4	247.2	872.0	399.2	395.4	81.6	612.1	596.4	263.9	222.4	42.3	332.3	16.2
2nd quarter	279.4	251.4	896.9	409.2	409.0	82.7	606.3	590.9	261.4	219.9	42.3	329.2	15.9
3rd quarter	275.9	250.0	905.9	417.4	409.6	83.6	600.1	584.9	257.2	216.6	41.4	327.5	15.7
4th quarter	289.6	254.9	931.3	428.0	422.7	85.2	593.0	578.0	250.2	207.6	43.1	327.7	15.3

[2] Excludes software "embedded," or bundled, in computers and other equipment.
[3] Excludes software development.

Table 5-4. Chain-Type Quantity Indexes for Private Fixed Investment by Type

(Index numbers, 2012 = 100, quarterly data are seasonally adjusted.)   NIPA Table 5.3.3

Year and quarter	Total gross private fixed investment	Nonresidential											
		Total	Structures						Equipment				
			Total	Commercial and health care	Manufacturing	Power and communication	Mining exploration, shafts, and wells	Other nonresidential structures	Total	Information processing equipment			Industrial equipment
										Total	Computers and peripheral equipment[1]	Other information processing	
1960	13.9	8.8	46.3	47.3	60.9	36.1	42.0	62.1	5.4	0.2	0.0	2.3	28.7
1961	13.9	8.8	46.9	54.0	59.6	33.8	42.8	60.6	5.3	0.2	0.0	2.5	27.2
1962	15.1	9.5	49.1	60.2	60.2	34.0	44.9	61.9	5.9	0.3	0.0	2.7	28.6
1963	16.3	10.1	49.7	58.3	60.4	36.3	42.1	65.5	6.4	0.3	0.0	2.8	30.9
1964	17.8	11.2	54.8	63.4	72.6	39.5	45.1	71.8	7.2	0.3	0.0	3.0	34.9
1965	19.7	13.0	63.6	74.3	100.4	43.7	44.6	80.7	8.5	0.4	0.0	3.4	41.1
1966	20.9	14.6	67.9	72.5	124.8	49.8	42.2	83.8	9.8	0.5	0.0	4.0	47.3
1967	20.7	14.6	66.2	69.7	109.8	54.5	40.4	82.0	9.7	0.6	0.0	4.0	47.5
1968	22.1	15.3	67.1	76.3	105.7	60.6	40.6	72.4	10.3	0.6	0.0	4.1	46.5
1969	23.4	16.4	70.7	88.0	111.2	60.6	42.3	71.8	11.1	0.7	0.0	4.8	49.6
1970	23.0	16.2	70.9	89.2	107.3	66.3	39.8	67.5	10.9	0.8	0.0	5.2	50.3
1971	24.5	16.2	69.8	97.5	88.6	66.9	36.9	63.3	11.0	0.8	0.0	5.2	46.2
1972	27.3	17.6	72.0	106.5	77.4	69.7	39.5	64.0	12.4	1.0	0.0	5.5	49.9
1973	29.7	19.9	77.8	111.2	95.9	74.2	42.2	69.0	14.7	1.1	0.0	6.7	58.9
1974	28.0	20.1	76.2	104.3	109.3	69.5	50.2	63.1	15.0	1.3	0.0	7.4	63.5
1975	25.3	18.3	68.2	81.3	104.8	64.0	58.6	55.6	13.4	1.3	0.0	7.1	53.8
1976	27.8	19.3	69.8	80.2	96.8	69.8	63.2	58.4	14.3	1.4	0.0	7.8	54.1
1977	31.5	21.4	72.7	84.4	99.4	68.7	71.4	61.6	16.5	1.8	0.0	9.4	57.3
1978	35.2	24.4	83.2	96.2	132.6	72.0	81.4	70.0	19.0	2.3	0.1	11.2	63.8
1979	37.2	26.8	93.7	114.7	162.9	74.7	86.7	76.9	20.5	2.8	0.1	12.6	68.3
1980	35.0	26.8	99.2	126.1	136.2	75.4	121.9	74.7	19.6	3.3	0.1	13.5	65.4
1981	35.9	28.5	107.2	136.5	154.7	77.4	141.7	72.3	20.3	3.8	0.2	14.3	64.4
1982	33.7	27.6	105.4	148.7	149.3	73.1	131.0	68.2	19.0	4.0	0.3	14.4	58.8
1983	36.3	27.5	94.0	144.3	108.0	62.8	108.4	72.0	19.9	4.6	0.4	15.1	54.1
1984	42.1	32.1	107.1	175.6	111.8	64.2	123.3	84.4	23.8	5.9	0.7	17.2	61.9
1985	44.4	34.2	114.7	202.8	125.0	63.5	109.8	92.8	25.1	6.5	0.9	17.8	64.8
1986	45.2	33.6	102.2	188.4	105.0	67.8	64.4	87.2	25.4	7.0	1.0	18.5	64.5
1987	45.5	33.6	99.2	181.5	102.4	60.3	62.9	92.4	25.5	7.3	1.3	17.7	62.9
1988	47.0	35.3	99.9	187.0	108.0	56.8	69.0	86.2	27.2	7.9	1.4	19.0	66.1
1989	48.5	37.3	101.9	184.6	129.4	59.0	62.4	87.9	28.6	8.7	1.7	19.8	70.7
1990	47.8	37.7	103.4	177.3	146.6	54.9	72.1	93.3	28.0	8.8	1.7	20.3	66.6
1991	45.3	36.2	91.9	137.7	135.0	65.2	70.5	81.0	26.7	9.0	1.9	20.2	62.3
1992	47.8	37.3	86.3	128.0	124.0	69.2	58.4	75.0	28.3	10.3	2.6	21.4	63.4
1993	51.5	40.1	86.1	133.2	97.4	66.0	67.8	81.0	31.9	11.7	3.3	23.1	68.8
1994	55.7	43.2	87.7	139.4	115.3	59.8	65.9	79.2	35.8	13.3	4.1	25.2	75.0
1995	59.1	47.4	93.2	147.6	135.9	61.0	56.0	88.7	40.1	16.1	6.1	27.1	82.3
1996	64.3	51.7	98.5	159.0	142.9	52.5	59.6	104.3	44.0	19.4	8.9	29.7	85.3
1997	69.8	57.3	105.8	175.7	136.5	50.6	70.2	118.3	48.9	23.4	12.9	32.3	87.2
1998	77.0	63.5	111.5	186.7	141.1	59.4	61.4	126.5	55.3	28.9	18.7	36.3	91.0
1999	84.0	69.8	112.0	193.5	117.7	70.1	51.9	128.4	62.1	35.3	26.5	40.5	91.5
2000	89.9	76.3	121.1	206.4	121.4	80.8	64.8	134.1	68.2	42.6	32.2	48.8	99.4
2001	88.7	74.6	119.6	196.4	118.0	83.4	76.4	129.5	65.3	42.1	33.0	47.3	92.1
2002	85.6	69.5	98.8	166.4	68.9	81.3	55.9	110.9	61.8	40.3	34.6	43.2	85.9
2003	89.4	71.2	95.2	155.8	63.8	74.9	63.9	108.2	64.5	45.4	38.9	48.8	86.3
2004	95.7	75.2	95.2	159.7	65.7	62.6	67.8	111.4	70.0	51.2	44.1	54.8	84.7
2005	102.7	81.0	97.2	158.3	74.6	64.8	74.2	108.3	77.3	56.6	50.0	60.0	91.9
2006	105.3	87.5	104.6	167.9	79.6	68.8	80.1	121.0	84.7	65.5	62.5	66.9	99.7
2007	104.0	93.6	118.6	184.6	94.1	95.8	78.8	142.6	88.0	74.6	71.4	76.1	102.9
2008	98.0	94.1	126.3	177.9	117.9	105.4	84.6	161.4	83.8	78.7	77.0	79.4	98.6
2009	82.3	80.5	102.7	123.3	123.9	106.8	60.4	133.2	66.1	74.7	78.3	73.1	77.0
2010	84.9	84.1	86.1	92.7	90.2	89.6	70.8	98.6	79.4	87.3	90.9	85.7	76.9
2011	90.9	91.4	88.5	92.3	86.8	82.6	89.4	90.4	90.1	91.5	90.7	91.9	92.3
2012	100.0	100.0	100.0	100.0	100.0	100.0	100.0	100.0	100.0	100.0	100.0	100.0	100.0
2013	105.6	104.1	101.3	103.6	104.2	95.6	101.6	103.3	104.7	106.2	99.5	109.3	98.6
2014	112.5	111.6	112.4	116.9	117.7	109.9	109.7	111.9	112.0	111.8	99.4	117.6	102.5
2015	116.3	113.7	109.0	129.3	157.5	106.2	77.7	122.3	115.6	119.1	100.2	128.2	102.7
2016	118.5	114.5	103.5	152.1	149.4	106.4	44.4	130.8	114.1	125.4	99.7	138.0	101.6
2017	123.5	119.5	108.4	157.0	126.8	105.5	62.5	133.0	119.5	137.8	108.5	152.2	108.0
2018	129.1	127.1	112.8	156.9	119.0	104.8	77.6	136.5	127.6	150.5	119.3	165.8	112.9
2016													
1st quarter	117.3	112.8	99.4	143.3	150.9	98.3	45.3	124.1	114.6	122.7	100.5	133.5	100.7
2nd quarter	118.1	113.9	101.8	148.3	152.4	103.8	41.3	132.8	113.9	123.5	99.4	135.2	101.6
3rd quarter	119.1	115.5	106.2	157.4	151.0	111.1	44.4	133.3	114.0	126.9	98.2	141.0	101.3
4th quarter	119.7	115.7	106.8	159.6	143.5	112.6	46.4	133.0	114.1	128.6	100.7	142.3	102.7
2017													
1st quarter	122.0	117.6	108.7	158.5	134.0	112.9	55.4	133.7	115.9	131.7	103.2	145.8	104.6
2nd quarter	122.8	118.8	109.2	158.2	128.5	106.7	63.1	133.0	118.3	136.2	108.5	149.8	107.6
3rd quarter	123.3	119.5	107.1	155.1	122.6	102.8	63.4	131.6	120.1	139.0	112.4	152.0	108.9
4th quarter	125.9	122.0	108.4	156.0	122.0	99.7	68.1	133.7	123.8	144.1	109.9	161.1	111.0
2018													
1st quarter	127.6	124.6	111.6	161.2	120.2	105.1	71.2	134.5	125.8	147.7	117.5	162.5	112.5
2nd quarter	129.2	127.0	114.5	159.0	115.1	107.9	80.8	136.4	126.9	149.1	120.5	162.9	111.4
3rd quarter	129.4	127.6	113.9	155.6	121.0	105.3	80.4	137.6	127.8	153.0	120.6	168.8	113.0
4th quarter	130.3	129.1	111.3	151.6	119.6	100.8	78.0	137.5	130.1	152.2	118.5	168.9	114.8

[1]See notes and definitions.

Table 5-4. Chain-Type Quantity Indexes for Private Fixed Investment by Type—*Continued*

(Index numbers, 2012 = 100, quarterly data are seasonally adjusted.)　　　　　NIPA Table 5.3.3

Year and quarter	Nonresidential—*Continued*						Residential						
	Equipment—*Continued*		Intellectual property				Total	Residential structures					Residential equipment
	Transportation equipment	Other nonresidential equipment	Total	Software [2]	Research and development [3]	Entertainment, literary, and artistic originals		Total	Permanent site			Other residential structures	
									Total	Single family	Multifamily		
1960	16.5	18.9	3.5	0.0	8.0	17.5	49.9	51.7	97.3	92.6	114.5	25.8	8.6
1961	15.6	18.7	3.9	0.0	8.5	21.1	50.1	51.9	96.8	87.7	144.3	26.4	8.6
1962	19.2	20.0	4.0	0.0	9.2	20.2	54.9	56.9	110.4	93.5	209.8	26.7	9.1
1963	18.6	23.2	4.4	0.1	9.9	21.7	61.3	63.6	125.2	99.9	281.6	28.8	10.2
1964	21.0	25.8	4.6	0.1	10.6	20.6	65.0	67.4	133.7	109.6	279.7	30.0	10.9
1965	26.3	28.6	5.2	0.1	11.9	22.9	63.4	65.5	127.8	107.1	251.4	30.3	12.2
1966	29.0	32.2	5.9	0.2	13.3	25.0	58.0	59.9	112.1	95.4	210.8	30.2	12.5
1967	28.0	30.4	6.4	0.3	14.7	24.8	56.5	58.3	106.9	93.5	183.1	30.6	12.8
1968	33.7	30.8	6.9	0.3	15.9	25.9	64.2	66.1	125.6	102.7	264.5	32.3	15.3
1969	35.0	32.8	7.2	0.4	16.9	24.5	66.2	68.0	129.2	97.6	329.1	33.3	17.4
1970	28.7	34.3	7.2	0.5	17.0	21.4	62.7	64.2	116.9	84.8	322.1	34.1	18.9
1971	31.0	34.2	7.2	0.6	16.9	21.1	79.4	81.6	157.5	117.7	409.9	38.6	21.3
1972	36.1	38.7	7.8	0.7	17.7	22.8	93.2	95.7	190.6	139.9	513.2	42.3	25.5
1973	43.4	45.2	8.1	0.7	18.9	21.8	92.6	94.8	189.7	136.8	527.6	41.4	28.6
1974	39.2	44.7	8.4	0.9	19.0	22.1	74.5	75.8	137.3	105.0	339.7	40.8	29.2
1975	33.8	41.3	8.5	1.0	18.7	21.4	65.5	66.5	105.1	95.9	151.5	44.1	27.2
1976	37.6	41.7	9.4	1.1	19.9	27.2	80.0	81.6	138.1	133.5	146.8	49.1	28.6
1977	45.7	46.9	10.0	1.1	20.9	30.3	96.4	98.5	178.2	171.5	194.6	53.2	31.3
1978	50.5	51.7	10.7	1.4	22.6	29.8	102.8	105.0	186.4	177.1	219.2	58.7	34.0
1979	52.7	52.7	11.9	1.7	24.2	34.2	98.9	100.8	174.4	157.0	267.8	58.7	37.3
1980	43.0	47.3	12.5	2.0	25.9	31.0	78.2	79.3	122.5	103.2	239.7	54.1	37.3
1981	41.8	40.0	13.9	2.3	27.5	37.4	71.8	72.6	113.0	93.9	230.6	49.1	37.3
1982	37.1	42.1	14.8	2.7	29.0	37.7	58.8	59.1	88.1	72.0	187.9	42.2	35.8
1983	41.8	40.5	15.9	3.2	30.6	39.7	83.5	84.6	143.9	125.0	257.3	50.6	40.0
1984	49.7	45.9	18.1	4.0	33.7	44.0	95.8	97.1	169.1	145.1	314.5	56.1	44.3
1985	51.7	46.3	19.7	4.7	36.6	44.1	98.0	99.2	166.9	143.8	306.4	60.4	47.8
1986	49.7	45.5	21.1	5.3	38.1	48.7	110.1	111.6	186.3	164.6	315.4	68.7	51.7
1987	47.3	46.8	21.9	6.1	38.8	47.9	112.3	113.8	188.8	177.9	248.9	70.6	53.8
1988	50.0	49.0	23.5	7.0	40.9	48.4	111.3	112.7	182.1	175.7	212.4	72.5	55.7
1989	45.1	54.7	26.2	9.0	43.7	50.6	107.7	108.9	177.6	170.7	211.8	69.3	56.0
1990	44.5	52.7	28.4	10.5	46.5	52.4	98.6	99.5	160.4	155.1	184.8	64.3	54.3
1991	43.3	45.0	30.3	11.3	49.9	52.9	89.8	90.5	138.1	135.9	142.4	62.5	52.2
1992	44.3	44.8	32.0	13.0	50.8	55.5	102.2	103.3	161.0	165.1	121.0	69.5	53.8
1993	52.8	49.0	33.4	14.7	50.6	58.5	110.5	111.8	171.9	180.8	97.6	76.6	54.8
1994	61.8	53.0	34.7	16.1	50.9	61.6	120.4	122.1	192.4	200.1	126.1	81.1	55.2
1995	66.5	55.8	37.1	17.6	53.7	66.7	116.3	117.8	179.7	181.6	156.3	81.5	55.9
1996	68.7	57.9	41.3	20.6	58.8	71.5	125.9	127.7	196.6	198.4	172.8	87.2	55.6
1997	73.7	63.5	46.7	26.9	62.8	73.6	128.9	130.8	197.7	198.8	186.1	91.4	55.2
1998	79.7	70.0	51.8	32.6	67.1	77.1	140.0	142.2	216.5	219.4	184.1	98.5	59.1
1999	94.0	71.9	58.6	40.6	72.4	79.4	148.8	151.0	232.5	235.6	198.7	103.2	65.3
2000	92.1	75.9	64.1	45.8	78.3	84.1	149.8	151.9	234.4	238.2	195.3	103.4	69.4
2001	83.4	75.3	64.8	46.3	79.9	81.5	151.2	153.2	235.8	238.5	203.8	104.7	70.7
2002	75.4	74.7	64.3	46.1	77.9	85.6	160.3	162.6	245.3	247.8	214.5	113.8	72.8
2003	69.8	79.8	66.8	49.1	78.5	92.0	174.9	177.4	269.3	274.2	220.2	123.3	77.3
2004	79.5	85.6	70.0	54.9	79.2	93.4	192.3	195.1	300.5	307.1	237.3	133.1	86.2
2005	89.2	93.9	75.2	60.1	83.5	101.2	204.9	207.9	321.7	327.8	260.9	141.1	90.0
2006	95.9	97.8	79.5	63.8	89.2	102.3	189.5	192.1	293.6	295.8	262.6	132.4	91.7
2007	91.8	96.6	84.5	70.2	94.5	99.8	154.1	155.8	218.5	214.7	232.6	119.0	89.1
2008	72.0	91.9	87.7	76.0	97.2	94.1	116.8	117.6	145.1	134.9	204.5	101.5	86.3
2009	33.7	73.3	87.3	78.2	94.9	90.4	91.5	91.8	87.0	79.7	130.1	94.5	80.1
2010	65.7	82.5	89.7	81.2	95.2	99.0	88.7	88.7	83.5	86.6	65.3	91.7	87.9
2011	84.5	91.4	95.3	90.1	99.2	98.2	88.5	88.4	80.6	82.6	68.2	92.9	95.4
2012	100.0	100.0	100.0	100.0	100.0	100.0	100.0	100.0	100.0	100.0	100.0	100.0	100.0
2013	110.8	102.2	105.4	105.6	106.5	100.2	112.4	112.5	124.5	122.6	136.0	105.5	107.9
2014	123.1	110.3	110.5	112.2	110.7	103.6	116.7	116.7	136.1	130.1	172.3	105.4	117.3
2015	136.2	103.0	114.5	117.5	113.5	107.5	128.5	128.6	152.2	145.0	195.6	114.9	127.4
2016	128.7	97.6	123.5	127.2	123.4	110.6	136.8	136.8	161.6	152.5	216.3	122.5	138.3
2017	122.3	105.3	128.1	137.4	123.9	113.4	141.6	141.5	167.8	162.6	200.3	126.3	148.2
2018	130.1	111.2	137.5	152.0	130.5	118.7	139.5	139.4	167.1	164.1	187.7	123.3	149.5
2016													
1st quarter	133.7	98.5	120.8	123.8	120.8	109.3	137.3	137.3	164.2	155.1	218.4	121.7	135.3
2nd quarter	130.5	97.1	123.5	126.0	124.7	109.1	136.6	136.6	162.5	153.2	217.7	121.6	136.2
3rd quarter	126.7	97.4	124.9	128.5	125.0	111.4	135.7	135.6	158.3	148.3	217.2	122.6	139.9
4th quarter	123.7	97.4	124.9	130.5	123.2	112.6	137.8	137.8	161.6	153.2	211.8	124.0	141.7
2017													
1st quarter	122.3	100.6	126.9	134.0	124.3	112.8	141.8	141.7	165.2	158.1	208.6	128.1	145.5
2nd quarter	120.6	104.2	126.9	135.7	123.1	112.8	141.0	140.9	167.0	161.4	201.6	125.7	146.2
3rd quarter	121.0	106.7	128.5	139.0	123.6	113.2	140.2	140.1	168.0	164.0	194.3	123.9	148.7
4th quarter	125.3	109.6	129.9	141.0	124.7	114.9	143.6	143.5	171.0	167.1	196.8	127.5	152.4
2018													
1st quarter	127.9	109.5	133.0	146.8	126.1	116.2	141.7	141.5	170.8	168.4	187.8	124.5	153.3
2nd quarter	129.8	111.4	136.8	150.4	130.5	117.9	140.3	140.2	169.2	166.6	187.8	123.3	150.6
3rd quarter	128.2	110.8	138.2	153.4	130.7	119.2	138.9	138.8	166.5	164.1	183.8	122.7	148.6
4th quarter	134.5	113.0	142.0	157.3	134.9	121.4	137.3	137.1	161.9	157.2	191.5	122.8	145.5

[2] Excludes software "embedded," or bundled, in computers and other equipment.
[3] Excludes software development.

Table 5-5A. Current-Cost Net Stock of Fixed Assets: Recent Data

(Billions of dollars, year-end estimates.)

Year	Total	Private Total	Private Nonresidential Total	Private Nonresidential Equipment	Private Nonresidential Structures	Private Nonresidential Intellectual property products	Private Residential	Government Total	Government Nonresidential Total	Government Nonresidential Equipment	Government Nonresidential Structures	Government Nonresidential Intellectual property products	Government Residential	Federal	State and local
1950	872.1	655.8	335.9	98.8	224.2	12.8	319.9	216.3	210.0	38.4	163.8	7.7	6.3	106.6	109.6
1951	957.4	713.9	366.2	109.1	243.6	13.6	347.7	243.5	235.7	42.4	184.6	8.8	7.7	120.3	123.2
1952	1 013.0	751.9	386.3	114.9	256.8	14.6	365.6	261.1	253.9	49.0	195.3	9.6	7.2	130.5	130.6
1953	1 051.3	783.3	403.7	124.4	263.1	16.2	379.5	268.1	260.5	56.1	193.8	10.5	7.6	138.7	129.4
1954	1 100.0	815.2	414.9	128.2	269.1	17.7	400.3	284.8	274.5	62.8	200.1	11.6	10.3	149.6	135.2
1955	1 197.2	888.8	457.0	141.2	296.3	19.6	431.7	308.4	301.1	68.3	219.5	13.3	7.3	158.3	150.0
1956	1 301.6	958.5	505.5	158.6	324.9	22.1	452.9	343.1	334.5	72.4	246.2	15.9	8.6	173.1	170.0
1957	1 369.5	1 008.9	540.6	172.9	343.3	24.4	468.3	360.6	351.6	74.7	257.9	19.0	9.0	182.4	178.2
1958	1 415.1	1 034.0	551.3	178.0	346.9	26.4	482.6	381.1	371.3	76.2	273.0	22.1	9.8	191.4	189.7
1959	1 471.2	1 078.1	573.9	186.6	358.4	29.0	504.1	393.1	382.4	80.5	277.0	25.0	10.7	198.0	195.1
1960	1 520.5	1 111.0	586.9	192.7	363.0	31.1	524.1	409.6	398.2	83.5	286.7	28.1	11.3	205.2	204.4
1961	1 580.2	1 147.6	604.3	196.1	374.6	33.7	543.3	432.6	420.5	87.5	301.4	31.6	12.1	215.7	216.8
1962	1 652.8	1 190.4	626.8	203.7	387.2	35.9	563.6	462.4	449.5	94.6	319.6	35.3	12.9	230.3	232.2
1963	1 717.5	1 228.7	650.6	212.3	399.5	38.9	578.1	488.8	475.6	96.7	338.3	40.6	13.2	240.7	248.1
1964	1 829.7	1 314.0	689.3	224.9	422.5	41.9	624.6	515.7	501.8	98.9	356.6	46.3	14.0	251.0	264.8
1965	1 954.3	1 402.9	738.8	242.5	450.6	45.8	664.0	551.5	536.8	100.6	384.0	52.2	14.7	263.0	288.5
1966	2 119.9	1 522.1	804.3	270.0	483.7	50.5	717.8	597.8	582.0	104.3	418.6	59.1	15.7	278.9	318.9
1967	2 285.4	1 636.6	871.2	296.2	518.7	56.3	765.3	648.8	632.3	110.2	455.6	66.6	16.5	299.8	349.0
1968	2 511.4	1 804.7	957.9	326.5	568.5	62.8	846.9	706.6	688.1	114.1	499.0	75.1	18.5	319.6	387.1
1969	2 744.0	1 962.9	1 055.8	360.1	625.2	70.5	907.1	781.1	760.5	117.7	558.0	84.8	20.6	343.0	438.0
1970	2 990.8	2 121.0	1 161.7	394.8	689.4	77.5	959.3	869.8	847.8	123.2	629.9	94.7	21.9	370.2	499.6
1971	3 294.5	2 352.7	1 275.1	422.5	770.1	82.5	1 077.6	941.9	917.3	123.3	691.3	102.6	24.6	391.2	550.7
1972	3 621.8	2 594.0	1 385.7	455.1	841.9	88.7	1 208.2	1 027.9	1 000.3	127.7	761.8	110.8	27.5	424.7	603.1
1973	4 110.6	2 946.9	1 560.5	507.1	955.0	98.4	1 386.4	1 163.7	1 132.6	135.9	872.9	123.8	31.1	470.8	692.9
1974	4 887.7	3 468.3	1 893.8	625.5	1 156.4	111.9	1 574.5	1 419.4	1 384.7	149.1	1 096.7	138.9	34.7	539.7	879.7
1975	5 277.3	3 786.4	2 083.8	716.2	1 246.2	121.5	1 702.6	1 490.9	1 452.8	165.0	1 138.2	149.6	38.0	572.1	918.8
1976	5 747.3	4 168.9	2 280.1	792.5	1 354.7	132.9	1 888.7	1 578.5	1 536.0	180.7	1 194.5	160.8	42.5	619.5	958.9
1977	6 415.5	4 735.7	2 532.5	891.6	1 495.1	145.8	2 203.2	1 679.7	1 630.9	198.3	1 258.6	174.1	48.9	656.0	1 023.7
1978	7 261.1	5 412.9	2 870.3	1 018.0	1 689.2	163.2	2 542.5	1 848.2	1 791.7	222.5	1 378.7	190.6	56.5	722.1	1 126.1
1979	8 360.7	6 264.5	3 313.9	1 181.9	1 945.2	186.8	2 950.5	2 096.2	2 028.9	236.7	1 578.9	213.3	67.3	800.2	1 296.0
1980	9 512.5	7 118.0	3 800.4	1 369.9	2 216.6	213.8	3 317.6	2 394.5	2 320.8	260.3	1 819.6	240.9	73.7	888.6	1 505.9
1981	10 490.4	7 860.3	4 300.0	1 528.5	2 526.7	244.9	3 560.2	2 630.1	2 549.3	286.7	1 992.5	270.1	80.8	960.0	1 670.1
1982	11 083.9	8 297.0	4 588.9	1 619.9	2 694.9	274.1	3 708.0	2 787.0	2 701.3	315.1	2 092.4	293.8	85.7	1 024.5	1 762.5
1983	11 468.9	8 599.6	4 744.7	1 670.5	2 770.2	304.1	3 854.8	2 869.3	2 773.3	345.2	2 110.8	317.3	96.0	1 084.6	1 784.7
1984	12 140.1	9 112.5	5 041.0	1 758.8	2 943.8	338.5	4 071.5	3 027.5	2 928.0	404.3	2 182.7	341.0	99.5	1 182.2	1 845.4
1985	12 759.0	9 619.0	5 332.4	1 858.0	3 101.8	372.6	4 286.6	3 140.0	3 039.6	400.2	2 274.6	364.7	100.4	1 210.4	1 929.6
1986	13 548.2	10 230.6	5 603.6	1 964.7	3 232.5	406.4	4 627.0	3 317.6	3 213.7	413.1	2 415.0	385.6	103.9	1 260.1	2 057.5
1987	14 342.5	10 843.5	5 919.1	2 055.3	3 415.1	448.7	4 924.4	3 499.0	3 385.6	431.0	2 538.2	416.3	113.4	1 318.2	2 180.8
1988	15 253.0	11 560.9	6 335.9	2 178.1	3 641.8	516.1	5 224.9	3 692.1	3 563.5	466.3	2 648.8	448.4	128.6	1 405.8	2 286.3
1989	16 120.6	12 230.0	6 726.7	2 300.6	3 858.7	567.4	5 503.2	3 890.7	3 753.2	498.4	2 780.9	474.0	137.4	1 477.7	2 413.0
1990	16 885.5	12 802.8	7 100.5	2 423.7	4 055.4	621.4	5 702.3	4 082.8	3 940.9	534.7	2 907.9	498.2	141.9	1 541.0	2 541.8
1991	17 305.0	13 090.9	7 274.6	2 482.2	4 116.7	675.6	5 816.3	4 214.1	4 071.3	559.1	2 994.6	517.6	142.8	1 592.5	2 621.5
1992	18 034.2	13 643.6	7 521.7	2 544.3	4 251.0	726.4	6 121.9	4 390.6	4 239.5	585.5	3 120.0	534.0	151.1	1 652.1	2 738.4
1993	18 937.5	14 358.0	7 874.3	2 642.4	4 458.5	773.4	6 483.7	4 579.6	4 417.8	608.9	3 262.5	546.4	161.8	1 704.8	2 874.8
1994	20 066.8	15 242.3	8 312.5	2 784.9	4 696.8	830.7	6 929.8	4 824.5	4 652.0	634.4	3 452.8	564.8	172.6	1 770.4	3 054.1
1995	21 042.0	15 993.5	8 765.6	2 959.2	4 913.2	893.2	7 228.0	5 048.5	4 869.7	641.7	3 648.7	579.2	178.8	1 811.0	3 237.5
1996	22 046.6	16 812.9	9 197.5	3 104.1	5 136.2	957.2	7 615.4	5 233.7	5 047.1	633.9	3 826.7	586.4	186.6	1 833.8	3 399.9
1997	23 219.0	17 752.7	9 723.7	3 235.2	5 445.7	1 042.8	8 029.0	5 466.3	5 271.2	629.5	4 043.3	598.5	195.1	1 864.3	3 602.0
1998	24 525.4	18 828.3	10 279.7	3 384.7	5 759.2	1 135.8	8 548.6	5 697.1	5 492.1	639.3	4 240.2	612.6	205.0	1 902.5	3 794.6
1999	26 101.0	20 085.2	10 907.6	3 578.7	6 063.8	1 265.0	9 177.6	6 015.7	5 798.0	660.8	4 503.0	634.2	217.8	1 967.3	4 048.4
2000	27 823.6	21 482.6	11 672.4	3 805.2	6 468.4	1 398.7	9 810.2	6 341.0	6 111.4	655.2	4 799.9	656.3	229.6	2 003.8	4 337.2
2001	29 376.5	22 772.5	12 255.5	3 913.4	6 880.7	1 461.4	10 517.0	6 604.0	6 359.8	650.9	5 040.4	668.5	244.2	2 026.4	4 577.6
2002	30 805.4	23 906.8	12 684.4	3 959.3	7 212.5	1 512.5	11 222.4	6 898.7	6 641.1	666.8	5 285.2	689.1	257.6	2 077.6	4 821.1
2003	32 468.3	25 270.5	13 109.5	4 000.4	7 514.3	1 594.9	12 161.0	7 197.7	6 922.0	685.7	5 510.1	726.2	275.7	2 149.8	5 047.9
2004	35 787.7	27 811.3	14 227.1	4 225.9	8 334.0	1 667.2	13 584.3	7 976.3	7 673.3	711.2	6 199.5	762.6	303.0	2 289.1	5 687.2
2005	39 374.9	30 662.1	15 511.7	4 423.0	9 311.4	1 777.3	15 150.4	8 712.8	8 380.8	735.3	6 842.1	803.4	332.0	2 432.9	6 279.9
2006	42 596.0	32 986.8	16 793.6	4 729.2	10 180.9	1 883.6	16 193.2	9 609.2	9 262.1	768.0	7 654.8	839.3	347.0	2 579.5	7 029.7
2007	44 509.5	34 154.6	17 747.4	4 952.1	10 772.1	2 023.2	16 407.2	10 354.9	10 010.1	803.6	8 316.4	890.1	344.8	2 709.8	7 645.1
2008	46 019.3	34 981.3	18 885.6	5 200.2	11 557.2	2 128.2	16 095.7	11 038.0	10 702.1	855.2	8 919.1	927.8	335.9	2 828.7	8 209.3
2009	45 191.4	34 101.1	18 343.1	5 120.2	11 029.7	2 193.2	15 758.0	11 090.2	10 760.4	885.7	8 918.4	956.4	329.8	2 838.4	8 251.8
2010	46 099.4	34 582.2	18 799.2	5 224.0	11 286.3	2 288.9	15 783.0	11 517.2	11 184.7	923.4	9 249.1	1 012.2	332.5	2 955.8	8 561.4
2011	47 690.3	35 557.7	19 665.1	5 459.7	11 799.6	2 405.8	15 892.7	12 132.5	11 795.5	954.8	9 800.5	1 040.2	337.0	3 061.3	9 071.2
2012	49 215.9	36 693.1	20 340.8	5 699.6	12 126.7	2 514.4	16 352.4	12 522.8	12 175.1	969.0	10 142.0	1 064.1	347.6	3 126.7	9 396.0
2013	51 642.8	38 699.6	21 205.3	5 897.2	12 641.7	2 666.5	17 494.3	12 943.2	12 572.0	979.9	10 492.1	1 100.0	371.1	3 204.8	9 738.4
2014	53 969.7	40 732.9	22 166.5	6 133.3	13 248.1	2 785.1	18 566.4	13 236.8	12 852.3	990.2	10 750.9	1 111.2	384.5	3 244.0	9 992.9
2015	55 079.2	41 651.8	22 637.3	6 298.4	13 453.4	2 885.5	19 014.5	13 427.4	13 035.0	988.1	10 932.4	1 114.5	392.4	3 243.6	10 183.8
2016	57 251.2	43 442.8	23 210.5	6 447.1	13 732.6	3 030.9	20 232.3	13 808.3	13 392.9	997.4	11 256.6	1 138.8	415.5	3 301.7	10 506.7
2017	59 580.2	45 272.4	24 119.6	6 679.0	14 222.1	3 218.4	21 152.8	14 307.9	13 879.4	1 009.2	11 697.4	1 172.8	428.5	3 370.6	10 937.3

Table 5-5B. Current-Cost Net Stock of Fixed Assets: Historical Data

(Billions of dollars, year-end estimates.)

Year	Total	Private						Government							
		Total	Nonresidential				Resi-dential	Total	Nonresidential				Resi-dential	Federal	State and local
			Total	Equip-ment	Struc-tures	Intellect-ual property products			Total	Equip-ment	Struc-tures	Intellect-ual property products			
1925	261.7	223.4	127.0	31.1	94.0	1.9	96.4	38.3	38.3	2.6	35.6	0.1	0.0	8.9	29.4
1926	270.5	231.6	131.5	32.8	96.7	2.0	100.0	38.9	38.9	2.6	36.2	0.1	0.0	8.8	30.1
1927	277.2	236.9	134.2	33.5	98.5	2.1	102.7	40.3	40.3	2.6	37.7	0.1	0.0	8.6	31.7
1928	289.2	248.6	137.8	34.2	101.3	2.3	110.8	40.7	40.7	2.5	38.0	0.1	0.0	8.3	32.4
1929	294.4	251.7	137.1	34.3	100.3	2.5	114.6	42.7	42.7	2.5	40.0	0.2	0.0	8.9	33.8
1930	281.7	239.4	130.2	32.7	95.0	2.5	109.2	42.3	42.3	2.4	39.7	0.2	0.0	8.4	33.9
1931	245.5	205.4	116.2	29.7	84.0	2.5	89.2	40.0	40.0	2.4	37.4	0.3	0.0	7.8	32.3
1932	222.3	187.6	107.6	26.7	78.6	2.4	79.9	34.7	34.7	2.3	32.1	0.3	0.0	7.5	27.2
1933	236.3	196.5	109.3	26.4	80.5	2.4	87.2	39.8	39.8	2.3	37.1	0.4	0.0	8.5	31.3
1934	246.0	198.6	110.6	26.5	81.6	2.6	87.9	47.4	47.4	2.5	44.5	0.4	0.0	9.4	38.0
1935	247.3	199.5	110.3	25.9	81.6	2.8	89.2	47.7	47.7	2.6	44.6	0.4	0.0	10.5	37.2
1936	275.1	220.4	121.4	27.8	90.4	3.1	99.1	54.7	54.5	2.7	51.3	0.5	0.1	12.1	42.5
1937	289.9	231.7	126.5	29.9	93.1	3.4	105.2	58.1	57.9	3.0	54.4	0.5	0.3	13.4	44.7
1938	292.7	232.0	125.2	29.9	91.6	3.7	106.8	60.7	60.4	3.1	56.7	0.6	0.3	14.2	46.5
1939	298.2	235.6	125.5	30.5	91.0	4.0	110.1	62.6	62.2	3.3	58.3	0.6	0.4	14.8	47.7
1940	320.5	254.4	133.6	32.8	96.5	4.3	120.8	66.1	65.6	3.7	61.3	0.6	0.6	16.8	49.3
1941	365.5	282.8	151.0	37.6	108.2	5.1	131.9	82.7	81.6	7.1	73.6	0.9	1.1	27.1	55.6
1942	424.9	302.3	160.5	38.5	115.9	6.0	141.8	122.5	120.6	21.0	98.3	1.3	1.9	57.5	65.0
1943	478.7	317.1	163.3	38.9	117.6	6.8	153.8	161.6	158.6	46.5	109.9	2.2	3.0	92.5	69.1
1944	513.4	329.3	165.2	38.8	110.0	7.5	164.1	184.0	180.7	67.4	109.8	3.5	3.3	117.0	67.1
1945	554.9	353.0	180.2	44.2	127.9	8.1	172.8	201.9	198.3	78.5	115.5	4.3	3.6	135.7	66.2
1946	639.7	434.0	221.0	53.3	158.6	9.1	213.0	205.7	201.4	70.7	125.7	5.0	4.3	132.7	73.0
1947	734.2	516.6	265.7	65.5	189.6	10.6	250.9	217.6	212.0	60.0	146.4	5.5	5.6	124.2	93.3
1948	780.0	562.2	291.6	80.3	199.9	11.3	270.6	217.9	212.9	50.2	156.6	6.2	5.0	116.2	101.7
1949	785.1	581.1	296.7	84.6	200.4	11.7	284.4	204.0	198.9	42.4	149.7	6.8	5.1	106.7	97.2

Table 5-6A. Chain-Type Quantity Indexes for Net Stock of Fixed Assets: Recent Data

(Index numbers, 2012 = 100.)

Year	Total	Private						Government							
		Total	Nonresidential				Resi-dential	Total	Nonresidential				Resi-dential	Federal	State and local
			Total	Equip-ment	Struc-tures	Intellect-ual property products			Total	Equip-ment	Struc-tures	Intellect-ual property products			
1950	17.3	16.4	14.5	8.3	24.0	3.1	18.9	20.5	20.6	28.2	22.3	4.7	17.1	34.7	14.5
1951	18.0	17.0	15.0	8.8	24.6	3.2	19.6	21.4	21.4	30.3	23.0	5.0	18.3	36.2	15.1
1952	18.7	17.5	15.5	9.2	25.1	3.4	20.4	22.5	22.6	34.2	23.8	5.4	19.7	38.9	15.6
1953	19.4	18.2	16.0	9.7	25.8	3.7	21.1	23.8	23.9	38.4	24.7	5.9	20.9	41.7	16.2
1954	20.2	18.8	16.5	10.0	26.5	4.0	21.9	24.9	25.0	40.5	25.7	6.4	21.5	43.5	17.0
1955	21.0	19.6	17.1	10.5	27.2	4.3	22.9	25.8	25.9	41.3	26.8	7.1	21.9	44.5	17.9
1956	21.8	20.3	17.8	11.0	28.1	4.6	23.8	26.8	26.9	41.8	27.8	8.3	22.5	45.7	18.8
1957	22.5	21.0	18.4	11.4	28.9	5.0	24.5	27.8	27.9	42.5	28.8	9.6	23.5	46.9	19.7
1958	23.2	21.6	18.8	11.6	29.6	5.3	25.3	29.0	29.1	43.2	30.0	10.8	25.4	48.4	20.8
1959	24.1	22.3	19.3	11.9	30.4	5.6	26.3	30.3	30.4	45.0	31.2	12.1	27.5	50.3	21.9
1960	24.9	23.0	19.9	12.2	31.2	6.0	27.3	31.5	31.6	46.3	32.4	13.5	29.1	52.0	23.0
1961	25.8	23.7	20.5	12.5	32.0	6.4	28.2	33.0	33.1	48.3	33.7	15.0	31.0	54.1	24.1
1962	26.7	24.5	21.1	12.9	32.9	6.8	29.2	34.5	34.6	50.4	35.0	16.8	33.1	56.5	25.3
1963	27.8	25.4	21.8	13.4	33.8	7.2	30.3	36.1	36.1	51.6	36.4	19.2	34.2	58.5	26.7
1964	28.9	26.4	22.7	14.1	34.9	7.6	31.6	37.6	37.7	52.4	37.9	21.8	35.3	60.4	28.1
1965	30.1	27.5	23.8	15.1	36.2	8.2	32.7	39.2	39.2	52.5	39.4	24.4	36.6	62.0	29.6
1966	31.5	28.7	25.2	16.4	37.6	8.8	33.7	40.9	41.0	53.2	41.1	27.5	37.9	64.1	31.3
1967	32.7	29.8	26.4	17.4	39.0	9.5	34.6	42.8	42.8	54.4	42.8	30.5	39.4	66.1	33.0
1968	34.0	31.0	27.6	18.6	40.3	10.1	35.7	44.4	44.5	54.4	44.5	33.4	40.9	67.4	34.8
1969	35.3	32.3	28.9	19.8	41.7	10.8	36.8	45.8	45.9	53.9	46.0	36.0	42.7	68.3	36.5
1970	36.4	33.4	30.0	20.8	43.1	11.3	37.8	47.0	47.0	53.2	47.3	37.8	44.6	68.7	38.0
1971	37.5	34.5	31.1	21.6	44.4	11.6	39.3	47.9	47.9	51.0	48.6	39.3	46.4	68.3	39.3
1972	38.8	36.0	32.3	22.8	45.7	12.0	41.0	48.7	48.7	49.2	49.7	40.8	48.0	68.1	40.6
1973	40.2	37.5	33.7	24.4	47.2	12.5	42.7	49.5	49.5	47.7	50.8	42.1	49.5	68.0	41.8
1974	41.3	38.8	35.1	25.9	48.6	12.9	43.8	50.4	50.3	47.2	51.9	42.9	50.9	68.0	43.0
1975	42.2	39.7	36.0	26.8	49.6	13.2	44.7	51.2	51.1	46.8	52.8	43.7	52.6	68.0	44.1
1976	43.2	40.8	36.9	27.7	50.7	13.8	45.9	52.0	52.0	46.7	53.8	44.6	53.9	68.3	45.2
1977	44.5	42.2	38.2	29.1	51.8	14.4	47.6	52.8	52.7	46.7	54.6	45.7	55.1	68.6	46.2
1978	46.0	43.8	39.8	31.0	53.2	15.1	49.3	53.7	53.6	46.5	55.6	47.0	56.2	69.0	47.3
1979	47.6	45.6	41.6	33.0	54.9	16.1	50.9	54.7	54.6	46.9	56.7	48.4	57.2	69.6	48.5
1980	48.9	46.9	43.2	34.4	56.8	17.0	51.9	55.7	55.6	47.4	57.7	49.9	58.4	70.3	49.6
1981	50.1	48.3	44.9	35.8	59.0	18.2	52.7	56.6	56.5	48.2	58.5	51.5	59.9	71.2	50.5
1982	51.1	49.3	46.3	36.5	61.0	19.4	53.2	57.5	57.3	49.3	59.2	53.0	61.3	72.3	51.3
1983	52.2	50.4	47.5	37.2	62.5	20.6	54.3	58.4	58.3	51.1	59.8	54.7	63.0	73.7	52.0
1984	53.7	52.1	49.3	38.7	64.5	22.3	55.8	59.6	59.4	53.2	60.6	56.8	64.4	75.3	53.0
1985	55.4	53.8	51.2	40.2	66.8	24.1	57.2	61.0	60.9	56.0	61.6	59.5	66.2	77.4	54.2
1986	57.0	55.5	52.8	41.3	68.6	25.8	58.9	62.6	62.5	59.6	62.6	62.1	68.0	79.9	55.4
1987	58.6	57.0	54.2	42.2	70.3	27.5	60.7	64.3	64.1	63.3	63.8	64.9	70.0	82.5	56.7
1988	60.2	58.6	55.7	43.3	71.9	29.2	62.4	65.8	65.6	65.9	64.9	67.1	71.7	84.2	58.1
1989	61.7	60.1	57.2	44.5	73.5	31.2	63.8	67.2	67.0	68.7	66.0	68.8	73.3	85.6	59.5
1990	63.1	61.5	58.7	45.4	75.3	33.3	65.1	68.7	68.5	71.6	67.2	70.4	74.9	87.0	61.1
1991	64.2	62.5	59.8	45.9	76.5	35.5	66.0	70.1	69.9	73.7	68.5	71.3	76.4	87.9	62.7
1992	65.3	63.6	60.8	46.5	77.3	37.6	67.2	71.4	71.1	75.4	69.8	72.0	77.9	88.5	64.2
1993	66.6	64.9	62.1	47.9	78.2	39.5	68.6	72.4	72.2	76.1	71.1	72.3	79.3	88.7	65.7
1994	68.0	66.5	63.6	49.8	79.1	41.3	70.3	73.3	73.1	76.0	72.3	72.4	80.4	88.4	67.1
1995	69.5	68.2	65.4	52.1	80.2	43.3	71.8	74.4	74.1	76.0	73.6	72.4	81.8	88.2	68.7
1996	71.4	70.2	67.6	54.8	81.6	45.9	73.6	75.5	75.3	76.0	75.1	72.5	83.2	88.4	70.3
1997	73.3	72.4	70.1	57.8	83.2	49.3	75.4	76.6	76.3	75.5	76.6	72.8	84.6	87.9	72.0
1998	75.5	74.9	72.9	61.4	84.9	53.0	77.5	77.8	77.5	75.4	78.1	73.3	85.9	87.5	73.9
1999	77.9	77.6	75.9	65.5	86.6	57.3	79.7	79.2	78.9	75.8	79.7	73.9	87.1	87.3	75.9
2000	80.3	80.4	79.2	69.8	88.6	61.9	81.9	80.6	80.3	76.1	81.3	74.7	88.2	87.1	78.0
2001	82.5	82.7	81.7	72.6	90.3	65.5	84.0	82.1	81.8	76.4	83.1	76.0	89.3	87.0	80.2
2002	84.3	84.5	83.2	74.4	91.3	68.1	86.3	83.9	83.7	77.6	85.0	77.9	90.6	87.6	82.5
2003	86.3	86.6	84.7	76.3	92.2	70.7	89.0	85.8	85.6	78.7	86.9	80.2	91.7	88.5	84.8
2004	88.5	88.8	86.3	78.8	92.9	73.3	91.8	87.6	87.4	80.3	88.7	82.7	92.9	89.6	86.9
2005	90.6	91.0	88.1	82.0	93.6	76.5	94.6	89.2	89.0	82.3	90.1	85.4	93.8	90.7	88.6
2006	92.9	93.5	90.4	86.0	94.7	80.0	97.3	90.9	90.8	84.8	91.7	88.1	94.7	92.0	90.6
2007	95.0	95.7	92.9	90.0	96.3	83.9	99.1	92.7	92.6	87.8	93.3	90.6	95.5	93.3	92.5
2008	96.5	97.2	95.2	92.6	98.0	87.7	99.7	94.5	94.4	91.4	94.9	92.9	96.6	94.8	94.4
2009	97.2	97.6	95.8	91.9	98.8	90.5	99.8	96.2	96.1	94.5	96.4	95.0	97.6	96.4	96.1
2010	98.0	98.1	96.7	93.3	99.0	93.1	99.8	97.8	97.7	97.1	97.9	97.0	98.8	98.1	97.7
2011	98.9	98.9	98.1	96.1	99.4	96.5	99.8	99.1	99.1	98.8	99.1	98.7	99.7	99.3	99.0
2012	100.0	100.0	100.0	100.0	100.0	100.0	100.0	100.0	100.0	100.0	100.0	100.0	100.0	100.0	100.0
2013	101.2	101.4	102.0	104.0	100.8	103.8	100.7	100.7	100.8	100.2	100.8	101.0	100.0	100.0	101.0
2014	102.6	103.1	104.4	108.5	101.9	107.7	101.4	101.4	101.4	100.0	101.5	101.5	100.0	99.8	101.9
2015	104.1	104.7	106.6	112.7	102.9	111.5	102.5	102.1	102.2	100.0	102.4	102.1	100.2	99.5	103.0
2016	105.5	106.3	108.5	116.0	103.6	116.2	103.7	102.9	103.0	100.4	103.2	102.7	100.6	99.3	104.1
2017	106.7	107.8	110.5	119.7	104.4	120.9	104.6	103.6	103.7	101.2	103.9	103.7	100.9	99.5	105.0

Table 5-6B. Chain-Type Quantity Indexes for Net Stock of Fixed Assets: Historical Data

(Index numbers, 2012 = 100.)

Year	Total	Private						Government							
		Total	Nonresidential				Resi-dential	Total	Nonresidential				Resi-dential	Federal	State and local
			Total	Equip-ment	Struc-tures	Intellect-ual property products			Total	Equip-ment	Struc-tures	Intellect-ual property products			
1925	10.6	11.5	10.6	4.8	19.8	0.7	13.1	5.5	5.7	2.5	8.0	0.10	0.00	4.5	6.7
1926	11.0	12.0	10.9	5.0	20.4	0.7	13.7	5.8	6.0	2.4	8.3	0.10	0.00	4.4	7.0
1927	11.4	12.4	11.2	5.1	21.0	0.8	14.2	6.0	6.2	2.4	8.7	0.10	0.00	4.3	7.5
1928	11.8	12.8	11.5	5.2	21.6	0.8	14.7	6.3	6.5	2.4	9.2	0.10	0.00	4.3	7.9
1929	12.1	13.2	11.9	5.4	22.2	0.9	15.0	6.6	6.8	2.5	9.6	0.10	0.00	4.3	8.4
1930	12.4	13.3	12.1	5.4	22.7	1.0	15.0	7.0	7.2	2.4	10.2	0.20	0.00	4.3	9.0
1931	12.5	13.3	12.0	5.2	22.8	1.1	15.1	7.3	7.6	2.5	10.7	0.30	0.10	4.4	9.6
1932	12.4	13.2	11.8	4.9	22.6	1.1	15.0	7.6	7.9	2.4	11.2	0.30	0.10	4.6	10.0
1933	12.4	13.0	11.6	4.7	22.4	1.2	14.9	7.9	8.1	2.4	11.5	0.40	0.10	4.8	10.2
1934	12.4	12.9	11.4	4.5	22.2	1.2	14.8	8.1	8.4	2.4	11.9	0.40	0.10	5.1	10.4
1935	12.4	12.9	11.3	4.5	22.0	1.3	14.8	8.4	8.7	2.6	12.3	0.40	0.20	5.6	10.7
1936	12.6	12.9	11.4	4.6	22.0	1.4	14.8	8.9	9.2	2.6	13.1	0.50	0.70	6.1	11.2
1937	12.8	13.0	11.6	4.8	22.1	1.5	14.9	9.3	9.6	2.7	13.6	0.50	1.30	6.5	11.6
1938	12.9	13.1	11.6	4.7	22.0	1.7	15.0	9.8	10.0	2.9	14.3	0.50	1.50	6.9	12.1
1939	13.2	13.2	11.6	4.8	22.0	1.8	15.2	10.3	10.5	3.1	15.0	0.60	1.90	7.3	12.8
1940	13.4	13.3	11.8	5.0	22.1	1.9	15.4	10.8	11.0	3.2	15.7	0.60	2.80	7.9	13.2
1941	13.9	13.6	12.0	5.2	22.2	2.1	15.7	12.3	12.5	6.1	17.0	0.90	5.20	11.9	13.5
1942	14.6	13.5	11.9	5.2	22.0	2.2	15.7	16.3	16.5	19.2	19.8	1.20	8.10	24.0	13.5
1943	15.2	13.4	11.8	5.1	21.8	2.3	15.6	20.8	21.0	45.0	20.9	1.70	12.00	38.7	13.4
1944	15.8	13.4	11.8	5.1	21.6	2.4	15.6	24.4	24.8	70.1	21.2	2.40	13.00	51.0	13.2
1945	16.1	13.5	12.0	5.4	21.6	2.5	15.5	25.8	26.2	79.2	21.4	3.00	13.40	55.8	13.1
1946	16.1	13.9	12.4	5.9	22.1	2.6	15.9	23.9	24.2	64.6	21.2	3.40	14.90	49.7	13.1
1947	16.2	14.5	13.0	6.7	22.5	2.8	16.5	22.3	22.5	50.9	21.1	3.70	15.30	43.8	13.3
1948	16.5	15.1	13.6	7.4	23.0	2.9	17.3	21.1	21.3	39.7	21.3	4.00	15.50	39.2	13.5
1949	16.8	15.7	14.0	7.8	23.4	3.0	17.9	20.8	20.9	33.9	21.8	4.30	16.40	36.9	14.0

Table 5-7. Capital Expenditures

(Millions of dollars, except totals are shown in billions.)

Capital expenditures	All companies												
	2005	2006	2007	2008	2009	2010	2011	2012	2013	2014	2015	2016	2017
TOTAL (billions)	1 144.8	1 309.9	1 354.7	1 374.2	1 090.7	1 105.7	1 243.0	1 424.2	1 491.3	1 597.9	1 642.0	1 574.8	1 682.9
Structures	401 653	488 701	525 273	562 381	449 545	428 713	471 779	570 489	581 836	643 590	642 859	599 707	666 393
New	365 938	448 861	480 839	522 999	422 780	394 517	442 745	534 696	549 359	606 980	589 689	543 758	617 327
Used	35 715	39 840	44 434	39 382	26 765	34 196	29 034	35 793	32 477	36 610	53 171	55 948	49 066
Equipment	743 130	821 238	829 455	811 779	641 149	676 989	771 177	853 661	909 477	954 261	999 122	975 138	1 016 490
New	701 247	777 059	790 407	765 279	606 576	639 214	730 033	800 519	856 012	898 581	941 297	924 954	959 116
Used	41 884	44 179	39 048	46 501	34 572	37 775	41 144	53 142	53 466	55 680	57 825	50 184	57 377
Not distributed as structures or equipment	0	0	0	0	0	0	...	0	0	0	0	0	0
CAPITALIZED COMPUTER SOFTWARE [1]	...	...	...	...	...	...	...	...	...	...	...	...	...
Prepackaged	...	...	...	...	...	...	...	...	...	...	...	...	...
Vendor-customized	...	...	...	...	...	...	...	...	...	...	...	...	...
Internally-developed	...	...	...	...	...	...	...	...	...	...	...	...	...
CAPITAL LEASE AND CAPITALIZED INTEREST EXPENSES [1]													
Capital leases	18 103	24 442	20 210	20 169	17 410	15 780	20 509	26 087	26 252	31 114	35 350	36 219	36 943

Capital expenditures	Companies with employees												
	2005	2006	2007	2008	2009	2010	2011	2012	2013	2014	2015	2016	2017
TOTAL (billions)	1 062.6	1 217.1	1 270.5	1 294.5	1 015.3	1 036.2	1 169.6	1 334.9	1 400.9	1 506.6	1 548.1	1 479.4	1 581.8
Structures	368 791	453 893	490 779	529 393	414 051	395 531	440 893	533 115	546 056	608 698	600 766	552 637	612 776
New	341 223	420 090	457 233	500 474	395 022	366 853	414 958	501 420	518 430	577 743	557 641	508 523	576 937
Used	27 568	33 802	33 546	28 919	19 030	28 678	25 935	31 695	27 625	30 955	43 126	44 114	35 839
Equipment	693 745	763 215	779 744	765 098	601 270	640 631	728 710	801 820	854 827	897 885	947 291	926 802	969 063
New	664 648	734 160	750 353	728 322	577 051	612 441	697 766	759 632	813 972	853 888	900 519	887 884	923 313
Used	29 096	29 055	29 391	36 776	24 219	28 190	30 944	42 189	40 855	43 996	46 772	38 918	45 750
Not distributed as structures or equipment	0	0	0	0	...	0	0	0	0	0	0	0	0
CAPITALIZED COMPUTER SOFTWARE [1]	49 149	58 522	63 116	72 241	...	63 780	71 068	87 474	89 893	90 023	93 985	100 110	102 431
Prepackaged	17 630	21 181	21 777	26 260	...	21 749	23 242	27 868	28 986	27 901	26 759	27 296	27 297
Vendor-customized	13 876	16 912	17 990	19 259	...	17 264	20 346	23 507	24 207	23 638	25 547	26 003	25 894
Internally-developed	17 643	20 433	23 350	26 723	...	24 768	27 480	36 099	36 701	38 484	41 679	46 811	49 239
CAPITAL LEASE AND CAPITALIZED INTEREST EXPENSES [1]													
Capital leases	17 640	23 923	19 432	19 422	...	15 212	20 145	25 301	25 550	30 400	34 570	35 408	36 172

Capital expenditures	Companies without employees												
	2005	2006	2007	2008	2009	2010	2011	2012	2013	2014	2015	2016	2017
TOTAL (billions)	82.2	92.8	84.2	79.7	75.4	69.5	73.4	89.2	90.4	91.3	93.9	95.4	101.0
Structures	32 862	34 809	34 494	32 988	35 493	33 182	30 886	37 374	35 780	34 892	42 093	47 070	53 617
New	24 715	28 771	23 606	22 525	27 758	27 664	27 787	33 276	30 929	29 237	32 048	35 236	40 390
Used	8 146	6 038	10 888	10 463	7 735	5 518	3 099	4 098	4 851	5 655	10 045	11 834	13 227
Equipment	49 386	58 023	49 711	46 681	39 878	36 357	42 467	51 840	54 650	56 376	51 831	48 336	47 429
New	36 598	42 899	40 054	36 957	29 525	26 773	32 267	40 887	42 040	44 693	40 778	37 070	35 802
Used	12 787	15 124	9 657	9 724	10 353	9 585	10 200	10 953	12 610	11 684	11 053	11 266	11 627
Not distributed as structures or equipment	0	0	0	0	0	0	0	0	0	0	0	0	0
CAPITALIZED COMPUTER SOFTWARE [1]	...	...	...	...	...	...	...	...	...	...	...	...	...
Prepackaged	...	...	...	...	...	...	...	...	...	...	...	...	...
Vendor-customized	...	...	...	...	...	...	...	...	...	...	...	...	...
Internally-developed	...	...	...	...	...	...	...	...	...	...	...	...	...
CAPITAL LEASE AND CAPITALIZED INTEREST EXPENSES [1]													
Capital leases	463	519	778	747	577	568	365	786	702	714	781	812	771

[1] Included in structures and equipment data shown above.
. . . = Not available.

Table 5-8. Capital Expenditures for Structures and Equipment for Companies with Employees by Major Industry Sector

(Millions of dollars.)

Year and type of expenditure	Total	Forestry, fishing, and agricultural services (113–115)	Mining (21)	Utilities (22)	Construction (23)	Manufacturing (31–33) Total	Durable goods industries (321, 327, 33)	Nondurable goods industries (31, 322–326)	Wholesale trade (42)	Retail trade (44–45)	Transportation and warehousing (48–49)	Information (51)
2000												
Total expenditures	1 089 862	1 488	42 522	61 302	25 049	214 827	133 786	81 041	33 579	69 791	59 851	160 177
Structures, total	338 120	139	28 620	29 472	2 803	39 434	21 228	18 207	8 923	32 037	13 457	41 502
New	309 541	134	25 500	29 258	2 583	36 643	19 748	16 895	8 364	30 413	13 190	40 062
Used	28 579	5	3 120	214	220	2 791	1 480	1 312	559	1 624	267	1 440
Equipment, total	751 742	1 350	13 902	31 830	22 245	175 393	112 558	62 835	24 656	37 754	46 394	118 675
New	718 227	1 086	12 854	27 937	17 788	169 454	108 703	60 751	23 610	36 428	43 455	117 835
Used	33 515	264	1 048	3 893	4 458	5 939	3 856	2 083	1 046	1 326	2 938	841
2001												
Total expenditures	1 052 344	1 532	51 278	82 823	24 802	192 835	118 875	73 959	29 981	66 917	57 756	144 793
Structures, total	346 221	226	32 678	38 093	3 859	39 815	22 032	17 784	6 932	30 010	16 594	41 742
New	323 871	149	31 825	36 504	3 389	38 001	20 701	17 301	5 357	29 118	14 479	41 384
Used	22 349	77	853	1 588	470	1 814	1 331	483	1 575	892	2 116	358
Equipment, total	706 123	1 306	18 600	44 731	20 943	153 019	96 844	56 176	23 049	36 906	41 161	103 051
New	679 090	1 091	17 567	42 939	17 432	148 397	94 251	54 145	20 757	35 074	38 521	102 410
Used	27 033	215	1 033	1 792	3 511	4 623	2 592	2 030	2 292	1 833	2 640	641
2002												
Total expenditures	917 490	1 910	42 467	65 502	24 773	157 243	84 062	73 181	26 789	59 316	47 124	88 156
Structures, total	325 168	184	30 685	29 893	1 890	32 643	15 133	17 510	5 885	26 286	14 498	33 607
New	299 941	118	29 775	29 008	1 254	31 022	14 396	16 626	5 447	25 051	13 870	33 472
Used	25 227	66	910	886	456	1 622	737	885	438	1 234	628	135
Equipment, total	592 321	1 726	11 783	35 609	23 063	124 600	68 929	55 671	20 904	33 030	32 626	54 550
New	564 218	1 319	10 262	34 816	19 257	118 621	66 112	52 510	18 562	31 157	29 178	54 247
Used	28 103	407	1 520	793	3 806	5 978	2 817	3 161	2 342	1 873	3 447	303
2003												
Total expenditures	886 846	1 894	50 548	54 569	23 159	149 065	80 226	68 839	26 014	65 868	44 460	80 524
Structures, total	314 021	202	36 617	24 841	1 676	31 108	13 330	17 778	5 615	29 675	13 005	30 765
New	281 892	177	35 897	24 580	1 424	29 315	12 631	16 685	4 921	27 393	11 779	30 406
Used	32 128	25	720	261	251	1 793	700	1 093	694	2 282	1 226	358
Equipment, total	572 825	1 692	13 931	29 729	21 484	117 956	66 895	51 061	20 399	36 193	31 454	49 759
New	540 611	1 267	12 135	29 044	16 170	112 102	62 810	49 292	19 457	32 162	26 786	47 857
Used	32 214	425	1 796	685	5 313	5 855	4 086	1 769	942	4 031	4 668	1 902
2004												
Total expenditures	953 171	2 081	51 253	50 409	28 627	156 651	85 119	71 532	32 314	72 170	46 054	83 488
Structures, total	335 405	324	34 564	24 398	4 511	31 823	13 606	18 217	7 133	33 308	13 992	28 636
New	300 371	309	33 583	23 626	4 167	30 016	12 818	17 198	6 555	31 486	13 018	26 253
Used	35 034	15	982	772	345	1 807	788	1 019	578	1 822	975	2 384
Equipment, total	617 766	1 757	16 689	26 011	24 115	124 828	71 513	53 315	25 181	38 862	32 062	54 852
New	588 110	1 507	15 415	25 724	18 939	120 481	68 904	51 576	21 888	36 965	28 472	53 120
Used	29 656	250	1 274	286	5 176	4 347	2 609	1 738	3 293	1 897	3 590	1 732
2005												
Total expenditures	1 062 536	2 702	66 746	58 032	30 072	165 634	92 180	73 455	40 578	73 531	56 926	91 373
Structures, total	368 791	344	46 433	24 186	2 544	34 132	14 735	19 397	9 184	34 119	17 855	31 977
New	341 223	283	45 655	23 485	2 247	32 564	14 033	18 531	8 830	33 360	16 954	31 716
Used	27 568	61	777	701	297	1 569	703	866	355	759	901	262
Equipment, total	693 745	2 358	20 313	33 847	27 528	131 502	77 444	54 058	31 394	39 412	39 072	59 396
New	664 648	2 016	18 495	33 083	22 082	126 387	73 889	52 498	28 224	38 301	34 953	59 071
Used	29 096	341	1 818	764	5 446	5 115	3 555	1 560	3 169	1 111	4 119	325
2006												
Total expenditures	1 217 107	2 672	99 309	69 757	30 257	192 364	106 843	85 521	36 600	86 735	68 021	104 373
Structures, total	453 893	391	68 662	30 587	2 556	41 617	17 515	24 103	10 375	43 188	20 852	31 947
New	420 090	316	67 322	29 294	2 217	39 419	16 243	23 176	9 956	41 985	19 765	31 621
Used	33 802	75	1 340	1 293	338	2 198	1 272	926	419	1 203	1 087	326
Equipment, total	763 215	2 281	30 647	39 170	27 701	150 747	89 328	61 419	26 226	43 547	47 168	72 425
New	734 160	1 846	28 813	37 617	23 276	146 551	86 637	59 914	24 366	41 943	41 258	71 830
Used	29 055	435	1 833	1 553	4 425	4 196	2 692	1 504	1 860	1 604	5 911	595
2007												
Total expenditures	1 270 522	2 149	120 681	85 354	36 692	197 298	107 664	89 633	30 776	82 511	67 351	106 084
Structures, total	490 779	469	85 242	40 178	3 529	42 458	17 879	24 579	7 526	41 527	23 712	29 081
New	457 233	320	83 206	37 647	2 704	41 247	17 431	23 816	7 091	40 471	22 808	28 304
Used	33 546	149	2 036	2 531	824	1 211	447	763	436	1 056	904	777
Equipment, total	779 744	1 681	35 440	45 176	33 164	154 840	89 786	65 054	23 250	40 983	43 639	77 003
New	750 353	1 368	33 095	44 107	27 325	150 333	87 020	63 313	21 690	39 668	39 238	76 143
Used	29 391	313	2 344	1 069	5 838	4 507	2 766	1 741	1 560	1 316	4 400	861
2008												
Total expenditures	1 294 491	2 337	149 272	98 668	40 838	213 117	103 022	110 095	32 370	73 234	79 617	103 327
Structures, total	529 393	421	109 683	43 515	11 129	49 346	18 896	30 450	8 387	36 147	30 078	27 376
New	500 474	417	105 064	41 746	10 525	48 184	18 168	30 015	7 958	35 342	29 214	27 080
Used	28 919	4	4 618	1 769	603	1 162	728	434	430	805	863	296
Equipment, total	765 098	1 917	39 590	55 154	29 709	163 771	84 125	79 645	23 982	37 087	49 539	75 951
New	728 322	1 610	35 928	53 486	22 404	158 104	80 690	77 414	22 554	36 328	41 877	75 342
Used	36 776	307	3 662	1 668	7 305	5 667	3 435	2 232	1 429	759	7 662	609

Table 5-8. Capital Expenditures for Structures and Equipment for Companies with Employees by Major Industry Sector—Continued

(Millions of dollars.)

Year and type of expenditure	Finance and insurance (52)	Real estate and rental and leasing (53)	Professional, scientific, and technical services (54)	Management of companies and enterprises (55)	Administrative and support and waste management (56)	Educational services (61)	Health care and social assistance (62)	Arts, entertainment, and recreation (71)	Accommodation and food services (72)	Other services, except public administration (81)	Structure and equipment expenditures serving multiple industries
2000											
Total expenditures	133 684	92 456	34 055	5 054	17 506	18 223	52 166	19 125	26 307	21 125	1 572
Structures, total	23 010	24 815	8 141	1 570	4 032	13 699	26 868	12 245	13 873	13 274	206
New	20 298	17 793	7 470	955	3 504	12 965	23 999	11 627	12 879	11 705	200
Used	2 712	7 022	671	615	528	735	2 869	618	993	1 569	6
Equipment, total	110 675	67 641	25 914	3 484	13 475	4 523	25 299	6 880	12 434	7 852	1 366
New	109 678	62 175	24 847	3 403	12 723	4 338	24 407	6 161	11 501	7 192	1 357
Used	997	5 466	1 067	81	752	186	892	719	933	659	10
2001											
Total expenditures	131 105	82 674	30 464	3 035	15 785	17 377	52 932	14 974	21 365	29 006	911
Structures, total	22 744	20 489	7 258	933	3 527	12 852	27 030	8 998	12 248	20 031	163
New	19 571	17 325	6 793	869	3 367	11 860	25 241	8 157	11 402	18 918	162
Used	3 173	3 164	465	64	160	991	1 789	841	846	1 112	0
Equipment, total	108 361	62 185	23 206	2 102	12 258	4 525	25 902	5 976	9 117	8 976	749
New	107 268	60 295	22 330	2 019	11 644	4 238	24 573	5 590	7 921	8 300	725
Used	1 093	1 891	876	83	613	287	1 329	386	1 196	676	24
2002											
Total expenditures	128 444	94 529	25 864	3 430	14 719	19 532	59 311	13 169	22 409	21 269	1 532
Structures, total	24 308	35 579	7 129	933	3 276	14 655	30 291	7 758	12 157	13 261	250
New	19 748	30 227	6 424	913	2 948	13 601	27 273	7 332	10 848	11 363	248
Used	4 739	5 352	706	21	328	1 055	3 018	425	1 309	1 899	2
Equipment, total	103 956	58 949	18 735	2 497	11 443	4 876	29 021	5 412	10 252	8 007	1 282
New	103 421	56 847	18 021	2 481	10 585	4 690	28 196	5 132	9 290	6 858	1 276
Used	535	2 102	714	16	857	186	825	280	962	1 149	6
2003											
Total expenditures	120 787	87 952	24 703	3 298	16 612	16 667	61 151	11 029	21 036	26 035	1 476
Structures, total	26 200	25 028	5 314	925	3 976	11 984	30 996	6 800	10 568	18 518	209
New	17 908	16 446	4 671	869	3 213	11 569	28 885	6 532	9 417	16 288	202
Used	8 292	8 583	643	56	763	415	2 111	268	1 151	2 230	7
Equipment, total	94 587	62 923	19 389	2 373	12 636	4 683	30 155	4 229	10 468	7 517	1 267
New	94 205	61 253	18 675	2 368	11 374	4 569	29 497	4 038	9 684	6 706	1 263
Used	383	1 671	714	5	1 262	114	658	192	783	811	4
2004											
Total expenditures	153 629	91 606	26 688	2 825	17 455	18 919	64 561	12 165	20 641	19 701	1 572
Structures, total	43 919	27 277	6 007	860	2 567	13 728	32 608	7 360	9 860	12 278	321
New	30 216	21 610	5 714	798	2 309	12 781	30 668	7 196	9 126	10 867	307
Used	13 703	5 667	293	62	259	947	1 939	164	734	1 411	13
Equipment, total	109 710	64 329	20 681	1 965	14 888	5 190	31 953	4 804	10 781	7 423	1 252
New	109 244	61 947	20 081	1 931	12 692	4 965	31 280	4 677	10 373	6 788	1 248
Used	466	2 382	600	34	2 196	225	673	128	408	635	3
2005											
Total expenditures	161 389	103 022	33 066	2 809	18 194	17 484	73 825	14 165	30 718	20 105	2 163
Structures, total	39 383	24 791	8 717	857	3 051	12 711	39 089	9 242	17 679	12 036	460
New	31 023	17 341	7 633	795	2 759	11 913	37 493	8 805	16 567	11 350	452
Used	8 360	7 450	1 084	62	292	798	1 597	436	1 112	686	8
Equipment, total	122 005	78 231	24 350	1 951	15 143	4 773	34 736	4 924	13 039	8 069	1 703
New	121 511	76 894	23 887	1 917	13 523	4 597	34 110	4 757	11 950	7 209	1 681
Used	494	1 337	463	34	1 620	176	626	166	1 089	861	22
2006											
Total expenditures	163 069	132 073	30 284	3 306	19 231	22 615	75 296	17 156	36 217	25 959	1 813
Structures, total	41 326	40 794	6 971	875	3 613	17 537	41 197	11 733	22 585	16 621	467
New	34 028	30 240	6 375	799	3 485	16 203	37 765	11 326	21 774	15 754	446
Used	7 298	10 554	596	76	128	1 334	3 433	406	811	866	21
Equipment, total	121 743	91 280	23 313	2 432	15 618	5 078	34 099	5 424	13 632	9 339	1 346
New	121 157	88 957	22 867	2 188	14 932	4 983	33 508	5 121	13 161	8 463	1 322
Used	586	2 323	446	244	686	95	591	303	472	875	24
2007											
Total expenditures	172 894	117 969	31 804	4 542	18 167	23 238	84 160	18 769	38 021	29 680	2 380
Structures, total	45 159	41 222	7 453	1 472	3 859	17 910	45 361	12 589	22 211	19 165	656
New	36 507	34 026	6 944	1 404	3 677	17 322	43 522	11 894	21 226	16 264	650
Used	8 652	7 196	509	69	182	588	1 839	695	984	2 901	6
Equipment, total	127 735	76 747	24 351	3 070	14 308	5 328	38 799	6 180	15 810	10 515	1 725
New	126 763	74 735	23 931	2 943	13 599	5 253	38 258	5 916	14 671	9 593	1 724
Used	972	2 012	420	127	709	75	541	264	1 139	922	1
2008											
Total expenditures	132 913	106 910	32 980	4 567	16 552	27 426	90 248	17 109	40 519	28 312	4 175
Structures, total	25 058	45 929	8 836	1 304	4 133	21 698	50 105	11 594	24 947	18 673	1 035
New	22 640	36 532	8 481	1 241	3 789	20 632	47 205	11 224	24 271	17 899	1 030
Used	2 418	9 397	355	63	344	1 065	2 900	370	675	775	6
Equipment, total	107 855	60 981	24 144	3 264	12 420	5 728	40 143	5 515	15 572	9 638	3 139
New	107 108	59 117	23 349	3 146	11 687	5 672	39 161	5 340	14 012	8 974	3 124
Used	747	1 864	794	117	733	56	981	175	1 560	664	15

Table 5-8. Capital Expenditures for Structures and Equipment for Companies with Employees by Major Industry Sector—*Continued*

(Millions of dollars.)

Year and type of expenditure	Total	Forestry, fishing, and agricultural services (113–115)	Mining (21)	Utilities (22)	Construction (23)	Manufacturing (31–33) Total	Durable goods industries (321, 327, 33)	Nondurable goods industries (31, 322–326)	Wholesale trade (42)	Retail trade (44–45)	Transportation and warehousing (48–49)	Information (51)
2009												
Total expenditures	1 015 322	2 168	100 564	103 024	19 751	155 153	76 039	79 114	25 252	58 428	55 702	88 373
Structures, total	414 051	460	72 255	45 973	4 556	35 735	13 054	22 680	5 485	28 205	22 088	21 764
New	395 022	453	69 942	45 168	4 215	34 537	12 273	22 264	5 131	27 220	21 128	21 410
Used	19 030	7	2 313	805	341	1 198	782	416	354	985	961	354
Equipment, total	601 270	1 708	28 308	57 051	15 195	119 418	62 985	56 434	19 767	30 223	33 614	66 609
New	577 051	1 428	26 790	54 848	11 815	115 470	60 699	54 772	18 512	29 063	30 182	65 879
Used	24 219	279	1 518	2 204	3 380	3 948	2 286	1 662	1 255	1 160	3 432	731
2010												
Total expenditures	1 036 153	3 255	115 749	94 462	17 856	160 798	86 570	74 227	31 075	65 252	58 952	97 150
Structures, total	396 398	679	85 221	43 896	2 633	31 166	14 099	17 068	6 879	29 169	23 401	21 771
New	367 759	671	82 147	42 853	2 361	30 069	13 492	16 576	5 709	27 608	22 622	20 696
Used	28 639	8	3 074	1 044	272	1 098	606	491	1 170	1 561	778	1 074
Equipment, total	639 755	2 576	30 528	50 566	15 223	129 631	72 472	57 160	24 196	36 084	35 551	75 379
New	611 573	1 979	28 820	48 675	11 377	125 911	70 692	55 219	21 756	34 457	29 912	74 657
Used	28 182	597	1 708	1 891	3 846	3 720	1 780	1 940	2 440	1 626	5 639	723
2011												
Total expenditures	1 169 604	3 063	165 693	98 047	21 778	192 441	110 151	82 290	35 745	68 131	72 722	100 057
Structures, total	440 893	529	124 508	46 729	2 867	36 339	17 394	18 946	7 988	27 332	29 026	20 152
New	414 958	518	120 421	45 882	2 590	34 737	16 655	18 081	7 512	26 419	28 148	19 734
Used	25 935	11	4 086	847	277	1 602	738	864	476	912	878	418
Equipment, total	728 710	2 534	41 185	51 319	18 910	156 102	92 758	63 344	27 757	40 799	43 696	79 905
New	697 766	2 050	39 373	48 843	13 895	151 094	89 389	61 705	26 699	39 575	37 612	79 686
Used	30 944	484	1 813	2 476	5 015	5 009	3 369	1 639	1 058	1 224	6 084	210
2012												
Total expenditures	1 334 421	3 149	196 653	124 958	23 555	203 119	113 299	89 820	40 852	77 567	81 792	106 537
Structures, total	533 042	496	150 653	70 940	1 975	43 080	20 181	22 900	9 367	33 443	32 397	23 263
New	501 764	452	146 020	69 578	1 821	41 765	19 563	22 202	8 324	30 978	31 393	22 844
Used	31 279	44	4 633	1 362	154	1 316	618	697	1 043	2 464	1 003	419
Equipment, total	801 378	2 653	46 000	54 017	21 580	160 039	93 118	66 921	31 485	44 124	49 395	83 274
New	759 355	2 266	43 985	50 358	16 446	153 572	89 830	63 742	29 297	42 466	42 663	82 762
Used	42 023	387	2 015	3 660	5 134	6 467	3 288	3 179	2 187	1 658	6 732	512
2013												
Total expenditures	1 400 883	2 965	202 213	111 310	27 562	221 345	121 494	99 850	37 505	77 520	92 616	123 881
Structures, total	546 056	492	152 441	58 051	2 545	46 594	20 303	26 291	9 659	33 865	37 645	33 297
New	518 430	474	148 917	56 353	2 102	44 100	19 536	24 564	9 223	33 160	36 454	32 663
Used	27 625	18	3 523	1 698	442	2 494	767	1 727	435	705	1 191	634
Equipment, total	854 827	2 473	49 773	53 260	25 018	174 751	101 191	73 560	27 846	43 655	54 971	90 584
New	813 972	2 124	47 372	49 935	20 641	169 310	98 239	71 071	26 704	42 561	47 951	90 180
Used	40 855	349	2 401	3 324	4 377	5 441	2 952	2 489	1 142	1 094	7 020	404
2014												
Total expenditures	1 506 582	3 985	230 776	118 895	30 277	231 089	124 797	106 292	44 758	82 402	111 010	132 049
Structures, total	608 698	952	187 348	63 844	3 815	50 481	19 520	30 962	12 288	35 564	40 557	33 315
New	577 743	792	183 822	61 741	2 873	49 175	18 898	30 276	10 679	34 099	39 136	32 785
Used	30 955	160	3 526	2 103	942	1 307	621	685	1 609	1 466	1 421	530
Equipment, total	897 885	3 033	43 428	55 052	26 462	180 608	105 277	75 331	32 470	46 838	70 453	98 734
New	853 888	2 488	39 552	51 676	19 642	174 845	101 761	73 084	31 021	45 233	62 902	98 307
Used	43 996	545	3 876	3 375	6 820	5 763	3 516	2 247	1 449	1 605	7 551	426
2015												
Total expenditures	1 548 057	3 322	174 069	130 514	33 276	245 123	127 714	117 409	42 378	85 963	116 633	132 671
Structures, total	600 766	598	135 552	71 277	2 600	54 716	22 845	31 871	12 624	37 005	42 490	30 161
New	557 641	571	133 903	68 610	2 232	53 804	22 515	31 288	12 514	34 581	41 325	29 929
Used	43 126	26	1 649	2 667	368	913	330	583	110	2 423	1 164	233
Equipment, total	947 291	2 724	38 517	59 237	30 676	190 407	104 869	85 537	29 754	48 959	74 144	102 510
New	900 519	2 392	35 869	55 598	24 385	184 904	101 952	82 951	27 848	47 578	65 280	101 595
Used	46 772	333	2 649	3 639	6 291	5 503	2 917	2 586	1 906	1 381	8 864	915
2016												
Total expenditures	1 479 439	4 561	92 641	133 455	35 979	243 595	122 898	120 698	43 768	86 931	109 650	142 911
Structures, total	552 637	559	75 865	73 533	4 302	57 171	22 759	34 412	13 290	39 098	35 717	32 303
New	508 523	. . .	74 049	69 539	3 284	55 628	22 361	33 267	12 957	38 103	35 120	31 541
Used	44 114	. . .	1 816	3 993	1 018	1 543	398	1 145	333	995	597	763
Equipment, total	926 802	4 002	16 776	59 922	31 676	186 425	100 139	86 286	30 478	47 833	73 933	110 608
New	887 884	. . .	14 698	57 256	23 993	181 333	97 266	84 067	28 730	46 090	66 228	110 145
Used	38 918	. . .	2 078	2 666	7 683	5 092	2 873	2 219	1 748	1 743	7 705	462
2017												
Total expenditures	1 581 840	4 503	132 563	133 689	34 991	248 304	130 490	117 814	43 371	91 828	110 165	158 584
Structures, total	612 776	757	106 522	66 500	3 251	58 298	24 221	34 077	14 115	39 469	42 048	34 388
New	576 937	743	104 167	66 093	2 891	57 068	23 475	33 593	13 208	38 209	41 090	34 374
Used	35 839	14	2 356	407	360	1 230	746	484	907	1 259	958	14
Equipment, total	969 063	3 746	26 041	67 189	31 741	190 007	106 270	83 737	29 256	52 360	68 117	124 196
New	923 313	3 224	23 470	59 754	24 327	184 696	102 824	81 872	27 514	51 097	60 590	123 846
Used	45 750	522	2 571	7 435	7 413	5 311	3 445	1 866	1 742	1 263	7 527	350

. . . = Not available.

Table 5-8. Capital Expenditures for Structures and Equipment for Companies with Employees by Major Industry Sector—*Continued*

(Millions of dollars.)

Year and type of expenditure	Finance and insurance (52)	Real estate and rental and leasing (53)	Professional, scientific, and technical services (54)	Management of companies and enterprises (55)	Administrative and support and waste management (56)	Educational services (61)	Health care and social assistance (62)	Arts, entertainment, and recreation (71)	Accommodation and food services (72)	Other services, except public administration (81)	Structure and equipment expenditures serving multiple industries
2009											
Total expenditures	99 466	72 902	28 163	4 719	19 234	28 018	79 370	16 265	26 439	29 296	3 034
Structures, total	21 825	26 564	6 669	1 371	5 876	22 388	44 375	11 035	14 735	22 054	641
New	20 419	22 350	5 975	1 334	5 361	21 577	42 188	10 804	13 607	21 564	639
Used	1 406	4 214	694	36	515	810	2 187	231	1 128	490	2
Equipment, total	77 641	46 339	21 494	3 349	13 359	5 630	34 994	5 230	11 704	7 243	2 393
New	77 169	44 594	20 980	3 311	12 471	5 403	34 271	5 073	10 858	6 545	2 389
Used	473	1 745	514	38	888	227	723	157	846	698	4
2010											
Total expenditures	103 093	81 282	28 203	4 946	16 873	23 368	78 381	12 121	19 915	20 951	2 471
Structures, total	16 209	32 375	5 706	1 567	3 552	17 673	42 861	7 709	9 765	13 605	562
New	13 653	23 085	5 365	1 517	3 466	16 678	40 113	7 551	9 051	11 987	558
Used	2 556	9 290	341	50	86	995	2 748	158	714	1 618	4
Equipment, total	86 884	48 908	22 497	3 379	13 321	5 695	35 520	4 412	10 151	7 346	1 909
New	86 401	47 593	21 889	3 359	12 626	5 553	34 808	4 164	8 856	6 879	1 902
Used	483	1 315	608	20	695	142	713	248	1 294	467	7
2011											
Total expenditures	109 229	91 124	28 109	5 449	18 423	21 019	83 114	11 430	24 833	16 064	3 133
Structures, total	14 405	33 540	6 733	1 368	3 536	15 037	42 910	6 589	12 021	8 675	608
New	12 238	24 764	6 340	1 308	3 422	14 444	40 534	. . .	11 352	7 867	. . .
Used	2 167	8 776	393	60	114	593	2 376	. . .	669	807	. . .
Equipment, total	94 823	57 583	21 376	4 081	14 887	5 982	40 204	4 841	12 811	7 389	2 524
New	94 551	55 163	20 822	4 020	14 340	5 910	39 342	. . .	10 892	7 031	. . .
Used	272	2 420	554	61	546	72	862	. . .	1 920	358	. . .
2012											
Total expenditures	130 168	115 652	31 605	6 916	18 811	21 629	88 860	13 296	28 802	16 789	3 711
Structures, total	17 749	42 034	7 536	2 555	3 578	15 721	45 981	8 055	14 154	9 078	988
New	14 141	31 344	7 098	2 429	3 452	15 449	43 773	8 031	13 359	8 537	977
Used	3 608	10 690	438	126	126	272	2 208	24	794	541	12
Equipment, total	112 419	73 618	24 069	4 361	15 233	5 908	42 879	5 241	14 649	7 711	2 722
New	107 539	70 375	23 183	4 290	14 059	5 806	41 973	5 052	13 471	7 110	2 683
Used	4 880	3 243	886	71	1 174	102	905	189	1 178	601	40
2013											
Total expenditures	137 824	114 181	35 655	6 147	21 760	22 619	94 181	14 981	34 532	18 615	3 472
Structures, total	15 458	39 359	8 571	1 577	4 123	16 583	49 364	8 263	16 304	10 999	868
New	13 167	31 813	8 346	1 523	3 995	15 417	47 088	7 969	14 303	10 505	860
Used	2 291	7 547	225	54	129	1 166	2 276	294	2 000	494	8
Equipment, total	122 366	74 821	27 084	4 570	17 637	6 036	44 817	6 718	18 228	7 616	2 604
New	117 827	69 780	26 419	4 526	16 709	5 891	43 564	6 221	16 509	7 166	2 582
Used	4 539	5 041	665	44	927	145	1 253	497	1 719	450	22
2014											
Total expenditures	153 260	121 919	30 383	5 366	22 568	25 823	89 011	19 550	29 836	20 304	3 321
Structures, total	17 843	44 677	7 228	1 237	4 112	19 710	47 481	12 153	14 181	11 075	837
New	16 127	33 225	6 603	1 214	3 984	18 871	45 661	11 987	13 568	10 574	828
Used	1 716	11 452	625	24	128	839	1 819	167	614	500	8
Equipment, total	135 417	77 241	23 156	4 128	18 456	6 114	41 531	7 397	15 654	9 230	2 484
New	130 479	74 946	22 303	4 093	17 451	5 863	40 180	7 007	14 760	8 681	2 461
Used	4 938	2 296	853	36	1 005	251	1 351	390	895	549	23
2015											
Total expenditures	164 594	151 875	33 327	5 180	26 183	33 176	93 755	16 142	33 066	23 195	3 613
Structures, total	19 787	62 188	8 393	1 047	6 547	25 906	49 234	8 822	17 331	13 092	1 395
New	18 137	37 404	8 179	1 045	5 159	25 213	46 504	8 760	16 490	11 886	1 392
Used	1 651	24 783	215	2	1 388	693	2 730	62	841	1 206	3
Equipment, total	144 807	89 688	24 933	4 133	19 636	7 270	44 521	7 321	15 734	10 103	2 218
New	139 004	86 649	23 755	4 117	17 758	7 155	43 423	7 155	14 407	9 447	2 199
Used	5 803	3 038	1 179	16	1 878	115	1 098	165	1 327	656	19
2016											
Total expenditures	161 653	150 686	31 728	6 161	27 672	30 424	93 588	22 544	29 855	27 249	4 389
Structures, total	19 485	62 906	8 736	1 618	6 867	23 999	48 238	13 534	16 795	16 794	1 827
New	17 927	38 222	8 450	1 610	6 251	23 329	46 199	13 358	15 837	14 881	. . .
Used	1 558	24 684	285	8	616	670	2 039	177	958	1 913	. . .
Equipment, total	142 168	87 779	22 992	4 543	20 805	6 425	45 350	9 010	13 059	10 456	2 562
New	141 604	84 182	22 367	4 532	19 753	6 324	44 441	8 670	12 164	9 380	. . .
Used	564	3 597	625	11	1 052	101	908	340	895	1 075	. . .
2017											
Total expenditures	167 468	158 309	37 545	7 554	26 541	36 679	104 889	22 443	35 829	22 015	4 567
Structures, total	22 423	73 020	11 713	1 674	5 263	29 847	56 575	13 968	18 292	13 185	1 469
New	20 516	54 315	11 449	1 645	5 093	27 820	54 052	13 821	16 720	12 255	1 408
Used	1 907	18 705	264	29	170	2 026	2 523	147	1 572	931	61
Equipment, total	145 045	85 289	25 832	5 880	21 278	6 833	48 314	8 475	17 537	8 830	3 098
New	144 856	81 127	25 106	5 734	19 001	6 700	47 221	7 917	16 434	7 612	3 088
Used	189	4 162	726	146	2 277	133	1 093	558	1 103	1 218	10

. . . = Not available.

NOTES AND DEFINITIONS, CHAPTER 5

TABLES 5-1 THROUGH 5-4

Gross Saving and Investment Accounts

SOURCE: U.S. DEPARTMENT OF COMMERCE, BUREAU OF ECONOMIC ANALYSIS (BEA)

All of the data in these tables are from the July 26, 2019 National Income and Product Accounts NIPA), publication. Explanation of NIPA data are described in the Notes and Definitions to Chapter 1. All quarterly series are shown at seasonally adjusted annual rates. Current and constant dollar values are in billions of dollars. Constant dollar values are in 2012 dollars. Indexes of quantity are based on the average for the year 2012, set to equal 100.

The 2013 revision included a major expansion of the investment accounts. Expenditures for research and development are now recognized as fixed investment. Before 2013, they were treated as if they were costs of producing this year's output. R&D spending by business, government, and nonprofit institutions serving households (NPISHs) is now counted as fixed investment. It is depreciated (with the estimated depreciation added to capital consumption allowances), so that there can be either positive or negative net investment in R&D. A similar treatment is now given to expenditures by private enterprises for the creation of entertainment, literary, and artistic originals. Finally, an expanded set of ownership transfer costs for residential fixed assets is recognized as fixed investment.

Results from the 2019 Annual Update of NIPA revised data between 2014 and 2018. Other revisions were incorporated include the Census Bureau's annual retail sales, construction, manufacturing plus others.

Definitions: Table 5-1

Gross saving is saving before the deduction of allowances for the consumption of fixed capital. It represents the amount of saving available to finance gross investment. *Net saving* is gross saving less allowances for fixed capital consumption. It represents the amount of saving available for financing expansion of the capital stock, and comprises net private saving (the sum of personal saving, undistributed corporate profits, and wage accruals less disbursements) and the net saving of federal, state, and local governments.

Personal saving is derived by subtracting personal outlays from disposable personal income. (See Chapter 4 for more information.) It is the current net saving of individuals (including proprietors of unincorporated businesses), nonprofit institutions that primarily serve individuals, life insurance carriers, retirement funds, private noninsured welfare funds, and private trust funds.

Conceptually, personal saving may also be viewed as the sum for all persons (including institutions as previously defined) of the net acquisition of financial assets and the change in physical assets, less the sum of net borrowing and consumption of fixed capital. In either case, it is defined to exclude capital gains. That is, it excludes profits on the increase in the value of homes, securities, and other property—whether realized or unrealized—and therefore includes the noncorporate inventory valuation adjustment and the capital consumption adjustment (IVA and CCAdj, respectively). (See notes and definitions to Chapter 1.)

The net saving of *Domestic corporate business* is corporate profits after tax less dividends plus the corporate IVA and corporate CCAdj. (See notes and definitions for Chapter 1.)

Government net saving was formerly called "current surplus or deficit (-) of general government." (See Chapter 6 for further detail from the government accounts.) Where current receipts of government exceed current expenditures, government has a current surplus (indicated by a positive value) and saving is made available to finance investment by government or other sectors— for example, by the repayment of debt, which can free up funds for private investment. Where current expenditures exceed current receipts, there is a government deficit (indicated by a negative value) and government must borrow, drawing on funds that would otherwise be available for private investment. In these accounts, current expenditures are defined to include a charge for the consumption of fixed capital.

Consumption of fixed capital is an accounting charge for the using-up of private and government fixed capital, including software, located in the United States. It is based on studies of prices of used equipment and structures in resale markets. As of the 2013 revision, it also includes estimated charges for the using-up of the research and development and other intellectual property capital now defined as investment spending.

For general government and nonprofit institutions that primarily serve individuals, consumption of fixed capital is recorded in government consumption expenditures and in personal consumption expenditures (PCE), respectively, and taken to be the value of the current services of the fixed capital assets owned and used by these entities and the estimated using-up of R&D and other intellectual property.

Private consumption of fixed capital consists of tax-return-based depreciation charges for corporations and nonfarm proprietorships and historical-cost depreciation (calculated by the Bureau of Economic Analysis [BEA] using a geometric pattern of price declines) for farm proprietorships, rental income of persons, and nonprofit institutions, minus the capital consumption adjustments. (In other words, in the NIPA treatment of saving, the amount of the CCAdj is taken out of book depreciation and added to income and profits—a reallocation from one form of gross saving to another.) It also includes the charges for the using-up of private R&D and other intellectual property, as described above.

Gross private domestic investment consists of gross private fixed investment and change in private inventories. (See the notes and definitions for Chapter 1.)

Gross government investment consists of federal, state, and local general government and government enterprise expenditures for fixed assets (structures, equipment, and intellectual property). Government inventory investment is included in government consumption expenditures. For further detail, see Chapter 6.

Capital account transactions, net are the net cash or in-kind transfers between the United States and the rest of the world that are linked to the acquisition or disposition of assets rather than the purchase or sale of currently-produced goods and services. When positive, it represents a net transfer from the United States to the rest of the world; when negative, it represents a net transfer to the United States from the rest of the world. This is a definitional category that was introduced in the 1999 revision of the NIPAs. Estimates are available only from 1982 forward. With the new treatment of disaster losses and disaster insurance introduced in the 2009 revision, this line will include disaster-related insurance payouts to the rest of the world less what is received from the rest of the world.

Net lending or net borrowing (-), NIPAs is equal to the international balance on current account as measured in the NIPAs (see Chapter 7) less capital account transactions, net. When positive, this represents net investment by the United States in the rest of the world; when negative, it represents net borrowing by the United States from the rest of the world. For data before 1982, net lending or net borrowing equals the NIPA balance on current account, because estimates of capital account transactions are not available.

By definition, gross national saving must equal the sum of gross domestic investment, capital account transactions, and net international lending (where net international borrowing appears as negative lending). In practice, due to differences in measurement, these two aggregates differ by the *statistical discrepancy* calculated in the product and income accounts. (See Chapter 1.) Gross saving is therefore equal to the sum of gross domestic investment, capital account transactions, and net international lending minus the statistical discrepancy. Where the statistical discrepancy is negative, it means that the sum of measured investment, capital transactions, and net international lending has fallen short of measured saving.

Net domestic investment is gross domestic investment minus consumption of fixed capital, calculated by the editors from the data shown in the table.

Gross national income is national income plus the consumption of fixed capital. (See Chapter 1 for further information.) This is a new concept introduced in the 2003 revision. It is conceptually equal to gross national product, but differs by the statistical discrepancy. Gross national income is an appropriate denominator for the national saving ratios. Saving was previously shown as a percentage of gross national product; in the revision, it is instead shown as a percentage of the income-side equivalent of gross national product. Since saving is measured as a residual from income, it is appropriate to involve consistent measurements—and consistent imperfections in those measurements—in both the numerator and the denominator of the fraction.

Definitions: Tables 5-2 through 5-4

Gross private fixed investment comprises both nonresidential and residential fixed investment. It consists of purchases of fixed assets, which are commodities that will be used in a production process for more than one year, including replacements and additions to the capital stock, and intellectual property, including software, research and development, and entertainment, literary, and artistic originals. It is "gross" in the sense that it is measured before a deduction for consumption of fixed capital. It covers investment by private businesses and nonprofit institutions in the United States, regardless of whether the investment is owned by U.S. residents. It does not include purchases of the same types of equipment, structures, or intellectual property by government agencies, which are included in government gross investment. It also does not include investment by U.S. residents in other countries.

Gross nonresidential fixed investment consists of structures, equipment, and intellectual property that are not related to personal residences.

Nonresidential structures consists of new construction, brokers' commissions on sales of structures, and net purchases (purchases less sales) of used structures by private business and by nonprofit institutions from government agencies. New construction includes hotels, motels, and mining exploration, shafts, and wells.

Other nonresidential structures consists primarily of religious, educational, vocational, lodging, railroads, farm, and amusement and recreational structures, net purchases of used structures, and brokers' commissions on the sale of structures.

Nonresidential equipment consists of private business purchases—on capital account—of new machinery, equipment, and vehicles; dealers' margins on sales of used equipment; and net purchases (purchases less sales) of used equipment from government agencies, persons, and the rest of the world. (However, it does not include the personal-use portion of equipment purchased for both business and personal use. This is included in PCE.)

Computers have displayed phenomenal growth in numbers and power not easily represented by index numbers or constant-dollar estimates at the scale shown in these tables. Zero entries shown in the quantity indexes for early years actually represent very small quantities. Because of rapid growth in computing power and declines in its price, constant-dollar measures are deemed not meaningful and are not calculated by BEA.

Other information processing includes communication equipment, nonmedical instruments, medical equipment and instruments, photocopy and related equipment, and office and accounting equipment.

Other nonresidential equipment consists primarily of furniture and fixtures, agricultural machinery, construction machinery, mining and oilfield machinery, service industry machinery, and electrical equipment not elsewhere classified.

Intellectual property comprises information processing software, research and development, and entertainment, literary, and artistic originals.

Software excludes the value of software "embedded," or bundled, in computers and other equipment, which is instead included in the value of that equipment.

Research and development consists of expenditures for both purchased and own-account R&D by businesses, NPISHs, and general governments. Government R&D expenditures are treated as investment regardless of whether the R&D is protected or made freely available to the public. Investment is measured as the sum of production costs.

Entertainment, literary, and other artistic originals include theatrical movies, long-lived television programs, books, music, and other miscellaneous entertainment.

Residential private fixed investment consists of both *structures* and residential producers' durable *equipment*—that is, equipment owned by landlords and rented to tenants. Investment in *structures* consists of new units, improvements to existing units, manufactured homes, brokers' commissions and other ownership transfer costs on the sale of residential property, and net purchases (purchases less sales) of used structures from government agencies.

Other residential structures consists primarily of manufactured homes, improvements, dormitories, net purchases of used structures, and brokers' commissions on the sale of residential structures.

Real gross private investment (Table 5-3) and *chain-type quantity indexes for private fixed investment* (Table 5-4) are defined and explained in the notes and definitions to Chapter 1. The chained-dollar (2012) estimates in Table 5-3 are constructed by applying the changes in the chain-type quantity indexes, as shown in Table 5-4, to the 2012 current-dollar values. Thus, they do not contain any information about time trends that is not already present in the quantity indexes.

In Table 5-4, the user may wish to distinguish between the use of the "not available" symbol (...) and the publication of zero values (0.0). The "not available" values shown for computers and software mean that BEA has no separate estimates of their values; they are included in total "information processing equipment and software." The zeroes indicate quantities so small relative to the 2012 base that they round to zero, but they do exist and are included in the higher-level aggregates.

TABLES 5-5 AND 5-6

Current-Cost Net Stock of Fixed Assets; Chain-Type Quantity Indexes for Net Stock of Fixed Assets

SOURCE: U.S. DEPARTMENT OF COMMERCE, BUREAU OF ECONOMIC ANALYSIS (BEA)

The Bureau of Economic Analysis (BEA) calculates annual, end-of-year measurements, integrated with the national income and product accounts (NIPAs), of the level of the stock of fixed assets in the U.S. economy, or what is commonly called the "capital stock." (The fixed investment component of the GDP is a flow, or the increment of new capital goods into the capital stock.) Data on consumer stocks of durable goods are also included in the accounts, but are not shown here. Historical data are available back to 1901, with detailed estimates of net stocks, depreciation, and investment by type and by NAICS (North American Industry Classification System) industry. From this data system, *Business Statistics* presents time series data on the net stock of fixed assets valued in current dollars and also as constant-dollar quantity indexes.

The expanded definition of investment introduced in the 2013 comprehensive revision of the NIPAs, explained at the beginning of the Notes and Definitions to this chapter, was not incorporated in the fixed assets accounts until October 2013, so this edition of *Business Statistics* now publishes for the first time fixed assets data including the new NIPA categories.

Definitions and methods

The definitions of capital stock categories are now the same as the fixed investment categories listed earlier in these Notes and Definitions.

The values of fixed capital and depreciation typically reported by businesses are inadequate for economic analysis and are not typically used in these measures. In business reports, capital is generally valued at historical costs—each year's capital acquisition in the prices of the year acquired—and the totals thus represent a mixture of pricing bases. Reported depreciation is generally based on historical cost and on depreciation rates allowable by federal income tax law, rather than on a realistic rate of economic depreciation.

In these data, the *net stock of fixed assets* is measured by a perpetual inventory method. In other words, net stock at any given time is the cumulative value of past gross investment less the cumulative value of past depreciation (measured by "consumption of fixed capital," the component of the NIPAs that is subtracted from GDP in order to yield net domestic product) and also less damages from disasters and war losses that exceed normal depreciation (such as Hurricane Katrina and the terrorist attacks of September 11, 2001).

The initial calculations using this perpetual inventory method are performed in real terms for each type of asset. They are then aggregated to higher levels using an annual-weighted Fisher-type index. (See the definition of *real or chained-dollar estimates* in the notes and definitions for Chapter 1.) This provides the *chain-type quantity indexes* shown in Table 5-6. Growth rates in these indexes measure real growth in the capital stock.

The real values are then converted to a *current-cost* basis to yield the values shown in Table 5-5. They are converted by multiplying the real values by the appropriate price index for the period under consideration. A major use of the current-cost net stock figures is comparison with the value of output in that year; for example, the current-cost net stock of fixed assets for the total economy divided by the current-dollar value of GDP yields a capital-output ratio for the entire economy. Growth rates in current-cost values will reflect both the real growth measured by the quantity indexes and the increase in the value at current prices of the existing stock.

Data availability and references

Full historical data are available on the BEA website at www.bea.gov.

TABLES 5-7 AND 5-8

Annual Capital Expenditures

SOURCE: U.S. DEPARTMENT OF COMMERCE, CENSUS BUREAU

These data are from the Census Bureau's Annual Capital Expenditures Survey (ACES). The survey provides detailed information on capital investment in new and used structures and equipment by nonfarm businesses.

The survey is based on a stratified random sample of approximately 45,000 companies with employees and 30,000 non-employer businesses (businesses with an owner but no employees). For companies with employees, the Census Bureau reports data for 132 separate industry categories from the North American Industry Classification System (NAICS). Major exclusions are foreign operations of U.S. businesses, businesses in U.S. territories, government operations (including the U.S. Postal Service, agricultural production companies and private households.

Table 5-8 shows these data for the major NAICS sectors. Total capital expenditures, with no industry detail, are reported for the nonemployer businesses and are shown in Table 5-7, where they can be compared with the totals for companies with employees. The 1999 ACES was the first to use NAICS, providing data for the years 1998 forward on that basis.

Definitions

Capital expenditures include all capitalized costs during the year for both new and used structures and equipment, including software, that were chargeable to fixed asset accounts for which depreciation or amortization accounts are ordinarily maintained. For projects lasting longer than one year, this definition includes gross additions to construction-in-progress accounts, even if the asset was not in use and not yet depreciated. For *capital leases*, the company using the asset (lessee) is asked to include the cost or present value of the leased assets in the year in which the lease was entered. Also included in capital expenditures are capitalized leasehold improvements and capitalized interest charges on loans used to finance capital projects.

Structures consist of the capitalized costs of buildings and other structures and all necessary expenditures to acquire, construct, and prepare the structures. The costs of any machinery and equipment that is integral to or built-in features of the structures are classified as structures. Also included are major additions and alterations to existing structures and capitalized repairs and improvements to buildings.

New structures include new buildings and other structures not previously owned, as well as buildings and other structures that have been previously owned but not used or occupied.

Used structures are buildings and other structures that have been previously owned and occupied.

Equipment includes machinery, furniture and fixtures, computers, and vehicles used in the production and distribution of goods and services. Expenditures for machinery and equipment that is housed in structures and can be removed or replaced without significantly altering the structure are classified as equipment.

New equipment consists of machinery and equipment purchased new, as well as equipment produced in the company for the company's own use.

Used equipment is secondhand machinery and equipment.

Capital leases consist of new assets acquired under capital lease arrangements entered into during the year. Capital leases are defined by the criteria in the Financial Accounting Standards (FASB) Number 13.

Capitalized computer software consists of costs of materials and services directly related to the development or acquisition of software; payroll and payroll-related costs for employees directly associated with software development; and interest cost incurred while developing the software. Capitalized computer software is defined by the criteria in Statement of Position 98-1, Accounting for the Costs of Computer Software Developed or Obtained for Internal Use.

Prepackaged software is purchased off-the-shelf through retailers or other mass-market outlets for internal use by the company and includes the cost of licensing fees and service/maintenance agreements.

Vendor-customized software is externally developed by vendors and customized for the company's use.

Internally-developed software is developed by the company's employees for internal use and includes loaded payroll (salaries, wages, benefits, and bonuses related to all software development activities).

Data availability and references

Current and historical data and references are available on the Census Bureau Web site at https://www.census.gov/programs-surveys/aces.html.

CHAPTER 6: GOVERNMENT

SECTION 6A: FEDERAL GOVERNMENT IN THE NATIONAL INCOME AND PRODUCT ACCOUNTS

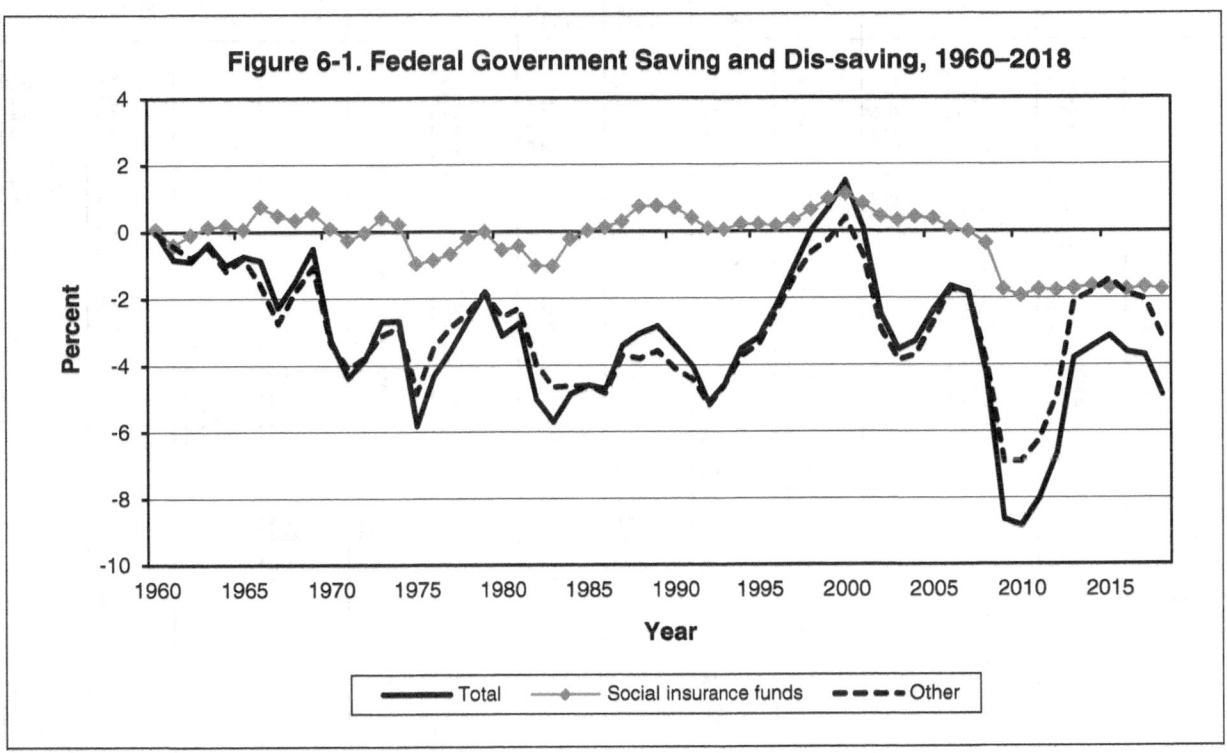

Figure 6-1. Federal Government Saving and Dis-saving, 1960–2018

- In the recovery and economic expansion of the 2000s, the federal budget never got back to a surplus such as those achieved in 1999 and 2000. With the onset of the most severe recession of the postwar period beginning in December 2007, deficits grew rapidly, due both to the "automatic stabilizers"; such as unemployment insurance, built into the budget structure and to anti-recession programs, such as the American Recovery and Reinvestment Act of 2009. (Table 6-1)

- In calendar 2009, the NIPA budget deficit peaked at 10.7 percent of GDP, including $251.7 billion deficit in social insurance funds. This was not as great as World War II deficits but exceeded in scale the deficits of the 1930s. By 2015, the deficit shrank to less than a third from its peak and was 2.6 percent of GDP, the smallest since 2007. However, the deficit has risen the last three years from 3.3 percent of GDP in 2016 to 3.9 percent of GDP in 2018. (Tables 6-1, 1-1, and 6-14B)

- The main contributors to the rise in the deficit from 2007 to 2010 were increasing social benefits to persons, such as Social Security and unemployment insurance, and a drop in non-social-insurance tax receipts. Federal consumption spending and grants to state and local governments also rose. The deficit shrinkage from 2010 to 2015 was powered mainly by a resurgence in non-social-insurance tax receipts while the increase in the deficit in recent years is due to tax cuts. (Table 6-1)

- Total federal government current expenditures have increased from 17.2 percent of gross domestic product (GDP) in 1960 to 24.1 percent in 2009, falling back to 21.9 percent in 2018. The composition of current expenditures changed significantly over the last several decades. "Consumption" spending on defense and nondefense programs fell from significantly while benefits to persons, such as Social Security and Medicare, increased significantly. (Tables 6-1, 1-1 and 6-3)

Table 6-1. Federal Government Current Receipts and Expenditures

(National income and product accounts, calendar years, billions of dollars, quarterly data are at seasonally adjusted annual rates.)

NIPA Table 3.2

Year and quarter	Total current receipts	Current tax receipts							Contributions for government social insurance	Income receipts on assets			Current transfer receipts	Current surplus of government enterprises
		Total tax receipts [1]	Personal current taxes	Taxes on production and imports		Taxes on corporate income		Taxes from the rest of the world		Interest receipts	Non-federal reserve bank dividends	Rents and royalties		
				Excise taxes	Customs duties	Federal Reserve banks	Other corporate taxes							
1960	93.6	75.6	41.8	12.0	1.1	0.9	20.6	0.1	16.0	1.3	...	0.2	0.3	-0.7
1961	95.1	76.8	42.7	12.1	1.0	0.7	20.8	0.1	16.6	1.4	...	0.2	0.4	-0.9
1962	103.2	82.5	46.5	12.9	1.2	0.8	21.7	0.1	18.6	1.6	...	0.2	0.4	-0.9
1963	111.3	87.7	49.1	13.5	1.2	0.9	23.7	0.2	21.1	1.7	...	0.2	0.5	-0.7
1964	111.3	86.2	46.0	14.1	1.3	1.6	24.6	0.2	21.8	1.7	...	0.2	0.6	-0.8
1965	120.4	94.3	51.1	13.8	1.6	1.3	27.6	0.2	22.7	1.8	...	0.2	0.9	-0.9
1966	137.4	103.1	58.6	12.6	1.9	1.6	29.8	0.2	30.6	2.0	...	0.2	1.0	-1.2
1967	146.3	107.9	64.4	13.3	1.9	1.9	28.1	0.2	34.1	2.3	...	0.3	1.0	-1.3
1968	170.6	127.2	76.4	14.6	2.3	2.5	33.6	0.3	37.9	2.7	...	0.3	1.0	-1.0
1969	191.8	143.0	91.7	15.4	2.4	3.0	33.0	0.4	43.3	2.5	...	0.3	1.0	-1.3
1970	185.1	134.4	88.9	15.6	2.5	3.5	27.1	0.4	45.5	2.8	...	0.4	1.0	-2.5
1971	190.7	135.3	85.8	15.9	3.1	3.4	30.1	0.4	50.3	3.1	...	0.5	0.9	-2.8
1972	219.0	155.0	102.8	15.5	3.0	3.2	33.4	0.4	58.3	3.3	...	0.5	1.1	-2.4
1973	249.2	168.7	109.6	16.5	3.3	4.3	38.9	0.4	74.5	3.4	...	0.7	1.0	-3.4
1974	278.5	186.5	126.5	16.4	3.7	5.6	39.6	0.4	84.1	3.6	...	0.9	1.1	-3.2
1975	276.8	181.5	120.7	16.2	5.9	5.4	38.2	0.5	88.1	4.3	...	1.0	1.1	-4.4
1976	322.6	212.4	141.6	16.8	4.6	5.9	48.7	0.7	99.8	5.2	...	1.1	1.2	-2.9
1977	363.9	241.6	162.5	17.3	5.4	5.9	55.7	0.7	111.1	5.8	...	1.3	1.6	-3.4
1978	423.8	279.8	189.2	18.2	7.1	7.0	64.4	1.0	128.7	7.4	...	1.5	2.2	-2.9
1979	487.0	316.9	224.9	18.2	7.5	9.3	65.1	1.1	149.8	9.2	...	2.1	2.5	-2.7
1980	533.7	344.5	250.6	26.5	7.2	11.7	58.6	1.6	163.6	11.3	...	3.0	3.1	-3.6
1981	621.1	394.3	291.2	41.3	8.6	14.0	51.7	1.5	193.0	14.8	...	4.4	3.1	-2.6
1982	618.7	371.8	295.6	32.3	8.6	15.2	33.8	1.4	206.0	18.3	...	5.1	4.4	-2.1
1983	644.8	379.5	286.8	35.3	9.1	14.2	47.1	1.2	223.1	20.4	...	4.3	5.2	-1.9
1984	711.2	409.7	301.9	35.4	11.9	16.1	59.2	1.3	254.1	22.8	...	4.9	6.3	-2.7
1985	775.7	442.9	336.5	33.9	12.2	17.8	58.5	1.9	277.9	25.6	...	4.6	8.6	-1.6
1986	817.9	462.0	350.6	30.0	13.7	17.8	66.0	1.7	298.9	28.8	...	3.3	7.6	-0.6
1987	899.5	526.4	393.0	30.4	15.5	17.7	85.4	2.0	317.4	25.1	...	3.1	10.2	-0.5
1988	962.4	549.8	403.8	33.4	16.4	17.4	93.8	2.4	354.8	27.4	...	2.8	9.6	0.7
1989	1 042.5	601.1	453.1	32.3	17.5	21.6	95.6	2.7	378.0	25.9	...	3.0	11.8	1.1
1990	1 087.6	620.6	472.1	33.4	17.5	23.6	94.5	3.0	402.0	26.9	...	3.5	13.3	-2.3
1991	1 107.8	617.1	463.6	44.9	16.8	20.8	89.2	2.6	420.6	26.2	...	3.8	17.2	2.1
1992	1 154.4	645.4	477.5	45.0	18.3	16.8	102.0	2.6	444.0	22.2	...	3.5	18.4	4.1
1993	1 231.0	699.3	507.7	46.5	19.8	16.0	122.5	2.7	465.5	22.9	...	3.8	20.2	3.5
1994	1 329.3	763.5	545.1	57.5	21.4	20.5	136.3	3.1	496.2	20.1	...	3.6	21.6	4.0
1995	1 417.4	825.7	590.3	55.7	19.8	23.4	155.9	3.9	521.9	20.9	...	3.0	17.5	5.0
1996	1 536.3	917.0	668.4	53.6	19.2	20.1	170.5	5.2	545.4	22.6	...	4.1	22.7	4.5
1997	1 667.4	1 015.1	749.8	58.2	19.6	20.7	182.3	5.1	579.4	21.0	...	4.5	20.1	6.6
1998	1 789.8	1 095.3	831.2	60.2	19.6	26.6	177.7	5.7	617.4	17.4	...	4.2	21.5	7.6
1999	1 906.6	1 174.5	897.4	63.3	19.2	25.4	187.6	6.1	654.8	17.2	...	4.0	23.4	7.2
2000	2 068.4	1 288.5	999.6	65.1	21.1	25.3	194.1	7.6	698.6	19.4	...	6.3	24.9	5.3
2001	2 032.2	1 226.8	996.0	63.7	20.6	27.1	137.6	7.9	723.3	18.1	...	8.4	25.7	2.8
2002	1 870.8	1 053.2	832.3	65.9	19.9	24.5	126.0	8.1	739.4	15.4	...	6.5	25.0	6.8
2003	1 895.6	1 053.9	778.6	67.7	21.5	22.0	175.8	9.3	763.3	16.6	...	7.0	26.2	6.6
2004	2 027.7	1 140.6	803.0	70.9	23.3	18.1	232.2	10.1	809.0	16.6	0.1	8.7	29.8	4.8
2005	2 304.4	1 367.8	937.2	72.7	25.3	21.5	319.5	11.7	853.4	17.1	0.2	9.7	33.7	0.9
2006	2 538.3	1 534.8	1 055.7	71.5	26.7	29.1	366.0	13.9	905.7	18.6	0.3	9.9	38.1	1.9
2007	2 667.8	1 607.7	1 170.6	64.8	28.8	34.6	328.2	14.2	947.3	22.1	0.2	11.1	42.7	2.1
2008	2 580.7	1 489.5	1 176.3	63.7	29.2	31.7	202.0	17.2	974.5	19.8	0.6	13.8	49.7	1.1
2009	2 239.5	1 123.7	866.2	67.3	23.1	47.4	153.0	13.1	950.7	23.8	18.7	7.0	67.2	0.9
2010	2 444.0	1 273.6	943.2	67.2	28.6	79.3	219.4	14.2	970.9	29.5	17.0	8.0	68.1	-2.4
2011	2 572.8	1 478.4	1 130.3	75.7	31.9	75.4	224.0	15.6	903.2	26.3	18.8	9.9	67.1	-6.3
2012	2 700.3	1 573.0	1 165.8	80.6	33.5	88.4	274.7	17.3	938.0	21.4	21.1	10.2	56.1	-7.8
2013	3 139.0	1 744.9	1 302.3	89.0	35.5	79.6	298.4	18.7	1 091.8	22.9	131.3	9.3	69.3	-10.1
2014	3 292.0	1 900.1	1 403.1	97.8	37.4	96.9	339.6	21.0	1 140.1	23.8	40.7	10.3	87.3	-7.1
2015	3 446.0	2 023.1	1 530.6	101.0	38.1	110.4	329.1	23.0	1 190.8	26.4	16.0	6.7	76.1	-3.5
2016	3 460.3	2 019.4	1 546.5	97.7	37.5	91.5	311.9	24.6	1 224.9	26.7	14.8	5.0	79.6	-1.6
2017	3 526.4	2 019.2	1 613.1	89.0	38.5	80.6	251.5	25.9	1 283.8	28.1	23.2	6.1	84.6	1.0
2018	3 497.7	1 956.1	1 620.2	106.3	53.3	65.3	147.4	27.7	1 339.4	31.3	13.7	8.3	86.0	-2.5
2016														
1st quarter	3 439.4	1 995.9	1 524.4	98.3	39.0	101.0	308.1	25.0	1 210.7	26.7	18.6	5.1	82.9	-1.4
2nd quarter	3 440.1	2 003.3	1 535.1	97.3	36.6	101.0	308.9	24.1	1 217.5	26.6	3.9	5.0	85.4	-2.6
3rd quarter	3 472.7	2 039.0	1 552.7	97.8	37.3	90.8	326.0	24.0	1 228.8	26.6	15.4	4.6	69.1	-1.7
4th quarter	3 489.1	2 039.3	1 574.0	97.5	37.2	73.1	304.5	25.0	1 242.7	27.0	21.4	5.3	81.0	-0.6
2017														
1st quarter	3 532.2	1 987.6	1 581.6	86.7	37.6	88.4	255.6	24.9	1 264.6	27.3	40.0	5.5	117.9	0.8
2nd quarter	3 496.2	2 003.7	1 601.8	88.7	38.0	87.2	248.5	25.6	1 276.5	27.8	20.2	5.9	73.5	1.3
3rd quarter	3 535.8	2 042.9	1 622.1	90.1	38.2	73.7	265.4	26.0	1 288.7	28.2	20.6	6.0	74.6	1.2
4th quarter	3 541.5	2 042.4	1 647.0	90.5	40.2	72.8	236.5	26.9	1 305.5	28.9	11.8	6.8	72.6	0.6
2018														
1st quarter	3 446.9	1 921.5	1 605.3	107.2	41.8	78.3	139.2	26.9	1 327.3	29.9	0.2	7.4	82.2	0.2
2nd quarter	3 469.3	1 943.5	1 613.5	106.5	45.4	67.1	149.2	27.7	1 333.9	30.5	3.9	7.7	84.0	-1.3
3rd quarter	3 545.4	1 971.4	1 628.3	106.0	52.5	60.7	156.2	27.2	1 345.1	31.8	24.4	7.7	107.5	-3.2
4th quarter	3 529.0	1 987.9	1 633.9	105.7	73.5	55.2	144.9	28.9	1 351.4	33.0	26.3	10.4	70.3	-5.6

[1] Does not include "contributions for government social insurance" (the taxes that fund Social Security, Medicare, and unemployment insurance), which are shown separately in column 9.
. . . = Not available.

Table 6-1. Federal Government Current Receipts and Expenditures—*Continued*

(National income and product accounts, calendar years, billions of dollars, quarterly data are at seasonally adjusted annual rates.)

NIPA Table 3.2

Year and quarter	Total current expenditures	Consumption expenditures	Government social benefits		Other current transfer payments		Interest payments			Subsidies	Net federal government saving, NIPA (surplus + / deficit -)		
			Benefits to persons	Benefits to the rest of the world	Grants-in-aid to state and local governments	Payments to the rest of the world (net)	Total	To persons and business	To the rest of the world		Total	Social insurance funds	Other
1960	93.4	51.0	19.9	0.2	3.8	3.5	13.9	13.5	0.3	1.1	0.2	0.4	-0.2
1961	99.8	52.7	23.1	0.3	4.3	3.5	13.9	13.6	0.3	2.0	-4.7	-2.3	-2.4
1962	108.6	59.0	23.5	0.3	4.7	3.6	15.1	14.8	0.3	2.3	-5.4	-0.6	-4.8
1963	113.5	61.2	24.6	0.3	5.2	3.6	16.3	15.9	0.4	2.2	-2.2	0.8	-3.0
1964	118.2	63.1	25.2	0.3	6.0	3.4	17.5	17.0	0.5	2.7	-6.9	1.2	-8.1
1965	125.9	66.5	27.3	0.5	6.6	3.6	18.6	18.1	0.5	3.0	-5.5	0.4	-5.9
1966	144.4	76.5	29.9	0.5	9.4	4.0	20.3	19.7	0.5	3.9	-7.0	6.0	-13.0
1967	165.8	88.1	36.5	0.6	10.9	4.0	21.9	21.3	0.6	3.8	-19.5	4.1	-23.6
1968	184.3	96.8	41.9	0.6	11.8	4.5	24.6	23.9	0.7	4.1	-13.8	3.2	-16.9
1969	197.0	100.5	45.8	0.6	13.7	4.5	27.5	26.7	0.8	4.5	-5.1	5.7	-10.8
1970	219.9	102.3	55.6	0.7	18.3	4.8	33.3	32.4	0.9	4.8	-34.8	1.0	-35.8
1971	241.6	106.7	66.1	0.8	22.1	6.0	35.2	33.5	1.8	4.6	-50.9	-3.0	-47.9
1972	268.0	112.3	72.9	1.0	30.5	7.2	37.6	35.1	2.6	6.6	-49.0	-0.6	-48.4
1973	287.6	114.5	84.5	1.2	33.5	5.4	43.3	39.6	3.7	5.1	-38.3	5.9	-44.2
1974	319.8	123.2	103.3	1.3	34.9	6.0	48.0	43.9	4.0	3.2	-41.3	3.2	-44.5
1975	374.8	133.9	132.3	2.0	43.6	6.1	52.5	48.1	4.3	4.3	-97.9	-16.3	-81.6
1976	403.5	140.0	143.5	2.5	49.1	4.5	58.9	54.6	4.3	4.9	-80.9	-16.0	-64.9
1977	437.3	151.1	152.4	2.6	54.8	4.2	65.3	60.0	5.2	6.9	-73.4	-14.1	-59.4
1978	485.9	164.2	162.7	2.7	63.5	5.0	79.1	70.9	8.2	8.7	-62.0	-4.7	-57.3
1979	534.4	178.8	183.0	3.0	64.0	5.8	91.5	80.9	10.6	8.2	-47.4	-0.1	-47.3
1980	622.5	204.6	220.3	3.5	69.7	7.3	107.5	95.7	11.9	9.4	-88.8	-15.8	-73.0
1981	709.1	233.6	250.7	4.3	69.4	6.7	133.4	117.2	16.3	11.1	-88.1	-14.3	-73.8
1982	786.0	258.4	281.9	4.1	66.3	8.1	152.7	134.7	18.0	14.6	-167.4	-34.4	-132.9
1983	851.0	280.0	303.6	3.6	67.9	8.9	167.0	148.9	18.1	20.9	-207.2	-37.9	-169.3
1984	907.7	296.2	309.7	3.7	72.3	11.2	193.8	173.7	20.2	20.7	-196.5	-9.4	-187.1
1985	975.0	320.4	325.9	4.0	76.2	13.8	213.7	191.6	22.2	21.0	-199.2	0.8	-200.1
1986	1 033.8	339.8	344.3	4.4	82.4	14.2	224.2	200.9	23.3	24.6	-215.9	5.6	-221.5
1987	1 065.2	350.3	357.2	4.3	78.4	12.7	232.3	208.2	24.1	30.0	-165.7	14.4	-180.2
1988	1 122.4	363.5	378.4	4.6	85.7	13.2	247.9	219.2	28.8	29.2	-160.0	39.0	-198.9
1989	1 201.8	382.9	411.7	5.1	91.8	13.5	269.7	235.1	34.5	27.1	-159.4	42.6	-201.9
1990	1 290.9	404.2	447.0	6.2	104.4	13.5	289.0	253.0	36.1	26.6	-203.3	42.3	-245.6
1991	1 356.2	426.1	494.0	6.3	124.0	-25.6	304.4	268.3	36.1	27.1	-248.4	24.6	-273.0
1992	1 488.9	432.1	551.8	6.2	141.7	20.6	306.9	272.6	34.3	29.7	-334.5	5.3	-339.8
1993	1 544.6	429.5	583.7	6.2	155.7	21.7	311.4	277.3	34.2	36.3	-313.5	2.8	-316.3
1994	1 585.0	429.3	609.0	6.8	166.8	19.9	320.8	284.5	36.4	32.2	-255.6	15.3	-270.9
1995	1 659.5	429.0	647.1	6.8	174.5	15.4	352.3	305.2	47.1	34.5	-242.1	15.5	-257.6
1996	1 715.7	429.4	682.1	7.6	181.5	20.4	360.0	303.2	56.8	34.9	-179.4	12.8	-192.2
1997	1 759.4	439.6	707.9	7.9	188.1	17.0	365.6	295.9	69.7	33.4	-92.0	28.8	-120.8
1998	1 788.4	437.8	722.1	8.2	200.8	18.0	365.6	293.5	72.1	35.9	1.4	58.2	-56.8
1999	1 839.7	456.5	739.9	8.5	219.2	18.7	352.2	285.4	66.7	44.8	66.9	92.4	-25.6
2000	1 912.9	476.3	773.4	8.7	233.1	22.4	353.7	289.3	64.4	45.3	155.5	114.2	41.3
2001	2 018.2	506.2	840.8	9.4	261.3	18.3	331.2	274.1	57.1	51.1	14.0	88.0	-74.0
2002	2 142.3	560.8	918.0	9.9	288.7	23.1	301.3	252.0	49.3	40.5	-271.5	48.8	-320.3
2003	2 299.7	631.1	967.5	10.4	321.7	28.3	291.7	244.3	47.4	49.0	-404.1	37.5	-441.6
2004	2 428.6	683.3	1 019.5	11.0	332.3	30.4	306.0	251.2	54.8	46.0	-400.9	50.8	-451.7
2005	2 610.3	725.4	1 084.4	11.6	343.5	40.5	344.4	276.4	68.0	60.5	-305.9	48.0	-353.9
2006	2 765.9	765.1	1 189.1	12.4	341.0	35.0	372.2	286.3	85.9	51.1	-227.6	11.8	-239.4
2007	2 933.9	800.5	1 263.3	13.3	359.1	42.3	408.0	308.9	99.0	47.5	-266.1	-3.2	-262.9
2008	3 211.8	878.9	1 463.2	15.4	371.2	45.1	388.4	291.3	97.0	49.6	-631.1	-53.6	-577.5
2009	3 488.4	935.6	1 614.5	16.0	458.1	52.7	354.5	265.7	88.8	56.9	-1 248.9	-251.7	-997.2
2010	3 769.1	1 000.7	1 757.5	16.5	505.2	53.5	381.5	288.9	92.6	54.2	-1 325.1	-290.6	-1 034.0
2011	3 814.7	1 003.3	1 779.5	17.0	472.5	57.6	425.4	328.0	97.5	59.5	-1 242.0	-271.5	-970.5
2012	3 779.0	999.3	1 781.8	18.0	444.4	55.3	422.6	326.7	95.9	57.6	-1 078.6	-287.7	-791.0
2013	3 776.9	956.9	1 821.5	18.9	450.1	53.9	416.3	319.8	96.5	59.2	-637.9	-288.6	-349.3
2014	3 896.3	951.2	1 881.1	19.5	495.0	52.8	439.1	345.5	93.6	57.6	-604.3	-287.2	-317.1
2015	4 016.0	956.3	1 967.5	20.4	533.2	52.7	429.3	335.2	94.0	56.7	-570.1	-308.9	-261.2
2016	4 137.4	968.6	2 020.9	20.9	556.9	54.7	454.1	357.8	96.3	61.3	-677.0	-327.6	-349.5
2017	4 251.1	992.6	2 087.9	21.8	559.8	52.6	475.9	371.6	104.4	60.6	-724.7	-327.2	-397.6
2018	4 507.4	1 056.9	2 181.7	22.5	582.9	58.9	540.7	420.0	120.7	63.8	-1 009.8	-359.1	-650.6
2016													
1st quarter	4 083.9	960.9	2 003.8	20.6	537.7	61.4	439.4	343.8	95.6	60.2	-644.5	-321.2	-323.3
2nd quarter	4 114.9	961.3	2 015.2	20.8	553.2	49.6	452.8	356.4	96.4	61.8	-674.8	-328.0	-346.8
3rd quarter	4 159.9	972.8	2 025.3	21.3	564.2	55.1	458.6	361.8	96.8	62.6	-687.2	-329.9	-357.4
4th quarter	4 190.8	979.3	2 039.4	21.0	572.4	52.8	465.5	369.2	96.3	60.4	-701.6	-331.2	-370.4
2017													
1st quarter	4 217.2	978.2	2 066.9	21.5	558.3	53.0	479.9	382.4	97.5	59.3	-685.0	-324.8	-360.2
2nd quarter	4 195.4	987.2	2 078.3	21.7	545.5	49.3	455.2	353.9	101.3	58.1	-699.2	-325.7	-373.5
3rd quarter	4 242.9	993.0	2 094.3	21.8	564.3	48.2	458.7	351.7	107.0	62.6	-707.1	-328.9	-378.1
4th quarter	4 349.1	1 011.9	2 111.9	22.1	571.2	59.8	509.9	398.2	111.7	62.3	-807.6	-329.2	-478.4
2018													
1st quarter	4 423.2	1 033.9	2 158.1	22.1	578.9	50.4	521.1	404.8	116.3	58.8	-976.3	-339.1	-637.2
2nd quarter	4 483.1	1 052.5	2 171.5	22.5	582.6	65.4	530.6	410.7	119.9	58.0	-1 013.8	-350.9	-662.9
3rd quarter	4 526.8	1 068.1	2 186.7	22.7	585.7	56.2	549.5	426.4	123.1	57.8	-981.3	-362.6	-618.8
4th quarter	4 596.6	1 073.1	2 210.5	22.7	584.6	63.6	561.4	437.9	123.5	80.7	-1 067.6	-383.9	-683.7

Table 6-2. Federal Government Consumption Expenditures and Gross Investment

(National income and product accounts, calendar years, billions of dollars, quarterly data are at seasonally adjusted annual rates.)

NIPA Tables 3.9.5, 3.10.5

Year and quarter	Total	Federal government consumption expenditures and gross investment											
		Consumption expenditures					Gross investment						
		Total	Compensation of general government employees	Consumption of general government fixed capital	Intermediate goods and services purchased [1]	Less: Own-account investment and sales to other sectors	Total	National defense			Nondefense		
								Structures	Equipment	Intellectual property	Structures	Equipment	Intellectual property
1960	72.9	51.0	26.6	15.0	12.0	2.7	21.9	2.2	10.1	5.9	1.7	0.3	1.7
1961	77.4	52.7	27.7	15.9	11.9	2.8	24.6	2.4	11.5	6.5	1.9	0.3	2.1
1962	85.5	59.0	29.5	17.0	15.6	3.1	26.5	2.0	12.5	6.7	2.1	0.3	2.9
1963	87.9	61.2	30.8	18.2	15.6	3.4	26.7	1.6	10.9	6.9	2.3	0.4	4.5
1964	90.3	63.1	33.0	19.0	14.9	3.8	27.1	1.3	10.1	6.9	2.5	0.6	5.7
1965	93.2	66.5	34.8	19.8	16.0	4.2	26.7	1.1	8.9	6.7	2.8	0.5	6.8
1966	106.6	76.5	39.8	20.8	20.6	4.7	30.1	1.3	10.4	7.0	2.8	0.6	7.9
1967	120.0	88.1	43.8	22.2	26.5	4.5	31.9	1.2	12.2	7.9	2.2	0.6	7.8
1968	128.0	96.8	48.3	23.9	29.3	4.6	31.1	1.2	10.8	7.9	2.1	0.5	8.6
1969	131.2	100.5	51.4	25.5	28.7	5.1	30.7	1.5	9.7	8.2	1.9	0.6	8.8
1970	132.8	102.3	55.2	27.3	25.3	5.4	30.5	1.3	9.6	8.1	2.1	0.7	8.8
1971	134.5	106.7	58.8	28.7	25.0	5.8	27.8	1.8	5.5	8.4	2.5	0.8	8.9
1972	141.6	112.3	62.6	29.7	26.9	6.9	29.3	1.8	5.3	9.2	2.7	1.1	9.2
1973	146.2	114.5	64.6	31.8	25.8	7.7	31.7	2.1	6.0	9.7	3.1	1.1	9.6
1974	158.8	123.2	67.8	34.7	29.2	8.6	35.7	2.2	8.2	10.0	3.4	1.5	10.4
1975	173.7	133.9	72.5	38.0	31.8	8.4	39.9	2.3	9.9	10.5	4.1	1.7	11.5
1976	184.8	140.0	76.5	40.8	31.3	8.5	44.8	2.1	12.1	11.3	4.6	1.8	12.7
1977	200.3	151.1	80.9	44.3	35.6	9.7	49.2	2.4	13.3	12.3	5.0	2.3	14.0
1978	218.9	164.2	87.1	48.6	39.3	10.8	54.7	2.5	14.3	13.3	6.1	2.7	15.9
1979	240.6	178.8	92.0	53.4	45.8	12.4	61.7	2.5	17.5	14.9	6.3	2.7	17.9
1980	274.9	204.6	100.8	59.1	58.4	13.6	70.2	3.2	19.8	17.2	7.1	3.2	19.8
1981	314.0	233.6	112.2	66.0	69.6	14.3	80.4	3.2	24.1	20.9	7.7	3.3	21.2
1982	348.3	258.4	122.0	73.7	77.6	14.9	90.0	4.0	29.1	25.2	6.8	3.9	21.1
1983	382.4	280.0	127.7	79.4	88.8	15.9	102.3	4.8	35.2	29.8	6.7	4.3	21.6
1984	411.8	296.2	136.5	88.7	88.3	17.4	115.6	4.9	41.1	35.0	7.0	4.7	22.8
1985	452.9	320.4	145.0	95.0	99.1	18.7	132.5	6.2	47.9	41.3	7.3	5.1	24.6
1986	481.7	339.8	149.1	101.5	108.7	19.6	141.9	6.8	52.7	43.9	8.0	4.8	25.7
1987	502.8	350.3	153.1	108.0	110.2	20.9	152.4	7.7	55.0	48.1	9.0	5.2	27.4
1988	511.4	363.5	161.9	116.6	107.4	22.5	148.0	7.4	49.6	48.5	6.8	6.1	29.6
1989	534.1	382.9	168.5	124.3	114.0	23.9	151.2	6.4	51.5	47.3	6.9	7.2	31.9
1990	562.4	404.2	176.0	130.7	123.1	25.6	158.2	6.1	54.6	46.6	8.0	8.1	34.8
1991	582.9	426.1	186.5	137.0	129.4	26.8	156.8	4.6	53.4	43.3	9.2	8.9	37.4
1992	588.5	432.1	189.6	140.0	129.4	26.9	156.4	5.2	50.7	41.3	10.3	10.0	38.9
1993	580.2	429.5	188.0	143.4	125.1	27.2	150.7	5.3	44.7	39.2	11.2	9.9	40.4
1994	574.7	429.3	186.1	146.6	126.1	29.3	145.4	5.8	41.9	38.1	10.2	7.5	41.9
1995	576.7	429.0	183.3	149.6	123.2	27.3	147.7	6.7	39.8	38.4	10.8	8.6	43.4
1996	579.2	429.4	182.2	149.6	125.2	27.6	149.8	6.3	39.9	38.9	11.3	9.7	43.8
1997	583.3	439.6	183.5	150.1	131.6	25.6	143.7	6.1	33.6	39.2	9.9	10.0	44.9
1998	585.5	437.8	185.8	150.9	127.7	26.6	147.7	5.8	34.1	39.4	10.8	10.2	47.5
1999	611.3	456.5	191.7	153.3	137.1	25.6	154.9	5.4	36.8	39.3	10.7	11.9	50.7
2000	633.7	476.3	202.4	157.9	144.2	28.2	157.4	5.4	37.8	40.2	8.4	10.8	54.8
2001	670.1	506.2	211.2	159.1	165.5	29.6	163.8	5.3	39.9	42.5	8.1	10.0	58.1
2002	743.0	560.8	236.3	161.5	194.4	31.5	182.2	6.1	46.4	47.9	10.0	11.9	59.8
2003	826.3	631.1	267.7	167.0	229.9	33.5	195.2	7.1	50.0	53.0	10.3	12.3	62.5
2004	891.7	683.3	284.3	175.1	258.6	34.6	208.4	6.9	55.7	58.3	9.6	13.3	64.5
2005	947.5	725.4	303.5	184.9	276.3	39.2	222.0	7.2	61.4	62.9	8.1	15.2	67.3
2006	1 000.7	765.1	314.5	195.4	295.2	40.0	235.6	7.8	66.0	65.7	9.4	17.4	69.2
2007	1 050.5	800.5	329.1	206.7	304.8	40.1	250.0	10.0	71.8	69.4	11.4	16.1	71.4
2008	1 150.6	878.9	349.6	219.5	352.2	42.3	271.7	13.7	82.6	70.8	11.4	17.5	75.7
2009	1 218.2	935.6	375.8	227.2	376.2	43.6	282.6	17.1	86.6	69.6	12.0	17.8	79.5
2010	1 297.9	1 000.7	402.8	237.2	407.1	46.5	297.2	16.7	89.3	70.2	16.0	19.2	85.9
2011	1 298.9	1 003.3	411.6	248.1	394.4	50.9	295.7	13.3	87.3	71.3	16.7	18.4	88.5
2012	1 286.5	999.3	409.1	254.6	382.5	47.0	287.3	8.1	85.2	70.5	14.5	18.7	90.2
2013	1 226.6	956.9	399.8	258.6	345.5	46.9	269.6	6.5	78.8	67.7	11.6	16.2	88.8
2014	1 215.0	951.2	404.2	262.7	333.4	49.0	263.8	5.3	74.7	64.8	12.0	17.4	89.6
2015	1 221.5	956.3	411.2	263.6	330.9	49.4	265.3	4.3	72.6	65.9	12.6	19.0	91.0
2016	1 234.1	968.6	422.5	263.8	334.0	51.8	265.5	3.4	72.2	64.5	12.3	19.5	93.7
2017	1 269.3	992.6	432.9	268.8	343.8	52.9	276.8	3.8	76.8	65.7	12.5	20.2	97.8
2018	1 347.3	1 056.9	451.6	277.3	380.1	52.2	290.4	3.7	83.7	68.8	12.3	21.8	100.0
2016													
1st quarter	1 227.5	960.9	418.2	262.1	331.4	50.7	266.7	3.6	74.4	65.0	12.2	19.2	92.3
2nd quarter	1 226.2	961.3	421.4	263.2	328.5	51.7	264.9	3.2	72.2	64.5	12.5	19.3	93.2
3rd quarter	1 237.5	972.8	424.3	264.2	337.6	53.3	264.7	3.3	71.5	64.3	12.2	19.4	94.1
4th quarter	1 245.2	979.3	426.1	265.9	338.7	51.3	265.9	3.3	70.7	64.4	12.4	19.9	95.2
2017													
1st quarter	1 248.4	978.2	428.7	266.7	334.3	51.4	270.2	3.8	72.7	64.9	12.3	19.9	96.7
2nd quarter	1 263.6	987.2	430.2	267.8	342.3	53.1	276.4	4.1	76.9	65.4	12.5	19.8	97.7
3rd quarter	1 270.2	993.0	433.9	269.4	342.9	53.1	277.2	3.5	77.3	65.9	12.1	20.3	98.1
4th quarter	1 295.1	1 011.9	438.8	271.3	355.6	53.8	283.2	3.8	80.4	66.4	13.2	20.7	98.7
2018													
1st quarter	1 318.2	1 033.9	444.0	274.1	366.7	51.0	284.3	3.6	81.1	67.2	12.0	21.6	98.8
2nd quarter	1 340.4	1 052.5	449.2	276.5	379.2	52.4	287.9	3.6	82.9	68.1	12.1	21.8	99.4
3rd quarter	1 358.6	1 068.1	454.6	278.7	385.8	51.1	290.6	3.7	82.8	69.2	12.5	22.1	100.2
4th quarter	1 371.8	1 073.1	458.6	280.0	388.5	54.1	298.7	3.9	88.3	70.5	12.7	21.8	101.6

[1] Includes general government intermediate inputs for goods and services sold to other sectors and for own-account investment.

Table 6-3. Federal Government Defense and Nondefense Consumption Expenditures by Type

(National income and product accounts, calendar years, billions of dollars, quarterly data are at seasonally adjusted annual rates.)

NIPA Table 3.10.5

Year and quarter	Defense consumption expenditures [1]						Nondefense consumption expenditures [1]					
	Total	Compensation of general government employees	Consumption of general government fixed capital	Intermediate goods and services purchased [2]			Total	Compensation of general government employees	Consumption of general government fixed capital	Intermediate goods and services purchased [2]		
				Durable goods	Nondurable goods	Services				Durable goods	Nondurable goods	Services
1960	42.6	21.1	13.8	4.4	1.7	3.5	8.4	5.6	1.3	0.0	0.5	1.9
1961	44.2	21.7	14.5	3.6	2.2	4.0	8.6	6.1	1.4	0.1	-0.3	2.3
1962	48.6	22.9	15.4	4.6	2.8	4.7	10.5	6.5	1.6	0.1	1.1	2.4
1963	50.5	23.6	16.3	4.7	2.5	5.2	10.7	7.2	1.9	0.1	0.6	2.6
1964	51.4	25.2	16.7	4.0	2.8	4.7	11.7	7.8	2.3	0.1	0.3	3.0
1965	53.9	26.4	17.1	4.2	3.2	5.1	12.6	8.3	2.8	0.1	0.4	3.1
1966	63.8	30.8	17.5	6.1	4.7	7.0	12.7	9.0	3.3	0.1	-0.7	3.3
1967	73.7	34.2	18.4	6.2	7.3	9.7	14.3	9.7	3.9	0.1	0.1	3.1
1968	81.4	37.6	19.4	7.4	8.4	10.5	15.4	10.7	4.5	0.1	0.8	2.0
1969	82.6	40.0	20.3	6.4	7.6	10.5	17.9	11.5	5.2	0.1	1.4	2.6
1970	81.8	41.9	21.3	6.1	5.4	9.6	20.6	13.3	6.0	0.1	0.5	3.6
1971	82.3	43.4	21.9	4.6	4.4	10.6	24.3	15.5	6.7	0.1	0.9	4.4
1972	84.4	45.6	22.4	5.5	4.7	9.5	27.9	17.0	7.4	0.2	1.2	5.8
1973	84.9	46.1	23.6	5.3	4.3	9.3	29.7	18.4	8.2	0.2	1.0	5.8
1974	89.5	47.7	25.3	5.0	5.2	10.7	33.6	20.1	9.4	0.3	1.4	6.7
1975	95.3	50.2	27.3	5.8	5.1	10.9	38.5	22.3	10.6	0.3	2.1	7.6
1976	99.3	51.8	29.2	5.7	4.4	11.7	40.7	24.7	11.6	0.3	2.1	7.1
1977	106.5	53.9	31.5	7.6	4.5	12.7	44.7	27.0	12.8	0.4	2.5	7.8
1978	115.1	57.6	34.4	9.0	4.9	13.3	49.0	29.5	14.3	0.5	3.2	8.4
1979	125.7	60.9	37.3	10.8	6.2	14.9	53.1	31.1	16.1	0.6	3.3	10.0
1980	143.2	66.7	40.6	12.3	10.0	18.7	61.4	34.0	18.5	0.8	5.5	11.2
1981	165.1	76.3	45.0	15.7	11.8	22.0	68.4	35.9	21.1	0.8	8.7	10.5
1982	187.2	84.8	50.4	19.1	11.5	28.1	71.2	37.2	23.3	0.8	7.5	10.7
1983	202.3	89.3	55.2	24.6	11.3	28.9	77.7	38.4	24.2	0.8	10.1	13.1
1984	217.7	95.9	62.3	26.6	10.4	30.7	78.5	40.7	26.4	0.9	5.8	13.9
1985	234.1	102.4	67.2	29.0	9.9	34.7	86.3	42.6	27.8	0.9	8.9	15.7
1986	249.0	106.2	72.6	31.7	10.2	38.3	90.8	42.9	28.9	0.9	11.7	16.1
1987	261.6	109.5	77.7	33.6	10.2	40.9	88.7	43.6	30.2	1.0	6.2	18.3
1988	276.6	113.8	84.4	33.4	10.6	45.4	86.8	48.1	32.2	1.1	-0.8	17.7
1989	286.1	117.9	90.1	32.0	10.8	46.5	96.8	50.6	34.2	1.2	5.1	18.2
1990	297.7	121.2	94.5	31.6	11.0	51.8	106.5	54.8	36.2	1.4	5.0	22.2
1991	312.7	126.8	98.5	31.0	10.7	58.1	113.4	59.7	38.6	1.5	5.9	22.3
1992	309.3	126.8	99.8	28.4	9.4	57.6	122.8	62.8	40.2	1.6	6.5	25.9
1993	302.4	121.0	101.3	26.4	8.5	58.1	127.1	67.1	42.1	1.6	7.6	23.0
1994	296.3	117.4	102.7	22.9	7.6	60.2	133.0	68.7	43.9	1.5	6.1	27.7
1995	292.2	114.1	103.3	20.9	6.3	59.8	136.8	69.2	46.4	1.4	7.5	27.3
1996	292.3	112.0	101.7	20.8	7.6	62.5	137.1	70.2	47.9	1.5	6.6	26.0
1997	293.0	111.4	100.4	20.9	7.6	64.6	146.6	72.0	49.7	1.6	8.5	28.5
1998	289.5	110.8	99.7	20.9	7.0	62.6	148.3	75.1	51.3	1.5	9.4	26.3
1999	301.8	112.7	99.4	22.2	8.2	70.9	154.7	79.0	53.9	1.7	7.2	27.0
2000	309.2	116.9	100.7	22.1	10.4	71.4	167.2	85.5	57.2	1.8	9.3	29.1
2001	325.5	123.2	99.8	22.3	10.3	84.2	180.7	88.0	59.3	1.9	12.2	34.7
2002	358.5	139.3	100.4	23.4	11.4	100.0	202.3	97.0	61.1	2.2	12.6	44.7
2003	411.2	160.8	103.5	26.0	13.5	124.6	219.0	106.9	63.5	2.2	15.4	48.2
2004	448.9	172.2	108.8	28.5	17.1	139.8	234.4	112.1	66.3	2.5	17.1	53.7
2005	478.0	186.0	115.2	29.8	21.0	145.3	247.5	117.5	69.7	2.7	19.0	58.6
2006	501.3	193.0	122.1	32.8	22.3	152.5	263.8	121.5	73.3	3.0	20.3	64.4
2007	528.1	201.5	129.6	37.0	24.0	157.9	272.4	127.6	77.1	3.0	20.2	62.8
2008	583.2	214.3	138.3	43.2	30.3	180.0	295.7	135.2	81.2	3.3	23.6	71.9
2009	614.3	229.2	143.5	47.1	24.5	193.0	321.3	146.6	83.7	3.5	26.6	81.4
2010	651.8	244.0	149.5	47.4	26.8	207.8	348.9	158.8	87.7	3.9	28.4	92.7
2011	662.0	250.8	156.3	46.2	33.0	201.8	341.3	160.8	91.8	3.7	24.8	84.9
2012	650.3	248.0	159.6	45.2	31.9	191.6	348.9	161.1	95.0	3.7	24.0	86.2
2013	611.2	239.6	160.5	39.9	28.2	169.4	345.7	160.2	98.1	3.5	22.6	81.9
2014	598.7	239.1	160.9	36.5	26.0	162.8	352.5	165.0	101.8	3.6	22.0	82.5
2015	587.4	239.0	159.6	35.6	21.0	158.8	368.9	172.2	104.0	3.8	23.4	88.2
2016	588.3	242.5	157.6	35.3	20.6	159.8	380.3	180.0	106.2	4.0	23.7	90.6
2017	600.0	246.6	158.5	37.3	22.8	162.9	392.6	186.3	110.3	4.0	24.5	92.3
2018	637.4	258.7	162.0	39.5	26.4	178.4	419.5	193.0	115.3	4.5	28.2	103.1
2016												
1st quarter	584.6	241.2	157.2	35.5	19.3	158.7	376.2	177.0	104.9	4.0	23.7	90.2
2nd quarter	582.4	242.2	157.5	34.2	20.5	155.6	378.9	179.2	105.7	4.0	23.7	90.6
3rd quarter	592.3	243.2	157.7	35.4	21.1	162.6	380.5	181.2	106.5	4.0	23.8	90.7
4th quarter	593.8	243.5	158.2	36.2	21.6	162.4	385.5	182.6	107.7	4.0	23.7	90.9
2017												
1st quarter	590.8	244.4	158.0	36.7	22.9	156.8	387.4	184.3	108.7	4.0	23.8	90.0
2nd quarter	599.8	244.9	158.1	38.2	22.1	164.6	387.4	185.3	109.7	3.9	23.6	89.7
3rd quarter	599.6	247.2	158.5	35.8	22.2	164.0	393.5	186.7	110.8	4.0	24.4	92.4
4th quarter	609.7	250.0	159.3	38.3	24.0	166.1	402.2	188.7	112.0	4.2	26.0	97.0
2018												
1st quarter	618.1	253.5	160.5	37.6	25.5	168.8	415.8	190.5	113.6	4.4	27.7	102.7
2nd quarter	634.9	257.1	161.6	39.9	25.7	178.2	417.7	192.2	114.9	4.4	28.1	102.9
3rd quarter	644.9	260.8	162.8	41.6	26.7	180.5	423.2	193.8	116.0	4.5	28.6	103.9
4th quarter	651.8	263.3	163.2	39.1	27.7	185.9	421.3	195.3	116.8	4.5	28.4	103.0

[1] Excludes government sales to other sectors and government own-account investment (construction and software).
[2] Includes general government intermediate inputs for goods and services sold to other sectors and for own-account investment.

Table 6-4. National Defense Consumption Expenditures and Gross Investment: Selected Detail

(National income and product accounts, calendar years, billions of dollars, quarterly data are at seasonally adjusted annual rates.)

NIPA Table 3.11.5

Year and quarter	Compensation of general government employees		Intermediate goods and services purchased [1]						Defense gross investment				
			Durable goods	Nondurable goods		Services			Aircraft	Missiles	Ships	Software	Research and development
	Military	Civilian	Aircraft	Petroleum products	Ammunition	Installation support	Weapons support	Personnel support					
1975	32.0	18.2	2.1	2.9	1.1	4.8	1.7	1.6	3.6	1.3	2.2	0.5	10.0
1976	32.6	19.2	2.0	2.5	0.6	5.3	1.9	1.7	3.4	1.4	2.4	0.5	10.8
1977	33.7	20.2	3.3	2.4	0.8	5.8	1.9	1.9	3.7	1.2	3.1	0.6	11.7
1978	35.5	22.2	3.5	2.5	1.0	5.9	2.1	2.1	3.7	1.1	3.9	0.7	12.6
1979	37.4	23.5	4.8	3.6	1.2	6.6	2.7	2.1	4.8	1.8	4.3	0.8	14.0
1980	41.5	25.2	5.6	6.8	1.4	8.3	3.9	2.3	6.1	2.3	4.1	1.0	16.2
1981	48.8	27.6	7.7	7.7	1.6	9.6	4.8	2.9	7.5	2.8	5.1	1.3	19.6
1982	54.9	29.9	10.2	6.8	2.1	12.6	6.3	3.6	8.4	3.4	6.2	1.4	23.7
1983	57.8	31.5	13.6	6.4	2.5	12.4	7.0	3.9	10.1	4.6	7.1	1.7	28.1
1984	62.0	33.9	14.0	5.9	2.2	13.8	7.2	3.7	10.9	5.7	8.0	2.2	32.8
1985	66.2	36.1	15.4	5.8	1.3	15.1	8.4	5.0	13.4	6.6	9.0	2.7	38.6
1986	69.4	36.8	17.2	3.6	3.6	16.0	9.4	6.2	17.9	7.9	8.9	3.2	40.7
1987	72.2	37.3	18.2	3.9	2.8	17.7	9.2	6.8	17.6	8.7	8.8	3.6	44.5
1988	74.8	39.0	17.9	3.5	3.5	18.3	10.3	9.5	13.5	7.8	8.6	4.3	44.3
1989	76.5	41.4	16.4	4.2	3.1	18.0	10.5	10.2	12.2	8.8	10.0	4.8	42.5
1990	78.5	42.7	14.8	5.3	2.8	21.4	11.6	10.1	12.0	11.2	10.8	5.2	41.4
1991	82.4	44.4	13.6	4.7	2.7	22.9	10.0	9.5	9.2	10.8	10.2	5.4	37.9
1992	80.9	45.9	12.2	3.5	2.6	23.1	9.4	13.7	8.3	10.6	10.1	5.6	35.6
1993	75.1	45.9	10.7	3.2	2.5	25.2	8.8	14.7	9.3	7.9	8.7	5.4	33.7
1994	71.7	45.7	9.2	3.0	1.8	26.2	9.4	16.3	10.5	5.7	8.1	5.3	32.8
1995	69.5	44.6	8.9	2.8	1.2	24.9	9.5	17.0	9.0	4.7	8.0	5.3	33.1
1996	67.7	44.3	8.8	3.4	1.4	25.6	9.0	19.3	9.2	4.1	6.8	5.5	33.4
1997	67.8	43.6	9.4	2.9	1.7	24.7	10.5	21.3	5.8	2.9	6.1	5.6	33.6
1998	68.0	42.8	9.9	2.1	1.9	23.5	10.0	20.6	5.8	3.3	6.4	5.9	33.5
1999	69.8	42.9	10.5	2.6	1.9	24.5	11.3	26.0	7.0	2.9	6.7	5.9	33.3
2000	72.9	44.1	9.8	4.1	1.8	24.7	11.7	26.1	7.8	2.7	6.6	6.0	34.2
2001	78.4	44.8	9.8	4.2	2.1	27.1	14.8	32.4	8.5	3.3	7.2	5.8	36.7
2002	89.7	49.6	9.8	4.6	2.4	30.6	18.0	41.3	9.6	3.3	8.4	5.6	42.3
2003	107.4	53.4	11.4	5.3	2.7	35.8	22.6	49.5	9.2	3.5	9.3	5.7	47.3
2004	113.9	58.3	12.0	7.0	3.6	36.8	23.8	62.4	11.1	4.0	9.8	5.8	52.5
2005	123.3	62.7	10.8	10.1	4.0	36.0	26.5	66.9	13.5	4.0	9.6	6.1	56.8
2006	128.1	64.9	11.1	11.4	4.1	38.3	27.8	71.4	13.6	4.5	10.5	6.4	59.3
2007	133.4	68.1	11.5	12.2	4.2	38.4	29.0	70.7	12.9	4.3	10.3	6.8	62.5
2008	143.0	71.3	12.9	17.6	4.3	41.6	31.3	86.5	13.5	4.3	11.0	7.4	63.4
2009	152.5	76.8	14.9	10.5	4.2	44.2	32.8	93.4	13.8	5.1	11.1	7.5	62.1
2010	159.1	84.9	16.2	13.7	4.2	46.5	37.4	99.3	16.7	5.6	11.8	8.1	62.1
2011	160.9	89.9	18.0	19.2	4.2	44.5	35.6	97.1	20.2	5.1	11.7	8.7	62.6
2012	159.1	88.9	19.3	18.3	4.3	41.9	33.6	97.3	20.2	6.9	12.0	9.0	61.5
2013	153.2	86.3	17.9	15.0	3.6	35.5	28.9	89.9	21.7	6.4	12.5	9.1	58.5
2014	150.0	89.1	16.3	13.3	3.1	39.2	31.1	78.0	19.2	6.6	13.3	9.4	55.4
2015	147.4	91.6	15.8	8.3	2.8	36.6	32.2	76.8	17.6	6.6	13.5	9.4	56.5
2016	148.0	94.5	15.8	6.8	3.1	38.3	32.2	76.1	16.4	4.8	14.4	10.1	54.5
2017	148.6	98.1	16.8	7.9	3.6	39.8	33.3	76.3	18.3	4.7	14.1	10.7	55.0
2018	156.1	102.6	17.0	10.2	5.1	40.8	35.8	87.8	20.3	4.5	15.5	11.7	57.1
2012													
1st quarter	160.2	89.3	19.3	19.0	4.4	43.5	36.1	96.3	19.3	6.8	11.3	8.8	61.9
2nd quarter	159.5	89.0	18.7	19.2	4.4	42.0	33.8	94.3	20.4	7.0	12.4	8.9	61.6
3rd quarter	158.8	88.9	19.5	19.6	4.2	42.3	34.3	101.3	22.5	6.6	11.5	9.0	61.3
4th quarter	157.9	88.4	19.5	15.4	4.0	39.9	30.3	97.2	18.5	7.0	12.7	9.4	61.1
2013													
1st quarter	154.3	88.0	17.8	16.3	3.8	36.4	28.9	94.7	21.9	5.2	11.6	9.4	59.9
2nd quarter	153.6	87.4	18.0	15.1	3.7	35.6	29.3	94.7	19.4	6.5	12.6	9.0	59.1
3rd quarter	152.9	81.8	18.4	14.3	3.5	34.3	28.2	89.5	21.1	7.9	12.8	9.0	58.1
4th quarter	152.2	88.1	17.5	14.1	3.4	35.9	29.2	80.7	24.3	6.2	13.2	9.1	57.0
2014													
1st quarter	151.3	87.6	17.2	14.7	3.3	37.8	30.3	79.1	17.3	6.1	12.4	9.2	55.9
2nd quarter	150.7	88.7	16.7	13.5	3.2	39.2	30.4	74.1	20.6	6.4	14.0	9.3	55.2
3rd quarter	149.6	89.6	15.2	13.2	3.0	40.8	33.0	87.2	19.3	6.6	13.6	9.5	55.1
4th quarter	148.5	90.6	16.1	11.7	2.9	39.0	30.9	71.4	19.5	7.4	13.0	9.4	55.4
2015													
1st quarter	147.5	90.8	16.0	8.9	2.8	37.3	31.8	77.1	15.6	6.5	13.8	9.4	56.3
2nd quarter	147.1	91.2	15.3	8.9	2.8	37.0	33.1	77.1	19.2	6.3	13.2	9.4	56.8
3rd quarter	147.3	92.0	15.6	8.1	2.8	35.8	31.9	76.3	18.1	5.8	13.7	9.4	56.7
4th quarter	147.8	92.5	16.3	7.3	2.9	36.4	31.8	76.9	17.6	7.7	13.5	9.5	56.2
2016													
1st quarter	148.2	93.0	16.3	6.0	2.9	37.1	32.1	76.4	18.7	5.1	14.6	9.8	55.1
2nd quarter	148.3	94.0	14.6	6.6	3.0	37.9	31.5	73.0	16.6	5.6	13.8	10.0	54.5
3rd quarter	148.2	95.0	15.6	7.1	3.2	38.8	32.8	77.8	15.2	5.0	14.7	10.2	54.1
4th quarter	147.6	95.9	16.5	7.4	3.3	39.3	32.5	77.2	15.3	3.7	14.5	10.3	54.1
2017													
1st quarter	147.6	96.8	16.7	8.3	3.4	38.9	32.1	72.5	17.6	5.1	12.0	10.4	54.5
2nd quarter	147.6	97.4	18.3	7.2	3.6	39.9	33.4	78.0	18.4	4.8	14.9	10.6	54.8
3rd quarter	148.7	98.4	15.5	7.2	3.6	40.4	34.2	75.9	18.1	4.7	14.4	10.7	55.2
4th quarter	150.4	99.6	16.7	9.0	3.7	40.0	33.7	78.8	19.1	4.2	15.2	10.9	55.6
2018													
1st quarter	152.9	100.6	15.7	10.2	4.0	39.7	33.9	81.4	17.7	4.7	15.5	11.2	55.9
2nd quarter	154.9	102.2	17.5	10.2	4.6	40.9	35.7	87.7	19.5	4.1	15.8	11.6	56.5
3rd quarter	157.4	103.4	18.9	10.3	5.4	40.9	36.3	89.3	19.8	4.1	15.0	11.9	57.4
4th quarter	159.0	104.3	16.0	10.3	6.4	41.7	37.3	92.7	24.0	5.3	15.7	12.1	58.4

[1] Includes general government intermediate inputs for goods and services sold to other sectors and for own-account investment.

Table 6-5. Federal Government Output, Lending and Borrowing, and Net Investment

(National income and product accounts, calendar years, billions of dollars, quarterly data are at seasonally adjusted annual rates.)

NIPA Tables 3.2, 3.10.5

Year and quarter	Output						Net lending (net borrowing -)							
	Gross		Value added		Intermediate goods and services purchased [1]		Net saving, current (surplus +, deficit -)	Plus: capital transfer receipts	Minus			Plus: Consumption of fixed capital	Equals: Net lending (borrowing -)	Net investment
	Defense	Non-defense	Defense	Non-defense	Defense	Non-defense			Gross investment	Capital transfer payments	Net purchases of non-produced assets			
1960	44.4	9.3	34.8	6.8	9.6	2.4	0.2	1.8	21.9	2.6	0.5	15.2	-7.9	6.7
1961	46.0	9.6	36.2	7.5	9.8	2.1	-4.7	2.0	24.6	2.9	0.5	16.0	-14.7	8.6
1962	50.4	11.7	38.3	8.2	12.1	3.5	-5.4	2.1	26.5	3.1	0.6	17.2	-16.3	9.3
1963	52.3	12.3	39.9	9.1	12.4	3.2	-2.2	2.2	26.7	3.6	0.5	18.4	-12.3	8.3
1964	53.3	13.6	41.9	10.1	11.4	3.5	-6.9	2.6	27.1	4.1	0.6	19.2	-16.9	7.9
1965	55.9	14.7	43.5	11.1	12.4	3.6	-5.5	2.8	26.7	4.0	0.5	20.0	-13.9	6.7
1966	66.1	15.1	48.3	12.4	17.9	2.7	-7.0	3.0	30.1	4.4	0.6	21.0	-18.0	9.1
1967	75.7	16.8	52.5	13.5	23.2	3.3	-19.5	3.1	31.9	4.3	-0.2	22.5	-30.0	9.4
1968	83.4	18.0	57.0	15.1	26.4	2.9	-13.8	3.1	31.1	6.0	-0.9	24.2	-22.7	6.9
1969	84.9	20.7	60.3	16.6	24.6	4.1	-5.1	3.6	30.7	5.9	0.1	25.8	-12.5	4.9
1970	84.2	23.5	63.2	19.3	21.0	4.3	-34.8	3.7	30.5	5.3	-0.3	27.6	-39.0	2.9
1971	84.9	27.6	65.3	22.2	19.6	5.4	-50.9	4.6	27.8	5.9	-0.4	29.1	-50.6	-1.3
1972	87.6	31.6	68.0	24.4	19.7	7.2	-49.0	5.4	29.3	6.1	-0.7	30.2	-48.1	-0.9
1973	88.6	33.6	69.7	26.6	18.9	6.9	-38.3	5.1	31.7	6.0	-3.2	32.4	-35.3	-0.7
1974	93.8	37.8	73.0	29.5	20.8	8.3	-41.3	4.8	35.7	7.9	-5.7	35.4	-39.1	0.3
1975	99.4	42.9	77.6	32.9	21.8	10.0	-97.9	4.9	39.9	9.7	-0.4	38.7	-103.4	1.2
1976	102.8	45.8	81.0	36.2	21.8	9.5	-80.9	5.6	44.8	10.6	-2.4	41.7	-86.6	3.1
1977	110.3	50.5	85.4	39.8	24.8	10.8	-73.4	7.2	49.2	11.2	-1.4	45.3	-80.0	3.9
1978	119.2	55.8	92.0	43.8	27.2	12.1	-62.0	5.2	54.7	12.0	-0.6	49.7	-73.2	5.0
1979	130.1	61.1	98.2	47.2	31.9	13.9	-47.4	5.5	61.7	14.5	-2.8	54.6	-60.7	7.1
1980	148.2	70.0	107.3	52.5	40.9	17.5	-88.8	6.5	70.2	16.7	-4.1	60.4	-104.8	9.8
1981	170.9	77.0	121.3	57.0	49.6	20.1	-88.1	6.9	80.4	15.7	-5.5	67.5	-104.3	12.9
1982	193.9	79.5	135.2	60.5	58.7	19.0	-167.4	7.5	90.0	14.7	-3.7	75.4	-185.4	14.6
1983	209.3	86.6	144.4	62.6	64.8	24.0	-207.2	5.8	102.3	15.6	-4.9	81.1	-233.3	21.2
1984	225.9	87.6	158.2	67.1	67.7	20.5	-196.5	6.0	115.6	17.8	-4.0	90.6	-229.4	25.0
1985	243.2	95.8	169.6	70.4	73.6	25.4	-199.2	6.4	132.5	19.6	-1.2	97.0	-246.8	35.5
1986	258.9	100.4	178.8	71.8	80.1	28.6	-215.9	7.0	141.9	20.1	-3.0	103.6	-264.3	38.3
1987	272.0	99.3	187.2	73.8	84.8	25.5	-165.7	7.2	152.4	19.1	-0.4	110.1	-219.5	42.3
1988	287.6	98.3	198.2	80.3	89.4	18.0	-160.0	7.6	148.0	19.8	-0.2	118.9	-201.0	29.1
1989	297.4	109.3	208.0	84.8	89.4	24.5	-159.4	8.9	151.2	20.2	-0.7	126.9	-194.3	24.3
1990	310.1	119.6	215.7	91.0	94.5	28.6	-203.3	11.6	158.2	28.2	-0.8	133.5	-243.8	24.7
1991	324.9	128.0	225.2	98.3	99.7	29.7	-248.4	11.0	156.8	26.5	0.1	140.1	-280.8	16.7
1992	322.0	137.0	226.6	103.0	95.4	34.0	-334.5	11.3	156.4	22.6	0.2	143.3	-359.2	13.1
1993	315.3	141.3	222.3	109.2	92.9	32.2	-313.5	12.9	150.7	24.3	0.2	147.0	-328.7	3.7
1994	310.8	147.9	220.0	112.6	90.7	35.3	-255.6	15.1	145.4	26.0	0.1	150.3	-261.7	-4.9
1995	304.4	151.8	217.4	115.6	87.0	36.2	-242.1	14.9	147.7	27.9	-7.9	153.5	-241.3	-5.8
1996	304.7	152.2	213.7	118.1	91.0	34.2	-179.4	17.5	149.8	28.4	-4.8	153.7	-181.7	-3.9
1997	304.9	160.3	211.8	121.7	93.0	38.6	-92.0	20.6	143.7	29.2	-8.8	154.3	-81.2	-10.6
1998	301.0	163.5	210.4	126.4	90.5	37.2	1.4	25.2	147.7	28.9	-6.0	155.4	11.4	-7.7
1999	313.3	168.8	212.0	133.0	101.3	35.8	66.9	28.8	154.9	36.6	-1.2	158.2	63.6	-3.3
2000	321.6	183.0	217.6	142.7	104.0	40.3	155.5	28.1	157.4	37.0	-0.6	163.1	152.9	-5.7
2001	339.8	196.0	223.0	147.3	116.8	48.8	14.0	28.0	163.8	42.5	-1.5	164.4	1.5	-0.6
2002	374.6	217.7	239.7	158.1	134.9	59.6	-271.5	25.3	182.2	49.7	-0.3	166.8	-310.9	15.4
2003	428.3	236.3	264.3	170.4	164.1	65.9	-404.1	22.0	195.2	63.1	-0.8	172.2	-467.2	23.0
2004	466.3	251.6	281.0	178.4	185.4	73.3	-400.9	24.6	208.4	64.0	-0.6	180.5	-467.5	27.9
2005	497.2	267.5	301.2	187.2	196.0	80.3	-305.9	25.0	222.0	85.3	-1.8	190.5	-396.0	31.5
2006	522.6	282.5	315.1	194.8	207.5	87.7	-227.6	27.8	235.6	71.0	-14.2	201.3	-290.9	34.3
2007	550.0	290.6	331.1	204.6	218.9	86.0	-266.1	26.5	250.0	79.4	-3.3	212.8	-353.0	37.2
2008	606.1	315.2	352.6	216.4	253.5	98.7	-631.1	28.3	271.7	145.9	-20.4	225.9	-774.2	45.8
2009	637.3	341.8	372.7	230.3	264.6	111.5	-1 248.9	20.6	282.6	206.9	-8.9	233.6	-1 475.3	49.0
2010	675.6	371.5	393.5	246.5	282.1	125.0	-1 325.1	15.1	297.2	141.4	-1.0	243.7	-1 504.0	53.5
2011	688.1	366.0	407.1	252.6	281.0	113.4	-1 242.0	9.6	295.7	123.4	-0.9	254.9	-1 395.6	40.8
2012	676.3	369.9	407.6	256.1	268.7	113.9	-1 078.6	14.1	287.3	99.0	-2.0	261.6	-1 187.3	25.7
2013	637.6	366.3	400.1	258.3	237.5	108.0	-637.9	20.9	269.6	79.0	-2.5	265.9	-697.3	3.7
2014	625.3	374.9	400.0	266.8	225.3	108.1	-604.3	18.8	263.8	77.2	-2.6	270.2	-653.7	-6.4
2015	614.1	391.5	398.7	276.1	215.4	115.4	-570.1	20.2	265.3	71.6	-30.8	271.4	-584.6	-6.1
2016	615.9	404.5	400.1	286.2	215.7	118.3	-677.0	20.1	265.5	73.9	-8.6	271.8	-716.0	-6.3
2017	628.1	417.4	405.1	296.6	223.0	120.8	-724.7	273.2	276.8	84.9	-2.2	277.1	-534.0	-0.3
2018	665.0	444.0	420.7	308.3	244.3	135.8	-1 009.8	22.9	290.4	75.6	-0.9	286.1	-1 065.8	4.3
2016														
1st quarter	611.8	399.8	398.3	281.9	213.5	117.9	-644.5	19.6	266.7	72.7	-0.1	269.9	-694.3	-3.2
2nd quarter	609.9	403.2	399.7	284.9	210.2	118.3	-674.8	20.0	264.9	72.5	-32.6	271.1	-688.5	-6.2
3rd quarter	620.0	406.1	400.8	287.7	219.1	118.4	-687.2	19.9	264.7	73.8	-1.5	272.2	-732.1	-7.5
4th quarter	621.8	408.8	401.7	290.3	220.1	118.6	-701.6	20.8	265.9	76.4	-0.3	273.9	-748.9	-8.0
2017														
1st quarter	618.8	410.8	402.4	293.0	216.5	117.8	-685.0	22.1	270.2	76.3	-0.5	274.9	-734.0	-4.7
2nd quarter	628.0	412.3	403.0	295.0	225.0	117.3	-699.2	22.8	276.4	77.2	-0.5	276.1	-753.5	0.3
3rd quarter	627.7	418.4	405.7	297.6	222.1	120.8	-707.1	23.2	277.2	112.1	-7.3	277.7	-788.1	-0.5
4th quarter	637.8	427.9	409.3	300.8	228.4	127.2	-807.6	1 024.5	283.2	74.0	-0.3	279.8	139.8	3.4
2018														
1st quarter	645.9	439.0	414.0	304.1	231.9	134.8	-976.3	24.1	284.3	74.0	-0.3	282.7	-1 027.5	1.6
2nd quarter	662.5	442.4	418.7	307.0	243.8	135.4	-1 013.8	22.5	287.9	73.8	-0.4	285.2	-1 067.3	2.7
3rd quarter	672.4	446.7	423.6	309.8	248.8	137.0	-981.3	22.5	290.6	79.6	-0.3	287.6	-1 041.1	3.0
4th quarter	679.3	448.0	426.5	312.1	252.7	135.8	-1 067.6	22.3	298.7	74.9	-2.4	289.0	-1 127.5	9.7

[1] Includes general government intermediate inputs for goods and services sold to other sectors and for own-account investment.

Table 6-6. Chain-Type Quantity Indexes for Federal Government Defense and Nondefense Consumption Expenditures and Gross Investment

(Seasonally adjusted, 2012 = 100.)

NIPA Tables 3.9.3, 3.10.3

Year and quarter	Defense consumption expenditures [1]						Defense gross investment	Nondefense consumption expenditures [1]						Nondefense gross investment
	Total	Compensation of general government employees	Consumption of general government fixed capital	Intermediate goods and services purchased [2]				Total	Compensation of general government employees	Consumption of general government fixed capital	Intermediate goods and services purchased [2]			
				Durable goods	Non-durable goods	Services					Durable goods	Non-durable goods excluding CCC inventory change	Services	
1960	60.2	118.5	49.5	39.1	64.2	18.7	45.7	20.9	48.3	6.6	1.9	6.8	15.8	12.7
1961	61.9	121.1	51.9	32.1	81.9	20.8	50.5	20.5	49.5	7.3	2.6	10.5	18.9	15.0
1962	67.2	126.3	54.7	39.6	107.2	23.7	51.9	24.4	52.1	8.2	3.1	15.7	18.7	18.5
1963	67.5	124.0	56.9	39.9	95.8	24.7	47.5	24.3	55.2	9.5	3.5	15.8	19.7	24.7
1964	66.1	124.1	58.3	33.7	109.1	20.4	44.6	25.3	56.5	11.3	4.5	15.9	22.2	29.6
1965	67.1	124.2	59.0	35.1	123.2	20.9	40.4	26.1	57.5	13.4	4.7	16.8	21.8	33.8
1966	76.2	136.9	59.8	51.0	175.7	26.8	44.9	25.3	59.8	15.9	5.0	17.0	22.6	37.8
1967	85.7	149.0	61.4	50.4	269.3	34.2	50.4	28.0	62.8	18.2	4.9	19.0	20.6	34.8
1968	89.6	151.0	63.0	58.3	309.3	35.4	46.0	28.3	64.2	20.5	3.2	12.6	12.6	35.5
1969	86.5	151.2	63.3	48.6	274.6	33.1	42.9	30.9	64.8	22.6	3.2	13.1	15.3	34.2
1970	79.4	140.2	62.9	43.5	191.2	30.7	39.2	32.2	65.7	24.4	3.9	16.5	21.0	32.8
1971	73.5	129.5	61.2	31.4	156.4	31.2	30.7	35.0	67.8	26.0	4.8	20.3	24.5	32.7
1972	68.8	119.7	59.0	38.3	159.1	26.8	27.6	38.1	69.8	27.3	7.9	24.8	32.2	33.8
1973	64.1	113.6	57.2	35.0	118.3	24.7	28.8	38.2	70.0	28.7	7.2	23.8	30.6	34.0
1974	62.3	111.9	55.9	30.4	102.4	26.0	31.0	40.7	73.5	30.0	7.7	25.1	31.7	34.1
1975	60.9	110.4	54.9	32.2	84.8	23.9	32.3	42.5	74.8	31.2	8.1	22.8	32.0	35.1
1976	59.5	108.2	54.3	29.1	67.5	23.9	34.4	42.3	78.5	32.6	8.5	22.6	26.8	37.4
1977	59.8	107.2	54.1	36.5	64.5	23.8	35.4	44.0	80.2	34.1	11.1	22.3	29.2	39.5
1978	60.2	108.0	53.9	40.1	65.4	23.0	35.7	45.7	82.3	35.9	12.8	32.9	30.0	43.6
1979	60.6	107.1	53.9	43.7	67.4	23.5	39.1	46.5	82.2	37.8	15.0	32.3	32.8	44.1
1980	62.1	107.7	54.4	46.0	73.7	26.0	41.9	49.7	84.4	39.7	16.9	31.4	33.4	45.2
1981	65.1	111.1	55.5	54.1	78.4	28.1	46.5	50.9	81.9	41.4	15.8	58.7	28.4	44.5
1982	68.6	113.6	57.3	59.4	78.1	33.7	52.5	49.9	80.4	42.8	14.6	43.8	26.9	41.3
1983	71.5	115.4	60.1	70.9	81.4	34.0	60.5	53.4	81.6	44.1	15.6	54.8	32.5	41.2
1984	73.2	117.0	63.7	72.0	77.0	35.5	68.1	51.7	81.6	45.4	16.9	57.0	33.4	42.4
1985	76.8	119.2	68.5	77.7	75.4	39.1	80.4	54.7	81.8	46.6	17.4	49.6	36.2	44.7
1986	80.8	119.6	73.9	83.7	95.8	41.7	89.5	56.5	80.5	47.9	17.2	42.3	36.0	45.8
1987	83.6	120.7	79.3	88.9	93.9	42.6	97.0	55.1	82.0	49.3	19.6	49.0	40.5	48.8
1988	85.3	119.2	83.8	91.2	91.2	44.8	90.6	51.9	84.1	50.7	20.7	48.8	38.2	48.2
1989	85.8	119.2	86.9	87.7	88.2	44.8	88.3	56.2	84.5	52.2	23.4	43.6	37.8	50.6
1990	86.1	118.5	89.1	84.5	78.3	47.5	88.3	60.2	88.7	53.8	26.7	46.4	44.9	54.5
1991	87.4	118.0	90.1	81.1	79.4	51.2	81.7	60.6	88.5	55.7	28.4	40.6	44.5	58.1
1992	83.6	111.3	90.0	72.8	73.6	49.5	77.0	64.1	90.1	57.3	30.9	50.5	50.1	61.6
1993	80.8	106.5	88.9	65.9	67.6	48.9	68.9	63.6	90.0	58.9	30.1	55.0	42.9	62.7
1994	77.2	101.0	87.1	56.5	61.5	49.4	64.4	64.2	87.5	60.3	28.3	46.2	50.4	59.7
1995	74.1	95.5	85.0	51.5	50.3	47.6	61.8	63.6	84.0	61.7	27.6	52.1	48.6	61.0
1996	72.7	91.4	83.0	51.1	55.5	48.7	61.3	62.3	81.8	63.3	31.4	45.0	45.8	62.9
1997	71.6	88.1	81.2	51.3	56.0	49.4	57.0	65.1	81.3	65.0	33.8	54.9	48.8	62.4
1998	69.7	85.3	79.4	51.7	57.8	47.0	57.1	64.6	82.2	66.8	34.0	59.7	44.7	65.9
1999	70.7	83.2	77.9	55.0	64.2	52.1	57.7	65.5	82.4	69.1	38.1	44.4	45.3	69.9
2000	69.7	82.6	76.7	54.4	67.3	51.1	58.3	68.1	84.9	71.5	41.9	50.5	47.5	69.0
2001	71.6	83.1	75.7	54.7	69.6	58.0	61.6	72.0	84.9	73.5	45.1	67.1	54.4	70.6
2002	75.5	85.3	75.7	57.3	80.0	67.4	70.8	77.2	86.4	75.6	54.9	73.3	68.2	75.8
2003	81.7	88.2	76.9	62.9	87.3	81.0	76.6	80.4	88.6	77.5	56.5	88.0	71.8	78.3
2004	85.7	89.9	78.9	68.3	97.9	88.0	83.0	82.3	87.8	79.5	62.6	98.5	77.3	79.2
2005	86.7	90.2	81.5	70.0	99.0	88.8	88.5	83.5	87.7	81.7	68.9	99.7	81.8	80.4
2006	87.4	88.9	84.5	75.4	96.3	90.3	92.6	86.4	87.8	84.4	77.2	98.4	87.0	84.2
2007	89.0	88.6	87.6	84.8	98.3	90.8	98.6	86.3	88.4	87.1	79.0	95.6	82.0	85.4
2008	94.8	91.9	91.2	97.9	100.1	100.2	106.6	91.2	91.7	89.6	88.6	106.3	90.7	88.5
2009	100.8	97.8	94.5	106.1	105.9	107.1	110.4	98.0	96.5	92.3	96.7	123.2	102.1	92.6
2010	104.4	101.0	97.3	106.7	102.3	113.4	111.1	103.3	100.7	94.8	106.5	129.8	113.1	100.6
2011	102.9	102.0	99.2	103.0	104.3	106.7	105.9	98.5	99.5	97.5	99.8	106.9	100.0	101.0
2012	100.0	100.0	100.0	100.0	100.0	100.0	100.0	100.0	100.0	100.0	100.0	100.0	100.0	100.0
2013	93.4	96.5	99.7	87.6	88.9	87.2	93.0	97.4	97.1	102.3	94.7	92.3	93.9	93.7
2014	90.1	94.5	98.5	79.7	83.3	82.4	86.9	97.2	97.0	104.3	97.9	90.2	93.0	94.1
2015	88.1	92.2	97.0	77.3	84.4	80.3	85.5	100.4	98.1	106.1	104.0	97.4	99.8	96.5
2016	87.8	92.1	95.6	76.9	91.0	80.1	84.0	102.2	100.1	107.9	108.0	100.4	101.9	98.6
2017	87.9	92.2	94.7	80.9	92.4	79.9	87.0	102.6	99.9	109.9	108.1	101.0	101.6	100.8
2018	90.4	93.1	94.6	85.0	97.6	85.1	91.4	105.8	99.2	111.9	118.1	110.9	109.9	101.3
2016														
1st quarter	88.0	92.1	96.1	77.3	90.7	80.1	86.1	102.1	99.6	107.2	107.2	101.5	102.3	97.7
2nd quarter	87.0	92.1	95.8	74.3	90.4	78.0	84.1	102.2	100.1	107.7	108.0	100.3	102.1	98.5
3rd quarter	88.2	92.1	95.5	77.1	91.3	81.2	83.3	102.0	100.4	108.2	108.2	100.1	101.7	98.7
4th quarter	88.1	92.0	95.2	78.8	91.5	81.0	82.6	102.7	100.3	108.7	108.5	99.9	101.6	99.6
2017														
1st quarter	87.1	92.0	95.0	79.7	93.7	77.6	84.3	102.4	100.4	109.2	106.6	99.1	99.8	100.3
2nd quarter	88.2	91.9	94.8	82.8	91.8	81.0	87.3	101.7	99.9	109.7	105.6	98.3	99.0	100.8
3rd quarter	87.7	92.2	94.6	77.8	90.5	80.4	87.2	102.6	99.7	110.2	107.9	100.7	101.8	100.7
4th quarter	88.5	92.6	94.6	83.1	93.6	80.7	89.2	103.8	99.5	110.6	112.3	106.1	105.7	101.7
2018														
1st quarter	88.6	92.5	94.5	81.2	95.0	81.3	89.4	106.1	99.4	111.1	117.6	111.1	110.9	100.8
2nd quarter	90.3	93.0	94.5	85.9	95.2	85.0	90.7	105.8	99.7	111.6	117.8	110.4	109.6	100.8
3rd quarter	91.1	93.4	94.6	89.3	98.3	85.8	90.9	106.5	99.8	112.2	118.9	111.4	110.0	101.6
4th quarter	91.6	93.5	94.7	83.6	101.7	88.1	94.7	104.6	97.9	112.8	118.2	110.4	108.9	102.3

[1] Excludes government sales to other sectors and government own-account investment (construction and software).
[2] Includes general government intermediate inputs for goods and services sold to other sectors and for own-account investment.

Table 6-7. Chain-Type Quantity Indexes for National Defense Consumption Expenditures and Gross Investment: Selected Detail

(Seasonally adjusted, 2012 = 100.) NIPA Table 3.11.3

Year and quarter	Defense consumption expenditures								Defense gross investment				
	Compensation of general government employees		Intermediate goods and services purchased [1]						Aircraft	Missiles	Ships	Software	Research and development
			Durable goods	Nondurable goods		Services							
	Military	Civilian	Aircraft	Petroleum products	Ammunition	Installation support	Weapons support	Personnel support					
1975	109.0	111.8	28.1	149.7	74.4	45.7	19.9	10.2	23.1	20.1	69.5	6.8	52.2
1976	106.3	110.5	23.6	123.7	37.4	47.1	21.4	9.6	21.6	17.1	71.3	7.1	53.9
1977	105.4	109.4	36.0	105.9	51.1	46.4	20.4	9.9	22.5	13.4	83.2	7.3	55.5
1978	104.4	113.3	36.8	105.0	59.0	43.4	20.4	9.8	21.3	11.0	93.6	8.6	56.2
1979	102.8	113.8	45.9	106.7	65.6	45.4	24.4	8.7	26.7	22.2	96.3	10.3	58.3
1980	103.8	113.6	49.9	118.3	65.8	51.4	30.8	8.6	31.9	32.0	84.7	11.7	61.4
1981	107.2	116.8	64.3	116.3	74.3	55.8	33.6	9.6	36.9	37.1	96.5	14.0	68.2
1982	109.1	120.6	75.0	109.6	91.6	68.1	41.4	11.3	37.1	47.9	112.6	15.3	76.8
1983	111.0	122.1	91.4	117.2	107.4	65.1	43.3	12.4	42.1	60.6	125.1	17.8	87.1
1984	112.3	124.2	86.5	116.9	90.9	70.5	43.9	11.5	45.3	72.9	132.5	23.5	97.5
1985	114.0	127.5	92.9	119.1	54.2	74.7	49.5	14.9	61.8	83.1	145.6	28.7	112.3
1986	115.0	126.8	102.7	119.6	145.9	74.7	54.5	17.5	94.2	105.3	140.5	34.1	116.6
1987	116.4	127.2	109.9	122.0	110.7	79.4	52.6	17.7	108.1	119.8	136.5	38.1	125.4
1988	115.5	124.4	113.7	103.4	127.8	79.0	56.6	21.0	88.9	109.1	129.5	44.2	119.5
1989	115.0	125.6	107.2	113.2	106.9	77.0	55.3	21.3	81.3	124.9	143.7	50.6	111.1
1990	115.3	122.7	94.2	111.9	96.4	87.4	59.3	19.6	75.0	162.2	151.6	54.8	105.8
1991	116.6	118.8	84.2	110.6	90.9	92.0	48.7	17.5	53.9	163.1	136.0	56.3	93.8
1992	106.7	118.2	73.4	92.4	86.3	90.5	43.7	24.0	47.2	161.3	131.4	60.8	86.2
1993	101.2	115.0	63.7	89.0	85.7	98.1	39.3	25.0	50.8	117.2	110.9	58.0	80.1
1994	96.4	108.3	54.0	91.6	59.0	99.0	41.0	26.8	50.8	86.0	99.8	57.1	76.1
1995	91.4	101.9	51.5	83.0	37.8	91.1	40.7	27.0	41.1	72.5	93.9	56.0	73.6
1996	88.1	96.5	51.1	83.7	44.2	92.7	37.8	20.5	40.0	64.9	79.3	58.9	73.1
1997	85.7	91.6	54.6	75.4	54.3	89.1	42.7	32.0	27.8	48.0	70.4	60.5	71.9
1998	83.8	87.4	57.9	73.5	61.5	83.7	40.5	30.2	28.0	54.1	74.4	64.6	70.7
1999	82.3	84.3	61.2	79.7	62.6	85.8	44.7	36.9	29.1	49.0	78.8	64.4	68.9
2000	83.0	81.5	57.0	73.5	59.5	85.3	44.9	35.5	33.9	46.2	75.0	63.6	68.6
2001	84.7	80.2	56.2	87.0	68.0	90.7	55.5	42.4	39.6	58.0	82.2	60.2	73.4
2002	87.8	80.7	56.6	103.2	80.5	100.3	66.9	52.5	47.3	58.0	95.1	58.7	84.3
2003	92.6	80.5	64.3	99.5	87.4	112.2	82.3	61.2	45.0	59.8	103.3	60.4	92.8
2004	93.7	83.2	66.2	103.0	110.7	110.3	84.4	75.3	55.4	67.9	105.0	62.6	100.0
2005	93.1	85.0	58.5	100.5	117.7	103.1	91.3	78.8	69.5	65.8	100.0	65.8	106.5
2006	91.0	85.2	58.9	99.2	112.9	105.7	93.6	81.9	70.7	74.1	105.7	68.9	109.0
2007	90.2	85.7	61.0	98.3	111.0	102.0	95.7	78.3	67.4	70.3	99.4	73.3	112.1
2008	94.2	87.8	68.0	99.0	105.1	106.6	100.7	94.1	68.2	68.1	102.3	78.6	110.6
2009	101.5	91.3	78.2	99.8	107.2	114.2	103.5	100.2	69.1	79.1	101.1	80.3	108.5
2010	103.1	97.2	85.0	102.0	105.3	116.7	116.3	105.2	82.7	85.6	104.6	88.6	106.1
2011	102.7	100.7	94.2	104.8	99.7	106.9	108.0	101.2	97.5	75.4	98.7	95.5	103.6
2012	100.0	100.0	100.0	100.0	100.0	100.0	100.0	100.0	100.0	100.0	100.0	100.0	100.0
2013	97.5	94.7	92.6	83.8	82.7	83.5	84.8	91.2	110.1	92.8	104.7	101.6	94.3
2014	94.4	94.6	83.8	77.5	70.7	90.2	89.8	78.0	97.5	96.5	108.7	104.9	87.3
2015	91.0	94.2	80.8	79.9	64.8	86.0	91.4	75.9	93.5	95.5	110.9	106.6	87.8
2016	90.4	94.8	80.7	82.8	73.2	89.5	90.5	74.0	87.2	71.0	118.0	114.7	84.8
2017	90.5	94.9	85.9	77.0	82.3	90.2	92.2	72.7	98.8	70.9	114.7	122.9	83.4
2018	91.6	95.6	86.8	79.6	113.6	88.5	97.0	81.9	109.9	68.8	125.0	135.8	83.7
2012													
1st quarter	101.4	101.1	100.5	101.8	103.4	103.9	108.0	99.5	95.1	96.7	94.5	96.9	101.3
2nd quarter	100.3	100.7	97.2	106.6	102.5	100.1	100.8	97.2	99.8	102.6	103.7	98.4	100.2
3rd quarter	99.5	99.8	101.2	109.9	100.0	101.0	101.6	104.0	111.3	97.7	95.6	99.8	99.3
4th quarter	98.8	98.4	101.2	81.7	94.2	95.0	89.6	99.3	93.8	103.0	106.3	104.9	99.2
2013													
1st quarter	98.1	98.3	92.4	86.3	88.0	85.8	85.2	96.5	111.6	75.9	97.0	104.8	97.5
2nd quarter	97.8	97.2	92.8	87.7	84.1	83.7	86.2	96.2	99.0	94.4	105.7	99.8	95.5
3rd quarter	97.3	91.1	95.0	81.1	80.6	80.3	82.7	90.6	107.9	112.5	106.8	100.5	93.5
4th quarter	96.7	92.2	90.1	80.1	78.1	84.0	85.1	81.6	122.0	88.4	109.1	101.2	90.6
2014													
1st quarter	96.0	94.7	88.6	81.2	75.1	87.4	87.9	79.6	86.4	88.1	101.9	102.6	88.2
2nd quarter	95.2	94.6	86.0	75.2	72.4	90.0	87.8	74.3	104.4	93.2	114.9	104.3	87.2
3rd quarter	93.8	94.4	78.0	75.4	68.9	93.3	94.9	87.1	98.5	96.4	111.6	106.8	86.7
4th quarter	92.5	94.5	82.5	78.2	66.6	90.3	88.4	71.0	100.9	108.3	106.4	106.0	87.3
2015													
1st quarter	91.6	94.5	81.6	81.7	64.6	87.9	90.6	76.5	80.7	94.5	113.4	106.0	87.8
2nd quarter	90.9	94.1	78.5	77.0	63.6	86.6	94.3	76.3	103.0	91.9	107.9	106.2	88.0
3rd quarter	90.7	94.1	80.0	79.1	64.4	83.7	90.6	75.3	96.2	85.8	111.6	106.4	87.9
4th quarter	90.7	94.2	83.2	81.9	66.5	85.7	90.1	75.6	94.3	110.0	110.6	107.6	87.6
2016													
1st quarter	90.7	94.4	83.6	88.1	68.8	88.0	90.5	74.8	99.5	76.0	119.7	111.6	86.6
2nd quarter	90.5	94.7	74.8	81.5	71.8	88.5	88.5	71.1	88.8	80.5	113.1	113.8	85.1
3rd quarter	90.4	95.0	79.8	81.2	74.7	90.1	91.9	75.4	79.8	72.7	120.5	116.1	84.2
4th quarter	90.2	95.1	84.5	80.5	77.6	91.6	90.8	74.7	80.7	55.0	118.6	117.4	83.2
2017													
1st quarter	90.1	95.1	85.4	82.6	79.6	89.6	89.4	69.5	94.3	76.2	97.8	119.7	83.4
2nd quarter	90.1	94.8	93.3	75.6	81.8	90.8	92.4	74.5	99.8	72.4	121.2	121.8	83.6
3rd quarter	90.6	94.9	79.4	71.0	83.5	91.2	94.4	72.3	98.2	71.9	116.7	124.0	83.4
4th quarter	91.2	94.9	85.6	78.9	84.5	89.2	92.4	74.5	102.7	63.1	123.1	126.0	83.3
2018													
1st quarter	91.2	94.7	80.3	80.8	89.4	87.2	92.6	76.5	96.5	70.8	125.6	130.3	82.7
2nd quarter	91.4	95.7	89.2	79.4	101.4	88.3	96.9	82.0	107.2	60.3	127.5	134.6	83.1
3rd quarter	91.8	96.1	96.2	79.2	120.1	88.2	98.1	83.0	107.3	62.9	120.8	137.8	83.8
4th quarter	92.0	96.0	81.4	79.1	143.4	90.2	100.4	85.9	128.6	81.3	126.2	140.6	85.3

[1] Includes general government intermediate inputs for goods and services sold to other sectors and for own-account investment.

SECTION 6B: STATE AND LOCAL GOVERNMENT IN THE NATIONAL INCOME AND PRODUCT ACCOUNTS

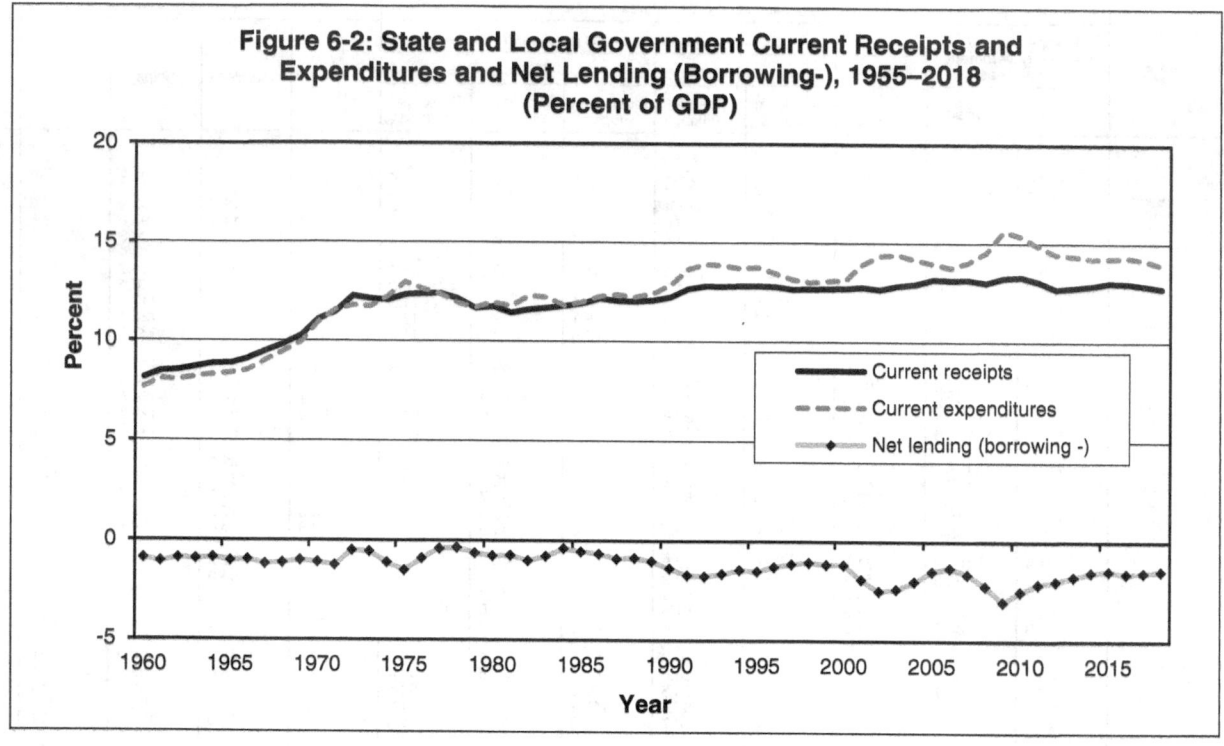

Figure 6-2: State and Local Government Current Receipts and Expenditures and Net Lending (Borrowing-), 1955–2018 (Percent of GDP)

- Both current receipts and current spending of state and local governments have increased as a share of gross domestic product (GDP) over the postwar period, but the growth slowed markedly after the early 1970s. (Tables 6-8 and 1-1)

- State and local governments have consistently been net borrowers in capital markets, as can be seen by the negative values in Figure 6-2. Although they are usually constrained to balance their statutory budgets, they have consistently made net investments and been able to borrow for that purpose. In addition, as revealed by the accrual accounting adopted in the NIPA accounts last year, they have often incurred hidden deficits by failing to fund commitments for public employee retirement benefits. (Table 6-10)

- A larger gap between current expenditures and current receipts appeared during the slow-growth years 2001–2003, and a still-greater one during the Great Recession beginning in late 2007. Generally, in both of those periods of rising deficits, receipts barely kept pace with the economy, while general ("consumption") expenditures, social benefits, and interest payments all fueled higher spending. (Table 6-8)

- After 2009, aggregate state and local spending and borrowing declined as the economy continued to struggle after the Great Recession. (Table 6-8)

Table 6-8. State and Local Government Current Receipts and Expenditures

(National income and product accounts, calendar years, billions of dollars, quarterly data are at seasonally adjusted annual rates.)

NIPA Table 3.3

Year and quarter	Total current receipts	Current tax receipts							Taxes on corporate income	Contributions for government social insurance	Income receipts on assets		
		Total tax receipts	Personal current taxes		Taxes on production and imports						Interest receipts	Dividends	Rents and royalties
			Income taxes	Other personal taxes	Sales taxes	Property taxes	Other taxes on prod. and imports						
1960	44.2	37.0	2.5	1.7	5.3	16.2	3.1	1.2	0.5	1.0	...	0.3
1961	47.7	39.7	2.8	1.8	5.8	17.6	3.2	1.3	0.5	1.1	...	0.4
1962	51.6	42.8	3.2	1.8	6.4	19.0	3.3	1.5	0.5	1.1	...	0.4
1963	55.5	45.8	3.4	2.0	6.9	20.2	3.5	1.7	0.6	1.2	...	0.4
1964	60.7	49.8	4.0	2.1	7.7	21.7	3.7	1.8	0.7	1.5	...	0.4
1965	65.9	53.9	4.4	2.2	8.6	23.2	3.9	2.0	0.8	1.8	...	0.4
1966	74.2	58.8	5.4	2.4	9.8	24.5	4.3	2.2	0.8	2.1	...	0.5
1967	81.7	64.0	6.1	2.5	10.7	27.0	4.4	2.6	0.9	2.4	...	0.6
1968	92.6	73.4	7.8	2.7	13.1	29.9	4.6	3.3	0.9	2.8	...	0.7
1969	104.5	82.5	9.8	3.0	15.1	32.8	4.7	3.6	1.0	3.6	...	0.8
1970	119.1	91.3	10.9	3.3	17.0	36.7	5.0	3.7	1.1	4.3	...	0.8
1971	133.7	101.7	12.4	3.5	19.1	40.4	5.7	4.3	1.2	4.6	...	0.9
1972	157.1	115.6	17.2	3.7	21.8	43.2	6.4	5.3	1.3	4.9	...	1.0
1973	173.0	126.3	18.9	4.0	24.5	46.4	7.0	6.0	1.5	6.6	0.0	1.1
1974	186.6	136.0	20.4	4.2	28.0	49.0	7.7	6.7	1.7	8.9	0.0	1.3
1975	208.0	147.4	22.5	4.4	30.3	53.4	8.1	7.3	1.8	9.8	0.0	1.3
1976	232.2	165.7	26.3	4.8	34.4	58.2	9.0	9.6	2.2	9.0	0.0	1.3
1977	258.3	183.7	30.4	5.0	39.0	63.2	9.7	11.4	2.8	10.4	0.0	1.3
1978	285.8	198.2	35.0	5.5	44.2	63.7	10.9	12.1	3.4	13.3	0.1	1.3
1979	306.3	212.0	38.2	5.8	49.5	64.4	12.7	13.6	3.9	18.1	0.1	1.9
1980	335.9	230.0	42.6	6.0	50.0	68.8	15.0	14.5	3.6	23.1	0.1	3.1
1981	367.5	255.8	47.9	6.7	58.8	77.1	17.9	15.4	3.9	28.5	0.1	3.3
1982	388.5	273.2	51.9	7.3	62.0	85.3	18.5	14.0	4.0	33.1	0.2	3.5
1983	425.3	300.9	58.3	7.8	70.4	91.9	19.4	15.9	4.1	37.0	0.2	4.3
1984	476.1	337.3	67.5	8.6	80.6	99.7	21.8	18.8	4.7	42.6	0.2	4.9
1985	517.5	363.7	72.1	9.2	87.9	107.5	23.5	20.2	4.9	50.1	0.2	5.4
1986	557.4	389.5	77.4	9.8	93.7	116.2	23.7	22.7	6.0	52.6	0.2	6.2
1987	585.5	422.1	86.0	10.6	100.9	126.4	24.9	23.9	7.2	53.0	0.2	5.3
1988	630.4	452.8	90.6	11.5	110.0	136.5	25.7	26.0	8.4	56.6	0.2	4.4
1989	681.4	488.0	102.3	12.4	117.4	149.9	26.9	24.2	9.0	62.3	0.2	4.1
1990	730.1	519.1	109.6	13.0	125.6	161.5	28.2	22.5	10.0	64.1	0.2	4.2
1991	779.9	544.3	111.7	13.6	128.0	176.1	28.6	23.6	11.6	62.4	0.3	4.5
1992	836.1	579.8	120.4	14.9	136.0	184.7	31.1	24.4	13.1	58.8	0.5	4.8
1993	878.0	604.7	126.2	14.8	143.7	187.3	33.1	26.9	14.1	56.2	0.6	4.5
1994	935.1	644.2	132.2	15.8	155.4	199.4	35.4	30.0	14.5	58.4	0.8	4.5
1995	981.0	672.1	141.7	16.4	164.4	202.6	37.0	31.7	13.6	63.8	1.0	4.5
1996	1 033.7	709.6	152.3	16.3	174.4	212.4	39.3	33.0	12.5	68.0	1.4	4.6
1997	1 086.7	749.9	164.7	17.3	184.2	223.5	41.6	34.1	10.8	72.0	1.5	4.8
1998	1 149.6	794.9	183.0	18.2	194.9	231.0	43.9	34.9	10.4	75.2	1.6	4.6
1999	1 222.7	840.4	195.5	19.0	208.6	242.8	45.8	35.8	9.8	78.9	1.5	5.1
2000	1 304.1	893.2	217.4	19.4	221.4	254.7	49.8	35.2	10.8	86.6	1.4	6.3
2001	1 353.4	914.3	223.3	19.7	223.1	268.0	52.6	28.9	13.7	81.6	1.4	6.5
2002	1 386.2	923.9	199.3	20.6	222.5	289.2	56.3	30.9	15.8	70.7	1.6	6.6
2003	1 472.0	974.3	202.8	22.1	233.3	306.9	63.0	34.0	19.9	64.8	1.8	7.6
2004	1 580.3	1 060.3	221.4	24.3	252.0	326.1	77.4	41.7	24.7	66.8	2.3	8.5
2005	1 718.5	1 173.2	249.1	26.2	272.8	351.2	91.7	54.9	24.6	77.3	2.5	9.8
2006	1 816.2	1 258.2	274.2	26.9	293.9	375.2	92.1	59.2	21.5	95.0	3.0	10.5
2007	1 902.1	1 321.7	294.3	27.3	301.4	403.7	92.9	57.9	18.8	104.6	3.4	11.7
2008	1 915.5	1 334.1	304.1	26.8	301.1	415.7	94.7	47.4	18.7	91.4	3.4	12.2
2009	1 915.2	1 265.8	258.9	26.9	279.6	439.6	72.3	44.5	18.6	73.5	2.9	11.2
2010	1 994.4	1 306.4	265.8	28.3	295.1	438.6	77.8	46.1	17.8	69.0	3.0	11.4
2011	2 030.4	1 366.4	294.1	28.8	311.3	439.2	83.9	48.4	17.9	67.1	3.4	12.2
2012	2 056.3	1 414.7	313.1	30.0	324.4	444.3	86.7	50.7	17.2	65.3	4.1	12.5
2013	2 145.6	1 490.6	343.1	30.4	343.9	456.8	91.6	53.9	17.7	65.2	4.7	12.5
2014	2 257.4	1 541.9	349.4	31.5	361.7	475.0	93.0	56.5	18.7	66.4	5.3	12.7
2015	2 375.3	1 600.1	374.3	32.8	375.5	490.3	90.1	56.2	19.2	67.3	5.6	11.6
2016	2 431.9	1 639.4	375.9	33.6	382.7	513.5	91.3	53.5	20.1	69.3	5.6	11.1
2017	2 515.1	1 722.9	397.9	34.8	394.7	545.8	96.6	54.5	20.8	72.0	5.9	11.4
2018	2 623.0	1 796.8	420.9	36.5	411.9	562.0	102.7	58.4	22.2	73.6	6.1	11.7
2016												
1st quarter	2 380.7	1 609.8	364.2	33.4	379.5	502.5	89.7	53.6	19.8	68.3	5.8	11.1
2nd quarter	2 413.8	1 628.9	376.8	33.4	380.1	509.1	90.1	52.5	20.1	68.9	5.5	11.1
3rd quarter	2 451.9	1 656.2	383.5	33.5	386.5	516.7	92.5	53.4	20.3	69.7	5.6	11.1
4th quarter	2 481.3	1 662.6	379.3	34.1	384.5	525.6	93.0	54.5	20.4	70.3	5.7	11.1
2017												
1st quarter	2 478.7	1 689.8	385.4	34.5	388.8	535.5	95.6	54.6	20.4	71.1	5.7	11.2
2nd quarter	2 471.3	1 694.7	379.7	34.5	391.4	543.6	96.1	53.3	20.6	71.8	5.8	11.3
3rd quarter	2 520.8	1 723.9	392.4	35.3	397.5	549.9	96.0	54.3	20.8	72.3	6.0	11.4
4th quarter	2 589.8	1 783.2	433.9	35.0	401.2	554.4	98.7	55.7	21.3	72.9	6.1	11.5
2018												
1st quarter	2 607.3	1 790.1	433.1	36.5	404.6	557.0	101.3	54.8	21.8	72.8	5.9	11.6
2nd quarter	2 622.4	1 798.8	422.2	36.1	414.8	560.0	102.1	59.8	22.1	73.2	6.2	11.7
3rd quarter	2 629.9	1 800.9	421.6	36.6	412.0	563.5	103.4	60.1	22.4	73.7	6.2	11.7
4th quarter	2 632.2	1 797.4	406.8	36.8	416.2	567.4	103.9	59.0	22.5	74.5	6.2	11.8

. . . = Not available.

Table 6-9. State and Local Government Consumption Expenditures and Gross Investment

(National income and product accounts, calendar years, billions of dollars, quarterly data are at seasonally adjusted annual rates.)

NIPA Tables 3.9.5, 3.10.5

Year and quarter	Total	State and local government consumption expenditures and gross investment								Gross investment			
		Consumption expenditures [1]					Less						
			Compensation of general government employees	Consumption of general government fixed capital	Intermediate goods and services purchased [2]			Sales to other sectors					
		Total				Own-account investment	Total [3]	Tuition and related educational charges	Health and hospital charges	Total	Structures	Equipment	Intellectual property
1960	47.6	33.5	25.9	3.6	8.9	0.9	3.9	0.4	1.0	14.1	12.7	1.2	0.2
1961	51.9	36.6	28.5	3.8	9.6	1.0	4.3	0.5	1.0	15.3	13.8	1.2	0.2
1962	54.8	38.7	30.6	4.1	10.1	1.1	4.9	0.6	1.3	16.1	14.5	1.3	0.2
1963	59.3	41.5	33.2	4.4	10.8	1.3	5.6	0.7	1.3	17.8	16.0	1.5	0.3
1964	64.5	45.2	36.2	4.7	11.9	1.3	6.3	0.8	1.5	19.3	17.2	1.7	0.3
1965	70.9	49.7	39.8	5.1	13.4	1.4	7.2	1.0	1.8	21.2	19.0	1.8	0.4
1966	78.9	55.4	44.6	5.7	14.9	1.5	8.2	1.2	2.1	23.5	21.0	2.0	0.4
1967	87.0	61.3	49.5	6.2	16.4	1.6	9.3	1.4	2.5	25.7	23.0	2.2	0.5
1968	97.6	69.4	56.3	6.9	18.6	1.8	10.7	1.6	3.1	28.1	25.2	2.3	0.6
1969	107.8	79.0	63.2	7.8	21.8	1.9	11.8	1.9	3.5	28.8	25.6	2.6	0.7
1970	119.8	90.4	71.5	8.8	25.3	2.1	13.1	2.4	3.8	29.3	25.8	2.8	0.8
1971	133.0	102.3	80.5	9.8	29.1	2.2	14.9	2.9	4.6	30.7	27.0	2.9	0.8
1972	144.5	113.2	89.5	10.7	32.0	2.3	16.7	3.2	5.6	31.3	27.1	3.3	0.9
1973	158.6	124.6	98.5	11.9	35.3	2.4	18.7	3.7	6.6	34.0	29.1	3.8	1.0
1974	182.5	142.0	108.1	14.7	42.6	2.8	20.5	4.0	7.4	40.5	34.7	4.6	1.2
1975	207.4	162.8	121.3	16.7	50.5	3.0	22.6	4.3	8.5	44.6	38.1	5.1	1.3
1976	219.4	174.5	130.6	17.4	55.0	3.0	25.6	4.7	9.9	44.9	38.1	5.3	1.5
1977	234.0	190.0	142.1	18.4	60.8	3.1	28.3	5.2	10.9	44.0	36.9	5.4	1.6
1978	257.4	206.6	155.7	19.9	66.6	3.5	32.0	5.8	12.7	50.8	42.8	6.1	1.9
1979	284.2	225.9	170.6	22.2	73.9	4.3	36.6	6.4	15.2	58.4	49.0	7.1	2.2
1980	314.7	249.0	187.9	25.7	81.6	4.9	41.2	7.2	17.3	65.7	55.1	8.1	2.6
1981	340.4	273.5	205.9	29.3	91.2	5.3	47.6	8.3	21.0	66.9	55.4	8.5	3.0
1982	363.1	296.1	223.6	32.0	100.0	5.7	53.8	9.4	24.5	67.0	54.2	9.4	3.4
1983	384.2	315.4	239.6	33.0	109.0	6.1	60.1	10.6	27.8	68.8	54.2	10.8	3.8
1984	416.1	338.5	257.8	34.1	118.4	7.0	64.7	11.7	29.4	77.6	60.5	12.7	4.4
1985	457.6	370.3	281.7	35.9	131.4	7.9	70.8	12.7	32.0	87.3	67.6	14.8	5.0
1986	494.4	398.2	304.2	38.6	141.7	8.8	77.4	13.8	34.8	96.2	74.2	16.4	5.6
1987	528.7	426.6	327.8	41.6	150.0	9.5	83.4	15.0	37.0	102.2	78.8	17.2	6.2
1988	567.4	457.0	354.4	44.5	159.8	10.4	91.2	16.6	40.3	110.4	84.8	18.6	7.0
1989	617.8	498.5	388.4	47.7	175.8	11.9	101.4	18.4	45.0	119.2	88.7	21.9	8.6
1990	676.2	544.0	426.5	51.6	190.5	13.1	111.6	20.3	50.0	132.2	98.5	23.9	9.7
1991	716.0	578.7	457.2	55.1	204.7	14.1	124.1	22.6	57.0	137.3	103.2	23.5	10.6
1992	756.0	616.3	492.6	57.6	218.9	14.4	138.4	25.4	65.3	139.7	104.2	24.0	11.4
1993	784.8	643.6	516.1	60.8	234.0	14.8	152.5	27.4	73.1	141.2	104.5	24.9	11.9
1994	827.6	678.8	541.9	64.3	251.5	15.4	163.5	29.3	78.8	148.8	108.7	28.1	12.0
1995	872.7	712.8	565.4	68.6	270.6	16.1	175.6	31.1	85.0	160.0	117.3	30.5	12.3
1996	913.7	743.5	586.6	72.0	285.8	16.8	184.0	32.9	86.6	170.2	126.8	30.5	12.8
1997	963.8	781.0	613.5	75.8	304.2	18.3	194.3	35.4	89.9	182.9	136.6	32.0	14.3
1998	1 026.1	829.8	646.3	79.9	326.7	19.4	203.7	37.9	93.4	196.3	146.1	34.7	15.5
1999	1 109.0	895.4	688.1	85.1	357.3	20.7	214.4	40.5	96.7	213.6	159.7	37.4	16.5
2000	1 193.1	961.7	732.5	91.5	392.1	22.6	231.8	43.6	104.0	231.5	174.6	38.8	18.1
2001	1 279.2	1 031.2	785.4	96.9	425.2	25.0	251.4	46.1	116.2	248.0	190.6	38.4	19.1
2002	1 345.7	1 084.2	829.1	101.4	444.3	26.0	264.6	49.3	120.7	261.5	204.1	38.0	19.4
2003	1 384.9	1 115.9	864.9	105.2	451.0	26.7	278.6	53.4	124.8	269.1	211.1	37.6	20.4
2004	1 447.1	1 169.4	907.6	111.8	472.3	27.8	294.4	56.7	130.9	277.8	219.1	37.2	21.5
2005	1 528.5	1 237.3	944.9	122.1	507.7	29.7	307.8	60.9	132.3	291.3	231.5	37.1	22.7
2006	1 623.5	1 308.3	989.8	132.2	542.6	32.2	324.2	64.0	141.0	315.3	251.4	39.1	24.7
2007	1 740.3	1 398.3	1 045.8	145.4	578.3	34.3	336.9	65.9	149.3	342.0	271.4	44.0	26.6
2008	1 831.4	1 473.5	1 095.9	156.2	605.3	36.1	347.8	67.8	158.4	357.9	284.5	45.1	28.3
2009	1 855.3	1 495.0	1 117.9	162.6	610.5	36.3	359.7	69.7	168.7	360.3	288.9	43.1	28.3
2010	1 856.7	1 509.5	1 141.1	165.0	612.1	35.6	373.1	71.6	176.1	347.3	279.8	39.6	27.9
2011	1 849.4	1 508.5	1 141.9	171.2	616.8	35.9	385.6	73.6	182.4	341.0	273.4	38.6	29.0
2012	1 850.5	1 516.7	1 147.1	178.4	622.6	35.8	395.5	76.1	188.5	333.7	265.0	39.2	29.6
2013	1 905.8	1 575.1	1 197.8	183.2	638.9	36.4	408.4	78.9	197.6	330.8	259.3	39.9	31.5
2014	1 953.0	1 614.2	1 232.5	188.7	656.2	37.3	425.8	81.3	209.0	338.8	266.1	39.9	32.8
2015	2 015.7	1 660.0	1 275.4	192.0	677.2	39.0	445.6	84.4	220.5	355.8	280.2	41.3	34.3
2016	2 072.6	1 702.8	1 303.6	195.5	710.5	40.6	466.1	89.1	231.5	369.7	290.2	44.0	35.5
2017	2 142.7	1 764.6	1 338.9	202.5	753.8	41.8	488.7	94.4	244.4	378.0	294.9	46.2	36.9
2018	2 244.2	1 847.4	1 387.6	213.0	805.2	44.1	514.2	99.1	259.8	396.8	309.3	47.9	39.6
2016													
1st quarter	2 046.3	1 675.5	1 292.4	193.1	687.6	39.6	458.1	87.1	227.2	370.8	291.7	43.9	35.2
2nd quarter	2 065.2	1 696.6	1 298.8	195.5	706.4	40.7	463.4	88.4	230.0	368.6	288.9	44.4	35.3
3rd quarter	2 080.0	1 713.3	1 308.6	195.9	718.7	41.1	468.8	89.8	233.0	366.7	287.0	44.2	35.6
4th quarter	2 098.7	1 726.0	1 314.8	197.3	729.1	41.2	474.0	91.0	235.8	372.7	293.2	43.6	35.9
2017													
1st quarter	2 120.3	1 744.3	1 324.1	199.4	741.4	41.4	479.2	92.4	238.9	376.1	293.6	46.2	36.3
2nd quarter	2 129.3	1 751.3	1 332.4	201.4	744.2	41.4	485.3	93.7	242.3	377.9	294.9	46.4	36.6
3rd quarter	2 145.2	1 768.4	1 343.4	203.8	754.1	41.8	491.1	94.9	245.8	376.8	293.8	45.8	37.2
4th quarter	2 175.9	1 794.5	1 355.6	205.5	775.4	42.8	499.2	96.5	250.7	381.4	297.3	46.5	37.6
2018													
1st quarter	2 203.2	1 814.9	1 367.3	208.1	788.9	43.3	506.3	97.7	254.9	388.3	303.0	46.6	38.7
2nd quarter	2 236.7	1 837.9	1 379.6	212.2	802.1	44.0	512.0	98.7	258.5	398.8	312.0	47.3	39.5
3rd quarter	2 263.9	1 860.6	1 396.5	214.4	811.0	44.5	516.9	99.6	261.5	403.3	314.7	48.7	40.0
4th quarter	2 273.0	1 876.3	1 406.9	217.1	818.7	44.7	521.6	100.5	264.1	396.7	307.5	48.9	40.3

[1] Excludes government sales to other sectors and government own-account investment (construction and software).
[2] Includes general government intermediate inputs for goods and services sold to other sectors and for own-account investment.
[3] Includes components not shown separately.

Table 6-10. State and Local Government Output, Lending and Borrowing, and Net Investment

(National income and product accounts, calendar years, billions of dollars, quarterly data are at seasonally adjusted annual rates.)

NIPA Tables 3.3, 3.10.5

Year and quarter	Output			Net lending (net borrowing -)								Net investment
	Gross	Value added	Intermediate goods and services purchased [1]	Net saving, current (surplus +, deficit -)	Plus: Capital transfer receipts	Minus			Plus: consumption of fixed capital	Equals: Net lending (borrowing -)		
						Gross investment	Capital transfer payments	Net purchases of nonproduced assets				
1960	38.4	29.5	8.9	2.5	3.0	14.1	...	0.9	4.5	-4.9		9.6
1961	41.9	32.3	9.6	2.2	3.3	15.3	...	1.0	4.8	-6.0		10.5
1962	44.7	34.6	10.1	3.1	3.5	16.1	...	1.1	5.1	-5.4		11.0
1963	48.4	37.6	10.8	3.3	4.1	17.8	...	1.2	5.5	-6.0		12.3
1964	52.8	40.9	11.9	3.9	4.7	19.3	...	1.3	5.9	-6.0		13.4
1965	58.3	44.9	13.4	3.7	4.7	21.2	...	1.3	6.4	-7.6		14.8
1966	65.1	50.2	14.9	4.7	5.1	23.5	...	1.4	7.1	-7.9		16.4
1967	72.2	55.8	16.4	4.0	5.1	25.7	...	1.4	7.8	-10.2		17.9
1968	81.9	63.3	18.6	3.5	6.8	28.1	...	1.4	8.6	-10.6		19.5
1969	92.7	70.9	21.8	3.1	6.8	28.8	...	1.0	9.6	-10.2		19.2
1970	105.6	80.3	25.3	1.4	6.2	29.3	...	1.1	10.9	-11.8		18.4
1971	119.4	90.3	29.1	-1.3	7.0	30.7	...	1.6	12.2	-14.4		18.5
1972	132.2	100.2	32.0	6.1	7.3	31.3	...	1.7	13.3	-6.3		18.0
1973	145.7	110.4	35.3	5.6	7.3	34.0	...	1.7	14.8	-8.0		19.2
1974	165.3	122.8	42.6	-2.3	9.2	40.5	...	1.9	18.4	-17.2		22.1
1975	188.5	138.0	50.5	-10.7	11.0	44.6	...	1.9	21.0	-25.3		23.6
1976	203.0	148.0	55.0	-4.4	12.0	44.9	...	1.7	22.0	-17.0		22.9
1977	221.4	160.6	60.8	0.5	13.1	44.0	...	1.6	23.4	-8.5		20.6
1978	242.1	175.5	66.6	4.9	13.7	50.8	...	1.8	25.4	-8.6		25.4
1979	266.7	192.8	73.9	-1.2	16.2	58.4	...	2.0	28.6	-16.8		29.8
1980	295.1	213.5	81.6	-5.9	18.6	65.7	...	2.2	33.1	-22.1		32.6
1981	326.4	235.2	91.2	-10.2	17.8	66.9	...	2.2	37.8	-23.8		29.1
1982	355.6	255.6	100.0	-22.8	16.9	67.0	...	2.2	41.2	-33.8		25.8
1983	381.6	272.6	109.0	-18.4	18.0	68.8	...	2.2	42.7	-28.8		26.1
1984	410.2	291.9	118.4	-0.2	20.1	77.6	...	2.6	44.3	-16.0		33.3
1985	449.0	317.6	131.4	-2.4	22.0	87.3	...	3.1	46.8	-24.1		40.5
1986	484.4	342.7	141.7	-4.0	23.0	96.2	...	3.7	50.1	-30.7		46.1
1987	519.5	369.4	150.0	-14.4	22.3	102.2	...	4.2	54.0	-44.4		48.2
1988	558.6	398.8	159.8	-11.3	23.1	110.4	...	4.3	57.7	-45.3		52.7
1989	611.8	436.1	175.8	-19.3	23.4	119.2	0.0	4.9	61.8	-58.3		57.4
1990	668.7	478.2	190.5	-36.2	25.0	132.2	0.0	5.7	66.6	-82.5		65.6
1991	716.9	512.2	204.7	-60.1	25.8	137.3	0.0	5.8	70.8	-106.6		66.5
1992	769.2	550.2	218.9	-71.0	26.9	139.7	0.0	5.9	74.1	-115.5		65.6
1993	810.9	576.9	234.0	-72.5	28.2	141.2	0.0	5.9	78.3	-113.1		62.9
1994	857.7	606.2	251.5	-63.9	29.9	148.8	0.0	6.2	82.8	-106.3		66.0
1995	904.5	633.9	270.6	-70.4	32.4	160.0	0.0	6.6	88.1	-116.5		71.9
1996	944.4	658.6	285.8	-53.8	33.8	170.2	0.0	6.0	92.5	-103.6		77.7
1997	993.5	689.3	304.2	-42.0	35.2	182.9	0.0	5.8	97.3	-98.2		85.6
1998	1 052.9	726.2	326.7	-30.1	35.9	196.3	0.0	7.6	102.2	-95.9		94.1
1999	1 130.5	773.1	357.3	-38.9	40.0	213.6	0.0	8.6	108.6	-112.6		105.0
2000	1 216.1	824.0	392.1	-40.6	44.2	231.5	0.0	8.6	116.6	-119.9		114.9
2001	1 307.5	882.3	425.2	-119.0	51.7	248.0	0.0	10.1	123.4	-202.1		124.6
2002	1 374.8	930.6	444.3	-182.9	52.9	261.5	0.0	11.2	129.3	-273.3		132.2
2003	1 421.2	970.1	451.0	-181.3	52.3	269.1	0.0	11.4	134.9	-274.6		134.2
2004	1 491.6	1 019.3	472.3	-149.0	52.5	277.8	4.5	11.3	144.2	-245.8		133.6
2005	1 574.8	1 067.0	507.7	-102.8	56.9	291.3	6.4	10.0	157.9	-195.7		133.4
2006	1 664.6	1 122.1	542.6	-85.0	57.9	315.3	0.0	11.2	171.0	-182.5		144.3
2007	1 769.5	1 191.2	578.3	-129.3	59.8	342.0	0.0	13.9	187.5	-237.9		154.5
2008	1 857.6	1 252.1	605.3	-220.9	63.1	357.9	0.0	13.3	201.2	-327.8		156.7
2009	1 891.0	1 280.5	610.5	-341.3	68.0	360.3	0.0	12.6	209.1	-437.1		151.2
2010	1 918.2	1 306.1	612.1	-307.5	76.9	347.3	0.0	12.0	213.4	-376.4		133.9
2011	1 930.0	1 313.1	616.8	-275.1	74.1	341.0	0.0	11.5	222.2	-331.2		118.8
2012	1 948.1	1 325.5	622.6	-282.8	74.4	333.7	0.0	10.9	232.0	-321.1		101.7
2013	2 019.8	1 380.9	638.9	-265.4	71.6	330.8	0.0	10.4	238.8	-296.1		92.0
2014	2 077.3	1 421.1	656.2	-238.0	70.6	338.8	0.0	10.6	246.3	-270.4		92.5
2015	2 144.6	1 467.4	677.2	-220.3	69.1	355.8	0.0	12.1	251.4	-267.7		104.4
2016	2 209.5	1 499.1	710.5	-246.8	72.4	369.7	0.0	13.3	256.6	-300.8		113.1
2017	2 295.2	1 541.4	753.8	-248.1	73.4	378.0	1.2	13.7	266.1	-301.6		111.9
2018	2 405.7	1 600.5	805.2	-239.2	73.3	396.8	0.0	14.5	279.6	-297.6		117.2
2016												
1st quarter	2 173.1	1 485.5	687.6	-253.4	72.2	370.8	0.0	13.0	253.3	-311.8		117.5
2nd quarter	2 200.7	1 494.2	706.4	-253.0	72.3	368.6	0.0	13.2	256.3	-306.3		112.3
3rd quarter	2 223.2	1 504.5	718.7	-243.0	73.2	366.7	0.0	13.4	257.4	-292.6		109.3
4th quarter	2 241.2	1 512.1	729.1	-237.6	72.0	372.7	0.0	13.5	259.4	-292.4		113.3
2017												
1st quarter	2 264.9	1 523.5	741.4	-258.1	72.7	376.1	0.0	13.6	262.1	-313.0		114.0
2nd quarter	2 278.0	1 533.8	744.2	-269.7	74.2	377.9	0.0	13.7	264.6	-322.4		113.3
3rd quarter	2 301.3	1 547.2	754.1	-255.6	76.0	376.8	4.9	13.8	267.7	-307.4		109.1
4th quarter	2 336.5	1 561.1	775.4	-208.9	70.8	381.4	0.0	13.9	270.0	-263.4		111.4
2018												
1st quarter	2 364.4	1 575.5	788.9	-216.3	71.0	388.3	0.0	14.1	273.4	-274.3		114.9
2nd quarter	2 393.9	1 591.8	802.1	-233.3	71.1	398.8	0.0	14.3	278.5	-296.9		120.3
3rd quarter	2 422.0	1 610.9	811.0	-250.1	77.6	403.3	0.0	14.6	281.5	-308.9		121.8
4th quarter	2 442.6	1 624.0	818.7	-256.8	73.3	396.7	0.0	14.9	284.9	-310.1		111.8

[1] Includes general government intermediate inputs for goods and services sold to other sectors and for own-account investment.
. . . = Not available.

Table 6-11. Chain-Type Quantity Indexes for State and Local Government Consumption Expenditures and Gross Investment

(Seasonally adjusted, 2012 = 100.) NIPA Tables 3.9.3, 3.10.3

Year and quarter	Total	Consumption expenditures [1] Total	Compensation of general government employees	Consumption of general government fixed capital	Intermediate goods and services purchased [2]	Less: Own-account investment	Sales to other sectors Total [3]	Tuition and related educational charges	Health and hospital charges	Gross investment Total	Structures	Equipment	Intellectual property
1960	27.7	25.9	34.1	13.8	11.6	21.8	13.9	16.8	10.1	33.8	52.2	6.0	2.4
1961	29.4	27.3	35.9	14.6	12.4	23.0	15.0	18.4	10.3	36.5	56.6	6.2	2.7
1962	30.3	28.0	37.1	15.4	12.9	24.6	16.8	20.8	12.0	37.9	58.4	6.6	3.0
1963	32.1	29.3	39.0	16.4	13.7	27.7	18.7	23.8	12.5	41.2	63.2	7.4	3.5
1964	34.2	31.2	41.3	17.4	15.0	28.0	20.6	27.8	13.5	44.2	67.3	8.4	3.9
1965	36.5	33.2	43.8	18.5	16.5	29.2	23.0	32.3	15.4	47.3	72.0	9.0	4.4
1966	38.7	35.2	46.4	19.7	17.9	31.2	25.1	36.6	17.6	50.4	76.5	9.8	5.2
1967	40.7	36.7	48.2	21.0	19.3	31.4	27.7	40.8	20.0	53.6	81.3	10.2	5.8
1968	43.1	39.2	51.1	22.3	21.3	33.2	30.1	45.9	23.3	56.0	85.0	10.6	6.2
1969	44.6	41.7	53.3	23.6	23.6	33.7	31.3	50.5	24.1	53.8	80.5	11.4	6.9
1970	45.8	44.2	55.7	24.8	26.0	34.6	32.6	58.8	24.6	50.5	74.4	11.8	7.6
1971	47.3	46.5	57.9	25.9	28.5	35.1	35.4	66.2	28.6	49.1	72.0	11.9	7.9
1972	48.3	48.2	59.9	26.8	30.2	34.7	37.9	70.3	32.8	47.6	68.5	13.3	8.3
1973	49.7	49.9	62.0	27.8	31.2	34.6	39.5	75.8	36.4	47.9	67.9	14.9	8.8
1974	51.5	52.1	64.3	28.8	32.6	36.2	39.3	75.6	37.5	48.2	67.6	16.2	9.1
1975	53.4	54.7	66.2	29.8	35.2	35.1	39.9	75.7	38.6	47.9	67.1	15.7	9.8
1976	53.8	55.4	67.0	30.7	36.3	32.9	42.2	78.1	41.0	47.2	66.1	15.4	10.4
1977	54.1	56.5	67.9	31.4	37.6	31.5	43.6	81.3	41.7	44.5	61.7	15.0	10.8
1978	55.8	57.7	69.5	32.2	38.6	33.9	45.8	85.8	44.6	48.6	67.7	15.7	12.0
1979	56.6	58.0	70.7	33.0	38.6	38.0	48.0	88.2	48.4	51.0	70.6	17.0	13.2
1980	56.5	57.8	71.6	33.9	37.2	39.6	48.8	91.1	48.8	51.2	70.2	18.1	14.0
1981	55.4	57.5	71.3	34.7	37.4	38.7	50.7	93.1	51.9	47.0	63.2	17.7	14.9
1982	55.4	58.5	71.7	35.4	39.1	39.4	51.9	92.9	53.4	44.2	58.0	18.6	15.9
1983	56.2	59.2	71.2	36.1	41.8	39.8	53.8	95.7	54.8	44.9	57.4	21.2	17.3
1984	58.3	60.5	71.4	37.1	43.9	43.3	53.9	95.6	53.5	50.3	63.9	24.8	19.1
1985	61.6	63.2	73.4	38.4	47.6	47.1	55.8	95.5	54.8	55.8	70.1	28.7	21.5
1986	64.7	66.0	75.5	40.0	51.5	51.4	57.8	96.3	56.2	59.7	74.4	31.5	23.8
1987	66.1	67.4	76.7	41.6	52.7	53.5	59.0	96.9	56.2	61.3	76.0	32.7	25.6
1988	68.7	69.8	79.3	43.4	54.3	56.0	60.2	99.9	56.2	64.5	79.3	34.9	28.4
1989	71.4	72.3	81.6	45.6	57.0	60.9	61.9	102.9	56.7	67.9	80.9	40.3	33.7
1990	74.3	74.7	83.8	48.1	58.8	63.9	63.2	105.2	57.3	73.1	86.9	43.3	37.5
1991	75.9	76.2	84.5	50.4	61.8	66.4	65.3	106.9	59.7	74.8	89.7	41.8	40.3
1992	77.5	78.0	85.7	52.6	64.8	65.8	68.2	108.9	63.2	75.8	90.1	42.6	43.8
1993	78.5	79.3	86.6	54.7	67.7	65.7	71.4	107.6	66.9	75.2	88.1	44.0	44.8
1994	80.7	81.5	87.9	56.7	71.5	66.7	73.8	107.7	69.6	77.5	89.1	49.6	45.2
1995	82.9	83.5	89.5	58.7	74.4	67.5	76.3	108.1	72.7	80.8	92.1	54.4	45.0
1996	84.9	85.0	90.6	60.9	76.4	69.3	77.8	108.4	72.4	84.3	96.5	55.7	47.4
1997	87.6	87.3	92.2	63.4	80.2	74.1	80.4	110.9	74.1	89.0	100.3	60.9	53.0
1998	91.5	90.9	93.9	66.3	86.5	77.7	82.8	114.0	76.3	94.3	104.1	69.4	58.1
1999	95.1	94.0	95.1	69.5	92.5	80.6	85.0	117.4	77.6	100.4	109.6	77.8	61.6
2000	97.6	95.9	96.8	72.8	96.2	84.9	89.3	121.4	81.4	105.4	114.7	82.2	65.9
2001	101.2	99.2	98.8	76.0	103.0	91.7	94.1	122.1	88.2	110.6	121.1	83.9	68.8
2002	104.2	101.9	100.6	78.9	107.0	93.9	96.2	122.2	88.6	114.7	126.1	85.6	70.2
2003	103.9	101.2	100.9	81.2	104.2	94.2	96.8	122.1	87.3	116.3	127.5	86.4	74.1
2004	103.8	101.3	101.0	83.6	104.3	95.2	97.6	118.3	87.6	115.0	125.1	86.3	78.3
2005	103.8	101.9	101.4	86.0	104.7	98.3	98.1	118.3	85.6	112.2	120.4	87.7	82.2
2006	104.9	102.9	102.1	88.4	106.4	103.6	99.2	116.6	87.7	114.2	120.5	95.5	88.6
2007	106.7	104.8	103.3	91.2	108.0	106.9	99.0	112.9	89.9	115.5	118.9	109.5	94.1
2008	106.9	105.0	104.4	93.9	105.0	108.4	98.1	109.3	92.7	115.9	118.3	113.3	97.7
2009	108.9	107.7	104.5	96.0	111.2	108.2	98.7	106.1	96.0	114.7	117.5	109.0	98.0
2010	106.0	105.0	103.2	97.6	106.9	104.0	99.5	103.6	97.5	110.4	113.4	101.4	96.1
2011	102.3	101.6	101.7	98.9	101.2	102.5	100.1	101.3	99.0	105.5	107.1	99.5	98.7
2012	100.0	100.0	100.0	100.0	100.0	100.0	100.0	100.0	100.0	100.0	100.0	100.0	100.0
2013	99.7	100.3	99.7	101.1	101.4	99.3	100.8	99.5	102.8	97.3	95.7	101.7	105.6
2014	99.9	100.4	99.6	102.1	103.0	99.7	103.0	98.9	107.4	97.5	95.8	100.3	109.3
2015	103.1	103.4	100.2	103.2	112.0	104.0	106.4	99.1	112.3	101.4	99.6	103.3	115.3
2016	105.8	106.1	101.3	104.6	118.7	107.6	109.7	101.9	116.5	104.6	102.1	110.5	119.5
2017	106.5	106.9	102.0	106.1	121.2	108.2	112.9	105.8	120.9	104.6	101.0	115.4	123.1
2018	107.5	107.9	103.0	107.9	123.7	110.7	116.0	108.7	125.6	105.9	101.5	118.4	130.4
2016 1st quarter	105.4	105.3	100.8	104.0	116.9	105.5	108.5	100.5	115.1	105.8	103.7	110.4	118.6
2nd quarter	105.5	105.8	101.1	104.4	118.4	108.0	109.3	101.2	116.1	104.2	101.6	111.2	119.1
3rd quarter	106.0	106.5	101.6	104.7	119.5	108.7	110.1	102.5	116.9	103.6	100.8	110.8	119.9
4th quarter	106.3	106.6	101.7	105.1	120.1	108.3	110.8	103.5	117.9	104.8	102.5	109.5	120.5
2017 1st quarter	106.4	106.7	101.7	105.5	120.5	108.1	111.6	104.4	119.1	105.1	101.8	115.7	121.5
2nd quarter	106.5	106.8	101.9	105.9	120.9	107.6	112.4	105.3	120.3	105.0	101.6	115.9	122.1
3rd quarter	106.4	107.0	102.1	106.3	121.4	107.7	113.3	106.2	121.6	103.7	100.0	114.1	124.0
4th quarter	106.7	107.2	102.3	106.7	122.2	109.3	114.2	107.4	122.7	104.5	100.7	115.9	124.8
2018 1st quarter	107.1	107.4	102.6	107.2	122.8	109.7	115.2	108.3	124.1	105.4	101.4	116.0	127.9
2nd quarter	107.5	107.7	102.8	107.7	123.4	110.7	115.9	108.9	125.3	106.7	102.7	117.4	129.8
3rd quarter	108.0	108.1	103.2	108.1	124.0	111.1	116.4	108.9	126.2	107.1	102.7	120.0	131.2
4th quarter	107.6	108.3	103.2	108.6	124.4	111.2	116.6	108.9	126.9	104.3	99.2	120.2	132.6

[1]Excludes government sales to other sectors and government own-account investment (construction and software).
[2]Includes general government intermediate inputs for goods and services sold to other sectors and for own-account investment.
[3]Includes components not shown separately.

Table 6-12. State Government Current Receipts and Expenditures

(National income and product accounts, calendar years, billions of dollars.)

NIPA Table 3.20

Year	Total [1]	Current tax receipts Total	Personal current taxes Total [1]	Personal current taxes Income taxes	Taxes on production and imports Total	Taxes on production and imports Sales taxes	Taxes on production and imports Property taxes	Taxes on production and imports Other	Taxes on corporate income	Contributions for government social insurance	Income receipts on assets Total [1]	Income receipts on assets Interest receipts	Income receipts on assets Rents and royalties
1959	21.6	16.7	3.1	2.0	12.5	10.0	0.5	2.0	1.1	0.4	0.5	0.3	0.2
1960	23.3	18.1	3.4	2.3	13.5	10.8	0.5	2.2	1.2	0.5	0.5	0.4	0.2
1961	25.0	19.3	3.7	2.5	14.4	11.6	0.5	2.3	1.3	0.5	0.6	0.4	0.2
1962	27.1	21.0	4.0	2.8	15.4	12.6	0.6	2.3	1.5	0.5	0.6	0.4	0.2
1963	29.1	22.4	4.3	3.1	16.4	13.4	0.6	2.4	1.6	0.6	0.6	0.4	0.2
1964	31.8	24.5	4.9	3.6	17.7	14.5	0.6	2.6	1.8	0.7	0.7	0.4	0.2
1965	35.1	26.9	5.4	3.9	19.6	16.1	0.7	2.8	1.9	0.8	0.8	0.5	0.3
1966	41.5	30.2	6.4	4.8	21.7	18.0	0.7	3.0	2.2	0.8	0.9	0.6	0.3
1967	45.6	32.7	7.0	5.3	23.3	19.4	0.7	3.1	2.5	0.9	1.1	0.8	0.3
1968	53.8	38.6	8.8	6.9	26.7	22.8	0.8	3.2	3.1	0.9	1.7	1.4	0.3
1969	61.6	44.0	10.8	8.6	29.9	25.7	0.9	3.3	3.4	1.0	2.1	1.8	0.3
1970	69.1	48.1	12.0	9.6	32.7	28.2	0.9	3.6	3.5	1.1	2.5	2.2	0.3
1971	78.2	53.8	13.4	11.0	36.3	31.4	1.0	3.9	4.0	1.2	2.7	2.3	0.4
1972	95.0	63.5	17.9	15.2	40.7	35.2	1.1	4.4	5.0	1.3	2.9	2.5	0.4
1973	103.1	70.2	19.8	16.8	44.7	38.8	1.2	4.7	5.7	1.5	3.9	3.3	0.5
1974	112.1	75.9	21.1	18.0	48.4	42.0	1.1	5.3	6.3	1.7	5.0	4.4	0.6
1975	125.8	81.9	23.2	19.9	51.8	44.7	1.4	5.6	6.9	1.8	5.6	5.0	0.6
1976	141.7	93.8	27.0	23.4	57.7	49.9	1.5	6.3	9.1	2.2	6.3	4.7	0.6
1977	158.1	105.1	30.9	27.2	63.4	55.0	1.5	6.9	10.8	2.8	6.1	5.4	0.6
1978	177.2	117.4	35.6	31.6	70.3	60.8	1.9	7.6	11.5	3.4	7.4	6.7	0.6
1979	194.8	128.9	38.9	34.6	77.1	66.7	2.3	8.0	12.0	3.0	10.0	9.2	1.1
1980	216.0	140.9	43.7	39.1	83.4	70.0	2.6	10.8	13.7	3.6	13.4	11.2	2.0
1981	236.4	155.6	48.5	43.6	92.7	76.5	2.7	13.6	14.5	3.9	15.6	13.2	2.2
1982	244.7	162.3	52.3	47.0	97.0	80.3	2.8	13.9	13.1	4.0	17.5	15.3	2.1
1983	269.3	180.5	58.9	53.2	106.8	90.0	3.0	13.8	14.9	4.1	19.5	17.2	2.2
1984	303.7	205.4	68.2	61.9	119.7	100.9	3.4	15.5	17.4	4.7	22.4	19.8	2.5
1985	328.6	220.9	73.2	66.1	129.0	109.0	3.5	16.5	18.7	4.9	26.1	23.4	2.6
1986	354.0	234.0	78.3	70.7	134.9	115.8	3.6	15.6	20.7	6.0	27.5	24.6	2.7
1987	375.0	253.7	87.5	79.1	144.3	124.6	3.7	16.0	21.8	7.2	28.4	25.6	2.6
1988	403.1	269.4	90.4	81.5	155.2	135.1	3.8	16.3	23.8	8.4	30.8	28.0	2.7
1989	435.2	288.4	102.4	92.9	163.5	142.3	4.3	17.0	22.4	9.0	32.8	30.2	2.4
1990	469.2	305.6	109.6	99.6	175.4	152.5	4.6	18.3	20.5	10.0	33.9	31.4	2.3
1991	504.1	314.2	111.8	101.4	180.8	157.5	4.9	18.3	21.6	11.6	34.1	31.2	2.6
1992	551.4	338.8	120.6	109.0	195.9	169.7	6.1	20.1	22.2	13.1	33.6	30.4	2.7
1993	584.8	357.1	126.3	114.9	206.3	179.4	5.9	21.0	24.5	14.1	32.4	29.4	2.5
1994	621.6	380.4	132.1	120.2	220.8	191.7	7.0	22.1	27.5	14.5	33.9	30.6	2.5
1995	653.2	401.5	140.9	128.4	231.3	200.5	7.2	23.6	29.2	13.6	36.9	33.4	2.6
1996	686.2	425.9	150.9	138.6	245.0	211.8	8.2	25.0	29.9	12.5	39.4	35.3	2.7
1997	719.6	449.1	162.7	149.7	255.4	221.3	8.1	26.0	31.0	10.8	41.8	37.5	2.7
1998	764.1	480.7	180.4	166.8	268.8	233.3	8.5	27.0	31.6	10.4	43.3	39.2	2.4
1999	812.5	507.9	192.6	178.4	283.0	245.9	9.1	27.9	32.3	9.8	46.6	42.5	2.6
2000	867.7	540.1	213.5	199.2	294.8	255.5	7.8	31.6	31.7	10.8	49.3	44.3	3.6
2001	904.5	546.8	220.1	205.7	300.8	259.9	8.1	32.8	25.9	13.7	46.5	41.4	3.7
2002	917.8	533.7	199.1	184.3	309.6	268.0	7.7	33.9	25.0	15.8	42.4	37.2	3.7
2003	970.5	558.8	202.4	186.4	328.9	282.0	9.1	37.8	27.5	19.9	42.0	35.9	4.4
2004	1 048.7	606.4	220.3	203.4	352.3	300.5	8.3	43.5	33.8	24.7	44.4	37.4	5.1
2005	1 136.5	676.3	246.5	229.0	385.3	325.6	9.1	50.6	44.5	24.6	50.6	42.7	6.0
2006	1 193.9	728.1	271.0	252.5	409.1	346.6	9.7	52.7	48.0	21.5	59.2	50.7	6.5
2007	1 247.9	759.7	288.9	269.8	424.0	359.0	9.8	55.2	46.9	18.9	65.0	55.3	7.5
2008	1 264.0	763.6	299.8	280.9	425.4	354.3	9.7	61.3	38.4	18.7	60.1	49.6	7.9
2009	1 267.1	688.2	256.6	236.3	393.2	335.6	10.4	47.2	38.4	18.6	52.2	43.0	7.0
2010	1 336.1	720.0	264.7	242.2	415.5	354.0	10.7	50.1	39.8	18.1	51.3	42.4	6.6
2011	1 357.8	773.2	293.2	270.5	437.6	369.9	9.8	57.9	42.4	18.3	50.7	41.8	6.7
2012	1 372.3	804.9	313.8	290.4	448.3	379.2	9.7	59.3	42.9	17.5	50.1	40.8	6.9

[1] Includes components not shown separately.

Table 6-12. State Government Current Receipts and Expenditures—*Continued*

(National income and product accounts, calendar years, billions of dollars.)　　　　NIPA Table 3.20

Year	Current receipts—*Continued* Current transfer receipts					Current expenditures					Net state government saving, NIPA (surplus + / deficit -)		
	Total	Federal grants-in-aid	Local grants-in-aid	From business, net	From persons	Total [1]	Consumption expenditures	Government social benefits to persons	Grants-in-aid to local governments	Interest payments	Total	Social insurance funds	Other
1959	3.6	3.2	0.2	0.0	0.1	20.9	8.2	3.6	7.8	1.2	0.7	0.0	0.7
1960	3.7	3.3	0.2	0.0	0.1	22.9	8.9	3.8	8.8	1.3	0.4	0.0	0.4
1961	4.2	3.7	0.3	0.0	0.1	25.0	9.6	4.1	9.8	1.5	0.0	0.0	0.0
1962	4.6	4.1	0.3	0.1	0.2	27.0	10.2	4.4	10.8	1.6	0.1	0.0	0.1
1963	5.0	4.5	0.3	0.1	0.2	29.4	11.0	4.7	11.9	1.7	-0.3	0.0	-0.3
1964	5.5	4.9	0.3	0.1	0.2	31.9	11.8	5.1	13.0	1.8	-0.1	0.0	-0.1
1965	6.1	5.5	0.3	0.1	0.3	35.6	13.0	5.5	15.0	2.0	-0.4	0.1	-0.5
1966	8.9	8.2	0.4	0.1	0.3	40.7	14.6	6.4	17.4	2.2	0.8	0.1	0.6
1967	10.3	9.4	0.5	0.1	0.3	46.8	16.7	7.7	19.9	2.3	-1.2	0.1	-1.3
1968	11.8	10.7	0.6	0.1	0.3	54.6	19.2	9.5	23.0	2.7	-0.8	0.1	-1.0
1969	13.7	12.4	0.8	0.1	0.4	62.8	22.4	10.8	26.4	2.9	-1.2	0.2	-1.3
1970	16.6	15.2	0.9	0.1	0.4	72.8	25.8	12.9	30.4	3.4	-3.7	0.2	-3.9
1971	19.9	18.4	1.0	0.1	0.4	82.9	28.8	15.3	34.3	4.2	-4.7	0.2	-5.0
1972	26.4	24.6	1.1	0.2	0.5	92.9	31.5	17.5	38.5	5.0	2.2	0.3	1.9
1973	26.6	24.5	1.2	0.2	0.7	103.3	35.3	19.4	42.6	5.4	-0.1	0.3	-0.4
1974	28.6	26.3	1.3	0.2	0.8	116.5	42.1	19.9	47.3	6.4	-4.4	0.4	-4.8
1975	35.4	32.4	1.7	0.2	1.0	134.8	49.1	24.2	52.8	7.5	-9.0	0.5	-9.5
1976	39.1	35.4	2.3	0.3	1.2	145.6	52.8	26.9	57.0	7.7	-4.0	0.6	-4.6
1977	42.8	38.8	2.4	0.3	1.4	158.0	57.5	29.1	61.6	8.5	0.1	1.0	-0.9
1978	47.4	43.1	2.5	0.3	1.5	174.2	63.1	32.0	68.2	9.3	3.0	1.5	1.5
1979	50.1	45.9	2.2	0.4	1.6	193.8	70.0	35.4	76.3	10.3	1.0	1.8	-0.8
1980	56.5	52.1	2.3	0.4	1.7	217.8	78.8	41.3	84.6	11.1	-1.8	1.3	-3.2
1981	59.7	54.6	2.7	0.5	2.0	239.8	86.9	46.7	91.3	12.6	-3.4	1.3	-4.7
1982	58.7	52.7	3.1	0.6	2.3	257.0	93.7	51.0	95.3	14.6	-12.4	1.2	-13.6
1983	62.4	55.0	4.2	0.6	2.6	274.7	99.5	55.9	99.9	16.7	-5.5	1.2	-6.7
1984	67.7	59.0	5.0	0.8	3.0	297.5	106.9	59.8	109.8	17.9	6.2	1.4	4.8
1985	72.2	62.6	5.2	0.9	3.6	327.9	117.4	65.2	121.2	20.8	0.6	1.3	-0.7
1986	81.5	69.3	5.3	2.8	4.1	352.3	124.9	71.4	130.1	22.2	1.7	1.9	-0.1
1987	80.3	69.4	5.5	0.9	4.5	374.8	132.7	77.4	139.5	20.9	0.2	2.2	-2.0
1988	88.0	76.2	5.6	1.1	5.2	402.7	142.4	84.4	149.8	21.2	0.4	2.5	-2.1
1989	98.1	85.1	5.8	1.3	5.9	440.8	153.9	94.4	162.6	24.7	-5.6	2.3	-7.9
1990	112.4	97.6	6.2	1.6	6.9	480.7	167.2	111.0	173.1	23.8	-11.5	2.0	-13.4
1991	136.7	117.1	7.6	2.0	10.0	532.2	175.4	137.6	186.1	27.3	-28.1	2.4	-30.5
1992	158.0	134.3	9.0	2.6	12.1	580.8	183.9	159.5	201.0	30.4	-29.4	3.1	-32.6
1993	172.7	147.6	10.1	2.9	12.1	614.6	193.1	173.6	212.9	29.1	-29.8	4.2	-34.0
1994	183.7	156.9	11.0	3.4	12.4	649.8	204.5	184.4	226.2	28.4	-28.1	4.6	-32.8
1995	191.3	164.0	11.1	4.1	12.2	685.0	213.6	194.8	239.0	30.8	-31.8	4.0	-35.8
1996	197.9	168.9	12.0	5.0	12.0	708.1	219.7	202.5	250.3	27.9	-21.8	2.8	-24.7
1997	207.2	174.1	13.5	6.6	13.0	735.5	231.4	206.8	263.0	25.7	-15.9	1.2	-17.1
1998	219.3	182.9	13.3	9.8	13.2	774.0	248.3	214.6	282.4	19.8	-9.9	1.7	-11.6
1999	238.2	199.1	12.9	11.4	14.9	836.1	272.6	230.3	306.6	17.1	-23.6	1.7	-25.3
2000	258.8	212.6	14.0	15.0	17.2	896.0	292.8	248.4	331.0	13.5	-28.4	2.0	-30.3
2001	289.7	238.9	14.9	16.1	19.8	984.0	314.4	281.0	350.7	20.2	-79.5	2.6	-82.1
2002	317.7	263.5	15.5	17.2	21.5	1 037.5	320.3	307.2	367.5	31.5	-119.7	1.4	-121.1
2003	340.5	285.1	16.4	16.6	22.3	1 086.6	319.2	326.2	383.5	47.5	-116.1	3.2	-119.4
2004	363.4	303.4	17.6	17.7	24.6	1 139.7	327.1	355.8	401.6	44.5	-91.0	7.3	-98.3
2005	374.1	312.2	16.6	18.4	26.9	1 195.2	347.4	375.2	413.9	46.8	-58.7	7.2	-65.9
2006	375.4	314.2	15.6	18.2	28.0	1 239.7	361.2	371.3	441.7	52.3	-45.8	4.6	-50.5
2007	397.4	332.2	17.7	19.0	29.0	1 324.9	384.6	398.5	469.2	51.8	-77.0	2.7	-79.7
2008	415.6	346.6	17.7	21.1	30.7	1 383.9	407.7	418.6	486.1	54.1	-119.9	1.7	-121.6
2009	501.2	432.8	16.5	21.7	30.7	1 455.0	411.7	453.0	486.3	87.3	-187.9	2.2	-190.1
2010	539.5	473.3	16.6	20.1	30.1	1 487.1	411.2	481.5	488.8	88.4	-151.0	3.2	-154.2
2011	507.6	441.0	16.6	19.7	31.3	1 515.1	425.5	489.9	494.6	88.4	-157.2	4.2	-161.4
2012	491.2	424.0	16.8	18.8	32.4	1 579.2	459.7	501.9	502.8	97.8	-206.9	3.9	-210.9

[1] Includes components not shown separately.

Table 6-13. Local Government Current Receipts and Expenditures

(National income and product accounts, calendar years, billions of dollars.)

NIPA Table 3.21

Year	Current receipts Total 1	Current tax receipts Total	Personal current taxes Total 1	Income taxes	Taxes on production and imports Total	Sales taxes	Property taxes	Other	Taxes on corporate income	Contributions for government social insurance	Income receipts on assets Total 1	Interest receipts	Rents and royalties
1959	27.0	17.1	0.8	0.2	16.4	1.2	14.3	0.9	0.0	...	0.7	0.5	0.1
1960	30.0	18.8	0.8	0.3	18.0	1.3	15.7	0.9	0.0	...	0.8	0.7	0.1
1961	32.9	20.3	0.9	0.3	19.4	1.4	17.0	1.0	0.0	...	0.9	0.7	0.2
1962	35.6	21.8	1.0	0.3	20.8	1.5	18.4	1.0	0.0	...	1.0	0.8	0.2
1963	38.7	23.4	1.1	0.4	22.3	1.6	19.7	1.0	0.0	...	1.0	0.9	0.2
1964	42.3	25.3	1.2	0.5	24.1	1.9	21.1	1.1	0.0	...	1.3	1.1	0.2
1965	46.1	27.0	1.2	0.5	25.7	2.1	22.5	1.1	0.0	...	1.4	1.2	0.2
1966	50.5	28.5	1.4	0.6	27.1	2.0	23.8	1.3	0.0	...	1.7	1.4	0.3
1967	56.5	31.3	1.6	0.8	29.5	1.9	26.2	1.3	0.2	...	2.0	1.6	0.3
1968	62.6	34.8	1.7	0.9	32.8	2.3	29.1	1.4	0.3	...	1.7	1.4	0.4
1969	70.1	38.4	2.0	1.1	36.1	2.9	31.9	1.4	0.3	...	2.2	1.8	0.4
1970	81.4	43.2	2.3	1.3	40.6	3.5	35.7	1.5	0.2	...	2.6	2.2	0.5
1971	90.9	47.9	2.5	1.5	45.2	4.0	39.5	1.8	0.3	...	2.8	2.3	0.5
1972	101.9	52.0	3.0	2.0	48.8	4.6	42.2	2.0	0.3	...	3.0	2.4	0.6
1973	114.0	56.1	3.0	2.0	52.7	5.2	45.2	2.3	0.3	...	3.9	3.3	0.6
1974	123.8	60.2	3.4	2.3	56.4	6.1	47.9	2.4	0.3	...	5.2	4.5	0.7
1975	137.7	65.5	3.7	2.5	61.4	7.0	51.9	2.5	0.4	...	5.6	4.9	0.7
1976	150.8	71.9	4.1	2.8	67.3	7.9	56.7	2.7	0.5	...	5.1	4.4	0.7
1977	165.3	78.6	4.5	3.2	73.6	9.0	61.7	2.9	0.6	...	5.6	4.9	0.7
1978	180.5	80.8	4.9	3.4	75.3	10.2	61.8	3.3	0.6	...	7.2	6.5	0.7
1979	191.3	83.1	5.1	3.6	77.3	11.5	62.1	3.7	0.6	...	9.8	8.9	0.9
1980	208.3	89.1	5.1	3.5	83.3	12.8	66.2	4.2	0.7	...	12.9	11.8	1.1
1981	226.9	100.2	6.2	4.3	93.0	14.3	74.4	4.3	1.0	...	16.4	15.3	1.1
1982	244.4	110.8	6.9	4.9	103.0	15.9	82.5	4.6	1.0	...	19.2	17.8	1.4
1983	262.5	120.4	7.2	5.1	112.1	17.7	88.9	5.5	1.0	...	21.9	19.8	2.1
1984	289.9	131.9	7.8	5.6	122.7	20.1	96.3	6.3	1.4	...	25.2	22.8	2.4
1985	318.4	142.8	8.2	6.0	133.1	22.1	104.0	7.0	1.5	...	29.5	26.7	2.7
1986	342.5	155.6	8.9	6.8	144.7	24.1	112.6	8.1	2.0	...	31.4	27.9	3.5
1987	360.1	168.5	9.1	6.8	157.3	25.7	122.7	8.9	2.1	...	30.1	27.4	2.7
1988	387.9	183.3	11.7	9.1	169.4	27.2	132.7	9.4	2.3	...	30.4	28.6	1.7
1989	420.0	199.6	12.3	9.4	185.6	30.1	145.6	10.0	1.8	...	33.8	32.1	1.7
1990	446.2	213.6	12.9	10.0	198.7	31.8	157.0	9.9	2.0	...	34.5	32.7	1.8
1991	475.9	230.1	13.5	10.3	214.5	33.2	171.1	10.2	2.1	...	33.2	31.2	1.9
1992	501.1	241.0	14.7	11.4	224.2	34.6	178.6	11.0	2.1	...	30.4	28.4	2.0
1993	522.2	247.7	14.8	11.3	230.5	37.0	181.3	12.1	2.4	...	28.7	26.6	2.1
1994	556.9	263.8	15.9	12.0	245.4	39.7	192.4	13.3	2.5	...	29.6	27.6	2.0
1995	584.6	270.7	17.2	13.3	251.0	42.2	195.3	13.4	2.5	...	32.2	30.3	2.0
1996	617.4	283.7	17.8	13.7	262.9	44.4	204.2	14.4	3.1	...	34.3	32.4	2.0
1997	652.4	300.8	19.3	14.9	278.4	47.4	215.5	15.6	3.1	...	36.3	34.3	2.1
1998	690.4	314.2	20.9	16.2	290.0	50.6	222.5	16.9	3.4	...	38.0	35.8	2.2
1999	739.5	332.5	21.9	17.1	307.1	55.6	233.7	17.8	3.5	...	38.6	36.1	2.5
2000	791.8	353.1	23.2	18.1	326.5	61.3	246.9	18.3	3.5	...	44.6	42.0	2.7
2001	825.4	367.5	22.9	17.6	341.6	61.9	260.0	19.8	3.0	...	42.6	39.8	2.8
2002	865.1	394.8	22.7	17.1	366.2	63.1	281.6	21.5	5.9	...	36.2	33.2	2.9
2003	915.7	419.8	24.4	18.3	389.0	66.8	297.9	24.2	6.5	...	31.9	28.6	3.2
2004	957.8	451.5	27.1	20.3	416.4	70.5	317.9	28.0	7.9	...	32.4	29.0	3.4
2005	1 016.4	490.2	30.0	22.5	449.8	76.8	342.3	30.7	10.4	...	38.0	34.2	3.8
2006	1 089.2	526.4	31.5	23.6	483.6	85.2	365.3	33.1	11.3	...	48.0	44.0	3.9
2007	1 154.8	561.6	34.6	26.5	516.0	90.0	394.1	32.0	11.0	...	52.9	48.9	4.0
2008	1 164.4	565.3	33.8	26.1	522.6	90.4	404.0	28.1	9.0	...	45.6	41.4	4.2
2009	1 171.2	580.0	31.2	23.3	541.6	88.3	424.7	28.6	7.2	...	35.7	31.4	4.2
2010	1 184.5	585.7	33.0	25.0	544.9	91.2	424.4	29.3	7.8	...	31.3	26.7	4.6
2011	1 200.4	593.1	33.8	25.7	550.9	93.8	427.1	30.1	8.4	...	29.3	24.5	4.8
2012	1 204.4	600.3	35.1	26.9	556.7	95.7	430.3	30.7	8.5	...	28.4	23.4	5.0

1Includes components not shown separately.
. . . = Not available.

Table 6-13. Local Government Current Receipts and Expenditures—*Continued*

(National income and product accounts, calendar years, billions of dollars.)   **NIPA Table 3.21**

| Year | Current receipts—Continued | | | | | Current expenditures | | | | | Net local government saving, NIPA (surplus + / deficit -) | | |
| | Current transfer receipts | | | | | | | | | | | | |
	Total	Federal grants-in-aid	State grants-in-aid	From business, net	From persons	Total [1]	Consumption expenditures	Government social benefits to persons	Grants-in-aid to state governments	Interest payments	Total	Social insurance funds	Other
1959	8.4	0.4	7.8	0.1	0.1	25.5	23.0	0.7	0.2	1.5	1.5	...	1.5
1960	9.6	0.5	8.8	0.1	0.2	27.9	25.2	0.8	0.2	1.7	2.2	...	2.2
1961	10.8	0.6	9.8	0.2	0.3	30.7	27.7	0.9	0.3	1.8	2.2	...	2.2
1962	11.9	0.6	10.8	0.2	0.3	32.3	29.1	0.9	0.3	1.9	3.3	...	3.3
1963	13.1	0.8	11.9	0.2	0.3	34.7	31.3	1.0	0.3	2.1	4.0	...	4.0
1964	14.6	1.1	13.0	0.2	0.3	37.9	34.2	1.0	0.3	2.3	4.4	...	4.4
1965	16.6	1.1	15.0	0.3	0.3	41.6	37.6	1.1	0.3	2.5	4.5	...	4.5
1966	19.3	1.2	17.4	0.2	0.4	46.1	41.7	1.3	0.4	2.7	4.4	...	4.4
1967	22.4	1.5	19.9	0.4	0.6	50.6	45.6	1.6	0.5	2.9	5.9	...	5.9
1968	25.2	1.1	23.0	0.4	0.7	57.5	51.5	2.0	0.6	3.3	5.1	...	5.1
1969	28.8	1.3	26.4	0.4	0.7	64.9	58.1	2.4	0.8	3.6	5.3	...	5.3
1970	34.8	3.2	30.4	0.4	0.8	74.7	66.3	3.2	0.9	4.3	6.7	...	6.7
1971	39.5	3.8	34.3	0.5	0.9	85.6	75.4	4.0	1.0	5.2	5.3	...	5.3
1972	46.0	5.9	38.5	0.5	1.1	95.6	83.9	4.5	1.1	6.1	6.3	...	6.3
1973	53.2	8.9	42.6	0.7	1.0	104.3	91.5	4.7	1.2	6.8	9.8	...	9.8
1974	57.9	8.7	47.3	0.9	1.1	117.0	102.4	5.4	1.3	7.9	6.8	...	6.8
1975	66.4	11.2	52.8	1.0	1.4	134.3	116.5	6.6	1.7	9.2	3.5	...	3.5
1976	73.5	13.6	57.0	1.1	1.7	144.0	124.4	7.3	2.3	9.8	6.8	...	6.8
1977	80.9	16.0	61.6	1.3	1.9	156.9	135.5	8.0	2.4	10.7	8.4	...	8.4
1978	92.3	20.4	68.2	1.5	2.2	170.1	146.8	8.7	2.5	11.7	10.4	...	10.4
1979	98.8	18.1	76.3	1.8	2.6	184.0	159.4	8.9	2.2	13.2	7.2	...	7.2
1980	107.3	17.6	84.6	2.0	3.0	201.1	174.0	9.9	2.3	14.5	7.2	...	7.2
1981	112.2	14.8	91.3	2.4	3.7	219.4	189.2	10.4	2.7	16.7	7.5	...	7.5
1982	115.6	13.5	95.3	2.7	4.1	237.1	203.7	10.2	3.1	19.6	7.3	...	7.3
1983	120.4	13.0	99.9	3.0	4.5	255.8	217.2	11.0	4.2	22.8	6.7	...	6.7
1984	131.7	13.3	109.8	3.4	5.2	275.2	233.0	11.4	5.0	25.3	14.7	...	14.7
1985	144.0	13.6	121.2	3.5	5.7	298.8	254.4	12.1	5.2	26.4	19.6	...	19.6
1986	153.7	13.2	130.1	3.9	6.5	323.8	275.0	12.9	5.3	29.6	18.7	...	18.7
1987	159.2	9.0	139.5	4.0	6.7	348.3	295.6	13.4	5.5	32.6	11.9	...	11.9
1988	170.5	9.5	149.8	4.3	6.9	371.3	316.5	14.1	5.6	33.8	16.6	...	16.6
1989	181.9	6.8	162.6	5.1	7.5	407.0	347.4	15.0	5.8	37.6	13.0	...	13.0
1990	193.4	6.7	173.1	5.5	8.0	441.2	379.0	16.6	6.2	38.0	5.0	...	5.0
1991	207.7	7.0	186.1	5.9	8.7	472.9	404.7	18.9	7.6	40.3	3.0	...	3.0
1992	224.8	7.4	201.0	6.7	9.8	509.2	435.2	20.5	9.0	43.2	-8.1	...	-8.1
1993	240.0	8.1	212.9	7.7	11.4	529.6	453.4	21.6	10.1	43.1	-7.4	...	-7.4
1994	257.6	9.9	226.2	8.6	12.9	555.8	477.7	22.4	11.0	43.3	1.2	...	1.2
1995	273.2	10.5	239.0	9.4	14.3	584.3	503.2	22.9	11.1	45.8	0.3	...	0.3
1996	288.7	12.6	250.3	10.2	15.7	608.0	528.4	21.8	12.0	44.5	9.4	...	9.4
1997	305.4	14.0	263.0	11.1	17.3	634.8	554.5	20.8	13.5	44.5	17.6	...	17.6
1998	330.0	17.9	282.4	11.8	17.9	665.1	587.4	21.1	13.3	41.7	25.3	...	25.3
1999	358.7	20.1	306.6	12.7	19.3	706.6	629.4	22.1	12.9	40.7	32.9	...	32.9
2000	385.8	20.5	331.0	13.5	20.8	753.6	676.3	23.0	14.0	38.6	38.3	...	38.3
2001	409.1	22.4	350.7	13.7	22.3	811.2	725.4	24.0	14.9	45.3	14.2	...	14.2
2002	430.7	23.7	367.5	15.4	24.1	866.4	770.8	25.8	15.5	52.8	-1.2	...	-1.2
2003	464.4	36.6	383.5	17.2	27.1	921.0	808.0	27.3	16.4	67.7	-5.3	...	-5.3
2004	478.3	28.8	401.6	18.8	29.1	975.2	860.5	29.2	17.6	65.9	-17.4	...	-17.4
2005	492.8	31.1	413.9	18.1	29.6	1 024.3	909.2	31.4	16.6	65.0	-7.9	...	-7.9
2006	520.8	26.6	441.7	20.6	31.3	1 082.8	964.8	32.6	15.6	67.8	6.4	...	6.4
2007	550.4	26.8	469.2	21.5	32.5	1 150.6	1 026.8	34.8	17.7	69.7	4.3	...	4.3
2008	566.0	24.4	486.1	22.4	32.6	1 209.6	1 081.0	36.8	17.7	73.0	-45.2	...	-45.2
2009	567.7	25.3	486.3	22.3	33.3	1 255.2	1 096.7	39.6	16.5	101.4	-84.0	...	-84.0
2010	577.8	31.9	488.8	23.3	33.1	1 270.8	1 107.1	42.3	16.6	103.7	-86.3	...	-86.3
2011	585.4	31.4	494.6	24.4	33.9	1 256.2	1 092.0	42.1	16.6	104.5	-55.9	...	-55.9
2012	581.0	19.2	502.8	23.1	35.1	1 250.2	1 076.7	42.4	16.8	113.2	-45.8	...	-45.8

[1]Includes components not shown separately.
. . . = Not available.

SECTION 6C: FEDERAL GOVERNMENT BUDGET ACCOUNTS

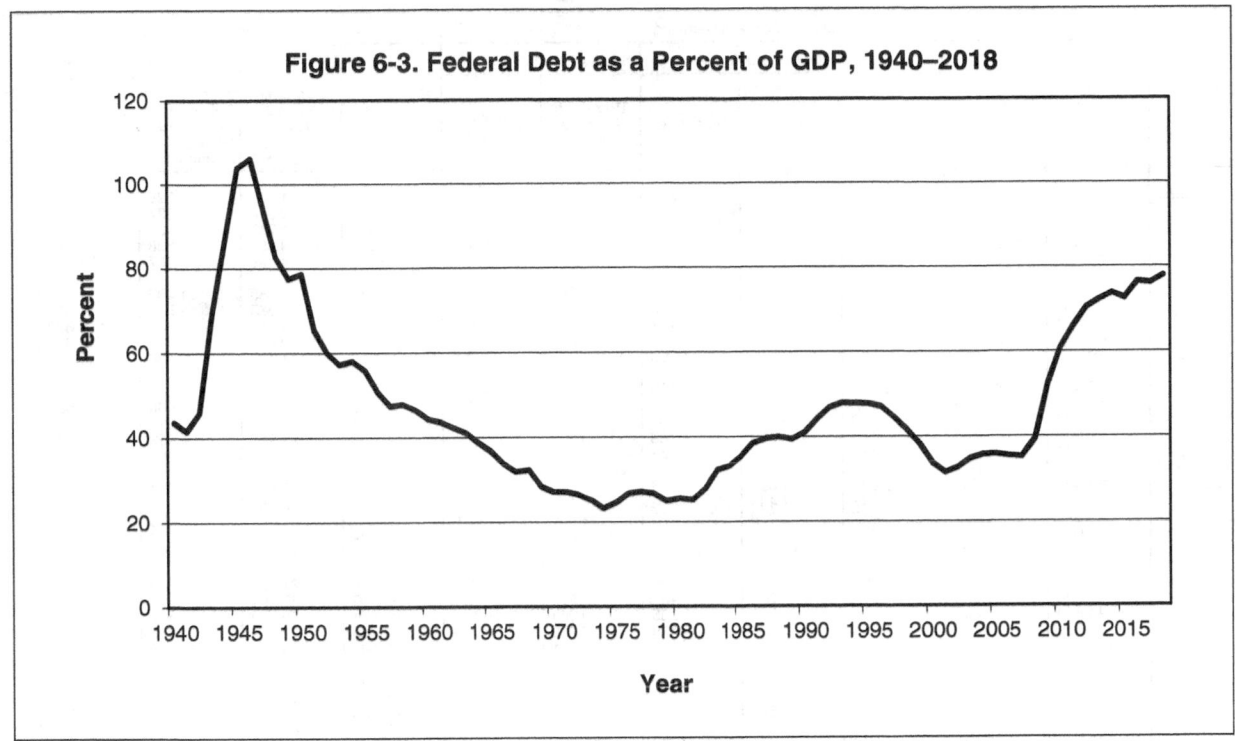

Figure 6-3. Federal Debt as a Percent of GDP, 1940–2018

- The "debt held by the public" is generally thought to be a more significant measure of the burden of the federal debt on credit markets than the gross debt, because it nets out the intragovernmental debt of the Social Security and other U.S. government trust funds. The ratio of debt held by the public to GDP was reduced in the late 1990s, began to creep back upward after 2001, and then rose rapidly beginning in 2008, as the recession reduced revenues and increased spending, and taxes were reduced further. The spending increases were a result of both automatic stabilizer programs, such as unemployment insurance, and an unprecedented level of spending for stimulus and financial rescue programs. At the end of fiscal year 2018, the debt held by the public was 77.8 percent of GDP, an increase from 76.1 percent in 2017. This represented the highest debt/GDP ratio since 1950. (Tables 6-14 and 6-15)

- At the end of 2017, foreign residents—including central banks—held nearly $6.3 trillion of Treasury debt, 39.5 percent of the nearly $15.8 trillion total debt held by the public. (Table 6-15)

Table 6-14A. Federal Government Receipts and Outlays by Fiscal Year [1]

(Budget accounts, millions of dollars.)

Year	Total receipts, net	Total outlays, net	Budget surplus or deficit (-)			Sources of financing, total		Individual income taxes	Corporate income taxes	Social insurance taxes and contributions		
			Total	On-budget	Off-budget	Borrowing from the public	Other financing			Employment taxes and contributions	Unemployment insurance	Other retirement contributions
1940	6 548	9 468	-2 920	-3 484	564	. . .	. . .	892	1 197	725	1 015	45
1941	8 712	13 653	-4 941	-5 594	653	5 451	-510	1 314	2 124	827	1 056	57
1942	14 634	35 137	-20 503	-21 333	830	19 530	973	3 263	4 719	1 064	1 299	89
1943	24 001	78 555	-54 554	-55 595	1 041	60 013	-5 459	6 505	9 557	1 338	1 477	229
1944	43 747	91 304	-47 557	-48 735	1 178	57 030	-9 473	19 705	14 838	1 557	1 644	272
1945	45 159	92 712	-47 553	-48 720	1 167	50 386	-2 833	18 372	15 988	1 592	1 568	291
1946	39 296	55 232	-15 936	-16 964	1 028	6 679	9 257	16 098	11 883	1 517	1 316	282
1947	38 514	34 496	4 018	2 861	1 157	-17 522	13 504	17 935	8 615	1 835	1 329	259
1948	41 560	29 764	11 796	10 548	1 248	-8 069	-3 727	19 315	9 678	2 168	1 343	239
1949	39 415	38 835	580	-684	1 263	-1 948	1 368	15 552	11 192	2 246	1 205	330
1950	39 443	42 562	-3 119	-4 702	1 583	4 701	-1 582	15 755	10 449	2 648	1 332	358
1951	51 616	45 514	6 102	4 259	1 843	-4 697	-1 405	21 616	14 101	3 688	1 609	377
1952	66 167	67 686	-1 519	-3 383	1 864	432	1 087	27 934	21 226	4 315	1 712	418
1953	69 608	76 101	-6 493	-8 259	1 766	3 625	2 868	29 816	21 238	4 722	1 675	423
1954	69 701	70 855	-1 154	-2 831	1 677	6 116	-4 962	29 542	21 101	5 192	1 561	455
1955	65 451	68 444	-2 993	-4 091	1 098	2 117	876	28 747	17 861	5 981	1 449	431
1956	74 587	70 640	3 947	2 494	1 452	-4 460	513	32 188	20 880	7 059	1 690	571
1957	79 990	76 578	3 412	2 639	773	-2 836	-576	35 620	21 167	7 405	1 950	642
1958	79 636	82 405	-2 769	-3 315	546	7 016	-4 247	34 724	20 074	8 624	1 933	682
1959	79 249	92 098	-12 849	-12 149	-700	8 365	4 484	36 719	17 309	8 821	2 131	770
1960	92 492	92 191	301	510	-209	2 139	-2 440	40 715	21 494	11 248	2 667	768
1961	94 388	97 723	-3 335	-3 766	431	1 517	1 818	41 338	20 954	12 679	2 903	857
1962	99 676	106 821	-7 146	-5 881	-1 265	9 653	-2 507	45 571	20 523	12 835	3 337	875
1963	106 560	111 316	-4 756	-3 966	-789	5 968	-1 212	47 588	21 579	14 746	4 112	946
1964	112 613	118 528	-5 915	-6 546	632	2 871	3 044	48 697	23 493	16 959	3 997	1 007
1965	116 817	118 228	-1 411	-1 605	194	3 929	-2 518	48 792	25 461	17 358	3 803	1 081
1966	130 835	134 532	-3 698	-3 068	-630	2 936	762	55 446	30 073	20 662	3 755	1 129
1967	148 822	157 464	-8 643	-12 620	3 978	2 912	5 731	61 526	33 971	27 823	3 575	1 221
1968	152 973	178 134	-25 161	-27 742	2 581	22 919	2 242	68 726	28 665	29 224	3 346	1 354
1969	186 882	183 640	3 242	-507	3 749	-11 437	8 195	87 249	36 678	34 236	3 328	1 451
1970	192 807	195 649	-2 842	-8 694	5 852	5 090	-2 248	90 412	32 829	39 133	3 464	1 765
1971	187 139	210 172	-23 033	-26 052	3 019	19 839	3 194	86 230	26 785	41 699	3 674	1 952
1972	207 309	230 681	-23 373	-26 068	2 695	19 340	4 033	94 737	32 166	46 120	4 357	2 097
1973	230 799	245 707	-14 908	-15 246	338	18 533	-3 625	103 246	36 153	54 876	6 051	2 187
1974	263 224	269 359	-6 135	-7 198	1 063	2 789	3 346	118 952	38 620	65 888	6 837	2 347
1975	279 090	332 332	-53 242	-54 148	906	51 001	2 241	122 386	40 621	75 199	6 771	2 565
1976	298 060	371 792	-73 732	-69 427	-4 306	82 704	-8 972	131 603	41 409	79 901	8 054	2 814
TQ	81 232	95 975	-14 744	-14 065	-679	18 105	-3 361	38 801	8 460	21 801	2 698	720
1977	355 559	409 218	-53 659	-49 933	-3 726	53 595	64	157 626	54 892	92 199	11 312	2 974
1978	399 561	458 746	-59 185	-55 416	-3 770	58 022	1 163	180 988	59 952	10 388	13 850	3 237
1979	463 302	504 028	-40 726	-39 633	-1 093	33 180	7 546	217 841	65 677	12 005	15 387	3 494
1980	517 112	590 941	-73 830	-73 141	-689	71 617	2 213	244 069	64 600	13 874	15 336	3 719
1981	599 272	678 241	-78 968	-73 859	-5 109	77 487	1 481	285 917	61 137	16 297	15 763	3 984
1982	617 766	745 743	-127 977	-120 593	-7 384	135 165	-7 188	297 744	49 207	18 068	16 600	4 212
1983	600 562	808 364	-207 802	-207 692	-110	212 693	-4 891	288 938	37 022	18 576	18 799	4 429
1984	666 438	851 805	-185 367	-185 269	-98	169 707	15 660	298 415	56 893	20 965	25 138	4 580
1985	734 037	946 344	-212 308	-221 529	9 222	200 285	12 023	334 531	61 331	23 464	25 758	4 759
1986	769 155	990 382	-221 227	-237 915	16 688	233 363	-12 136	348 959	63 143	25 506	24 098	4 742
1987	854 287	1 004 017	-149 730	-168 357	18 627	149 130	600	392 557	83 926	27 302	25 575	4 715
1988	909 238	1 064 416	-155 178	-192 265	37 087	161 863	-6 685	401 181	94 508	30 509	24 584	4 658
1989	991 104	1 143 743	-152 639	-205 393	52 754	139 100	13 539	445 690	10 329	33 285	22 011	4 546
1990	1 031 958	1 252 993	-221 036	-277 626	56 590	220 842	194	466 884	93 507	35 389	21 635	4 522
1991	1 054 988	1 324 226	-269 238	-321 435	52 198	277 441	-8 203	467 827	98 086	37 052	20 922	4 568
1992	1 091 208	1 381 529	-290 321	-340 408	50 087	310 738	-20 417	475 964	10 027	38 549	23 410	4 788
1993	1 154 334	1 409 386	-255 051	-300 398	45 347	248 659	6 392	509 680	11 752	39 693	26 556	4 805
1994	1 258 566	1 461 752	-203 186	-258 840	55 654	184 669	18 517	543 055	14 038	42 881	28 004	4 661
1995	1 351 790	1 515 742	-163 952	-226 367	62 415	171 313	-7 361	590 244	15 700	45 104	28 878	4 550
1996	1 453 053	1 560 484	-107 431	-174 019	66 588	129 695	-22 264	656 417	17 182	47 636	28 584	4 469
1997	1 579 232	1 601 116	-21 884	-103 248	81 364	38 271	-16 387	737 466	18 229	50 675	28 202	4 418
1998	1 721 728	1 652 458	69 270	-29 925	99 195	-51 245	-18 025	828 586	18 867	54 001	27 484	4 333
1999	1 827 452	1 701 842	125 610	1 920	123 690	-88 736	-36 874	879 480	18 468	58 088	26 480	4 473
2000	2 025 191	1 788 950	236 241	86 422	149 819	-222 559	-13 682	100 446	20 728	62 045	27 640	4 761
2001	1 991 082	1 862 846	128 236	-32 445	160 681	-90 189	-38 047	994 339	15 107	66 144	27 812	4 713
2002	1 853 136	2 010 894	-157 758	-317 417	159 659	220 812	-63 054	858 345	14 804	66 854	27 619	4 594
2003	1 782 314	2 159 899	-377 585	-538 418	160 833	373 016	4 569	793 699	131 778	674 981	33 366	4 631
2004	1 880 114	2 292 841	-412 727	-567 961	155 234	382 101	30 626	808 959	189 371	689 360	39 453	4 594
2005	2 153 611	2 471 957	-318 346	-493 611	175 265	296 668	21 678	927 222	278 282	747 664	42 002	4 459
2006	2 406 869	2 655 050	-248 181	-434 494	186 313	236 760	11 421	1 043 908	353 915	790 043	43 420	4 358
2007	2 567 985	2 728 686	-160 701	-342 153	181 452	206 157	-45 456	1 163 472	370 243	824 258	41 091	4 258
2008	2 523 991	2 982 544	-458 553	-641 848	183 295	767 921	-309 368	1 145 747	304 346	856 459	39 527	4 169
2009	2 104 989	3 517 677	-1 412 688	-1 549 681	136 993	1 741 657	-328 969	915 308	138 229	848 885	37 889	4 143
2010	2 162 706	3 457 079	-1 294 373	-1 371 378	77 005	1 474 175	-179 802	898 549	191 437	815 894	44 823	4 097
2011	2 303 466	3 603 065	-1 299 599	-1 366 781	67 182	1 109 305	190 294	1 091 473	181 085	758 516	56 241	4 035
2012	2 449 990	3 526 563	-1 076 573	-1 138 486	61 913	1 152 944	-76 371	1 132 206	242 289	774 927	66 647	3 740
2013	2 775 106	3 454 681	-679 775	-719 238	39 463	701 582	-21 807	1 316 405	273 506	887 445	56 811	3 564
2014	3 021 491	3 506 284	-484 793	-514 305	29 512	797 186	-312 393	1 394 568	320 731	965 029	54 957	3 472
2015	3 249 887	3 691 847	-441 960	-469 255	27 295	336 793	105 167	1 540 802	343 797	101 042	51 178	3 652
2016	3 267 961	3 852 612	-584 652	-620 158	35 507	1 050 932	-466 281	1 546 075	299 571	106 230	48 856	3 904
2017	3 316 182	3 981 628	-665 446	-714 863	49 417	497 815	167 631	1 587 120	297 048	111 189	45 808	4 192
2018	3 329 904	4 109 042	-779 138	-785 313	6 175	1 084 146	-305 008	1 683 538	204 733	112 115	45 042	4 504

[1]Fiscal years through 1976 are from July 1 through June 30. Beginning with October 1976 (fiscal year 1977), fiscal years are from October 1 through September 30. The period from July 1 through September 30, 1976, is a separate fiscal period known as the transition quarter (TQ) and is not included in any fiscal year.
. . . = Not available.

Table 6-14A. Federal Government Receipts and Outlays by Fiscal Year [1]—*Continued*

(Budget accounts, millions of dollars.)

Year	Receipts by source—Continued					Outlays by function						
	Excise taxes	Estate and gift taxes	Customs duties and fees	Miscellaneous receipts		National defense	International affairs	General science, space, and technology	Energy	Natural resources and environment	Agriculture	Commerce and housing credit
				Federal reserve deposits	All other							
1940	1 977	353	331	. . .	14	1 660	51	0	88	997	369	550
1941	2 552	403	365	. . .	14	6 435	145	0	91	817	339	398
1942	3 399	420	369	. . .	11	25 658	968	4	156	819	344	1 521
1943	4 096	441	308	. . .	50	66 699	1 286	1	116	726	343	2 151
1944	4 759	507	417	. . .	48	79 143	1 449	48	65	642	1 275	624
1945	6 265	637	341	. . .	105	82 965	1 913	111	25	455	1 635	-2 630
1946	6 998	668	424	. . .	109	42 681	1 935	34	41	482	610	-1 857
1947	7 211	771	477	15	69	12 808	5 791	5	18	700	814	-923
1948	7 356	890	403	100	68	9 105	4 566	1	202	780	69	306
1949	7 502	780	367	187	54	13 150	6 052	48	341	1 080	1 924	800
1950	7 550	698	407	192	55	13 724	4 673	55	327	1 308	2 049	1 035
1951	8 648	708	609	189	72	23 566	3 647	51	383	1 310	-323	1 228
1952	8 852	818	533	278	81	46 089	2 691	49	474	1 233	176	1 278
1953	9 877	881	596	298	81	52 802	2 119	49	425	1 289	2 253	910
1954	9 945	934	542	341	88	49 266	1 596	46	432	1 007	1 817	-184
1955	9 131	924	585	251	90	42 729	2 223	74	325	940	3 514	92
1956	9 929	1 161	682	287	140	42 523	2 414	79	174	870	3 486	506
1957	9 055	1 365	735	434	139	45 430	3 147	122	240	1 098	2 288	1 424
1958	8 612	1 393	782	664	123	46 815	3 364	141	348	1 407	2 411	930
1959	8 504	1 333	925	491	171	49 015	3 144	294	382	1 632	4 509	1 933
1960	9 137	1 606	1 105	1 093	119	48 130	2 988	599	464	1 559	2 623	1 618
1961	9 063	1 896	982	788	130	49 601	3 184	1 042	510	1 779	2 641	1 203
1962	9 585	2 016	1 142	718	125	52 345	5 639	1 723	604	2 044	3 562	1 424
1963	9 915	2 167	1 205	828	194	53 400	5 308	3 051	530	2 251	4 384	62
1964	10 211	2 394	1 252	947	139	54 757	4 945	4 897	572	2 364	4 609	418
1965	10 911	2 716	1 442	1 372	222	50 620	5 273	5 823	699	2 531	3 954	1 157
1966	9 145	3 066	1 767	1 713	163	58 111	5 580	6 717	612	2 719	2 447	3 245
1967	9 278	2 978	1 901	1 805	302	71 417	5 566	6 233	782	2 869	2 990	3 979
1968	9 700	3 051	2 038	2 091	400	81 926	5 301	5 524	1 037	2 988	4 544	4 280
1969	10 585	3 491	2 319	2 662	247	82 497	4 600	5 020	1 010	2 900	5 826	-119
1970	10 352	3 644	2 430	3 266	158	81 692	4 330	4 511	997	3 065	5 166	2 112
1971	10 510	3 735	2 591	3 533	325	78 872	4 159	4 182	1 035	3 915	4 290	2 366
1972	9 506	5 436	3 287	3 252	380	79 174	4 781	4 175	1 296	4 241	5 227	2 222
1973	9 836	4 917	3 188	3 495	425	76 681	4 149	4 032	1 237	4 775	4 821	931
1974	9 743	5 035	3 334	4 845	523	79 347	5 710	3 980	1 303	5 697	2 194	4 705
1975	9 400	4 611	3 676	5 777	935	86 509	7 097	3 991	2 916	7 346	2 997	9 947
1976	10 612	5 216	4 074	5 451	2 576	89 619	6 433	4 373	4 204	8 184	3 109	7 619
TQ	2 520	1 455	1 212	1 500	111	22 269	2 458	1 162	1 129	2 524	972	931
1977	9 648	7 327	5 150	5 908	623	97 241	6 353	4 736	5 770	10 032	6 734	3 093
1978	10 054	5 285	6 573	6 641	778	10 449	7 482	4 926	7 991	10 983	11 301	6 254
1979	9 808	5 411	7 439	8 327	925	11 634	7 459	5 234	9 179	12 135	11 176	4 686
1980	15 563	6 389	7 174	11 767	981	13 399	12 714	5 831	10 156	13 858	8 774	9 390
1981	34 128	6 787	8 083	12 834	956	15 751	13 104	6 468	15 166	13 568	11 241	8 206
1982	28 670	7 991	8 854	15 186	975	18 530	12 300	7 199	13 527	12 998	15 866	6 256
1983	24 086	6 053	8 655	14 492	1 108	20 990	11 848	7 934	9 353	12 672	22 807	6 681
1984	22 279	6 010	11 370	15 684	1 328	22 741	15 869	8 311	7 073	12 586	13 477	6 959
1985	19 097	6 422	12 079	17 059	1 460	25 274	16 169	8 622	5 608	13 345	25 427	4 337
1986	16 053	6 958	13 327	18 374	1 574	27 337	14 146	8 962	4 690	13 628	31 319	5 058
1987	14 844	7 493	15 085	16 817	2 635	28 199	11 645	9 200	4 072	13 355	26 466	6 434
1988	16 185	7 594	16 198	17 163	3 031	29 036	10 466	10 820	2 296	14 601	17 088	19 163
1989	13 147	8 745	16 334	19 604	3 639	30 355	9 583	12 821	2 705	16 169	16 698	29 709
1990	15 591	11 500	16 707	24 319	3 647	29 932	13 758	14 426	3 341	17 055	11 637	67 599
1991	18 275	11 138	15 949	19 158	4 412	27 328	15 846	16 092	2 436	18 544	14 886	76 270
1992	21 836	11 143	17 359	22 920	4 293	29 834	16 090	16 389	4 499	20 001	14 922	10 918
1993	24 522	12 577	18 802	14 908	4 491	29 110	17 218	17 006	4 319	20 224	20 081	-2 185
1994	31 226	15 225	20 099	18 023	5 081	28 164	17 067	16 189	5 218	21 000	14 795	-4 228
1995	26 941	14 763	19 301	23 378	5 143	27 206	16 429	16 692	4 936	21 889	9 671	-1 780
1996	25 447	17 189	18 670	20 477	5 048	26 574	13 487	16 684	2 839	21 503	9 035	-1 047
1997	27 831	19 845	17 928	19 636	5 769	27 050	15 173	17 136	1 475	21 201	8 889	-1 464
1998	21 665	24 076	18 297	24 540	8 048	26 819	13 054	18 172	1 270	22 278	12 077	1 007
1999	19 293	27 782	18 336	25 917	9 010	27 476	15 239	18 084	911	23 943	22 879	2 641
2000	22 692	29 010	19 914	32 293	10 506	29 436	17 213	18 594	-761	25 003	36 458	3 207
2001	24 286	28 400	19 369	26 124	11 576	30 473	16 485	19 753	9	25 532	26 252	5 731
2002	24 017	26 507	18 602	23 683	10 206	34 845	22 315	20 734	475	29 426	21 965	-407
2003	23 804	21 959	19 862	21 878	12 636	40 473	21 199	20 831	-725	29 667	22 496	727
2004	24 566	24 831	21 083	19 652	12 956	45 581	26 870	23 029	-147	30 694	15 439	5 265
2005	22 547	24 764	23 379	19 297	13 448	495 294	34 565	23 597	440	27 983	26 565	7 566
2006	22 460	27 877	24 810	29 945	14 632	521 820	29 499	23 584	785	33 025	25 969	6 187
2007	11 076	26 044	26 010	32 043	15 497	551 258	28 482	24 407	-852	31 721	17 662	487
2008	15 726	28 844	27 568	33 598	16 399	616 066	28 857	26 773	631	31 820	18 387	27 870
2009	13 854	23 482	22 453	34 318	17 799	661 012	37 529	28 417	4 755	35 573	22 237	291 535
2010	18 256	18 885	25 298	75 845	20 969	693 485	45 195	30 100	11 618	43 667	21 356	-82 316
2011	18 904	7 399	29 519	82 546	20 271	705 554	45 685	29 466	12 174	45 473	20 662	-12 564
2012	20 359	13 973	30 307	81 957	24 883	677 852	36 802	29 060	14 858	41 631	17 791	40 647
2013	28 330	18 912	31 815	75 767	26 874	633 446	46 464	28 908	11 042	38 145	29 678	-83 198
2014	34 240	19 300	33 926	99 235	36 905	603 457	46 879	28 570	5 270	36 171	24 386	-94 861
2015	37 759	19 232	35 041	96 468	51 011	589 659	52 040	29 412	6 838	36 033	18 500	-37 905
2016	33 991	21 354	34 838	115 672	40 360	593 372	45 306	30 174	3 719	39 082	18 342	-34 077
2017	21 191	22 768	34 574	81 287	47 665	598 722	46 309	30 394	3 856	37 896	18 870	-26 685
2018	30 067	22 983	41 299	70 750	40 914	631 161	48 972	31 534	2 169	39 140	21 787	-9 470

[1] Fiscal years through 1976 are from July 1 through June 30. Beginning with October 1976 (fiscal year 1977), fiscal years are from October 1 through September 30. The period from July 1 through September 30, 1976, is a separate fiscal period known as the transition quarter (TQ) and is not included in any fiscal year.
. . . = Not available.

Table 6-14A. Federal Government Receipts and Outlays by Fiscal Year [1]—*Continued*

(Budget accounts, millions of dollars.)

Year	Transportation	Community and regional development	Education, employment, and social services	Health	Medicare	Income security	Social Security	Veterans benefits and services	Administration of justice	General government	Net interest
1940	392	285	1 972	55	0	1 514	28	570	81	274	899
1941	353	123	1 592	60	0	1 855	91	560	92	306	943
1942	1 283	113	1 062	71	0	1 828	137	501	117	397	1 052
1943	3 220	219	375	92	0	1 739	177	276	154	673	1 529
1944	3 901	238	160	174	0	1 503	217	-126	192	900	2 219
1945	3 654	243	134	211	0	1 137	267	110	178	581	3 112
1946	1 970	200	85	201	0	2 384	358	2 465	176	825	4 111
1947	1 130	302	102	177	0	2 820	466	6 344	176	1 114	4 204
1948	787	78	191	162	0	2 499	558	6 457	170	1 045	4 341
1949	916	-33	178	197	0	3 174	657	6 599	184	824	4 523
1950	967	30	241	268	0	4 097	781	8 834	193	986	4 812
1951	956	47	235	323	0	3 352	1 565	5 526	218	1 097	4 665
1952	1 124	73	339	347	0	3 655	2 063	5 341	267	1 163	4 701
1953	1 264	117	441	336	0	3 823	2 717	4 519	243	1 209	5 156
1954	1 229	100	370	307	0	4 434	3 352	4 613	257	799	4 811
1955	1 246	129	445	291	0	5 071	4 427	4 675	256	651	4 850
1956	1 450	92	591	359	0	4 734	5 478	4 891	302	1 201	5 079
1957	1 662	135	590	479	0	5 427	6 661	5 005	303	1 360	5 354
1958	2 334	169	643	541	0	7 535	8 219	5 350	325	655	5 604
1959	3 655	211	789	685	0	8 239	9 737	5 443	356	926	5 762
1960	4 126	224	968	795	0	7 378	11 602	5 441	366	1 184	6 947
1961	3 987	275	1 063	913	0	9 683	12 474	5 705	400	1 354	6 716
1962	4 290	469	1 241	1 198	0	9 207	14 365	5 619	429	1 049	6 889
1963	4 596	574	1 458	1 451	0	9 311	15 788	5 514	465	1 230	7 740
1964	5 242	933	1 555	1 788	0	9 657	16 620	5 675	489	1 518	8 199
1965	5 763	1 114	2 140	1 791	0	9 469	17 460	5 716	536	1 499	8 591
1966	5 730	1 105	4 363	2 543	64	9 678	20 694	5 916	564	1 603	9 386
1967	5 936	1 108	6 453	3 351	2 748	10 261	21 725	6 735	618	1 719	10 268
1968	6 316	1 382	7 634	4 390	4 649	11 816	23 854	7 032	659	1 757	11 090
1969	6 526	1 552	7 548	5 162	5 695	13 076	27 298	7 631	766	1 939	12 699
1970	7 008	2 392	8 634	5 907	6 213	15 655	30 270	8 669	959	2 320	14 380
1971	8 052	2 917	9 849	6 843	6 622	22 946	35 872	9 768	1 307	2 442	14 841
1972	8 392	3 423	12 529	8 674	7 479	27 650	40 157	10 720	1 684	2 960	15 478
1973	9 066	4 605	12 744	9 356	8 052	28 278	49 090	12 003	2 174	9 774	17 349
1974	9 172	4 229	12 455	10 733	9 639	33 714	55 867	13 374	2 505	10 032	21 449
1975	10 918	4 322	16 022	12 930	12 875	50 176	64 658	16 584	3 028	10 374	23 244
1976	13 739	5 442	18 910	15 734	15 834	60 799	73 899	18 419	3 430	9 706	26 727
TQ	3 358	1 569	5 169	3 924	4 264	14 985	19 763	3 960	918	3 878	6 949
1977	14 829	7 021	21 104	17 302	19 345	61 060	85 061	18 022	3 701	12 791	29 901
1978	15 521	11 841	26 706	18 524	22 768	61 509	93 861	18 961	3 923	11 961	35 458
1979	18 079	10 480	30 218	20 494	26 495	66 382	10 407	19 914	4 286	12 241	42 633
1980	21 329	11 252	31 835	23 169	32 090	86 565	11 854	21 169	4 702	12 975	52 533
1981	23 379	10 568	33 146	26 866	39 149	10 030	13 958	22 973	4 908	11 373	68 766
1982	20 625	8 347	26 609	27 445	46 567	10 815	15 596	23 938	4 842	10 861	85 032
1983	21 334	7 564	26 194	28 641	52 588	12 304	17 072	24 824	5 246	11 181	89 808
1984	23 669	7 673	26 916	30 417	57 540	11 340	17 822	25 575	5 811	11 746	11 110
1985	25 838	7 676	28 589	33 541	65 822	12 903	18 862	26 251	6 426	11 515	12 947
1986	28 113	7 233	29 773	35 933	70 164	12 068	19 875	26 314	6 735	12 491	13 601
1987	26 222	5 049	28 918	39 964	75 120	12 413	20 735	26 729	7 715	7 487	13 861
1988	27 272	5 293	30 928	44 483	78 878	13 043	21 934	29 367	9 397	9 399	15 180
1989	27 608	5 362	35 325	48 380	84 964	13 758	23 254	30 003	9 644	9 317	16 898
1990	29 485	8 531	37 167	57 699	98 102	14 883	24 862	29 034	10 185	10 462	18 434
1991	31 099	6 810	41 231	71 168	10 448	17 263	26 901	31 275	12 486	11 568	19 444
1992	33 332	6 836	42 735	89 486	11 902	19 973	28 758	34 037	14 650	12 883	19 934
1993	35 004	9 146	47 374	99 401	13 055	21 013	30 458	35 642	15 193	12 944	19 871
1994	38 066	10 620	43 281	10 710	14 474	21 729	31 956	37 559	15 516	11 159	20 293
1995	39 350	10 746	51 020	11 539	15 985	22 380	33 584	37 862	16 508	13 799	23 213
1996	39 565	10 741	48 311	11 936	17 422	22 974	34 967	36 956	17 898	11 755	24 105
1997	40 767	11 049	48 972	12 383	19 001	23 503	36 525	39 283	20 617	12 547	24 398
1998	40 343	9 771	50 512	13 142	19 282	23 775	37 921	41 741	23 359	15 544	24 111
1999	42 532	11 865	50 605	14 104	19 044	24 247	39 003	43 155	26 536	15 363	22 975
2000	46 853	10 623	53 764	15 450	19 711	25 372	40 942	46 989	28 499	13 013	22 294
2001	54 447	11 773	57 094	17 223	21 738	26 977	43 295	44 974	30 201	14 358	20 616
2002	61 833	12 981	70 566	19 649	23 085	31 272	45 598	50 929	35 061	16 951	17 094
2003	67 069	18 850	82 587	219 541	249 433	334 632	474 680	56 984	35 340	23 164	15 307
2004	64 627	15 820	87 974	240 122	269 360	333 059	495 548	59 746	45 576	22 338	16 024
2005	67 894	26 262	97 555	250 548	298 638	345 847	523 305	70 120	40 019	16 997	18 398
2006	70 244	54 465	118 482	252 739	329 868	352 477	548 549	69 811	41 016	18 177	22 660
2007	72 905	29 567	91 656	266 382	375 407	365 975	586 153	72 818	42 362	17 425	23 710
2008	77 616	23 952	91 287	280 599	390 758	431 313	617 027	84 653	48 097	20 323	25 275
2009	84 289	27 676	79 749	334 335	430 093	533 224	682 963	95 429	52 581	22 017	18 690
2010	91 972	23 894	128 598	369 068	451 636	622 210	706 737	108 384	54 383	23 014	19 619
2011	92 966	23 883	101 233	372 504	485 653	597 349	730 811	127 189	56 056	27 476	22 996
2012	93 019	25 132	90 823	346 742	471 793	541 344	773 290	124 595	56 277	28 035	22 040
2013	91 673	32 335	72 808	358 316	497 826	536 511	813 551	138 938	52 601	27 737	22 088
2014	91 915	20 670	90 615	409 449	511 688	513 644	850 533	149 616	50 457	26 913	22 895
2015	89 533	20 669	122 061	482 231	546 202	508 843	887 753	159 738	51 906	20 956	22 318
2016	92 566	20 140	109 737	511 297	594 536	514 139	916 067	174 516	55 768	23 146	24 003
2017	93 552	24 907	143 976	533 129	597 307	503 484	944 878	176 543	57 944	23 821	26 255
2018	92 785	42 159	95 516	551 216	588 706	495 318	987 791	178 856	60 418	23 878	324 975

[1] Fiscal years through 1976 are from July 1 through June 30. Beginning with October 1976 (fiscal year 1977), fiscal years are from October 1 through September 30. The period from July 1 through September 30, 1976, is a separate fiscal period known as the transition quarter (TQ) and is not included in any fiscal year.
. . . = Not available.

Table 6-14B. The Federal Budget and GDP

(Billions of dollars; percent.)

Year	Fiscal year GDP	Billions of dollars						Percent of GDP					
		Receipts			Outlays		Total budget surplus or deficit	Receipts			Outlays		Total budget surplus or deficit
		Total	Individual income taxes	Corporate income taxes	Total	National defense		Total	Individual income taxes	Corporate income taxes	Total	National defense	
1950	278.7	39.4	15.8	10.4	42.6	13.7	-3.1	14.2	5.7	3.7	15.1	4.9	-1.7
1951	327.0	51.6	21.6	14.1	45.5	23.6	6.1	15.8	6.6	4.3	13.5	7.2	1.3
1952	357.1	66.2	27.9	21.2	67.7	46.1	-1.5	18.5	7.8	5.9	18.5	12.9	-0.9
1953	382.0	69.6	29.8	21.2	76.1	52.8	-6.5	18.2	7.8	5.6	19.3	13.8	-2.2
1954	387.2	69.7	29.5	21.1	70.9	49.3	-1.2	18.0	7.6	5.5	17.5	12.7	-0.7
1955	406.3	65.5	28.7	17.9	68.4	42.7	-3.0	16.1	7.1	4.4	15.9	10.5	-1.0
1956	438.2	74.6	32.2	20.9	70.6	42.5	3.9	17.0	7.3	4.8	15.0	9.7	0.6
1957	463.4	80.0	35.6	21.2	76.6	45.4	3.4	17.3	7.7	4.6	15.2	9.8	0.6
1958	473.5	79.6	34.7	20.1	82.4	46.8	-2.8	16.8	7.3	4.2	15.8	9.9	-0.7
1959	504.6	79.2	36.7	17.3	92.1	49.0	-12.8	15.7	7.3	3.4	16.5	9.7	-2.4
1960	534.3	92.5	40.7	21.5	92.2	48.1	0.3	17.3	7.6	4.0	15.2	9.0	0.1
1961	546.6	94.4	41.3	21.0	97.7	49.6	-3.3	17.3	7.6	3.8	15.7	9.1	-0.7
1962	585.7	99.7	45.6	20.5	106.8	52.3	-7.1	17.0	7.8	3.5	15.9	8.9	-1.0
1963	618.2	106.6	47.6	21.6	111.3	53.4	-4.8	17.2	7.7	3.5	15.6	8.6	-0.6
1964	661.7	112.6	48.7	23.5	118.5	54.8	-5.9	17.0	7.4	3.6	15.5	8.3	-1.0
1965	709.3	116.8	48.8	25.5	118.2	50.6	-1.4	16.5	6.9	3.6	14.3	7.1	-0.2
1966	780.5	130.8	55.4	30.1	134.5	58.1	-3.7	16.8	7.1	3.9	14.7	7.4	-0.4
1967	836.5	148.8	61.5	34.0	157.5	71.4	-8.6	17.8	7.4	4.1	16.4	8.5	-1.5
1968	897.6	153.0	68.7	28.7	178.1	81.9	-25.2	17.0	7.7	3.2	17.4	9.1	-3.1
1969	980.3	186.9	87.2	36.7	183.6	82.5	3.2	19.1	8.9	3.7	16.2	8.4	-0.1
1970	1 040.7	192.8	90.4	32.8	195.6	81.7	-2.8	18.4	8.6	3.1	16.1	7.8	-0.8
1971	1 116.6	187.1	86.2	26.8	210.2	78.9	-23.0	16.8	7.7	2.4	15.9	7.1	-2.3
1972	1 216.2	207.3	94.7	32.2	230.7	79.2	-23.4	17.0	7.8	2.6	15.9	6.5	-2.1
1973	1 352.7	230.8	103.2	36.2	245.7	76.7	-14.9	17.1	7.6	2.7	14.8	5.7	-1.1
1974	1 482.8	263.2	119.0	38.6	269.4	79.3	-6.1	17.8	8.0	2.6	14.6	5.4	-0.5
1975	1 606.9	279.1	122.4	40.6	332.3	86.5	-53.2	17.4	7.6	2.5	16.9	5.4	-3.4
1976	1 786.1	298.1	131.6	41.4	371.8	89.6	-73.7	16.7	7.4	2.3	16.9	5.0	-3.9
1977	2 024.3	355.6	157.6	54.9	409.2	97.2	-53.7	17.6	7.8	2.7	16.2	4.8	-2.5
1978	2 273.4	399.6	181.0	60.0	458.7	104.5	-59.2	17.6	8.0	2.6	16.3	4.6	-2.4
1979	2 565.6	463.3	217.8	65.7	504.0	116.3	-40.7	18.1	8.5	2.6	15.8	4.5	-1.5
1980	2 791.9	517.1	244.1	64.6	590.9	134.0	-73.8	18.5	8.7	2.3	17.1	4.8	-2.6
1981	3 133.2	599.3	285.9	61.1	678.2	157.5	-79.0	19.1	9.1	2.0	17.3	5.0	-2.4
1982	3 313.4	617.8	297.7	49.2	745.7	185.3	-128.0	18.6	9.0	1.5	18.0	5.6	-3.6
1983	3 536.0	600.6	288.9	37.0	808.4	209.9	-207.8	17.0	8.2	1.0	18.7	5.9	-5.9
1984	3 949.2	666.4	298.4	56.9	851.8	227.4	-185.4	16.9	7.6	1.4	17.4	5.8	-4.7
1985	4 265.1	734.0	334.5	61.3	946.3	252.7	-212.3	17.2	7.8	1.4	18.0	5.9	-5.2
1986	4 526.2	769.2	349.0	63.1	990.4	273.4	-221.2	17.0	7.7	1.4	17.8	6.0	-5.3
1987	4 767.6	854.3	392.6	83.9	1 004.0	282.0	-149.7	17.9	8.2	1.8	17.0	5.9	-3.5
1988	5 138.6	909.2	401.2	94.5	1 064.4	290.4	-155.2	17.7	7.8	1.8	16.7	5.7	-3.7
1989	5 554.7	991.1	445.7	103.3	1 143.7	303.6	-152.6	17.8	8.0	1.9	16.8	5.5	-3.7
1990	5 898.8	1 032.0	466.9	93.5	1 253.0	299.3	-221.0	17.5	7.9	1.6	17.4	5.1	-4.7
1991	6 093.2	1 055.0	467.8	98.1	1 324.2	273.3	-269.2	17.3	7.7	1.6	17.8	4.5	-5.3
1992	6 416.2	1 091.2	476.0	100.3	1 381.5	298.3	-290.3	17.0	7.4	1.6	17.6	4.6	-5.3
1993	6 775.3	1 154.3	509.7	117.5	1 409.4	291.1	-255.1	17.0	7.5	1.7	16.9	4.3	-4.4
1994	7 176.8	1 258.6	543.1	140.4	1 461.8	281.6	-203.2	17.5	7.6	2.0	16.5	3.9	-3.6
1995	7 560.4	1 351.8	590.2	157.0	1 515.7	272.1	-164.0	17.9	7.8	2.1	16.2	3.6	-3.0
1996	7 951.3	1 453.1	656.4	171.8	1 560.5	265.7	-107.4	18.3	8.3	2.2	15.8	3.3	-2.2
1997	8 451.0	1 579.2	737.5	182.3	1 601.1	270.5	-21.9	18.7	8.7	2.2	15.3	3.2	-1.2
1998	8 930.8	1 721.7	828.6	188.7	1 652.5	268.2	69.3	19.3	9.3	2.1	15.0	3.0	-0.3
1999	9 479.4	1 827.5	879.5	184.7	1 701.8	274.8	125.6	19.3	9.3	1.9	14.6	2.9	. . .
2000	10 117.4	2 025.2	1 004.5	207.3	1 789.0	294.4	236.2	20.0	9.9	2.0	14.4	2.9	0.9
2001	10 526.5	1 991.1	994.3	151.1	1 862.8	304.7	128.2	18.9	9.4	1.4	14.4	2.9	-0.3
2002	10 833.6	1 853.1	858.3	148.0	2 010.9	348.5	-157.8	17.1	7.9	1.4	15.3	3.2	-2.9
2003	11 283.8	1 782.3	793.7	131.8	2 159.9	404.7	-377.6	15.8	7.0	1.2	15.9	3.6	-4.8
2004	12 025.4	1 880.1	809.0	189.4	2 292.8	455.8	-412.7	15.6	6.7	1.6	15.9	3.8	-4.7
2005	12 834.2	2 153.6	927.2	278.3	2 472.0	495.3	-318.3	16.8	7.2	2.2	16.1	3.9	-3.8
2006	13 638.4	2 406.9	1 043.9	353.9	2 655.1	521.8	-248.2	17.6	7.7	2.6	16.4	3.8	-3.2
2007	14 290.8	2 568.0	1 163.5	370.2	2 728.7	551.3	-160.7	18.0	8.1	2.6	15.9	3.9	-2.4
2008	14 743.3	2 524.0	1 145.7	304.3	2 982.5	616.1	-458.6	17.1	7.8	2.1	17.0	4.2	-4.4
2009	14 431.8	2 105.0	915.3	138.2	3 517.7	661.0	-1 412.7	14.6	6.3	1.0	20.8	4.6	-10.7
2010	14 838.8	2 162.7	898.5	191.4	3 457.1	693.5	-1 294.4	14.6	6.1	1.3	19.6	4.7	-9.2
2011	15 403.7	2 303.5	1 091.5	181.1	3 603.1	705.6	-1 299.6	15.0	7.1	1.2	20.2	4.6	-8.9
2012	16 056.4	2 450.0	1 132.2	242.3	3 526.6	677.9	-1 076.6	15.3	7.1	1.5	18.8	4.2	-7.1
2013	16 603.8	2 775.1	1 316.4	273.5	3 454.9	633.4	-679.8	16.7	7.9	1.6	17.0	3.8	-4.3
2014	17 332.9	3 021.5	1 394.6	320.7	3 506.3	603.5	-484.8	17.4	8.0	1.9	16.2	3.5	-3.0
2015	18 090.3	3 249.9	1 540.8	343.8	3 691.8	589.7	-442.0	18.0	8.5	1.9	16.3	3.3	-2.6
2016	18 551.0	3 268.0	1 546.1	299.6	3 852.6	593.4	-584.7	17.6	8.3	1.6	16.6	3.2	-3.3
2017	19 272.2	3 316.2	1 587.1	297.0	3 981.6	598.7	-665.4	17.2	8.2	1.5	16.5	3.1	-3.7
2018	20 235.9	3 329.9	1 683.5	204.7	4 109.0	631.2	-779.1	16.5	8.3	1.0	16.1	3.1	-3.9

. . . = Not available.

Table 6-15. Federal Government Debt by Fiscal Year

(Billions of dollars, except as noted.)

Year	Federal government debt held by the public at end of fiscal year		Gross federal debt at end of fiscal year held by:						
	Debt held by the public	Debt/GDP ratio (percent)	Total	Social Security funds [1]	Other U.S. government accounts	Federal Reserve System	Private investors		
							Total	Foreign residents	Domestic investors
1940	43	43.6	51	2	6	2	40	...	...
1941	48	41.5	58	2	7	2	46	...	...
1942	68	45.9	79	3	8	3	65	...	...
1943	128	69.2	143	4	11	7	121	...	...
1944	185	86.4	204	5	14	15	170	...	...
1945	235	103.9	260	7	18	22	213	...	...
1946	242	106.1	271	8	21	24	218	...	...
1947	224	93.9	257	9	24	22	202	...	...
1948	216	82.6	252	10	26	21	195	...	...
1949	214	77.5	253	11	27	19	195	...	...
1950	219	78.6	257	13	25	18	201	...	...
1951	214	65.5	255	15	26	23	191	...	...
1952	215	60.1	259	17	28	23	192	...	...
1953	218	57.2	266	18	29	25	194	...	...
1954	224	58.0	271	20	26	25	199	...	...
1955	227	55.8	274	21	27	24	203	...	...
1956	222	50.7	273	23	28	24	198	...	...
1957	219	47.3	272	23	30	23	196	...	...
1958	226	47.8	280	24	29	25	201	...	...
1959	235	46.5	287	23	30	26	209	...	...
1960	237	44.3	291	23	31	27	210	...	...
1961	238	43.6	293	23	31	27	211	...	...
1962	248	42.3	303	22	33	30	218	...	...
1963	254	41.1	310	21	35	32	222	...	...
1964	257	38.8	316	22	37	35	222	...	...
1965	261	36.8	322	22	39	39	222	12	210
1966	264	33.8	328	22	43	42	222	12	210
1967	267	31.9	340	26	48	47	220	11	209
1968	290	32.3	369	28	51	52	237	11	226
1969	278	28.4	366	32	56	54	224	10	214
1970	283	27.1	381	38	60	58	225	14	211
1971	303	27.1	408	41	64	66	238	32	206
1972	322	26.5	436	44	70	71	251	49	202
1973	341	25.2	466	44	81	75	266	59	207
1974	344	23.2	484	46	94	81	263	57	206
1975	395	24.6	542	48	99	85	310	66	244
1976	477	26.7	629	45	107	95	383	70	313
1977	549	27.1	706	40	118	105	444	96	348
1978	607	26.7	777	35	134	115	492	121	371
1979	640	25.0	829	33	156	116	525	120	405
1980	712	25.5	909	32	165	121	591	122	469
1981	789	25.2	995	27	178	124	665	131	534
1982	925	27.9	1 137	19	193	134	790	141	649
1983	1 137	32.2	1 372	32	202	156	982	160	822
1984	1 307	33.1	1 565	32	225	155	1 152	176	976
1985	1 507	35.3	1 817	40	270	170	1 337	223	1 114
1986	1 741	38.5	2 121	46	334	191	1 550	266	1 284
1987	1 890	39.6	2 346	65	391	212	1 678	280	1 398
1988	2 052	39.9	2 601	104	445	229	1 822	346	1 476
1989	2 191	39.4	2 868	157	520	220	1 971	395	1 576
1990	2 412	40.9	3 206	215	580	234	2 177	464	1 713
1991	2 689	44.1	3 598	268	641	259	2 430	506	1 924
1992	3 000	46.8	4 002	319	683	296	2 703	563	2 140
1993	3 248	47.9	4 351	366	737	326	2 923	619	2 304
1994	3 433	47.8	4 643	423	788	355	3 078	682	2 396
1995	3 604	47.7	4 921	483	833	374	3 230	820	2 410
1996	3 734	47.0	5 181	550	898	391	3 343	993	2 350
1997	3 772	44.6	5 369	631	966	425	3 348	1 231	2 117
1998	3 721	41.7	5 478	730	1 027	458	3 263	1 224	2 039
1999	3 632	38.3	5 606	855	1 118	497	3 136	1 281	1 855
2000	3 410	33.7	5 629	1 007	1 212	511	2 898	1 039	1 859
2001	3 320	31.5	5 770	1 170	1 280	534	2 785	1 006	1 779
2002	3 540	32.7	6 198	1 329	1 329	604	2 936	1 201	1 735
2003	3 913	34.7	6 760	1 485	1 362	656	3 257	1 454	1 803
2004	4 296	35.7	7 355	1 635	1 424	700	3 595	1 799	1 796
2005	4 592	35.8	7 905	1 809	1 504	736	3 856	1 930	1 926
2006	4 829	35.4	8 451	1 994	1 628	769	4 060	2 025	2 035
2007	5 035	35.2	8 951	2 181	1 735	780	4 255	2 235	2 020
2008	5 803	39.4	9 986	2 366	1 817	491	5 312	2 802	2 510
2009	7 545	52.3	11 876	2 504	1 827	769	6 776	3 571	3 205
2010	9 019	60.8	13 529	2 585	1 924	812	8 207	4 316	3 891
2011	10 128	65.8	14 764	2 653	1 983	1 665	8 464	4 912	3 551
2012	11 281	70.3	16 051	2 718	2 052	1 645	9 636	5 476	4 160
2013	11 983	72.2	16 719	2 756	1 981	2 072	9 910	5 653	4 258
2014	12 780	73.7	17 794	2 783	2 232	2 452	10 328	6 069	4 259
2015	13 117	72.5	18 120	2 808	2 195	2 462	10 655	6 104	4 551
2016	14 168	76.4	19 539	2 842	2 529	2 463	11 704	6 156	5 548
2017	14 665	76.1	20 206	2 890	2 651	2 465	12 200	6 302	5 898
2018	15 750	77.8	21 462	2 894	2 818	2 313	13 436	6 226	7 211

[1] Sum of old age and survivors insurance fund and disability insurance trust fund.
. . . = Not available.

SECTION 6D: GOVERNMENT OUTPUT AND EMPLOYMENT

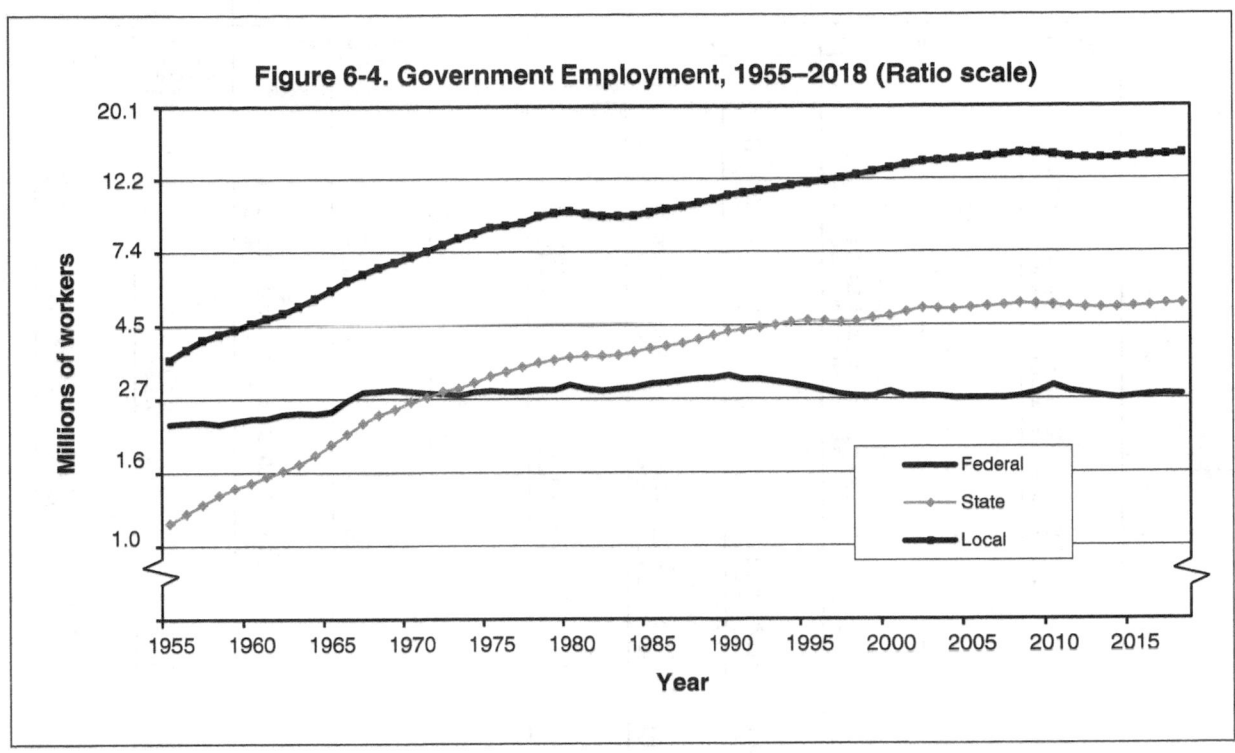

Figure 6-4. Government Employment, 1955–2018 (Ratio scale)

- From 1955 to 2018, total reported employment of civilians by all levels of government has increased 219.7 percent. Federal employment rose 21.8 percent; state government employment rose 343.2 percent, with 54.3 percent of this increase due to education. However, budget cuts hit all levels of government employment. For example, the Department of Defense lost 29 percent of its employees between 1955 and 2018. (Tables 6-17)

- Contracting and outsourcing of federal government spending can be seen in comparisons between government purchases and government value added. From 1960 to 2018, the volume (measured by the quantity index) of "intermediate" goods and services purchased for nondefense purposes rose 904 percent, compared with the 298 percent rise in nondefense value added by government itself in such activities. (Real value added comprises the input of government employees and government-owned capital, valued in constant dollars). For defense, government value added rose 12.4 percent over that period, while purchases of goods and services rose 203.5 percent in real terms. (Table 6-16)

- A similar process has occurred at state and local governments, where real value added increased 236.4 percent while goods and services purchased rose 966.4 percent from 1960 to 2018. (Table 6-16)

Table 6-16. Chain-Type Quantity Indexes for Government Output

(Seasonally adjusted, 2012 = 100.)

NIPA Table 3.10.3

Year and quarter	Federal government									State and local government		
	Gross output of general government			Value added			Intermediate goods and services purchased [1]			Gross output of general government	Value added	Intermediate goods and services purchased [1]
	Total	Defense	Nondefense	Total	Defense	Nondefense	Total	Defense	Nondefense			
1960	45.1	59.3	21.3	60.6	83.4	26.1	22.3	28.4	11.0	23.7	30.8	11.6
1961	46.1	60.8	21.2	62.6	86.1	27.1	21.8	28.6	9.5	25.0	32.4	12.4
1962	50.7	65.9	25.1	65.7	90.1	28.9	28.0	34.7	15.3	25.9	33.5	12.9
1963	51.1	66.1	25.9	67.0	90.6	31.2	27.4	34.5	13.9	27.4	35.3	13.7
1964	50.9	64.9	27.1	68.2	91.5	33.0	25.2	30.9	14.3	29.2	37.4	15.0
1965	52.0	66.0	28.3	69.3	92.0	34.9	26.3	32.6	14.5	31.3	39.7	16.5
1966	57.4	74.9	28.0	74.0	98.2	37.5	32.4	44.8	10.4	33.3	42.0	17.9
1967	63.9	83.5	30.7	79.1	104.7	40.4	40.1	55.8	12.5	35.0	43.8	19.3
1968	66.2	87.2	31.0	81.0	106.5	42.4	42.9	61.6	10.4	37.4	46.5	21.3
1969	65.5	84.4	33.6	81.8	106.8	44.0	40.1	54.9	13.9	39.7	48.6	23.6
1970	61.7	77.6	34.6	79.2	101.4	45.5	35.0	46.5	14.1	42.0	50.7	26.0
1971	59.2	71.9	37.3	76.3	95.3	47.4	32.9	41.0	17.4	44.2	52.8	28.5
1972	58.0	68.0	40.7	73.4	89.3	49.1	34.6	40.1	23.1	46.0	54.7	30.2
1973	55.3	63.8	40.6	71.4	85.4	49.9	30.8	35.1	21.5	47.6	56.6	31.2
1974	55.3	62.3	43.0	71.4	83.9	52.3	30.6	33.5	23.8	49.4	58.7	32.6
1975	54.8	60.5	44.6	71.1	82.6	53.6	29.8	31.1	25.7	51.6	60.4	35.2
1976	53.7	58.8	44.8	71.3	81.2	56.2	27.0	28.9	22.2	52.5	61.2	36.3
1977	54.7	59.1	46.7	71.6	80.6	57.9	28.8	30.4	24.3	53.6	62.2	37.6
1978	55.6	59.5	48.8	72.6	80.9	59.9	29.7	30.9	26.1	54.9	63.7	38.6
1979	56.4	59.9	50.1	72.7	80.5	60.9	31.3	32.3	27.8	55.6	64.8	38.6
1980	58.4	61.5	52.7	73.9	81.1	63.0	34.2	35.1	30.9	55.6	65.7	37.2
1981	60.5	64.5	53.3	75.1	83.3	62.8	37.3	38.8	32.7	55.7	65.7	37.4
1982	62.3	67.9	52.1	76.5	85.5	62.8	39.5	43.5	29.7	56.7	66.2	39.1
1983	65.4	70.8	55.6	78.4	87.8	64.1	43.9	46.8	36.1	57.6	65.9	41.8
1984	66.1	72.7	54.0	80.5	90.7	64.8	42.8	47.5	31.1	58.7	66.3	43.9
1985	69.5	76.4	57.0	83.0	94.3	65.6	47.2	51.1	37.1	61.3	68.2	47.6
1986	72.8	80.4	58.9	85.0	97.4	65.7	52.0	55.9	41.8	64.0	70.2	51.5
1987	74.3	83.2	57.9	87.7	100.9	67.2	51.7	57.8	36.9	65.4	71.6	52.7
1988	74.5	84.9	55.2	89.4	102.5	69.0	49.9	59.6	27.1	67.5	74.1	54.3
1989	76.2	85.3	59.5	90.8	104.1	70.0	52.1	58.5	37.1	70.0	76.3	57.0
1990	77.9	85.9	63.3	92.4	104.8	73.0	53.8	58.7	42.1	72.1	78.6	58.8
1991	78.8	86.8	64.2	92.9	105.2	73.9	55.4	60.5	43.0	73.8	79.7	61.8
1992	77.5	83.3	67.1	91.4	101.7	75.5	54.4	57.0	48.3	75.7	81.0	64.8
1993	75.4	80.5	66.3	89.8	98.7	76.3	51.5	54.5	44.3	77.4	82.1	67.7
1994	73.7	77.4	67.0	87.4	95.1	75.6	50.9	52.2	47.7	79.6	83.6	71.5
1995	71.1	73.8	66.2	84.6	91.2	74.5	48.7	49.0	47.9	81.7	85.2	74.4
1996	69.7	72.5	64.9	82.5	88.2	74.1	48.5	50.1	44.6	83.2	86.5	76.4
1997	69.6	71.2	66.7	81.2	85.6	74.6	50.2	50.6	49.1	85.6	88.2	80.2
1998	68.5	69.3	67.0	80.3	83.2	75.9	48.6	49.2	47.3	89.0	90.1	86.5
1999	69.1	70.2	67.1	79.7	81.4	77.1	51.3	54.0	44.9	91.9	91.6	92.5
2000	69.6	69.4	70.0	80.1	80.5	79.6	52.0	53.3	48.9	94.4	93.6	96.2
2001	72.2	71.5	73.4	80.3	80.3	80.4	58.3	58.6	57.6	98.0	95.8	103.0
2002	76.5	75.6	78.2	81.8	81.5	82.2	67.4	66.7	69.0	100.5	97.7	107.0
2003	81.5	81.6	81.3	83.9	83.7	84.3	77.2	78.4	74.4	100.1	98.3	104.2
2004	84.7	85.5	83.2	85.1	85.5	84.6	83.8	85.4	80.1	100.4	98.6	104.3
2005	86.1	86.6	85.1	86.2	86.7	85.4	85.8	86.4	84.4	101.0	99.4	104.7
2006	87.4	87.6	87.2	86.9	87.1	86.5	88.4	88.2	88.9	102.1	100.2	106.4
2007	88.3	89.1	86.9	88.1	88.2	87.9	88.8	90.5	84.7	103.6	101.6	108.0
2008	93.7	94.8	91.7	91.4	91.6	91.0	97.8	99.7	93.4	103.6	103.0	105.0
2009	99.8	100.6	98.4	95.9	96.5	94.9	106.6	106.7	106.5	105.8	103.4	111.2
2010	104.0	104.1	103.9	99.2	99.6	98.5	112.5	111.0	116.2	103.9	102.5	106.9
2011	101.7	102.9	99.6	100.1	100.9	98.8	104.5	105.8	101.5	101.3	101.3	101.2
2012	100.0	100.0	100.0	100.0	100.0	100.0	100.0	100.0	100.0	100.0	100.0	100.0
2013	95.0	93.7	97.4	98.2	97.8	99.0	89.3	87.5	93.7	100.3	99.9	101.4
2014	93.0	90.5	97.5	97.5	96.1	99.7	85.2	82.1	92.5	100.9	99.9	103.0
2015	92.8	88.6	100.5	96.8	94.1	101.0	86.0	80.2	99.5	104.0	100.6	112.0
2016	93.4	88.3	102.6	97.2	93.5	103.0	86.9	80.6	101.7	106.8	101.7	118.7
2017	93.6	88.4	103.0	97.3	93.2	103.5	87.4	81.3	101.7	108.1	102.6	121.2
2018	96.0	90.7	105.7	97.7	93.7	103.8	93.5	86.2	110.4	109.6	103.6	123.7
2016												
1st quarter	93.4	88.5	102.3	97.1	93.7	102.4	87.1	80.6	102.2	106.0	101.2	116.9
2nd quarter	92.9	87.6	102.5	97.2	93.5	102.9	85.6	78.6	101.9	106.6	101.5	118.4
3rd quarter	93.7	88.7	102.7	97.3	93.5	103.3	87.5	81.5	101.5	107.2	102.0	119.5
4th quarter	93.7	88.6	102.7	97.2	93.3	103.4	87.5	81.6	101.4	107.5	102.1	120.1
2017												
1st quarter	93.0	87.7	102.5	97.3	93.2	103.6	85.6	79.4	99.9	107.7	102.2	120.5
2nd quarter	93.5	88.7	102.2	97.1	93.0	103.5	87.3	82.3	99.1	107.9	102.4	120.9
3rd quarter	93.5	88.3	103.0	97.2	93.2	103.5	87.2	80.9	101.8	108.3	102.7	121.4
4th quarter	94.4	89.0	104.2	97.4	93.4	103.6	89.5	82.4	106.0	108.7	102.9	122.2
2018												
1st quarter	95.0	89.0	105.8	97.4	93.3	103.6	91.1	82.6	111.1	109.1	103.2	122.8
2nd quarter	96.0	90.6	105.7	97.7	93.6	104.0	93.3	86.2	110.1	109.4	103.4	123.4
3rd quarter	96.6	91.3	106.1	98.0	93.9	104.3	94.5	87.6	110.7	109.9	103.8	124.0
4th quarter	96.5	91.8	105.1	97.6	94.0	103.3	94.9	88.6	109.6	110.1	103.9	124.4

[1] Includes general government intermediate inputs for goods and services sold to other sectors and for own-account investment.

Table 6-17. Government Employment

(Calendar years, payroll employment in thousands.)

Year and month	Total government employment	Federal			State			Local		
		Total	Department of Defense	Postal Service	Total	Education	State government hospitals	Total	Education	Local government hospitals
1950	6 120	2 023	533	516	...	...	...	...	...	...
1951	6 502	2 415	797	521	...	...	...	...	...	...
1952	6 727	2 539	868	542	...	...	...	...	...	...
1953	6 758	2 418	818	530	...	...	...	...	...	...
1954	6 858	2 295	744	533	...	...	...	...	...	...
1955	7 021	2 295	744	534	1 168	308	...	3 558	1 751	...
1956	7 386	2 318	749	539	1 249	334	...	3 819	1 884	...
1957	7 724	2 326	729	555	1 328	363	...	4 071	2 026	...
1958	7 946	2 298	695	567	1 415	389	...	4 232	2 115	...
1959	8 192	2 342	699	578	1 484	420	...	4 366	2 198	...
1960	8 464	2 381	681	591	1 536	448	...	4 547	2 314	...
1961	8 706	2 391	683	601	1 607	474	...	4 708	2 411	...
1962	9 004	2 455	697	601	1 669	511	...	4 881	2 522	...
1963	9 341	2 473	687	603	1 747	557	...	5 121	2 674	...
1964	9 711	2 463	676	604	1 856	609	...	5 392	2 839	...
1965	10 191	2 495	679	619	1 996	679	...	5 700	3 031	...
1966	10 910	2 690	741	686	2 141	775	...	6 080	3 297	...
1967	11 525	2 852	802	719	2 302	873	...	6 371	3 490	...
1968	11 972	2 871	801	729	2 442	958	...	6 660	3 649	...
1969	12 330	2 893	815	737	2 533	1 042	...	6 904	3 785	...
1970	12 687	2 865	756	741	2 664	1 104	...	7 158	3 912	...
1971	13 012	2 828	731	731	2 747	1 149	...	7 437	4 001	...
1972	13 465	2 815	720	703	2 859	1 188	459	7 790	4 262	467
1973	13 862	2 794	696	698	2 923	1 205	472	8 146	4 433	477
1974	14 303	2 858	698	710	3 039	1 267	483	8 407	4 584	483
1975	14 820	2 882	704	699	3 179	1 323	503	8 758	4 722	489
1976	15 001	2 863	693	676	3 273	1 371	518	8 865	4 786	492
1977	15 258	2 859	676	657	3 377	1 385	538	9 023	4 859	494
1978	15 812	2 893	661	660	3 474	1 367	541	9 446	4 958	535
1979	16 068	2 894	649	673	3 541	1 378	538	9 633	4 989	571
1980	16 375	3 000	645	673	3 610	1 398	530	9 765	5 090	604
1981	16 180	2 922	655	675	3 640	1 420	515	9 619	5 095	622
1982	15 982	2 884	690	684	3 640	1 433	494	9 458	5 049	635
1983	16 011	2 915	699	685	3 662	1 450	471	9 434	5 020	644
1984	16 159	2 943	716	706	3 734	1 488	459	9 482	5 076	623
1985	16 533	3 014	738	750	3 832	1 540	449	9 687	5 221	608
1986	16 838	3 044	736	792	3 893	1 561	438	9 901	5 358	601
1987	17 156	3 089	736	815	3 967	1 586	439	10 100	5 469	606
1988	17 540	3 124	719	835	4 076	1 621	446	10 339	5 590	619
1989	17 927	3 136	735	838	4 182	1 668	442	10 609	5 740	632
1990	18 415	3 196	722	825	4 305	1 730	426	10 914	5 902	646
1991	18 545	3 110	702	813	4 355	1 768	417	11 081	5 994	653
1992	18 787	3 111	702	800	4 408	1 799	419	11 267	6 076	665
1993	18 989	3 063	670	793	4 488	1 834	414	11 438	6 206	673
1994	19 275	3 018	657	821	4 576	1 882	407	11 682	6 329	673
1995	19 432	2 949	627	850	4 635	1 919	395	11 849	6 453	669
1996	19 539	2 877	597	867	4 606	1 911	376	12 056	6 592	648
1997	19 664	2 806	588	866	4 582	1 904	360	12 276	6 759	632
1998	19 909	2 772	550	881	4 612	1 922	346	12 525	6 921	630
1999	20 307	2 769	525	890	4 709	1 983	344	12 829	7 120	626
2000	20 790	2 865	510	880	4 786	2 031	343	13 139	7 294	622
2001	21 118	2 764	504	873	4 905	2 113	345	13 449	7 479	628
2002	21 513	2 766	499	842	5 029	2 243	349	13 718	7 654	642
2003	21 583	2 761	486	809	5 002	2 255	348	13 820	7 709	651
2004	21 621	2 730	473	782	4 982	2 238	348	13 909	7 765	656
2005	21 804	2 732	488	774	5 032	2 260	350	14 041	7 856	655
2006	21 974	2 732	493	770	5 075	2 293	357	14 167	7 913	647
2007	22 218	2 734	491	769	5 122	2 318	360	14 362	7 987	654
2008	22 509	2 762	496	747	5 177	2 354	361	14 571	8 084	659
2009	22 555	2 832	519	703	5 169	2 360	359	14 554	8 079	661
2010	22 490	2 977	545	659	5 137	2 373	354	14 376	8 013	650
2011	22 086	2 859	559	631	5 078	2 374	348	14 150	7 873	648
2012	21 920	2 820	552	611	5 055	2 389	349	14 045	7 778	647
2013	21 853	2 769	537	595	5 046	2 393	350	14 037	7 777	646
2014	21 882	2 733	524	593	5 050	2 389	347	14 098	7 815	642
2015	22 029	2 757	526	597	5 077	2 401	354	14 195	7 871	648
2016	22 224	2 795	529	609	5 110	2 429	367	14 319	7 905	658
2017	22 350	2 805	527	615	5 165	2 479	376	14 379	7 922	667
2018	22 449	2 796	528	609	5 176	2 486	381	14 477	7 963	675

. . . = Not available.

NOTES AND DEFINITIONS, CHAPTER 6

TABLES 6-1 THROUGH 6-11 AND 6-16

Federal, State, and Local Government in the National Income and Product Accounts

SOURCE: U.S. DEPARTMENT OF COMMERCE, BUREAU OF ECONOMIC ANALYSIS (BEA)

These data are from the national income and product accounts (NIPAs), as redefined in the 2018 comprehensive NIPA revision and updated in the 2019 annual revision. For general information about the NIPAs see the notes and definitions for Chapter 1.

The 2019 annual update of the National Income and Product Accounts was released on July 26, 2019. Data for 2014 through 2018 were revised. The reference year remained 2012. The updated statistics largely reflect the incorporation of newly available and revised source data and improvements to existing methodologies.

The framework for the government accounts

In an earlier major revision in 2003, a new framework was introduced for government consumption expenditures—federal, state, and local—that explicitly recognizes the services produced by general government. Governments serve several functions in the economy. Three of these functions are recognized in the NIPAs: the production of nonmarket services; the consumption of these services, as the value of services provided to the general public is treated as government consumption expenditures; and the provision of transfer payments. These functions are financed through taxation, through contributions to social insurance funds, and by borrowing in the world's capital markets.

In this framework, the value of the government services produced and consumed (most of which are not sold in the market) is measured as the sum of the costs of the three major inputs: compensation of government employees, consumption of fixed capital (CFC), and intermediate goods and services purchased. The purchase from the private sector of goods and services by government, classified as final sales to government before the 2003 revision, was reclassified as intermediate purchases.

The value of government final purchases of consumption expenditures and gross investment, which constitutes the contribution of government demand to the gross domestic product (GDP), was not changed by this reclassification, since the previous definition of that contribution was the sum of compensation, CFC, and goods and services purchased. However, the distribution of GDP by type of product was changed—final sales of goods were reduced by the amount of goods purchased by government, and final sales of services were increased by the same amount.

In addition to this change in the conceptual framework, a number of the categories of government receipts and expenditures were redefined to make more precise distinctions. For example, items that used to be called "nontax payments" and included with taxes are now classified as transfer or fee payments and not included in taxes.

Finally, the concept previously known as "current surplus or deficit (-), national income and product accounts" was renamed "net government saving." This recognizes, in part, the role of government in the capital markets. When government runs a current surplus, net government saving is positive and funds are made available (for example, by retiring outstanding government debt) to finance investment—both private-sector capital spending and government investment. By the same token, when government runs a current deficit, or "dis-saves," it must borrow funds that would otherwise be available to finance investment.

However, this definition of net government saving does not give a complete picture of governments' role in capital markets, because it is based on current receipts and expenditures alone and does not include government investment activity.

In the NIPAs, the capital spending of all levels of government is treated the same way as the accounts treat private investment spending. A depreciation, or more precisely "consumption of fixed capital" (CFC), entry for existing capital is calculated, using estimated replacement costs and realistic depreciation rates. In the government accounts this CFC value is entered as one element of current expenditures and output. Capital spending is excluded from government current expenditures but appears in the account for "net lending or borrowing (-)."

The basic concept expressed in the net lending section of the NIPAs is that when CFC exceeds actual investment expenditures, governments have a positive net cash flow and can lend (or repay debt); if gross investment exceeds CFC, government must borrow to finance the difference, indicating negative net cash flow and requiring borrowing. (As will be seen below in the definitions, capital transfer and purchase accounts also enter into the calculation of net lending.)

The federal *budget* accounts (see Tables 6-1 and 6-2) do not draw a distinction between current and capital spending. The budget accounts of individual state and local governments typically separate capital from current spending and allow capital spending to be financed by borrowing—even when deficit financing of current spending is constitutionally forbidden. However, neither federal nor state and local government budget accounts typically show depreciation as a current expense in the way that is standard to private-sector accounting or in the way adopted in the NIPAs.

Notes on the data

Government receipts and expenditures data are derived from the U.S. government accounts and from Census Bureau censuses and surveys of the finances of state and local governments. However, BEA makes a number of adjustments to the data to convert them from fiscal year to calendar year and quarter bases and to agree with the concepts of national income accounting. Data are converted from the cash basis usually found in financial statements to the timing bases required for the NIPAs. In the NIPAs, receipts from businesses are generally on an accrual basis, purchases of goods and services are recorded when delivered, and receipts from and transfer payments to persons are on a cash basis. The federal receipts and expenditure data from the NIPAs in Table 6-1 therefore differ from the federal receipts and outlay data in Table 6-14. Among other differences, the latter are by fiscal year and are on a modified cash basis.

The NIPA data on government receipts and expenditures record transactions of governments (federal, state, and local) with other U.S. residents and foreigners. Each entry in the government receipts and expenditures account has a corresponding entry elsewhere in the NIPAs. Thus, for example, the sum of personal current taxes received by federal and state and local governments (Tables 6-1 and 6-8) is equal to personal current taxes paid, as shown in personal income (Table 4-1).

Definitions: general

In the 2003 revision of the NIPAs, several items appear separately that were previously treated as "negative expenditures" and netted against other items on the expenditures side. This grossing-up of the accounts raises both receipts and outlays and has no effect on net saving. Grossing-up has been applied to taxes from the rest of the world, interest receipts (back to 1960 for the federal government and back to 1946 for state and local governments), dividends, and subsidies and the current surplus of government enterprises (back to 1959).

Definitions: current receipts

Current tax receipts includes personal current taxes, taxes on production and imports, taxes on corporate income and (for the federal government only) taxes from the rest of the world. The category *total tax receipts* does not include *contributions for government social insurance,* which are the taxes levied to finance Social Security, unemployment insurance, and Medicare, and are included in *receipts* in the budget accounts (Table 6-14). Analysts using NIPA data to analyze tax burdens as they are usually understood should add *contributions for government social insurance* to *tax receipts* for this purpose.

Personal current taxes is personal tax payments from residents of the United States that are not chargeable to business expense. Personal taxes consist of taxes on income, including on realized net capital gains, and on personal property. Personal contributions for social insurance are not included in this category. As of the 1999 revisions, estate and gift taxes are classified as capital transfers and are no longer included in personal current taxes. However, estate and gift taxes continue to be included in federal government receipts in Table 6-14.

Taxes on production and imports in the case of the federal government consists of *excise taxes* and *customs duties.* In the case of state and local governments, these taxes include *sales taxes, property taxes* (including residential real estate taxes), and *Other* taxes such as motor vehicle licenses, severance taxes, and special assessments. Before the 2003 revision, all of these taxes were components of "indirect business tax and nontax liabilities."

Taxes on corporate income covers federal, state, and local government taxes on all corporate income subject to taxes. This taxable income includes capital gains and other income excluded from NIPA profits. The taxes are measured on an accrual basis, net of applicable tax credits. Federal corporate income tax receipts in the NIPAs include, but show separately, receipts from *Federal Reserve Banks* that represent the return of the surplus earnings of the Federal Reserve system. (In the budget accounts in Table 6-14, these are classified not as corporate taxes but as the major component of "Miscellaneous receipts.")

Contributions for social insurance includes employer and personal contributions for Social Security, Medicare, unemployment insurance, and other government social insurance programs. As of the 1999 revisions, contributions to government employee retirement plans are no longer included in this category; these plans are now treated the same as private pension plans.

Income receipts on assets consists of *interest, dividends* and *rents and royalties.*

Interest receipts (1960 to the present for federal government; 1946 to the present for state and local governments) consists of monetary and imputed interest received on loans and investments. In the NIPAs, this no longer includes interest received by government employee retirement plans, which is now credited to personal income. However, such interest received is still deducted from interest paid in the budget accounts that are shown in Table 6-14. Before the indicated years, receipts are deducted from aggregate interest payments in the NIPAs. Hence, they are not shown as receipts, and net interest is presented on the expenditure side. In the federal budget accounts in Table 6-14, net interest (total interest expenditures minus interest receipts) is the interest "expenditure" concept used throughout the period covered.

Current transfer receipts include receipts in categories other than those specified above from persons and business. In the case of state and local government accounts (Table 6-8), it also includes *federal grants-in-aid,* a component of federal expenditures. Receipts from *business* and *persons* were previously included with income taxes in "tax and nontax payments." They consist of federal deposit insurance premiums and other nontaxes (largely fines and regulatory and inspection fees), state and local fines and other nontaxes (largely donations and tobacco settlements), and net insurance settlements paid to governments as policyholders.

The *current surplus of government enterprises* is the current operating revenue and subsidies received from other levels of government by such enterprises less their current expenses. No deduction is made for depreciation charges or net interest paid. Before 1959, this category of receipts is treated as a deduction from subsidies. In the federal NIPA accounts before 1959, there is no entry shown for the current surplus on the receipts side, and on the expenditure side, there is an entry for subsidies, which is net of the current surplus. (Subsidies are usually a larger amount than the enterprise surplus in the federal accounts.) In the state and local NIPA accounts before 1959, there is an entry for the surplus on the receipts side, which is net of subsidies. (Subsidies are typically smaller than the enterprise surplus in state and local finance.)

Definitions: consumption expenditures, saving, and gross investment

Government consumption expenditures is expenditures by governments (federal or state and local) on services for current consumption. It includes *compensation of general government employees* (including employer contributions to government employee retirement plans, as of the 1999 revision); an allowance for *consumption of general government fixed capital (CFC)*, including R&D and software (depreciation); and *intermediate goods and services purchased.* (See the general discussion above for an explanation.) The estimated value of *own-account investment*—investment goods, including R&D and software, produced by government resources and purchased inputs—is subtracted here, and added to *government gross investment. Sales to other sectors*—primarily tuition payments received from individuals for higher education and charges for medical care to individuals—are also deducted, since these are counted elsewhere in the accounts, as (for example) personal consumption expenditures for those categories.

Government social benefits consists of payments to individuals for which the individuals do not render current services. Examples are Social Security benefits, Medicare, Medicaid, unemployment benefits, and public assistance. Retirement payments to retired government employees from their pension plans are no longer included in this category.

Government social benefits to persons consists of payments to persons residing in the United States (with a corresponding entry of an equal amount in the personal income receipts accounts). Government social benefits to the *rest of the world* appear only in the federal government account, and are transfers, mainly retirement benefits, to former residents of the United States.

Other current transfer payments (federal account only) includes *grants-in-aid to state and local governments* and *grants to the rest of the world*—military and nonmilitary grants to foreign governments.

Federal grants-in-aid comprises net payments from federal to state and local governments that are made to help finance programs such as health (Medicaid), public assistance (the old Aid to Families with Dependent Children and the new Temporary Assistance for Needy Families), and education. Investment grants to state and local governments for highways, transit, air transportation, and water treatment plants are now classified as capital transfers and are no longer included in this category. However, such investment grants continue to be included as federal government outlays in Table 6-14.

Interest payments is monetary interest paid to U.S. and foreign persons and businesses and to foreign governments for public debt and other financial obligations. As noted above, from 1960 forward for the federal government and from 1946 forward for state and local governments, this represents gross total (not net) interest payments. Before those dates in the NIPAs, and throughout the federal budget accounts presented in Table 6-14, net instead of aggregate interest is shown; that is, gross total interest paid less interest received.

Subsidies are monetary grants paid by government to business, including to government enterprises at another level of government. Subsidies no longer include federal maritime construction subsidies, which are now classified as a capital transfer. For years prior to 1959, subsidies continue to be presented net of the *current surplus of government enterprises*, because detailed data to separate the series are not available for this period. See the entry for *current surplus of government enterprises*, described above, for explanation of the pre-1959 treatment of this item in the federal accounts in *Business Statistics*, which differs from the treatment in the state and local accounts.

Net saving, NIPA (surplus+/deficit-), is the sum of current receipts less the sum of current expenditures. This is shown separately for *social insurance funds* (which, in the case of the federal government, include Social Security and other trust funds) and *other* (all other government).

Gross government investment consists of general government and government enterprise expenditures for fixed assets—*structures*, *equipment*, and *intellectual property* (R&D and software). The expenditures include the compensation of government employees producing the assets and the purchase of goods and services as intermediate inputs to be incorporated in the assets. Government inventory investment is included in government consumption expenditures.

Capital consumption. Consumption of fixed capital (CFC; economic depreciation) is included in government consumption expenditures as a partial measure of the value of the services of general government fixed assets, including structures, equipment, R&D, and software.

Definitions: output, lending and borrowing, and net investment

In Tables 6-5 and 6-10, current-dollar values of gross output and value added of government are presented, as described in the general discussion above. *Gross output* of government is the sum of the *intermediate goods and services purchased* by government and the value added by government as a producing industry. *Value added* consists of compensation of general government employees and consumption of general government fixed capital. Since this depreciation allowance is the only entry on the product side of the accounts measuring the output associated with such capital, a zero <u>net</u> return on these assets is implicitly assumed.

Gross output minus own-account investment and sales to other sectors (see the previous description) yields *government consumption expenditures*, which represents the contribution of government consumption spending to final demands in GDP.

Net lending (net borrowing) consists of current *net saving* as defined above, plus the *consumption of fixed capital* (CFC, from the current expenditure account), minus *gross investment*, plus *capital transfer receipts*, and minus *capital transfer payments* and *net purchases of non-produced assets*. (If this definition sounds somehow counter-intuitive, it should be remembered that "subtraction" in this context means "increasing the deficit [negative value] which must be financed.)

Capital transfer receipts and *payments* include grants between levels of government, or between government and the private sector, associated with acquisition or disposal of assets rather than with current consumption expenditures. Examples are federal grants to state and local government for highways, transit, air transportation, and water treatment plants; federal shipbuilding subsidies and other subsidies to businesses; and lump-sum payments to amortize the unfunded liability of the Uniformed Services Retiree Health Care Fund. Government capital transfer receipts include estate and gift taxes, which are no longer included in personal tax receipts. (Federal estate and gift taxes are shown in Table 6-14, where they are reported on a fiscal year basis, and are similar in order of magnitude to the calendar year values for federal capital transfer receipts in Table 6-5.)

Non-produced assets are land and radio spectrum. Unusually large negative entries for net purchases of non-produced assets in some recent years result from the negative purchase—that is, the sale—of spectrum.

The values of *net investment* shown in these tables are calculated by the editor, as gross investment minus the consumption of fixed capital.

Definitions: chain-type quantity indexes

Chain-type quantity indexes represent changes over time in real values, removing the effects of inflation. Indexes for key categories in the government expenditure accounts, as well as for government gross output, value added, and intermediate goods and services purchased, use the chain formula described in the notes and definitions for Chapter 1 and are expressed as index numbers, with the average for the year 2012 equal to 100.

Data availability

The most recent data are published each month in the *Survey of Current Business*. Current and historical data may be obtained from the BEA Web site at <http://www.bea.gov> and the STAT-USA subscription Web site at <http://www.stat-usa.gov>.

References

See the references regarding the 2018 comprehensive revision of the NIPAs in the notes and definitions for Chapter 1. NIPA concepts and their differences from the budget estimates are discussed in "NIPA Estimates of the Federal Sector and the Federal Budget Estimates," *Survey of Current Business*, March 2007, p. 11.

For information about the classification of government expenditures into current consumption and gross investment, first undertaken in the 1996 comprehensive revisions, see the *Survey of Current Business* article, "Preview of the Comprehensive Revision of the National Income and Product Accounts: Recognition of Government Investment and Incorporation of a New Methodology for Calculating Depreciation" September 1995. Other sources of information about the NIPAs are listed in the notes and definitions for Chapter 1.

TABLES 6-12 AND 6-13

State Government Current Receipts and Expenditures; Local Government Current Receipts and Expenditures

SOURCE: BUREAU OF ECONOMIC ANALYSIS (BEA)

In the standard presentation of the state and local sector of the national income and product accounts (NIPAs) such as in Tables 6-8 through 6-11 above, state and local governments are combined. Annual measures for aggregate state governments and aggregate local governments are also available on the BEA Web site. These measures are shown in Tables 6-12 and 6-13. The definitions are the same as in the other NIPA tables described above.

The data shown here now reflect the 2013 comprehensive revision, affecting some concepts, that has been incorporated in the quarterly and annual data elsewhere in this chapter and this volume.

Two categories not shown in Table 6-8 appear in each table, detailing inter-governmental flows that are consolidated in Table 6-8. State government receipts include not only *federal grants-in-aid* but also *local grants-in-aid*, and state government expenditures include *grants-in-aid to local governments*. Local government receipts include not only *federal grants-in-aid* but also *state grants-in-aid*, and local government expenditures include *grants-in-aid to state governments*. To make room for these columns, the components *current surplus of government enterprises* and *subsidies* are not shown, though they are included in total current receipts and total current expenditures respectively.

These measures are described in "Receipts and Expenditures of State Governments and of Local Governments," *Survey of Current Business*, October 2005. Data back to 1959 are available on the BEA Web site at <http://www.bea.gov>.

TABLES 6-14A AND 6-14B

Federal Government Receipts and Outlays by Fiscal Year; The Federal Budget and GDP

SOURCE: U.S. OFFICE OF MANAGEMENT AND BUDGET

These data on federal government receipts and outlays are on a modified cash basis and are from the *Budget of the United States Government: Historical Tables*. The data are by federal fiscal years, which are defined as July 1 through June 30 through 1976 and October 1 through September 30 for 1977 and subsequent years. They are identified by the year of the final fiscal quarter—i.e., the year from July 1, 1975, through June 30, 1976, is identified as Fiscal 1976, and the year from October 1, 1976, through June 30, 1977, is identified as Fiscal Year 1977. The period July 1 through September 30, 1976, is a separate fiscal period known as the transition quarter (TQ) and is not included in any fiscal year.

There are numerous differences in both timing and definition between these estimates and the NIPA estimates in Tables 6-1 through 6-7. See the notes and definitions for those tables for the definitional differences that were introduced with the 1999 comprehensive revision of the NIPAs.

Definitions

The definitions for these tables are not affected by the 2013, 2003, or 1999 changes in the government sectors of the NIPAs.

Table references will be given indicating the source of each item in the *Historical Tables*; for example, "HT Table 1.1."

Receipts consist of gifts and of taxes or other compulsory payments to the government. Other types of payments to the government are netted against outlays. (HT Table 1.1)

Outlays occur when the federal government liquidates an obligation through a cash payment or when interest accrues on public debt issues. Beginning with the data for 1992, outlays include the subsidy cost of direct and guaranteed loans made. Before 1992, the costs and repayments associated with such loans are recorded on a cash basis. As noted previously, various types of nontax receipts are netted against cash outlays. These accounts do not distinguish between investment outlays and current consumption and do not include allowances for depreciation. (HT Table 1.1)

The *total budget surplus (deficit-)* is receipts minus outlays. It is sometimes presented, including in Table 6-14, as composed of two components, *on-budget and off-budget*. By law, two government programs that are included in the federal receipts and outlays totals are "off-budget"—old-age, survivors, and disability insurance (Social Security) and the Postal Service. The former accounts for nearly all of the off-budget activity. The *surplus (deficit-)* not accounted for by these two programs is the on-budget surplus or deficit. The *on-budget deficit* is based on an arbitrary distinction and is not customarily referred to as the *budget deficit*, a term reserved in customary use for the total deficit. (HT Table 1.1)

Sources of financing is the means by which the total deficit is financed or the surplus is distributed. By definition, sources of financing sum to the total deficit or surplus with the sign reversed. The principal source is *borrowing from the public*, shown as a positive number, that is, the increase in the debt held by the public. (This entry is calculated by the editors as the change in the debt held by the public as shown in HT Table 7.1.) When there is a budget surplus, as in fiscal years 1998 to 2001, this provides resources for debt reduction, which is indicated by a minus sign in this column. *Other financing* includes drawdown (or buildup, shown here with a minus sign) in Treasury cash balances, seigniorage on coins, direct and guaranteed loan account cash transactions, and miscellaneous other transactions. Large negative "other financing" in 2008 through 2010 resulted from direct loans and asset purchases under TARP and other emergency financial procedures; in other words, the government borrowed from the public to undertake these loans and asset purchases, but they are not included in budget expenditures or the budget deficit. *Other financing* is calculated by the editors, by reversing the sign of the surplus/deficit—so that the deficit becomes a positive number—and subtracting the value of borrowing from the public.

Some of the categories of *receipts by source* are self-explanatory. *Employment taxes and contributions* includes taxes for old-age, survivors, and disability insurance (Social Security), hospital insurance (Medicare), and railroad retirement funds. *Other retirement contributions* includes the employee share of payments for retirement pensions, mainly those for federal employees. *Excise taxes* includes federal taxes on alcohol, tobacco, telephone service, and transportation fuels, as well as taxes funding smaller programs such as black lung disability and vaccine injury compensation. *Miscellaneous receipts* includes deposits of earnings by the Federal Reserve system and all other receipts. (HT Tables 2.1, 2.4, and 2.5)

Outlays by function presents outlays according to the major purpose of the spending. Functional classifications cut across departmental and agency lines. Most categories of offsetting receipts are netted against cash outlays in the appropriate function, which explains how recorded outlays in *energy* and *commerce and housing credit* (which, as its name suggests, includes loan programs) can be negative. There is also a category of "undistributed offsetting receipts" (not shown), always with a negative sign, that includes proceeds from the sale or lease of assets and payments from federal agencies to federal retirement funds and to the Social Security and Medicare trust funds. Note that *Social Security* is recorded separately from other *income security* outlays, and *Medicare* separately from other *health* outlays. (HT Table 3.1) For further explanation, consult the *Budget of the United States Government.*

In order to provide authoritative comparisons of these budget values with the overall size of the economy, a special calculation of gross domestic product by fiscal year is supplied to the Office of Management and Budget by the BEA, and is shown in Table 6-14B. That table also displays selected budget aggregates in dollar values and as a percent of GDP. (HT Tables 1.2, 1.3, 2.3, and 3.1)

Data availability and references

Definitions, budget concepts, and historical data are from *Budget of the United States Government for Fiscal Year 2015: Historical Tables*, available on the Office of Management and Budget Web site at <http://www.whitehouse.gov/omb/Budget/historicals>

Similarly defined data for the latest month, the year-ago month, and the current and year-ago fiscal year to date are published in the *Monthly Treasury Statement* prepared by the Financial Management Service, U.S. Department of the Treasury. For those who need up-to-date budget information, this publication is available on the Financial Management Service Web site at <http://www.fms.treas.gov>. As these monthly figures are never revised to agree with the final annual data, they are not published in this volume.

TABLE 6-15

Federal Government Debt by Fiscal Year

SOURCE: U.S. OFFICE OF MANAGEMENT AND BUDGET

Debt outstanding at the end of each fiscal year is from the *Budget of the United States Government*. Most securities are recorded at sales price plus amortized discount or less amortized premium.

Definitions

Federal government debt held by the public consists of all federal debt held outside the federal government accounts—by individuals, financial institutions (including the Federal Reserve Banks), and foreign individuals, businesses, and central banks. It does not include federal debt held by federal government trust funds such as the Social Security trust fund. The level and change of the ratio of this debt to the value of gross domestic product (GDP) provide proportional measures of the impact of federal borrowing on credit markets. (HT Table 7.1) This measure of debt held by the public is very similar in concept and scope to the total federal government credit market debt outstanding in the flow-of-funds accounts, shown in *Business Statistics* in Table 12-5; however, it is not identical, being priced somewhat differently, and is shown here in Chapter 6 on a fiscal year rather than calendar year basis.

Gross federal debt—total. This is the total debt owed by the U.S. Treasury. It includes a small amount of matured debt. (HT Table 7.1)

Debt held by Social Security funds is the sum of the end-year trust fund balances for old age and survivors insurance and disability insurance. The separate disability trust fund begins in 1957. (HT Table 13.1)

Debt held by other U.S. government accounts is calculated by the editors by subtracting the Social Security debt holdings from the total debt held by federal government accounts, which is shown in HT Table 7.1. It includes the balances in all the other trust funds, including the Medicare funds, federal employee retirement funds, and the highway trust fund.

Debt held by the Federal Reserve System is the total value of Treasury securities held by the 12 Federal Reserve Banks, which are acquired in open market operations that carry out monetary policy. (HT Table 7.1)

Debt held by private investors is calculated by subtracting the Federal Reserve debt from the total debt held by the public, and is shown as "Debt Held by the Public: Other" in HT Table 7.1.

Debt held by foreign residents is based on surveys by the Treasury Department. (Table 5-7, *Budget of the United States for Fiscal Year 2015: Analytical Perspectives.*)

Debt held by domestic investors is calculated by the editors by subtracting the foreign-held debt from the total debt held by private investors.

The "debt subject to statutory limitation," not shown here, is close to the gross federal debt in concept and size, but there are some relatively minor definitional differences specified by law. The debt limit can only be changed by an Act of Congress. For information about the debt subject to limit and other debt subjects, see the latest *Budget of the United States Government: Historical Statistics* and *Analytical Perspectives*.

Data availability and references

For the end-of-fiscal-year data, see *Historical Tables* and *Analytical Perspectives* in the *Budget of the United States Government*, which is available on the OMB Web site at <http://www.whitehouse.gov/omb/Budget>. In the *Analytical Perspectives,* debt analysis and data are included in the "Economic and Budget Analysis" section.

Recent quarterly data, measured on a somewhat different basis, are found in the *Treasury Bulletin* in the chapter on "Ownership of Federal Securities (OFS)," in Tables OFS-1 and OFS-2. The *Treasury Bulletin* can be accessed on the Internet at <http://www.fms.treas.gov/bulletin>. Holdings by Social Security funds are also available in the *Bulletin* in the chapter on "Federal Debt," Table FD-3. The Disability Fund is listed separately from the Old-Age and Survivors Fund.

TABLE 6-17

Government Employment

SOURCE: U.S. DEPARTMENT OF LABOR, BUREAU OF LABOR STATISTICS (SEE NOTES AND DEFINITIONS FOR TABLE 10-8).

Government payroll employment includes federal, state, and local activities such as legislative, executive, and judicial functions, as well as all government-owned and government-operated business enterprises, establishments, and institutions (arsenals, navy yards, hospitals, etc.), and government force account construction. The figures relate to civilian employment only. The Bureau of Labor Statistics (BLS) considers regular fulltime teachers (private and governmental) to be employed during the summer vacation period, regardless of whether they are specifically paid in those months.

Employment in federal government establishments reflects employee counts as of the pay period containing the 12th of the month. Federal government employment excludes employees of the Central Intelligence Agency and the National Security Agency.

CHAPTER 7: U.S. FOREIGN TRADE AND FINANCE

SECTION 7A: FOREIGN TRANSACTIONS IN THE NATIONAL INCOME AND PRODUCT ACCOUNTS

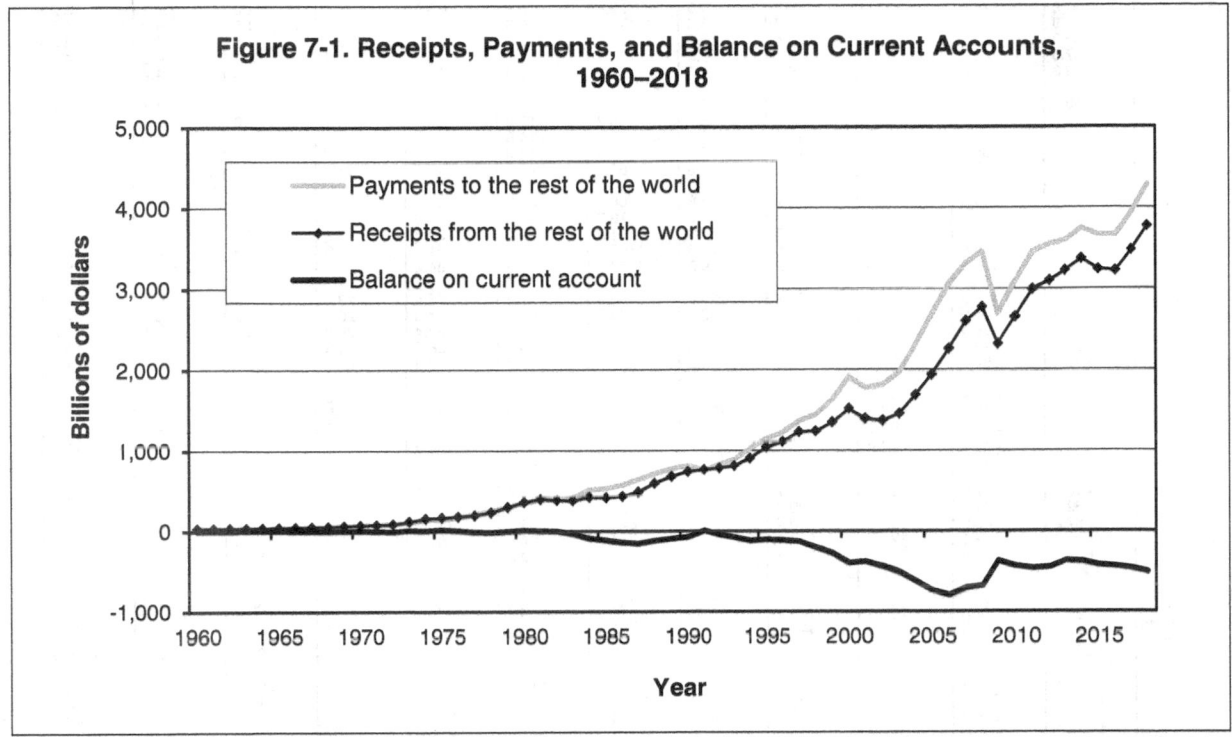

Figure 7-1. Receipts, Payments, and Balance on Current Accounts, 1960–2018

- Beginning in 1983, chronic current-account deficits in U.S. transactions with the rest of the world have emerged and increased, requiring increasing net capital inflows from abroad to finance them. The only current-account surplus since then occurred in 1991, when payments from other countries financed most of the cost of the first Gulf War. In 2006, the NIPA deficit reached a record $803.9 billion, 5.8 percent of GDP. From 2006 to 2013, the current account deficit declined by half. However, from 2013 to 2018, it has increased by 37.2 percent. (Tables 7-1 and 1-1A)

- The NIPA current-account deficit of $506.7 billion in 2018 was 6.4 percent higher than in 2017. This was the fifth consecutive year that the deficit increased. (Table 7-1)

- Chain-type quantity indexes measure the real changes in the flow or volume of goods, especially since these products can be measured by barrels, tons, British thermal unit, pounds, etc. In contrast, the consumer price index (CPI) in Chapter 8 tracks the fluctuation in prices. In 2018, the quantity of exports of goods and services increased slightly by 0.2 percent between the first quarter and the fourth quarter while imports of goods and services increased 3.0 percent. (Table 7-2)

- The term "services" refers to an expanding range of economic activities, such as audiovisual, construction, computer and related services, express delivery, telecommunications, e-commerce, financial, professional (such as accounting and legal services), retail and wholesaling, transportation, and tourism. Because measures of service trade are not anchored in any observation of physical movement, they are dependent on definitions of residence. (Table 7-4)

Table 7-1. Foreign Transactions in the National Income and Product Accounts

(Billions of dollars, quarterly data are at seasonally adjusted annual rates.)

NIPA Table 4.1

Year and quarter	Current receipts from the rest of the world					Current payments to the rest of the world						Balance on current account, NIPAs	Net lending or net borrowing (-), NIPAs
	Total	Exports of goods and services			Income receipts	Total	Imports of goods and services			Income payments	Current taxes and transfer payments, net		
		Goods [1]		Services [1]			Goods [1]		Services [1]				
		Durable	Nondurable				Durable	Nondurable					
1960	31.9	20.5	9.2	6.6	4.9	28.8	6.4	8.9	7.6	1.8	4.1	3.2	3.2
1961	32.9	20.9	9.4	6.7	5.3	28.7	6.0	9.0	7.6	1.8	4.2	4.2	4.2
1962	35.0	21.7	9.7	7.4	5.9	31.2	6.9	9.9	8.1	1.8	4.4	3.8	3.8
1963	37.6	23.3	10.7	7.7	6.5	32.7	7.4	10.3	8.4	2.1	4.5	4.9	4.9
1964	42.3	26.8	12.1	8.3	7.2	34.8	8.4	11.0	8.7	2.3	4.4	7.5	7.5
1965	45.0	28.0	12.1	9.2	7.9	38.9	10.4	11.8	9.3	2.6	4.7	6.2	6.2
1966	49.0	31.1	13.4	9.8	8.1	45.2	13.3	13.1	10.7	3.0	5.1	3.8	3.8
1967	52.2	32.5	13.9	10.9	8.7	48.7	14.5	13.2	12.2	3.3	5.5	3.5	3.5
1968	58.0	35.7	14.5	12.2	10.1	56.5	18.8	15.1	12.6	4.0	5.9	1.5	1.5
1969	63.7	38.7	14.8	13.2	11.8	62.1	20.8	16.1	13.7	5.7	5.9	1.6	1.6
1970	72.5	45.0	17.7	14.7	12.8	68.8	22.8	18.1	14.9	6.4	6.6	3.7	3.7
1971	77.0	46.2	18.3	16.8	14.0	76.7	26.6	19.9	15.8	6.4	7.9	0.3	0.3
1972	87.1	52.6	21.2	18.3	16.3	91.2	33.3	23.6	17.3	7.7	9.2	-4.0	-4.1
1973	118.8	75.8	33.5	19.5	23.5	109.9	40.8	31.0	19.3	10.9	7.9	8.9	8.8
1974	156.5	103.5	45.8	23.2	29.8	150.5	50.3	54.2	22.9	14.3	8.7	6.0	5.9
1975	166.7	112.5	47.3	26.2	28.0	146.9	45.6	53.4	23.7	15.0	9.1	19.8	19.8
1976	181.9	121.5	50.6	28.0	32.4	174.8	56.8	67.8	26.5	15.5	8.1	7.1	7.0
1977	196.5	128.4	54.0	30.9	37.2	207.5	69.2	83.4	29.8	16.9	8.1	-10.9	-11.0
1978	233.1	149.9	62.4	37.0	46.3	245.8	89.0	88.4	34.8	24.7	8.8	-12.6	-12.7
1979	298.5	187.3	77.6	42.9	68.3	299.6	100.4	112.3	39.9	36.4	10.6	-1.2	-1.3
1980	359.9	230.4	94.8	50.3	79.1	351.4	111.9	136.6	45.3	44.9	12.6	8.5	8.4
1981	397.3	245.2	102.0	60.0	92.0	393.9	126.0	141.8	49.9	59.1	17.0	3.4	3.3
1982	384.2	222.6	94.1	60.7	101.0	387.5	125.1	125.4	52.6	64.5	19.8	-3.3	-3.4
1983	378.9	214.0	90.3	62.9	101.9	413.9	147.3	125.4	56.0	64.8	20.5	-35.1	-35.2
1984	424.2	231.3	96.1	71.1	121.9	514.3	192.4	143.9	68.8	85.6	23.6	-90.1	-90.2
1985	415.9	227.5	87.6	75.7	112.7	530.2	204.2	139.1	73.9	87.3	25.7	-114.3	-114.5
1986	432.3	231.4	86.1	89.6	111.3	575.0	238.8	131.2	82.9	94.4	27.8	-142.7	-142.8
1987	487.2	265.6	98.6	98.4	123.3	641.3	264.2	150.6	93.9	105.8	26.8	-154.1	-154.2
1988	596.7	332.1	120.1	112.5	152.1	712.4	294.8	157.3	101.9	129.5	29.0	-115.7	-115.9
1989	682.0	374.8	132.2	129.5	177.7	774.3	310.3	174.5	106.2	152.9	30.4	-92.4	-92.7
1990	740.7	403.3	138.0	148.6	188.8	815.6	314.5	193.5	121.7	154.2	31.7	-74.9	-82.3
1991	763.3	430.1	144.6	164.8	168.4	755.4	315.5	185.2	122.8	136.8	-4.9	7.9	2.6
1992	785.1	455.3	151.1	177.7	152.1	830.7	346.7	198.2	122.9	121.0	41.9	-45.6	-44.3
1993	810.4	467.7	149.9	187.1	155.6	889.8	386.4	206.4	127.2	124.4	45.4	-79.4	-80.2
1994	905.5	518.4	164.6	202.6	184.5	1 021.1	454.0	222.8	136.6	161.6	46.1	-115.6	-116.9
1995	1 042.6	592.4	193.8	220.4	229.8	1 148.5	510.8	246.7	145.1	201.9	44.1	-105.9	-106.3
1996	1 114.0	628.8	202.4	238.8	246.4	1 229.0	533.4	274.1	156.5	215.5	49.5	-115.0	-115.2
1997	1 233.9	699.9	211.5	253.9	280.1	1 364.0	588.5	297.1	170.1	256.8	51.4	-130.1	-130.6
1998	1 239.8	692.6	200.2	260.4	286.8	1 445.1	637.7	293.0	184.9	269.4	60.0	-205.3	-205.6
1999	1 350.9	711.7	203.0	281.1	320.2	1 629.3	715.8	335.4	197.4	294.7	86.0	-278.4	-282.8
2000	1 518.0	795.9	223.2	300.3	380.6	1 914.4	822.2	427.9	221.2	345.6	97.6	-396.4	-396.8
2001	1 394.1	741.2	217.2	283.4	324.1	1 777.0	756.2	417.6	218.8	275.3	109.1	-383.0	-370.0
2002	1 370.4	709.0	218.3	289.7	314.8	1 813.6	771.7	422.7	229.8	269.6	119.9	-443.2	-443.7
2003	1 456.1	737.1	237.4	299.1	353.8	1 969.4	803.9	487.4	248.0	295.4	134.6	-513.2	-515.3
2004	1 689.3	830.0	266.5	347.7	446.9	2 314.5	934.5	572.8	289.4	368.8	149.0	-625.2	-622.4
2005	1 941.5	921.9	294.5	383.3	566.0	2 678.8	1 023.9	691.6	311.0	488.1	164.3	-737.3	-724.5
2006	2 259.9	1 044.9	332.2	427.7	712.0	3 061.7	1 129.5	766.3	347.8	661.5	156.7	-801.9	-803.9
2007	2 603.0	1 161.3	380.9	499.6	866.6	3 313.7	1 174.4	825.3	379.6	757.6	176.9	-710.8	-710.7
2008	2 775.8	1 292.5	466.4	544.5	848.8	3 458.9	1 160.6	983.7	415.9	694.2	204.6	-683.2	-677.8
2009	2 321.5	1 058.4	390.1	523.6	647.8	2 693.6	891.4	694.0	393.1	505.8	209.3	-372.1	-372.7
2010	2 657.2	1 272.4	474.6	573.8	715.2	3 093.9	1 104.5	840.3	415.4	519.5	214.2	-436.7	-437.4
2011	2 996.3	1 462.3	570.6	640.7	789.2	3 461.8	1 236.6	1 003.9	441.9	552.8	226.6	-465.6	-467.2
2012	3 104.3	1 521.6	583.9	669.7	799.7	3 552.4	1 326.4	975.0	458.5	567.4	225.2	-448.1	-441.6
2013	3 228.0	1 559.2	603.8	714.2	823.4	3 596.5	1 357.6	938.7	467.8	592.7	239.6	-368.5	-369.4
2014	3 371.1	1 615.0	622.0	756.7	853.5	3 746.7	1 450.3	941.3	487.8	612.5	254.8	-375.6	-376.0
2015	3 240.3	1 494.6	540.5	772.2	837.7	3 664.4	1 488.9	799.1	498.6	613.1	264.7	-424.1	-424.5
2016	3 224.6	1 444.0	518.5	776.6	861.7	3 665.9	1 464.9	756.1	518.3	643.5	283.0	-441.4	-441.9
2017	3 478.6	1 538.4	578.5	957.9	925.5	3 945.2	1 566.5	813.3	552.3	714.6	298.5	-466.6	-476.0
2018	3 771.8	1 661.3	655.5	848.9	1 106.2	4 281.3	1 664.3	906.3	577.9	838.3	294.5	-509.5	-506.7
2016													
1st quarter	3 129.1	1 405.1	487.6	759.8	826.4	3 594.8	1 445.7	731.7	509.6	624.9	282.8	-465.7	-466.3
2nd quarter	3 210.9	1 433.6	510.1	774.4	861.8	3 623.7	1 448.9	743.3	511.2	648.0	272.2	-412.8	-413.2
3rd quarter	3 256.7	1 466.7	539.3	787.7	860.1	3 693.0	1 470.4	761.3	522.4	655.3	283.6	-436.4	-437.1
4th quarter	3 301.6	1 470.7	536.9	784.5	898.4	3 752.2	1 494.8	788.3	530.1	645.7	293.3	-450.6	-451.0
2017													
1st quarter	3 376.0	1 503.0	566.2	800.3	898.4	3 826.4	1 516.0	821.5	536.7	665.2	287.0	-450.4	-451.0
2nd quarter	3 388.7	1 508.7	564.8	804.5	924.9	3 904.6	1 550.5	799.0	547.4	708.4	299.2	-515.8	-516.6
3rd quarter	3 521.0	1 535.4	568.8	824.7	982.1	3 944.9	1 566.7	786.6	557.4	725.9	308.3	-423.9	-459.7
4th quarter	3 628.6	1 606.4	614.2	844.0	1 026.2	4 104.9	1 632.7	846.1	567.6	758.9	299.6	-476.2	-476.8
2018													
1st quarter	3 694.9	1 626.4	615.6	850.2	1 070.5	4 172.8	1 644.9	891.6	569.1	789.5	277.7	-477.9	-478.4
2nd quarter	3 810.0	1 697.6	681.3	846.0	1 111.4	4 254.3	1 641.1	901.7	569.3	845.8	296.6	-444.3	-444.8
3rd quarter	3 786.0	1 661.3	666.3	849.0	1 116.0	4 316.1	1 682.8	919.2	579.6	843.6	290.8	-530.1	-528.4
4th quarter	3 796.6	1 659.9	658.8	850.6	1 127.0	4 382.1	1 688.5	912.8	593.4	874.4	313.0	-585.5	-575.1

[1]Exports and imports of certain goods, primarily military equipment purchased and sold by the federal government, are included in services. Beginning with 1986, repairs and alterations of equipment are reclassified from goods to services.

Table 7-2. Chain-Type Quantity Indexes for Exports and Imports of Goods and Services

(Index numbers, 2012 = 100; quarterly indexes are seasonally adjusted.) NIPA Table 4.2.3

Year and quarter	Exports of goods and services					Imports of goods and services				
	Total	Goods [1]			Services [1]	Total	Goods [1]			Services [1]
		Total	Durable	Nondurable			Total	Durable	Nondurable	
1975	12.82	12.81	9.84	21.95	12.68	10.79	9.80	4.81	24.48	16.54
1976	13.38	13.47	10.02	24.13	12.81	12.90	12.02	5.90	30.04	17.68
1977	13.70	13.73	10.05	25.13	13.39	14.31	13.49	6.67	33.46	18.56
1978	15.14	15.16	11.14	27.60	14.89	15.55	14.70	7.71	34.55	19.89
1979	16.64	16.76	12.55	29.71	15.95	15.80	14.95	7.89	34.95	20.16
1980	18.44	18.82	14.04	33.55	16.62	14.75	13.84	7.95	30.09	19.71
1981	18.66	18.70	13.59	34.58	18.23	15.14	14.12	8.61	29.18	20.84
1982	17.23	17.11	11.89	33.70	17.43	14.95	13.76	8.70	27.48	21.95
1983	16.79	16.56	11.60	32.22	17.40	16.83	15.63	10.49	29.27	23.73
1984	18.15	17.73	12.76	33.23	19.44	20.93	19.41	13.96	33.33	29.68
1985	18.76	18.35	13.76	32.36	19.99	22.29	20.63	15.36	33.73	31.94
1986	20.20	19.34	14.58	33.77	22.86	24.19	22.74	16.78	37.71	32.29
1987	22.41	21.70	16.78	36.32	24.59	25.62	23.80	17.52	39.61	36.10
1988	26.04	25.56	20.65	39.65	27.51	26.63	24.76	18.26	41.13	37.32
1989	29.05	28.48	23.47	42.63	30.81	27.80	25.83	19.07	42.80	39.13
1990	31.62	30.92	26.01	44.55	33.75	28.80	26.58	19.57	44.24	41.69
1991	33.71	33.00	27.90	47.03	35.90	28.76	26.70	19.67	44.37	40.60
1992	36.04	35.48	30.13	50.15	37.83	30.77	29.22	21.74	47.76	39.51
1993	37.22	36.63	31.69	49.83	39.09	33.43	32.14	24.31	50.97	40.56
1994	40.51	40.14	35.48	52.14	41.82	37.42	36.43	28.33	54.76	42.72
1995	44.68	44.81	40.51	55.52	44.66	40.41	39.72	31.58	57.01	44.01
1996	48.33	48.80	45.18	57.51	47.46	43.93	43.44	34.96	60.83	46.31
1997	54.09	55.87	53.36	61.46	49.96	49.84	49.70	40.73	67.15	50.33
1998	55.35	57.08	55.06	61.35	51.34	55.67	55.59	46.09	73.29	55.81
1999	58.11	59.47	57.61	63.21	54.99	61.96	62.74	52.77	80.24	57.77
2000	62.95	65.37	64.61	66.07	57.37	69.94	70.88	60.73	87.30	64.97
2001	59.31	61.23	59.16	65.47	54.89	67.98	68.49	56.87	89.47	65.18
2002	58.28	59.19	55.96	66.58	56.22	70.45	71.00	59.23	91.94	67.48
2003	59.56	60.93	57.85	67.89	56.42	73.92	75.13	62.67	97.30	67.72
2004	65.31	66.27	64.41	70.27	63.13	82.35	83.72	71.70	103.85	75.30
2005	69.97	71.34	70.75	72.50	66.84	87.72	89.61	78.18	108.00	78.21
2006	76.51	78.41	78.86	77.33	72.17	93.54	95.20	85.62	109.69	85.12
2007	83.16	83.87	84.96	81.47	81.52	95.87	97.16	88.09	110.66	89.26
2008	87.87	88.75	89.05	87.80	85.82	93.74	93.89	85.24	106.74	92.76
2009	80.49	78.23	74.07	85.61	85.55	81.48	79.52	68.03	97.26	90.72
2010	90.26	89.95	87.25	94.49	90.97	92.17	91.80	84.11	102.71	93.97
2011	96.70	96.30	95.31	97.89	97.63	97.37	97.44	92.51	104.22	96.98
2012	100.00	100.00	100.00	100.00	100.00	100.00	100.00	100.00	100.00	100.00
2013	103.58	103.18	102.39	104.46	104.46	101.54	101.76	104.49	98.06	100.46
2014	107.94	107.96	106.79	109.86	107.92	106.62	107.42	113.76	98.85	102.75
2015	108.45	107.59	104.44	112.97	110.25	112.26	113.52	121.24	102.98	106.32
2016	108.44	108.18	103.30	116.96	109.06	114.49	115.17	122.98	104.51	111.00
2017	112.21	112.38	106.55	123.00	112.03	119.88	120.67	131.91	104.66	115.89
2018	115.59	117.17	110.22	129.84	112.79	125.12	126.71	139.44	108.53	117.77
2012										
1st quarter	98.96	99.04	100.33	96.96	98.79	99.61	99.66	99.46	99.98	99.31
2nd quarter	100.04	100.22	99.90	100.74	99.63	100.10	100.09	100.02	100.19	100.13
3rd quarter	100.56	100.86	100.09	102.10	99.89	100.63	100.65	100.22	101.24	100.53
4th quarter	100.43	99.88	99.68	100.20	101.69	99.66	99.59	100.31	98.59	100.02
2013										
1st quarter	101.58	100.64	100.52	100.82	103.69	99.96	100.04	101.58	97.93	99.55
2nd quarter	102.80	102.25	103.05	100.96	104.03	101.40	101.60	103.78	98.63	100.42
3rd quarter	103.45	102.90	101.91	104.50	104.65	102.13	102.40	105.39	98.34	100.80
4th quarter	106.48	106.94	104.08	111.56	105.45	102.69	103.02	107.20	97.34	101.08
2014										
1st quarter	105.72	105.09	104.13	106.65	107.09	103.85	104.45	108.58	98.82	100.97
2nd quarter	108.01	107.89	106.56	110.03	108.29	106.49	107.31	113.79	98.56	102.56
3rd quarter	108.49	108.97	108.21	110.19	107.49	106.77	107.59	114.61	98.10	102.80
4th quarter	109.54	109.90	108.25	112.58	108.81	109.36	110.35	118.06	99.90	104.65
2015										
1st quarter	108.32	107.26	104.78	111.42	110.53	111.23	112.60	119.98	102.63	104.80
2nd quarter	109.36	108.92	105.46	114.90	110.32	111.99	113.43	120.91	103.28	105.27
3rd quarter	108.28	107.64	104.40	113.19	109.62	112.91	114.13	122.11	103.17	107.14
4th quarter	107.85	106.52	103.11	112.38	110.54	112.91	113.91	121.94	102.85	108.08
2016										
1st quarter	107.02	106.75	102.26	114.79	107.68	113.15	113.87	120.99	104.34	109.50
2nd quarter	108.06	107.47	102.90	115.63	109.32	113.37	114.17	121.40	104.45	109.41
3rd quarter	109.68	109.36	103.63	119.83	110.42	114.68	115.20	123.54	103.66	111.82
4th quarter	108.98	109.14	104.43	117.59	108.82	116.77	117.45	126.00	105.60	113.26
2017										
1st quarter	110.60	110.72	104.36	122.36	110.51	117.94	118.64	128.03	105.46	114.33
2nd quarter	111.03	111.33	105.06	122.80	110.61	118.95	119.63	130.78	103.75	115.44
3rd quarter	112.24	111.93	107.29	120.27	112.97	119.32	119.89	131.79	102.86	116.26
4th quarter	114.97	115.55	109.49	126.58	114.03	123.30	124.52	137.06	106.57	117.54
2018										
1st quarter	115.18	115.89	111.18	124.45	114.02	123.49	124.95	137.37	107.21	116.71
2nd quarter	116.82	119.22	111.12	134.00	112.46	123.57	125.20	137.03	108.29	116.12
3rd quarter	114.97	116.40	108.86	130.19	112.44	126.16	127.98	141.06	109.32	117.85
4th quarter	115.39	117.15	109.74	130.70	112.25	127.24	128.71	142.31	109.31	120.40

[1] Exports and imports of certain goods, primarily military equipment purchased and sold by the federal government, are included in services. Beginning with 1986, repairs and alterations of equipment are reclassified from goods to services.

Table 7-3. Chain-Type Price Indexes for Exports and Imports of Goods and Services

(Index numbers, 2012 = 100; quarterly indexes are seasonally adjusted.)　　　　　　　　　　　　　　　　　　　　　NIPA Table 4.2.4

Year and quarter	Exports of goods and services					Imports of goods and services				
	Total	Goods [1]			Services [1]	Total	Goods [1]			Services [1]
		Total	Durable	Nondurable			Total	Durable	Nondurable	
1975	49.39	57.70	31.30	28.79	30.88	41.23	43.88	30.16	15.43	31.29
1976	51.01	59.27	31.27	28.75	32.67	42.47	45.05	29.82	16.12	32.71
1977	53.09	61.47	32.55	29.94	34.50	46.21	49.18	32.56	17.46	35.03
1978	56.32	65.00	34.05	31.33	37.08	49.47	52.45	34.98	17.77	38.22
1979	63.10	73.43	41.28	37.94	40.13	57.93	61.85	41.76	23.94	43.16
1980	69.50	80.46	46.38	42.62	45.24	72.17	78.06	50.46	36.47	50.09
1981	74.65	86.18	48.09	44.20	49.18	76.07	82.43	51.15	40.08	52.26
1982	75.01	85.47	46.62	42.87	51.97	73.51	79.11	49.25	37.26	52.30
1983	75.31	84.98	45.31	41.64	54.04	70.75	75.80	46.63	34.01	51.44
1984	76.02	85.74	46.72	42.94	54.60	70.14	75.28	46.12	33.82	50.56
1985	73.75	81.46	44.85	40.84	56.57	67.84	72.32	43.75	31.76	50.50
1986	72.52	78.67	45.60	37.81	58.50	67.83	70.70	44.98	21.80	55.97
1987	74.12	80.43	49.70	42.55	59.75	71.94	75.73	48.64	24.68	56.76
1988	77.92	85.38	55.83	46.82	61.08	75.38	79.34	57.70	23.42	59.53
1989	79.21	86.48	57.30	47.50	62.77	77.02	81.56	60.98	26.25	59.22
1990	79.66	85.71	57.02	48.31	65.75	79.23	83.06	58.40	29.35	63.66
1991	80.55	85.67	56.51	46.73	68.56	78.57	81.49	56.70	26.37	65.98
1992	80.15	84.34	57.43	44.34	70.17	78.64	81.04	56.77	25.48	67.84
1993	80.28	83.91	60.97	43.24	71.46	78.03	80.13	56.81	24.06	68.41
1994	81.21	84.87	63.22	46.82	72.32	78.77	80.72	58.83	23.49	69.75
1995	83.03	86.89	67.71	54.68	73.68	80.92	82.87	63.38	25.98	71.93
1996	81.92	84.68	65.37	51.82	75.14	79.51	80.77	62.48	28.29	73.73
1997	80.48	82.34	64.78	51.75	75.90	76.75	77.43	63.60	27.22	73.72
1998	78.57	79.74	62.12	48.62	75.74	72.62	72.76	60.61	21.76	72.28
1999	77.97	78.66	60.50	48.31	76.33	73.02	72.80	60.73	24.61	74.54
2000	79.47	80.02	61.99	53.72	78.16	76.22	76.64	64.11	34.93	74.26
2001	78.84	79.56	60.96	51.82	77.11	74.22	74.46	61.96	31.85	73.21
2002	78.20	78.73	61.12	50.18	76.95	73.24	73.10	59.65	30.87	74.27
2003	79.40	79.51	63.82	54.57	79.17	75.45	74.69	61.17	36.57	79.87
2004	82.28	82.32	72.10	60.65	82.24	79.06	78.23	71.96	43.29	83.82
2005	85.13	84.93	78.57	68.84	85.63	83.70	83.18	75.90	55.60	86.72
2006	87.84	87.59	88.36	74.56	88.49	86.91	86.53	83.18	64.09	89.13
2007	91.14	91.00	94.11	80.56	91.50	89.92	89.43	89.50	70.07	92.77
2008	95.41	95.71	97.00	92.74	94.75	98.96	99.23	98.81	95.73	97.80
2009	89.69	88.92	82.75	73.38	91.39	87.99	86.63	83.50	62.43	94.51
2010	93.35	92.97	92.26	85.13	94.20	92.78	92.05	95.39	77.82	96.42
2011	99.24	99.79	102.08	102.30	98.00	99.83	99.91	104.84	98.81	99.39
2012	100.00	100.00	100.00	100.00	100.00	100.00	100.00	100.00	100.00	100.00
2013	100.17	99.31	96.95	98.35	102.10	98.64	98.05	97.83	96.30	101.58
2014	100.27	98.31	95.52	96.50	104.71	97.85	96.74	97.34	93.55	103.56
2015	95.39	91.30	89.76	76.78	104.58	89.95	87.58	88.77	60.98	102.29
2016	93.46	87.73	86.34	68.30	106.33	86.70	83.80	84.72	50.66	101.86
2017	95.85	89.96	89.78	75.95	109.08	88.62	85.69	92.35	58.94	103.94
2018	99.10	93.19	93.02	84.92	112.39	91.18	88.15	98.98	69.56	107.03
2012										
1st quarter	99.82	100.02	101.06	102.33	99.35	101.12	101.43	101.39	104.48	99.56
2nd quarter	100.00	99.99	100.56	100.81	100.01	100.36	100.46	100.52	101.26	99.86
3rd quarter	99.92	99.91	98.79	98.43	99.93	98.75	98.52	98.39	95.18	99.88
4th quarter	100.26	100.07	99.60	98.43	100.71	99.77	99.59	99.70	99.09	100.70
2013										
1st quarter	100.60	100.23	98.73	99.47	101.43	99.90	99.66	100.33	99.81	101.12
2nd quarter	99.91	99.12	97.09	97.29	101.70	98.61	98.10	97.94	95.75	101.18
3rd quarter	100.12	99.15	95.63	98.50	102.28	98.11	97.45	96.09	95.24	101.47
4th quarter	100.05	98.75	96.36	98.16	102.99	97.92	97.01	96.96	94.41	102.53
2014										
1st quarter	100.82	99.50	96.22	101.25	103.80	99.14	98.34	97.03	99.87	103.22
2nd quarter	100.85	99.16	95.87	98.09	104.66	98.47	97.45	96.96	96.40	103.70
3rd quarter	100.60	98.52	95.82	97.04	105.30	97.86	96.68	98.45	93.26	103.89
4th quarter	98.81	96.05	94.16	89.63	105.07	95.94	94.49	96.91	84.68	103.42
2015										
1st quarter	96.36	92.85	92.30	79.25	104.28	91.84	89.81	93.64	66.66	102.40
2nd quarter	96.16	92.39	91.00	80.51	104.66	90.70	88.42	89.98	63.45	102.60
3rd quarter	95.26	90.95	88.78	76.44	104.98	89.63	87.18	86.93	60.69	102.41
4th quarter	93.76	89.02	86.94	70.94	104.40	87.62	84.92	84.52	53.12	101.73
2016										
1st quarter	92.32	86.51	85.31	63.76	105.36	86.05	83.09	82.43	47.00	101.51
2nd quarter	93.25	87.68	85.99	67.41	105.79	86.41	83.44	83.97	49.10	101.93
3rd quarter	93.80	88.15	86.57	69.61	106.52	87.03	84.18	86.17	52.27	101.91
4th quarter	94.44	88.57	87.52	72.43	107.65	87.30	84.47	86.29	54.27	102.09
2017										
1st quarter	95.05	89.23	89.33	74.40	108.15	88.31	85.62	89.42	59.94	102.40
2nd quarter	95.09	89.08	89.09	73.47	108.63	88.25	85.35	90.97	57.60	103.43
3rd quarter	95.97	90.17	89.44	76.38	109.02	88.39	85.30	92.97	56.26	104.59
4th quarter	97.28	91.38	91.24	79.54	110.52	89.53	86.50	96.04	61.96	105.35
2018										
1st quarter	98.13	92.24	92.85	81.88	111.35	91.12	88.21	98.35	68.87	106.36
2nd quarter	99.36	93.58	94.29	85.22	112.32	91.25	88.25	100.17	69.16	106.94
3rd quarter	99.64	93.80	92.81	87.50	112.75	91.38	88.34	99.27	70.88	107.29
4th quarter	99.28	93.13	92.12	85.07	113.14	90.97	87.82	98.14	69.34	107.51

[1] Exports and imports of certain goods, primarily military equipment purchased and sold by the federal government, are included in services. Beginning with 1986, repairs and alterations of equipment are reclassified from goods to services.

Table 7-4. Exports and Imports of Selected NIPA Types of Product

(Billions of dollars, quarterly data are at seasonally adjusted annual rates.)

NIPA Table 4.2.5

| Year and quarter | Exports | | | | | | | Imports | | | | | | | |
| | Goods | | | | | Services | | Goods | | | | | | Services | |
	Foods, feeds, and beverages	Industrial supplies and materials	Capital goods, except auto-motive	Auto-motive vehicles, engines, and parts	Consumer goods, except food and auto-motive	Travel	Other business services	Foods, feeds, and beverages	Industrial supplies and materials, except petroleum and products	Petroleum and products	Capital goods, except auto-motive	Auto-motive vehicles, engines, and parts	Consumer goods, except food and auto-motive	Travel	Other business services
1970	5.9	13.8	14.7	3.9	2.8	2.3	1.0	6.1	15.2	2.9	4.0	5.7	7.4	4.0	0.6
1971	6.1	12.6	15.4	4.7	2.9	2.5	1.3	6.4	17.2	3.7	4.3	7.6	8.4	4.4	0.7
1972	7.5	13.9	16.9	5.5	3.6	2.8	1.5	7.3	20.6	4.7	5.9	9.0	11.1	5.0	0.8
1973	15.2	19.7	22.0	7.0	4.8	3.4	1.7	9.1	27.6	8.4	8.3	10.7	12.9	5.5	0.9
1974	18.6	29.9	30.9	8.8	6.4	4.0	3.0	10.6	53.6	26.6	9.8	12.4	14.4	6.0	1.9
1975	19.2	29.3	36.6	10.8	6.6	4.7	3.7	9.6	50.6	27.0	10.2	12.1	13.2	6.4	2.3
1976	19.8	31.6	39.1	12.2	8.0	5.7	4.5	11.5	63.1	34.6	12.3	16.8	17.2	6.9	2.9
1977	19.7	33.2	39.8	13.5	8.9	6.2	4.9	14.0	78.4	45.0	14.0	19.4	21.8	7.5	3.2
1978	25.7	38.4	47.5	15.2	11.4	7.2	6.2	15.8	81.9	42.6	19.3	25.0	29.4	8.5	3.9
1979	30.5	53.3	60.2	17.9	14.0	8.4	7.3	18.0	105.4	60.4	24.6	26.6	31.3	9.4	4.6
1980	36.3	68.0	76.3	17.4	17.8	10.6	8.6	18.6	126.9	79.5	31.6	28.3	34.3	10.4	5.1
1981	38.8	65.7	84.2	19.7	17.7	15.6	10.5	18.6	130.4	78.4	37.1	31.0	38.4	11.8	6.0
1982	32.2	61.8	76.5	17.2	16.1	15.5	13.8	17.5	107.3	62.0	38.4	34.3	39.7	12.8	7.0
1983	32.1	57.1	71.7	18.5	14.9	14.5	14.1	18.8	106.2	55.1	43.7	43.0	47.3	13.6	6.8
1984	32.2	61.9	77.0	22.4	15.1	21.3	14.5	21.9	120.7	58.1	60.4	56.5	61.1	23.4	7.7
1985	24.6	59.4	79.3	24.9	14.6	22.4	14.8	21.8	110.6	51.4	61.3	64.9	66.3	25.1	8.8
1986	23.5	59.0	82.8	25.1	16.7	25.5	23.3	24.4	96.7	34.3	72.0	78.1	79.4	26.5	13.3
1987	25.2	67.4	92.7	27.6	20.3	29.0	24.2	24.8	109.0	42.9	85.1	85.2	88.8	29.9	16.7
1988	33.8	84.2	119.1	33.4	27.0	35.2	25.7	24.9	116.3	39.6	102.2	87.9	96.4	32.8	17.8
1989	36.3	05.4	136.9	05.1	35.9	43.0	30.3	24.9	129.7	50.9	112.3	87.4	103.6	34.2	19.2
1990	35.1	101.9	153.0	36.2	43.5	50.9	32.7	26.4	140.5	62.3	116.4	88.2	105.0	38.2	22.4
1991	35.7	106.2	166.6	39.9	46.6	56.7	39.9	26.2	127.4	51.7	121.1	85.5	107.7	36.2	25.9
1992	40.3	105.2	176.4	46.9	51.2	64.0	41.1	27.6	134.0	51.6	134.8	91.5	122.4	39.6	24.3
1993	40.5	102.9	182.7	51.6	54.5	67.9	43.5	27.9	140.2	51.5	153.2	102.1	133.7	41.9	26.6
1994	42.4	115.4	205.7	57.5	59.7	69.4	49.9	31.0	156.2	51.3	185.0	118.1	145.9	45.1	30.3
1995	50.8	141.0	234.4	61.4	64.2	74.9	53.6	33.2	175.6	56.0	222.1	123.7	159.7	46.4	33.7
1996	56.0	141.3	254.0	64.4	69.3	81.8	61.3	35.7	197.3	72.7	228.4	128.7	172.5	49.7	38.1
1997	52.0	153.0	295.8	73.4	77.0	86.2	71.2	39.7	205.5	71.8	253.6	139.4	195.2	53.8	41.4
1998	46.8	143.3	299.8	72.5	79.4	85.1	78.1	41.3	193.1	50.9	269.8	148.6	218.5	58.5	45.6
1999	46.0	145.7	311.2	75.3	80.9	92.3	79.8	44.1	221.9	72.1	296.1	178.2	243.7	59.6	56.2
2000	47.9	171.1	357.0	80.4	89.3	100.2	83.5	46.5	301.3	126.1	347.7	195.0	284.6	65.8	61.6
2001	49.4	159.5	321.7	75.4	88.3	86.7	88.2	47.2	276.8	109.4	299.2	188.7	287.1	60.7	66.9
2002	49.6	157.3	290.4	78.9	84.3	81.9	95.4	50.3	270.6	109.3	284.9	202.8	311.3	59.9	74.1
2003	55.0	173.3	293.7	80.6	89.9	80.3	102.3	56.5	317.8	140.4	297.6	209.2	338.4	61.9	79.6
2004	56.6	206.3	327.5	89.2	103.2	92.4	118.8	63.0	418.3	189.9	346.1	227.3	378.1	74.0	89.9
2005	59.0	236.8	358.4	98.4	115.2	101.5	128.9	69.1	531.4	263.2	382.8	238.7	412.7	80.0	95.8
2006	66.0	279.1	404.0	107.3	129.0	105.1	151.4	76.1	610.1	316.7	422.6	256.0	447.6	84.2	126.6
2007	84.3	316.3	433.0	121.3	145.9	119.0	184.8	83.0	644.5	346.7	449.1	258.5	479.8	89.2	149.3
2008	108.3	386.9	457.7	121.5	161.2	133.8	202.9	90.4	794.6	476.1	458.7	233.2	485.7	92.5	174.0
2009	93.9	293.5	391.5	81.7	149.3	119.9	211.7	82.9	464.4	267.7	374.1	159.2	429.9	81.4	178.5
2010	107.7	388.6	447.8	112.0	164.9	137.0	227.4	92.5	603.2	353.6	450.4	225.6	485.1	86.6	183.6
2011	126.2	485.3	494.2	133.0	174.7	150.9	251.6	108.3	755.0	462.1	513.4	255.2	515.9	89.7	197.3
2012	133.0	483.2	527.5	146.2	181.0	161.6	263.6	111.1	723.2	434.3	551.8	298.5	518.8	100.3	200.2
2013	136.2	492.4	534.8	152.7	188.1	177.5	286.3	116.0	679.0	387.8	559.0	309.6	532.9	98.1	208.1
2014	143.7	500.7	551.8	159.8	198.4	191.9	309.0	126.8	669.9	353.6	598.8	329.5	558.7	105.7	214.7
2015	127.7	418.1	539.8	151.9	197.3	206.9	315.8	128.8	488.1	197.2	607.2	350.0	596.4	114.5	218.3
2016	130.5	387.6	520.0	150.4	193.3	206.7	323.9	131.0	437.4	159.6	593.6	350.8	584.9	123.5	222.6
2017	132.7	456.2	533.5	157.9	197.2	210.7	353.3	138.8	505.9	199.6	642.9	359.2	603.6	134.9	231.8
2018	133.2	534.6	563.2	158.8	205.5	214.7	369.5	148.4	577.3	240.4	695.9	373.1	649.1	144.5	235.6
2013															
1st quarter	136.0	482.7	526.3	151.1	182.6	174.0	281.4	115.1	699.1	404.1	553.4	296.4	529.6	98.0	203.0
2nd quarter	125.7	478.6	539.1	154.3	193.4	175.9	282.9	116.5	677.0	384.6	551.3	308.0	533.6	96.5	208.0
3rd quarter	134.1	493.0	534.5	151.8	187.0	178.2	287.0	115.3	676.5	389.2	561.5	315.6	530.7	97.6	210.2
4th quarter	148.9	515.4	539.2	153.4	189.4	181.8	294.1	117.2	663.5	373.1	569.7	318.3	537.6	100.4	211.3
2014															
1st quarter	151.0	496.0	543.1	152.4	192.3	187.2	303.2	120.6	702.8	393.3	577.3	312.5	545.3	103.3	210.5
2nd quarter	146.5	511.2	550.9	160.3	200.7	190.2	313.2	129.1	679.3	363.1	597.2	332.9	563.7	105.9	214.9
3rd quarter	134.3	516.0	556.6	165.1	200.3	194.1	306.9	128.3	661.1	343.8	607.8	336.0	553.6	105.6	214.4
4th quarter	143.1	479.5	556.4	161.5	200.2	196.2	312.9	129.3	636.3	314.3	612.7	336.6	572.3	107.9	218.9
2015															
1st quarter	134.9	430.7	548.4	148.7	200.2	201.5	322.0	128.9	534.2	224.9	616.5	334.7	592.1	111.6	215.4
2nd quarter	130.7	443.0	546.3	152.0	196.2	207.8	314.3	130.8	500.1	206.4	612.2	354.2	598.9	113.5	216.3
3rd quarter	123.7	415.8	535.3	154.8	196.8	209.3	309.7	128.8	482.5	195.9	601.8	357.5	602.1	114.7	221.2
4th quarter	121.6	383.1	529.2	152.0	196.1	209.1	317.1	126.5	435.8	161.8	598.2	353.8	592.5	118.4	220.2
2016															
1st quarter	119.0	365.0	521.9	147.8	191.2	207.0	312.2	129.9	404.3	138.7	583.4	348.7	586.0	122.1	222.2
2nd quarter	121.9	386.3	520.3	151.7	191.7	203.7	322.4	129.2	421.7	150.6	595.5	346.8	582.3	121.0	220.4
3rd quarter	146.7	395.6	515.4	152.4	196.0	208.6	328.9	130.5	451.0	168.3	594.4	352.3	580.9	123.4	221.9
4th quarter	134.5	403.4	522.5	149.7	194.1	207.3	332.0	134.4	472.6	180.9	601.1	355.7	590.4	127.7	226.0
2017															
1st quarter	132.6	440.1	522.5	157.9	196.3	206.2	345.1	134.5	511.5	218.1	613.4	358.8	592.0	131.2	224.7
2nd quarter	132.2	440.9	522.6	156.4	197.6	207.7	346.2	137.7	495.7	190.7	631.1	359.0	600.5	133.8	230.0
3rd quarter	134.7	446.5	540.1	156.4	195.6	213.0	356.0	140.5	485.2	179.3	651.1	357.9	590.5	135.5	235.9
4th quarter	131.4	497.5	548.7	160.8	199.2	215.8	365.9	142.5	531.0	210.2	675.9	361.0	631.3	139.0	236.5
2018															
1st quarter	128.3	505.6	556.0	167.0	204.7	217.6	365.0	147.3	565.1	235.8	681.7	369.4	652.6	142.1	233.2
2nd quarter	149.1	544.7	566.5	161.6	205.1	215.9	363.9	148.4	579.3	244.3	696.3	363.1	631.3	141.7	232.4
3rd quarter	135.3	544.0	561.3	155.5	204.5	212.1	373.8	148.4	594.3	256.8	705.3	375.9	646.3	144.6	237.1
4th quarter	120.0	544.3	568.9	151.3	207.7	213.2	375.2	149.4	570.6	224.8	700.4	384.1	666.2	149.5	239.7

Table 7-5. Chain-Type Quantity Indexes for Exports and Imports of Selected NIPA Types of Product

(Index numbers, 2012 = 100; quarterly indexes are seasonally adjusted.) NIPA Table 4.2.3

Year and quarter	Exports							Imports							
	Goods					Services		Goods						Services	
	Foods, feeds, and beverages	Industrial supplies and materials	Capital goods, except auto-motive	Auto-motive vehicles, engines, and parts	Consumer goods, except food and auto-motive	Travel	Other business services	Foods, feeds, and beverages	Industrial supplies and materials, except petroleum and products	Petroleum and products	Capital goods, except auto-motive	Auto-motive vehicles, engines, and parts	Consumer goods, except food and auto-motive	Travel	Other business services
1975	33.04	20.42	5.71	24.41	8.73	14.12	3.14	26.37	38.82	53.59	0.86	14.20	4.67	17.59	2.24
1976	37.49	22.05	5.63	25.52	9.85	16.08	3.63	30.01	47.02	64.61	1.05	18.81	6.02	18.70	2.69
1977	37.46	22.24	5.54	25.70	10.65	16.12	3.74	29.71	53.87	77.73	1.12	19.93	7.27	19.27	2.82
1978	46.28	24.58	6.32	26.39	12.30	17.31	4.43	33.36	54.44	73.49	1.44	21.56	8.93	19.91	3.29
1979	49.09	28.14	7.59	26.55	12.93	18.56	4.94	34.24	53.80	74.34	1.77	20.85	9.00	19.42	3.73
1980	55.34	31.94	8.86	22.36	15.67	20.73	5.28	29.80	44.81	60.04	2.08	20.66	9.07	19.40	3.81
1981	56.50	29.79	8.90	22.11	15.02	27.59	6.03	31.01	42.66	52.60	2.45	19.99	9.93	21.14	4.23
1982	53.09	28.89	7.87	18.12	13.52	25.46	7.47	31.61	37.47	45.24	2.67	21.40	10.38	25.08	4.78
1983	50.75	27.49	7.59	18.74	12.37	22.61	7.26	34.30	40.27	44.78	3.23	26.19	12.49	28.41	4.43
1984	49.38	28.89	8.40	22.12	12.30	31.42	7.24	38.86	46.09	47.30	4.82	33.65	15.77	51.35	4.95
1985	42.73	29.06	9.24	24.07	11.94	31.45	7.12	40.54	44.88	44.41	5.43	37.71	17.21	56.65	5.53
1986	44.08	30.36	10.08	23.67	13.21	34.69	10.89	41.50	50.46	54.91	6.17	40.82	19.07	52.72	8.05
1987	47.18	31.11	11.64	25.51	15.54	37.63	10.90	42.54	51.02	57.35	7.04	41.89	19.69	62.40	9.23
1988	51.85	35.08	14.79	30.28	19.85	43.89	11.54	40.85	53.31	63.44	8.12	40.84	19.98	64.19	9.84
1989	53.93	39.01	17.16	31.13	25.45	51.86	13.47	41.77	54.07	68.60	9.18	39.80	20.91	66.21	11.51
1990	55.64	41.31	19.85	31.17	29.86	58.22	14.02	43.37	54.80	69.33	9.92	39.85	20.54	70.73	12.49
1991	56.70	44.12	21.72	33.43	30.93	60.95	16.48	41.40	54.13	69.08	10.66	37.16	20.89	64.24	13.94
1992	64.27	44.94	23.79	38.54	33.32	67.09	16.66	43.71	58.33	72.20	12.35	39.07	23.07	66.25	12.98
1993	63.95	43.75	25.35	42.02	35.01	70.21	17.42	44.23	63.52	79.09	14.41	42.93	24.98	69.55	13.76
1994	64.61	46.02	29.25	46.44	38.25	71.06	19.84	45.24	71.13	84.44	17.67	48.14	27.07	72.08	15.54
1995	71.38	49.64	34.86	48.95	40.54	75.52	20.94	46.38	72.95	82.91	21.75	48.97	29.20	72.86	17.11
1996	70.24	52.16	40.40	50.76	43.24	80.41	23.59	51.01	77.78	89.90	25.56	50.63	31.44	75.76	18.76
1997	70.68	56.71	49.52	57.46	47.69	82.71	27.14	56.17	83.04	93.62	32.08	54.73	36.01	82.34	20.10
1998	69.96	56.08	51.63	56.67	49.14	80.46	29.96	60.29	91.15	100.07	36.90	58.26	40.87	93.17	22.42
1999	72.00	57.85	54.48	58.46	50.24	85.03	31.51	66.60	96.94	107.56	42.35	69.39	45.90	93.93	27.67
2000	75.60	62.97	62.68	61.74	55.25	88.07	33.71	70.95	102.23	111.48	50.88	75.06	54.04	107.08	31.55
2001	77.63	60.41	56.46	57.70	54.82	75.56	38.00	74.14	101.22	114.76	45.10	72.60	54.92	98.84	36.55
2002	76.26	60.79	51.87	60.10	52.62	71.19	42.25	78.05	102.32	112.09	44.42	77.79	60.21	93.33	40.82
2003	77.57	62.43	54.51	60.95	55.77	67.77	44.97	84.10	105.76	119.57	48.60	79.81	65.88	86.11	43.31
2004	72.50	66.50	61.35	66.91	63.44	74.63	51.18	89.09	117.79	127.47	57.57	85.29	73.49	94.41	48.23
2005	76.76	68.21	67.34	72.98	70.06	78.18	53.87	92.32	122.99	130.04	64.35	88.62	80.03	97.54	50.55
2006	82.83	73.24	76.12	78.59	77.46	77.37	61.73	98.02	124.14	127.59	72.12	94.66	87.15	98.67	66.03
2007	90.10	77.21	81.21	87.83	85.76	84.29	73.32	99.29	120.43	124.57	77.01	94.64	92.96	96.36	76.79
2008	95.64	85.26	85.93	86.90	92.77	90.34	79.35	97.91	114.04	120.29	78.95	83.26	92.48	92.04	88.35
2009	91.84	79.71	73.86	58.17	85.76	83.02	85.57	92.50	96.93	110.55	66.16	56.41	82.93	84.47	92.10
2010	101.26	92.13	85.10	79.30	93.71	91.70	90.14	94.59	102.84	111.09	81.19	79.41	94.51	88.22	94.56
2011	100.79	98.23	93.82	92.63	97.97	96.05	96.73	96.41	104.38	108.99	92.74	87.15	99.34	89.09	99.61
2012	100.00	100.00	100.00	100.00	100.00	100.00	100.00	100.00	100.00	100.00	100.00	100.00	100.00	100.00	100.00
2013	102.11	104.03	101.31	103.88	105.07	108.45	105.63	103.93	97.16	94.12	102.76	104.12	105.19	96.37	102.51
2014	110.29	107.68	104.92	108.12	112.28	114.96	110.61	109.31	98.11	90.37	111.61	111.49	112.62	103.00	103.75
2015	112.40	107.67	103.60	102.87	114.08	124.95	111.26	113.69	99.55	91.43	116.44	120.85	123.07	116.07	104.59
2016	118.80	109.54	101.56	102.67	114.27	123.74	110.49	115.31	102.45	96.46	117.67	122.05	123.33	127.56	104.80
2017	119.39	118.28	104.13	107.50	118.18	123.72	116.99	117.93	104.17	96.11	128.81	125.45	128.93	137.58	106.80
2018	119.24	126.54	108.89	107.34	121.99	123.26	117.81	127.12	104.10	91.41	139.63	130.29	139.00	144.97	105.85
2013															
1st quarter	98.87	100.64	99.34	102.87	101.33	106.49	105.06	104.48	96.74	94.28	101.01	98.78	102.93	96.56	100.43
2nd quarter	93.28	101.87	102.20	105.07	108.05	108.00	104.82	104.43	97.27	95.21	101.26	103.33	104.85	94.71	102.92
3rd quarter	100.95	104.42	101.18	103.35	104.78	108.52	105.61	102.77	98.02	94.70	103.40	106.72	105.30	95.99	103.45
4th quarter	115.36	109.20	102.52	104.23	106.13	110.78	107.02	104.03	96.61	92.27	105.37	107.65	107.67	98.20	103.23
2014															
1st quarter	115.57	102.75	103.27	103.46	108.62	113.14	109.29	104.86	97.97	93.18	107.14	105.41	109.19	100.65	102.18
2nd quarter	107.77	108.45	104.56	108.45	113.63	113.68	112.62	111.74	97.31	89.00	111.13	112.14	113.30	102.63	104.08
3rd quarter	102.75	110.37	105.92	111.53	113.02	115.62	109.39	110.50	96.75	87.66	113.37	113.97	111.69	102.65	103.39
4th quarter	115.06	109.16	105.95	109.04	113.87	117.41	111.13	110.15	100.41	91.64	114.82	114.43	116.33	106.08	105.34
2015															
1st quarter	113.12	107.55	104.51	100.68	115.19	122.30	114.27	111.47	100.78	91.76	116.65	115.00	121.14	112.25	103.71
2nd quarter	115.15	109.95	104.68	102.87	113.36	125.55	110.46	114.91	98.87	90.33	116.83	122.30	123.03	114.67	103.60
3rd quarter	110.32	107.83	102.97	104.81	113.78	125.83	108.37	113.98	99.44	91.52	115.94	123.54	124.63	116.52	105.61
4th quarter	111.00	105.33	102.26	103.12	113.98	126.11	111.95	114.39	99.11	92.11	116.32	122.58	123.48	120.82	105.43
2016															
1st quarter	110.37	108.60	101.50	100.69	112.98	125.05	107.72	117.51	100.44	94.97	114.72	121.48	122.40	125.63	105.37
2nd quarter	109.88	110.30	101.27	103.45	113.31	122.35	111.57	116.86	101.19	94.80	117.61	120.54	122.36	124.08	104.06
3rd quarter	131.74	110.25	100.91	104.13	115.45	124.80	111.68	112.04	102.96	97.82	118.20	122.67	123.03	127.40	104.20
4th quarter	123.19	109.01	102.55	102.40	115.33	122.75	111.98	114.85	105.22	98.24	120.15	123.51	125.52	133.12	105.56
2017															
1st quarter	120.11	116.00	102.33	107.64	118.64	121.44	115.75	116.59	105.35	101.44	123.06	125.29	126.02	137.06	104.50
2nd quarter	120.13	117.37	102.21	106.55	118.70	122.59	114.99	117.09	104.18	96.27	126.64	125.19	128.04	136.91	106.17
3rd quarter	120.52	115.49	105.41	106.37	116.66	125.51	117.70	117.69	102.72	92.74	130.28	125.02	126.21	136.15	108.54
4th quarter	116.82	124.26	106.57	109.45	118.69	125.32	119.51	120.36	104.43	93.99	135.28	126.28	135.47	140.21	107.99
2018															
1st quarter	113.85	123.04	108.08	113.11	121.53	126.22	117.62	123.88	102.83	90.54	136.47	128.66	139.32	141.91	105.59
2nd quarter	130.80	128.17	109.52	109.06	121.59	123.79	116.39	126.91	104.45	93.00	139.40	126.79	134.95	142.20	104.60
3rd quarter	123.89	125.98	108.24	104.95	121.36	121.26	118.64	129.84	105.72	94.39	141.60	131.42	138.03	145.14	106.16
4th quarter	108.43	128.98	109.73	102.23	123.47	121.74	118.58	127.83	103.39	87.71	141.04	134.29	143.69	150.61	107.03

SECTION 7B: EXPORTS AND IMPORTS

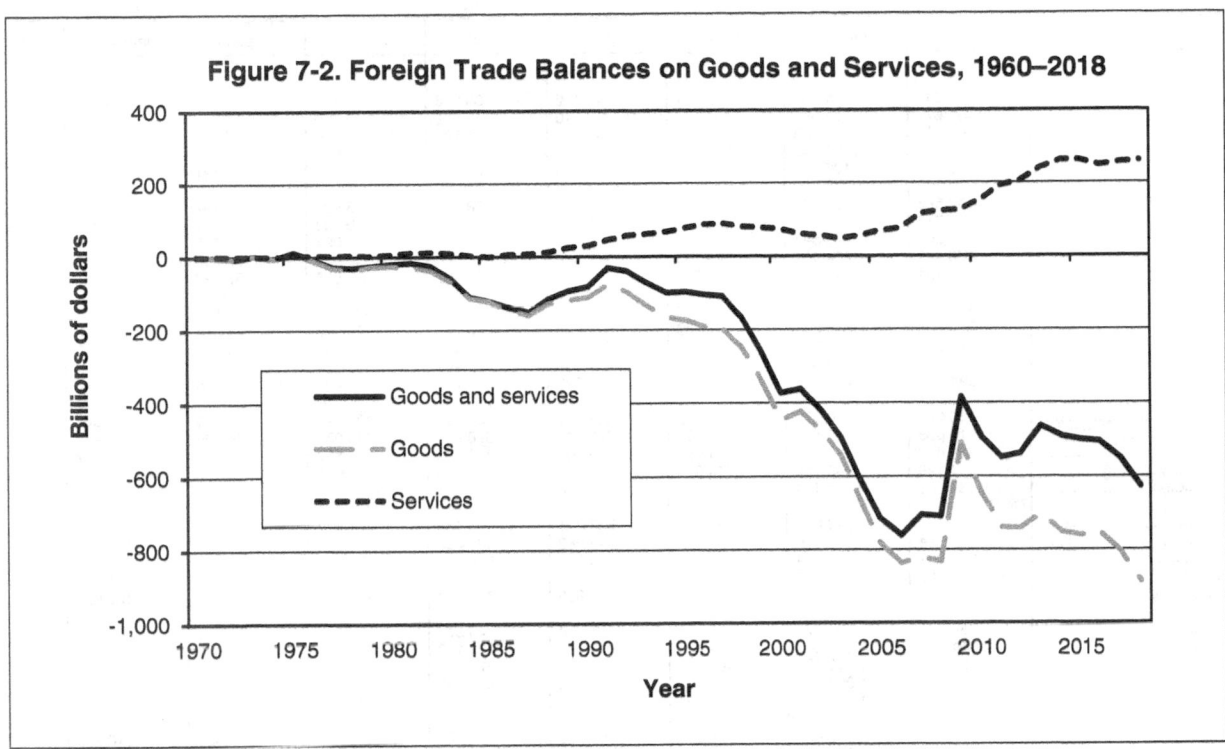

Figure 7-2. Foreign Trade Balances on Goods and Services, 1960–2018

- U.S. imports of goods and services exceeded exports by a record $762 billion in 2006. This negative "trade balance" (to differentiate it from the more comprehensive "current account deficit" including income payments and receipts that is shown in earlier pages) diminished to $384 billion in 2009 but was back up to $628 billion in 2018. (Table 7-6)

- Canada and Mexico are principal trading partners of the United States. In 2018, U.S. goods exports to those two countries brought in $565 billion, compared with $318 billion in exports to the European Union and $195 billion in exports to China and Japan. (Table 7-10)

- U.S. goods imports from Canada and Mexico in 2018 amounted to $665 billion, more than the $487 billion imported from the European Union but less than the $682 billion from China and Japan. (Table 7-11)

- Imports may decline due to proposed or imposed imports tariffs on various products. Countries have begun imposing or threatening to impose tariffs include China, the European Union, representing 28 nations, and Canada. The impact on U.S. trade is unknown but as retaliatory tariffs increase, the impact on the global economy could grow. (Table 7-11)

- The value of imports from China rapidly increased 8,646 percent between 1985 to 2008 but fell from $338 billion in 2008 to $296 billion in 2009—the lowest amount since then. Imports from China grew 82.1 percent from 2009 to 2018. (Table 7-11)

- The continual and growing U.S. surplus in services is in contrast with the goods deficit, though it offsets only about a third of it. Revised categories permit a more detailed look at the composition of the total services surplus, which was $260 billion in 2018. The largest component, $81 billion, was financial services which includes commercial banks but also investment banks and hedge funds managers. The second largest service sector was $73 billion for "Charges for the use of intellectual property" (formerly known as "royalties and license fees"). A close third at nearly $70 billion, was "travel for all purposes including education," which includes the spending of foreigners here for tourism and education. The U.S. runs a deficit on insurance services. (Tables 7-12A and 7-13A)

Table 7-6. U.S. Exports and Imports of Goods and Services

(Balance of payments basis; millions of dollars, seasonally adjusted.)

Year and month	Goods and services			Goods			Services		
	Exports	Imports	Balance	Exports	Imports	Balance	Exports	Imports	Balance
1970	56 640	54 386	2 254	42 469	39 866	2 603	14 171	14 520	-349
1971	59 677	60 979	-1 302	43 319	45 579	-2 260	16 358	15 400	958
1972	67 222	72 665	-5 443	49 381	55 797	-6 416	17 841	16 868	973
1973	91 242	89 342	1 900	71 410	70 499	911	19 832	18 843	989
1974	120 897	125 190	-4 293	98 306	103 811	-5 505	22 591	21 379	1 212
1975	132 585	120 181	12 404	107 088	98 185	8 903	25 497	21 996	3 501
1976	142 716	148 798	-6 082	114 745	124 228	-9 483	27 971	24 570	3 401
1977	152 301	179 547	-27 246	120 816	151 907	-31 091	31 485	27 640	3 845
1978	178 428	208 191	-29 763	142 075	176 002	-33 927	36 353	32 189	4 164
1979	224 131	248 696	-24 565	184 439	212 007	-27 568	39 692	36 689	3 003
1980	271 834	291 241	-19 407	224 250	249 750	-25 500	47 584	41 491	6 093
1981	294 398	310 570	-16 172	237 044	265 067	-28 023	57 354	45 503	11 851
1982	275 236	299 391	-24 156	211 157	247 642	-36 485	64 079	51 749	12 329
1983	266 106	323 874	-57 767	201 799	268 901	-67 102	64 307	54 973	9 335
1984	291 094	400 166	-109 072	219 926	332 418	-112 492	71 168	67 748	3 420
1985	289 070	410 950	-121 880	215 915	338 088	-122 173	73 155	72 862	294
1986	310 033	448 572	-138 538	223 344	368 425	-145 081	86 689	80 147	6 543
1987	348 869	500 552	-151 684	250 208	409 765	-159 557	98 661	90 787	7 874
1988	431 149	545 715	-114 566	320 230	447 189	-126 959	110 919	98 526	12 393
1989	487 003	580 144	-93 141	359 916	477 665	-117 749	127 087	102 479	24 607
1990	535 233	616 097	-80 864	387 401	498 438	-111 037	147 832	117 659	30 173
1991	578 344	609 479	-31 135	414 083	491 020	-76 937	164 261	118 459	45 802
1992	616 882	656 094	-39 212	439 631	536 528	-96 897	177 251	119 566	57 685
1993	642 863	713 174	-70 311	456 943	589 394	-132 451	185 920	123 780	62 141
1994	703 254	801 747	-98 493	502 859	668 690	-165 831	200 395	133 057	67 338
1995	794 387	890 771	-96 384	575 204	749 374	-174 170	219 183	141 397	77 786
1996	851 602	955 667	-104 065	612 113	803 113	-191 000	239 489	152 554	86 935
1997	934 453	1 042 726	-108 273	678 366	876 794	-198 428	256 087	165 932	90 155
1998	933 174	1 099 314	-166 140	670 416	918 637	-248 221	262 758	180 677	82 081
1999	969 867	1 228 485	-258 617	698 524	1 035 592	-337 068	271 343	192 893	78 450
2000	1 075 321	1 447 837	-372 517	784 940	1 231 722	-446 783	290 381	216 115	74 266
2001	1 005 654	1 367 165	-361 511	731 331	1 153 701	-422 370	274 323	213 465	60 858
2002	978 706	1 397 660	-418 955	698 036	1 173 281	-475 245	280 670	224 379	56 290
2003	1 020 418	1 514 308	-493 890	730 446	1 272 089	-541 643	289 972	242 219	47 754
2004	1 161 549	1 771 433	-609 883	823 584	1 488 349	-664 766	337 966	283 083	54 882
2005	1 286 022	2 000 267	-714 245	913 016	1 695 820	-782 804	373 006	304 448	68 558
2006	1 457 642	2 219 358	-761 716	1 040 905	1 878 194	-837 289	416 738	341 165	75 573
2007	1 653 548	2 358 922	-705 375	1 165 151	1 986 347	-821 196	488 396	372 575	115 821
2008	1 841 612	2 550 339	-708 726	1 308 795	2 141 287	-832 492	532 817	409 052	123 765
2009	1 583 053	1 966 827	-383 774	1 070 331	1 580 025	-509 694	512 722	386 801	125 920
2010	1 853 606	2 348 263	-494 658	1 290 273	1 938 950	-648 678	563 333	409 313	154 020
2011	2 127 021	2 675 646	-548 625	1 499 240	2 239 886	-740 646	627 781	435 761	192 020
2012	2 218 989	2 755 762	-536 773	1 562 578	2 303 749	-741 171	656 411	452 013	204 398
2013	2 293 457	2 755 334	-461 876	1 592 002	2 294 247	-702 244	701 455	461 087	240 368
2014	2 375 905	2 866 241	-490 336	1 633 986	2 385 480	-751 494	741 919	480 761	261 157
2015	2 263 907	2 764 352	-500 445	1 510 757	2 272 612	-761 855	753 150	491 740	261 410
2016	2 208 072	2 712 866	-504 794	1 455 704	2 208 211	-752 507	752 368	504 654	247 714
2017	2 352 546	2 902 669	-550 123	1 553 589	2 358 789	-805 200	798 957	543 880	255 077
2018	2 501 310	3 128 989	-627 679	1 674 330	2 561 667	-887 337	826 980	567 322	259 658
2017									
January	192 191	238 607	-46 416	127 301	194 586	-67 285	64 890	44 021	20 869
February	192 601	235 704	-43 103	127 345	191 670	-64 325	65 256	44 034	21 222
March	192 314	236 845	-44 531	127 034	192 619	-65 585	65 280	44 226	21 054
April	191 562	238 946	-47 384	126 413	194 542	-68 129	65 149	44 404	20 745
May	192 223	238 908	-46 685	126 887	193 815	-66 928	65 336	45 093	20 243
June	194 260	239 868	-45 608	128 376	194 544	-66 168	65 884	45 324	20 560
July	194 747	238 908	-44 161	128 228	193 416	-65 188	66 519	45 492	21 027
August	195 565	239 254	-43 689	128 617	193 706	-65 089	66 948	45 548	21 400
September	198 166	241 737	-43 571	130 283	195 589	-65 306	67 883	46 148	21 735
October	199 315	244 793	-45 478	131 100	198 362	-67 262	68 215	46 431	21 784
November	202 903	252 024	-49 121	134 236	205 493	-71 257	68 667	46 531	22 136
December	206 700	257 076	-50 376	137 770	210 448	-72 678	68 930	46 628	22 302
2018									
January	202 575	254 688	-52 113	133 614	208 472	-74 858	68 961	46 216	22 745
February	205 608	259 425	-53 817	136 377	212 245	-75 868	69 231	47 180	22 051
March	209 937	257 114	-47 177	140 742	210 731	-69 989	69 195	46 383	22 812
April	208 883	257 102	-48 219	140 364	210 657	-70 293	68 519	46 445	22 074
May	213 341	257 693	-44 352	144 552	211 222	-66 670	68 789	46 471	22 318
June	210 968	258 398	-47 430	142 172	211 606	-69 434	68 796	46 792	22 004
July	208 734	261 175	-52 441	139 943	214 078	-74 135	68 791	47 097	21 694
August	207 758	262 647	-54 889	138 857	215 353	-76 496	68 901	47 294	21 607
September	209 747	265 841	-56 094	140 745	218 015	-77 270	69 002	47 826	21 176
October	210 124	266 817	-56 693	141 259	218 563	-77 304	68 865	48 254	20 611
November	207 976	261 623	-53 647	139 124	213 234	-74 110	68 852	48 389	20 463
December	205 661	266 468	-60 807	136 581	217 491	-80 910	69 080	48 977	20 103

Table 7-7. U.S. Exports of Goods by End-Use and Advanced Technology Categories

(Census basis, except as noted; billions of dollars; seasonally adjusted, except as noted.)

Year and month	Total exports of goods			Principal end-use category							Advanced technology products [1]
	Total, balance of payments basis	Net adjustments	Total, Census basis	Foods, feeds, and beverages	Industrial supplies and materials		Capital goods, except automotive	Automotive vehicles, engines, and parts	Consumer goods, except automotive	Other goods	
					Total	Petroleum and products					
1980	224.25	3.55	220.70	36.28	72.09	3.57	76.28	17.44	17.75	...	...
1981	237.04	3.31	233.74	38.84	70.19	4.56	84.17	19.69	17.70	...	...
1982	211.16	-1.12	212.28	32.20	64.05	6.87	76.50	17.23	16.13	...	...
1983	201.80	0.09	201.71	32.09	58.94	5.59	71.66	18.46	14.93	...	...
1984	219.93	1.18	218.74	32.20	64.12	5.43	77.01	22.42	15.09	...	...
1985	215.92	3.29	212.62	24.57	61.16	5.71	79.32	24.95	14.59	...	...
1986	223.34	-3.13	226.47	23.52	64.72	4.43	82.82	25.10	16.73	...	...
1987	250.21	-3.70	253.90	25.23	70.05	4.63	92.71	27.58	20.31	...	...
1988	320.23	-3.11	323.34	33.77	90.02	4.48	119.10	33.40	26.98	...	...
1989	359.92	-3.08	363.00	36.34	98.36	6.46	136.94	35.05	36.01	...	...
1990	387.40	-5.57	392.97	35.18	105.55	8.36	153.07	36.07	43.60	20.73	...
1991	414.08	-7.77	421.85	35.79	109.69	8.40	166.72	39.72	46.65	23.66	...
1992	439.63	-8.54	448.17	40.34	109.59	7.62	176.50	46.71	51.31	24.39	...
1993	456.94	-7.92	464.86	40.59	111.89	7.49	182.85	51.35	54.56	23.89	...
1994	502.86	-9.77	512.63	41.96	121.55	6.97	205.82	57.31	59.86	26.50	...
1995	575.20	-9.54	584.74	50.47	146.37	8.10	234.46	61.26	64.31	28.72	...
1996	612.11	-12.96	625.08	55.53	147.98	9.63	253.99	64.24	70.11	33.85	...
1997	678.37	-10.82	689.18	51.51	158.32	10.42	295.87	73.30	77.96	33.51	...
1998	670.42	-11.72	682.14	46.40	148.31	8.08	299.87	72.39	80.29	35.44	...
1999	683.97	-11.83	695.80	45.98	147.52	8.62	310.79	75.26	80.92	35.32	200.28
2000	771.99	-9.92	781.92	47.87	172.62	12.01	356.93	80.36	89.38	34.77	227.40
2001	718.71	-10.39	729.10	49.41	160.10	10.64	321.71	75.44	88.33	34.11	199.63
2002	682.42	-10.68	693.10	49.62	156.81	10.34	290.44	78.94	84.36	32.04	170.57
2003	710.42	-11.36	724.77	55.03	173.04	12.69	293.67	80.63	89.91	32.49	180.21
2004	807.52	-11.26	818.78	56.57	203.96	17.08	331.56	89.21	103.08	34.40	201.56
2005	894.63	-11.35	905.98	58.90	233.05	22.66	363.32	98.41	115.29	36.96	216.82
2006	1 015.81	-10.16	1 025.97	65.96	276.05	31.57	404.03	107.26	129.08	43.59	247.07
2007	1 138.38	-9.82	1 148.20	84.26	316.38	37.76	433.02	121.26	145.98	47.30	264.88
2008	1 307.50	20.06	1 287.44	108.35	388.03	67.18	457.66	121.45	161.28	50.67	270.13
2009	1 069.73	13.69	1 056.04	93.91	296.51	49.18	391.24	81.72	149.46	43.22	244.71
2010	1 288.80	10.30	1 278.50	107.72	391.66	70.83	447.54	112.01	165.23	54.34	273.31
2011	1 495.85	15.56	1 480.29	126.25	501.08	113.71	493.96	133.04	175.30	52.89	287.72
2012	1 562.58	16.76	1 545.82	133.05	483.23	142.02	527.46	146.16	180.99	53.85	304.98
2013	1 592.00	13.48	1 578.52	136.16	492.42	137.42	534.76	152.66	188.09	53.56	319.73
2014	1 633.99	12.11	1 621.87	143.72	505.81	145.24	551.49	159.81	199.00	62.04	338.48
2015	1 510.76	7.66	1 503.10	127.74	426.32	97.94	539.50	151.92	197.84	59.79	343.13
2016	1 457.39	5.93	1 451.46	130.52	397.27	88.85	519.74	150.40	193.66	59.88	345.24
2017	1 553.59	7.12	1 546.47	132.74	464.70	125.18	533.21	157.87	197.69	60.27	353.57
2018	1 674.33	8.34	1 665.99	133.18	541.74	175.30	562.92	158.84	205.98	63.34	368.36
2016											
January	116.98	0.36	116.62	9.50	31.24	6.26	42.97	12.34	16.07	4.09	25.47
February	118.90	0.35	118.55	9.95	31.03	6.02	43.27	12.99	16.72	4.74	25.55
March	117.99	0.45	117.55	9.55	31.02	6.67	43.60	12.27	15.70	5.46	31.55
April	119.77	0.50	119.28	9.92	32.41	7.12	43.41	12.86	15.78	4.81	28.47
May	120.38	0.52	119.86	10.09	32.69	8.22	43.01	12.65	15.73	5.29	27.87
June	120.79	0.49	120.30	10.77	32.64	8.12	43.37	12.33	16.07	5.22	31.06
July	121.99	0.53	121.46	13.10	32.57	7.48	43.06	12.53	16.03	4.72	27.97
August	123.85	0.55	123.30	13.36	34.04	7.35	42.34	12.91	16.16	4.73	27.76
September	124.54	0.51	124.03	11.83	34.66	7.74	43.58	12.52	16.79	4.89	29.55
October	123.59	0.53	123.06	10.89	33.60	7.55	43.76	12.46	15.99	5.66	30.50
November	121.73	0.56	121.17	10.79	34.80	7.94	42.15	12.16	16.37	4.94	27.54
December	126.89	0.60	126.29	10.82	35.74	8.21	45.07	12.30	16.43	5.72	32.14
2017											
January	127.30	0.58	126.72	11.06	37.65	9.16	42.97	12.34	16.07	4.09	26.04
February	127.35	0.62	126.72	10.32	38.34	10.17	43.27	12.99	16.72	4.74	25.37
March	127.03	0.61	126.42	11.01	37.05	9.11	43.60	12.27	15.70	5.46	31.21
April	126.41	0.53	125.88	11.61	37.51	9.99	43.41	12.86	15.78	4.81	28.59
May	126.89	0.51	126.38	10.88	37.36	9.60	43.01	12.65	15.73	5.29	28.56
June	128.38	0.53	127.85	11.54	37.59	9.80	43.37	12.33	16.07	5.22	30.57
July	128.23	0.57	127.66	11.95	37.45	10.23	43.06	12.53	16.03	4.72	29.19
August	128.62	0.59	128.03	11.49	36.50	9.05	42.34	12.91	16.16	4.73	30.45
September	130.28	0.49	129.79	11.62	38.39	10.29	43.58	12.52	16.79	4.89	30.23
October	131.10	0.61	130.49	10.26	40.94	12.17	43.76	12.46	15.99	5.66	29.62
November	134.24	0.76	133.48	10.40	41.27	12.28	42.15	12.16	16.37	4.94	30.30
December	137.77	0.72	137.05	10.85	42.81	12.49	45.07	12.30	16.43	5.72	33.43
2018											
January	133.61	0.74	132.87	10.45	41.07	11.64	45.27	13.50	17.68	4.90	27.44
February	136.38	0.73	135.64	10.57	43.09	12.04	46.08	14.27	16.58	5.06	26.01
March	140.74	0.67	140.07	11.07	45.05	13.83	47.60	13.98	17.08	5.30	34.90
April	140.36	0.72	139.65	11.49	45.82	14.43	46.17	13.87	17.22	5.09	28.52
May	144.55	0.80	143.75	13.09	45.35	15.13	48.08	13.59	17.66	6.00	32.28
June	142.17	0.67	141.50	12.70	46.61	15.79	47.31	12.94	16.48	5.47	32.87
July	139.94	0.71	139.24	12.02	46.86	16.15	46.32	13.03	16.15	4.86	28.13
August	138.86	0.68	138.18	11.32	44.65	14.43	46.61	12.85	17.52	5.22	29.87
September	140.75	0.67	140.07	10.49	46.66	15.40	47.33	12.99	17.60	5.02	31.44
October	141.26	0.70	140.56	10.00	47.26	16.79	47.21	12.80	17.78	5.52	32.28
November	139.12	0.67	138.45	10.06	45.32	15.46	48.10	12.55	17.10	5.33	31.99
December	136.58	0.58	136.00	9.93	44.02	14.24	46.85	12.48	17.14	5.58	32.61

[1] Not seasonally adjusted.
... = Not available.

Table 7-8. U.S. Imports of Goods by End-Use and Advanced Technology Categories

(Census basis, except as noted; billions of dollars; seasonally adjusted, except as noted.)

Year and month	Total imports of goods			Principal end-use category							Advanced technology products [1]
	Total, balance of payments basis	Net adjustments	Total, Census basis	Foods, feeds, and beverages	Industrial supplies and materials		Capital goods, except automotive	Automotive vehicles, engines, and parts	Consumer goods, except automotive	Other goods	
					Total	Petroleum and products					
1980	249.75	4.23	245.52	18.55	124.96	...	30.72	28.13	34.22	...	...
1981	265.07	3.76	261.31	18.53	131.10	...	36.86	30.80	38.30	...	...
1982	247.64	3.70	243.94	17.47	107.82	...	38.22	34.26	39.66	...	...
1983	268.90	7.18	261.72	18.56	105.63	...	42.61	42.04	46.59	...	...
1984	332.42	1.91	330.51	21.92	122.72	...	60.15	56.77	61.19	...	...
1985	338.09	1.71	336.38	21.89	112.48	...	60.81	65.21	66.43	...	...
1986	368.43	2.75	365.67	24.40	101.37	...	71.86	78.25	79.43	...	...
1987	409.77	3.48	406.28	24.81	110.67	...	84.77	85.17	88.82	...	...
1988	447.19	5.26	441.93	24.93	118.06	...	101.79	87.95	96.42	...	...
1989	477.37	3.72	473.65	25.08	132.40	...	112.45	87.38	102.26	...	...
1990	498.34	2.36	495.98	26.65	143.41	62.16	116.04	87.69	105.29	16.09	...
1991	490.98	2.53	488.45	26.21	131.38	51.78	120.80	84.94	107.78	15.94	...
1992	536.58	3.92	532.67	27.61	138.64	51.60	134.25	91.79	122.66	17.71	71.87
1993	589.44	8.78	580.66	27.87	145.61	51.50	152.37	102.42	134.02	18.39	81.23
1994	668.70	5.44	663.26	30.96	162.11	51.28	184.37	118.27	146.27	21.27	98.12
1995	749.37	5.83	743.54	33.18	181.85	56.16	221.43	123.80	159.90	23.39	124.79
1996	803.11	7.92	795.19	35.71	204.48	72.75	228.07	128.94	172.00	26.10	130.36
1997	876.40	6.70	869.70	39.69	213.77	71.77	253.28	139.81	193.81	29.34	147.28
1998	917.64	5.74	911.90	41.24	200.14	50.90	269.45	148.68	217.00	35.39	156.75
1999	1 035.60	10.98	1 024.62	43.60	221.39	67.81	295.72	178.96	241.91	43.04	181.18
2000	1 231.72	13.70	1 218.02	45.98	298.98	120.28	347.03	195.88	281.83	48.33	222.08
2001	1 153.70	12.70	1 141.00	46.64	273.87	103.59	297.99	189.78	284.29	48.42	195.18
2002	1 173.28	11.91	1 161.37	49.69	267.69	103.51	283.32	203.74	307.84	49.08	195.15
2003	1 272.10	14.98	1 257.12	55.83	313.82	133.10	295.87	210.14	333.88	47.59	207.03
2004	1 488.35	18.65	1 469.70	62.14	412.77	180.46	343.58	228.20	372.94	50.11	238.28
2005	1 695.82	22.36	1 673.46	68.09	523.77	251.86	379.33	239.45	407.24	55.57	259.74
2006	1 878.19	24.25	1 853.94	74.94	601.99	302.43	418.26	256.63	442.64	59.49	290.76
2007	1 986.35	29.39	1 956.96	81.68	634.75	330.98	444.51	256.67	474.55	64.81	326.81
2008	2 141.29	37.65	2 103.64	89.00	779.48	453.28	453.74	231.24	481.64	68.54	331.15
2009	1 580.03	20.40	1 559.63	81.62	462.38	253.69	370.48	157.65	427.33	60.17	300.89
2010	1 938.95	25.09	1 913.86	91.75	603.10	336.11	449.39	225.10	483.23	61.30	354.25
2011	2 239.89	31.93	2 207.95	107.48	755.78	439.34	510.80	254.62	514.11	65.17	386.44
2012	2 303.75	27.45	2 276.30	110.27	730.64	415.17	548.71	297.78	516.93	71.95	396.23
2013	2 294.25	26.26	2 267.99	115.15	681.53	369.68	555.65	308.80	531.67	75.20	401.14
2014	2 385.48	29.12	2 356.36	125.88	667.02	334.00	594.14	328.64	557.09	83.59	422.13
2015	2 273.25	24.44	2 248.81	127.80	485.97	98.09	602.45	349.16	594.18	89.25	434.99
2016	2 207.20	20.41	2 186.79	130.01	443.27	146.61	589.70	349.90	583.10	90.80	429.17
2017	2 358.79	18.91	2 339.88	137.82	507.09	186.49	639.90	358.30	601.50	95.30	464.60
2018	2 561.67	20.86	2 540.81	147.40	575.62	225.34	692.60	372.20	646.80	106.20	497.40
2016											
January	180.75	1.63	179.12	10.66	34.66	11.44	48.26	29.97	48.39	7.18	30.48
February	184.65	1.68	182.97	11.08	34.33	10.63	49.14	29.34	51.83	7.26	30.70
March	175.98	1.72	174.26	10.54	33.25	10.30	47.52	28.61	46.61	7.73	34.87
April	178.71	1.68	177.03	10.72	33.99	10.58	49.40	28.63	46.86	7.43	33.36
May	182.11	1.84	180.26	10.74	36.34	11.24	48.68	28.98	48.08	7.43	34.73
June	186.03	1.93	184.10	10.58	38.14	12.99	49.57	28.61	49.68	7.52	38.40
July	184.95	1.83	183.13	10.68	38.91	12.62	48.99	28.58	48.58	7.39	34.75
August	185.31	1.74	183.57	10.88	38.00	12.92	49.93	28.89	48.06	7.81	37.96
September	184.18	1.54	182.64	10.88	38.04	12.87	48.78	29.81	47.44	7.69	37.35
October	185.91	1.56	184.35	10.90	37.62	13.18	49.63	29.14	49.26	7.80	39.23
November	188.48	1.60	186.88	11.16	39.75	14.06	49.70	29.16	49.27	7.85	41.33
December	191.16	1.65	189.51	11.25	40.28	13.82	50.38	30.39	49.50	7.71	36.02
2017											
January	194.59	1.71	192.88	11.14	41.70	16.76	51.00	31.24	52.00	7.72	34.53
February	191.67	1.68	190.00	11.27	43.93	18.15	50.96	29.04	48.46	7.62	29.96
March	192.62	1.72	190.90	11.05	43.39	17.39	50.51	30.53	48.87	7.05	36.94
April	194.54	1.55	192.99	11.41	41.94	15.28	51.45	29.85	50.80	7.77	34.65
May	193.82	1.56	192.26	11.35	42.01	15.62	52.72	29.13	49.34	8.00	37.97
June	194.54	1.58	192.96	11.42	41.07	14.11	52.79	30.16	48.63	8.28	39.61
July	193.42	1.54	191.87	11.62	40.29	13.10	54.07	29.32	48.64	7.93	37.39
August	193.71	1.58	192.13	11.54	39.79	13.77	53.57	30.00	48.72	7.82	39.47
September	195.59	1.48	194.11	11.74	40.82	14.08	55.08	29.46	49.08	7.72	40.61
October	198.36	1.43	196.93	11.76	42.60	15.28	54.85	29.52	49.88	8.84	43.40
November	205.49	1.52	203.97	11.64	44.74	16.82	56.41	29.86	52.28	8.47	46.08
December	210.45	1.57	208.88	11.89	45.31	15.77	57.25	30.92	55.50	8.41	43.83
2018											
January	208.47	1.70	206.78	11.86	46.98	18.73	55.72	30.47	53.48	8.28	38.91
February	212.25	1.59	210.66	12.40	47.15	18.36	57.29	30.88	54.99	7.94	33.89
March	210.73	1.58	209.15	12.27	47.21	18.36	56.65	30.78	54.17	8.07	40.78
April	210.66	1.70	208.95	12.28	47.86	18.83	57.26	30.23	52.25	9.07	38.02
May	211.22	1.84	209.38	12.36	48.00	18.97	58.56	29.97	51.89	8.62	42.43
June	211.61	1.64	209.96	12.19	48.55	19.49	57.45	30.36	53.03	8.39	41.52
July	214.08	1.78	212.30	12.42	49.08	20.09	58.02	30.85	52.94	8.98	41.37
August	215.35	1.91	213.45	12.30	49.44	20.23	57.74	31.62	53.34	9.00	42.22
September	218.02	1.71	216.31	12.17	49.18	20.00	59.72	31.26	54.73	9.24	44.04
October	218.56	1.85	216.71	12.32	49.12	19.52	57.06	31.79	56.49	9.93	46.68
November	213.23	1.87	211.37	12.23	46.40	16.80	57.55	32.02	53.71	9.46	44.67
December	217.49	1.70	215.79	12.57	46.67	15.96	59.60	31.97	55.79	9.20	41.98

[1] Not seasonally adjusted.
. . . = Not available.

Table 7-9. U.S. Exports and Imports of Goods by Principal End-Use Category in Constant Dollars

(Census basis; billions of 2012 chain-weighted dollars, except as noted; seasonally adjusted.)

Year and month	Exports							Imports						
	Total	Foods, feeds, and beverages	Industrial supplies and materials	Capital goods, except automotive	Automotive vehicles, engines, and parts	Consumer goods, except automotive	Other goods	Total	Foods, feeds, and beverages	Industrial supplies and materials	Capital goods, except automotive	Automotive vehicles, engines, and parts	Consumer goods, except automotive	Other goods
1995	712.38	101.87	253.24	196.62	70.95	76.73	42.73	928.46	50.57	551.80	132.86	144.68	166.43	28.41
1996	787.89	98.62	268.78	233.39	73.84	82.35	49.89	1 025.33	55.99	585.42	160.26	149.62	178.25	31.71
1997	893.07	98.34	290.85	285.92	83.32	90.44	49.54	1 167.72	61.36	625.78	200.68	161.75	202.94	35.87
1998	914.15	97.73	289.74	298.56	81.32	93.97	53.65	1 305.30	66.02	685.91	230.18	171.86	230.06	44.21
1999	943.54	101.08	291.84	314.72	84.21	94.63	53.61	1 462.44	72.10	696.85	264.02	205.26	258.33	53.94
2000	1 047.54	106.71	321.11	364.20	89.05	103.87	51.60	1 656.42	76.78	740.31	316.80	223.15	303.86	59.66
2001	982.79	109.54	306.65	328.47	83.25	102.71	50.69	1 601.50	79.75	736.67	279.63	216.30	308.87	60.23
2002	909.96	107.07	301.64	280.75	86.77	98.41	47.18	1 631.30	85.02	737.94	258.51	231.54	337.43	60.92
2003	933.89	108.87	314.83	286.68	88.18	103.78	45.28	1 718.52	92.87	767.45	274.39	237.52	366.47	58.04
2004	1 016.04	100.50	336.09	322.28	96.86	118.15	45.95	1 921.28	98.59	862.16	323.53	254.00	406.44	59.42
2005	1 088.65	105.18	346.06	352.41	105.73	130.19	47.33	2 058.36	102.24	906.53	360.64	263.89	439.58	63.74
2006	1 198.66	112.76	372.59	396.62	114.04	143.15	53.39	2 182.04	108.21	910.77	403.10	281.53	474.87	66.65
2007	1 293.52	119.92	397.85	426.95	127.59	157.91	55.12	2 223.92	109.13	883.13	429.06	278.75	501.57	70.64
2008	1 385.46	126.88	445.88	453.35	126.38	170.78	55.79	2 158.81	106.70	842.41	438.11	245.30	495.69	70.96
2009	1 198.93	123.67	396.47	388.40	84.63	157.41	50.08	1 820.18	99.87	705.51	363.71	166.20	441.07	64.98
2010	1 379.61	135.65	454.60	445.18	115.49	170.24	59.44	2 091.57	103.66	746.18	447.24	235.88	498.29	64.38
2011	1 486.80	132.97	494.94	493.40	135.10	177.18	53.20	2 221.30	106.79	758.64	508.41	259.44	521.70	65.37
2012	1 545.82	133.05	501.18	527.18	146.16	181.68	56.58	2 276.27	110.27	730.64	548.71	297.78	516.92	71.95
2013	1 589.07	132.72	523.08	533.01	151.88	190.81	58.61	2 301.79	113.99	706.16	563.42	309.75	532.57	75.79
2014	1 648.18	141.79	532.96	550.08	158.11	204.04	63.01	2 405.58	119.39	707.26	607.30	331.35	555.61	84.03
2015	1 636.51	144.92	531.77	541.31	150.43	207.29	64.84	2 519.22	123.48	709.91	630.14	358.82	597.23	92.50
2016	1 638.19	156.86	531.84	528.74	150.12	207.17	67.54	2 540.94	125.25	736.71	633.49	362.41	588.25	95.64
2017	1 707.81	158.40	578.79	540.17	157.23	213.98	66.34	2 642.64	128.14	743.97	688.10	372.41	606.83	99.15
2018	1 780.90	158.89	618.19	566.67	156.97	220.46	67.23	2 786.64	137.50	748.94	740.72	386.76	649.98	109.12
2016														
January	132.87	11.80	43.35	43.70	12.07	17.13	5.06	207.58	10.39	59.55	51.23	30.76	47.78	7.66
February	135.85	12.39	44.00	44.18	12.56	17.42	5.54	213.90	10.86	60.54	52.19	30.24	51.81	7.60
March	134.78	11.95	43.63	44.32	12.18	16.79	6.14	206.04	10.59	59.12	51.20	29.16	47.10	8.70
April	136.13	11.84	45.13	44.09	12.90	17.12	5.45	208.63	10.55	59.18	53.28	29.84	47.46	7.90
May	135.29	12.06	44.29	43.85	12.62	16.88	5.93	210.29	10.67	61.64	51.96	30.00	48.43	7.90
June	134.51	12.30	42.82	44.06	12.30	17.44	5.72	214.42	10.60	62.73	53.27	29.65	50.52	7.90
July	135.70	14.17	42.75	43.87	12.43	17.42	5.26	211.84	10.19	63.01	52.45	29.90	49.17	7.77
August	139.06	15.27	45.78	43.23	12.92	17.28	5.29	212.62	10.10	61.83	53.94	29.96	48.95	8.07
September	139.47	14.25	45.90	44.15	12.73	17.57	5.44	211.44	10.09	61.66	52.55	31.20	48.18	8.02
October	137.98	13.75	43.61	44.79	12.69	17.08	6.23	212.13	10.20	60.72	53.59	30.12	49.36	8.00
November	135.81	13.54	44.95	42.81	12.22	17.37	5.49	215.36	10.36	63.49	53.82	30.39	49.74	8.04
December	140.77	13.54	45.63	45.70	12.53	17.69	6.00	216.71	10.65	63.24	54.03	31.20	49.75	8.09
2017														
January	141.16	13.55	47.45	44.59	13.28	17.76	5.10	219.83	10.73	62.46	55.12	31.93	51.18	8.11
February	140.75	12.90	48.07	43.89	13.11	17.98	5.45	214.73	10.53	63.11	54.76	29.91	48.77	8.07
March	140.30	13.24	46.92	44.19	12.96	17.99	5.56	215.52	10.55	62.00	55.12	31.17	48.94	7.78
April	139.55	13.20	47.24	44.19	12.56	17.44	5.57	218.22	10.70	60.84	55.76	31.02	51.14	8.17
May	141.23	13.05	48.53	43.81	12.85	18.22	5.55	218.33	10.63	62.67	56.36	30.41	49.98	8.31
June	142.86	13.59	48.33	44.32	13.56	18.09	5.63	219.45	10.44	62.25	56.96	31.47	49.74	8.44
July	142.06	13.44	48.04	45.36	12.79	17.37	5.68	218.74	10.67	62.03	57.62	30.75	49.28	8.24
August	141.15	13.11	45.95	45.98	13.05	17.95	5.33	217.98	10.63	60.73	57.59	31.08	49.46	7.98
September	141.96	13.43	47.19	45.46	13.05	17.49	5.80	218.88	10.64	60.45	58.62	30.94	49.58	8.01
October	142.91	12.64	50.81	44.45	12.79	17.52	5.57	221.43	10.69	62.09	58.98	30.59	49.96	8.76
November	145.10	13.03	48.94	46.40	13.61	18.10	5.49	227.51	10.83	62.76	60.41	31.29	53.00	8.54
December	148.79	13.24	51.32	47.55	13.62	18.07	5.60	232.01	11.09	62.59	60.79	31.86	55.80	8.74
2018														
January	143.36	12.44	48.10	45.63	13.40	18.95	5.26	227.60	10.98	61.82	59.67	31.57	54.04	8.53
February	146.06	12.59	50.12	46.55	14.12	17.74	5.41	230.52	11.22	61.30	61.20	31.98	55.38	8.15
March	150.65	12.83	52.56	48.11	13.82	18.27	5.66	229.61	11.28	62.04	60.46	31.95	54.49	8.29
April	149.54	13.42	52.91	46.49	13.70	18.39	5.42	229.52	11.38	63.05	61.07	31.37	52.53	9.31
May	152.65	15.18	51.22	48.38	13.41	18.87	6.32	229.28	11.41	62.42	62.50	31.14	52.10	8.84
June	150.08	14.88	52.31	47.56	12.77	17.65	5.76	230.08	11.58	62.56	61.42	31.58	53.34	8.62
July	148.18	14.78	52.42	46.58	12.87	17.28	5.14	232.34	12.00	62.82	62.06	32.08	53.07	9.24
August	147.11	13.77	50.15	46.89	12.69	18.75	5.52	233.94	11.79	63.71	61.81	32.90	53.48	9.27
September	149.14	12.89	52.29	47.58	12.82	18.85	5.30	236.81	11.43	63.37	63.94	32.52	54.89	9.51
October	148.71	12.33	51.98	47.45	12.64	19.03	5.79	236.61	11.24	62.97	61.06	33.07	56.73	10.20
November	148.16	12.21	51.68	48.35	12.41	18.31	5.66	231.87	11.44	60.26	61.64	33.33	53.94	9.72
December	147.27	11.57	52.45	47.09	12.33	18.37	6.00	238.45	11.76	62.64	63.90	33.27	56.01	9.44

Table 7-10. U.S. Exports of Goods by Selected Regions and Countries

(Census f.a.s. basis; millions of dollars, not seasonally adjusted.)

Year and month	Total, all countries	Selected regions [1]			Selected countries				
		European Union, 15 countries	European Union, 28 countries	OPEC	Brazil	Canada	China	Japan	Mexico
1980	. . .	53 679	. . .	. . .	4 344	40 331	3 755	20 790	15 145
1981	. . .	52 363	. . .	. . .	3 798	44 602	3 603	21 823	17 789
1982	. . .	47 932	. . .	. . .	3 423	37 887	2 912	20 966	11 817
1983	. . .	44 311	. . .	. . .	2 557	43 345	2 173	21 894	9 082
1984	. . .	46 976	. . .	. . .	2 640	51 777	3 004	23 575	11 992
1985	. . .	48 994	. . .	12 478	3 140	47 251	3 856	22 631	13 635
1986	. . .	53 154	. . .	10 844	3 885	45 333	3 106	26 882	12 392
1987	254 122	. . .	. . .	11 057	4 040	59 814	3 497	28 249	14 582
1988	322 426	. . .	. . .	13 994	4 266	71 622	5 022	37 725	20 629
1989	363 812	. . .	. . .	13 196	4 804	78 809	5 755	44 494	24 982
1990	393 592	. . .	. . .	13 703	5 048	83 674	4 806	48 580	28 279
1991	421 730	. . .	. . .	19 054	6 148	85 150	6 278	48 125	33 277
1992	448 165	. . .	. . .	21 960	5 751	90 594	7 419	47 813	40 592
1993	465 090	. . .	. . .	19 500	6 058	100 444	8 763	47 892	41 581
1994	512 625	. . .	. . .	17 868	8 102	114 439	9 282	53 488	50 844
1995	584 740	. . .	. . .	19 533	11 439	127 226	11 754	64 343	46 292
1996	625 073	. . .	. . .	22 275	12 718	134 210	11 993	67 607	56 792
1997	689 180	. . .	143 931	25 525	15 915	151 767	12 862	65 549	71 389
1998	682 139	. . .	151 967	25 154	15 142	156 604	14 241	57 831	78 773
1999	695 797	. . .	154 825	20 166	13 203	166 600	13 111	57 466	86 909
2000	781 918	. . .	168 181	19 078	15 321	178 941	16 185	64 924	111 349
2001	729 101	. . .	161 931	20 052	15 879	163 424	19 182	57 452	101 297
2002	693 101	. . .	146 621	18 812	12 376	160 923	22 128	51 449	97 470
2003	724 771	. . .	155 170	17 279	11 211	169 924	28 368	52 004	97 412
2004	814 875	. . .	171 230	22 256	13 886	189 880	34 428	53 569	110 731
2005	901 082	. . .	185 166	31 641	15 372	211 899	41 192	54 681	120 248
2006	1 025 967	. . .	211 887	38 658	18 887	230 656	53 673	58 459	133 722
2007	1 148 199	. . .	244 166	48 485	24 172	248 888	62 937	61 160	135 918
2008	1 287 442	. . .	271 810	64 880	32 299	261 150	69 733	65 142	151 220
2009	1 056 043	. . .	220 599	49 857	26 095	204 658	69 497	51 134	128 892
2010	1 278 495	. . .	239 591	54 226	35 418	249 256	91 911	60 472	163 665
2011	1 482 508	. . .	269 069	64 809	43 019	281 292	104 122	65 800	198 289
2012	1 545 821	. . .	265 373	81 727	43 771	292 651	110 517	69 976	215 875
2013	1 578 517	. . .	262 095	84 730	44 106	300 755	121 746	65 237	225 954
2014	1 621 874	. . .	276 274	82 533	42 432	312 817	123 657	66 892	241 007
2015	1 503 328	. . .	271 911	72 878	31 641	280 855	115 873	62 388	236 460
2016	1 451 460	. . .	269 650	71 053	30 193	266 734	115 595	63 247	230 229
2017	1 546 473	. . .	283 256	59 336	37 331	282 473	129 798	67 585	243 508
2018	1 665 992	. . .	318 376	59 348	39 560	299 769	120 148	75 229	265 443
2016									
January	116 617	. . .	20 375	5 174	2 074	19 700	8 209	4 656	18 045
February	118 553	. . .	22 548	6 672	1 985	20 904	8 081	4 976	18 131
March	117 548	. . .	24 532	7 023	2 329	23 253	8 926	5 373	19 295
April	119 275	. . .	22 988	5 751	2 252	23 350	8 680	4 690	19 362
May	119 858	. . .	22 046	5 594	2 625	23 075	8 542	5 225	18 991
June	120 298	. . .	23 874	5 507	2 471	24 254	8 846	5 164	19 456
July	121 462	. . .	21 438	5 519	2 604	20 898	9 130	5 168	18 347
August	123 298	. . .	21 837	6 052	2 967	23 265	9 373	5 464	20 034
September	124 033	. . .	23 208	5 721	2 628	22 661	9 521	5 617	19 720
October	123 056	. . .	22 986	5 230	2 654	22 601	12 600	5 793	20 209
November	121 174	. . .	20 793	5 620	3 045	21 668	12 044	5 470	19 567
December	126 287	. . .	23 026	7 188	2 559	21 104	11 645	5 653	19 071
2017									
January	126 721	. . .	21 220	4 043	2 695	20 870	9 956	4 986	19 641
February	126 721	. . .	22 889	4 515	2 642	21 144	9 740	5 283	18 218
March	126 420	. . .	25 794	5 580	2 831	24 964	9 720	5 804	21 035
April	125 882	. . .	22 878	4 886	2 828	22 737	9 807	5 952	18 933
May	126 379	. . .	23 759	5 102	3 243	24 904	9 880	5 144	19 907
June	127 845	. . .	23 657	5 216	3 035	25 172	9 719	5 576	21 400
July	127 663	. . .	21 469	4 943	3 166	21 795	9 955	5 737	19 841
August	128 030	. . .	23 382	4 267	3 268	24 521	10 825	5 390	20 879
September	129 794	. . .	24 277	5 442	3 297	24 096	10 896	5 949	20 161
October	130 485	. . .	25 612	4 562	3 835	23 908	12 964	5 603	22 120
November	133 480	. . .	23 541	4 879	3 119	25 108	12 707	5 832	21 734
December	137 052	. . .	24 780	5 901	3 371	23 255	13 630	6 330	19 639
2018									
January	132 872	. . .	23 377	3 538	3 043	22 636	9 903	5 688	21 419
February	135 644	. . .	24 912	4 191	2 642	23 668	9 760	5 389	20 370
March	140 069	. . .	30 013	5 250	3 281	27 263	12 652	6 495	21 971
April	139 649	. . .	26 744	4 493	3 212	25 820	10 504	5 838	22 507
May	143 753	. . .	27 970	5 771	3 242	27 062	10 428	6 123	23 005
June	141 502	. . .	28 123	5 177	3 328	26 353	10 860	6 386	22 203
July	139 238	. . .	23 861	4 174	3 381	23 712	10 135	6 379	22 847
August	138 176	. . .	25 604	5 251	3 665	25 588	9 286	6 339	22 492
September	140 073	. . .	27 018	4 973	3 354	24 576	9 730	6 429	21 544
October	140 562	. . .	28 042	5 147	3 940	26 197	9 140	6 534	24 647
November	138 452	. . .	26 878	5 829	2 979	24 549	8 606	6 832	22 959
December	136 001	. . .	25 833	5 554	3 225	22 345	9 145	6 797	19 479

[1]See notes and definitions for definitions of regions.
. . . = Not available.

Table 7-11. U.S. Imports of Goods by Selected Regions and Countries

(Census Customs basis; millions of dollars, not seasonally adjusted.)

| Year and month | Total, all countries | Selected regions [1] | | | Selected countries | | | | |
		European Union, 15 countries	European Union, 28 countries	OPEC	Brazil	Canada	China	Japan	Mexico
1980	...	35 958	...	...	...	...	...	...	...
1981	...	41 624	...	...	...	...	...	...	...
1982	...	42 509	...	...	...	...	...	...	...
1983	...	43 892	...	...	...	...	...	...	...
1984	...	57 360	...	...	...	...	...	...	...
1985	...	67 822	...	22 801	7 526	69 006	3 862	68 783	19 132
1986	...	75 736	...	19 751	6 813	68 253	4 771	81 911	17 302
1987	406 241	81 188	...	23 952	7 865	71 085	6 294	84 575	20 271
1988	440 952	84 939	...	22 962	9 294	81 398	8 511	89 519	23 260
1989	473 211	85 153	...	30 611	8 410	87 953	11 990	93 553	27 162
1990	495 311	91 868	...	38 052	7 898	91 380	15 237	89 684	30 157
1991	488 453	86 481	...	32 644	6 717	91 064	18 969	91 511	31 130
1992	532 662	93 993	...	33 200	7 609	98 630	25 728	97 414	35 211
1993	580 656	97 941	...	31 739	7 479	111 216	31 540	107 246	39 918
1994	663 252	110 875	...	31 685	8 683	128 406	38 787	119 156	49 494
1995	743 545	131 871	...	35 606	8 833	144 370	45 543	123 479	62 100
1996	795 287	142 947	...	44 285	8 773	155 893	51 513	115 187	74 297
1997	869 703	157 528	160 896	44 025	9 625	167 234	62 558	121 663	85 938
1998	911 897	176 380	180 550	33 925	10 102	173 256	71 169	121 845	94 629
1999	1 024 616	195 227	200 053	41 977	11 314	198 711	81 788	130 864	109 721
2000	1 218 023	220 019	226 901	67 090	13 853	230 838	100 018	146 479	135 926
2001	1 140 998	220 057	226 568	59 754	14 466	216 268	102 278	126 473	131 338
2002	1 161 366	225 771	232 313	53 245	15 781	209 088	125 193	121 429	134 616
2003	1 257 121	244 826	253 042	68 344	17 910	221 595	152 436	118 037	138 060
2004	1 469 705	272 439	281 959	92 099	21 160	256 360	196 682	129 805	155 902
2005	1 673 454	298 879	309 628	127 169	24 436	290 384	243 470	138 004	170 109
2006	1 853 938	319 590	330 482	150 758	26 367	302 438	287 774	148 181	198 253
2007	1 956 961	...	354 409	165 651	25 644	317 057	321 443	145 463	210 714
2008	2 103 641	...	367 617	242 579	30 453	339 491	337 773	139 262	215 942
2009	1 559 625	...	281 801	111 602	20 070	226 248	296 374	95 804	176 654
2010	1 913 857	...	319 264	149 893	23 958	277 637	364 953	120 552	229 986
2011	2 207 954	...	368 464	191 470	31 737	315 325	399 371	128 928	262 874
2012	2 276 302	...	381 755	180 752	32 123	324 263	425 619	146 432	277 594
2013	2 267 987	...	387 510	152 688	27 541	332 504	440 430	138 575	280 556
2014	2 385 489	...	420 609	132 370	30 021	349 286	468 475	134 505	295 730
2015	2 272 868	...	427 810	66 241	27 474	296 305	483 202	131 445	296 433
2016	2 187 805	...	416 331	77 551	26 043	277 720	462 420	132 000	293 501
2017	2 341 963	...	434 902	72 168	29 450	299 090	505 220	136 418	312 809
2018	2 540 806	...	487 037	80 215	31 104	318 824	539 676	142 425	346 101
2016									
January	177 648	...	29 271	5 582	1 682	22 268	37 126	9 586	22 338
February	181 907	...	32 396	4 951	1 651	21 813	36 067	10 307	23 122
March	174 726	...	37 564	6 161	2 241	23 179	29 812	12 084	24 692
April	177 414	...	34 996	5 462	1 871	22 220	32 920	10 855	24 986
May	179 920	...	35 520	6 271	2 050	23 022	37 514	9 872	24 761
June	184 253	...	36 692	7 018	2 197	24 231	38 539	10 994	24 810
July	183 229	...	33 882	7 126	2 109	21 609	39 439	11 276	22 945
August	184 165	...	35 725	7 134	2 879	24 217	43 222	11 428	25 545
September	183 339	...	33 482	6 781	2 421	23 749	42 021	10 391	25 122
October	184 019	...	36 041	6 697	2 329	23 848	43 798	11 693	26 359
November	187 014	...	35 583	7 316	2 150	24 410	42 603	11 331	25 358
December	189 151	...	35 179	7 051	2 463	23 153	39 359	12 184	23 464
2017									
January	192 880	...	32 653	6 834	2 244	24 391	41 339	10 493	23 461
February	189 995	...	32 382	6 037	2 009	23 386	32 788	9 869	23 810
March	190 904	...	36 890	6 999	2 403	25 999	34 163	13 019	27 925
April	192 994	...	35 427	5 963	2 334	24 155	37 443	11 201	25 021
May	192 259	...	36 516	6 486	2 464	26 289	41 761	10 951	27 037
June	192 963	...	36 216	6 170	2 612	25 893	42 261	11 187	27 158
July	191 873	...	34 949	5 400	2 473	22 803	43 566	11 456	24 519
August	192 127	...	35 780	5 729	2 892	24 907	45 788	11 958	26 931
September	194 111	...	35 639	4 778	2 396	24 212	45 409	10 821	25 737
October	196 928	...	39 409	6 005	2 491	25 655	48 138	12 021	28 578
November	203 973	...	38 358	6 139	2 794	26 132	48 108	11 514	27 488
December	208 878	...	40 683	5 628	2 339	25 268	44 456	11 928	25 144
2018									
January	206 776	...	36 868	6 480	2 767	26 201	45 766	11 282	25 930
February	210 658	...	36 939	5 577	1 984	24 013	39 021	10 811	25 927
March	209 153	...	41 828	6 050	2 353	26 898	38 328	12 852	29 958
April	208 954	...	41 438	7 689	2 524	26 526	38 304	12 062	28 174
May	209 384	...	41 066	6 494	2 338	28 223	43 966	11 661	29 432
June	209 964	...	40 038	6 884	2 641	28 094	44 612	11 506	29 558
July	212 295	...	41 529	7 659	2 742	26 835	47 121	11 834	28 214
August	213 447	...	41 246	6 558	3 251	28 014	47 869	12 338	31 099
September	216 309	...	37 680	6 854	2 621	26 315	50 015	10 315	29 170
October	216 710	...	45 392	7 536	3 105	28 058	52 202	12 790	31 861
November	211 365	...	42 043	6 604	2 339	25 691	46 501	12 639	29 661
December	215 791	...	40 971	5 829	2 440	23 956	45 972	12 334	27 116

[1]See notes and definitions for definitions of regions.
... = Not available.

Table 7-12A. U.S. Exports of Services: Recent Data

(Balance of payments basis, millions of dollars, seasonally adjusted.)

Year and month	Total services	Maintenance and repair services, n.i.e.	Transport	Travel for all purposes including education	Insurance services	Financial services	Charges for the use of intellectual property n.i.e	Telecommunications, computer and information services	Other business services	Government goods and services, n.i.e.
1999	271 343	3 812	43 218	92 338	3 052	19 433	47 731	12 287	40 976	8 495
2000	290 381	4 686	45 758	100 187	3 631	22 117	51 808	12 215	40 497	9 481
2001	274 323	5 575	41 716	86 733	3 424	21 899	49 489	12 829	44 146	8 514
2002	280 670	5 769	41 912	81 869	4 415	24 496	53 859	12 451	47 996	7 903
2003	289 972	5 458	41 446	80 332	5 974	27 840	56 813	14 061	48 775	9 274
2004	337 966	5 342	47 723	92 387	7 314	36 389	67 094	14 962	54 398	12 357
2005	373 006	7 218	52 622	101 470	7 566	39 878	74 448	15 515	58 302	15 989
2006	416 738	7 673	57 462	105 140	9 445	47 882	83 549	17 184	68 619	19 783
2007	488 396	9 062	65 824	119 037	10 841	61 376	97 803	20 192	82 382	21 879
2008	532 817	10 019	74 973	133 761	13 403	63 027	102 125	23 119	92 738	19 652
2009	512 722	12 077	62 189	119 902	14 586	64 437	98 406	23 816	95 984	21 324
2010	563 333	13 860	71 656	137 010	14 397	72 348	107 521	25 038	101 029	20 474
2011	627 781	14 279	79 830	150 867	15 114	78 271	123 333	29 171	112 568	24 348
2012	654 850	15 115	83 592	161 249	16 534	76 605	125 492	32 103	119 892	24 267
2013	701 455	18 568	86 776	177 484	16 696	95 131	128 034	34 419	121 530	22 816
2014	743 257	22 132	90 701	191 325	17 312	107 712	129 890	35 044	128 817	20 325
2015	753 150	23 406	87 609	205 418	16 229	102 595	124 442	35 664	136 622	21 165
2016	758 446	25 132	84 749	206 650	16 819	99 074	124 387	38 245	144 614	18 777
2017	798 957	26 880	88 836	210 655	18 015	109 203	126 523	42 001	157 190	19 653
2018	826 980	30 968	92 852	214 680	17 466	112 015	128 748	43 196	165 821	21 235
2016										
January	62 053	2 074	7 115	17 361	1 354	7 844	10 103	3 168	11 591	1 443
February	61 587	2 051	6 897	17 207	1 342	7 817	10 090	3 168	11 578	1 436
March	61 891	1 981	6 973	17 188	1 343	7 859	10 184	3 178	11 695	1 490
April	62 391	1 992	6 978	16 976	1 357	7 959	10 385	3 198	11 942	1 604
May	63 017	2 048	7 006	16 938	1 375	8 118	10 518	3 202	12 123	1 690
June	63 683	2 150	7 061	17 000	1 396	8 320	10 581	3 188	12 237	1 750
July	63 985	2 192	6 994	17 336	1 421	8 235	10 583	3 158	12 284	1 782
August	64 186	2 143	7 146	17 369	1 439	8 380	10 542	3 148	12 279	1 741
September	64 169	2 224	7 080	17 456	1 450	8 489	10 463	3 157	12 222	1 627
October	63 528	2 104	7 035	17 302	1 454	8 542	10 350	3 187	12 114	1 441
November	63 756	2 055	7 136	17 276	1 450	8 788	10 293	3 224	12 167	1 367
December	64 198	2 118	7 328	17 240	1 437	8 722	10 296	3 269	12 382	1 407
2017										
January	64 890	2 303	7 249	17 255	1 415	8 695	10 353	3 321	12 758	1 541
February	65 256	2 351	7 202	17 259	1 410	8 691	10 403	3 359	12 954	1 629
March	65 280	2 369	7 148	17 027	1 423	8 872	10 440	3 382	12 968	1 652
April	65 149	2 334	7 127	17 293	1 453	8 676	10 465	3 390	12 801	1 610
May	65 336	2 260	7 180	17 263	1 487	8 915	10 473	3 412	12 747	1 598
June	65 884	2 187	7 354	17 375	1 524	9 105	10 465	3 449	12 807	1 617
July	66 519	2 103	7 564	17 614	1 565	9 086	10 441	3 500	12 981	1 666
August	66 948	2 123	7 363	17 681	1 583	9 339	10 467	3 554	13 148	1 690
September	67 883	2 133	7 687	17 944	1 580	9 391	10 542	3 610	13 307	1 689
October	68 215	2 194	7 625	17 847	1 555	9 527	10 677	3 668	13 459	1 664
November	68 667	2 219	7 636	18 016	1 526	9 531	10 819	3 688	13 582	1 650
December	68 930	2 304	7 703	18 080	1 495	9 376	10 978	3 669	13 677	1 647
2018										
January	68 961	2 356	7 620	18 042	1 460	9 326	11 147	3 611	13 744	1 655
February	69 231	2 426	7 818	18 135	1 432	9 270	11 178	3 564	13 720	1 688
March	69 195	2 522	7 830	18 229	1 412	9 263	11 061	3 527	13 604	1 746
April	68 519	2 529	7 814	17 922	1 398	9 330	10 797	3 502	13 398	1 829
May	68 789	2 539	7 920	18 077	1 406	9 463	10 638	3 495	13 388	1 862
June	68 796	2 513	7 870	17 965	1 435	9 505	10 582	3 508	13 573	1 844
July	68 791	2 558	7 745	17 748	1 485	9 354	10 632	3 539	13 955	1 777
August	68 901	2 641	7 585	17 654	1 510	9 382	10 640	3 582	14 164	1 742
September	69 002	2 658	7 767	17 611	1 511	9 267	10 607	3 639	14 201	1 741
October	68 865	2 711	7 686	17 633	1 488	9 264	10 537	3 708	14 066	1 772
November	68 852	2 731	7 632	17 731	1 471	9 264	10 482	3 752	14 001	1 789
December	69 080	2 784	7 565	17 932	1 458	9 327	10 448	3 770	14 007	1 789

n.i.e = Not included elsewhere.

Table 7-12B. U.S. Exports of Services: Historical Data

(Balance of payments basis, millions of dollars, seasonally adjusted.)

Year	Total	Travel	Passenger fares	Other transportation	Royalties and license fees	Other private services (financial, professional, etc.)	Transfers under U.S. military sales contracts [1]	U.S. government miscellaneous services
1960	6 290	919	175	1 607	837	570	2 030	153
1961	6 295	947	183	1 620	906	607	1 867	164
1962	6 941	957	191	1 764	1 056	585	2 193	195
1963	7 348	1 015	205	1 898	1 162	613	2 219	236
1964	7 840	1 207	241	2 076	1 314	651	2 086	265
1965	8 824	1 380	271	2 175	1 534	714	2 465	285
1966	9 616	1 590	317	2 333	1 516	814	2 721	326
1967	10 667	1 646	371	2 426	1 747	951	3 191	336
1968	11 917	1 775	411	2 548	1 867	1 024	3 939	353
1969	12 806	2 043	450	2 652	2 019	1 160	4 138	343
1970	14 171	2 331	544	3 125	2 331	1 294	4 214	332
1971	16 358	2 534	615	3 299	2 545	1 546	5 472	347
1972	17 841	2 817	699	3 579	2 770	1 764	5 856	357
1973	19 832	3 412	975	4 465	3 225	1 985	5 369	401
1974	22 591	4 032	1 104	5 697	3 821	2 321	5 197	419
1975	25 497	4 697	1 039	5 840	4 300	2 920	6 256	446
1976	27 971	5 742	1 229	6 747	4 353	3 584	5 826	489
1977	31 485	6 150	1 366	7 090	4 920	3 848	7 554	557
1978	36 353	7 183	1 603	8 136	5 885	4 717	8 209	620
1979	39 692	8 441	2 156	9 971	6 184	5 439	6 981	520
1980	47 584	10 588	2 591	11 618	7 085	6 276	9 029	398
1981	57 354	12 913	3 111	12 560	7 284	10 250	10 720	517
1982	64 079	12 393	3 174	12 317	5 603	17 444	12 572	576
1983	64 307	10 947	3 610	12 590	5 778	18 192	12 524	666
1004	71 168	17 177	4 067	13 809	6 177	19 255	9 969	714
1985	73 155	17 762	4 411	14 674	6 678	20 035	8 718	878
1986	86 689	20 385	5 582	15 438	8 113	28 027	8 549	595
1987	98 661	23 563	7 003	17 027	10 174	29 263	11 106	526
1988	110 919	29 434	8 976	19 311	12 139	31 111	9 284	664
1989	127 087	36 205	10 657	20 526	13 818	36 729	8 564	587
1990	147 832	43 007	15 298	22 042	16 634	40 251	9 932	668
1991	164 261	48 385	15 854	22 631	17 819	47 748	11 135	690
1992	177 252	54 742	16 618	21 531	20 841	50 292	12 387	841
1993	185 920	57 875	16 528	21 958	21 695	53 510	13 471	883
1994	200 395	58 417	16 997	23 754	26 712	60 841	12 787	887
1995	219 183	63 395	18 909	26 081	30 289	65 048	14 643	818
1996	239 489	69 809	20 422	26 074	32 470	73 340	16 446	928
1997	256 087	73 426	20 868	27 006	33 228	83 929	16 675	955
1998	262 758	71 325	20 098	25 604	35 626	91 774	17 405	926
1999	268 790	75 161	19 425	23 792	47 731	96 812	5 211	657

Note: The category "Royalties and license fees" is the earlier terminology for the category called "Charges for the use of intellectual property n.i.e" in the revised and updated data shown in Tables 7-15A/7-16A. Other categories differ in both title and content; see the Notes and Definitions.

[1]Contains goods that cannot be separately identified.

Table 7-13A. U.S. Imports of Services: Recent Data

(Balance of payments basis, millions of dollars, seasonally adjusted.)

Year and month	Total services	Maintenance and repair services, n.i.e.	Transport	Travel for all purposes including education	Insurance services	Financial services	Charges for the use of intellectual property n.i.e	Telecommunications, computer and information services	Other business services	Government goods and services, n.i.e.
1999	192 893	1 278	49 620	59 592	9 389	8 280	13 302	13 332	23 887	14 212
2000	216 115	2 569	57 606	65 787	11 284	10 936	16 606	12 397	24 414	14 516
2001	213 465	1 999	53 840	60 730	16 706	10 157	16 661	12 421	25 629	15 322
2002	224 379	2 217	51 491	59 942	21 927	8 963	19 493	11 721	29 274	19 353
2003	242 219	2 246	57 863	61 884	25 233	8 948	19 259	13 063	30 103	23 619
2004	283 083	2 395	69 158	74 024	29 089	11 156	23 691	14 210	33 065	26 296
2005	304 448	3 015	75 643	79 988	28 710	12 126	25 577	15 975	35 960	27 454
2006	341 165	4 583	77 962	84 206	39 382	14 733	25 038	19 776	48 130	27 353
2007	372 575	5 209	79 326	89 235	47 517	19 197	26 479	22 384	54 968	28 260
2008	409 052	5 742	83 988	92 545	58 913	17 218	29 623	24 655	67 488	28 880
2009	386 801	5 938	64 133	81 421	63 801	14 415	31 297	25 784	68 553	31 460
2010	409 313	6 909	74 628	86 623	61 478	15 502	32 551	29 015	70 646	31 960
2011	435 761	8 236	81 377	89 700	55 654	17 368	36 087	32 756	83 289	31 293
2012	450 360	7 970	85 029	100 317	53 203	16 975	39 502	32 156	87 347	27 861
2013	461 087	7 420	90 634	98 120	53 420	21 545	38 860	35 034	90 714	25 341
2014	481 264	7 521	94 160	105 529	51 824	24 906	42 208	36 313	94 568	24 236
2015	491 740	9 010	97 061	114 723	47 822	25 740	39 858	36 270	99 665	21 592
2016	511 627	8 764	96 982	123 549	50 144	25 710	46 987	37 418	100 570	21 503
2017	543 880	8 400	101 756	134 868	50 599	28 957	53 440	39 628	104 185	22 047
2018	567 322	8 718	108 202	144 463	42 485	31 298	56 117	41 190	111 874	22 975
2016										
January	41 747	704	7 875	10 107	4 033	2 102	3 565	3 129	8 497	1 737
February	42 150	699	8 070	10 222	4 064	2 128	3 572	3 147	8 502	1 746
March	41 898	690	7 800	10 185	4 069	2 147	3 611	3 121	8 508	1 767
April	42 008	687	8 085	10 025	4 049	2 115	3 684	3 052	8 513	1 798
May	42 040	685	8 070	10 088	4 052	2 082	3 746	3 025	8 478	1 814
June	42 125	712	8 017	10 143	4 078	2 121	3 796	3 042	8 401	1 815
July	42 261	728	8 032	10 228	4 126	2 126	3 835	3 102	8 284	1 801
August	43 810	761	8 141	10 292	4 197	2 150	5 111	3 139	8 226	1 794
September	42 844	776	8 153	10 338	4 292	2 179	3 931	3 154	8 226	1 794
October	43 513	773	8 280	10 662	4 410	2 166	3 988	3 146	8 285	1 802
November	43 492	782	8 170	10 569	4 428	2 204	4 045	3 162	8 320	1 812
December	43 738	766	8 289	10 691	4 347	2 189	4 102	3 201	8 329	1 823
2017										
January	44 021	735	8 428	10 909	4 166	2 209	4 159	3 265	8 314	1 836
February	44 034	706	8 374	10 940	4 084	2 251	4 216	3 297	8 327	1 841
March	44 226	711	8 362	10 943	4 100	2 336	4 270	3 298	8 368	1 838
April	44 404	710	8 360	10 983	4 215	2 281	4 323	3 267	8 439	1 826
May	45 093	718	8 495	11 223	4 295	2 401	4 357	3 253	8 528	1 823
June	45 324	707	8 488	11 240	4 342	2 458	4 370	3 255	8 635	1 829
July	45 492	689	8 519	11 242	4 354	2 446	4 364	3 273	8 761	1 844
August	45 548	661	8 326	11 285	4 351	2 501	4 430	3 295	8 851	1 849
September	46 148	692	8 590	11 351	4 333	2 542	4 569	3 321	8 903	1 846
October	46 431	685	8 579	11 462	4 300	2 521	4 782	3 350	8 919	1 834
November	46 531	702	8 541	11 558	4 157	2 525	4 844	3 371	8 998	1 834
December	46 628	683	8 693	11 733	3 905	2 486	4 756	3 384	9 142	1 845
2018										
January	46 216	689	8 607	11 683	3 543	2 570	4 517	3 387	9 350	1 869
February	47 180	690	8 752	11 792	3 346	2 557	5 342	3 381	9 432	1 887
March	46 383	691	8 748	12 049	3 315	2 589	4 340	3 364	9 389	1 898
April	46 445	681	8 935	11 820	3 449	2 622	4 479	3 337	9 221	1 903
May	46 471	687	8 927	11 796	3 515	2 616	4 511	3 334	9 174	1 910
June	46 792	707	8 888	11 810	3 515	2 634	4 713	3 356	9 249	1 921
July	47 097	742	8 922	11 898	3 447	2 581	4 725	3 403	9 445	1 934
August	47 294	768	8 984	12 065	3 453	2 612	4 507	3 448	9 517	1 939
September	47 826	776	9 319	12 185	3 533	2 583	4 539	3 491	9 463	1 936
October	48 254	772	9 328	12 250	3 685	2 670	4 806	3 533	9 284	1 925
November	48 389	733	9 237	12 470	3 802	2 654	4 818	3 566	9 185	1 923
December	48 977	782	9 555	12 643	3 882	2 608	4 821	3 591	9 166	1 931

n.i.e = Not included elsewhere.

Table 7-13B. U.S. Imports of Services: Historical Data

(Balance of payments basis, millions of dollars, seasonally adjusted.)

Year	Total	Travel	Passenger fares	Other transportation	Royalties and license fees	Other private services (financial, professional, etc.)	Direct defense expenditures [1]	U.S. government miscellaneous services
1960	7 674	1 750	513	1 402	74	593	3 087	254
1961	7 671	1 785	506	1 437	89	588	2 998	268
1962	8 092	1 939	567	1 558	100	528	3 105	296
1963	8 362	2 114	612	1 701	112	493	2 961	370
1964	8 619	2 211	642	1 817	127	527	2 880	415
1965	9 111	2 438	717	1 951	135	461	2 952	457
1966	10 494	2 657	753	2 161	140	506	3 764	513
1967	11 863	3 207	829	2 157	166	565	4 378	561
1968	12 302	3 030	885	2 367	186	668	4 535	631
1969	13 322	3 373	1 080	2 455	221	751	4 856	586
1970	14 520	3 980	1 215	2 843	224	827	4 855	576
1971	15 400	4 373	1 290	3 130	241	956	4 819	592
1972	16 868	5 042	1 596	3 520	294	1 043	4 784	589
1973	18 843	5 526	1 790	4 694	385	1 180	4 629	640
1974	21 379	5 980	2 095	5 942	346	1 262	5 032	722
1975	21 996	6 417	2 263	5 708	472	1 551	4 795	789
1976	24 570	6 856	2 568	6 852	482	2 006	4 895	911
1977	27 640	7 451	2 748	7 972	504	2 190	5 823	951
1978	32 189	8 475	2 896	9 124	671	2 573	7 352	1 099
1979	36 689	9 413	3 184	10 906	831	2 822	8 294	1 239
1980	41 491	10 397	3 607	11 790	724	2 909	10 851	1 214
1981	45 503	11 479	4 487	12 474	650	3 562	11 564	1 287
1982	51 749	12 394	4 772	11 710	795	8 159	12 460	1 460
1983	54 973	13 149	6 003	12 222	943	9 001	13 087	1 568
1984	67 748	22 913	5 735	14 843	1 168	9 040	12 516	1 534
1985	72 862	24 558	6 444	15 643	1 170	10 203	13 108	1 735
1986	80 147	25 913	6 505	17 766	1 401	13 146	13 730	1 686
1987	90 787	29 310	7 283	19 010	1 857	16 485	14 950	1 893
1988	98 526	32 114	7 729	20 891	2 601	17 667	15 604	1 921
1989	102 479	33 416	8 249	22 172	2 528	18 930	15 313	1 871
1990	117 659	37 349	10 531	24 966	3 135	22 229	17 531	1 919
1991	118 459	35 322	10 012	24 975	4 035	25 590	16 409	2 116
1992	119 566	38 552	10 603	23 767	5 161	25 386	13 835	2 263
1993	123 779	40 713	11 410	24 524	5 032	27 760	12 086	2 255
1994	133 057	43 782	13 062	26 019	5 852	31 565	10 217	2 560
1995	141 397	44 916	14 663	27 034	6 919	35 199	10 043	2 623
1996	152 554	48 078	15 809	27 403	7 837	39 679	11 061	2 687
1997	165 932	52 051	18 138	28 959	9 161	43 154	11 707	2 762
1998	180 677	56 483	19 971	30 363	11 235	47 591	12 185	2 849
1999	195 172	59 332	20 946	31 494	13 302	55 885	11 849	2 364

Note: The category "Royalties and license fees" is the earlier terminology for the category called "Charges for the use of intellectual property n.i.e" in the revised and updated data shown in Tables 7-15A/7-16A. Other categories differ in both title and content; see the Notes and Definitions.

[1]Contains goods that cannot be separately identified.

Table 7-14. U.S. Export and Import Price Indexes by End-Use Category

(2000 = 100, not seasonally adjusted.)

Year and month	Exports			Imports		
	All commodities	Agricultural	Nonagricultural	All commodities	Petroleum [1]	Nonpetroleum
1990	95.5	105.1	94.4	94.0	75.5	97.1
1991	96.3	103.4	95.4	94.2	67.3	98.7
1992	96.3	102.5	95.7	94.9	62.9	100.0
1993	96.9	104.4	96.2	94.6	57.7	100.6
1994	98.9	109.4	98.0	96.2	54.3	103.2
1995	103.9	119.0	102.5	100.6	59.8	107.2
1996	104.5	132.6	101.6	101.6	71.1	106.4
1997	103.1	120.6	101.3	99.1	66.0	104.1
1998	99.7	108.8	98.8	93.1	44.8	100.4
1999	98.4	101.1	98.2	93.9	60.1	99.0
2000	100.0	100.0	100.0	100.0	100.0	100.0
2001	99.2	101.2	99.0	96.5	82.8	98.5
2002	98.2	103.2	97.8	94.1	85.3	96.2
2003	99.7	112.3	98.8	96.9	103.2	97.3
2004	103.6	123.4	102.1	102.3	134.6	99.8
2005	106.9	121.0	105.9	110.0	185.1	102.5
2006	110.7	125.8	109.6	115.4	223.3	104.2
2007	116.1	150.9	113.6	120.2	249.1	107.0
2008	123.1	183.5	118.8	134.1	343.2	112.7
2009	117.4	160.0	114.3	118.6	219.9	108.0
2010	123.1	172.6	119.6	126.8	282.2	111.0
2011	133.0	211.0	127.5	140.6	385.1	115.9
2012	133.5	216.1	127.6	141.0	384.0	116.3
2013	133.0	219.7	126.7	139.5	374.2	115.6
2014	132.3	213.8	126.3	138.0	353.2	115.7
2015	123.9	185.4	119.3	123.9	190.8	112.5
2016	119.9	175.5	115.7	119.8	153.1	110.8
2017	122.8	178.2	118.6	123.3	193.9	112.0
2018	126.9	179.3	122.9	127.1	236.6	113.5
2015						
January	126.1	199.0	120.7	126.0	196.0	114.3
February	125.9	194.6	120.9	125.5	193.3	114.0
March	125.9	191.4	121.1	125.3	197.6	113.5
April	125.1	189.7	120.3	125.1	202.4	113.0
May	125.7	187.3	121.0	126.5	226.1	112.9
June	125.3	184.3	120.9	126.6	229.7	112.8
July	124.8	186.1	120.2	125.4	215.3	112.5
August	123.0	181.5	118.6	123.2	186.2	112.1
September	122.3	178.9	118.0	121.9	168.7	111.9
October	122.0	179.0	117.7	121.5	168.3	111.5
November	121.1	177.3	116.9	120.8	161.5	111.2
December	119.8	175.7	115.6	119.3	144.6	110.8
2016						
January	118.7	173.5	114.6	117.8	119.8	110.7
February	118.2	174.7	113.9	117.2	111.2	110.5
March	118.1	171.0	114.1	117.7	123.2	110.4
April	118.7	172.4	114.6	118.5	136.4	110.4
May	120.0	177.5	115.7	119.9	155.4	110.8
June	120.9	181.9	116.3	120.7	173.1	110.5
July	121.1	181.3	116.6	120.8	167.8	111.0
August	120.1	175.1	115.9	120.5	160.9	111.1
September	120.5	173.3	116.5	120.6	162.3	111.2
October	120.7	174.2	116.7	121.2	174.4	111.1
November	120.8	175.4	116.7	121.1	171.5	111.1
December	121.3	175.1	117.2	121.6	181.6	111.1
2017						
January	121.6	175.8	117.4	122.3	192.0	111.2
February	122.0	178.3	117.7	122.7	193.6	111.5
March	122.1	180.2	117.7	122.5	188.6	111.6
April	122.4	180.3	118.0	122.8	187.2	111.9
May	121.7	177.4	117.5	122.7	185.8	111.9
June	121.6	174.9	117.5	122.4	178.7	112.0
July	122.2	178.1	117.9	122.2	177.6	111.9
August	122.9	178.1	118.7	122.9	185.7	112.1
September	123.9	176.9	119.8	123.9	197.7	112.5
October	124.0	180.4	119.7	124.1	198.6	112.5
November	124.6	179.1	120.5	125.3	217.5	112.7
December	124.7	178.5	120.5	125.5	223.6	112.6
2018						
January	125.6	178.5	121.5	126.5	229.6	113.3
February	125.8	179.5	121.7	126.8	227.1	113.7
March	126.3	185.2	121.8	126.5	221.6	113.7
April	126.9	183.1	122.6	127.1	231.2	113.8
May	127.8	186.0	123.3	128.2	248.4	113.9
June	128.0	184.2	123.7	128.2	255.4	113.5
July	127.4	174.7	123.8	128.1	256.8	113.4
August	127.3	175.4	123.6	127.6	251.5	113.1
September	127.3	173.2	123.8	127.7	251.5	113.2
October	127.9	173.0	124.4	128.3	258.9	113.4
November	126.9	176.0	123.1	126.2	220.2	113.5
December	126.1	182.6	121.7	124.4	186.4	113.7

[1] Petroleum and petroleum products.

NOTES AND DEFINITIONS, CHAPTER 7

This chapter presents data from two different data systems on international flows of goods, services, income payments, and financial transactions as they affect the U.S. economy. Tables 7-1 through 7-5 present data on the value, quantities, and prices of foreign transactions in the national income and product accounts (NIPAs). The Census Bureau source data for exports and imports of goods and services are presented in greater detail in Tables 7-6 through 7-13. Table 7-14 shows selected summary values for export and import price indexes compiled by the Bureau of Labor Statistics (BLS).

Due to a few differences in concept, scope, and definitions, the aggregate values of international transactions in the NIPAs (shown in Tables 7-1 and 7-4) are not exactly equal to the values for similar concepts in the Census values that serve as their sources, shown in Tables 7-6 through 7-14. The principal sources of differences are as follows:

- The NIPAs cover only the 50 states and the District of Columbia. Census data include the U.S. territories and Puerto Rico as part of the U.S. economy.

- Gold is treated differently.

- Services without payment by financial intermediaries except life insurance carriers (imputed interest) is treated differently.

A reconciliation of the two sets of international accounts is published from time to time in the *Survey of Current Business*.

The conventions of presentation no longer differ between the two sets of international accounts. In all the tables that follow, as in their source documents, values of imports of goods and services and of all other transactions that result in a payment to the rest of the world—income payments to foreigners, net transfers to foreigners, and net acquisition of assets from abroad—are presented normally as positive. Where a minus sign does appear, it means that the usual direction of flow is reversed.

The 2019 annual update of the National Income and Product Accounts was released on July 26, 2019. Data for 2014 through 2018 were revised. The reference year remained 2012. The updated statistics largely reflect the incorporation of newly available and revised source data and improvements to existing methodologies.

TABLES 7-1 AND 7-4

Foreign Transactions in the National Income and Product Accounts

SOURCE: U.S. DEPARTMENT OF COMMERCE, BUREAU OF ECONOMIC ANALYSIS

See the notes and definitions to Chapter 1 for an overview of the national income and product accounts (NIPAs).

Definitions

In accordance with the split between current and capital account, there are now two NIPA measures of the balance of international transactions.

The *balance on current account, national income and product accounts* is *current receipts from the rest of the world* minus *current payments to the rest of the world*. A negative value indicates that current payments exceed current receipts.

Net lending or net borrowing (-), national income and product accounts is equal to the balance on current account less capital account transactions with the rest of the world (net). Capital account transactions with the rest of the world (net) is not shown separately in Table 7-1 (for space reasons) but can be calculated from that table as the difference between net lending/borrowing and the current account balance. (A similar measure, a component of the ITAs, is shown explicitly in Table 7-6.) Capital account transactions with the rest of the world are cash or in-kind transfers linked to the acquisition or disposition of an existing, nonproduced, nonfinancial asset. In contrast, the current account is limited to flows associated with current production of goods and services.

Net lending or net borrowing provides an indirect measure of the net acquisition of foreign assets by U.S. residents less the net acquisition of U.S. assets by foreign residents. See Table 7-1 and its notes and definitions for a more extensive discussion of the relationship between the balances on current and capital account and international asset flows.

Current receipts from the rest of the world is *exports of goods and services* plus *income receipts*.

Current payments to the rest of the world is *imports of goods and services* plus *income payments* plus *current taxes and transfer payments (net)*.

Exports and imports of goods and services. Goods, in general, are products that can be stored or inventoried. *Services*, in general, are products that cannot be stored and are consumed at the place and time of their purchase. Goods imports include expenditures abroad by U.S. residents, except for travel. Services include foreign travel by U.S. residents, expenditures in the United States by foreign travelers, and exports and imports of certain goods—primarily military equipment purchased and sold by the federal government. See the following paragraph for the definition of *travel*.

Table 7-4 shows values for selected sectors of total goods and services; the sectors shown will not add to the total because of omitted items. In the case of goods, a miscellaneous "other" category is not shown. In the case of services, only two components are shown in this table. One is *travel* (for all purposes including education), which does not include passenger fares but includes as exports spending by foreign tourists and students in the United States, and includes as imports all other spending abroad by tourists and students from the United States. The other component shown here is a category called *other business services*, which includes the professional and financial services (for example, computer services) that have accounted for a large part of the long-term growth in the service category. The remaining components of total services are transport (of both persons and goods), charges for the use of intellectual property not elsewhere classified, government goods and services not elsewhere classified, and a miscellaneous, smaller *other* category. They are shown separately in Tables 7-12 and 7-13.

Income receipts and payments. Income receipts—receipts from abroad of factor (labor or capital) income by U.S. residents—are analogous to exports and are combined with them to yield total *current receipts from the rest of the world. Income payments* by U.S. entities of factor income to entities abroad are analogous to imports.

Current taxes and transfer payments (net) consists of net payments between the United States and abroad that do not involve payment for the services of the labor or capital factors of production, purchase of currently-produced goods and services, or transfer of an existing asset. It includes net flows from persons, government, and business. The types of payments included are personal remittances from U.S. residents to the rest of the world, net of remittances from foreigners to U.S. residents; government grants; and transfer payments from businesses. Only the net payment to the rest of the world is shown. It usually appears in these accounts as a positive value, with transfers from the United States to abroad exceeding the reverse flow. An exception came in 1991, when U.S. allies in the Gulf War reimbursed the United States for the cost of the war, causing the only current-account surplus since 1981. This resulted in net payments to the United States from the rest of the world and appears as a negative entry in the net transfer payments column of the NIPA accounts.

TABLES 7-2, 7-3 AND 7-5

Chain-Type Quantity and Price Indexes for NIPA Foreign Transactions

These indexes represent the separation of the current-dollar values in Tables 7-1 and 7-4 into their real quantity and price trends components. (See the notes and definitions to Chapter 1 for a general explanation of chained-dollar estimates of real output and prices.) As those notes explain, quantity indexes are shown instead of constant-dollar estimates, because BEA no longer publishes its real output estimates before 1999 in any detail in the constant-dollar form. Therefore, quantity indexes are the only comprehensive source of information about longer-term trends in real volumes.

TABLES 7-6 THROUGH 7-13

Exports and Imports of Goods and Services

SOURCES: U.S. DEPARTMENT OF COMMERCE, CENSUS BUREAU AND BUREAU OF ECONOMIC ANALYSIS

These tables present the source data used to build up the aggregate measures of goods and services flows shown in Tables 7-1 through 7-5. These data are compiled and published monthly, making trends evident before the publication of the quarterly aggregate estimates. They also provide more detail than the quarterly aggregates.

Monthly and annual data on exports and imports of *goods* are compiled by the Census Bureau from documents collected by the U.S. Customs Service. The Bureau of Economic Analysis (BEA) makes certain adjustments to these data (as described below) to place the estimates on a *balance of payments* basis—a basis consistent with the national and international accounts.

Data on exports and imports of *services* are prepared by BEA from a variety of sources. Monthly data on services are available from January 1992. Annual and quarterly data for earlier years are available as part of the international transactions accounts. Current data on goods and services are available each month in a joint Census Bureau-BEA press release.

In the case of some of the detailed breakdowns of exports and imports, such as by end-use categories, monthly data may not sum exactly to annual totals. This is due to later revisions, which are made only to annual data and are not allocated to monthly data.

In addition, monthly and annual data on exports and imports of goods for individual countries and various country groupings do not reflect subsequent revisions of annual total data. These country data are compiled by the Census Bureau for all countries, although this volume includes only a selection of the most significant countries and areas. The full set of data can be accessed on the Census Web site at <http://www.census.gov>.

Definitions: Goods

Goods: Census basis. The Census basis goods data are compiled from documents collected by the U.S. Customs Service. They reflect the movement of goods between foreign countries and the 50 states, the District of Columbia, Puerto Rico, the U.S. Virgin Islands, and U.S. Foreign Trade Zones. They include government and nongovernment shipments of goods, and exclude shipments between the United States and its territories and possessions; transactions with U.S. military, diplomatic, and consular installations abroad; U.S. goods returned to the United States by its armed forces; personal and household effects of travelers; and intransit shipments. The general import values reflect the total arrival of merchandise from foreign countries that immediately enters consumption channels, warehouses, or Foreign Trade Zones.

For *imports*, the value reported is the U.S. Customs Service appraised value of merchandise (generally, the price paid for merchandise for export to the United States). Import duties, freight, insurance, and other charges incurred in bringing merchandise to the United States are excluded.

Exports are valued at the f.a.s. (free alongside ship) value of merchandise at the U.S. port of export, based on the transaction price including inland freight, insurance, and other charges incurred in placing the merchandise alongside the carrier at the U.S. port of exportation.

Goods: balance of payments (BOP) basis. Goods on a Census basis are adjusted by BEA to goods on a BOP basis to bring the data in line with the concepts and definitions used to prepare the international and national accounts. In general, the adjustments include changes in ownership that occur without goods passing into or out of the customs territory of the United States. These adjustments are necessary to supplement coverage of the Census basis data, to eliminate duplication of transactions recorded elsewhere in the international accounts, and to value transactions according to a standard definition. With the June 4, 2013 release, improved BOP adjustments involving U.S. military transactions were introduced for data beginning with 1999.

The *export* adjustments include the following: (1) The deduction of *U.S. military sales contracts.* The Census Bureau has included these contracts in the goods data, but BEA includes them in the service category "Transfers Under U.S. Military Sales Contracts." BEA's source material for these contracts is more comprehensive but does not distinguish between goods and services. (2) The addition of *private gift parcels* mailed to foreigners by individuals through the U.S. Postal Service. Only commercial shipments are covered in Census goods exports. (3) The addition to *nonmonetary gold exports* of gold purchased by foreign official agencies from private dealers in the United States and held at the Federal Reserve Bank of New York. The Census data include only gold that leaves the customs territory. (4) *Smaller adjustments* includes deductions for repairs of goods, exposed motion picture film, and military grant aid, and additions for sales of fish in U.S. territorial waters, exports of electricity to Mexico, and vessels and oil rigs that change ownership without export documents being filed.

The *import* adjustments include the following: (1) On *inland freight in Canada,* the customs value for imports for certain Canadian goods is the point of origin in Canada. BEA makes an addition for the inland freight charges of transporting these Canadian goods to the U.S. border. (2) An addition is made to *nonmonetary gold imports* for gold sold by foreign official agencies to private purchasers out of stock held at the Federal Reserve Bank of New

York. The Census Bureau data include only gold that enters the customs territory. (3) A deduction is made for *imports by U.S. military agencies.* The Census Bureau has included these contracts in the goods data, but BEA includes them in the service category "Direct Defense Expenditures." BEA's source material is more comprehensive but does not distinguish between goods and services. (4) *Smaller adjustments* includes deductions for repairs of goods and for exposed motion picture film and additions for imported electricity from Mexico, conversion of vessels for commercial use, and repairs to U.S. vessels abroad.

Definitions: Services

These statistics are estimates of service transactions between foreign countries and the 50 states, the District of Columbia, Puerto Rico, the U.S. Virgin Islands, and other U.S. territories and possessions. Transactions with U.S. military, diplomatic, and consular installations abroad are excluded because they are considered to be part of the U.S. economy.

In the 2014 comprehensive revision of NIPAs, new categories were introduced in the services reports, beginning with January 1999. The 9 new categories, shown in Tables 7-12A and 7-13A, are as follows:

Maintenance and repair services n.i.e. (not included elsewhere) were previously included in "other private services."

Transport includes the previous categories of "passenger fares" and "other transportation."

Travel for all purposes including education is broader than the previous "travel" category and now includes business travel, expenditures by border, seasonal, and other short-term workers, and personal travel including that related to health and education. As before, it consists of expenditures for food, lodging, recreation, gifts, and other items incidental to a foreign visit but not the actual transportation expenses.

Insurance services and *financial services* were previously included in other private services.

Charges for the use of intellectual property n.i.e. was previously named "royalties and license fees."

Telecommunications, computer, and information services and *other business services* were previously included in "other private services."

Government goods and services n.i.e. replaces the government categories in the previous system.

Earlier services definitions

In Tables 7-12B and 7-13B, showing data before 1999, services are classified in the broad categories described below. For six of these categories, the definitions are the same for imports and exports. For the seventh, the export category is "Transfers under U.S. Military Sales Contracts," while for imports, the category is "Direct Defense Expenditures."

Travel includes purchases of services and goods by U.S. travelers abroad and by foreign visitors to the United States. A traveler is defined as a person who stays for a period of less than one year in a country where the person is not a resident. Included are expenditures for food, lodging, recreation, gifts, and other items incidental to a foreign visit. Not included are the international costs of the travel itself, which are covered in *passenger fares* (see below).

Passenger fares consists of fares paid by residents of one country to residents in other countries. Receipts consist of fares received by U.S. carriers from foreign residents for travel between the United States and foreign countries and between two foreign points. Payments consist of fares paid by U.S. residents to foreign carriers for travel between the United States and foreign countries.

Other transportation includes charges for the transportation of goods by ocean, air, waterway, pipeline, and rail carriers to and from the United States. Included are freight charges, operating expenses that transportation companies incur in foreign ports, and payments for vessel charter and aircraft and freight car rentals.

Royalties and license fees consists of transactions with foreign residents involving intangible assets and proprietary rights, such as the use of patents, techniques, processes, formulas, designs, know-how, trademarks, copyrights, franchises, and manufacturing rights. The term *royalties* generally refers to payments for the utilization of copyrights or trademarks, and the term *license fees* generally refers to payments for the use of patents or industrial processes.

Other private services includes transactions with "affiliated" foreigners for which no identification by type is available and transactions with unaffiliated foreigners.

The term "affiliated" refers to a direct investment relationship, which exists when a U.S. person has ownership or control (directly or indirectly) of 10 percent or more of a foreign business enterprise, or when a foreign person has a similar interest in a U.S. enterprise.

Transactions with "unaffiliated" foreigners in this "other private services" category consist of education services, financial services, insurance services, telecommunications services, and business, professional, and technical services. Included in the last group are advertising services; computer and data processing services; database and other information services; research, development, and testing services; management, consulting, and public relations services; legal services; construction, engineering, architectural, and mining services; industrial engineering services; installation, maintenance, and repair of equipment; and other services, including medical services and film and tape rental.

The insurance component of "other private services" was measured before the July 2003 revision as premiums less actual losses paid or recovered. Furthermore, catastrophic losses were entered immediately when the loss occurred, rather than when the insurance claim was actually paid out. This led to sharp swings for any month in which catastrophic losses occurred, such as Hurricane Katrina in August 2005 or the September 11, 2001, terror attacks. In the accounts as revised in July 2003 and presented here, insurance services are now measured as premiums less "normal" losses. Normal losses consist of a measure of expected regularly occurring losses based on six years of past experience plus an additional allowance for catastrophic loss. Catastrophic losses, when they occur, are added in equal increments to the estimate of regularly occurring losses over the 20 years following the occurrence. As adoption of this methodology introduces a difference between actual and normal losses, an amount equal to the difference is entered in the international accounts as a capital account transaction.

Transfers under U.S. military sales contracts (exports only) includes exports in which U.S. government military agencies participate. This category has included both goods, such as equipment, and services, such as repair services and training, that could not be separately identified. In the 2010 and 2013 annual revisions, more precise breakdowns between goods and services in this category were introduced and incorporated as revised estimates of services and of the BOP goods adjustment, beginning with the data for 2007. Transfers of goods and services under U.S. military grant programs are included.

Direct defense expenditures (imports only) consists of expenditures incurred by U.S. military agencies abroad, including expenditures by U.S. personnel, payments of wages to foreign residents, construction expenditures, payments for foreign contractual services, and procurement of foreign goods. Improved breakdowns between goods and services are now incorporated beginning with statistics for 1999.

U.S. government miscellaneous services includes transactions of U.S. government nonmilitary agencies with foreign residents. Most of these transactions involve the provision of services to, or purchases of services from, foreigners. Transfers of some goods are also included.

Services estimates are based on quarterly, annual, and benchmark surveys and partial information generated from monthly reports. Service transactions are estimated at market prices. Estimates are seasonally adjusted when statistically significant seasonal patterns are present.

Definitions: Area groupings

The Census trade statistics for groups of countries present "groups as they were at the time of reporting", which means that as multinational organizations such as the European Union expand, the statistics for trade with that group include the added country only beginning with the year it entered the group. The European Union was expanded from 15 to 28 nations as of July 1, 2013, and data for that year are available for both the new and the old group; to provide historical perspective, they are both shown here, along with the historical data back to 1974 for the original group.

The *European Union* (EU) now includes 28 countries: Austria, Belgium, Bulgaria, Croatia, Cyprus, Czech Republic, Denmark, Estonia, Finland, France, Germany, Greece, Hungary, Ireland, Italy, Latvia, Lithuania, Luxembourg, Malta, Netherlands, Poland, Portugal, Romania, Slovakia, Slovenia, Spain, Sweden, and the United Kingdom. The last country to join was Croatia in July 2013.

In a referendum on June 23, 2016, a majority of British voters supported leaving the EU. On March 29, 2017, the United Kingdom government invoked Article 50 of the Treaty on European Union. The European Union (Withdrawal) Act 2018 declares "exit day" was to be March 29, 2019. However, "Brexit" day has changed over the last two years. As of the date of this publication, the United Kingdom new government has set a deadline of October 31 for either a negotiated exit plan or an exit without negotiated plans.

The *Euro area* now includes Austria, Belgium, Cyprus, Estonia, Finland, France, Germany, Greece, Ireland, Italy, Luxembourg, Malta, the Netherlands, Portugal, Slovakia, Slovenia, and Spain. See the notes and definitions in Table 13-2 for further information about the euro.

With the June 3, 2016 International Trade and Services Annual Revision for 2015, the area grouping *Asian Newly Industrialized Countries (NICS)* has been removed from all relevant exhibits and publications and are listed separately as Hong Kong, South Korea, Singapore, and Taiwan. This group was initially called the *Four Asian Tigers.*

The *Organization of Petroleum Exporting Countries (OPEC)* currently consists of Algeria, Angola, Ecuador, Equatorial Guinea, Gabon, Iran, Iraq, Kuwait, Libya, Nigeria, Qatar, Republic of Congo, Saudi Arabia, the United Arab Emirates, and Venezuela.

Notes on the data

U.S./Canada data exchange and substitution. The data for U.S. exports to Canada are derived from import data compiled by Canada. The use of Canada's import data to produce U.S. export data requires several alignments in order to compare the two series.

- *Coverage*: Canadian imports are based on country of origin. U.S. goods shipped from a third country are included, but U.S. exports exclude these foreign shipments. U.S. export coverage also excludes certain Canadian postal shipments.

- *Valuation*: Canadian imports are valued at their point of origin in the United States. However, U.S. exports are valued at the

port of exit in the United States and include inland freight charges, making the U.S. export value slightly larger. Canada requires inland freight to be reported.

- *Re-exports*: U.S. exports include re-exports of foreign goods. Again, the aggregate U.S. export figure is slightly larger.

- *Exchange Rate*: Average monthly exchange rates are applied to convert the published data to U.S. currency.

End-use categories and seasonal adjustment of trade in goods. Goods are initially classified under the Harmonized System, which describes and measures the characteristics of goods traded. Combining trade into approximately 140 export and 140 import enduse categories makes it possible to examine goods according to their principal uses. These categories are used as the basis for computing the seasonal and working-day adjusted data. Adjusted data are then summed to the six end-use aggregates for publication.

The seasonal adjustment procedure is based on a model that estimates the monthly movements as percentages above or below the general level of each end-use commodity series (unlike other methods that redistribute the actual series values over the calendar year). Imports of petroleum and petroleum products are adjusted for the length of the month.

Data availability

Data are released monthly in a joint Census Bureau-BEA press release (FT-900), which is published about six weeks after the end of the month to which the data pertain. The release and historical data are available on the Census Bureau Web site at <https://www.census.gov/foreign-trade/index.html >.

Revisions

Data for recent years are normally revised annually. In some cases, revisions to annual totals are not distributed to monthly data; therefore, monthly data may not sum to the revised total shown. Data on trade in services may be subject to extensive revision as part of BEA's annual revision of the international transactions accounts (ITAs), usually released in July.

References

See the references for Table 7-6.

TABLE 7-14

Export and Import Price Indexes

SOURCE: U.S. DEPARTMENT OF LABOR, BUREAU OF LABOR STATISTICS

The International Price Program of the Bureau of Labor Statistics (BLS) collects price data for nonmilitary goods traded between the United States and the rest of the world and for selected transportation services in international markets. BLS aggregates the goods price data into export and import price indexes. Summary values of these price indexes for goods are presented in *Business Statistics.* For product and locality detail on international prices for both goods and services, see the *Handbook of U.S. Labor Statistics,* also published by Bernan Press.

Definitions

The *export* price index provides a measure of price change for all goods sold by U.S. residents (businesses and individuals located within the geographic boundaries of the United States, whether or not owned by U.S. citizens) to foreign buyers.

The *import* price index provides a measure of price change for goods purchased from other countries by U.S. residents.

Notes on the data

Published index series use a base year of 2000 = 100.

The product universe for both the import and export indexes includes raw materials, agricultural products, and manufactures. Price data are primarily collected by mail questionnaire, and directly from the exporter or importer in all but a few cases.

To the greatest extent possible, the data refer to prices at the U.S. border for exports and at either the foreign border or the U.S. border for imports. For nearly all products, the prices refer to transactions completed during the first week of the month and represent the actual price for which the product was bought or sold, including discounts, allowances, and rebates.

For the export price indexes, the preferred pricing basis is f.a.s. (free alongside ship) U.S. port of exportation. Where necessary, adjustments are made to reported prices to place them on this basis. An attempt is made to collect two prices for imports: f.o.b.

(free on board) at the port of exportation and c.i.f. (cost, insurance, and freight) at the U.S. port of importation. Adjustments are made to account for changes in product characteristics in order to obtain a pure measure of price change.

The indexes are weighted indexes of the Laspeyres type. (See "General Notes" at the beginning of this volume for further explanation.) The values assigned to each weight category are based on trade value figures compiled by the Census Bureau. They are reweighted annually, with a two-year lag (as concurrent value data are not available) in revisions.

The merchandise price indexes are published using three different classification systems: the Harmonized System, the Bureau of Economic Analysis End-Use System, and the Standard International Trade Classification (SITC) system. The aggregate indexes shown here are from the End-Use System.

Data availability

Indexes are published monthly in a press release and a more detailed report. Indexes are published for detailed product categories, as well as for all commodities. Aggregate import indexes by country or region of origin also are available, as are indexes for selected categories of internationally traded services. Additional information is available from the Division of International Prices in the Bureau of Labor Statistics. Complete historical data are available on the BLS Web site at <http://www.bls.gov>.

REFERENCES

The indexes are described in "BLS to Produce Monthly Indexes of Export and Import Prices," *Monthly Labor Review* (December 1988), and Chapter 15, "International Price Indexes," *BLS Handbook of Methods* Bulletin 2490 (April 1997).

CHAPTER 8: PRICES

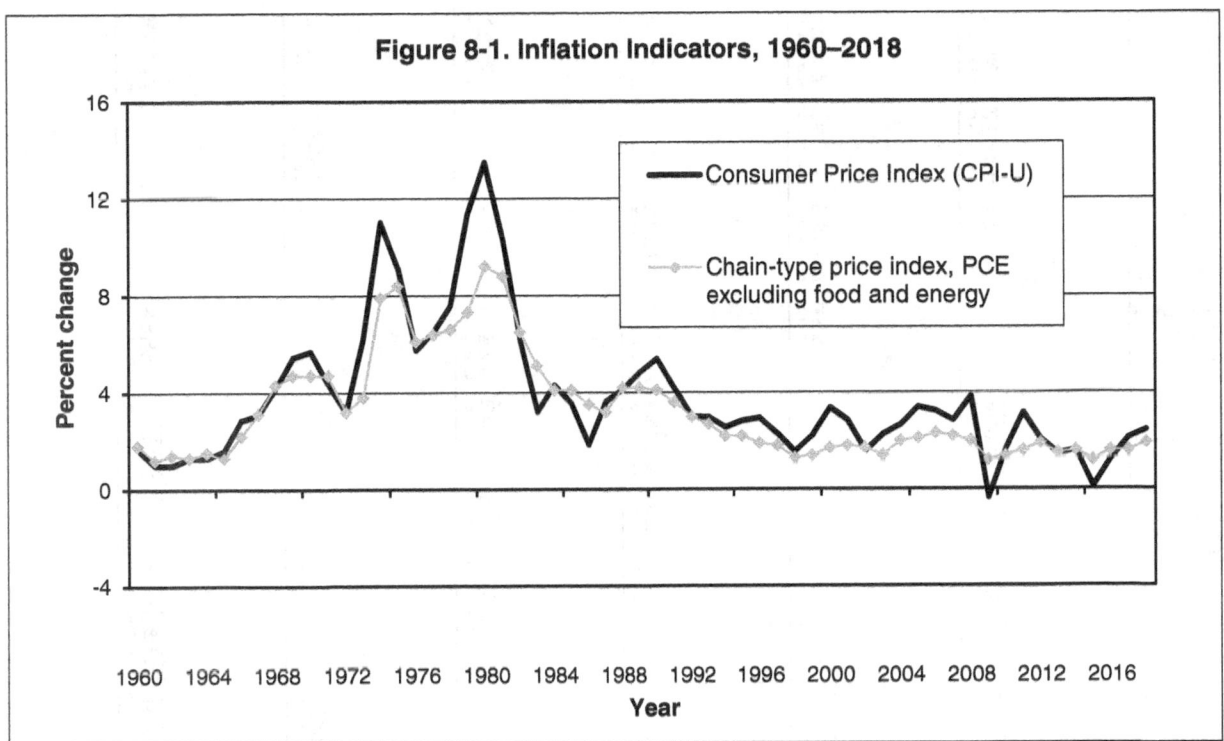

Figure 8-1. Inflation Indicators, 1960–2018

- Figure 8-1 shows annual rates of change in the Consumer Price Index for All Urban Consumers (CPI-U), the most widely used measure of the general price level. It also shows changes in the chain-type price index for personal consumption expenditures (PCE) excluding food and energy, which provides one widely used measure of the "core" or underlying rate of inflation. (Tables 8-1, 8-4, and 1-6)

- Sharp swings in energy prices caused the all-items CPI to drop 0.4 percent in 2009—the first year-to-year deflation registered since 1955—and swing back up to 3.2 percent inflation in 2011. Inflation measured by CPI-U dropped back to 0.1 percent in 2015 then grew by 2.1 percent in 2017 and 2.4 percent in 2018. (Table 8-4)

- The consumer price index (CPI) measures a basket of goods that represents all the types of products and services consumers buy. Changes in the index measures the rate of inflation. With both the increase in interest rates by the Federal Reserve and tariffs being imposed on items such steel and aluminum, the costs of items in the basket will rise.

- With inflation nearing 3 percent, an increase in wages of 2 percent actually ends up being a loss of 1 percent. Real wages, income, GDP, etc. mean inflation has been factored out. With real statistics, comparisons can be made across many years.

- The Producer Price Index (PPI) measures prices at the point of production, rather than at the consumer level. For most of its history, the PPI has measured goods only. The PPI for all commodities fell in 2015 after increasing each year for the previous five years and declined again in 2016 before increasing 4.4 percent in 2017 and 2018. The PPI for "unprocessed nonfood materials less energy," also known (in the old goods-only system) as "crude nonfood materials less energy") is a sensitive leading indicator, not of aggregate prices, but of general business conditions. (Tables 8-7)

Table 8-1A. Summary Consumer Price Indexes: Recent Data

(Seasonally adjusted.)

Year and month	Urban wage earners and clerical workers (CPI-W), all items	Consumer Price Index, 1982–1984 = 100					
		All urban consumers (CPI-U)					
		All items	Food	Energy	All items less food and energy	Commod-ities	Services
1965	31.7	31.5	32.2	22.9	32.7	35.2	26.6
1966	32.6	32.4	33.8	23.3	33.5	36.1	27.6
1967	33.6	33.4	34.1	23.8	34.7	36.8	28.8
1968	35.0	34.8	35.3	24.2	36.3	38.1	30.3
1969	36.9	36.7	37.1	24.8	38.4	39.9	32.4
1970	39.0	38.8	39.2	25.5	40.8	41.7	35.0
1971	40.7	40.5	40.4	26.5	42.7	43.2	37.0
1972	42.1	41.8	42.1	27.2	44.0	44.5	38.4
1973	44.7	44.4	48.2	29.4	45.6	47.8	40.1
1974	49.6	49.3	55.1	38.1	49.4	53.5	43.8
1975	54.1	53.8	59.8	42.1	53.9	58.2	48.0
1976	57.2	56.9	61.6	45.1	57.4	60.7	52.0
1977	60.9	60.6	65.5	49.4	61.0	64.2	56.0
1978	65.6	65.2	72.0	52.5	65.5	68.8	60.8
1979	73.1	72.6	79.9	65.7	71.9	76.6	67.5
1980	82.9	82.4	86.8	86.0	80.8	86.0	77.9
1981	91.4	90.9	93.6	97.7	89.2	93.2	88.1
1982	96.9	96.5	97.4	99.2	95.8	97.0	96.0
1983	99.8	99.6	99.4	99.9	99.6	99.8	99.4
1984	103.3	103.9	103.2	100.9	104.6	103.2	104.6
1985	106.9	107.6	105.6	101.6	109.1	105.4	109.9
1986	108.6	109.6	109.0	88.2	113.5	104.4	115.4
1987	112.5	113.6	113.5	88.6	118.2	107.7	120.2
1988	117.0	118.3	118.2	89.3	123.4	111.5	125.7
1989	122.6	124.0	125.1	94.3	129.0	116.7	131.9
1990	129.0	130.7	132.4	102.1	135.5	122.8	139.2
1991	134.3	136.2	136.3	102.5	142.1	126.6	146.3
1992	138.2	140.3	137.9	103.0	147.3	129.1	152.0
1993	142.1	144.5	140.9	104.2	152.2	131.5	157.9
1994	145.6	148.2	144.3	104.6	156.5	133.8	163.1
1995	149.8	152.4	148.4	105.2	161.2	136.4	168.7
1996	154.1	156.9	153.3	110.1	165.6	139.9	174.1
1997	157.6	160.5	157.3	111.5	169.5	141.8	179.4
1998	159.7	163.0	160.7	102.9	173.4	141.9	184.2
1999	163.2	166.6	164.1	106.6	177.0	144.4	188.8
2000	168.9	172.2	167.8	124.6	181.3	149.2	195.3
2001	173.5	177.1	173.1	129.3	186.1	150.7	203.4
2002	175.9	179.9	176.2	121.7	190.5	149.7	209.8
2003	179.8	184.0	180.0	136.5	193.2	151.2	216.5
2004	184.5	188.9	186.2	151.4	196.6	154.7	222.8
2005	191.0	195.3	190.7	177.1	200.9	160.2	230.1
2006	197.1	201.6	195.2	196.9	205.9	164.0	238.9
2007	202.8	207.3	202.9	207.7	210.7	167.5	246.8
2008	211.1	215.3	214.1	236.7	215.6	174.8	255.5
2009	209.6	214.5	218.0	193.1	219.2	169.7	259.2
2010	214.0	218.1	219.6	211.4	221.3	174.6	261.3
2011	221.6	224.9	227.8	243.9	225.0	183.9	265.8
2012	226.2	229.6	233.8	246.1	229.8	187.6	271.4
2013	229.3	233.0	237.0	244.4	233.8	187.7	277.9
2014	232.8	236.7	242.7	243.6	237.9	187.9	285.1
2015	231.8	237.0	247.2	202.9	242.2	181.7	291.7
2016	234.1	240.0	247.9	189.5	247.6	179.2	299.9
2017	239.1	245.1	250.1	204.5	252.2	181.2	308.1
2018	245.1	251.1	253.6	219.9	257.6	184.6	316.6
2017							
January	237.9	243.8	247.9	206.0	250.6	181.8	304.9
February	238.0	244.0	248.6	203.2	251.1	181.3	305.8
March	237.7	243.7	249.2	201.5	250.9	181.0	305.6
April	237.9	244.1	249.6	202.4	251.1	181.0	306.2
May	237.8	244.0	250.1	198.6	251.3	180.1	306.9
June	238.0	244.2	249.9	198.3	251.7	180.0	307.5
July	238.1	244.4	250.4	197.3	252.0	179.8	308.0
August	239.2	245.3	250.5	202.3	252.6	180.7	309.0
September	240.4	246.4	250.8	211.1	253.0	182.3	309.6
October	240.5	246.6	251.1	207.8	253.5	181.7	310.5
November	241.4	247.3	251.1	213.1	253.9	182.6	311.1
December	242.0	247.9	251.5	214.1	254.4	183.0	311.9
2018							
January	243.0	248.9	252.0	217.5	255.2	184.0	312.8
February	243.5	249.4	252.1	219.0	255.7	184.2	313.6
March	243.6	249.5	252.5	215.8	256.1	183.7	314.3
April	244.0	250.0	253.0	217.7	256.4	184.1	314.9
May	244.7	250.6	253.2	221.0	256.9	184.6	315.7
June	245.2	251.1	253.5	222.4	257.3	185.1	316.2
July	245.7	251.6	253.9	222.3	257.9	185.2	317.0
August	246.0	251.9	254.1	223.3	258.1	185.2	317.6
September	246.0	252.0	254.2	221.1	258.5	184.8	318.2
October	246.9	252.8	254.1	225.6	259.0	185.7	318.9
November	246.7	252.8	254.6	219.3	259.6	184.9	319.6
December	246.4	252.7	255.5	213.6	260.1	183.9	320.6

Table 8-1B. Summary Consumer Price Indexes: Historical Data, 1946 to Date

(Seasonally adjusted.)

Year and month	Urban wage earners and clerical workers (CPI-W), all items	Consumer Price Index, 1982–1984 = 100					
		All urban consumers (CPI-U)					
		All items	Food	Energy	All items less food and energy	Commod-ities	Services
1946	19.6	19.5	19.8	. . .	. . .	22.9	14.1
1947	22.5	22.3	24.1	. . .	. . .	27.6	14.7
1948	24.2	24.1	26.1	. . .	. . .	29.6	15.6
1949	24.0	23.8	25.0	. . .	. . .	28.8	16.4
1950	24.2	24.1	25.4	. . .	. . .	29.0	16.9
1951	26.1	26.0	28.2	. . .	. . .	31.6	17.8
1952	26.7	26.5	28.7	. . .	. . .	32.0	18.6
1953	26.9	26.7	28.3	. . .	. . .	31.9	19.4
1954	27.0	26.9	28.2	. . .	. . .	31.6	20.0
1955	26.9	26.8	27.8	. . .	. . .	31.3	20.4
1956	27.3	27.2	28.0	. . .	. . .	31.6	20.9
1957	28.3	28.1	28.9	21.5	28.9	32.6	21.8
1958	29.1	28.9	30.2	21.5	29.6	33.3	22.6
1959	29.3	29.1	29.7	21.9	30.2	33.3	23.3
1960	29.8	29.6	30.0	22.4	30.6	33.6	24.1
1961	30.1	29.9	30.4	22.5	31.0	33.8	24.5
1962	30.4	30.2	30.6	22.6	31.4	34.1	25.0
1963	30.8	30.6	31.1	22.6	31.8	34.4	25.5
1964	31.2	31.0	31.5	22.5	32.3	34.8	26.0
1947							
January	21.6	21.5	22.8	. . .	. . .	. . .	. . .
February	21.7	21.6	23.1	. . .	. . .	. . .	. . .
March	22.1	22.0	23.8	. . .	. . .	. . .	. . .
April	22.1	22.0	23.5	. . .	. . .	. . .	. . .
May	22.1	22.0	23.4	. . .	. . .	. . .	. . .
June	22.2	22.1	23.5	. . .	. . .	. . .	. . .
July	22.4	22.2	23.8	. . .	. . .	. . .	. . .
August	22.5	22.4	24.1	. . .	. . .	. . .	. . .
September	23.0	22.8	24.8	. . .	. . .	. . .	. . .
October	23.0	22.9	24.9	. . .	. . .	. . .	. . .
November	23.2	23.1	25.2	. . .	. . .	. . .	. . .
December	23.5	23.4	25.7	. . .	. . .	. . .	. . .
1948							
January	23.8	23.7	26.1	. . .	. . .	. . .	. . .
February	23.8	23.7	25.9	. . .	. . .	. . .	. . .
March	23.6	23.5	25.3	. . .	. . .	. . .	. . .
April	24.0	23.8	26.0	. . .	. . .	. . .	. . .
May	24.1	24.0	26.3	. . .	. . .	. . .	. . .
June	24.3	24.2	26.5	. . .	. . .	. . .	. . .
July	24.5	24.4	26.7	. . .	. . .	. . .	. . .
August	24.6	24.4	26.5	. . .	. . .	. . .	. . .
September	24.5	24.4	26.3	. . .	. . .	. . .	. . .
October	24.4	24.3	26.1	. . .	. . .	. . .	. . .
November	24.3	24.2	25.7	. . .	. . .	. . .	. . .
December	24.2	24.1	25.5	. . .	. . .	. . .	. . .
1949							
January	24.2	24.0	25.4	. . .	. . .	. . .	. . .
February	24.1	23.9	25.3	. . .	. . .	. . .	. . .
March	24.0	23.9	25.3	. . .	. . .	. . .	. . .
April	24.1	23.9	25.3	. . .	. . .	. . .	. . .
May	24.0	23.9	25.2	. . .	. . .	. . .	. . .
June	24.1	23.9	25.3	. . .	. . .	. . .	. . .
July	23.8	23.7	24.8	. . .	. . .	. . .	. . .
August	23.8	23.7	24.8	. . .	. . .	. . .	. . .
September	23.9	23.8	25.0	. . .	. . .	. . .	. . .
October	23.8	23.7	24.8	. . .	. . .	. . .	. . .
November	23.8	23.7	24.8	. . .	. . .	. . .	. . .
December	23.8	23.6	24.5	. . .	. . .	. . .	. . .
1950							
January	23.6	23.5	24.3	. . .	. . .	. . .	. . .
February	23.7	23.6	24.7	. . .	. . .	. . .	. . .
March	23.8	23.6	24.6	. . .	. . .	. . .	. . .
April	23.8	23.7	24.6	. . .	. . .	. . .	. . .
May	23.9	23.8	24.8	. . .	. . .	. . .	. . .
June	24.0	23.9	25.1	. . .	. . .	. . .	. . .
July	24.2	24.1	25.6	. . .	. . .	. . .	. . .
August	24.3	24.2	25.8	. . .	. . .	. . .	. . .
September	24.5	24.3	25.8	. . .	. . .	. . .	. . .
October	24.6	24.5	26.0	. . .	. . .	. . .	. . .
November	24.7	24.6	26.1	. . .	. . .	. . .	. . .
December	25.1	25.0	26.9	. . .	. . .	. . .	. . .

. . . = Not available.

Table 8-1B. Summary Consumer Price Indexes: Historical Data, 1946 to Date—*Continued*

(Seasonally adjusted.)

| Year and month | Urban wage earners and clerical workers (CPI-W), all items | Consumer Price Index, 1982–1984 = 100 | | | | | |
| | | All urban consumers (CPI-U) | | | | | |
		All items	Food	Energy	All items less food and energy	Commod-ities	Services
1951							
January	25.5	25.4	27.6	. . .	. . .	. . .	. . .
February	26.0	25.8	28.5	. . .	. . .	. . .	. . .
March	26.0	25.9	28.4	. . .	. . .	. . .	. . .
April	26.1	25.9	28.2	. . .	. . .	. . .	. . .
May	26.1	26.0	28.3	. . .	. . .	. . .	. . .
June	26.1	25.9	28.0	. . .	. . .	. . .	. . .
July	26.1	25.9	27.9	. . .	. . .	. . .	. . .
August	26.0	25.9	27.8	. . .	. . .	. . .	. . .
September	26.2	26.0	27.9	. . .	. . .	. . .	. . .
October	26.3	26.2	28.4	. . .	. . .	. . .	. . .
November	26.5	26.3	28.6	. . .	. . .	. . .	. . .
December	26.6	26.5	28.9	. . .	. . .	. . .	. . .
1952							
January	26.6	26.5	28.9	. . .	. . .	. . .	. . .
February	26.6	26.4	28.6	. . .	. . .	. . .	. . .
March	26.5	26.4	28.5	. . .	. . .	. . .	. . .
April	26.6	26.5	28.7	. . .	. . .	. . .	. . .
May	26.6	26.5	28.7	. . .	. . .	. . .	. . .
June	26.7	26.5	28.6	. . .	. . .	. . .	. . .
July	26.8	26.7	28.9	. . .	. . .	. . .	. . .
August	26.8	26.7	28.9	. . .	. . .	. . .	. . .
September	26.8	26.6	28.7	. . .	. . .	. . .	. . .
October	26.8	26.7	28.8	. . .	. . .	. . .	. . .
November	26.8	26.7	28.8	. . .	. . .	. . .	. . .
December	26.9	26.7	28.6	. . .	. . .	. . .	. . .
1953							
January	26.8	26.6	28.4	. . .	. . .	. . .	. . .
February	26.7	26.6	28.3	. . .	. . .	. . .	. . .
March	26.8	26.6	28.3	. . .	. . .	. . .	. . .
April	26.8	26.7	28.1	. . .	. . .	. . .	. . .
May	26.9	26.7	28.2	. . .	. . .	. . .	. . .
June	26.9	26.8	28.4	. . .	. . .	. . .	. . .
July	26.9	26.8	28.2	. . .	. . .	. . .	. . .
August	27.0	26.9	28.3	. . .	. . .	. . .	. . .
September	27.0	26.9	28.4	. . .	. . .	. . .	. . .
October	27.1	27.0	28.4	. . .	. . .	. . .	. . .
November	27.0	26.9	28.1	. . .	. . .	. . .	. . .
December	27.0	26.9	28.3	. . .	. . .	. . .	. . .
1954							
January	27.1	26.9	28.5	. . .	. . .	. . .	. . .
February	27.1	27.0	28.5	. . .	. . .	. . .	. . .
March	27.1	26.9	28.4	. . .	. . .	. . .	. . .
April	27.0	26.9	28.4	. . .	. . .	. . .	. . .
May	27.1	26.9	28.4	. . .	. . .	. . .	. . .
June	27.1	26.9	28.4	. . .	. . .	. . .	. . .
July	27.0	26.9	28.4	. . .	. . .	. . .	. . .
August	27.0	26.9	28.3	. . .	. . .	. . .	. . .
September	27.0	26.8	28.0	. . .	. . .	. . .	. . .
October	26.9	26.7	27.9	. . .	. . .	. . .	. . .
November	26.9	26.8	27.9	. . .	. . .	. . .	. . .
December	26.9	26.8	27.8	. . .	. . .	. . .	. . .
1955							
January	26.9	26.8	27.8	. . .	. . .	. . .	. . .
February	27.0	26.8	28.0	. . .	. . .	. . .	. . .
March	26.9	26.8	28.0	. . .	. . .	. . .	. . .
April	26.9	26.8	28.0	. . .	. . .	. . .	. . .
May	26.9	26.8	27.9	. . .	. . .	. . .	. . .
June	26.9	26.7	27.7	. . .	. . .	. . .	. . .
July	26.9	26.8	27.7	. . .	. . .	. . .	. . .
August	26.9	26.7	27.6	. . .	. . .	. . .	. . .
September	27.0	26.9	27.8	. . .	. . .	. . .	. . .
October	27.0	26.8	27.7	. . .	. . .	. . .	. . .
November	27.0	26.9	27.6	. . .	. . .	. . .	. . .
December	27.0	26.9	27.6	. . .	. . .	. . .	. . .
1956							
January	27.0	26.8	27.5	. . .	. . .	31.2	20.7
February	27.0	26.9	27.5	. . .	. . .	31.1	20.7
March	27.1	26.9	27.5	. . .	. . .	31.2	20.7
April	27.1	26.9	27.6	. . .	. . .	31.3	20.8
May	27.2	27.0	27.8	. . .	. . .	31.4	20.8
June	27.3	27.2	28.1	. . .	. . .	31.5	20.9
July	27.4	27.3	28.4	. . .	. . .	31.8	20.9
August	27.5	27.3	28.2	. . .	. . .	31.7	21.0
September	27.5	27.4	28.2	. . .	. . .	31.8	21.1
October	27.7	27.5	28.3	. . .	. . .	32.0	21.1
November	27.7	27.5	28.4	. . .	. . .	32.0	21.2
December	27.8	27.6	28.5	. . .	. . .	32.1	21.3

. . . = Not available.

Table 8-1B. Summary Consumer Price Indexes: Historical Data, 1946 to Date—*Continued*

(Seasonally adjusted.)

| Year and month | Urban wage earners and clerical workers (CPI-W), all items | Consumer Price Index, 1982–1984 = 100 | | | | | |
| | | All urban consumers (CPI-U) | | | | | |
		All items	Food	Energy	All items less food and energy	Commod-ities	Services
1957							
January	27.8	27.7	28.4	21.3	28.5	32.1	21.4
February	28.0	27.8	28.7	21.4	28.6	32.3	21.4
March	28.0	27.9	28.6	21.5	28.7	32.3	21.6
April	28.1	27.9	28.6	21.6	28.8	32.4	21.6
May	28.2	28.0	28.7	21.6	28.8	32.4	21.7
June	28.3	28.1	28.9	21.6	28.9	32.5	21.8
July	28.4	28.2	29.1	21.5	29.0	32.6	21.8
August	28.4	28.3	29.4	21.4	29.0	32.8	21.9
September	28.5	28.3	29.2	21.4	29.1	32.8	22.0
October	28.5	28.3	29.2	21.4	29.2	32.7	22.1
November	28.6	28.4	29.2	21.5	29.3	32.8	22.2
December	28.6	28.5	29.2	21.5	29.3	32.9	22.2
1958							
January	28.8	28.6	29.8	21.6	29.3	33.1	22.3
February	28.9	28.7	29.9	21.3	29.4	33.2	22.4
March	29.0	28.9	30.5	21.4	29.5	33.4	22.4
April	29.1	28.9	30.6	21.4	29.5	33.5	22.5
May	29.1	28.9	30.5	21.5	29.5	33.5	22.6
June	29.1	28.9	30.3	21.5	29.6	33.4	22.6
July	29.1	28.9	30.2	21.6	29.6	33.3	22.7
August	29.1	28.9	30.1	21.7	29.6	33.3	22.7
September	29.1	28.9	30.0	21.7	29.7	33.3	22.8
October	29.1	28.9	30.0	21.7	29.7	33.2	22.8
November	29.1	29.0	30.0	21.4	29.8	33.0	22.8
December	29.1	29.0	29.9	21.4	29.9	33.3	22.8
1959							
January	29.2	29.0	30.0	21.4	29.9	33.3	22.9
February	29.2	29.0	29.8	21.6	29.9	33.3	23.0
March	29.1	29.0	29.7	21.7	30.0	33.2	23.0
April	29.1	29.0	29.5	21.8	30.0	33.2	23.1
May	29.2	29.0	29.5	21.8	30.1	33.3	23.2
June	29.3	29.1	29.7	21.9	30.2	33.3	23.2
July	29.3	29.2	29.6	21.8	30.2	33.3	23.3
August	29.3	29.2	29.6	21.9	30.2	33.3	23.4
September	29.4	29.3	29.7	21.9	30.3	33.5	23.5
October	29.5	29.4	29.7	22.2	30.4	33.5	23.6
November	29.5	29.4	29.7	22.2	30.4	33.5	23.6
December	29.6	29.4	29.6	22.3	30.5	33.5	23.7
1960							
January	29.5	29.4	29.6	22.3	30.5	33.5	23.7
February	29.6	29.4	29.5	22.2	30.6	33.4	23.8
March	29.6	29.4	29.6	22.3	30.6	33.5	23.9
April	29.7	29.5	30.0	22.4	30.6	33.6	23.9
May	29.7	29.6	30.0	22.3	30.6	33.6	24.0
June	29.8	29.6	30.0	22.4	30.7	33.6	24.0
July	29.7	29.6	29.9	22.5	30.6	33.5	24.1
August	29.8	29.6	30.0	22.5	30.6	33.6	24.1
September	29.8	29.6	30.1	22.6	30.6	33.6	24.2
October	29.9	29.8	30.3	22.5	30.8	33.7	24.2
November	30.0	29.8	30.5	22.7	30.8	33.8	24.3
December	30.0	29.8	30.5	22.6	30.7	33.9	24.3
1961							
January	30.0	29.8	30.5	22.7	30.8	33.8	24.4
February	30.0	29.8	30.5	22.6	30.8	33.8	24.4
March	30.0	29.8	30.5	22.6	30.9	33.8	24.4
April	30.0	29.8	30.4	22.2	30.9	33.8	24.5
May	30.0	29.8	30.3	22.4	30.9	33.8	24.5
June	30.0	29.8	30.2	22.5	31.0	33.8	24.5
July	30.1	29.9	30.3	22.5	31.0	33.9	24.5
August	30.1	29.9	30.3	22.5	31.1	33.9	24.6
September	30.2	30.0	30.3	22.6	31.1	33.9	24.6
October	30.2	30.0	30.3	22.4	31.1	33.9	24.7
November	30.2	30.0	30.3	22.5	31.2	33.8	24.7
December	30.2	30.0	30.3	22.4	31.2	33.9	24.8
1962							
January	30.2	30.0	30.4	22.4	31.2	33.9	24.8
February	30.3	30.1	30.5	22.6	31.2	34.0	24.8
March	30.4	30.2	30.6	22.4	31.3	34.0	24.9
April	30.4	30.2	30.7	22.7	31.3	34.1	24.9
May	30.4	30.2	30.6	22.7	31.4	34.1	25.0
June	30.4	30.2	30.5	22.5	31.4	34.1	25.0
July	30.4	30.2	30.4	22.3	31.4	34.0	25.1
August	30.5	30.3	30.6	22.4	31.5	34.1	25.1
September	30.6	30.4	30.9	22.8	31.5	34.3	25.1
October	30.6	30.4	30.8	22.7	31.5	34.2	25.1
November	30.6	30.4	30.9	22.7	31.5	34.3	25.2
December	30.6	30.4	30.7	22.8	31.6	34.2	25.2

Table 8-1B. Summary Consumer Price Indexes: Historical Data, 1946 to Date—*Continued*

(Seasonally adjusted.)

Year and month	Urban wage earners and clerical workers (CPI-W), all items	Consumer Price Index, 1982–1984 = 100 — All urban consumers (CPI-U)					
		All items	Food	Energy	All items less food and energy	Commod-ities	Services
1963							
January	30.6	30.4	31.0	22.8	31.5	34.3	25.3
February	30.7	30.5	31.1	22.7	31.6	34.3	25.3
March	30.7	30.5	31.0	22.7	31.7	34.3	25.3
April	30.7	30.5	30.9	22.6	31.7	34.3	25.4
May	30.7	30.5	30.9	22.6	31.7	34.3	25.4
June	30.8	30.6	31.0	22.5	31.8	34.4	25.5
July	30.9	30.7	31.2	22.7	31.8	34.5	25.5
August	30.9	30.8	31.2	22.6	31.9	34.6	25.6
September	30.9	30.7	31.1	22.5	31.9	34.5	25.6
October	30.9	30.8	31.0	22.7	32.0	34.5	25.6
November	31.0	30.8	31.2	22.6	32.0	34.6	25.7
December	31.1	30.9	31.3	22.6	32.1	34.7	25.8
1964							
January	31.1	30.9	31.4	22.8	32.2	34.7	25.8
February	31.1	30.9	31.4	22.2	32.2	34.7	25.8
March	31.1	30.9	31.4	22.6	32.2	34.7	25.8
April	31.1	31.0	31.4	22.5	32.2	34.7	25.9
May	31.2	31.0	31.4	22.5	32.2	34.7	25.9
June	31.2	31.0	31.4	22.6	32.3	34.7	26.0
July	31.2	31.0	31.5	22.5	32.3	34.7	26.0
August	31.2	31.1	31.4	22.6	32.3	34.7	26.0
September	31.3	31.1	31.6	22.5	32.3	34.8	26.0
October	31.3	31.1	31.6	22.5	32.4	34.8	26.1
November	31.4	31.2	31.7	22.5	32.5	34.9	26.2
December	31.4	31.3	31.7	22.6	32.5	35.0	26.2
1965							
January	31.5	31.3	31.6	22.8	32.6	35.0	26.3
February	31.5	31.3	31.5	22.7	32.6	34.9	26.4
March	31.5	31.3	31.7	22.6	32.6	35.0	26.4
April	31.6	31.4	31.8	22.9	32.7	35.0	26.5
May	31.7	31.5	32.1	23.0	32.7	35.1	26.5
June	31.8	31.6	32.6	23.1	32.7	35.3	26.5
July	31.8	31.6	32.5	23.0	32.7	35.3	26.6
August	31.7	31.6	32.4	23.0	32.7	35.2	26.6
September	31.8	31.6	32.3	23.1	32.8	35.2	26.7
October	31.8	31.7	32.5	23.0	32.8	35.3	26.8
November	31.9	31.8	32.6	23.1	32.9	35.4	26.9
December	32.0	31.9	32.8	23.1	33.0	35.5	26.9
1966							
January	32.1	31.9	33.0	23.1	33.0	35.6	27.0
February	32.3	32.1	33.5	23.2	33.1	35.8	27.0
March	32.4	32.2	33.8	23.2	33.1	35.9	27.1
April	32.5	32.3	33.8	23.2	33.3	36.0	27.3
May	32.5	32.4	33.7	23.2	33.4	36.0	27.4
June	32.6	32.4	33.7	23.3	33.5	36.0	27.5
July	32.6	32.5	33.5	23.4	33.6	36.1	27.7
August	32.8	32.7	34.0	23.3	33.7	36.2	27.7
September	32.9	32.8	34.1	23.4	33.8	36.4	27.9
October	33.0	32.9	34.2	23.4	34.0	36.4	28.0
November	33.1	32.9	34.1	23.5	34.0	36.4	28.2
December	33.1	32.9	34.0	23.5	34.1	36.4	28.2
1967							
January	33.1	32.9	33.9	23.6	34.2	36.4	28.3
February	33.2	33.0	33.8	23.7	34.2	36.4	28.4
March	33.2	33.0	33.8	23.6	34.3	36.4	28.5
April	33.3	33.1	33.7	23.9	34.4	36.4	28.6
May	33.3	33.1	33.7	23.9	34.5	36.5	28.6
June	33.5	33.3	34.0	23.8	34.6	36.6	28.8
July	33.6	33.4	34.1	23.8	34.7	36.8	28.8
August	33.7	33.5	34.3	23.9	34.9	37.0	28.9
September	33.8	33.6	34.3	24.0	35.0	37.0	29.0
October	33.9	33.7	34.4	23.9	35.1	37.1	29.2
November	34.0	33.9	34.5	24.0	35.2	37.2	29.2
December	34.1	34.0	34.6	23.9	35.4	37.4	29.4
1968							
January	34.3	34.1	34.6	24.0	35.5	37.5	29.5
February	34.4	34.2	34.8	24.1	35.7	37.6	29.6
March	34.5	34.3	34.9	24.1	35.8	37.7	29.8
April	34.6	34.4	35.0	24.0	35.9	37.8	29.9
May	34.7	34.5	35.1	24.1	36.0	37.8	30.0
June	34.9	34.7	35.2	24.2	36.2	38.0	30.2
July	35.0	34.9	35.3	24.2	36.4	38.1	30.4
August	35.2	35.0	35.4	24.3	36.5	38.3	30.6
September	35.3	35.1	35.6	24.3	36.7	38.4	30.7
October	35.5	35.3	35.9	24.3	36.9	38.6	30.9
November	35.6	35.4	35.9	24.4	37.1	38.7	31.0
December	35.8	35.6	36.0	24.3	37.2	38.8	31.2

Table 8-1B. Summary Consumer Price Indexes: Historical Data, 1946 to Date—*Continued*

(Seasonally adjusted.)

Year and month	Urban wage earners and clerical workers (CPI-W), all items	Consumer Price Index, 1982–1984 = 100					
		All urban consumers (CPI-U)					
		All items	Food	Energy	All items less food and energy	Commod-ities	Services
1969							
January	35.9	35.7	36.1	24.4	37.3	38.9	31.4
February	36.0	35.8	36.1	24.4	37.6	39.0	31.5
March	36.3	36.1	36.2	24.7	37.8	39.3	31.8
April	36.5	36.3	36.4	24.9	38.1	39.4	32.0
May	36.6	36.4	36.6	24.8	38.1	39.5	32.2
June	36.8	36.6	37.0	25.0	38.3	39.8	32.3
July	37.0	36.8	37.3	24.9	38.5	39.9	32.5
August	37.1	36.9	37.5	24.9	38.7	40.1	32.7
September	37.3	37.1	37.7	25.0	38.9	40.2	33.0
October	37.5	37.3	37.8	25.0	39.1	40.4	33.1
November	37.7	37.5	38.2	25.0	39.2	40.6	33.3
December	37.9	37.7	38.6	25.1	39.4	40.8	33.5
1970							
January	38.1	37.9	38.7	25.1	39.6	41.0	33.8
February	38.3	38.1	38.9	25.1	39.8	41.2	34.0
March	38.5	38.3	38.9	25.0	40.1	41.2	34.4
April	38.7	38.5	39.0	25.5	40.4	41.4	34.6
May	38.8	38.6	39.2	25.4	40.5	41.5	34.8
June	39.0	38.8	39.2	25.3	40.8	41.6	35.0
July	39.1	38.9	39.2	25.5	40.9	41.7	35.2
August	39.2	39.0	39.2	25.4	41.1	41.8	35.4
September	39.4	39.2	39.4	25.6	41.3	42.0	35.6
October	39.6	39.4	39.5	25.9	41.5	42.2	35.8
November	39.8	39.6	39.5	26.0	41.0	42.3	36.0
December	40.0	39.8	39.5	26.2	42.0	42.5	36.2
1971							
January	40.1	39.9	39.4	26.3	42.1	42.5	36.4
February	40.2	39.9	39.5	26.2	42.2	42.6	36.5
March	40.2	40.0	39.8	26.2	42.2	42.7	36.5
April	40.4	40.1	40.1	26.1	42.4	42.9	36.6
May	40.5	40.3	40.3	26.2	42.6	43.1	36.7
June	40.7	40.5	40.5	26.3	42.8	43.2	37.0
July	40.9	40.6	40.6	26.3	42.9	43.3	37.1
August	41.0	40.7	40.6	26.8	43.0	43.4	37.3
September	41.0	40.8	40.6	26.9	43.0	43.4	37.4
October	41.1	40.9	40.7	27.0	43.1	43.5	37.5
November	41.2	41.0	40.9	26.9	43.2	43.5	37.6
December	41.4	41.1	41.3	27.0	43.3	43.8	37.7
1972							
January	41.5	41.2	41.1	27.0	43.5	43.8	37.9
February	41.6	41.4	41.7	26.8	43.6	44.0	38.0
March	41.7	41.4	41.6	26.9	43.6	44.0	38.1
April	41.7	41.5	41.6	26.9	43.8	44.1	38.2
May	41.8	41.6	41.7	27.0	43.9	44.2	38.3
June	41.9	41.7	41.9	27.0	44.0	44.3	38.4
July	42.1	41.8	42.1	27.1	44.1	44.5	38.5
August	42.2	41.9	42.2	27.3	44.3	44.5	38.6
September	42.3	42.1	42.5	27.6	44.3	44.8	38.7
October	42.5	42.2	42.8	27.7	44.4	44.9	38.8
November	42.6	42.4	43.0	27.9	44.4	45.1	38.9
December	42.8	42.5	43.2	27.8	44.6	45.2	39.0
1973							
January	43.0	42.7	44.0	27.9	44.6	45.5	39.1
February	43.2	43.0	44.6	28.2	44.8	45.9	39.2
March	43.6	43.4	45.8	28.3	45.0	46.4	39.4
April	43.9	43.7	46.5	28.6	45.1	46.8	39.5
May	44.2	43.9	47.1	28.8	45.3	47.2	39.6
June	44.4	44.2	47.6	29.2	45.4	47.5	39.8
July	44.5	44.2	47.7	29.2	45.5	47.5	39.9
August	45.3	45.0	50.5	29.4	45.7	48.7	40.2
September	45.4	45.2	50.4	29.4	46.0	48.7	40.5
October	45.8	45.6	50.7	30.3	46.3	49.0	41.0
November	46.2	45.9	51.4	31.5	46.5	49.5	41.3
December	46.5	46.3	51.9	32.5	46.7	49.9	41.5
1974							
January	47.0	46.8	52.5	34.1	46.9	50.5	41.8
February	47.6	47.3	53.6	35.4	47.2	51.3	42.0
March	48.1	47.8	54.2	36.9	47.6	51.9	42.4
April	48.3	48.1	54.1	37.6	47.9	52.1	42.6
May	48.8	48.6	54.5	38.3	48.5	52.7	43.1
June	49.2	49.0	54.5	38.6	49.0	53.1	43.5
July	49.6	49.3	54.3	38.9	49.5	53.3	44.0
August	50.2	49.9	55.1	39.2	50.2	54.1	44.5
September	50.9	50.6	56.2	39.3	50.7	54.8	45.0
October	51.3	51.0	56.8	39.2	51.2	55.3	45.4
November	51.8	51.5	57.5	39.4	51.6	55.8	45.8
December	52.2	51.9	58.2	39.6	52.0	56.3	46.2

Table 8-1B. Summary Consumer Price Indexes: Historical Data, 1946 to Date—*Continued*

(Seasonally adjusted.)

Year and month	Urban wage earners and clerical workers (CPI-W), all items	Consumer Price Index, 1982–1984 = 100 All urban consumers (CPI-U)					
		All items	Food	Energy	All items less food and energy	Commod-ities	Services
1975							
January	52.6	52.3	58.4	40.0	52.3	56.7	46.5
February	52.9	52.6	58.5	40.3	52.8	56.9	46.9
March	53.1	52.8	58.4	40.6	53.0	57.1	47.0
April	53.3	53.0	58.3	41.0	53.3	57.2	47.3
May	53.4	53.1	58.6	41.3	53.5	57.5	47.5
June	53.8	53.5	59.2	41.7	53.8	57.9	47.8
July	54.3	54.0	60.3	42.5	54.0	58.6	48.0
August	54.5	54.2	60.3	42.8	54.2	58.7	48.3
September	54.9	54.6	60.7	43.2	54.5	59.0	48.7
October	55.2	54.9	61.3	43.5	54.8	59.4	49.0
November	55.6	55.3	61.7	43.9	55.2	59.7	49.6
December	55.9	55.6	62.1	44.1	55.5	59.9	49.9
1976							
January	56.2	55.8	61.9	44.5	55.9	60.0	50.5
February	56.2	55.9	61.3	44.4	56.2	59.9	50.8
March	56.3	56.0	60.9	44.1	56.5	59.8	51.1
April	56.5	56.1	60.9	43.9	56.7	59.9	51.3
May	56.7	56.4	61.1	44.1	57.0	60.2	51.4
June	57.0	56.7	61.3	44.4	57.2	60.4	51.7
July	57.3	57.0	61.6	44.8	57.6	60.7	52.1
August	57.6	57.3	61.8	45.2	57.9	61.0	52.4
September	57.9	57.6	62.1	45.7	58.2	61.3	52.8
October	58.2	57.9	62.4	46.1	58.5	61.6	53.1
November	58.4	58.1	62.3	46.8	58.7	61.7	53.4
December	58.7	58.4	62.5	47.5	58.9	62.0	53.7
1977							
January	59.1	58.7	62.7	48.1	59.3	62.3	54.1
February	59.6	59.3	63.9	48.1	59.7	63.0	54.4
March	59.9	59.6	64.2	48.4	60.0	63.2	54.8
April	60.3	60.0	65.0	48.6	60.3	63.7	55.2
May	60.6	60.2	65.3	48.9	60.6	63.9	55.4
June	60.9	60.5	65.7	48.9	61.0	64.2	55.8
July	61.2	60.8	65.9	49.1	61.2	64.4	56.3
August	61.4	61.1	66.2	49.5	61.5	64.6	56.6
September	61.7	61.3	66.4	49.8	61.8	64.8	56.9
October	61.9	61.6	66.6	50.5	62.0	65.0	57.2
November	62.3	62.0	67.1	51.3	62.3	65.5	57.6
December	62.6	62.3	67.4	51.6	62.7	65.7	57.9
1978							
January	63.0	62.7	67.9	51.1	63.1	66.1	58.3
February	63.3	63.0	68.6	50.6	63.4	66.4	58.7
March	63.8	63.4	69.5	51.0	63.8	66.8	59.1
April	64.2	63.9	70.6	51.4	64.3	67.4	59.6
May	64.8	64.5	71.6	51.7	64.7	68.0	60.0
June	65.3	65.0	72.7	51.9	65.2	68.6	60.5
July	65.8	65.5	73.0	52.1	65.6	69.1	61.0
August	66.2	65.9	73.3	52.6	66.1	69.4	61.5
September	66.7	66.5	73.6	53.2	66.7	70.0	62.1
October	67.4	67.1	74.2	54.1	67.2	70.6	62.6
November	67.8	67.5	74.7	54.9	67.6	71.1	63.1
December	68.3	67.9	75.1	55.9	68.0	71.6	63.3
1979							
January	68.8	68.5	76.4	55.8	68.5	72.2	63.8
February	69.6	69.2	77.7	55.9	69.2	72.9	64.4
March	70.3	69.9	78.4	57.4	69.8	73.8	64.9
April	71.1	70.6	79.0	59.5	70.3	74.7	65.5
May	71.9	71.4	79.7	62.0	70.8	75.6	66.2
June	72.7	72.2	80.0	64.7	71.3	76.4	66.8
July	73.5	73.0	80.5	67.3	71.9	77.2	67.6
August	74.2	73.7	80.4	69.7	72.7	77.9	68.5
September	75.0	74.4	80.9	71.9	73.3	78.6	69.2
October	75.7	75.2	81.5	73.5	74.0	79.3	70.1
November	76.5	76.0	82.0	74.8	74.8	80.0	71.1
December	77.3	76.9	82.8	76.8	75.7	80.8	72.0
1980							
January	78.5	78.0	83.3	79.1	76.7	82.0	73.1
February	79.4	79.0	83.4	81.9	77.5	82.8	74.1
March	80.6	80.1	84.1	84.5	78.6	83.9	75.4
April	81.4	80.9	84.7	85.4	79.5	84.4	76.6
May	82.2	81.7	85.2	86.4	80.1	84.9	77.6
June	83.0	82.5	85.7	86.5	81.0	85.3	79.0
July	83.1	82.6	86.6	86.7	80.8	85.9	78.5
August	83.7	83.2	88.0	87.2	81.3	86.9	78.5
September	84.4	83.9	89.1	87.5	82.1	87.8	79.0
October	85.3	84.7	89.8	88.0	83.0	88.5	80.0
November	86.2	85.6	90.8	88.8	83.9	89.2	81.1
December	87.0	86.4	91.3	90.7	84.9	89.7	82.2

Table 8-1B. Summary Consumer Price Indexes: Historical Data, 1946 to Date—*Continued*

(Seasonally adjusted.)

Year and month	Urban wage earners and clerical workers (CPI-W), all items	Consumer Price Index, 1982–1984 = 100					
		All urban consumers (CPI-U)					
		All items	Food	Energy	All items less food and energy	Commod- ities	Services
1981							
January	87.7	87.2	91.6	92.1	85.4	90.4	83.0
February	88.6	88.0	92.1	95.2	85.9	91.4	83.7
March	89.1	88.6	92.6	97.4	86.4	91.9	84.4
April	89.6	89.1	92.8	97.6	87.0	92.0	85.3
May	90.2	89.7	92.8	97.9	87.8	92.4	86.4
June	90.9	90.5	93.2	97.3	88.6	92.9	87.5
July	92.0	91.5	93.9	97.3	89.8	93.6	88.9
August	92.7	92.2	94.4	97.8	90.7	94.0	89.9
September	93.5	93.1	94.8	98.6	91.8	94.6	91.2
October	93.8	93.4	95.0	99.2	92.1	94.7	91.7
November	94.2	93.8	95.1	100.5	92.5	94.9	92.5
December	94.5	94.1	95.3	101.5	93.0	95.1	93.0
1982							
January	94.8	94.4	95.6	100.6	93.3	95.2	93.5
February	95.1	94.7	96.3	98.0	93.8	95.4	93.9
March	95.0	94.7	96.2	96.6	93.9	95.3	94.0
April	95.3	95.0	96.4	94.2	94.7	95.1	94.9
May	96.1	95.9	97.2	95.7	95.4	96.0	95.7
June	97.3	97.0	98.1	98.4	96.1	97.4	96.5
July	97.8	97.5	98.2	99.3	96.7	97.9	97.0
August	98.1	97.7	98.0	99.8	97.1	97.9	97.6
September	98.1	97.7	98.2	100.3	97.2	97.8	97.6
October	98.5	98.1	98.2	101.7	97.5	98.2	97.9
November	98.4	98.0	98.2	102.5	07.0	98.3	97.7
December	98.1	07.7	98.2	102.8	97.2	98.3	96.9
1983							
January	98.2	97.9	08.1	99.6	97.6	98.3	97.5
February	98.2	98.0	98.2	97.7	98.0	98.1	97.9
March	98.5	98.1	98.8	96.8	98.2	98.2	98.1
April	99.1	98.8	99.2	98.9	98.6	98.9	98.7
May	99.5	99.2	99.5	100.4	98.9	99.5	98.9
June	99.7	99.4	99.6	100.6	99.2	99.8	99.2
July	100.0	99.8	99.6	100.9	99.8	100.2	99.6
August	100.5	100.1	99.7	101.2	100.1	100.5	99.8
September	100.7	100.4	100.0	101.0	100.5	100.7	100.2
October	101.0	100.8	100.3	100.8	101.0	101.0	100.7
November	101.2	101.1	100.3	100.5	101.5	101.1	101.3
December	101.3	101.4	100.6	100.0	101.8	101.2	101.6
1984							
January	101.8	102.1	102.0	100.2	102.5	101.9	102.1
February	102.0	102.6	102.7	101.4	102.8	102.4	102.6
March	102.0	102.9	102.9	101.4	103.2	102.6	103.0
April	102.2	103.3	102.9	101.7	103.7	102.9	103.5
May	102.5	103.5	102.7	101.6	104.1	103.0	103.9
June	102.7	103.7	103.1	100.8	104.5	103.1	104.2
July	103.2	104.1	103.3	100.5	105.0	103.2	104.9
August	104.1	104.4	103.9	100.1	105.4	103.4	105.4
September	104.5	104.7	103.8	100.6	105.8	103.6	105.9
October	104.7	105.1	104.0	101.1	106.2	103.9	106.3
November	104.8	105.3	104.1	100.8	106.4	103.9	106.7
December	104.9	105.5	104.5	100.1	106.8	103.9	107.1
1985							
January	105.2	105.7	104.7	100.3	107.1	104.1	107.4
February	105.7	106.3	105.2	100.3	107.7	104.7	107.9
March	106.1	106.8	105.5	101.3	108.1	105.1	108.4
April	106.4	107.0	105.4	102.3	108.4	105.4	108.7
May	106.6	107.2	105.2	102.2	108.8	105.2	109.4
June	106.9	107.5	105.5	102.2	109.1	105.3	109.8
July	107.0	107.7	105.5	102.2	109.4	105.3	110.3
August	107.1	107.9	105.6	101.2	109.8	105.2	110.7
September	107.3	108.1	105.8	101.2	110.0	105.4	111.0
October	107.7	108.5	105.8	101.2	110.5	105.6	111.5
November	108.2	109.0	106.5	101.8	111.1	106.1	112.1
December	108.7	109.5	107.3	102.4	111.4	106.6	112.5
1986							
January	109.1	109.9	107.5	102.6	111.9	106.9	113.1
February	108.8	109.7	107.3	99.5	112.2	106.0	113.5
March	108.1	109.1	107.5	92.6	112.5	104.5	114.1
April	107.7	108.7	107.7	87.2	112.9	103.3	114.6
May	107.8	109.0	108.2	87.2	113.1	103.5	114.8
June	108.3	109.4	108.3	88.8	113.4	103.8	115.5
July	108.3	109.5	109.1	85.6	113.8	103.7	115.7
August	108.4	109.6	110.1	83.6	114.2	103.6	116.1
September	108.8	110.0	110.2	84.4	114.6	104.0	116.5
October	108.9	110.2	110.5	82.8	115.0	104.0	116.9
November	109.2	110.4	111.1	82.1	115.3	104.2	117.2
December	109.5	110.8	111.4	82.5	115.6	104.5	117.5

Table 8-1B. Summary Consumer Price Indexes: Historical Data, 1946 to Date—*Continued*

(Seasonally adjusted.)

Year and month	Urban wage earners and clerical workers (CPI-W), all items	Consumer Price Index, 1982–1984 = 100 — All urban consumers (CPI-U)					
		All items	Food	Energy	All items less food and energy	Commod-ities	Services
1987							
January	110.2	111.4	111.8	85.4	115.9	105.5	117.9
February	110.7	111.8	112.2	87.4	116.2	106.1	118.3
March	111.1	112.2	112.4	87.6	116.6	106.5	118.6
April	111.6	112.7	112.6	87.6	117.3	106.9	119.2
May	111.9	113.0	113.2	87.1	117.7	107.2	119.6
June	112.4	113.5	113.9	88.5	117.9	107.7	120.0
July	112.7	113.8	113.7	89.2	118.3	108.0	120.3
August	113.2	114.3	113.9	90.5	118.7	108.5	120.9
September	113.6	114.7	114.3	90.3	119.2	108.8	121.4
October	113.9	115.0	114.5	89.6	119.8	109.0	121.8
November	114.2	115.4	114.5	90.0	120.1	109.3	122.2
December	114.4	115.6	115.1	89.5	120.4	109.3	122.6
1988							
January	114.7	116.0	115.6	88.8	120.9	109.5	123.0
February	114.9	116.2	115.6	88.7	121.2	109.5	123.5
March	115.2	116.5	115.8	88.4	121.7	109.8	123.9
April	115.8	117.2	116.4	88.8	122.3	110.5	124.4
May	116.2	117.5	116.9	88.5	122.7	110.7	124.8
June	116.6	118.0	117.6	88.9	123.2	111.2	125.4
July	117.3	118.5	118.8	89.4	123.6	111.9	125.8
August	117.7	119.0	119.4	90.1	124.0	112.2	126.4
September	118.2	119.5	120.1	89.8	124.7	112.8	126.9
October	118.6	119.9	120.3	89.8	125.2	113.0	127.5
November	118.9	120.3	120.5	89.8	125.6	113.3	127.9
December	119.3	120.7	121.1	89.6	126.0	113.5	128.4
1989							
January	119.9	121.2	121.6	90.3	126.5	114.1	128.9
February	120.3	121.6	122.5	90.8	126.9	114.5	129.4
March	120.9	122.2	123.2	91.8	127.4	115.1	130.0
April	121.9	123.1	123.9	96.6	127.8	116.5	130.5
May	122.5	123.7	124.7	97.4	128.3	117.1	131.1
June	122.8	124.1	125.1	96.9	128.8	117.2	131.6
July	123.2	124.5	125.6	96.7	129.2	117.3	132.3
August	123.2	124.5	125.9	94.9	129.5	117.0	132.8
September	123.4	124.8	126.3	93.8	129.9	117.2	133.1
October	123.9	125.4	126.8	94.4	130.6	117.8	133.7
November	124.4	125.9	127.4	93.9	131.1	118.1	134.3
December	124.9	126.3	127.8	94.2	131.6	118.4	134.9
1990							
January	126.1	127.5	129.7	98.9	132.1	120.2	135.4
February	126.6	128.0	130.8	98.2	132.7	120.7	136.0
March	127.0	128.6	131.0	97.6	133.5	120.9	136.8
April	127.3	128.9	130.8	97.5	134.0	121.0	137.4
May	127.5	129.1	131.1	96.7	134.4	121.0	137.9
June	128.2	129.9	132.1	97.3	135.1	121.6	138.8
July	128.8	130.5	132.8	97.1	135.8	122.0	139.6
August	129.9	131.6	133.2	101.6	136.6	123.2	140.6
September	130.9	132.5	133.6	106.5	137.1	124.5	141.1
October	131.7	133.4	134.1	110.8	137.6	125.8	141.6
November	132.1	133.7	134.5	111.2	138.0	126.0	142.2
December	132.5	134.2	134.6	111.0	138.6	126.3	142.7
1991							
January	132.9	134.7	135.0	108.5	139.5	126.3	143.7
February	132.9	134.8	135.1	104.5	140.2	125.9	144.4
March	133.0	134.8	135.3	101.9	140.5	125.5	144.7
April	133.3	135.1	136.1	101.2	140.9	126.0	144.9
May	133.8	135.6	136.6	102.1	141.3	126.4	145.4
June	134.1	136.0	137.4	101.1	141.8	126.7	145.9
July	134.3	136.2	136.7	100.7	142.3	126.6	146.5
August	134.6	136.6	136.2	101.1	142.9	126.8	146.9
September	135.0	137.0	136.4	101.5	143.4	127.0	147.6
October	135.2	137.2	136.2	101.6	143.7	127.0	148.0
November	135.8	137.8	136.7	102.4	144.2	127.6	148.5
December	136.2	138.2	137.0	103.1	144.7	127.9	149.2
1992							
January	136.2	138.3	136.6	101.5	145.1	127.6	149.6
February	136.5	138.6	137.1	101.2	145.4	127.8	149.9
March	136.9	139.1	137.6	101.2	145.9	128.2	150.4
April	137.2	139.4	137.5	101.4	146.3	128.3	150.9
May	137.5	139.7	137.2	102.0	146.8	128.6	151.3
June	138.0	140.1	137.6	103.3	147.1	129.1	151.7
July	138.4	140.5	137.4	103.7	147.6	129.3	152.2
August	138.7	140.8	138.4	103.5	147.9	129.6	152.6
September	139.0	141.1	138.9	103.6	148.1	129.9	152.9
October	139.5	141.7	138.9	104.3	148.8	130.2	153.7
November	139.8	142.1	138.7	105.1	149.2	130.3	154.3
December	140.1	142.3	138.8	105.3	149.6	130.5	154.7

Table 8-1B. Summary Consumer Price Indexes: Historical Data, 1946 to Date—*Continued*

(Seasonally adjusted.)

Year and month	Urban wage earners and clerical workers (CPI-W), all items	Consumer Price Index, 1982–1984 = 100					
		All urban consumers (CPI-U)					
		All items	Food	Energy	All items less food and energy	Commod-ities	Services
1993							
January	140.5	142.8	139.1	105.0	150.1	130.7	155.3
February	140.8	143.1	139.6	104.3	150.6	131.1	155.6
March	141.0	143.3	139.6	104.9	150.8	131.1	156.0
April	141.4	143.8	140.0	104.9	151.4	131.4	156.7
May	141.8	144.2	141.0	104.3	151.8	131.6	157.3
June	142.0	144.3	140.6	103.9	152.1	131.3	157.8
July	142.1	144.5	140.6	103.4	152.3	131.3	158.1
August	142.4	144.8	141.1	103.4	152.8	131.6	158.6
September	142.5	145.0	141.4	103.0	152.9	131.3	159.0
October	143.2	145.6	142.0	105.3	153.4	132.2	159.4
November	143.4	146.0	142.3	104.4	153.9	132.4	159.9
December	143.7	146.3	142.8	103.7	154.3	132.4	160.5
1994							
January	143.8	146.3	142.9	102.8	154.5	132.3	160.8
February	144.0	146.7	142.7	104.1	154.8	132.4	161.4
March	144.3	147.1	142.7	104.3	155.3	132.5	162.0
April	144.5	147.2	143.0	103.7	155.5	132.6	162.2
May	144.8	147.5	143.3	102.8	155.9	132.9	162.4
June	145.3	147.9	143.8	103.1	156.4	133.5	162.8
July	145.9	148.4	144.6	104.5	156.7	134.1	163.1
August	146.5	149.0	145.1	106.7	157.1	134.7	163.8
September	146.8	149.3	145.3	106.1	157.5	134.8	164.1
October	146.9	149.4	145.3	105.7	157.8	134.8	164.5
November	147.3	149.8	145.6	106.1	158.2	135.0	165.0
December	147.6	150.1	146.0	105.9	158.3	135.4	165.2
1995							
January	148.0	150.5	146.7	105.7	159.0	135.4	166.0
February	148.4	150.9	147.3	105.8	159.4	135.6	166.5
March	148.6	151.2	147.1	105.5	159.9	135.6	167.1
April	149.2	151.8	148.1	105.6	160.4	136.2	167.7
May	149.5	152.1	148.2	105.8	160.7	136.4	168.1
June	149.8	152.4	148.3	106.7	161.1	136.6	168.5
July	149.9	152.6	148.5	105.8	161.4	136.6	168.9
August	150.2	152.9	148.6	105.6	161.8	136.8	169.3
September	150.4	153.1	149.1	104.1	162.2	136.8	169.7
October	150.8	153.5	149.5	104.4	162.7	137.1	170.3
November	150.9	153.7	149.6	103.4	163.0	137.0	170.7
December	151.3	153.9	149.9	104.4	163.1	137.3	170.9
1996							
January	152.0	154.7	150.4	106.9	163.7	138.2	171.5
February	152.3	155.0	150.8	107.1	164.0	138.3	172.0
March	152.9	155.5	151.4	108.3	164.4	139.0	172.4
April	153.4	156.1	152.0	111.2	164.6	139.6	172.8
May	153.8	156.4	151.9	112.0	165.0	139.7	173.4
June	154.0	156.7	152.9	110.3	165.4	139.8	173.8
July	154.3	157.0	153.4	110.1	165.7	139.8	174.4
August	154.5	157.2	153.9	109.7	166.0	139.8	174.9
September	154.9	157.7	154.6	109.8	166.5	140.3	175.4
October	155.4	158.2	155.5	110.5	166.8	140.8	175.8
November	155.9	158.7	156.1	111.8	167.2	141.4	176.3
December	156.3	159.1	156.3	113.9	167.4	141.7	176.7
1997							
January	156.6	159.4	155.9	115.2	167.8	141.8	177.3
February	156.9	159.7	156.5	115.0	168.1	142.1	177.6
March	156.9	159.8	156.6	113.0	168.4	141.8	178.0
April	157.0	159.9	156.5	111.0	168.9	141.6	178.4
May	157.0	159.9	156.6	108.8	169.2	141.4	178.7
June	157.3	160.2	156.9	110.0	169.4	141.5	179.2
July	157.4	160.4	157.2	109.1	169.7	141.4	179.7
August	157.8	160.8	157.7	110.9	169.8	141.9	179.9
September	158.2	161.2	158.0	112.4	170.2	142.2	180.4
October	158.4	161.5	158.3	111.8	170.6	142.2	180.9
November	158.6	161.7	158.6	111.4	170.8	142.1	181.4
December	158.6	161.8	158.7	109.8	171.2	142.1	181.7
1998							
January	158.8	162.0	159.5	107.5	171.6	142.0	182.1
February	158.7	162.0	159.4	105.1	171.9	141.8	182.3
March	158.7	162.0	159.7	103.3	172.2	141.4	182.8
April	158.8	162.2	159.7	102.4	172.5	141.3	183.3
May	159.3	162.6	160.3	103.2	172.9	141.7	183.7
June	159.5	162.8	160.2	103.6	173.2	141.8	184.0
July	159.8	163.2	160.6	103.3	173.5	142.1	184.3
August	160.0	163.4	161.0	102.1	174.0	142.2	184.7
September	160.1	163.5	161.1	101.3	174.2	142.0	185.1
October	160.5	163.9	162.0	101.5	174.4	142.3	185.5
November	160.7	164.1	162.2	101.1	174.8	142.2	186.0
December	161.1	164.4	162.4	100.1	175.4	142.5	186.4

Table 8-1B. Summary Consumer Price Indexes: Historical Data, 1946 to Date—*Continued*

(Seasonally adjusted.)

Year and month	Urban wage earners and clerical workers (CPI-W), all items	Consumer Price Index, 1982–1984 = 100					
		All urban consumers (CPI-U)					
		All items	Food	Energy	All items less food and energy	Commod-ities	Services
1999							
January	161.4	164.7	163.0	99.7	175.6	142.8	186.6
February	161.3	164.7	163.3	99.2	175.6	142.4	186.9
March	161.4	164.8	163.3	100.4	175.7	142.4	187.4
April	162.4	165.9	163.5	105.5	176.3	144.0	187.9
May	162.6	166.0	163.8	104.9	176.5	143.9	188.1
June	162.6	166.0	163.7	104.5	176.6	143.8	188.3
July	163.3	166.7	163.9	106.7	177.1	144.5	188.9
August	163.8	167.1	164.2	109.5	177.3	145.0	189.3
September	164.6	167.8	164.6	111.8	177.8	145.9	189.8
October	164.9	168.1	165.0	112.0	178.1	146.1	190.2
November	165.1	168.4	165.3	111.5	178.4	145.9	190.9
December	165.6	168.8	165.5	113.8	178.7	146.5	191.2
2000							
January	166.0	169.3	165.6	115.0	179.3	146.7	192.0
February	166.7	170.0	166.2	118.8	179.4	147.6	192.5
March	167.8	171.0	166.5	124.3	180.0	149.1	193.1
April	167.6	170.9	166.7	120.9	180.3	148.5	193.5
May	167.9	171.2	167.3	120.0	180.7	148.5	194.0
June	169.0	172.2	167.4	126.8	181.1	149.6	194.9
July	169.5	172.7	168.3	127.3	181.5	149.8	195.7
August	169.3	172.7	168.7	123.8	181.9	149.2	196.3
September	170.3	173.6	168.9	129.2	182.3	150.4	196.9
October	170.5	173.9	169.0	129.6	182.6	150.1	197.7
November	170.8	174.2	169.2	129.2	183.1	150.3	198.1
December	171.2	174.6	170.0	130.1	183.3	150.4	198.8
2001							
January	172.2	175.6	170.3	135.0	183.9	150.6	200.6
February	172.5	176.0	171.2	134.1	184.4	150.8	201.2
March	172.6	176.1	171.7	131.7	184.7	150.5	201.6
April	173.0	176.4	172.1	132.3	185.1	150.9	202.0
May	174.0	177.3	172.4	138.4	185.3	151.9	202.7
June	174.2	177.7	173.1	136.9	186.0	151.9	203.5
July	173.8	177.4	173.6	129.6	186.4	150.9	203.8
August	173.8	177.4	174.0	127.3	186.7	150.4	204.4
September	174.7	178.1	174.2	130.9	187.1	151.6	204.5
October	173.9	177.6	174.8	122.9	187.4	150.3	204.8
November	173.7	177.5	174.9	116.9	188.1	149.2	205.6
December	173.4	177.4	174.7	113.9	188.4	148.3	206.1
2002							
January	173.7	177.7	175.3	114.2	188.7	148.3	206.8
February	173.9	178.0	175.7	113.5	189.1	148.3	207.5
March	174.5	178.5	176.1	117.6	189.2	149.0	207.9
April	175.4	179.3	176.4	121.5	189.7	150.0	208.5
May	175.5	179.5	175.8	121.9	190.0	149.7	209.0
June	175.7	179.6	175.9	121.6	190.2	149.7	209.3
July	176.1	180.0	176.1	122.4	190.5	149.9	209.9
August	176.6	180.5	176.1	123.1	191.1	150.2	210.6
September	176.8	180.8	176.5	123.7	191.3	150.2	211.1
October	177.2	181.2	176.4	126.9	191.5	150.5	211.7
November	177.5	181.5	176.9	126.7	191.9	150.5	212.4
December	177.7	181.8	177.1	127.1	192.1	150.3	213.0
2003							
January	178.6	182.6	177.1	133.5	192.4	151.3	213.7
February	179.7	183.6	178.1	140.8	192.5	152.7	214.2
March	180.1	183.9	178.4	143.9	192.5	152.6	215.0
April	179.2	183.2	178.5	136.5	192.5	151.1	215.1
May	178.7	182.9	178.8	129.4	192.9	149.4	216.0
June	178.9	183.1	179.7	129.8	193.0	149.7	216.2
July	179.4	183.7	179.8	132.2	193.4	150.3	216.8
August	180.3	184.5	180.5	137.8	193.6	151.4	217.2
September	180.9	185.1	180.9	142.9	193.7	152.1	217.8
October	180.6	184.9	181.6	137.8	194.0	151.1	218.4
November	180.6	185.0	182.6	136.9	194.0	151.1	218.6
December	181.0	185.5	183.5	138.8	194.2	151.6	219.0
2004							
January	181.9	186.3	183.4	143.9	194.6	152.5	219.7
February	182.4	186.7	183.9	145.8	194.9	153.1	220.1
March	182.7	187.1	184.2	145.1	195.5	153.3	220.8
April	182.8	187.4	184.6	143.4	195.9	153.0	221.4
May	183.8	188.2	186.1	147.6	196.2	154.2	221.9
June	184.4	188.9	186.4	151.5	196.6	154.9	222.7
July	184.6	189.1	186.8	151.1	196.8	154.6	223.2
August	184.7	189.2	187.0	151.5	196.9	154.4	223.7
September	185.3	189.8	186.9	152.9	197.5	154.9	224.2
October	186.4	190.8	187.8	158.6	197.9	156.7	224.6
November	187.4	191.7	188.4	163.8	198.3	157.8	225.4
December	187.4	191.7	188.4	162.3	198.6	157.4	225.8

Table 8-1B. Summary Consumer Price Indexes: Historical Data, 1946 to Date—*Continued*

(Seasonally adjusted.)

| Year and month | Urban wage earners and clerical workers (CPI-W), all items | Consumer Price Index, 1982–1984 = 100 | | | | | |
| | | All urban consumers (CPI-U) | | | | | |
		All items	Food	Energy	All items less food and energy	Commod-ities	Services
2005							
January	187.2	191.6	188.7	157.2	199.0	156.8	226.2
February	188.0	192.4	188.6	161.8	199.4	157.5	226.9
March	188.6	193.1	189.1	163.5	200.1	157.9	228.0
April	189.3	193.7	190.4	166.7	200.2	158.6	228.5
May	189.1	193.6	190.6	163.3	200.5	158.1	228.9
June	189.3	193.7	190.5	163.7	200.6	158.0	229.1
July	190.6	194.9	190.9	172.8	200.9	159.6	229.9
August	192.0	196.1	191.1	183.5	201.1	161.5	230.4
September	195.1	198.8	191.5	208.2	201.3	166.0	231.4
October	195.2	199.1	192.0	205.9	202.0	164.8	233.1
November	193.7	198.1	192.6	191.0	202.5	161.8	234.2
December	193.7	198.1	192.9	187.6	202.8	161.5	234.4
2006							
January	195.1	199.3	193.6	196.6	203.2	162.9	235.5
February	195.0	199.4	193.7	194.1	203.6	162.5	235.9
March	195.3	199.7	194.0	192.0	204.3	162.6	236.5
April	196.4	200.7	193.8	198.5	204.8	164.1	237.1
May	197.0	201.3	194.1	199.8	205.4	164.6	237.8
June	197.4	201.8	194.7	199.8	205.9	164.8	238.5
July	198.6	202.9	195.2	207.9	206.3	166.4	239.2
August	199.5	203.8	195.7	211.9	206.8	167.2	239.9
September	198.3	202.8	196.3	198.1	207.2	164.6	240.7
October	197.1	201.9	196.9	184.1	207.6	162.5	241.0
November	197.2	202.0	197.0	184.1	207.8	162.0	241.7
December	198.4	203.1	197.1	192.7	208.1	163.4	242.4
2007							
January	198.6	203.4	198.4	190.3	208.6	163.3	243.2
February	199.4	204.2	199.7	192.3	209.1	164.1	244.1
March	200.7	205.3	200.4	200.2	209.4	165.6	244.7
April	201.3	205.9	201.0	203.3	209.7	166.2	245.3
May	202.3	206.8	201.7	208.6	210.1	167.3	245.9
June	202.7	207.2	202.6	209.8	210.4	167.5	246.6
July	203.0	207.6	203.2	209.6	210.8	167.8	247.1
August	203.0	207.7	204.0	206.4	211.1	167.5	247.5
September	204.0	208.5	205.0	210.7	211.6	168.5	248.2
October	204.7	209.2	205.7	212.4	212.1	169.0	249.1
November	206.5	210.8	206.5	223.8	212.7	171.5	249.8
December	207.1	211.4	206.9	225.6	213.2	172.0	250.5
2008							
January	207.9	212.2	208.1	226.8	213.8	172.7	251.3
February	208.4	212.6	208.9	229.7	213.9	173.0	251.8
March	209.2	213.4	209.3	233.3	214.4	173.7	252.9
April	209.8	214.0	211.1	234.8	214.6	174.1	253.6
May	211.1	215.2	212.0	243.9	214.9	175.5	254.7
June	213.7	217.5	213.3	262.1	215.4	178.8	255.8
July	215.5	219.1	215.4	271.1	216.0	180.7	257.1
August	215.0	218.7	216.5	262.6	216.4	179.6	257.6
September	215.1	218.9	217.8	260.1	216.7	179.8	257.6
October	212.7	217.0	218.7	238.1	216.8	175.8	257.7
November	207.9	213.1	219.1	195.2	216.9	168.0	257.8
December	205.9	211.4	219.1	176.6	216.9	164.5	258.0
2009							
January	206.5	212.0	219.2	178.8	217.3	165.3	258.4
February	207.4	212.8	219.0	184.9	217.8	166.8	258.7
March	207.1	212.6	218.5	178.8	218.3	166.2	258.7
April	207.4	212.7	218.2	177.1	218.7	166.5	258.7
May	207.8	213.0	217.8	179.7	218.9	167.2	258.6
June	209.9	214.7	217.8	196.9	219.2	170.7	258.7
July	210.0	214.7	217.4	195.6	219.3	170.5	258.7
August	210.9	215.5	217.4	201.5	219.5	171.3	259.3
September	211.3	215.9	217.3	202.8	219.9	171.9	259.6
October	211.9	216.5	217.5	204.6	220.5	172.7	260.0
November	212.7	217.1	217.6	210.3	220.6	173.8	260.2
December	213.0	217.3	217.9	210.3	220.8	174.2	260.3
2010							
January	213.4	217.5	218.5	212.8	220.6	174.9	259.8
February	213.3	217.3	218.6	209.6	220.7	174.3	260.0
March	213.3	217.4	219.0	209.3	220.8	174.1	260.3
April	213.2	217.4	219.2	209.2	220.8	173.9	260.7
May	212.9	217.3	219.3	206.6	221.0	173.4	260.9
June	212.9	217.2	219.3	203.8	221.2	172.9	261.2
July	213.5	217.6	219.2	206.9	221.4	173.4	261.5
August	213.9	217.9	219.5	208.8	221.5	173.8	261.7
September	214.3	218.3	220.2	209.8	221.7	174.4	261.9
October	215.1	219.0	220.5	216.7	221.8	175.7	262.1
November	215.5	219.6	220.9	219.5	222.1	176.5	262.4
December	216.9	220.5	221.2	227.1	222.3	177.9	262.7

Table 8-1B. Summary Consumer Price Indexes: Historical Data, 1946 to Date—*Continued*

(Seasonally adjusted.)

Year and month	Urban wage earners and clerical workers (CPI-W), all items	Consumer Price Index, 1982–1984 = 100 — All urban consumers (CPI-U)					
		All items	Food	Energy	All items less food and energy	Commod-ities	Services
2011							
January	217.5	221.2	222.5	229.3	222.8	178.9	263.2
February	218.2	221.9	223.5	232.1	223.2	179.8	263.7
March	219.5	223.0	225.2	240.1	223.5	181.8	264.1
April	220.8	224.1	226.1	248.0	223.7	183.5	264.5
May	221.6	224.8	226.9	250.7	224.2	184.5	264.9
June	221.5	224.8	227.5	245.5	224.7	184.1	265.3
July	222.1	225.4	228.5	246.2	225.2	184.6	266.0
August	222.8	226.1	229.6	246.9	225.9	185.3	266.7
September	223.4	226.6	230.7	248.6	226.1	185.8	267.1
October	223.5	226.8	230.9	246.7	226.5	185.7	267.5
November	224.0	227.2	231.1	247.6	226.9	186.2	267.8
December	223.9	227.2	231.6	243.4	227.4	185.7	268.5
2012							
January	224.5	227.8	232.2	244.9	227.9	186.5	268.9
February	225.1	228.3	232.1	248.9	228.0	187.4	269.0
March	225.7	228.8	232.6	249.7	228.5	187.9	269.6
April	226.0	229.2	233.0	249.7	228.9	188.1	270.1
May	225.4	228.7	233.2	241.8	229.2	186.8	270.4
June	225.0	228.5	233.7	235.9	229.6	185.8	271.0
July	224.9	228.6	233.8	233.6	230.0	185.5	271.4
August	226.4	229.9	234.2	245.0	230.2	187.5	272.1
September	227.7	231.0	234.4	253.0	230.7	189.1	272.7
October	228.4	231.6	234.8	256.0	231.0	189.8	273.2
November	227.8	231.2	235.3	248.8	231.3	188.5	273.8
December	227.7	231.2	235.7	244.7	231.7	187.7	274.4
2013							
January	228.1	231.7	236.0	245.0	232.2	188.0	275.1
February	229.6	232.9	236.0	255.7	232.6	189.9	275.8
March	228.8	232.3	236.2	246.6	232.8	188.1	276.2
April	228.1	231.8	236.7	240.5	232.8	186.8	276.5
May	228.3	231.9	236.5	240.5	233.0	186.5	277.0
June	228.8	232.4	237.0	242.7	233.4	187.1	277.4
July	229.2	232.9	237.2	243.0	233.9	187.3	278.1
August	229.8	233.5	237.5	244.8	234.3	187.8	278.7
September	229.8	233.5	237.5	242.7	234.7	187.4	279.4
October	229.9	233.7	237.8	242.0	234.9	187.2	279.8
November	230.4	234.1	237.9	242.7	235.4	187.5	280.3
December	231.0	234.7	238.2	245.7	235.8	188.2	280.9
2014							
January	231.6	235.3	238.5	250.3	236.0	188.5	281.7
February	231.9	235.5	239.4	249.9	236.2	188.4	282.3
March	232.3	236.0	240.4	250.0	236.6	188.1	283.5
April	232.8	236.5	241.2	249.9	237.1	188.8	283.8
May	233.1	236.9	242.4	249.2	237.5	188.8	284.7
June	233.4	237.2	242.6	249.7	237.8	189.0	285.1
July	233.6	237.5	243.2	248.7	238.2	188.9	285.7
August	233.5	237.5	243.8	245.7	238.4	188.5	286.0
September	233.4	237.5	244.5	241.6	238.8	188.1	286.4
October	233.2	237.4	244.9	237.1	239.2	187.5	286.9
November	232.6	237.0	245.4	229.0	239.5	186.1	287.4
December	231.5	236.3	246.2	218.5	239.6	184.0	288.0
2015							
January	229.4	234.7	246.2	199.5	239.8	180.2	288.6
February	230.1	235.2	246.5	202.1	240.1	180.9	289.0
March	231.0	236.0	246.1	206.1	240.7	182.0	289.5
April	231.0	236.2	246.0	202.9	241.3	181.5	290.2
May	232.0	237.0	246.3	209.1	241.6	182.8	290.5
June	232.7	237.7	247.0	212.5	242.0	183.5	291.3
July	233.1	238.1	247.2	212.3	242.5	183.5	292.0
August	233.0	238.0	247.7	208.9	242.8	183.0	292.4
September	232.1	237.5	248.5	197.3	243.3	181.2	293.1
October	232.3	237.8	248.7	196.0	243.8	181.0	293.9
November	232.5	238.0	248.5	194.4	244.3	180.6	294.7
December	232.2	237.8	248.2	190.3	244.6	179.8	295.1
2016							
January	232.1	237.8	248.2	186.1	245.1	179.1	295.8
February	231.5	237.5	248.6	176.4	245.7	177.7	296.5
March	232.2	238.0	248.1	181.1	246.0	178.2	297.1
April	233.0	238.8	248.3	185.4	246.5	179.0	297.9
May	233.6	239.5	248.0	188.4	247.0	179.3	298.8
June	234.4	240.2	247.8	193.1	247.4	179.9	299.6
July	234.2	240.2	247.7	190.1	247.8	179.0	300.4
August	234.6	240.6	247.7	189.8	248.4	179.0	301.3
September	235.0	241.1	247.7	191.8	248.7	179.3	302.0
October	235.7	241.7	247.8	195.8	249.1	180.0	302.5
November	236.0	242.0	247.6	195.5	249.6	179.8	303.4
December	236.8	242.8	247.6	200.3	250.0	180.6	304.1

Table 8-1C. Consumer and Producer Price Indexes: Historical Data, 1913–1949

(Not seasonally adjusted.)

| Year and month | Consumer price indexes, 1982–1984 =100 | | | | Producer Price Indexes for goods, 1982 = 100 | | | |
| | All urban consumers (CPI-U) | | Urban wage earners and clerical workers (CPI-W) | | All commodities | | Farm products | Industrial commodities |
	Index	Percent change	Index	Percent change	Index	Percent change		
1913	9.9	...	10.0	...	12.0	...	18.0	11.9
1914	10.0	1.0	10.1	1.0	11.8	-1.7	17.9	11.3
1915	10.1	1.0	10.2	1.0	12.0	1.7	18.0	11.6
1916	10.9	7.9	11.0	7.8	14.7	22.5	21.3	15.0
1917	12.8	17.4	12.9	17.3	20.2	37.4	32.6	19.5
1918	15.1	18.0	15.1	17.1	22.6	11.9	37.4	21.1
1919	17.3	14.6	17.4	15.2	23.9	5.8	39.8	22.0
1920	20.0	15.6	20.1	15.5	26.6	11.3	38.0	27.4
1921	17.9	-10.5	18.0	-10.4	16.8	-36.8	22.3	17.8
1922	16.8	-6.1	16.9	-6.1	16.7	-0.6	23.7	17.4
1923	17.1	1.8	17.2	1.8	17.3	3.6	24.9	17.8
1924	17.1	0.0	17.2	0.0	16.9	-2.3	25.2	17.0
1925	17.5	2.3	17.6	2.3	17.8	5.3	27.7	17.5
1926	17.7	1.1	17.8	1.1	17.2	-3.4	25.3	17.0
1927	17.4	-1.7	17.5	-1.7	16.5	-4.1	25.1	16.0
1928	17.1	-1.7	17.2	-1.7	16.7	1.2	26.7	15.8
1929	17.1	0.0	17.2	0.0	16.4	-1.8	26.4	15.6
1930	16.7	-2.3	16.8	-2.3	14.9	-9.1	22.4	14.5
1931	15.2	-9.0	15.3	-8.9	12.6	-15.4	16.4	12.8
1932	13.7	-9.9	13.7	-10.5	11.2	-11.1	12.2	11.9
1933	13.0	-5.1	13.0	-5.1	11.4	1.8	13.0	12.1
1934	13.4	3.1	13.5	3.8	12.9	13.2	16.5	13.3
1935	13.7	2.2	13.8	2.2	13.8	7.0	19.8	13.3
1936	13.9	1.5	13.9	0.7	13.0	0.7	20.4	13.5
1937	14.4	3.6	14.4	3.6	14.9	7.2	21.8	14.5
1938	14.1	-2.1	14.2	-1.4	13.5	-9.4	17.3	13.9
1939	13.9	-1.4	14.0	-1.4	13.3	-1.5	16.5	13.9
1940	14.0	0.7	14.1	0.7	13.5	1.5	17.1	14.1
1941	14.7	5.0	14.8	5.0	15.1	11.9	20.8	15.1
1942	16.3	10.9	16.4	10.8	17.0	12.6	26.7	16.2
1943	17.3	6.1	17.4	6.1	17.8	4.7	30.9	16.5
1944	17.6	1.7	17.7	1.7	17.9	0.6	31.2	16.7
1945	18.0	2.3	18.1	2.3	18.2	1.7	32.4	17.0
1946	19.5	8.3	19.6	8.3	20.8	14.3	37.5	18.6
1947	22.3	14.4	22.5	14.8	25.6	23.1	45.1	22.7
1948	24.1	8.1	24.2	7.6	27.7	8.2	48.5	24.6
1949	23.8	-1.2	24.0	-0.8	26.3	-5.1	41.9	24.1
1913								
January	9.8	...	9.9	...	12.1	...	17.6	12.3
February	9.8	0.0	9.8	-1.0	12.0	-0.8	17.5	12.2
March	9.8	0.0	9.8	0.0	12.0	0.0	17.6	12.1
April	9.8	0.0	9.9	1.0	12.0	0.0	17.5	12.0
May	9.7	-1.0	9.8	-1.0	11.9	-0.8	17.4	11.9
June	9.8	1.0	9.8	0.0	11.9	0.0	17.6	11.8
July	9.9	1.0	9.9	1.0	12.0	0.8	18.1	11.8
August	9.9	0.0	10.0	1.0	12.0	0.0	18.2	11.7
September	10.0	1.0	10.0	0.0	12.2	1.7	18.8	11.8
October	10.0	0.0	10.1	1.0	12.2	0.0	18.8	11.8
November	10.1	1.0	10.1	0.0	12.1	-0.8	18.9	11.7
December	10.0	-1.0	10.1	0.0	11.9	-1.7	18.5	11.6
1914								
January	10.0	0.0	10.1	0.0	11.8	-0.8	18.4	11.5
February	9.9	-1.0	10.0	-1.0	11.8	0.0	18.3	11.5
March	9.9	0.0	10.0	0.0	11.7	-0.8	18.2	11.5
April	9.8	-1.0	9.9	-1.0	11.7	0.0	18.0	11.5
May	9.9	1.0	9.9	0.0	11.6	-0.9	18.0	11.4
June	9.9	0.0	10.0	1.0	11.6	0.0	18.1	11.3
July	10.0	1.0	10.1	1.0	11.6	0.0	18.0	11.2
August	10.2	2.0	10.2	1.0	12.0	3.4	18.3	11.2
September	10.2	0.0	10.3	1.0	12.1	0.8	17.9	11.3
October	10.1	-1.0	10.2	-1.0	11.7	-3.3	17.2	11.0
November	10.2	1.0	10.2	0.0	11.7	0.0	17.6	11.0
December	10.1	-1.0	10.2	0.0	11.6	-0.9	17.4	11.0
1915								
January	10.1	0.0	10.2	0.0	11.8	1.7	18.1	11.1
February	10.0	-1.0	10.1	-1.0	11.8	0.0	18.4	11.0
March	9.9	-1.0	10.0	-1.0	11.8	0.0	17.9	11.0
April	10.0	1.0	10.1	1.0	11.8	0.0	18.2	11.1
May	10.1	1.0	10.1	0.0	11.9	0.8	18.2	11.2
June	10.1	0.0	10.2	1.0	11.8	-0.8	17.7	11.4
July	10.1	0.0	10.2	0.0	11.9	0.8	18.1	11.5
August	10.1	0.0	10.2	0.0	11.8	-0.8	17.9	11.5
September	10.1	0.0	10.2	0.0	11.8	0.0	17.5	11.7
October	10.2	1.0	10.3	1.0	12.1	2.5	18.1	11.9
November	10.3	1.0	10.4	1.0	12.3	1.7	18.0	12.3
December	10.3	0.0	10.4	0.0	12.8	4.1	18.4	12.9

... = Not available.

Table 8-1C. Consumer and Producer Price Indexes: Historical Data, 1913–1949—*Continued*

(Not seasonally adjusted.)

Year and month	Consumer price indexes, 1982–1984 =100				Producer Price Indexes for goods, 1982 = 100			
	All urban consumers (CPI-U)		Urban wage earners and clerical workers (CPI-W)		All commodities		Farm products	Industrial commodities
	Index	Percent change	Index	Percent change	Index	Percent change		
1916								
January	10.4	1.0	10.5	1.0	13.3	3.9	19.4	13.6
February	10.4	0.0	10.5	0.0	13.5	1.5	19.4	14.0
March	10.5	1.0	10.6	1.0	13.9	3.0	19.4	14.4
April	10.6	1.0	10.7	0.9	14.1	1.4	19.6	14.6
May	10.7	0.9	10.7	0.0	14.2	0.7	19.8	14.7
June	10.8	0.9	10.9	1.9	14.3	0.7	19.7	14.8
July	10.8	0.0	10.9	0.0	14.4	0.7	20.3	14.6
August	10.9	0.9	11.0	0.9	14.7	2.1	21.7	14.6
September	11.1	1.8	11.2	1.8	15.0	2.0	22.6	14.8
October	11.3	1.8	11.3	0.9	15.7	4.7	23.7	15.5
November	11.5	1.8	11.5	1.8	16.8	7.0	25.3	16.8
December	11.6	0.9	11.6	0.9	17.1	1.8	25.0	17.7
1917								
January	11.7	0.9	11.8	1.7	17.6	2.9	26.2	18.3
February	12.0	2.6	12.0	1.7	18.0	2.3	27.2	18.6
March	12.0	0.0	12.1	0.8	18.5	2.8	28.6	18.8
April	12.6	5.0	12.6	4.1	19.7	6.5	31.6	19.0
May	12.8	1.6	12.9	2.4	20.8	5.6	33.6	19.8
June	13.0	1.6	13.0	0.8	21.0	1.0	33.8	20.4
July	12.8	-1.5	12.9	-0.8	21.2	1.0	34.0	20.7
August	13.0	1.6	13.1	1.6	21.5	1.4	34.6	20.7
September	13.3	2.3	13.3	1.5	21.3	-0.9	34.3	20.1
October	13.5	1.5	13.6	2.3	21.1	-0.9	35.2	19.1
November	13.5	0.0	13.6	0.0	21.2	0.5	36.0	19.2
December	13.7	1.5	13.8	1.5	21.2	0.0	35.6	19.5
1918								
January	14.0	2.2	14.0	1.4	21.6	1.9	37.0	19.9
February	14.1	0.7	14.2	1.4	21.1	-2.3	37.1	19.0
March	14.0	-0.7	14.1	-0.7	21.8	3.3	37.3	20.2
April	14.2	1.4	14.3	1.4	22.1	1.4	36.6	20.7
May	14.5	2.1	14.5	1.4	22.1	0.0	35.4	21.0
June	14.7	1.4	14.8	2.1	22.2	0.5	35.4	21.3
July	15.1	2.7	15.2	2.7	22.7	2.3	37.0	21.5
August	15.4	2.0	15.4	1.3	23.2	2.2	38.6	21.7
September	15.7	1.9	15.8	2.6	23.7	2.2	39.6	22.1
October	16.0	1.9	16.1	1.9	23.5	-0.8	38.2	22.1
November	16.3	1.9	16.3	1.2	23.5	0.0	38.0	22.1
December	16.5	1.2	16.6	1.8	23.5	0.0	38.1	21.8
1919								
January	16.5	0.0	16.6	0.0	23.2	-1.3	38.9	21.1
February	16.2	-1.8	16.2	-2.4	22.4	-3.4	37.5	20.6
March	16.4	1.2	16.5	1.9	22.6	0.9	38.5	20.1
April	16.7	1.8	16.8	1.8	22.9	1.3	40.0	20.0
May	16.9	1.2	17.0	1.2	23.3	1.7	40.9	20.2
June	16.9	0.0	17.0	0.0	23.4	0.4	39.6	21.1
July	17.4	3.0	17.5	2.9	24.3	3.8	41.5	22.1
August	17.7	1.7	17.8	1.7	24.9	2.5	41.3	23.1
September	17.8	0.6	17.9	0.6	24.3	-2.4	38.7	23.2
October	18.1	1.7	18.2	1.7	24.4	0.4	38.5	23.5
November	18.5	2.2	18.6	2.2	24.9	2.0	40.3	23.8
December	18.9	2.2	19.0	2.2	26.0	4.4	41.8	24.7
1920								
January	19.3	2.1	19.4	2.1	27.2	4.6	43.0	26.1
February	19.5	1.0	19.6	1.0	27.1	-0.4	41.2	27.1
March	19.7	1.0	19.8	1.0	27.3	0.7	41.5	27.7
April	20.3	3.0	20.4	3.0	28.5	4.4	42.5	28.7
May	20.6	1.5	20.7	1.5	28.8	1.1	42.8	29.0
June	20.9	1.5	21.0	1.4	28.7	-0.3	42.3	29.0
July	20.8	-0.5	20.9	-0.5	28.6	-0.3	40.5	29.5
August	20.3	-2.4	20.4	-2.4	27.8	-2.8	37.8	29.7
September	20.0	-1.5	20.1	-1.5	26.8	-3.6	36.4	28.5
October	19.9	-0.5	20.0	-0.5	24.9	-7.1	32.2	26.9
November	19.8	-0.5	19.9	-0.5	23.0	-7.6	30.0	24.5
December	19.4	-2.0	19.5	-2.0	20.8	-9.6	26.4	22.7
1921								
January	19.0	-2.1	19.1	-2.1	19.6	-5.8	25.6	21.1
February	18.4	-3.2	18.5	-3.1	18.1	-7.7	23.4	19.4
March	18.3	-0.5	18.4	-0.5	17.7	-2.2	22.7	18.7
April	18.1	-1.1	18.2	-1.1	17.0	-4.0	20.9	18.4
May	17.7	-2.2	17.8	-2.2	16.6	-2.4	21.0	17.9
June	17.6	-0.6	17.7	-0.6	16.1	-3.0	20.3	17.4
July	17.7	0.6	17.8	0.6	16.1	0.0	21.8	16.9
August	17.7	0.0	17.8	0.0	16.1	0.0	22.5	16.5
September	17.5	-1.1	17.6	-1.1	16.1	0.0	22.7	16.5
October	17.5	0.0	17.6	0.0	16.2	0.6	22.7	17.0
November	17.4	-0.6	17.5	-0.6	16.2	0.0	22.1	17.3
December	17.3	-0.6	17.4	-0.6	16.0	-1.2	22.2	17.0

Table 8-1C. Consumer and Producer Price Indexes: Historical Data, 1913–1949—*Continued*

(Not seasonally adjusted.)

| Year and month | Consumer price indexes, 1982–1984 =100 | | | | Producer Price Indexes for goods, 1982 = 100 | | | |
| | All urban consumers (CPI-U) | | Urban wage earners and clerical workers (CPI-W) | | All commodities | | Farm products | Industrial commodities |
	Index	Percent change	Index	Percent change	Index	Percent change		
1922								
January	16.9	-2.3	17.0	-2.3	15.7	-1.9	22.2	16.7
February	16.9	0.0	17.0	0.0	16.0	1.9	24.0	16.6
March	16.7	-1.2	16.8	-1.2	16.0	0.0	23.6	16.5
April	16.7	0.0	16.8	0.0	16.1	0.6	23.4	16.6
May	16.7	0.0	16.8	0.0	16.6	3.1	23.8	17.4
June	16.7	0.0	16.8	0.0	16.6	0.0	23.4	17.4
July	16.8	0.6	16.9	0.6	17.1	3.0	24.1	18.1
August	16.6	-1.2	16.7	-1.2	17.0	-0.6	23.0	18.2
September	16.6	0.0	16.7	0.0	17.1	0.6	23.3	18.2
October	16.7	0.6	16.8	0.6	17.2	0.6	23.8	17.9
November	16.8	0.6	16.9	0.6	17.3	0.6	24.7	17.7
December	16.9	0.6	17.0	0.6	17.3	0.0	25.0	17.7
1923								
January	16.8	-0.6	16.9	-0.6	17.6	1.7	25.1	18.2
February	16.8	0.0	16.9	0.0	17.8	1.1	25.3	18.6
March	16.8	0.0	16.9	0.0	18.0	1.1	25.3	18.8
April	16.9	0.6	17.0	0.6	17.9	-0.6	24.8	18.7
May	16.9	0.0	17.0	0.0	17.5	-2.2	24.4	18.3
June	17.0	0.6	17.1	0.6	17.3	-1.1	24.2	17.9
July	17.2	1.2	17.3	1.2	17.0	-1.7	23.7	17.6
August	17.1	-0.6	17.2	-0.6	16.9	-0.6	24.2	17.4
September	17.2	0.6	17.3	0.6	17.2	1.8	25.3	17.3
October	17.3	0.6	17.4	0.6	17.1	-0.6	25.4	17.1
November	17.3	0.0	17.4	0.0	17.0	-0.0	25.7	16.9
December	17.3	0.0	17.4	0.0	16.9	-0.6	25.5	16.9
1924								
January	17.3	0.0	17.4	0.0	17.2	1.8	25.6	17.4
February	17.2	-0.6	17.3	-0.6	17.2	0.0	25.0	17.6
March	17.1	-0.6	17.2	-0.6	17.0	-1.2	24.2	17.5
April	17.0	-0.6	17.1	-0.6	16.7	-1.8	24.6	17.3
May	17.0	0.0	17.1	0.0	16.5	-1.2	24.0	17.0
June	17.0	0.0	17.1	0.0	16.4	-0.6	23.8	16.7
July	17.1	0.6	17.2	0.6	16.5	0.6	24.9	16.6
August	17.0	-0.6	17.1	-0.6	16.7	1.2	25.7	16.6
September	17.1	0.6	17.2	0.6	16.7	0.0	25.3	16.6
October	17.2	0.6	17.3	0.6	16.9	1.2	26.0	16.6
November	17.2	0.0	17.3	0.0	17.1	1.2	26.2	16.8
December	17.3	0.6	17.4	0.6	17.5	2.3	27.3	17.1
1925								
January	17.3	0.0	17.4	0.0	17.7	1.1	28.7	17.3
February	17.2	-0.6	17.3	-0.6	17.9	1.1	28.4	17.7
March	17.3	0.6	17.4	0.6	17.9	0.0	28.5	17.5
April	17.2	-0.6	17.3	-0.6	17.5	-2.2	27.1	17.2
May	17.3	0.6	17.4	0.6	17.5	0.0	27.1	17.3
June	17.5	1.2	17.6	1.1	17.7	1.1	27.6	17.5
July	17.7	1.1	17.8	1.1	18.0	1.7	28.3	17.6
August	17.7	0.0	17.8	0.0	17.9	-0.6	28.1	17.4
September	17.7	0.0	17.8	0.0	17.8	-0.6	27.7	17.4
October	17.7	0.0	17.8	0.0	17.8	0.0	27.0	17.5
November	18.0	1.7	18.1	1.7	18.0	1.1	27.3	17.6
December	17.9	-0.6	18.0	-0.6	17.8	-1.1	26.6	17.6
1926								
January	17.9	0.0	18.0	0.0	17.8	0.0	27.1	17.5
February	17.9	0.0	18.0	0.0	17.6	-1.1	26.5	17.3
March	17.8	-0.6	17.9	-0.6	17.3	-1.7	25.7	17.1
April	17.9	0.6	18.0	0.6	17.3	0.0	26.0	17.0
May	17.8	-0.6	17.9	-0.6	17.3	0.0	25.8	17.0
June	17.7	-0.6	17.8	-0.6	17.3	0.0	25.5	17.0
July	17.5	-1.1	17.6	-1.1	17.1	-1.2	24.9	16.9
August	17.4	-0.6	17.5	-0.6	17.1	0.0	24.6	16.9
September	17.5	0.6	17.6	0.6	17.2	0.6	25.1	16.9
October	17.6	0.6	17.7	0.6	17.1	-0.6	24.7	16.9
November	17.7	0.6	17.8	0.6	17.0	-0.6	23.9	16.9
December	17.7	0.0	17.8	0.0	16.9	-0.6	24.0	16.7
1927								
January	17.5	-1.1	17.6	-1.1	16.4	-3.0	24.3	16.4
February	17.4	-0.6	17.5	-0.6	16.6	1.2	24.1	16.3
March	17.3	-0.6	17.4	-0.6	16.5	-0.6	23.8	16.1
April	17.3	0.0	17.4	0.0	16.3	-1.2	23.8	15.9
May	17.4	0.6	17.5	0.6	16.2	-0.6	24.3	15.9
June	17.6	1.1	17.7	1.1	16.2	0.0	24.3	15.9
July	17.3	-1.7	17.4	-1.7	16.2	0.0	24.6	15.9
August	17.2	-0.6	17.3	-0.6	16.4	1.2	25.8	15.9
September	17.3	0.6	17.4	0.6	16.6	1.2	26.7	16.0
October	17.4	0.6	17.5	0.6	16.7	0.6	26.5	15.9
November	17.3	-0.6	17.4	-0.6	16.6	-0.6	26.3	15.8
December	17.3	0.0	17.4	0.0	16.6	0.0	26.3	15.9

Table 8-1C. Consumer and Producer Price Indexes: Historical Data, 1913–1949—*Continued*

(Not seasonally adjusted.)

Year and month	Consumer price indexes, 1982–1984 =100				Producer Price Indexes for goods, 1982 = 100			
	All urban consumers (CPI-U)		Urban wage earners and clerical workers (CPI-W)		All commodities		Farm products	Industrial commodities
	Index	Percent change	Index	Percent change	Index	Percent change		
1928								
January	17.3	0.0	17.4	0.0	16.6	0.0	26.8	15.8
February	17.1	-1.2	17.2	-1.1	16.5	-0.6	26.4	15.8
March	17.1	0.0	17.2	0.0	16.5	0.0	26.1	15.8
April	17.1	0.0	17.2	0.0	16.7	1.2	27.1	15.8
May	17.2	0.6	17.3	0.6	16.8	0.6	27.7	15.8
June	17.1	-0.6	17.2	-0.6	16.7	-0.6	26.9	15.8
July	17.1	0.0	17.2	0.0	16.8	0.6	27.4	15.8
August	17.1	0.0	17.2	0.0	16.8	0.0	27.0	15.8
September	17.3	1.2	17.4	1.2	17.0	1.2	27.5	15.8
October	17.2	-0.6	17.3	-0.6	16.7	-1.8	26.1	15.8
November	17.2	0.0	17.3	0.0	16.5	-1.2	25.7	15.8
December	17.1	-0.6	17.2	-0.6	16.5	0.0	26.2	15.8
1929								
January	17.1	0.0	17.2	0.0	16.5	0.0	26.7	15.7
February	17.1	0.0	17.2	0.0	16.4	-0.6	26.6	15.6
March	17.0	-0.6	17.1	-0.6	16.6	1.2	27.1	15.7
April	16.9	-0.6	17.0	-0.6	16.5	-0.6	26.5	15.6
May	17.0	0.6	17.1	0.6	16.3	-1.2	25.8	15.6
June	17.1	0.6	17.2	0.6	16.4	0.6	26.1	15.6
July	17.3	1.2	17.4	1.2	16.6	1.2	27.1	15.6
August	17.3	0.0	17.4	0.0	16.6	0.0	27.1	15.5
September	17.3	0.0	17.4	0.0	16.6	0.0	26.9	15.6
October	17.3	0.0	17.4	0.0	16.4	-1.2	26.2	15.6
November	17.3	0.0	17.4	0.0	16.1	-1.8	25.5	15.5
December	17.2	-0.6	17.3	-0.6	16.1	0.0	25.7	15.4
1930								
January	17.1	-0.6	17.2	-0.6	15.9	-1.2	25.5	15.2
February	17.0	-0.6	17.1	-0.6	15.7	-1.3	24.7	15.1
March	16.9	-0.6	17.0	-0.6	15.5	-1.3	23.9	15.0
April	17.0	0.6	17.1	0.6	15.5	0.0	24.2	15.0
May	16.9	-0.6	17.0	-0.6	15.3	-1.3	23.5	14.9
June	16.8	-0.6	16.9	-0.6	15.0	-2.0	22.5	14.6
July	16.6	-1.2	16.7	-1.2	14.5	-3.3	21.0	14.4
August	16.5	-0.6	16.6	-0.6	14.5	0.0	21.4	14.2
September	16.6	0.6	16.7	0.6	14.5	0.0	21.5	14.2
October	16.5	-0.6	16.6	-0.6	14.3	-1.4	20.8	14.0
November	16.4	-0.6	16.5	-0.6	14.0	-2.1	20.0	13.8
December	16.1	-1.8	16.2	-1.8	13.7	-2.1	19.0	13.6
1931								
January	15.9	-1.2	16.0	-1.2	13.5	-1.5	18.4	13.4
February	15.7	-1.3	15.7	-1.9	13.2	-2.2	17.7	13.3
March	15.6	-0.6	15.6	-0.6	13.1	-0.8	17.8	13.1
April	15.5	-0.6	15.5	-0.6	12.9	-1.5	17.7	12.9
May	15.3	-1.3	15.4	-0.6	12.6	-2.3	16.9	12.8
June	15.1	-1.3	15.2	-1.3	12.4	-1.6	16.5	12.6
July	15.1	0.0	15.2	0.0	12.4	0.0	16.4	12.6
August	15.1	0.0	15.1	-0.7	12.4	0.0	16.1	12.6
September	15.0	-0.7	15.1	0.0	12.3	-0.8	15.3	12.6
October	14.9	-0.7	15.0	-0.7	12.1	-1.6	14.8	12.4
November	14.7	-1.3	14.8	-1.3	12.1	0.0	14.8	12.5
December	14.6	-0.7	14.7	-0.7	11.8	-2.5	14.1	12.3
1932								
January	14.3	-2.1	14.4	-2.0	11.6	-1.7	13.3	12.2
February	14.1	-1.4	14.2	-1.4	11.4	-1.7	12.8	12.1
March	14.0	-0.7	14.1	-0.7	11.4	0.0	12.7	12.0
April	13.9	-0.7	14.0	-0.7	11.3	-0.9	12.4	12.0
May	13.7	-1.4	13.8	-1.4	11.1	-1.8	11.8	11.9
June	13.6	-0.7	13.7	-0.7	11.0	-0.9	11.6	11.9
July	13.6	0.0	13.7	0.0	11.1	0.9	12.1	11.8
August	13.5	-0.7	13.5	-1.5	11.2	0.9	12.4	11.9
September	13.4	-0.7	13.5	0.0	11.3	0.9	12.4	11.9
October	13.3	-0.7	13.4	-0.7	11.1	-1.8	11.8	11.9
November	13.2	-0.8	13.3	-0.7	11.0	-0.9	11.8	11.9
December	13.1	-0.8	13.2	-0.8	10.8	-1.8	11.1	11.7
1933								
January	12.9	-1.5	13.0	-1.5	10.5	-2.8	10.8	11.4
February	12.7	-1.6	12.8	-1.5	10.3	-1.9	10.3	11.2
March	12.6	-0.8	12.7	-0.8	10.4	1.0	10.8	11.2
April	12.6	0.0	12.6	-0.8	10.4	0.0	11.2	11.1
May	12.6	0.0	12.7	0.8	10.8	3.8	12.7	11.3
June	12.7	0.8	12.8	0.8	11.2	3.7	13.4	11.7
July	13.1	3.1	13.2	3.1	11.9	6.3	15.2	12.3
August	13.2	0.8	13.3	0.8	12.0	0.8	14.6	12.6
September	13.2	0.0	13.3	0.0	12.2	1.7	14.4	13.0
October	13.2	0.0	13.3	0.0	12.3	0.8	14.1	13.1
November	13.2	0.0	13.3	0.0	12.3	0.0	14.3	13.1
December	13.2	0.0	13.2	-0.8	12.2	-0.8	14.0	13.2

Table 8-1C. Consumer and Producer Price Indexes: Historical Data, 1913–1949—*Continued*

(Not seasonally adjusted.)

Year and month	Consumer price indexes, 1982–1984 =100				Producer Price Indexes for goods, 1982 = 100			
	All urban consumers (CPI-U)		Urban wage earners and clerical workers (CPI-W)		All commodities		Farm products	Industrial commodities
	Index	Percent change	Index	Percent change	Index	Percent change		
1934								
January	13.2	0.0	13.3	0.8	12.4	1.6	14.8	13.3
February	13.3	0.8	13.4	0.8	12.7	2.4	15.5	13.4
March	13.3	0.0	13.4	0.0	12.7	0.0	15.5	13.4
April	13.3	0.0	13.4	0.0	12.7	0.0	15.1	13.4
May	13.3	0.0	13.4	0.0	12.7	0.0	15.1	13.4
June	13.4	0.8	13.4	0.0	12.9	1.6	16.0	13.3
July	13.4	0.0	13.4	0.0	12.9	0.0	16.3	13.3
August	13.4	0.0	13.5	0.7	13.2	2.3	17.6	13.3
September	13.6	1.5	13.7	1.5	13.4	1.5	18.5	13.3
October	13.5	-0.7	13.6	-0.7	13.2	-1.5	17.8	13.3
November	13.5	0.0	13.5	-0.7	13.2	0.0	17.8	13.3
December	13.4	-0.7	13.5	0.0	13.3	0.8	18.2	13.3
1935								
January	13.6	1.5	13.7	1.5	13.6	2.3	19.6	13.2
February	13.7	0.7	13.8	0.7	13.7	0.7	20.0	13.2
March	13.7	0.0	13.8	0.0	13.7	0.0	19.7	13.2
April	13.8	0.7	13.9	0.7	13.8	0.7	20.3	13.1
May	13.8	0.0	13.8	-0.7	13.8	0.0	20.3	13.2
June	13.7	-0.7	13.8	0.0	13.8	0.0	19.8	13.3
July	13.7	0.0	13.7	-0.7	13.7	-0.7	19.5	13.3
August	13.7	0.0	13.7	0.0	13.9	1.5	20.0	13.3
September	13.7	0.0	13.8	0.7	13.9	0.0	20.1	13.2
October	13.7	0.0	13.8	0.0	13.9	0.0	19.7	13.3
November	10.0	0.7	13.9	0.7	13.9	0.0	19.6	13.4
December	13.8	0.0	13.9	0.0	14.0	0.7	19.8	13.4
1936								
January	13.8	0.0	13.9	0.0	13.9	-0.7	19.7	13.4
February	13.8	0.0	13.8	-0.7	13.9	0.0	20.1	13.4
March	13.7	-0.7	13.8	0.0	13.7	-1.4	19.3	13.4
April	13.7	0.0	13.8	0.0	13.7	0.0	19.4	13.4
May	13.7	0.0	13.8	0.0	13.5	-1.5	19.0	13.4
June	13.8	0.7	13.9	0.7	13.7	1.5	19.7	13.4
July	13.9	0.7	14.0	0.7	13.9	1.5	20.5	13.5
August	14.0	0.7	14.1	0.7	14.0	0.7	21.2	13.5
September	14.0	0.0	14.1	0.0	14.0	0.0	21.2	13.5
October	14.0	0.0	14.1	0.0	14.0	0.0	21.2	13.6
November	14.0	0.0	14.1	0.0	14.2	1.4	21.5	13.8
December	14.0	0.0	14.1	0.0	14.5	2.1	22.4	14.0
1937								
January	14.1	0.7	14.2	0.7	14.8	2.1	23.1	14.2
February	14.1	0.0	14.2	0.0	14.9	0.7	23.1	14.3
March	14.2	0.7	14.3	0.7	15.1	1.3	23.8	14.5
April	14.3	0.7	14.4	0.7	15.2	0.7	23.3	14.7
May	14.4	0.7	14.4	0.0	15.1	-0.7	22.7	14.7
June	14.4	0.0	14.5	0.7	15.0	-0.7	22.3	14.6
July	14.5	0.7	14.5	0.0	15.2	1.3	22.6	14.7
August	14.5	0.0	14.6	0.7	15.1	-0.7	21.8	14.6
September	14.6	0.7	14.7	0.7	15.1	0.0	21.7	14.6
October	14.6	0.0	14.6	-0.7	14.7	-2.6	20.3	14.5
November	14.5	-0.7	14.5	-0.7	14.4	-2.0	19.1	14.3
December	14.4	-0.7	14.5	0.0	14.1	-2.1	18.4	14.2
1938								
January	14.2	-1.4	14.3	-1.4	14.0	-0.7	18.1	14.2
February	14.1	-0.7	14.2	-0.7	13.8	-1.4	17.6	14.1
March	14.1	0.0	14.2	0.0	13.7	-0.7	17.7	14.1
April	14.2	0.7	14.2	0.0	13.5	-1.5	17.2	14.0
May	14.1	-0.7	14.2	0.0	13.5	0.0	17.0	13.9
June	14.1	0.0	14.2	0.0	13.5	0.0	17.3	13.8
July	14.1	0.0	14.2	0.0	13.6	0.7	17.5	13.9
August	14.1	0.0	14.2	0.0	13.4	-1.5	17.0	13.9
September	14.1	0.0	14.2	0.0	13.5	0.7	17.2	13.9
October	14.0	-0.7	14.1	-0.7	13.4	-0.7	16.8	13.8
November	14.0	0.0	14.1	0.0	13.4	0.0	17.1	13.7
December	14.0	0.0	14.1	0.0	13.3	-0.7	17.0	13.6
1939								
January	14.0	0.0	14.0	-0.7	13.3	0.0	17.0	13.6
February	13.9	-0.7	14.0	0.0	13.3	0.0	16.9	13.6
March	13.9	0.0	13.9	-0.7	13.2	-0.8	16.6	13.7
April	13.8	-0.7	13.9	0.0	13.1	-0.8	16.1	13.7
May	13.8	0.0	13.9	0.0	13.1	0.0	16.1	13.7
June	13.8	0.0	13.9	0.0	13.0	-0.8	15.7	13.6
July	13.8	0.0	13.9	0.0	13.0	0.0	15.8	13.6
August	13.8	0.0	13.9	0.0	12.9	-0.8	15.4	13.6
September	14.1	2.2	14.2	2.2	13.6	5.4	17.3	14.0
October	14.0	-0.7	14.1	-0.7	13.7	0.7	16.9	14.2
November	14.0	0.0	14.1	0.0	13.6	-0.7	17.0	14.3
December	14.0	0.0	14.0	-0.7	13.7	0.7	17.1	14.3

Table 8-1C. Consumer and Producer Price Indexes: Historical Data, 1913–1949—*Continued*

(Not seasonally adjusted.)

Year and month	Consumer price indexes, 1982–1984 =100				Producer Price Indexes for goods, 1982 = 100			
	All urban consumers (CPI-U)		Urban wage earners and clerical workers (CPI-W)		All commodities		Farm products	Industrial commodities
	Index	Percent change	Index	Percent change	Index	Percent change		
1940								
January	13.9	-0.7	14.0	0.0	13.7	0.0	17.4	14.3
February	14.0	0.7	14.1	0.7	13.6	-0.7	17.3	14.2
March	14.0	0.0	14.1	0.0	13.5	-0.7	17.1	14.1
April	14.0	0.0	14.1	0.0	13.5	0.0	17.5	14.0
May	14.0	0.0	14.1	0.0	13.5	0.0	17.1	14.0
June	14.1	0.7	14.1	0.0	13.4	-0.7	16.7	14.0
July	14.0	-0.7	14.1	0.0	13.4	0.0	16.8	14.0
August	14.0	0.0	14.1	0.0	13.4	0.0	16.5	14.0
September	14.0	0.0	14.1	0.0	13.4	0.0	16.7	14.0
October	14.0	0.0	14.1	0.0	13.6	1.5	16.8	14.2
November	14.0	0.0	14.1	0.0	13.7	0.7	17.2	14.3
December	14.1	0.7	14.2	0.7	13.8	0.7	17.6	14.3
1941								
January	14.1	0.0	14.2	0.0	13.9	0.7	18.1	14.3
February	14.1	0.0	14.2	0.0	13.9	0.0	17.7	14.3
March	14.2	0.7	14.2	0.0	14.0	0.7	18.1	14.4
April	14.3	0.7	14.4	1.4	14.4	2.9	18.8	14.6
May	14.4	0.7	14.5	0.7	14.6	1.4	19.3	14.9
June	14.7	2.1	14.7	1.4	15.0	2.7	20.7	15.1
July	14.7	0.0	14.8	0.7	15.3	2.0	21.7	15.2
August	14.9	1.4	14.9	0.7	15.6	2.0	22.1	15.5
September	15.1	1.3	15.2	2.0	15.8	1.3	23.0	15.6
October	15.3	1.3	15.4	1.3	15.9	0.6	22.7	15.9
November	15.4	0.7	15.5	0.6	15.9	0.0	22.9	15.9
December	15.5	0.6	15.5	0.0	16.2	1.9	23.9	15.9
1942								
January	15.7	1.3	15.7	1.3	16.5	1.9	25.5	16.1
February	15.8	0.6	15.9	1.3	16.7	1.2	25.6	16.1
March	16.0	1.3	16.1	1.3	16.8	0.6	26.0	16.2
April	16.1	0.6	16.2	0.6	17.0	1.2	26.4	16.2
May	16.3	1.2	16.3	0.6	17.0	0.0	26.3	16.3
June	16.3	0.0	16.4	0.6	17.0	0.0	26.3	16.3
July	16.4	0.6	16.5	0.6	17.0	0.0	26.6	16.3
August	16.5	0.6	16.6	0.6	17.1	0.6	26.8	16.2
September	16.5	0.0	16.6	0.0	17.2	0.6	27.2	16.2
October	16.7	1.2	16.8	1.2	17.2	0.0	27.5	16.2
November	16.8	0.6	16.9	0.6	17.3	0.6	27.9	16.3
December	16.9	0.6	17.0	0.6	17.4	0.6	28.7	16.3
1943								
January	16.9	0.0	17.0	0.0	17.5	0.6	29.5	16.4
February	16.9	0.0	17.0	0.0	17.7	1.1	30.0	16.4
March	17.2	1.8	17.3	1.8	17.8	0.6	31.0	16.4
April	17.4	1.2	17.5	1.2	17.9	0.6	31.2	16.5
May	17.5	0.6	17.6	0.6	17.9	0.0	31.7	16.5
June	17.5	0.0	17.6	0.0	17.9	0.0	31.9	16.5
July	17.4	-0.6	17.5	-0.6	17.8	-0.6	31.5	16.5
August	17.3	-0.6	17.4	-0.6	17.8	0.0	31.2	16.5
September	17.4	0.6	17.5	0.6	17.8	0.0	31.0	16.5
October	17.4	0.0	17.5	0.0	17.8	0.0	30.9	16.5
November	17.4	0.0	17.5	0.0	17.7	-0.6	30.6	16.6
December	17.4	0.0	17.5	0.0	17.8	0.6	30.7	16.6
1944								
January	17.4	0.0	17.5	0.0	17.8	0.0	30.7	16.6
February	17.4	0.0	17.5	0.0	17.8	0.0	30.9	16.7
March	17.4	0.0	17.5	0.0	17.9	0.6	31.2	16.7
April	17.5	0.6	17.6	0.6	17.9	0.0	31.1	16.7
May	17.5	0.0	17.6	0.0	17.9	0.0	31.0	16.7
June	17.6	0.6	17.7	0.6	18.0	0.6	31.5	16.7
July	17.7	0.6	17.8	0.6	17.9	-0.6	31.3	16.7
August	17.7	0.0	17.8	0.0	17.9	0.0	30.9	16.8
September	17.7	0.0	17.8	0.0	17.9	0.0	31.0	16.8
October	17.7	0.0	17.8	0.0	17.9	0.0	31.2	16.8
November	17.7	0.0	17.8	0.0	18.0	0.6	31.4	16.8
December	17.8	0.6	17.9	0.6	18.0	0.0	31.6	16.8
1945								
January	17.8	0.0	17.9	0.0	18.1	0.6	31.9	16.8
February	17.8	0.0	17.9	0.0	18.1	0.0	32.1	16.9
March	17.8	0.0	17.9	0.0	18.1	0.0	32.1	16.9
April	17.8	0.0	17.9	0.0	18.2	0.6	32.5	16.9
May	17.9	0.6	18.0	0.6	18.3	0.5	32.8	16.9
June	18.1	1.1	18.2	1.1	18.3	0.0	32.9	16.9
July	18.1	0.0	18.2	0.0	18.3	0.0	32.5	17.0
August	18.1	0.0	18.2	0.0	18.2	-0.5	32.0	17.0
September	18.1	0.0	18.2	0.0	18.1	-0.5	31.4	17.0
October	18.1	0.0	18.2	0.0	18.2	0.6	32.1	17.0
November	18.1	0.0	18.2	0.0	18.4	1.1	33.1	17.0
December	18.2	0.6	18.3	0.5	18.4	0.0	33.2	17.1

Table 8-1C. Consumer and Producer Price Indexes: Historical Data, 1913–1949—*Continued*

(Not seasonally adjusted.)

Year and month	Consumer price indexes, 1982–1984 =100				Producer Price Indexes for goods, 1982 = 100			
	All urban consumers (CPI-U)		Urban wage earners and clerical workers (CPI-W)		All commodities		Farm products	Industrial commodities
	Index	Percent change	Index	Percent change	Index	Percent change		
1946								
January	18.2	0.0	18.3	0.0	18.4	0.0	32.7	17.1
February	18.1	-0.5	18.2	-0.5	18.5	0.5	33.0	17.2
March	18.3	1.1	18.4	1.1	18.8	1.6	33.6	17.4
April	18.4	0.5	18.5	0.5	19.0	1.1	34.1	17.5
May	18.5	0.5	18.6	0.5	19.1	0.5	34.7	17.7
June	18.7	1.1	18.8	1.1	19.4	1.6	35.4	18.0
July	19.8	5.9	19.9	5.9	21.5	10.8	39.6	18.6
August	20.2	2.0	20.3	2.0	22.2	3.3	40.6	19.0
September	20.4	1.0	20.5	1.0	21.4	-3.6	39.0	19.1
October	20.8	2.0	20.9	2.0	23.1	7.9	41.7	19.7
November	21.3	2.4	21.5	2.9	24.1	4.3	42.8	20.6
December	21.5	0.9	21.6	0.5	24.3	0.8	42.4	21.2
1947								
January	21.5	0.0	21.6	0.0	24.5	0.8	41.6	21.8
February	21.5	0.0	21.6	0.0	24.7	0.8	42.6	22.0
March	21.9	1.9	22.1	2.3	25.3	2.4	45.5	22.3
April	21.9	0.0	22.1	0.0	25.1	-0.8	44.0	22.4
May	21.9	0.0	22.0	-0.5	25.0	-0.4	43.6	22.3
June	22.0	0.5	22.2	0.9	25.0	0.0	43.8	22.4
July	22.2	0.9	22.4	0.9	25.3	1.2	44.3	22.5
August	22.5	1.4	22.6	0.9	25.6	1.2	44.8	22.8
September	23.0	2.2	23.1	2.2	26.1	2.0	46.6	23.1
October	23.0	0.0	23.1	0.0	26.4	1.1	47.4	23.3
November	23.1	0.4	23.3	0.0	26.7	1.1	47.7	23.6
December	23.4	1.3	23.6	1.3	27.2	1.9	50.1	23.9
1948								
January	23.7	1.3	23.8	0.8	27.7	1.8	51.2	24.3
February	23.5	-0.8	23.6	-0.8	27.2	-1.8	47.7	24.1
March	23.4	-0.4	23.6	0.0	27.2	0.0	47.6	24.1
April	23.8	1.7	23.9	1.3	27.4	0.7	48.3	24.3
May	23.9	0.4	24.1	0.8	27.5	0.4	49.4	24.3
June	24.1	0.8	24.2	0.4	27.7	0.7	50.4	24.4
July	24.4	1.2	24.5	1.2	28.0	1.1	50.2	24.6
August	24.5	0.4	24.6	0.4	28.2	0.7	49.6	24.9
September	24.5	0.0	24.6	0.0	28.1	-0.4	48.8	25.0
October	24.4	-0.4	24.5	-0.4	27.8	-1.1	46.9	25.0
November	24.2	-0.8	24.4	-0.4	27.8	0.0	46.3	25.1
December	24.1	-0.4	24.2	-0.8	27.6	-0.7	45.2	25.1
1949								
January	24.0	-0.4	24.2	0.0	27.3	-1.1	43.8	24.9
February	23.8	-0.8	23.9	-1.2	26.8	-1.8	42.0	24.7
March	23.8	0.0	24.0	0.4	26.8	0.0	42.7	24.6
April	23.9	0.4	24.0	0.0	26.5	-1.1	42.7	24.3
May	23.8	-0.4	24.0	0.0	26.3	-0.8	42.7	24.0
June	23.9	0.4	24.0	0.0	26.0	-1.1	41.8	23.8
July	23.7	-0.8	23.8	-0.8	26.0	0.0	41.7	23.7
August	23.8	0.4	23.9	0.4	26.0	0.0	41.7	23.8
September	23.9	0.4	24.0	0.4	26.1	0.4	41.8	23.8
October	23.7	-0.8	23.9	-0.4	26.0	-0.4	41.0	23.8
November	23.8	0.4	23.9	0.0	26.0	0.0	40.9	23.8
December	23.6	-0.8	23.8	-0.4	25.9	-0.4	40.3	23.8

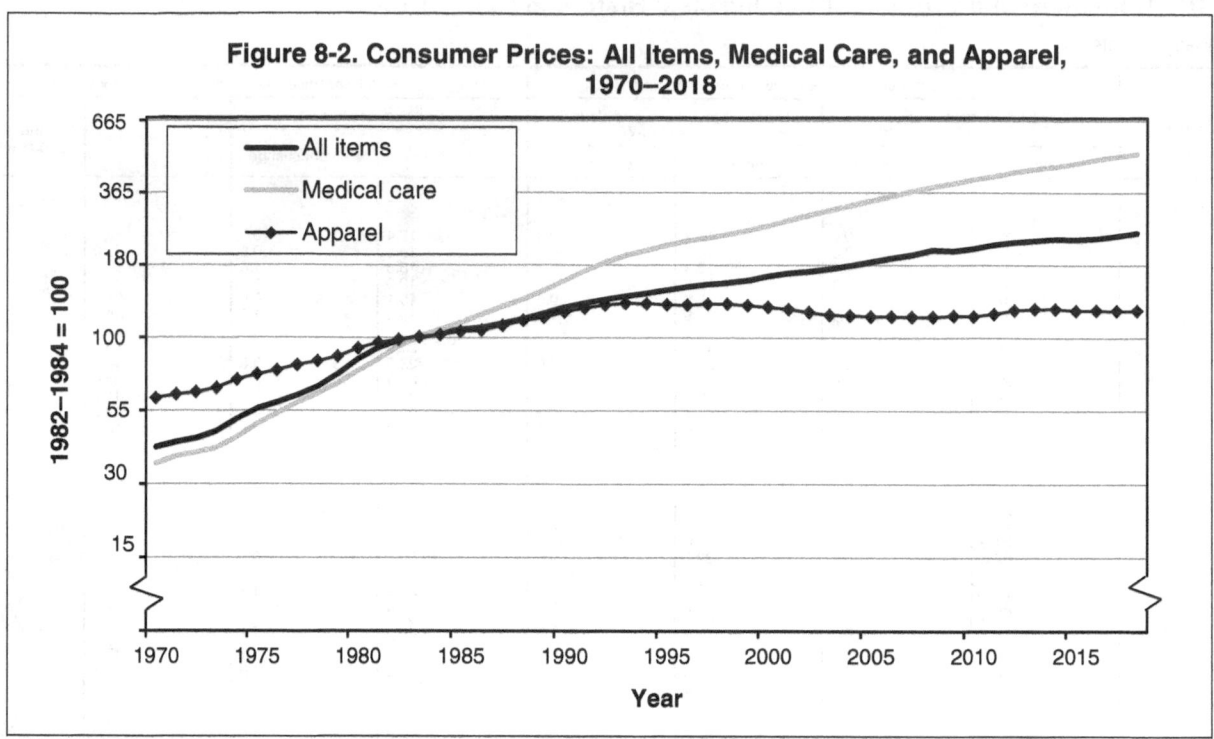

Figure 8-2. Consumer Prices: All Items, Medical Care, and Apparel, 1970–2018

- Figure 8-2 charts two components of the Consumer Price Index for All Urban Consumers (CPI-U) along with the all-items total. Since all three indexes have the base years 1982–1984, they converge around 100 in those years. However, over the entire postwar period, the trends of the two components are very different. (Table 8-2)

- Apparel has been one of the areas most subject to international competition, and the apparel index shows far less growth than the overall average of prices.

- Medical care, on the other hand, has little price competition from producers in other countries. It is often paid for by third-party insurers, both government and private, rather than directly by consumers. Furthermore, it is characterized by trend growth in demand, due to rising income and expectations and to technological progress. All of these economic factors cause medical care prices to rise faster than the general price level.

- Medical care has arguably been overstated in the CPI due to the difficulties of making quality adjustments. Quality adjustments in the indexes have been much improved in recent years, although such improvements are not retroactively introduced into the official CPIs. But even since 1997, when a major improvement was introduced into the hospital cost component of the CPI, measured medical care prices have increased at a much higher rate than the CPI. (Table 8-2)

- A broader system of PPIs including services and construction as well as goods was introduced in 2014, featuring a more detailed representation by stages of production as well. These new measures only go back to November 2009—not enough time yet to determine what kind of signals about future inflation they will contain. (Tables 8-5 and 8-6)

Table 8-2. Consumer Price Indexes and Purchasing Power of the Dollar

(1982–1984 = 100, seasonally adjusted, except as noted.)

Year and month	CPI-U, All items — Not seasonally adjusted	Seasonally adjusted — Index	Seasonally adjusted — Percent change from previous period	Total	Total food	Food at home — Total	Cereals and bakery products	Meats, poultry, fish, and eggs	Dairy and related products[1]	Fruits and vegetables	Non-alcoholic beverages	Other food at home	Food away from home [1]	Alcoholic beverages
1970	38.8	38.8	5.7	40.1	39.2	39.9	37.1	44.6	44.7	37.8	27.1	32.9	37.5	52.1
1971	40.5	40.5	4.4	41.4	40.4	40.9	38.8	44.1	46.1	39.7	28.1	34.3	39.4	54.2
1972	41.8	41.8	3.2	43.1	42.1	42.7	39.0	48.0	46.8	41.6	28.0	34.6	41.0	55.4
1973	44.4	44.4	6.2	48.8	48.2	49.7	43.5	60.9	51.2	47.4	30.1	36.7	44.2	56.8
1974	49.3	49.3	11.0	55.5	55.1	57.1	56.5	62.2	60.7	55.2	35.9	47.8	49.8	61.1
1975	53.8	53.8	9.1	60.2	59.8	61.8	62.9	67.0	62.6	56.9	41.3	55.4	54.5	65.9
1976	56.9	56.9	5.8	62.1	61.6	63.1	61.5	68.0	67.7	58.4	49.4	56.4	58.2	68.1
1977	60.6	60.6	6.5	65.8	65.5	66.8	62.5	67.4	69.5	63.8	74.4	68.4	62.6	70.0
1978	65.2	65.2	7.6	72.2	72.0	73.8	68.1	77.6	74.2	70.9	78.7	73.6	68.3	74.1
1979	72.6	72.6	11.3	79.9	79.9	81.8	74.9	89.0	82.8	76.6	82.6	79.0	75.9	79.9
1980	82.4	82.4	13.5	86.7	86.8	88.4	83.9	92.0	90.9	82.1	91.4	88.4	83.4	86.4
1981	90.9	90.9	10.3	93.5	93.6	94.8	92.3	96.0	97.4	92.0	95.3	94.9	90.9	92.5
1982	96.5	96.5	6.2	97.3	97.4	98.1	96.5	00.6	98.8	97.0	97.9	97.3	95.8	96.7
1983	99.6	99.6	3.2	99.5	99.4	99.1	99.6	99.2	100.0	97.3	99.8	99.5	100.0	100.4
1984	103.9	103.9	4.3	103.2	103.2	102.8	103.9	101.3	101.3	105.7	102.3	103.1	104.2	103.0
1985	107.6	107.6	3.6	105.6	105.6	104.3	107.9	100.1	103.2	108.4	104.3	105.7	108.3	106.4
1986	109.6	109.6	1.9	109.1	109.0	107.3	110.9	104.5	103.3	109.4	110.4	109.4	112.5	111.1
1987	113.6	113.6	3.6	113.5	113.5	111.9	114.8	110.5	105.9	119.1	107.5	110.5	117.0	114.1
1988	118.3	118.3	4.1	118.2	118.2	116.6	122.1	114.3	108.4	128.1	107.5	113.1	121.8	118.6
1989	124.0	124.0	4.8	124.9	125.1	124.2	132.4	121.3	115.6	138.0	111.3	119.1	127.4	123.5
1990	130.7	130.7	5.4	132.1	132.4	132.3	140.0	130.0	126.5	149.0	113.5	123.4	133.4	129.3
1991	136.2	136.2	4.2	136.8	136.3	135.8	145.8	132.6	125.1	155.8	114.1	127.3	137.9	142.8
1992	140.3	140.3	3.0	138.7	137.9	136.8	151.5	130.9	128.5	155.4	114.3	128.8	140.7	147.3
1993	144.5	144.5	3.0	141.6	140.9	140.1	156.6	135.5	129.4	159.0	114.6	130.5	143.2	149.6
1994	148.2	148.2	2.6	144.9	144.3	144.1	163.0	137.2	131.7	165.0	123.2	135.6	145.7	151.5
1995	152.4	152.4	2.8	148.9	148.4	148.8	167.5	138.8	132.8	177.7	131.7	140.8	149.0	153.9
1996	156.9	156.9	3.0	153.7	153.3	154.3	174.0	144.8	142.1	183.9	128.6	142.9	152.7	158.5
1997	160.5	160.5	2.3	157.7	157.3	158.1	177.6	148.5	145.5	187.5	133.4	147.3	157.0	162.8
1998	163.0	163.0	1.6	161.1	160.7	161.1	181.1	147.3	150.8	198.2	133.0	150.8	161.1	165.7
1999	166.6	166.6	2.2	164.6	164.1	164.2	185.0	147.9	159.6	203.1	134.3	153.5	165.1	169.7
2000	172.2	172.2	3.4	168.4	167.8	167.9	188.3	154.5	160.7	204.6	137.8	155.6	169.0	174.7
2001	177.1	177.1	2.8	173.6	173.1	173.4	193.8	161.3	167.1	212.2	139.2	159.6	173.9	179.3
2002	179.9	179.9	1.6	176.8	176.2	175.6	198.0	162.1	168.1	220.9	139.2	160.8	178.3	183.6
2003	184.0	184.0	2.3	180.5	180.0	179.4	202.8	169.3	167.9	225.9	139.8	162.6	182.1	187.2
2004	188.9	188.9	2.7	186.6	186.2	186.2	206.0	181.7	180.2	232.7	140.4	164.9	187.5	192.1
2005	195.3	195.3	3.4	191.2	190.7	189.8	209.0	184.7	182.4	241.4	144.4	167.0	193.4	195.9
2006	201.6	201.6	3.2	195.7	195.2	193.1	212.8	186.6	181.4	252.9	147.4	169.6	199.4	200.7
2007	207.3	207.3	2.8	203.3	202.9	201.2	222.1	195.6	194.8	262.6	153.4	173.3	206.7	207.0
2008	215.3	215.3	3.8	214.2	214.1	214.1	244.9	204.7	210.4	278.9	160.0	184.2	215.8	214.5
2009	214.5	214.5	-0.4	218.2	218.0	215.1	252.6	203.8	197.0	272.9	163.0	191.2	223.3	220.8
2010	218.1	218.1	1.6	220.0	219.6	215.8	250.4	207.7	199.2	273.5	161.6	191.1	226.1	223.3
2011	224.9	224.9	3.2	227.9	227.8	226.2	260.3	223.2	212.7	284.7	166.8	197.4	231.4	226.7
2012	229.6	229.6	2.1	233.7	233.8	231.8	267.7	231.0	217.3	282.8	168.6	204.8	238.0	230.8
2013	233.0	233.0	1.5	237.0	237.0	233.9	270.4	236.0	217.6	290.0	166.9	204.8	243.1	234.6
2014	236.7	236.7	1.6	242.4	242.7	239.5	271.1	253.0	225.3	294.4	166.0	206.2	249.0	237.3
2015	237.0	237.0	0.1	246.8	247.2	242.3	274.1	260.3	222.4	293.8	167.9	209.3	256.1	239.5
2016	240.0	240.0	1.3	247.7	247.9	239.1	273.1	247.7	217.3	296.3	167.3	209.5	262.7	242.5
2017	245.1	245.1	-0.2	249.8	250.1	238.6	271.7	245.8	217.5	295.7	167.6	209.9	268.6	245.1
2018	251.1	251.1	-0.2	253.3	253.6	239.7	272.8	248.9	216.5	297.8	167.6	210.2	275.9	249.1
2017														
January	242.8	243.8	2.5	247.7	247.9	236.8	272.2	243.9	218.0	287.4	166.0	209.5	266.1	244.2
February	243.6	244.0	2.7	248.4	248.6	237.7	271.8	244.3	219.9	290.1	168.4	209.0	266.6	244.0
March	243.8	243.7	2.4	249.0	249.2	238.4	271.9	244.7	218.3	294.5	167.9	210.0	267.1	244.5
April	244.5	244.1	2.2	249.4	249.6	238.6	271.8	243.2	217.8	299.1	167.6	209.9	267.7	245.0
May	244.7	244.0	1.9	249.9	250.1	239.2	272.4	244.4	218.1	298.1	169.4	210.2	268.1	244.8
June	245.0	244.2	1.7	249.7	249.9	238.8	271.8	245.6	217.3	296.9	168.1	209.8	268.2	245.0
July	244.8	244.4	1.8	250.1	250.4	239.2	271.4	247.2	217.8	297.4	167.9	209.9	268.6	245.4
August	245.5	245.3	2.0	250.3	250.5	238.8	272.0	246.5	217.1	296.8	167.3	209.8	269.5	245.8
September	246.8	246.4	2.2	250.6	250.8	238.8	272.0	246.2	216.4	296.9	167.5	210.1	270.4	246.2
October	246.7	246.6	2.0	250.8	251.1	239.1	271.3	247.7	216.2	297.9	167.4	209.8	270.7	246.3
November	246.7	247.3	2.2	250.9	251.1	238.8	270.6	247.1	216.5	296.9	166.8	210.2	271.2	246.8
December	246.5	247.9	2.1	251.3	251.5	239.1	270.9	249.2	216.2	296.3	166.5	210.2	271.8	246.9
2018														
January	247.9	248.9	2.1	251.8	252.0	239.2	271.5	248.7	216.2	297.2	166.6	210.3	272.8	247.0
February	249.0	249.4	2.2	251.9	252.1	239.0	271.4	248.4	215.7	296.4	166.7	210.2	273.4	247.5
March	249.6	249.5	2.4	252.2	252.5	239.3	272.1	249.8	216.2	295.6	167.2	210.0	273.7	247.8
April	250.5	250.0	2.4	252.8	253.0	239.8	271.8	251.6	216.8	297.8	166.3	209.9	274.4	248.3
May	251.6	250.6	2.7	253.0	253.2	239.5	271.9	250.0	216.7	297.0	167.2	209.7	275.3	248.3
June	252.0	251.1	2.8	253.3	253.5	239.6	273.0	248.5	218.1	297.6	167.3	210.0	276.1	249.2
July	252.0	251.6	2.9	253.7	253.9	240.0	272.8	249.4	216.9	300.0	167.4	210.0	276.1	249.2
August	252.1	251.9	2.7	253.8	254.1	240.0	272.8	249.3	217.0	299.3	167.7	210.1	276.6	249.3
September	252.4	252.0	2.3	254.0	254.2	239.8	273.8	247.6	216.6	298.2	168.5	210.5	277.3	250.2
October	252.9	252.8	2.5	254.0	254.1	239.5	272.8	247.4	216.0	297.0	168.9	210.5	277.5	250.4
November	252.0	252.8	2.2	254.5	254.6	239.8	274.3	248.1	215.6	296.6	168.5	211.0	278.3	251.0
December	251.2	252.7	1.9	255.3	255.5	240.5	275.3	248.2	216.0	300.8	168.9	210.5	279.4	251.2

[1]Not seasonally adjusted.

Table 8-2. Consumer Price Indexes and Purchasing Power of the Dollar—*Continued*

(1982–1984 = 100, seasonally adjusted, except as noted.)

Year and month	CPI-U, Housing												
	Total	Shelter						Fuels and utilities				Household furnishings and operations	
		Total	Rent of shelter [2]	Rent of primary residence	Lodging away from home [3]	Owners' equivalent rent of primary residence [2]	Tenants' and household insurance [1,3]	Total	Fuels		Water and sewer and trash collection services [3]	Total	Household operations [1,3]
									Fuel oil and other fuels	Energy services			
1970	36.4	35.5	...	46.5	...	...	...	29.1	17.0	25.4	...	46.8	...
1971	38.0	37.0	...	48.7	...	...	...	31.1	18.2	27.1	...	48.6	...
1972	39.4	38.7	...	50.4	...	...	...	32.5	18.3	28.5	...	49.7	...
1973	41.2	40.5	...	52.5	...	...	...	34.3	21.1	29.9	...	51.1	...
1974	45.8	44.4	...	55.2	...	...	...	40.7	33.2	34.5	...	56.8	...
1975	50.7	48.8	...	58.0	...	...	...	45.4	36.4	40.1	...	63.4	...
1976	53.8	51.5	...	61.1	...	...	...	49.4	38.8	44.7	...	67.3	...
1977	57.4	54.9	...	64.8	...	...	...	54.7	43.9	50.5	...	70.4	...
1978	62.4	60.5	...	69.3	...	...	...	58.5	46.2	55.0	...	74.7	...
1979	70.1	68.9	...	74.3	...	...	...	64.8	62.4	61.0	...	79.9	...
1980	81.1	81.0	...	80.9	...	...	...	75.4	86.1	71.4	...	86.3	...
1981	90.4	90.5	...	87.9	...	...	...	86.4	104.6	81.9	...	93.0	...
1982	96.9	96.9	...	94.6	...	...	...	94.9	103.4	93.2	...	98.0	...
1983	99.5	99.1	102.7	100.1	...	102.5	...	100.2	97.2	101.5	...	100.2	...
1984	103.6	104.0	107.7	105.3	...	107.3	...	104.8	99.4	105.4	...	101.9	...
1985	107.7	109.8	113.9	111.8	...	113.2	...	106.5	95.9	107.1	...	103.8	...
1986	110.9	115.8	120.2	118.3	...	119.4	...	104.1	77.6	105.7	...	105.2	...
1987	114.2	121.3	125.9	123.1	...	124.8	...	103.0	77.9	103.8	...	107.1	...
1988	118.5	127.1	132.0	127.8	...	131.1	...	104.4	78.1	104.6	...	109.4	...
1989	123.0	132.8	138.0	132.8	...	137.4	...	107.8	81.7	107.5	...	111.2	...
1990	128.5	140.0	145.5	138.4	...	144.8	...	111.6	99.3	109.3	...	113.3	...
1991	133.6	146.3	152.1	143.3	...	150.4	...	115.3	94.6	112.6	...	116.0	...
1992	137.5	151.2	157.3	146.9	...	155.5	...	117.8	90.7	114.8	...	118.0	...
1993	141.2	155.7	162.0	150.3	...	160.5	...	121.3	90.3	118.5	...	119.3	...
1994	144.8	160.5	167.0	154.0	...	165.8	...	122.8	88.8	119.2	...	121.0	...
1995	148.5	165.7	172.4	157.8	...	171.3	...	123.7	88.1	119.2	...	123.0	...
1996	152.8	171.0	178.0	162.0	...	176.8	...	127.5	99.2	122.1	...	124.7	...
1997	156.8	176.3	183.4	166.7	...	181.9	...	130.8	99.8	125.1	...	125.4	...
1998	160.4	182.1	189.6	172.1	109.0	187.8	99.8	128.5	90.0	121.2	101.6	126.6	101.5
1999	163.9	187.3	195.0	177.5	112.3	192.9	101.3	128.8	91.4	120.9	104.0	126.7	104.5
2000	169.6	193.4	201.3	183.9	117.5	198.7	103.7	137.9	129.7	128.0	106.5	128.2	110.5
2001	176.4	200.6	208.9	192.1	118.6	206.3	106.2	150.2	129.3	142.4	109.6	129.1	115.6
2002	180.3	208.1	216.7	199.7	118.3	214.7	108.7	143.6	115.5	134.4	113.0	128.3	119.0
2003	184.8	213.1	221.9	205.5	119.3	219.9	114.8	154.5	139.5	145.0	117.2	126.1	121.8
2004	189.5	218.8	227.9	211.0	125.9	224.9	116.2	161.9	160.5	150.6	124.0	125.5	125.0
2005	195.7	224.4	233.7	217.3	130.3	230.2	117.6	179.0	208.6	166.5	130.3	126.1	130.3
2006	203.2	232.1	241.9	225.1	136.0	238.2	116.5	194.7	234.9	182.1	136.8	127.0	136.6
2007	209.6	240.6	250.8	234.7	142.8	246.2	117.0	200.6	251.5	186.3	143.7	126.9	140.6
2008	216.3	246.7	257.2	243.3	143.7	252.4	118.8	220.0	334.4	202.2	152.1	127.8	147.5
2009	217.1	249.4	259.9	248.8	134.2	256.6	121.5	210.7	239.8	193.6	161.1	128.7	150.3
2010	216.3	248.4	258.8	249.4	133.7	256.6	125.7	214.2	275.1	192.9	170.9	125.5	150.3
2011	219.1	251.6	262.2	253.6	137.4	259.6	127.4	220.4	337.1	194.4	179.6	124.9	151.8
2012	222.7	257.1	267.8	260.4	140.5	264.8	131.3	219.0	335.9	189.7	189.3	125.7	155.2
2013	227.4	263.1	274.0	267.7	142.4	270.7	135.4	225.2	332.0	194.8	197.6	124.8	157.6
2014	233.2	270.5	281.8	276.2	148.5	277.8	141.9	234.6	338.9	203.4	204.9	123.1	161.6
2015	238.1	278.8	290.4	286.0	153.0	285.8	146.4	230.1	256.2	198.7	214.0	122.6	166.9
2016	244.0	288.2	300.3	296.8	158.1	295.4	147.7	228.9	226.3	196.1	221.7	121.6	171.6
2017	251.2	297.8	310.3	308.1	159.4	305.1	148.8	237.3	254.4	202.7	229.1	120.7	176.3
2018	258.5	307.7	320.7	319.3	161.1	315.2	150.7	241.6	293.0	203.8	237.1	121.6	186.0
2017													
January	248.2	293.7	306.0	303.1	160.4	300.9	147.8	233.9	251.1	199.7	226.2	121.3	174.4
February	248.8	294.5	306.8	304.0	160.8	301.7	147.8	234.9	250.4	200.7	226.9	121.1	174.4
March	249.1	295.0	307.4	304.8	157.4	302.3	148.1	234.6	249.3	200.3	227.3	121.1	174.5
April	249.8	295.8	308.2	305.7	159.7	303.0	149.1	236.4	250.5	202.1	227.8	120.9	174.7
May	250.4	296.5	308.9	306.8	159.5	303.6	149.2	237.4	246.6	203.3	228.4	120.9	175.5
June	251.0	297.3	309.8	307.8	158.6	304.5	149.3	237.9	244.4	203.9	228.8	120.7	175.4
July	251.2	297.8	310.3	308.6	153.7	305.3	149.4	237.7	241.5	203.7	229.2	120.5	176.6
August	252.1	299.0	311.5	309.7	158.5	306.3	149.4	237.9	249.9	203.4	229.6	120.7	177.0
September	252.6	299.8	312.4	310.4	160.1	307.1	149.3	238.3	259.5	203.3	230.2	120.4	177.0
October	253.3	300.8	313.4	311.3	163.0	308.0	148.9	238.7	261.2	203.5	231.0	120.4	178.3
November	253.8	301.4	314.0	312.2	161.1	308.7	148.6	239.8	272.0	204.1	231.8	120.3	178.8
December	254.5	302.3	315.1	313.3	160.6	309.7	148.7	240.3	275.8	204.4	232.3	120.2	178.7
2018													
January	255.1	303.1	315.9	314.4	158.5	310.6	149.1	240.1	292.5	203.4	232.7	120.6	181.3
February	255.8	303.7	316.5	315.1	158.4	311.2	149.3	242.0	286.2	205.6	233.4	120.9	184.1
March	256.5	304.8	317.7	315.9	161.3	312.2	149.3	241.6	285.4	205.0	233.9	121.0	184.4
April	257.2	305.8	318.8	317.0	162.0	313.2	149.4	241.4	289.1	204.5	234.8	121.6	184.7
May	257.8	306.8	319.9	317.9	165.9	314.0	150.6	241.6	289.9	204.4	235.8	121.2	185.7
June	258.0	307.3	320.4	318.8	161.1	314.8	150.9	240.7	294.5	202.9	236.7	121.1	186.2
July	258.7	308.2	321.3	319.8	161.6	315.7	151.8	240.1	295.7	202.0	237.3	121.5	186.9
August	259.4	309.0	322.2	320.8	162.2	316.5	152.0	241.2	300.3	202.8	238.3	121.6	186.9
September	259.6	309.6	322.7	321.7	161.0	317.1	151.8	239.7	299.4	201.0	238.6	121.8	186.8
October	260.4	310.3	323.3	322.4	159.2	318.0	151.5	242.3	305.5	203.6	239.4	122.2	186.7
November	261.2	311.2	324.3	323.5	159.0	318.9	151.7	243.1	298.4	204.0	242.1	122.4	187.2
December	262.0	312.0	325.2	324.2	161.8	319.6	151.5	245.1	279.8	207.0	242.7	122.7	190.9

[1] Not seasonally adjusted.
[2] December 1982 = 100.
[3] December 1997 = 100.
. . . = Not available.

Table 8-2. Consumer Price Indexes and Purchasing Power of the Dollar—*Continued*

(1982–1984 = 100, seasonally adjusted, except as noted.)

Year and month	CPI-U, Apparel					CPI-U, Transportation								
	Total	Men's and boys' apparel	Women's and girls' apparel	Infants' and toddlers' apparel	Footwear	Total	Private transportation						Motor vehicle parts and equipment	Motor vehicle maintenance and repair [1]
							Total	New and used motor vehicles			Motor fuel			
								Total [3]	New vehicles	Used cars and trucks	Total	Gasoline (all types)		
1970	59.2	62.2	71.8	39.2	56.8	37.5	37.5	...	53.1	31.2	27.9	27.9	...	36.6
1971	61.1	63.9	74.4	40.0	58.6	39.5	39.4	...	55.3	33.0	28.1	28.1	...	39.3
1972	62.3	64.7	76.2	41.1	60.3	39.9	39.7	...	54.8	33.1	28.4	28.4	...	41.1
1973	64.6	67.1	78.8	42.5	62.8	41.2	41.0	...	54.8	35.2	31.2	31.2	...	43.2
1974	69.4	72.4	83.5	54.2	66.6	45.8	46.2	...	58.0	36.7	42.2	42.2	...	47.6
1975	72.5	75.5	85.5	64.5	69.6	50.1	50.6	...	63.0	43.8	45.1	45.1	...	53.7
1976	75.2	78.1	87.9	68.0	72.3	55.1	55.6	...	67.0	50.3	47.0	47.0	...	57.6
1977	78.6	81.7	90.6	74.6	75.7	59.0	59.7	...	70.5	54.7	49.7	49.7	...	61.9
1978	81.4	83.5	92.4	77.4	79.0	61.7	62.5	...	75.9	55.8	51.8	51.8	77.6	67.0
1979	84.9	85.4	94.0	79.0	85.3	70.5	71.7	...	81.9	60.2	70.1	70.2	85.1	73.7
1980	90.9	89.4	96.0	85.5	91.8	83.1	84.2	...	88.5	62.3	97.4	97.5	95.3	81.5
1981	95.3	94.2	97.5	92.9	96.7	93.2	93.8	...	93.9	76.9	108.5	108.5	101.0	89.2
1982	97.8	97.6	98.5	96.3	99.1	97.0	97.1	...	97.5	88.8	102.8	102.8	103.6	96.0
1983	100.2	100.3	100.2	101.1	99.8	99.3	99.3	...	99.9	98.7	99.4	99.4	100.7	100.3
1984	102.1	102.1	101.3	102.6	101.1	103.7	103.6	...	102.6	112.5	97.9	97.8	95.6	103.8
1985	105.0	105.0	104.9	107.2	102.3	106.4	106.2	...	106.1	113.7	98.7	98.6	95.9	106.8
1986	105.9	106.2	104.0	111.8	101.9	102.3	101.2	...	110.6	108.8	77.1	77.0	95.4	110.3
1987	110.6	109.1	110.4	112.1	105.1	105.4	104.2	...	114.4	113.1	80.2	80.1	96.1	114.8
1988	115.4	113.4	114.9	116.4	109.9	108.7	107.6	...	116.5	118.0	80.9	80.8	97.9	119.7
1989	118.6	117.0	116.4	119.1	114.4	114.1	112.9	...	119.2	120.4	88.5	88.5	100.2	124.9
1990	124.1	120.4	122.6	125.8	117.4	120.5	118.8	...	121.4	117.6	101.2	101.0	100.9	130.1
1991	128.7	124.2	127.6	128.9	120.9	123.8	121.9	...	126.0	118.1	99.4	99.2	102.2	136.0
1992	131.9	126.5	130.4	129.3	125.0	126.5	124.6	...	129.2	123.2	99.0	99.0	103.1	141.3
1993	133.7	127.5	132.6	127.1	125.9	130.4	127.5	91.8	132.7	133.9	98.0	97.7	101.6	145.9
1994	133.4	126.4	130.9	128.1	126.0	134.3	131.4	95.5	137.6	141.7	98.5	98.2	101.4	150.2
1995	132.0	126.2	126.9	127.2	125.4	139.1	136.3	99.4	141.0	156.5	100.0	99.8	102.1	154.0
1996	131.7	127.7	124.7	129.7	126.6	143.0	140.0	101.0	143.7	157.0	106.3	105.9	102.2	158.4
1997	132.9	130.1	126.1	129.0	127.6	144.3	141.0	100.5	144.3	151.1	106.2	105.8	101.9	162.7
1998	133.0	131.8	126.0	126.1	128.0	141.6	137.9	100.1	143.4	150.6	92.2	91.6	101.1	167.1
1999	131.3	131.1	123.3	129.0	125.7	144.4	140.5	100.1	142.9	152.0	100.7	100.1	100.5	171.9
2000	129.6	129.7	121.5	130.6	123.8	153.3	149.1	100.8	142.8	155.8	129.3	128.6	101.5	177.3
2001	127.3	125.7	119.3	129.2	123.0	154.3	150.0	101.3	142.1	158.7	124.7	124.0	104.8	183.5
2002	124.0	121.7	115.8	126.4	121.4	152.9	148.8	99.2	140.0	152.0	116.6	116.0	106.9	190.2
2003	120.9	118.0	113.1	122.1	119.6	157.6	153.6	96.5	137.9	142.9	135.8	135.1	107.8	195.6
2004	120.4	117.5	113.0	118.5	119.3	163.1	159.4	94.2	137.1	133.3	160.4	159.7	108.7	200.2
2005	119.5	116.1	110.8	116.7	122.6	173.9	170.2	95.6	137.9	139.4	195.7	194.7	111.9	206.9
2006	119.5	114.1	110.7	116.5	123.5	180.9	177.0	95.6	137.6	140.0	221.0	219.9	117.3	215.6
2007	119.0	112.4	110.3	113.9	122.4	184.7	180.8	94.3	136.3	135.7	239.1	238.0	121.6	223.0
2008	118.9	113.0	107.5	113.8	124.2	195.5	191.0	93.3	134.2	134.0	279.7	277.5	128.7	233.9
2009	120.1	113.6	108.1	114.5	126.9	179.3	174.8	93.5	135.6	127.0	202.0	201.6	134.1	243.3
2010	119.5	111.9	107.1	114.2	128.0	193.4	188.7	97.1	138.0	143.1	239.2	238.6	137.0	248.0
2011	122.1	114.7	109.2	113.6	128.5	212.4	207.6	99.8	141.9	149.0	302.6	301.7	143.9	253.1
2012	126.3	119.5	113.0	119.7	131.8	217.3	212.8	100.6	144.2	150.3	312.7	311.5	148.6	257.6
2013	127.4	121.6	113.3	116.5	135.0	217.4	212.4	100.9	145.8	149.9	303.9	302.6	146.4	261.6
2014	127.5	120.6	114.4	117.6	135.5	215.9	211.0	100.8	146.3	149.1	292.4	290.9	144.8	266.0
2015	125.9	119.6	111.2	119.7	136.8	199.1	193.7	100.8	147.1	147.1	213.1	212.0	144.2	270.7
2016	126.0	118.8	111.2	115.9	137.3	194.9	189.5	100.2	147.4	143.5	188.4	187.6	143.6	275.4
2017	125.6	117.2	111.0	114.9	136.7	201.6	196.6	98.9	147.0	138.3	212.7	211.8	143.0	280.8
2018	125.7	118.1	110.4	120.3	136.1	210.7	206.4	99.1	146.3	138.4	241.9	240.6	143.7	286.4
2017														
January	126.6	118.7	111.8	115.0	138.5	202.7	197.7	99.8	148.5	140.4	219.6	218.8	143.0	279.5
February	127.1	120.7	111.8	115.8	137.3	201.2	196.0	99.6	148.2	139.6	212.1	211.3	143.3	279.8
March	126.4	118.0	112.4	113.6	137.3	200.5	195.4	99.2	147.8	139.0	209.0	208.1	143.8	279.6
April	125.8	116.0	111.5	114.8	137.6	200.5	195.3	99.2	147.8	138.5	208.8	207.9	143.9	280.2
May	125.1	116.7	110.2	113.4	137.3	198.6	193.4	99.1	147.6	138.2	199.3	198.3	143.3	280.4
June	125.2	116.1	110.5	113.3	138.1	198.2	193.2	98.9	147.2	137.5	198.0	197.1	143.4	279.3
July	125.3	116.1	111.5	113.0	136.3	197.7	192.6	98.3	146.3	136.6	196.3	195.4	143.6	279.6
August	125.7	117.0	111.8	113.7	135.9	200.2	195.3	98.2	146.2	136.7	207.3	206.3	143.4	280.5
September	125.7	117.0	110.9	119.7	136.2	204.4	199.7	98.2	145.9	137.1	226.3	225.4	143.2	282.2
October	125.6	117.4	111.2	116.9	135.6	203.0	198.2	98.3	145.6	137.9	218.6	217.6	142.2	283.3
November	124.8	116.5	110.0	114.5	135.5	205.8	201.2	98.7	146.0	138.8	229.2	228.3	141.5	282.5
December	124.4	116.8	109.3	115.1	135.2	206.9	202.3	99.2	146.8	139.5	230.8	229.8	142.0	283.1
2018														
January	125.8	117.5	111.8	116.0	135.4	208.9	204.5	99.3	146.7	139.3	237.4	236.3	142.6	283.3
February	127.3	119.8	113.0	116.7	136.8	209.5	205.1	99.1	146.0	139.0	238.4	237.3	142.8	284.0
March	126.7	118.4	111.1	118.9	138.3	208.4	203.9	99.2	146.0	139.4	232.6	231.4	143.3	283.7
April	126.7	118.4	112.0	120.1	136.8	208.5	204.2	98.3	145.4	137.2	236.8	235.6	142.8	284.4
May	126.8	117.3	112.9	122.3	136.2	210.1	205.9	98.3	146.0	136.0	243.5	242.3	143.7	284.9
June	125.9	117.4	111.2	123.8	136.2	211.4	207.4	98.8	146.5	136.8	247.6	246.3	143.8	285.8
July	125.5	117.8	109.9	121.5	136.9	212.2	208.0	99.2	146.6	137.7	248.2	246.9	143.8	286.1
August	123.9	115.3	108.5	119.7	136.3	212.7	208.5	99.3	146.6	138.3	249.4	248.1	143.9	286.9
September	125.0	118.0	109.8	121.0	135.2	211.7	207.5	98.6	146.6	135.3	246.6	245.3	143.7	288.5
October	125.3	120.0	108.9	121.5	134.9	213.9	209.8	99.3	146.4	138.7	253.1	251.9	144.0	289.1
November	124.5	118.6	108.4	120.7	135.0	211.8	207.7	100.2	146.4	142.2	240.2	238.7	144.5	290.1
December	124.6	118.8	108.2	121.8	135.1	208.6	204.4	100.1	146.4	141.5	226.4	224.9	145.1	289.7

[1] Not seasonally adjusted.
[3] December 1997 = 100.
. . . = Not available.

Table 8-2. Consumer Price Indexes and Purchasing Power of the Dollar—*Continued*

(1982–1984 = 100, seasonally adjusted, except as noted.)

| Year and month | CPI-U, Transportation—Continued | | CPI-U, Medical care | | | | | CPI-U, Recreation | | CPI-U, Education and communication | | | |
| | Public transportation | Transportation services | Medical care, total | Medical care commodities | Medical care services | | | Total [3] | Video and audio [3] | Total [3] | Education | | |
					Total	Professional services	Hospital and related services				Total [3]	Educational books and supplies	Tuition, other school fees, and childcare
1970	35.2	40.2	34.0	46.5	32.3	37.0	...	...	...	...	...	38.8	...
1971	37.8	43.4	36.1	47.3	34.7	39.4	...	...	...	...	...	41.4	...
1972	39.3	44.4	37.3	47.4	35.9	40.8	...	...	...	...	...	44.2	...
1973	39.7	44.7	38.8	47.5	37.5	42.2	...	...	...	...	...	45.6	...
1974	40.6	46.3	42.4	49.2	41.4	45.8	...	...	...	...	...	47.2	...
1975	43.5	49.8	47.5	53.3	46.6	50.8	...	...	...	...	...	50.3	...
1976	47.8	56.9	52.0	56.5	51.3	55.5	...	...	...	...	...	53.7	...
1977	50.0	61.5	57.0	60.2	56.4	60.0	...	...	...	...	...	56.9	...
1978	51.5	64.4	61.8	64.4	61.2	64.5	55.1	...	...	...	...	61.6	59.8
1979	54.9	69.5	67.5	69.0	67.2	70.1	61.0	...	...	...	...	65.7	64.7
1980	69.0	79.2	74.9	75.4	74.8	77.9	69.2	...	...	...	...	71.4	71.2
1981	85.6	88.6	82.9	83.7	82.8	85.9	79.1	...	...	...	...	80.3	79.9
1982	94.9	96.1	92.5	92.3	92.6	93.2	90.3	...	...	...	...	91.0	90.5
1983	99.5	99.1	100.6	100.2	100.7	99.8	100.5	...	...	...	...	100.3	99.7
1984	105.7	104.8	106.8	107.5	106.7	107.0	109.2	...	...	...	...	108.7	109.8
1985	110.5	110.0	113.5	115.2	113.2	113.5	116.1	...	...	...	...	118.2	119.7
1986	117.0	116.3	122.0	122.8	121.9	120.8	123.1	...	...	...	...	128.1	129.6
1987	121.1	121.9	130.1	131.0	130.0	128.8	131.6	...	...	...	...	138.1	140.0
1988	123.3	128.0	138.6	139.9	138.3	137.5	143.9	...	...	...	...	148.1	151.0
1989	129.5	135.6	149.3	150.8	148.9	146.4	160.5	...	...	...	...	158.0	162.7
1990	142.6	144.2	162.8	163.4	162.7	156.1	178.0	...	...	...	...	171.3	175.7
1991	148.9	151.2	177.0	176.8	177.1	165.7	196.1	...	...	...	...	180.3	191.4
1992	151.4	155.7	190.1	188.1	190.5	175.8	214.0	...	...	...	...	190.3	208.5
1993	167.0	162.9	201.4	195.0	202.9	184.7	231.9	90.7	96.5	85.5	78.4	197.6	225.3
1994	172.0	168.6	211.0	200.7	213.4	192.5	245.6	92.7	95.4	88.8	83.3	205.5	239.8
1995	175.9	175.9	220.5	204.5	224.2	201.0	257.8	94.5	95.1	92.2	88.0	214.4	253.8
1996	181.9	180.5	228.2	210.4	232.4	208.3	269.5	97.4	96.6	95.3	92.7	226.9	267.1
1997	186.7	185.0	234.6	215.3	239.1	215.4	278.4	99.6	99.4	98.4	97.3	238.4	280.4
1998	190.3	187.9	242.1	221.8	246.8	222.2	287.5	101.1	101.1	100.3	102.1	250.8	294.2
1999	197.7	190.7	250.6	230.7	255.1	229.2	299.5	102.0	100.7	101.2	107.0	261.7	308.4
2000	209.6	196.1	260.8	238.1	266.0	237.7	317.3	103.3	101.0	102.5	112.5	279.9	324.0
2001	210.6	201.9	272.8	247.6	278.8	246.5	338.3	104.9	101.5	105.2	118.5	295.9	341.1
2002	207.4	209.1	285.6	256.4	292.9	253.9	367.8	106.2	102.8	107.9	126.0	317.6	362.1
2003	209.3	216.3	297.1	262.8	306.0	261.2	394.8	107.5	103.6	109.8	134.4	335.4	386.7
2004	209.1	220.6	310.1	269.3	321.3	271.5	417.9	108.6	104.2	111.6	143.7	351.0	414.3
2005	217.3	225.7	323.2	276.0	336.7	281.7	439.9	109.4	104.2	113.7	152.7	365.6	440.9
2006	226.6	230.8	336.2	285.9	350.6	289.3	468.1	110.9	104.6	116.8	162.1	388.9	468.1
2007	230.0	233.7	351.1	290.0	369.3	300.8	498.9	111.4	102.9	119.6	171.4	420.4	494.1
2008	250.5	244.1	364.1	296.0	384.9	311.0	534.0	113.3	102.6	123.6	181.3	450.2	522.1
2009	236.3	251.0	375.6	305.1	397.3	319.4	567.9	114.3	101.3	127.4	190.9	482.1	549.0
2010	251.4	259.8	388.4	314.7	411.2	328.2	607.7	113.3	99.1	129.9	199.3	505.6	573.2
2011	269.4	268.0	400.3	324.1	423.8	335.7	641.5	113.4	98.4	131.5	207.8	529.5	597.2
2012	271.4	272.9	414.9	333.6	440.3	342.0	672.1	114.7	99.4	133.8	216.3	562.6	621.0
2013	278.9	280.0	425.1	335.1	454.0	349.5	701.3	115.3	99.7	135.9	224.5	594.7	643.7
2014	276.4	285.3	435.3	343.4	464.8	355.2	733.8	115.5	99.8	137.5	231.9	615.4	664.8
2015	268.7	291.0	446.8	354.6	476.2	361.5	761.9	115.9	99.6	138.2	240.5	648.2	688.8
2016	265.4	299.4	463.7	366.8	494.8	371.5	795.1	117.0	100.9	139.1	247.5	678.8	708.1
2017	263.1	309.9	475.3	377.0	506.8	375.1	831.7	118.5	104.2	136.5	253.2	689.5	724.7
2018	258.8	321.8	484.7	381.4	517.8	378.4	866.9	119.1	104.2	136.8	258.8	696.1	741.3
2017													
January	265.4	305.1	471.9	374.9	503.0	377.0	811.9	117.6	102.3	139.0	250.8	690.4	717.4
February	266.1	306.3	472.7	374.8	504.1	376.8	816.0	118.2	102.9	138.8	251.5	696.7	719.0
March	266.5	307.2	473.1	375.2	504.5	376.6	818.9	118.2	103.3	136.4	252.0	699.7	720.4
April	266.7	307.5	472.8	373.5	504.7	373.6	825.1	118.3	103.7	136.0	252.3	691.1	721.8
May	264.3	308.5	472.6	373.0	504.6	374.1	826.3	118.4	104.1	135.9	252.6	685.6	723.1
June	261.6	309.2	474.1	376.2	505.5	373.1	831.7	118.2	104.2	135.9	253.3	685.5	725.1
July	262.3	310.2	476.2	380.4	506.7	373.4	834.3	118.6	104.5	135.8	253.6	692.6	725.6
August	260.7	311.0	477.2	380.1	508.3	374.9	836.9	118.8	104.9	135.6	253.2	687.9	724.9
September	260.9	311.9	477.2	377.4	509.2	375.4	840.0	119.0	105.3	135.8	254.0	680.5	727.5
October	261.5	312.9	478.1	377.3	510.4	375.8	844.5	119.0	105.2	136.1	254.5	685.2	728.9
November	261.0	314.1	478.4	379.3	510.1	375.4	846.5	118.9	104.9	136.3	255.2	693.6	730.4
December	260.7	315.3	479.7	382.3	510.8	375.8	849.1	118.9	104.7	136.5	255.4	685.7	731.4
2018													
January	260.1	317.4	481.2	381.6	513.1	376.1	857.6	118.9	104.7	136.5	255.8	687.2	732.6
February	260.1	320.0	481.1	380.9	513.2	377.1	854.7	118.9	104.4	136.3	256.3	693.7	733.8
March	259.9	320.1	482.5	380.5	515.2	378.7	858.9	119.0	104.2	136.2	256.6	696.2	734.5
April	258.7	319.9	483.2	380.9	516.0	378.4	859.9	118.6	103.8	136.1	257.0	693.8	735.9
May	258.0	320.4	483.9	382.9	516.2	378.3	863.0	118.7	103.8	136.6	257.9	714.7	737.5
June	254.9	320.7	485.8	385.1	518.0	378.9	868.5	118.9	103.7	136.9	258.4	703.6	739.5
July	259.0	322.5	485.2	382.1	518.3	378.7	870.3	119.0	103.7	137.2	259.1	704.2	741.5
August	260.0	322.9	484.6	381.1	517.8	378.0	870.9	118.9	104.0	137.3	260.0	695.1	744.6
September	260.0	324.4	485.5	380.1	519.3	378.6	871.6	119.3	104.5	137.5	260.5	693.5	746.3
October	259.5	324.7	486.3	379.9	520.4	378.9	871.9	119.2	104.4	137.3	261.0	690.7	747.9
November	258.5	324.6	488.1	381.7	522.2	379.2	876.1	119.6	104.5	136.6	261.5	687.7	749.5
December	256.9	324.4	489.3	380.3	524.3	379.6	880.2	120.2	104.6	136.7	262.0	693.6	750.9

[3]December 1997 = 100.
. . . = Not available.

Table 8-2. Consumer Price Indexes and Purchasing Power of the Dollar—*Continued*

(1982–1984 = 100, seasonally adjusted except as noted.)

Year and month	CPI-U, Education and communication—*Continued*					CPI-U, Other goods and services					CPI-W, All items, not seasonally adjusted	Purchasing power of the dollar, CPI-U, 1982–1984 = $1.00, not seasonally adjusted
	Communication					Total	Tobacco and smoking products [1]	Personal care				
	Total [3]	Information and information processing						Total	Personal care products [1]	Personal care services [1]		
		Total [3]	Telephone services [1,3]	Information technology, hardware, and services								
				Total [1,4]	Personal computers and peripheral equipment [1,5]							
1970	...	...	...	...	...	40.9	43.1	43.5	42.7	44.2	39.0	257.400
1971	...	...	...	...	...	42.9	44.9	44.9	44.0	45.7	40.7	246.600
1972	...	...	...	...	...	44.7	47.4	46.0	45.2	46.8	42.1	239.100
1973	...	...	...	...	...	46.4	48.7	48.1	46.4	49.7	44.7	225.100
1974	...	...	...	...	...	49.8	51.1	52.8	51.5	53.9	49.6	202.900
1975	...	...	...	...	...	53.9	54.7	57.9	58.0	57.7	54.1	185.900
1976	...	...	...	...	...	57.0	57.0	61.7	61.3	61.9	57.2	175.700
1977	...	...	...	...	...	60.4	59.8	65.7	64.7	66.4	60.9	164.900
1978	...	...	...	...	...	64.3	63.0	69.9	68.2	71.3	65.6	153.200
1979	...	...	...	...	...	68.9	66.8	75.2	72.9	77.2	73.1	138.000
1980	...	...	...	...	...	75.2	72.0	81.9	79.6	83.7	82.9	121.500
1981	...	...	...	...	...	82.6	77.8	89.1	87.8	90.2	91.4	109.800
1982	...	...	...	...	...	91.1	86.5	95.4	95.1	95.7	96.9	103.500
1983	...	...	...	...	...	101.1	103.4	100.3	100.7	100.0	99.8	100.300
1984	...	...	...	...	...	107.9	110.1	104.3	104.2	104.4	103.3	96.100
1985	...	...	...	...	...	114.5	116.7	108.3	107.6	108.9	106.9	92.800
1986	...	...	...	...	...	121.4	124.7	111.9	111.3	112.5	108.6	91.300
1987	...	...	...	...	...	128.5	133.6	115.1	113.9	116.2	112.5	88.000
1988	...	...	...	...	...	137.0	145.8	119.4	118.1	120.7	117.0	84.600
1989	...	...	...	06.3	...	147.7	164.4	125.0	123.2	126.8	122.6	80.700
1990	...	...	...	93.5	...	159.0	181.5	130.4	128.2	132.8	129.0	76.600
1991	...	...	...	88.6	...	171.6	202.7	134.9	132.8	137.0	134.3	73.400
1992	...	...	...	83.7	...	183.3	219.8	138.3	136.5	140.0	138.2	71.300
1993	96.7	97.7	...	78.8	...	192.9	228.4	141.5	139.0	144.0	142.1	69.200
1994	97.6	98.6	...	72.0	...	198.5	220.0	144.6	141.5	147.9	145.6	67.500
1995	98.8	98.7	...	63.8	...	206.9	225.7	147.1	143.1	151.5	149.8	65.600
1996	99.6	99.5	...	57.2	...	215.4	232.8	150.1	144.3	156.6	154.1	63.800
1997	100.3	100.4	...	50.1	...	224.8	243.7	152.7	144.2	162.4	157.6	62.300
1998	98.7	98.5	100.7	39.9	875.1	237.7	274.8	156.7	148.3	166.0	159.7	61.400
1999	96.0	95.5	100.1	30.5	598.7	258.3	355.8	161.1	151.8	171.4	163.2	60.000
2000	93.6	92.8	98.5	25.9	459.9	271.1	394.9	165.6	153.7	178.1	168.9	58.100
2001	93.3	92.3	99.3	21.3	330.1	282.6	425.2	170.5	155.1	184.3	173.5	56.500
2002	92.3	90.8	99.7	18.3	248.4	293.2	461.5	174.7	154.7	188.4	175.9	55.600
2003	89.7	87.8	98.3	16.1	196.9	298.7	469.0	178.0	153.5	193.2	179.8	54.400
2004	86.7	84.6	95.8	14.8	171.2	304.7	478.0	181.7	153.9	197.6	184.5	53.000
2005	84.7	82.6	94.9	13.6	143.2	313.4	502.8	185.6	154.4	203.9	191.0	51.200
2006	84.1	81.7	95.8	12.5	120.9	321.7	519.9	190.2	155.8	209.7	197.1	49.600
2007	83.4	80.7	98.2	10.6	108.4	333.3	554.2	195.6	158.3	216.6	202.8	48.200
2008	84.2	81.4	100.5	10.1	94.9	345.4	588.7	201.3	159.3	223.7	211.1	46.500
2009	85.0	81.9	102.4	9.7	82.3	368.6	730.3	204.6	162.6	227.6	209.6	46.600
2010	84.7	81.5	102.4	9.4	76.4	381.3	807.3	206.6	161.1	229.6	214.0	45.900
2011	83.3	80.0	101.2	9.0	68.9	387.2	834.8	208.6	160.5	230.8	221.6	44.500
2012	83.1	79.5	101.7	8.7	62.3	394.4	853.5	212.1	162.2	234.2	226.2	43.600
2013	82.6	78.9	101.6	8.5	56.8	401.0	876.8	215.0	161.8	238.8	229.3	42.900
2014	82.1	78.2	101.1	8.4	52.6	408.1	903.3	218.0	163.4	242.0	232.8	42.200
2015	80.2	76.4	99.3	8.1	47.9	414.9	930.8	220.8	163.3	247.2	231.8	42.200
2016	79.2	75.4	98.8	7.8	44.4	423.1	963.4	224.3	163.1	252.9	234.1	41.700
2017	75.0	71.1	91.8	7.6	42.6	432.6	1 022.8	227.0	161.7	257.4	239.1	40.800
2018	73.9	70.0	90.4	7.5	40.8	442.3	1 063.5	231.1	161.5	264.2	245.1	39.800
2017												
January	78.2	74.4	97.4	7.7	43.6	427.4	982.4	226.0	162.4	255.6	237.9	41.200
February	77.8	74.0	96.8	7.7	43.5	428.0	985.4	226.3	162.5	255.8	238.0	41.100
March	75.2	71.4	92.0	7.7	42.9	428.5	989.8	226.3	161.6	256.5	237.7	41.000
April	74.7	70.8	90.9	7.7	42.6	432.3	1 026.9	226.6	161.9	256.6	237.9	40.900
May	74.6	70.7	90.7	7.6	42.2	432.1	1 028.6	226.4	161.0	256.6	237.8	40.900
June	74.3	70.5	90.2	7.7	42.5	433.2	1 029.4	227.0	161.6	256.5	238.0	40.800
July	74.2	70.3	90.1	7.6	42.9	434.0	1 030.6	227.5	161.6	257.5	238.1	40.900
August	74.1	70.2	90.1	7.6	42.6	433.4	1 030.7	227.1	160.7	258.8	239.2	40.700
September	74.1	70.2	90.3	7.6	42.2	434.4	1 035.6	227.5	161.5	258.5	240.4	40.500
October	74.2	70.3	90.6	7.6	42.2	435.9	1 043.7	227.9	161.7	258.7	240.5	40.500
November	74.3	70.4	90.9	7.5	42.0	436.1	1 045.1	228.0	162.2	258.4	241.4	40.500
December	74.4	70.5	91.1	7.5	41.6	435.7	1 044.9	227.7	161.3	258.9	242.0	40.600
2018												
January	74.3	70.5	91.0	7.5	41.5	437.5	1 047.7	228.8	161.9	260.2	243.0	40.300
February	74.0	70.1	90.6	7.5	41.4	438.4	1 050.0	229.2	162.1	260.6	243.5	40.200
March	73.9	70.0	90.6	7.4	41.3	439.5	1 049.5	230.0	162.0	261.4	243.6	40.100
April	73.7	69.8	90.6	7.4	41.2	442.5	1 056.6	231.6	161.9	262.6	244.0	39.900
May	73.9	70.0	90.7	7.4	41.2	442.8	1 060.8	231.5	161.7	263.1	244.7	39.700
June	74.1	70.2	90.7	7.5	41.0	442.9	1 061.2	231.6	161.1	265.1	245.2	39.700
July	74.2	70.3	90.9	7.5	40.2	443.2	1 063.7	231.7	161.3	265.3	245.7	39.700
August	74.1	70.2	90.5	7.5	40.7	443.1	1 066.5	231.4	161.3	265.1	246.0	39.700
September	74.2	70.2	90.6	7.5	41.1	443.8	1 070.1	231.7	161.5	265.3	246.0	39.600
October	73.9	70.0	90.4	7.5	40.4	443.8	1 075.4	231.4	161.1	266.8	246.9	39.500
November	73.0	69.1	89.0	7.4	39.8	445.1	1 078.8	232.0	160.8	267.0	246.7	39.700
December	73.1	69.2	88.9	7.5	40.3	444.9	1 080.6	231.8	160.7	268.2	246.4	39.800

[1] Not seasonally adjusted.
[3] December 1997 = 100.
[4] December 1988 = 100.
[5] December 2007 = 100.
... = Not available.

Table 8-3. Alternative Measures of Total and Core Consumer Prices: Index Levels

(Various bases; monthly data seasonally adjusted, except as noted.)

Year and month	CPIs, all items					CPIs, all items less food and energy		
	CPI-U, 1982–1984 = 100	CPI-W, 1982–1984 = 100	CPI-U-X1, 1982–1984 = 100	CPI-U-RS, Dec. 1977 = 100, not seasonally adjusted	C-CPI-U, Dec. 1999 = 100, not seasonally adjusted	CPI-U, 1982–1984 = 100	CPI-U-RS, Dec. 1977 = 100, not seasonally adjusted	C-CPI-U, Dec. 1999 = 100, not seasonally adjusted
1975	53.8	54.1	56.2	...	...	53.9	...	...
1976	56.9	57.2	59.4	...	...	57.4	...	...
1977	60.6	60.9	63.2	...	...	61.0	...	...
1978	65.2	65.6	67.5	104.4	...	65.5	103.6	...
1979	72.6	73.1	74.0	114.3	...	71.9	111.0	...
1980	82.4	82.9	82.3	127.1	...	80.8	120.9	...
1981	90.9	91.4	90.1	139.1	...	89.2	132.2	...
1982	96.5	96.9	95.6	147.5	...	95.8	142.4	...
1983	99.6	99.8	99.6	153.8	...	99.6	150.4	...
1984	103.9	103.3	103.9	160.2	...	104.6	157.9	...
1985	107.6	106.9	107.6	165.7	...	109.1	164.8	...
1986	109.6	108.6	109.6	168.6	...	113.5	171.4	...
1987	113.6	112.5	113.6	174.4	...	118.2	178.1	...
1988	118.3	117.0	118.3	180.7	...	123.4	185.2	...
1989	124.0	122.6	124.0	188.6	...	129.0	192.6	...
1990	130.7	129.0	130.7	197.9	...	135.5	201.4	...
1991	136.2	134.3	136.2	205.1	...	142.1	209.9	...
1992	140.3	138.2	140.3	210.2	...	147.3	216.4	...
1993	144.5	142.1	144.5	215.5	...	152.2	222.5	...
1994	148.2	145.6	148.2	220.0	...	156.5	227.7	...
1995	152.4	149.8	152.4	225.3	...	161.2	233.4	...
1996	156.9	154.1	156.9	231.3	...	165.6	239.1	...
1997	160.5	157.6	160.5	236.3	...	169.5	244.4	...
1998	163.0	159.7	163.0	239.5	...	173.4	249.6	...
1999	166.6	163.2	166.6	244.6	...	177.0	254.6	...
2000	172.2	168.9	172.2	252.9	102.0	181.3	260.9	101.4
2001	177.1	173.5	177.1	260.1	104.3	186.1	267.9	103.5
2002	179.9	175.9	179.9	264.2	105.6	190.5	274.1	105.4
2003	184.0	179.8	184.0	270.2	107.8	193.2	278.1	106.6
2004	188.9	184.5	188.9	277.5	110.5	196.6	283.1	108.4
2005	195.3	191.0	195.3	286.9	113.7	200.9	289.2	110.4
2006	201.6	197.1	201.6	296.2	117.0	205.9	296.5	112.9
2007	207.3	202.8	207.3	304.6	120.0	210.7	303.4	115.0
2008	215.3	211.1	215.3	316.3	124.4	215.6	310.3	117.3
2009	214.5	209.6	214.5	315.2	123.9	219.2	315.6	119.1
2010	218.1	214.0	218.1	320.4	125.6	221.3	318.7	120.0
2011	224.9	221.6	224.9	330.5	129.5	225.0	324.0	121.9
2012	229.6	226.2	229.6	337.5	132.0	229.8	331.0	124.3
2013	233.0	229.3	233.0	342.5	...	233.8	337.0	...
2014	236.7	232.8	236.7	348.3	...	237.9	343.1	...
2015	237.0	231.8	237.0	348.9	...	242.2	349.7	...
2016	240.0	234.1	240.0	353.4	...	247.6	357.5	...
2017	245.1	239.1	245.1	361.0	...	252.2	364.2	...
2018	251.1	245.1	251.1	369.8	...	257.6	372.0	...
2017								
January	243.8	237.9	243.8	359.0	138.0	250.6	361.9	133.3
February	244.0	238.0	244.0	359.3	138.4	251.1	362.6	133.8
March	243.7	237.7	243.7	358.9	138.5	250.9	362.3	133.8
April	244.1	237.9	244.1	359.4	138.8	251.1	362.6	134.0
May	244.0	237.8	244.0	359.3	138.9	251.3	362.9	134.1
June	244.2	238.0	244.2	359.6	139.0	251.7	363.5	134.1
July	244.4	238.1	244.4	359.9	138.8	252.0	363.9	133.9
August	245.3	239.2	245.3	361.2	139.1	252.6	364.8	134.2
September	246.4	240.4	246.4	362.9	139.9	253.0	365.3	134.4
October	246.6	240.5	246.6	363.1	139.8	253.5	366.1	134.8
November	247.3	241.4	247.3	364.2	139.7	253.9	366.6	134.6
December	247.9	242.0	247.9	365.1	139.5	254.4	367.5	134.5
2018								
January	248.9	243.0	248.9	366.5	140.2	255.2	368.6	135.1
February	249.4	243.5	249.4	367.2	140.8	255.7	369.2	135.6
March	249.5	243.6	249.5	367.4	141.1	256.1	369.9	136.0
April	250.0	244.0	250.0	368.1	141.7	[1]256.4	370.3	136.3
May	250.6	244.7	250.6	369.1	142.1	[1]256.9	371.0	136.5
June	251.1	245.2	251.1	369.8	142.3	[1]257.3	371.6	136.6
July	251.6	245.7	251.6	370.5	142.3	[1]257.9	372.4	136.6
August	251.9	246.0	251.9	370.9	142.4	[1]258.1	372.7	136.7
September	252.0	246.0	252.0	371.1	142.6	[1]258.5	373.3	136.9
October	252.8	246.9	252.8	372.3	142.8	259.0	374.0	137.2
November	252.8	246.7	252.8	372.2	142.3	259.6	374.9	137.2
December	252.7	246.4	252.7	372.2	141.8	260.1	375.6	137.2

[1]Interim values.
. . . = Not available.

Table 8-4. Alternative Measures of Total and Core Consumer Prices: Inflation Rates

(Percent changes from year earlier, except as noted; monthly data seasonally adjusted, except as noted.)

Year and month	CPIs, all items					CPIs, all items less food and energy		
	CPI-U, 1982–1984 = 100	CPI-W, 1982–1984 = 100	CPI-U-X1, 1982–1984 = 100	CPI-U-RS, Dec. 1977 = 100, not seasonally adjusted	C-CPI-U, Dec. 1999 = 100, not seasonally adjusted	CPI-U, 1982–1984 = 100	CPI-U-RS, Dec. 1977 = 100, not seasonally adjusted	C-CPI-U, Dec. 1999 = 100, not seasonally adjusted
1960	1.7	1.7	1.9	...	...	1.3	...	...
1961	1.0	1.0	0.9	...	...	1.3	...	...
1962	1.0	1.0	0.9	...	...	1.3	...	...
1963	1.3	1.3	1.5	...	...	1.3	...	...
1964	1.3	1.3	1.2	...	...	1.6	...	...
1965	1.6	1.6	1.5	...	...	1.2	...	...
1966	2.9	2.8	2.9	...	...	2.4	...	...
1967	3.1	3.1	3.1	...	...	3.6	...	...
1968	4.2	4.2	3.9	...	...	4.6	...	...
1969	5.5	5.4	4.5	...	...	5.8	...	...
1970	5.7	5.7	4.8	...	...	6.3	...	...
1971	4.4	4.4	4.4	...	...	4.7	...	...
1972	3.2	3.4	3.0	...	...	3.0	...	...
1973	6.2	6.2	6.3	...	...	3.6	...	...
1974	11.0	11.0	10.0	...	...	8.3	...	...
1975	9.1	9.1	8.3	...	...	9.1	...	...
1976	5.8	5.7	5.7	...	...	6.5	...	...
1977	6.5	6.5	6.4	...	...	6.3	...	...
1978	7.6	7.7	6.8	...	...	7.4	...	...
1979	11.3	11.4	9.6	9.5	...	9.8	7.1	...
1980	13.5	13.4	11.2	11.2	...	12.4	8.9	...
1981	10.3	10.3	9.5	9.4	...	10.4	9.3	...
1982	6.2	6.0	6.1	6.0	...	7.4	7.7	...
1983	3.2	3.0	4.2	4.3	...	4.0	5.6	...
1984	4.3	3.5	4.3	4.2	...	5.0	5.0	...
1985	3.6	3.5	3.6	3.4	...	4.3	4.4	...
1986	1.9	1.6	1.9	1.8	...	4.0	4.0	...
1987	3.6	3.6	3.6	3.4	...	4.1	3.9	...
1988	4.1	4.0	4.1	3.6	...	4.4	4.0	...
1989	4.8	4.8	4.8	4.4	...	4.5	4.0	...
1990	5.4	5.2	5.4	4.9	...	5.0	4.6	...
1991	4.2	4.1	4.2	3.6	...	4.9	4.2	...
1992	3.0	2.9	3.0	2.5	...	3.7	3.1	...
1993	3.0	2.8	3.0	2.5	...	3.3	2.8	...
1994	2.6	2.5	2.6	2.1	...	2.8	2.3	...
1995	2.8	2.9	2.8	2.4	...	3.0	2.5	...
1996	3.0	2.9	3.0	2.7	...	2.7	2.4	...
1997	2.3	2.3	2.3	2.2	...	2.4	2.2	...
1998	1.6	1.3	1.6	1.4	...	2.3	2.1	...
1999	2.2	2.2	2.2	2.1	...	2.1	2.0	...
2000	3.4	3.5	3.4	3.4	...	2.4	2.5	...
2001	2.8	2.7	2.8	2.8	2.3	2.6	2.7	2.1
2002	1.6	1.4	1.6	1.6	1.2	2.4	2.3	1.8
2003	2.3	2.2	2.3	2.3	2.1	1.4	1.5	1.1
2004	2.7	2.6	2.7	2.7	2.5	1.8	1.8	1.7
2005	3.4	3.5	3.4	3.4	2.9	2.2	2.2	1.8
2006	3.2	3.2	3.2	3.2	2.9	2.5	2.5	2.3
2007	2.8	2.9	2.8	2.8	2.6	2.3	2.3	1.9
2008	3.8	4.1	3.8	3.8	3.7	2.3	2.3	2.0
2009	-0.4	-0.7	-0.4	-0.3	-0.4	1.7	1.7	1.5
2010	1.6	2.1	1.6	1.6	1.4	1.0	1.0	0.8
2011	3.2	3.6	3.2	3.2	3.1	1.7	1.7	1.6
2012	2.1	2.1	2.1	2.1	1.9	2.1	2.2	2.0
2013	1.5	1.4	1.5	1.5	...	[1]1.8	1.8	...
2014	1.6	1.5	1.6	1.7	...	[1]1.7	1.8	...
2015	0.1	-0.4	0.1	0.2	...	[1]1.8	1.9	...
2016	1.3	1.0	1.3	1.3	...	[1]2.2	2.2	...
2017	2.1	2.1	2.1	2.2	...	[1]1.8	1.9	...
2018	2.4	2.5	2.4	2.4	...	[1]2.1	2.1	...
2018								
January	2.1	2.2	2.1	2.1	1.6	[1]1.8	1.9	1.3
February	2.2	2.3	2.2	2.2	1.7	[1]1.8	1.8	1.3
March	2.4	2.5	2.4	2.4	1.9	[1]2.1	2.1	1.7
April	2.4	2.5	2.4	2.4	2.1	[1]2.1	2.1	1.7
May	2.7	2.9	2.7	2.7	2.3	[1]2.2	2.2	1.8
June	2.8	3.1	2.8	2.8	2.4	[1]2.2	2.2	1.8
July	2.9	3.2	2.9	2.9	2.6	[1]2.3	2.3	2.0
August	2.7	2.9	2.7	2.7	2.3	[1]2.2	2.2	1.9
September	2.3	2.3	2.3	2.3	1.9	[1]2.2	2.2	1.8
October	2.5	2.7	2.5	2.5	2.1	[1]2.2	2.2	1.8
November	2.2	2.2	2.2	2.2	1.9	[1]2.3	2.3	1.9
December	1.9	1.8	1.9	1.9	1.6	[1]2.2	2.2	2.0

[1]Interim values.
. . . = Not available.

Table 8-5. Producer Price Indexes for Goods and Services: Final Demand

(November 2009 =100, except as noted.)

Year and month	Final demand	Final demand goods	Final demand foods	Final demand energy goods	Final demand goods less food and energy Total	Finished goods less foods and energy (1982 =100) Total	Finished goods less foods and energy (1982 =100) Private capital equipment	Final demand services	Final demand trade services	Final demand transportation and warehousing services	Final demand services less trade, transportation, and warehousing	Final demand construction	Final demand less food and energy (April 2010 =100)
NOT SEASONALLY ADJUSTED													
2010	101.8	102.8	103.7	107.2	101.4	173.6	157.3	101.3	101.7	103.2	100.9	100.3	. . .
2011	105.7	109.9	112.5	126.2	104.9	177.8	159.7	103.4	104.0	110.0	102.5	102.5	102.7
2012	107.7	111.7	115.9	126.3	106.8	182.4	162.8	105.4	106.7	114.2	103.9	105.5	104.7
2013	109.1	112.6	117.8	125.3	107.9	185.1	164.2	107.1	108.2	115.3	105.8	107.5	106.2
2014	110.9	114.0	121.6	124.2	109.5	188.6	166.4	109.0	110.2	117.7	107.5	110.6	108.1
2015	109.9	109.1	118.4	98.6	109.9	192.3	168.5	110.0	111.6	115.3	108.7	112.7	108.9
2016	110.4	107.6	115.1	90.4	110.7	195.3	169.3	111.5	113.1	113.5	110.6	114.0	110.2
2017	113.0	111.2	116.5	99.8	113.2	198.9	170.9	113.5	114.8	115.9	112.8	116.5	112.3
2018	116.2	115.0	116.7	110.0	116.0	203.4	173.6	116.5	116.9	122.0	115.8	121.2	115.3
2015													
January	109.7	109.0	120.6	95.7	110.1	191.3	168.0	109.8	111.9	116.8	108.0	112.0	108.8
February	109.5	108.9	118.8	96.8	110.0	191.8	168.5	109.5	110.8	115.8	108.2	112.1	108.5
March	109.8	109.2	118.1	99.7	109.9	191.8	168.5	109.8	111.4	116.4	108.3	112.1	108.7
April	109.8	108.9	117.5	98.6	109.8	191.6	168.4	110.0	111.4	115.9	108.6	112.1	108.8
May	110.2	110.4	118.9	104.8	109.9	191.7	168.4	109.7	111.1	115.7	108.5	112.2	108.7
June	110.6	111.1	119.8	106.9	110.2	192.5	168.3	110.0	111.2	116.5	108.8	112.2	109.0
July	110.8	110.9	119.0	106.6	110.2	192.7	168.5	110.4	112.0	116.4	109.0	112.8	109.3
August	110.5	110.1	119.3	103.2	109.9	192.4	168.3	110.3	111.6	115.3	109.2	112.9	109.1
September	109.9	108.7	118.5	97.0	109.7	192.4	168.2	110.1	112.0	113.7	108.9	112.9	108.9
October	109.8	108.1	117.3	94.0	109.9	193.3	169.0	110.3	112.4	114.0	108.9	113.4	109.1
November	109.4	107.5	117.0	92.0	109.7	193.2	169.0	110.1	112.0	114.0	108.7	113.8	108.9
December	109.1	106.7	115.8	88.3	109.7	193.4	169.0	110.0	111.9	113.4	108.8	113.9	108.9
2016													
January	109.7	106.4	116.4	85.1	110.1	194.5	169.1	111.1	113.1	114.1	109.8	113.4	109.7
February	109.6	105.9	116.4	81.9	110.2	194.7	169.2	111.2	113.4	113.5	110.0	113.4	109.9
March	109.7	106.1	115.1	84.3	110.2	194.6	169.1	111.2	113.0	114.2	110.2	113.3	109.9
April	110.0	106.7	115.0	86.5	110.5	194.7	169.1	111.3	113.2	113.7	110.3	114.4	110.0
May	110.2	107.6	115.9	90.2	110.6	194.7	169.2	111.2	113.2	112.8	110.1	114.4	110.0
June	110.8	108.7	116.9	94.5	110.8	195.3	169.3	111.6	113.7	113.6	110.5	114.4	110.3
July	110.8	108.5	115.9	94.5	110.8	195.1	169.0	111.6	112.7	114.2	110.9	113.6	110.3
August	110.5	108.0	114.4	92.9	110.8	195.1	168.8	111.4	111.9	113.7	111.1	113.7	110.2
September	110.6	108.3	114.8	94.2	110.7	195.1	168.7	111.6	112.9	111.7	111.0	113.7	110.2
October	111.0	108.5	113.5	94.5	111.2	196.3	169.7	112.1	113.9	113.0	111.2	114.6	110.7
November	110.8	108.0	112.9	92.2	111.3	196.3	169.7	112.0	113.6	113.5	111.1	114.6	110.7
December	111.0	108.7	113.9	93.9	111.6	196.7	170.1	111.8	113.0	114.7	111.0	114.5	110.7
2017													
January	111.6	109.7	113.9	96.8	112.3	197.8	170.4	112.3	113.6	115.2	111.5	114.9	111.2
February	111.8	110.1	114.8	97.6	112.5	197.9	170.6	112.3	112.9	115.4	111.8	114.9	111.3
March	112.1	110.4	115.8	97.2	112.8	198.1	170.8	112.7	114.0	114.9	111.9	115.0	111.6
April	112.7	111.1	117.5	98.7	113.1	198.6	170.9	113.2	114.5	115.6	112.4	115.5	112.1
May	112.7	110.7	117.1	97.3	113.0	198.4	170.6	113.4	115.1	115.0	112.5	115.7	112.2
June	112.9	111.1	118.2	98.1	113.1	198.7	170.7	113.5	114.4	116.1	112.9	115.8	112.3
July	113.0	111.0	117.8	98.2	113.0	198.6	170.5	113.7	114.5	115.7	113.2	117.3	112.4
August	113.2	111.4	117.0	100.6	113.1	198.6	170.5	113.9	115.0	115.6	113.2	117.4	112.6
September	113.5	112.0	116.6	104.3	113.1	198.4	170.5	113.9	115.3	115.6	113.1	117.3	112.6
October	114.1	112.0	116.4	102.0	113.8	200.2	171.4	114.8	116.9	116.9	113.6	118.0	113.4
November	114.1	112.5	116.5	103.7	114.0	200.5	171.7	114.6	116.0	117.9	113.6	117.9	113.3
December	113.8	112.5	116.2	103.4	114.1	200.6	171.8	114.2	115.0	117.3	113.5	118.0	113.1
2018													
January	114.5	113.2	116.1	105.2	114.7	201.4	171.9	114.8	115.2	117.9	114.4	119.0	113.7
February	114.9	113.5	115.5	106.4	115.0	201.8	172.2	115.3	115.8	119.0	114.8	119.0	114.1
March	115.4	114.0	118.0	105.6	115.4	202.1	172.5	115.8	116.0	120.5	115.3	119.1	114.6
April	115.7	114.4	116.8	108.0	115.6	202.4	172.6	116.0	116.5	120.8	115.2	120.4	114.8
May	116.2	115.5	117.8	112.8	115.8	202.6	172.9	116.1	116.7	120.7	115.4	120.5	114.9
June	116.6	116.0	116.9	115.3	116.0	202.9	173.2	116.6	117.2	122.6	115.7	120.7	115.3
July	116.8	115.9	116.3	114.9	116.2	203.3	173.5	116.8	116.8	122.9	116.3	121.2	115.5
August	116.6	115.8	115.7	114.6	116.3	203.7	174.0	116.7	116.4	122.3	116.3	121.4	115.5
September	116.6	115.8	115.2	114.7	116.3	203.9	174.3	116.7	116.3	122.7	116.3	121.4	115.5
October	117.6	116.3	115.7	114.9	116.9	205.2	175.3	117.8	118.8	124.6	116.7	123.8	116.5
November	117.1	115.4	117.0	108.0	117.1	205.7	175.6	117.6	118.3	125.4	116.5	124.0	116.4
December	116.8	114.3	119.4	100.2	117.1	205.8	175.7	117.6	118.6	124.9	116.4	124.1	116.4

. . . = Not available.

Table 8-5. Producer Price Indexes for Goods and Services: Final Demand—*Continued*

(November 2009 =100, except as noted.)

Year and month	Final demand	Final demand goods						Final demand services				Final demand construction	Final demand less food and energy (April 2010 =100)
		Final demand goods	Final demand foods	Final demand energy goods	Final demand goods less food and energy			Final demand services	Final demand trade services	Final demand transportation and warehousing services	Final demand services less trade, transportation, and warehousing		
					Total	Finished goods less foods and energy (1982 =100)							
						Total	Private capital equipment						
SEASONALLY ADJUSTED													
2015													
January	110.1	109.5	121.5	98.4	109.9	190.7	167.7	110.1	112.6	116.7	108.1	112.0	108.9
February	109.6	109.3	119.5	99.1	109.8	191.3	168.2	109.5	110.8	115.7	108.1	112.1	108.5
March	109.8	109.5	118.5	101.0	109.7	191.5	168.3	109.6	111.1	115.7	108.2	112.1	108.5
April	109.6	108.9	117.7	98.4	109.8	191.6	168.3	109.8	110.9	115.5	108.6	112.1	108.7
May	110.1	110.0	118.3	103.4	109.9	191.8	168.5	109.8	111.1	115.9	108.5	112.2	108.7
June	110.4	110.4	118.7	104.2	110.2	192.7	168.5	110.0	111.4	115.8	108.7	112.2	109.0
July	110.5	110.3	118.4	103.4	110.3	193.0	168.9	110.3	112.2	115.8	108.8	112.8	109.2
August	110.3	109.7	118.7	100.8	110.1	193.0	168.8	110.3	111.9	115.0	109.1	112.9	109.2
September	109.8	108.4	118.3	94.6	110.1	193.2	168.9	110.2	112.0	114.8	108.9	112.9	109.1
October	109.6	108.1	117.4	94.0	109.8	193.0	168.7	110.0	111.6	114.5	108.8	113.9	108.9
November	109.7	108.0	117.3	94.2	109.7	193.1	168.7	110.2	111.9	114.6	108.9	113.8	109.0
December	109.6	107.3	116.3	90.9	100.0	193.4	168.7	110.4	112.3	113.8	109.2	113.9	109.2
2016													
January	109.9	106.8	117.1	87.3	109.9	193.9	168.8	111.3	113.8	113.9	109.8	113.4	109.8
February	109.7	106.1	117.0	83.5	110.0	194.2	168.9	111.2	113.5	113.5	110.0	113.4	109.8
March	109.6	106.3	115.5	85.6	110.1	194.3	168.9	111.0	112.7	113.4	110.0	113.3	109.7
April	109.9	106.7	115.2	86.6	110.4	194.6	169.0	111.2	112.8	113.2	110.3	114.4	109.9
May	110.1	107.4	115.4	89.3	110.6	194.9	169.2	111.2	113.1	113.0	110.2	114.4	110.0
June	110.6	108.1	115.8	92.2	110.8	195.4	169.5	111.6	113.8	113.0	110.4	114.4	110.3
July	110.5	108.0	115.3	92.0	110.9	195.4	169.3	111.5	112.9	113.7	110.7	113.6	110.2
August	110.3	107.7	113.8	90.8	111.0	195.7	169.3	111.4	112.2	113.6	110.9	113.7	110.2
September	110.6	108.0	114.7	91.8	111.0	195.8	169.4	111.7	112.8	112.7	111.1	113.7	110.4
October	110.9	108.5	113.7	94.5	111.2	196.1	169.4	111.8	113.1	113.4	111.1	114.6	110.6
November	111.1	108.4	113.2	94.0	111.4	196.3	169.4	112.1	113.5	113.9	111.3	114.6	110.8
December	111.4	109.3	114.4	96.6	111.7	196.7	169.8	112.2	113.4	115.1	111.4	114.5	111.0
2017													
January	111.9	110.2	114.6	99.5	112.1	197.1	170.0	112.6	114.2	115.1	111.6	114.9	111.3
February	111.9	110.5	115.4	99.7	112.3	197.4	170.3	112.3	113.0	115.3	111.7	114.9	111.2
March	112.1	110.7	116.1	99.0	112.7	197.8	170.6	112.5	113.8	114.2	111.8	115.0	111.4
April	112.6	111.1	117.7	98.8	113.0	198.5	170.8	113.0	114.1	115.1	112.4	115.5	111.9
May	112.6	110.4	116.6	96.4	113.0	198.6	170.7	113.4	115.0	115.2	112.6	115.7	112.2
June	112.6	110.5	117.1	95.9	113.0	198.8	170.9	113.5	114.5	115.5	112.8	115.8	112.2
July	112.7	110.5	117.2	95.7	113.1	198.9	170.9	113.5	114.7	115.3	112.9	117.3	112.3
August	113.1	111.0	116.4	98.5	113.3	199.2	171.0	113.8	115.3	115.7	113.0	117.4	112.6
September	113.4	111.7	116.5	101.6	113.4	199.2	171.1	114.0	115.2	116.7	113.2	117.3	112.7
October	113.9	112.0	116.7	101.8	113.8	200.0	171.1	114.5	116.2	117.3	113.5	118.0	113.2
November	114.3	112.9	116.9	105.5	114.0	200.5	171.4	114.7	116.0	118.3	113.8	117.9	113.4
December	114.3	113.2	116.6	106.6	114.2	200.6	171.5	114.6	115.4	117.7	113.9	118.0	113.4
2018													
January	114.8	113.7	116.7	108.3	114.5	200.9	171.6	115.0	115.7	117.8	114.5	119.0	113.8
February	115.1	113.9	116.1	108.7	114.9	201.3	172.0	115.3	115.8	119.0	114.7	119.0	114.1
March	115.4	114.3	118.3	107.7	115.2	201.8	172.3	115.6	115.8	119.6	115.1	119.1	114.4
April	115.5	114.3	117.0	108.0	115.5	202.3	172.5	115.8	116.1	120.2	115.2	120.4	114.6
May	116.0	115.2	117.3	111.4	115.8	202.7	172.9	116.1	116.7	120.9	115.4	120.5	114.9
June	116.4	115.3	115.9	112.6	116.0	203.1	173.4	116.5	117.3	121.9	115.7	120.7	115.3
July	116.5	115.4	115.8	112.1	116.3	203.7	173.8	116.6	117.0	122.5	115.9	121.2	115.4
August	116.5	115.5	115.2	112.2	116.5	204.2	174.5	116.6	116.6	122.5	116.1	121.4	115.5
September	116.6	115.4	115.1	111.4	116.6	204.6	174.8	116.8	116.2	124.1	116.4	121.4	115.7
October	117.5	116.3	116.1	114.4	116.9	205.1	175.0	117.7	118.2	125.0	116.7	123.8	116.4
November	117.4	115.7	117.3	109.7	117.1	205.6	175.3	117.8	118.3	125.6	116.8	124.0	116.5
December	117.3	115.0	119.8	103.3	117.1	205.8	175.5	118.0	119.1	125.4	116.8	124.1	116.7

Table 8-6. Producer Price Indexes for Goods and Services: Intermediate Demand

(November 2009 = 100, except as noted.)

Year and month	Intermediate demand by commodity type						Intermediate demand by production flow (Prices of inputs to indicated stage of production)			
	Processed goods (1982=100)	Unprocessed goods (1982=100)	Services	Construction	Processed materials less foods and energy (1982=100)	Unprocessed nonfood materials less energy (1982 =100)	Stage 4 intermediate demand	Stage 3 intermediate demand	Stage 2 intermediate demand	Stage 1 intermediate demand
NOT SEASONALLY ADJUSTED										
2010	183.4	212.2	101.1	101.1	180.8	329.1	102.0	104.2	103.5	106.2
2011	199.9	249.4	103.2	102.3	192.0	390.4	106.9	111.7	112.9	115.7
2012	200.7	241.4	105.3	103.8	192.6	369.6	108.6	112.7	110.7	114.7
2013	200.8	246.7	107.2	105.7	193.8	351.2	109.9	114.1	112.8	114.8
2014	201.9	249.3	108.9	108.0	195.2	345.7	111.2	116.9	112.2	116.1
2015	188.0	189.1	110.2	110.1	189.4	296.0	109.9	110.5	98.9	106.8
2016	182.2	173.4	112.1	111.7	186.9	288.0	109.8	107.3	98.3	104.2
2017	190.7	190.8	115.0	114.1	193.3	324.1	113.0	111.9	103.4	110.4
2018	200.9	200.1	118.6	116.2	201.8	340.7	117.0	116.9	108.5	116.7
2015										
January	189.7	199.7	109.8	109.2	191.6	325.5	110.0	111.1	100.1	108.8
February	189.2	193.4	109.6	109.2	191.1	309.5	109.8	110.1	99.8	107.7
March	189.3	193.9	110.0	109.4	190.6	306.0	110.0	111.1	99.6	107.8
April	188.3	195.5	110.6	109.5	190.4	306.3	110.1	111.0	99.8	107.6
May	190.2	202.5	110.4	110.0	190.5	305.6	110.5	112.8	100.7	108.3
June	191.6	203.3	110.5	110.3	190.5	306.3	110.8	113.1	101.0	109.1
July	191.2	196.1	110.7	110.4	190.1	303.0	110.8	112.5	100.3	109.0
August	189.6	187.6	110.9	110.4	189.2	288.3	110.6	111.8	99.0	107.7
September	186.8	182.4	110.2	110.5	188.0	286.8	109.9	109.6	97.9	105.8
October	185.1	179.5	109.7	110.6	187.4	278.8	109.3	108.5	97.5	104.6
November	183.4	171.3	109.8	110.7	186.8	268.6	109.0	107.6	96.1	103.4
December	181.3	164.1	109.9	110.7	186.1	267.2	108.6	106.2	95.6	102.0
2016										
January	179.6	164.7	111.5	110.8	185.6	266.6	109.0	106.7	95.6	102.0
February	178.3	161.7	111.7	110.9	185.1	269.2	108.9	106.0	95.1	101.6
March	178.8	166.0	111.8	110.9	185.3	276.9	109.0	106.5	95.6	102.2
April	179.7	170.1	111.8	111.0	185.9	287.1	109.2	106.7	96.5	103.2
May	181.6	176.6	111.6	111.3	186.5	297.0	109.5	107.4	97.8	104.3
June	183.8	181.3	112.1	111.3	186.9	296.8	110.2	108.5	99.1	105.6
July	184.2	181.2	112.6	111.4	187.0	295.8	110.4	108.6	99.9	105.6
August	183.9	175.4	112.3	112.1	187.5	295.4	110.2	107.7	99.2	105.0
September	184.4	173.5	112.2	112.4	187.5	289.1	110.2	107.4	99.3	105.0
October	184.0	171.6	112.3	112.5	187.8	285.2	110.0	106.6	100.2	104.7
November	183.6	173.0	112.7	112.8	188.4	292.6	110.1	107.4	99.7	105.2
December	184.6	185.4	112.8	113.2	189.0	303.9	110.4	108.3	101.7	106.5
2017										
January	186.7	193.4	113.4	113.5	189.8	314.4	111.2	109.7	103.2	107.8
February	188.0	192.8	114.0	113.5	191.3	318.4	111.8	110.6	103.2	108.7
March	188.3	188.3	114.0	113.7	192.3	325.3	111.8	111.1	101.8	109.1
April	189.4	191.9	114.6	113.8	193.2	323.5	112.5	111.4	103.2	109.6
May	189.9	191.6	114.6	114.0	193.2	321.6	112.7	111.8	102.7	109.5
June	190.8	191.8	115.2	114.2	193.3	323.4	113.2	112.4	102.8	110.1
July	190.8	192.0	115.1	114.3	193.0	325.5	113.2	112.2	102.9	110.3
August	191.5	187.6	115.5	114.3	193.5	328.2	113.5	112.1	102.9	111.1
September	192.7	186.8	115.3	114.4	194.0	332.5	113.6	112.3	103.3	111.4
October	192.9	186.2	116.0	114.6	194.8	323.7	113.8	112.5	104.0	111.6
November	193.5	192.0	116.3	114.6	195.7	323.1	114.1	113.5	104.9	112.3
December	193.9	194.7	116.1	114.7	195.8	329.8	114.1	113.5	105.5	112.6
2018										
January	195.4	199.7	117.0	115.3	197.0	341.9	114.9	114.4	106.7	114.0
February	197.0	203.0	117.5	115.3	198.2	342.8	115.4	115.4	107.7	114.7
March	197.3	198.2	117.7	115.5	199.5	347.2	115.8	115.7	106.5	115.0
April	198.5	199.9	118.0	115.6	200.1	349.6	116.2	116.1	107.4	115.7
May	201.7	205.7	118.2	115.9	201.6	349.3	116.9	117.7	108.9	117.0
June	203.4	204.2	118.6	116.0	202.5	350.8	117.5	118.5	108.8	118.0
July	203.6	204.8	119.0	116.1	203.1	342.4	117.7	118.4	109.7	118.2
August	203.8	194.6	119.0	116.5	203.7	332.3	117.6	117.3	109.0	117.8
September	204.3	194.8	119.1	116.5	204.4	328.2	117.7	117.5	109.4	117.8
October	204.6	200.3	119.8	117.0	204.7	330.6	118.1	118.7	110.2	118.6
November	202.2	194.4	119.9	117.2	204.0	333.5	117.8	117.6	108.4	117.7
December	199.4	202.0	119.7	117.4	202.7	339.2	117.6	115.8	109.3	116.1

Table 8-6. Producer Price Indexes for Goods and Services: Intermediate Demand—*Continued*

(November 2009 = 100, except as noted.)

Year and month	Intermediate demand by commodity type						Intermediate demand by production flow (Prices of inputs to indicated stage of production)			
	Processed goods (1982=100)	Unprocessed goods (1982=100)	Services	Construction	Processed materials less foods and energy (1982=100)	Unprocessed nonfood materials less energy (1982=100)	Stage 4 intermediate demand	Stage 3 intermediate demand	Stage 2 intermediate demand	Stage 1 intermediate demand
SEASONALLY ADJUSTED										
2015										
January	190.8	200.7	109.8	109.2	191.8	326.1	110.2	111.7	100.2	109.3
February	190.0	194.3	109.5	109.2	191.1	309.3	109.9	110.6	99.8	108.0
March	189.8	192.2	110.0	109.4	190.5	304.8	110.1	111.0	99.4	107.9
April	188.5	193.5	110.4	109.5	190.1	304.3	110.1	110.7	99.6	107.5
May	189.7	199.7	110.3	110.0	190.1	303.5	110.4	112.1	100.5	108.1
June	190.3	200.8	110.3	110.3	190.2	304.5	110.4	112.3	100.7	108.4
July	189.8	194.4	110.7	110.4	190.0	302.4	110.5	111.8	100.0	108.3
August	188.5	187.9	110.8	110.4	189.1	289.3	110.3	111.5	98.9	107.2
September	185.9	184.3	110.3	110.5	188.1	288.2	109.7	109.6	98.1	105.6
October	185.2	182.0	109.9	110.6	187.7	280.8	109.5	108.9	97.7	104.8
November	184.5	173.1	109.9	110.7	187.1	270.5	109.3	108.2	96.5	103.9
December	182.4	165.9	110.3	110.7	186.6	268.2	109.1	107.1	95.9	102.7
2016										
January	180.5	165.7	111.4	110.8	185.8	267.1	109.2	107.1	95.8	102.4
February	178.9	162.5	111.6	110.9	185.2	268.7	109.0	106.3	95.1	101.7
March	179.2	164.5	111.7	110.9	185.1	275.8	109.1	106.4	95.4	102.3
April	180.0	168.6	111.6	111.0	185.7	285.3	109.2	106.5	96.3	103.1
May	181.3	174.0	111.6	111.3	186.2	294.8	109.4	106.9	97.6	104.1
June	182.6	178.3	111.9	111.3	186.6	294.9	109.8	107.7	98.8	104.9
July	183.0	179.1	112.5	111.4	186.9	295.3	110.1	107.9	99.6	105.1
August	182.9	175.8	112.2	112.1	187.4	296.4	109.9	107.3	99.2	104.5
September	183.6	175.4	112.3	112.4	187.7	290.6	110.0	107.4	99.5	104.8
October	184.2	174.3	112.4	112.5	188.0	287.1	110.2	107.0	100.4	104.9
November	184.6	174.9	112.8	112.8	188.7	294.5	110.4	107.9	100.0	105.7
December	185.8	187.3	113.2	113.2	189.4	305.0	110.9	109.2	102.0	107.2
2017										
January	187.7	194.4	113.4	113.5	190.0	314.9	111.4	110.2	103.3	108.3
February	188.7	193.5	113.8	113.5	191.3	317.8	111.8	110.9	103.2	108.9
March	188.8	186.7	113.9	113.7	192.1	324.1	112.0	111.2	101.6	109.2
April	189.7	190.3	114.4	113.8	192.9	321.7	112.5	111.1	103.0	109.6
May	189.5	188.7	114.6	114.0	192.8	319.3	112.5	111.2	102.5	109.3
June	189.6	188.6	115.0	114.2	193.1	321.5	112.8	111.5	102.6	109.5
July	189.5	189.6	115.1	114.3	192.9	325.2	112.8	111.4	102.7	109.7
August	190.5	188.0	115.3	114.3	193.5	329.2	113.2	111.8	102.9	110.6
September	191.8	189.1	115.4	114.4	194.2	334.2	113.4	112.3	103.5	111.2
October	193.0	189.4	116.2	114.6	195.0	325.7	114.0	113.0	104.2	111.9
November	194.6	194.3	116.4	114.6	196.0	325.0	114.5	114.1	105.2	112.7
December	195.2	196.7	116.6	114.7	196.3	330.9	114.6	114.6	105.9	113.4
2018										
January	196.7	200.8	117.0	115.3	197.2	342.6	115.2	115.1	106.8	114.6
February	197.9	203.4	117.3	115.3	198.3	342.1	115.5	115.8	107.7	115.0
March	198.0	196.4	117.6	115.5	199.3	345.8	115.9	115.7	106.3	115.1
April	198.9	198.3	117.9	115.6	199.8	347.7	116.2	115.8	107.3	115.7
May	201.1	202.9	118.2	115.9	201.3	346.6	116.7	117.0	108.7	116.7
June	202.0	200.9	118.4	116.0	202.3	348.8	117.1	117.4	108.6	117.2
July	202.2	202.7	118.9	116.1	203.0	342.2	117.4	117.5	109.5	117.5
August	202.6	195.1	118.9	116.5	203.7	333.4	117.4	117.0	109.0	117.2
September	203.1	196.9	119.2	116.5	204.5	330.0	117.6	117.4	109.6	117.4
October	204.6	203.4	119.9	117.0	204.8	332.8	118.3	119.1	110.3	118.7
November	203.3	196.7	120.0	117.2	204.2	335.5	118.1	118.2	108.7	118.2
December	200.8	203.9	120.1	117.4	203.1	340.3	118.1	116.8	109.6	117.0

Table 8-7. Producer Price Indexes by Major Commodity Groups

(1982 = 100.)

Year and month	All com-modities	Farm products	Pro-cessed foods and feeds	Industrial commodities													
				Total	Textile products and apparel	Hides, leather, and related products	Fuels and related products and power	Chemi-cals and allied products	Rubber and plastics products	Lumber and wood products	Pulp, paper, and allied products	Metals and metal products	Machin-ery and equip-ment	Fur-niture and house-hold durables	Non-metallic mineral products	Trans-portation equip-ment	Miscel-laneous products
1950	27.3	44.0	33.2	25.0	50.2	32.9	12.6	30.4	35.6	31.4	25.7	22.0	22.6	40.9	23.5	. . .	28.6
1951	30.4	51.2	36.9	27.6	56.0	37.7	13.0	34.8	43.7	34.1	30.5	24.5	25.3	44.4	25.0	. . .	30.3
1952	29.6	48.4	36.4	26.9	50.5	30.5	13.0	33.0	39.6	33.2	29.7	24.5	25.3	43.5	25.0	. . .	30.2
1953	29.2	43.8	34.8	27.2	49.3	31.0	13.4	33.4	36.9	33.1	29.6	25.3	25.9	44.0	26.0	. . .	31.0
1954	29.3	43.2	35.4	27.2	48.2	29.5	13.2	33.8	37.5	32.5	29.6	25.5	26.3	44.9	26.6	. . .	31.3
1955	29.3	40.5	33.8	27.8	48.2	29.4	13.2	33.7	42.4	34.1	30.4	27.2	27.2	45.1	27.3	. . .	31.3
1956	30.3	40.0	33.8	29.1	48.2	31.2	13.6	33.9	43.0	34.6	32.4	29.6	29.3	46.3	28.5	. . .	31.7
1957	31.2	41.1	34.8	29.9	48.3	31.2	14.3	34.6	42.8	32.8	33.0	30.2	31.4	47.5	29.6	. . .	32.6
1958	31.6	42.9	36.5	30.0	47.4	31.6	13.7	34.9	42.8	32.5	33.4	30.0	32.1	47.9	29.9	. . .	33.3
1959	31.7	40.2	35.6	30.5	48.1	35.9	13.7	34.8	42.6	34.7	33.7	30.6	32.8	48.0	30.3	. . .	33.4
1960	31.7	40.1	35.6	30.5	48.6	34.6	13.9	34.8	42.7	33.5	34.0	30.6	33.0	47.8	30.4	. . .	33.6
1961	31.6	39.7	36.2	30.4	47.8	34.9	14.0	34.5	41.1	32.0	33.0	30.5	33.0	47.5	30.5	. . .	33.7
1962	31.7	40.4	36.5	30.4	48.2	35.3	14.0	33.9	39.9	32.2	33.4	30.2	33.0	47.2	30.5	. . .	33.9
1963	31.6	39.6	36.8	30.3	48.2	34.3	13.9	33.5	40.1	32.8	33.1	30.3	33.1	46.9	30.3	. . .	34.2
1964	31.6	39.0	36.7	30.5	48.5	34.4	13.5	33.6	39.6	33.5	33.0	31.1	33.3	47.1	30.4	. . .	34.4
1965	32.3	40.7	38.0	30.9	48.8	35.9	13.8	33.9	39.7	33.7	33.3	32.0	33.7	46.8	30.4	. . .	34.7
1966	33.3	43.7	40.2	31.5	48.9	39.4	14.1	34.0	40.5	35.2	34.2	32.8	34.7	47.4	30.7	. . .	35.3
1967	33.4	41.3	39.8	32.0	48.9	38.1	14.4	34.2	41.4	35.1	34.6	33.2	35.9	48.3	31.2	. . .	36.2
1968	34.2	42.3	40.6	32.8	50.7	39.3	14.3	34.1	42.8	39.8	35.0	34.0	37.0	49.7	32.4	. . .	37.0
1969	35.6	45.0	42.7	33.9	51.8	41.5	14.6	34.2	43.6	44.0	36.0	36.0	38.2	50.7	33.6	40.4	38.1
1970	36.9	45.8	44.6	35.2	52.4	42.0	15.3	35.0	44.9	39.9	37.5	38.7	40.0	51.9	35.3	41.9	39.8
1971	38.1	46.6	45.5	36.5	53.3	43.4	16.6	35.6	45.2	44.7	38.1	39.4	41.4	53.1	38.2	44.2	40.8
1972	39.8	51.6	48.0	37.8	55.5	50.0	17.1	35.6	45.3	50.7	39.3	40.9	42.3	53.8	39.4	45.5	41.5
1973	45.0	72.7	58.9	40.3	60.5	54.5	19.4	37.6	46.6	62.2	42.3	44.0	43.7	55.7	40.7	46.1	43.3
1974	53.5	77.4	68.0	49.2	68.0	55.2	30.1	50.2	56.4	64.5	52.5	57.0	50.0	61.8	47.8	50.3	48.1
1975	58.4	77.0	72.6	54.9	67.4	56.5	35.4	62.0	62.2	62.1	59.0	61.5	57.9	67.5	54.4	56.7	53.4
1976	61.1	78.8	70.8	58.4	72.4	63.9	38.3	64.0	66.0	72.2	62.1	65.0	61.3	70.3	58.2	60.5	55.6
1977	64.9	79.4	74.0	62.5	75.3	68.3	43.6	65.9	69.4	83.0	64.6	69.3	65.2	73.2	62.6	64.6	59.4
1978	69.9	87.7	80.6	67.0	78.1	76.1	46.5	68.0	72.4	96.9	67.7	75.3	70.3	77.5	69.6	69.5	66.7
1979	78.7	99.6	88.5	75.7	82.5	96.1	58.9	76.0	80.5	105.5	75.9	86.0	76.7	82.8	77.6	75.3	75.5
1980	89.8	102.9	95.9	88.0	89.7	94.7	82.8	89.0	90.1	101.5	86.3	95.0	86.0	90.7	88.4	82.9	93.6
1981	98.0	105.2	98.9	97.4	97.6	99.3	100.2	98.4	96.4	102.8	94.8	99.6	94.4	95.9	96.7	94.3	96.1
1982	100.0	100.0	100.0	100.0	100.0	100.0	100.0	100.0	100.0	100.0	100.0	100.0	100.0	100.0	100.0	100.0	100.0
1983	101.3	102.4	101.8	101.1	100.3	103.2	95.9	100.3	100.8	107.9	103.3	101.8	102.7	103.4	101.6	102.8	104.8
1984	103.7	105.5	105.4	103.3	102.7	109.0	94.8	102.9	102.3	108.0	110.3	104.8	105.1	105.7	105.4	105.2	107.0
1985	103.2	95.1	103.5	103.7	102.9	108.9	91.4	103.7	101.9	106.6	113.3	104.4	107.2	107.1	108.6	107.9	109.4
1986	100.2	92.9	105.4	100.0	103.2	113.0	69.8	102.6	101.9	107.2	116.1	103.2	108.8	108.2	110.0	110.5	111.6
1987	102.8	95.5	107.9	102.6	105.1	120.4	70.2	106.4	103.0	112.8	121.8	107.1	110.4	109.9	110.0	112.5	114.9
1988	106.9	104.9	112.7	106.3	109.2	131.4	66.7	116.3	109.3	118.9	130.4	118.7	113.2	113.1	111.2	114.3	120.2
1989	112.2	110.9	117.8	111.6	112.3	136.3	72.9	123.0	112.6	126.7	137.8	124.1	117.4	116.9	112.6	117.7	126.5
1990	116.3	112.2	121.9	115.8	115.0	141.7	82.3	123.6	113.6	129.7	141.2	122.9	120.7	119.2	114.7	121.5	134.2
1991	116.5	105.7	121.9	116.5	116.3	138.9	81.2	125.6	115.1	132.1	142.9	120.2	123.0	121.2	117.2	126.4	140.8
1992	117.2	103.6	122.1	117.4	117.8	140.4	80.4	125.9	115.1	146.6	145.2	119.2	123.4	122.2	117.3	130.4	145.3
1993	118.9	107.1	124.0	119.0	118.0	143.7	80.0	128.2	116.0	174.0	147.3	119.2	124.0	123.7	120.0	133.7	145.4
1994	120.4	106.3	125.5	120.7	118.3	148.5	77.8	132.1	117.6	180.0	152.5	124.8	125.1	126.1	124.2	137.2	141.9
1995	124.7	107.4	127.0	125.5	120.8	153.7	78.0	142.5	124.3	178.1	172.2	134.5	126.6	128.2	129.0	139.7	145.4
1996	127.7	122.4	133.3	127.3	122.4	150.5	85.8	142.1	123.8	176.1	168.7	131.0	126.5	130.4	131.0	141.7	147.7
1997	127.6	112.9	134.0	127.7	122.6	154.2	86.1	143.6	123.2	183.8	167.9	131.8	125.9	130.8	133.2	141.6	150.9
1998	124.4	104.6	131.6	124.8	122.9	148.0	75.3	143.9	122.6	179.1	171.7	127.8	124.9	131.3	135.4	141.2	156.0
1999	125.5	98.4	131.1	126.5	121.1	146.0	80.5	144.2	122.5	183.6	174.1	124.6	124.3	131.7	138.9	141.8	166.6
2000	132.7	99.5	133.1	134.8	121.4	151.5	103.5	151.0	125.5	178.2	183.7	128.1	124.0	132.6	142.5	143.8	170.8
2001	134.2	103.8	137.3	135.7	121.3	158.4	105.3	151.8	127.2	174.4	184.8	125.4	123.7	133.2	144.3	145.2	181.3
2002	131.1	99.0	136.2	132.4	119.9	157.6	93.2	151.9	126.8	173.3	185.9	125.9	122.9	133.5	146.2	144.6	182.4
2003	138.1	111.5	143.4	139.1	119.8	162.3	112.9	161.8	130.1	177.4	190.0	129.2	121.9	133.9	148.2	145.7	179.6
2004	146.7	123.3	151.2	147.6	121.0	164.5	126.9	174.4	133.8	195.6	195.7	149.6	122.1	135.1	153.2	148.6	183.2
2005	157.4	118.5	153.1	160.2	122.8	165.4	156.4	192.0	136.3	196.5	206.0	160.8	123.7	139.4	164.2	151.0	195.1
2006	164.7	117.0	153.8	168.8	124.5	168.4	166.7	205.8	153.8	194.4	209.8	181.6	126.2	142.6	179.9	152.6	205.6
2007	172.6	143.4	165.1	175.1	125.8	173.6	177.6	214.8	155.0	192.4	216.9	193.5	127.3	144.7	186.2	155.0	210.3
2008	189.6	161.3	180.5	192.3	128.9	173.1	214.6	245.5	165.9	191.3	226.8	213.0	129.7	148.9	197.1	158.6	216.6
2009	172.9	134.6	176.2	174.8	129.5	157.0	158.7	229.4	165.2	182.8	225.6	186.8	131.3	153.1	202.4	162.2	217.5
2010	184.7	151.0	182.3	187.0	131.7	181.4	185.8	246.6	170.7	192.7	236.9	207.6	131.1	153.2	201.8	163.4	221.5
2011	201.0	186.7	197.5	202.0	141.7	199.9	215.9	275.1	182.7	194.7	245.1	225.9	132.7	156.4	205.0	166.1	229.2
2012	202.2	192.5	205.2	202.1	142.2	202.3	212.1	276.6	186.9	201.6	244.2	219.9	134.2	160.6	211.0	169.8	235.6
2013	203.4	195.3	208.3	203.0	143.4	217.9	211.8	279.2	189.0	214.9	248.8	213.5	135.2	161.1	216.9	171.8	239.5
2014	205.3	197.4	216.5	204.1	145.5	229.1	209.8	280.9	190.2	224.2	250.5	215.0	136.2	163.2	223.7	174.1	243.0
2015	190.4	173.8	209.1	188.8	144.1	210.4	160.5	266.0	187.0	221.9	248.8	200.3	136.9	164.7	228.9	176.5	247.4
2016	185.4	157.0	203.5	184.6	143.3	195.1	145.9	265.1	184.6	222.7	247.7	194.3	136.9	165.1	233.3	177.4	252.1
2017	193.5	161.8	205.4	193.7	145.1	192.1	163.7	280.8	189.0	230.4	254.6	207.8	137.9	166.9	238.6	179.0	257.6
2018	202.0	160.9	206.1	203.7	149.3	180.4	181.6	295.1	195.1	243.9	260.0	223.6	140.3	171.6	247.3	181.2	264.0

. . . = Not available.

NOTES AND DEFINITIONS, CHAPTER 8

This chapter presents price indexes (Consumer and Producer) from two major price collection systems, both conducted by the U.S. Bureau of Labor Statistics. Effective on February 26, 2015, the Bureau of Labor Statistics began utilizing a new estimation system for the Consumer Price Index. The new estimation system, the first major improvement to the existing system in over 25 years. This system is redesigned, state-of-the-art with improved flexibility and review capabilities.

In 2014 the Producer Price Index system was expanded to include services and construction as well as goods. A new index for "Final Demand" for goods, services and construction, with more than double the scope, is now presented alongside the familiar Finished Goods index. The expanded system is only available beginning in November 2009, and the historical Producer Price Indexes for goods—with data going back to 1913—are also maintained and carried forward. See the Notes and Definitions for Tables 8-5 through 8-7, below, for further information.

TABLES 8-1 THROUGH 8-4

Consumer Price Indexes

SOURCES: U.S. DEPARTMENT OF LABOR, BUREAU OF LABOR STATISTICS (BLS) AND U.S. DEPARTMENT OF COMMERCE, BUREAU OF ECONOMIC ANALYSIS (BEA)

The Consumer Price Index (CPI), which is compiled by the Bureau of Labor Statistics (BLS), was originally conceived as a statistical measure of the average change in the cost to consumers of a market basket of goods and services purchased by urban wage earners and clerical workers. In 1978, its scope was broadened to also provide a measure of the change in the cost of the average market basket for all urban consumers. There was still a demand for a wage-earner index, so both versions have been calculated and published since then. The most commonly cited measure in this system is the Consumer Price Index for All Urban Consumers (CPI-U). The wage-earner alternative, used according to law for calculating cost of living adjustments in many government programs, including Social Security, and also used by choice of the contracting parties in wage agreements, is called the Consumer Price Index for Urban Wage Earners and Clerical Workers (CPI-W). Both are presented by the BLS back to 1913, and reproduced in *Business Statistics*; however, the movements (percent changes) in the two indexes before 1978 are identical and are based on the wage-earner market basket. (The pre-1978 levels of the two indexes differ because of differences between 1978 and the reference base period.)

These CPIs have typically been called "cost-of-living" indexes, even though the original fixed market basket concept does not correspond to economists' definition of a cost-of-living index, which is the cost of maintaining a constant standard of living or level of satisfaction. In recent years, the concept measured in practice in the CPI has developed into something intended to be closer to the theoretical definition of a cost-of-living index. In addition, a new variation of the CPI—the Chained Consumer Price Index for All Urban Consumers (C-CPI-U)—is intended to provide an even closer approximation.

The reference base for the total BLS Consumer Price Index and most of its components is currently 1982–1984 = 100. However, new products that have been introduced into the index since January 1982 are shown on later reference bases, as is the entire C-CPI-U.

Price indexes for personal consumption expenditures (PCE) are calculated and published by the Bureau of Economic Analysis (BEA) as a part of the national income and product accounts (NIPAs). (See Chapters 1 and 4 and their notes and definitions.) Monthly, quarterly, and annual values are all available. The reference base for these indexes is the average in the NIPA base year, 2009. These indexes differ in a number of other respects from the CPIs, and are often emphasized by the Federal Reserve in its analyses of the nation's economy. The Federal Reserve is mandated to maximize employment (see Chapter 10) and maintain price stability as measured by the Consumer Price Index (CPI). The target inflation rate is 2 percent. Four important NIPA aggregate price indexes are shown in Tables 8-3 and 8-4 for convenient comparison with the CPIs. See the definitions for those tables for more information.

The CPI-U and the CPI-W

The *CPI-U*, which is displayed in Tables 8-1 through 8-4 and also provides all of the component category sub-indexes shown in Table 8-2, uses the consumption patterns for all urban consumers, who currently comprise about 89 percent of the population.

A slightly different index that is widely used for adjusting wages and government benefits is the *CPI-W*, of which the all-items total is shown in each of Tables 8-1 through 8-4. It represents the buying habits of only urban wage earners and clerical workers, who currently comprise about 29 percent of the population. The weights are derived from the same Consumer Expenditure Surveys (CES) used for the CPI-U weights, and are changed on the same schedule. However, they include only consumers from the specified categories instead of all urban consumers.

Beginning with February 2016, the weights in both indexes are based on consumer expenditures in the 2014–2015 period. In January 2012, the weights in both indexes are based on consumer expenditures in the 2009–2010 period. From January 2010 to December 2011, 2007–2008 weights were used; from January 2008 to December 2009, 2005–2006 weights; from January 2006 to December 2007, 2003–2004 weights; from January 2004 to December 2005, 2001–2002 weights; from January 2002 to

December 2003, 1999–2000 weights; and from January 1998 to December 2001, 1993–1995 weights. The weights will continue to be updated at two-year intervals, with new weights introduced in the January indexes of each even-numbered year. Previously, new weights were introduced only at the time of a major revision, which translated into a lag of a decade or more.

Specifically, the CPI weights for 1964 through 1977 were derived from reported expenditures of a sample of wageearner and clericalworker families and individuals in 1960–1961 and adjusted for price changes between the survey dates and 1963. Weights for the 1978–1986 period were derived from a survey undertaken during the 1972–1974 period and adjusted for price change between the survey dates and December 1977. For 1987 through 1997, the spending patterns reflected in the CPI were derived from a survey undertaken during the 1982–1984 period. The reported expenditures were adjusted for price change between the survey dates and December 1986.

The CPI was overhauled and updated in the latest major revision, which took effect in January 1998. In addition, new products and improved methods are regularly introduced into the index, usually in January.

The latest change in methods was the introduction of a geometric mean formula for calculating many of the basic components of the index. Beginning with the index for January 1999, this formula is used for categories comprising approximately 61 percent of total consumer spending. The new formula allows for the possibility that some consumers may react to changing relative prices within a category by substituting items whose relative prices have declined for products whose relative prices have risen, while maintaining their overall level of satisfaction. The geometric mean formula is not used for a few categories in which consumer substitution in the short term is not feasible, currently housing rent and utilities.

The CPI-U was introduced in 1978. Before that time, only CPI-W data were available. The movements of the CPI-U before 1978 are therefore based on the changes in the CPI-W. However, the index levels are different because the two indexes differed in the 1982–1984 base period.

Because the official CPI-U and CPI-W are so widely used in "escalation"—the calculation of cost-of-living adjustments for wages and other private contracts, and for government payments and tax parameters—these price indexes are not retrospectively revised to incorporate new information and methods. (An exception is occasionally made for outright error, which happened in September 2000 and affected the data for January through August of that year.) Instead, the new information and methods of calculation are introduced in the current index and affect future index changes only. In Tables 8-3 and 8-4, PCE indexes, which are subject to routine revision, and special CPI indexes that have been retrospectively revised are presented. These indexes can be used by researchers to provide more consistent historical information.

Notes on the CPI data

The CPI is based on prices of food, clothing, shelter, fuel, utilities, transportation, medical care, and other goods and services that people buy for daytoday living. The quantity and quality of these priced items are kept essentially constant between revisions to ensure that only price changes will be measured. All taxes directly associated with the purchase and use of these items, such as sales and property taxes, are included in the index; the effects of income and payroll tax changes are not included.

Data are collected on about 83,400 individually defined goods and services from about 22,000 retail and service establishments and about 6,000 housing units in 75 urban areas across the country. These data are used to develop the U.S. city average.

Periodic major revisions of the indexes update the content and weights of the market basket of goods and services; update the statistical sample of urban areas, outlets, and unique items used in calculating the CPI; and improve the statistical methods used. In addition, retail outlets and items are resampled on a rotating 5-year basis. Adjustments for changing quality are made at times of major product changes, such as the annual auto model changeover. Other methodological changes are introduced from time to time.

The Consumer Expenditure Survey (CES) provides the weights—that is, the relative importance—used to combine the individual price changes into subtotals and totals. This survey is composed of two separate surveys: an interview survey and a diary survey, both of which are conducted by the Census Bureau for BLS. Each expenditure reported in the two surveys is classified into a series of detailed categories, which are then combined into expenditure classes and ultimately into major expenditure groups. CPI data as of 1998 are grouped into eight such categories: (1) food and beverages, (2) housing, (3) apparel, (4) transportation, (5) medical care, (6) entertainment (7) education and communication, and (8) other goods and services.

Seasonally adjusted national CPI indexes are published for selected series for which there is a significant seasonal pattern of price change. The factors currently in use were derived by the X13ARIMA-SEATS seasonal adjustment method. Some series with extreme or sharp movements are seasonally adjusted using Intervention Analysis Seasonal Adjustment. Seasonally adjusted indexes and seasonal factors for the preceding five years are updated annually based on data through the previous December. Due to these revisions, BLS advises against the use of seasonally adjusted data for escalation. Detailed descriptions of seasonal adjustment procedures are available upon request from BLS.

BLS estimates the "standard error"—the error due to collecting data from a sample instead of the universe—of the one-month percent change in the not-seasonally-adjusted U.S. all-items index in 2013 was 0.03 percentage point. This means that for the 2013 median 0.12 percent change in the All Items CPI-U, BLS is 95 percent confident that the actual percent change based on all retail prices would fall between 0.06 and 0.18 percent. Monthly percent changes in the seasonally adjusted indexes, in contrast, may be revised by 0.1 percentage point when reviewed at the end of the year, and occasionally by even more.

CPI Definitions

Definitions of the major CPI groupings were modified beginning with the data for January 1998. These modifications were carried back to 1993. The following definitions are the current definitions currently used for the CPI components.

The *food and beverage index* includes both food at home and food away from home (restaurant meals and other food bought and eaten away from home).

The *housing index* measures changes in rental costs and expenses connected with the acquisition and operation of a home. The CPIU, beginning with data for January 1983, and the CPIW, beginning with data for January 1985, reflect a change in the methodology used to compute the homeownership component. A rental equivalence measure replaced an assetprice approach, which included purchase prices and interest costs. The intent of the change was to separate shelter costs from the investment component of homeownership, so that the index would only reflect the cost of the shelter services provided by owner-occupied homes. In addition to measures of the cost of shelter, the housing category includes insurance, fuel, utilities, and household furnishings and operations.

The *apparel index* includes the purchase of apparel and footwear.

The *private transportation index* includes prices paid by urban consumers for such items as new and used automobiles and other vehicles, gasoline, motor oil, tires, repairs and maintenance, insurance, registration fees, driver's licenses, parking fees, and the like. Auto finance charges, like mortgage interest payments, are considered to be a cost of asset acquisition, not of current consumption. Therefore, they are no longer included in the CPI. City bus, streetcar, subway, taxicab, intercity bus, airplane, and railroad coach fares are some of the components of the *public transportation index*.

The *medical care index* includes prices for professional medical services, hospital and related services, prescription and nonprescription drugs, and other medical care commodities. The weight for the portion of health insurance premiums that is used to cover the costs of these medical goods and services is distributed among the items; the weight for the portion of health insurance costs attributable to administrative expenses and profits of insurance providers constitutes a separate health insurance item. Effective with the January 1997 data, the method of calculating the hospital cost component was changed from the pricing of individual commodities and services to a more comprehensive cost-of-treatment approach.

Recreation includes components formerly listed in housing, apparel, entertainment, and "other goods and services."

Education and communication is a new group including components formerly categorized in housing and "other goods and services," such as college tuition and books, telephone and internet services, and computers.

Other goods and services now includes tobacco, personal care, and miscellaneous.

TABLE 8-2

Purchasing Power of the Dollar

SOURCE: U.S. DEPARTMENT OF LABOR, BUREAU OF LABOR STATISTICS (BLS)

The purchasing power of the dollar measures changes in the quantity of goods and services a dollar will buy at a particular date compared with a selected base date. It must be defined in terms of the following: (1) the specific commodities and services that are to be purchased with the dollar; (2) the market level (producer, retail, etc.) at which they are purchased; and (3) the dates for which the comparison is to be made. Thus, the purchasing power of the dollar for a selected period, compared with another period, may be measured in terms of a single commodity or a large group of commodities such as all goods and services purchased by consumers at retail.

The purchasing power of the dollar is computed by dividing the price index number for the base period by the price index number for the comparison date and expressing the result in dollars and cents. The base period is the period in which the price index equals 100; the average purchasing power in that base period—1982–1984 in the case of the measure shown here—is therefore $1.00.

Purchasing power estimates in terms of both the CPI-U and the CPI-W are calculated by BLS, based on indexes not adjusted for seasonal variation, and published in the CPI press release. The CPI-U version is shown here.

Alternative price measures in Tables 8-3 and 8-4

Table 8-3 shows the all-items CPI-U and CPI-W, along with a number of other indexes that various analysts of price trends have preferred as measures of the price level. Table 8-4 shows the inflation rates (percent changes in price levels) implied by each of the indexes in Table 8-3.

As food and energy prices are volatile and frequently determined by forces separate from monetary aggregate demand pressures, many analysts prefer an index of prices excluding those components. Indexes *excluding food and energy* are known as *core* indexes, and inflation rates calculated from them are known as *core inflation rates*.

The *CPI-U-X1* is a special experimental version of the CPI that researchers have used to provide a more historically consistent series. As explained above, the official CPI-U treated homeownership on an asset-price basis until January 1983. It then changed to a rental equivalence method. The CPI-U-X1 incorporates a rental equivalence approach to homeowners' costs for the years 1967–1982 as well. It is rebased to the December 1982 value of the CPI-U (1982–1984 = 100); thus, it is identical to the CPI-U in December 1982 and all subsequent periods, as can be seen in Table 8-3. For this reason, it is not updated or published in the CPI news release or on the BLS Web site. We continue to present it here because it provides the only available data before 1978 on changes in an improved and more consistent CPI.

The *CPI-E* is an experimental re-weighting of components of the CPI-U to represent price change for the goods and services purchased by Americans age 62 years and over, who accounted for 16.5 percent of the total number of urban consumer units in the 2001–2002 CES.

BLS does not consider the CPI-E to be an ideal measure of price change for older Americans. Because the sample is small, the sampling error in the weights is greater than the error in the all-urban index. The products and outlets sampled are those characteristic of the general urban population rather than older residents. In addition, senior discounts are not included in the prices collected. Such discounts are included—appropriately— in the weights, which are based on the expenditures reported by the older consumers' households. Therefore, such discounts are only a problem if they do not move proportionately to general prices.

The *CPI-U-RS* is a "research series" CPI that retroactively incorporates estimates of the effects of most of the methodological changes implemented since 1978, including the rental equivalence method, new or improved quality adjustments, and improvement of formulas to eliminate bias and allow for some consumer substitution within categories. This index is calculated from 1977 onward. Its reference base is December 1977 = 100. Thus, although it generally shows less <u>increase</u> than the official index, its current <u>levels</u> are considerably higher because the earlier reference base period had lower prices. Unlike the official CPIs and the CPI-U-X1, its historical values will be revised each time a significant change is made in the calculation of the current index. This index is not seasonally adjusted and is not included in the CPI news release. It is available on the BLS Web site, along with an explanation and background material. The CPI-U-RS is used by BLS in the calculation of historical trends in real compensation per hour in its Productivity and Costs system; see Table 9-3 and its notes and definitions. It is also now used by the Census Bureau to convert household incomes into constant dollars, as seen in Chapter 3. And it is used by the editor in some analytical calculations in this volume.

The *C-CPI-U* (Chained Consumer Price Index for All Urban Consumers) is a new, supplemental index that has been published in the monthly CPI news release since August 2002. It is available only from December 1999 to date and is calculated with the base December 1999 = 100; it is not seasonally adjusted. It is designed to be a still-closer approximation to a true cost-of-living index than the CPI-U and the CPI-W, in that it assumes that consumers substitute between as well as within categories as relative prices change, in order to maintain a fixed basket of "consumer satisfaction."

The C-CPI-U is technically a "superlative" index, using a method known as the "Tornqvist formula" to incorporate the composition of consumer spending in the current period as well as in the earlier base period. (All of the other Consumer Price Indexes use a "Laspeyres" formula; see the General Notes at the beginning of this volume.) As it requires consumer expenditure data for the current as well as the earlier period, its final version can only be calculated after the expenditure data become available—about two years later—and is approximated in more recent periods by making more extensive use of the geometric mean formula (see above). With the release of January 2014 data, the indexes for 2012 were revised to their final form, and the initial indexes for 2013 were revised to "interim" levels, shown in Tables 8-3 and 8-4.

Personal consumption expenditure (PCE) chain-type price indexes are calculated by the Bureau of Economic Analysis (BEA) in the framework of the national income and product accounts (NIPAs). (See the notes and definitions for Chapters 1 and 4.) The scope of NIPA PCE is broader than the scope of the CPI. PCE includes the rural as well as the urban population and also covers the consumption spending of nonprofit entities. The CPI includes only consumer out-of-pocket cash spending, whereas PCE includes expenditures financed by government and private insurance, particularly in the medical care area. For this reason, there is a large difference between the relatively small weight of medical care spending in the CPI and the markedly greater percentage of PCE that is accounted for by total medical care spending. Housing, on the other hand, has a somewhat smaller weight in PCE while all non-housing components have a higher weight. The reason for this is that the CES—the survey on which the CPI weights are based—tends to report housing expenditures accurately and somewhat underestimate other spending. This suggests that the weight of housing relative to all other products may be overestimated in the CPI but measured more correctly in the PCE price index.

PCE chain-type indexes use the expenditure weights of both the earlier and the later period to determine the aggregate price change between the two periods. (See the notes and definitions for Chapter 1, as well as the General Notes on index number formulas.) Thus, they are subject to revision as improved data on the composition of consumption spending become available, and in this respect resemble the C-CPI-U.

For a large share of PCE, the price movements for basic individual spending categories are determined by CPI components.

Hence, the differences between the rates of change in the aggregate CPI and PCE indexes are largely the result of the different weights, but also reflect some alternative methodologies and the previously mentioned differences in scope.

Market-based PCE indexes are based on household expenditures for which there are observable price measures. They exclude most implicit prices (for example, the services furnished without payment by financial intermediaries) and they exclude items not deflated by a detailed component of either the Consumer Price Index (CPI) or the Producer Price Index (PPI). This means that the price observations that make up these new aggregate measures are all based on observed market transactions, making them "market-based price indexes." The imputed rent for owner-occupied housing is included in the market-based price index, since it is based on observed rentals of comparable homes. Household insurance premiums are also included in the market-based index, since they are deflated by the CPI for tenants' and household insurance. Excluded are services furnished without payment by financial intermediaries, most insurance purchases, expenses of NPISHs (nonprofit institutions serving households), legal gambling (illegal gambling is excluded from all measures), margins on used light motor vehicles, and expenditures by U.S. residents working and traveling abroad. Also excluded are medical, hospitalization, and income loss insurance; expense of handling life insurance; motor vehicle insurance; and workers' compensation.

The *inflation rates* shown in Table 8-4 are percent changes in the price indexes introduced in Table 8-3. For annual indexes, the rate is the percent change from the previous year. For monthly indexes, the rate is the percent change from the same month a year earlier. To give an indication of the longer-run implications of these different price indicators, comparisons of compound annual inflation rates, calculated by the editor, are also shown for the 1978–2013, 1983–2013, 2000–2013 and 2013–2015 periods, using the growth rate formula presented in the article at the beginning of this volume.

Data availability and references

The CPI-U, CPI-W, and C-CPI-U are initially issued in a press release two to three weeks after the end of the month for which the data were collected. This release and detailed and complete current and historical data on the CPI and its variants and components, along with extensive documentation, are available on the BLS Web site at <http://www.bls.gov/cpi>. Seasonal factors and seasonally adjusted indexes are revised once a year with the issuance of the January index.

Information available on the BLS Web site includes "Common Misconceptions about the Consumer Price Index: Questions and Answers;" another fact sheet on frequently asked questions; a fact sheet on seasonal adjustment; Chapter 17 of the *BLS Handbook of Methods*, entitled "The Consumer Price Index"; a section entitled "Note on Chained Consumer Price Index for All Urban Consumers"; and a number of explanatory CPI fact sheets on specific subjects.

As previously indicated, the CPI-U-X1 is not currently published because its recent values are identical to the CPI-U. The CPI-E is presented in articles in the CPI section of the BLS Web site, the most recent of which is "Experimental Consumer Price Index for Americans 62 Years of Age and Older, 1998-2005"; recent values are available by request from BLS. The CPI-U-RS is updated each month in a report entitled "CPI Research Series Using Current Methods" on the site. In both cases, the reports describe the indexes and provide references.

The monthly PCE indexes are included in the personal income report issued by BEA, which is published near the end of the following month. These indexes are revised month-by-month to reflect new information and annually to reflect the annual and quinquennial benchmarking of the NIPAs. The complete historical record can be found on the BEA Web site at <http://www.bea.gov> in the Personal Income and Outlays section of the NIPA tables, in tables entitled "Price Indexes for Personal Consumption Expenditures by Major Type of Product."

Two special editions of the *Monthly Labor Review* were devoted to CPI issues. The December 1996 issue describes the subsequently implemented 1997 and 1998 revisions in a series of articles, and the December 1993 issue, entitled *The Anatomy of Price Change*, includes the following articles: "The Consumer Price Index: Underlying Concepts and Caveats"; "Basic Components of the CPI: Estimation of Price Changes"; "The Commodity Substitution Effect in CPI Data, 1982–1991"; and "Quality Adjustment of Price Indexes."

The new formula for calculating basic components is described in "Incorporating a Geometric Mean Formula into the CPI," *Monthly Labor Review* (October 1998). For a detailed discussion of the treatment of homeownership, see "Changing the Homeownership Component of the Consumer Price Index to Rental Equivalence," *CPI Detailed Report* (January 1983).

For a comprehensive professional review of CPI concepts and methodology, see Charles Schultze and Christopher Mackie, ed., *At What Price? Conceptualizing and Measuring Cost-of-Living and Price Indexes* (Washington, DC: National Academy Press, 2001). Earlier references include: "Using Survey Data to Assess Bias in the Consumer Price Index," *Monthly Labor Review* (April 1998); Joel Popkin, "Improving the CPI: The Record and Suggested Next Steps," *Business Economics*, Vol. XXXII, No. 3 (July 1997), pages 42–47; *Measurement Issues in the Consumer Price Index* (Bureau of Labor Statistics, U.S. Department of Labor, June 1997); *Toward a More Accurate Measure of the Cost of Living* (Final Report to the Senate Finance Committee from the Advisory Commission to Study the Consumer Price Index, December 4, 1996)—also known as the "Boskin Commission" report; and *Government Price Statistics* (U.S. Congress Joint Economic Committee, 87th Congress, 1st Session, January 24, 1961)—also known as the "Stigler Committee" report.

For an explanation of the differences between the CPI-U and the PCE index, see Clinton P. McCully, Brian C. Moyer, and Kenneth J. Stewart, "Comparing the Consumer Price Index and the Personal Consumption Expenditures Price Index," *Survey of Current Business,* November 2007, pp. 26-33.

TABLES 8-1 AND 8-5 THROUGH 8-7

Producer Price Indexes

SOURCE: U.S. DEPARTMENT OF LABOR, BUREAU OF LABOR STATISTICS (BLS)

The Bureau of Labor Statistics first compiled and published a "Wholesale Price Index," measuring prices of various goods from the perspective of the seller, in 1902; it covered the years 1890-1901. Over the years this index was broadened and refined. In a major revamping in 1978, it was structured into an input-output format, differentiated between stages of processing, and renamed the Producer Price Index. The index for "Finished Goods," from the stage-of-processing system, became the featured "headline" number. Still, this index only covered goods.

New FD-ID system (Tables 8-5 and 8-6)

Over subsequent years the Bureau instituted collection of an increasing number of prices for services as well. All prices are collected within an industry input-output structure, with one of the most important purposes being to measure real output quantities by deflating industry outputs and inputs. In the new aggregation system introduced in February 2014, prices for goods, services, and construction have been integrated into a set of Final Demand-Intermediate Demand (FD-ID) indexes, organized by class of buyer, degree of fabrication, and stage of production. Its principal components are calculated beginning in November 2009 with that month as the reference base (i.e. November 2009 = 100). Components of the previous goods PPI system continue to be calculated on the base 1982 = 100 as well.

In the new FD-ID system, all products, including services and construction, are referred to as "commodities." In the old PPI system and in the CPI system, the term "commodities" refers to goods only; note, for example, how the CPI components are aggregated into summary measures of "commodities" [goods] and services, displaying different trends and behaviors. Unless and until these terminologies are reconciled, users need to be mindful of which system they are operating in.

More than 100,000 producer price quotations from over 25,000 establishment are collected each month; more than 10,000 indexes for individual products and groups of products are released each month. The prices are organized into three sets of PPIs: the FD-ID indexes; commodity indexes; and indexes for the net output of industries and their products.

Final demand. The final demand portion of the FD-ID structure measures price change for commodities sold for personal consumption, capital investment, government, and export. *Final demand trade services* measures changes in margins received by wholesalers and retailers. The other main components are *final demand goods, final demand transportation and warehousing services, other final demand services,* and *final demand construction.* The total index for final demand is now featured by BLS in its press release as its "headline number," replacing the previously featured Finished Goods index.

Intermediate demand. This portion of the system tracks price changes for goods, services, and construction products sold to businesses as inputs to production and not as capital investment. There are two parallel treatments.

The first treatment organizes intermediate demand commodities by type, including the distinction between *processed goods* and *unprocessed goods,* which is carried over from the older PPI goods system.

The second system organizes intermediate demand commodities into production stages, with the explicit goal of developing a forward-flow model of production and price change. For each production stage, the intermediate demand index measures the prices of inputs to the included industries. Many commodities, for example energy products, are sold to both final demand and various intermediate stages of production; their weight is allocated proportionately to all the consuming stages of production and their price movements are proportionately represented in each stage.

Industries assigned to *Stage 4* primarily produce final demand commodities. The intermediate demand index for Stage 4 measures inputs to those industries, such as motor vehicle parts, beef and veal, and long distance motor carrying.

Industries assigned to *Stage 3* produce the output sold to Stage 4, and therefore include motor vehicle parts and slaughtering. Examples of inputs to Stage 3 included in their intermediate demand index include slaughter steers and heifers, industrial electric power, steel plates, and temporary help services.

Industries assigned to *Stage 2* include petroleum refining, electricity, and insurance agencies. The goods and services that they purchase and that are included in their intermediate demand index include crude oil and business loans.

Industries assigned to *Stage 1* include oil and gas extraction, paper mills, and advertising services. Inputs to those industries include electric power and solid waste collection. According to BLS, "It should be noted that all inputs purchased by stage 1 industries are by definition produced either within stage 1 or by later stages of processing, leaving stage 1 less useful for price transmission analysis."

Commodity indexes

Aggregation by *commodity* organizes the price measures by similarity of product or end use, disregarding industry of origin.

For the original Wholesale Price Index (goods only), this was the principal means of aggregation; in Table 8-1C, producer price indexes are shown for *farm products, industrial commodities*, and *"all commodities"* back to the year 1913. Table 8-7 shows 16 goods commodity groups from 1950 to date. Whatever their imperfections, these measures provide the longest historical perspective for producer prices.

The major goods commodity groups—particularly the totals for all commodities and industrial commodities—came under criticism in the energy price crisis of the early 1970s, for aggregating successive stages of processing of the same product (e.g. crude oil) and thus exaggerating price trends. As a result, the stage-of-processing grouping was introduced in 1978, and the finished goods components, in total and minus food and energy, became the headline Producer Price Indexes.

Stage of processing indexes

In the PPI for goods as reconstructed in 1978 and shown in Table 8-1B, the three major indexes are: (1) *finished goods*, or goods that will not undergo further processing and are ready for sale to the ultimate user (such as automobiles, meats, apparel, and machine tools, and also unprocessed foods such as eggs and fresh vegetables, that are ready for the consumer); (2) *processed goods for intermediate demand* (formerly *intermediate materials, supplies, and components*), or goods that have been processed but require further processing before they become finished goods (such as steel mill products, cotton yarns, lumber, and flour), as well as physically complete goods that are purchased by business firms as inputs for their operations (such as diesel fuel and paper boxes); and (3) *unprocessed goods for intermediate demand* (formerly *crude materials for further processing*), or goods entering the market for the first time that have not been manufactured or fabricated and are not sold directly to consumers (such as ores, scrap metals, crude petroleum, raw cotton, and livestock).

Notes on the data

The probability sample used for calculating the PPI provides more than 100,000 price quotations per month. To the greatest extent possible, prices used in calculating the PPI represent prices received by domestic producers in the first important commercial transaction for each commodity. These indexes attempt to measure only price changes (changes in receipts per unit of measurement not influenced by changes in quality, quantity sold, terms of sale, or level of distribution). Transaction prices are sought instead of list or book prices. Price data are collected monthly via Internet, mail, and fax. Prices are obtained directly from producing companies on a voluntary and confidential basis. Prices are generally reported for the Tuesday of the week containing the 13th day of the month. Samples are updated regularly.

The BLS revises the PPI weighting structure when data from economic censuses become available. Beginning with data for January 2012, the weights used to construct the PPI reflect 2007 shipments values as measured by the 2007 Economic Censuses. Data for January 2007 through December 2011 used 2002 values; 2002 through 2006, 1997 shipments values; 1996 through 2001, 1992 values; 1992 through 1995, 1987 values; 1987 through 1991, 1982 values; 1976 through 1986, 1972 values; and 1967 through 1975, 1963 values.

Price indexes are available in unadjusted form and are also adjusted for seasonal variation using the X-12-ARIMA method. Since January 1988, BLS has also used X-12-ARIMA Intervention Analysis Seasonal Adjustment for a small number of series to remove unusual values that might distort seasonal patterns before calculating the seasonal adjustment factors. Seasonal factors for the PPI are revised annually to take into account the most recent 12 months of data. Seasonally adjusted data for the previous 5 years are subject to these annual revisions.

The newer FD-ID indexes are only available beginning in November 2009. Hence, they only regard, so far, the recovery and expansion phase of a particularly severe business cycle; the editor suggests that up to now, both seasonal adjustments and broader inferences about relative price behavior are necessarily based on a short and possibly unrepresentative time period. Partly for that reason, we present both unadjusted and seasonally adjusted indexes in Tables 8-5 and 8-6.

Data availability and references

The indexes are initially issued in a press release two to three weeks after the end of the month for which the data were collected. Data are subsequently published in greater detail in a monthly BLS publication, *PPI Detailed Report*. Each month, data for the fourth previous month (both unadjusted and seasonally adjusted) are revised to reflect late reports and corrections.

The press release, the *PPI Detailed Report*, detailed and complete current and historical data, and extensive documentation are available at <http://www.bls.gov/ppi>. The items available on this Web site include Chapter 14 of the *BLS Handbook of Methods*, "Producer Price Indexes"; a selection of *Monthly Labor Review* articles on the PPI; and fact sheets on a number of issues and index components.

CHAPTER 9: EMPLOYMENT COSTS, PRODUCTIVITY, AND PROFITS

SECTION 9A: EMPLOYMENT COST INDEXES

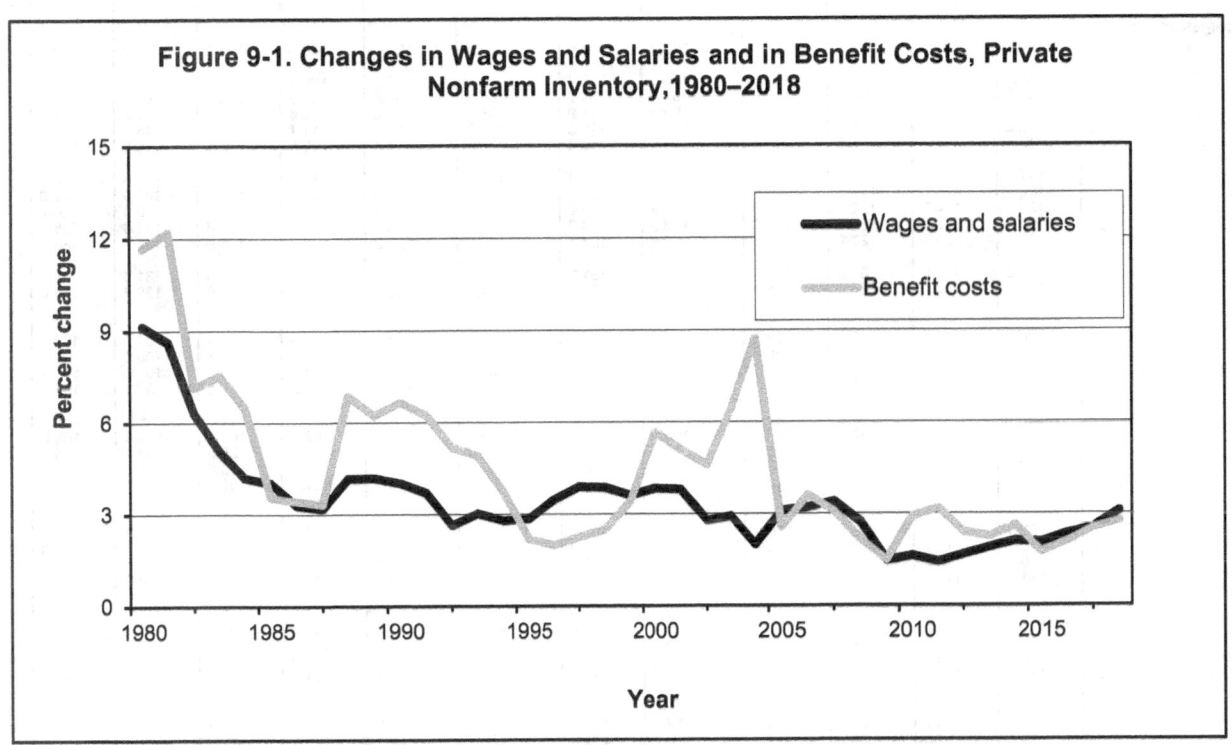

Figure 9-1. Changes in Wages and Salaries and in Benefit Costs, Private Nonfarm Inventory, 1980–2018

- The Employment Cost Index measures average increases in the dollar value of hourly compensation for nonfarm wage and salary workers, holding the composition of employment constant so as to reflect only the aggregate of changes affecting individual defined jobs. (The analogy is to the Consumer Price Index, which holds the market basket constant in any individual time period so as to reflect only an aggregate of price changes affecting individual defined products.) This makes the ECI the most accurate indicator of changes in price for a typical unit of labor.

- From 2006 to 2018, total compensation for all civilian workers rose 30.7 percent. Compensation includes wages, salaries, and benefits, but not stock options. Benefits rose even more, at 33.3 percent. Much of this rise in benefits was due to medical care costs. (Tables 9-1, 9-2, and 8-2)

- The ECI reports separately the rates of increase for wages and salaries and for benefits such as health insurance, as shown in Figure 9-1. (Tables 9-1 and 9-2)

Table 9-1. Employment Cost Indexes, NAICS Basis

(December 2005 [not seasonally adjusted] = 100; annual values are for December, not seasonally adjusted; quarterly values, seasonally adjusted, except as noted.)

Year and month	All civilian workers [1]	State and local government workers	Private industry workers									Union [2]	Non-union [2]
				By occupational group					By industry				
			All private industry workers	Management, professional, and related	Sales and office	Natural resources, construction, and maintenance	Production, transportation, and material moving	Service occupations	Goods-producing industries		Service-providing		
									Total	Manufacturing			
TOTAL COMPENSATION													
2006	103.3	104.1	103.2	103.5	102.9	103.6	102.3	103.1	102.5	101.8	103.4	103.0	103.2
2007	106.7	108.4	106.3	106.8	106.1	106.7	104.5	107.0	105.0	103.8	106.7	105.1	106.5
2008	109.5	111.6	108.9	109.9	107.9	109.6	106.9	109.8	107.5	105.9	109.4	108.0	109.1
2009	111.0	114.2	110.2	110.7	109.2	111.2	108.9	111.8	108.6	107.0	110.8	111.1	110.1
2010	113.2	116.2	112.5	113.0	111.6	113.3	111.5	113.5	111.1	110.0	113.0	114.8	112.1
2011	115.5	117.7	115.0	115.4	114.2	115.8	114.2	115.4	113.8	113.1	115.3	117.9	114.5
2012	117.7	119.9	117.1	117.7	116.3	117.8	116.0	117.4	115.6	114.9	117.6	120.5	116.6
2013	120.0	122.2	119.4	120.2	119.0	120.1	118.0	119.0	117.7	117.0	120.0	122.6	119.0
2014	122.7	124.7	122.2	123.0	121.9	123.2	120.6	121.1	120.3	119.8	122.8	126.7	121.5
2015	125.1	127.8	124.5	125.3	123.9	124.9	123.7	123.2	123.2	122.8	124.9	128.7	123.8
2016	127.9	130.9	127.2	127.4	126.8	127.7	127.0	127.2	125.8	125.5	127.7	130.6	126.6
2017	131.2	134.2	130.5	130.5	130.1	131.2	131.0	130.9	128.9	128.9	131.0	134.5	129.9
2018	135.0	137.7	134.4	133.8	135.0	134.5	134.7	135.6	131.9	131.6	135.2	138.8	133.7
2007													
March	104.2	105.1	103.9	104.5	103.8	104.1	102.5	104.4	102.9	101.9	104.3	102.7	104.2
June	105.1	106.2	104.8	105.5	104.5	104.9	103.3	105.2	103.8	102.8	105.2	103.9	105.1
September	105.9	107.2	105.6	106.3	105.2	105.8	103.9	106.3	104.4	103.2	106.0	104.4	105.9
December	106.8	108.3	106.5	107.1	106.2	106.8	104.7	107.2	105.2	104.0	106.9	105.1	106.5
2008													
March	107.6	109.0	107.2	108.0	106.8	107.7	105.5	107.8	106.0	104.6	107.6	105.9	107.5
June	108.3	109.8	108.0	108.8	107.4	108.2	106.0	108.7	106.7	105.0	108.4	106.7	108.3
September	109.1	110.9	108.6	109.6	107.8	108.9	106.5	109.4	107.2	105.6	109.1	107.4	108.9
December	109.6	111.6	109.1	110.2	108.0	109.7	107.0	109.9	107.7	106.1	109.5	108.0	109.1
2009													
March	109.9	112.4	109.3	110.2	108.1	109.9	107.7	110.6	107.9	106.4	109.8	109.1	109.4
June	110.2	113.2	109.5	110.4	108.2	110.3	108.1	111.0	108.1	106.6	110.0	109.8	109.6
September	110.7	113.5	109.9	110.6	108.8	110.7	108.6	111.6	108.3	106.8	110.5	110.5	109.9
December	111.1	114.1	110.4	111.0	109.3	111.3	109.0	112.0	108.8	107.3	110.9	111.1	110.1
2010													
March	111.8	114.5	111.1	111.7	109.9	112.3	109.9	112.3	109.7	108.4	111.5	112.8	110.9
June	112.3	115.1	111.6	112.1	110.6	112.6	110.4	112.6	110.2	109.0	112.1	113.7	111.4
September	112.8	115.5	112.1	112.7	111.1	112.9	111.2	113.3	110.9	109.9	112.5	114.6	111.8
December	113.3	116.1	112.6	113.3	111.7	113.4	111.6	113.6	111.2	110.2	113.1	114.8	112.1
2011													
March	114.0	116.7	113.3	114.0	112.2	113.9	112.2	114.5	112.0	111.3	113.8	115.6	113.0
June	114.7	117.0	114.2	114.7	113.2	114.8	113.5	114.7	113.2	112.6	114.5	117.1	113.8
September	115.1	117.3	114.6	115.1	113.7	115.4	113.7	115.0	113.3	112.8	115.0	117.4	114.2
December	115.6	117.7	115.1	115.7	114.3	115.9	114.3	115.5	113.9	113.3	115.5	117.9	114.5
2012													
March	116.3	118.3	115.7	116.3	115.1	116.4	114.5	116.0	114.1	113.4	116.3	118.3	115.3
June	116.8	118.9	116.3	117.0	115.7	116.9	115.0	116.5	114.6	113.9	116.9	119.3	116.0
September	117.3	119.5	116.8	117.3	116.2	117.5	115.6	116.8	115.2	114.6	117.3	120.2	116.3
December	117.8	119.9	117.2	118.0	116.5	118.0	116.1	117.4	115.7	115.1	117.7	120.5	116.6
2013													
March	118.4	120.5	117.9	118.5	117.3	118.7	116.7	117.8	116.4	115.7	118.3	121.5	117.3
June	119.0	121.0	118.5	119.3	117.7	119.0	117.1	118.3	116.9	116.2	119.0	122.1	118.0
September	119.5	121.4	119.0	119.9	118.4	119.7	117.5	118.4	117.4	116.7	119.5	122.5	118.5
December	120.1	122.2	119.6	120.4	119.2	120.3	118.1	119.1	117.8	117.2	120.2	122.6	119.0
2014													
March	120.5	122.8	119.9	120.7	119.3	120.9	118.8	119.0	118.6	118.0	120.4	123.5	119.4
June	121.4	123.5	120.9	121.9	120.3	121.9	119.5	119.6	119.1	118.6	121.5	125.0	120.4
September	122.2	124.1	121.7	122.7	121.1	122.6	120.2	120.5	119.8	119.3	122.3	125.8	121.1
December	122.8	124.7	122.4	123.1	122.1	123.4	120.8	121.2	120.4	120.0	123.0	126.7	121.5
2015													
March	123.6	125.4	123.2	123.7	123.3	123.7	121.6	121.8	121.1	120.8	123.8	127.4	122.5
June	123.8	126.2	123.2	124.1	122.6	124.0	122.4	122.1	121.8	121.6	123.7	127.5	122.7
September	124.6	126.9	124.0	124.8	123.5	124.5	123.0	122.6	122.4	122.1	124.5	128.0	123.4
December	125.3	127.8	124.6	125.4	124.3	125.1	123.8	123.3	123.3	122.9	125.1	128.7	123.8
2016													
March	126.0	128.5	125.3	126.0	125.0	125.8	124.6	124.2	123.8	123.6	125.8	129.6	124.7
June	126.7	129.2	126.1	126.3	126.5	126.5	125.5	124.5	124.6	124.4	125.8	130.0	125.7
September	127.4	130.2	126.7	127.1	126.6	126.9	126.3	126.5	125.1	125.1	127.3	130.4	126.3
December	128.1	130.9	127.4	127.6	127.2	127.9	127.2	127.3	125.8	125.7	127.9	130.6	126.6
2017													
March	129.0	131.8	128.2	128.4	128.0	128.6	128.2	128.3	126.5	126.3	128.8	131.9	127.8
June	129.7	132.5	129.0	129.4	128.6	129.5	129.0	129.2	127.2	127.0	129.6	132.7	128.6
September	130.6	133.3	130.0	130.0	129.8	130.5	130.4	130.0	128.3	128.3	130.5	133.7	129.5
December	131.4	134.2	130.7	130.7	130.5	131.3	131.2	131.1	129.0	129.0	131.3	134.5	129.9
2018													
March	132.4	134.8	131.9	131.7	131.9	131.9	132.3	132.4	129.9	129.9	132.5	135.4	131.4
June	133.3	135.8	132.7	132.5	133.0	133.2	133.2	133.4	130.8	130.7	133.4	137.4	132.2
September	134.3	136.8	133.7	133.3	134.3	133.7	134.0	134.6	131.2	130.9	134.5	137.8	133.2
December	135.2	137.8	134.6	134.0	135.5	134.6	134.9	135.8	131.9	131.7	135.4	138.8	133.7

[1] Excludes farm workers, private household workers, and federal government employees.
[2] Not seasonally adjusted.

Table 9-1. Employment Cost Indexes, NAICS Basis—*Continued*

(December 2005 [not seasonally adjusted] = 100; annual values are for December, not seasonally adjusted; quarterly values, seasonally adjusted, except as noted.)

Year and month	All civilian workers [1]	State and local government workers	All private industry workers	By occupational group					By industry			Union [2]	Non-union [2]
				Management, professional, and related	Sales and office	Natural resources, construction, and maintenance	Production, transportation, and material moving	Service occupations	Goods-producing industries Total	Goods-producing industries Manufacturing	Service-providing		
WAGES AND SALARIES													
2006	103.2	103.5	103.2	103.6	103.0	103.4	102.4	102.9	102.9	102.3	103.3	102.3	103.3
2007	106.7	107.1	106.6	107.2	106.2	107.1	105.0	107.1	106.0	104.9	106.8	104.7	106.9
2008	109.6	110.4	109.4	110.5	108.0	110.5	107.8	110.1	109.0	107.7	109.6	108.1	109.6
2009	111.2	112.5	110.8	111.5	109.4	112.0	109.6	112.3	110.0	108.9	111.1	110.9	110.9
2010	113.0	113.8	112.8	113.7	111.5	113.3	111.3	113.5	111.6	110.7	113.1	112.9	112.7
2011	114.6	114.9	114.6	115.5	113.6	115.4	112.8	115.1	113.5	112.7	114.9	114.9	114.6
2012	116.5	116.2	116.6	117.7	115.8	116.7	115.1	116.8	115.4	114.8	117.0	117.4	116.5
2013	118.7	117.5	119.0	120.2	118.4	118.8	117.2	118.3	117.6	117.2	119.4	119.8	118.9
2014	121.2	119.4	121.6	122.8	121.3	121.4	119.9	120.7	120.1	119.8	122.1	123.1	121.5
2015	123.7	121.6	124.2	125.6	123.5	123.5	122.8	122.8	123.2	123.0	124.5	125.5	124.0
2016	126.6	124.1	127.1	128.1	126.2	126.7	126.6	127.1	126.2	126.2	127.4	127.3	127.1
2017	129.8	126.7	130.6	131.1	129.8	130.4	130.7	131.3	129.3	129.3	131.0	130.9	130.6
2018	133.8	129.7	134.7	134.6	135.0	133.4	135.1	136.4	133.0	132.9	135.2	134.6	134.7
2007													
March	104.3	104.2	104.3	104.9	104.0	104.3	103.2	104.6	103.9	103.2	104.4	102.8	104.5
June	105.1	105.0	105.1	105.7	104.7	105.1	103.8	105.4	104.6	103.8	105.2	103.7	105.3
September	105.9	106.0	105.9	106.6	105.2	106.1	104.4	106.4	105.4	104.4	106.1	104.4	106.2
December	106.8	107.0	106.7	107.4	106.2	107.1	105.0	107.2	106.1	105.1	106.9	104.7	106.9
2008													
March	107.6	107.8	107.6	108.4	106.9	108.2	106.0	107.8	107.1	105.9	107.7	105.5	107.9
June	108.5	108.6	108.4	109.2	107.5	108.9	106.8	108.9	107.9	106.7	108.6	106.7	108.7
September	109.2	109.7	109.0	110.1	107.9	109.7	107.4	109.6	108.5	107.3	109.2	107.4	109.4
December	109.7	110.3	109.5	110.7	108.0	110.5	107.9	110.2	109.1	107.9	109.7	108.1	109.6
2009													
March	110.0	111.0	109.8	110.9	108.1	110.7	108.3	110.9	109.2	108.0	110.0	108.8	110.0
June	110.4	111.7	110.1	111.1	108.2	111.1	108.8	111.3	109.4	108.3	110.2	109.6	110.2
September	110.8	111.9	110.5	111.3	108.9	111.5	109.3	112.0	109.7	108.6	110.7	110.2	110.6
December	111.2	112.4	111.0	111.8	109.5	112.1	109.7	112.4	110.2	109.1	111.2	110.9	110.9
2010													
March	111.7	112.8	111.4	112.3	109.8	112.6	109.9	112.5	110.5	109.3	111.7	111.5	111.4
June	112.1	113.2	111.9	112.8	110.5	112.8	110.3	112.8	110.9	109.9	112.2	112.1	111.9
September	112.5	113.3	112.3	113.4	110.8	113.0	111.0	113.2	111.4	110.5	112.6	112.7	112.4
December	113.0	113.8	112.9	113.9	111.5	113.3	111.3	113.5	111.7	110.9	113.2	112.9	112.7
2011													
March	113.4	114.1	113.2	114.4	111.8	113.8	111.6	114.2	112.2	111.5	113.6	113.6	113.2
June	113.9	114.4	113.7	114.8	112.5	114.5	112.0	114.3	112.7	111.9	114.1	114.0	113.8
September	114.3	114.5	114.2	115.2	113.1	115.0	112.4	114.6	113.1	112.5	114.6	114.6	114.3
December	114.7	114.8	114.7	115.7	113.7	115.4	112.9	115.1	113.6	113.0	115.0	114.9	114.6
2012													
March	115.3	115.2	115.3	116.3	114.5	115.7	113.7	115.4	114.0	113.5	115.7	115.6	115.2
June	115.8	115.6	115.8	116.9	115.0	116.0	114.0	115.9	114.4	113.9	116.2	116.2	115.9
September	116.2	115.8	116.3	117.2	115.7	116.4	114.6	116.2	115.0	114.5	116.7	116.9	116.3
December	116.6	116.1	116.7	117.8	115.9	116.8	115.2	116.8	115.5	115.0	117.1	117.4	116.5
2013													
March	117.2	116.4	117.4	118.5	116.7	117.3	115.8	117.2	116.1	115.6	117.7	118.4	117.2
June	117.7	116.7	118.0	119.3	117.1	117.6	116.2	117.7	116.7	116.3	118.3	119.0	117.9
September	118.2	116.9	118.4	119.7	117.7	118.4	116.6	117.6	117.3	116.8	118.8	119.6	118.4
December	118.8	117.4	119.1	120.3	118.6	118.9	117.3	118.3	117.7	117.4	119.5	119.8	118.9
2014													
March	119.1	117.8	119.4	120.6	118.6	119.5	118.0	118.4	118.2	118.0	119.7	120.5	119.2
June	119.8	118.2	120.2	121.7	119.5	120.0	118.7	119.0	118.9	118.7	120.6	121.2	120.2
September	120.7	118.8	121.1	122.5	120.3	120.7	119.5	120.1	119.5	119.3	121.5	122.1	121.0
December	121.3	119.4	121.8	122.8	121.5	121.5	120.0	120.8	120.2	120.0	122.2	123.1	121.5
2015													
March	122.1	120.0	122.6	123.4	123.2	121.8	120.7	121.5	120.9	120.8	123.1	123.7	122.4
June	122.4	120.6	122.8	124.2	122.1	122.6	121.5	121.6	121.7	121.6	123.1	124.5	122.7
September	123.1	121.0	123.6	125.1	123.1	123.0	122.1	122.2	122.4	122.3	124.0	124.8	123.6
December	123.8	121.5	124.3	125.7	123.8	123.6	123.0	122.9	123.3	123.2	124.7	125.5	124.0
2016													
March	124.5	122.1	125.1	126.4	124.4	124.5	123.8	123.8	124.0	123.9	125.4	126.4	125.0
June	125.4	122.7	126.0	126.8	126.1	125.4	125.0	125.2	124.8	124.8	126.3	126.8	126.0
September	126.0	123.4	126.6	127.7	126.0	125.8	125.8	126.4	125.4	125.7	127.0	127.2	126.6
December	126.7	124.0	127.3	128.1	126.6	126.8	126.8	127.3	126.3	126.4	127.6	127.3	127.1
2017													
March	127.6	124.8	128.2	128.9	127.6	127.7	127.9	128.4	127.1	127.0	128.6	128.6	128.2
June	128.3	125.3	129.0	129.9	128.1	128.6	128.8	129.3	127.8	127.8	129.4	129.2	129.1
September	129.2	125.9	130.0	130.5	129.4	129.6	130.0	130.2	128.7	128.7	130.3	129.9	130.0
December	130.0	126.5	130.8	131.2	130.2	130.5	131.0	131.5	129.5	129.5	131.2	130.9	130.6
2018													
March	131.1	127.1	132.0	132.3	131.7	131.1	132.2	132.7	130.4	130.4	132.4	131.6	132.1
June	131.8	127.8	132.8	132.9	132.7	131.9	133.2	134.0	131.3	131.2	133.2	132.7	132.9
September	133.0	128.7	133.9	134.0	134.2	132.5	134.3	135.3	132.2	132.0	134.4	133.5	134.1
December	133.9	129.5	134.9	134.7	135.4	133.4	135.4	136.6	133.1	133.0	135.4	134.6	134.7

[1] Excludes farm workers, private household workers, and federal government employees.
[2] Not seasonally adjusted.

Table 9-1. Employment Cost Indexes, NAICS Basis—Continued

(December 2005 [not seasonally adjusted] = 100; annual values are for December, not seasonally adjusted; quarterly values, seasonally adjusted, except as noted.)

Year and month	All civilian workers [1]	State and local government workers	Private industry workers — All private industry workers	By occupational group — Management, professional, and related	By occupational group — Sales and office	By occupational group — Natural resources, construction, and maintenance	By occupational group — Production, transportation, and material moving	By occupational group — Service occupations	By industry — Goods-producing industries — Total	By industry — Goods-producing industries — Manufacturing	By industry — Service-providing	Union [2]	Nonunion [2]
TOTAL BENEFITS													
2006	103.6	105.2	103.1	103.4	102.9	104.0	102.0	103.6	101.7	100.8	103.7	104.2	102.9
2007	106.8	111.0	105.6	106.0	106.0	105.9	103.7	106.7	103.2	101.7	106.6	105.8	105.6
2008	109.1	114.2	107.7	108.5	107.8	107.7	105.1	108.8	104.7	102.5	108.9	107.8	107.6
2009	110.7	117.7	108.7	108.8	108.7	109.5	107.4	110.5	105.8	103.6	109.9	111.4	108.2
2010	113.9	121.1	111.9	111.2	111.8	113.2	112.0	113.5	110.1	108.8	112.6	117.9	110.6
2011	117.5	123.6	115.9	115.2	115.5	116.8	117.0	116.4	114.4	113.9	116.4	122.8	114.4
2012	120.3	127.8	118.2	117.9	117.6	120.2	118.0	119.2	116.0	115.0	119.1	125.6	116.7
2013	123.0	132.0	120.5	120.2	120.5	122.9	119.5	121.0	118.0	116.6	121.5	127.3	119.1
2014	126.2	135.8	123.5	123.4	123.4	127.1	122.1	122.2	120.7	119.8	124.6	132.6	121.7
2015	128.4	140.6	125.1	124.5	125.0	127.9	125.4	124.3	123.1	122.5	125.9	133.9	123.3
2016	131.1	145.0	127.3	126.0	128.4	129.9	127.9	127.1	124.9	124.3	128.3	136.1	125.5
2017	134.4	149.7	130.2	129.1	130.8	132.9	131.4	129.7	128.0	128.0	131.2	140.6	128.2
2018	138.1	154.4	133.6	132.1	135.3	136.9	133.8	132.8	129.6	129.1	135.1	145.7	131.2
2007													
March	104.0	107.1	103.1	103.5	103.3	103.5	101.1	104.0	100.9	99.4	104.0	102.4	103.4
June	105.2	108.7	104.2	104.8	104.2	104.5	102.3	105.0	102.1	100.8	105.1	104.1	104.3
September	106.0	109.7	105.0	105.5	105.2	105.2	102.7	106.0	102.4	100.9	106.0	104.3	105.1
December	107.0	110.9	105.9	106.4	106.1	106.3	103.9	107.0	103.4	101.9	106.9	105.8	105.6
2008													
March	107.5	111.4	106.4	107.1	106.5	106.6	104.4	107.4	104.0	102.2	107.4	106.6	106.5
June	108.1	112.4	106.9	107.8	106.9	106.7	104.4	108.3	104.3	102.0	108.0	106.6	107.1
September	108.8	113.3	107.5	108.5	107.6	107.4	104.8	108.7	104.6	102.4	108.7	107.2	107.6
December	109.3	114.1	107.9	109.0	108.0	108.0	105.3	109.2	105.0	102.8	109.1	107.8	107.6
2009													
March	109.6	115.3	108.1	108.5	108.0	108.3	106.4	109.5	105.4	103.4	109.2	109.5	107.9
June	109.9	116.2	108.2	108.7	108.0	108.5	106.7	109.8	105.5	103.4	109.3	110.3	108.0
September	110.4	116.8	108.6	108.9	108.6	109.1	107.1	110.4	105.6	103.4	109.9	110.9	108.2
December	110.9	117.7	109.1	109.2	108.9	109.8	107.5	110.9	106.2	104.0	110.2	111.4	108.2
2010													
March	112.0	118.1	110.3	110.0	110.1	111.6	109.9	111.5	108.4	106.6	111.1	114.8	109.5
June	112.7	119.1	110.9	110.3	110.9	112.1	110.7	112.3	108.9	107.4	111.7	116.2	110.0
September	113.5	120.2	111.6	110.9	111.6	112.8	111.7	113.3	110.0	108.7	112.3	117.6	110.4
December	114.1	121.2	112.1	111.7	112.1	113.5	112.2	113.8	110.2	108.9	112.9	117.9	110.6
2011													
March	115.4	122.0	113.6	113.2	113.3	114.2	113.5	115.2	111.7	111.0	114.4	119.0	112.6
June	116.8	122.5	115.2	114.6	114.9	115.6	116.4	115.9	114.1	113.9	115.7	122.3	113.9
September	117.1	123.2	115.4	114.7	115.3	116.1	116.3	116.0	113.8	113.4	116.0	122.0	114.0
December	117.7	123.7	116.1	115.6	115.8	117.1	117.1	116.7	114.5	114.0	116.7	122.8	114.4
2012													
March	118.5	124.7	116.8	116.6	116.6	117.9	116.1	117.8	114.2	113.2	117.8	122.9	115.6
June	119.3	125.8	117.5	117.1	117.4	118.8	117.0	118.2	114.8	113.9	118.5	124.3	116.2
September	119.9	127.1	117.9	117.7	117.4	119.9	117.6	118.8	115.7	114.7	118.8	125.5	116.4
December	120.5	127.9	118.5	118.3	117.9	120.6	118.1	119.4	116.1	115.1	119.4	125.6	116.7
2013													
March	121.3	129.1	119.1	118.5	118.9	121.6	118.7	119.7	117.0	115.7	120.0	126.6	117.7
June	121.9	129.9	119.6	119.3	119.3	122.0	119.0	120.3	117.3	116.1	120.6	127.3	118.3
September	122.6	130.9	120.3	120.2	120.1	122.6	119.1	120.8	117.6	116.4	121.4	127.1	118.9
December	123.2	132.1	120.8	120.7	120.9	123.2	119.7	121.2	118.1	116.7	121.9	127.3	119.1
2014													
March	123.8	133.0	121.3	120.9	121.2	124.0	120.5	120.9	119.2	118.1	122.2	128.5	119.9
June	125.1	134.2	122.5	122.4	122.5	126.0	120.9	121.1	119.4	118.2	123.8	131.3	120.9
September	125.7	134.8	123.2	123.1	123.0	126.7	121.6	121.7	120.3	119.3	124.3	131.9	121.4
December	126.4	135.8	123.8	123.8	123.8	127.4	122.3	122.4	120.8	119.9	125.0	132.6	121.7
2015													
March	127.1	136.7	124.5	124.6	123.7	127.7	123.5	122.9	121.4	120.8	125.7	133.4	122.7
June	127.2	137.8	124.2	123.9	123.9	126.9	124.2	123.3	122.1	121.5	125.1	132.3	122.7
September	127.8	138.9	124.8	124.3	124.6	127.5	124.9	123.9	122.3	121.6	125.8	133.2	123.0
December	128.7	140.5	125.3	124.9	125.5	128.2	125.5	124.5	123.2	122.5	126.3	133.9	123.3
2016													
March	129.3	141.5	125.9	124.9	126.7	128.6	126.3	125.2	123.6	123.1	126.9	135.0	124.2
June	129.9	142.4	126.3	125.2	127.4	128.9	126.7	126.0	124.2	123.7	127.3	135.3	124.7
September	130.7	144.0	127.0	125.8	128.2	129.4	127.4	126.7	124.5	124.0	128.0	135.6	125.3
December	131.3	144.8	127.5	126.4	128.8	130.1	128.0	127.3	124.9	124.4	128.6	136.1	125.5
2017													
March	132.2	146.0	128.3	127.2	129.2	130.6	128.7	128.0	125.4	125.0	129.5	137.4	126.6
June	133.0	147.0	129.1	128.0	129.9	131.5	129.5	128.6	126.0	125.6	130.3	138.4	127.4
September	134.0	148.2	130.0	129.0	130.7	132.3	131.0	129.3	127.5	127.6	131.0	139.9	128.0
December	134.7	149.6	130.5	129.6	131.3	133.1	131.6	129.9	128.0	128.1	131.6	140.6	128.2
2018													
March	135.7	150.3	131.5	130.4	132.3	133.6	132.6	131.0	129.0	129.1	132.6	141.7	129.6
June	136.9	151.7	132.7	131.4	133.8	136.0	133.3	131.5	129.7	129.7	133.9	145.1	130.5
September	137.5	152.8	133.2	131.7	134.7	136.4	133.4	132.4	129.2	128.8	134.7	144.9	130.9
December	138.4	154.2	133.9	132.6	135.8	137.1	134.0	133.0	129.6	129.2	135.5	145.7	131.2

[1] Excludes farm workers, private household workers, and federal government employees.
[2] Not seasonally adjusted.

Table 9-2. Employment Cost Indexes, SIC Basis

(Not seasonally adjusted, December 2005 = 100; annual values are for December.)

Year	All civilian workers [1,2]	State and local government workers [2]	All private industry workers [2]	Private industry workers excluding sales occupations	Production and nonsupervisory occupations	White-collar occupations	Blue-collar occupations	Service occupations [2]	Goods-producing: Total [2]	Construction [2]	Manufacturing [2]	Service-providing: Total [2]	Transportation and utilities	Wholesale trade	Retail trade	Finance, insurance, and real estate [2]	Services
TOTAL COMPENSATION																	
1979	...	...	32.8	32.5	...	31.2	35.0	33.9	33.7	...	33.1	32.0	...	...	...	...	...
1980	...	...	35.9	35.9	...	34.2	38.5	37.1	37.0	...	36.4	35.1	...	...	...	...	...
1981	39.0	36.8	39.5	39.4	40.1	37.6	42.2	40.5	40.7	...	40.0	38.6	...	...	...	...	...
1982	41.5	39.4	42.0	42.0	42.8	40.1	44.7	43.9	43.2	...	42.4	41.1	...	...	...	...	...
1983	43.9	41.8	44.4	44.4	45.2	42.7	47.0	46.3	45.3	...	44.6	43.8	...	...	...	...	...
1984	46.2	44.6	46.6	46.7	47.3	44.8	49.0	49.4	47.4	...	46.9	46.0	...	...	...	...	...
1985	48.2	47.1	48.4	48.3	49.1	47.0	50.5	50.9	49.0	50.8	48.4	48.1	50.6	...	52.4	44.7	46.1
1986	49.9	49.6	49.9	49.9	50.5	48.6	51.9	52.4	50.5	52.3	50.0	49.6	51.8	48.5	53.5	46.1	48.1
1987	51.7	51.8	51.6	51.7	52.2	50.4	53.5	53.7	52.1	54.2	51.5	51.4	53.3	50.4	54.8	47.0	50.5
1988	54.2	54.7	54.1	54.1	54.8	52.9	55.7	55.9	54.4	56.5	53.8	54.0	54.9	52.5	58.0	50.0	53.5
1989	56.9	58.1	56.7	56.5	57.6	55.7	58.2	59.0	56.7	59.0	56.3	56.8	56.9	57.1	59.9	52.7	56.4
1990	59.7	61.5	59.3	59.3	60.1	58.4	60.7	61.8	59.4	60.9	59.1	59.4	59.1	58.2	62.5	54.9	59.9
1991	62.3	63.7	61.9	62.0	62.7	61.0	63.4	64.7	62.1	63.3	61.9	61.8	61.7	60.7	65.2	57.2	62.5
1992	64.4	66.0	64.1	64.2	64.9	63.1	65.6	66.7	64.5	65.6	64.3	63.9	63.9	62.5	66.9	57.9	65.2
1993	66.7	67.9	66.4	66.6	67.3	65.4	68.1	68.8	67.0	67.1	66.9	66.2	66.1	64.4	68.9	60.5	67.5
1994	68.7	69.9	68.5	68.6	69.2	67.5	70.0	70.8	69.0	69.6	69.0	68.1	68.7	66.4	70.8	61.8	69.4
1995	70.6	72.0	70.2	70.4	71.0	69.4	71.7	72.1	70.7	71.1	70.8	70.0	71.2	69.4	72.3	64.0	70.9
1996	72.6	73.9	72.4	72.4	73.1	71.7	73.6	74.2	72.7	72.9	72.9	72.3	73.4	71.5	75.1	65.5	73.1
1997	75.0	75.6	74.9	74.9	75.4	74.4	75.5	77.2	74.5	74.8	74.6	75.1	75.5	73.8	77.7	69.9	75.9
1998	77.6	77.8	77.5	77.2	78.1	77.3	77.6	79.4	76.5	77.4	76.6	78.0	78.4	78.0	80.0	74.1	78.2
1999	80.2	80.5	80.2	80.0	80.4	79.9	80.2	82.1	79.1	79.9	79.2	80.6	80.1	81.1	83.0	77.1	80.9
2000	83.6	82.9	83.6	83.6	84.0	83.6	83.6	85.3	82.6	84.6	82.3	84.2	83.5	84.4	86.4	81.0	84.5
2001	87.0	86.4	87.1	87.0	87.4	87.1	86.7	89.1	85.7	88.2	85.3	87.8	87.5	87.2	90.3	83.9	88.3
2002	90.0	89.9	90.0	89.9	90.2	89.9	89.8	92.0	88.9	91.0	88.5	90.5	91.0	91.1	91.9	87.6	90.7
2003	93.5	92.9	93.6	93.6	93.6	93.6	93.4	94.9	92.4	94.1	92.2	94.2	94.0	94.0	94.9	94.1	94.0
2004	96.9	96.1	97.1	97.2	97.2	96.9	97.5	97.7	96.8	96.4	96.7	97.3	97.6	96.5	97.1	96.7	97.5
2005	100.0	100.0	100.0	100.0	100.0	100.0	100.0	100.0	100.0	100.0	100.0	100.0	100.0	100.0	100.0	100.0	100.0
WAGES AND SALARIES																	
1979	...	...	36.1	36.1	36.8	34.1	39.4	37.9	38.2	41.4	37.5	34.9	39.1	33.7	39.7	32.8	31.7
1980	...	...	39.4	39.3	40.3	37.1	43.1	41.0	41.8	45.0	41.0	38.0	43.5	37.1	42.5	35.2	34.5
1981	42.3	40.1	42.8	42.9	43.9	40.4	46.8	44.4	45.4	49.0	44.5	41.4	47.1	40.0	45.6	38.8	38.1
1982	45.0	42.7	45.5	45.6	46.7	43.1	49.4	48.2	48.0	51.5	47.0	44.2	50.5	42.5	47.5	41.3	41.2
1983	47.3	45.0	47.8	47.9	48.9	45.7	51.3	50.4	49.9	53.0	49.0	46.7	53.0	45.1	49.5	44.3	43.9
1984	49.4	47.7	49.8	50.1	50.8	47.6	53.1	53.5	51.8	53.7	51.2	48.7	54.8	47.6	52.0	43.9	46.7
1985	51.5	50.3	51.8	51.9	52.9	50.0	55.0	54.8	53.6	55.3	53.0	51.0	56.9	49.7	54.5	47.9	48.4
1986	53.3	53.0	53.5	53.6	54.3	51.7	56.4	56.4	55.3	56.7	54.8	52.5	57.9	51.5	55.7	49.2	50.3
1987	55.2	55.2	55.2	55.5	56.0	53.6	58.1	57.6	57.1	58.6	56.6	54.3	59.1	53.6	57.2	49.8	53.0
1988	57.5	57.9	57.5	57.5	58.4	56.1	59.9	60.1	58.9	60.7	58.3	56.9	60.6	55.6	60.1	52.9	55.6
1989	60.1	61.0	59.9	59.8	61.0	58.7	62.0	62.3	61.2	62.8	60.6	59.5	62.2	60.7	62.0	55.7	58.3
1990	62.6	64.2	62.3	62.4	63.2	61.2	64.2	64.8	63.4	64.1	63.1	61.8	64.3	61.2	64.2	57.6	61.6
1991	64.9	66.4	64.6	64.7	65.4	63.5	66.4	67.4	65.8	66.0	65.6	64.1	66.6	63.6	66.6	59.6	63.8
1992	66.6	68.4	66.3	66.5	67.2	65.2	68.1	68.8	67.6	67.3	67.6	65.7	68.7	65.5	68.2	59.5	66.0
1993	68.7	70.2	68.3	68.5	69.2	67.4	70.0	70.3	69.6	68.6	69.7	67.8	70.9	67.1	70.2	62.1	68.0
1994	70.6	72.4	70.2	70.5	71.1	69.3	72.0	72.0	71.7	70.8	71.8	69.6	73.5	69.1	71.9	62.8	70.0
1995	72.7	74.7	72.2	72.5	73.0	71.3	74.1	74.0	73.7	72.5	73.9	71.7	76.0	72.4	73.6	65.1	71.7
1996	75.1	76.8	74.7	74.9	75.5	73.8	76.3	76.6	76.0	74.6	76.3	74.2	78.1	74.7	76.8	67.2	74.2
1997	77.9	78.9	77.6	77.7	78.3	77.0	78.8	79.9	78.3	77.1	78.6	77.4	80.7	77.0	79.7	71.8	77.5
1998	80.8	81.3	80.6	80.4	81.4	80.3	81.3	82.4	81.1	79.9	81.3	80.5	83.0	81.5	82.2	76.9	80.1
1999	83.6	84.2	83.5	83.4	83.8	83.1	84.0	85.1	83.8	82.5	84.1	83.4	84.8	84.5	85.2	79.8	83.0
2000	86.7	87.0	86.7	86.7	87.1	86.4	87.1	88.3	87.1	86.9	87.1	86.6	87.5	87.4	88.6	83.4	86.3
2001	90.0	90.2	90.0	90.0	90.4	89.6	90.5	91.8	90.2	90.4	90.2	89.9	91.7	89.3	91.9	85.8	90.0
2002	92.6	93.1	92.4	92.5	92.6	92.0	93.0	94.1	92.9	92.8	93.0	92.3	94.7	92.8	93.2	89.4	92.0
2003	95.2	95.0	95.2	95.4	95.1	95.2	95.2	96.2	95.1	95.1	95.2	95.3	96.2	95.3	95.5	95.9	94.8
2004	97.5	97.0	97.5	97.8	97.5	97.5	97.6	97.9	97.4	97.0	97.5	97.7	98.6	96.6	97.2	97.7	97.8
2005	100.0	100.0	100.0	100.0	100.0	100.0	100.0	100.0	100.0	100.0	100.0	100.0	100.0	100.0	100.0	100.0	100.0

[1] Excludes farm workers, private household workers, and federal government employees.
[2] Roughly continuous and comparable with new NAICS-based series. See notes and definitions for more information.
. . . = Not available.

Table 9-2. Employment Cost Indexes, SIC Basis—*Continued*

(Not seasonally adjusted, December 2005 = 100; annual values are for December.)

Year	All civilian workers [1,2]	State and local government workers [2]	Private industry workers															
			All private industry workers [2]	Private industry workers excluding sales occupations	By occupational group				By industry division								
					Production and nonsupervisory occupations	White-collar occupations	Blue-collar occupations	Service occupations [2]	Goods-producing industries			Service-providing industries					
									Total [2]	Construction [2]	Manufacturing [2]	Total [2]	Transportation and utilities	Wholesale trade	Retail trade	Finance, insurance, and real estate [2]	Services
TOTAL BENEFITS																	
1979	...	...	25.7	...	...	24.7	27.4	...	26.0	...	25.8	25.4	...	...	...	...	...
1980	...	...	28.7	...	...	27.7	30.4	...	28.8	...	28.5	28.6	...	...	...	...	...
1981	31.8	...	32.2	...	...	31.1	34.0	...	32.4	...	32.1	31.9	...	...	...	...	...
1982	34.2	...	34.5	...	...	33.3	36.5	...	34.8	...	34.4	34.1	...	...	...	...	...
1983	36.8	...	37.1	...	...	35.8	39.1	...	37.2	...	36.9	36.8	...	...	...	...	...
1984	39.3	...	39.5	...	...	38.3	41.4	...	39.6	...	39.3	39.4	...	...	...	...	...
1985	40.8	...	40.9	...	...	40.0	42.5	41.2	40.8	...	40.4	40.9	...	...	...	...	...
1986	42.4	...	42.3	...	...	41.4	43.8	42.9	42.0	...	41.6	42.5	...	...	...	...	...
1987	44.0	...	43.7	...	...	42.9	45.3	43.9	43.2	...	42.7	44.2	...	...	...	...	...
1988	47.0	...	46.7	...	...	45.6	48.6	47.4	46.3	...	46.0	47.1	...	...	...	...	...
1989	50.1	52.2	49.6	...	...	48.6	51.2	50.5	48.8	...	48.7	50.2	...	...	...	...	...
1990	53.5	55.8	52.9	...	...	52.0	54.4	53.8	52.3	...	52.1	53.4	...	...	...	...	...
1991	56.5	58.0	56.2	...	...	55.2	57.8	57.7	55.5	...	55.2	56.7	...	...	...	...	...
1992	59.5	61.1	59.1	...	...	57.8	61.0	61.0	58.7	...	58.3	59.4	...	...	...	...	...
1993	62.2	62.9	62.0	...	...	60.5	64.4	64.4	62.0	...	61.8	62.0	...	...	...	...	...
1994	64.4	64.6	64.3	...	...	63.2	66.2	66.0	64.1	...	63.9	64.4	...	...	...	...	...
1995	65.8	66.3	65.7	...	...	64.8	67.2	66.6	65.2	...	65.0	66.0	...	...	...	...	...
1996	67.1	67.8	67.0	...	...	66.2	68.4	67.3	66.4	...	66.5	67.3	...	...	...	...	...
1997	68.5	68.6	68.5	...	...	68.0	69.4	69.6	67.3	...	67.4	69.2	...	...	...	...	...
1998	70.3	70.7	70.2	...	...	69.9	70.7	70.9	68.1	...	67.9	71.4	...	...	...	...	...
1999	72.6	72.7	72.6	...	...	72.3	73.0	73.4	70.5	...	70.3	73.8	...	...	...	...	...
2000	76.2	74.4	76.7	...	...	76.5	76.9	76.6	74.3	...	73.6	78.1	...	...	...	...	...
2001	80.2	78.5	80.6	...	...	81.1	79.5	81.3	77.3	...	76.3	82.5	...	...	...	...	...
2002	84.2	83.3	84.4	...	...	84.6	83.8	85.7	81.3	...	80.4	86.1	...	...	...	...	...
2003	89.5	88.4	89.8	...	...	89.7	89.8	91.3	87.4	...	86.7	91.2	...	...	...	...	...
2004	95.7	94.3	96.0	...	...	95.3	97.3	97.1	95.7	...	95.3	96.2	...	...	...	...	...
2005	100.0	100.0	100.0	...	...	100.0	100.0	100.0	100.0	...	100.0	100.0	...	...	...	...	...

[1]Excludes farm workers, private household workers, and federal government employees.
[2]Roughly continuous and comparable with new NAICS-based series. See notes and definitions for more information.
... = Not available.

SECTION 9B: PRODUCTIVITY AND RELATED DATA

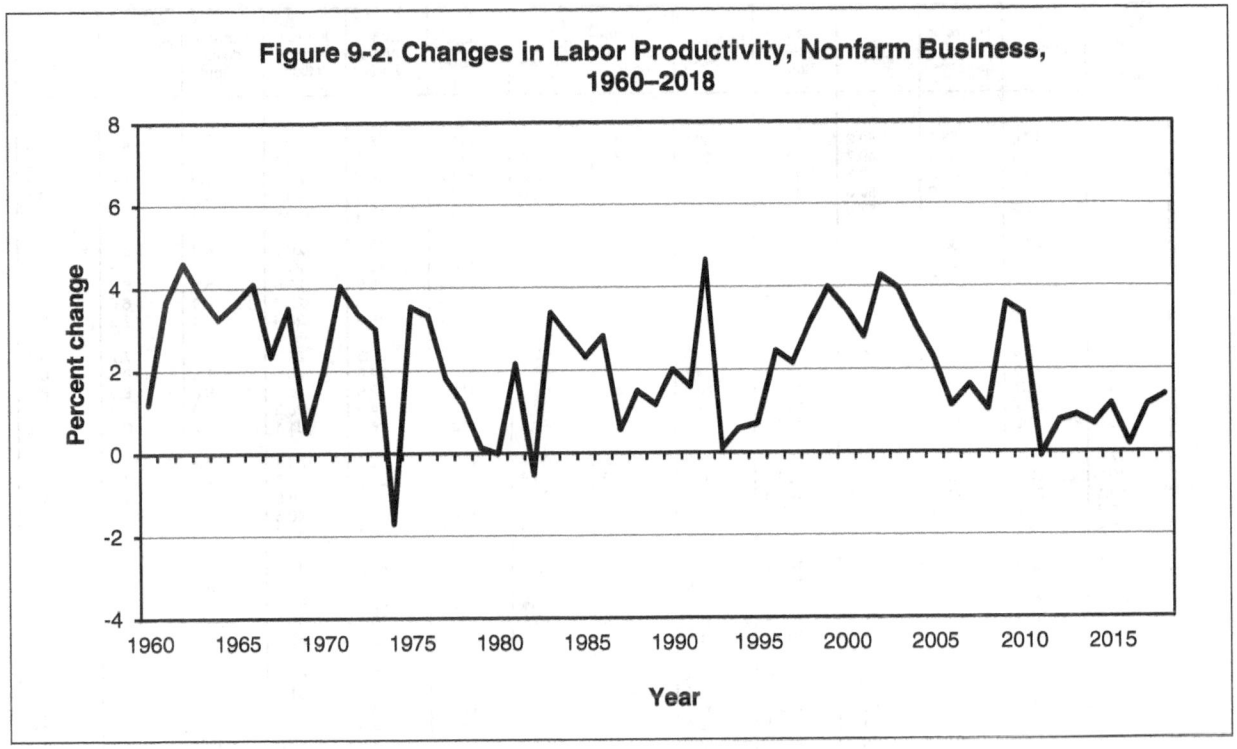

Figure 9-2. Changes in Labor Productivity, Nonfarm Business, 1960–2018

- As Figure 9-2 demonstrates, the rate of change in U.S. nonfarm labor productivity has often been quite variable from year to year. Up through the 1980s, productivity tended to decline (growth rates were less than zero) in recession years but rebound sharply in recovery. One explanation of this was that firms expected that declines in demand would be temporary, and therefore held on to their experienced workers in order to be prepared for the recovery. (Table 9-3A and B)

- In the 1990s and 2000s, productivity growth rates were still variable, but output per hour no longer declined even in a severe recession. This suggests a somewhat greater tendency to regard negative demand trends as lasting. (Table 9-3A)

- Productivity growth slowed between 1973 and 1989 when compared with growth decades earlier. From 1989 to 2018, productivity growth increased or remained stable each year except between 2010 and 2011 when it declined by 0.1 percent. (Table 9-3A and B)

Table 9-3A. Productivity and Related Data: Recent Data

(2012 = 100, seasonally adjusted.)

Year and quarter	Business sector														
	Output per hour of all persons	Output	Hours of all persons	Employ-ment	Average weekly hours	Unit labor costs	Compen-sation per hour	Real hourly compen-sation	Labor share	Unit nonlabor payments	Implicit price deflator	Current dollar output	Compen-sation	Nonlabor payments	Output per job
1960	31.0	17.6	56.6	50.2	112.9	23.0	7.1	50.6	115.7	15.9	19.9	3.5	4.0	2.8	35.0
1961	32.2	17.9	55.8	49.7	112.4	23.1	7.4	52.0	115.1	16.2	20.1	3.6	4.1	2.9	36.1
1962	33.6	19.1	56.8	50.3	113.0	23.0	7.8	53.8	113.8	16.7	20.2	3.9	4.4	3.2	38.0
1963	34.9	20.0	57.2	50.6	113.1	23.0	8.0	55.0	112.8	17.0	20.4	4.1	4.6	3.4	39.5
1964	36.1	21.2	58.9	51.5	114.4	23.1	8.3	56.3	112.0	17.4	20.6	4.4	4.9	3.7	41.2
1965	37.4	22.7	60.9	53.0	114.9	23.1	8.6	57.5	110.4	18.1	20.9	4.8	5.3	4.1	42.9
1966	38.9	24.3	62.4	54.6	114.5	23.7	9.2	59.6	110.5	18.6	21.4	5.2	5.8	4.5	44.5
1967	39.8	24.8	62.3	55.3	112.7	24.5	9.7	61.1	111.1	18.9	22.0	5.5	6.1	4.7	44.8
1968	41.2	26.0	63.2	56.4	112.1	25.5	10.5	63.2	111.4	19.5	22.9	6.0	6.6	5.1	46.2
1969	41.4	26.8	64.8	58.2	111.5	27.1	11.2	64.2	113.4	19.8	23.9	6.4	7.3	5.3	46.2
1970	42.2	26.8	63.5	58.0	109.6	28.6	12.1	65.2	114.6	20.3	24.9	6.7	7.7	5.4	46.3
1971	44.0	27.9	63.4	58.1	109.1	29.1	12.8	66.3	112.0	22.0	26.0	7.2	8.1	6.1	47.9
1972	45.4	29.7	65.3	59.8	109.2	30.0	13.6	68.3	111.4	22.9	26.9	8.0	8.9	6.8	49.6
1973	46.8	31.7	67.8	62.4	108.7	31.4	14.7	69.4	111.0	24.3	28.3	9.0	10.0	7.7	50.8
1974	46.0	31.2	67.9	63.4	107.3	34.9	16.1	68.3	112.4	26.1	31.1	9.7	10.9	8.2	49.3
1975	47.6	31.0	65.0	61.5	105.8	37.3	17.8	69.3	109.6	29.9	34.1	10.5	11.6	9.2	50.4
1976	49.2	33.0	67.1	63.4	106.0	39.0	19.2	70.8	108.9	31.8	35.8	11.8	12.9	10.5	52.2
1977	50.1	34.9	69.7	66.1	105.5	41.4	20.7	71.8	109.0	33.6	38.0	13.3	14.5	11.7	52.8
1978	50.7	37.2	73.3	69.7	105.1	44.3	22.5	72.7	109.2	35.8	40.6	15.1	16.5	13.3	53.3
1979	50.8	38.5	75.8	72.4	104.7	48.6	24.7	72.8	110.4	38.1	44.0	16.9	18.7	14.7	53.1
1980	50.8	38.1	75.1	72.6	103.5	53.8	27.3	72.5	112.3	40.3	47.9	18.3	20.5	15.4	52.5
1981	51.9	39.2	75.7	73.4	103.2	57.6	29.9	72.5	110.2	45.5	52.3	20.5	22.6	17.9	53.5
1982	51.6	38.1	73.9	72.2	102.5	62.2	32.1	73.5	112.6	46.3	55.3	21.1	23.7	17.7	52.8
1983	53.4	40.1	75.2	72.8	103.4	62.8	33.5	73.6	109.7	50.1	57.3	23.0	25.2	20.1	55.1
1984	54.9	43.7	79.6	76.5	104.2	63.8	35.0	73.8	108.3	52.6	58.9	25.7	27.9	23.0	57.1
1985	56.2	45.7	81.5	78.4	104.0	65.5	36.8	75.0	108.5	53.9	60.4	27.6	30.0	24.6	58.4
1986	57.7	47.4	82.1	79.7	103.1	67.4	38.9	77.8	110.0	53.4	61.3	29.0	31.9	25.3	59.5
1987	58.1	49.1	84.6	81.8	103.4	69.5	40.4	78.1	111.4	53.3	62.4	30.6	34.1	26.2	60.0
1988	58.9	51.2	86.9	84.3	103.1	72.1	42.5	79.4	112.0	54.5	64.4	33.0	36.9	27.9	60.7
1989	59.6	53.2	89.2	86.2	103.6	73.4	43.8	78.3	110.0	58.2	66.8	35.5	39.0	30.9	61.7
1990	60.8	54.0	88.8	86.7	102.5	76.5	46.5	79.3	110.9	59.3	69.0	37.3	41.3	32.0	62.3
1991	61.8	53.7	86.9	85.4	101.9	78.8	48.7	80.1	111.0	60.9	71.0	38.1	42.3	32.7	62.9
1992	64.7	56.0	86.5	84.9	102.0	79.9	51.7	83.0	110.8	62.1	72.1	40.4	44.7	34.8	65.9
1993	64.7	57.6	88.9	86.7	102.6	81.0	52.4	82.1	109.8	64.5	73.8	42.5	46.6	37.1	66.4
1994	65.1	60.3	92.7	89.7	103.4	81.1	52.8	81.0	108.0	67.4	75.1	45.3	48.9	40.7	67.3
1995	65.6	62.2	94.8	92.2	103.0	82.5	54.1	81.0	107.9	68.7	76.5	47.5	51.3	42.7	67.4
1996	67.2	65.1	96.9	94.2	102.9	83.4	56.0	81.8	107.4	70.3	77.7	50.5	54.3	45.7	69.0
1997	68.6	68.5	99.8	96.8	103.2	84.9	58.3	83.2	107.8	71.0	78.8	54.0	58.2	48.6	70.8
1998	70.8	72.0	101.8	98.8	103.1	87.2	61.7	87.0	110.0	69.0	79.3	57.1	62.8	49.7	72.9
1999	73.6	76.1	103.4	100.4	103.1	87.9	64.7	89.3	110.2	69.3	79.8	60.7	66.9	52.7	75.8
2000	76.1	79.8	104.8	102.1	102.7	90.9	69.2	92.4	112.3	68.2	81.0	64.6	72.6	54.4	78.1
2001	78.3	80.4	102.7	101.4	101.4	92.5	72.4	93.9	112.4	69.2	82.3	66.1	74.3	55.6	79.3
2002	81.6	81.8	100.2	99.2	101.2	90.7	74.0	94.5	109.4	72.9	82.9	67.8	74.2	59.6	82.5
2003	84.8	84.5	99.6	99.0	100.7	90.5	76.8	95.9	107.9	75.5	83.9	70.9	76.5	63.7	85.3
2004	87.4	88.1	100.8	100.3	100.6	92.0	80.4	97.7	106.9	78.5	86.1	75.8	81.0	69.1	87.8
2005	89.3	91.5	102.5	102.1	100.4	93.2	83.3	98.0	105.1	82.9	88.7	81.2	85.3	75.9	89.6
2006	90.4	94.6	104.7	104.1	100.7	95.7	86.5	98.6	105.1	85.2	91.1	86.2	90.6	80.6	90.9
2007	91.8	96.8	105.4	105.0	100.5	98.4	90.4	100.1	105.6	86.5	93.2	90.2	95.2	83.7	92.2
2008	92.8	95.8	103.3	103.5	99.8	100.0	92.8	99.0	105.7	87.8	94.7	90.7	95.8	84.1	92.6
2009	96.1	92.3	96.0	97.7	98.3	97.4	93.6	100.2	102.6	91.7	94.9	87.5	89.8	84.6	94.5
2010	99.3	95.2	95.9	96.5	99.3	95.9	95.3	100.4	99.9	96.1	96.0	91.4	91.3	91.5	98.6
2011	99.2	97.1	97.8	98.1	99.7	98.1	97.3	99.4	99.9	98.3	98.2	95.3	95.2	95.4	99.0
2012	100.0	100.0	100.0	100.0	100.0	100.0	100.0	100.0	100.0	100.0	100.0	100.0	100.0	100.0	100.0
2013	100.9	102.4	101.5	101.7	99.9	100.6	101.5	100.0	99.1	102.6	101.5	103.9	103.0	105.1	100.8
2014	101.6	105.6	103.9	103.8	100.1	102.5	104.1	100.9	99.4	104.0	103.2	108.9	108.2	109.8	101.7
2015	102.8	109.3	106.3	106.3	100.1	104.2	107.1	103.6	100.4	103.2	103.8	113.4	113.9	112.7	102.8
2016	103.0	111.1	107.9	108.2	99.7	105.2	108.3	103.4	100.5	104.1	104.7	116.3	116.8	115.6	102.7
2017	104.1	114.0	109.5	109.8	99.7	107.5	112.0	104.7	101.0	105.1	106.5	121.4	122.6	119.8	103.8
2018	105.6	117.9	111.7	111.9	99.8	108.9	115.0	105.0	100.1	108.7	108.8	128.3	128.4	128.2	105.3
2016															
1st quarter	102.7	110.3	107.4	107.6	99.9	104.6	107.5	103.6	100.9	102.4	103.7	114.4	115.4	113.0	102.6
2nd quarter	102.8	110.9	107.9	108.1	99.8	104.7	107.7	103.1	100.3	104.0	104.4	115.8	116.2	115.4	102.6
3rd quarter	103.3	111.6	108.1	108.3	99.8	104.9	108.3	103.2	100.1	104.6	104.8	116.9	117.1	116.8	103.1
4th quarter	103.9	112.3	108.1	108.7	99.5	105.6	109.7	103.8	100.4	104.7	105.2	118.2	118.6	117.6	103.3
2017															
1st quarter	104.0	113.1	108.7	109.2	99.5	106.4	110.7	104.0	100.7	104.7	105.6	119.4	120.3	118.4	103.5
2nd quarter	104.2	113.8	109.2	109.5	99.8	106.8	111.2	104.5	100.8	104.8	105.9	120.5	121.5	119.2	103.9
3rd quarter	105.0	114.9	109.4	109.8	99.6	107.2	112.6	105.2	100.7	105.5	106.5	122.3	123.2	121.2	104.6
4th quarter	105.1	116.1	110.5	110.7	99.8	108.3	113.8	105.5	101.1	105.7	107.2	124.4	125.7	122.7	104.9
2018															
1st quarter	105.4	116.9	110.9	111.2	99.7	109.2	115.1	105.9	101.4	105.7	107.7	125.9	127.7	123.6	105.1
2nd quarter	106.0	118.0	111.3	111.4	99.9	108.8	115.4	105.5	100.1	108.5	108.7	128.3	128.4	128.0	106.0
3rd quarter	106.3	119.0	112.0	112.2	99.9	109.3	116.1	105.7	100.2	108.8	109.1	129.8	130.1	129.5	106.1
4th quarter	106.3	119.4	112.4	112.8	99.6	109.4	116.2	105.4	100.0	109.4	109.4	130.6	130.6	130.6	105.8

Table 9-3A. Productivity and Related Data: Recent Data—*Continued*

(2012 = 100, seasonally adjusted.)

Year and quarter	Output per hour of all persons	Output	Hours of all persons	Employ-ment	Average weekly hours	Unit labor costs	Compen-sation per hour	Real hourly compen-sation	Labor share	Unit nonlabor payments	Implicit price deflator	Current dollar output	Compen-sation	Nonlabor payments	Output per job
								Nonfarm business sector							
1960	33.1	17.3	52.5	45.9	114.4	22.6	7.5	53.0	115.9	15.5	19.5	3.4	3.9	2.7	37.8
1961	34.2	17.7	51.9	45.5	114.1	22.7	7.7	54.2	115.0	15.8	19.7	3.5	4.0	2.8	38.9
1962	35.7	18.9	53.0	46.4	114.4	22.5	8.0	55.8	113.4	16.4	19.9	3.8	4.3	3.1	40.8
1963	36.9	19.8	53.6	46.9	114.4	22.5	8.3	57.0	112.5	16.8	20.0	4.0	4.5	3.3	42.2
1964	38.0	21.1	55.6	48.1	115.9	22.6	8.6	58.0	111.4	17.3	20.3	4.3	4.8	3.7	44.0
1965	39.2	22.6	57.8	49.7	116.3	22.6	8.9	59.0	110.1	17.9	20.5	4.7	5.1	4.0	45.5
1966	40.6	24.3	59.8	51.7	115.8	23.1	9.4	60.7	110.0	18.3	21.0	5.1	5.6	4.4	46.9
1967	41.4	24.7	59.8	52.6	113.8	24.0	9.9	62.3	110.8	18.6	21.7	5.4	5.9	4.6	47.0
1968	42.8	26.0	60.8	53.8	113.2	24.9	10.7	64.3	110.8	19.3	22.5	5.9	6.5	5.0	48.4
1969	42.9	26.8	62.6	55.7	112.4	26.6	11.4	65.2	113.1	19.5	23.5	6.3	7.1	5.2	48.2
1970	43.5	26.8	61.6	55.8	110.5	28.0	12.2	65.9	114.2	20.0	24.5	6.6	7.5	5.4	48.1
1971	45.2	27.8	61.5	55.9	110.0	28.6	13.0	67.0	111.9	21.7	25.6	7.1	8.0	6.0	49.7
1972	46.8	29.7	63.4	57.6	110.1	29.5	13.8	69.2	111.7	22.4	26.4	7.8	8.7	6.6	51.5
1973	48.3	31.8	66.0	60.2	109.7	30.8	14.8	70.1	112.6	22.8	27.3	8.7	9.8	7.3	52.9
1974	47.5	31.4	66.1	61.2	108.1	34.2	16.3	69.1	113.6	24.8	30.2	9.5	10.7	7.8	51.3
1975	48.8	30.9	63.3	59.4	106.6	36.8	18.0	70.0	110.4	28.8	33.4	10.3	11.4	8.9	51.9
1976	50.5	33.1	65.5	61.4	106.7	38.4	19.4	71.3	109.0	31.0	35.2	11.6	12.7	10.3	53.8
1977	51.3	35.0	68.1	64.2	106.1	40.8	20.9	72.5	109.2	32.9	37.4	13.1	14.3	11.5	54.4
1978	52.1	37.3	71.6	67.8	105.8	43.7	22.7	73.5	109.7	34.8	39.8	14.8	16.3	13.0	55.0
1979	51.9	38.6	74.2	70.7	105.1	47.9	24.9	73.5	111.1	36.9	43.1	16.6	18.5	14.2	54.5
1980	51.9	38.2	73.6	70.9	103.9	53.1	27.6	73.3	112.4	39.6	47.2	18.1	20.3	15.1	53.9
1981	52.7	39.1	74.2	71.7	103.6	57.4	30.2	73.4	110.8	44.5	51.8	20.2	22.4	17.4	54.5
1982	52.3	37.9	72.5	70.6	102.8	62.1	32.5	74.3	113.0	45.7	55.0	20.8	23.5	17.3	53.7
1983	54.4	40.3	74.0	71.3	103.9	62.3	33.9	74.4	109.5	49.8	56.9	22.9	25.1	20.1	56.5
1984	55.6	43.7	78.5	75.1	104.6	63.6	35.4	74.5	108.7	51.8	58.5	25.5	27.8	22.6	58.1
1985	56.6	45.6	80.5	77.2	104.3	65.5	37.1	75.6	108.8	53.4	60.2	27.5	29.9	24.3	59.0
1986	58.3	47.3	81.1	78.6	103.3	67.3	39.2	78.5	110.2	53.0	61.1	28.9	31.8	25.1	60.2
1987	58.6	49.0	83.6	80.8	103.6	69.4	40.7	78.8	111.6	52.8	62.2	30.5	34.0	25.9	60.7
1988	59.6	51.3	86.0	83.4	103.2	71.8	42.8	80.0	112.0	54.1	64.1	32.9	36.8	27.7	61.5
1989	60.2	53.1	88.3	85.2	103.7	73.2	44.1	78.8	110.2	57.6	66.5	35.3	38.9	30.6	62.3
1990	61.2	53.9	88.2	85.9	102.8	76.4	46.7	79.7	111.1	58.8	68.7	37.1	41.2	31.7	62.8
1991	62.2	53.6	86.2	84.5	102.2	78.7	48.9	80.5	111.0	60.7	70.9	38.0	42.2	32.5	63.5
1992	65.0	55.8	85.9	84.0	102.3	80.0	52.0	83.5	111.0	61.8	72.1	40.2	44.6	34.5	66.4
1993	65.0	57.5	88.4	86.0	102.9	80.9	52.6	82.4	109.7	64.5	73.8	42.4	46.5	37.1	66.8
1994	65.5	60.1	91.8	88.8	103.5	81.1	53.1	81.5	108.0	67.3	75.1	45.2	48.8	40.5	67.7
1995	66.2	62.2	94.0	91.4	102.9	82.2	54.5	81.6	107.6	68.9	76.5	47.6	51.2	42.9	68.1
1996	67.6	65.0	96.2	93.5	102.9	83.3	56.3	82.2	107.5	70.0	77.5	50.4	54.2	45.5	69.5
1997	68.9	68.4	99.2	96.2	103.3	84.9	58.5	83.6	107.7	71.0	78.9	53.9	58.1	48.5	71.1
1998	71.0	72.0	101.3	98.3	103.2	87.2	61.9	87.3	109.8	69.3	79.4	57.1	62.7	49.9	73.2
1999	73.7	76.1	103.2	100.1	103.1	87.9	64.8	89.4	109.8	69.8	80.0	60.9	66.8	53.1	76.0
2000	76.2	79.7	104.6	101.9	102.7	91.0	69.3	92.5	112.0	68.7	81.3	64.8	72.5	54.7	78.2
2001	78.3	80.3	102.6	101.3	101.4	92.5	72.4	93.9	112.0	69.7	82.6	66.3	74.2	56.0	79.3
2002	81.6	81.7	100.0	99.0	101.1	90.7	74.0	94.6	108.9	73.7	83.3	68.0	74.1	60.2	82.5
2003	84.8	84.3	99.5	98.9	100.6	90.6	76.8	95.9	107.6	75.9	84.2	71.0	76.4	64.0	85.2
2004	87.2	87.9	100.8	100.3	100.6	92.1	80.3	97.7	106.8	78.5	86.2	75.8	80.9	69.0	87.6
2005	89.1	91.3	102.5	102.2	100.4	93.4	83.2	97.9	104.9	83.4	89.1	81.3	85.3	76.2	89.4
2006	90.1	94.4	104.8	104.2	100.7	95.9	86.4	98.5	104.8	85.9	91.6	86.5	90.6	81.1	90.7
2007	91.7	96.7	105.5	105.1	100.5	98.4	90.2	99.9	105.3	86.9	93.4	90.4	95.2	84.1	92.0
2008	92.6	95.7	103.3	103.6	99.8	100.1	92.7	98.9	105.5	88.1	94.9	90.8	95.8	84.3	92.4
2009	95.9	92.0	96.0	97.7	98.2	97.5	93.5	100.2	102.3	92.5	95.4	87.8	89.8	85.2	94.2
2010	99.2	95.0	95.8	96.5	99.3	96.1	95.3	100.4	99.8	96.6	96.3	91.5	91.4	91.8	98.5
2011	99.1	96.9	97.8	98.0	99.7	98.2	97.4	99.5	100.1	98.1	98.2	95.2	95.2	95.1	98.9
2012	100.0	100.0	100.0	100.0	100.0	100.0	100.0	100.0	100.0	100.0	100.0	100.0	100.0	100.0	100.0
2013	100.5	102.2	101.7	101.8	99.9	100.8	101.3	99.8	99.3	102.3	101.5	103.7	103.0	104.6	100.4
2014	101.3	105.4	104.0	103.9	100.1	102.8	104.1	100.9	99.5	104.0	103.3	108.9	108.3	109.6	101.4
2015	102.6	109.0	106.2	106.2	100.0	104.6	107.3	103.8	100.4	103.6	104.2	113.6	114.0	113.0	102.6
2016	102.8	110.8	107.8	108.1	99.7	105.5	108.5	103.6	100.3	104.9	105.3	116.6	116.9	116.3	102.5
2017	104.0	113.8	109.4	109.8	99.6	107.9	112.2	104.9	100.8	105.9	107.0	121.8	122.7	120.5	103.6
2018	105.4	117.8	111.8	112.0	99.8	109.2	115.1	105.0	105.0	99.8	109.7	109.4	128.8	128.6	129.1	105.2
2016															
1st quarter	102.6	110.1	107.3	107.5	99.8	105.0	107.7	103.8	100.7	103.2	104.2	114.7	115.6	113.6	102.4
2nd quarter	102.7	110.6	107.7	108.0	99.7	105.2	108.0	103.4	100.1	104.9	105.0	116.2	116.3	116.0	102.4
3rd quarter	103.1	111.3	107.9	108.2	99.7	105.3	108.6	103.4	99.9	105.5	105.4	117.3	117.2	117.5	102.8
4th quarter	103.6	112.0	108.2	108.8	99.4	106.0	109.7	103.9	100.1	105.7	105.9	118.6	118.7	118.5	103.0
2017															
1st quarter	103.8	112.7	108.6	109.2	99.4	106.8	110.9	104.2	100.6	105.4	106.2	119.7	120.4	118.8	103.2
2nd quarter	103.9	113.5	109.2	109.4	99.8	107.2	111.4	104.6	100.7	105.5	106.5	120.8	121.7	119.7	103.7
3rd quarter	104.7	114.7	109.5	109.9	99.6	107.6	112.7	105.2	100.5	106.3	107.0	122.7	123.4	121.9	104.3
4th quarter	104.9	115.8	110.4	110.7	99.7	108.7	114.1	105.7	100.9	106.5	107.8	124.8	125.9	123.4	104.6
2018															
1st quarter	105.2	116.7	110.9	111.3	99.7	109.6	115.2	106.0	101.2	106.6	108.3	126.3	127.9	124.4	104.9
2nd quarter	105.7	117.8	111.5	111.6	99.9	109.1	115.3	105.5	99.9	109.4	109.3	128.7	128.6	128.9	105.6
3rd quarter	106.0	118.8	112.1	112.2	99.9	109.6	116.1	105.7	99.9	109.9	109.7	130.4	130.2	130.6	105.9
4th quarter	106.0	119.2	112.4	112.8	99.6	109.8	116.4	105.5	99.7	110.4	110.0	131.2	130.8	131.6	105.6

Table 9-3B. Productivity and Related Data: Historical Data

(2012 = 100, seasonally adjusted.)

Year and quarter	Business sector								Nonfarm business sector							
	Output per hour of all persons	Output	Hours of all persons	Compensation per hour	Real compensation per hour	Unit labor costs	Unit nonlabor payments	Implicit price deflator	Output per hour of all persons	Output	Hours of all persons	Compensation per hour	Real compensation per hour	Unit labor costs	Unit nonlabor payments	Implicit price deflator
1947	20.4	11.2	54.7	3.6	33.9	17.7	11.8	15.1	20.4	11.2	54.7	3.6	33.9	16.4	11.3	14.2
1948	21.3	11.7	55.1	3.9	34.0	18.4	13.0	16.0	21.3	11.7	55.1	3.9	34.0	17.4	12.2	15.1
1949	21.8	11.6	53.3	4.0	34.9	18.2	12.9	15.9	21.8	11.6	53.3	4.0	34.9	17.3	12.5	15.2
1950	23.6	12.7	54.1	4.2	36.9	18.0	13.5	16.1	23.6	12.7	54.1	4.2	36.9	17.2	13.1	15.4
1951	24.3	13.6	55.8	4.7	37.5	19.2	14.9	17.3	24.3	13.6	55.8	4.7	37.5	18.2	14.1	16.4
1952	25.0	14.0	55.9	4.9	39.1	19.7	14.6	17.5	25.0	14.0	55.9	4.9	39.1	18.9	14.0	16.8
1953	25.9	14.7	56.6	5.3	41.2	20.2	14.3	17.6	25.9	14.7	56.6	5.3	41.2	19.5	14.0	17.1
1954	26.5	14.5	54.7	5.4	42.3	20.4	14.2	17.7	26.5	14.5	54.7	5.4	42.3	19.7	14.0	17.2
1955	27.6	15.7	56.8	5.6	43.5	20.1	15.2	18.0	27.6	15.7	56.8	5.6	43.5	19.6	14.9	17.5
1956	27.7	16.0	57.6	5.9	45.7	21.4	14.9	18.5	27.7	16.0	57.6	5.9	45.7	20.9	14.6	18.2
1957	28.6	16.3	56.8	6.3	47.1	22.1	15.4	19.1	28.6	16.3	56.8	6.3	47.1	21.5	15.1	18.7
1958	29.4	16.0	54.3	6.6	47.8	22.4	15.7	19.5	29.4	16.0	54.3	6.6	47.8	21.9	15.3	19.0
1959	30.5	17.2	56.5	6.9	49.4	22.5	16.1	19.7	30.5	17.2	56.5	6.9	49.4	22.0	15.9	19.3
1947																
1st quarter	20.4	10.8	46.3	3.7	35.9	16.0	11.6	14.6	23.3	10.8	46.3	3.7	35.9	16.0	10.8	13.7
2nd quarter	20.5	11.0	46.3	3.8	36.2	16.0	11.5	14.8	23.8	11.0	46.3	3.8	36.2	16.0	11.2	13.9
3rd quarter	20.3	10.7	46.4	3.9	36.5	17.0	11.8	15.2	23.1	10.7	46.4	3.9	36.5	17.0	11.4	14.5
4th quarter	20.5	11.3	46.9	4.0	36.3	16.7	12.3	15.7	24.0	11.3	46.9	4.0	36.3	16.7	11.7	14.5
1948																
1st quarter	21.0	11.4	47.2	4.1	36.4	17.0	12.8	15.8	24.1	11.4	47.2	4.1	36.4	17.0	11.8	14.7
2nd quarter	21.4	11.3	47.1	4.2	36.3	17.3	13.2	15.9	24.1	11.3	47.1	4.2	36.3	17.3	12.0	15.0
3rd quarter	21.3	11.4	47.5	4.2	36.4	17.6	13.2	16.2	24.1	11.4	47.5	4.2	36.4	17.6	12.3	15.3
4th quarter	21.5	11.4	47.0	4.3	37.1	17.7	12.9	16.2	24.2	11.4	47.0	4.3	37.1	17.7	12.6	15.5
1949																
1st quarter	21.4	11.3	46.2	4.3	37.7	17.6	13.0	16.1	24.5	11.3	46.2	4.3	37.7	17.6	12.5	15.4
2nd quarter	21.5	11.2	45.5	4.3	37.8	17.5	12.8	15.9	24.7	11.2	45.5	4.3	37.8	17.5	12.3	15.2
3rd quarter	22.1	11.4	45.0	4.3	38.2	17.1	13.0	15.8	25.3	11.4	45.0	4.3	38.2	17.1	12.7	15.2
4th quarter	22.1	11.3	44.8	4.3	38.2	17.1	12.7	15.8	25.2	11.3	44.8	4.3	38.2	17.1	12.5	15.1
1950																
1st quarter	23.1	11.7	45.1	4.4	39.3	17.0	12.9	15.7	26.0	11.7	45.1	4.4	39.3	17.0	12.8	15.2
2nd quarter	23.4	12.2	46.3	4.5	39.9	17.2	13.0	15.8	26.3	12.2	46.3	4.5	39.9	17.2	12.8	15.2
3rd quarter	23.8	12.8	47.7	4.6	39.9	17.1	13.8	16.2	26.9	12.8	47.7	4.6	39.9	17.1	13.2	15.4
4th quarter	23.9	13.0	48.2	4.7	40.1	17.5	14.3	16.5	27.0	13.0	48.2	4.7	40.1	17.5	13.4	15.7
1951																
1st quarter	23.9	13.2	49.0	4.8	39.4	17.9	14.9	17.1	27.0	13.2	49.0	4.8	39.4	17.9	13.9	16.2
2nd quarter	24.0	13.3	49.4	4.9	39.9	18.4	14.7	17.3	26.9	13.3	49.4	4.9	39.9	18.4	13.8	16.4
3rd quarter	24.7	13.4	48.9	5.0	40.4	18.2	15.0	17.3	27.5	13.4	48.9	5.0	40.4	18.2	14.3	16.5
4th quarter	24.6	13.5	48.9	5.1	40.6	18.5	15.0	17.5	27.6	13.5	48.9	5.1	40.6	18.5	14.2	16.6
1952																
1st quarter	24.7	13.6	49.2	5.1	40.8	18.6	14.8	17.4	27.7	13.6	49.2	5.1	40.8	18.6	14.1	16.6
2nd quarter	24.9	13.6	49.0	5.2	41.0	18.7	14.5	17.4	27.7	13.6	49.0	5.2	41.0	18.7	13.9	16.6
3rd quarter	25.0	13.6	49.4	5.3	41.3	19.1	14.6	17.6	27.5	13.6	49.4	5.3	41.3	19.1	13.9	16.8
4th quarter	25.4	14.3	50.7	5.4	42.2	19.1	14.5	17.6	28.1	14.3	50.7	5.4	42.2	19.1	14.1	16.9
1953																
1st quarter	25.8	14.5	51.2	5.4	42.9	19.2	14.4	17.6	28.4	14.5	51.2	5.4	42.9	19.2	14.0	17.0
2nd quarter	26.0	14.6	51.3	5.5	43.2	19.4	14.4	17.6	28.4	14.6	51.3	5.5	43.2	19.4	14.1	17.1
3rd quarter	26.0	14.5	50.8	5.6	43.5	19.5	14.2	17.7	28.5	14.5	50.8	5.6	43.5	19.5	14.1	17.1
4th quarter	25.9	14.2	50.0	5.6	43.8	19.8	14.2	17.7	28.4	14.2	50.0	5.6	43.8	19.8	13.8	17.1
1954																
1st quarter	25.9	14.1	49.3	5.7	44.1	19.9	14.1	17.7	28.5	14.1	49.3	5.7	44.1	19.9	13.7	17.2
2nd quarter	26.3	14.1	49.0	5.7	44.2	19.8	14.0	17.7	28.7	14.1	49.0	5.7	44.2	19.8	13.9	17.2
3rd quarter	26.7	14.3	48.8	5.7	44.7	19.6	14.4	17.7	29.2	14.3	48.8	5.7	44.7	19.6	14.1	17.2
4th quarter	27.1	14.6	49.4	5.8	45.3	19.5	14.4	17.7	29.6	14.6	49.4	5.8	45.3	19.5	14.3	17.3
1955																
1st quarter	27.5	15.1	50.2	5.8	45.5	19.3	15.2	17.8	30.1	15.1	50.2	5.8	45.5	19.3	14.8	17.4
2nd quarter	27.8	15.4	50.8	5.9	45.9	19.4	15.1	17.8	30.2	15.4	50.8	5.9	45.9	19.4	14.8	17.4
3rd quarter	27.7	15.6	51.4	6.0	46.7	19.7	15.2	18.0	30.4	15.6	51.4	6.0	46.7	19.7	14.9	17.6
4th quarter	27.5	15.7	52.0	6.0	47.0	19.9	15.3	18.2	30.2	15.7	52.0	6.0	47.0	19.9	15.0	17.8
1956																
1st quarter	27.4	15.6	52.3	6.1	47.7	20.5	14.9	18.3	29.9	15.6	52.3	6.1	47.7	20.5	14.6	17.9
2nd quarter	27.6	15.7	52.4	6.2	48.3	20.8	14.7	18.4	30.0	15.7	52.4	6.2	48.3	20.8	14.5	18.0
3rd quarter	27.6	15.7	52.2	6.3	48.6	21.1	15.0	18.7	30.0	15.7	52.2	6.3	48.6	21.1	14.6	18.3
4th quarter	28.2	15.9	52.5	6.4	49.0	21.2	15.1	18.7	30.3	15.9	52.5	6.4	49.0	21.2	14.7	18.4
1957																
1st quarter	28.4	16.1	52.5	6.5	49.3	21.3	15.3	19.0	30.7	16.1	52.5	6.5	49.3	21.3	15.1	18.6
2nd quarter	28.4	16.0	52.4	6.6	49.3	21.6	15.3	19.1	30.6	16.0	52.4	6.6	49.3	21.6	15.0	18.7
3rd quarter	28.7	16.2	52.2	6.7	49.6	21.6	15.6	19.2	31.0	16.2	52.2	6.7	49.6	21.6	15.2	18.8
4th quarter	29.0	15.9	51.1	6.8	50.0	21.8	15.3	19.2	31.1	15.9	51.1	6.8	50.0	21.8	15.0	18.8
1958																
1st quarter	28.8	15.3	49.9	6.8	49.4	22.1	15.3	19.4	30.7	15.3	49.9	6.8	49.4	22.1	14.8	18.9
2nd quarter	29.2	15.4	49.2	6.8	49.5	21.9	15.7	19.4	31.3	15.4	49.2	6.8	49.5	21.9	15.2	18.9
3rd quarter	29.7	15.9	49.8	7.0	50.7	22.0	15.7	19.5	31.9	15.9	49.8	7.0	50.7	22.0	15.3	19.0
4th quarter	30.0	16.4	50.5	7.0	50.9	21.7	16.2	19.6	32.4	16.4	50.5	7.0	50.9	21.7	15.8	19.1
1959																
1st quarter	30.3	16.7	51.5	7.1	51.3	21.9	16.0	19.6	32.4	16.7	51.5	7.1	51.3	21.9	15.8	19.2
2nd quarter	30.4	17.2	52.5	7.1	51.6	21.8	16.2	19.6	32.7	17.2	52.5	7.1	51.6	21.8	15.9	19.3
3rd quarter	30.6	17.2	52.3	7.2	51.6	21.9	16.2	19.7	32.8	17.2	52.3	7.2	51.6	21.9	16.0	19.3
4th quarter	30.6	17.1	52.4	7.3	51.8	22.2	16.0	19.8	32.7	17.1	52.4	7.3	51.8	22.2	15.8	19.4

Table 9-3B. Productivity and Related Data: Historical Data—*Continued*

(2012 = 100, seasonally adjusted.)

Year and quarter	Business sector								Nonfarm business sector							
	Output per hour of all persons	Output	Hours of all persons	Compensation per hour	Real compensation per hour	Unit labor costs	Unit nonlabor payments	Implicit price deflator	Output per hour of all persons	Output	Hours of all persons	Compensation per hour	Real compensation per hour	Unit labor costs	Unit nonlabor payments	Implicit price deflator
1960																
1st quarter	31.5	17.6	52.7	7.4	52.9	22.2	16.2	19.8	33.5	17.6	52.7	7.4	52.9	22.2	15.9	19.5
2nd quarter	30.8	17.4	52.8	7.5	52.9	22.7	15.8	19.9	32.9	17.4	52.8	7.5	52.9	22.7	15.4	19.5
3rd quarter	31.0	17.4	52.5	7.5	53.2	22.7	16.0	19.9	33.1	17.4	52.5	7.5	53.2	22.7	15.5	19.6
4th quarter	30.7	17.0	52.0	7.5	53.1	23.1	15.6	20.0	32.7	17.0	52.0	7.5	53.1	23.1	15.2	19.6
1961																
1st quarter	31.1	17.1	51.7	7.6	53.5	23.0	15.7	20.0	33.1	17.1	51.7	7.6	53.5	23.0	15.3	19.7
2nd quarter	32.1	17.5	51.6	7.7	54.2	22.7	16.1	20.0	34.0	17.5	51.6	7.7	54.2	22.7	15.8	19.7
3rd quarter	32.5	17.9	51.9	7.8	54.4	22.5	16.4	20.1	34.6	17.9	51.9	7.8	54.4	22.5	16.1	19.7
4th quarter	32.8	18.3	52.5	7.8	54.7	22.5	16.5	20.1	34.9	18.3	52.5	7.8	54.7	22.5	16.1	19.7
1962																
1st quarter	33.2	18.7	52.7	8.0	55.4	22.4	16.7	20.2	35.4	18.7	52.7	8.0	55.4	22.4	16.4	19.8
2nd quarter	33.3	18.8	53.4	8.0	55.5	22.7	16.5	20.2	35.3	18.8	53.4	8.0	55.5	22.7	16.2	19.9
3rd quarter	33.9	19.1	53.2	8.1	55.8	22.5	16.8	20.3	35.9	19.1	53.2	8.1	55.8	22.5	16.5	19.9
4th quarter	34.2	19.1	52.9	8.1	56.1	22.5	16.7	20.3	36.1	19.1	52.9	8.1	56.1	22.5	16.5	19.9
1963																
1st quarter	34.3	19.3	53.2	8.2	56.5	22.7	16.8	20.3	36.3	19.3	53.2	8.2	56.5	22.7	16.5	20.0
2nd quarter	34.5	19.6	53.6	8.3	56.7	22.6	16.9	20.3	36.5	19.6	53.6	8.3	56.7	22.6	16.6	20.0
3rd quarter	35.4	20.1	53.7	8.3	56.9	22.3	17.2	20.4	37.4	20.1	53.7	8.3	56.9	22.3	17.0	20.0
4th quarter	35.4	20.2	54.1	8.4	57.4	22.6	17.3	20.5	37.4	20.2	54.1	8.4	57.4	22.6	16.9	20.1
1964																
1st quarter	35.8	20.8	55.2	8.4	57.1	22.4	17.5	20.5	37.7	20.8	55.2	8.4	57.1	22.4	17.3	20.2
2nd quarter	35.9	21.0	55.4	8.5	57.7	22.5	17.4	20.6	37.9	21.0	55.4	8.5	57.7	22.5	17.3	20.2
3rd quarter	36.3	21.4	55.8	8.6	58.3	22.5	17.5	20.6	38.3	21.4	55.8	8.6	58.3	22.5	17.4	20.3
4th quarter	36.1	21.4	56.3	8.7	58.4	22.9	17.3	20.7	37.9	21.4	56.3	8.7	58.4	22.9	17.1	20.4
1965																
1st quarter	36.7	22.0	57.2	8.7	58.5	22.7	17.8	20.8	38.5	22.0	57.2	8.7	58.5	22.7	17.6	20.5
2nd quarter	36.7	22.3	57.8	8.8	58.5	22.7	17.8	20.9	38.6	22.3	57.8	8.8	58.5	22.7	17.6	20.5
3rd quarter	37.7	22.8	57.9	8.9	58.9	22.5	18.3	20.9	39.4	22.8	57.9	8.9	58.9	22.5	18.0	20.6
4th quarter	38.2	23.4	58.4	9.0	59.5	22.5	18.6	21.1	40.1	23.4	58.4	9.0	59.5	22.5	18.3	20.7
1966																
1st quarter	38.9	24.1	59.3	9.2	60.0	22.6	18.7	21.2	40.7	24.1	59.3	9.2	60.0	22.6	18.3	20.7
2nd quarter	38.7	24.2	59.9	9.3	60.4	23.1	18.4	21.3	40.4	24.2	59.9	9.3	60.4	23.1	18.1	20.9
3rd quarter	38.8	24.4	60.2	9.4	60.7	23.3	18.5	21.5	40.5	24.4	60.2	9.4	60.7	23.3	18.2	21.1
4th quarter	39.1	24.5	60.0	9.6	61.0	23.5	18.7	21.7	40.8	24.5	60.0	9.6	61.0	23.5	18.5	21.3
1967																
1st quarter	39.5	24.6	59.9	9.7	61.7	23.6	18.8	21.8	41.1	24.6	59.9	9.7	61.7	23.6	18.6	21.4
2nd quarter	39.8	24.6	59.6	9.9	62.3	23.9	18.7	21.9	41.2	24.6	59.6	9.9	62.3	23.9	18.5	21.6
3rd quarter	39.9	24.8	59.8	10.0	62.6	24.1	18.9	22.1	41.5	24.8	59.8	10.0	62.6	24.1	18.6	21.7
4th quarter	40.0	25.0	60.0	10.1	62.7	24.4	19.1	22.3	41.6	25.0	60.0	10.1	62.7	24.4	18.8	21.9
1968																
1st quarter	40.9	25.6	60.1	10.4	63.8	24.5	19.4	22.5	42.5	25.6	60.1	10.4	63.8	24.5	19.2	22.2
2nd quarter	41.3	26.0	60.7	10.6	64.2	24.6	19.7	22.8	42.9	26.0	60.7	10.6	64.2	24.6	19.5	22.4
3rd quarter	41.3	26.2	61.1	10.7	64.4	25.0	19.5	22.9	42.9	26.2	61.1	10.7	64.4	25.0	19.3	22.6
4th quarter	41.3	26.3	61.5	11.0	64.8	25.6	19.5	23.3	42.8	26.3	61.5	11.0	64.8	25.6	19.3	22.9
1969																
1st quarter	41.4	26.8	62.1	11.1	64.9	25.7	20.0	23.5	43.2	26.8	62.1	11.1	64.9	25.7	19.8	23.1
2nd quarter	41.4	26.8	62.7	11.3	65.0	26.4	19.8	23.8	42.8	26.8	62.7	11.3	65.0	26.4	19.5	23.4
3rd quarter	41.5	27.0	63.0	11.5	65.2	26.8	19.8	24.0	42.9	27.0	63.0	11.5	65.2	26.8	19.5	23.6
4th quarter	41.4	26.8	62.8	11.7	65.5	27.5	19.6	24.3	42.7	26.8	62.8	11.7	65.5	27.5	19.2	23.9
1970																
1st quarter	41.6	26.8	62.5	11.9	65.6	27.9	19.6	24.6	42.8	26.8	62.5	11.9	65.6	27.9	19.3	24.1
2nd quarter	42.0	26.8	61.7	12.1	65.7	27.9	20.3	24.9	43.4	26.8	61.7	12.1	65.7	27.9	20.0	24.5
3rd quarter	42.8	27.1	61.4	12.3	66.1	27.9	20.6	25.0	44.1	27.1	61.4	12.3	66.1	27.9	20.3	24.6
4th quarter	42.5	26.6	60.9	12.4	65.8	28.5	20.6	25.3	43.7	26.6	60.9	12.4	65.8	28.5	20.4	25.0
1971																
1st quarter	43.7	27.5	61.2	12.7	66.6	28.2	21.6	25.6	45.0	27.5	61.2	12.7	66.6	28.2	21.3	25.2
2nd quarter	43.8	27.7	61.4	12.9	67.0	28.5	21.9	25.9	45.1	27.7	61.4	12.9	67.0	28.5	21.6	25.5
3rd quarter	44.3	28.0	61.4	13.1	67.3	28.7	22.3	26.1	45.6	28.0	61.4	13.1	67.3	28.7	22.0	25.8
4th quarter	43.9	28.0	62.1	13.1	67.2	29.1	22.0	26.3	45.2	28.0	62.1	13.1	67.2	29.1	21.6	25.9
1972																
1st quarter	44.4	28.8	62.8	13.5	68.5	29.5	22.2	26.6	45.9	28.8	62.8	13.5	68.5	29.5	21.9	26.2
2nd quarter	45.5	29.6	63.2	13.7	68.9	29.2	22.9	26.7	46.8	29.6	63.2	13.7	68.9	29.2	22.5	26.3
3rd quarter	45.6	29.9	63.6	13.9	69.2	29.4	23.2	27.0	47.0	29.9	63.6	13.9	69.2	29.4	22.5	26.4
4th quarter	46.1	30.5	64.2	14.1	69.8	29.7	23.5	27.3	47.4	30.5	64.2	14.1	69.8	29.7	22.5	26.6
1973																
1st quarter	47.0	31.6	65.2	14.5	70.6	29.8	23.7	27.5	48.5	31.6	65.2	14.5	70.6	29.8	22.7	26.8
2nd quarter	47.1	32.0	65.9	14.7	70.1	30.3	24.2	27.9	48.5	32.0	65.9	14.7	70.1	30.3	22.9	27.1
3rd quarter	46.4	31.9	66.3	14.9	69.9	31.1	24.3	28.5	48.1	31.9	66.3	14.9	69.9	31.1	22.6	27.4
4th quarter	46.6	31.9	66.7	15.2	69.6	31.9	25.0	29.1	47.8	31.9	66.7	15.2	69.6	31.9	23.0	28.0
1974																
1st quarter	46.0	31.6	66.4	15.6	69.1	32.7	25.1	29.8	47.7	31.6	66.4	15.6	69.1	32.7	23.5	28.7
2nd quarter	46.1	31.6	66.6	16.0	69.0	33.7	25.7	30.5	47.5	31.6	66.6	16.0	69.0	33.7	24.6	29.7
3rd quarter	45.7	31.2	66.3	16.5	69.2	35.0	26.1	31.5	47.1	31.2	66.3	16.5	69.2	35.0	25.0	30.7
4th quarter	46.1	31.0	65.2	16.9	68.9	35.6	27.6	32.5	47.5	31.0	65.2	16.9	68.9	35.6	26.3	31.6

Table 9-3B. Productivity and Related Data: Historical Data—*Continued*

(2012 = 100, seasonally adjusted.)

Year and quarter	Business sector								Nonfarm business sector							
	Output per hour of all persons	Output	Hours of all persons	Compensation per hour	Real compensation per hour	Unit labor costs	Unit nonlabor payments	Implicit price deflator	Output per hour of all persons	Output	Hours of all persons	Compensation per hour	Real compensation per hour	Unit labor costs	Unit nonlabor payments	Implicit price deflator
1975																
1st quarter	46.8	30.2	63.2	17.4	69.6	36.4	28.5	33.3	47.9	30.2	63.2	17.4	69.6	36.4	27.6	32.6
2nd quarter	47.5	30.4	62.6	17.8	70.2	36.6	29.3	33.7	48.6	30.4	62.6	17.8	70.2	36.6	28.5	33.1
3rd quarter	48.0	31.1	63.2	18.1	70.0	36.8	30.6	34.3	49.2	31.1	63.2	18.1	70.0	36.8	29.4	33.6
4th quarter	48.2	31.7	64.2	18.4	69.9	37.4	31.0	34.9	49.3	31.7	64.2	18.4	69.9	37.4	29.8	34.1
1976																
1st quarter	48.8	32.6	65.3	18.8	70.4	37.6	31.3	35.2	50.0	32.6	65.3	18.8	70.4	37.6	30.5	34.5
2nd quarter	49.2	33.0	65.3	19.1	71.1	37.9	31.6	35.5	50.5	33.0	65.3	19.1	71.1	37.9	30.9	34.9
3rd quarter	49.2	33.2	65.6	19.5	71.5	38.6	31.8	36.0	50.6	33.2	65.6	19.5	71.5	38.6	31.1	35.3
4th quarter	49.5	33.5	65.9	19.9	71.9	39.3	32.3	36.6	50.7	33.5	65.9	19.9	71.9	39.3	31.7	36.0
1977																
1st quarter	49.8	34.0	66.6	20.3	72.0	39.8	32.9	37.2	51.1	34.0	66.6	20.3	72.0	39.8	32.2	36.5
2nd quarter	49.9	34.8	67.9	20.7	72.2	40.5	33.3	37.7	51.3	34.8	67.9	20.7	72.2	40.5	32.7	37.1
3rd quarter	50.6	35.6	68.6	21.1	72.6	40.8	33.8	38.1	51.8	35.6	68.6	21.1	72.6	40.8	33.4	37.6
4th quarter	50.0	35.4	69.3	21.5	72.8	42.1	34.1	38.9	51.1	35.4	69.3	21.5	72.8	42.1	33.3	38.2
1978																
1st quarter	49.8	35.6	69.6	22.1	73.6	43.2	33.8	39.4	51.2	35.6	69.6	22.1	73.6	43.2	32.9	38.7
2nd quarter	50.8	37.4	71.7	22.5	73.2	43.0	35.7	40.2	52.2	37.4	71.7	22.5	73.2	43.0	34.6	39.4
3rd quarter	51.0	37.7	72.3	22.9	73.3	43.8	36.3	40.9	52.2	37.7	72.3	22.9	73.3	43.8	35.3	40.1
4th quarter	51.1	38.4	73.1	23.5	73.7	44.6	37.1	41.7	52.5	38.4	73.1	23.5	73.7	44.6	36.1	40.9
1979																
1st quarter	50.9	38.4	73.7	24.1	73.9	46.2	36.9	42.5	52.1	38.4	73.7	24.1	73.9	46.2	35.6	41.6
2nd quarter	50.8	38.4	73.9	24.6	73.6	47.4	38.0	43.6	52.0	38.4	73.9	24.6	73.6	47.4	36.8	42.8
3rd quarter	50.8	38.7	74.6	25.2	73.3	48.5	38.6	44.5	51.9	38.7	74.6	25.2	73.3	48.5	37.4	43.7
4th quarter	50.7	38.8	74.7	25.8	73.3	49.7	38.8	45.3	51.8	38.8	74.7	25.8	73.3	49.7	37.7	44.5
1980																
1st quarter	51.0	38.8	74.6	26.5	73.0	50.9	39.5	46.3	52.1	38.8	74.6	26.5	73.0	50.9	38.8	45.6
2nd quarter	50.5	37.7	73.2	27.2	73.2	52.8	39.4	47.3	51.6	37.7	73.2	27.2	73.2	52.8	39.3	46.9
3rd quarter	50.6	37.7	72.8	27.9	73.3	54.0	40.3	48.4	51.8	37.7	72.8	27.9	73.3	54.0	39.5	47.7
4th quarter	51.1	38.6	73.8	28.7	73.5	54.8	42.0	49.6	52.3	38.6	73.8	28.7	73.5	54.8	40.8	48.7
1981																
1st quarter	52.0	39.4	74.3	29.4	73.4	55.4	44.6	50.9	53.1	39.4	74.3	29.4	73.4	55.4	43.6	50.2
2nd quarter	51.6	39.0	74.3	29.9	73.2	57.1	44.8	51.9	52.4	39.0	74.3	29.9	73.2	57.1	43.7	51.3
3rd quarter	52.3	39.3	74.3	30.6	73.4	57.8	46.4	52.8	52.9	39.3	74.3	30.6	73.4	57.8	45.2	52.3
4th quarter	51.7	38.7	73.9	31.1	73.3	59.4	46.1	53.6	52.4	38.7	73.9	31.1	73.3	59.4	45.4	53.3
1982																
1st quarter	51.4	37.9	72.7	31.9	74.4	61.2	45.2	54.3	52.1	37.9	72.7	31.9	74.4	61.2	44.5	53.9
2nd quarter	51.5	38.1	73.0	32.2	74.2	61.7	46.1	54.9	52.1	38.1	73.0	32.2	74.2	61.7	45.5	54.6
3rd quarter	51.5	37.9	72.5	32.7	74.1	62.5	46.7	55.7	52.2	37.9	72.5	32.7	74.1	62.5	45.9	55.3
4th quarter	52.0	37.8	71.8	33.1	74.3	62.9	47.4	56.2	52.7	37.8	71.8	33.1	74.3	62.9	46.9	55.9
1983																
1st quarter	52.4	38.5	72.2	33.5	74.7	62.7	48.4	56.6	53.4	38.5	72.2	33.5	74.7	62.7	47.6	56.2
2nd quarter	53.3	39.8	73.1	33.8	74.5	62.0	49.7	57.0	54.5	39.8	73.1	33.8	74.5	62.0	49.2	56.4
3rd quarter	53.6	40.8	74.6	34.0	74.2	62.1	50.9	57.5	54.8	40.8	74.6	34.0	74.2	62.1	51.0	57.3
4th quarter	54.0	41.8	76.0	34.4	74.3	62.4	51.3	57.9	55.0	41.8	76.0	34.4	74.3	62.4	51.3	57.6
1984																
1st quarter	54.3	42.8	77.4	34.8	74.2	62.9	51.9	58.3	55.3	42.8	77.4	34.8	74.2	62.9	51.0	57.7
2nd quarter	54.8	43.6	78.4	35.1	74.2	63.2	52.6	58.7	55.6	43.6	78.4	35.1	74.2	63.2	51.8	58.3
3rd quarter	55.1	44.0	78.8	35.6	74.7	63.8	52.8	59.1	55.8	44.0	78.8	35.6	74.7	63.8	52.1	58.7
4th quarter	55.2	44.3	79.3	35.9	74.7	64.3	53.0	59.4	55.9	44.3	79.3	35.9	74.7	64.3	52.2	59.0
1985																
1st quarter	55.5	44.8	79.9	36.4	75.0	64.9	53.7	60.0	56.1	44.8	79.9	36.4	75.0	64.9	53.0	59.7
2nd quarter	55.7	45.2	80.4	36.7	75.1	65.3	53.8	60.3	56.2	45.2	80.4	36.7	75.1	65.3	53.3	60.1
3rd quarter	56.6	45.9	80.7	37.3	75.7	65.4	54.4	60.6	57.0	45.9	80.7	37.3	75.7	65.4	54.1	60.5
4th quarter	56.8	46.4	81.1	38.0	76.4	66.4	53.6	60.8	57.2	46.4	81.1	38.0	76.4	66.4	53.1	60.6
1986																
1st quarter	57.3	46.8	80.9	38.5	77.1	66.6	53.9	61.0	57.9	46.8	80.9	38.5	77.1	66.6	53.6	61.0
2nd quarter	57.7	47.1	80.7	39.0	78.4	66.8	53.7	61.2	58.3	47.1	80.7	39.0	78.4	66.8	53.4	61.0
3rd quarter	58.0	47.5	81.1	39.5	78.9	67.4	53.5	61.3	58.6	47.5	81.1	39.5	78.9	67.4	52.9	61.1
4th quarter	57.9	47.8	81.7	40.0	79.5	68.4	52.6	61.5	58.5	47.8	81.7	40.0	79.5	68.4	52.1	61.3
1987																
1st quarter	57.6	48.2	82.7	40.2	78.9	69.0	52.6	61.8	58.2	48.2	82.7	40.2	78.9	69.0	52.1	61.7
2nd quarter	58.0	48.8	83.2	40.5	78.7	69.1	53.3	62.2	58.6	48.8	83.2	40.5	78.7	69.1	52.8	62.0
3rd quarter	58.1	49.1	83.8	40.9	78.7	69.8	53.5	62.7	58.6	49.1	83.8	40.9	78.7	69.8	53.1	62.5
4th quarter	58.6	50.1	84.7	41.4	79.0	69.9	53.8	62.9	59.2	50.1	84.7	41.4	79.0	69.9	53.3	62.7
1988																
1st quarter	58.7	50.3	84.7	42.1	79.8	70.9	53.7	63.5	59.4	50.3	84.7	42.1	79.8	70.9	53.1	63.1
2nd quarter	58.8	51.2	85.9	42.6	80.0	71.5	53.9	64.0	59.5	51.2	85.9	42.6	80.0	71.5	53.5	63.7
3rd quarter	59.0	51.4	86.2	43.1	80.0	72.2	54.8	64.8	59.7	51.4	86.2	43.1	80.0	72.2	54.3	64.5
4th quarter	59.1	52.1	87.1	43.4	79.9	72.6	55.5	65.3	59.8	52.1	87.1	43.4	79.9	72.6	55.6	65.2
1989																
1st quarter	59.3	52.7	88.0	43.6	79.3	72.8	56.9	65.9	59.9	52.7	88.0	43.6	79.3	72.8	56.1	65.5
2nd quarter	59.6	53.0	88.3	43.7	78.5	72.9	58.4	66.6	60.1	53.0	88.3	43.7	78.5	72.9	57.9	66.3
3rd quarter	59.8	53.4	88.5	44.1	78.6	73.2	58.9	67.0	60.3	53.4	88.5	44.1	78.6	73.2	58.6	66.8
4th quarter	59.9	53.4	88.5	44.8	79.0	74.2	58.5	67.4	60.4	53.4	88.5	44.8	79.0	74.2	58.0	67.2

Table 9-3B. Productivity and Related Data: Historical Data—*Continued*

(2012 = 100, seasonally adjusted.)

Year and quarter	Business sector								Nonfarm business sector							
	Output per hour of all persons	Output	Hours of all persons	Compensation per hour	Real compensation per hour	Unit labor costs	Unit nonlabor payments	Implicit price deflator	Output per hour of all persons	Output	Hours of all persons	Compensation per hour	Real compensation per hour	Unit labor costs	Unit nonlabor payments	Implicit price deflator
1990																
1st quarter	60.5	54.1	88.8	45.6	79.3	74.9	59.2	68.1	60.9	54.1	88.8	45.6	79.3	74.9	58.5	67.8
2nd quarter	60.9	54.2	88.4	46.5	80.1	75.9	59.3	68.8	61.3	54.2	88.4	46.5	80.1	75.9	58.8	68.5
3rd quarter	61.3	54.1	87.9	47.2	79.9	76.6	59.7	69.3	61.5	54.1	87.9	47.2	79.9	76.6	59.2	69.0
4th quarter	60.6	53.3	87.5	47.5	79.3	78.0	59.1	69.8	60.9	53.3	87.5	47.5	79.3	78.0	58.8	69.7
1991																
1st quarter	60.7	53.0	86.7	47.8	79.3	78.2	60.2	70.4	61.1	53.0	86.7	47.8	79.3	78.2	60.0	70.3
2nd quarter	61.7	53.5	86.2	48.7	80.5	78.5	60.8	70.8	62.1	53.5	86.2	48.7	80.5	78.5	60.6	70.7
3rd quarter	62.2	53.9	86.1	49.3	80.9	78.7	61.4	71.2	62.6	53.9	86.1	49.3	80.9	78.7	61.3	71.2
4th quarter	62.5	54.0	85.9	50.0	81.4	79.4	61.3	71.5	62.9	54.0	85.9	50.0	81.4	79.4	61.0	71.4
1992																
1st quarter	63.9	54.8	85.4	51.2	83.1	79.9	61.1	71.6	64.2	54.8	85.4	51.2	83.1	79.9	60.7	71.5
2nd quarter	64.4	55.5	85.7	51.7	83.3	79.9	61.8	71.9	64.7	55.5	85.7	51.7	83.3	79.9	61.3	71.8
3rd quarter	65.0	56.1	85.9	52.3	83.7	80.1	62.3	72.2	65.3	56.1	85.9	52.3	83.7	80.1	61.8	72.2
4th quarter	65.4	56.8	86.4	52.6	83.6	80.1	63.4	72.7	65.7	56.8	86.4	52.6	83.6	80.1	63.2	72.7
1993																
1st quarter	63.9	56.8	87.2	52.2	82.5	80.2	64.1	73.2	65.1	56.8	87.2	52.2	82.5	80.2	64.2	73.2
2nd quarter	64.4	57.1	88.2	52.6	82.5	81.3	63.8	73.6	64.7	57.1	88.2	52.6	82.5	81.3	63.8	73.7
3rd quarter	65.0	57.7	88.7	52.6	82.3	80.9	64.7	74.0	65.0	57.7	88.7	52.6	82.3	80.9	64.7	73.9
4th quarter	65.4	58.4	89.5	53.0	82.3	81.2	65.5	74.4	65.3	58.4	89.5	53.0	82.3	81.2	65.2	74.3
1994																
1st quarter	65.2	59.0	90.1	52.7	81.6	80.5	67.0	74.6	65.5	59.0	90.1	52.7	81.6	80.5	66.8	74.5
2nd quarter	65.1	60.0	91.6	53.1	81.8	81.1	66.9	74.9	65.5	60.0	91.6	53.1	81.8	81.1	66.8	74.9
3rd quarter	64.8	60.4	92.6	53.1	81.2	81.5	67.3	75.3	65.1	60.4	92.6	53.1	81.2	81.5	67.3	75.3
4th quarter	65.4	61.2	93.0	53.5	81.4	81.3	68.3	75.6	65.8	61.2	93.0	53.5	81.4	81.3	68.5	75.7
1995																
1st quarter	65.3	61.6	93.5	53.9	81.4	81.7	68.2	76.0	65.9	61.6	93.5	53.9	81.4	81.7	68.6	76.0
2nd quarter	65.5	61.8	93.3	54.3	81.5	82.0	68.6	76.3	66.2	61.8	93.3	54.3	81.5	82.0	69.0	76.4
3rd quarter	65.5	62.5	94.4	54.6	81.5	82.5	68.8	76.6	66.2	62.5	94.4	54.6	81.5	82.5	69.0	76.6
4th quarter	66.0	63.0	94.6	55.0	81.8	82.7	69.1	76.9	66.6	63.0	94.6	55.0	81.8	82.7	69.1	76.8
1996																
1st quarter	66.5	63.5	94.8	55.6	82.1	83.1	69.5	77.2	67.0	63.5	94.8	55.6	82.1	83.1	69.2	77.0
2nd quarter	67.2	64.7	95.7	56.2	82.1	83.0	70.5	77.6	67.6	64.7	95.7	56.2	82.1	83.0	69.9	77.3
3rd quarter	67.5	65.5	96.5	56.6	82.4	83.4	70.3	77.7	67.9	65.5	96.5	56.6	82.4	83.4	70.1	77.6
4th quarter	67.6	66.3	97.6	56.9	82.2	83.8	70.7	78.1	67.9	66.3	97.6	56.9	82.2	83.8	70.5	78.0
1997																
1st quarter	67.4	66.7	98.4	57.6	82.6	84.9	70.4	78.6	67.8	66.7	98.4	57.6	82.6	84.9	70.1	78.5
2nd quarter	68.5	68.0	98.9	58.1	83.2	84.5	71.2	78.6	68.8	68.0	98.9	58.1	83.2	84.5	71.3	78.7
3rd quarter	69.1	69.0	99.5	58.7	83.7	84.6	71.7	78.9	69.4	69.0	99.5	58.7	83.7	84.6	71.8	79.0
4th quarter	69.5	69.7	99.9	59.7	84.7	85.7	70.6	79.1	69.7	69.7	99.9	59.7	84.7	85.7	70.7	79.2
1998																
1st quarter	69.9	70.5	100.6	60.7	86.0	86.6	69.5	79.1	70.1	70.5	100.6	60.7	86.0	86.6	69.6	79.2
2nd quarter	70.2	71.3	101.1	61.5	86.9	87.2	68.7	79.2	70.5	71.3	101.1	61.5	86.9	87.2	69.0	79.3
3rd quarter	71.2	72.3	101.2	62.5	87.9	87.5	68.9	79.4	71.5	72.3	101.2	62.5	87.9	87.5	69.2	79.5
4th quarter	71.8	73.8	102.5	62.9	88.1	87.3	69.1	79.4	72.0	73.8	102.5	62.9	88.1	87.3	69.3	79.5
1999																
1st quarter	72.9	74.6	102.3	64.0	89.3	87.7	68.9	79.5	73.0	74.6	102.3	64.0	89.3	87.7	69.2	79.6
2nd quarter	73.0	75.3	102.9	64.2	89.0	87.8	69.0	79.6	73.1	75.3	102.9	64.2	89.0	87.8	69.5	79.9
3rd quarter	73.7	76.5	103.5	64.8	89.1	87.8	69.4	79.8	73.8	76.5	103.5	64.8	89.1	87.8	70.1	80.1
4th quarter	74.8	77.9	104.0	66.1	90.2	88.2	69.7	80.1	75.0	77.9	104.0	66.1	90.2	88.2	70.4	80.5
2000																
1st quarter	74.7	78.2	104.5	68.5	92.5	91.4	66.3	80.4	74.9	78.2	104.5	68.5	92.5	91.4	66.8	80.7
2nd quarter	76.3	80.0	104.7	68.6	92.0	89.8	69.3	80.8	76.4	80.0	104.7	68.6	92.0	89.8	69.7	81.1
3rd quarter	76.3	80.0	104.8	70.0	92.9	91.7	67.9	81.2	76.3	80.0	104.8	70.0	92.9	91.7	68.4	81.5
4th quarter	77.2	80.5	104.3	70.3	92.8	91.2	69.2	81.5	77.2	80.5	104.3	70.3	92.8	91.2	69.7	81.8
2001																
1st quarter	76.9	80.1	104.3	71.9	94.0	93.7	66.9	81.9	76.8	80.1	104.3	71.9	94.0	93.7	67.4	82.2
2nd quarter	78.2	80.7	103.2	72.2	93.7	92.3	69.3	82.3	78.2	80.7	103.2	72.2	93.7	92.3	69.9	82.6
3rd quarter	78.5	80.1	102.0	72.3	93.5	92.1	70.0	82.5	78.5	80.1	102.0	72.3	93.5	92.1	70.4	82.7
4th quarter	79.6	80.2	100.9	73.0	94.5	91.8	70.6	82.5	79.5	80.2	100.9	73.0	94.5	91.8	71.3	82.9
2002																
1st quarter	81.0	81.3	100.0	73.3	94.6	90.2	72.5	82.6	81.3	81.3	100.0	73.3	94.6	90.2	73.2	82.8
2nd quarter	81.3	81.6	100.3	74.0	94.7	90.9	72.3	82.7	81.4	81.6	100.3	74.0	94.7	90.9	73.2	83.2
3rd quarter	82.0	81.9	99.9	74.3	94.7	90.7	73.2	83.0	82.0	81.9	99.9	74.3	94.7	90.7	73.9	83.4
4th quarter	82.0	82.0	100.0	74.5	94.4	90.9	73.6	83.3	82.0	82.0	100.0	74.5	94.4	90.9	74.3	83.7
2003																
1st quarter	82.9	82.4	99.5	75.0	94.0	90.5	74.8	83.6	82.8	82.4	99.5	75.0	94.0	90.5	75.5	84.0
2nd quarter	84.2	83.2	99.2	76.3	95.7	91.0	74.6	83.7	83.9	83.2	99.2	76.3	95.7	91.0	75.1	84.1
3rd quarter	85.8	85.2	99.4	77.4	96.5	90.3	76.1	84.1	85.8	85.2	99.4	77.4	96.5	90.3	76.5	84.3
4th quarter	86.5	86.5	99.8	78.5	97.4	90.7	76.3	84.4	86.6	86.5	99.8	78.5	97.4	90.7	76.5	84.5
2004																
1st quarter	86.5	86.5	100.3	78.4	96.4	90.8	78.2	85.2	86.3	86.5	100.3	78.4	96.4	90.8	78.3	85.4
2nd quarter	87.1	87.4	100.4	79.9	97.5	91.7	78.3	85.8	87.1	87.4	100.4	79.9	97.5	91.7	78.2	85.8
3rd quarter	87.6	88.3	101.0	81.3	98.6	92.9	77.8	86.2	87.5	88.3	101.0	81.3	98.6	92.9	77.9	86.4
4th quarter	88.3	89.2	101.5	81.7	98.0	92.9	79.6	87.0	87.9	89.2	101.5	81.7	98.0	92.9	79.8	87.2

Table 9-3B. Productivity and Related Data: Historical Data—*Continued*

(2012 = 100, seasonally adjusted.)

Year and quarter	Business sector								Nonfarm business sector							
	Output per hour of all persons	Output	Hours of all persons	Compensation per hour	Real compensation per hour	Unit labor costs	Unit nonlabor payments	Implicit price deflator	Output per hour of all persons	Output	Hours of all persons	Compensation per hour	Real compensation per hour	Unit labor costs	Unit nonlabor payments	Implicit price deflator
2005																
1st quarter	89.2	90.4	101.7	82.2	98.2	92.5	81.5	87.6	88.9	90.4	101.7	82.2	98.2	92.5	82.0	87.9
2nd quarter	88.8	90.8	102.3	82.7	98.1	93.3	82.1	88.3	88.7	90.8	102.3	82.7	98.1	93.3	82.6	88.6
3rd quarter	89.6	91.7	102.6	83.7	97.9	93.7	83.5	89.1	89.4	91.7	102.6	83.7	97.9	93.7	84.1	89.5
4th quarter	89.8	92.4	103.2	84.3	97.6	94.2	84.4	89.8	89.5	92.4	103.2	84.3	97.6	94.2	85.1	90.2
2006																
1st quarter	90.5	93.9	104.2	86.0	99.0	95.4	84.1	90.3	90.1	93.9	104.2	86.0	99.0	95.4	84.8	90.8
2nd quarter	90.3	94.1	104.5	86.0	98.1	95.5	85.5	91.0	90.0	94.1	104.5	86.0	98.1	95.5	86.3	91.5
3rd quarter	90.0	94.3	105.1	86.1	97.4	96.0	86.1	91.5	89.7	94.3	105.1	86.1	97.4	96.0	86.8	92.0
4th quarter	90.6	95.4	105.4	87.7	99.5	96.9	85.1	91.6	90.5	95.4	105.4	87.7	99.5	96.9	85.7	92.0
2007																
1st quarter	90.9	95.7	105.5	89.8	101.0	99.0	84.3	92.5	90.7	95.7	105.5	89.8	101.0	99.0	84.7	92.8
2nd quarter	91.4	96.5	105.9	89.8	99.8	98.5	86.1	93.1	91.1	96.5	105.9	89.8	99.8	98.5	86.6	93.3
3rd quarter	92.3	97.1	105.5	90.1	99.5	97.9	87.8	93.5	92.0	97.1	105.5	90.1	99.5	97.9	88.2	93.7
4th quarter	92.8	97.6	105.3	91.1	99.4	98.2	88.0	93.8	92.7	97.6	105.3	91.1	99.4	98.2	88.2	93.9
2008																
1st quarter	92.1	96.6	104.9	92.0	99.3	100.0	86.3	94.0	92.0	96.6	104.9	92.0	99.3	100.0	86.4	94.1
2nd quarter	93.1	97.0	104.4	92.1	98.1	99.1	88.5	94.4	92.9	97.0	104.4	92.1	98.1	99.1	88.7	94.6
3rd quarter	93.3	96.2	103.3	92.9	97.5	99.8	89.2	95.1	93.1	96.2	103.3	92.9	97.5	99.8	89.6	95.3
4th quarter	92.7	93.2	100.7	93.9	100.9	101.5	87.2	95.1	92.5	93.2	100.7	93.9	100.9	101.5	87.7	95.5
2009																
1st quarter	93.6	91.7	98.1	91.4	98.9	97.9	91.8	95.1	93.4	91.7	98.1	91.4	98.9	97.9	92.8	95.7
2nd quarter	95.5	91.5	96.0	93.7	100.8	98.3	90.3	94.7	95.3	91.5	96.0	93.7	100.8	98.3	91.1	95.2
3rd quarter	97.1	91.8	94.8	94.3	100.6	97.4	91.5	94.7	96.8	91.8	94.8	94.3	100.6	97.4	92.4	95.2
4th quarter	98.4	93.1	94.9	94.8	100.4	96.6	93.2	95.0	98.1	93.1	94.9	94.8	100.4	96.6	93.8	95.4
2010																
1st quarter	98.8	93.6	94.9	94.2	99.6	95.5	95.5	95.4	98.6	93.6	94.9	94.2	99.6	95.5	96.2	95.8
2nd quarter	99.0	94.6	95.7	95.3	100.7	96.3	95.4	95.8	98.9	94.6	95.7	95.3	100.7	96.3	96.1	96.2
3rd quarter	99.6	95.6	96.2	95.7	100.8	96.2	96.1	96.0	99.4	95.6	96.2	95.7	100.8	96.2	96.6	96.4
4th quarter	99.9	96.4	96.6	96.1	100.5	96.4	97.2	96.6	99.7	96.4	96.6	96.1	100.5	96.4	97.4	96.8
2011																
1st quarter	99.1	95.9	96.8	97.8	101.2	98.8	95.6	97.3	99.0	95.9	96.8	97.8	101.2	98.8	95.3	97.3
2nd quarter	99.3	96.9	97.6	97.3	99.5	98.0	98.0	97.9	99.3	96.9	97.6	97.3	99.5	98.0	97.9	97.9
3rd quarter	98.9	96.9	98.0	97.9	99.5	99.1	98.2	98.6	98.8	96.9	98.0	97.9	99.5	99.1	97.9	98.6
4th quarter	99.6	98.2	98.7	96.6	97.7	97.1	101.3	98.8	99.5	98.2	98.7	96.6	97.7	97.1	101.2	98.9
2012																
1st quarter	99.9	99.4	99.5	98.9	99.5	99.0	99.7	99.3	99.9	99.4	99.5	98.9	99.5	99.0	99.7	99.3
2nd quarter	100.3	100.0	99.7	99.4	99.8	99.1	100.6	99.7	100.3	100.0	99.7	99.4	99.8	99.1	100.6	99.8
3rd quarter	100.0	100.2	100.1	99.5	99.4	99.4	101.3	100.3	100.1	100.2	100.1	99.5	99.4	99.4	101.3	100.3
4th quarter	99.8	100.4	100.7	102.1	101.4	102.4	98.4	100.7	99.7	100.4	100.7	102.1	101.4	102.4	98.3	100.6
2013																
1st quarter	100.6	101.4	101.1	100.7	99.5	100.4	101.9	101.0	100.3	101.4	101.1	100.7	99.5	100.4	101.4	100.8
2nd quarter	100.5	101.4	101.4	101.4	100.3	101.4	101.1	101.2	100.0	101.4	101.4	101.4	100.3	101.4	100.7	101.1
3rd quarter	100.9	102.4	102.0	101.2	99.5	100.7	103.1	101.6	100.5	102.4	102.0	101.2	99.5	100.7	102.8	101.6
4th quarter	101.6	103.6	102.3	101.9	99.9	100.7	104.3	102.1	101.2	103.6	102.3	101.9	99.9	100.7	104.4	102.3
2014																
1st quarter	100.7	103.2	102.8	104.2	101.4	103.8	100.9	102.4	100.4	103.2	102.8	104.2	101.4	103.8	100.9	102.6
2nd quarter	101.5	104.8	103.6	103.4	100.1	102.2	104.6	103.1	101.2	104.8	103.6	103.4	100.1	102.2	104.5	103.2
3rd quarter	102.3	106.5	104.2	103.9	100.4	101.8	106.1	103.5	102.1	106.5	104.2	103.9	100.4	101.8	106.2	103.7
4th quarter	101.7	107.0	105.4	105.0	101.7	103.4	104.2	103.6	101.6	107.0	105.4	105.0	101.7	103.4	104.2	103.8
2015																
1st quarter	102.6	108.5	105.8	106.3	103.5	103.6	102.6	103.2	102.5	108.3	105.6	106.5	103.8	104.0	103.1	103.6
2nd quarter	103.1	109.5	106.2	107.2	103.7	104.0	103.4	103.7	102.9	109.2	106.2	107.4	103.9	104.3	103.8	104.1
3rd quarter	103.4	109.8	106.2	107.7	103.8	104.1	103.9	104.0	103.2	109.6	106.1	107.8	103.9	104.5	104.4	104.4
4th quarter	102.5	109.7	107.1	107.4	103.5	104.8	102.7	103.9	102.3	109.4	107.0	107.6	103.7	105.2	103.2	104.4
2016																
1st quarter	102.7	110.3	107.4	107.5	103.6	104.6	102.4	103.7	102.6	110.1	107.3	107.7	103.8	105.0	103.2	104.2
2nd quarter	102.8	110.9	107.9	107.7	103.1	104.7	104.0	104.4	102.7	110.6	107.7	108.0	103.4	105.2	104.9	105.0
3rd quarter	103.3	111.6	108.1	108.3	103.2	104.9	104.6	104.8	103.1	111.3	107.9	108.6	103.4	105.3	105.5	105.4
4th quarter	103.9	112.3	108.1	109.7	103.8	105.6	104.7	105.2	103.6	112.0	108.2	109.7	103.9	106.0	105.7	105.9
2017																
1st quarter	104.0	113.1	108.7	110.7	104.0	106.4	104.7	105.6	103.8	112.7	108.6	110.9	104.2	106.8	105.4	106.2
2nd quarter	104.2	113.8	109.2	111.2	104.5	106.8	104.8	105.9	103.9	113.5	109.2	111.4	104.6	107.2	105.5	106.5
3rd quarter	105.0	114.9	109.4	112.6	105.2	107.2	105.5	106.5	104.7	114.7	109.5	112.7	105.2	107.6	106.3	107.0
4th quarter	105.1	116.1	110.5	113.8	105.5	108.3	105.7	107.2	104.9	115.8	110.4	114.1	105.7	108.7	106.5	107.8
2018																
1st quarter	105.4	116.9	110.9	115.1	105.9	109.2	105.7	107.7	105.2	116.7	110.9	115.2	106.0	109.6	106.6	108.3
2nd quarter	106.0	118.0	111.3	115.4	105.8	108.5	108.5	108.7	105.7	117.8	111.5	115.3	105.5	109.1	109.4	109.3
3rd quarter	106.3	119.0	112.0	116.1	105.7	109.3	108.8	109.1	106.0	118.8	112.1	116.1	105.7	109.6	109.9	109.7
4th quarter	106.3	119.4	112.4	116.2	105.4	109.4	109.4	109.4	106.0	119.2	112.4	116.4	105.5	109.8	110.4	110.0

Table 9-4. Corporate Profits with Inventory Valuation Adjustment by Industry Group, NAICS Basis

(Billions of dollars, quarterly data are at seasonally adjusted annual rates.)   NIPA Table 6.16D

Year and quarter	Total	Domestic industries			Nonfinancial								
		Financial					Manufacturing						
										Durable goods			
		Total	Federal Reserve banks	Other financial	Total	Utilities	Total	Fabricated metal products	Machinery	Computer and electronic products	Electrical equipment, appliances, and components	Motor vehicles, bodies and trailers, and parts	Other durable goods
2000	729.8	584.1	31.2	118.5	434.4	24.3	175.6	15.6	9.4	20.5	6.9	4.0	25.8
2001	697.1	528.3	28.9	166.1	333.3	22.5	75.1	9.2	2.2	-29.1	1.1	-6.8	14.0
2002	797.4	640.6	23.5	241.9	375.3	10.5	78.3	9.9	2.9	-25.0	-0.8	-3.1	19.5
2003	955.7	796.7	20.0	282.7	494.0	13.2	123.9	8.9	2.8	-6.1	2.7	7.9	15.5
2004	1 217.5	1 022.4	20.0	326.0	676.3	21.1	186.2	12.6	8.3	1.2	0.4	-4.6	35.2
2005	1 629.2	1 403.4	26.5	383.0	993.9	32.4	279.7	19.0	16.4	15.5	-0.6	1.6	61.4
2006	1 812.2	1 572.5	33.8	379.3	1 159.4	55.2	352.9	19.3	20.8	28.5	11.7	-6.9	67.6
2007	1 708.3	1 370.5	36.0	264.2	1 070.3	49.6	321.1	21.4	23.4	24.3	-0.7	-16.3	67.1
2008	1 344.5	954.3	35.1	59.5	859.7	30.4	240.0	15.7	18.0	27.7	5.0	-39.2	39.5
2009	1 470.1	1 121.3	47.3	315.3	758.7	23.4	164.7	11.8	9.6	27.1	9.2	-54.8	33.0
2010	1 786.4	1 400.6	71.6	334.3	994.8	30.6	281.8	15.4	17.3	48.4	10.1	-10.9	43.4
2011	1 750.2	1 337.7	76.0	302.4	959.3	10.2	296.0	16.5	25.8	37.8	4.9	-0.3	49.7
2012	2 144.7	1 739.3	71.7	410.6	1 256.9	13.8	403.0	24.0	33.5	52.9	12.0	23.0	60.1
2013	2 165.9	1 767.1	79.7	351.1	1 336.3	28.3	446.9	25.3	36.5	58.7	20.4	21.5	66.6
2014	2 266.6	1 861.7	103.5	379.6	1 378.6	32.8	458.7	24.1	35.5	60.3	14.2	32.0	68.1
2015	2 190.0	1 787.5	100.7	347.4	1 339.4	20.1	424.8	24.9	24.3	68.3	24.0	26.7	66.6
2016	2 116.5	1 704.6	92.0	364.8	1 247.8	0.4	392.2	20.7	19.0	50.2	5.2	29.0	64.0
2017	2 084.1	1 630.0	78.3	335.2	1 216.5	11.6	315.5	20.4	18.3	44.2	9.6	15.9	64.9
2018	2 011.9	1 510.3	63.6	341.4	1 105.3	-4.0	283.7	19.3	12.0	43.0	6.1	-0.3	51.9
2017													
1st quarter	2 128.9	1 692.3	89.3	320.5	1 282.5	13.5	306.5	20.1	21.2	35.6	7.7	24.2	63.5
2nd quarter	2 151.4	1 728.1	80.2	336.8	1 311.1	14.2	337.1	23.4	20.5	44.6	12.2	18.7	73.0
3rd quarter	2 171.5	1 703.8	71.9	369.0	1 262.9	11.7	348.8	20.5	18.7	50.1	10.7	15.4	66.2
4th quarter	1 884.5	1 395.8	71.8	314.5	1 009.5	6.8	269.6	17.5	12.7	46.4	7.8	5.4	57.1
2018													
1st quarter	1 979.9	1 472.1	70.0	343.3	1 058.8	1.7	246.0	18.8	11.2	32.7	10.9	-1.0	48.1
2nd quarter	1 991.5	1 496.5	65.6	352.8	1 078.2	-1.6	287.0	18.0	11.3	46.1	7.6	-1.1	55.4
3rd quarter	2 045.0	1 533.4	61.9	335.5	1 136.1	-5.4	298.9	20.2	12.4	49.7	5.0	4.4	52.3
4th quarter	2 031.3	1 539.1	56.8	334.0	1 148.2	-10.7	303.0	20.1	13.0	43.7	0.9	-3.5	51.8

Year and quarter	Domestic industries—Continued											Rest of the world, net
	Nonfinancial—Continued											
	Manufacturing—Continued											
	Nondurable goods					Wholesale trade	Retail trade	Transportation and warehousing	Information	Other nonfinancial	
	Total	Food and beverage and tobacco products	Petroleum and coal products	Chemical products	Other nondurable goods							
2000	93.3	25.8	25.9	25.1	16.6	59.5	51.3	9.5	-11.9	126.1		145.7
2001	84.5	28.5	27.5	22.0	6.6	51.1	71.3	-0.7	-26.4	140.2		168.8
2002	74.9	26.7	4.8	30.7	12.8	53.5	83.3	-6.5	5.0	151.2		156.8
2003	92.3	25.6	26.7	31.8	8.1	56.6	87.9	4.4	28.1	179.9		158.9
2004	133.0	26.2	50.8	39.4	16.6	72.7	94.0	12.0	61.6	228.8		195.1
2005	166.3	29.2	80.0	36.2	20.9	96.0	123.3	28.4	100.7	333.5		225.7
2006	211.8	34.6	82.0	68.8	26.4	105.0	133.6	40.8	115.2	356.8		239.7
2007	201.9	31.3	76.1	70.8	23.7	102.8	119.4	23.3	120.5	333.6		337.8
2008	173.3	31.5	85.0	51.5	5.3	92.7	82.2	29.3	98.8	286.3		390.2
2009	128.9	45.0	10.4	55.2	18.2	88.9	107.9	21.7	87.0	265.1		348.8
2010	158.1	45.1	26.2	63.4	23.5	99.3	115.9	44.6	102.3	320.4		385.8
2011	161.6	40.0	47.5	53.2	20.9	97.2	115.1	30.6	95.7	314.5		412.6
2012	197.5	44.7	57.7	63.5	31.5	137.9	155.7	54.4	112.0	380.1		405.4
2013	217.8	55.6	53.9	75.0	33.3	146.4	153.3	45.2	137.6	378.6		398.8
2014	224.5	58.3	65.5	72.2	28.4	150.6	157.3	55.7	126.6	397.0		404.9
2015	190.0	69.0	17.1	65.6	38.3	152.0	169.3	61.0	135.6	376.5		402.5
2016	139.9	68.4	-30.6	64.6	37.6	126.6	170.5	63.9	157.4	387.8		411.9
2017	142.3	60.4	-7.5	62.2	27.1	124.2	156.9	58.2	141.0	409.1		454.1
2018	151.7	46.9	19.4	58.5	26.9	108.9	133.1	45.0	121.7	416.9		501.7
2017												
1st quarter	134.2	58.0	-14.1	58.6	31.8	132.7	174.5	63.2	158.0	434.3		436.6
2nd quarter	144.8	60.4	-13.0	66.7	30.7	140.0	168.1	67.5	145.6	438.6		423.3
3rd quarter	167.3	69.0	1.0	70.6	26.7	127.8	161.9	59.4	151.2	402.0		467.6
4th quarter	122.7	54.2	-3.7	52.9	19.3	96.4	123.2	42.8	109.2	361.5		488.7
2018												
1st quarter	125.4	47.1	4.2	48.4	25.7	109.3	137.7	42.9	123.9	397.2		507.7
2nd quarter	149.7	50.8	11.9	59.7	27.3	92.3	122.8	39.9	127.3	410.4		495.0
3rd quarter	154.9	50.1	16.5	60.5	27.8	110.9	141.8	43.5	124.3	422.1		511.6
4th quarter	176.9	39.6	45.2	65.4	26.8	122.9	130.0	53.6	111.4	438.1		492.3

NOTES AND DEFINITIONS, CHAPTER 9

GENERAL NOTE ON DATA ON COMPENSATION PER HOUR

This chapter includes two data series with similar names—the Employment Cost Index for total compensation and the index of compensation per hour—that often display different behavior. Both are compiled and published by the Bureau of Labor Statistics (BLS), but the definitions, sources, and methods of compilation are different. Users should be aware of these differences and of the consequent differences in the appropriate uses and interpretations for each of the two series.

The *Employment Cost Index (ECI)* (Tables 9-1 and 9-2) measures changes in hourly compensation for "all civilian workers", which is not quite as broad as it sounds, as it excludes federal government workers, farm workers, and private household workers. Indexes are also published for subgroups including state and local workers, "all private industry" (again excluding farm and private household workers), and a number of industry and occupational subgroups.

The ECI is calculated and published separately for *total compensation* and for the two components of hourly compensation, *wages and salaries* and the employer cost of employee *benefits*. It is constructed by analogy with the Consumer Price Index (CPI); that is, it <u>holds the composition of employment constant</u> in order to isolate hourly compensation trends that take place for individual occupations, which are then aggregated, using fixed relative importance weights. The ECI is based on a sample survey and may be revised from time to time, due to updated classification, weighting, and seasonal adjustments. However, it is not subject to major benchmark revision of the underlying wage, salary, and benefit rate observations. By design, it excludes any representation of employee stock options. As it is based on a sample survey, the ECI is measured "from the bottom up," aggregating from individual employers' reports to higher levels. The ECI is frequently and appropriately used as the best available measure of the general trend of wages and of the extent of inflationary pressure exerted on prices by labor costs.

The *compensation per hour* component of the report on "Productivity and Costs" (Table 9-3) is calculated and published for total compensation in total business, nonfarm business, nonfinancial corporations, and manufacturing. The nonfarm business category is similar in scope to the "all private industry" category in the ECI. The measures in Table 9-3, however, are compiled "from the top down," starting with aggregate estimates of compensation and hours, then dividing the former by the latter. Compensation per hour <u>is affected by changes in the composition of employment</u>. If the composition of employment shifts toward a larger proportion of higher-paid employees and/or industries, compensation per hour will rise even if there is no increase in hourly compensation for <u>any</u> individual worker.

In addition, *compensation per hour* includes the value of exercised stock options as expensed by companies. Also included are other transitory payments, many of which may be of little relevance to the typical worker or to ongoing production costs. These values are not reported immediately. Instead, they are incorporated when later, more comprehensive reports are received. This process can lead to dramatic revisions in compensation per hour and in the unit labor costs index, which is based on compensation. For example, the fourth-quarter 2004 increase in compensation per hour in nonfarm business was initially reported at an annual rate of 3.1 percent. Four months later, the reported rate for the same time period was 10.2 percent. The rate of increase from a year earlier was revised from 3.6 to 5.9 percent. According to then-Federal Reserve Chairman Alan Greenspan, in testimony before the Joint Economic Committee on June 9, 2005, this reflected "a large but apparently transitory surge in bonuses and the proceeds of stock option exercises," not a potentially inflationary acceleration in the rate of labor compensation increase. More recently, it was reported in August 2009 that "first-quarter unit labor costs were revised to negative 2.7% from positive 3%." (*Wall Street Journal,* August 12, 2009, p. A2.)

These characteristics suggest that *compensation per hour* should not be considered a reliable or appropriate indicator of wage or compensation trends for typical workers. It is useful in conjunction with the productivity series, because aggregate productivity is subject to the same composition shifts—higher-productivity industries also tend to have higher-paid employees. Hence, the measure of *unit labor costs* (derived by dividing compensation per hour by output per hour in this system) is not distorted when the composition of output shifts toward higher-productivity industries. The shift affects the numerator and denominator of the ratio similarly. However, both compensation and unit labor costs can still be distorted by transitory payments, such as those discussed above.

There are other, probably less important differences between the two measures. Compensation per hour refers to the entire quarter, while the ECI is observed in the terminal month of each quarter. Tips and other forms of compensation not provided by employers are included in hourly compensation but not in the ECI. The ECI excludes persons working for token wages, business owners and others who set their own wage, and family workers who do not earn a market wage; all of these workers are included in the productivity and cost accounts. Hourly compensation excludes employees of nonprofit institutions serving individuals—about 10 percent of private workers (mostly in education and medical care) who are within the scope of the ECI. Hourly compensation measures include an estimate for the unincorporated self-employed, who are assumed to earn the same hourly compensation as other employees in the sector. Implicitly, unpaid family workers also are included in the hourly compensation measures with the assumption that their hourly compensation is zero. Both of these groups are also excluded from the ECI.

TABLES 9-1 AND 9-2

Employment Cost Indexes

SOURCE: U.S. DEPARTMENT OF LABOR, BUREAU OF LABOR STATISTICS (BLS)

The Employment Cost Index (ECI) is a quarterly measure of the change in the cost of labor, independent of the influence of employment shifts among occupations and industries. It uses a fixed market basket of labor—similar in concept to the Consumer Price Index's fixed market basket of goods and services—to measure changes over time in employer costs of employing labor. Data are quarterly in all cases and are reported for the final month of each quarter. These measures are expressed as indexes, with the not-seasonally-adjusted value for December 2005 set at 100.

Care should be used in comparing the ECI with other data sets. The "all private industry" category in the ECI excludes farm and household workers (it is sometimes, and more precisely, called "private nonfarm industry"), and the "all civilian workers" category excludes federal government, farm, and household workers.

The data for 1979 through 2016, which are presented in Table 9-2 and are the official ECI measures for that time period, were based on the 1987 Standard Industrial Classification (SIC) and 2010 Occupational Classification System (OCS).

Currently the ECI is compiled based on the 2012 North American Industry Classification System (NAICS) and the 2010 Standard Occupational Classification Manual (SOC). These data, along with comparable data for 2004 through 2016, are shown in Table 9-1.

For certain broad categories shown in this volume—indicated by footnote 2 in Table 9-2—the old SIC categories are roughly comparable and continuous with the data for 2006 and subsequent years. (However, they differ slightly on overlap dates, and should be "linked" if a continuous time series is desired; see the article at the beginning of this volume.) Many of the new industry and occupational categories are not continuous with the old series shown here, and some of the old categories are not being continued because BLS finds them obsolete and no longer meaningful.

Definitions

Total compensation comprises wages, salaries, and the employer's costs for employee benefits. Excluded from wages and salaries and employee benefits are the value of stock option exercises and items such as payment-in-kind, free room and board, and tips.

Wages and salaries consists of straight-time earnings per hour before payroll deductions, including production bonuses, incentive earnings, commissions, and cost-of-living adjustments. These wage rates exclude premium pay for overtime and for work on weekends and holidays, shift differentials, and nonproduction

bonuses such as lump-sum payments provided in lieu of wage increases. According to BLS, wages and salaries are about 70 percent of total compensation.

Benefits includes the cost to employers for paid leave—vacations, holidays, sick leave, and other leave; for supplemental pay—premium pay for work in addition to the regular work schedule (such as overtime, weekends, and holidays), shift differentials, and nonproduction bonuses (such as referral bonuses and lump-sum payments provided in lieu of wage increases); for insurance benefits—life, health, short-term disability, and long-term disability; for retirement and savings benefits—defined benefit and defined contribution plans; and for legally required benefits—Social Security, Medicare, federal and state unemployment insurance, and workers' compensation. Severance pay and supplemental unemployment benefit (SUB) plans are included in the data through December 2005 but were dropped beginning with the March 2006 data. The combined cost of these two benefits accounted for less than one-tenth of one percent of compensation, and according to BLS, dropping these benefits has had virtually no impact on the index. According to BLS, benefit costs are about 30 percent of total compensation.

Civilian workers are private industry workers, as defined below, and workers in state and local government. Federal workers are not included.

Private industry workers are paid workers in private industry excluding farms and private households. To be included in the ECI, employees in occupations must receive cash payments from the establishment for services performed and the establishment must pay the employer's portion of Medicare taxes on that individual's wages. Major exclusions from the survey are the self-employed, individuals who set their own pay (for example, proprietors, owners, major stockholders, and partners in unincorporated firms), volunteers, unpaid workers, family members being paid token wages, individuals receiving long-term disability compensation, and U.S. citizens working overseas.

Private industry workers excluding incentive paid occupations is a new category introduced in the 2006 revision to eliminate the quarter-to-quarter variability related to the way workers (for example, salespersons working on commission) are paid. (The category *private industry workers excluding sales occupations* was intended to serve a similar purpose in the previous SIC-based classification system, but was much less accurate in separately identifying workers with highly variable compensation.)

Goods-producing industries include mining, construction, and manufacturing.

Service-providing industries include the following NAICS industries: wholesale trade; retail trade; transportation and warehousing; utilities; information; finance and insurance; real estate and rental and leasing; professional, scientific, and technical services; management of companies and enterprises; administrative and support and waste management and remediation services; education services; health care and social assistance; arts, entertainment, and recreation; accommodation and food services; and other services, except public administration.

Notes on the data

Employee benefit costs are calculated as cents per hour worked.

The 2018 data were collected from probability samples of approximately 26,500 occupational observations in about 6,400 sample establishments in private industry, and approximately 7900 occupations within about 1,400 establishments in state and local governments. The private industry sample is rotated over approximately five years.

Currently, the sample establishments are classified in industry categories based on the NAICS. Within an establishment, specific job categories are selected and classified into approximately 800 occupational classifications according to the SOC. Similar procedures were followed under the previous classification systems. Data are collected each quarter for the pay periods including the 12th day of March, June, September, and December.

Aggregate indexes are calculated using fixed employment weights. Beginning with December 2013, estimates are based on 2012 employment weight, using Occupational Employment counts. From March 2006 through June 2018, ECI weights were based on fixed employment counts for 2012 from the BLS Occupational Employment Statistics survey. Employment data from the 2012 Survey were introduced in December 2013. ECI measures were based on 1990 employment counts from March 1995 through December 2005 and 1980 census employment counts from June 1986 through December 1994. Prior to June 1986, they were based on 1970 census employment counts.

Use of fixed weights ensures that changes in the indexes reflect only changes in hourly compensation, not employment shifts among industries or occupations with different levels of wages and compensation. This feature distinguishes the ECI from other compensation series such as average hourly earnings (see Chapter 10 and its notes and definitions) and the compensation per hour component of the productivity series (see Table 9-3 and its notes and definitions, and the general note above), each of which is affected by such employment shifts.

Data availability

Data for wages and salaries for the private nonfarm economy are available beginning with the data for 1975; data for compensation begin with the 1980 data. The series for state and local government and for the civilian nonfarm economy begin with the 1981 data. All series are available on the BLS Web site at <http://www.bls.gov>.

Wage and salary change and compensation cost change data also are available by major occupational and industry groups, as well as by region and collective bargaining status. Information on wage and salary change is available from 1975 to the present for most of these series. Compensation cost change data are available from 1980 to the present for most series. For 10 occupational and industry series, benefit cost change data are available from the early 1980s to the present. For state and local governments and the civilian economy (state and local governments plus private industry), wage and salary change and compensation cost change data are available for major occupational and industry series. BLS provides data for all these series from June 1981 to the present.

Updates are available about four weeks after the end of the reference quarter. Reference quarters end in March, June, September, and December. Seasonal adjustment factors and seasonally adjusted indexes are released in late April.

References

Explanatory notes, including references, are included in a Technical Note in each quarter's ECI news release, and can be found on the BLS Web site, in the PDF version of the release.

More detailed information on the ECI is available from Chapter 8, "National compensation measures," (www.bls.gov/opub/hom/pdf/homch8.pdf) from the BLS Handbook of Methods, and several articles published in the *Monthly Labor Review* and *Compensation and Working Conditions*. The articles and other descriptive pieces are available at www.bls.gov/ect/#publications, by calling (202) 691-6199, or sending e-mail to NCSinfo@bls.gov.

TABLE 9-3A AND 9-3B

Productivity and Related Data

The indexes of productivity and costs for the entire postwar period, 1947 through 2018, have been revised to reflect the recent comprehensive revisions of the National Income and Product Accounts (NIPAs).

Productivity measures relate real physical output to real input. They encompass a family of measures that includes single-factor input measures, such as output per unit of labor input or output per unit of capital input, as well as measures of multifactor productivity—that is, output per unit of combined labor and capital inputs. The indexes published in this book are indexes of labor productivity expressed in terms of output per hour of labor input. (A larger group of BLS productivity measures can be found in Bernan Press's *Handbook of U.S. Labor Statistics*.) All data are presented as indexes with a base of 2012 = 100.

Definitions

Current dollar output is the current-dollar value of the goods and services produced in the specified sector. *Output* is the constant-dollar value of the same goods and services, derived as current-dollar output divided by the sector *implicit price deflator.* All these data are derived from components of the NIPAs; see Notes on the data below, and for general information on the NIPAs, see the Notes and Definitions to Chapter 1.

Output per hour of all persons (labor productivity) is the value of goods and services in constant dollars produced per hour of labor input. By definition, nonfinancial corporations include no self-employed persons. Productivity in this sector is expressed as *output per hour of all employees. Output per person* is output divided by the index for employment instead of by the index for hours of all persons.

Compensation is the total value of the wages and salaries of employees, plus employers' contributions for social insurance and private benefit plans, plus wages, salaries, and supplementary payments for the self-employed. *Compensation per hour* is compensation divided by *hours of all persons.* Included in compensation is the value of exercised stock options that companies report as a charge against earnings. Stock option values are reported with a delay; consequently, recent values are estimated based on extrapolation. They are revised to actual values when the data become available; sometimes the revisions are very large. The labor compensation of proprietors cannot be explicitly identified and must be estimated. This is done by assuming that proprietors have the same hourly compensation as employees in the same sector. The quarterly labor productivity and cost measures do not contain estimates of compensation for unpaid family workers.

Real compensation per hour is compensation per hour deflated by the Consumer Price Index Research Series (CPI-U-RS) for the period 1978 through 2018 (For current quarters, before the CPI-U-RS becomes available, changes in the CPI-U are used.) Changes in the Consumer Price Index for Urban Wage Earners and Clerical Workers (CPI-W) are used for data before 1978, as there was no CPI-U (or CPIU-RS) for that period. See the Notes and Definitions to Chapter 8 for explanation of the CPI-U, the CPI-W, and the CPI-U-RS.

Unit labor costs are the current-dollar labor costs expended in the production of a unit of output. They are derived by dividing compensation by output.

Unit nonlabor payments include profits, depreciation, interest, rental income of persons, and indirect taxes per unit of output. They are computed by subtracting current-dollar compensation of all persons from current-dollar value of output, providing the data for the index of total *nonlabor payments,* which is then divided by output.

Unit nonlabor costs are available for nonfinancial corporations only. They contain all the components of unit nonlabor payments except unit profits (and rental income of persons, which is zero by definition for nonfinancial corporations).

Unit profits, the other component of unit nonlabor payments, are also only available for nonfinancial corporations.

Hours of all persons (labor input) consists of the total hours at work (*employment* multiplied by *average weekly hours*) of payroll workers, self-employed persons, and unpaid family workers. For the nonfinancial corporations data, there are no self-employed persons; the data represent *employee hours.*

Labor share is the total dollar amount of labor compensation divided by the total current-dollar value of output.

Notes on the data

Output for the business sector is equal to constant-dollar gross domestic product minus: the rental value of owner-occupied dwellings, the output of nonprofit institutions, the output of paid employees of private households, and general government output. The measures are derived from national income and product account (NIPA) data supplied by the U.S. Department of Commerce's Bureau of Economic Analysis (BEA). For manufacturing, BLS produces annual estimates of sectoral output. Quarterly manufacturing output indexes derived from the Federal Reserve Board of Governors' monthly indexes of industrial production (see Chapter 2) are adjusted to these annual measures by the BLS, and are also used to project the quarterly values in the current period.

Nonfinancial corporate output consists of business output minus unincorporated businesses and those corporations classified as offices of bank holding companies, offices of other holding companies, or offices in the finance and insurance sector.. Unit profits and unit nonlabor costs can be calculated separately for this sector and are shown in this table.

Compensation and hours data are developed from BLS and BEA data. The primary source for hours and employment is BLS's Current Employment Statistics (CES) program (see the notes and definitions for Chapter 10). Other data sources are the Current Population Survey (CPS) and the National Compensation Survey (NCS). Weekly paid hours are adjusted to hours at work using the NCS. For paid employees, hours at work differ from hours paid, in that they exclude paid vacation and holidays, paid sick leave, and other paid personal or administrative leave.

Although the labor productivity measures relate output to labor input, they do not measure the contribution of labor or any other specific factor of production. Instead, they reflect the joint effect of many influences, including changes in technology; capital investment; level of output; utilization of capacity, energy, and materials; the organization of production; managerial skill; and the characteristics and efforts of the work force.

Revisions

Data for recent years are revised frequently to take account of revisions in the output and labor input measures that underlie the estimates. Customarily, all revisions to source data are reflected in the release following the source data revision.

Data availability

Series are available quarterly and annually. Quarterly measures are based entirely on seasonally adjusted data. For some detailed manufacturing series (not shown here), only annual averages are available. Productivity indexes are published early in the second and third months of each quarter, reflecting new data for preceding quarters. Complete historical data are available on the BLS Web site at <http://www.bls.gov>.

BLS also publishes productivity estimates for a number of individual industries. A release entitled "Productivity and Costs by Industry" is available on the BLS Web site at <http://www.bls.gov>.

References

Further information is available in the Technical Notes and footnotes on the most current monthly release, available on the Web site, and from the following sources: Chapter 10, "Productivity Measures: Business Sector and Major Subsectors," *BLS Handbook of Methods*—Bulletin 2490 (April 1997) and the following

Monthly Labor Review articles: "Alternative Measures of Supervisory Employee Hours and Productivity Growth" (April 2004); "Possible Measurement Bias in Aggregate Productivity Growth" (February 1999); "Improvements to the Quarterly Productivity Measures" (October 1995); "Hours of Work: A New Base for BLS Productivity Statistics" (February 1990); and "New Sector Definitions for Productivity Series" (October 1976).

TABLE 9-4

Corporate Profits with Inventory Valuation Adjustment by Industry Group

SOURCE: U.S. DEPARTMENT OF COMMERCE, BUREAU OF ECONOMIC ANALYSIS

These profits measures are derived from the NIPAs and classified according to the NAICS. See the notes and definitions to Chapter 1 for definitions, data availability, and references. Note that this industry breakdown of profits incorporates the inventory valuation adjustment (IVA), which eliminates any capital gain element in profits arising from changes in the prices at which inventories are valued, but does <u>not</u> incorporate the capital consumption adjustment (CCAdj), which adjusts historical costs of fixed capital to replacement costs and uses actual rather than tax-based service lives. This is because the CCAdj is calculated by BEA at an aggregate level, whereas the IVA is calculated at an industry level.

CHAPTER 10: EMPLOYMENT, HOURS, AND EARNINGS

SECTION 10A: LABOR FORCE, EMPLOYMENT, AND UNEMPLOYMENT

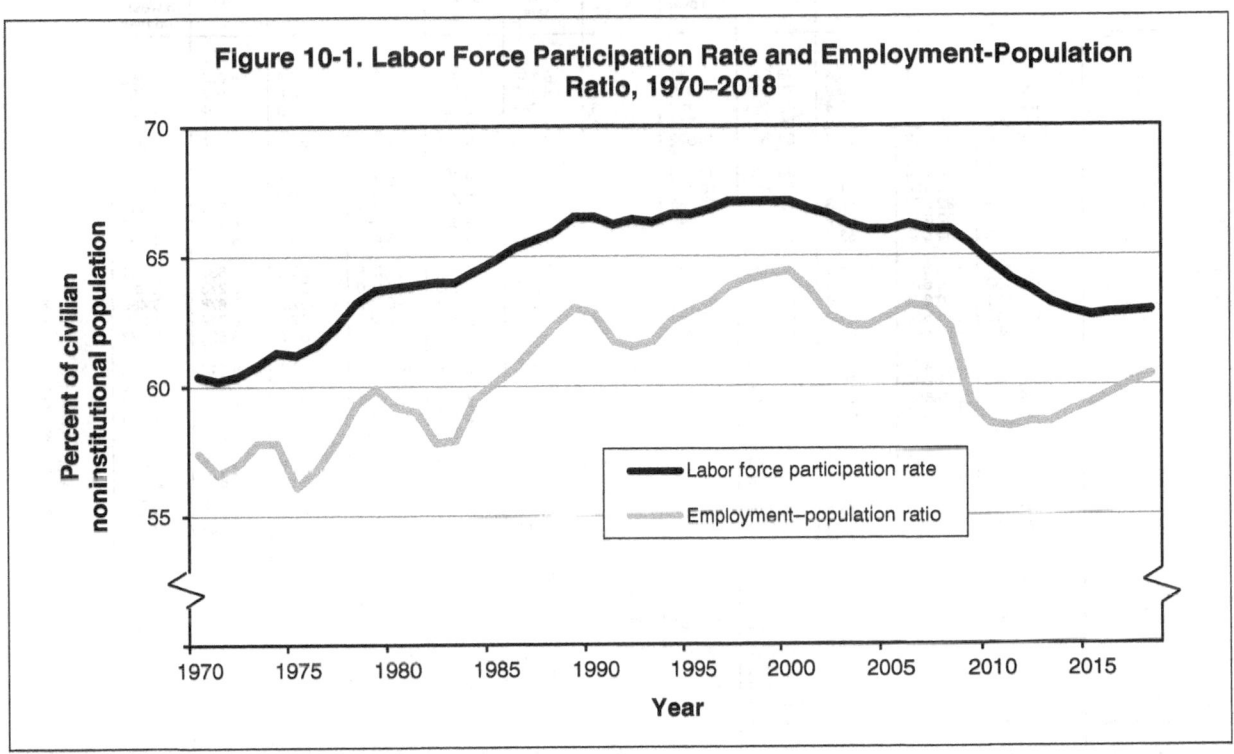

Figure 10-1. Labor Force Participation Rate and Employment-Population Ratio, 1970–2018

- The National Bureau of Economic Research (NBER) determined that the Great Recession endured from the fourth quarter of 2007 to the second quarter of 2009, lasting seven quarters. From 2007 to 2009, the number of unemployed people more than doubled from nearly 7.1 million to over 14.2 million and the unemployment rate grew from 4.6 percent to 9.3 percent. (Table 10-1A)

- Labor force statistics cover the working-age population to determine the size of the civilian labor force, employment, unemployment, wages, and more. Two measurements that are not as common are employment/population rate and labor force participation rate. These economic metrics provide information to gauge the health of the U.S. job market. (Table 10-1 and 10-2)

- The employment-population ratio plateaued at 64.4 percent in 2000 and fell to 58.4 percent in 2011. This ratio is influenced by population demographics such as the baby boomers starting to retire and younger individuals choosing to stay longer in school. Labor participation rates peaked between 1997 and 2000 at 67.1 percent but declined or remained the same each year from 2001 through 2015 except in 2006 when it increased 0.2 percent. After remaining steady at 62.8 percent in 2017, it increased 0.1 percent in 2018. (Table 10-1)

- Women began in earnest joining the workforce in the 1970s. In 1970, employed women 20 years and over made up 32.6 percent of the labor force. By 2018, they made up 43.5 percent of the labor force. (Table 10-1A)

- Since the end of the Vietnam War, government employment has steadily dropped. Between 1965 and 1975 government share of total nonfarm payroll employment grew from 16.7 percent of total nonfarm payroll employment to 19.2 percent of total nonfarm payroll employment. By 2018, government share of total employment fell to 15.1 percent. Total government employment includes federal state and local employment. (Table 10-8)

Table 10-1A. Summary Labor Force, Employment, and Unemployment: Recent Data

(Thousands of persons, percent, seasonally adjusted, except as noted.)

Year and month	Civilian noninsti-tutional population [1]	Civilian labor force		Employment, thousands of persons						Employ-ment-popul-ation ratio, percent	Unemployment		Rate (percent)
		Thous-ands of persons	Participa-tion rate (percent)	Total	By age and sex			By industry			Thousands of persons		
					Men, 20 years and over	Women, 20 years and over	Both sexes, 16 to 19 years	Agri-cultural	Nonagri-cultural		Total	Unem-ployed 15 weeks and over	
1970	137 085	82 771	60.4	78 678	45 581	26 952	6 144	3 463	75 215	57.4	4 093	663	4.9
1971	140 216	84 382	60.2	79 367	45 912	27 246	6 209	3 394	75 972	56.6	5 016	1 187	5.9
1972	144 126	87 034	60.4	82 153	47 130	28 276	6 746	3 484	78 669	57.0	4 882	1 167	5.6
1973	147 096	89 429	60.8	85 064	48 310	29 484	7 271	3 470	81 594	57.8	4 365	826	4.9
1974	150 120	91 949	61.3	86 794	48 922	30 424	7 448	3 515	83 279	57.8	5 156	955	5.6
1975	153 153	93 775	61.2	85 846	48 018	30 726	7 103	3 408	82 438	56.1	7 929	2 505	8.5
1976	156 150	96 158	61.6	88 752	49 190	32 226	7 336	3 331	85 421	56.8	7 406	2 366	7.7
1977	159 033	99 009	62.3	92 017	50 555	33 775	7 688	3 283	88 734	57.9	6 991	1 942	7.1
1978	161 910	102 251	63.2	96 048	52 143	35 836	8 070	3 387	92 661	59.3	6 202	1 414	6.1
1979	164 863	104 962	63.7	98 824	53 308	37 434	8 083	3 347	95 477	59.9	6 137	1 241	5.8
1980	167 745	106 940	63.8	99 303	53 101	38 492	7 710	3 364	95 938	59.2	7 637	1 871	7.1
1981	170 130	108 670	63.9	100 397	53 582	39 590	7 225	3 368	97 030	59.0	8 273	2 285	7.6
1982	172 271	110 204	64.0	99 526	52 891	40 086	6 549	3 401	96 125	57.8	10 678	3 485	9.7
1983	174 215	111 550	64.0	100 834	53 487	41 004	6 342	3 383	97 450	57.9	10 717	4 210	9.6
1984	176 383	113 544	64.4	105 005	55 769	42 793	6 444	3 321	101 685	59.5	8 539	2 737	7.5
1985	178 206	115 461	64.8	107 150	56 562	44 154	6 434	3 179	103 971	60.1	8 312	2 305	7.2
1986	180 587	117 834	65.3	109 597	57 569	45 556	6 472	3 163	106 434	60.7	8 237	2 232	7.0
1987	182 753	119 865	65.6	112 440	58 726	47 074	6 641	3 208	109 232	61.5	7 425	1 983	6.2
1988	184 613	121 669	65.9	114 968	59 781	48 383	6 805	3 169	111 800	62.3	6 701	1 610	5.5
1989	186 393	123 869	66.5	117 342	60 837	49 745	6 750	3 199	114 142	63.0	6 528	1 375	5.3
1990	189 164	125 840	66.5	118 793	61 678	50 535	6 610	3 223	115 570	62.8	7 047	1 525	5.6
1991	190 925	126 346	66.2	117 718	61 178	50 634	5 906	3 269	114 449	61.7	8 628	2 357	6.8
1992	192 805	128 105	66.4	118 492	61 496	51 328	5 669	3 247	115 245	61.5	9 613	3 408	7.5
1993	194 838	129 200	66.3	120 259	62 355	52 099	6 035	3 115	117 144	61.7	8 940	3 094	6.9
1994	196 814	131 056	66.6	123 060	63 294	53 606	6 162	3 409	119 651	62.5	7 996	2 860	6.1
1995	198 584	132 304	66.6	124 900	64 085	54 396	6 419	3 440	121 460	62.9	7 404	2 363	5.6
1996	200 591	133 943	66.8	126 708	64 897	55 311	6 500	3 443	123 264	63.2	7 236	2 316	5.4
1997	203 133	136 297	67.1	129 558	66 284	56 613	6 661	3 399	126 159	63.8	6 739	2 062	4.9
1998	205 220	137 673	67.1	131 463	67 135	57 278	7 051	3 378	128 085	64.1	6 210	1 637	4.5
1999	207 753	139 368	67.1	133 488	67 761	58 555	7 172	3 281	130 207	64.3	5 880	1 480	4.2
2000	212 577	142 583	67.1	136 891	69 634	60 067	7 190	2 464	134 427	64.4	5 692	1 318	4.0
2001	215 092	143 734	66.8	136 933	69 776	60 417	6 740	2 299	134 635	63.7	6 801	1 752	4.7
2002	217 570	144 863	66.6	136 485	69 734	60 420	6 332	2 311	134 174	62.7	8 378	2 904	5.8
2003	221 168	146 510	66.2	137 736	70 415	61 402	5 919	2 275	135 461	62.3	8 774	3 378	6.0
2004	223 357	147 401	66.0	139 252	71 572	61 773	5 907	2 232	137 020	62.3	8 149	3 072	5.5
2005	226 082	149 320	66.0	141 730	73 050	62 702	5 978	2 197	139 532	62.7	7 591	2 619	5.1
2006	228 815	151 428	66.2	144 427	74 431	63 834	6 163	2 206	142 221	63.1	7 001	2 266	4.6
2007	231 867	153 124	66.0	146 047	75 337	64 799	5 911	2 095	143 952	63.0	7 078	2 303	4.6
2008	233 788	154 287	66.0	145 362	74 750	65 039	5 573	2 168	143 194	62.2	8 924	3 188	5.8
2009	235 801	154 142	65.4	139 877	71 341	63 699	4 837	2 103	137 775	59.3	14 265	7 272	9.3
2010	237 830	153 889	64.7	139 064	71 230	63 456	4 378	2 206	136 858	58.5	14 825	8 786	9.6
2011	239 618	153 617	64.1	139 869	72 182	63 360	4 327	2 254	137 615	58.4	13 747	8 077	8.9
2012	243 284	154 975	63.7	142 469	73 403	64 640	4 426	2 186	140 283	58.6	12 506	6 996	8.1
2013	245 679	155 389	63.2	143 929	74 176	65 295	4 458	2 130	141 799	58.6	11 460	6 117	7.4
2014	247 947	155 922	62.9	146 305	75 471	66 287	4 548	2 237	144 068	59.0	9 617	4 714	6.2
2015	250 801	157 130	62.7	148 834	76 776	67 323	4 737	2 422	146 411	59.3	8 296	3 595	5.3
2016	253 538	159 187	62.8	151 436	78 084	66 387	4 965	2 460	148 976	59.7	7 751	3 163	4.9
2017	255 079	160 310	62.8	153 337	78 919	69 343	5 074	2 454	150 883	60.1	6 973	2 704	4.4
2018	257 791	162 075	62.9	155 761	80 211	70 424	5 126	2 425	153 337	60.4	6 314	2 268	3.9
2017													
January	254 082	159 693	62.9	152 128	78 440	68 633	5 055	2 411	149 709	59.9	7 565	3 021	4.7
February	254 246	159 854	62.9	152 417	78 439	68 971	5 007	2 437	149 939	59.9	7 437	2 820	4.7
March	254 414	160 036	62.9	152 958	78 472	69 343	5 143	2 503	150 260	60.1	7 078	2 768	4.4
April	254 588	160 169	62.9	153 150	78 807	69 239	5 104	2 682	150 432	60.2	7 019	2 717	4.4
May	254 767	159 910	62.8	152 920	78 748	69 134	5 037	2 501	150 397	60.0	6 991	2 794	4.4
June	254 957	160 124	62.8	153 176	78 755	69 250	5 171	2 466	150 816	60.1	6 948	2 642	4.3
July	255 151	160 383	62.9	153 456	78 863	69 529	5 065	2 349	151 073	60.1	6 927	2 740	4.3
August	255 357	160 706	62.9	153 591	78 972	69 511	5 111	2 378	151 312	60.1	7 115	2 786	4.4
September	255 562	161 190	63.1	154 399	79 453	69 694	5 252	2 286	152 143	60.4	6 791	2 684	4.2
October	255 766	160 436	62.7	153 847	79 278	69 545	5 025	2 487	151 353	60.2	6 588	2 494	4.1
November	255 949	160 626	62.8	153 945	79 344	69 670	4 931	2 461	151 562	60.1	6 682	2 566	4.2
December	256 109	160 636	62.7	154 065	79 493	69 587	4 985	2 512	151 628	60.2	6 572	2 404	4.1
2018													
January	256 780	161 123	62.7	154 482	79 719	69 620	5 143	2 480	152 030	60.2	6 641	2 387	4.1
February	256 934	161 900	63.0	155 213	80 186	69 849	5 178	2 450	152 695	60.4	6 687	2 336	4.1
March	257 097	161 646	62.9	155 160	80 091	69 946	5 123	2 331	152 664	60.4	6 486	2 237	4.0
April	257 272	161 551	62.8	155 216	80 108	70 033	5 074	2 312	152 860	60.3	6 335	2 330	3.9
May	257 454	161 667	62.8	155 539	80 299	70 161	5 079	2 353	153 127	60.4	6 128	2 164	3.8
June	257 642	162 129	62.9	155 592	80 006	70 455	5 131	2 363	153 267	60.4	6 537	2 329	4.0
July	257 843	162 209	62.9	155 964	80 217	70 622	5 125	2 493	153 425	60.5	6 245	2 377	3.9
August	258 066	161 802	62.7	155 604	80 149	70 563	4 892	2 346	153 376	60.3	6 197	2 247	3.8
September	258 290	162 055	62.7	156 069	80 251	70 710	5 108	2 478	153 634	60.4	5 986	2 240	3.7
October	258 514	162 694	62.9	156 582	80 388	70 935	5 258	2 418	154 135	60.6	6 112	2 229	3.8
November	258 708	162 821	62.9	156 803	80 633	70 949	5 221	2 556	154 297	60.6	6 018	2 124	3.7
December	258 888	163 240	63.1	156 945	80 501	71 218	5 226	2 522	154 520	60.6	6 294	2 203	3.9

[1] Not seasonally adjusted.

Table 10-1B. Summary Labor Force, Employment, and Unemployment: Historical Data

(Thousands of persons, percent, seasonally adjusted, except as noted.)

Year and month	Civilian noninstitutional population [1]	Civilian labor force		Employment, thousands of persons						Employment-population ratio, percent	Unemployment		
		Thousands of persons	Participation rate (percent)	Total	Men, 20 years and over	Women, 20 years and over	Both sexes, 16 to 19 years	Agricultural	Nonagricultural		Thousands of persons		Rate (percent) [2]
						By age and sex		By industry			Total	Unemployed 15 weeks and over	

14 Years and Over

Year and month													
1929	...	49 180	...	47 630	...	...	...	10 450	37 180	...	1 550	...	3.2
1930	...	49 820	...	45 480	...	...	...	10 340	35 140	...	4 340	...	8.7
1931	...	50 420	...	42 400	...	...	...	10 290	32 110	...	8 020	...	15.9 (15.3)
1932	...	51 000	...	38 940	...	...	...	10 170	28 770	...	12 060	...	23.6 (22.5)
1933	...	51 590	...	38 760	...	...	...	10 090	28 670	...	12 830	...	24.9 (20.6)
1934	...	52 230	...	40 890	...	...	...	9 900	30 990	...	11 340	...	21.7 (16.0)
1935	...	52 870	...	42 260	...	...	...	10 110	32 150	...	10 610	...	20.1 (14.2)
1936	...	53 440	...	44 410	...	...	...	10 000	34 410	...	9 030	...	16.9 (9.9)
1937	...	54 000	...	46 300	...	...	...	9 820	36 480	...	7 700	...	14.3 (9.1)
1938	...	54 610	...	44 220	...	...	...	9 690	34 530	...	10 390	...	19.0 (12.5)
1939	...	55 230	...	45 750	...	...	...	9 610	36 140	...	9 480	...	17.2 (11.3)
1940	99 840	55 640	55.7	47 520	...	...	...	9 540	37 980	47.6	8 120	...	14.6 (9.5)
1941	99 900	55 910	56.0	50 350	...	...	...	9 100	41 250	50.4	5 560	...	9.9 (6.0)
1942	98 640	56 410	57.2	53 750	...	...	...	9 250	44 500	54.5	2 660	...	4.7 (3.1)
1943	94 640	55 540	58.7	54 470	...	...	...	9 080	45 390	57.6	1 070	...	1.9 (1.8)
1944	93 220	54 630	58.6	53 960	...	...	...	8 950	45 010	57.9	670	...	1.2
1945	94 090	53 860	57.2	52 820	...	...	...	8 580	44 240	56.1	1 040	...	1.9
1946	103 070	57 520	55.8	55 250	...	...	...	8 320	46 930	53.6	2 270	...	3.9
1947	106 018	60 168	56.8	57 812	...	...	...	8 256	49 557	54.5	2 356	...	3.9

16 Years and Over

Year and month													
1947	101 827	59 350	58.3	57 038	...	...	...	7 890	49 148	56.0	2 311	...	3.9
1948	103 068	60 621	58.8	58 343	39 382	14 936	4 026	7 629	50 714	56.6	2 276	300	3.8
1949	103 994	61 286	58.9	57 651	38 803	15 137	3 712	7 658	49 993	55.4	3 637	684	5.9
1950	104 995	62 208	59.2	58 918	39 394	15 824	3 703	7 160	51 758	56.1	3 288	782	5.3
1951	104 621	62 017	59.2	59 961	39 626	16 570	3 767	6 726	53 235	57.3	2 055	303	3.3
1952	105 231	62 138	59.0	60 250	39 578	16 958	3 719	6 500	53 749	57.3	1 883	232	3.0
1953	107 056	63 015	58.9	61 179	40 296	17 164	3 720	6 260	54 919	57.1	1 834	210	2.9
1954	108 321	63 643	58.8	60 109	39 634	17 000	3 475	6 205	53 904	55.5	3 532	812	5.5
1955	109 683	65 023	59.3	62 170	40 526	18 002	3 642	6 450	55 722	56.7	2 852	702	4.4
1956	110 954	66 552	60.0	63 799	41 216	18 767	3 818	6 283	57 514	57.5	2 750	533	4.1
1957	112 265	66 929	59.6	64 071	41 239	19 052	3 778	5 947	58 123	57.1	2 859	560	4.3
1958	113 727	67 639	59.5	63 036	40 411	19 043	3 582	5 586	57 450	55.4	4 602	1 452	6.8
1959	115 329	68 369	59.3	64 630	41 267	19 524	3 838	5 565	59 065	56.0	3 740	1 040	5.5
1960	117 245	69 628	59.4	65 778	41 543	20 105	4 129	5 458	60 318	56.1	3 852	957	5.5
1961	118 771	70 459	59.3	65 746	41 342	20 296	4 108	5 200	60 546	55.4	4 714	1 532	6.7
1962	120 153	70 614	58.8	66 702	41 815	20 693	4 195	4 944	61 759	55.5	3 911	1 119	5.5
1963	122 416	71 833	58.7	67 762	42 251	21 257	4 255	4 687	63 076	55.4	4 070	1 088	5.7
1964	124 485	73 091	58.7	69 305	42 886	21 903	4 516	4 523	64 782	55.7	3 786	973	5.2

1948

Year and month													
January	102 603	60 095	58.6	58 061	39 386	14 556	4 119	8 077	49 984	56.6	2 034	311	3.4
February	102 698	60 524	58.9	58 196	39 480	14 621	4 095	7 696	50 500	56.7	2 328	283	3.8
March	102 771	60 070	58.5	57 671	39 098	14 481	4 092	7 333	50 338	56.1	2 399	292	4.0
April	102 831	60 677	59.0	58 291	39 157	15 001	4 133	7 557	50 734	56.7	2 386	324	3.9
May	102 923	59 972	58.3	57 854	39 139	14 712	4 003	7 141	50 713	56.2	2 118	329	3.5
June	102 992	60 957	59.2	58 743	39 392	15 213	4 138	7 591	51 152	57.0	2 214	322	3.6
July	103 216	61 181	59.3	58 968	39 607	15 348	4 013	7 602	51 366	57.1	2 213	295	3.6
August	103 240	60 806	58.9	58 456	39 510	14 994	3 952	7 562	50 894	56.6	2 350	332	3.9
September	103 291	60 815	58.9	58 513	39 324	15 207	3 982	7 865	50 648	56.6	2 302	298	3.8
October	103 361	60 646	58.7	58 387	39 522	14 956	3 909	7 626	50 761	56.5	2 259	324	3.7
November	103 424	60 702	58.7	58 417	39 459	15 054	3 904	7 624	50 793	56.5	2 285	282	3.8
December	103 468	61 169	59.1	58 740	39 539	15 137	4 064	7 984	50 756	56.8	2 429	305	4.0

1949

Year and month													
January	103 529	60 771	58.7	58 175	39 233	14 991	3 951	7 790	50 385	56.2	2 596	315	4.3
February	103 559	61 057	59.0	58 208	39 117	15 117	3 974	8 022	50 186	56.2	2 849	374	4.7
March	103 665	61 073	58.9	58 043	39 015	15 069	3 959	8 008	50 035	56.0	3 030	414	5.0
April	103 739	61 007	58.8	57 747	38 993	14 978	3 776	7 911	49 836	55.7	3 260	483	5.3
May	103 845	61 259	59.0	57 552	38 701	15 066	3 785	8 067	49 485	55.4	3 707	602	6.1
June	103 930	60 948	58.6	57 172	38 632	15 003	3 537	7 802	49 370	55.0	3 776	705	6.2
July	104 042	61 301	58.9	57 190	38 405	15 244	3 541	8 021	49 169	55.0	4 111	848	6.7
August	104 121	61 590	59.2	57 397	38 610	15 181	3 606	7 604	49 793	55.1	4 193	917	6.8
September	104 219	61 633	59.1	57 584	38 744	15 129	3 711	7 297	50 287	55.3	4 049	973	6.6
October	104 338	62 185	59.6	57 269	38 394	15 260	3 615	6 814	50 455	54.9	4 916	1 000	7.9
November	104 421	62 005	59.4	58 009	38 860	15 422	3 727	7 497	50 512	55.6	3 996	1 056	6.4
December	104 524	61 908	59.2	57 845	38 908	15 300	3 637	7 379	50 466	55.3	4 063	961	6.6

1950

Year and month													
January	104 619	61 661	58.9	57 635	38 780	15 255	3 600	7 065	50 570	55.1	4 026	947	6.5
February	104 737	61 687	58.9	57 751	38 818	15 339	3 594	7 057	50 694	55.1	3 936	947	6.4
March	104 844	61 604	58.8	57 728	38 851	15 366	3 511	7 116	50 612	55.1	3 876	912	6.3
April	104 943	62 158	59.2	58 583	39 100	15 831	3 652	7 264	51 319	55.8	3 575	920	5.8
May	105 014	62 083	59.1	58 649	39 416	15 628	3 605	7 277	51 372	55.8	3 434	890	5.5
June	105 104	62 419	59.4	59 052	39 476	15 953	3 623	7 285	51 767	56.2	3 367	868	5.4
July	105 194	62 121	59.1	59 001	39 517	15 793	3 691	7 126	51 875	56.1	3 120	769	5.0
August	105 282	62 596	59.5	59 797	39 879	16 124	3 794	7 248	52 549	56.8	2 799	633	4.5
September	105 269	62 349	59.2	59 575	39 865	15 902	3 808	6 992	52 583	56.6	2 774	648	4.4
October	105 096	62 428	59.4	59 803	39 737	16 175	3 891	7 371	52 432	56.9	2 625	545	4.2
November	104 979	62 286	59.3	59 697	39 668	16 195	3 834	7 163	52 534	56.9	2 589	507	4.2
December	104 872	62 068	59.2	59 429	39 536	16 149	3 744	6 760	52 669	56.7	2 639	482	4.3

[1] Not seasonally adjusted.
[2] In 1930 through 1943, the official BLS data count persons on work relief as unemployed. The unemployment rates for those years shown in parentheses count persons on work relief as employed, which is more consistent with the postwar practice. See notes and definitions.
. . . = Not available.

Table 10-1B. Summary Labor Force, Employment, and Unemployment: Historical Data—*Continued*

(Thousands of persons, percent, seasonally adjusted, except as noted.)

Year and month	Civilian noninsti-tutional popu-lation [1]	Civilian labor force — Thousands of persons	Civilian labor force — Participa-tion rate (percent)	Employment — Total	By age and sex — Men, 20 years and over	By age and sex — Women, 20 years and over	By age and sex — Both sexes, 16 to 19 years	By industry — Agri-cultural	By industry — Nonagri-cultural	Employ-ment-population ratio, percent	Unemployment — Thousands of persons — Total	Unemployment — Thousands of persons — Unem-ployed 15 weeks and over	Unemployment — Rate (percent) [2]
1951													
January	104 844	61 941	59.1	59 636	39 595	16 279	3 762	6 828	52 808	56.9	2 305	438	3.7
February	104 604	61 778	59.1	59 661	39 695	16 257	3 709	6 738	52 923	57.0	2 117	386	3.4
March	104 629	62 526	59.8	60 401	40 013	16 557	3 831	6 858	53 543	57.7	2 125	355	3.4
April	104 541	61 808	59.1	59 889	39 804	16 426	3 659	6 722	53 167	57.3	1 919	294	3.1
May	104 491	62 044	59.4	60 188	39 752	16 581	3 855	6 752	53 436	57.6	1 856	269	3.0
June	104 488	61 615	59.0	59 620	39 538	16 368	3 714	6 529	53 091	57.1	1 995	258	3.2
July	104 504	62 106	59.4	60 156	39 483	16 898	3 775	6 601	53 555	57.6	1 950	260	3.1
August	104 536	61 927	59.2	59 994	39 508	16 665	3 821	6 790	53 204	57.4	1 933	249	3.1
September	104 588	61 780	59.1	59 713	39 416	16 504	3 793	6 558	53 155	57.1	2 067	223	3.3
October	104 690	62 204	59.4	60 010	39 555	16 674	3 781	6 636	53 374	57.3	2 194	269	3.5
November	104 740	62 014	59.2	59 836	39 504	16 669	3 663	6 699	53 137	57.1	2 178	316	3.5
December	104 810	62 457	59.6	60 497	39 691	16 946	3 860	7 065	53 432	57.7	1 960	269	3.1
1952													
January	104 862	62 432	59.5	60 460	39 714	17 001	3 745	7 148	53 312	57.7	1 972	282	3.2
February	104 868	62 419	59.5	60 462	39 772	16 935	3 755	7 020	53 442	57.7	1 957	248	3.1
March	104 860	61 721	58.9	59 908	39 580	16 627	3 701	6 468	53 440	57.1	1 813	234	2.9
April	104 906	61 720	58.8	59 909	39 542	16 659	3 708	6 525	53 384	57.1	1 811	242	2.9
May	104 996	62 058	59.1	60 195	39 588	16 844	3 763	6 334	53 861	57.3	1 863	219	3.0
June	105 118	62 103	59.1	60 219	39 558	16 837	3 824	6 529	53 690	57.3	1 884	210	3.0
July	105 246	61 962	58.9	59 971	39 496	16 778	3 697	6 334	53 637	57.0	1 991	194	3.2
August	105 346	61 877	58.7	59 790	39 289	16 867	3 634	6 174	53 616	56.8	2 087	211	3.4
September	105 436	62 457	59.2	60 521	39 386	17 477	3 658	6 537	53 984	57.4	1 936	249	3.1
October	105 591	61 971	58.7	60 132	39 451	17 032	3 649	6 363	53 769	56.9	1 839	230	3.0
November	105 706	62 491	59.1	60 748	39 549	17 450	3 749	6 509	54 239	57.5	1 743	216	2.8
December	105 812	62 621	59.2	60 954	40 011	17 181	3 762	6 361	54 593	57.6	1 667	238	2.7
1953													
January	106 594	63 439	59.5	61 600	40 256	17 482	3 862	6 642	54 958	57.8	1 839	268	2.9
February	106 678	63 520	59.5	61 884	40 546	17 321	4 017	6 463	55 421	58.0	1 636	208	2.6
March	106 744	63 657	59.6	62 010	40 648	17 397	3 965	6 420	55 590	58.1	1 647	213	2.6
April	106 826	63 167	59.1	61 444	40 346	17 242	3 856	6 362	55 082	57.5	1 723	180	2.7
May	106 910	62 615	58.6	61 019	40 323	16 983	3 713	5 937	55 082	57.1	1 596	176	2.5
June	106 978	63 063	58.9	61 456	40 358	17 301	3 797	6 361	55 095	57.4	1 607	213	2.5
July	107 034	63 057	58.9	61 397	40 378	17 341	3 678	6 267	55 130	57.4	1 660	168	2.6
August	107 132	62 816	58.6	61 151	40 352	17 108	3 691	6 319	54 832	57.1	1 665	177	2.7
September	107 253	62 727	58.5	60 906	40 192	17 063	3 651	6 198	54 708	56.8	1 821	178	2.9
October	107 383	62 867	58.5	60 893	40 155	17 236	3 502	6 096	54 797	56.7	1 974	190	3.1
November	107 504	62 949	58.6	60 738	40 163	16 974	3 601	6 345	54 393	56.5	2 211	259	3.5
December	107 623	62 795	58.3	59 977	39 885	16 599	3 493	5 929	54 048	55.7	2 818	309	4.5
1954													
January	107 763	63 101	58.6	60 024	39 834	16 574	3 616	6 073	53 951	55.7	3 077	372	4.9
February	107 880	63 994	59.3	60 663	39 899	17 162	3 602	6 590	54 073	56.2	3 331	532	5.2
March	107 987	63 793	59.1	60 186	39 497	17 022	3 667	6 395	53 791	55.7	3 607	765	5.7
April	108 080	63 934	59.2	60 185	39 613	17 015	3 557	6 142	54 043	55.7	3 749	774	5.9
May	108 184	63 675	58.9	59 908	39 467	16 975	3 466	6 210	53 698	55.4	3 767	879	5.9
June	108 267	63 343	58.5	59 792	39 476	16 894	3 422	6 162	53 630	55.2	3 551	880	5.6
July	108 344	63 302	58.4	59 643	39 467	16 777	3 399	6 222	53 421	55.0	3 659	932	5.8
August	108 440	63 707	58.7	59 853	39 582	16 868	3 403	6 087	53 766	55.2	3 854	1 002	6.0
September	108 546	64 209	59.2	60 282	39 702	17 133	3 447	6 453	53 829	55.5	3 927	1 017	6.1
October	108 668	63 936	58.8	60 270	39 618	17 209	3 443	6 242	54 028	55.5	3 666	1 009	5.7
November	108 798	63 759	58.6	60 357	39 745	17 213	3 399	5 934	54 423	55.5	3 402	975	5.3
December	108 892	63 312	58.1	60 116	39 763	17 121	3 232	5 848	54 268	55.2	3 196	827	5.0
1955													
January	109 059	63 910	58.6	60 753	39 937	17 375	3 441	6 113	54 640	55.7	3 157	882	4.9
February	109 078	63 696	58.4	60 727	39 964	17 413	3 350	5 854	54 873	55.7	2 969	826	4.7
March	109 254	63 882	58.5	60 964	40 111	17 415	3 438	6 242	54 722	55.8	2 918	816	4.6
April	109 377	64 564	59.0	61 515	40 120	17 867	3 528	6 363	55 152	56.2	3 049	811	4.7
May	109 544	64 381	58.8	61 634	40 410	17 665	3 559	6 327	55 307	56.3	2 747	734	4.3
June	109 680	64 482	58.8	61 781	40 444	17 837	3 500	6 243	55 538	56.3	2 701	668	4.2
July	109 792	65 145	59.3	62 513	40 751	18 123	3 639	6 438	56 075	56.9	2 632	640	4.0
August	109 882	65 581	59.7	62 797	40 747	18 377	3 673	6 575	56 222	57.1	2 784	535	4.2
September	109 977	65 628	59.7	62 950	40 920	18 285	3 745	6 819	56 131	57.2	2 678	558	4.1
October	110 085	65 821	59.8	62 991	40 858	18 327	3 806	6 728	56 263	57.2	2 830	572	4.3
November	110 177	66 037	59.9	63 257	40 936	18 422	3 899	6 655	56 602	57.4	2 780	564	4.2
December	110 296	66 445	60.2	63 684	41 063	18 630	3 991	6 653	57 031	57.7	2 761	581	4.2
1956													
January	110 390	66 419	60.2	63 753	41 203	18 691	3 859	6 590	57 163	57.8	2 666	561	4.0
February	110 478	66 124	59.9	63 518	41 175	18 582	3 761	6 457	57 061	57.5	2 606	545	3.9
March	110 582	66 175	59.8	63 411	41 199	18 496	3 716	6 221	57 190	57.3	2 764	521	4.2
April	110 650	66 264	59.9	63 614	41 289	18 629	3 696	6 460	57 154	57.5	2 650	476	4.0
May	110 810	66 722	60.2	63 861	41 166	18 844	3 851	6 375	57 486	57.6	2 861	506	4.3
June	110 903	66 702	60.1	63 820	41 196	18 748	3 876	6 335	57 485	57.5	2 882	516	4.3
July	111 019	66 752	60.1	63 800	41 216	18 718	3 866	6 320	57 480	57.5	2 952	523	4.4
August	111 099	66 673	60.0	63 972	41 265	18 864	3 843	6 280	57 692	57.6	2 701	543	4.1
September	111 222	66 714	60.0	64 079	41 221	19 019	3 839	6 375	57 704	57.6	2 635	577	3.9
October	111 335	66 546	59.8	63 975	41 261	18 928	3 786	6 137	57 838	57.5	2 571	530	3.9
November	111 432	66 657	59.8	63 796	41 208	18 846	3 742	5 997	57 799	57.3	2 861	575	4.3
December	111 526	66 700	59.8	63 910	41 192	18 859	3 859	5 806	58 104	57.3	2 790	567	4.2

[1] Not seasonally adjusted.
[2] In 1930 through 1943, the official BLS data count persons on work relief as unemployed. The unemployment rates for those years shown in parentheses count persons on work relief as employed, which is more consistent with the postwar practice. See notes and definitions.

Table 10-1B. Summary Labor Force, Employment, and Unemployment: Historical Data—*Continued*

(Thousands of persons, percent, seasonally adjusted, except as noted.)

Year and month	Civilian noninstitutional population [1]	Civilian labor force		Employment, thousands of persons						Employment-population ratio, percent	Unemployment		Rate (percent) [2]
		Thousands of persons	Participation rate (percent)	Total	Men, 20 years and over	Women, 20 years and over	Both sexes, 16 to 19 years	Agricultural	Nonagricultural		Total	Unemployed 15 weeks and over	
					By age and sex			By industry			Thousands of persons		
1957													
January	111 626	66 428	59.5	63 632	41 168	18 740	3 724	5 790	57 842	57.0	2 796	509	4.2
February	111 711	66 879	59.9	64 257	41 341	19 115	3 801	6 125	58 132	57.5	2 622	530	3.9
March	111 824	66 913	59.8	64 404	41 500	19 066	3 838	5 963	58 441	57.6	2 509	514	3.7
April	111 933	66 647	59.5	64 047	41 345	18 937	3 765	5 836	58 211	57.2	2 600	516	3.9
May	112 031	66 695	59.5	63 985	41 334	18 897	3 754	5 999	57 986	57.1	2 710	538	4.1
June	112 172	67 052	59.8	64 196	41 411	18 973	3 812	6 002	58 194	57.2	2 856	526	4.3
July	112 317	67 336	60.0	64 540	41 472	19 262	3 806	6 401	58 139	57.5	2 796	535	4.2
August	112 421	66 706	59.3	63 959	41 243	19 020	3 696	5 898	58 061	56.9	2 747	542	4.1
September	112 554	67 064	59.6	64 121	41 213	19 116	3 792	5 728	58 393	57.0	2 943	559	4.4
October	112 710	67 066	59.5	64 046	41 069	19 160	3 817	5 875	58 171	56.8	3 020	650	4.5
November	112 874	67 123	59.5	63 669	40 853	19 082	3 734	5 686	57 983	56.4	3 454	674	5.1
December	113 013	67 398	59.6	63 922	40 884	19 285	3 753	6 037	57 885	56.6	3 476	731	5.2
1958													
January	113 138	67 095	59.3	63 220	40 617	19 035	3 568	5 831	57 389	55.9	3 875	879	5.8
February	113 234	67 201	59.3	62 898	40 336	18 951	3 611	5 654	57 244	55.5	4 303	1 005	6.4
March	113 337	67 223	59.3	62 731	40 180	18 968	3 583	5 561	57 170	55.3	4 492	1 128	6.7
April	113 415	67 647	59.6	62 631	40 129	18 969	3 533	5 602	57 029	55.2	5 016	1 387	7.4
May	113 534	67 895	59.8	62 874	40 253	18 978	3 643	5 647	57 227	55.4	5 021	1 493	7.4
June	113 647	67 674	59.5	62 730	40 208	19 008	3 514	5 510	57 220	55.2	4 944	1 677	7.3
July	113 727	67 824	59.6	62 745	40 270	19 039	3 436	5 525	57 220	55.2	5 079	1 796	7.5
August	113 835	68 037	59.8	63 012	40 343	19 103	3 566	5 673	57 300	55.4	5 025	1 888	7.4
September	113 977	68 002	59.7	63 181	40 564	19 033	3 584	5 453	57 728	55.4	4 821	1 795	7.1
October	114 138	68 045	59.6	63 475	40 699	19 091	3 685	5 563	57 912	55.6	4 570	1 708	6.7
November	114 283	67 658	59.2	63 470	40 684	19 157	3 629	5 571	57 899	55.5	4 188	1 570	6.2
December	114 429	67 740	59.2	63 549	40 666	19 170	3 713	5 521	58 028	55.5	4 191	1 490	6.2
1959													
January	114 582	67 936	59.3	63 868	40 769	19 292	3 807	5 481	58 387	55.7	4 068	1 396	6.0
February	114 712	67 649	59.0	63 684	40 699	19 167	3 818	5 429	58 255	55.5	3 965	1 277	5.9
March	114 849	68 068	59.3	64 267	41 079	19 379	3 809	5 677	58 590	56.0	3 801	1 210	5.6
April	114 986	68 339	59.4	64 768	41 419	19 498	3 851	5 893	58 875	56.3	3 571	1 039	5.2
May	115 144	68 178	59.2	64 699	41 355	19 565	3 779	5 792	58 907	56.2	3 479	965	5.1
June	115 287	68 278	59.2	64 849	41 387	19 658	3 804	5 712	59 137	56.3	3 429	963	5.0
July	115 429	68 539	59.4	65 011	41 596	19 595	3 820	5 564	59 447	56.3	3 528	889	5.1
August	115 555	68 432	59.2	64 844	41 485	19 568	3 791	5 442	59 402	56.1	3 588	889	5.2
September	115 668	68 545	59.3	64 770	41 351	19 531	3 888	5 447	59 323	56.0	3 775	895	5.5
October	115 798	68 821	59.4	64 911	41 362	19 702	3 847	5 355	59 556	56.1	3 910	883	5.7
November	115 916	68 533	59.1	64 530	41 062	19 594	3 874	5 480	59 050	55.7	4 003	982	5.8
December	116 040	68 994	59.5	65 341	41 651	19 717	3 973	5 458	59 883	56.3	3 653	920	5.3
1960													
January	116 594	68 962	59.1	65 347	41 637	19 686	4 024	5 458	59 889	56.0	3 615	915	5.2
February	116 702	68 949	59.1	65 620	41 729	19 765	4 126	5 443	60 177	56.2	3 329	841	4.8
March	116 827	68 399	58.5	64 673	41 320	19 388	3 965	4 959	59 714	55.4	3 726	959	5.4
April	116 910	69 579	59.5	65 959	41 641	20 110	4 208	5 471	60 488	56.4	3 620	896	5.2
May	117 033	69 626	59.5	66 057	41 668	20 186	4 203	5 359	60 698	56.4	3 569	797	5.1
June	117 167	69 934	59.7	66 168	41 553	20 290	4 325	5 416	60 752	56.5	3 766	854	5.4
July	117 281	69 745	59.5	65 909	41 490	20 257	4 162	5 542	60 367	56.2	3 836	921	5.5
August	117 431	69 841	59.5	65 895	41 503	20 316	4 076	5 520	60 375	56.1	3 946	927	5.6
September	117 521	70 151	59.7	66 267	41 604	20 493	4 170	5 755	60 512	56.4	3 884	982	5.5
October	117 643	69 884	59.4	65 632	41 464	20 076	4 092	5 436	60 196	55.8	4 252	1 189	6.1
November	117 829	70 439	59.8	66 109	41 543	20 384	4 182	5 513	60 596	56.1	4 330	1 223	6.1
December	118 001	70 395	59.7	65 778	41 416	20 332	4 030	5 622	60 156	55.7	4 617	1 142	6.6
1961													
January	118 155	70 447	59.6	65 776	41 363	20 325	4 088	5 422	60 354	55.7	4 671	1 328	6.6
February	118 250	70 420	59.6	65 588	41 177	20 392	4 019	5 472	60 116	55.5	4 832	1 416	6.9
March	118 358	70 703	59.7	65 850	41 273	20 459	4 118	5 406	60 444	55.6	4 853	1 463	6.9
April	118 503	70 267	59.3	65 374	41 206	20 145	4 023	5 037	60 337	55.2	4 893	1 598	7.0
May	118 638	70 452	59.4	65 449	41 139	20 261	4 049	5 099	60 350	55.2	5 003	1 686	7.1
June	118 767	70 878	59.7	65 993	41 349	20 446	4 198	5 220	60 773	55.6	4 885	1 651	6.9
July	118 889	70 536	59.3	65 608	41 245	20 252	4 111	5 153	60 455	55.2	4 928	1 830	7.0
August	119 006	70 534	59.3	65 852	41 362	20 279	4 211	5 366	60 486	55.3	4 682	1 649	6.6
September	119 107	70 217	59.0	65 541	41 400	20 112	4 029	5 021	60 520	55.0	4 676	1 531	6.7
October	119 202	70 492	59.1	65 919	41 509	20 338	4 072	5 203	60 716	55.3	4 573	1 481	6.5
November	119 153	70 376	59.1	66 081	41 556	20 330	4 195	5 090	60 991	55.5	4 295	1 388	6.1
December	119 214	70 077	58.8	65 900	41 534	20 287	4 079	4 992	60 908	55.3	4 177	1 361	6.0
1962													
January	119 300	70 189	58.8	66 108	41 547	20 501	4 060	5 094	61 014	55.4	4 081	1 235	5.8
February	119 360	70 409	59.0	66 538	41 745	20 693	4 100	5 289	61 249	55.7	3 871	1 244	5.5
March	119 476	70 414	58.9	66 493	41 696	20 567	4 230	5 157	61 336	55.7	3 921	1 162	5.6
April	119 702	70 278	58.7	66 372	41 647	20 567	4 158	5 009	61 363	55.4	3 906	1 122	5.6
May	119 813	70 551	58.9	66 688	41 847	20 558	4 283	4 964	61 724	55.7	3 863	1 134	5.5
June	119 943	70 514	58.8	66 670	41 761	20 547	4 362	4 943	61 727	55.6	3 844	1 079	5.5
July	120 128	70 302	58.5	66 483	41 671	20 592	4 220	4 840	61 643	55.3	3 819	1 049	5.4
August	120 323	70 981	59.0	66 968	41 900	20 841	4 227	4 866	62 102	55.7	4 013	1 081	5.7
September	120 653	71 153	59.0	67 192	42 020	20 982	4 190	4 867	62 325	55.7	3 961	1 096	5.6
October	120 856	70 917	58.7	67 114	42 086	20 856	4 172	4 816	62 298	55.5	3 803	1 022	5.4
November	121 045	70 871	58.5	66 847	41 985	20 794	4 068	4 831	62 016	55.2	4 024	1 051	5.7
December	121 236	70 854	58.4	66 947	41 934	20 831	4 182	4 647	62 300	55.2	3 907	1 068	5.5

[1]Not seasonally adjusted.
[2]In 1930 through 1943, the official BLS data count persons on work relief as unemployed. The unemployment rates for those years shown in parentheses count persons on work relief as employed, which is more consistent with the postwar practice. See notes and definitions.

Table 10-1B. Summary Labor Force, Employment, and Unemployment: Historical Data—*Continued*

(Thousands of persons, percent, seasonally adjusted, except as noted.)

Year and month	Civilian noninsti- tutional popu- lation [1]	Civilian labor force		Employment, thousands of persons						Employ- ment- population ratio, percent	Unemployment		
		Thousands of persons	Participa- tion rate (percent)	Total	By age and sex			By industry			Thousands of persons		Rate (percent) [2]
					Men, 20 years and over	Women, 20 years and over	Both sexes, 16 to 19 years	Agri- cultural	Nonagri- cultural		Total	Unem- ployed 15 weeks and over	
1963													
January	121 463	71 146	58.6	67 072	41 938	20 933	4 201	4 882	62 190	55.2	4 074	1 122	5.7
February	121 633	71 262	58.6	67 024	41 876	21 046	4 102	4 652	62 372	55.1	4 238	1 137	5.9
March	121 824	71 423	58.6	67 351	42 047	21 162	4 142	4 696	62 655	55.3	4 072	1 087	5.7
April	121 986	71 697	58.8	67 642	42 131	21 281	4 230	4 670	62 972	55.5	4 055	1 071	5.7
May	122 162	71 832	58.8	67 615	42 145	21 225	4 245	4 729	62 886	55.3	4 217	1 157	5.9
June	122 352	71 626	58.5	67 649	42 268	21 185	4 196	4 642	63 007	55.3	3 977	1 067	5.6
July	122 521	71 956	58.7	67 905	42 427	21 268	4 210	4 694	63 211	55.4	4 051	1 070	5.6
August	122 667	71 786	58.5	67 908	42 400	21 185	4 323	4 604	63 304	55.4	3 878	1 114	5.4
September	122 821	72 131	58.7	68 174	42 500	21 317	4 357	4 650	63 524	55.5	3 957	1 069	5.5
October	123 014	72 281	58.8	68 294	42 437	21 456	4 401	4 702	63 592	55.5	3 987	1 071	5.5
November	123 192	72 418	58.8	68 267	42 415	21 553	4 299	4 694	63 573	55.4	4 151	1 054	5.7
December	123 360	72 188	58.5	68 213	42 427	21 481	4 305	4 629	63 584	55.3	3 975	1 007	5.5
1964													
January	123 560	72 356	58.6	68 327	42 510	21 462	4 355	4 603	63 724	55.3	4 029	1 057	5.6
February	123 707	72 683	58.8	68 751	42 579	21 652	4 520	4 563	64 188	55.6	3 932	1 015	5.4
March	123 857	72 713	58.7	68 763	42 600	21 685	4 478	4 366	64 397	55.5	3 950	1 039	5.4
April	124 019	73 274	59.1	69 356	42 885	22 110	4 361	4 414	64 942	55.9	3 918	934	5.3
May	124 204	73 395	59.1	69 631	43 025	22 103	4 503	4 603	65 028	56.1	3 764	975	5.1
June	124 386	73 032	58.7	69 218	42 760	21 995	4 463	4 556	64 662	55.6	3 814	1 047	5.2
July	124 567	73 007	58.6	69 399	42 998	21 846	4 555	4 591	64 808	55.7	3 608	1 002	4.9
August	124 731	73 118	58.6	69 463	42 963	22 002	4 498	4 573	64 890	55.7	3 655	934	5.0
September	124 920	73 290	58.7	69 578	43 009	21 863	4 706	4 619	64 959	55.7	3 712	917	5.1
October	125 108	73 308	58.6	69 582	43 023	21 984	4 575	4 550	65 032	55.6	3 726	903	5.1
November	125 291	73 286	58.5	69 735	43 171	21 954	4 610	4 496	65 239	55.7	3 551	922	4.8
December	125 468	73 465	58.6	69 814	43 109	22 136	4 569	4 322	65 492	55.6	3 651	873	5.0
1965													
January	125 647	73 569	58.6	69 997	43 237	22 282	4 478	4 271	65 726	55.7	3 572	793	4.9
February	125 810	73 857	58.7	70 127	43 279	22 276	4 572	4 322	65 805	55.7	3 730	919	5.1
March	125 985	73 949	58.7	70 439	43 370	22 373	4 696	4 318	66 121	55.9	3 510	796	4.7
April	126 155	74 228	58.8	70 633	43 397	22 416	4 820	4 424	66 209	56.0	3 595	796	4.8
May	126 320	74 466	59.0	71 034	43 579	22 494	4 961	4 724	66 310	56.2	3 432	736	4.6
June	126 499	74 412	58.8	71 025	43 487	22 759	4 779	4 444	66 581	56.1	3 387	786	4.6
July	126 573	74 761	59.1	71 460	43 489	22 841	5 130	4 390	67 070	56.5	3 301	683	4.4
August	126 756	74 616	58.9	71 362	43 447	22 783	5 132	4 355	67 007	56.3	3 254	733	4.4
September	126 906	74 502	58.7	71 286	43 371	22 684	5 231	4 271	67 015	56.2	3 216	732	4.3
October	127 043	74 838	58.9	71 695	43 461	22 819	5 415	4 418	67 277	56.4	3 143	672	4.2
November	127 171	74 797	58.8	71 724	43 447	22 829	5 448	4 093	67 631	56.4	3 073	645	4.1
December	127 294	75 093	59.0	72 062	43 513	22 983	5 566	4 159	67 903	56.6	3 031	659	4.0
1966													
January	127 394	75 186	59.0	72 198	43 495	23 098	5 605	4 077	68 121	56.7	2 988	623	4.0
February	127 514	74 954	58.8	72 134	43 528	23 089	5 517	4 078	68 056	56.6	2 820	594	3.8
March	127 626	75 075	58.8	72 188	43 576	23 109	5 503	4 069	68 119	56.6	2 887	583	3.8
April	127 744	75 338	59.0	72 510	43 679	23 227	5 604	4 108	68 402	56.8	2 828	575	3.8
May	127 879	75 447	59.0	72 497	43 710	23 282	5 505	3 930	68 567	56.7	2 950	534	3.9
June	127 983	75 647	59.1	72 775	43 662	23 359	5 754	3 967	68 808	56.9	2 872	475	3.8
July	128 102	75 736	59.1	72 860	43 574	23 422	5 864	3 920	68 940	56.9	2 876	427	3.8
August	128 240	76 046	59.3	73 146	43 636	23 605	5 905	3 921	69 225	57.0	2 900	464	3.8
September	128 359	76 056	59.3	73 258	43 718	23 881	5 659	3 952	69 306	57.1	2 798	488	3.7
October	128 494	76 199	59.3	73 401	43 776	23 881	5 744	3 912	69 489	57.1	2 798	494	3.7
November	128 627	76 610	59.6	73 840	43 804	24 130	5 906	3 945	69 895	57.4	2 770	464	3.6
December	128 730	76 641	59.5	73 729	43 820	24 025	5 884	3 906	69 823	57.3	2 912	488	3.8
1967													
January	128 909	76 639	59.5	73 671	44 029	23 872	5 770	3 890	69 781	57.1	2 968	489	3.9
February	129 032	76 521	59.3	73 606	43 997	23 919	5 690	3 723	69 883	57.0	2 915	459	3.8
March	129 190	76 328	59.1	73 439	43 922	23 832	5 685	3 757	69 682	56.8	2 889	436	3.8
April	129 344	76 777	59.4	73 882	44 061	24 161	5 660	3 748	70 134	57.1	2 895	428	3.8
May	129 515	76 773	59.3	73 844	44 100	24 172	5 572	3 658	70 186	57.0	2 929	417	3.8
June	129 722	77 270	59.6	74 278	44 230	24 303	5 745	3 689	70 589	57.3	2 992	422	3.9
July	129 918	77 464	59.6	74 520	44 364	24 416	5 740	3 833	70 687	57.4	2 944	412	3.8
August	130 187	77 712	59.7	74 767	44 410	24 600	5 757	3 963	70 804	57.4	2 945	441	3.8
September	130 392	77 812	59.7	74 854	44 535	24 683	5 636	3 851	71 003	57.4	2 958	448	3.8
October	130 582	78 194	59.9	75 051	44 610	24 802	5 639	4 008	71 043	57.5	3 143	472	4.0
November	130 754	78 191	59.8	75 125	44 625	24 914	5 586	3 933	71 192	57.5	3 066	490	3.9
December	130 936	78 491	59.9	75 473	44 719	25 104	5 650	4 076	71 397	57.6	3 018	485	3.8
1968													
January	131 112	77 578	59.2	74 700	44 606	24 581	5 513	3 908	70 792	57.0	2 878	503	3.7
February	131 277	78 230	59.6	75 229	44 659	24 881	5 689	3 959	71 270	57.3	3 001	468	3.8
March	131 412	78 256	59.6	75 379	44 663	25 019	5 697	3 904	71 475	57.4	2 877	447	3.7
April	131 553	78 270	59.5	75 561	44 753	25 072	5 736	3 875	71 686	57.4	2 709	393	3.5
May	131 712	78 847	59.9	76 107	44 841	25 513	5 753	3 814	72 293	57.8	2 740	395	3.5
June	131 872	79 120	60.0	76 182	44 914	25 466	5 802	3 806	72 376	57.8	2 938	405	3.7
July	132 053	78 970	59.8	76 087	44 935	25 347	5 805	3 820	72 267	57.6	2 883	426	3.7
August	132 251	78 811	59.6	76 043	44 897	25 201	5 945	3 736	72 307	57.5	2 768	393	3.5
September	132 446	78 858	59.5	76 172	44 893	25 445	5 834	3 758	72 414	57.5	2 686	375	3.4
October	132 617	78 913	59.5	76 224	44 884	25 475	5 865	3 741	72 483	57.5	2 689	386	3.4
November	132 903	79 209	59.6	76 494	44 996	25 674	5 824	3 758	72 736	57.6	2 715	357	3.4
December	133 120	79 463	59.7	76 778	45 262	25 712	5 804	3 746	73 032	57.7	2 685	351	3.4

[1]Not seasonally adjusted.
[2]In 1930 through 1943, the official BLS data count persons on work relief as unemployed. The unemployment rates for those years shown in parentheses count persons on work relief as employed, which is more consistent with the postwar practice. See notes and definitions.

Table 10-1B. Summary Labor Force, Employment, and Unemployment: Historical Data—*Continued*

(Thousands of persons, percent, seasonally adjusted, except as noted.)

Year and month	Civilian noninstitutional population [1]	Civilian labor force		Employment, thousands of persons						Employment-population ratio, percent	Unemployment		
		Thousands of persons	Participation rate (percent)	Total	By age and sex			By industry			Thousands of persons		Rate (percent) [2]
					Men, 20 years and over	Women, 20 years and over	Both sexes, 16 to 19 years	Agricultural	Nonagricultural		Total	Unemployed 15 weeks and over	
1969													
January	133 324	79 523	59.6	76 805	45 154	25 777	5 874	3 704	73 101	57.6	2 718	339	3.4
February	133 465	80 019	60.0	77 327	45 339	26 092	5 896	3 770	73 557	57.9	2 692	358	3.4
March	133 639	80 079	59.9	77 367	45 305	26 115	5 947	3 668	73 699	57.9	2 712	353	3.4
April	133 821	80 281	60.0	77 523	45 262	26 233	6 028	3 629	73 894	57.9	2 758	386	3.4
May	134 027	80 125	59.8	77 412	45 278	26 283	5 851	3 706	73 706	57.8	2 713	387	3.4
June	134 213	80 696	60.1	77 880	45 313	26 429	6 138	3 663	74 217	58.0	2 816	368	3.5
July	134 414	80 827	60.1	77 959	45 305	26 516	6 138	3 548	74 411	58.0	2 868	377	3.5
August	134 597	81 106	60.3	78 250	45 513	26 556	6 181	3 613	74 637	58.1	2 856	373	3.5
September	134 774	81 290	60.3	78 250	45 447	26 572	6 231	3 551	74 699	58.1	3 040	391	3.7
October	135 012	81 494	60.4	78 445	45 488	26 658	6 299	3 517	74 928	58.1	3 049	374	3.7
November	135 239	81 397	60.2	78 541	45 505	26 652	6 384	3 477	75 064	58.1	2 856	392	3.5
December	135 489	81 624	60.2	78 740	45 577	26 832	6 331	3 409	75 331	58.1	2 884	413	3.5
1970													
January	135 713	81 981	60.4	78 780	45 654	26 908	6 218	3 422	75 358	58.0	3 201	431	3.9
February	135 957	82 151	60.4	78 698	45 627	26 828	6 243	3 439	75 259	57.9	3 453	470	4.2
March	136 179	82 498	60.6	78 863	45 668	26 933	6 262	3 499	75 364	57.9	3 635	534	4.4
April	136 416	82 727	60.6	78 930	45 679	27 114	6 137	3 568	75 362	57.9	3 797	602	4.6
May	136 686	82 483	60.3	78 564	45 666	26 739	6 159	3 547	75 017	57.5	3 919	591	4.8
June	136 928	82 484	60.2	78 413	45 554	26 904	5 955	3 555	74 858	57.3	4 071	657	4.9
July	137 196	82 901	60.4	78 726	45 516	27 083	6 127	3 517	75 209	57.4	4 175	662	5.0
August	137 455	82 880	60.3	78 624	45 495	27 011	6 118	3 418	75 206	57.2	4 256	705	5.1
September	137 717	82 954	60.2	78 498	45 535	26 704	6 179	3 451	75 047	57.0	4 456	788	5.4
October	137 988	83 070	60.4	78 685	45 508	27 058	6 119	3 337	75 348	57.0	4 591	771	5.5
November	138 264	83 548	60.4	78 650	45 540	27 020	6 090	3 372	75 278	56.9	4 808	871	5.9
December	138 529	83 670	60.4	78 594	45 466	27 038	6 090	3 380	75 214	56.7	5 076	1 102	6.1
1971													
January	138 795	83 850	60.4	78 864	45 527	27 173	6 164	3 393	75 471	56.8	4 986	1 113	5.9
February	139 021	83 603	60.1	78 700	45 455	27 040	6 205	3 288	75 412	56.6	4 903	1 068	5.9
March	139 285	83 575	60.0	78 588	45 520	26 967	6 101	3 356	75 232	56.4	4 987	1 098	6.0
April	139 566	83 946	60.1	78 987	45 789	26 984	6 214	3 574	75 413	56.6	4 959	1 149	5.9
May	139 826	84 135	60.2	79 139	45 917	27 056	6 166	3 449	75 690	56.6	4 996	1 173	5.9
June	140 090	83 706	59.8	78 757	45 879	27 013	5 865	3 334	75 423	56.2	4 949	1 167	5.9
July	140 343	84 340	60.1	79 305	46 000	27 054	6 251	3 386	75 919	56.5	5 035	1 251	6.0
August	140 596	84 673	60.2	79 539	46 041	27 171	6 327	3 395	76 144	56.6	5 134	1 261	6.1
September	140 869	84 731	60.1	79 689	46 090	27 390	6 209	3 367	76 322	56.6	5 042	1 239	6.0
October	141 146	84 872	60.1	79 918	46 132	27 538	6 248	3 405	76 513	56.6	4 954	1 268	5.8
November	141 393	85 458	60.4	80 297	46 209	27 721	6 367	3 410	76 887	56.8	5 161	1 277	6.0
December	141 666	85 625	60.4	80 471	46 280	27 791	6 400	3 371	77 100	56.8	5 154	1 283	6.0
1972													
January	142 736	85 978	60.2	80 959	46 471	27 956	6 532	3 366	77 593	56.7	5 019	1 257	5.8
February	143 017	86 036	60.2	81 108	46 600	28 016	6 492	3 358	77 750	56.7	4 928	1 292	5.7
March	143 263	86 611	60.5	81 573	46 821	28 126	6 626	3 438	78 135	56.9	5 038	1 232	5.8
April	143 483	86 614	60.4	81 655	46 863	28 114	6 678	3 382	78 273	56.9	4 959	1 203	5.7
May	143 760	86 809	60.4	81 887	46 950	28 184	6 753	3 412	78 475	57.0	4 922	1 168	5.7
June	144 033	87 006	60.4	82 083	47 147	28 175	6 761	3 402	78 681	57.0	4 923	1 141	5.7
July	144 285	87 143	60.4	82 230	47 244	28 225	6 761	3 461	78 769	57.0	4 913	1 154	5.6
August	144 522	87 517	60.6	82 578	47 321	28 382	6 875	3 603	78 975	57.1	4 939	1 156	5.6
September	144 761	87 392	60.4	82 543	47 394	28 417	6 732	3 568	78 975	57.0	4 849	1 131	5.5
October	144 988	87 491	60.3	82 616	47 354	28 438	6 824	3 634	78 982	57.0	4 875	1 123	5.6
November	145 211	87 592	60.3	82 990	47 529	28 567	6 894	3 517	79 473	57.2	4 602	1 040	5.3
December	145 446	87 943	60.5	83 400	47 747	28 698	6 955	3 596	79 804	57.3	4 543	1 006	5.2
1973													
January	145 720	87 487	60.0	83 161	47 701	28 596	6 864	3 456	79 705	57.1	4 326	947	4.9
February	145 943	88 364	60.5	83 912	47 884	28 995	7 033	3 415	80 497	57.5	4 452	894	5.0
March	146 230	88 846	60.8	84 452	48 117	29 110	7 225	3 469	80 983	57.8	4 394	889	4.9
April	146 459	89 018	60.8	84 559	48 098	29 304	7 157	3 407	81 152	57.7	4 459	809	5.0
May	146 719	88 977	60.6	84 648	48 068	29 432	7 148	3 376	81 272	57.7	4 329	816	4.9
June	146 981	89 548	60.9	85 185	48 244	29 505	7 436	3 509	81 676	58.0	4 363	779	4.9
July	147 233	89 604	60.9	85 299	48 452	29 592	7 255	3 540	81 759	57.9	4 305	756	4.8
August	147 471	89 509	60.7	85 204	48 353	29 578	7 273	3 425	81 779	57.8	4 305	788	4.8
September	147 731	89 838	60.8	85 488	48 408	29 710	7 370	3 342	82 146	57.9	4 350	785	4.8
October	147 980	90 131	60.9	85 987	48 631	29 885	7 471	3 424	82 563	58.1	4 144	793	4.6
November	148 219	90 716	61.2	86 320	48 764	30 071	7 485	3 593	82 727	58.2	4 396	832	4.8
December	148 479	90 890	61.2	86 401	48 902	29 991	7 508	3 658	82 743	58.2	4 489	767	4.9
1974													
January	148 753	91 199	61.3	86 555	49 107	29 893	7 555	3 756	82 799	58.2	4 644	799	5.1
February	148 982	91 485	61.4	86 754	49 057	30 146	7 551	3 824	82 930	58.2	4 731	829	5.2
March	149 225	91 453	61.3	86 819	48 986	30 293	7 540	3 726	83 093	58.2	4 634	849	5.1
April	149 478	91 287	61.1	86 669	48 853	30 376	7 440	3 582	83 087	58.0	4 618	889	5.1
May	149 750	91 596	61.2	86 891	49 039	30 424	7 428	3 529	83 362	58.0	4 705	880	5.1
June	150 012	91 868	61.2	86 941	48 946	30 512	7 483	3 386	83 555	58.0	4 927	926	5.4
July	150 248	92 212	61.4	87 149	48 883	30 869	7 397	3 436	83 713	58.0	5 063	924	5.5
August	150 493	92 059	61.2	87 037	48 950	30 662	7 425	3 429	83 608	57.8	5 022	960	5.5
September	150 753	92 488	61.4	87 051	48 978	30 569	7 504	3 460	83 591	57.7	5 437	1 021	5.9
October	151 009	92 518	61.3	86 995	48 959	30 570	7 466	3 431	83 564	57.6	5 523	1 072	6.0
November	151 256	92 766	61.3	86 626	48 833	30 424	7 369	3 405	83 221	57.3	6 140	1 128	6.6
December	151 494	92 780	61.2	86 144	48 458	30 431	7 255	3 361	82 783	56.9	6 636	1 326	7.2

[1] Not seasonally adjusted.
[2] In 1930 through 1943, the official BLS data count persons on work relief as unemployed. The unemployment rates for those years shown in parentheses count persons on work relief as employed, which is more consistent with the postwar practice. See notes and definitions.

Table 10-1B. Summary Labor Force, Employment, and Unemployment: Historical Data—Continued

(Thousands of persons, percent, seasonally adjusted, except as noted.)

Year and month	Civilian noninstitutional population [1]	Civilian labor force		Employment, thousands of persons						Employment-population ratio, percent	Unemployment		
		Thousands of persons	Participation rate (percent)	Total	By age and sex			By industry			Thousands of persons		Rate (percent) [2]
					Men, 20 years and over	Women, 20 years and over	Both sexes, 16 to 19 years	Agricultural	Nonagricultural		Total	Unemployed 15 weeks and over	
1975													
January	151 755	93 128	61.4	85 627	48 086	30 343	7 198	3 401	82 226	56.4	7 501	1 555	8.1
February	151 990	92 776	61.0	85 256	47 927	30 215	7 114	3 361	81 895	56.1	7 520	1 841	8.1
March	152 217	93 165	61.2	85 187	47 776	30 334	7 077	3 358	81 829	56.0	7 978	2 074	8.6
April	152 443	93 399	61.3	85 189	47 759	30 410	7 020	3 315	81 874	55.9	8 210	2 442	8.8
May	152 704	93 884	61.5	85 451	47 835	30 483	7 133	3 560	81 891	56.0	8 433	2 643	9.0
June	152 976	93 575	61.2	85 355	47 754	30 618	6 983	3 368	81 987	55.8	8 220	2 843	8.8
July	153 309	94 021	61.3	85 894	48 050	30 794	7 050	3 457	82 437	56.0	8 127	2 943	8.6
August	153 580	94 162	61.3	86 234	48 239	30 966	7 029	3 429	82 805	56.1	7 928	2 862	8.4
September	153 848	94 202	61.2	86 279	48 126	30 979	7 174	3 508	82 771	56.1	7 923	2 906	8.4
October	154 082	94 267	61.2	86 370	48 165	31 121	7 084	3 397	82 973	56.1	7 897	2 689	8.4
November	154 338	94 250	61.1	86 456	48 203	31 135	7 118	3 331	83 125	56.0	7 794	2 789	8.3
December	154 589	94 409	61.1	86 665	48 266	31 268	7 131	3 259	83 406	56.1	7 744	2 868	8.2
1976													
January	154 853	94 934	61.3	87 400	48 592	31 595	8 967	3 387	84 013	56.4	7 534	2 713	7.9
February	155 066	94 998	61.3	87 672	48 721	31 680	8 981	3 304	84 368	56.5	7 326	2 519	7.7
March	155 306	95 215	61.3	87 985	48 836	31 842	9 007	3 296	84 689	56.7	7 230	2 441	7.6
April	155 529	95 746	61.6	88 416	49 097	31 951	9 151	3 438	84 978	56.8	7 330	2 210	7.7
May	155 765	95 847	61.5	88 794	49 193	32 147	9 155	3 367	85 427	57.0	7 053	2 115	7.4
June	156 026	95 885	61.5	88 563	49 010	32 267	8 943	3 310	85 253	56.8	7 322	2 332	7.6
July	156 276	96 583	61.8	89 093	49 236	32 334	9 204	3 358	85 735	57.0	7 490	2 316	7.8
August	156 525	96 741	61.8	89 223	49 417	32 437	9 168	3 380	85 843	57.0	7 518	2 378	7.8
September	156 779	96 553	61.6	89 173	49 485	32 390	8 968	3 278	85 895	56.9	7 380	2 296	7.6
October	156 993	96 704	61.6	89 274	49 524	32 412	9 054	3 316	85 958	56.9	7 430	2 292	7.7
November	157 235	97 254	61.9	89 634	49 561	32 753	9 055	3 263	86 371	57.0	7 620	2 354	7.8
December	157 438	97 348	61.8	89 803	49 599	32 914	9 011	3 251	86 552	57.0	7 545	2 375	7.8
1977													
January	157 688	97 208	61.6	89 928	49 738	32 872	9 025	3 185	86 743	57.0	7 280	2 200	7.5
February	157 913	97 785	61.9	90 342	49 838	32 997	9 198	3 222	87 120	57.2	7 443	2 174	7.6
March	158 131	98 115	62.0	90 808	50 031	33 246	9 257	3 212	87 596	57.4	7 307	2 057	7.4
April	158 371	98 330	62.1	91 271	50 185	33 470	9 289	3 313	87 958	57.6	7 059	1 936	7.2
May	158 657	98 665	62.2	91 754	50 280	33 851	9 279	3 432	88 322	57.8	6 911	1 928	7.0
June	158 928	99 093	62.4	91 959	50 544	33 678	9 525	3 340	88 619	57.9	7 134	1 918	7.2
July	159 185	98 913	62.1	92 084	50 597	33 749	9 377	3 247	88 837	57.8	6 829	1 907	6.9
August	159 430	99 366	62.3	92 441	50 745	33 809	9 550	3 260	89 181	58.0	6 925	1 836	7.0
September	159 674	99 453	62.3	92 702	50 825	34 218	9 340	3 201	89 501	58.1	6 751	1 853	6.8
October	159 915	99 815	62.4	93 052	51 046	34 187	9 441	3 272	89 780	58.2	6 763	1 789	6.8
November	160 129	100 576	62.8	93 761	51 316	34 536	9 551	3 375	90 386	58.6	6 815	1 804	6.8
December	160 376	100 491	62.7	94 105	51 492	34 668	9 406	3 320	90 785	58.7	6 386	1 717	6.4
1978													
January	160 617	100 873	62.8	94 384	51 542	34 948	9 473	3 434	90 950	58.8	6 489	1 643	6.4
February	160 831	100 837	62.7	94 519	51 578	35 118	9 448	3 320	91 199	58.8	6 318	1 584	6.3
March	161 038	101 092	62.8	94 755	51 635	35 310	9 441	3 351	91 404	58.8	6 337	1 531	6.3
April	161 263	101 574	63.0	95 394	51 912	35 546	9 518	3 349	92 045	59.2	6 180	1 502	6.1
May	161 518	101 896	63.1	95 769	52 050	35 597	9 668	3 325	92 444	59.3	6 127	1 420	6.0
June	161 794	102 371	63.3	96 343	52 240	35 828	9 781	3 483	92 860	59.5	6 028	1 352	5.9
July	162 034	102 399	63.2	96 090	52 190	35 764	9 749	3 441	92 649	59.3	6 309	1 373	6.2
August	162 259	102 511	63.2	96 431	52 228	35 856	9 903	3 401	93 030	59.4	6 080	1 242	5.9
September	162 502	102 795	63.3	96 670	52 284	36 274	9 700	3 400	93 270	59.5	6 125	1 308	6.0
October	162 783	103 080	63.3	97 133	52 448	36 525	9 727	3 409	93 724	59.7	5 947	1 319	5.8
November	163 017	103 562	63.5	97 485	52 802	36 559	9 704	3 284	94 201	59.8	6 077	1 242	5.9
December	163 272	103 809	63.6	97 581	52 807	36 686	9 708	3 396	94 185	59.8	6 228	1 269	6.0
1979													
January	163 516	104 057	63.6	97 948	53 072	36 697	9 749	3 305	94 643	59.9	6 109	1 250	5.9
February	163 726	104 502	63.8	98 329	53 233	36 904	9 762	3 373	94 956	60.1	6 173	1 297	5.9
March	164 027	104 589	63.8	98 480	53 120	37 159	9 751	3 368	95 112	60.0	6 109	1 365	5.8
April	164 162	104 172	63.5	98 103	53 085	36 944	9 652	3 291	94 812	59.8	6 069	1 272	5.8
May	164 459	104 171	63.3	98 331	53 178	37 134	9 553	3 272	95 059	59.8	5 840	1 239	5.6
June	164 720	104 638	63.5	98 679	53 309	37 221	9 664	3 331	95 348	59.9	5 959	1 171	5.7
July	164 970	105 002	63.6	99 006	53 384	37 514	9 606	3 335	95 671	60.0	5 996	1 123	5.7
August	165 198	105 096	63.6	98 776	53 336	37 548	9 456	3 374	95 402	59.8	6 320	1 203	6.0
September	165 431	105 530	63.8	99 340	53 510	37 798	9 623	3 371	95 969	60.0	6 190	1 172	5.9
October	165 813	105 700	63.7	99 404	53 478	37 931	9 574	3 325	96 079	59.9	6 296	1 219	6.0
November	166 051	105 812	63.7	99 574	53 435	38 065	9 599	3 436	96 138	60.0	6 238	1 239	5.9
December	166 300	106 258	63.9	99 933	53 555	38 259	9 690	3 400	96 533	60.1	6 325	1 277	6.0
1980													
January	166 544	106 562	64.0	99 879	53 501	38 367	9 590	3 316	96 563	60.0	6 683	1 353	6.3
February	166 759	106 697	64.0	99 995	53 686	38 389	9 501	3 397	96 598	60.0	6 702	1 358	6.3
March	166 984	106 442	63.7	99 713	53 353	38 406	9 500	3 418	96 295	59.7	6 729	1 457	6.3
April	167 197	106 591	63.8	99 233	53 035	38 427	9 272	3 326	95 907	59.4	7 358	1 694	6.9
May	167 407	106 929	63.9	98 945	52 915	38 335	9 457	3 382	95 563	59.1	7 984	1 740	7.5
June	167 643	106 780	63.7	98 682	52 712	38 312	9 438	3 296	95 386	58.9	8 098	1 760	7.6
July	167 932	107 159	63.8	98 796	52 733	38 374	9 499	3 319	95 477	58.8	8 363	1 995	7.8
August	168 103	107 105	63.7	98 824	52 815	38 511	9 247	3 234	95 590	58.8	8 281	2 162	7.7
September	168 297	107 098	63.6	99 077	52 866	38 595	9 289	3 443	95 634	58.9	8 021	2 309	7.5
October	168 503	107 405	63.7	99 317	53 094	38 620	9 319	3 372	95 945	58.9	8 088	2 306	7.5
November	168 695	107 568	63.8	99 545	53 210	38 795	9 246	3 396	96 149	59.0	8 023	2 329	7.5
December	168 883	107 352	63.6	99 634	53 333	38 737	9 175	3 492	96 142	59.0	7 718	2 406	7.2

[1] Not seasonally adjusted.
[2] In 1930 through 1943, the official BLS data count persons on work relief as unemployed. The unemployment rates for those years shown in parentheses count persons on work relief as employed, which is more consistent with the postwar practice. See notes and definitions.

Table 10-1B. Summary Labor Force, Employment, and Unemployment: Historical Data—*Continued*

(Thousands of persons, percent, seasonally adjusted, except as noted.)

Year and month	Civilian noninsti-tutional popu-lation [1]	Civilian labor force		Employment, thousands of persons						Employ-ment-population ratio, percent	Unemployment		
					By age and sex			By industry			Thousands of persons		Rate (percent) [2]
		Thousands of persons	Participa-tion rate (percent)	Total	Men, 20 years and over	Women, 20 years and over	Both sexes, 16 to 19 years	Agri-cultural	Nonagri-cultural		Total	Unem-ployed 15 weeks and over	
1981													
January	169 104	108 026	63.9	99 955	53 392	39 042	9 300	3 429	96 526	59.1	8 071	2 389	7.5
February	169 280	108 242	63.9	100 191	53 445	39 280	9 257	3 345	96 846	59.2	8 051	2 344	7.4
March	169 453	108 553	64.1	100 571	53 662	39 464	9 212	3 365	97 206	59.4	7 982	2 276	7.4
April	169 641	108 925	64.2	101 056	53 886	39 628	9 289	3 529	97 527	59.6	7 869	2 231	7.2
May	169 829	109 222	64.3	101 048	53 879	39 759	9 160	3 369	97 679	59.5	8 174	2 221	7.5
June	170 042	108 396	63.7	100 298	53 576	39 682	8 780	3 334	96 964	59.0	8 098	2 250	7.5
July	170 246	108 556	63.8	100 693	53 814	39 683	8 839	3 296	97 397	59.1	7 863	2 166	7.2
August	170 399	108 725	63.8	100 689	53 718	39 723	8 931	3 379	97 310	59.1	8 036	2 241	7.4
September	170 593	108 294	63.5	100 064	53 625	39 342	8 835	3 361	96 703	58.7	8 230	2 261	7.6
October	170 809	109 024	63.8	100 378	53 482	39 843	8 851	3 412	96 966	58.8	8 646	2 303	7.9
November	170 996	109 236	63.9	100 207	53 335	39 908	8 852	3 415	96 792	58.6	9 029	2 345	8.3
December	171 166	108 912	63.6	99 645	53 149	39 708	8 602	3 227	96 418	58.2	9 267	2 374	8.5
1982													
January	171 335	109 089	63.7	99 692	53 103	39 821	8 676	3 393	96 299	58.2	9 397	2 409	8.6
February	171 489	109 467	63.8	99 762	53 172	39 859	8 697	3 375	96 387	58.2	9 705	2 758	8.9
March	171 667	109 567	63.8	99 672	53 054	39 936	8 550	3 372	96 300	58.1	9 895	2 965	9.0
April	171 844	109 820	63.9	99 576	53 081	39 848	8 605	3 351	96 225	57.9	10 244	3 086	9.3
May	172 026	110 451	64.2	100 116	53 234	40 121	8 753	3 434	96 682	58.2	10 335	3 276	9.4
June	172 190	110 081	63.9	99 543	52 933	40 219	8 293	3 331	96 212	57.8	10 538	3 451	9.6
July	172 364	110 342	64.0	99 493	52 896	40 228	8 380	3 402	96 091	57.7	10 849	3 555	9.8
August	172 511	110 514	64.1	99 633	52 797	40 336	8 514	3 408	96 225	57.8	10 001	3 696	9.8
September	172 690	110 721	64.1	99 504	52 700	40 275	8 469	3 385	96 119	57.6	11 217	3 889	10.1
October	172 881	110 744	64.1	99 215	52 624	40 105	8 499	3 489	95 726	57.4	11 529	4 185	10.4
November	173 058	111 050	64.2	99 112	52 537	40 111	8 520	3 510	95 602	57.3	11 938	4 485	10.8
December	173 199	111 083	64.1	99 032	52 497	40 164	8 397	3 414	95 618	57.2	12 051	4 662	10.8
1983													
January	173 354	110 695	63.9	99 161	52 487	40 268	8 335	3 439	95 722	57.2	11 534	4 668	10.4
February	173 505	110 634	63.8	99 089	52 453	40 336	8 159	3 382	95 707	57.1	11 545	4 641	10.4
March	173 656	110 587	63.7	99 179	52 615	40 368	8 098	3 360	95 819	57.1	11 408	4 612	10.3
April	173 794	110 828	63.8	99 560	52 814	40 542	8 094	3 341	96 219	57.3	11 268	4 370	10.2
May	173 953	110 790	63.7	99 642	52 922	40 538	8 006	3 328	96 314	57.3	11 154	4 538	10.1
June	174 125	111 879	64.3	100 633	53 515	40 695	8 448	3 462	97 171	57.8	11 246	4 470	10.1
July	174 306	111 756	64.1	101 208	53 835	41 041	8 207	3 481	97 727	58.1	10 548	4 329	9.4
August	174 440	112 231	64.3	101 608	53 837	41 314	8 379	3 502	98 106	58.2	10 623	4 070	9.5
September	174 602	112 298	64.3	102 016	53 983	41 650	8 147	3 347	98 669	58.4	10 282	3 854	9.2
October	174 779	111 926	64.0	102 039	54 146	41 597	8 010	3 303	98 736	58.4	9 887	3 648	8.8
November	174 951	112 228	64.1	102 729	54 499	41 788	8 075	3 291	99 438	58.7	9 499	3 535	8.5
December	175 121	112 327	64.1	102 996	54 662	41 852	8 089	3 332	99 664	58.8	9 331	3 379	8.3
1984													
January	175 533	112 209	63.9	103 201	54 975	41 812	7 965	3 293	99 908	58.8	9 008	3 254	8.0
February	175 679	112 615	64.1	103 824	55 213	42 196	7 958	3 353	100 471	59.1	8 791	2 991	7.8
March	175 824	112 713	64.1	103 967	55 281	42 328	7 926	3 233	100 734	59.1	8 746	2 881	7.8
April	175 969	113 098	64.3	104 336	55 373	42 512	7 988	3 291	101 045	59.3	8 762	2 858	7.7
May	176 123	113 649	64.5	105 193	55 661	43 071	7 944	3 343	101 850	59.7	8 456	2 884	7.4
June	176 284	113 817	64.6	105 591	55 996	42 944	8 131	3 383	102 208	59.9	8 226	2 612	7.2
July	176 440	113 972	64.6	105 435	55 921	42 979	8 045	3 344	102 091	59.8	8 537	2 638	7.5
August	176 583	113 682	64.4	105 163	55 930	42 885	7 809	3 286	101 877	59.6	8 519	2 604	7.5
September	176 763	113 857	64.4	105 490	56 095	42 967	7 951	3 393	102 097	59.7	8 367	2 538	7.3
October	176 956	114 019	64.4	105 638	56 183	43 052	7 862	3 194	102 444	59.7	8 381	2 526	7.4
November	177 135	114 170	64.5	105 972	56 274	43 244	7 844	3 394	102 578	59.8	8 198	2 438	7.2
December	177 306	114 581	64.6	106 223	56 313	43 472	7 933	3 385	102 838	59.9	8 358	2 401	7.3
1985													
January	177 384	114 725	64.7	106 302	56 184	43 589	8 036	3 317	102 985	59.9	8 423	2 284	7.3
February	177 516	114 876	64.7	106 555	56 216	43 787	8 020	3 317	103 238	60.0	8 321	2 389	7.2
March	177 667	115 328	64.9	106 989	56 356	44 035	8 062	3 250	103 739	60.2	8 339	2 394	7.2
April	177 799	115 331	64.9	106 936	56 374	44 000	7 958	3 306	103 630	60.1	8 395	2 393	7.3
May	177 944	115 234	64.8	106 932	56 531	43 905	7 966	3 280	103 652	60.1	8 302	2 292	7.2
June	178 096	114 965	64.6	106 505	56 288	43 958	7 680	3 161	103 344	59.8	8 460	2 310	7.4
July	178 263	115 320	64.7	106 807	56 435	43 975	8 013	3 143	103 664	59.9	8 513	2 329	7.4
August	178 405	115 291	64.6	107 095	56 655	44 110	7 714	3 121	103 974	60.0	8 196	2 258	7.1
September	178 572	115 905	64.9	107 657	56 845	44 395	7 818	3 064	104 593	60.3	8 248	2 242	7.1
October	178 770	116 145	65.0	107 847	56 969	44 565	7 889	3 051	104 796	60.3	8 298	2 295	7.1
November	178 940	116 135	64.9	108 007	56 972	44 617	7 852	3 062	104 945	60.4	8 128	2 207	7.0
December	179 112	116 354	65.0	108 216	56 995	44 889	7 826	3 141	105 075	60.4	8 138	2 208	7.0
1986													
January	179 670	116 682	64.9	108 887	57 637	44 944	7 704	3 287	105 600	60.6	7 795	2 089	6.7
February	179 821	116 882	65.0	108 480	57 269	44 804	7 895	3 083	105 397	60.3	8 402	2 308	7.2
March	179 985	117 220	65.1	108 837	57 353	44 960	7 977	3 200	105 637	60.5	8 383	2 261	7.2
April	180 148	117 316	65.1	108 952	57 358	45 081	8 058	3 153	105 799	60.5	8 364	2 162	7.1
May	180 311	117 528	65.2	109 089	57 287	45 289	7 997	3 150	105 939	60.5	8 439	2 232	7.2
June	180 503	118 084	65.4	109 576	57 471	45 621	8 024	3 193	106 383	60.7	8 508	2 320	7.2
July	180 682	118 129	65.4	109 810	57 514	45 837	7 914	3 141	106 669	60.8	8 319	2 269	7.0
August	180 828	118 150	65.3	110 015	57 597	45 926	7 920	3 082	106 933	60.8	8 135	2 276	6.9
September	180 997	118 395	65.4	110 085	57 630	45 972	7 945	3 171	106 914	60.8	8 310	2 318	7.0
October	181 186	118 516	65.4	110 273	57 660	46 046	7 977	3 128	107 145	60.9	8 243	2 188	7.0
November	181 363	118 634	65.4	110 475	57 941	46 070	7 891	3 220	107 255	60.9	8 159	2 202	6.9
December	181 547	118 611	65.3	110 728	58 185	46 132	7 769	3 148	107 580	61.0	7 883	2 161	6.6

[1] Not seasonally adjusted.
[2] In 1930 through 1943, the official BLS data count persons on work relief as unemployed. The unemployment rates for those years shown in parentheses count persons on work relief as employed, which is more consistent with the postwar practice. See notes and definitions.

Table 10-1B. Summary Labor Force, Employment, and Unemployment: Historical Data—*Continued*

(Thousands of persons, percent, seasonally adjusted, except as noted.)

Year and month	Civilian noninsti- tutional popu- lation [1]	Civilian labor force		Employment, thousands of persons						Employ- ment- population ratio, percent	Unemployment		
					By age and sex			By industry			Thousands of persons		
		Thousands of persons	Participa- tion rate (percent)	Total	Men, 20 years and over	Women, 20 years and over	Both sexes, 16 to 19 years	Agri- cultural	Nonagri- cultural		Total	Unem- ployed 15 weeks and over	Rate (percent) [2]
1987													
January	181 827	118 845	65.4	110 953	58 264	46 219	7 861	3 143	107 810	61.0	7 892	2 168	6.6
February	181 998	119 122	65.5	111 257	58 279	46 444	7 969	3 208	108 049	61.1	7 865	2 117	6.6
March	182 179	119 270	65.5	111 408	58 362	46 549	7 913	3 214	108 194	61.2	7 862	2 070	6.6
April	182 344	119 336	65.4	111 794	58 503	46 746	7 916	3 246	108 548	61.3	7 542	2 091	6.3
May	182 533	120 008	65.7	112 434	58 713	47 052	8 075	3 345	109 089	61.6	7 574	2 104	6.3
June	182 703	119 644	65.5	112 246	58 581	47 102	7 857	3 216	109 030	61.4	7 398	2 087	6.2
July	182 885	119 902	65.6	112 634	58 740	47 229	7 911	3 235	109 399	61.6	7 268	1 921	6.1
August	183 002	120 318	65.7	113 057	58 810	47 322	8 232	3 112	109 945	61.8	7 261	1 878	6.0
September	183 161	120 011	65.5	112 909	58 964	47 285	7 945	3 189	109 720	61.6	7 102	1 866	5.9
October	183 311	120 509	65.7	113 282	59 073	47 533	8 074	3 219	110 063	61.8	7 227	1 794	6.0
November	183 470	120 540	65.7	113 505	59 210	47 622	8 002	3 145	110 360	61.9	7 035	1 797	5.8
December	183 620	120 729	65.7	113 793	59 217	47 781	8 092	3 213	110 580	62.0	6 936	1 767	5.7
1988													
January	183 822	120 969	65.8	114 016	59 346	47 862	8 110	3 247	110 769	62.0	6 953	1 714	5.7
February	183 969	121 156	65.9	114 227	59 535	47 919	8 025	3 201	111 026	62.1	6 929	1 738	5.7
March	184 111	120 913	65.7	114 037	59 393	48 090	7 854	3 169	110 868	61.9	6 876	1 744	5.7
April	184 232	121 251	65.8	114 650	59 832	48 147	7 937	3 224	111 426	62.2	6 601	1 563	5.4
May	184 374	121 071	65.7	114 292	59 644	47 946	7 912	3 121	111 171	62.0	6 779	1 647	5.6
June	184 562	121 473	65.8	114 927	59 751	48 146	8 193	3 111	111 816	62.3	6 546	1 531	5.4
July	184 729	121 665	65.9	115 060	59 888	48 186	8 198	3 060	112 000	62.3	6 605	1 601	5.4
August	184 830	122 125	66.1	115 282	59 877	48 467	8 201	3 119	112 163	62.4	6 843	1 639	5.6
September	184 962	121 960	65.9	115 356	59 980	48 511	8 124	3 165	112 191	62.4	6 604	1 569	5.4
October	185 114	122 206	66.0	115 638	60 023	48 859	7 961	3 231	112 407	62.5	6 568	1 562	5.4
November	185 244	122 637	66.2	116 100	60 042	49 254	7 904	3 241	112 859	62.7	6 537	1 468	5.3
December	185 402	122 622	66.1	116 104	60 059	49 257	7 966	3 194	112 910	62.6	6 518	1 490	5.3
1989													
January	185 644	123 390	66.5	116 708	60 477	49 529	8 019	3 287	113 421	62.9	6 682	1 480	5.4
February	185 777	123 135	66.3	116 776	60 588	49 497	7 871	3 234	113 542	62.9	6 359	1 304	5.2
March	185 897	123 227	66.3	117 022	60 795	49 503	7 809	3 198	113 824	62.9	6 205	1 353	5.0
April	186 024	123 565	66.4	117 097	60 764	49 565	7 929	3 162	113 935	62.9	6 468	1 397	5.2
May	186 181	123 474	66.3	117 099	60 795	49 583	7 890	3 125	113 974	62.9	6 375	1 348	5.2
June	186 329	123 995	66.5	117 418	61 054	49 542	8 094	3 068	114 350	63.0	6 577	1 300	5.3
July	186 483	123 967	66.5	117 472	60 947	49 693	7 961	3 227	114 245	63.0	6 495	1 435	5.2
August	186 598	124 166	66.5	117 655	60 915	49 804	8 126	3 284	114 371	63.1	6 511	1 302	5.2
September	186 726	123 944	66.4	117 354	60 668	50 015	7 870	3 219	114 135	62.8	6 590	1 360	5.3
October	186 871	124 211	66.5	117 581	60 958	49 871	7 940	3 215	114 366	62.9	6 630	1 392	5.3
November	187 017	124 637	66.6	117 912	60 958	50 221	7 964	3 132	114 780	63.0	6 725	1 418	5.4
December	187 165	124 497	66.5	117 830	61 068	50 116	7 851	3 188	114 642	63.0	6 667	1 375	5.4
1990													
January	188 413	125 833	66.8	119 081	61 742	50 436	8 103	3 210	115 871	63.2	6 752	1 412	5.4
February	188 516	125 710	66.7	119 059	61 805	50 438	8 015	3 188	115 871	63.2	6 651	1 350	5.3
March	188 630	125 801	66.7	119 203	61 832	50 463	8 061	3 260	115 943	63.2	6 598	1 331	5.2
April	188 778	125 649	66.6	118 852	61 579	50 457	7 987	3 231	115 621	63.0	6 797	1 376	5.4
May	188 913	125 893	66.6	119 151	61 778	50 646	7 911	3 266	115 885	63.1	6 742	1 415	5.4
June	189 058	125 573	66.4	118 983	61 762	50 550	7 783	3 245	115 738	62.9	6 590	1 436	5.2
July	189 188	125 732	66.5	118 810	61 683	50 514	7 777	3 192	115 618	62.8	6 922	1 534	5.5
August	189 342	125 990	66.5	118 802	61 715	50 635	7 708	3 197	115 605	62.7	7 188	1 607	5.7
September	189 528	125 892	66.4	118 524	61 608	50 587	7 575	3 206	115 318	62.5	7 368	1 695	5.9
October	189 710	125 995	66.4	118 536	61 606	50 616	7 565	3 270	115 266	62.5	7 459	1 689	5.9
November	189 872	126 070	66.4	118 306	61 545	50 541	7 506	3 189	115 117	62.3	7 764	1 831	6.2
December	190 017	126 142	66.4	118 241	61 506	50 530	7 516	3 245	114 996	62.2	7 901	1 804	6.3
1991													
January	190 163	125 955	66.2	117 940	61 383	50 472	7 478	3 208	114 732	62.0	8 015	1 866	6.4
February	190 271	126 020	66.2	117 755	61 117	50 523	7 405	3 270	114 485	61.9	8 265	1 955	6.6
March	190 381	126 238	66.3	117 652	61 144	50 422	7 447	3 177	114 475	61.8	8 586	2 137	6.8
April	190 517	126 548	66.4	118 109	61 280	50 760	7 381	3 241	114 868	62.0	8 439	2 206	6.7
May	190 650	126 176	66.2	117 440	61 052	50 457	7 303	3 275	114 165	61.6	8 736	2 252	6.9
June	190 800	126 331	66.2	117 639	61 147	50 585	7 248	3 300	114 339	61.7	8 692	2 533	6.9
July	190 946	126 154	66.1	117 568	61 179	50 636	7 138	3 319	114 249	61.6	8 586	2 388	6.8
August	191 116	126 150	66.0	117 484	61 122	50 601	7 105	3 313	114 171	61.5	8 666	2 460	6.9
September	191 302	126 650	66.2	117 928	61 279	50 864	7 123	3 319	114 609	61.6	8 722	2 497	6.9
October	191 497	126 642	66.1	117 800	61 174	50 811	7 185	3 289	114 511	61.5	8 842	2 638	7.0
November	191 657	126 701	66.1	117 770	61 201	50 759	7 169	3 296	114 474	61.4	8 931	2 718	7.0
December	191 798	126 664	66.0	117 466	61 074	50 728	7 104	3 146	114 320	61.2	9 198	2 892	7.3
1992													
January	191 953	127 261	66.3	117 978	61 116	51 095	7 138	3 155	114 823	61.5	9 283	3 060	7.3
February	192 067	127 207	66.2	117 753	61 062	51 033	7 083	3 239	114 514	61.3	9 454	3 182	7.4
March	192 204	127 604	66.4	118 144	61 363	51 204	6 998	3 236	114 908	61.5	9 460	3 196	7.4
April	192 354	127 841	66.5	118 426	61 468	51 323	6 910	3 245	115 181	61.6	9 415	3 130	7.4
May	192 503	128 119	66.6	118 375	61 513	51 245	7 028	3 213	115 162	61.5	9 744	3 444	7.6
June	192 663	128 459	66.7	118 419	61 537	51 383	7 137	3 297	115 122	61.5	10 040	3 758	7.8
July	192 826	128 563	66.7	118 713	61 641	51 458	7 087	3 285	115 428	61.6	9 850	3 614	7.7
August	193 018	128 613	66.6	118 826	61 681	51 386	7 194	3 279	115 547	61.6	9 787	3 579	7.6
September	193 229	128 501	66.5	118 720	61 663	51 359	7 216	3 274	115 446	61.4	9 781	3 504	7.6
October	193 442	128 026	66.2	118 628	61 550	51 373	6 985	3 254	115 374	61.3	9 398	3 505	7.3
November	193 621	128 441	66.3	118 876	61 644	51 535	7 164	3 207	115 669	61.4	9 565	3 397	7.4
December	193 784	128 554	66.3	118 997	61 721	51 524	7 174	3 259	115 738	61.4	9 557	3 651	7.4

[1]Not seasonally adjusted.
[2]In 1930 through 1943, the official BLS data count persons on work relief as unemployed. The unemployment rates for those years shown in parentheses count persons on work relief as employed, which is more consistent with the postwar practice. See notes and definitions.

Table 10-1B. Summary Labor Force, Employment, and Unemployment: Historical Data—*Continued*

(Thousands of persons, percent, seasonally adjusted, except as noted.)

| Year and month | Civilian noninstitutional population [1] | Civilian labor force | | Employment, thousands of persons | | | | | | Employment-population ratio, percent | Unemployment | | |
| | | Thousands of persons | Participation rate (percent) | Total | By age and sex | | | By industry | | | Thousands of persons | | Rate (percent) [2] |
					Men, 20 years and over	Women, 20 years and over	Both sexes, 16 to 19 years	Agricultural	Nonagricultural		Total	Unemployed 15 weeks and over	
1993													
January	193 962	128 400	66.2	119 075	61 895	51 505	7 089	3 222	115 853	61.4	9 325	3 346	7.3
February	194 108	128 458	66.2	119 275	61 963	51 573	7 144	3 125	116 150	61.4	9 183	3 190	7.1
March	194 248	128 598	66.2	119 542	62 007	51 808	7 132	3 119	116 423	61.5	9 056	3 115	7.0
April	194 398	128 584	66.1	119 474	62 032	51 732	7 091	3 074	116 400	61.5	9 110	3 014	7.1
May	194 549	129 264	66.4	120 115	62 309	51 996	7 244	3 100	117 015	61.7	9 149	3 101	7.1
June	194 719	129 411	66.5	120 290	62 409	52 183	7 114	3 108	117 182	61.8	9 121	3 141	7.0
July	194 882	129 397	66.4	120 467	62 497	52 088	7 205	3 126	117 341	61.8	8 930	3 046	6.9
August	195 063	129 619	66.4	120 856	62 634	52 294	7 264	3 026	117 830	62.0	8 763	3 026	6.8
September	195 259	129 268	66.2	120 554	62 437	52 241	7 180	3 174	117 380	61.7	8 714	3 042	6.7
October	195 444	129 573	66.3	120 823	62 614	52 379	7 171	3 084	117 739	61.8	8 750	3 029	6.8
November	195 625	129 711	66.3	121 169	62 732	52 531	7 248	3 157	118 012	61.9	8 542	2 986	6.6
December	195 794	129 941	66.4	121 464	62 760	52 813	7 178	3 116	118 348	62.0	8 477	2 968	6.5
1994													
January	195 953	130 596	66.6	121 966	62 798	53 052	7 486	3 302	118 664	62.2	8 630	3 060	6.6
February	196 090	130 669	66.6	122 086	62 708	53 266	7 457	3 339	118 747	62.3	8 583	3 118	6.6
March	196 213	130 400	66.5	121 930	62 780	53 099	7 381	3 354	118 576	62.1	8 470	3 055	6.5
April	196 363	130 621	66.5	122 290	62 906	53 274	7 551	3 428	118 862	62.3	8 331	2 921	6.4
May	196 510	130 779	66.6	122 864	63 116	53 624	7 466	3 409	119 455	62.5	7 915	2 836	6.1
June	196 693	130 561	66.4	122 634	63 041	53 393	7 527	3 299	119 335	62.3	7 927	2 735	6.1
July	196 859	130 652	66.4	122 706	63 034	53 531	7 453	3 333	119 373	62.3	7 946	2 822	6.1
August	197 043	131 275	66.6	123 312	63 204	53 744	7 019	3 451	119 891	62.6	7 933	2 750	6.0
September	197 248	131 421	66.6	123 687	63 631	53 991	7 352	3 430	120 257	62.7	7 734	2 746	5.9
October	197 430	131 744	66.7	124 112	63 818	54 071	7 543	3 490	120 622	62.9	7 632	2 955	5.8
November	197 607	131 891	66.7	124 516	64 080	54 168	7 423	3 574	120 942	63.0	7 375	2 666	5.6
December	197 765	131 951	66.7	124 721	64 359	54 054	7 599	3 577	121 144	63.1	7 230	2 488	5.5
1995													
January	197 753	132 038	66.8	124 663	64 185	54 087	7 650	3 519	121 144	63.0	7 375	2 396	5.6
February	197 886	132 115	66.8	124 928	64 378	54 226	7 659	3 620	121 308	63.1	7 187	2 345	5.4
March	198 007	132 108	66.7	124 955	64 321	54 141	7 742	3 634	121 321	63.1	7 153	2 287	5.4
April	198 148	132 590	66.9	124 945	64 165	54 366	7 774	3 566	121 379	63.1	7 645	2 473	5.8
May	198 286	131 851	66.5	124 421	63 829	54 272	7 664	3 349	121 072	62.7	7 430	2 577	5.6
June	198 452	131 949	66.5	124 522	63 992	54 020	7 850	3 461	121 061	62.7	7 427	2 266	5.6
July	198 615	132 343	66.6	124 816	63 962	54 476	7 798	3 379	121 437	62.8	7 527	2 311	5.7
August	198 801	132 336	66.6	124 852	63 875	54 434	7 910	3 374	121 478	62.8	7 484	2 391	5.7
September	199 005	132 611	66.6	125 133	64 179	54 507	7 825	3 285	121 848	62.9	7 478	2 306	5.6
October	199 192	132 716	66.6	125 388	64 272	54 692	7 774	3 438	121 950	62.9	7 328	2 272	5.5
November	199 355	132 614	66.5	125 188	63 931	54 850	7 765	3 338	121 850	62.8	7 426	2 339	5.6
December	199 508	132 511	66.4	125 088	64 041	54 674	7 768	3 352	121 736	62.7	7 423	2 331	5.6
1996													
January	199 634	132 616	66.4	125 125	64 180	54 580	7 733	3 483	121 642	62.7	7 491	2 371	5.6
February	199 772	132 952	66.6	125 639	64 398	54 844	7 688	3 547	122 092	62.9	7 313	2 307	5.5
March	199 921	133 180	66.6	125 862	64 506	54 994	7 673	3 489	122 373	63.0	7 318	2 454	5.5
April	200 101	133 409	66.7	125 994	64 481	55 067	7 774	3 406	122 588	63.0	7 415	2 455	5.6
May	200 278	133 667	66.7	126 244	64 683	55 034	7 841	3 473	122 771	63.0	7 423	2 403	5.6
June	200 459	133 697	66.7	126 602	64 940	55 177	7 739	3 424	123 178	63.2	7 095	2 355	5.3
July	200 641	134 284	66.9	126 947	65 068	55 362	7 859	3 433	123 514	63.3	7 337	2 297	5.5
August	200 847	134 054	66.7	127 172	65 216	55 525	7 731	3 395	123 777	63.3	6 882	2 267	5.1
September	201 060	134 515	66.9	127 536	65 169	55 669	7 935	3 448	124 088	63.4	6 979	2 220	5.2
October	201 273	134 921	67.0	127 890	65 460	55 750	7 980	3 463	124 427	63.5	7 031	2 268	5.2
November	201 463	135 007	67.0	127 771	65 320	55 896	7 875	3 356	124 415	63.4	7 236	2 159	5.4
December	201 636	135 113	67.0	127 860	65 435	55 849	7 883	3 445	124 415	63.4	7 253	2 124	5.4
1997													
January	202 285	135 456	67.0	128 298	65 679	56 024	7 930	3 449	124 849	63.4	7 158	2 162	5.3
February	202 388	135 400	66.9	128 298	65 758	55 955	7 948	3 353	124 945	63.4	7 102	2 140	5.2
March	202 513	135 891	67.1	128 891	65 974	56 270	7 951	3 419	125 472	63.6	7 000	2 110	5.2
April	202 674	136 016	67.1	129 143	66 092	56 347	7 974	3 462	125 681	63.7	6 873	2 176	5.1
May	202 832	136 119	67.1	129 464	66 328	56 446	7 964	3 437	126 027	63.8	6 655	2 121	4.9
June	203 000	136 211	67.1	129 412	66 308	56 573	7 849	3 409	126 003	63.7	6 799	2 085	5.0
July	203 166	136 477	67.2	129 822	66 422	56 785	7 975	3 422	126 400	63.9	6 655	2 119	4.9
August	203 364	136 618	67.2	130 010	66 508	56 852	7 922	3 359	126 651	63.9	6 608	2 004	4.8
September	203 570	136 675	67.1	130 019	66 483	56 931	7 873	3 392	126 627	63.9	6 656	2 074	4.9
October	203 767	136 633	67.1	130 179	66 511	56 982	7 876	3 312	126 867	63.9	6 454	1 950	4.7
November	203 941	136 961	67.2	130 653	66 765	57 039	8 042	3 386	127 267	64.1	6 308	1 817	4.6
December	204 098	137 155	67.2	130 679	66 643	57 219	7 923	3 405	127 274	64.0	6 476	1 901	4.7
1998													
January	204 238	137 095	67.1	130 726	66 750	56 941	8 171	3 299	127 389	64.0	6 368	1 833	4.6
February	204 400	137 112	67.1	130 807	66 856	56 992	8 137	3 284	127 522	64.0	6 306	1 809	4.6
March	204 546	137 236	67.1	130 814	66 721	57 080	8 235	3 146	127 650	64.0	6 422	1 772	4.7
April	204 731	137 150	67.0	131 209	67 151	57 074	8 071	3 334	127 852	64.1	5 941	1 476	4.3
May	204 899	137 372	67.0	131 325	67 164	57 155	8 228	3 360	127 959	64.1	6 047	1 490	4.4
June	205 085	137 455	67.0	131 244	67 054	57 156	8 268	3 380	127 874	64.0	6 212	1 613	4.5
July	205 270	137 588	67.0	131 329	67 119	57 192	8 221	3 455	127 913	64.0	6 259	1 577	4.5
August	205 479	137 570	67.0	131 390	66 985	57 332	8 289	3 509	127 970	63.9	6 179	1 626	4.5
September	205 699	138 286	67.2	131 986	67 254	57 520	8 489	3 500	128 399	64.2	6 300	1 688	4.6
October	205 919	138 279	67.2	131 999	67 433	57 529	8 347	3 593	128 389	64.1	6 280	1 582	4.5
November	206 104	138 381	67.1	132 280	67 591	57 638	8 269	3 375	128 897	64.2	6 100	1 590	4.4
December	206 270	138 634	67.2	132 602	67 548	57 840	8 340	3 246	129 320	64.3	6 032	1 559	4.4

[1] Not seasonally adjusted.
[2] In 1930 through 1943, the official BLS data count persons on work relief as unemployed. The unemployment rates for those years shown in parentheses count persons on work relief as employed, which is more consistent with the postwar practice. See notes and definitions.

Table 10-1B. Summary Labor Force, Employment, and Unemployment: Historical Data—Continued

(Thousands of persons, percent, seasonally adjusted, except as noted.)

Year and month	Civilian noninsti-tutional popu-lation [1]	Civilian labor force Thousands of persons	Civilian labor force Participa-tion rate (percent)	Employment Total	Men, 20 years and over	Women, 20 years and over	Both sexes, 16 to 19 years	Agri-cultural	Nonagri-cultural	Employ-ment-population ratio, percent	Unemployment Total	Unem-ployed 15 weeks and over	Rate (percent) [2]
1999													
January	206 719	139 003	67.2	133 027	67 679	58 256	8 367	3 233	129 802	64.4	5 976	1 490	4.3
February	206 873	138 967	67.2	132 856	67 498	58 129	8 399	3 246	129 647	64.2	6 111	1 551	4.4
March	207 036	138 730	67.0	132 947	67 660	58 132	8 343	3 238	129 656	64.2	5 783	1 472	4.2
April	207 236	138 959	67.1	132 955	67 542	58 260	8 334	3 336	129 615	64.2	6 004	1 480	4.3
May	207 427	139 107	67.1	133 311	67 539	58 440	8 454	3 335	129 937	64.3	5 796	1 505	4.2
June	207 632	139 329	67.1	133 378	67 700	58 641	8 175	3 386	129 982	64.2	5 951	1 624	4.3
July	207 828	139 439	67.1	133 414	67 731	58 490	8 306	3 346	130 146	64.2	6 025	1 513	4.3
August	208 038	139 430	67.0	133 591	67 768	58 707	8 208	3 234	130 366	64.2	5 838	1 455	4.2
September	208 265	139 622	67.0	133 707	67 882	58 735	8 323	3 173	130 434	64.2	5 915	1 449	4.2
October	208 483	139 771	67.0	133 993	67 840	58 921	8 394	3 229	130 758	64.3	5 778	1 438	4.1
November	208 666	140 025	67.1	134 309	68 094	59 018	8 358	3 343	130 989	64.4	5 716	1 378	4.1
December	208 832	140 177	67.1	134 523	68 217	59 056	8 370	3 260	131 257	64.4	5 653	1 375	4.0
2000													
January	211 410	142 267	67.3	136 559	69 419	59 842	8 360	2 613	133 863	64.6	5 708	1 380	4.0
February	211 576	142 456	67.3	136 598	69 505	59 887	8 359	2 731	133 912	64.6	5 858	1 300	4.1
March	211 772	142 434	67.3	136 701	69 482	59 977	8 355	2 579	134 022	64.6	5 733	1 312	4.0
April	212 018	142 751	67.3	137 270	69 519	60 358	8 458	2 505	134 806	64.7	5 481	1 261	3.8
May	212 242	142 388	67.1	136 630	69 399	59 951	8 344	2 480	134 144	64.4	5 758	1 325	4.0
June	212 466	142 591	67.1	136 940	69 629	60 027	8 308	2 445	134 528	64.5	5 651	1 242	4.0
July	212 677	142 278	66.9	136 531	69 525	60 011	8 078	2 408	134 196	64.2	5 747	1 343	4.0
August	212 916	142 514	66.9	136 662	69 823	59 719	8 280	2 433	134 311	64.2	5 853	1 394	4.1
September	213 163	142 518	66.9	136 893	69 700	60 083	8 176	2 384	134 489	64.2	5 625	1 290	3.9
October	213 405	142 622	66.8	137 088	69 762	60 238	8 124	2 319	134 808	64.2	5 534	1 337	3.9
November	213 540	142 962	66.9	137 322	69 910	60 269	8 211	2 330	134 921	64.3	5 639	1 315	3.9
December	213 736	143 248	67.0	137 614	69 939	60 503	8 258	2 389	135 194	64.4	5 634	1 329	3.9
2001													
January	213 888	143 800	67.2	137 778	70 064	60 609	8 243	2 360	135 304	64.4	6 023	1 372	4.2
February	214 110	143 701	67.1	137 612	69 959	60 615	8 154	2 370	135 291	64.3	6 089	1 491	4.2
March	214 305	143 924	67.2	137 783	69 881	60 902	8 120	2 350	135 372	64.3	6 141	1 521	4.3
April	214 525	143 569	66.9	137 299	69 916	60 523	7 970	2 336	135 036	64.0	6 271	1 499	4.4
May	214 732	143 318	66.7	137 092	69 865	60 509	7 760	2 353	134 735	63.8	6 226	1 502	4.3
June	214 950	143 357	66.7	136 873	69 690	60 371	7 942	2 082	134 755	63.7	6 484	1 532	4.5
July	215 180	143 654	66.8	137 071	69 808	60 480	7 929	2 295	134 858	63.7	6 583	1 653	4.6
August	215 420	143 284	66.5	136 241	69 585	60 301	7 531	2 305	133 944	63.2	7 042	1 861	4.9
September	215 665	143 989	66.8	136 846	69 933	60 265	7 836	2 322	134 558	63.5	7 142	1 950	5.0
October	215 903	144 086	66.7	136 392	69 621	60 168	7 858	2 327	134 098	63.2	7 694	2 082	5.3
November	216 117	144 240	66.7	136 238	69 444	60 174	7 873	2 203	133 955	63.0	8 003	2 318	5.5
December	216 315	144 305	66.7	136 047	69 551	60 095	7 712	2 293	133 751	62.9	8 258	2 444	5.7
2002													
January	216 506	143 883	66.5	135 701	69 308	60 032	7 619	2 385	133 233	62.7	8 182	2 578	5.7
February	216 663	144 653	66.8	136 438	69 534	60 479	7 650	2 397	134 127	63.0	8 215	2 608	5.7
March	216 823	144 481	66.6	136 177	69 480	60 190	7 802	2 368	133 816	62.8	8 304	2 719	5.7
April	217 006	144 725	66.7	136 126	69 574	60 204	7 621	2 371	133 833	62.7	8 599	2 852	5.9
May	217 198	144 938	66.7	136 539	69 981	60 226	7 588	2 260	134 278	62.9	8 399	2 967	5.8
June	217 407	144 808	66.6	136 415	69 769	60 297	7 617	2 161	134 135	62.7	8 393	3 023	5.8
July	217 630	144 803	66.5	136 413	69 806	60 290	7 591	2 324	134 107	62.7	8 390	2 966	5.8
August	217 866	145 009	66.6	136 705	69 937	60 560	7 477	2 127	134 593	62.7	8 304	2 887	5.7
September	218 107	145 552	66.7	137 302	70 207	60 679	7 667	2 285	135 102	63.0	8 251	2 971	5.7
October	218 340	145 314	66.6	137 008	69 948	60 663	7 537	2 471	134 580	62.7	8 307	3 042	5.7
November	218 548	145 041	66.4	136 521	69 615	60 697	7 487	2 261	134 171	62.5	8 520	3 062	5.9
December	218 741	145 066	66.3	136 426	69 620	60 667	7 385	2 352	134 071	62.4	8 640	3 271	6.0
2003													
January	219 897	145 937	66.4	137 417	69 919	61 406	7 358	2 337	135 045	62.5	8 520	3 166	5.8
February	220 114	146 100	66.4	137 482	70 262	61 159	7 322	2 234	135 306	62.5	8 618	3 161	5.9
March	220 317	146 022	66.3	137 434	70 243	61 317	7 143	2 263	135 232	62.4	8 588	3 161	5.9
April	220 540	146 474	66.4	137 633	70 311	61 374	7 223	2 150	135 561	62.4	8 842	3 348	6.0
May	220 768	146 500	66.4	137 544	70 215	61 393	7 228	2 183	135 370	62.3	8 957	3 318	6.1
June	221 014	147 056	66.5	137 790	70 159	61 747	7 268	2 185	135 419	62.3	9 266	3 552	6.3
July	221 252	146 485	66.2	137 474	70 190	61 437	7 150	2 187	135 242	62.1	9 011	3 633	6.2
August	221 507	146 445	66.1	137 549	70 237	61 442	7 035	2 313	135 200	62.1	8 896	3 557	6.1
September	221 779	146 530	66.1	137 609	70 631	61 119	7 112	2 349	135 355	62.0	8 921	3 486	6.1
October	222 039	146 716	66.1	137 984	70 685	61 451	7 060	2 479	135 571	62.1	8 732	3 451	6.0
November	222 279	147 000	66.1	138 424	70 935	61 494	7 112	2 373	136 003	62.3	8 576	3 420	5.8
December	222 509	146 729	65.9	138 411	71 170	61 396	6 978	2 243	136 145	62.2	8 317	3 366	5.7
2004													
January	222 161	146 842	66.1	138 472	71 318	61 183	7 194	2 196	136 228	62.3	8 370	3 364	5.7
February	222 357	146 709	66.0	138 542	71 122	61 514	7 077	2 210	136 362	62.3	8 167	3 248	5.6
March	222 550	146 944	66.0	138 453	71 155	61 534	6 930	2 180	136 302	62.2	8 491	3 314	5.8
April	222 757	146 850	65.9	138 680	71 121	61 647	7 093	2 241	136 474	62.3	8 170	2 971	5.6
May	222 967	147 065	66.0	138 852	71 180	61 762	7 126	2 300	136 556	62.3	8 212	3 103	5.6
June	223 196	147 460	66.1	139 174	71 562	61 793	7 007	2 237	136 748	62.4	8 286	3 130	5.6
July	223 422	147 692	66.1	139 556	71 780	61 884	7 166	2 222	137 354	62.5	8 136	2 918	5.5
August	223 677	147 564	66.0	139 573	71 808	61 836	7 123	2 333	137 230	62.4	7 990	2 846	5.4
September	223 941	147 415	65.8	139 487	71 744	61 859	7 056	2 251	137 323	62.3	7 927	2 910	5.4
October	224 192	147 793	65.9	139 732	71 860	61 942	7 180	2 216	137 598	62.3	8 061	3 041	5.5
November	224 422	148 162	66.0	140 231	72 110	62 088	7 216	2 206	137 978	62.5	7 932	2 960	5.4
December	224 640	148 059	65.9	140 125	72 058	62 136	7 202	2 171	137 947	62.4	7 934	2 927	5.4

[1] Not seasonally adjusted.
[2] In 1930 through 1943, the official BLS data count persons on work relief as unemployed. The unemployment rates for those years shown in parentheses count persons on work relief as employed, which is more consistent with the postwar practice. See notes and definitions.

Table 10-1B. Summary Labor Force, Employment, and Unemployment: Historical Data—*Continued*

(Thousands of persons, percent, seasonally adjusted, except as noted.)

Year and month	Civilian noninstitutional population [1]	Civilian labor force		Employment, thousands of persons						Employment-population ratio, percent	Unemployment		
					By age and sex			By industry			Thousands of persons		
		Thousands of persons	Participation rate (percent)	Total	Men, 20 years and over	Women, 20 years and over	Both sexes, 16 to 19 years	Agricultural	Nonagricultural		Total	Unemployed 15 weeks and over	Rate (percent) [2]
2005													
January	224 837	148 029	65.8	140 245	72 063	62 260	7 071	2 112	138 111	62.4	7 784	2 851	5.3
February	225 041	148 364	65.9	140 385	72 299	62 253	7 073	2 129	138 267	62.4	7 980	2 896	5.4
March	225 236	148 391	65.9	140 654	72 478	62 222	7 185	2 180	138 462	62.4	7 737	2 817	5.2
April	225 441	148 926	66.1	141 254	72 860	62 483	7 190	2 248	138 992	62.7	7 672	2 678	5.2
May	225 670	149 261	66.1	141 609	73 133	62 557	7 201	2 225	139 379	62.8	7 651	2 683	5.1
June	225 911	149 238	66.1	141 714	73 223	62 516	7 140	2 302	139 276	62.7	7 524	2 405	5.0
July	226 153	149 432	66.1	142 026	73 337	62 691	7 150	2 308	139 789	62.8	7 406	2 449	5.0
August	226 421	149 779	66.2	142 434	73 513	62 844	7 240	2 183	140 277	62.9	7 345	2 569	4.9
September	226 693	149 954	66.1	142 401	73 333	63 035	7 141	2 181	140 276	62.8	7 553	2 537	5.0
October	226 959	150 001	66.1	142 548	73 445	63 127	7 124	2 191	140 435	62.8	7 453	2 492	5.0
November	227 204	150 065	66.0	142 499	73 341	63 127	7 265	2 174	140 299	62.7	7 566	2 486	5.0
December	227 425	150 030	66.0	142 752	73 467	63 209	7 139	2 094	140 635	62.8	7 279	2 429	4.9
2006													
January	227 553	150 214	66.0	143 150	73 892	63 151	7 197	2 164	140 933	62.9	7 064	2 270	4.7
February	227 763	150 641	66.1	143 457	73 972	63 304	7 298	2 178	141 254	63.0	7 184	2 546	4.8
March	227 975	150 813	66.2	143 741	74 228	63 353	7 342	2 153	141 604	63.1	7 072	2 373	4.7
April	228 199	150 881	66.1	143 761	74 204	63 413	7 198	2 249	141 388	63.0	7 120	2 353	4.7
May	228 428	151 069	66.1	144 089	74 229	63 656	7 215	2 194	141 859	63.1	6 980	2 303	4.6
June	228 671	151 354	66.2	144 353	74 261	63 866	7 393	2 256	142 019	63.1	7 001	2 127	4.6
July	228 912	151 377	66.1	144 202	74 011	64 030	7 324	2 278	142 066	63.0	7 175	2 289	4.7
August	229 167	151 716	66.2	144 625	74 384	64 139	7 266	2 240	142 440	63.1	7 091	2 293	4.7
September	229 420	151 662	66.1	144 815	74 866	63 922	7 196	2 182	142 661	63.1	6 847	2 231	4.5
October	229 675	152 041	66.2	145 314	74 833	64 298	7 289	2 183	143 222	63.3	6 727	2 062	4.4
November	229 905	152 406	66.3	145 534	74 962	64 328	7 331	2 164	143 350	63.3	6 872	2 159	4.5
December	230 108	152 732	66.4	145 970	75 232	64 518	7 287	2 233	143 716	63.4	6 762	2 083	4.4
2007													
January	230 650	153 144	66.4	146 028	75 238	64 621	7 240	2 214	143 785	63.3	7 116	2 156	4.6
February	230 834	152 983	66.3	146 057	75 239	64 750	7 133	2 295	143 741	63.3	6 927	2 211	4.5
March	231 034	153 051	66.2	146 320	75 386	64 928	7 061	2 178	144 153	63.3	6 731	2 255	4.4
April	231 253	152 435	65.9	145 586	75 342	64 366	6 985	2 069	143 385	63.0	6 850	2 281	4.5
May	231 480	152 670	66.0	145 903	75 386	64 724	6 888	2 081	143 773	63.0	6 766	2 231	4.4
June	231 713	153 041	66.0	146 063	75 327	64 788	7 109	1 946	144 112	63.0	6 979	2 281	4.6
July	231 958	153 054	66.0	145 905	75 247	64 766	6 953	2 016	144 041	62.9	7 149	2 364	4.7
August	232 211	152 749	65.8	145 682	75 195	64 838	6 714	1 863	143 865	62.7	7 067	2 333	4.6
September	232 461	153 414	66.0	146 244	75 273	65 086	7 001	2 095	144 148	62.9	7 170	2 355	4.7
October	232 715	153 183	65.8	145 946	75 146	64 842	7 039	2 121	143 926	62.7	7 237	2 300	4.7
November	232 939	153 835	66.0	146 595	75 683	64 985	7 071	2 148	144 413	62.9	7 240	2 366	4.7
December	233 156	153 918	66.0	146 273	75 510	64 910	7 032	2 219	144 018	62.7	7 645	2 501	5.0
2008													
January	232 616	154 063	66.2	146 378	75 551	65 058	7 015	2 205	144 168	62.9	7 685	2 540	5.0
February	232 809	153 653	66.0	146 156	75 450	65 021	6 820	2 190	143 943	62.8	7 497	2 446	4.9
March	232 995	153 908	66.1	146 086	75 294	65 074	6 813	2 172	143 871	62.7	7 822	2 491	5.1
April	233 198	153 769	65.9	146 132	75 182	65 107	6 947	2 115	143 874	62.7	7 637	2 687	5.0
May	233 405	154 303	66.1	145 908	74 953	65 137	7 182	2 113	143 737	62.5	8 395	2 780	5.4
June	233 627	154 313	66.1	145 737	74 975	65 169	6 925	2 127	143 629	62.4	8 575	2 913	5.6
July	233 864	154 469	66.1	145 532	74 950	65 103	6 909	2 134	143 519	62.2	8 937	3 114	5.8
August	234 107	154 641	66.1	145 203	74 658	65 008	6 801	2 148	143 086	62.0	9 438	3 420	6.1
September	234 360	154 570	66.0	145 076	74 456	65 078	6 850	2 231	142 796	61.9	9 494	3 626	6.1
October	234 612	154 876	66.0	144 802	74 263	65 103	6 793	2 199	142 689	61.7	10 074	3 990	6.5
November	234 828	154 639	65.9	144 100	73 943	64 885	6 612	2 207	141 962	61.4	10 538	3 923	6.8
December	235 035	154 655	65.8	143 369	73 311	64 813	6 601	2 209	141 139	61.0	11 286	4 547	7.3
2009													
January	234 739	154 210	65.7	142 152	72 724	64 209	5 218	2 147	140 029	60.6	12 058	4 764	7.8
February	234 913	154 538	65.8	141 640	72 260	64 195	5 185	2 130	139 493	60.3	12 898	5 455	8.3
March	235 086	154 133	65.6	140 707	71 629	64 027	5 051	2 031	138 664	59.9	13 426	5 886	8.7
April	235 271	154 509	65.7	140 656	71 599	64 021	5 036	2 146	138 437	59.8	13 853	6 385	9.0
May	235 452	154 747	65.7	140 248	71 410	63 807	5 031	2 145	138 036	59.6	14 499	7 022	9.4
June	235 655	154 716	65.7	140 009	71 326	63 747	4 937	2 165	137 841	59.4	14 707	7 837	9.5
July	235 870	154 502	65.5	139 901	71 273	63 761	4 867	2 126	137 819	59.3	14 601	7 839	9.5
August	236 087	154 307	65.4	139 492	71 111	63 633	4 749	2 088	137 371	59.1	14 814	7 828	9.6
September	236 322	153 827	65.1	138 818	70 844	63 324	4 649	2 033	136 734	58.7	15 009	8 402	9.8
October	236 550	153 784	65.0	138 432	70 713	63 275	4 445	2 039	136 483	58.5	15 352	8 648	10.0
November	236 743	153 878	65.0	138 659	70 774	63 422	4 463	2 098	136 640	58.6	15 219	8 755	9.9
December	236 924	153 111	64.6	138 013	70 479	63 086	4 449	2 087	135 904	58.3	15 098	8 835	9.9
2010													
January	236 832	153 484	64.8	138 438	70 509	63 498	4 430	2 128	136 384	58.5	15 046	8 894	9.8
February	236 998	153 694	64.9	138 581	70 603	63 482	4 497	2 308	136 327	58.5	15 113	8 922	9.8
March	237 159	153 954	64.9	138 751	70 877	63 378	4 497	2 211	136 579	58.5	15 202	9 087	9.9
April	237 329	154 622	65.2	139 297	71 352	63 415	4 530	2 295	136 958	58.7	15 325	9 130	9.9
May	237 499	154 091	64.9	139 241	71 374	63 435	4 433	2 202	136 984	58.6	14 849	8 897	9.6
June	237 690	153 616	64.6	139 141	71 345	63 555	4 241	2 144	136 936	58.5	14 474	8 907	9.4
July	237 890	153 691	64.6	139 179	71 448	63 420	4 311	2 174	136 927	58.5	14 512	8 714	9.4
August	238 099	154 086	64.7	139 438	71 608	63 426	4 404	2 165	137 193	58.6	14 648	8 421	9.5
September	238 322	153 975	64.6	139 396	71 588	63 541	4 267	2 157	137 220	58.5	14 579	8 454	9.5
October	238 530	153 635	64.4	139 119	71 434	63 391	4 294	2 331	136 893	58.3	14 516	8 613	9.4
November	238 715	154 125	64.6	139 044	71 151	63 503	4 390	2 184	136 915	58.2	15 081	8 696	9.8
December	238 889	153 650	64.3	139 301	71 482	63 515	4 304	2 178	137 123	58.3	14 348	8 549	9.3

[1] Not seasonally adjusted.
[2] In 1930 through 1943, the official BLS data count persons on work relief as unemployed. The unemployment rates for those years shown in parentheses count persons on work relief as employed, which is more consistent with the postwar practice. See notes and definitions.

Table 10-1B. Summary Labor Force, Employment, and Unemployment: Historical Data—*Continued*

(Thousands of persons, percent, seasonally adjusted, except as noted.)

Year and month	Civilian noninsti-tutional popu-lation [1]	Civilian labor force		Employment, thousands of persons						Employ-ment-population ratio, percent	Unemployment		
		Thousands of persons	Participa-tion rate (percent)	Total	By age and sex			By industry			Thousands of persons		Rate (percent) [2]
					Men, 20 years and over	Women, 20 years and over	Both sexes, 16 to 19 years	Agri-cultural	Nonagri-cultural		Total	Unem-ployed 15 weeks and over	
2011													
January	238 704	153 263	64.2	139 250	71 546	63 370	4 334	2 264	137 022	58.3	14 013	8 393	9.1
February	238 851	153 214	64.1	139 394	71 819	63 250	4 326	2 264	137 160	58.4	13 820	8 175	9.0
March	239 000	153 376	64.2	139 639	71 824	63 464	4 351	2 259	137 434	58.4	13 737	8 166	9.0
April	239 146	153 543	64.2	139 586	71 926	63 374	4 287	2 150	137 420	58.4	13 957	8 016	9.1
May	239 313	153 479	64.1	139 624	72 099	63 266	4 259	2 240	137 356	58.3	13 855	8 205	9.0
June	239 489	153 346	64.0	139 384	71 970	63 135	4 278	2 262	137 061	58.2	13 962	8 143	9.1
July	239 671	153 288	64.0	139 524	71 984	63 309	4 232	2 233	137 204	58.2	13 763	8 177	9.0
August	239 871	153 760	64.1	139 942	72 231	63 341	4 370	2 357	137 507	58.3	13 818	8 247	9.0
September	240 071	154 131	64.2	140 183	72 410	63 398	4 375	2 225	137 963	58.4	13 948	8 316	9.0
October	240 269	153 961	64.1	140 368	72 435	63 560	4 373	2 197	138 285	58.4	13 594	7 802	8.8
November	240 441	154 128	64.1	140 826	72 926	63 511	4 388	2 240	138 571	58.6	13 302	7 706	8.6
December	240 584	153 995	64.0	140 902	73 084	63 430	4 388	2 347	138 568	58.6	13 093	7 545	8.5
2012													
January	242 269	154 381	63.7	141 584	73 055	64 148	4 382	2 203	139 382	58.4	12 797	7 433	8.3
February	242 435	154 671	63.8	141 858	73 088	64 369	4 400	2 191	139 718	58.5	12 813	7 293	8.3
March	242 604	154 749	63.8	142 036	73 144	64 533	4 359	2 236	139 800	58.5	12 713	7 176	8.2
April	242 784	154 545	63.7	141 899	73 080	64 459	4 359	2 206	139 682	58.4	12 646	7 072	8.2
May	242 966	154 866	63.7	142 206	73 218	64 594	4 394	2 299	139 915	58.5	12 660	7 091	8.2
June	243 155	155 083	63.8	142 391	73 249	64 627	4 515	2 233	140 160	58.6	12 692	7 227	8.2
July	243 354	154 948	63.7	142 292	73 265	64 491	4 536	2 230	140 013	58.5	12 656	6 983	8.2
August	243 566	154 763	63.5	142 291	73 240	64 647	4 404	2 117	140 107	58.4	12 471	6 889	8.1
September	243 772	155 160	63.6	143 044	73 704	64 925	4 415	2 164	140 912	58.7	12 115	6 775	7.8
October	243 983	155 554	63.8	143 431	73 981	64 995	4 455	2 157	141 387	58.8	12 124	6 784	7.8
November	244 174	155 338	63.6	143 333	73 907	64 966	4 460	2 113	141 224	58.7	12 005	6 550	7.7
December	244 350	155 628	63.7	143 330	74 024	64 930	4 376	2 062	141 284	58.7	12 298	6 629	7.9
2013													
January	244 663	155 763	63.7	143 292	74 011	64 758	4 523	2 052	141 124	58.6	12 471	6 572	8.0
February	244 828	155 312	63.4	143 362	74 128	64 879	4 354	2 067	141 299	58.6	11 950	6 457	7.7
March	244 995	155 005	63.3	143 316	74 107	64 873	4 337	1 991	141 192	58.5	11 689	6 393	7.5
April	245 175	155 394	63.4	143 635	74 082	65 205	4 348	2 041	141 557	58.6	11 760	6 421	7.6
May	245 363	155 536	63.4	143 882	74 111	65 342	4 429	2 102	141 758	58.6	11 654	6 333	7.5
June	245 552	155 749	63.4	143 999	74 203	65 318	4 477	2 106	141 896	58.6	11 751	6 231	7.5
July	245 756	155 599	63.3	144 264	74 254	65 511	4 499	2 209	142 079	58.7	11 335	6 049	7.3
August	245 959	155 605	63.3	144 326	74 085	65 766	4 474	2 230	142 154	58.7	11 279	5 973	7.2
September	246 168	155 687	63.2	144 418	74 211	65 617	4 590	2 190	142 304	58.7	11 270	5 929	7.2
October	246 381	154 673	62.8	143 537	73 882	65 215	4 440	2 182	141 411	58.3	11 136	5 761	7.2
November	246 567	155 265	63.0	144 479	74 540	65 404	4 535	2 121	142 463	58.6	10 787	5 737	6.9
December	246 745	155 182	62.9	144 778	74 628	65 675	4 474	2 232	142 561	58.7	10 404	5 539	6.7
2014													
January	246 915	155 352	62.9	145 150	74 799	65 956	4 396	2 171	142 852	58.8	10 202	5 304	6.6
February	247 085	155 483	62.9	145 134	74 680	66 193	4 261	2 135	143 024	58.7	10 349	5 404	6.7
March	247 258	156 028	63.1	145 648	75 157	66 017	4 474	2 084	143 393	58.9	10 380	5 415	6.7
April	247 439	155 369	62.8	145 667	75 057	66 144	4 467	2 140	143 512	58.9	9 702	4 981	6.2
May	247 622	155 684	62.9	145 825	75 112	66 179	4 534	2 046	143 778	58.9	9 859	4 785	6.3
June	247 814	155 707	62.8	146 247	75 481	66 308	4 458	2 162	144 084	59.0	9 460	4 619	6.1
July	248 023	156 007	62.9	146 399	75 636	66 218	4 545	2 199	144 188	59.0	9 608	4 613	6.2
August	248 229	156 130	62.9	146 530	75 745	66 259	4 527	2 305	144 303	59.0	9 599	4 464	6.1
September	248 446	156 040	62.8	146 778	76 021	66 223	4 534	2 387	144 502	59.1	9 262	4 386	5.9
October	248 657	156 417	62.9	147 427	76 056	66 559	4 812	2 425	144 989	59.3	8 990	4 292	5.7
November	248 844	156 494	62.9	147 404	75 837	66 753	4 815	2 397	145 113	59.2	9 090	4 228	5.8
December	249 027	156 332	62.8	147 615	76 209	66 639	4 767	2 379	145 229	59.3	8 717	4 069	5.6
2015													
January	249 723	157 053	62.9	148 150	76 363	67 071	4 715	2 431	145 621	59.3	8 903	4 156	5.7
February	249 899	156 663	62.7	148 053	76 415	66 895	4 743	2 416	145 639	59.2	8 610	3 992	5.5
March	250 080	156 626	62.6	148 122	76 506	66 861	4 754	2 479	145 460	59.2	8 504	3 787	5.4
April	250 266	157 017	62.7	148 491	76 703	67 026	4 762	2 388	146 069	59.3	8 526	3 650	5.4
May	250 455	157 616	62.9	148 802	76 794	67 222	4 786	2 370	146 413	59.4	8 814	3 749	5.6
June	250 663	157 014	62.6	148 765	76 772	67 370	4 624	2 556	146 254	59.3	8 249	3 559	5.3
July	250 876	157 008	62.6	148 815	76 928	67 213	4 674	2 406	146 364	59.3	8 194	3 387	5.2
August	251 096	157 165	62.6	149 175	76 977	67 522	4 676	2 379	146 849	59.4	7 990	3 456	5.1
September	251 325	156 745	62.4	148 853	76 872	67 290	4 691	2 373	146 594	59.2	7 892	3 334	5.0
October	251 541	157 188	62.5	149 270	76 958	67 558	4 753	2 430	146 848	59.3	7 918	3 344	5.0
November	251 747	157 502	62.6	149 506	76 869	67 868	4 770	2 411	147 220	59.4	7 995	3 349	5.1
December	251 936	158 080	62.7	150 164	77 270	67 972	4 922	2 433	147 724	59.6	7 916	3 361	5.0
2016													
January	252 397	158 371	62.7	150 622	77 611	68 091	4 920	2 390	148 160	59.7	7 749	3 198	4.9
February	252 577	158 705	62.8	150 934	77 844	68 125	4 965	2 454	148 444	59.8	7 771	3 286	4.9
March	252 768	159 079	62.9	151 146	77 777	68 272	4 897	2 555	148 375	59.8	7 932	3 394	5.0
April	252 969	158 891	62.8	150 963	77 941	68 089	4 933	2 572	148 377	59.7	7 928	3 351	5.0
May	253 174	158 700	62.7	151 074	77 880	68 256	4 938	2 556	148 511	59.7	7 626	3 018	4.8
June	253 397	158 899	62.7	151 104	78 145	68 150	4 809	2 514	148 673	59.6	7 795	3 167	4.9
July	253 620	159 150	62.8	151 450	78 086	68 400	4 964	2 423	149 006	59.7	7 700	3 181	4.8
August	253 854	159 582	62.9	151 766	78 256	68 430	5 080	2 564	149 285	59.8	7 817	3 061	4.9
September	254 091	159 810	62.9	151 877	78 240	68 599	5 037	2 432	149 514	59.8	7 933	3 113	5.0
October	254 321	159 768	62.8	151 949	78 271	68 681	4 998	2 330	149 610	59.7	7 819	3 138	4.9
November	254 540	159 629	62.7	152 150	78 387	68 713	5 049	2 394	149 839	59.8	7 480	2 960	4.7
December	254 742	159 779	62.7	152 276	78 449	68 816	5 011	2 323	149 947	59.8	7 503	3 065	4.7

[1] Not seasonally adjusted.
[2] In 1930 through 1943, the official BLS data count persons on work relief as unemployed. The unemployment rates for those years shown in parentheses count persons on work relief as employed, which is more consistent with the postwar practice. See notes and definitions.

Table 10-2. Labor Force and Employment by Major Age and Sex Groups

(Thousands of persons, percent, seasonally adjusted.)

Year and month	Civilian labor force (thousands)			Participation rate (percent)			Employment (thousands)			Employment-population ratio, percent		
	Men, 20 years and over	Women, 20 years and over	Both sexes, 16 to 19 years	Men, 20 years and over	Women, 20 years and over	Both sexes, 16 to 19 years	Men, 20 years and over	Women, 20 years and over	Both sexes, 16 to 19 years	Men, 20 years and over	Women, 20 years and over	Both sexes, 16 to 19 years
1970	47 220	28 301	7 249	82.6	43.3	49.9	45 581	26 952	6 144	79.7	41.2	42.3
1971	48 009	28 904	7 470	82.1	43.3	49.7	45 912	27 246	6 208	78.5	40.9	41.3
1972	49 079	29 901	8 054	81.6	43.7	51.9	47 130	28 276	6 746	78.4	41.3	43.5
1973	49 932	30 991	8 507	81.3	44.4	53.7	48 310	29 484	7 271	78.6	42.2	45.9
1974	50 879	32 201	8 871	81.0	45.3	54.8	48 922	30 424	7 448	77.9	42.8	46.0
1975	51 494	33 410	8 870	80.3	46.0	54.0	48 018	30 726	7 104	74.8	42.3	43.3
1976	52 288	34 814	9 056	79.8	47.0	54.5	49 190	32 226	7 336	75.1	43.5	44.2
1977	53 348	36 310	9 351	79.7	48.1	56.0	50 555	33 775	7 688	75.6	44.8	46.1
1978	54 471	38 128	9 652	79.8	49.6	57.8	52 143	35 836	8 070	76.4	46.6	48.3
1979	55 615	39 708	9 638	79.8	50.6	57.9	53 308	37 434	8 083	76.5	47.7	48.5
1980	56 455	41 106	9 378	79.4	51.3	56.7	53 101	38 492	7 710	74.6	48.1	46.6
1981	57 197	42 485	8 988	79.0	52.1	55.4	53 582	39 590	7 225	74.0	48.6	44.6
1982	57 980	43 699	8 526	78.7	52.7	54.1	52 891	40 086	6 549	71.8	48.4	41.5
1983	58 744	44 636	8 171	78.5	53.1	53.5	53 487	41 004	6 342	71.4	48.8	41.5
1984	59 701	45 900	7 943	78.3	53.7	53.9	55 769	42 793	6 444	73.2	50.1	43.7
1985	60 277	47 283	7 901	78.1	54.7	54.5	56 562	44 154	6 434	73.3	51.0	44.4
1986	61 320	48 589	7 926	78.1	55.5	54.7	57 569	45 556	6 472	73.3	52.0	44.6
1987	62 095	49 783	7 988	78.0	56.2	54.7	58 726	47 074	6 640	73.8	53.1	45.5
1988	62 768	50 870	8 031	77.9	56.8	55.3	59 781	48 383	6 805	74.2	54.0	46.8
1989	63 704	52 212	7 954	78.1	57.7	55.9	60 837	49 745	6 759	74.5	54.9	47.5
1990	64 916	53 131	7 792	78.2	58.0	53.7	61 678	50 535	6 581	74.3	55.2	45.3
1991	65 374	53 708	7 265	77.7	57.9	51.6	61 178	50 634	5 906	72.7	54.6	42.0
1992	66 213	54 796	7 096	77.7	58.5	51.0	61 496	51 328	5 669	72.1	54.8	41.0
1993	66 642	55 388	7 170	77.3	58.5	51.5	62 355	52 099	5 805	72.3	55.0	41.7
1994	66 921	56 655	7 481	76.8	59.3	52.7	63 294	53 606	6 161	72.6	56.2	43.4
1995	67 324	57 215	7 765	76.7	59.4	53.5	64 085	54 396	6 419	73.0	56.5	44.2
1996	68 044	58 094	7 806	76.8	59.9	52.3	64 897	55 311	6 500	73.2	57.0	43.5
1997	69 166	59 198	7 932	77.0	60.5	51.6	66 284	56 613	6 661	73.7	57.8	43.4
1998	69 715	59 702	8 256	76.8	60.4	52.8	67 135	57 278	7 051	73.9	58.0	45.1
1999	70 194	60 840	8 333	76.7	60.7	52.0	67 761	58 555	7 172	74.0	58.5	44.7
2000	72 010	62 301	8 271	76.7	60.6	52.0	69 634	60 067	7 189	74.2	58.4	45.2
2001	72 816	63 016	7 902	76.5	60.6	49.6	69 776	60 417	6 740	73.3	58.1	42.3
2002	73 630	63 648	7 585	76.3	60.5	47.4	69 734	60 420	6 332	72.3	57.5	39.6
2003	74 623	64 716	7 170	75.9	60.6	44.5	70 415	61 402	5 919	71.7	57.5	36.8
2004	75 364	64 923	7 114	75.8	60.3	43.9	71 572	61 773	5 907	71.9	57.4	36.4
2005	76 443	65 714	7 164	75.8	60.4	43.7	73 050	62 702	5 978	72.4	57.6	36.5
2006	77 562	66 585	7 281	75.9	60.5	43.7	74 431	63 834	6 162	72.9	58.0	36.9
2007	78 596	67 516	7 012	75.9	60.6	41.3	75 337	64 799	5 911	72.8	58.2	34.8
2008	79 047	68 382	6 858	75.7	60.9	40.2	74 750	65 039	5 573	71.6	57.9	32.6
2009	78 897	68 856	6 390	74.8	60.8	37.5	71 341	63 699	4 837	67.6	56.2	28.4
2010	78 994	68 990	5 906	74.1	60.3	34.9	71 230	63 456	4 378	66.8	55.5	25.9
2011	79 080	68 810	5 727	73.4	59.8	34.1	72 182	63 360	4 327	67.0	55.0	25.8
2012	79 387	69 765	5 823	73.0	59.3	34.3	73 403	64 640	4 426	67.5	55.0	26.1
2013	79 744	69 860	5 785	72.5	58.8	34.5	74 176	65 295	4 458	67.4	54.9	26.6
2014	80 056	70 212	5 654	71.9	58.5	34.0	75 471	66 287	4 548	67.8	55.2	27.3
2015	80 735	70 695	5 700	71.7	58.2	34.3	76 776	67 323	4 734	68.1	55.4	28.5
2016	81 759	71 538	5 889	71.7	58.3	35.2	78 074	68 387	4 965	68.5	55.7	29.7
2017	82 198	72 293	5 904	71.6	58.1	35.2	78 920	69 343	5 074	68.7	56.1	30.3
2018	83 183	72 997	5 890	71.6	58.5	35.1	80 212	70 422	5 130	69.0	56.4	30.6
2017												
January	82 005	71 761	5 927	71.7	58.3	35.4	78 440	68 633	5 055	68.6	55.8	30.2
February	81 955	72 032	5 867	71.7	58.5	35.1	78 439	68 971	5 007	68.6	56.0	29.9
March	81 906	72 173	5 957	71.6	58.6	35.6	78 472	69 343	5 143	68.6	56.3	30.7
April	81 983	72 195	5 991	71.6	58.6	35.8	78 807	69 239	5 104	68.8	56.2	30.5
May	81 962	72 086	5 862	71.5	58.4	35.0	78 748	69 134	5 037	68.7	56.0	30.1
June	82 012	72 137	5 974	71.5	58.4	35.7	78 755	69 250	5 171	68.6	56.1	30.9
July	82 126	72 417	5 839	71.5	58.6	34.8	78 863	69 529	5 065	68.7	56.3	30.2
August	82 389	72 380	5 936	71.7	58.5	35.4	78 972	69 508	5 111	68.7	56.2	30.5
September	82 639	72 512	6 039	71.8	58.6	36.0	79 453	69 694	5 252	69.1	56.3	31.3
October	82 402	72 203	5 831	71.6	58.3	34.8	79 278	69 545	5 025	68.9	56.1	30.0
November	82 447	72 327	5 853	71.6	58.3	34.9	79 344	69 670	4 931	68.9	56.2	29.4
December	82 594	72 272	5 771	71.6	58.3	34.4	79 493	69 587	4 985	68.9	56.1	29.7
2018												
January	82 915	72 238	5 970	71.7	58.1	35.6	79 719	69 620	5 143	68.9	56.0	30.6
February	83 258	72 595	6 047	71.9	58.4	36.0	80 186	69 849	5 178	69.3	56.1	30.9
March	83 149	72 580	5 916	71.8	58.3	35.3	80 091	69 946	5 123	69.1	56.2	30.5
April	83 163	72 558	5 829	71.7	58.2	34.8	80 108	70 033	5 074	69.1	56.2	30.3
May	83 257	72 590	5 820	71.8	58.2	34.7	80 299	70 161	5 079	69.2	56.3	30.3
June	83 103	73 155	5 871	71.6	58.6	35.0	80 006	70 455	5 131	68.9	56.5	30.6
July	83 019	73 295	5 896	71.4	58.7	35.2	80 217	70 622	5 125	69.0	56.6	30.6
August	83 044	73 153	5 604	71.4	58.5	33.4	80 149	70 563	4 892	68.9	56.5	29.2
September	83 104	73 107	5 843	71.4	58.4	34.9	80 251	70 710	5 108	68.9	56.5	30.5
October	83 277	73 442	5 974	71.5	58.7	35.6	80 388	70 935	5 258	69.0	56.7	31.4
November	83 408	73 478	5 935	71.5	58.6	35.4	80 633	70 949	5 221	69.1	56.6	31.1
December	83 500	73 769	5 971	71.5	58.8	35.6	80 501	71 218	5 226	69.0	56.8	31.2

Table 10-3. Employment by Type of Job

(Thousands of persons, seasonally adjusted, except as noted.)

Year and month	Agricultural	By class of worker								Multiple jobholders		Employed and at work part-time	
		Nonagricultural industries											
			Wage and salary				Self-employed (unincor-porated)	Unpaid family workers [1]	Total (thousands)	Percent of total employed	Economic reasons	Non-economic reasons
		Total	Total	Government	Private industries							
					Private house-holds [1]	Other private industries						
1970	3 463	75 215	69 491	12 431	...	...	5 221	502	...	...	2 446	9 999
1971	3 394	75 972	70 120	12 799	...	...	5 327	522	...	...	2 688	10 152
1972	3 484	78 669	72 785	13 393	...	...	5 365	519	...	...	2 648	10 612
1973	3 470	81 594	75 580	13 655	...	...	5 474	540	...	...	2 554	10 972
1974	3 515	83 279	77 094	14 124	...	...	5 697	489	...	...	2 988	11 153
1975	3 408	82 438	76 249	14 675	...	...	5 705	483	...	...	3 804	11 228
1976	3 331	85 421	79 175	15 132	...	...	5 783	464	...	...	3 607	11 607
1977	3 283	88 734	82 121	15 361	...	...	6 114	498	...	...	3 608	12 120
1978	3 387	92 661	85 753	15 525	...	...	6 429	479	...	...	3 516	12 650
1979	3 347	95 477	88 222	15 635	...	...	6 791	463	...	...	3 577	12 893
1980	3 364	95 938	88 525	15 912	...	...	7 000	413	...	...	4 321	13 067
1981	3 368	97 030	89 543	15 689	...	...	7 097	390	...	...	4 768	13 025
1982	3 401	96 125	88 462	15 516	...	...	7 262	401	...	...	6 170	12 953
1983	3 383	97 450	89 500	15 537	...	...	7 575	376	...	...	6 266	12 911
1984	3 321	101 685	93 565	15 770	...	...	7 785	335	...	...	5 744	13 169
1985	3 179	103 971	95 871	16 031	...	...	7 811	289	...	...	5 590	13 489
1986	3 163	106 434	98 299	16 342	...	...	7 881	255	...	...	5 588	13 935
1987	3 208	109 232	100 771	16 800	...	...	8 201	260	...	...	5 401	14 395
1988	3 169	111 800	103 021	17 114	...	...	8 519	260	...	...	5 206	14 963
1989	3 199	114 142	105 259	17 469	...	...	8 605	279	...	...	4 894	15 393
1990	3 223	115 570	106 598	17 769	...	...	8 719	253	...	...	5 204	15 341
1991	3 269	114 449	105 373	17 934	...	...	8 851	226	...	...	6 161	15 172
1992	3 247	115 245	106 437	18 136	...	...	8 575	233	...	...	6 520	14 918
1993	3 115	117 144	107 966	18 579	...	...	8 959	218	...	...	6 481	15 240
1994	3 409	119 651	110 517	18 293	...	...	9 003	131	7 260	5.9	4 625	17 638
1995	3 440	121 460	112 448	18 362	...	...	8 902	110	7 693	6.2	4 473	17 734
1996	3 443	123 264	114 171	18 217	...	...	8 971	122	7 832	6.2	4 315	17 770
1997	3 399	126 159	116 983	18 131	...	...	9 056	120	7 955	6.1	4 068	18 149
1998	3 378	128 085	119 019	18 383	...	...	8 962	103	7 926	6.0	3 665	18 530
1999	3 281	130 207	121 323	18 903	...	...	8 790	95	7 802	5.8	3 357	18 758
2000	2 464	134 427	125 114	19 248	718	105 148	9 205	108	7 604	5.6	3 227	18 814
2001	2 299	134 635	125 407	19 335	694	105 378	9 121	107	7 357	5.4	3 715	18 790
2002	2 311	134 174	125 156	19 636	757	104 764	8 923	95	7 291	5.3	4 213	18 843
2003	2 275	135 461	126 015	19 634	764	105 616	9 344	101	7 315	5.3	4 701	19 014
2004	2 232	137 020	127 463	19 983	779	106 701	9 467	90	7 473	5.4	4 567	19 380
2005	2 197	139 532	129 931	20 357	812	108 761	9 509	93	7 546	5.3	4 350	19 491
2006	2 206	142 221	132 449	20 337	803	111 309	9 685	87	7 576	5.2	4 162	19 591
2007	2 095	143 952	134 283	21 003	813	112 467	9 557	112	7 655	5.2	4 401	19 756
2008	2 168	143 194	133 882	21 258	805	111 819	9 219	93	7 620	5.2	5 875	19 343
2009	2 103	137 775	128 713	21 178	783	106 752	8 995	66	7 271	5.2	8 913	18 710
2010	2 206	136 858	127 914	21 003	667	106 244	8 860	84	6 878	4.9	8 874	18 251
2011	2 254	137 615	128 934	20 536	722	107 676	8 603	78	6 880	4.9	8 560	18 334
2012	2 186	140 283	131 452	20 360	738	110 355	8 749	81	6 943	4.9	8 122	18 806
2013	2 130	141 799	133 111	20 247	723	112 141	8 619	70	7 002	4.9	7 935	18 903
2014	2 237	144 068	135 402	20 135	820	114 456	8 602	64	7 146	4.9	7 213	19 489
2015	2 422	146 411	137 678	20 601	798	116 279	8 665	68	7 262	4.9	6 371	20 018
2016	2 460	148 975	140 161	20 630	724	118 807	8 715	65	7 531	5.0	5 843	20 680
2017	2 457	150 875	142 090	20 834	657	120 600	8 734	52	7 545	4.9	5 186	20 557
2018	2 425	153 336	144 326	20 942	777	122 607	8 941	69	7 769	5.0	6 788	19 370
2017												
January	2 411	149 709	141 138	20 814	731	119 618	8 487	47	7 511	4.9	5 753	20 609
February	2 437	149 939	141 097	20 804	689	119 526	8 808	52	7 748	5.1	5 603	20 812
March	2 503	150 260	141 600	20 848	698	120 065	8 663	53	7 961	5.2	5 455	20 677
April	2 682	150 432	141 887	20 642	680	120 623	8 512	53	7 636	5.0	5 279	20 425
May	2 501	150 397	141 870	20 751	655	120 394	8 572	53	7 570	5.0	5 234	21 091
June	2 466	150 816	141 947	20 882	630	120 358	8 840	67	7 638	5.0	5 266	20 828
July	2 349	151 073	142 454	21 028	686	120 767	8 512	54	7 542	4.9	5 281	21 181
August	2 378	151 312	142 486	21 064	670	120 808	8 683	69	7 317	4.8	5 237	21 222
September	2 286	152 143	143 126	20 964	583	121 571	8 855	42	7 393	4.8	5 179	20 936
October	2 487	151 353	142 315	20 759	571	121 084	8 958	33	7 195	4.7	4 912	20 866
November	2 461	151 562	142 556	20 742	608	121 210	8 951	46	7 389	4.8	4 866	21 061
December	2 512	151 628	142 677	20 731	687	121 245	8 971	57	7 636	5.0	4 986	21 163
2018												
January	2 480	152 030	142 968	20 867	701	121 402	9 014	47	7 860	5.1	4 982	20 978
February	2 450	152 695	143 741	21 012	738	121 928	8 954	56	7 843	5.1	5 115	21 120
March	2 331	152 664	143 668	20 973	781	121 932	8 983	50	7 605	4.9	4 969	21 439
April	2 312	152 860	143 815	21 175	780	121 895	9 001	70	7 667	4.9	4 952	21 295
May	2 353	153 127	144 115	20 992	773	122 316	9 013	61	7 416	4.8	4 920	21 134
June	2 363	153 267	144 456	20 932	769	122 667	8 728	93	7 672	4.9	4 736	21 336
July	2 493	153 425	144 411	20 876	800	122 722	8 898	66	8 064	5.2	4 588	21 525
August	2 346	153 376	144 380	20 785	782	122 806	8 872	79	7 924	5.1	4 368	21 803
September	2 478	153 634	144 480	20 753	741	122 968	8 957	96	7 717	4.9	4 656	21 404
October	2 418	154 135	145 071	21 186	769	123 239	8 949	88	7 873	5.0	4 630	21 448
November	2 556	154 297	145 313	21 054	811	123 512	8 889	69	7 749	4.9	4 781	20 909
December	2 522	154 520	145 478	20 677	879	123 904	9 031	52	7 866	5.0	4 657	21 234

[1]Not seasonally adjusted.
... = Not available.

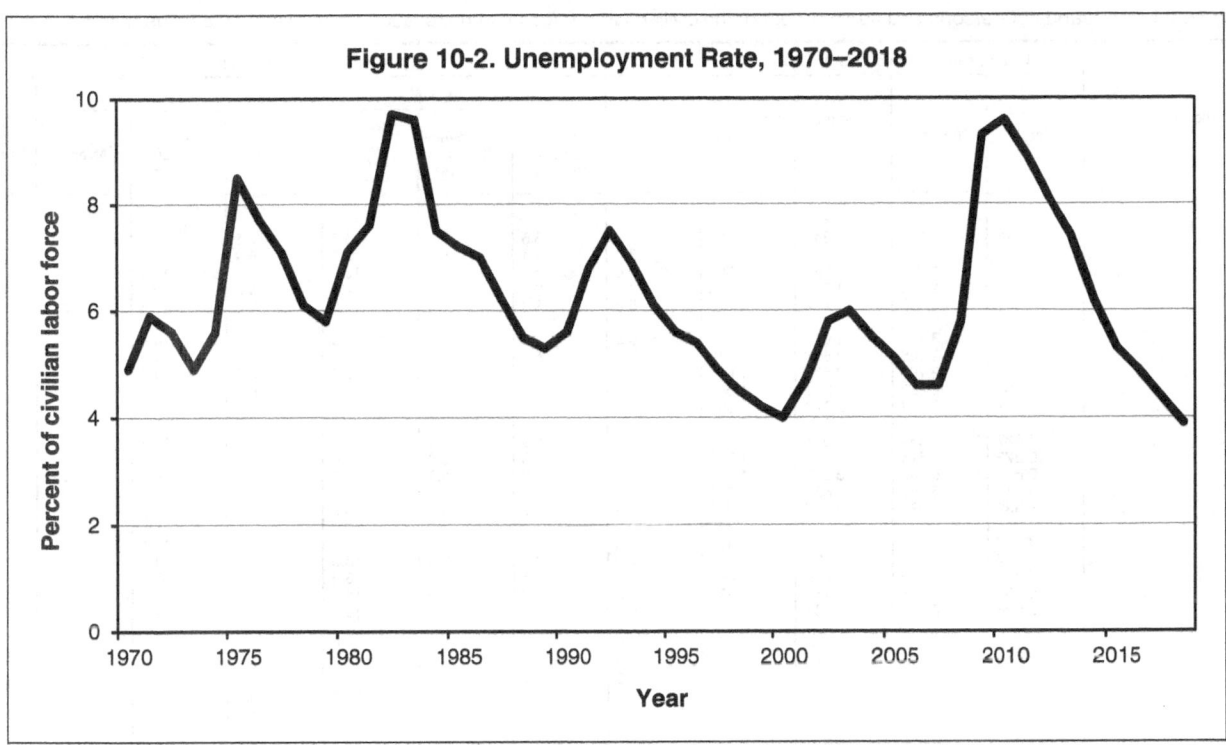

Figure 10-2. Unemployment Rate, 1970–2018

- The unemployment rate peaked at 10.0 percent in October 2009 (lagging, as it usually does, the business cycle trough, which was in June 2009). This was not quite as high as the postwar record of 10.8 percent, set in November and December 1982. Yet the employment losses of 2008–2009 were proportionately greater than the drop in 1981–1982 employment. The unemployment rate in September 2019 was 3.5 percent. (Tables 10-1A and B)

- Recovery from high unemployment during the Great Recession has been slow for most racial and ethnic groups. For example, Hispanics had an unemployment rate of 5.2 percent in 2006. This rate increased to 12.5 percent in 2010. It was not until 2017 that the unemployment rate dropped below the 2006 level to 5.1 percent. (Table 10-4)

- In 2010 and 2011, an unemployed worker was out of work for a median of 21.4 weeks. This is much higher than the median of 10.1 weeks in 1983 and 2003—both recession years. Median duration is more representative of the typical job-seeker than "average duration," which is distorted by the presence of a few extremely long-term unemployed and recently by a technical statistical problem. Even by this better measure, the rise in long-term unemployment has been large—and even understates the problem, since it does not include discouraged workers. (Table 10-5)

- Adding discouraged workers, workers marginally attached to the labor force, and persons on involuntary part time to get a comprehensive measure of the work gap, BLS calculates the "U-6" rate of combined unemployment and underutilization, which rose from 8.3 percent in the high-employment year 2007 to 16.7 percent in 2010. U-6 has since declined to 7.7 percent in 2018. (Table 10-5 and recent data).

- By July 2019, there were more than 18.7 million jobs than in February 2010 based on seasonally adjusted data from the Current Population Survey according to the Current Population Survey.

Table 10-4. Unemployment by Demographic Group

(Unemployment in thousands of persons and as a percent of the civilian labor force in group, seasonally adjusted, except as noted.)

Year and month	Unemployment (thousands of persons)				Unemployment rate (percent)							
	Total	Men, 20 years and over	Women, 20 years and over	Both sexes, 16 to 19 years	All civilian workers	By age and sex			By race			Hispanic or Latino ethnicity
						Men, 20 years and over	Women, 20 years and over	Both sexes, 16 to 19 years	White	Black or African American	Asian[1]	
1970	4 093	1 638	1 349	1 106	4.9	3.5	4.8	15.3	4.5	...	...	...
1971	5 016	2 097	1 658	1 262	5.9	4.4	5.7	16.9	5.4	...	...	...
1972	4 882	1 948	1 625	1 308	5.6	4.0	5.4	16.2	5.1	10.4	...	...
1973	4 365	1 624	1 507	1 235	4.9	3.3	4.9	14.5	4.3	9.4	...	7.5
1974	5 156	1 957	1 777	1 422	5.6	3.8	5.5	16.0	5.0	10.5	...	8.1
1975	7 929	3 476	2 684	1 767	8.5	6.8	8.0	19.9	7.8	14.8	...	12.2
1976	7 406	3 098	2 588	1 719	7.7	5.9	7.4	19.0	7.0	14.0	...	11.5
1977	6 991	2 794	2 535	1 663	7.1	5.2	7.0	17.8	6.2	14.0	...	10.1
1978	6 202	2 328	2 292	1 583	6.1	4.3	6.0	16.4	5.2	12.8	...	9.1
1979	6 137	2 308	2 276	1 555	5.8	4.2	5.7	16.1	5.1	12.3	...	8.3
1980	7 637	3 353	2 615	1 669	7.1	5.9	6.4	17.8	6.3	14.3	...	10.1
1981	8 273	3 615	2 895	1 763	7.6	6.3	6.8	19.6	6.7	15.6	...	10.4
1982	10 678	5 089	3 613	1 977	9.7	8.8	8.3	23.2	8.6	18.9	...	13.8
1983	10 717	5 257	3 632	1 829	9.6	8.9	8.1	22.4	8.4	19.5	...	13.7
1984	8 539	3 932	3 107	1 499	7.5	6.6	6.8	18.9	6.5	15.9	...	10.7
1985	8 312	3 715	3 129	1 468	7.2	6.2	6.6	18.6	6.2	15.1	...	10.5
1986	8 237	3 751	3 032	1 454	7.0	6.1	6.2	18.3	6.0	14.5	...	10.6
1987	7 425	3 369	2 709	1 347	6.2	5.4	5.4	16.9	5.3	13.0	...	8.8
1988	6 701	2 987	2 487	1 226	5.5	4.8	4.9	15.3	4.7	11.7	...	8.2
1989	6 528	2 867	2 467	1 194	5.3	4.5	4.7	15.0	4.5	11.4	...	8.0
1990	7 047	3 239	2 596	1 212	5.6	5.0	4.9	15.5	4.8	11.4	...	8.2
1991	8 628	4 195	3 074	1 359	6.8	6.4	5.7	18.7	6.1	12.5	...	10.0
1992	9 613	4 717	3 469	1 427	7.5	7.1	6.3	20.1	6.6	14.2	...	11.6
1993	8 940	4 287	3 288	1 365	6.9	6.4	5.9	19.0	6.1	13.0	...	10.8
1994	7 996	3 627	3 049	1 320	6.1	5.4	5.4	17.6	5.3	11.5	...	9.9
1995	7 404	3 239	2 819	1 346	5.6	4.8	4.9	17.3	4.9	10.4	...	9.3
1996	7 236	3 146	2 783	1 306	5.4	4.6	4.8	16.7	4.7	10.5	...	8.9
1997	6 739	2 882	2 585	1 271	4.9	4.2	4.4	16.0	4.2	10.0	...	7.7
1998	6 210	2 580	2 424	1 205	4.5	3.7	4.1	14.6	3.9	8.9	...	7.2
1999	5 880	2 433	2 285	1 162	4.2	3.5	3.8	13.9	3.7	8.0	...	6.4
2000	5 692	2 376	2 235	1 081	4.0	3.3	3.6	13.1	3.5	7.6	3.6	5.7
2001	6 801	3 040	2 599	1 162	4.7	4.2	4.1	14.7	4.2	8.6	4.5	6.6
2002	8 378	3 896	3 228	1 253	5.8	5.3	5.1	16.5	5.1	10.2	5.9	7.5
2003	8 774	4 209	3 314	1 251	6.0	5.6	5.1	17.5	5.2	10.8	6.0	7.7
2004	8 149	3 791	3 150	1 208	5.5	5.0	4.9	17.0	4.8	10.4	4.4	7.0
2005	7 591	3 392	3 013	1 186	5.1	4.4	4.6	16.6	4.4	10.0	4.0	6.0
2006	7 001	3 131	2 751	1 119	4.6	4.0	4.1	15.4	4.0	8.9	3.0	5.2
2007	7 078	3 259	2 718	1 101	4.6	4.1	4.0	15.7	4.1	8.3	3.2	5.6
2008	8 924	4 297	3 342	1 285	5.8	5.4	4.9	18.7	5.2	10.1	4.0	7.6
2009	14 265	7 555	5 157	1 552	9.3	9.6	7.5	24.3	8.5	14.8	7.3	12.1
2010	14 825	7 763	5 534	1 528	9.6	9.8	8.0	25.9	8.7	16.0	7.5	12.5
2011	13 747	6 898	5 450	1 400	8.9	8.7	7.9	24.4	7.9	15.8	7.0	11.5
2012	12 506	5 984	5 125	1 397	8.1	7.5	7.3	24.0	7.2	13.8	5.9	10.3
2013	11 460	5 568	4 565	1 327	7.4	7.0	6.5	22.9	6.5	13.1	5.2	9.1
2014	9 617	4 585	3 926	1 106	6.2	5.7	5.6	19.6	5.3	11.3	5.0	7.4
2015	8 296	3 959	3 371	966	5.3	4.9	4.8	16.9	4.6	9.6	3.8	6.6
2016	7 751	3 675	3 151	925	4.9	4.5	5.5	15.7	4.3	8.4	3.6	5.8
2017	6 973	3 220	2 865	831	4.4	4.0	4.0	14.1	3.8	7.5	3.4	5.1
2018	6 306	2 971	2 575	760	3.9	3.6	3.5	12.9	3.5	6.5	3.0	4.7
2017												
January	7 565	3 565	3 128	872	4.7	4.3	4.4	14.7	4.3	7.7	3.7	5.8
February	7 437	3 516	3 060	861	4.7	4.3	4.2	14.7	4.0	8.0	3.5	5.5
March	7 078	3 435	2 829	814	4.4	4.2	3.9	13.7	3.9	7.9	3.3	5.0
April	7 019	3 176	2 956	887	4.4	3.9	4.1	14.8	3.8	7.9	3.2	5.1
May	6 991	3 213	2 952	825	4.4	3.9	4.1	14.1	3.8	7.7	3.7	5.1
June	6 948	3 258	2 888	803	4.3	4.0	4.0	13.4	3.8	7.1	3.6	4.9
July	6 927	3 264	2 888	775	4.3	4.0	4.0	13.3	3.7	7.4	3.8	5.2
August	7 115	3 417	2 872	826	4.4	4.1	4.0	13.9	3.8	7.6	4.0	5.1
September	6 791	3 186	2 818	787	4.2	3.9	3.9	13.0	3.7	7.1	3.6	5.1
October	6 588	3 123	2 658	806	4.1	3.8	3.7	13.8	3.5	7.3	3.0	4.9
November	6 682	3 103	2 657	922	4.2	3.8	3.7	15.8	3.7	7.3	3.0	4.8
December	6 572	3 101	2 686	785	4.1	3.8	3.7	13.6	3.7	6.7	2.5	4.9
2018												
January	6 641	3 196	2 618	827	4.1	3.9	3.6	13.9	3.5	7.7	3.0	5.0
February	6 687	3 072	2 746	870	4.1	3.7	3.8	14.4	3.7	6.8	3.0	4.9
March	6 486	3 059	2 634	793	4.0	3.7	3.6	13.4	3.6	6.8	3.1	5.1
April	6 335	3 055	2 525	755	3.9	3.7	3.5	13.0	3.5	6.5	2.8	4.8
May	6 128	2 958	2 429	741	3.8	3.6	3.3	12.7	3.5	5.9	2.2	4.9
June	6 537	3 097	2 701	740	4.0	3.7	3.7	12.6	3.5	6.5	3.2	4.6
July	6 245	2 801	2 673	771	3.9	3.4	3.6	13.1	3.3	6.6	3.1	4.5
August	6 197	2 895	2 590	712	3.8	3.5	3.5	12.7	3.4	6.3	3.0	4.7
September	5 986	2 853	2 398	735	3.7	3.4	3.3	12.6	3.3	6.0	3.5	4.5
October	6 112	2 889	2 507	715	3.8	3.5	3.4	12.0	3.3	6.2	3.1	4.4
November	6 018	2 775	2 529	714	3.7	3.3	3.4	12.0	3.4	6.0	2.7	4.5
December	6 294	2 999	2 550	745	3.9	3.6	3.5	12.5	3.4	6.6	3.3	4.4

[1] Not seasonally adjusted.
. . . = Not available.

Table 10-5. Unemployment Rates and Related Data

(Seasonally adjusted, except as noted.)

Year and month	Unemployment rates by marital status (percent of labor force in group)			Unemployment rates by reason for unemployment (percent of total civilian labor force in group.)					Duration of unemployment		Alternative measures of labor under-utilization (percent)		
	Married men, spouse present	Married women, spouse present	Women who maintain families[1]	Total	Job losers and persons who completed temporary jobs	Job leavers	Reentrants	New entrants	Average (mean) weeks unemployed	Median weeks unemployed	Including discouraged workers (U-4)	Including all marginally attached workers (U-5)	Including marginally attached and under-employed (U-6)
1970	2.6	4.9	5.4	2.2	0.7	0.7	1.5	0.6	8.6	4.9	...	...	...
1971	3.2	5.7	7.3	2.8	0.7	0.7	1.7	0.7	11.3	6.3	...	...	...
1972	2.8	5.4	7.2	2.4	0.7	0.7	1.7	0.8	12.0	6.2	...	...	...
1973	2.3	4.7	7.1	1.9	0.8	0.8	1.5	0.7	10.0	5.2	...	...	...
1974	2.7	5.3	7.0	2.4	0.8	0.8	1.6	0.7	9.8	5.2	...	...	...
1975	5.1	7.9	10.0	4.7	0.9	0.9	2.0	0.9	14.2	8.4	...	...	...
1976	4.2	7.1	10.1	3.8	0.9	0.9	2.0	0.9	15.8	8.2	...	...	...
1977	3.6	6.5	9.4	3.2	0.9	0.9	2.0	1.0	14.3	7.0	...	...	...
1978	2.8	5.5	8.5	2.5	0.9	0.9	1.8	0.9	11.9	5.9	...	...	...
1979	2.8	5.1	8.3	2.5	0.8	0.8	1.7	0.8	10.8	5.4	...	...	...
1980	4.2	5.8	9.2	3.7	0.8	0.8	1.8	0.8	11.9	6.5	...	...	...
1981	4.3	6.0	10.4	3.9	0.8	0.8	1.9	0.9	13.7	6.9	...	...	...
1982	6.5	7.4	11.7	5.7	0.8	0.8	2.2	1.1	15.6	8.7	...	...	...
1983	6.5	7.0	12.2	5.6	0.7	0.7	2.2	1.1	20.0	10.1	...	...	...
1984	4.6	5.7	10.3	3.9	0.7	0.7	1.9	1.0	18.2	7.9	...	...	...
1985	4.3	5.6	10.4	3.6	0.8	0.8	2.0	0.9	15.6	6.8	...	...	...
1986	4.4	5.2	9.8	3.4	0.9	0.9	1.8	0.9	15.0	6.9	...	...	...
1987	3.9	4.3	9.2	3.0	0.8	0.8	1.6	0.8	14.5	6.5	...	...	...
1988	3.3	3.9	8.1	2.5	0.8	0.0	1.5	0.7	13.5	5.9	...	...	...
1989	3.0	3.7	8.1	2.4	0.8	0.8	1.5	0.5	11.9	4.8	...	...	...
1990	3.4	3.8	8.3	2.7	0.8	0.8	1.6	0.5	12.0	5.3	...	...	...
1991	4.4	4.5	9.3	3.7	0.8	0.8	1.7	0.6	13.7	6.8	...	...	...
1992	5.1	5.0	10.0	4.2	0.8	0.8	1.8	0.7	17.7	8.7	...	...	...
1993	4.4	4.6	9.7	3.8	0.8	0.8	1.7	0.7	18.0	8.3	...	...	...
1994	3.7	4.1	8.9	2.9	0.6	0.6	2.1	0.5	18.8	9.2	6.5	7.4	10.9
1995	3.3	3.9	8.0	2.6	0.6	0.6	1.9	0.4	16.6	8.3	5.9	6.7	10.1
1996	3.0	3.6	8.2	2.5	0.6	0.6	1.9	0.4	16.7	8.3	5.7	6.5	9.7
1997	2.7	3.1	8.1	2.2	0.6	0.6	1.7	0.4	15.8	8.0	5.2	5.9	8.9
1998	2.4	2.9	7.2	2.1	0.5	0.5	1.5	0.4	14.5	6.7	4.7	5.4	8.0
1999	2.2	2.7	6.4	1.9	0.6	0.6	1.4	0.3	13.4	6.4	4.4	5.0	7.4
2000	2.0	2.7	5.9	1.8	0.5	0.5	1.4	0.3	12.6	5.9	4.2	4.8	7.0
2001	2.7	3.1	6.6	2.4	0.6	0.6	1.4	0.3	13.1	6.8	4.9	5.6	8.1
2002	3.6	3.7	8.0	3.2	0.6	0.6	1.6	0.4	16.6	9.1	6.0	6.7	9.6
2003	3.8	3.7	8.5	3.3	0.6	0.6	1.7	0.4	19.2	10.1	6.3	7.0	10.1
2004	3.1	3.5	8.0	2.8	0.6	0.6	1.6	0.5	19.6	9.8	5.8	6.5	9.6
2005	2.8	3.3	7.8	2.5	0.6	0.6	1.6	0.4	18.4	8.9	5.4	6.1	8.9
2006	2.4	2.9	7.1	2.2	0.5	0.5	1.5	0.4	16.8	8.3	4.9	5.5	8.2
2007	2.5	2.8	6.5	2.3	0.5	0.5	1.4	0.4	16.8	8.5	4.9	5.5	8.3
2008	3.4	3.6	8.0	3.1	0.6	0.6	1.6	0.5	17.9	9.4	6.1	6.8	10.5
2009	6.6	5.5	11.5	5.9	0.6	0.6	2.1	0.7	24.4	15.1	9.7	10.5	16.2
2010	6.8	5.9	12.3	6.0	0.6	0.6	2.3	0.8	33.0	21.4	10.3	11.1	16.7
2011	5.8	5.6	12.4	5.3	0.6	0.6	2.2	0.8	39.3	21.4	9.5	10.4	15.9
2012	4.9	5.3	11.4	4.4	0.6	0.6	2.2	0.8	39.4	19.3	8.6	9.5	14.7
2013	4.3	4.6	10.2	3.9	0.6	0.6	2.1	0.8	36.5	17.0	7.9	8.8	13.8
2014	3.4	3.8	8.6	3.1	0.5	0.5	1.8	0.7	33.7	14.0	6.6	7.5	12.0
2015	2.8	3.1	7.4	2.6	0.5	0.5	1.6	0.6	29.2	11.6	5.7	6.4	10.4
2016	2.7	3.0	6.8	2.3	0.5	0.5	1.5	0.5	27.5	10.6	5.2	5.9	9.6
2017	2.4	2.7	6.2	2.1	0.5	0.5	1.3	0.4	25.0	10.0	4.6	5.3	8.5
2018	2.0	2.4	5.4	1.8	0.5	0.5	1.2	0.4	22.7	9.3	4.1	4.8	7.7
2017													
January	2.7	3.1	6.3	4.7	2.3	0.5	1.4	0.5	25.0	10.2	5.1	5.8	9.3
February	2.6	3.0	6.5	4.7	2.3	0.5	1.4	0.5	25.3	10.1	5.0	5.7	9.1
March	2.5	2.7	5.5	4.4	2.2	0.5	1.3	0.5	25.4	10.5	4.7	5.4	8.7
April	2.4	2.8	6.0	4.4	2.2	0.5	1.3	0.4	24.4	10.1	4.7	5.3	8.6
May	2.4	2.7	6.8	4.4	2.1	0.5	1.3	0.4	25.0	10.6	4.6	5.2	8.5
June	2.2	2.7	6.9	4.3	2.1	0.5	1.3	0.4	25.1	10.1	4.6	5.3	8.5
July	2.4	2.8	6.8	4.3	2.1	0.5	1.3	0.4	24.9	10.1	4.6	5.3	8.5
August	2.6	2.8	7.2	4.4	2.2	0.5	1.3	0.4	24.3	10.4	4.7	5.3	8.6
September	2.3	2.6	6.5	4.2	2.1	0.5	1.3	0.4	26.4	10.2	4.5	5.1	8.3
October	2.1	2.5	5.6	4.1	2.0	0.5	1.2	0.4	25.5	9.7	4.4	5.0	8.0
November	2.1	2.3	5.5	4.2	2.0	0.5	1.3	0.4	25.1	9.7	4.4	5.0	8.0
December	2.2	2.6	5.3	4.1	2.0	0.5	1.2	0.4	23.8	8.9	4.4	5.1	8.1
2018													
January	2.3	2.4	6.5	4.1	2.0	0.4	1.2	0.4	23.9	9.4	4.4	5.1	8.2
February	2.1	2.7	6.2	4.1	2.0	0.5	1.2	0.4	22.9	9.3	4.4	5.1	8.2
March	2.1	2.6	5.6	4.0	1.9	0.5	1.2	0.4	24.2	9.2	4.3	4.9	7.9
April	2.1	2.4	5.5	3.9	1.8	0.5	1.2	0.4	23.0	9.8	4.2	4.7	7.8
May	1.9	2.3	4.7	3.8	1.8	0.5	1.2	0.4	21.3	9.3	4.0	4.6	7.7
June	2.1	2.5	5.5	4.0	1.9	0.5	1.3	0.4	21.2	9.0	4.2	4.9	7.8
July	2.0	2.5	5.6	3.9	1.8	0.5	1.1	0.4	23.1	9.6	4.2	4.7	7.5
August	2.0	2.5	5.4	3.8	1.8	0.5	1.2	0.4	22.6	9.2	4.1	4.7	7.4
September	1.9	2.1	5.1	3.7	1.7	0.5	1.2	0.4	24.1	9.3	3.9	4.6	7.5
October	1.9	2.3	5.3	3.8	1.8	0.4	1.2	0.4	22.4	9.4	4.1	4.6	7.5
November	2.0	2.3	5.4	3.7	1.7	0.4	1.2	0.4	21.7	9.0	4.0	4.7	7.6
December	2.1	2.3	4.5	3.9	1.8	0.5	1.2	0.4	21.8	9.1	4.1	4.8	7.6

[1]Not seasonally adjusted.
... = Not available.

Table 10-6. Labor Force and Employment Estimates Smoothed for Population Adjustments (Unofficial)

(Thousands of persons, seasonally adjusted.)

Year	January	February	March	April	May	June	July	August	September	October	November	December
CIVILIAN LABOR FORCE												
1990	125 845	125 734	125 837	125 697	125 953	125 645	125 816	126 087	126 001	126 116	126 203	126 287
1991	126 112	126 189	126 420	126 742	126 382	126 549	126 384	126 392	126 905	126 909	126 981	126 956
1992	127 566	127 524	127 934	128 184	128 475	128 829	128 945	129 008	128 908	128 444	128 872	128 998
1993	128 856	128 926	129 079	129 077	129 772	129 932	129 931	130 166	129 826	130 145	130 296	130 539
1994	131 210	131 296	131 038	131 272	131 444	131 237	131 341	131 980	132 139	132 477	132 637	132 710
1995	132 811	132 901	132 906	133 404	132 673	132 784	133 193	133 199	133 489	133 607	133 517	133 426
1996	133 545	133 896	134 138	134 381	134 654	134 697	135 302	135 083	135 560	135 982	136 082	136 202
1997	136 560	136 517	137 025	137 164	137 281	137 387	137 668	137 824	137 894	137 865	138 209	138 418
1998	138 370	138 401	138 539	138 465	138 703	138 800	138 947	138 942	139 679	139 685	139 801	140 070
1999	140 456	140 433	140 207	140 452	140 615	140 852	140 977	140 981	141 189	141 353	141 623	141 790
2000	142 260	142 443	142 414	142 724	142 355	142 551	142 232	142 461	142 458	142 556	142 889	143 168
2001	143 713	143 607	143 823	143 462	143 204	143 237	143 527	143 150	143 848	143 938	144 085	144 143
2002	143 715	144 477	144 299	144 536	144 742	144 605	144 593	144 792	145 327	145 083	144 803	144 822
2003	145 054	145 193	145 091	145 517	145 519	146 047	145 456	145 392	145 453	145 614	145 871	145 579
2004	146 111	145 963	146 182	146 074	146 273	146 651	146 867	146 724	146 561	146 922	147 274	147 156
2005	147 161	147 480	147 493	148 010	148 329	148 292	148 470	148 800	148 960	148 992	149 041	148 992
2006	149 292	149 704	149 862	149 917	150 091	150 361	150 371	150 695	150 629	150 993	151 342	151 653
2007	151 885	151 711	151 764	151 138	151 356	151 709	151 708	151 390	152 035	151 791	152 422	152 490
2008	153 260	152 844	153 090	152 943	153 466	153 468	153 614	153 777	153 698	153 994	153 750	153 758
2009	153 759	154 082	153 674	154 044	154 277	154 242	154 025	153 826	153 344	153 297	153 386	152 618
2010	153 161	153 478	153 724	154 402	153 869	153 396	153 473	153 855	153 741	153 403	153 900	153 414
2011	153 485	153 572	153 701	153 870	153 835	153 694	153 629	154 051	154 392	154 278	154 416	154 276
2012	154 422	154 924	154 913	154 671	155 056	155 248	155 087	154 790	155 143	155 636	155 412	155 622
2013	156 059	155 635	155 271	155 653	155 820	156 026	155 903	155 806	155 832	154 903	155 600	155 348
2014	155 771	155 979	156 478	155 722	155 937	156 014	156 369	156 345	156 177	156 582	156 747	156 479
2015	156 842	156 588	156 595	156 854	157 254	156 798	156 927	156 908	156 606	156 916	157 148	157 743
2016	157 922	158 440	158 823	158 479	158 045	158 418	158 816	159 022	159 338	159 145	158 953	159 130
2017	160 013	160 296	160 537	160 487	160 037	160 527	160 784	160 919	161 407	160 698	160 864	160 931
CIVILIAN EMPLOYMENT, TOTAL												
1990	119 093	119 082	119 238	118 898	119 209	119 052	118 891	118 894	118 628	118 651	118 432	118 379
1991	118 089	117 915	117 823	118 293	117 634	117 845	117 785	117 712	118 169	118 052	118 033	117 740
1992	118 265	118 050	118 454	118 748	118 709	118 764	119 071	119 195	119 101	119 020	119 280	119 413
1993	119 503	119 715	119 995	119 938	120 594	120 781	120 970	121 373	121 081	121 363	121 722	122 031
1994	122 547	122 679	122 534	122 908	123 497	123 277	123 362	124 013	124 372	124 811	125 230	125 448
1995	125 402	125 681	125 720	125 722	125 207	125 321	125 629	125 677	125 972	126 241	126 052	125 963
1996	126 013	126 542	126 779	126 924	127 189	127 562	127 922	128 161	128 540	128 909	128 801	128 904
1997	129 358	129 370	129 981	130 247	130 584	130 544	130 970	131 172	131 194	131 368	131 859	131 898
1998	131 958	132 053	132 072	132 484	132 614	132 545	132 643	132 718	133 333	133 359	133 655	133 994
1999	134 436	134 276	134 381	134 402	134 775	134 855	134 905	135 097	135 227	135 529	135 862	136 092
2000	136 552	136 585	136 681	137 243	136 597	136 900	136 485	136 609	136 833	137 022	137 249	137 534
2001	137 691	137 519	137 683	137 193	136 979	136 754	136 945	136 109	136 707	136 246	136 086	135 889
2002	135 536	136 266	135 998	135 941	136 347	136 216	136 207	136 492	137 082	136 781	136 289	136 187
2003	136 580	136 622	136 552	136 727	136 616	136 838	136 501	136 553	136 590	136 940	137 354	137 318
2004	137 771	137 827	137 724	137 935	138 092	138 398	138 763	138 766	138 666	138 895	139 377	139 257
2005	139 407	139 533	139 786	140 369	140 708	140 798	141 094	141 486	141 439	141 571	141 508	141 746
2006	142 251	142 544	142 814	142 821	143 135	143 385	143 222	143 630	143 806	144 289	144 495	144 915
2007	144 807	144 821	145 068	144 326	144 626	144 770	144 599	144 364	144 907	144 597	145 226	144 892
2008	145 583	145 354	145 276	145 314	145 083	144 905	144 693	144 358	144 223	143 943	143 237	142 502
2009	141 691	141 176	140 242	140 187	139 776	139 534	139 422	139 010	138 334	137 946	138 168	137 520
2010	138 195	138 345	138 522	139 074	139 025	138 933	138 978	139 244	139 209	138 940	138 872	139 136
2011	139 569	139 724	139 981	139 938	139 987	139 757	139 909	140 339	140 591	140 788	141 258	141 345
2012	141 821	142 105	142 293	142 165	142 482	142 677	142 588	142 596	143 360	143 758	143 669	143 676
2013	143 516	143 592	143 552	143 878	144 132	144 255	144 526	144 595	144 693	143 816	144 766	145 072
2014	145 400	145 445	145 964	145 976	146 127	146 576	146 716	146 843	147 079	147 745	147 731	147 943
2015	148 135	148 122	148 198	148 528	148 811	148 830	148 854	149 160	148 835	149 279	149 512	150 161
2016	150 393	150 819	151 040	150 786	150 854	150 967	151 284	151 484	151 610	151 731	151 915	152 019
2017	152 355	152 794	153 351	153 452	153 185	153 547	153 812	153 775	154 634	154 158	154 232	154 340

Table 10-7. Insured Unemployment

(Averages of weekly data; thousands of persons, except as noted.)

Year and month	State programs, seasonally adjusted			Federal programs, not seasonally adjusted			
				Initial claims		Persons claiming benefits	
	Initial claims	Insured unemployment	Insured unemployment rate (percent) [1]	Federal employees	Newly discharged veterans	Federal employees	Newly discharged veterans
1980	488	3 365	3.9	. . .	. . .	. . .	. . .
1981	451	3 032	3.5	. . .	. . .	. . .	. . .
1982	586	4 094	4.7	. . .	. . .	. . .	. . .
1983	441	3 337	3.9	. . .	. . .	. . .	. . .
1984	374	2 452	2.8	. . .	. . .	. . .	. . .
1985	392	2 584	2.9	. . .	. . .	. . .	. . .
1986	378	2 632	2.8	2	2.5	21	17.5
1987	325	2 273	2.4	2	2.6	21	17.7
1988	309	2 075	2.1	2	2.7	23	18.1
1989	330	2 174	2.1	2	2.3	22	15.1
1990	385	2 539	2.4	2	2.5	24	18.4
1991	447	3 338	3.2	3	2.9	31	22.7
1992	409	3 208	3.1	3	5.0	32	60.6
1993	344	2 768	2.6	3	3.9	32	54.4
1994	340	2 667	2.5	3	3.0	32	37.5
1995	359	2 590	2.4	5	2.5	32	29.6
1996	352	2 552	2.3	7	2.1	29	24.2
1997	322	2 300	2.0	2	1.8	24	19.6
1998	317	2 213	1.9	2	1.4	19	15.6
1999	298	2 186	1.8	1	1.2	17	14.2
2000	299	2 112	1.7	2	1.0	19	12.5
2001	406	3 016	2.4	1	1.2	18	14.0
2002	404	3 570	2.8	1	1.2	18	16.2
2003	402	3 532	2.8	2	1.5	18	19.8
2004	342	2 930	2.3	2	1.9	18	26.9
2005	331	2 660	2.1	1	2.0	17	27.6
2006	312	2 456	1.9	1	2.0	15	26.3
2007	321	2 548	1.9	1	1.7	15	22.8
2008	418	3 337	2.5	1	1.6	15	22.2
2009	574	5 808	4.4	2	2.0	20	30.4
2010	459	4 544	3.6	3	2.6	28	39.2
2011	409	3 744	3.0	2	2.6	31	39.1
2012	374	3 319	2.6	1	2.6	20	39.7
2013	343	2 978	2.3	4	2.3	22	34.8
2014	308	2 590	2.2	1	1.8	16	26.0
2015	278	2 267	1.7	1	1.4	13	19.4
2016	263	2 136	1.6	1	1.4	11	13.8
2017	245	1 961	1.4	1	0.8	11	10.2
2018	220	1 756	1.2	1	0.6	9	7.4
2016							
January	280	2 221	1.6	1	1.2	15	16.4
February	270	2 210	1.6	4	4.7	14	15.4
March	267	2 184	1.6	1	1.0	12	14.5
April	266	2 158	1.6	1	1.0	10	13.8
May	276	2 165	1.6	1	1.1	8	13.5
June	267	2 150	1.6	1	1.1	7	13.1
July	260	2 141	1.6	1	1.1	8	13.2
August	262	2 143	1.6	1	1.1	8	13.3
September	249	2 097	1.5	1	1.1	9	13.3
October	255	2 060	1.5	1	1.1	10	13.4
November	249	2 039	1.5	1	0.9	13	12.9
December	252	2 069	1.5	1	0.9	14	12.9
2017							
January	248	2 055	1.5	1	0.9	14	12.4
February	242	2 046	1.5	1	0.8	14	11.8
March	253	2 017	1.5	1	0.8	13	11.2
April	244	1 979	1.4	1	0.8	11	10.7
May	237	1 920	1.4	1	0.8	9	10.6
June	243	1 952	1.4	1	0.8	8	10.2
July	244	1 960	1.4	1	0.7	8	10.1
August	238	1 948	1.4	1	0.8	9	9.9
September	265	1 931	1.4	1	0.7	8	9.4
October	235	1 903	1.4	2	0.8	10	8.9
November	241	1 916	1.4	1	0.7	12	8.7
December	239	1 905	1.3	1	0.7	15	8.7
2018							
January	236	1 911	1.4	1	0.7	14	8.6
February	219	1 874	1.3	1	0.6	10	6.5
March	223	1 841	1.3	1	0.6	11	8.1
April	219	1 808	1.3	1	0.6	9	7.7
May	220	1 743	1.2	1	0.8	8	7.7
June	222	1 734	1.2	1	0.7	7	7.6
July	215	1 746	1.2	1	0.6	7	7.5
August	215	1 723	1.2	1	0.6	8	7.6
September	212	1 678	1.2	1	0.6	7	7.3
October	216	1 654	1.2	1	0.6	7	7.1
November	225	1 675	1.2	1	0.5	8	6.8
December	223	1 700	1.2	2	0.5	11	6.9

[1] Insured unemployed as a percent of employment covered by state programs.

. . . = Not available.

SECTION 10B: PAYROLL EMPLOYMENT, HOURS, AND EARNINGS

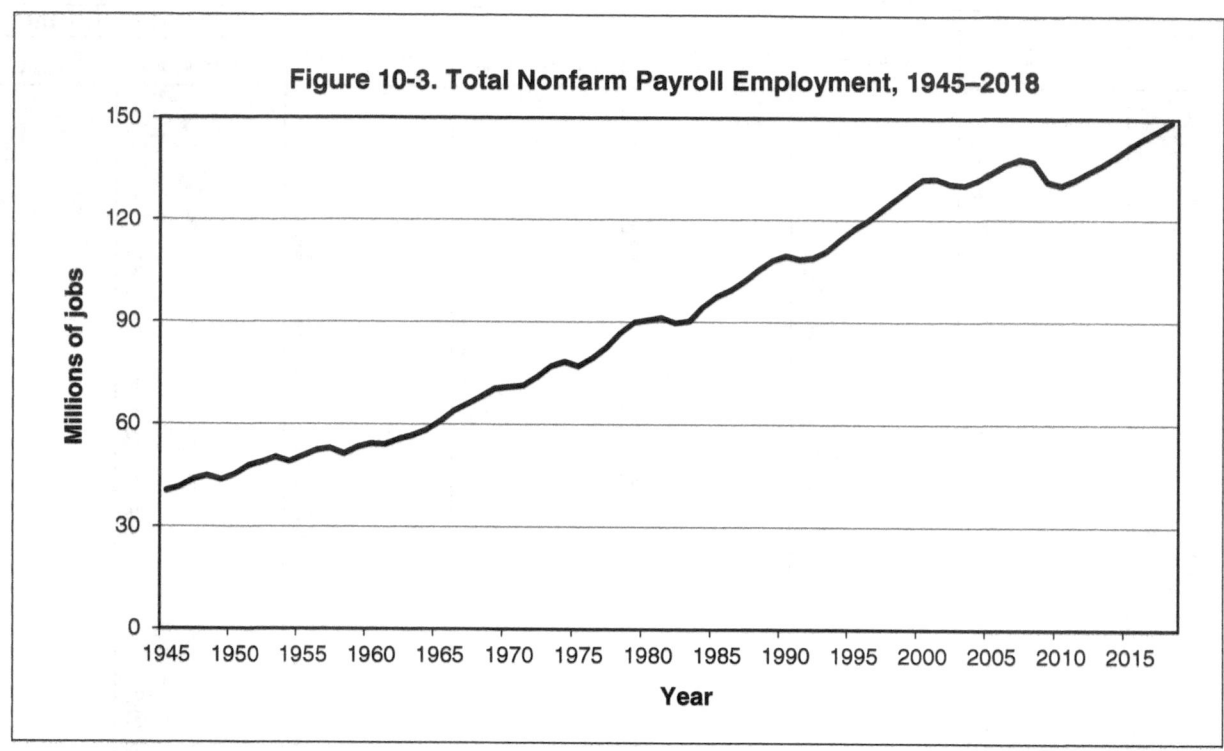

Figure 10-3. Total Nonfarm Payroll Employment, 1945–2018

- Nonfarm payroll employment peaked in January 2008 and plunged 6.3 percent, a postwar record, between then and its lowest point in February 2010. The total number of jobs lost by this measure was 8.7 million. Nonfarm payroll employment has gradually increased from 2010 through 2018. (Table 10-9)

- From 2001 to 2018, total nonfarm employment has increased 14.4 percent. Much of this growth has been in the private service providing industries such as leisure and hospitality, education and health services, and professional and business services. Government employment declined from 2010 to 2019. (Table 10-8)

- The diffusion index indicates the percentage of private industries in which employment is higher than six months earlier. Diffusion indexes measure the extent to which expansion or contraction have spread throughout the economy; indexes below 50 percent indicate recession. This index reached a low of 12.8 percent in April 2009—lower than the lowest points reached in any of the previous four recessions (1980, 1981–1982, 1990–1991, and 2001), and thus indicating the most widespread and pervasive decline in 30 years or more. Gains have been widespread during the recovery, with the vast majority of industries showing an increase. However, after reaching 79.5 percent in October 2014, it fell again to 61.8 in November 2015. By September 2018, the diffusion index peaked at 76.9, the largest amount since December 2014. (Tables 10-8 and recent data)

Table 10-8. Nonfarm Payroll Employment by NAICS Supersector: Recent Data

(Thousands; seasonally adjusted, except as noted.)

Year and month	Total	Private								Service-providing				
		Total	Goods-producing							Total	Private			
			Total	Mining and logging	Construction	Manufacturing					Total	Trade, transportation, and utilities		
						Total	Durable	Nondurable				Total	Wholesale trade	Retail trade
2005	134 051	112 247	22 190	628	7 336	14 227	8 956	5 271	111 861	90 057	25 910	5 706	15 285
2006	136 453	114 479	22 530	684	7 691	14 155	8 981	5 174	113 922	91 949	26 223	5 842	15 359
2007	137 999	115 781	22 233	724	7 630	13 879	8 808	5 071	115 766	93 548	26 573	5 948	15 526
2008	137 241	114 732	21 335	767	7 162	13 406	8 463	4 943	115 907	93 398	26 236	5 875	15 289
2009	131 313	108 758	18 558	694	6 016	11 847	7 284	4 564	112 755	90 201	24 850	5 521	14 528
2010	130 362	107 871	17 751	705	5 518	11 528	7 064	4 464	112 611	90 121	24 581	5 387	14 446
2011	131 932	109 845	18 047	788	5 533	11 726	7 273	4 453	113 884	91 798	25 008	5 475	14 674
2012	134 175	112 255	18 420	848	5 646	11 927	7 470	4 457	115 755	93 835	25 416	5 595	14 847
2013	136 381	114 529	18 738	863	5 856	12 020	7 548	4 472	117 643	95 791	25 801	5 660	15 085
2014	138 958	117 076	19 226	891	6 151	12 185	7 674	4 512	119 732	97 850	26 321	5 740	15 363
2015	141 843	119 814	19 610	813	6 461	12 336	7 765	4 571	122 233	100 204	26 824	5 780	15 611
2016	144 352	122 128	19 750	668	6 728	12 354	7 714	4 640	124 603	102 379	27 195	5 787	15 832
2017	146 624	124 275	20 084	676	6 969	12 439	7 741	4 699	126 540	104 191	27 409	5 814	15 846
2018	149 074	126 625	20 710	732	7 289	12 689	7 945	4 743	128 365	105 916	27 659	5 853	15 833
2014													
January	137 567	115 753	18 937	871	5 985	12 081	7 583	4 498	118 630	96 816	26 095	5 703	15 263
February	137 735	115 910	18 988	875	6 007	12 106	7 605	4 501	118 747	96 922	26 077	5 709	15 244
March	137 985	116 152	19 037	879	6 038	12 120	7 621	4 499	118 948	97 115	26 103	5 714	15 248
April	138 312	116 459	19 104	886	6 084	12 134	7 632	4 502	119 208	97 355	26 187	5 728	15 303
May	138 533	116 701	19 142	887	6 109	12 146	7 647	4 499	119 391	97 559	26 243	5 735	15 020
June	138 857	116 962	19 195	892	6 133	12 170	7 669	4 501	119 662	97 767	26 304	5 746	15 352
July	139 084	117 183	19 264	897	6 178	12 189	7 686	4 503	119 820	97 919	26 353	5 749	15 375
August	139 270	117 417	19 316	898	6 210	12 208	7 699	4 509	119 956	98 101	26 378	5 747	15 386
September	139 583	117 686	19 371	903	6 242	12 226	7 709	4 517	120 212	98 315	26 431	5 755	15 419
October	139 841	117 924	19 419	901	6 259	12 259	7 732	4 527	120 422	98 505	26 508	5 757	15 470
November	140 127	118 105	19 456	900	6 272	12 284	7 749	4 535	120 671	98 739	26 566	5 760	15 503
December	140 396	118 449	19 481	896	6 293	12 292	7 752	4 540	120 915	98 968	26 612	5 768	15 511
2015													
January	140 609	118 650	19 512	887	6 330	12 295	7 754	4 541	121 097	99 138	26 642	5 772	15 514
February	140 857	118 877	19 529	875	6 351	12 303	7 762	4 541	121 328	99 348	26 675	5 782	15 534
March	140 934	118 964	19 503	859	6 333	12 311	7 764	4 547	121 431	99 461	26 701	5 784	15 545
April	141 234	119 226	19 550	843	6 390	12 317	7 762	4 555	121 684	99 676	26 736	5 781	15 573
May	141 553	119 540	19 582	820	6 428	12 334	7 773	4 561	121 971	99 958	26 789	5 786	15 597
June	141 723	119 712	19 601	817	6 446	12 338	7 774	4 564	122 122	100 111	26 830	5 784	15 616
July	142 016	119 964	19 632	806	6 469	12 357	7 775	4 582	122 384	100 332	26 878	5 786	15 639
August	142 138	120 077	19 623	794	6 486	12 343	7 773	4 570	122 515	100 454	26 884	5 784	15 635
September	142 271	120 229	19 633	778	6 505	12 350	7 771	4 579	122 638	100 596	26 880	5 774	15 635
October	142 610	120 554	19 681	771	6 549	12 361	7 771	4 590	122 929	100 873	26 927	5 777	15 686
November	142 845	120 768	19 710	756	6 597	12 357	7 757	4 600	123 135	101 058	26 973	5 778	15 703
December	143 125	121 023	19 740	746	6 632	12 362	7 749	4 613	123 385	101 283	26 994	5 776	15 696
2016													
January	143 215	121 096	19 745	731	6 630	12 384	7 764	4 620	123 470	101 351	27 007	5 781	15 722
February	143 447	121 302	19 720	711	6 640	12 369	7 751	4 618	123 727	101 582	27 057	5 777	15 774
March	143 681	121 498	19 712	691	6 677	12 344	7 729	4 615	123 969	101 786	27 107	5 779	15 809
April	143 892	121 703	19 728	678	6 699	12 351	7 729	4 622	124 164	101 975	27 144	5 786	15 819
May	143 907	121 711	19 690	666	6 691	12 333	7 707	4 626	124 217	102 021	27 149	5 784	15 813
June	144 189	121 986	19 711	657	6 701	12 353	7 703	4 650	124 478	102 275	27 174	5 780	15 837
July	144 525	122 227	19 755	652	6 733	12 370	7 719	4 651	124 770	102 472	27 214	5 781	15 852
August	144 660	122 389	19 729	648	6 734	12 347	7 696	4 651	124 931	102 660	27 261	5 783	15 870
September	144 930	122 634	19 759	647	6 768	12 344	7 689	4 655	125 171	102 875	27 284	5 794	15 884
October	145 058	122 764	19 780	644	6 795	12 341	7 689	4 652	125 278	102 984	27 302	5 796	15 889
November	145 228	122 948	19 806	648	6 817	12 341	7 686	4 655	125 422	103 142	27 314	5 800	15 871
December	145 443	123 139	19 827	647	6 825	12 355	7 696	4 659	125 616	103 312	27 359	5 798	15 895
2017													
January	145 695	123 385	19 874	649	6 857	12 368	7 701	4 667	125 821	103 511	27 404	5 803	15 941
February	145 836	123 516	19 931	655	6 890	12 386	7 701	4 685	125 905	103 585	27 371	5 807	15 888
March	145 963	123 634	19 959	660	6 904	12 395	7 704	4 691	126 004	103 675	27 353	5 806	15 862
April	146 176	123 844	19 993	669	6 921	12 403	7 706	4 697	126 183	103 851	27 360	5 812	15 849
May	146 304	123 971	20 008	674	6 929	12 405	7 714	4 691	126 296	103 963	27 353	5 807	15 822
June	146 533	124 177	20 054	678	6 956	12 420	7 726	4 694	126 479	104 123	27 374	5 813	15 821
July	146 737	124 369	20 054	679	6 958	12 417	7 718	4 699	126 683	104 315	27 376	5 814	15 814
August	146 924	124 563	20 132	685	6 988	12 459	7 751	4 708	126 792	104 431	27 385	5 811	15 811
September	146 942	124 579	20 158	687	7 004	12 467	7 759	4 708	126 784	104 421	27 422	5 813	15 811
October	147 202	124 828	20 200	687	7 026	12 487	7 773	4 714	127 002	104 628	27 448	5 819	15 813
November	147 422	125 040	20 269	692	7 060	12 517	7 797	4 720	127 153	104 771	27 482	5 820	15 828
December	147 596	125 207	20 330	692	7 093	12 545	7 821	4 724	127 266	104 877	27 484	5 826	15 807
2018													
January	147 767	125 393	20 386	699	7 126	12 561	7 838	4 723	127 381	105 007	27 502	5 824	15 809
February	148 097	125 697	20 497	706	7 199	12 592	7 865	4 727	127 600	105 200	27 560	5 827	15 833
March	148 279	125 870	20 527	714	7 201	12 612	7 886	4 726	127 752	105 343	27 591	5 835	15 834
April	148 475	126 054	20 587	723	7 230	12 634	7 903	4 731	127 888	105 467	27 589	5 821	15 838
May	148 745	126 318	20 650	728	7 267	12 655	7 917	4 738	128 095	105 668	27 630	5 827	15 856
June	149 007	126 554	20 706	735	7 284	12 687	7 944	4 743	128 301	105 848	27 622	5 835	15 822
July	149 185	126 727	20 744	734	7 303	12 707	7 961	4 746	128 441	105 983	27 643	5 849	15 824
August	149 467	126 973	20 794	742	7 337	12 715	7 973	4 742	128 673	106 179	27 693	5 869	15 830
September	149 575	127 081	20 832	745	7 354	12 733	7 987	4 746	128 743	106 249	27 692	5 871	15 804
October	149 852	127 366	20 892	751	7 379	12 762	8 006	4 756	128 960	106 474	27 715	5 878	15 794
November	150 048	127 566	20 921	748	7 384	12 789	8 022	4 767	129 127	106 645	27 783	5 889	15 826
December	150 275	127 790	20 961	752	7 400	12 809	8 036	4 773	129 314	106 829	27 788	5 901	15 821

Table 10-8. Nonfarm Payroll Employment by NAICS Supersector: Recent Data—*Continued*

(Thousands; seasonally adjusted, except as noted.)

							Service-providing—*Continued*							Diffusion index, 6-month span, private nonfarm [2]
	Private—*Continued*						Government							
								Federal		State		Local		
Year and month	Information	Financial activities	Profes-sional and business services	Education and health services	Leisure and hospitality	Other services	Total	Total	Depart-ment of Defense [1]	Total	Education	Total	Education	
2005	3 061	8 197	17 003	17 676	12 816	5 395	21 804	2 732	488	5 032	2 260	14 041	7 856	64.5
2006	3 038	8 367	17 619	18 154	13 110	5 438	21 974	2 732	493	5 075	2 293	14 167	7 913	59.9
2007	3 032	8 348	17 998	18 676	13 427	5 494	22 218	2 734	491	5 122	2 318	14 362	7 987	57.2
2008	2 984	8 206	17 792	19 228	13 436	5 515	22 509	2 762	496	5 177	2 354	14 571	8 084	26.9
2009	2 804	7 838	16 634	19 630	13 077	5 367	22 555	2 832	519	5 169	2 360	14 554	8 079	19.2
2010	2 707	7 695	16 783	19 975	13 049	5 331	22 490	2 977	545	5 137	2 373	14 376	8 013	62.4
2011	2 674	7 697	17 389	20 318	13 353	5 360	22 086	2 859	559	5 078	2 374	14 150	7 873	73.8
2012	2 676	7 784	17 992	20 769	13 768	5 430	21 920	2 820	552	5 055	2 389	14 045	7 778	62.2
2013	2 706	7 886	18 575	21 086	14 254	5 483	21 853	2 769	537	5 046	2 393	14 037	7 777	63.0
2014	2 726	7 977	19 124	21 439	14 696	5 567	21 882	2 733	524	5 050	2 389	14 098	7 815	78.5
2015	2 750	8 123	19 695	22 029	15 160	5 622	22 029	2 757	526	5 077	2 401	14 195	7 871	64.5
2016	2 794	8 287	20 114	22 639	15 660	5 691	22 224	2 795	529	5 110	2 429	14 319	7 905	62.8
2017	2 814	8 451	20 508	23 188	16 051	5 770	22 350	2 805	527	5 165	2 479	14 379	7 922	69.4
2018	2 828	8 569	20 999	23 667	16 348	5 845	22 449	2 796	528	5 176	2 486	14 477	7 963	76.9
2014														
January	2 721	7 913	18 837	21 228	14 489	5 533	21 814	2 736	525	5 052	2 399	14 026	7 762	75.0
February	2 719	7 932	18 899	21 251	14 509	5 535	21 825	2 731	524	5 057	2 403	14 037	7 768	73.3
March	2 725	7 932	18 944	21 294	14 564	5 553	21 833	2 728	524	5 055	2 404	14 050	7 778	73.4
April	2 722	7 944	19 002	21 321	14 614	5 565	21 853	2 728	524	5 055	2 403	14 070	7 799	73.8
May	2 719	7 953	19 044	21 364	14 670	5 566	21 832	2 728	524	5 051	2 394	14 053	7 767	74.8
June	2 725	7 967	19 105	21 410	14 693	5 563	21 895	2 729	523	5 058	2 396	14 108	7 816	75.8
July	2 724	7 979	19 142	21 453	14 703	5 565	21 901	2 731	524	5 035	2 371	14 135	7 848	76.0
August	2 731	7 995	19 201	21 491	14 725	5 580	21 855	2 735	525	5 016	2 354	14 104	7 812	77.9
September	2 732	8 009	19 257	21 543	14 768	5 575	21 897	2 735	525	5 041	2 380	14 121	7 838	78.5
October	2 726	8 020	19 285	21 573	14 811	5 582	21 917	2 735	525	5 046	2 381	14 136	7 845	79.5
November	2 735	8 034	19 339	21 626	14 849	5 590	21 932	2 740	525	5 053	2 386	14 139	7 845	79.3
December	2 735	8 041	19 418	21 678	14 892	5 592	21 947	2 743	524	5 059	2 388	14 145	7 849	78.9
2015														
January	2 737	8 058	19 441	21 743	14 917	5 600	21 959	2 744	525	5 066	2 393	14 149	7 853	76.4
February	2 742	8 070	19 487	21 790	14 979	5 605	21 980	2 748	526	5 072	2 398	14 160	7 864	75.4
March	2 737	8 081	19 511	21 832	14 992	5 607	21 970	2 748	526	5 067	2 392	14 155	7 853	71.7
April	2 742	8 089	19 557	21 907	15 031	5 614	22 008	2 762	526	5 072	2 395	14 174	7 865	69.8
May	2 751	8 096	19 636	21 956	15 105	5 625	22 013	2 752	526	5 076	2 396	14 185	7 869	69.6
June	2 751	8 114	19 691	21 994	15 114	5 617	22 011	2 753	526	5 069	2 392	14 189	7 866	68.2
July	2 756	8 134	19 727	22 050	15 158	5 629	22 052	2 757	526	5 072	2 395	14 223	7 893	67.4
August	2 756	8 146	19 759	22 093	15 196	5 620	22 061	2 759	526	5 076	2 401	14 226	7 894	65.7
September	2 762	8 153	19 776	22 141	15 263	5 621	22 042	2 759	526	5 085	2 409	14 198	7 851	64.5
October	2 765	8 168	19 850	22 217	15 316	5 630	22 056	2 755	526	5 086	2 408	14 215	7 868	68.0
November	2 751	8 180	19 884	22 263	15 370	5 637	22 077	2 765	526	5 090	2 412	14 222	7 873	61.8
December	2 759	8 188	19 962	22 322	15 406	5 652	22 102	2 777	527	5 091	2 414	14 234	7 876	64.1
2016														
January	2 762	8 202	19 950	22 340	15 441	5 649	22 119	2 769	528	5 092	2 415	14 258	7 885	63.8
February	2 776	8 209	19 967	22 405	15 503	5 665	22 145	2 780	528	5 086	2 410	14 279	7 890	65.3
March	2 784	8 226	19 996	22 451	15 549	5 673	22 183	2 783	529	5 095	2 414	14 305	7 906	68.2
April	2 789	8 251	20 029	22 507	15 576	5 679	22 189	2 783	529	5 099	2 417	14 307	7 907	64.9
May	2 754	8 263	20 033	22 562	15 588	5 672	22 196	2 792	529	5 092	2 413	14 312	7 911	61.0
June	2 798	8 280	20 074	22 626	15 638	5 685	22 203	2 798	530	5 108	2 430	14 297	7 892	58.7
July	2 799	8 303	20 123	22 671	15 673	5 689	22 298	2 801	530	5 117	2 437	14 380	7 970	64.1
August	2 807	8 321	20 135	22 719	15 715	5 702	22 271	2 803	530	5 112	2 428	14 356	7 933	62.8
September	2 815	8 327	20 210	22 774	15 747	5 718	22 296	2 806	530	5 146	2 460	14 344	7 917	62.8
October	2 816	8 333	20 221	22 830	15 767	5 715	22 294	2 806	531	5 137	2 452	14 351	7 914	65.5
November	2 812	8 342	20 256	22 866	15 825	5 727	22 280	2 799	527	5 136	2 450	14 345	7 900	67.6
December	2 813	8 370	20 286	22 927	15 844	5 713	22 304	2 817	532	5 137	2 456	14 350	7 910	66.7
2017														
January	2 817	8 399	20 336	22 942	15 884	5 729	22 310	2 810	531	5 150	2 464	14 350	7 918	64.5
February	2 812	8 397	20 340	23 005	15 920	5 740	22 320	2 812	530	5 164	2 472	14 344	7 910	67.8
March	2 811	8 408	20 373	23 046	15 939	5 745	22 329	2 810	529	5 167	2 475	14 352	7 918	67.8
April	2 804	8 426	20 403	23 098	16 010	5 750	22 332	2 800	528	5 167	2 481	14 365	7 918	65.9
May	2 804	8 432	20 454	23 130	16 029	5 761	22 333	2 810	526	5 170	2 491	14 353	7 919	64.5
June	2 809	8 447	20 491	23 170	16 061	5 771	22 356	2 807	527	5 170	2 487	14 379	7 920	66.9
July	2 810	8 460	20 537	23 238	16 118	5 776	22 368	2 807	526	5 174	2 489	14 387	7 918	68.4
August	2 818	8 473	20 573	23 278	16 123	5 781	22 361	2 802	527	5 168	2 480	14 391	7 926	67.8
September	2 814	8 480	20 586	23 298	16 043	5 778	22 363	2 801	526	5 161	2 474	14 401	7 930	69.4
October	2 814	8 485	20 627	23 317	16 145	5 792	22 374	2 804	524	5 160	2 473	14 410	7 939	69.4
November	2 815	8 493	20 662	23 355	16 163	5 801	22 382	2 797	525	5 166	2 479	14 419	7 939	69.4
December	2 821	8 500	20 693	23 380	16 195	5 804	22 389	2 795	526	5 164	2 477	14 430	7 943	67.6
2018														
January	2 812	8 502	20 730	23 445	16 208	5 808	22 374	2 795	525	5 147	2 476	14 432	7 934	68.2
February	2 812	8 528	20 774	23 481	16 233	5 812	22 400	2 792	525	5 155	2 476	14 453	7 950	69.4
March	2 824	8 537	20 816	23 518	16 244	5 813	22 409	2 792	525	5 160	2 480	14 457	7 950	73.8
April	2 829	8 541	20 878	23 542	16 262	5 826	22 421	2 793	527	5 169	2 479	14 459	7 953	74.2
May	2 831	8 556	20 929	23 581	16 300	5 841	22 427	2 793	528	5 168	2 476	14 466	7 957	75.6
June	2 831	8 567	20 980	23 646	16 343	5 859	22 453	2 795	529	5 178	2 482	14 480	7 968	75.2
July	2 832	8 572	21 017	23 694	16 378	5 847	22 458	2 796	529	5 179	2 484	14 483	7 973	73.1
August	2 826	8 583	21 075	23 754	16 395	5 853	22 494	2 796	530	5 190	2 495	14 508	7 993	73.6
September	2 822	8 597	21 128	23 779	16 371	5 860	22 494	2 797	530	5 204	2 508	14 493	7 976	76.9
October	2 832	8 611	21 183	23 816	16 450	5 867	22 486	2 798	530	5 197	2 502	14 491	7 970	76.7
November	2 829	8 614	21 217	23 845	16 489	5 868	22 482	2 804	532	5 180	2 484	14 498	7 970	76.4
December	2 827	8 615	21 254	23 912	16 554	5 879	22 485	2 798	532	5 183	2 487	14 504	7 974	74.6

[1] Not seasonally adjusted.
[2] See notes and definitions for explanation. September value used to represent year.

Table 10-9. Employment, Hours, and Earnings, Total Nonfarm and Manufacturing, Historical Annual and Monthly

(Wage and salary workers on nonfarm payrolls, seasonally adjusted.)

Year and month	All wage and salary workers (thousands)					Production and nonsupervisory workers on private payrolls							
	Total	Private			Service-providing	Number (thousands)		Average hours per week		Average hourly earnings, dollars		Average weekly earnings, dollars	
		Total	Goods-producing			Total private	Manufac-turing	Total private	Manufac-turing	Total private	Manufac-turing	Total private	Manufac-turing
			Total	Manufac-turing									
1939	30 645	26 606	11 511	9 450	19 134	...	8 163	...	37.7	...	0.49	...	18.47
1940	32 407	28 156	12 378	10 099	20 029	...	8 737	...	38.2	...	0.53	...	20.25
1941	36 600	31 874	14 940	12 121	21 660	...	10 641	...	40.7	...	0.61	...	24.83
1942	40 213	34 621	17 275	14 030	22 938	...	12 447	...	43.2	...	0.74	...	31.97
1943	42 574	36 353	18 738	16 153	23 837	...	14 407	...	45.1	...	0.86	...	38.79
1944	42 006	35 819	17 981	15 903	24 026	...	14 031	...	45.4	...	0.91	...	41.31
1945	40 510	34 428	16 308	14 255	24 203	...	12 445	...	43.6	...	0.90	...	39.24
1946	41 759	36 054	16 122	13 513	25 637	...	11 781	...	40.4	...	0.95	...	38.38
1947	43 945	38 379	17 314	14 287	26 631	...	12 453	...	40.5	...	1.10	...	44.55
1948	44 954	39 213	17 579	14 324	27 376	...	12 383	...	40.1	...	1.20	...	48.12
1949	43 843	37 893	16 464	13 281	27 379	...	11 355	...	39.2	...	1.25	...	49.00
1950	45 287	39 167	17 343	14 013	27 945	...	12 032	...	40.6	...	1.32	...	53.59
1951	47 930	41 427	18 703	15 070	29 227	...	12 808	...	40.7	...	1.45	...	59.02
1952	48 909	42 182	18 928	15 291	29 981	...	12 797	...	40.8	...	1.53	...	62.42
1953	50 310	43 552	19 733	16 131	30 577	...	13 437	...	40.6	...	1.63	...	66.18
1954	49 093	42 235	18 515	15 002	30 578	...	12 300	...	39.7	...	1.66	...	65.90
1955	50 744	43 722	19 234	15 524	31 510	...	12 735	...	40.8	...	1.74	...	70.99
1956	52 473	45 087	19 799	15 858	32 674	...	12 869	...	40.5	...	1.84	...	74.52
1957	52 959	45 235	19 669	15 798	33 290	...	12 640	...	38.9	...	1.93	...	77.01
1958	51 426	43 480	18 319	14 656	33 107	...	11 532	...	39.2	...	1.99	...	78.01
1959	53 374	45 182	19 163	15 325	34 211	...	12 089	...	40.3	...	2.08	...	83.82
1960	54 296	45 832	19 182	15 438	35 114	...	12 074	...	39.8	...	2.15	...	85.57
1961	54 105	45 399	18 647	15 011	35 458	...	11 612	...	39.9	...	2.20	...	87.78
1962	55 659	46 655	19 203	15 498	36 455	...	11 986	...	40.5	...	2.27	...	91.94
1963	56 764	47 423	19 385	15 631	37 379	...	12 051	...	40.6	...	2.34	...	95.00
1964	58 391	48 680	19 733	15 888	38 658	40 575	12 298	38.5	40.8	2.53	2.41	97.41	98.33
1947													
January	43 535	37 916	17 213	14 328	26 322	...	12 445	...	40.5	...	1.03	...	41.72
February	43 557	37 951	17 200	14 278	26 357	...	12 487	...	40.4	...	1.04	...	42.02
March	43 607	38 019	17 196	14 259	26 411	...	12 518	...	40.3	...	1.06	...	42.72
April	43 499	37 941	17 178	14 240	26 321	...	12 531	...	40.5	...	1.06	...	42.93
May	43 638	38 087	17 176	14 189	26 462	...	12 449	...	40.4	...	1.08	...	43.63
June	43 810	38 286	17 253	14 200	26 557	...	12 389	...	40.4	...	1.10	...	44.44
July	43 743	38 219	17 106	14 076	26 637	...	12 265	...	40.4	...	1.10	...	44.44
August	43 960	38 441	17 280	14 200	26 680	...	12 370	...	40.0	...	1.11	...	44.40
September	44 203	38 664	17 398	14 315	26 805	...	12 430	...	40.5	...	1.11	...	44.96
October	44 411	38 845	17 499	14 393	26 912	...	12 464	...	40.5	...	1.13	...	45.77
November	44 484	38 899	17 517	14 414	26 967	...	12 494	...	40.5	...	1.14	...	46.17
December	44 581	38 976	17 563	14 428	27 018	...	12 526	...	40.8	...	1.16	...	47.33
1948													
January	44 679	39 055	17 625	14 438	27 054	...	12 520	...	40.5	...	1.16	...	46.98
February	44 533	38 918	17 447	14 339	27 086	...	12 437	...	40.2	...	1.16	...	46.63
March	44 683	39 060	17 544	14 364	27 139	...	12 489	...	40.4	...	1.16	...	46.86
April	44 379	38 736	17 302	14 183	27 077	...	12 304	...	40.1	...	1.17	...	46.92
May	44 796	39 115	17 508	14 235	27 288	...	12 344	...	40.2	...	1.18	...	47.44
June	45 034	39 298	17 633	14 318	27 401	...	12 402	...	40.3	...	1.19	...	47.96
July	45 160	39 386	17 649	14 359	27 511	...	12 417	...	40.2	...	1.21	...	48.64
August	45 178	39 387	17 655	14 353	27 523	...	12 397	...	40.2	...	1.23	...	49.45
September	45 294	39 489	17 741	14 441	27 553	...	12 450	...	40.1	...	1.24	...	49.72
October	45 245	39 416	17 683	14 390	27 562	...	12 364	...	39.8	...	1.25	...	49.75
November	45 192	39 323	17 599	14 292	27 593	...	12 302	...	39.8	...	1.26	...	50.15
December	45 032	39 144	17 417	14 086	27 615	...	12 127	...	39.6	...	1.26	...	49.90
1949													
January	44 668	38 774	17 170	13 867	27 498	...	11 905	...	39.4	...	1.26	...	49.64
February	44 497	38 604	17 019	13 734	27 478	...	11 792	...	39.4	...	1.26	...	49.64
March	44 240	38 325	16 848	13 581	27 392	...	11 654	...	39.1	...	1.26	...	49.27
April	44 236	38 288	16 685	13 439	27 551	...	11 517	...	38.8	...	1.25	...	48.50
May	43 984	38 022	16 492	13 269	27 492	...	11 352	...	38.9	...	1.25	...	48.63
June	43 739	37 783	16 351	13 178	27 388	...	11 266	...	39.0	...	1.26	...	49.14
July	43 531	37 569	16 222	13 067	27 309	...	11 168	...	39.2	...	1.26	...	49.39
August	43 624	37 639	16 327	13 158	27 297	...	11 251	...	39.2	...	1.25	...	49.00
September	43 780	37 790	16 403	13 225	27 377	...	11 284	...	39.4	...	1.25	...	49.25
October	42 942	36 972	15 739	12 891	27 203	...	10 940	...	39.6	...	1.24	...	49.10
November	43 242	37 292	16 040	12 882	27 202	...	10 964	...	39.1	...	1.24	...	48.48
December	43 522	37 570	16 217	13 062	27 305	...	11 173	...	39.4	...	1.25	...	49.25
1950													
January	43 526	37 592	16 255	13 161	27 271	...	11 258	...	39.6	...	1.27	...	50.29
February	43 297	37 371	16 035	13 169	27 262	...	11 262	...	39.7	...	1.26	...	50.02
March	43 954	37 876	16 482	13 290	27 472	...	11 362	...	39.7	...	1.28	...	50.82
April	44 382	38 288	16 718	13 471	27 664	...	11 528	...	40.3	...	1.29	...	51.99
May	44 718	38 675	17 080	13 780	27 638	...	11 855	...	40.3	...	1.30	...	52.39
June	45 083	39 061	17 288	13 923	27 795	...	11 979	...	40.6	...	1.30	...	52.78
July	45 454	39 364	17 464	14 072	27 990	...	12 107	...	40.9	...	1.31	...	53.58
August	46 192	40 005	17 917	14 461	28 275	...	12 476	...	41.3	...	1.33	...	54.93
September	46 438	40 210	18 040	14 561	28 398	...	12 519	...	40.8	...	1.33	...	54.26
October	46 706	40 457	18 249	14 737	28 457	...	12 659	...	41.1	...	1.36	...	55.90
November	46 776	40 514	18 288	14 762	28 488	...	12 682	...	41.0	...	1.37	...	56.17
December	46 861	40 547	18 283	14 782	28 578	...	12 710	...	40.9	...	1.39	...	56.85

. . . = Not available.

Table 10-9. Employment, Hours, and Earnings, Total Nonfarm and Manufacturing, Historical Annual and Monthly—Continued

(Wage and salary workers on nonfarm payrolls, seasonally adjusted.)

Year and month	All wage and salary workers (thousands)					Production and nonsupervisory workers on private payrolls							
	Total	Private				Number (thousands)		Average hours per week		Average hourly earnings, dollars		Average weekly earnings, dollars	
		Total	Goods-producing		Service-providing	Total private	Manufac-turing	Total private	Manufac-turing	Total private	Manufac-turing	Total private	Manufac-turing
			Total	Manufac-turing									
1951													
January	47 288	40 936	18 518	14 950	28 770	...	12 816	...	41.0	...	1.40	...	57.40
February	47 577	41 195	18 666	15 076	28 911	...	12 929	...	40.9	...	1.41	...	57.67
March	47 873	41 463	18 754	15 125	29 119	...	12 936	...	41.0	...	1.42	...	58.22
April	47 861	41 410	18 810	15 166	29 051	...	12 960	...	41.0	...	1.43	...	58.63
May	47 952	41 535	18 829	15 164	29 123	...	12 941	...	41.0	...	1.44	...	59.04
June	48 064	41 565	18 826	15 176	29 238	...	12 934	...	40.9	...	1.45	...	59.31
July	48 061	41 523	18 747	15 110	29 314	...	12 848	...	40.6	...	1.45	...	58.87
August	48 012	41 493	18 709	15 061	29 303	...	12 772	...	40.4	...	1.46	...	58.98
September	47 954	41 402	18 622	14 996	29 332	...	12 659	...	40.4	...	1.46	...	58.98
October	48 006	41 429	18 630	14 973	29 376	...	12 615	...	40.3	...	1.47	...	59.24
November	48 147	41 521	18 617	14 999	29 530	...	12 628	...	40.4	...	1.48	...	59.79
December	48 314	41 626	18 698	15 045	29 616	...	12 667	...	40.7	...	1.49	...	60.64
1952													
January	48 296	41 707	18 719	15 067	29 577	...	12 675	...	40.8	...	1.49	...	60.79
February	48 522	41 872	18 813	15 105	29 709	...	12 689	...	40.8	...	1.49	...	60.79
March	48 504	41 842	18 775	15 127	29 729	...	12 695	...	40.6	...	1.51	...	61.31
April	48 620	41 958	18 806	15 162	29 814	...	12 713	...	40.3	...	1.51	...	60.85
May	48 642	41 948	18 784	15 143	29 858	...	12 676	...	40.5	...	1.51	...	61.16
June	48 282	41 570	18 419	14 828	29 863	...	12 353	...	40.6	...	1.51	...	61.31
July	48 143	41 406	18 268	14 707	29 875	...	12 233	...	40.2	...	1.50	...	60.30
August	48 924	42 206	18 928	15 279	29 996	...	12 764	...	40.7	...	1.53	...	62.27
September	49 320	42 586	19 206	15 553	30 114	...	13 007	...	41.1	...	1.55	...	63.71
October	49 597	42 781	19 312	15 690	30 285	...	13 122	...	41.2	...	1.56	...	64.27
November	49 816	43 015	19 473	15 843	30 343	...	13 262	...	41.2	...	1.58	...	65.10
December	50 166	43 231	19 610	15 973	30 556	...	13 375	...	41.2	...	1.58	...	65.10
1953													
January	50 144	43 350	19 721	16 067	30 423	...	13 447	...	41.1	...	1.59	...	65.35
February	50 339	43 542	19 841	16 158	30 498	...	13 529	...	41.0	...	1.61	...	66.01
March	50 473	43 689	19 909	16 270	30 564	...	13 620	...	41.2	...	1.61	...	66.33
April	50 435	43 665	19 908	16 293	30 527	...	13 629	...	40.9	...	1.62	...	66.26
May	50 490	43 773	19 930	16 341	30 560	...	13 645	...	41.0	...	1.62	...	66.42
June	50 519	43 785	19 909	16 343	30 610	...	13 636	...	40.9	...	1.63	...	66.67
July	50 536	43 813	19 910	16 353	30 626	...	13 653	...	40.7	...	1.64	...	66.75
August	50 489	43 735	19 834	16 278	30 655	...	13 564	...	40.7	...	1.64	...	66.75
September	50 368	43 619	19 726	16 151	30 642	...	13 419	...	40.1	...	1.65	...	66.17
October	50 240	43 476	19 578	15 981	30 662	...	13 244	...	40.1	...	1.65	...	66.17
November	49 908	43 159	19 315	15 728	30 593	...	12 990	...	40.0	...	1.65	...	66.00
December	49 703	42 960	19 173	15 581	30 530	...	12 847	...	39.7	...	1.65	...	65.51
1954													
January	49 469	42 709	18 963	15 440	30 506	...	12 706	...	39.5	...	1.65	...	65.18
February	49 382	42 599	18 880	15 307	30 502	...	12 593	...	39.7	...	1.65	...	65.51
March	49 157	42 361	18 748	15 197	30 409	...	12 493	...	39.6	...	1.65	...	65.34
April	49 179	42 373	18 602	15 065	30 577	...	12 361	...	39.7	...	1.65	...	65.51
May	48 965	42 136	18 476	14 974	30 489	...	12 282	...	39.7	...	1.67	...	66.30
June	48 895	42 049	18 400	14 910	30 495	...	12 220	...	39.7	...	1.67	...	66.30
July	48 835	41 967	18 280	14 799	30 555	...	12 121	...	39.7	...	1.66	...	65.90
August	48 826	41 934	18 251	14 772	30 575	...	12 089	...	39.8	...	1.66	...	66.07
September	48 886	41 992	18 261	14 805	30 625	...	12 104	...	39.9	...	1.66	...	66.23
October	48 942	42 042	18 321	14 841	30 621	...	12 142	...	39.7	...	1.67	...	66.30
November	49 180	42 216	18 438	14 913	30 742	...	12 206	...	40.1	...	1.68	...	67.37
December	49 331	42 374	18 508	14 967	30 823	...	12 254	...	40.1	...	1.68	...	67.37
1955													
January	49 496	42 543	18 609	15 034	30 887	...	12 309	...	40.4	...	1.69	...	68.28
February	49 644	42 721	18 726	15 138	30 918	...	12 408	...	40.6	...	1.70	...	69.02
March	49 962	43 024	18 910	15 258	31 052	...	12 524	...	40.7	...	1.71	...	69.60
April	50 248	43 289	19 067	15 375	31 181	...	12 626	...	40.8	...	1.71	...	69.77
May	50 512	43 521	19 223	15 493	31 289	...	12 731	...	41.1	...	1.73	...	71.10
June	50 790	43 770	19 331	15 585	31 459	...	12 810	...	40.8	...	1.73	...	70.58
July	50 987	43 938	19 376	15 614	31 611	...	12 818	...	40.7	...	1.75	...	71.23
August	51 111	44 088	19 432	15 679	31 679	...	12 866	...	40.7	...	1.76	...	71.63
September	51 266	44 199	19 427	15 668	31 839	...	12 830	...	40.7	...	1.77	...	72.04
October	51 429	44 311	19 482	15 740	31 947	...	12 896	...	41.0	...	1.77	...	72.57
November	51 592	44 509	19 554	15 813	32 038	...	12 967	...	41.1	...	1.78	...	73.16
December	51 805	44 673	19 608	15 859	32 197	...	13 009	...	40.9	...	1.78	...	72.80
1956													
January	51 975	44 808	19 665	15 882	32 310	...	13 011	...	40.8	...	1.78	...	72.62
February	52 167	44 955	19 731	15 889	32 436	...	12 986	...	40.7	...	1.79	...	72.85
March	52 294	45 042	19 691	15 829	32 603	...	12 905	...	40.6	...	1.80	...	73.08
April	52 375	45 099	19 811	15 909	32 564	...	12 970	...	40.5	...	1.82	...	73.71
May	52 506	45 139	19 825	15 893	32 681	...	12 925	...	40.4	...	1.83	...	73.93
June	52 586	45 219	19 905	15 835	32 681	...	12 836	...	40.3	...	1.83	...	73.75
July	51 955	44 550	19 390	15 468	32 565	...	12 435	...	40.3	...	1.82	...	73.35
August	52 631	45 180	19 922	15 893	32 709	...	12 860	...	40.3	...	1.85	...	74.56
September	52 604	45 123	19 860	15 863	32 744	...	12 822	...	40.5	...	1.87	...	75.74
October	52 777	45 258	19 918	15 937	32 859	...	12 908	...	40.6	...	1.88	...	76.33
November	52 821	45 268	19 886	15 916	32 935	...	12 864	...	40.4	...	1.88	...	75.95
December	52 929	45 345	19 926	15 957	33 003	...	12 882	...	40.6	...	1.91	...	77.55

. . . = Not available.

Table 10-9. Employment, Hours, and Earnings, Total Nonfarm and Manufacturing, Historical Annual and Monthly—*Continued*

(Wage and salary workers on nonfarm payrolls, seasonally adjusted.)

Year and month	All wage and salary workers (thousands)					Production and nonsupervisory workers on private payrolls							
	Total	Private			Service-providing	Number (thousands)		Average hours per week		Average hourly earnings, dollars		Average weekly earnings, dollars	
		Total	Goods-producing			Total private	Manufac-turing	Total private	Manufac-turing	Total private	Manufac-turing	Total private	Manufac-turing
			Total	Manufac-turing									
1957													
January	52 887	45 267	19 833	15 970	33 054	...	12 881	...	40.4	...	1.90	...	76.76
February	53 097	45 451	19 933	15 998	33 164	...	12 885	...	40.5	...	1.91	...	77.36
March	53 156	45 484	19 936	15 994	33 220	...	12 855	...	40.4	...	1.92	...	77.57
April	53 238	45 537	19 887	15 970	33 351	...	12 811	...	40.0	...	1.91	...	76.40
May	53 150	45 437	19 834	15 931	33 316	...	12 762	...	40.0	...	1.92	...	76.80
June	53 067	45 365	19 777	15 873	33 290	...	12 697	...	40.0	...	1.92	...	76.80
July	53 123	45 369	19 735	15 854	33 388	...	12 669	...	40.0	...	1.93	...	77.20
August	53 128	45 371	19 728	15 867	33 400	...	12 673	...	40.0	...	1.94	...	77.60
September	52 934	45 185	19 545	15 710	33 389	...	12 528	...	39.7	...	1.95	...	77.42
October	52 763	44 995	19 421	15 599	33 342	...	12 437	...	39.4	...	1.96	...	77.22
November	52 558	44 789	19 260	15 466	33 298	...	12 301	...	39.2	...	1.96	...	76.83
December	52 384	44 538	19 111	15 332	33 273	...	12 170	...	39.1	...	1.95	...	76.25
1958													
January	52 076	44 255	18 902	15 130	33 174	...	11 969	...	38.9	...	1.95	...	75.86
February	51 576	43 744	18 529	14 908	33 047	...	11 757	...	38.7	...	1.95	...	75.47
March	51 299	43 451	18 335	14 670	32 964	...	11 532	...	38.8	...	1.96	...	76.05
April	51 027	43 159	18 120	14 506	32 907	...	11 373	...	38.9	...	1.96	...	76.24
May	50 914	43 020	18 008	14 414	32 906	...	11 294	...	38.9	...	1.97	...	76.63
June	50 914	42 988	17 984	14 408	32 930	...	11 300	...	39.1	...	1.98	...	77.42
July	51 039	43 067	18 038	14 450	33 001	...	11 349	...	39.3	...	1.98	...	77.81
August	51 233	43 221	18 147	14 524	33 086	...	11 412	...	39.5	...	2.01	...	79.40
September	51 506	43 490	18 331	14 650	33 175	...	11 556	...	39.5	...	2.00	...	79.00
October	51 485	43 454	18 218	14 503	33 267	...	11 394	...	39.6	...	2.00	...	79.20
November	51 944	43 916	18 610	14 827	33 334	...	11 702	...	39.9	...	2.03	...	81.00
December	52 085	43 985	18 592	14 877	33 493	...	11 743	...	39.9	...	2.04	...	81.40
1959													
January	52 478	44 373	18 796	14 998	33 682	...	11 849	...	40.2	...	2.04	...	82.01
February	52 688	44 572	18 890	15 115	33 798	...	11 950	...	40.3	...	2.05	...	82.62
March	53 014	44 882	19 069	15 259	33 945	...	12 078	...	40.4	...	2.07	...	83.63
April	53 321	45 179	19 269	15 385	34 052	...	12 185	...	40.5	...	2.08	...	84.24
May	53 550	45 397	19 378	15 487	34 172	...	12 277	...	40.7	...	2.08	...	84.66
June	53 681	45 538	19 462	15 554	34 219	...	12 330	...	40.6	...	2.09	...	84.85
July	53 804	45 631	19 529	15 623	34 275	...	12 366	...	40.3	...	2.09	...	84.23
August	53 336	45 155	19 049	15 202	34 287	...	11 936	...	40.4	...	2.07	...	83.63
September	53 428	45 189	19 052	15 254	34 376	...	11 984	...	40.4	...	2.08	...	84.03
October	53 358	45 093	18 925	15 158	34 433	...	11 864	...	40.1	...	2.07	...	83.01
November	53 634	45 350	19 108	15 300	34 526	...	11 991	...	39.9	...	2.07	...	82.59
December	54 174	45 806	19 425	15 573	34 749	...	12 254	...	40.3	...	2.11	...	85.03
1960													
January	54 274	45 967	19 491	15 687	34 783	...	12 362	...	40.6	...	2.13	...	86.48
February	54 513	46 187	19 605	15 765	34 908	...	12 434	...	40.3	...	2.14	...	86.24
March	54 454	45 929	19 373	15 707	35 081	...	12 362	...	40.0	...	2.14	...	85.60
April	54 813	46 279	19 446	15 654	35 367	...	12 299	...	40.0	...	2.14	...	85.60
May	54 475	46 043	19 374	15 575	35 101	...	12 218	...	40.1	...	2.14	...	85.81
June	54 348	45 916	19 240	15 466	35 108	...	12 102	...	39.9	...	2.14	...	85.39
July	54 306	45 864	19 170	15 413	35 136	...	12 047	...	39.9	...	2.14	...	85.39
August	54 272	45 800	19 105	15 360	35 167	...	11 986	...	39.7	...	2.15	...	85.36
September	54 227	45 733	19 057	15 330	35 170	...	11 956	...	39.4	...	2.16	...	85.10
October	54 142	45 640	18 952	15 231	35 190	...	11 846	...	39.7	...	2.16	...	85.75
November	53 961	45 445	18 799	15 112	35 162	...	11 726	...	39.3	...	2.15	...	84.50
December	53 742	45 145	18 548	14 947	35 194	...	11 556	...	38.4	...	2.16	...	82.94
1961													
January	53 683	45 119	18 508	14 863	35 175	...	11 473	...	39.3	...	2.16	...	84.89
February	53 557	44 970	18 418	14 801	35 139	...	11 414	...	39.4	...	2.16	...	85.10
March	53 659	45 048	18 438	14 802	35 221	...	11 410	...	39.5	...	2.16	...	85.32
April	53 627	44 998	18 432	14 825	35 195	...	11 444	...	39.5	...	2.18	...	86.11
May	53 786	45 122	18 523	14 932	35 263	...	11 544	...	39.7	...	2.19	...	86.94
June	53 977	45 289	18 618	14 981	35 359	...	11 593	...	40.0	...	2.20	...	88.00
July	54 123	45 399	18 640	15 029	35 483	...	11 639	...	40.0	...	2.21	...	88.40
August	54 298	45 534	18 725	15 093	35 573	...	11 701	...	40.1	...	2.22	...	89.02
September	54 388	45 592	18 730	15 080	35 658	...	11 679	...	39.5	...	2.20	...	86.90
October	54 522	45 717	18 805	15 143	35 717	...	11 731	...	40.3	...	2.23	...	89.87
November	54 742	45 930	18 927	15 259	35 815	...	11 842	...	40.7	...	2.23	...	90.76
December	54 872	46 036	18 981	15 309	35 891	...	11 872	...	40.4	...	2.24	...	90.50
1962													
January	54 891	46 040	18 936	15 322	35 955	...	11 865	...	40.0	...	2.25	...	90.00
February	55 188	46 310	19 109	15 411	36 079	...	11 950	...	40.4	...	2.26	...	91.30
March	55 275	46 374	19 109	15 451	36 166	...	11 970	...	40.6	...	2.26	...	91.76
April	55 602	46 680	19 258	15 524	36 344	...	12 034	...	40.6	...	2.27	...	92.16
May	55 628	46 670	19 253	15 513	36 375	...	12 011	...	40.6	...	2.27	...	92.16
June	55 644	46 644	19 186	15 518	36 458	...	12 006	...	40.5	...	2.26	...	91.53
July	55 746	46 720	19 248	15 522	36 498	...	12 004	...	40.5	...	2.27	...	91.94
August	55 838	46 775	19 251	15 517	36 587	...	11 990	...	40.5	...	2.28	...	92.34
September	55 978	46 889	19 305	15 568	36 673	...	12 033	...	40.5	...	2.28	...	92.34
October	56 041	46 927	19 301	15 569	36 740	...	12 034	...	40.3	...	2.29	...	92.29
November	56 056	46 911	19 260	15 530	36 796	...	11 977	...	40.5	...	2.29	...	92.75
December	56 028	46 902	19 219	15 520	36 809	...	11 961	...	40.3	...	2.29	...	92.29

. . . = Not available.

Table 10-9. Employment, Hours, and Earnings, Total Nonfarm and Manufacturing, Historical Annual and Monthly—Continued

(Wage and salary workers on nonfarm payrolls, seasonally adjusted.)

Year and month	All wage and salary workers (thousands)					Production and nonsupervisory workers on private payrolls							
	Total	Private			Service-providing	Number (thousands)		Average hours per week		Average hourly earnings, dollars		Average weekly earnings, dollars	
		Total	Goods-producing			Total private	Manufacturing	Total private	Manufacturing	Total private	Manufacturing	Total private	Manufacturing
			Total	Manufacturing									
1963													
January	56 115	46 911	19 257	15 545	36 858	...	11 974	...	40.5	...	2.30	...	93.15
February	56 230	46 999	19 228	15 542	37 002	...	11 965	...	40.5	...	2.31	...	93.56
March	56 320	47 075	19 233	15 564	37 087	...	11 990	...	40.5	...	2.32	...	93.96
April	56 580	47 316	19 343	15 602	37 237	...	12 029	...	40.5	...	2.32	...	93.96
May	56 616	47 328	19 399	15 641	37 217	...	12 066	...	40.5	...	2.33	...	94.37
June	56 659	47 357	19 371	15 624	37 288	...	12 050	...	40.7	...	2.34	...	95.24
July	56 794	47 460	19 423	15 646	37 371	...	12 080	...	40.6	...	2.35	...	95.41
August	56 910	47 542	19 437	15 644	37 473	...	12 060	...	40.6	...	2.34	...	95.00
September	57 078	47 661	19 483	15 674	37 595	...	12 088	...	40.6	...	2.36	...	95.82
October	57 283	47 804	19 517	15 714	37 766	...	12 131	...	40.7	...	2.36	...	96.05
November	57 255	47 771	19 456	15 675	37 799	...	12 072	...	40.7	...	2.37	...	96.46
December	57 361	47 864	19 493	15 712	37 868	...	12 108	...	40.6	...	2.38	...	96.63
1964													
January	57 487	47 925	19 406	15 715	38 081	39 914	12 132	38.2	40.1	2.50	2.38	95.50	95.44
February	57 753	48 172	19 570	15 742	38 183	40 123	12 165	38.5	40.7	2.50	2.38	96.25	96.87
March	57 897	48 286	19 587	15 770	38 310	40 171	12 190	38.5	40.6	2.51	2.38	96.64	96.63
April	57 922	48 278	19 593	15 785	38 329	40 209	12 211	38.6	40.7	2.52	2.40	97.27	97.68
May	58 089	48 419	19 630	15 812	38 459	40 331	12 231	38.6	40.8	2.52	2.40	97.27	97.92
June	58 219	48 550	19 682	15 839	38 537	40 447	12 255	38.6	40.8	2.53	2.41	97.66	98.33
July	58 412	48 735	19 740	15 887	38 672	40 624	12 309	38.6	40.8	2.54	2.41	98.04	98.33
August	58 619	48 887	19 810	15 948	38 809	40 771	12 354	38.5	40.9	2.55	2.42	98.18	98.98
September	58 903	49 117	19 943	16 073	38 960	41 028	12 479	38.5	40.9	2.56	2.44	98.56	99.80
October	58 793	48 948	19 723	15 821	39 070	40 829	12 224	38.5	40.7	2.55	2.40	98.18	97.68
November	59 218	49 339	20 026	16 096	39 192	41 158	12 477	38.6	41.0	2.56	2.42	98.82	99.22
December	59 421	49 524	20 111	16 176	39 310	41 308	12 551	38.7	41.2	2.58	2.44	99.85	100.53
1965													
January	59 582	49 645	20 173	16 245	39 409	41 454	12 603	38.7	41.3	2.58	2.45	99.85	101.19
February	59 800	49 826	20 216	16 291	39 584	41 584	12 644	38.7	41.3	2.59	2.46	100.23	101.60
March	60 003	49 993	20 292	16 353	39 711	41 675	12 701	38.7	41.3	2.61	2.47	101.01	102.01
April	60 259	50 208	20 317	16 418	39 942	41 885	12 752	38.7	41.2	2.60	2.47	100.62	101.76
May	60 491	50 397	20 444	16 477	40 047	42 044	12 792	38.8	41.3	2.62	2.49	101.66	102.84
June	60 690	50 562	20 522	16 554	40 168	42 183	12 854	38.6	41.2	2.63	2.49	101.52	102.59
July	60 965	50 764	20 611	16 669	40 354	42 364	12 962	38.6	41.2	2.63	2.49	101.52	102.59
August	61 228	50 957	20 726	16 732	40 502	42 536	12 994	38.5	41.1	2.64	2.50	101.64	102.75
September	61 490	51 152	20 808	16 802	40 682	42 726	13 046	38.6	41.1	2.65	2.51	102.29	103.16
October	61 719	51 341	20 895	16 864	40 824	42 871	13 100	38.5	41.2	2.66	2.52	102.41	103.82
November	61 996	51 560	21 021	16 962	40 975	43 045	13 175	38.6	41.3	2.67	2.52	103.06	104.08
December	62 322	51 823	21 151	17 051	41 171	43 271	13 243	38.6	41.3	2.67	2.53	103.06	104.49
1966													
January	62 529	51 988	21 214	17 143	41 315	43 400	13 301	38.6	41.5	2.68	2.54	103.45	105.41
February	62 796	52 185	21 315	17 288	41 481	43 551	13 426	38.7	41.7	2.69	2.56	104.10	106.75
March	63 192	52 500	21 515	17 400	41 677	43 802	13 508	38.7	41.6	2.70	2.56	104.49	106.50
April	63 437	52 678	21 568	17 517	41 869	43 959	13 598	38.7	41.8	2.71	2.58	104.88	107.84
May	63 712	52 891	21 675	17 625	42 037	44 138	13 678	38.5	41.5	2.72	2.58	104.72	107.07
June	64 111	53 209	21 846	17 733	42 265	44 390	13 753	38.5	41.4	2.73	2.58	105.11	106.81
July	64 301	53 327	21 872	17 760	42 429	44 484	13 758	38.4	41.2	2.74	2.60	105.22	107.12
August	64 507	53 501	21 972	17 882	42 535	44 597	13 843	38.4	41.4	2.75	2.61	105.60	108.05
September	64 643	53 580	21 948	17 886	42 695	44 659	13 841	38.3	41.2	2.76	2.63	105.71	108.36
October	64 854	53 727	21 991	17 956	42 863	44 789	13 906	38.4	41.3	2.77	2.64	106.37	109.03
November	65 019	53 816	21 988	17 981	43 031	44 835	13 918	38.3	41.2	2.78	2.65	106.47	109.18
December	65 199	53 943	22 008	17 998	43 191	44 915	13 906	38.2	40.9	2.78	2.64	106.20	107.98
1967													
January	65 407	54 092	22 057	18 033	43 350	45 051	13 925	38.3	41.1	2.80	2.65	107.24	108.92
February	65 429	54 076	21 987	17 978	43 442	44 967	13 853	37.9	40.4	2.81	2.67	106.50	107.87
March	65 530	54 133	21 919	17 940	43 611	44 990	13 797	37.9	40.5	2.81	2.67	106.50	108.14
April	65 466	54 031	21 842	17 878	43 624	44 872	13 712	37.8	40.4	2.82	2.68	106.60	108.27
May	65 620	54 146	21 779	17 832	43 841	44 961	13 664	37.8	40.4	2.83	2.69	106.97	108.68
June	65 750	54 216	21 761	17 812	43 989	45 003	13 632	37.8	40.4	2.85	2.69	107.73	108.68
July	65 888	54 344	21 772	17 784	44 116	45 105	13 602	37.8	40.5	2.86	2.71	108.11	109.76
August	66 143	54 553	21 887	17 905	44 256	45 267	13 681	37.8	40.6	2.88	2.73	108.86	110.84
September	66 164	54 541	21 775	17 794	44 389	45 236	13 559	37.8	40.6	2.88	2.73	108.86	110.84
October	66 225	54 583	21 779	17 800	44 446	45 279	13 587	37.8	40.6	2.89	2.74	109.24	111.24
November	66 703	55 008	21 996	17 985	44 707	45 702	13 777	37.9	40.6	2.91	2.75	110.29	111.65
December	66 900	55 165	22 037	18 025	44 863	45 799	13 782	37.7	40.7	2.92	2.77	110.08	112.74
1968													
January	66 804	55 010	21 917	18 040	44 887	45 655	13 798	37.6	40.4	2.94	2.81	110.54	113.52
February	67 215	55 396	22 117	18 054	45 098	45 980	13 793	37.8	40.8	2.95	2.82	111.51	115.06
March	67 295	55 453	22 119	18 067	45 176	46 040	13 803	37.7	40.8	2.97	2.84	111.97	115.87
April	67 556	55 678	22 207	18 131	45 349	46 239	13 858	37.6	40.3	2.99	2.85	112.42	114.86
May	67 652	55 747	22 255	18 190	45 397	46 267	13 900	37.7	40.9	3.00	2.87	113.10	117.38
June	67 905	55 918	22 264	18 228	45 641	46 402	13 921	37.8	40.9	3.01	2.88	113.78	117.79
July	68 126	56 108	22 329	18 265	45 797	46 562	13 953	37.7	40.8	3.03	2.89	114.23	117.91
August	68 330	56 288	22 350	18 254	45 980	46 670	13 903	37.7	40.7	3.04	2.89	114.61	117.62
September	68 484	56 417	22 390	18 252	46 094	46 791	13 914	37.7	40.9	3.07	2.92	115.74	119.43
October	68 721	56 620	22 419	18 293	46 302	46 991	13 974	37.7	41.0	3.08	2.94	116.12	120.54
November	68 984	56 877	22 512	18 346	46 472	47 245	14 028	37.5	40.9	3.10	2.96	116.25	121.06
December	69 248	57 103	22 617	18 410	46 631	47 386	14 044	37.5	40.7	3.11	2.97	116.63	120.88

. . . = Not available.

Table 10-9. Employment, Hours, and Earnings, Total Nonfarm and Manufacturing, Historical Annual and Monthly—Continued

(Wage and salary workers on nonfarm payrolls, seasonally adjusted.)

Year and month	All wage and salary workers (thousands)					Production and nonsupervisory workers on private payrolls							
	Total	Private			Service-providing	Number (thousands)		Average hours per week		Average hourly earnings, dollars		Average weekly earnings, dollars	
		Total	Goods-producing			Total private	Manufac-turing	Total private	Manufac-turing	Total private	Manufac-turing	Total private	Manufac-turing
			Total	Manufac-turing									
1969													
January	69 439	57 230	22 644	18 432	46 795	47 529	14 086	37.7	40.8	3.12	2.99	117.62	121.99
February	69 699	57 475	22 755	18 502	46 944	47 697	14 134	37.5	40.4	3.14	3.00	117.75	121.20
March	69 905	57 676	22 813	18 558	47 092	47 852	14 169	37.6	40.8	3.16	3.01	118.82	122.81
April	70 072	57 827	22 815	18 554	47 257	47 960	14 144	37.7	41.0	3.18	3.03	119.89	124.23
May	70 328	58 044	22 899	18 588	47 429	48 122	14 161	37.6	40.7	3.19	3.04	119.94	123.73
June	70 636	58 277	22 981	18 640	47 655	48 329	14 206	37.5	40.7	3.21	3.05	120.38	124.14
July	70 729	58 389	22 990	18 642	47 739	48 431	14 194	37.5	40.6	3.22	3.08	120.75	125.05
August	71 008	58 635	23 111	18 767	47 897	48 616	14 287	37.6	40.6	3.24	3.10	121.82	125.86
September	70 914	58 535	22 988	18 620	47 926	48 524	14 159	37.5	40.6	3.26	3.12	122.25	126.67
October	71 121	58 691	22 976	18 613	48 145	48 670	14 174	37.4	40.5	3.28	3.13	122.67	126.77
November	71 086	58 638	22 840	18 467	48 246	48 590	14 035	37.5	40.5	3.30	3.14	123.75	127.17
December	71 241	58 764	22 884	18 485	48 357	48 639	14 013	37.5	40.6	3.31	3.15	124.13	127.89
1970													
January	71 176	58 680	22 726	18 424	48 450	48 566	13 964	37.3	40.4	3.31	3.16	123.46	127.66
February	71 305	58 787	22 747	18 361	48 558	48 601	13 897	37.3	40.2	3.33	3.17	124.21	127.43
March	71 451	58 848	22 738	18 360	48 713	48 690	13 917	37.2	40.1	3.36	3.19	124.99	127.92
April	71 348	58 643	22 552	18 207	48 796	48 482	13 785	37.0	39.8	3.36	3.19	124.32	126.96
May	71 124	58 456	22 336	18 029	48 788	48 288	13 616	37.0	39.8	3.37	3.22	124.69	128.16
June	71 029	58 362	22 241	17 930	48 788	48 225	13 554	36.9	39.8	3.40	3.24	125.46	128.95
July	71 053	58 356	22 195	17 877	48 858	48 235	13 527	37.0	40.0	3.41	3.25	126.17	130.00
August	70 937	58 226	22 105	17 779	48 832	48 089	13 449	37.0	39.8	3.44	3.26	127.28	129.75
September	70 944	58 203	21 988	17 692	48 956	48 101	13 401	36.8	39.6	3.45	3.29	126.96	130.28
October	70 521	57 728	21 477	17 173	49 044	47 621	12 905	36.8	39.5	3.45	3.26	126.96	128.77
November	70 409	57 579	21 345	17 024	49 064	47 468	12 781	36.8	39.5	3.47	3.26	127.70	128.77
December	70 792	57 947	21 673	17 309	49 119	47 779	13 062	36.8	39.5	3.50	3.32	128.80	131.14
1971													
January	70 865	57 987	21 594	17 280	49 271	47 859	13 069	36.8	39.9	3.53	3.36	129.90	134.06
February	70 807	57 930	21 514	17 216	49 293	47 780	13 033	36.7	39.7	3.55	3.39	130.29	134.58
March	70 860	57 952	21 491	17 154	49 369	47 820	12 984	36.7	39.8	3.56	3.39	130.65	134.92
April	71 036	58 091	21 552	17 149	49 484	47 967	12 993	36.8	39.9	3.58	3.41	131.74	136.06
May	71 247	58 277	21 645	17 225	49 602	48 154	13 081	36.7	40.0	3.61	3.43	132.49	137.20
June	71 254	58 246	21 568	17 139	49 686	48 109	13 012	36.8	39.9	3.62	3.45	133.22	137.66
July	71 315	58 304	21 564	17 126	49 751	48 159	13 001	36.7	40.0	3.63	3.46	133.22	138.40
August	71 373	58 332	21 570	17 115	49 803	48 166	12 993	36.7	39.8	3.66	3.48	134.32	138.50
September	71 614	58 546	21 650	17 154	49 964	48 364	13 038	36.7	39.7	3.67	3.48	134.69	138.16
October	71 642	58 527	21 604	17 126	50 038	48 308	13 027	36.8	39.9	3.68	3.50	135.42	139.65
November	71 847	58 699	21 684	17 166	50 163	48 441	13 064	36.8	40.0	3.70	3.49	136.16	139.60
December	72 109	58 919	21 741	17 202	50 368	48 613	13 078	36.9	40.2	3.73	3.55	137.64	142.71
1972													
January	72 441	59 175	21 865	17 283	50 576	49 033	13 173	36.8	40.2	3.80	3.57	139.84	143.51
February	72 648	59 350	21 915	17 361	50 733	49 140	13 235	36.9	40.4	3.82	3.61	140.96	145.84
March	72 944	59 615	22 036	17 447	50 908	49 437	13 316	36.9	40.4	3.84	3.63	141.70	146.65
April	73 162	59 804	22 099	17 508	51 063	49 561	13 373	36.9	40.5	3.86	3.65	142.43	147.83
May	73 469	60 053	22 222	17 602	51 247	49 749	13 451	36.8	40.5	3.88	3.67	142.78	148.64
June	73 758	60 353	22 282	17 641	51 476	49 997	13 475	36.9	40.6	3.88	3.68	143.17	149.41
July	73 709	60 227	22 162	17 556	51 547	49 847	13 387	36.8	40.5	3.90	3.69	143.52	149.45
August	74 141	60 611	22 400	17 741	51 741	50 159	13 562	36.8	40.6	3.92	3.73	144.26	151.44
September	74 264	60 689	22 456	17 774	51 808	50 232	13 572	36.9	40.6	3.94	3.75	145.39	152.25
October	74 674	61 068	22 613	17 893	52 061	50 555	13 681	37.0	40.7	3.97	3.78	146.89	153.85
November	74 973	61 330	22 688	18 005	52 285	50 788	13 783	36.9	40.7	3.99	3.79	147.23	154.25
December	75 268	61 584	22 772	18 158	52 496	51 053	13 902	36.8	40.6	4.00	3.83	147.20	155.50
1973													
January	75 617	61 927	22 955	18 276	52 662	51 346	14 006	36.8	40.4	4.03	3.86	148.30	155.94
February	76 014	62 303	23 160	18 410	52 854	51 684	14 127	36.9	40.9	4.05	3.87	149.45	158.28
March	76 284	62 539	23 262	18 493	53 022	51 903	14 181	37.0	40.9	4.06	3.88	150.22	158.69
April	76 455	62 678	23 316	18 530	53 139	51 983	14 192	36.9	40.8	4.09	3.91	150.92	159.53
May	76 648	62 831	23 382	18 564	53 266	52 086	14 217	36.9	40.7	4.10	3.93	151.29	159.95
June	76 887	63 015	23 485	18 606	53 402	52 235	14 253	36.9	40.7	4.12	3.95	152.03	160.77
July	76 913	63 048	23 522	18 598	53 391	52 242	14 232	36.9	40.7	4.15	3.98	153.14	161.99
August	77 168	63 264	23 559	18 629	53 609	52 396	14 251	36.8	40.6	4.16	4.00	153.09	162.40
September	77 276	63 384	23 548	18 609	53 728	52 413	14 213	36.8	40.7	4.20	4.03	154.56	164.02
October	77 607	63 630	23 641	18 702	53 966	52 655	14 289	36.7	40.6	4.21	4.05	154.51	164.43
November	77 920	63 885	23 719	18 773	54 201	52 843	14 342	36.9	40.6	4.24	4.07	156.46	165.24
December	78 031	63 961	23 779	18 820	54 252	52 950	14 386	36.7	40.6	4.25	4.09	155.98	166.05
1974													
January	78 100	64 010	23 709	18 788	54 391	52 894	14 340	36.6	40.5	4.27	4.10	156.28	166.05
February	78 254	64 119	23 718	18 727	54 536	52 972	14 269	36.6	40.4	4.29	4.13	157.01	166.85
March	78 296	64 144	23 687	18 700	54 609	52 955	14 223	36.6	40.4	4.32	4.15	158.11	167.66
April	78 382	64 191	23 670	18 702	54 712	53 007	14 225	36.4	39.5	4.33	4.16	157.61	164.32
May	78 549	64 328	23 635	18 688	54 914	53 098	14 199	36.6	40.3	4.40	4.25	161.04	171.28
June	78 604	64 365	23 591	18 690	55 013	53 109	14 197	36.5	40.2	4.44	4.30	162.06	172.86
July	78 636	64 348	23 462	18 656	55 174	53 047	14 152	36.5	40.1	4.45	4.33	162.43	173.63
August	78 619	64 291	23 396	18 623	55 223	53 027	14 089	36.5	40.2	4.49	4.38	163.89	176.08
September	78 610	64 188	23 274	18 492	55 336	52 917	14 025	36.4	40.0	4.53	4.42	164.89	176.80
October	78 630	64 146	23 118	18 364	55 512	52 836	13 884	36.3	40.0	4.56	4.48	165.53	179.20
November	78 265	63 733	22 773	18 077	55 492	52 407	13 607	36.1	39.5	4.57	4.49	164.98	177.36
December	77 652	63 093	22 303	17 693	55 349	51 853	13 259	36.1	39.3	4.60	4.52	166.06	177.64

Table 10-9. Employment, Hours, and Earnings, Total Nonfarm and Manufacturing, Historical Annual and Monthly—*Continued*

(Wage and salary workers on nonfarm payrolls, seasonally adjusted.)

Year and month	All wage and salary workers (thousands)					Production and nonsupervisory workers on private payrolls							
	Total	Private			Service-providing	Number (thousands)		Average hours per week		Average hourly earnings, dollars		Average weekly earnings, dollars	
		Total	Goods-producing			Total private	Manufac-turing	Total private	Manufac-turing	Total private	Manufac-turing	Total private	Manufac-turing
			Total	Manufac-turing									
1975													
January	77 293	62 669	21 974	17 344	55 319	51 435	12 933	36.0	39.2	4.61	4.54	165.96	177.97
February	76 918	62 171	21 512	17 004	55 406	50 934	12 622	35.9	38.9	4.64	4.58	166.58	178.16
March	76 648	61 894	21 274	16 853	55 374	50 667	12 483	35.7	38.8	4.67	4.63	166.72	179.64
April	76 460	61 665	21 109	16 759	55 351	50 443	12 407	35.8	39.0	4.67	4.63	167.19	180.57
May	76 624	61 797	21 097	16 746	55 527	50 559	12 406	35.9	39.0	4.69	4.65	168.37	181.35
June	76 521	61 737	21 018	16 690	55 503	50 543	12 371	35.9	39.2	4.72	4.68	169.45	183.46
July	76 770	61 909	20 981	16 678	55 789	50 730	12 374	35.9	39.4	4.74	4.71	170.17	185.57
August	77 153	62 283	21 176	16 824	55 977	51 072	12 538	36.1	39.7	4.78	4.75	172.56	188.58
September	77 228	62 404	21 284	16 904	55 944	51 183	12 617	36.1	39.8	4.79	4.78	172.92	190.24
October	77 540	62 640	21 384	16 984	56 156	51 378	12 687	36.1	39.9	4.82	4.80	174.00	191.52
November	77 685	62 782	21 442	17 025	56 243	51 451	12 700	36.1	39.9	4.86	4.83	175.45	192.72
December	78 017	63 071	21 602	17 140	56 415	51 758	12 811	36.2	40.2	4.87	4.86	176.29	195.37
1976													
January	78 503	63 534	21 799	17 287	56 704	52 181	12 945	36.3	40.3	4.90	4.90	177.87	197.47
February	78 816	63 835	21 893	17 384	56 923	52 430	13 030	36.3	40.4	4.93	4.94	178.96	199.58
March	79 048	64 061	21 980	17 470	57 068	52 616	13 092	36.1	40.2	4.95	4.98	178.70	200.20
April	79 292	64 307	22 050	17 541	57 242	52 814	13 160	36.0	39.6	4.98	4.98	179.28	197.21
May	79 312	64 341	21 988	17 513	57 324	52 803	13 130	36.2	40.3	5.01	5.04	181.36	203.11
June	79 376	64 413	21 982	17 521	57 394	52 832	13 121	36.1	40.2	5.03	5.07	181.58	203.81
July	79 547	64 554	21 988	17 524	57 559	52 972	13 124	36.1	40.3	5.06	5.11	182.67	205.93
August	79 704	64 697	22 038	17 596	57 666	53 073	13 195	36.0	40.2	5.11	5.16	183.96	207.43
September	79 892	64 921	22 142	17 665	57 750	53 268	13 253	36.1	40.2	5.14	5.20	185.55	209.04
October	79 911	64 883	22 037	17 548	57 874	53 167	13 109	35.9	40.0	5.16	5.19	185.24	207.60
November	80 240	65 167	22 207	17 682	58 033	53 358	13 199	35.9	40.1	5.21	5.25	187.04	210.53
December	80 448	65 373	22 261	17 719	58 187	53 544	13 230	35.9	39.9	5.23	5.29	187.76	211.07
1977													
January	80 690	65 634	22 320	17 803	58 370	53 755	13 305	35.6	39.4	5.26	5.35	187.26	210.79
February	80 988	65 932	22 478	17 843	58 510	54 017	13 331	36.0	40.2	5.30	5.36	190.80	215.47
March	81 391	66 341	22 672	17 941	58 719	54 391	13 424	35.9	40.3	5.32	5.40	190.99	217.62
April	81 728	66 653	22 807	18 024	58 921	54 673	13 490	36.0	40.4	5.36	5.45	192.96	220.18
May	82 088	66 956	22 919	18 107	59 169	54 941	13 567	36.0	40.5	5.40	5.49	194.40	222.35
June	82 488	67 281	23 046	18 192	59 442	55 194	13 622	36.0	40.5	5.42	5.54	195.12	224.37
July	82 834	67 535	23 106	18 259	59 728	55 394	13 670	35.9	40.4	5.46	5.58	196.01	225.43
August	83 075	67 747	23 124	18 276	59 951	55 542	13 679	35.9	40.4	5.48	5.61	196.73	226.64
September	83 532	68 129	23 244	18 334	60 288	55 858	13 714	35.8	40.4	5.51	5.65	197.26	228.26
October	83 800	68 337	23 279	18 356	60 521	56 006	13 722	36.0	40.6	5.55	5.69	199.80	231.01
November	84 173	68 658	23 371	18 419	60 802	56 278	13 771	35.9	40.5	5.58	5.72	200.32	231.66
December	84 410	68 872	23 371	18 531	61 039	56 468	13 861	35.8	40.4	5.61	5.75	200.84	232.30
1978													
January	84 594	68 983	23 374	18 593	61 220	56 548	13 917	35.3	39.5	5.66	5.83	199.80	230.29
February	84 948	69 277	23 453	18 639	61 495	56 770	13 950	35.6	39.9	5.69	5.86	202.56	233.81
March	85 460	69 729	23 649	18 699	61 811	57 176	13 992	35.8	40.5	5.73	5.88	205.13	238.14
April	86 162	70 365	24 008	18 772	62 154	57 712	14 037	35.8	40.4	5.79	5.93	207.28	239.57
May	86 509	70 675	24 082	18 848	62 427	57 941	14 096	35.8	40.4	5.82	5.96	208.36	240.78
June	86 950	71 098	24 238	18 919	62 712	58 263	14 129	35.9	40.6	5.86	6.01	210.37	244.01
July	87 204	71 303	24 300	18 951	62 904	58 421	14 152	35.9	40.6	5.90	6.06	211.81	246.04
August	87 483	71 592	24 374	19 006	63 109	58 626	14 187	35.8	40.5	5.92	6.09	211.94	246.65
September	87 621	71 802	24 444	19 068	63 177	58 819	14 241	35.7	40.5	5.97	6.15	213.13	249.08
October	87 956	72 098	24 548	19 142	63 408	59 019	14 291	35.8	40.5	6.02	6.20	215.52	251.10
November	88 391	72 497	24 678	19 257	63 713	59 374	14 388	35.7	40.6	6.05	6.26	215.99	254.16
December	88 671	72 760	24 758	19 334	63 913	59 598	14 459	35.7	40.5	6.10	6.31	217.77	255.56
1979													
January	88 808	72 871	24 740	19 388	64 068	59 660	14 497	35.6	40.4	6.13	6.36	218.23	256.94
February	89 055	73 108	24 784	19 409	64 271	59 841	14 501	35.7	40.5	6.18	6.40	220.63	259.20
March	89 479	73 523	24 998	19 453	64 481	60 219	14 526	35.8	40.6	6.22	6.45	222.68	261.87
April	89 417	73 440	24 958	19 450	64 459	60 068	14 515	35.3	39.3	6.21	6.43	219.21	252.70
May	89 789	73 799	25 071	19 509	64 718	60 370	14 551	35.6	40.2	6.27	6.52	223.21	262.10
June	90 108	74 063	25 161	19 553	64 947	60 582	14 566	35.6	40.2	6.32	6.56	224.99	263.71
July	90 217	74 067	25 163	19 531	65 054	60 559	14 536	35.6	40.2	6.36	6.59	226.42	264.92
August	90 300	74 071	25 059	19 406	65 241	60 517	14 397	35.6	40.1	6.40	6.63	227.84	265.86
September	90 327	74 199	25 088	19 442	65 239	60 632	14 440	35.6	40.1	6.45	6.67	229.62	267.47
October	90 481	74 345	25 038	19 390	65 443	60 747	14 384	35.6	40.2	6.46	6.71	229.98	269.74
November	90 573	74 400	24 947	19 299	65 626	60 778	14 295	35.6	40.1	6.51	6.74	231.76	270.27
December	90 672	74 492	24 970	19 301	65 702	60 864	14 300	35.6	40.1	6.56	6.80	233.54	272.68
1980													
January	90 800	74 599	24 949	19 282	65 851	60 901	14 241	35.4	40.0	6.57	6.82	232.58	272.80
February	90 883	74 657	24 874	19 219	66 009	60 967	14 170	35.4	40.1	6.63	6.88	234.70	275.89
March	90 994	74 698	24 818	19 217	66 176	60 990	14 165	35.3	39.9	6.69	6.95	236.16	277.31
April	90 849	74 266	24 507	18 973	66 342	60 544	13 914	35.2	39.8	6.72	6.97	236.54	277.41
May	90 420	73 966	24 234	18 726	66 186	60 198	13 632	35.1	39.3	6.76	7.02	237.28	275.89
June	90 101	73 660	23 968	18 490	66 133	59 896	13 405	35.0	39.2	6.82	7.10	238.70	278.32
July	89 840	73 422	23 698	18 276	66 142	59 697	13 227	34.9	39.1	6.86	7.16	239.41	279.96
August	90 099	73 689	23 860	18 414	66 239	59 912	13 349	35.1	39.5	6.91	7.24	242.54	285.98
September	90 213	73 883	23 931	18 445	66 282	60 079	13 396	35.1	39.6	6.95	7.30	243.95	289.08
October	90 490	74 104	24 012	18 506	66 478	60 242	13 437	35.2	39.8	7.02	7.38	247.10	293.72
November	90 747	74 356	24 123	18 601	66 624	60 451	13 528	35.3	39.9	7.09	7.47	250.28	298.05
December	90 943	74 570	24 182	18 640	66 761	60 616	13 550	35.3	40.1	7.13	7.52	251.69	301.55

Table 10-9. Employment, Hours, and Earnings, Total Nonfarm and Manufacturing, Historical Annual and Monthly—Continued

(Wage and salary workers on nonfarm payrolls, seasonally adjusted.)

Year and month	All wage and salary workers (thousands)					Production and nonsupervisory workers on private payrolls							
	Total	Private			Service-providing	Number (thousands)		Average hours per week		Average hourly earnings, dollars		Average weekly earnings, dollars	
		Total	Goods-producing			Total private	Manufacturing	Total private	Manufacturing	Total private	Manufacturing	Total private	Manufacturing
			Total	Manufacturing									
1981													
January	91 033	74 673	24 152	18 639	66 881	60 716	13 545	35.4	40.1	7.18	7.58	254.17	303.96
February	91 105	74 759	24 118	18 613	66 987	60 741	13 518	35.2	39.8	7.23	7.62	254.50	303.28
March	91 210	74 918	24 203	18 647	67 007	60 883	13 550	35.3	40.0	7.29	7.68	257.34	307.20
April	91 283	75 023	24 151	18 711	67 132	60 978	13 594	35.3	40.1	7.33	7.76	258.75	311.18
May	91 296	75 098	24 148	18 766	67 148	60 981	13 633	35.3	40.2	7.37	7.81	260.16	313.96
June	91 490	75 331	24 290	18 789	67 200	61 140	13 632	35.2	40.0	7.41	7.85	260.83	314.00
July	91 601	75 426	24 302	18 785	67 299	61 225	13 629	35.2	39.9	7.45	7.89	262.24	314.81
August	91 565	75 455	24 258	18 748	67 307	61 221	13 573	35.2	40.0	7.52	7.97	264.70	318.80
September	91 477	75 446	24 210	18 712	67 267	61 238	13 565	35.0	39.6	7.56	8.03	264.60	317.99
October	91 380	75 311	24 051	18 566	67 329	61 071	13 399	35.0	39.6	7.59	8.06	265.65	319.18
November	91 171	75 093	23 875	18 409	67 296	60 823	13 235	35.1	39.4	7.63	8.08	267.81	318.35
December	90 895	74 822	23 656	18 223	67 239	60 529	13 033	34.9	39.2	7.64	8.09	266.64	317.13
1982													
January	90 565	74 524	23 362	18 047	67 203	60 218	12 874	34.1	37.3	7.72	8.26	263.25	308.10
February	90 563	74 552	23 361	17 981	67 202	60 284	12 831	35.1	39.6	7.72	8.21	270.97	325.12
March	90 434	74 410	23 214	17 857	67 220	60 150	12 727	34.9	39.1	7.75	8.24	270.48	322.18
April	90 150	74 140	22 996	17 683	67 154	59 877	12 566	34.8	39.1	7.77	8.28	270.40	323.75
May	90 107	74 104	22 884	17 588	67 223	59 851	12 506	34.8	39.1	7.84	8.33	272.83	325.70
June	89 865	73 849	22 643	17 430	67 222	59 599	12 365	34.8	39.2	7.85	8.37	273.18	328.10
July	89 521	73 631	22 434	17 278	67 087	59 417	12 261	34.8	39.2	7.90	8.40	274.92	329.28
August	89 363	73 433	22 268	17 160	67 095	59 212	12 153	34.7	39.0	7.94	8.43	275.52	328.77
September	89 183	73 260	22 146	17 074	67 037	59 077	12 104	34.8	39.0	7.94	8.45	276.31	329.55
October	88 907	72 951	21 879	16 853	67 028	58 762	11 880	34.6	38.9	7.96	8.44	275.42	328.32
November	88 786	72 809	21 736	16 722	67 050	58 629	11 766	34.6	39.0	7.98	8.46	276.11	329.94
December	88 771	72 790	21 688	16 690	67 083	58 610	11 746	34.7	39.0	8.01	8.49	277.95	331.11
1983													
January	88 990	72 967	21 757	16 705	67 233	58 826	11 783	34.9	39.3	8.06	8.52	281.29	334.84
February	88 917	72 913	21 676	16 706	67 241	58 800	11 794	34.5	39.3	8.09	8.59	279.11	337.59
March	89 090	73 085	21 649	16 711	67 441	58 971	11 817	34.7	39.6	8.10	8.59	281.07	340.16
April	89 364	73 374	21 729	16 794	67 635	59 213	11 894	34.8	39.7	8.13	8.61	282.92	341.82
May	89 644	73 639	21 829	16 885	67 815	59 457	11 987	34.9	40.0	8.17	8.64	285.13	345.60
June	90 021	74 001	21 949	16 960	68 072	59 818	12 054	34.9	40.1	8.19	8.66	285.83	347.27
July	90 437	74 426	22 103	17 059	68 334	60 199	12 155	34.9	40.3	8.22	8.71	286.88	351.01
August	90 129	74 113	22 207	17 118	67 922	59 832	12 200	34.9	40.3	8.20	8.71	286.18	351.01
September	91 247	75 205	22 381	17 255	68 866	60 859	12 318	35.0	40.6	8.27	8.76	289.45	355.66
October	91 520	75 534	22 546	17 367	68 974	61 109	12 408	35.2	40.6	8.31	8.80	292.51	357.28
November	91 875	75 878	22 698	17 479	69 177	61 398	12 503	35.1	40.6	8.31	8.84	291.68	358.90
December	92 230	76 222	22 803	17 551	69 427	61 688	12 553	35.1	40.5	8.33	8.87	292.38	359.24
1984													
January	92 673	76 663	22 942	17 630	69 731	61 924	12 617	35.1	40.6	8.38	8.91	294.14	361.75
February	93 157	77 132	23 146	17 728	70 011	62 343	12 703	35.3	41.1	8.38	8.92	295.81	366.61
March	93 429	77 399	23 209	17 806	70 220	62 530	12 768	35.1	40.7	8.41	8.96	295.19	364.67
April	93 792	77 717	23 305	17 872	70 487	62 817	12 814	35.2	40.8	8.45	8.98	297.44	366.38
May	94 098	77 995	23 389	17 916	70 709	63 028	12 840	35.1	40.7	8.45	8.99	296.60	365.89
June	94 479	78 352	23 497	17 967	70 982	63 310	12 871	35.1	40.6	8.48	9.03	297.65	366.62
July	94 789	78 617	23 571	18 013	71 218	63 528	12 901	35.1	40.6	8.51	9.05	298.70	367.43
August	95 032	78 808	23 608	18 034	71 424	63 667	12 906	35.0	40.5	8.51	9.09	297.85	368.15
September	95 344	79 089	23 617	18 019	71 727	63 889	12 880	35.1	40.5	8.55	9.12	300.11	369.36
October	95 629	79 355	23 626	18 024	72 003	64 101	12 868	34.9	40.5	8.55	9.15	298.40	370.58
November	95 982	79 671	23 639	18 016	72 343	64 347	12 846	35.0	40.4	8.57	9.19	299.95	371.28
December	96 107	79 825	23 673	18 023	72 434	64 464	12 848	35.1	40.5	8.61	9.22	302.21	373.41
1985													
January	96 372	80 036	23 672	18 009	72 700	64 678	12 833	34.9	40.3	8.61	9.27	300.49	373.58
February	96 503	80 154	23 621	17 966	72 882	64 776	12 784	34.8	40.1	8.64	9.29	300.67	372.53
March	96 842	80 447	23 661	17 939	73 181	65 025	12 758	34.9	40.4	8.66	9.32	302.23	376.53
April	97 038	80 608	23 644	17 886	73 394	65 134	12 701	34.9	40.5	8.69	9.35	303.28	378.68
May	97 312	80 838	23 632	17 855	73 680	65 328	12 673	34.9	40.4	8.70	9.37	303.63	378.55
June	97 459	80 961	23 592	17 819	73 867	65 403	12 635	34.9	40.5	8.74	9.39	305.03	380.30
July	97 648	81 028	23 549	17 776	74 099	65 449	12 596	34.8	40.4	8.74	9.42	304.15	380.57
August	97 840	81 221	23 546	17 756	74 294	65 631	12 593	34.8	40.6	8.76	9.43	304.85	382.86
September	98 045	81 407	23 528	17 718	74 517	65 771	12 556	34.8	40.6	8.80	9.44	306.24	383.26
October	98 233	81 579	23 529	17 708	74 704	65 939	12 556	34.8	40.7	8.79	9.46	305.89	385.02
November	98 443	81 769	23 520	17 697	74 923	66 093	12 545	34.8	40.7	8.81	9.49	306.59	386.24
December	98 609	81 915	23 518	17 693	75 091	66 222	12 550	34.9	40.9	8.87	9.55	309.56	390.60
1986													
January	98 732	82 017	23 530	17 686	75 202	66 317	12 546	34.9	40.7	8.85	9.53	308.87	387.87
February	98 847	82 088	23 485	17 663	75 362	66 382	12 530	34.8	40.7	8.88	9.56	309.02	389.09
March	98 934	82 179	23 428	17 624	75 506	66 427	12 498	34.8	40.7	8.89	9.58	309.37	389.91
April	99 121	82 356	23 427	17 616	75 694	66 556	12 495	34.6	40.5	8.88	9.56	307.25	387.18
May	99 248	82 458	23 349	17 593	75 899	66 636	12 474	34.7	40.7	8.90	9.59	308.83	390.31
June	99 155	82 376	23 263	17 530	75 892	66 560	12 424	34.6	40.7	8.92	9.58	308.63	389.91
July	99 473	82 694	23 235	17 497	76 238	66 830	12 389	34.6	40.6	8.91	9.60	308.29	389.76
August	99 588	82 788	23 225	17 489	76 363	66 929	12 399	34.7	40.7	8.94	9.61	310.22	391.13
September	99 934	83 024	23 216	17 498	76 718	67 136	12 411	34.6	40.7	8.94	9.60	309.32	390.72
October	100 121	83 152	23 208	17 477	76 913	67 234	12 396	34.6	40.6	8.96	9.62	310.02	390.57
November	100 308	83 303	23 204	17 472	77 104	67 372	12 407	34.7	40.7	9.01	9.64	312.65	392.35
December	100 509	83 488	23 237	17 478	77 272	67 524	12 425	34.6	40.8	9.01	9.66	311.75	394.13

Table 10-9. Employment, Hours, and Earnings, Total Nonfarm and Manufacturing, Historical Annual and Monthly—*Continued*

(Wage and salary workers on nonfarm payrolls, seasonally adjusted.)

Year and month	All wage and salary workers (thousands)					Production and nonsupervisory workers on private payrolls							
	Total	Private	Goods-producing		Service-providing	Number (thousands)		Average hours per week		Average hourly earnings, dollars		Average weekly earnings, dollars	
		Total	Total	Manufac-turing		Total private	Manufac-turing	Total private	Manufac-turing	Total private	Manufac-turing	Total private	Manufac-turing
1987													
January	100 678	83 633	23 232	17 465	77 446	67 639	12 405	34.7	40.8	9.02	9.67	312.99	394.54
February	100 919	83 883	23 296	17 499	77 623	67 857	12 438	34.8	41.2	9.05	9.69	314.94	399.23
March	101 164	84 100	23 307	17 507	77 857	68 017	12 446	34.8	41.0	9.07	9.71	315.64	398.11
April	101 499	84 390	23 342	17 525	78 157	68 261	12 465	34.7	40.8	9.08	9.71	315.08	396.17
May	101 728	84 616	23 390	17 542	78 338	68 470	12 481	34.8	41.0	9.11	9.73	317.03	398.93
June	101 900	84 776	23 390	17 537	78 510	68 580	12 482	34.7	40.9	9.11	9.74	316.12	398.37
July	102 247	85 087	23 455	17 593	78 792	68 820	12 521	34.7	41.0	9.12	9.74	316.46	399.34
August	102 420	85 248	23 506	17 630	78 914	68 945	12 560	34.8	40.9	9.18	9.80	319.46	400.82
September	102 647	85 512	23 566	17 691	79 081	69 165	12 614	34.7	40.8	9.19	9.85	318.89	401.88
October	103 138	85 869	23 655	17 729	79 483	69 447	12 637	34.8	41.1	9.22	9.84	320.86	404.42
November	103 372	86 073	23 711	17 775	79 661	69 626	12 678	34.8	41.0	9.27	9.87	322.60	404.67
December	103 661	86 314	23 772	17 809	79 889	69 860	12 707	34.6	41.0	9.28	9.89	321.09	405.49
1988													
January	103 753	86 388	23 668	17 790	80 085	69 859	12 684	34.6	41.1	9.29	9.91	321.43	407.30
February	104 214	86 825	23 769	17 823	80 445	70 237	12 706	34.7	41.1	9.29	9.92	322.36	407.71
March	104 489	87 042	23 824	17 844	80 665	70 400	12 712	34.5	40.9	9.31	9.94	321.20	406.55
April	104 732	87 280	23 880	17 874	80 852	70 608	12 733	34.6	41.0	9.36	9.99	323.86	409.59
May	104 962	87 481	23 896	17 892	81 066	70 751	12 747	34.6	41.0	9.41	10.02	325.59	410.82
June	105 326	87 811	23 951	17 916	81 375	71 033	12 767	34.6	41.1	9.41	10.04	325.59	412.64
July	105 550	88 056	23 966	17 926	81 584	71 237	12 774	34.6	41.1	9.45	10.05	326.97	413.06
August	105 674	88 130	23 926	17 891	81 748	71 291	12 752	34.5	40.9	9.46	10.07	326.37	411.86
September	106 013	88 379	23 942	17 914	82 071	71 488	12 764	34.5	41.0	9.51	10.12	328.10	414.92
October	106 276	88 606	23 987	17 966	82 289	71 684	12 812	34.7	41.1	9.56	10.16	331.73	417.58
November	106 617	88 871	24 030	18 003	82 587	71 908	12 851	34.6	41.1	9.58	10.19	331.47	418.81
December	106 898	89 162	24 054	18 025	82 844	72 183	12 864	34.5	40.9	9.60	10.20	331.20	417.18
1989													
January	107 161	89 387	24 097	18 057	83 064	72 390	12 882	34.7	41.1	9.65	10.23	334.86	420.45
February	107 427	89 615	24 080	18 055	83 347	72 561	12 880	34.5	41.2	9.68	10.26	333.96	422.71
March	107 621	89 799	24 069	18 060	83 552	72 698	12 878	34.5	41.1	9.70	10.29	334.65	422.92
April	107 791	89 951	24 100	18 055	83 691	72 823	12 867	34.6	41.1	9.75	10.28	337.35	422.51
May	107 913	90 037	24 089	18 040	83 824	72 892	12 852	34.4	41.0	9.74	10.30	335.06	422.30
June	108 027	90 115	24 052	18 013	83 975	72 945	12 822	34.4	40.9	9.77	10.33	336.09	422.50
July	108 069	90 164	24 027	17 980	84 042	72 976	12 790	34.5	40.9	9.83	10.36	339.14	423.72
August	108 120	90 131	24 048	17 964	84 072	72 943	12 790	34.5	40.9	9.83	10.39	339.14	424.95
September	108 369	90 342	24 000	17 922	84 369	73 118	12 745	34.4	40.8	9.87	10.41	339.53	424.73
October	108 476	90 443	23 997	17 895	84 479	73 215	12 723	34.6	40.8	9.92	10.43	343.23	425.54
November	108 752	90 695	24 009	17 886	84 743	73 432	12 713	34.4	40.7	9.93	10.44	341.59	424.91
December	108 836	90 761	23 949	17 881	84 887	73 508	12 705	34.3	40.5	9.98	10.49	342.31	424.85
1990													
January	109 197	91 046	23 982	17 797	85 215	73 753	12 738	34.4	40.5	10.02	10.52	344.69	426.06
February	109 435	91 258	24 071	17 893	85 364	73 944	12 848	34.4	40.6	10.07	10.64	346.41	431.98
March	109 644	91 350	24 023	17 868	85 621	73 996	12 821	34.4	40.6	10.11	10.70	347.78	434.42
April	109 688	91 311	23 966	17 845	85 722	73 963	12 804	34.3	40.5	10.12	10.68	347.12	432.54
May	109 839	91 240	23 888	17 797	85 951	73 860	12 755	34.3	40.6	10.16	10.74	348.49	436.04
June	109 857	91 301	23 849	17 776	86 008	73 858	12 739	34.4	40.7	10.20	10.78	350.88	438.75
July	109 824	91 264	23 746	17 704	86 078	73 808	12 673	34.2	40.6	10.22	10.81	349.52	438.89
August	109 617	91 160	23 648	17 649	85 969	73 745	12 624	34.3	40.5	10.24	10.80	351.23	437.40
September	109 520	91 083	23 571	17 609	85 949	73 633	12 596	34.2	40.5	10.28	10.86	351.58	439.83
October	109 367	90 924	23 473	17 577	85 894	73 505	12 577	34.1	40.4	10.30	10.92	351.23	441.17
November	109 213	90 763	23 283	17 428	85 930	73 343	12 439	34.2	40.2	10.32	10.88	352.94	437.38
December	109 167	90 699	23 203	17 395	85 964	73 274	12 415	34.2	40.3	10.34	10.93	353.63	440.48
1991													
January	109 056	90 582	23 061	17 330	85 995	73 158	12 351	34.1	40.2	10.38	10.97	353.96	440.99
February	108 735	90 253	22 900	17 211	85 835	72 862	12 242	34.1	40.1	10.39	10.97	354.30	439.90
March	108 576	90 088	22 779	17 140	85 797	72 705	12 191	34.1	40.0	10.41	11.00	354.98	440.00
April	108 367	89 882	22 687	17 093	85 680	72 536	12 157	34.0	40.1	10.46	11.05	355.64	443.11
May	108 251	89 753	22 618	17 070	85 633	72 452	12 148	34.1	40.2	10.49	11.09	357.71	445.82
June	108 337	89 776	22 571	17 044	85 766	72 456	12 135	34.1	40.5	10.51	11.13	358.39	450.77
July	108 292	89 694	22 507	17 015	85 785	72 425	12 130	34.1	40.5	10.54	11.18	359.41	452.79
August	108 313	89 746	22 493	17 025	85 820	72 487	12 153	34.1	40.6	10.56	11.18	360.10	453.91
September	108 338	89 795	22 466	17 010	85 872	72 499	12 140	34.1	40.6	10.58	11.21	360.78	455.13
October	108 357	89 764	22 418	16 999	85 939	72 477	12 138	34.1	40.6	10.59	11.25	361.12	456.75
November	108 297	89 670	22 316	16 961	85 981	72 397	12 102	34.1	40.7	10.61	11.26	361.80	458.28
December	108 330	89 689	22 274	16 916	86 056	72 441	12 075	34.1	40.7	10.64	11.27	362.82	458.69
1992													
January	108 372	89 684	22 213	16 839	86 159	72 482	12 013	34.1	40.6	10.65	11.24	363.17	456.34
February	108 313	89 624	22 142	16 829	86 171	72 461	12 015	34.1	40.7	10.67	11.30	363.85	459.91
March	108 368	89 653	22 127	16 805	86 241	72 484	12 006	34.1	40.7	10.69	11.32	364.53	460.72
April	108 523	89 784	22 112	16 831	86 391	72 631	12 029	34.3	40.9	10.72	11.36	367.70	464.62
May	108 652	89 899	22 135	16 835	86 517	72 743	12 046	34.3	40.9	10.74	11.39	368.38	465.85
June	108 719	89 957	22 097	16 826	86 622	72 785	12 042	34.2	40.8	10.77	11.41	368.33	465.53
July	108 794	89 977	22 074	16 819	86 720	72 807	12 049	34.2	40.8	10.79	11.43	369.02	466.34
August	108 929	90 046	22 045	16 783	86 884	72 880	12 020	34.2	40.8	10.82	11.46	370.04	467.57
September	108 963	90 134	22 020	16 761	86 943	72 982	12 002	34.3	40.7	10.83	11.45	371.47	466.02
October	109 144	90 316	22 029	16 751	87 115	73 147	11 999	34.2	40.8	10.85	11.47	371.07	467.98
November	109 277	90 436	22 042	16 758	87 235	73 298	12 012	34.2	40.8	10.87	11.49	371.75	469.94
December	109 499	90 621	22 076	16 769	87 423	73 472	12 031	34.3	40.9	10.89	11.51	373.53	470.76

Table 10-9. Employment, Hours, and Earnings, Total Nonfarm and Manufacturing, Historical Annual and Monthly—Continued

(Wage and salary workers on nonfarm payrolls, seasonally adjusted.)

Year and month	All wage and salary workers (thousands)					Production and nonsupervisory workers on private payrolls							
	Total	Private				Number (thousands)		Average hours per week		Average hourly earnings, dollars		Average weekly earnings, dollars	
		Total	Goods-producing		Service-providing	Total private	Manufacturing	Total private	Manufacturing	Total private	Manufacturing	Total private	Manufacturing
			Total	Manufacturing									
1993													
January	109 799	90 898	22 133	16 791	87 666	73 748	12 060	34.3	41.1	10.93	11.55	374.90	474.71
February	110 047	91 145	22 188	16 805	87 859	74 002	12 072	34.3	41.1	10.95	11.57	375.59	475.53
March	109 999	91 092	22 142	16 795	87 857	73 927	12 074	34.1	40.8	10.99	11.58	374.76	472.46
April	110 302	91 364	22 131	16 772	88 171	74 147	12 056	34.4	41.4	10.99	11.63	378.06	481.48
May	110 573	91 622	22 189	16 766	88 384	74 403	12 056	34.3	41.0	11.01	11.65	377.64	477.65
June	110 752	91 783	22 165	16 742	88 587	74 509	12 039	34.3	40.9	11.03	11.68	378.33	477.71
July	111 057	91 997	22 183	16 739	88 874	74 703	12 043	34.4	41.1	11.05	11.69	380.12	480.46
August	111 211	92 183	22 203	16 741	89 008	74 878	12 050	34.3	41.1	11.07	11.72	379.70	481.69
September	111 452	92 411	22 252	16 769	89 200	75 076	12 080	34.4	41.3	11.10	11.76	381.84	485.69
October	111 737	92 695	22 306	16 778	89 431	75 319	12 092	34.4	41.3	11.13	11.79	382.87	486.93
November	111 990	92 922	22 347	16 800	89 643	75 540	12 119	34.4	41.3	11.15	11.83	383.56	488.58
December	112 319	93 210	22 413	16 815	89 906	75 792	12 140	34.4	41.4	11.18	11.88	384.59	491.83
1994													
January	112 601	93 454	22 465	16 855	90 136	75 988	12 181	34.4	41.4	11.20	11.90	385.28	492.66
February	112 785	93 635	22 451	16 862	90 334	76 163	12 198	34.2	40.9	11.25	11.97	384.75	489.57
March	113 248	94 058	22 550	16 897	90 698	76 536	12 234	34.5	41.7	11.25	11.95	388.13	498.32
April	113 592	94 369	22 641	16 933	90 951	76 827	12 276	34.5	41.8	11.27	11.96	388.82	499.93
May	113 928	94 664	22 704	16 962	91 224	77 103	12 304	34.5	41.8	11.29	11.98	389.51	500.76
June	114 242	94 968	22 764	17 010	91 478	77 354	12 351	34.5	41.8	11.31	12.00	000.20	501.60
July	114 613	95 312	22 807	17 026	91 806	77 656	12 367	34.6	41.8	11.35	12.03	392.71	502.85
August	114 902	95 596	22 876	17 081	92 026	77 901	12 428	34.5	41.7	11.36	12.06	391.92	502.00
September	115 251	95 914	22 948	17 115	92 303	78 109	12 459	34.5	41.6	11.39	12.09	392.96	502.94
October	115 464	96 120	22 974	17 144	92 490	78 349	12 489	34.5	41.8	11.43	12.12	394.34	506.62
November	115 876	96 509	23 050	17 186	92 826	78 702	12 529	34.5	41.7	11.45	12.15	395.03	506.66
December	116 171	96 783	23 095	17 217	93 076	78 970	12 558	34.5	41.8	11.48	12.17	396.06	508.71
1995													
January	116 508	97 111	23 147	17 262	93 361	79 204	12 593	34.5	41.8	11.48	12.20	396.06	509.96
February	116 702	97 295	23 103	17 265	93 599	79 338	12 602	34.4	41.7	11.53	12.24	396.63	510.41
March	116 913	97 486	23 151	17 263	93 762	79 517	12 600	34.4	41.5	11.55	12.24	397.32	507.96
April	117 076	97 642	23 174	17 278	93 902	79 655	12 608	34.3	41.0	11.56	12.25	396.51	502.25
May	117 056	97 638	23 120	17 259	93 936	79 661	12 590	34.2	41.1	11.60	12.28	396.72	504.71
June	117 293	97 848	23 137	17 247	94 156	79 831	12 574	34.3	41.2	11.63	12.31	398.91	507.17
July	117 386	97 949	23 119	17 218	94 267	79 907	12 538	34.3	41.1	11.67	12.39	400.28	509.23
August	117 642	98 213	23 164	17 240	94 478	80 160	12 566	34.3	41.2	11.69	12.39	400.97	510.47
September	117 881	98 451	23 208	17 247	94 673	80 359	12 567	34.3	41.2	11.73	12.41	402.34	511.29
October	118 038	98 574	23 206	17 216	94 832	80 473	12 535	34.3	41.2	11.75	12.44	403.03	512.53
November	118 182	98 719	23 200	17 209	94 982	80 544	12 515	34.3	41.3	11.78	12.45	404.05	514.19
December	118 328	98 862	23 209	17 231	95 119	80 695	12 551	34.2	40.9	11.81	12.49	403.90	510.84
1996													
January	118 323	98 873	23 196	17 208	95 127	80 651	12 517	33.8	39.7	11.86	12.60	400.87	500.22
February	118 744	99 259	23 280	17 229	95 464	81 023	12 533	34.3	41.3	11.87	12.56	407.14	518.73
March	119 001	99 469	23 276	17 193	95 725	81 198	12 488	34.3	41.1	11.89	12.50	407.83	513.75
April	119 168	99 653	23 316	17 204	95 852	81 369	12 505	34.2	41.2	11.96	12.69	409.03	522.83
May	119 496	99 967	23 358	17 222	96 138	81 658	12 518	34.3	41.4	11.98	12.71	410.91	526.19
June	119 778	100 250	23 399	17 226	96 379	81 837	12 522	34.4	41.5	12.03	12.76	413.83	529.54
July	120 020	100 473	23 418	17 223	96 602	82 036	12 517	34.3	41.4	12.06	12.80	413.66	529.92
August	120 207	100 703	23 479	17 255	96 728	82 245	12 546	34.4	41.5	12.09	12.82	415.90	532.03
September	120 418	100 851	23 497	17 252	96 921	82 346	12 544	34.4	41.6	12.13	12.85	417.27	534.56
October	120 674	101 120	23 546	17 268	97 128	82 602	12 559	34.4	41.4	12.16	12.83	418.30	531.16
November	120 969	101 404	23 584	17 277	97 385	82 786	12 560	34.4	41.5	12.20	12.88	419.68	534.52
December	121 152	101 581	23 598	17 284	97 554	82 951	12 571	34.4	41.7	12.24	12.95	421.06	540.02
1997													
January	121 372	101 779	23 618	17 297	97 754	83 099	12 579	34.3	41.4	12.29	12.99	421.55	537.79
February	121 682	102 084	23 686	17 316	97 996	83 379	12 592	34.4	41.6	12.32	12.99	423.81	540.38
March	121 999	102 391	23 739	17 340	98 260	83 617	12 613	34.5	41.8	12.36	13.04	426.42	545.07
April	122 296	102 693	23 765	17 349	98 531	83 872	12 618	34.5	41.8	12.39	13.03	427.46	544.65
May	122 557	102 956	23 809	17 362	98 748	84 105	12 633	34.6	41.7	12.44	13.07	430.42	545.02
June	122 825	103 165	23 834	17 387	98 991	84 230	12 647	34.3	41.6	12.46	13.09	427.38	544.54
July	123 119	103 433	23 862	17 389	99 257	84 476	12 645	34.5	41.6	12.50	13.10	431.25	544.96
August	123 099	103 482	23 952	17 452	99 147	84 440	12 699	34.6	41.7	12.57	13.17	434.92	549.19
September	123 594	103 915	23 996	17 465	99 598	84 811	12 711	34.6	41.7	12.60	13.18	435.96	549.61
October	123 940	104 202	24 053	17 513	99 887	85 031	12 746	34.6	41.8	12.67	13.29	438.38	555.52
November	124 246	104 485	24 112	17 556	100 134	85 233	12 777	34.5	41.8	12.72	13.33	438.84	557.19
December	124 559	104 793	24 184	17 588	100 375	85 450	12 800	34.6	42.0	12.75	13.35	441.15	560.70
1998													
January	124 823	105 053	24 262	17 619	100 561	85 609	12 818	34.6	41.9	12.79	13.34	442.53	558.95
February	125 024	105 238	24 283	17 627	100 741	85 784	12 828	34.5	41.7	12.83	13.39	442.64	558.36
March	125 174	105 382	24 264	17 637	100 910	85 817	12 822	34.5	41.6	12.87	13.43	444.02	558.69
April	125 453	105 637	24 340	17 637	101 113	86 033	12 814	34.4	41.3	12.92	13.40	444.45	553.42
May	125 855	105 980	24 361	17 624	101 494	86 307	12 790	34.6	41.5	12.97	13.45	448.76	558.18
June	126 087	106 208	24 387	17 608	101 700	86 469	12 769	34.4	41.4	12.99	13.43	446.86	556.00
July	126 213	106 283	24 238	17 422	101 975	86 483	12 558	34.5	41.4	13.01	13.34	448.85	552.28
August	126 551	106 592	24 420	17 563	102 131	86 785	12 702	34.5	41.4	13.07	13.46	450.92	557.24
September	126 764	106 779	24 419	17 557	102 345	86 921	12 715	34.3	41.3	13.11	13.52	449.67	558.38
October	126 968	106 967	24 406	17 512	102 562	87 082	12 673	34.5	41.4	13.15	13.52	453.68	559.73
November	127 243	107 199	24 394	17 465	102 849	87 258	12 634	34.4	41.4	13.18	13.55	453.39	560.97
December	127 607	107 528	24 454	17 449	103 153	87 509	12 625	34.5	41.5	13.22	13.56	456.09	562.74

Table 10-9. Employment, Hours, and Earnings, Total Nonfarm and Manufacturing, Historical Annual and Monthly—*Continued*

(Wage and salary workers on nonfarm payrolls, seasonally adjusted.)

Year and month	All wage and salary workers (thousands)					Production and nonsupervisory workers on private payrolls								
	Total	Private			Service-providing	Number (thousands)		Average hours per week		Average hourly earnings, dollars		Average weekly earnings, dollars	
		Total	Goods-producing			Total private	Manufac-turing	Total private	Manufac-turing	Total private	Manufac-turing	Total private	Manufac-turing
			Total	Manufac-turing									
1999													
January	127 713	107 629	24 401	17 427	103 312	87 560	12 605	34.4	41.3	13.27	13.59	456.49	561.27
February	128 131	107 987	24 434	17 395	103 697	87 884	12 575	34.4	41.4	13.30	13.63	457.52	564.28
March	128 239	108 071	24 378	17 368	103 861	87 952	12 562	34.3	41.3	13.33	13.69	457.22	565.40
April	128 610	108 373	24 425	17 344	104 185	88 177	12 540	34.4	41.3	13.38	13.73	460.27	567.05
May	128 822	108 593	24 445	17 333	104 377	88 357	12 534	34.3	41.4	13.43	13.80	460.65	571.32
June	129 099	108 827	24 434	17 295	104 665	88 542	12 503	34.3	41.3	13.47	13.86	462.02	572.42
July	129 420	109 081	24 474	17 308	104 946	88 741	12 525	34.4	41.4	13.52	13.91	465.09	575.87
August	129 576	109 201	24 467	17 287	105 109	88 865	12 502	34.4	41.5	13.54	13.93	465.78	578.10
September	129 781	109 377	24 485	17 281	105 296	89 009	12 495	34.3	41.4	13.61	13.99	466.82	579.19
October	130 190	109 733	24 505	17 272	105 685	89 313	12 484	34.4	41.4	13.64	13.99	469.22	579.19
November	130 479	109 983	24 561	17 282	105 918	89 533	12 488	34.4	41.4	13.66	14.01	469.90	580.01
December	130 786	110 246	24 582	17 280	106 204	89 770	12 490	34.4	41.4	13.70	14.06	471.28	582.08
2000													
January	131 020	110 449	24 628	17 284	106 392	89 918	12 491	34.4	41.5	13.75	14.13	473.00	586.40
February	131 136	110 537	24 609	17 285	106 527	89 967	12 481	34.4	41.5	13.80	14.14	474.72	586.81
March	131 609	110 876	24 705	17 302	106 904	90 250	12 490	34.4	41.4	13.85	14.17	476.44	586.64
April	131 900	111 098	24 688	17 298	107 212	90 485	12 477	34.4	41.6	13.90	14.23	478.16	591.97
May	132 118	110 971	24 647	17 279	107 471	90 362	12 464	34.3	41.2	13.94	14.23	478.14	586.28
June	132 079	111 192	24 672	17 296	107 407	90 545	12 468	34.3	41.3	13.99	14.30	479.86	590.59
July	132 247	111 380	24 717	17 322	107 530	90 668	12 474	34.3	41.5	14.03	14.33	481.23	594.70
August	132 240	111 403	24 683	17 287	107 557	90 708	12 431	34.2	41.0	14.07	14.36	481.19	588.76
September	132 364	111 629	24 642	17 230	107 722	90 866	12 382	34.3	41.1	14.12	14.40	484.32	591.84
October	132 365	111 622	24 638	17 217	107 727	90 847	12 357	34.3	41.1	14.17	14.48	486.03	595.13
November	132 570	111 810	24 623	17 202	107 947	90 972	12 336	34.2	41.1	14.24	14.52	487.01	596.77
December	132 722	111 918	24 575	17 181	108 147	91 000	12 307	34.0	40.4	14.29	14.50	485.86	585.80
2001													
January	132 712	111 877	24 533	17 104	108 179	91 002	12 229	34.2	40.6	14.30	14.48	489.06	587.89
February	132 804	111 898	24 474	17 028	108 330	90 946	12 155	34.0	40.5	14.36	14.55	488.24	589.28
March	132 761	111 816	24 409	16 938	108 352	90 891	12 085	34.1	40.5	14.43	14.58	492.06	590.49
April	132 475	111 483	24 254	16 802	108 221	90 667	11 982	34.0	40.4	14.45	14.63	491.30	591.05
May	132 426	111 397	24 119	16 661	108 307	90 562	11 859	34.0	40.4	14.50	14.69	493.00	593.48
June	132 312	111 175	23 965	16 515	108 347	90 365	11 739	34.0	40.3	14.55	14.74	494.70	594.02
July	132 187	111 002	23 837	16 382	108 350	90 253	11 632	34.0	40.7	14.56	14.82	495.04	603.17
August	132 043	110 825	23 667	16 232	108 376	90 122	11 495	33.9	40.3	14.60	14.85	494.94	598.46
September	131 791	110 549	23 537	16 117	108 254	89 846	11 400	33.8	40.2	14.63	14.89	494.49	598.58
October	131 468	110 193	23 378	15 972	108 090	89 568	11 286	33.7	40.1	14.66	14.89	494.04	597.09
November	131 158	109 832	23 209	15 825	107 949	89 250	11 175	33.8	40.1	14.71	14.96	497.20	599.90
December	130 997	109 642	23 094	15 711	107 903	89 139	11 081	33.9	40.2	14.75	15.01	500.03	603.40
2002													
January	130 868	109 491	22 961	15 587	107 907	89 130	10 993	33.7	40.1	14.77	15.05	497.75	603.51
February	130 752	109 362	22 876	15 515	107 876	89 078	10 951	33.8	40.3	14.78	15.12	499.56	609.34
March	130 732	109 301	22 786	15 443	107 946	89 035	10 901	33.9	40.5	14.83	15.15	502.74	613.58
April	130 636	109 193	22 689	15 392	107 947	88 883	10 859	33.9	40.6	14.84	15.17	503.08	615.90
May	130 647	109 133	22 604	15 337	108 043	88 738	10 824	33.9	40.6	14.88	15.24	504.43	618.74
June	130 695	109 146	22 578	15 298	108 117	88 649	10 797	33.9	40.7	14.94	15.27	506.47	621.49
July	130 604	109 060	22 521	15 256	108 083	88 495	10 764	33.8	40.4	14.98	15.30	506.32	618.12
August	130 603	109 014	22 449	15 171	108 154	88 446	10 697	33.9	40.5	15.02	15.34	509.18	621.27
September	130 524	108 978	22 397	15 119	108 127	88 410	10 668	33.9	40.5	15.06	15.37	510.53	622.49
October	130 643	109 084	22 325	15 060	108 318	88 495	10 631	33.8	40.3	15.12	15.45	511.06	622.64
November	130 632	109 051	22 282	14 992	108 350	88 440	10 583	33.8	40.4	15.15	15.48	512.07	625.39
December	130 488	108 900	22 189	14 912	108 299	88 256	10 523	33.8	40.5	15.20	15.53	513.76	628.97
2003													
January	130 596	108 970	22 146	14 866	108 450	88 291	10 484	33.8	40.3	15.22	15.58	514.44	627.87
February	130 461	108 837	22 023	14 781	108 438	88 163	10 416	33.6	40.3	15.29	15.63	513.74	629.89
March	130 246	108 636	21 945	14 721	108 301	87 861	10 357	33.8	40.4	15.28	15.64	516.46	631.86
April	130 194	108 599	21 864	14 609	108 330	87 839	10 256	33.6	40.1	15.28	15.63	513.41	626.76
May	130 210	108 643	21 832	14 557	108 378	87 832	10 218	33.7	40.2	15.33	15.69	516.62	630.74
June	130 209	108 603	21 788	14 493	108 421	87 805	10 165	33.7	40.3	15.36	15.72	517.63	633.52
July	130 207	108 574	21 708	14 402	108 499	87 780	10 093	33.6	40.1	15.40	15.76	517.44	631.98
August	130 167	108 611	21 706	14 376	108 461	87 843	10 084	33.7	40.2	15.42	15.78	519.65	634.36
September	130 279	108 775	21 700	14 347	108 579	87 953	10 059	33.6	40.5	15.42	15.83	518.11	641.12
October	130 473	108 915	21 691	14 334	108 782	88 029	10 054	33.7	40.6	15.43	15.83	519.99	642.70
November	130 490	108 955	21 688	14 316	108 802	88 094	10 039	33.8	40.9	15.47	15.90	522.89	650.31
December	130 605	109 059	21 703	14 300	108 902	88 159	10 031	33.6	40.7	15.47	15.92	519.79	647.94
2004													
January	130 787	109 249	21 715	14 290	109 072	88 244	10 027	33.7	40.9	15.51	15.95	522.69	652.36
February	130 844	109 294	21 693	14 279	109 151	88 284	10 012	33.8	41.0	15.54	15.98	525.25	655.18
March	131 156	109 568	21 758	14 287	109 398	88 527	10 025	33.7	40.9	15.56	16.01	524.37	654.81
April	131 426	109 812	21 802	14 315	109 624	88 790	10 060	33.7	40.7	15.59	16.05	525.38	653.24
May	131 710	110 096	21 881	14 342	109 829	89 077	10 091	33.8	41.1	15.64	16.07	528.63	660.48
June	131 807	110 206	21 885	14 332	109 922	89 235	10 087	33.6	40.8	15.67	16.11	526.51	657.29
July	131 864	110 258	21 900	14 330	109 964	89 349	10 095	33.7	40.8	15.70	16.12	529.09	657.70
August	131 955	110 329	21 944	14 345	110 011	89 495	10 115	33.7	40.8	15.73	16.20	530.10	660.96
September	132 112	110 477	21 957	14 331	110 155	89 678	10 104	33.7	40.8	15.78	16.29	531.79	664.63
October	132 466	110 810	22 004	14 332	110 462	89 984	10 103	33.8	40.6	15.81	16.27	534.38	660.56
November	132 521	110 829	21 997	14 307	110 524	90 011	10 078	33.7	40.5	15.84	16.31	533.81	660.56
December	132 644	110 951	22 005	14 287	110 639	90 163	10 063	33.8	40.6	15.86	16.34	536.07	663.40

Table 10-9. Employment, Hours, and Earnings, Total Nonfarm and Manufacturing, Historical Annual and Monthly—*Continued*

(Wage and salary workers on nonfarm payrolls, seasonally adjusted.)

Year and month	All wage and salary workers (thousands)					Production and nonsupervisory workers on private payrolls							
	Total	Private				Number (thousands)		Average hours per week		Average hourly earnings, dollars		Average weekly earnings, dollars	
		Total	Goods-producing		Service-providing	Total private	Manufacturing	Total private	Manufacturing	Total private	Manufacturing	Total private	Manufacturing
			Total	Manufacturing									
2005													
January	132 791	111 056	21 958	14 257	110 833	90 254	10 045	33.7	40.7	15.90	16.38	535.83	666.67
February	133 050	111 306	22 036	14 273	111 014	90 503	10 052	33.8	40.6	15.93	16.44	538.43	667.46
March	133 172	111 432	22 066	14 269	111 106	90 665	10 056	33.7	40.4	15.97	16.44	538.19	664.18
April	133 536	111 782	22 136	14 250	111 400	90 996	10 048	33.8	40.4	16.01	16.46	541.14	664.98
May	133 706	111 925	22 172	14 256	111 534	91 117	10 061	33.7	40.4	16.04	16.53	540.55	667.81
June	133 957	112 194	22 185	14 227	111 772	91 387	10 047	33.7	40.4	16.08	16.54	541.90	668.22
July	134 314	112 457	22 205	14 226	112 109	91 619	10 039	33.7	40.5	16.15	16.57	544.26	671.09
August	134 517	112 654	22 228	14 203	112 289	91 810	10 043	33.7	40.5	16.17	16.63	544.93	673.52
September	134 583	112 738	22 226	14 175	112 357	91 916	10 046	33.8	40.7	16.19	16.59	547.22	675.21
October	134 673	112 844	22 292	14 192	112 381	92 053	10 073	33.8	41.0	16.29	16.69	550.60	684.29
November	135 012	113 153	22 357	14 187	112 655	92 378	10 093	33.8	40.9	16.30	16.68	550.94	682.21
December	135 168	113 289	22 376	14 193	112 792	92 531	10 112	33.8	40.8	16.36	16.68	552.97	680.54
2006													
January	135 446	113 599	22 467	14 210	112 979	92 843	10 153	33.9	41.0	16.42	16.70	556.64	684.70
February	135 753	113 875	22 535	14 209	113 218	93 131	10 164	33.8	41.1	16.48	16.70	557.02	686.37
March	136 063	114 160	22 572	14 214	113 491	93 429	10 174	33.8	41.1	16.54	16.72	559.05	687.19
April	136 221	114 302	22 631	14 226	113 590	93 600	10 190	33.9	41.3	16.64	16.75	564.10	691.78
May	136 261	114 335	22 597	14 203	113 664	93 672	10 179	33.8	41.2	16.65	16.77	562.77	690.92
June	136 342	114 420	22 598	14 213	113 744	93 743	10 190	33.9	41.2	16.72	16.78	566.81	691.34
July	136 538	114 565	22 590	14 188	113 948	93 876	10 177	33.9	41.4	16.78	16.78	568.84	694.69
August	136 713	114 702	22 572	14 159	114 141	94 029	10 158	33.8	41.2	16.83	16.83	568.85	693.40
September	136 860	114 778	22 537	14 125	114 323	94 077	10 120	33.8	41.1	16.87	16.83	570.21	691.71
October	136 870	114 802	22 456	14 075	114 414	94 130	10 071	33.9	41.1	16.94	16.91	574.27	695.00
November	137 082	114 999	22 408	14 041	114 674	94 307	10 047	33.8	41.0	16.98	16.92	573.92	693.72
December	137 268	115 180	22 405	14 015	114 863	94 507	10 041	33.9	41.1	17.04	16.99	577.66	698.29
2007													
January	137 493	115 398	22 439	14 008	115 054	94 698	10 036	33.8	41.0	17.09	17.01	577.64	697.41
February	137 573	115 442	22 334	13 997	115 239	94 734	10 035	33.7	40.9	17.15	17.05	577.96	697.35
March	137 810	115 661	22 391	13 970	115 419	94 973	10 011	33.9	41.3	17.22	17.10	583.76	706.23
April	137 860	115 685	22 350	13 945	115 510	95 036	10 006	33.8	41.3	17.27	17.20	583.73	710.36
May	138 012	115 819	22 323	13 929	115 689	95 212	10 007	33.8	41.2	17.34	17.23	586.09	709.88
June	138 088	115 881	22 323	13 911	115 765	95 319	9 996	33.9	41.4	17.42	17.29	590.54	715.81
July	138 055	115 884	22 277	13 889	115 778	95 411	9 990	33.8	41.3	17.46	17.29	590.15	714.08
August	138 032	115 806	22 165	13 828	115 867	95 356	9 944	33.8	41.3	17.50	17.35	591.50	716.56
September	138 114	115 835	22 093	13 790	116 021	95 457	9 933	33.9	41.3	17.56	17.37	595.28	717.38
October	138 190	115 893	22 056	13 764	116 134	95 563	9 911	33.8	41.2	17.58	17.36	594.20	715.23
November	138 299	115 965	22 015	13 757	116 284	95 633	9 920	33.8	41.3	17.63	17.44	595.89	720.27
December	138 409	116 033	21 976	13 746	116 433	95 729	9 924	33.8	41.1	17.69	17.44	597.92	716.78
2008													
January	138 422	116 034	21 947	13 725	116 475	95 743	9 914	33.7	41.1	17.74	17.50	597.84	719.25
February	138 340	115 923	21 897	13 696	116 443	95 656	9 888	33.7	41.2	17.80	17.57	599.86	723.88
March	138 292	115 849	21 820	13 659	116 472	95 622	9 866	33.8	41.3	17.88	17.63	604.34	728.12
April	138 056	115 606	21 681	13 599	116 375	95 429	9 810	33.7	41.1	17.92	17.63	603.90	724.59
May	137 872	115 389	21 599	13 564	116 273	95 236	9 779	33.7	41.0	17.98	17.69	605.93	725.29
June	137 706	115 189	21 483	13 504	116 223	95 054	9 721	33.7	41.0	18.04	17.75	607.95	727.75
July	137 508	114 940	21 360	13 430	116 148	94 864	9 655	33.6	41.0	18.10	17.80	608.16	729.80
August	137 229	114 662	21 250	13 358	115 979	94 630	9 586	33.7	40.9	18.18	17.80	612.67	728.02
September	136 769	114 232	21 101	13 275	115 668	94 296	9 504	33.6	40.5	18.20	17.83	611.52	722.12
October	136 288	113 739	20 895	13 147	115 393	93 883	9 384	33.5	40.5	18.25	17.90	611.38	724.95
November	135 561	113 001	20 623	13 034	114 938	93 183	9 290	33.4	40.1	18.32	17.95	611.89	719.80
December	134 857	112 301	20 322	12 850	114 535	92 615	9 137	33.3	39.8	18.38	18.00	612.05	716.40
2009													
January	134 074	111 495	19 889	12 561	114 185	91 908	8 889	33.2	39.7	18.40	18.01	610.88	715.00
February	133 332	110 756	19 575	12 380	113 757	91 297	8 741	33.3	39.6	18.45	18.09	614.39	716.36
March	132 529	109 969	19 227	12 208	113 302	90 627	8 592	33.1	39.3	18.50	18.13	612.35	712.51
April	131 835	109 158	18 894	12 030	112 941	89 925	8 453	33.1	39.5	18.52	18.16	613.01	717.32
May	131 491	108 874	18 655	11 862	112 836	89 709	8 311	33.0	39.3	18.53	18.13	611.49	712.51
June	131 026	108 450	18 422	11 726	112 604	89 322	8 203	33.0	39.6	18.56	18.17	612.48	719.53
July	130 685	108 164	18 278	11 668	112 407	89 122	8 172	33.1	39.9	18.59	18.29	615.33	729.77
August	130 501	107 964	18 151	11 626	112 350	88 949	8 148	33.1	40.0	18.67	18.32	617.98	732.80
September	130 259	107 808	18 043	11 591	112 216	88 832	8 133	33.1	40.0	18.70	18.42	618.97	736.80
October	130 061	107 537	17 915	11 538	112 146	88 587	8 100	33.0	40.2	18.74	18.38	618.42	738.88
November	130 073	107 540	17 869	11 509	112 204	88 621	8 079	33.2	40.5	18.81	18.42	624.49	746.01
December	129 804	107 322	17 792	11 475	112 012	88 554	8 050	33.2	40.6	18.84	18.42	625.49	747.85
2010													
January	129 807	107 316	17 707	11 460	112 100	88 518	8 034	33.3	40.8	18.89	18.43	629.04	751.94
February	129 715	107 239	17 627	11 453	112 088	88 438	8 032	33.1	40.5	18.92	18.47	626.25	748.04
March	129 895	107 377	17 672	11 453	112 223	88 578	8 027	33.3	41.0	18.92	18.47	630.04	757.27
April	130 132	107 563	17 729	11 489	112 403	88 718	8 050	33.4	41.1	18.96	18.51	633.26	760.76
May	130 666	107 670	17 742	11 525	112 924	88 781	8 078	33.4	41.4	19.01	18.58	634.93	769.21
June	130 530	107 790	17 763	11 545	112 767	88 883	8 100	33.3	41.0	19.03	18.57	633.70	761.37
July	130 442	107 873	17 777	11 561	112 665	88 960	8 106	33.4	41.1	19.04	18.63	635.94	765.69
August	130 437	108 017	17 793	11 553	112 644	89 059	8 099	33.5	41.2	19.11	18.65	640.19	768.38
September	130 373	108 126	17 787	11 563	112 586	89 155	8 098	33.5	41.4	19.11	18.71	640.19	774.59
October	130 642	108 345	17 801	11 562	112 841	89 320	8 096	33.6	41.2	19.20	18.73	645.12	771.68
November	130 765	108 478	17 827	11 585	112 938	89 450	8 103	33.5	41.3	19.21	18.76	643.54	774.79
December	130 839	108 573	17 796	11 595	113 043	89 564	8 117	33.5	41.3	19.22	18.80	643.87	776.44

Table 10-9. Employment, Hours, and Earnings, Total Nonfarm and Manufacturing, Historical Annual and Monthly—Continued

(Wage and salary workers on nonfarm payrolls, seasonally adjusted.)

Year and month	All wage and salary workers (thousands)					Production and nonsupervisory workers on private payrolls							
	Total	Private			Service-providing	Number (thousands)		Average hours per week		Average hourly earnings, dollars		Average weekly earnings, dollars	
		Total	Goods-producing			Total private	Manufac-turing	Total private	Manufac-turing	Total private	Manufac-turing	Total private	Manufac-turing
			Total	Manufac-turing									
2011													
January	130 859	108 601	17 784	11 621	113 075	89 531	8 133	33.4	41.1	19.32	18.89	645.29	776.38
February	131 072	108 857	17 844	11 654	113 228	89 734	8 160	33.5	41.4	19.31	18.91	646.89	782.87
March	131 304	109 112	17 906	11 675	113 398	89 973	8 183	33.5	41.4	19.31	18.90	646.89	782.46
April	131 625	109 441	17 957	11 704	113 668	90 235	8 212	33.6	41.4	19.37	18.91	650.83	782.87
May	131 720	109 591	18 007	11 713	113 713	90 377	8 218	33.6	41.5	19.41	18.92	652.18	785.18
June	131 955	109 791	18 045	11 727	113 910	90 547	8 228	33.6	41.3	19.42	18.90	652.51	780.57
July	132 016	109 967	18 094	11 746	113 922	90 715	8 244	33.7	41.4	19.47	18.93	656.14	783.70
August	132 138	110 121	18 120	11 764	114 018	90 848	8 259	33.5	41.3	19.48	18.90	652.58	780.57
September	132 374	110 391	18 167	11 769	114 207	91 080	8 266	33.6	41.3	19.49	18.91	654.86	780.98
October	132 578	110 580	18 188	11 780	114 390	91 249	8 274	33.7	41.5	19.56	18.98	659.17	787.67
November	132 710	110 739	18 187	11 770	114 523	91 441	8 268	33.6	41.5	19.56	18.96	657.22	786.84
December	132 914	110 960	18 244	11 802	114 670	91 660	8 295	33.7	41.6	19.56	19.00	659.17	790.40
2012													
January	133 269	111 323	18 304	11 838	114 965	91 945	8 322	33.8	41.8	19.58	19.02	661.80	795.04
February	133 531	111 584	18 334	11 860	115 197	92 220	8 352	33.7	41.7	19.59	19.00	660.18	792.30
March	133 769	111 826	18 372	11 898	115 397	92 390	8 386	33.7	41.6	19.65	19.02	662.21	791.23
April	133 852	111 921	18 384	11 916	115 468	92 488	8 399	33.7	41.7	19.69	19.09	663.55	796.05
May	133 951	112 040	18 386	11 927	115 565	92 618	8 411	33.6	41.6	19.68	19.04	661.25	792.06
June	134 023	112 093	18 411	11 936	115 612	92 670	8 412	33.7	41.7	19.72	19.08	664.56	795.64
July	134 176	112 263	18 447	11 964	115 729	92 791	8 438	33.7	41.7	19.75	19.11	665.58	796.89
August	134 346	112 430	18 459	11 960	115 887	92 931	8 429	33.6	41.5	19.74	19.05	663.26	790.58
September	134 535	112 610	18 461	11 954	116 074	93 079	8 419	33.7	41.5	19.79	19.06	666.92	790.99
October	134 693	112 790	18 472	11 961	116 221	93 263	8 426	33.6	41.5	19.79	19.09	664.94	792.24
November	134 851	112 968	18 477	11 950	116 374	93 397	8 411	33.7	41.5	19.84	19.13	668.61	793.90
December	135 088	113 201	18 532	11 960	116 556	93 602	8 415	33.8	41.7	19.89	19.13	672.28	797.72
2013													
January	135 283	113 414	18 580	11 983	116 703	93 670	8 422	33.6	41.7	19.94	19.16	669.98	798.97
February	135 562	113 681	18 638	11 996	116 924	93 876	8 421	33.8	41.8	19.99	19.21	675.66	802.98
March	135 698	113 828	18 660	11 999	117 038	94 008	8 409	33.8	41.9	20.02	19.21	676.68	804.90
April	135 890	114 021	18 653	12 000	117 237	94 163	8 414	33.7	41.8	20.05	19.23	675.69	803.81
May	136 114	114 250	18 691	12 000	117 423	94 317	8 402	33.7	41.8	20.06	19.24	676.02	804.23
June	136 295	114 455	18 724	12 004	117 571	94 496	8 398	33.7	41.8	20.12	19.27	678.04	805.49
July	136 400	114 585	18 705	11 984	117 695	94 650	8 383	33.5	41.7	20.14	19.26	674.69	803.14
August	136 642	114 811	18 756	12 014	117 886	94 827	8 406	33.6	41.9	20.18	19.31	678.05	809.09
September	136 831	114 996	18 810	12 032	118 021	94 970	8 426	33.7	41.9	20.23	19.33	681.75	809.93
October	137 056	115 226	18 857	12 056	118 199	95 192	8 450	33.6	41.9	20.25	19.35	680.40	810.77
November	137 323	115 480	18 911	12 079	118 412	95 418	8 464	33.7	41.9	20.31	19.40	684.45	812.86
December	137 390	115 570	18 881	12 083	118 509	95 509	8 468	33.6	41.9	20.33	19.46	683.09	815.37
2014													
January	137 567	115 753	18 937	12 081	118 630	95 602	8 468	33.5	41.5	20.40	19.46	683.40	807.59
February	137 735	115 910	18 988	12 106	118 747	95 773	8 497	33.4	41.6	20.50	19.49	684.70	810.78
March	137 985	116 152	19 037	12 120	118 948	96 010	8 510	33.7	42.0	20.48	19.53	690.18	820.26
April	138 312	116 459	19 104	12 134	119 208	96 232	8 525	33.7	41.9	20.51	19.48	691.19	816.21
May	138 533	116 701	19 142	12 146	119 391	96 454	8 532	33.7	42.1	20.55	19.51	692.54	821.37
June	138 857	116 962	19 195	12 170	119 662	96 662	8 558	33.7	42.1	20.57	19.54	693.21	822.63
July	139 084	117 183	19 264	12 189	119 820	96 834	8 580	33.7	42.0	20.61	19.59	694.56	822.78
August	139 272	117 417	19 316	12 208	119 956	96 997	8 594	33.7	42.0	20.67	19.60	696.58	823.20
September	139 583	117 686	19 371	12 226	120 212	97 183	8 598	33.7	42.2	20.68	19.59	696.92	826.70
October	139 841	117 924	19 419	12 259	120 422	97 353	8 623	33.7	42.2	20.71	19.65	697.93	829.23
November	140 127	118 195	19 456	12 284	120 671	97 539	8 646	33.8	42.2	20.76	19.65	701.69	829.23
December	140 396	118 449	19 481	12 292	120 915	97 736	8 652	33.8	42.1	20.72	19.63	700.34	826.42
2015													
January	140 609	118 650	19 512	12 295	121 097	97 840	8 648	33.7	42.0	20.81	19.65	701.30	825.30
February	140 857	118 877	19 529	12 303	121 328	98 056	8 657	33.7	42.0	20.84	19.73	702.31	828.66
March	140 934	118 964	19 503	12 311	121 431	98 162	8 661	33.7	41.9	20.88	19.77	703.66	828.36
April	141 234	119 226	19 550	12 317	121 684	98 304	8 661	33.6	41.8	20.92	19.80	702.91	827.64
May	141 553	119 540	19 582	12 334	121 971	98 597	8 682	33.6	41.8	20.99	19.86	705.26	830.15
June	141 723	119 712	19 601	12 338	122 122	98 732	8 691	33.7	41.8	21.01	19.89	708.04	831.40
July	142 016	119 964	19 632	12 357	122 384	98 868	8 701	33.7	41.8	21.03	19.95	708.71	833.91
August	142 138	120 077	19 623	12 343	122 515	98 982	8 690	33.7	41.8	21.08	19.99	710.40	835.58
September	142 271	120 229	19 633	12 350	122 638	99 099	8 690	33.6	41.7	21.11	20.07	709.30	836.92
October	142 610	120 554	19 681	12 361	122 929	99 358	8 708	33.7	41.8	21.20	20.06	714.44	838.51
November	142 845	120 768	19 710	12 357	123 135	99 516	8 697	33.7	41.8	21.20	20.06	714.44	838.51
December	143 125	121 023	19 740	12 362	123 385	99 722	8 698	33.8	41.8	21.24	20.12	717.91	841.02
2016													
January	143 215	121 096	19 745	12 384	123 470	99 753	8 709	33.7	41.8	21.31	20.16	718.15	842.69
February	143 447	121 302	19 720	12 369	123 727	99 923	8 696	33.6	41.8	21.36	20.20	717.70	844.36
March	143 681	121 498	19 712	12 344	123 969	100 100	8 680	33.6	41.7	21.40	20.27	719.04	845.26
April	143 892	121 703	19 728	12 351	124 164	100 200	8 673	33.6	41.8	21.46	20.39	721.06	852.30
May	143 907	121 711	19 690	12 333	124 217	100 215	8 651	33.6	41.8	21.48	20.41	721.73	853.14
June	144 189	121 986	19 711	12 353	124 478	100 432	8 675	33.6	41.8	21.53	20.48	723.41	856.06
July	144 525	122 227	19 755	12 370	124 770	100 640	8 686	33.6	42.0	21.58	20.49	725.09	860.58
August	144 660	122 389	19 729	12 347	124 931	100 763	8 661	33.6	41.8	21.61	20.56	726.10	859.41
September	144 930	122 634	19 759	12 344	125 171	100 932	8 653	33.6	41.9	21.63	20.54	726.77	860.63
October	145 058	122 764	19 780	12 341	125 278	101 054	8 647	33.6	42.0	21.71	20.58	729.46	864.36
November	145 228	122 948	19 806	12 341	125 422	101 251	8 650	33.6	41.8	21.73	20.61	730.13	863.56
December	145 443	123 139	19 827	12 355	125 616	101 457	8 662	33.5	41.9	21.77	20.62	729.30	863.98

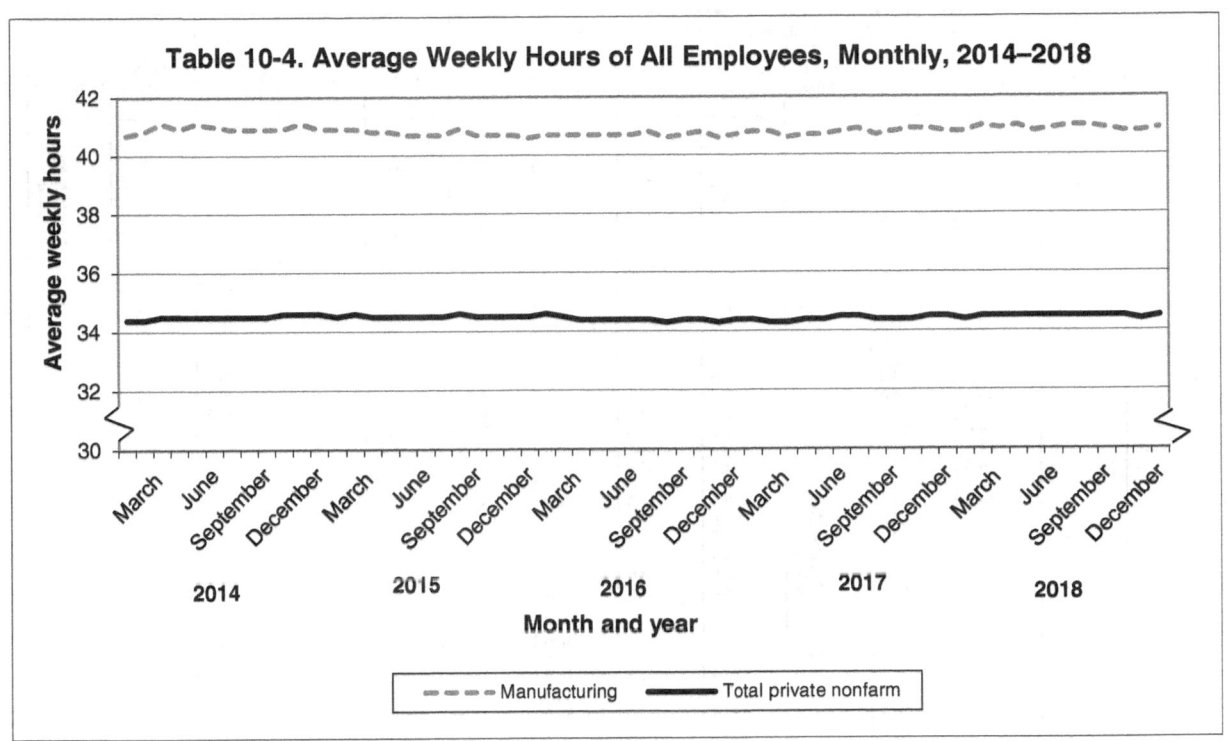

Table 10-4. Average Weekly Hours of All Employees, Monthly, 2014–2018

- The payroll survey now includes hours and earnings data for all nonfarm jobs, as well as the hours and earnings for production and nonsupervisory worker jobs that have been collected for many years. The newer, more comprehensive series are available beginning with March 2006. The new monthly data on the workweek for all nonfarm jobs and for manufacturing are shown in Figure 10-4. (Table 10-10 and Table 10-15)

- Firms experiencing increases in demand are likely to meet that demand, at first, by having their existing employees work longer hours. If the demand increase is sustained, they become more likely to hire more people. This is the basis for the long-time reputation of the workweek as a leading indicator for employment. As Figure 10-4 demonstrates, the all-employee workweek at manufacturing—the most cyclical sector—has been trending upward, though irregularly, from a low of 40.2 in 2010 to a high of 41.0 in 2014. By 2018, it had declined slightly to 40.9 hours. The workweek for all private nonfarm industry workers appears stuck around 34.5 hours per week. (Table 10-10 and recent data)

- Miners and loggers worked the longest hours per week in 2018 at 46.0 hours. Service-providing sectors such as leisure and hospitality worked the least amount of hours at 26.1 (Table 10-10)

- It is important to note that the "average" all-employee workweek of less than 34 1/2 hours reported in the payroll survey is not necessarily an accurate representation of how many hours per week a typical American worker works. This is because it is based on a count of jobs, as reported by employers. Suppose a full-time worker takes a second, "moonlighting" part-time job, newly created in another industry. This reduces the average workweek reported by the payroll survey, even though the individual is working longer hours than before. If a new labor force entrant takes a half-time job, it reduces the average workweek even if no other workers are working less. For measures of the workweek as reported by persons in the Current Population Survey, see Table 10-19, Figure 10-5, and the discussions of each.

Table 10-10. Average Weekly Hours of All Employees on Private Nonfarm Payrolls by NAICS Supersector

(Hours per week, seasonally adjusted.)

Year and month	Total private	Mining and logging	Construction	Manufacturing		Trade, transportation, and utilities			Information	Financial activities	Professional and business services	Education and health services	Leisure and hospitality	Other services
				Average weekly hours	Overtime hours	Total	Wholesale trade	Retail trade						
2010	34.1	43.4	37.8	40.2	3.0	34.2	38.1	31.3	36.5	36.9	35.4	32.7	25.7	31.6
2011	34.4	44.5	38.3	40.5	3.2	34.6	38.6	31.6	36.6	37.3	35.7	32.7	25.9	31.7
2012	34.5	44.0	38.7	40.7	3.3	34.6	38.7	31.7	36.6	37.4	36.0	32.8	26.1	31.6
2013	34.4	43.9	39.0	40.8	3.4	34.5	38.7	31.4	36.7	37.1	36.1	32.7	26.0	31.7
2014	34.5	34.5	39.0	41.0	3.5	34.5	38.9	31.3	36.8	37.3	36.3	32.7	26.2	31.8
2015	34.5	44.0	39.2	40.8	3.3	34.6	38.9	31.4	36.2	37.6	36.2	32.8	26.3	31.9
2016	34.4	43.4	39.0	40.7	3.3	34.3	38.9	31.0	36.0	37.5	36.1	32.9	26.1	31.9
2017	34.4	44.9	39.1	40.8	3.1	34.4	39.3	31.0	36.3	37.3	36.0	32.9	26.1	31.8
2018	34.5	46.0	39.2	40.9	3.5	34.5	39.0	31.0	36.1	37.6	36.1	33.0	26.1	35.9
2014														
January	34.4	44.1	38.6	40.7	3.5	34.4	38.8	31.1	36.6	37.1	36.1	32.7	26.1	31.7
February	34.4	45.1	38.5	40.8	3.4	34.3	38.7	31.1	36.8	37.2	36.1	32.7	26.1	31.7
March	34.5	45.7	39.1	41.1	3.5	34.4	38.9	31.2	36.9	37.3	36.3	32.7	26.2	31.8
April	34.5	44.6	39.1	40.9	3.5	34.5	38.8	31.4	36.8	37.1	36.2	32.7	26.2	31.8
May	34.5	44.5	39.0	41.1	3.5	34.5	38.9	31.3	36.8	37.3	36.3	32.7	26.1	31.8
June	34.5	44.9	39.1	41.0	3.5	34.4	38.9	31.2	36.6	37.2	36.2	32.7	26.1	31.7
July	34.5	44.5	39.2	40.9	3.5	34.5	38.9	31.3	36.7	37.2	36.2	32.7	26.2	31.7
August	34.5	44.9	39.2	40.9	3.4	34.5	38.9	31.3	36.6	37.2	36.2	32.8	26.2	31.8
September	34.5	44.5	39.0	40.9	3.5	34.5	38.9	31.3	36.7	37.3	36.2	32.8	26.2	31.8
October	34.6	45.0	39.1	40.9	3.4	34.6	38.9	31.4	36.7	37.4	36.3	32.8	26.2	31.8
November	34.6	44.9	39.0	41.1	3.5	34.6	38.9	31.4	36.7	37.3	36.3	32.8	26.2	31.8
December	34.6	44.9	39.2	40.9	3.6	34.6	38.9	31.4	36.3	37.4	36.3	32.7	26.3	31.8
2015														
January	34.5	44.7	39.0	40.9	3.5	34.5	38.9	31.3	36.4	37.4	36.2	32.8	26.3	31.8
February	34.6	44.2	39.3	40.9	3.4	34.6	38.9	31.4	36.4	37.4	36.2	32.8	26.4	31.9
March	34.5	43.9	39.0	40.8	3.3	34.6	38.7	31.4	36.4	37.5	36.2	32.8	26.3	31.9
April	34.5	43.8	38.9	40.8	3.4	34.5	38.8	31.3	36.3	37.6	36.0	32.8	26.2	31.8
May	34.5	43.8	39.0	40.7	3.4	34.7	38.9	31.5	36.3	37.6	36.2	32.8	26.2	31.8
June	34.5	43.7	39.2	40.7	3.4	34.6	38.8	31.4	36.3	37.7	36.2	32.8	26.3	31.8
July	34.5	43.9	39.0	40.7	3.4	34.6	38.9	31.4	36.3	37.6	36.2	32.8	26.2	31.9
August	34.6	43.8	39.2	40.9	3.3	34.7	38.8	31.5	36.3	37.6	36.2	32.9	26.2	31.9
September	34.5	43.9	38.8	40.7	3.2	34.8	38.8	31.7	36.0	37.7	36.1	32.8	26.3	31.8
October	34.5	43.8	39.6	40.7	3.3	34.6	38.9	31.4	36.0	37.7	36.2	32.8	26.3	31.9
November	34.5	43.9	39.1	40.7	3.2	34.6	38.9	31.4	36.0	37.6	36.1	32.8	26.2	31.9
December	34.5	44.3	39.7	40.6	3.3	34.6	38.9	31.3	36.0	37.7	36.2	32.8	26.2	31.9
2016														
January	34.6	43.5	39.3	40.7	3.3	34.6	38.9	31.3	36.2	37.7	36.3	32.9	26.2	31.9
February	34.5	43.1	39.2	40.7	3.3	34.5	38.9	31.2	36.0	37.7	36.1	32.8	26.2	31.9
March	34.4	42.8	38.7	40.7	3.3	34.3	38.8	31.0	36.0	37.6	36.1	32.8	26.1	31.9
April	34.4	43.0	39.1	40.7	3.3	34.4	38.9	31.1	36.0	37.6	36.1	32.8	26.1	31.9
May	34.4	43.1	39.1	40.7	3.3	34.3	38.8	31.0	36.0	37.4	36.1	32.8	26.1	31.9
June	34.4	43.0	39.1	40.7	3.2	34.3	38.8	31.0	36.0	37.5	36.0	32.9	26.1	32.0
July	34.4	43.3	39.2	40.8	3.3	34.4	38.9	31.1	36.1	37.6	36.1	32.9	26.1	32.0
August	34.3	43.6	38.8	40.6	3.3	34.2	38.9	30.8	35.9	37.5	36.0	32.9	26.0	31.9
September	34.4	43.7	38.8	40.7	3.3	34.2	38.9	30.8	35.8	37.5	36.1	32.9	26.1	31.9
October	34.4	43.9	39.2	40.8	3.3	34.2	38.9	30.8	35.9	37.3	36.1	32.9	26.0	32.0
November	34.3	43.6	39.1	40.6	3.2	34.2	38.9	30.8	36.0	37.5	36.0	32.9	26.1	31.9
December	34.4	43.7	38.9	40.7	3.3	34.4	38.9	31.1	36.1	37.4	36.0	32.9	26.0	31.9
2017														
January	34.4	43.9	39.1	40.8	3.3	34.3	38.9	30.9	36.4	37.3	36.1	32.9	26.0	31.8
February	34.3	44.1	39.0	40.8	3.3	34.2	38.9	30.8	36.2	37.4	36.0	32.9	25.9	31.8
March	34.3	44.7	38.7	40.6	0.2	34.3	38.9	30.9	36.3	37.3	36.0	32.9	26.0	31.9
April	34.4	45.0	39.2	40.7	3.2	34.5	39.0	31.2	36.3	37.5	36.0	32.9	26.1	31.9
May	34.4	44.9	39.2	40.7	3.3	34.4	39.0	31.0	36.3	37.4	36.0	32.9	26.0	31.8
June	34.5	44.9	39.1	40.8	3.3	34.4	39.1	31.0	36.3	37.6	36.1	32.9	26.1	31.9
July	34.5	45.3	39.1	40.9	3.3	34.4	39.1	31.0	36.3	37.5	36.1	32.9	26.1	31.8
August	34.4	44.9	39.0	40.7	3.3	34.4	39.1	30.9	36.3	37.5	36.0	32.9	26.0	31.8
September	34.4	45.2	38.8	40.8	3.3	34.4	40.8	30.9	36.2	37.5	36.0	32.9	26.0	31.7
October	34.4	45.2	39.0	40.9	3.5	34.4	40.0	31.0	36.3	37.5	36.0	32.9	26.2	31.7
November	34.5	45.5	39.1	40.9	3.5	34.7	39.2	31.3	36.0	37.6	36.1	32.9	26.1	31.7
December	34.5	45.7	39.4	40.8	3.5	34.5	39.3	31.1	36.2	36.0	36.0	33.0	26.2	31.8
2018														
January	34.4	45.2	39.1	40.8	3.5	34.5	39.0	31.1	35.8	37.6	35.9	32.9	26.0	37.4
February	34.5	45.9	39.3	41.0	3.7	34.5	39.0	31.1	36.0	37.6	36.2	33.0	26.1	37.6
March	34.5	45.7	39.3	40.9	3.6	34.5	39.1	31.1	36.0	37.5	36.2	32.9	26.1	37.2
April	34.5	45.9	39.4	41.0	3.7	34.5	39.0	31.0	36.1	37.6	36.1	33.0	26.0	36.0
May	34.5	46.1	39.5	40.8	3.4	34.5	39.1	31.0	36.0	37.6	36.2	33.0	26.1	34.4
June	34.5	46.4	39.2	40.9	3.5	34.6	39.1	31.2	35.8	37.7	36.2	33.0	26.1	35.8
July	34.5	45.9	39.4	41.0	3.5	34.6	39.0	31.1	36.0	37.5	36.2	32.9	26.1	35.8
August	34.5	46.1	39.2	41.0	3.5	34.5	39.1	30.9	36.1	37.6	36.1	33.0	26.1	35.4
September	34.5	46.0	39.1	40.9	3.5	34.4	39.0	30.9	36.3	37.5	36.1	32.9	26.0	35.3
October	34.5	46.1	38.8	40.8	3.5	34.4	38.9	30.8	36.2	37.6	36.2	32.9	26.1	35.4
November	34.4	45.9	38.7	40.8	3.5	34.5	39.0	30.9	36.1	37.6	36.1	32.9	25.9	35.3
December	34.5	46.2	39.5	40.9	3.5	34.3	39.0	30.6	36.3	37.6	36.1	33.0	26.0	35.0

Table 10-11. Indexes of Aggregate Weekly Hours of All Employees on Private Nonfarm Payrolls by NAICS Supersector

(2007 =100, seasonally adjusted.)

Year and month	Total private	Mining and logging	Construction	Manu-facturing	Trade, transportation, and utilities			Information	Financial activities	Profes-sional and business services	Education and health services	Leisure and hospitality	Other services
					Total	Wholesale trade	Retail trade						
2010	92.4	96.1	71.9	83.4	91.7	90.5	92.0	90.3	93.0	93.1	106.5	95.9	97.2
2011	94.7	110.2	73.1	85.4	94.2	93.3	94.1	89.4	93.9	97.4	108.6	98.7	98.0
2012	97.1	117.1	75.4	87.4	95.8	95.6	95.4	89.6	95.2	101.7	111.2	102.6	99.0
2013	99.0	119.2	78.7	88.3	97.0	96.7	96.2	90.6	95.9	105.1	112.6	105.7	100.3
2014	101.5	125.5	82.7	89.8	99.1	98.5	97.7	91.5	97.4	108.8	114.6	109.8	102.0
2015	103.9	112.4	87.1	90.5	101.3	99.2	99.6	91.1	100.1	111.8	118.0	113.7	103.3
2016	105.4	91.0	90.8	90.5	101.8	99.1	99.7	91.6	101.5	113.6	121.3	116.5	104.6
2017	107.4	96.0	94.1	91.3	102.8	100.2	99.8	93.1	103.8	116.0	124.5	119.5	105.7
2018	109.7	105.7	98.7	93.4	104.1	100.8	99.6	93.3	105.6	119.1	127.3	121.6	107.2
2014													
January	99.9	121.5	79.6	88.5	97.9	97.6	96.7	91.2	96.1	106.9	113.3	107.5	101.1
February	99.8	123.2	79.3	88.6	97.5	97.5	96.0	91.4	96.6	107.0	113.1	106.8	101.2
March	100.6	124.6	81.6	89.4	98.2	98.0	96.9	91.8	96.6	107.8	113.7	108.9	102.1
April	100.8	125.3	81.6	89.3	98.5	98.0	97.6	91.5	96.5	108.2	113.8	109.2	102.4
May	101.1	124.0	81.9	89.8	98.7	98.4	97.4	91.1	96.9	108.4	114.1	109.2	102.1
June	101.3	126.4	82.5	90.0	98.7	98.6	97.6	91.1	97.1	108.7	114.3	109.4	101.7
July	101.5	126.3	83.5	89.7	99.1	98.7	97.7	91.1	96.9	108.6	114.2	109.5	101.7
August	102.0	127.0	83.7	90.0	99.5	98.9	98.1	91.5	97.7	109.0	115.1	110.5	102.3
September	101.9	126.8	84.1	90.2	99.4	99.0	98.0	91.8	97.8	109.3	115.0	110.4	102.2
October	102.4	127.1	84.1	90.4	100.0	98.8	98.6	91.4	98.0	109.8	115.5	110.7	102.3
November	102.6	127.0	84.5	90.0	100.2	99.1	98.8	91.4	98.4	110.1	115.8	111.0	102.5
December	102.9	127.0	84.8	90.7	100.4	99.0	98.9	91.2	98.5	110.5	116.1	111.8	102.5
2015													
January	102.7	125.1	84.9	90.5	100.2	99.0	98.6	91.2	99.0	110.3	116.5	111.9	103.0
February	103.2	122.1	86.5	90.7	100.6	99.2	99.0	91.2	98.8	110.6	116.7	112.0	103.1
March	103.0	118.8	85.6	90.6	100.7	99.0	99.1	91.0	99.2	110.4	116.9	112.1	103.1
April	103.2	116.0	85.7	90.4	100.9	98.9	99.3	90.9	99.6	110.7	117.3	112.4	102.9
May	103.5	112.3	86.2	90.3	101.4	99.3	99.8	91.2	99.7	111.5	117.6	112.9	103.1
June	103.7	111.9	87.1	90.3	100.9	99.0	99.3	91.2	100.2	111.8	117.8	113.4	103.0
July	103.9	111.2	87.2	90.7	101.4	99.3	99.7	91.4	100.2	112.0	118.1	113.3	103.5
August	104.3	109.0	87.9	90.8	101.7	99.0	100.0	91.4	100.3	112.2	118.7	113.6	103.0
September	104.1	107.8	87.2	90.2	102.0	98.8	101.0	91.1	100.7	111.9	118.6	114.5	103.4
October	104.7	106.6	89.6	90.5	101.6	99.1	100.0	91.2	100.6	113.0	119.0	114.9	103.6
November	104.6	105.0	88.9	90.5	101.8	99.2	100.1	90.7	100.7	112.5	119.2	114.5	103.7
December	104.8	103.8	90.8	90.5	101.8	99.4	99.4	90.7	101.1	113.3	119.6	115.6	104.0
2016													
January	105.2	100.6	89.8	90.7	101.9	99.4	99.9	91.6	101.5	113.5	120.0	115.4	104.2
February	104.7	96.5	89.7	90.6	101.8	99.1	99.9	91.3	101.1	113.0	120.0	115.9	104.2
March	104.9	92.7	89.5	90.4	101.4	98.9	99.5	91.5	101.3	113.2	120.2	115.8	104.3
April	105.1	91.4	90.3	90.7	101.8	99.3	99.9	91.7	101.9	113.7	120.5	116.0	104.5
May	105.1	90.0	90.0	90.5	101.5	98.7	99.5	90.8	101.5	113.4	120.8	116.1	104.3
June	105.3	88.8	90.1	90.4	101.6	98.9	99.7	92.0	101.7	113.3	121.5	116.5	104.6
July	105.5	88.5	90.7	90.6	101.8	99.2	99.8	92.3	102.2	113.9	121.8	116.7	105.0
August	105.4	89.0	90.3	90.2	101.7	99.0	99.6	92.0	102.2	113.7	122.0	116.6	104.9
September	105.9	88.8	91.0	90.4	102.0	99.7	99.7	92.3	102.3	114.4	122.3	117.7	105.2
October	106.0	88.6	91.8	90.4	102.1	99.7	99.7	92.3	102.1	114.5	122.6	117.0	105.4
November	105.8	89.2	91.9	90.1	101.9	99.5	99.3	92.5	102.2	114.3	122.8	117.8	105.3
December	106.3	88.8	91.5	90.5	102.6	99.5	100.4	92.7	102.5	114.5	123.2	117.5	105.1
2017													
January	106.5	89.7	92.7	90.8	102.5	99.6	100.0	94.2	102.3	115.4	123.2	117.8	105.0
February	106.3	91.4	92.6	90.7	102.1	99.6	99.0	93.2	102.9	115.1	123.6	117.6	105.2
March	106.4	93.3	92.6	90.5	102.0	99.9	99.2	92.9	102.7	115.0	123.8	118.2	105.3
April	106.9	95.2	93.5	91.0	102.9	100.0	100.4	93.0	103.2	115.5	124.1	119.2	105.4
May	107.0	96.2	93.4	90.8	102.6	99.9	99.6	93.0	103.3	115.8	123.9	118.9	105.6
June	107.2	96.1	93.8	91.2	102.7	100.0	99.6	93.1	104.0	116.0	124.5	119.6	105.8
July	107.4	97.1	93.5	91.4	102.7	100.3	99.5	92.9	103.6	116.2	124.8	120.0	105.6
August	107.6	97.1	93.7	91.4	102.7	100.2	99.2	92.6	104.1	116.1	125.1	120.1	105.7
September	107.3	97.8	93.9	91.5	102.6	100.0	99.2	93.0	104.1	116.2	124.8	119.0	105.6
October	107.8	98.0	94.7	91.9	103.0	100.1	99.5	92.8	104.2	116.4	125.3	120.2	105.9
November	108.3	98.9	95.4	92.3	104.0	100.6	100.6	92.3	104.6	117.0	125.5	120.4	106.0
December	108.4	99.4	96.1	92.1	103.4	101.0	99.8	93.3	104.7	116.8	126.0	121.1	106.1
2018													
January	108.3	99.3	96.0	92.2	103.5	100.2	99.8	91.9	104.7	116.7	125.9	120.2	106.2
February	108.8	101.8	97.5	92.9	103.7	100.2	100.0	92.5	105.0	117.9	126.5	120.9	106.2
March	109.0	102.5	97.6	92.8	103.8	100.6	100.0	92.8	104.8	118.1	126.3	121.0	106.2
April	109.2	104.3	98.2	93.2	103.8	100.1	99.7	93.3	105.2	118.2	126.9	120.6	106.8
May	109.4	105.5	98.9	92.9	103.9	100.5	99.8	93.1	105.4	118.8	127.1	121.4	106.8
June	109.6	107.2	98.4	93.4	104.2	100.6	100.2	92.6	105.8	119.1	127.4	121.7	107.4
July	109.7	105.9	99.2	93.7	104.3	100.6	99.9	93.1	105.3	119.3	127.3	122.0	107.2
August	110.0	107.5	99.1	93.8	104.2	101.2	99.3	93.2	105.7	119.3	128.0	122.1	107.3
September	110.0	107.7	99.1	93.7	103.9	101.0	99.2	93.6	105.6	119.6	127.7	121.4	107.4
October	110.3	108.8	98.7	93.7	104.0	100.9	98.8	93.6	106.6	120.2	127.9	122.5	107.9
November	110.1	107.9	98.5	93.9	104.5	101.3	99.3	93.3	106.1	120.1	128.1	121.9	107.9
December	110.7	109.2	100.8	94.2	103.9	101.5	98.3	93.7	106.1	120.3	128.8	122.8	108.1

Table 10-12. Average Hourly Earnings of All Employees on Private Nonfarm Payrolls by NAICS Supersector

(Dollars, seasonally adjusted.)

Year and month	Total private	Mining and logging	Construc-tion	Manu-facturing	Trade, transportation, and utilities			Information	Financial activities	Profes-sional and business services	Education and health services	Leisure and hospitality	Other services
					Total	Wholesale trade	Retail trade						
2010	22.56	27.39	25.19	23.31	19.61	26.04	15.57	30.53	27.21	27.27	22.76	13.08	20.16
2011	23.03	28.10	25.41	23.69	20.00	26.28	15.87	31.59	27.91	27.79	23.42	13.23	20.50
2012	23.49	28.76	25.73	23.92	20.46	26.79	16.31	31.83	29.26	28.16	24.01	13.37	20.85
2013	23.96	29.72	26.12	24.35	20.93	27.54	16.64	32.91	30.15	28.58	24.42	13.50	21.40
2014	24.47	30.79	26.69	24.81	21.34	27.97	17.01	34.07	30.76	29.32	24.72	13.91	21.97
2015	25.02	31.15	27.37	25.25	21.79	28.53	17.52	35.14	31.52	30.11	25.24	14.32	22.48
2016	25.64	31.91	28.12	25.99	22.28	29.36	17.88	36.67	32.29	30.82	25.74	14.87	23.05
2017	26.33	32.05	28.90	26.59	22.74	29.92	18.18	38.30	33.24	31.67	26.32	15.47	23.85
2018	27.11	32.54	29.89	27.05	23.35	30.48	18.78	40.01	34.79	32.57	27.02	15.98	24.56
2014													
January	24.21	30.50	26.41	24.68	21.14	27.81	16.76	33.41	30.32	28.92	24.58	13.69	21.75
February	24.32	30.60	26.76	24.73	21.22	27.84	16.82	33.44	30.43	29.08	24.60	13.73	21.80
March	24.31	30.56	26.48	24.73	21.26	27.90	16.89	33.73	30.55	29.06	24.59	13.75	21.73
April	24.34	30.82	26.56	24.71	21.31	27.93	16.97	33.91	30.56	29.12	24.60	13.76	21.85
May	24.40	30.91	26.61	24.75	21.35	27.94	17.01	33.94	30.58	29.23	24.65	13.83	21.87
June	24.44	30.86	26.67	24.83	21.33	27.83	17.03	34.01	30.72	29.27	24.67	13.90	21.96
July	24.48	30.90	26.66	24.84	21.35	27.85	17.07	34.29	30.77	29.34	24.72	13.93	22.01
August	24.55	30.93	26.73	24.88	21.43	28.14	17.08	34.35	30.88	29.45	24.79	13.98	22.07
September	24.56	30.88	26.81	24.83	21.40	27.98	17.09	34.54	30.90	29.42	24.80	14.03	22.08
October	24.58	30.80	26.86	24.91	21.38	27.95	17.10	34.26	30.93	29.47	24.83	14.08	22.10
November	24.65	30.82	26.92	24.94	21.49	28.13	17.20	34.38	31.04	29.56	24.89	14.11	22.17
December	24.65	30.82	26.87	24.92	21.41	28.13	17.07	34.44	31.04	29.61	24.93	14.13	22.19
2015													
January	24.74	30.51	27.06	25.01	21.57	28.19	17.34	34.55	31.07	29.74	25.00	14.15	22.18
February	24.79	30.73	27.12	25.04	21.62	28.26	17.34	34.64	31.18	29.77	25.03	14.18	22.29
March	24.85	30.87	27.24	25.10	21.61	28.28	17.29	34.75	31.25	29.87	25.12	14.22	22.33
April	24.89	30.69	27.28	25.11	21.67	28.35	17.39	34.76	31.37	29.97	25.10	14.25	22.32
May	24.97	31.13	27.31	25.15	21.74	28.58	17.43	34.88	31.54	30.05	25.22	14.30	22.43
June	24.99	31.03	27.34	25.14	21.74	28.55	17.47	34.96	31.59	30.07	25.24	14.30	22.55
July	25.01	31.20	27.36	25.23	21.79	28.51	17.53	35.05	31.50	30.11	25.22	14.31	22.47
August	25.09	31.36	27.43	25.37	21.82	28.54	17.58	35.35	31.55	30.19	25.31	14.34	22.57
September	25.12	31.54	27.35	25.40	21.85	28.63	17.67	35.47	31.67	30.29	25.34	14.38	22.61
October	25.20	31.43	27.52	25.44	21.96	28.76	17.71	35.51	31.71	30.35	25.41	14.42	22.64
November	25.25	31.88	27.66	25.51	21.97	28.70	17.72	35.75	31.71	30.41	25.45	14.46	22.67
December	25.26	31.49	27.63	25.54	21.99	28.71	17.75	35.87	31.81	30.33	25.49	14.51	22.72
2016													
January	25.37	31.78	27.64	25.64	22.04	28.88	17.74	36.06	32.03	30.51	25.56	14.59	22.78
February	25.39	31.73	27.77	25.66	22.07	28.93	17.79	36.24	32.03	30.56	25.59	14.62	22.78
March	25.46	31.90	27.87	25.75	22.17	29.21	17.84	36.12	32.17	30.58	25.62	14.67	22.85
April	25.54	31.98	27.98	25.86	22.20	29.29	17.83	36.29	32.17	30.70	25.67	14.74	22.91
May	25.57	32.12	28.08	25.96	22.22	29.37	17.87	36.56	32.18	30.78	25.64	14.78	22.97
June	25.64	32.11	28.14	25.98	22.33	29.38	17.96	36.54	32.21	30.89	25.68	14.86	23.03
July	25.70	32.03	28.19	26.02	22.34	29.52	17.90	36.67	32.41	30.93	25.73	14.93	23.09
August	25.73	31.65	28.20	26.13	22.38	29.59	17.92	36.82	32.49	30.95	25.69	14.97	23.15
September	25.78	31.90	28.23	26.12	22.38	29.53	17.92	37.05	32.57	30.99	25.83	15.02	23.17
October	25.87	32.04	28.41	26.27	22.46	29.61	17.97	37.33	32.47	31.10	25.89	15.08	23.32
November	25.91	31.93	28.34	26.25	22.52	29.64	18.09	37.42	32.67	31.13	25.93	15.13	23.32
December	25.93	32.10	28.40	26.31	22.45	29.70	17.94	37.54	32.69	31.21	25.97	15.16	23.38
2017													
January	25.98	32.16	28.50	26.33	22.55	29.80	18.02	37.51	32.61	31.20	25.99	15.22	23.51
February	26.08	32.10	28.50	26.40	22.58	29.83	18.03	37.59	32.85	31.36	26.10	15.29	23.62
March	26.11	31.98	28.60	26.40	22.59	29.79	18.05	37.63	32.75	31.53	26.11	15.35	23.61
April	26.17	31.80	28.61	26.55	22.60	29.84	18.07	38.00	32.89	31.58	26.20	15.38	23.66
May	26.22	31.80	28.73	26.53	22.67	29.85	18.13	38.19	32.89	31.53	26.25	15.50	23.69
June	26.28	31.90	28.91	26.56	22.75	29.92	18.18	38.51	33.08	31.55	26.25	15.46	23.75
July	26.36	32.19	28.94	26.68	22.79	30.00	18.20	38.69	33.24	31.64	26.36	15.47	23.85
August	26.39	32.04	29.00	26.61	22.80	29.95	18.26	38.66	33.35	31.70	26.40	15.54	23.93
September	26.51	32.01	29.18	26.70	22.86	30.02	18.26	38.58	33.47	31.89	26.52	15.62	24.06
October	26.47	32.09	29.12	26.73	22.79	29.78	18.24	38.42	33.72	31.82	26.44	15.59	24.09
November	26.55	32.15	29.23	26.73	22.88	30.08	18.28	38.65	33.74	31.91	26.54	15.65	24.16
December	26.64	32.21	29.32	26.79	22.96	30.12	18.33	38.84	34.03	32.03	26.61	15.71	24.21
2018													
January	26.71	32.29	29.36	26.84	22.98	30.07	18.42	39.06	34.23	32.10	26.69	15.74	24.25
February	26.75	32.19	29.53	26.84	23.04	30.13	18.47	39.15	34.25	32.10	26.71	15.75	24.25
March	26.84	32.39	29.47	26.89	23.09	30.15	18.49	39.30	34.39	32.25	26.85	15.81	24.41
April	26.90	32.36	29.67	26.95	23.13	30.11	18.60	39.46	34.46	32.28	26.84	15.86	24.46
May	26.99	32.26	29.72	26.97	23.24	30.31	18.70	39.55	34.67	32.40	26.95	15.89	24.51
June	27.05	32.53	29.79	27.03	23.27	30.51	18.68	39.73	34.68	32.50	27.00	15.95	24.52
July	27.11	32.49	29.92	27.03	23.32	30.47	18.76	39.75	34.81	32.61	27.05	16.00	24.55
August	27.23	32.67	30.02	27.11	23.46	30.59	18.87	40.04	34.92	32.82	27.13	16.05	24.59
September	27.30	32.94	30.15	27.14	23.51	30.70	18.92	40.50	35.02	32.82	27.15	16.08	24.68
October	27.35	32.72	30.23	27.16	23.58	30.73	19.02	40.68	34.96	32.89	27.22	16.14	24.74
November	27.43	32.84	30.26	27.24	23.63	30.82	19.07	41.02	35.31	32.93	27.27	16.21	24.80
December	27.53	32.77	30.42	27.33	23.79	30.92	19.23	41.29	35.37	32.99	27.32	16.27	24.88

Table 10-13. Average Weekly Earnings of All Employees on Private Nonfarm Payrolls by NAICS Supersector

(Dollars, seasonally adjusted.)

Year and month	Total private	Mining and logging	Construction	Manu-facturing	Trade, transportation, and utilities			Information	Financial activities	Profes-sional and business services	Education and health services	Leisure and hospitality	Other services
					Total	Wholesale trade	Retail trade						
2010	769.63	1 189.32	952.78	937.34	671.38	991.59	488.08	1 114.65	1 004.14	964.63	743.41	336.83	637.65
2011	790.85	1 250.91	973.85	958.84	690.95	1 015.85	501.04	1 156.56	1 039.70	992.94	766.88	342.67	649.87
2012	809.57	1 263.98	997.01	973.96	707.71	1 038.30	516.10	1 166.45	1 093.00	1 014.67	787.43	349.12	659.49
2013	825.02	1 306.16	1 018.01	994.30	721.85	1 066.81	522.57	1 206.05	1 119.71	1 031.57	798.73	350.95	679.43
2014	844.91	1 380.77	1 040.85	1 016.42	736.63	1 088.33	532.88	1 252.44	1 146.22	1 063.77	809.11	364.07	698.41
2015	864.21	1 371.13	1 070.93	1 029.68	754.66	1 109.79	550.74	1 274.99	1 185.77	1 089.87	828.36	376.20	716.17
2016	881.20	1 383.16	1 100.94	1 057.77	764.95	1 140.05	554.27	1 315.74	1 207.93	1 110.34	844.80	387.80	734.90
2017	906.30	1 448.75	1 131.76	1 084.86	782.61	1 169.15	563.80	1 388.30	1 246.39	1 142.52	865.65	403.77	757.95
2018	936.06	1 495.98	1 174.46	1 107.47	805.61	1 190.39	582.00	1 444.58	1 309.68	1 178.42	889.97	416.68	781.44
2014													
January	832.82	1 354.20	1 019.43	1 004.48	727.22	1 079.03	522.91	1 226.15	1 124.87	1 046.90	803.77	355.94	689.48
February	834.18	1 370.88	1 024.91	1 006.51	727.85	1 077.41	521.42	1 230.59	1 132.00	1 049.79	801.96	354.23	691.06
March	838.70	1 378.26	1 038.02	1 013.93	733.47	1 085.31	528.66	1 244.64	1 136.46	1 054.88	804.09	360.25	693.19
April	839.73	1 386.90	1 033.18	1 010.64	735.20	1 083.68	532.86	1 247.89	1 133.78	1 057.06	804.42	360.51	697.02
May	841.80	1 375.50	1 035.13	1 017.23	736.58	1 086.87	532.41	1 245.60	1 137.58	1 061.05	806.06	360.96	695.47
June	843.18	1 391.79	1 040.13	1 020.51	733.75	1 082.59	533.04	1 244.77	1 142.78	1 062.50	806.71	362.79	696.13
July	844.56	1 384.32	1 045.07	1 015.96	736.58	1 083.37	534.29	1 255.01	1 141.57	1 062.11	805.87	363.57	697.72
August	849.43	1 391.85	1 045.14	1 020.08	741.48	1 097.46	536.31	1 260.65	1 151.82	1 066.09	813.11	367.67	701.83
September	847.32	1 380.34	1 048.27	1 018.03	738.30	1 091.22	534.92	1 271.07	1 152.57	1 065.00	810.96	367.59	702.14
October	850.47	1 382.92	1 047.54	1 021.31	739.75	1 087.26	536.94	1 257.34	1 153.69	1 069.76	814.42	368.90	702.79
November	852.89	1 383.82	1 052.57	1 025.03	743.55	1 097.07	540.08	1 258.31	1 160.90	1 073.03	816.39	369.68	705.01
December	852.89	1 389.98	1 050.62	1 021.72	740.79	1 094.26	536.00	1 257.06	1 160.90	1 074.84	817.70	371.62	705.64
2015													
January	853.53	1 369.90	1 052.63	1 022.91	744.17	1 096.59	542.74	1 261.08	1 165.13	1 076.59	820.00	372.15	707.54
February	857.73	1 364.41	1 071.24	1 026.64	748.05	1 099.31	544.48	1 260.90	1 166.13	1 077.67	820.98	371.52	711.05
March	857.33	1 358.28	1 067.81	1 026.59	747.71	1 097.26	542.91	1 264.90	1 171.88	1 078.31	823.94	372.56	712.33
April	858.71	1 344.22	1 061.19	1 024.49	749.78	1 099.98	546.05	1 261.79	1 179.51	1 081.92	823.28	373.35	709.78
May	861.47	1 357.27	1 062.36	1 023.61	754.38	1 111.76	549.05	1 266.14	1 185.90	1 087.81	827.22	374.66	713.27
June	862.16	1 352.91	1 071.73	1 023.20	750.03	1 107.74	546.81	1 269.05	1 190.94	1 088.53	827.87	376.09	717.09
July	862.85	1 369.68	1 069.78	1 029.38	753.93	1 109.04	550.44	1 272.32	1 184.40	1 089.98	827.22	374.92	716.79
August	868.11	1 370.43	1 078.00	1 037.63	757.15	1 107.35	553.77	1 283.21	1 186.28	1 092.88	832.70	375.71	717.73
September	866.64	1 390.91	1 063.92	1 031.24	760.38	1 110.84	561.91	1 280.47	1 193.96	1 093.47	831.15	378.19	721.26
October	871.92	1 382.92	1 092.54	1 035.41	759.82	1 118.76	556.09	1 281.91	1 192.30	1 101.71	833.45	379.25	722.22
November	871.13	1 409.10	1 081.51	1 038.26	760.16	1 116.43	556.41	1 290.58	1 192.30	1 097.80	834.76	377.41	723.17
December	871.47	1 395.01	1 096.91	1 039.48	760.85	1 119.69	553.80	1 291.32	1 199.24	1 097.95	836.07	381.61	724.77
2016													
January	877.80	1 391.96	1 086.25	1 043.55	762.58	1 126.32	555.26	1 308.98	1 210.73	1 107.51	840.92	382.26	728.96
February	873.42	1 370.74	1 088.58	1 044.36	761.42	1 125.38	555.05	1 304.64	1 204.33	1 103.22	839.35	383.04	726.68
March	875.82	1 362.13	1 084.14	1 048.03	760.43	1 133.35	553.04	1 300.32	1 209.59	1 103.94	840.34	382.89	728.92
April	878.58	1 371.94	1 094.02	1 055.09	763.68	1 139.38	554.51	1 306.44	1 212.81	1 111.34	841.98	384.71	730.83
May	879.61	1 381.16	1 095.12	1 059.17	762.15	1 136.62	553.97	1 319.82	1 206.75	1 111.16	840.99	385.76	732.74
June	882.02	1 380.73	1 097.46	1 057.39	765.92	1 139.94	556.76	1 315.44	1 207.88	1 112.04	844.87	387.85	734.66
July	884.08	1 383.70	1 102.23	1 059.01	766.26	1 148.33	554.90	1 323.79	1 218.62	1 116.57	846.52	389.67	738.88
August	882.54	1 383.11	1 096.98	1 060.88	765.40	1 148.09	553.73	1 321.84	1 218.38	1 114.20	845.20	389.22	738.49
September	886.83	1 394.03	1 100.97	1 063.08	767.63	1 151.67	553.73	1 330.10	1 221.38	1 118.74	849.81	393.52	739.12
October	889.93	1 403.35	1 113.67	1 069.19	770.38	1 154.79	555.27	1 340.15	1 214.38	1 122.71	851.78	392.08	746.24
November	888.71	1 398.53	1 108.09	1 065.75	770.18	1 153.00	557.17	1 347.12	1 221.86	1 120.68	853.10	394.89	743.91
December	891.99	1 402.77	1 104.76	1 070.82	772.28	1 155.33	557.93	1 355.19	1 222.61	1 123.56	854.41	394.16	745.82
2017													
January	893.71	1 415.04	1 117.20	1 074.26	773.47	1 159.22	556.82	1 372.87	1 213.09	1 129.44	855.07	395.72	747.62
February	894.54	1 425.24	1 111.50	1 074.48	772.24	1 160.39	553.52	1 364.52	1 228.59	1 132.10	858.69	396.01	751.12
March	895.57	1 439.10	1 112.54	1 071.84	772.58	1 161.81	555.94	1 362.21	1 221.58	1 135.08	859.02	399.10	750.80
April	900.25	1 440.54	1 121.51	1 083.24	779.70	1 163.76	563.78	1 379.40	1 230.09	1 140.04	861.98	401.42	752.39
May	901.97	1 443.72	1 123.34	1 079.77	779.85	1 164.15	562.03	1 386.30	1 230.09	1 138.23	861.00	403.00	753.34
June	904.03	1 438.69	1 130.38	1 083.65	782.60	1 166.88	563.58	1 397.91	1 243.81	1 138.96	863.63	403.51	755.25
July	906.78	1 464.65	1 128.66	1 091.21	783.98	1 173.00	564.20	1 400.58	1 243.18	1 142.20	867.24	403.77	756.05
August	907.82	1 445.00	1 128.10	1 085.69	784.32	1 171.05	564.23	1 391.76	1 250.63	1 141.20	868.56	405.59	758.58
September	909.29	1 450.05	1 135.10	1 089.36	784.10	1 170.78	564.23	1 396.60	1 255.13	1 148.04	869.86	406.12	762.70
October	910.57	1 456.89	1 138.59	1 093.26	783.98	1 161.42	565.44	1 386.96	1 264.50	1 145.52	869.88	406.90	763.65
November	915.98	1 462.83	1 145.82	1 095.93	793.94	1 179.14	572.16	1 387.54	1 268.62	1 151.95	873.17	408.47	765.87
December	919.08	1 472.00	1 152.28	1 093.03	792.12	1 183.72	570.06	1 406.01	1 279.53	1 153.08	878.13	411.60	767.46
2018													
January	918.82	1 459.51	1 147.98	1 095.07	792.81	1 172.73	572.86	1 398.35	1 287.05	1 152.39	878.10	409.24	768.73
February	922.88	1 477.52	1 160.53	1 100.44	794.88	1 175.07	574.42	1 409.40	1 287.80	1 162.02	881.43	411.08	768.73
March	925.98	1 480.22	1 158.17	1 099.80	796.61	1 178.87	575.04	1 414.80	1 289.63	1 167.45	883.37	412.64	773.80
April	928.05	1 485.32	1 169.00	1 104.95	797.99	1 174.29	576.60	1 424.51	1 295.70	1 165.31	885.72	412.36	777.83
May	931.16	1 487.19	1 173.94	1 100.38	801.78	1 185.12	579.70	1 423.80	1 303.59	1 172.88	889.35	414.73	776.97
June	933.23	1 509.39	1 167.77	1 105.53	805.14	1 192.94	582.82	1 422.33	1 307.44	1 176.50	891.00	416.30	779.74
July	935.30	1 491.29	1 178.85	1 108.23	806.87	1 188.33	583.44	1 431.00	1 305.38	1 180.48	889.95	417.60	780.69
August	939.44	1 506.09	1 176.78	1 111.51	809.37	1 196.07	583.08	1 445.44	1 312.99	1 184.80	895.29	418.91	781.96
September	941.85	1 515.24	1 178.87	1 110.03	808.74	1 197.30	584.63	1 470.15	1 313.25	1 184.80	893.24	418.08	784.82
October	943.58	1 508.39	1 172.92	1 108.13	811.15	1 195.40	585.82	1 472.62	1 321.49	1 190.62	895.54	421.25	789.21
November	943.59	1 507.36	1 171.06	1 111.39	815.24	1 201.98	589.26	1 480.82	1 327.66	1 188.77	897.18	419.84	791.12
December	949.79	1 513.97	1 201.59	1 117.80	816.00	1 205.88	588.44	1 498.83	1 329.91	1 190.94	901.56	423.02	793.67

Table 10-14. Production and Nonsupervisory Workers on Private Nonfarm Payrolls by NAICS Supersector

(Thousands, seasonally adjusted.)

Year and month	Total private	Mining and logging	Construc-tion	Manu-facturing	Trade, transportation, and utilities			Information	Financial activities	Profes-sional and business services	Education and health services	Leisure and hospitality	Other services
					Total	Wholesale trade	Retail trade						
1965	42 302	523	2 906	12 905	10 702	...	...	1 268	2 434	3 515	3 443	3 443	1 161
1966	44 292	517	2 977	13 703	11 095	...	...	1 334	2 492	3 715	3 623	3 607	1 230
1967	45 185	501	2 903	13 714	11 369	...	...	1 365	2 585	3 890	3 818	3 734	1 306
1968	46 519	491	2 986	13 908	11 688	...	...	1 394	2 700	4 067	4 008	3 898	1 379
1969	48 246	501	3 177	14 147	12 152	...	...	1 438	2 841	4 252	4 196	4 089	1 452
1970	48 180	496	3 158	13 490	12 388	...	...	1 422	2 922	4 321	4 305	4 185	1 494
1971	48 151	474	3 238	13 034	12 502	...	...	1 392	2 978	4 354	4 372	4 286	1 521
1972	49 971	494	3 425	13 497	12 954	2 920	7 257	1 437	3 066	4 518	4 531	4 467	1 583
1973	52 235	502	3 576	14 227	13 437	3 041	7 551	1 504	3 164	4 748	4 747	4 664	1 666
1974	52 846	550	3 469	14 040	13 700	3 148	7 673	1 516	3 217	4 907	4 941	4 766	1 740
1975	51 010	581	2 990	12 576	13 578	3 121	7 714	1 416	3 227	4 939	5 088	4 821	1 795
1976	52 916	606	2 999	13 127	14 038	3 212	8 048	1 459	3 300	5 153	5 309	5 046	1 880
1977	55 207	636	3 209	13 591	14 579	3 323	8 396	1 514	3 452	5 404	5 561	5 284	1 978
1978	58 188	658	3 544	14 150	15 329	3 509	8 861	1 586	3 645	5 717	5 874	5 588	2 099
1979	60 404	737	3 760	14 458	15 843	3 667	9 113	1 650	3 825	5 993	6 158	5 772	2 209
1980	60 377	785	3 623	13 667	15 907	3 708	9 158	1 626	3 957	6 197	6 447	5 850	2 318
1981	60 967	861	3 469	13 492	16 004	3 753	9 238	1 633	4 052	6 396	6 700	5 944	2 414
1982	59 475	834	3 208	12 315	15 821	3 662	9 254	1 564	4 055	6 421	6 822	5 976	2 458
1983	60 018	698	3 240	12 121	15 999	3 639	9 494	1 502	4 128	6 581	7 045	6 161	2 542
1984	63 333	714	3 614	12 821	16 797	3 821	9 964	1 631	4 289	6 918	7 385	6 491	2 672
1985	65 456	686	3 868	12 648	17 427	3 935	10 399	1 660	4 476	7 258	7 789	6 817	2 827
1986	66 825	577	3 984	12 449	17 769	3 941	10 704	1 663	4 698	7 532	8 130	7 066	2 957
1987	68 725	541	4 088	12 537	18 196	3 989	10 986	1 717	4 861	7 859	8 513	7 310	3 104
1988	71 059	545	4 199	12 765	18 771	4 132	11 306	1 775	4 894	8 256	8 985	7 587	3 280
1989	72 960	526	4 257	12 805	19 230	4 235	11 565	1 807	4 931	8 648	9 465	7 833	3 459
1990	73 721	538	4 115	12 669	19 032	4 198	11 308	1 866	4 973	8 889	9 784	8 299	3 555
1991	72 567	515	3 674	12 164	18 640	4 122	11 008	1 871	4 914	8 748	10 257	8 247	3 539
1992	72 854	478	3 546	12 020	18 506	4 071	10 931	1 871	4 924	8 971	10 607	8 406	3 526
1993	74 671	462	3 704	12 070	18 752	4 072	11 104	1 896	5 084	9 451	10 961	8 667	3 623
1994	77 478	461	3 973	12 361	19 392	4 196	11 502	1 928	5 220	10 078	11 397	8 979	3 689
1995	79 943	458	4 113	12 567	19 984	4 361	11 841	2 007	5 199	10 645	11 829	9 330	3 812
1996	81 888	461	4 325	12 532	20 325	4 423	12 057	2 096	5 322	11 161	12 195	9 565	3 907
1997	84 313	479	4 546	12 673	20 698	4 523	12 274	2 181	5 482	11 896	12 566	9 780	4 013
1998	86 516	473	4 807	12 729	21 059	4 605	12 440	2 217	5 692	12 566	12 903	9 947	4 124
1999	88 647	438	5 105	12 524	21 576	4 673	12 772	2 351	5 818	13 184	13 217	10 216	4 219
2000	90 543	446	5 295	12 428	21 965	4 686	13 040	2 502	5 819	13 790	13 487	10 516	4 296
2001	90 202	457	5 332	11 677	21 709	4 555	12 952	2 531	5 888	13 588	13 986	10 662	4 373
2002	88 645	436	5 196	10 768	21 337	4 474	12 774	2 398	5 964	13 049	14 471	10 576	4 449
2003	87 938	420	5 123	10 189	21 078	4 396	12 655	2 347	6 052	12 911	14 726	10 666	4 426
2004	89 212	440	5 309	10 072	21 319	4 444	12 788	2 371	6 052	13 287	14 984	10 955	4 425
2005	91 401	473	5 611	10 060	21 830	4 584	13 030	2 386	6 127	13 854	15 359	11 263	4 438
2006	93 727	519	5 903	10 137	22 166	4 724	13 110	2 399	6 312	14 446	15 783	11 568	4 494
2007	95 203	547	5 883	9 975	22 546	4 851	13 317	2 403	6 365	14 784	16 261	11 861	4 578
2008	94 610	574	5 521	9 629	22 337	4 822	13 134	2 388	6 320	14 585	16 777	11 873	4 606
2009	89 556	510	4 567	8 322	21 116	4 506	12 472	2 240	6 066	13 520	17 166	11 560	4 488
2010	88 875	525	4 172	8 077	20 874	4 378	12 425	2 170	5 942	13 699	17 452	11 507	4 458
2011	90 537	594	4 184	8 228	21 234	4 443	12 647	2 148	5 900	14 251	17 736	11 772	4 491
2012	92 715	641	4 246	8 400	21 617	4 562	12 793	2 164	5 986	14 802	18 165	12 154	4 541
2013	94 580	639	4 401	8 412	21 882	4 632	12 921	2 178	6 064	15 351	18 519	12 579	4 558
2014	96 794	658	4 631	8 568	22 281	4 707	13 113	2 221	6 157	15 795	18 859	12 981	4 642
2015	98 783	592	4 866	8 683	22 629	4 702	13 264	2 226	6 278	16 133	19 337	13 360	4 678
2016	100 531	476	5 062	8 666	22 882	4 696	13 424	2 235	429	16 479	19 839	13 749	4 715
2017	102 331	504	5 186	8 728	23 121	4 742	13 491	2 205	6 569	16 875	20 320	14 047	4 771
2018	104 310	544	5 437	8 899	23 386	4 698	13 521	2 275	6 638	17 121	20 788	14 384	4 839
2017													
January	101 567	474	5 156	8 676	23 032	4 717	13 501	2 220	6 529	16 759	20 094	13 902	4 734
February	101 784	482	5 193	8 689	23 044	4 722	13 494	2 213	6 530	16 801	20 158	13 932	4 742
March	101 826	488	5 186	8 695	23 017	4 727	13 462	2 207	6 531	16 845	20 177	13 937	4 743
April	101 921	497	5 171	8 706	23 019	4 732	13 456	2 199	6 538	16 856	20 217	13 968	4 750
May	101 987	503	5 165	8 710	23 010	4 731	13 448	2 194	6 544	16 870	20 258	13 947	4 759
June	102 222	512	5 170	8 714	23 061	4 744	13 466	2 200	6 569	16 921	20 291	14 017	4 767
July	102 339	512	5 158	8 701	23 073	4 746	13 469	2 196	6 578	16 952	20 340	14 059	4 770
August	102 489	515	5 172	8 735	23 087	4 757	13 475	2 193	6 587	16 978	20 386	14 061	4 775
September	102 493	518	5 178	8 740	23 227	4 757	13 521	2 186	6 595	16 986	20 398	13 971	4 779
October	102 980	513	5 175	8 763	23 262	4 750	13 527	2 189	6 608	16 846	20 474	14 245	4 805
November	103 119	511	5 255	8 793	23 307	4 757	13 550	2 235	6 609	16 840	20 511	14 250	4 808
December	103 239	517	5 255	8 809	23 308	4 767	13 528	2 231	6 611	16 840	20 538	14 277	4 814
2018													
January	103 304	518	5 330	8 822	23 240	4 673	13 493	2 263	6 597	16 870	20 595	14 264	4 805
February	103 549	523	5 376	8 837	23 300	4 680	13 533	2 260	6 611	16 915	20 632	14 285	4 810
March	103 703	530	5 379	8 847	23 330	4 686	13 536	2 269	6 618	16 949	20 667	14 302	4 812
April	103 853	536	5 396	8 864	23 334	4 671	13 543	2 276	6 619	17 007	20 682	14 316	4 823
May	104 103	540	5 430	8 870	23 384	4 678	13 570	2 277	6 627	17 073	20 718	14 350	4 834
June	104 294	548	5 441	8 900	23 367	4 681	13 532	2 278	6 634	17 116	20 779	14 386	4 845
July	104 418	546	5 454	8 909	23 389	4 696	13 524	2 276	6 630	17 153	20 807	14 413	4 841
August	104 624	553	5 467	8 914	23 425	4 711	13 524	2 278	6 646	17 199	20 866	14 428	4 848
September	104 660	552	5 475	8 923	23 433	4 712	13 508	2 272	6 659	17 234	20 883	14 375	4 854
October	104 921	562	5 492	8 944	23 445	4 716	13 488	2 282	6 669	17 294	20 914	14 459	4 860
November	105 051	558	5 491	8 966	23 495	4 730	13 508	2 284	6 668	17 306	20 931	14 489	4 863
December	105 244	559	5 517	8 987	23 488	4 744	13 497	2 285	6 672	17 340	20 984	14 541	4 871

. . . = Not available.

Table 10-15. Average Weekly Hours of Production and Nonsupervisory Workers on Private Nonfarm Payrolls by NAICS Supersector

(Hours per week, seasonally adjusted.)

Year and month	Total private	Mining and logging	Construc-tion	Manufacturing Average weekly hours	Manufacturing Overtime hours	Trade, transportation, and utilities Total	Trade, transportation, and utilities Wholesale trade	Trade, transportation, and utilities Retail trade	Informa-tion	Financial activities	Profes-sional and business services	Education and health services	Leisure and hospitality	Other services
1965	38.6	43.7	37.9	41.2	3.6	39.6	...	...	38.3	37.1	37.3	35.2	32.5	36.1
1966	38.5	44.1	38.1	41.4	3.9	39.1	...	...	38.3	37.2	37.0	34.9	31.9	35.8
1967	37.9	43.9	38.1	40.6	3.3	38.5	...	...	37.6	36.9	36.6	34.5	31.3	35.4
1968	37.7	44.0	37.8	40.7	3.6	38.2	...	...	37.6	36.8	36.3	34.1	30.8	35.0
1969	37.5	44.3	38.4	40.6	3.6	37.9	...	...	37.6	36.9	36.3	34.1	30.4	35.0
1970	37.0	43.9	37.8	39.8	2.9	37.6	...	...	37.2	36.6	35.9	33.8	30.0	34.7
1971	36.7	43.7	37.6	39.9	2.9	37.4	...	...	37.0	36.4	35.5	33.3	29.9	34.2
1972	36.9	44.1	37.0	40.6	3.4	37.4	39.8	35.1	37.3	36.4	35.5	33.3	29.7	34.2
1973	36.9	43.8	37.2	40.7	3.8	37.2	39.6	34.8	37.3	36.4	35.5	33.3	29.4	34.1
1974	36.4	43.7	37.1	40.0	3.2	36.8	39.2	34.3	37.0	36.3	35.3	33.1	29.1	33.9
1975	36.0	43.7	36.9	39.5	2.6	36.4	39.1	34.0	36.6	36.2	35.1	33.0	28.8	33.8
1976	36.1	44.2	37.3	40.1	3.1	36.3	39.1	33.8	36.7	36.2	34.9	32.7	28.5	33.6
1977	35.9	44.7	37.0	40.3	3.4	36.0	39.2	33.3	36.8	36.2	34.7	32.5	28.1	33.4
1978	35.8	44.9	37.3	40.4	3.6	35.6	39.2	32.7	36.8	36.1	34.6	32.3	27.7	33.2
1979	35.6	44.7	37.5	40.2	3.3	35.4	39.2	32.4	36.6	35.9	34.4	32.2	27.4	33.0
1980	35.2	44.9	37.5	39.6	2.8	35.0	38.8	31.9	36.4	36.0	34.3	32.1	27.0	33.0
1981	35.2	45.1	37.4	39.8	2.8	35.0	38.9	31.9	36.3	36.0	34.3	32.1	26.9	33.0
1982	34.7	44.1	37.2	38.9	2.3	34.6	38.7	31.7	35.8	36.0	34.2	32.1	26.8	33.0
1983	34.9	43.9	37.6	40.1	2.9	34.6	38.8	31.6	36.2	35.9	34.4	32.1	26.8	33.0
1984	35.1	44.6	38.2	40.6	3.4	34.7	38.9	31.6	36.6	36.2	34.3	32.0	26.7	32.9
1985	34.9	44.6	38.2	40.5	3.3	34.4	38.8	31.2	36.5	36.1	34.2	31.9	26.4	32.8
1986	34.7	43.6	37.9	40.7	3.4	34.1	38.7	31.0	36.4	36.1	34.3	32.0	26.2	32.9
1987	34.7	43.5	38.2	40.9	3.7	34.1	38.5	31.0	36.5	36.0	34.3	32.0	26.3	32.8
1988	34.6	43.3	38.2	41.0	3.8	33.8	38.5	30.9	36.1	35.6	34.2	32.0	26.3	32.9
1989	34.5	44.1	38.3	40.9	3.8	33.8	38.4	30.7	36.1	35.6	34.2	32.0	26.1	32.9
1990	34.3	45.0	38.3	40.5	3.9	33.7	38.4	30.6	35.8	35.5	34.2	31.9	26.0	32.8
1991	34.1	45.3	38.1	40.4	3.8	33.6	38.4	30.4	35.6	35.4	34.0	31.9	25.6	32.7
1992	34.2	44.6	38.0	40.7	4.0	33.8	38.5	30.7	35.8	35.6	34.0	32.0	25.7	32.6
1993	34.3	44.9	38.4	41.1	4.4	34.1	38.5	30.7	36.0	35.5	34.0	32.0	25.9	32.6
1994	34.5	45.3	38.8	41.7	5.0	34.3	38.8	30.9	36.0	35.5	34.1	32.0	26.0	32.7
1995	34.3	45.3	38.8	41.3	4.7	34.1	38.6	30.8	36.0	35.5	34.0	32.0	25.9	32.6
1996	34.3	46.0	38.9	41.3	4.8	34.1	38.6	30.7	36.3	35.5	34.1	31.9	25.9	32.5
1997	34.5	46.2	38.9	41.7	5.1	34.3	38.8	30.9	36.3	35.8	34.3	32.2	26.1	32.7
1998	34.5	44.9	38.8	41.4	4.9	34.2	38.6	30.9	36.6	36.0	34.3	32.2	26.2	32.6
1999	34.3	44.2	39.0	41.4	4.9	33.9	38.6	30.8	36.7	35.8	34.4	32.1	26.1	32.5
2000	34.3	44.4	39.2	41.3	4.7	33.8	38.8	30.7	36.8	35.9	34.5	32.2	26.1	32.5
2001	34.0	44.6	38.7	40.3	4.0	33.5	38.4	30.7	36.9	35.8	34.2	32.3	25.8	32.3
2002	33.9	43.2	38.4	40.5	4.2	33.6	38.0	30.9	36.5	35.6	34.2	32.4	25.8	32.1
2003	33.7	43.6	38.4	40.4	4.2	33.6	37.9	30.9	36.2	35.6	34.1	32.3	25.6	31.4
2004	33.7	44.5	38.3	40.8	4.6	33.5	37.8	30.7	36.3	35.6	34.2	32.4	25.7	31.0
2005	33.8	45.6	38.6	40.7	4.6	33.4	37.7	30.6	36.5	36.0	34.2	32.6	25.7	30.9
2006	33.9	45.6	39.0	41.1	4.4	33.4	38.0	30.5	36.6	35.8	34.6	32.5	25.7	30.9
2007	33.8	45.9	39.0	41.2	4.2	33.3	38.2	30.2	36.5	35.9	34.8	32.6	25.5	30.9
2008	33.6	45.1	38.5	40.8	3.7	33.2	38.2	30.0	36.7	35.9	34.8	32.4	25.2	30.8
2009	33.1	43.2	37.6	39.8	2.9	32.9	37.6	29.9	36.6	36.1	34.7	32.2	24.8	30.5
2010	33.4	44.6	38.4	41.1	3.8	33.3	37.9	30.2	36.3	36.2	35.1	32.1	24.8	30.7
2011	33.6	46.7	39.0	41.4	4.1	33.7	38.5	30.5	36.2	36.4	35.2	32.2	24.8	30.8
2012	33.7	46.6	39.3	41.7	4.2	33.8	38.7	30.6	36.0	36.8	35.3	32.3	25.0	30.7
2013	33.7	45.9	39.6	41.8	4.3	33.7	38.7	30.2	35.9	36.7	35.3	32.1	25.0	30.7
2014	33.7	47.4	39.6	42.0	4.5	33.6	38.6	30.0	35.9	36.8	35.6	32.0	25.1	30.7
2015	33.7	45.8	39.6	41.8	4.3	33.7	38.6	30.1	35.7	37.1	35.5	32.1	25.1	30.7
2016	33.6	45.3	39.7	41.9	4.3	33.5	38.6	29.7	35.6	36.9	35.4	32.2	24.9	30.8
2017	31.2	43.4	39.8	41.9	4.4	33.8	38.9	30.2	33.4	36.9	35.4	32.2	24.9	30.8
2018	33.7	46.8	39.8	42.1	4.5	33.9	38.9	30.4	35.6	37.0	35.3	32.2	24.9	30.8
2017														
January	33.6	45.5	39.5	41.9	3.9	33.6	38.8	29.9	35.9	36.9	35.4	32.0	24.8	30.8
February	33.3	45.6	38.7	41.6	4.2	33.2	38.5	29.5	35.5	36.7	35.0	32.1	24.4	30.6
March	33.3	45.3	38.7	41.6	4.2	33.2	38.4	29.6	35.3	36.6	35.0	32.0	24.7	30.6
April	33.7	46.0	39.9	41.9	4.2	33.8	38.8	30.3	35.9	37.0	35.5	32.1	24.9	30.8
May	33.6	45.9	39.9	41.9	4.3	33.8	38.8	30.2	35.8	37.0	35.4	32.1	24.8	30.7
June	3.7	45.6	39.7	41.9	4.3	33.8	38.8	30.2	35.8	37.0	35.5	32.2	24.9	30.8
July	34.0	46.3	40.3	41.7	4.4	34.2	39.3	30.7	36.5	37.4	35.9	32.4	25.6	31.1
August	33.8	40.6	40.6	42.1	4.4	34.0	38.9	30.5	6.5	36.7	35.4	32.2	25.2	30.9
September	33.6	40.1	40.1	42.0	4.4	33.8	38.9	30.2	35.7	36.6	35.2	32.2	24.4	30.6
October	34.0	40.3	40.3	42.2	4.4	34.1	39.6	30.4	36.3	37.5	35.9	32.4	25.1	30.9
November	33.7	39.6	39.6	41.9	4.7	34.0	39.2	30.4	35.6	36.9	35.4	32.3	24.9	30.7
December	33.7	40.0	40.0	41.7	5.0	33.9	39.1	30.3	35.8	36.9	35.3	32.2	25.0	30.8
2018														
January	33.6	46.1	39.7	41.9	4.5	33.9	38.9	30.4	35.5	37.0	35.1	32.2	24.8	30.6
February	33.8	47.0	39.8	42.2	4.7	33.9	38.9	30.3	35.7	37.0	35.4	32.3	25.0	30.7
March	33.7	46.8	40.0	42.2	4.6	33.9	39.0	30.4	35.9	37.0	35.3	32.2	24.9	30.7
April	33.8	47.0	40.0	42.4	4.7	33.9	38.9	30.3	35.9	37.0	35.3	32.3	24.9	30.8
May	33.8	47.4	40.3	42.0	4.5	34.0	39.0	30.4	35.5	37.0	35.3	32.3	24.9	30.8
June	33.8	47.8	39.9	42.1	4.5	34.0	39.0	30.5	35.4	37.0	35.4	32.3	25.0	30.8
July	33.8	46.7	40.1	42.2	4.5	34.1	38.9	30.5	35.6	37.1	35.4	32.2	24.9	30.7
August	33.8	47.0	39.8	42.2	4.5	34.0	39.0	30.4	35.6	37.1	35.4	32.2	24.9	30.9
September	33.7	46.4	39.6	42.1	4.5	33.9	38.9	30.4	35.7	37.0	35.3	32.2	24.7	30.8
October	33.7	46.4	39.4	42.1	4.5	33.9	38.8	30.4	35.5	37.1	35.4	32.2	24.8	30.9
November	33.7	46.1	39.3	42.0	4.5	33.9	38.9	30.2	35.5	36.9	35.3	32.2	24.8	30.9
December	33.7	46.7	39.9	42.0	4.5	33.7	38.8	30.1	35.6	36.9	35.4	32.2	24.8	30.9

. . . = Not available.

Table 10-16. Indexes of Aggregate Weekly Hours of Production and Nonsupervisory Workers on Private Nonfarm Payrolls by NAICS Supersector

(2002 = 100, seasonally adjusted.)

Year and month	Total private	Mining and logging	Construc-tion	Manu-facturing	Trade, transportation, and utilities			Information	Financial activities	Profes-sional and business services	Education and health services	Leisure and hospitality	Other services
					Total	Wholesale trade	Retail trade						
1965	54.4	121.6	55.2	122.1	59.1	...	...	55.5	42.6	29.5	25.9	41.0	29.4
1966	56.8	121.1	56.8	130.2	60.5	...	...	58.3	43.6	30.9	27.0	42.2	30.9
1967	57.0	117.0	55.4	127.8	61.1	...	...	58.6	44.9	31.9	28.1	42.9	32.4
1968	58.4	114.9	56.5	130.1	62.3	...	...	59.9	46.8	33.1	29.2	44.0	33.9
1969	60.4	118.1	61.0	131.9	64.3	...	...	61.8	49.4	34.6	30.5	45.6	35.6
1970	59.4	115.7	59.8	123.3	64.9	...	...	60.4	50.3	34.8	31.0	46.1	36.3
1971	59.0	109.9	61.0	119.3	65.1	...	...	58.7	51.1	34.7	31.1	46.9	36.5
1972	61.4	115.6	63.4	125.7	67.5	68.4	64.4	61.1	52.5	36.0	32.2	48.5	37.9
1973	64.2	117.0	66.7	132.9	69.6	71.0	66.4	64.1	54.2	37.8	33.7	50.3	39.9
1974	64.2	127.7	64.5	129.0	70.2	72.7	66.7	64.0	54.9	38.8	34.9	50.8	41.4
1975	61.1	134.9	55.2	113.9	69.0	71.8	66.4	59.1	55.0	38.9	35.8	50.9	42.6
1976	63.6	142.4	56.0	120.8	71.0	73.9	68.8	61.2	56.2	40.4	37.1	52.8	44.3
1977	66.1	151.1	59.4	125.8	73.1	76.7	70.7	63.5	58.8	42.1	38.6	54.4	46.3
1978	69.3	156.9	66.2	131.3	76.2	81.0	73.4	66.6	62.0	44.3	40.5	56.6	48.8
1979	71.6	175.2	70.6	133.3	78.3	84.6	74.7	69.0	64.7	46.3	42.3	57.9	51.2
1980	70.8	187.2	68.0	124.4	77.7	84.8	74.0	67.5	67.0	47.7	44.1	57.8	53.6
1981	71.4	206.2	64.9	123.1	78.0	86.0	74.5	67.7	68.6	49.2	45.9	58.7	55.8
1982	68.8	195.5	59.7	109.9	76.4	83.4	74.2	64.0	68.6	49.3	46.7	58.7	56.8
1983	69.8	162.8	61.0	111.6	77.2	83.3	76.0	62.1	69.8	50.7	48.3	60.4	58.9
1984	74.1	169.3	69.1	119.6	81.2	87.5	79.8	68.0	73.0	53.3	50.5	63.6	61.7
1985	76.0	162.7	73.9	117.5	83.5	89.9	82.2	69.1	76.1	55.7	53.1	66.0	65.0
1986	77.3	133.5	75.5	116.3	84.5	89.8	83.9	69.1	79.9	57.9	55.4	67.9	68.1
1987	79.5	125.2	78.1	117.8	86.5	90.5	86.3	71.5	82.3	60.4	58.1	70.5	71.5
1988	81.9	125.6	80.4	120.3	88.6	93.6	88.5	73.1	82.0	63.4	61.4	73.0	75.6
1989	83.9	123.2	81.7	120.3	90.5	95.9	90.0	74.4	82.6	66.4	64.7	74.9	79.7
1990	84.2	128.6	78.8	117.7	89.6	94.9	87.5	76.2	83.1	68.1	66.7	78.9	81.8
1991	82.4	123.8	70.1	112.8	87.5	93.3	84.8	76.1	82.0	66.7	69.8	77.3	81.1
1992	83.0	113.3	67.5	112.4	87.3	92.4	85.1	76.5	82.4	68.4	72.5	79.2	80.6
1993	85.3	110.3	71.3	113.9	89.1	92.4	86.3	78.0	84.9	72.0	74.9	82.1	82.8
1994	89.1	111.0	77.3	118.3	92.7	95.8	89.8	79.2	87.2	77.0	77.9	85.5	84.5
1995	91.5	110.3	79.9	119.0	95.1	99.2	92.3	82.5	86.9	81.2	80.8	88.5	87.0
1996	93.6	112.7	84.3	118.8	96.6	100.7	93.7	86.9	89.1	85.3	83.0	90.7	89.1
1997	97.0	117.6	88.6	121.4	98.9	103.4	95.9	90.4	92.3	91.5	86.4	93.4	91.9
1998	99.4	112.8	93.4	121.0	100.3	104.8	97.2	92.6	96.4	96.7	88.7	95.5	94.3
1999	101.5	102.9	99.7	119.0	101.9	106.2	99.5	98.5	98.1	101.6	90.6	97.8	96.2
2000	103.5	105.1	104.0	117.8	103.6	107.0	101.3	105.0	98.4	106.5	92.7	100.5	97.8
2001	102.1	108.3	103.2	108.1	101.5	102.9	100.5	106.7	99.3	104.0	96.5	100.6	99.0
2002	100.0	100.0	100.0	100.0	100.0	100.0	100.0	100.0	100.0	100.0	100.0	100.0	100.0
2003	98.7	97.4	98.4	94.5	98.6	98.0	98.9	97.0	101.3	98.7	101.6	100.0	97.4
2004	100.2	104.0	101.7	94.3	99.5	98.8	99.4	98.2	101.4	101.8	103.7	103.0	96.1
2005	102.8	114.7	108.3	93.9	101.6	101.8	100.8	99.4	103.7	106.4	106.8	106.2	96.2
2006	105.8	125.8	115.4	95.6	103.3	105.7	101.1	100.2	106.3	112.2	109.4	108.8	97.4
2007	107.3	133.5	114.7	94.4	104.7	109.2	101.8	100.1	107.6	115.4	113.0	110.8	99.3
2008	106.0	137.6	106.5	90.2	103.2	108.6	99.8	100.0	106.8	114.0	116.2	109.7	99.5
2009	98.8	117.2	85.9	76.1	96.7	99.9	94.3	93.5	103.1	105.3	117.9	105.1	96.0
2010	98.9	124.5	80.2	76.2	96.7	97.7	95.0	90.0	101.3	107.7	119.4	104.6	96.0
2011	101.4	147.4	81.7	78.3	99.6	100.7	97.7	88.9	101.2	112.5	122.0	106.9	96.8
2012	104.2	158.7	83.6	80.3	101.8	104.0	98.9	88.9	103.7	117.3	125.1	111.3	97.7
2013	106.1	155.7	87.2	80.8	102.6	105.6	98.7	89.3	104.6	121.4	126.8	115.1	98.2
2014	108.8	165.5	91.8	82.7	104.3	107.1	99.6	99.0	106.5	125.9	128.7	119.4	100.0
2015	110.9	144.1	96.5	83.3	106.2	106.8	100.9	90.8	109.6	128.3	132.3	122.7	100.7
2016	112.5	114.4	100.6	83.3	106.9	106.7	101.1	90.7	111.6	130.2	136.0	125.2	101.7
2017	114.7	214.8	103.1	83.9	109.1	108.8	179.7	90.4	114.2	132.6	139.6	127.7	102.9
2018	117.3	135.2	108.4	86.0	110.8	108.5	103.9	92.5	115.7	135.5	142.9	131.0	104.5
2017													
January	113.8	112.0	102.4	83.6	108.2	107.0	102.1	93.4	113.5	131.7	138.3	127.1	102.5
February	114.0	114.2	102.8	83.7	108.1	107.1	102.5	92.3	113.5	131.4	138.7	126.9	102.6
March	113.8	115.6	102.2	83.3	107.7	107.1	102.0	92.3	113.6	131.7	138.5	127.7	102.8
April	114.6	118.1	103.4	83.8	109.0	107.6	103.2	92.4	114.1	132.5	139.3	128.7	102.5
May	114.4	119.0	103.6	83.8	108.4	107.3	102.4	92.3	114.5	132.3	139.5	128.4	102.7
June	114.9	119.5	103.5	84.0	108.6	107.6	102.6	92.9	114.4	132.9	139.7	129.1	103.1
July	115.1	121.2	103.4	84.0	109.0	107.9	102.9	92.5	114.3	133.1	140.1	128.9	103.0
August	114.9	120.7	104.0	84.3	109.0	108.1	102.6	92.4	114.5	132.9	140.0	129.0	103.0
September	114.9	123.8	103.2	84.2	109.3	107.9	103.1	92.5	114.6	132.8	140.5	127.7	103.0
October	115.5	123.5	104.4	84.5	109.7	108.0	103.4	92.6	114.6	132.8	140.6	129.7	103.2
November	115.7	125.5	105.0	84.7	110.1	108.6	103.8	92.0	114.6	133.3	141.3	129.8	103.3
December	116.2	126.8	106.3	84.5	110.2	108.5	103.6	92.9	114.9	133.4	141.4	130.0	104.0
2018													
January	115.6	126.9	105.9	84.9	110.0	107.9	103.8	91.7	114.9	132.4	141.4	129.6	103.1
February	116.6	130.6	107.1	85.6	110.3	108.1	103.8	92.1	115.2	133.9	142.1	130.8	103.6
March	116.4	131.8	107.7	85.7	110.4	108.5	104.1	93.0	115.3	133.8	141.9	130.4	103.6
April	117.0	133.9	108.1	86.3	110.5	107.8	103.8	93.3	115.3	134.2	142.4	130.6	104.2
May	117.2	136.0	109.6	85.5	111.0	108.3	104.4	92.3	115.4	134.8	142.7	130.9	104.4
June	117.4	139.2	108.7	86.0	110.9	108.3	104.4	92.0	115.6	135.5	143.1	131.7	104.7
July	117.6	135.5	109.5	86.3	111.4	108.4	104.4	92.5	115.8	135.8	142.8	131.5	104.2
August	117.8	138.1	108.9	86.4	111.2	109.0	104.0	92.6	116.1	136.1	143.3	131.6	105.1
September	117.5	136.1	108.6	86.2	110.9	108.8	103.9	92.6	116.0	136.0	143.4	130.1	104.8
October	117.8	138.6	108.3	86.4	111.0	108.6	103.8	92.5	116.5	136.9	143.6	131.3	105.3
November	118.0	136.7	108.0	86.4	111.2	109.2	103.2	92.5	115.8	136.6	143.7	131.6	105.4
December	118.2	138.7	110.2	86.6	110.5	109.2	102.8	92.8	115.9	137.2	144.1	132.1	105.6

... = Not available.

Table 10-17. Average Hourly Earnings of Production and Nonsupervisory Workers on Private Nonfarm Payrolls by NAICS Supersector

(Dollars, seasonally adjusted.)

Year and month	Total private	Mining and logging	Construction	Manu-facturing	Trade, transportation, and utilities			Information	Financial activities	Professional and business services	Education and health services	Leisure and hospitality	Other services
					Total	Wholesale trade	Retail trade						
1965	2.63	2.87	3.23	2.49	2.94	...	...	4.47	2.38	3.28	2.12	1.17	1.25
1966	2.73	3.00	3.41	2.60	3.04	...	...	4.56	2.47	3.39	2.23	1.26	1.37
1967	2.85	3.14	3.63	2.71	3.15	...	...	4.68	2.58	3.51	2.36	1.37	1.49
1968	3.02	3.30	3.92	2.89	3.32	...	...	4.85	2.75	3.65	2.49	1.53	1.62
1969	3.22	3.54	4.30	3.07	3.48	...	...	5.05	2.92	3.84	2.68	1.69	1.81
1970	3.40	3.77	4.74	3.24	3.65	...	...	5.25	3.07	4.04	2.88	1.82	2.01
1971	3.63	3.99	5.17	3.45	3.86	...	...	5.53	3.23	4.26	3.11	1.95	2.24
1972	3.90	4.28	5.55	3.70	4.23	4.58	3.52	5.87	3.37	4.50	3.33	2.08	2.46
1973	4.14	4.59	5.89	3.97	4.46	4.80	3.69	6.17	3.55	4.72	3.54	2.20	2.67
1974	4.43	5.09	6.29	4.31	4.74	5.11	3.92	6.52	3.80	5.01	3.82	2.40	2.95
1975	4.73	5.68	6.78	4.71	5.01	5.45	4.14	6.92	4.08	5.29	4.09	2.58	3.21
1976	5.06	6.19	7.17	5.10	5.31	5.75	4.36	7.37	4.30	5.60	4.39	2.78	3.51
1977	5.44	6.70	7.56	5.55	5.67	6.12	4.65	7.84	4.58	5.95	4.72	3.03	3.84
1978	5.88	7.44	8.11	6.05	6.11	6.61	5.00	8.34	4.93	6.32	5.07	3.33	4.19
1979	6.34	8.20	8.71	6.57	6.56	7.12	5.34	8.86	5.31	6.71	5.44	3.63	4.56
1980	6.85	8.97	9.37	7.15	7.04	7.68	5.71	9.47	5.82	7.22	5.93	3.98	5.05
1981	7.44	9.89	10.24	7.87	7.55	8.28	6.09	10.21	6.34	7.80	6.49	4.36	5.61
1982	7.87	10.64	11.04	8.36	7.91	8.81	6.34	10.76	6.82	8.30	7.00	4.63	6.11
1983	8.20	11.14	11.36	8.70	8.23	9.27	6.60	11.18	7.32	8.70	7.39	4.89	6.51
1984	8.49	11.54	11.56	9.05	8.45	9.61	6.73	11.50	7.65	8.98	7.66	4.99	6.79
1985	8.74	11.87	11.75	9.40	8.60	9.88	6.83	11.81	7.97	9.28	7.97	5.10	7.10
1986	8.93	12.14	11.92	9.60	8.74	10.07	6.93	12.08	8.37	9.55	8.24	5.20	7.38
1987	9.14	12.17	12.15	9.77	8.92	10.32	7.02	12.30	8.73	9.05	0.50	5.00	7.00
1988	9.44	12.45	12.52	10.05	9.15	10.71	7.23	12.63	9.07	10.22	8.95	5.50	8.08
1989	9.80	12.90	12.98	10.35	9.46	11.12	7.46	12.99	9.54	10.69	9.45	5.76	8.58
1990	10.20	13.40	13.42	10.78	9.83	11.58	7.71	13.40	9.98	11.14	9.98	6.02	9.08
1991	10.51	13.82	13.65	11.13	10.08	11.95	7.89	13.90	10.43	11.50	10.48	6.22	9.39
1992	10.77	14.09	13.81	11.40	10.30	12.21	8.12	14.29	10.86	11.78	10.85	6.36	9.66
1993	11.05	14.12	14.04	11.70	10.55	12.57	8.36	14.86	11.38	11.96	11.19	6.48	9.90
1994	11.34	14.41	14.38	12.04	10.80	12.93	8.61	15.32	11.84	12.15	11.48	6.62	10.18
1995	11.65	14.78	14.73	12.34	11.10	13.34	8.85	15.68	12.31	12.53	11.78	6.79	10.51
1996	12.04	15.09	15.11	12.75	11.46	13.80	9.21	16.30	12.75	13.00	12.15	6.99	10.85
1997	12.51	15.57	15.67	13.14	11.90	14.41	9.59	17.14	13.28	13.57	12.53	7.32	11.29
1998	13.01	16.20	16.23	13.45	12.40	15.07	10.05	17.67	14.00	14.27	12.96	7.67	11.79
1999	13.49	16.33	16.80	13.85	12.82	15.62	10.45	18.40	14.55	14.85	13.40	7.96	12.26
2000	14.02	16.55	17.48	14.32	13.31	16.28	10.87	19.07	15.04	15.52	13.91	8.32	12.73
2001	14.54	17.00	18.00	14.76	13.70	16.77	11.29	19.80	15.65	16.33	14.59	8.57	13.27
2002	14.97	17.19	18.52	15.29	14.02	16.98	11.67	20.20	16.25	16.80	15.16	8.81	13.72
2003	15.37	17.56	18.95	15.74	14.34	17.36	11.90	21.01	17.21	17.21	15.57	9.00	13.84
2004	15.69	18.07	19.23	16.14	14.58	17.65	12.08	21.40	17.58	17.48	16.08	9.15	13.98
2005	16.12	18.72	19.46	16.56	14.92	18.16	12.36	22.06	17.98	18.08	16.63	9.38	14.34
2006	16.75	19.90	20.02	16.81	15.39	18.91	12.57	23.23	18.83	19.13	17.30	9.75	14.77
2007	17.42	20.97	20.95	17.26	15.78	19.59	12.75	23.96	19.67	20.15	18.01	10.41	15.42
2008	18.07	22.50	21.87	17.75	16.16	20.13	12.87	24.78	20.32	21.18	18.75	10.84	16.09
2009	18.61	23.29	22.66	18.24	16.48	20.84	13.01	25.45	20.90	22.35	19.35	11.12	16.59
2010	19.05	23.82	23.22	18.61	16.82	21.54	13.25	25.87	21.55	22.78	19.98	11.31	17.06
2011	19.44	24.50	23.65	18.93	17.15	21.97	13.51	26.62	21.93	23.12	20.63	11.45	17.32
2012	19.74	25.79	23.97	19.08	17.43	22.24	13.82	27.04	22.82	23.29	20.94	11.62	17.59
2013	20.13	26.81	24.22	19.30	17.75	22.63	14.02	27.99	23.87	23.69	21.30	11.78	18.00
2014	20.61	26.85	24.67	19.56	18.27	23.24	14.39	28.66	24.71	24.28	21.65	12.09	18.52
2015	21.03	26.48	25.50	19.91	18.67	23.63	14.82	29.05	25.34	24.79	22.09	12.41	19.01
2016	21.54	27.05	25.97	20.43	18.99	24.18	15.05	30.04	26.11	25.42	22.53	12.85	19.36
2017	22.13	27.66	26.44	20.95	19.35	24.75	15.11	30.69	26.52	26.09	22.96	13.45	19.98
2018	22.70	28.28	27.71	21.53	19.89	25.16	15.90	31.92	26.93	26.79	23.65	13.87	20.77
2017													
January	22.03	27.68	26.17	20.68	19.33	24.80	15.29	30.49	26.54	26.29	22.94	13.19	19.82
February	21.94	27.58	26.14	20.63	19.24	24.56	15.25	30.46	26.27	26.05	22.93	13.27	19.88
March	21.92	27.71	23.32	20.69	19.24	24.37	15.26	30.26	26.45	25.97	22.84	13.31	19.81
April	21.97	27.55	26.37	20.77	19.26	24.67	15.23	30.54	26.45	26.01	22.96	13.30	19.80
May	21.93	27.47	26.50	20.76	19.28	24.83	15.25	30.57	26.43	25.83	22.91	13.88	19.81
June	21.88	27.50	26.68	20.77	19.36	24.77	15.28	30.48	26.33	25.76	22.88	13.24	19.76
July	22.09	27.99	26.87	20.96	19.38	24.83	15.42	31.03	26.66	26.18	23.06	13.22	19.87
August	22.22	27.56	27.24	21.01	19.53	24.75	15.50	30.74	26.61	26.12	23.16	13.42	20.16
September	22.89	27.74	27.05	21.08	19.53	24.85	15.43	31.07	26.73	26.35	23.14	13.84	20.24
October	22.21	27.62	26.93	21.80	19.41	24.89	15.40	30.99	26.52	26.15	22.23	13.52	20.18
November	22.20	27.60	26.93	21.10	19.40	24.83	15.39	30.81	26.53	26.13	23.17	13.51	20.21
December	22.23	27.88	27.07	21.20	19.28	24.84	12.57	30.86	26.67	26.22	23.30	13.67	20.24
2018													
January	22.36	27.93	27.18	21.29	19.51	24.67	15.55	31.23	26.70	26.44	23.38	13.61	20.47
February	22.40	27.74	27.41	21.34	19.52	24.74	15.52	31.32	26.76	26.45	23.39	13.64	20.50
March	22.49	27.76	27.40	21.39	19.65	24.77	15.68	31.59	26.82	26.57	23.46	13.69	20.63
April	22.55	27.88	27.52	21.46	19.75	24.84	15.83	31.41	26.87	26.61	23.48	13.75	20.67
May	22.62	27.93	27.54	21.43	19.83	25.00	15.93	31.63	26.89	26.70	23.57	13.79	20.71
June	22.67	28.14	27.58	21.48	19.88	25.25	15.90	31.76	26.90	26.79	23.65	13.83	20.73
July	22.71	28.22	27.73	21.46	19.88	25.14	15.94	31.81	26.90	26.85	23.67	13.89	20.81
August	22.80	28.51	27.85	21.55	20.00	25.30	16.00	32.04	26.95	26.92	23.75	13.94	20.82
September	22.86	28.87	27.94	21.62	20.06	25.48	16.02	32.27	26.98	26.96	23.75	13.99	20.89
October	22.90	28.79	28.05	21.69	20.10	25.50	16.06	32.37	26.94	27.00	23.79	14.04	20.95
November	22.99	28.86	28.14	21.78	20.17	25.53	16.14	32.70	27.17	27.08	23.91	14.12	21.01
December	23.09	28.71	28.20	21.85	20.32	25.70	16.28	32.95	27.26	27.16	23.97	14.17	21.10

. . . = Not available.

Table 10-18. Average Weekly Earnings of Production and Nonsupervisory Workers on Private Nonfarm Payrolls by NAICS Supersector

(Dollars, seasonally adjusted.)

Year and month	Total private	Mining and logging	Construc-tion	Manu-facturing	Trade, transportation, and utilities			Information	Financial activities	Profes-sional and business services	Education and health services	Leisure and hospitality	Other services
					Total	Wholesale trade	Retail trade						
1965	101.51	125.48	122.35	102.69	116.36	. . .	. . .	171.24	88.50	122.30	74.84	38.06	45.30
1966	105.23	132.50	130.11	107.47	118.78	. . .	. . .	174.52	91.83	125.53	78.04	40.34	48.94
1967	108.07	137.76	138.55	109.85	121.57	. . .	. . .	176.09	95.35	128.34	81.34	43.06	52.69
1968	113.82	145.26	148.22	117.73	126.69	. . .	. . .	182.65	101.18	132.47	85.02	47.07	56.81
1969	120.70	157.11	164.88	124.74	132.13	. . .	. . .	190.18	107.91	139.37	91.46	51.29	63.44
1970	125.79	165.82	179.11	128.84	137.15	. . .	. . .	195.24	112.32	144.76	97.23	54.57	69.67
1971	133.22	174.18	194.70	137.59	144.11	. . .	. . .	204.50	117.60	151.51	103.57	58.39	76.58
1972	143.87	188.76	205.04	150.23	158.09	181.98	123.57	218.72	122.86	159.88	111.12	61.78	84.14
1973	152.59	201.43	219.32	161.59	165.61	190.33	128.16	230.48	129.32	167.40	117.85	64.92	91.06
1974	161.61	222.24	233.24	172.37	174.18	200.53	134.39	241.07	137.71	176.47	126.50	69.90	99.98
1975	170.29	248.38	249.70	185.69	182.61	212.85	140.99	253.21	147.75	185.81	134.83	74.49	108.67
1976	182.65	273.69	267.00	204.41	192.62	224.66	147.21	270.80	155.54	195.61	143.47	79.47	117.85
1977	195.58	299.46	279.61	224.01	203.87	239.57	154.51	288.29	165.47	206.78	153.38	85.20	128.12
1978	210.29	334.27	302.27	244.49	217.53	258.75	163.74	306.71	177.96	218.42	163.48	92.16	138.74
1979	225.69	366.58	326.55	263.99	232.36	278.74	173.18	324.51	190.72	231.16	174.80	99.47	150.47
1980	241.07	402.51	351.49	283.35	246.60	298.24	182.25	344.38	209.21	247.64	190.16	107.52	166.32
1981	261.53	446.10	382.75	312.74	263.93	322.08	193.97	370.36	228.21	267.46	208.06	117.49	184.77
1982	273.10	469.10	410.39	325.07	274.09	340.84	200.74	385.82	245.16	284.11	224.53	124.14	201.49
1983	286.43	488.89	427.00	348.89	284.76	359.98	208.81	404.89	262.83	298.74	237.40	130.82	214.87
1984	298.26	514.77	441.28	367.77	292.99	373.78	212.82	420.45	276.50	308.24	245.16	133.49	223.33
1985	304.62	529.62	448.35	380.60	295.81	383.22	213.38	430.86	287.66	317.41	254.19	134.71	232.81
1986	309.78	529.20	451.22	390.46	298.07	389.73	214.52	439.84	302.32	327.30	263.41	136.34	242.33
1987	317.39	529.72	463.78	400.00	304.10	397.39	218.05	450.77	314.01	337.57	273.49	139.45	252.58
1988	326.48	539.63	478.74	412.62	309.58	411.71	223.67	455.59	322.87	349.89	286.57	144.51	265.83
1989	338.34	568.55	497.43	423.58	319.40	427.40	229.14	468.77	339.46	365.93	302.70	150.29	282.09
1990	349.63	602.43	513.43	436.13	331.55	444.48	235.56	479.50	354.60	380.61	318.87	156.32	297.88
1991	358.46	625.46	520.41	449.83	339.10	459.17	240.13	495.14	369.57	391.06	334.07	159.20	306.75
1992	368.20	628.94	525.13	464.43	348.60	470.41	249.66	511.95	386.31	400.64	347.70	163.70	315.08
1993	378.89	634.77	539.81	480.89	359.51	484.46	256.89	535.19	403.68	406.20	358.44	167.54	322.69
1994	391.17	653.13	558.53	501.95	370.38	501.14	265.74	551.21	420.18	414.16	367.50	172.27	332.44
1995	400.04	670.40	571.57	509.26	378.79	515.14	272.63	564.92	437.26	426.54	376.94	175.74	342.36
1996	413.25	695.04	588.48	526.55	390.67	533.36	282.74	592.45	453.05	442.81	387.50	181.02	352.68
1997	431.86	720.07	609.48	548.26	407.63	559.39	295.94	622.37	475.03	465.51	403.63	190.66	368.63
1998	448.59	727.19	629.75	557.12	423.33	582.21	310.33	646.52	503.94	490.13	417.58	200.82	384.25
1999	463.15	721.77	655.11	573.29	434.31	602.77	321.71	675.47	520.74	510.99	429.98	208.05	398.77
2000	480.99	734.88	685.78	591.01	449.96	631.24	333.41	700.92	540.39	535.07	447.82	217.20	413.30
2001	493.74	757.96	695.86	595.19	459.53	643.45	346.16	731.18	560.46	557.84	471.36	220.73	428.64
2002	506.60	741.97	711.82	618.62	471.27	644.38	360.84	737.94	578.94	574.60	490.73	227.31	439.87
2003	517.82	765.94	727.00	636.03	481.14	657.29	367.18	760.84	611.82	587.02	503.27	230.49	434.41
2004	528.89	804.01	735.55	658.52	488.51	666.79	371.13	776.72	625.53	597.54	521.24	234.86	433.04
2005	544.05	853.87	750.37	673.30	498.46	685.00	377.58	805.11	646.48	618.71	541.78	241.36	443.40
2006	567.39	907.95	781.59	690.88	514.37	718.50	383.12	850.64	673.63	662.27	561.59	250.34	456.50
2007	589.27	962.63	816.23	711.53	525.91	748.94	385.00	874.45	706.52	700.82	586.41	265.54	477.06
2008	607.53	1 014.69	842.61	724.46	536.11	769.62	386.21	908.78	729.64	737.90	608.58	273.39	495.57
2009	616.01	1 006.67	851.76	726.12	541.88	784.49	388.57	931.08	754.90	775.81	622.73	275.95	506.26
2010	636.25	1 063.11	891.83	765.15	559.63	816.50	400.07	939.85	780.19	798.54	640.60	280.87	523.70
2011	653.19	1 144.64	921.84	784.29	577.71	845.44	412.09	964.85	798.68	813.37	665.05	283.82	532.63
2012	665.82	1 201.69	942.14	794.63	588.86	860.70	422.10	973.52	840.04	822.58	675.85	290.54	539.46
2013	677.67	1 230.00	958.67	807.51	597.51	875.82	423.00	1 005.49	875.22	837.27	683.48	294.34	553.12
2014	694.89	1 271.15	977.05	822.24	613.95	897.74	431.64	1 029.73	908.43	864.12	692.41	303.83	568.90
2015	708.90	1 211.91	998.02	832.05	628.76	911.41	445.52	1 038.10	939.46	878.71	709.25	311.32	583.54
2016	723.69	1 224.33	1 031.16	855.69	636.64	933.16	447.62	1 068.25	962.73	898.76	724.40	319.52	595.73
2017	742.48	1 267.91	969.01	872.52	654.05	962.29	462.06	1 097.75	981.24	923.51	741.48	331.08	614.38
2018	766.05	1 322.86	1 103.35	906.70	674.73	979.13	482.82	1 136.98	996.57	946.74	762.23	344.94	639.69
2017													
January	738.01	1 267.74	991.84	862.36	643.69	964.72	451.06	1 109.84	997.90	928.04	743.26	320.52	608.47
February	730.60	1 257.65	1 011.62	858.21	638.77	945.56	449.88	1 081.33	964.11	911.75	736.05	323.79	608.33
March	729.94	1 255.26	1 018.58	860.70	638.77	935.81	451.70	1 068.18	968.07	914.98	730.88	328.76	606.19
April	746.98	1 272.82	1 026.38	860.69	660.62	976.86	466.34	1 124.94	1 007.30	920.75	742.25	333.50	618.45
May	736.85	1 248.98	1 062.65	870.26	651.33	952.15	460.55	1 081.47	967.78	918.51	734.72	331.82	606.57
June	739.21	1 253.04	1 085.88	873.15	656.20	953.32	464.21	1 087.01	966.74	923.71	736.37	333.14	608.30
July	744.43	1 276.45	1 060.09	876.54	656.98	967.20	464.20	1 102.28	978.88	926.20	742.21	331.33	614.77
August	743.90	1 262.63	1 061.28	878.22	656.06	967.20	462.04	1 095.06	981.17	924.86	743.82	331.23	617.23
September	746.63	1 268.22	1 064.19	874.87	658.76	969.93	465.68	1 108.59	982.28	919.42	745.75	330.02	617.99
October	748.14	1 276.97	1 062.55	884.52	659.36	967.73	466.32	1 103.36	985.97	945.97	742.10	334.06	619.53
November	751.37	1 282.51	106.92	884.94	663.34	971.64	471.23	1 102.18	984.94	922.39	749.04	336.65	623.08
December	753.74	1 292.70	1 076.19	885.77	664.70	975.30	471.50	1 108.73	989.75	925.57	751.30	338.14	623.70
2018													
January	751.30	1 287.57	1 079.05	892.05	661.39	959.66	472.72	1 108.67	987.90	928.04	752.84	337.53	626.38
February	757.12	1 303.78	1 090.92	900.55	661.73	962.39	470.26	1 118.12	990.12	936.33	755.50	341.00	629.35
March	757.91	1 299.17	1 096.00	902.66	666.14	966.03	476.67	1 134.08	992.34	937.92	755.41	340.88	633.34
April	762.19	1 310.36	1 100.80	909.90	669.53	966.28	479.65	1 127.62	994.19	939.33	758.40	342.38	636.64
May	764.56	1 323.88	1 109.86	900.06	674.22	975.00	484.27	1 122.87	994.93	942.51	761.31	343.37	637.87
June	766.25	1 345.09	1 100.44	904.31	675.92	984.75	484.95	1 124.30	995.30	948.37	763.90	345.75	638.48
July	767.60	1 317.87	1 111.97	905.61	677.91	977.95	486.17	1 132.44	997.99	950.49	762.17	345.86	638.87
August	770.64	1 339.97	1 108.43	909.41	680.00	986.70	486.40	1 140.62	999.85	952.97	764.75	347.11	643.34
September	770.38	1 339.57	1 106.42	910.20	680.03	991.17	487.01	1 152.04	998.26	951.69	764.75	345.55	643.41
October	771.73	1 335.86	1 105.17	913.15	681.39	989.40	488.22	1 149.14	999.47	955.80	766.04	348.19	647.36
November	774.76	1 330.45	1 105.90	914.76	683.76	993.12	487.43	1 160.85	1 002.57	955.92	769.90	350.18	649.21
December	778.13	1 340.76	1 125.18	917.70	684.78	997.16	490.03	1 173.02	1 005.89	961.46	771.83	351.42	651.99

. . . = Not available.

SECTION 10C: OTHER SIGNIFICANT LABOR MARKET DATA

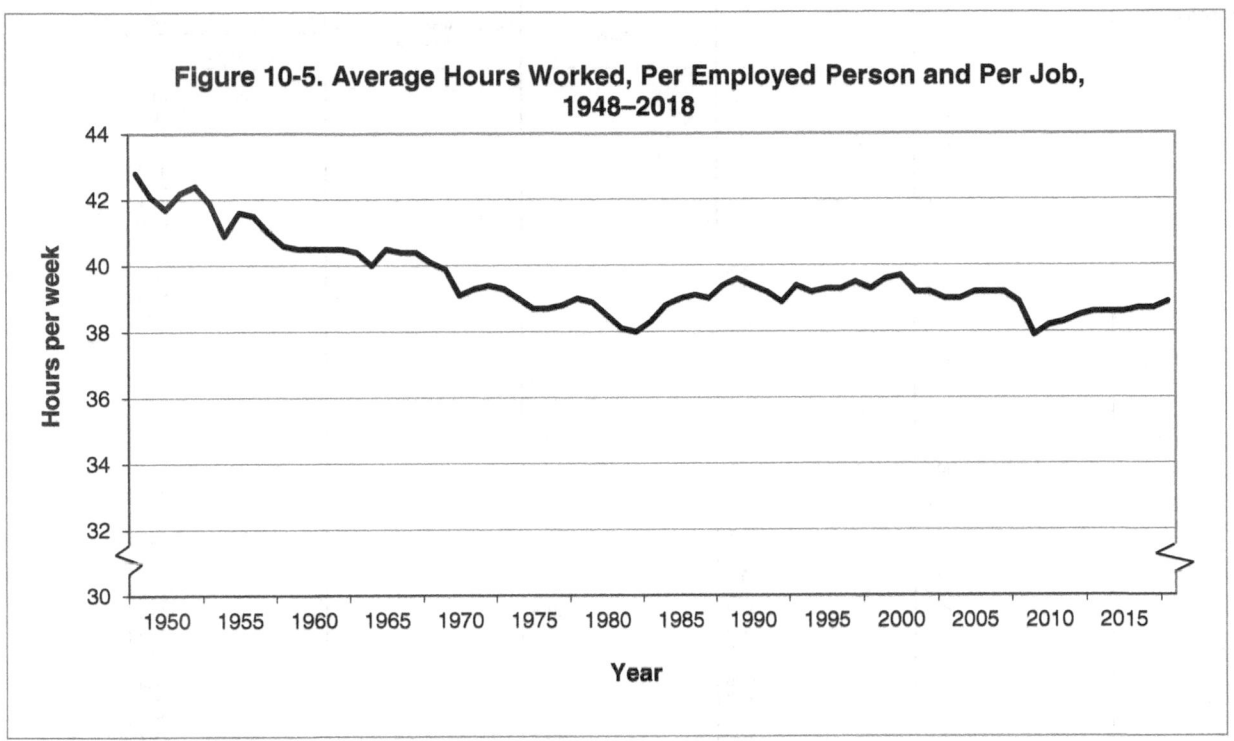

Figure 10-5. Average Hours Worked, Per Employed Person and Per Job, 1948–2018

- This section of *Business Statistics* includes indicators of hours and earnings from the Current Population Statistics (household) survey that are, for some purposes, more appropriate than similarly-named indicators produced by the payroll survey and shown in the previous section. It also includes annual employment data from the Bureau of Economic Analysis, produced as part of the National Income and Product Accounts (NIPA), that provide a consistent historical record of U.S. employment from 1929 through 2000.

- Figure 10-5 above shows annual averages of hours worked per week reported by all persons in the household survey employed in nonagricultural industries. Along with the recession dips expected in any measure of the average workweek, it displays a clear downward trend through the early 1970s, but no trend since then. The recession low in 2009 was slightly lower than the recession low in 1982. Hours in 2000 were the highest since 1969. However, since 2000, the number of hours at work has declined by almost one hour. The hours per week was unchanged between 2016–2017 at 38.7 and increased slightly in 2018 to 38.9 hours. (Table 10-19)

- Average weekly hours for nonfarm production and nonsupervisory workers are the workweeks reported by employers in the Current Employment Statistics (payroll) survey and they pertain to jobs, not to workers. (Table 10-15) This CES measure of "hours per job" represents many of the same workers as those in the "hours per employed person" measure. But it is increasingly pulled downward by the increasing prevalence of part-time jobs. It is not a good indicator of either the level or the trend of the amount of work being done each week by the typical American worker. (Table 10-19)

- A look at some of the other categories in Table 10-19 confirms the absence of any significant downtrend in the workers' workweek in recent years. However, there is a historical downtrend particularly for men. Men reported working an average of 41.7 hours at the pre-recession peak in 2007—two hours and six minutes below the peak of 43.8 hours in 1956. On the other hand, women reported working 36.4 hours in 2018—just 6 minutes below the peak in 1956. (Table 10-19)

Table 10-19. Hours at Work, Current Population Survey

(Hours per week.)

Year and month	Hours per person, all industries				Hours per person, nonagricultural industries	
	Total	Men	Women	Persons who usually work full-time	All workers	
					Total	Usually work full-time
1948	42.8	...	...	...	41.6	...
1949	42.1	...	...	...	40.9	...
1950	41.7	...	...	...	40.7	...
1951	42.2	...	...	...	41.3	...
1952	42.4	...	...	...	41.6	...
1953	41.9	...	...	...	41.1	...
1954	40.9	...	...	...	40.0	...
1955	41.6	...	...	...	40.9	...
1956	41.5	43.8	36.5	...	40.9	...
1957	41.0	43.4	36.1	...	40.5	...
1958	40.6	42.9	35.8	...	40.1	...
1959	40.5	42.8	35.6	...	40.0	...
1960	40.5	43.0	35.4	...	40.0	...
1961	40.5	43.0	35.3	...	40.1	...
1962	40.5	43.2	35.2	...	40.1	...
1963	40.4	43.2	35.1	...	40.1	...
1964	40.0	42.8	34.7	...	39.7	...
1965	40.5	43.3	35.1	...	40.2	...
1966	40.4	43.2	35.2	...	40.1	...
1967	40.4	43.3	35.2	...	40.0	...
1968	40.1	43.0	34.9	...	39.7	...
1969	39.9	42.9	34.9	...	39.5	...
1970	39.1	42.0	34.2	...	38.7	...
1971	39.3	42.2	34.3	...	38.9	...
1972	39.4	42.3	34.5	...	39.0	...
1973	39.3	42.4	34.4	...	39.0	...
1974	39.0	42.0	34.3	...	38.7	...
1975	38.7	41.6	34.1	...	38.4	...
1976	38.7	41.7	34.1	...	38.4	...
1977	38.8	41.9	34.2	...	38.5	...
1978	39.0	42.1	34.5	...	38.7	...
1979	38.9	42.0	34.5	...	38.6	...
1980	38.5	41.5	34.5	...	38.3	...
1981	38.1	41.1	34.1	...	37.9	...
1982	38.0	40.9	34.1	...	37.7	...
1983	38.3	41.2	34.5	...	38.1	...
1984	38.8	41.8	34.9	...	38.6	...
1985	39.0	42.0	35.2	...	38.9	...
1986	39.1	42.1	35.4	...	38.9	...
1987	39.0	42.0	35.3	...	38.8	...
1988	39.4	42.4	35.7	...	39.3	...
1989	39.6	42.6	35.8	...	39.4	...
1990	39.4	42.3	35.8	...	39.3	...
1991	39.2	42.0	35.8	...	39.1	...
1992	38.9	41.7	35.6	...	38.8	...
1993	39.4	42.2	36.0	...	39.3	...
1994	39.2	42.2	35.5	43.4	39.1	43.3
1995	39.3	42.3	35.6	43.4	39.2	43.2
1996	39.3	42.3	35.7	43.3	39.2	43.2
1997	39.5	42.4	36.0	43.4	39.4	43.3
1998	39.3	42.2	35.8	43.2	39.2	43.1
1999	39.6	42.4	36.2	43.4	39.5	43.3
2000	39.7	42.5	36.4	43.4	39.6	43.3
2001	39.2	41.9	36.1	42.9	39.2	42.8
2002	39.2	41.8	36.0	42.9	39.1	42.8
2003	39.0	41.7	35.9	42.9	39.0	42.7
2004	39.0	41.7	35.9	42.9	39.0	42.8
2005	39.2	41.8	36.1	42.9	39.1	42.8
2006	39.2	41.8	36.2	42.9	39.2	42.8
2007	39.2	41.7	36.1	42.8	39.1	42.7
2008	38.9	41.3	36.1	42.6	38.8	42.5
2009	37.9	40.2	35.3	41.9	37.8	41.8
2010	38.2	40.5	35.5	42.2	38.1	42.2
2011	38.3	40.6	35.6	42.4	38.2	42.3
2012	38.5	40.8	35.8	42.5	38.4	42.4
2013	38.6	40.9	36.0	42.6	38.5	42.5
2014	38.6	41.0	35.9	42.5	38.6	42.5
2015	38.6	40.9	35.9	42.4	38.5	42.3
2016	38.7	41.0	36.2	42.5	38.7	42.4
2017	38.7	40.9	36.2	42.4	38.6	42.3
2018	38.9	41.1	36.4	42.5	38.9	42.5

. . . = Not available.

Table 10-20. Median Usual Weekly Earnings of Full-Time Wage and Salary Workers

(Current dollars, [second quartile] except as noted; not seasonally adjusted.)

Year and quarter	Total, 16 years and over	Sex and age								Race and ethnicity			
		Men, 16 years and over				Women, 16 years and over				White	Black or African American	Asian	Hispanic or Latino ethnicity
		Total	16 to 24 years	25 to 54 years	55 years and over	Total	16 to 24 years	25 to 54 years	55 years and over				
1980	262	313	208	...	...	201	167	...	...	269	212	...	209
1981	284	340	218	...	...	219	180	...	...	291	235	...	223
1982	302	364	225	...	...	239	192	...	...	310	245	...	240
1983	313	379	223	...	...	252	198	...	...	320	261	...	250
1984	326	392	231	...	...	265	203	...	...	336	269	...	259
1985	344	407	241	...	...	277	211	...	...	356	277	...	270
1986	359	419	246	...	...	291	219	...	...	371	291	...	277
1987	374	434	257	...	...	303	227	...	...	384	301	...	285
1988	385	449	262	...	...	315	235	...	...	395	314	...	290
1989	399	468	271	...	...	328	246	...	...	409	319	...	298
1990	412	481	282	...	...	346	254	...	...	424	329	...	304
1991	426	493	285	...	...	366	266	...	...	442	348	...	312
1992	440	501	284	...	...	380	267	...	...	458	357	...	321
1993	459	510	288	...	...	393	273	...	...	475	369	...	331
1994	467	522	294	...	...	399	276	...	...	484	371	...	324
1995	479	538	303	...	...	406	275	...	...	494	383	...	329
1996	490	557	307	...	...	418	284	...	...	506	387	...	339
1997	503	579	317	...	...	431	292	...	...	519	400	...	351
1998	523	598	334	...	...	456	305	...	...	545	426	...	370
1999	549	618	356	...	...	473	324	...	...	573	445	...	385
2000	576	641	375	690	714	493	344	519	497	590	474	615	399
2001	596	670	391	717	742	512	353	547	519	610	491	000	417
2002	608	679	391	727	773	529	367	571	556	623	498	658	424
2003	620	695	398	736	797	552	371	584	584	636	514	693	440
2004	638	713	400	755	816	573	375	599	601	657	525	708	456
2005	651	722	409	763	829	585	381	610	619	672	520	753	471
2006	671	743	418	784	870	600	395	624	641	690	554	784	486
2007	695	766	443	809	902	614	409	644	657	716	569	830	503
2008	722	798	461	843	920	638	420	666	687	742	589	861	529
2009	739	819	458	857	944	657	424	683	706	757	601	880	541
2010	747	824	443	854	961	669	422	700	716	765	611	855	535
2011	756	832	455	864	976	684	421	713	738	775	615	866	549
2012	768	854	468	890	987	691	416	719	752	792	621	920	568
2013	776	860	479	891	999	706	423	733	766	802	629	942	578
2014	791	871	493	899	1 007	719	451	746	773	816	639	953	594
2015	809	895	510	921	1 054	726	450	757	774	835	641	993	604
2016	832	915	512	943	1 082	749	486	778	801	862	678	1 021	624
2017	860	941	547	977	1 084	770	499	801	843	890	682	1 043	655
2018	886	973	569	1 005	1 123	788	528	825	847	916	694	1 093	681
2011													
1st quarter	755	829	470	856	970	683	426	709	741	774	604	831	549
2nd quarter	753	825	446	863	965	689	417	714	744	770	623	872	565
3rd quarter	753	827	440	869	964	673	422	707	737	772	616	869	545
4th quarter	764	843	466	893	1 005	688	420	721	730	786	621	880	537
2012													
1st quarter	769	848	469	904	980	697	421	723	761	793	635	918	567
2nd quarter	771	865	460	918	991	689	409	722	748	792	637	930	576
3rd quarter	758	828	459	891	960	685	414	715	755	780	606	915	556
4th quarter	775	875	480	926	1 016	692	422	717	746	802	615	910	571
2013													
1st quarter	773	867	487	914	969	704	419	729	754	802	622	951	575
2nd quarter	776	860	479	913	1 018	707	422	735	769	799	634	973	572
3rd quarter	771	847	452	904	984	698	414	729	765	794	630	922	587
4th quarter	786	869	492	915	1 036	713	449	739	776	813	632	916	576
2014													
1st quarter	796	872	480	925	979	722	434	750	767	819	646	955	593
2nd quarter	780	857	481	914	993	716	449	744	768	802	649	954	588
3rd quarter	790	870	498	920	1 034	715	448	743	778	816	638	945	598
4th quarter	799	882	508	929	1 019	724	469	749	784	823	621	959	600
2015													
1st quarter	808	895	491	944	1 029	730	461	757	768	835	650	966	590
2nd quarter	801	866	497	942	1 032	726	444	753	782	829	647	965	601
3rd quarter	803	889	517	946	1 069	721	442	757	781	829	624	974	602
4th quarter	825	907	543	958	1 092	729	453	764	768	847	643	1 091	624
2016													
1st quarter	830	912	511	963	1 102	750	488	771	821	857	673	1 032	612
2nd quarter	824	909	505	967	1 049	744	470	774	785	854	677	1 021	618
3rd quarter	827	911	510	971	1 095	745	480	782	793	854	685	1 010	632
4th quarter	849	927	522	977	1 084	758	506	788	810	881	675	1 022	646
2017													
1st quarter	865	950	558	998	1 045	765	489	795	828	894	679	1 019	649
2nd quarter	859	934	524	995	1 046	780	496	811	859	886	689	1 103	657
3rd quarter	859	937	527	995	1 125	767	500	797	859	887	696	1 010	655
4th quarter	957	946	581	998	1 116	769	512	799	818	891	666	1 061	657
2018													
1st quarter	881	965	563	1 001	1 097	783	545	819	819	911	696	1 066	675
2nd quarter	876	959	528	991	1 117	780	511	811	868	907	683	1 083	674
3rd quarter	887	973	575	1 012	1 111	796	515	845	826	915	686	1 128	689
4th quarter	900	993	609	1 014	1 166	794	539	825	873	931	712	1 095	684

... = Not available.

NOTES AND DEFINITIONS, CHAPTER 10

General note on monthly employment data

This chapter presents data from two different data sets that measure employment monthly. Both are compiled and published by the Bureau of Labor Statistics (BLS), but each set has different characteristics. Users should be aware of these dissimilarities and the consequent differences in the appropriate uses and interpretations of data from the two systems.

One set of monthly employment estimates comes from the Current Population Survey (CPS), a large sample survey of approximately 60,000 U.S. households. The numbers in the sample are expanded to match the latest estimates of the total U.S. population. These are the most comprehensive estimates in their scope—that is, in the universe that they are designed to measure. These estimates represent all civilian workers, including the following groups that are excluded by definition from the other set of estimates: all farm workers; household workers (domestic servants); nonagricultural, nonincorporated self-employed workers; and nonagricultural unpaid family workers.

However, official CPS data are characterized by periodic discontinuities, which occur when new benchmarks for Census measures of the total population are introduced. These updates take place in a single month—usually January—and the official data for previous months are typically not modified to provide a smooth transition. Therefore, shorter-term comparisons (for a year or two or for a business cycle phase) will be misleading if such a discontinuity is included in the period. A recent example will illustrate. Beginning with January 2017, the updated controls decreased the estimated size of the civilian noninstitutional population 16 years and over by 800,000, the civilian labor force by 506,000, employment by 488,000, unemployment by 18,000, and the number of persons not in the labor force by 294,000. The total unemployment rate, the labor force participation rate, and the employment-population ratio were not affected. Such discontinuities occur throughout the history of the series.

> For users who would like to examine monthly CPS data in which these discontinuities have been smoothed, BLS now provides unofficial smoothed estimates of total labor force and total employment from January 1990 through December 2012 on its Web site, <http://www.bls.gov>. These two series are shown in Table 10-6.

The CPS is a count of persons employed, rather than a count of jobs. A person is counted as employed in this data set only once, no matter how many jobs he or she may hold. The CPS count is limited to persons 16 years of age and over.

The second set of employment estimates—the payroll survey—comes from a very large sample survey of about 142,000 business and government employers at approximately 689,000 worksites, the Current Employment Statistics (CES) survey. It is benchmarked annually to a survey of all employers. Benchmark data are introduced with a smooth adjustment back to the previous benchmark, thus preserving the continuity of the series and making it more appropriate for measurement of employment change over a year or two, a business cycle, or other short- to medium-length periods. This sample is much larger than the CPS sample, including about one-third of all nonfarm payroll employees, and consequently the threshold of statistical significance for changes is lower. The minimum size of the over-the-month change required to be statistically significant is about 115,000 jobs in the payroll survey, versus about 500,000 persons in the household survey.

Over recent years, benchmark revisions to payroll survey employment have ranged in magnitude from an upward revision of 0.6 percent, in 2006, to the March 2009 benchmark, which reduced the seasonally adjusted level of payroll employment in that month by 902,000 persons, or 0.7 percent. The most recent revision increased the level of employment in March 2018 increased employment by 135,000, or 0.1 percent. In absolute terms (that is, disregarding the sign), revisions have ranged from 0.1 to 0.7 percentage points over the latest 11 years and averaged 0.195 percentage points. Within that range, relatively large upward revisions have been seen for strong economies, and relatively large downward revisions for weak ones.

The scope of the CES survey is smaller than that of the CPS survey, as it is limited to wage and salary workers on nonfarm payrolls. There is also a significant definitional difference, because the CES survey is a count of jobs. Thus, a person with more than one nonfarm wage or salary job is counted as employed in each job. In addition, workers are not classified by age; as a result, there may be some workers younger than 16 years old in the job count.

Persons with a job but not at work (absent due to bad weather, work stoppages, personal reasons, and the like) are included in the household survey. However, they are excluded from the payroll survey if on leave without pay for the entire payroll period.

In addition to the differences in definitions and scope between the two series, there are also differences in sample design, collection methodology, and the sampling variability inherent in the surveys.

The payroll survey provides the most reliable and detailed information on employment by industry (for example, the data shown in Table 15-1). In addition, it provides data on weekly hours per job and hourly and weekly earnings per job.

The CPS employment estimates provide information not collected in the CES on employment by demographic characteristics, such as age, race, and Hispanic ethnicity; by education levels; and by occupation. A few of these tabulations are shown in *Business Statistics*. Many more breakdowns, in richer detail, can be found in the *Handbook of U.S. Labor Statistics*, also published by Bernan Press.

The differences between the two employment measurement systems are discussed in an article on the BLS Web site, "Employment from the BLS household and payroll surveys: summary of recent trends," which is updated monthly along with the release of the monthly data, and can be found under "Publications" on the CPS homepage at http://www.bls.gov. This article also includes references to further relevant studies.

Additional annual employment data

In addition to the CPS and CES monthly and annual data series, a third set of historical, annual-only estimates of employment is presented at the end of this chapter. These estimates of "full-time and part-time employees" by industry are compiled by the U.S. Department of Commerce, Bureau of Economic Analysis, as part of the National Income and Product Accounts (NIPAs). These estimates provide the most comprehensive and consistent employment estimates for the years before 1948.

Further detail on the historical and other characteristics of all of these employment series is presented below in the notes associated with specific data tables.

TABLES 10-1 THROUGH 10-5

Labor Force, Employment, and Unemployment

SOURCE: U.S. DEPARTMENT OF LABOR, BUREAU OF LABOR STATISTICS (BLS)

Labor force, employment, and unemployment data are derived from the Current Population Survey (CPS), a sample survey of households conducted each month by the Census Bureau for the Bureau of Labor Statistics (BLS). The data pertain to the U.S. civilian noninstitutional population age 16 years and over.

Due to changes in questionnaire design and survey methodology, data for 1994 and subsequent years are not fully comparable with data for 1993 and earlier years. Additionally, discontinuities in the reported number of persons in the population, and consequently in the estimated numbers of employed and unemployed persons and the number of persons in the labor force, are introduced whenever periodic updates are made to U.S. population estimates.

For example, population controls based on Census 2000 were introduced beginning with the data for January 2000. These data are therefore not comparable with data for December 1999 and earlier. Data for 1990 through 1999 incorporate 1990 census–based population controls and are not comparable with the

preceding years. An additional large population adjustment was introduced in January 2004, making the data from that time forward not comparable with data for December 2003 and earlier; further adjustments have been made in each subsequent January and other discontinuities have been introduced in various earlier years, usually with January data. See "Notes on the Data," below, for information on adjustments in other years, including the incorporation of the 2010 Census count beginning in January 2012.

For the most part, these population adjustments distort comparisons involving the <u>numbers of persons</u> in the population, labor force, and employment. They generally have negligible effects on the <u>percentages</u> that comprise the most important features of the CPS: the unemployment rates, the labor force participation rates, and the employment-population ratios.

BLS now makes available unofficial smoothed data for the total number of persons in the civilian labor force and the number of persons employed for 1990 through 2012, which introduce the population adjustments gradually within the period shown. These data are shown in Table 10-6.

Beginning with the data for January 2000, data classified by industry and occupation use the North American Industry Classification System (NAICS) and the 2000 Standard Occupational Classification System. This creates breaks in the time series between December 1999 and January 2000 for occupational and industry data at all levels of aggregation. Since the recent history is so short, most CPS industry and occupation data have been dropped from *Business Statistics* in favor of other important and economically meaningful data for which a longer consistent history can be supplied. However, detailed employment data by occupation and industry can be found in Bernan Press's *Handbook of U.S. Labor Statistics*.

Pre-1948 data

The Census Bureau began the survey of households that provides labor force data in 1942, and the Census of Population supplied data for 1940. For the years before that, annual data for labor force, employment, and unemployment are retrospective estimates. The 1929-1947 estimates were made by BLS. The 1890-1929 estimates shown in Table 10-1B were made by Stanley Lebergott. Both of those used trend interpolation of labor force participation between Census years, and the Census and Lebergott estimates are identical in 1929.

From 1929 to 1947, the data collected pertained to persons 14 years of age and over. From 1947 to the present the survey pertains to persons 16 years of age and over. Table 10-1C shows summary data for both age definitions in the overlap year, 1947. In that year, the unemployment rates are the same (3.9 percent) for both age definitions. However, the labor force participation rate and the employment/population ratio are higher when the

14- and 15-year-olds are excluded. Naturally, the raw numbers for population, labor force, and employment are smaller when the 14- and 15-year-olds are excluded.

The 1940 Census and the BLS data for 1931 through 1942 did not treat government work relief employment as employment; persons engaged in such work were counted as unemployed. In the labor force survey used today, anyone who worked for pay or profit is counted as employed (see the definitions below). Therefore, today's survey would count work relief as employment and not unemployment, and the BLS figures for the period before 1942 are not consistent with today's unemployment rates. Michael Darby calculated an alternative unemployment rate in which such workers are counted as employed, and this rate is shown in parentheses in Table 10-1B. BLS counts employment in nonprofit and private groups that provide relief under the "emergency and other relief services" industry (also referred to in some BLS data as the "community food and housing, and emergency and other relief services" industry). This employment is usually found in the "Other Services" of the NAICS supersector.

Race and ethnic origin

Data for two broad racial categories were made available beginning in 1954: *White* and *Black and other.* The latter included Asians and all other "nonwhite" races, and was discontinued after 2002. Data for *Blacks* only are available beginning with 1972; this category is now called *Black or African American.* Data for *Asians* are shown beginning with 2000. Persons in the remaining race categories—American Indian or Alaska Native, Native Hawaiian or Other Pacific Islanders, and persons who selected more than one race category when that became possible, beginning in 2003 (see below)—are included in the estimates of total employment and unemployment, but are not shown separately because their numbers are too small to yield quality estimates.

Hispanic or Latino ethnicity, previously labeled *Hispanic origin*, is not a racial category and is established in a survey question separate from the question about race. Persons of Hispanic or Latino ethnicity may be of any race.

In January 2003, changes that affected classification by race and Hispanic ethnicity were introduced. These changes caused discontinuities in race and ethnic group data between December 2002 and January 2003.

- Individuals in the sample are now asked whether they are of Hispanic ethnicity <u>before</u> being asked about their race. Prior to 2003, individuals were asked their ethnic origin <u>after</u> they were asked about their race. Furthermore, respondents are now asked directly if they are Spanish, Hispanic, or Latino. Previously, they were identified based on their or their ancestors' country of origin.

- Individuals in the sample are now allowed to choose more than one race category. Before 2003, they were required to select a single primary race. This change had no impact on

the size of the overall civilian noninstitutional population and labor force. It did reduce the population and labor force levels of Whites and Blacks beginning in January 2003, as individuals who reported more than one race are now excluded from those groups.

BLS has estimated, based on a special survey, that these changes reduced the population and labor force levels for Whites by about 950,000 and 730,000 persons, respectively, and for Blacks by about 320,000 and 240,000 persons, respectively, while having little or no impact on either of their unemployment rates. The changes did not affect the size of the Hispanic population or labor force, but they did cause an increase of about half a percentage point in the Hispanic unemployment rate.

Definitions

The employment status of the civilian population is surveyed each month with respect to a specific week in mid-month—not for the entire month. This is known as the "reference week." For a precise definition and explanation of the reference week, see Notes on the Data, which follows these definitions.

The *civilian noninstitutional population* comprises all civilians 16 years of age and over who are not inmates of penal or mental institutions, sanitariums, or homes for the aged, infirm, or needy.

Civilian employment includes those civilians who (1) worked for pay or profit at any time during the Sunday-through-Saturday week that includes the 12th day of the month (the reference week), or who worked for 15 hours or more as an unpaid worker in a family-operated enterprise; or (2) were temporarily absent from regular jobs because of vacation, illness, industrial dispute, bad weather, or similar reasons. Each employed person is counted only once; those who hold more than one job are counted as being in the job at which they worked. However, official CPS data are characterized by periodic discontinuities, which occur when new benchmarks for Census measures of the total population are introduced. These updates take place in a single month—usually January—and the official data for previous months are typically not modified to provide a smooth transition. Therefore, shorter-term comparisons (for a year or two or for a business cycle phase) will be misleading if such a discontinuity is included in the period. A recent example will illustrate. Beginning with January 2017, the updated controls decreased the estimated size of the civilian noninstitutional population 16 years and over by 800,000, the civilian labor force by 506,000, employment by 488,000, unemployment by 18,000, and the number of persons not in the labor force by 294,000. The total unemployment rate, the labor force participation rate, and the employment-population ratio were not affected. Such discontinuities occur throughout the history of the series the greatest number of hours during the reference week.

Unemployed persons are all civilians who were not employed (according to the above definition) during the reference week, but who were available for work—except for temporary illness—and who had made specific efforts to find employment sometime during the previous four weeks. Persons who did not look for work because they were on layoff are also counted as unemployed.

The *civilian labor force* comprises all civilians classified as employed or unemployed.

Civilians 16 years of age and over in the noninstitutional population who are not classified as employed or unemployed are defined as *not in the labor force*. This group includes those engaged in own-home housework; in school; unable to work because of longterm illness, retirement, or age; seasonal workers for whom the reference week fell in an "off" season (if not qualifying as unemployed by looking for a job); persons who became discouraged and gave up the search for work; and the voluntarily idle. Also included are those doing only incidental work (less than 15 hours) in a family-operated business during the reference week.

The civilian *labor force participation rate* represents the percentage of the civilian noninstitutional population (age 16 years and over) that is in the civilian labor force.

The *employment-population ratio* represents the percentage of the civilian noninstitutional population (age 16 years and over) that is employed. This is traditionally called a "ratio," although it is also traditionally expressed as a percent and therefore would be more appropriately called a "rate," as is the case with the labor force participation rate.

Employment is shown by *class of worker*, including a breakdown of total employment into *agricultural* and *nonagricultural* industries. Employment in *nonagricultural industries* includes *wage and salary workers*, the *self-employed*, and *unpaid family workers*.

Wage and salary workers receive wages, salaries, commissions, tips, and/or pay-in-kind. This category includes owners of self-owned incorporated businesses.

Self-employed workers are those who work for profit or for fees in their own business, profession, trade, or farm. This category includes only the unincorporated; a person whose business is incorporated is considered to be a wage and salary worker since he or she is a paid employee of a corporation, even if he or she is the corporation's president and sole employee. These categories are now labeled "Self-employed workers, unincorporated" to clarify this definition.

Wage and salary employment comprises *government* and *private industry* wage and salary workers. Domestic workers and other employees of *private households*, who are not included in the payroll employment series, are shown separately from *all other private industries*. The series for *government* and *other private industries* wage and salary workers are the closest in scope to similar categories in the payroll employment series.

Multiple jobholders are employed persons who, during the reference week, either had two or more jobs as a wage and salary worker, were self-employed and also held a wage and salary job, or worked as an unpaid family worker and also held a wage and salary job. This category does not include self-employed persons with multiple businesses or persons with multiple jobs as unpaid family workers. For purposes of industry and occupational classification, multiple jobholders are counted as being in the job at which they worked the greatest number of hours during the reference week.

Employed and at work part time does not include employed persons who were absent from their jobs during the entire reference week for reasons such as vacation, illness, or industrial dispute.

At work part time for economic reasons ("involuntary" part time) refers to individuals who worked 1 to 34 hours during the reference week because of slack work, unfavorable business conditions, an inability to find full-time work, or seasonal declines in demand. To be included in this category, workers must also indicate that they want and are available for full-time work.

At work part time for noneconomic reasons ("voluntary" part time) refers to persons who usually work part time and were at work for 1 to 34 hours during the reference week for reasons such as illness, other medical limitations, family obligations, education, retirement, Social Security limits on earnings, or working in an industry where the workweek is less than 35 hours. It also includes respondents who gave an economic reason but were not available for, or did not want, full-time work. At work part time for noneconomic reasons excludes persons who usually work full time, but who worked only 1 to 34 hours during the reference week for reasons such as holidays, illnesses, and bad weather.

The *long-term unemployed* are persons currently unemployed (searching or on layoff) who have been unemployed for 15 consecutive weeks or longer. If a person ceases to look for work for two weeks or more, or becomes temporarily employed, the continuity of longterm unemployment is broken. If he or she starts searching for work or is laid off again, the monthly CPS will record the length of his or her unemployment from the time the search recommenced or since the latest layoff.

The civilian *unemployment rate* is the number of unemployed as a percentage of the civilian labor force. The unemployment rates for groups within the civilian population (such as males age 20 years and over) are the number of unemployed in a group as a percent of that group's labor force.

Unemployment rates by reason provides a breakdown of the total unemployment rate. Each unemployed person is classified into one of four groups.

Job losers and persons who completed temporary jobs includes persons on temporary layoff, permanent job losers, and persons who completed temporary jobs and began looking for work after those jobs ended. These three categories are shown separately without seasonal adjustment in the BLS's "Employment Situation" news release and on its Web site. They are combined, under the title shown here, for the purpose of seasonal adjustment. This is the category of unemployment that responds most strongly to the business cycle.

Job leavers terminated their employment voluntarily and immediately began looking for work.

Reentrants are persons who previously worked, but were out of the labor force prior to beginning their current job search.

New entrants are persons searching for a first job who have never worked.

Each of these categories is expressed as a proportion of the entire civilian labor force, so that the sum of the four rates equals the unemployment rate for all civilian workers, except for possible discrepancies due to rounding or separate seasonal adjustment.

Median and average weeks unemployed are summary measures of the length of time that persons classified as unemployed have been looking for work. For persons on layoff, the duration represents the number of full weeks of the layoff. The *average (mean)* number of weeks is computed by aggregating all the weeks of unemployment experienced by all unemployed persons during their current spell of unemployment and dividing by the number of unemployed. The average can be distorted by what is called "top coding" because the length of unemployment is reported in ranges, including a top range of "over __ weeks", rather than exact numbers. See the paragraph below for further explanation. The *median* number of weeks unemployed is the number of weeks of unemployment experienced by the person at the midpoint of the distribution of all unemployed persons, as ranked by duration of unemployment, and is not distorted by top coding. Like medians in other economic time series, it is likely to be a better measure of typical experience.

Beginning with the data for January 2011 and phasing in through April 2011, respondents could report unemployment durations of up to 5 years; before that time, the "top code" was up to 2 years. This change causes a sharp increase in January 2011 and continued rises through April for *average weeks unemployed.* It did not affect total unemployment, total long-term unemployment, or the *median weeks unemployed.* Comparisons of the average weeks statistics on the old and new basis are available on the BLS Web site.

Alternative measures of labor underutilization are calculated by BLS and published in the monthly Employment Situation release. They measure alternative concepts of unused working capacity, and are numbered "U-1" through "U-6" in order of increasing breadth of the definition of underutilization.

"U-1" is persons unemployed 15 weeks or longer, as a percent of the civilian labor force. It is not shown in *Business Statistics.*

"U-2" is job losers and persons who completed temporary jobs, as a percent of the civilian labor force.

"U-3" is the official rate, described above.

"U-4" through "U-6" are shown in Table 10-5. They are based on additional labor force status questions, now included in the CPS survey, that were introduced beginning in 1994. U-4 and U-5 are increasingly broader rates of unemployment, while U-6 can be described as an "unemployment and underemployment rate."

"U-4" adds discouraged workers to unemployment and the labor force. Discouraged workers are persons not in the officially defined labor force who have given a job-market-related reason for not looking currently for a job—for example, they have not looked for a job because they believed that no jobs were available.

"U-5" adds both discouraged workers and all other "marginally attached" workers to unemployment and the labor force. "Marginally attached" workers are all persons who currently are neither working nor looking for work but indicate that they want and are available for a job and have looked for work some time in the recent past.

"U-6" adds persons employed part time for economic reasons, to the number of persons counted as unemployed in "U-5", with the same labor force definition as in "U-5." This statistic, which measures a combination of unemployment and underemployment, is often cited in media reports on the employment situation.

For more information, see "BLS introduced new range of alternative unemployment measures" in the October 1995 issue of the *Monthly Labor Review.*

Notes on the data

The CPS data are collected by trained interviewers from about 60,000 sample households selected to represent the U.S. civilian noninstitutional population. The sample size has fluctuated between 50,000 and 60,000 households. In January 1996, the size was reduced to 50,00 due to budgetary reasons. The sample size was increased back to 60,000 households, beginning with the data for July 2001, as part of a plan to meet the requirements of the State Children's Health Insurance Program legislation. The CPS provides data for other data series in addition to the BLS employment status data, such as household income and poverty) and health insurance (see Chapter 3.)

The employment status data are based on the activity or status reported for the calendar week, Sunday through Saturday, that includes the 12th day of the month (the reference week). This specification is chosen so as to avoid weeks that contain major holidays. Households are interviewed in the week following the reference week. Sample households are phased in and out of the sample on a rotating basis. Consequently, three-fourths of the sample is the same for any two consecutive months. One-half of the sample is the same as the sample in the same month a year earlier.

Data relating to 1994 and subsequent years are not strictly comparable with data for 1993 and earlier years because of the major 1994 redesign of the survey questionnaire and collection methodology. The redesign included new and revised questions for the classification of individuals as employed or unemployed, the collection of new data on multiple jobholding, a change in the definition of discouraged workers, and the implementation of a more completely automated data collection.

The 1994 redesign of the CPS was the most extensive since 1967. However, there are many other significant periods of year-to-year noncomparability in the labor force data. These typically result from the introduction of new decennial census data into the CPS estimation procedures, expansions of the sample, or other improvements made to increase the reliability of the estimates. Each change introduces a new discontinuity, usually between December of the previous year and January of the newly altered year. The discontinuities are usually minor or negligible with respect to figures expressed as nationwide percentages (such as the unemployment rate or the labor force participation rate), but can be significant with respect to levels (such as labor force and employment in thousands of persons). A list of the dates of the major discontinuities follows, with BLS estimates of their quantitative impact on the national totals. (There are likely to be larger impacts on population subgroups.) The discontinuities occur in January unless otherwise indicated. Note that some of the changes caused adjustments that were carried back to an earlier year.

- 2012: Reflecting the 2010 Census, the civilian noninstitutional population was increased by 1,510,000; the civilian labor force by 258,000; employment by 216,000; and unemployment by 42,000. BLS states, "Although the total unemployment rate was unaffected, the labor force participation rate and the employment-population ratio were each reduced by 0.3 percentage point. This was because the population increase was primarily among persons 55 and older and, to a lesser degree, persons 16 to 24 years of age. Both these age groups have lower levels of labor force participation than the general population."

- 2013: Updated population estimates increased the population by 138,000, the labor force by 136,000, and employment by 127,000. The rates and ratios were not affected.

- 2014: the smallest population control adjustment on record increased labor force by 24,000 and employment by 22,000, with no effect on rates or ratios.

- 2015: civilian population aged 16 or older increased by over one-half million, the labor force by 348,000, employment by 324,000 and unemployment by 24,000. The rates and ratios were not affected.

- 2016: the population adjustments rose once again by population 265,000, labor force 218,000, employment 206,000, unemployment 12,000 and not in labor force 47,000. The rates and ratios were not affected.

- 2017: the first year since 2012 the controls decreased. Population down by 800,000, labor force, 506,000, employment 488,000, unemployment 18,000 and those not in labor force 294,000. The rates and ratios were not affected.

For further information on these changes, see "Publications and Other Documentation" at http://www.bls.gov/cps.

The monthly labor force, employment, and unemployment data are seasonally adjusted by the X-12-ARIMA method. All seasonally adjusted civilian labor force and unemployment rate statistics, as well as major employment and unemployment estimates, are computed by aggregating independently adjusted series. For example, the seasonally adjusted level of total unemployment is the sum of the seasonally adjusted levels of unemployment for the four age/sex groups (men and women age 16 to 19 years, and men and women age 20 years and over). Seasonally adjusted employment is the sum of the seasonally adjusted levels of employment for the same four groups. The seasonally adjusted civilian labor force is the sum of all eight components. Finally, the seasonally adjusted civilian worker unemployment rate is calculated by taking total seasonally adjusted unemployment as a percent of the total seasonally adjusted civilian labor force.

To minimize subsequent revisions, BLS uses a concurrent technique that estimates factors for the latest month using the most recent data. Then, seasonal adjustment factors are fully revised at the end of each year to reflect recent experience. The revisions also affect the preceding four years. An article describing the seasonal adjustment methodology for the household survey data is available at http://www.bls.gov/cps/cpsrs2010.pdf.

Breakdowns other than the basic age/sex classification described above—such as the employment data by class of worker in Table 10-3—will not necessarily add to totals because of independent seasonal adjustment.

Data availability

Data for each month are usually released on the first Friday of the following month in the "Employment Situation" press release, which also includes data from the establishment survey (Tables 10-8 through 10-18 and Chapter 15). (The release date actually is determined by the timing of the survey week.) The press release and data are available on the BLS Web site at <http://www.bls. gov>. The *Monthly Labor Review*, also available online at the BLS Web site, features frequent articles analyzing developments in the labor force, employment, and unemployment.

Monthly and annual data on the current basis are available beginning with 1948. Historical unadjusted data are published in *Labor Force Statistics Derived from the Current Population Survey* (BLS Bulletin 2307). Historical seasonally adjusted data are available from BLS upon request. Complete historical data are available on the BLS Web site at <http://www.bls.gov/cps>.

Seasonal adjustment factors are revised each year for the five previous years, with the release of December data in early January. New population controls are introduced with the release of January data in early February.

BLS annual data for 1940 through 1947 are published on their Web site at <http://www.bls.gov>. The data for 1929 through 1939 were published in *Employment and Earnings,* May 1972, and in U.S. Commerce Department, Bureau of Economic Analysis, *Long-Term Economic Growth, 1860–1970,* June 1973, p. 163. The Lebergott estimates are also found in the latter volume.

The Darby alternative unemployment rate is found in Michael Darby, "Three-and-a-Half Million U.S. Employees Have Been Mislaid," *Journal of Political Economy,* February 1976, v. 84, no. 1. It is also displayed and discussed in Robert A. Margo, "Employment and Unemployment in the 1930s," *Journal of Economic Perspectives,* v. 7, no. 2, Spring 1993.

References

Comprehensive descriptive material can be found at <http://www.bls.gov/cps> under the "Publications and Other Documentation" section. Historical background on the CPS, as well as a description of the 1994 redesign, can be found in three articles from the September 1993 edition of *Monthly Labor Review*: "Why Is It Necessary to Change?"; "Redesigning the Questionnaire"; and "Evaluating Changes in the Estimates." The redesign is also described in the February 1994 issue of *Employment and Earnings.* See also Chapter 1, "Labor Force Data Derived from the Current Population Survey," *BLS Handbook of Methods,* Bulletin 2490 (April 1997).

TABLE 10-6

Labor Force and Employment Estimates Smoothed for Population Adjustments

SOURCE: U.S. DEPARTMENT OF LABOR, BUREAU OF LABOR STATISTICS

This table presents seasonally adjusted monthly estimates of total civilian labor force and total civilian employment in which discontinuities caused by the introduction of new population controls in the official series—as described above—have been smoothed (see box above). The method of smoothing is described in Marisa L. Di Natale, "Creating Comparability in CPS Employment Series," on the BLS Web site at <http://www.bls.gov/cps/cpscomp.pdf>. BLS notes that these series do not match the official estimates in BLS publications, which are also the data shown in all other tables in this volume.

Because the January 2014 population control adjustment was negligible in size (see above), no smoothing of the 2013 data was required. Hence, Table 10-6 was calculated only through the end of 2012, which can be considered continuous with the official data for 2013 and 2014. This series was discontinued in December 2017.

TABLE 10-7

Insured Unemployment

SOURCE: U.S. DEPARTMENT OF LABOR, EMPLOYMENT AND TRAINING ADMINISTRATION

Definitions

State programs of unemployment insurance cover operations of regular programs under state unemployment insurance laws. In 1976, the law was amended to extend coverage to include virtually all state and local government employees, as well as many agricultural and domestic workers. (This took effect on January 1, 1978.) Benefits under state programs are financed by taxes levied by the states on employers.

Federal programs are those directly financed by the federal government. They include unemployment benefits for *federal employees* (Unemployment Compensation for Federal Employees, or UCFE), *newly discharged veterans* (Unemployment Compensation for Ex-Service Members, or UCX) and *railroad retirement.*

UCX pays benefits, based on service, to veterans who were on active duty and honorably separated. In the case of both UCFE and UCX, state laws determine the benefit amounts, number of weeks benefits can be paid, and other eligibility conditions.

An *initial claim* is the first claim in a benefit year filed by a worker after losing his or her job, or the first claim filed at the beginning of a subsequent period of unemployment in the same benefit year. The initial claim establishes the starting date for any insured unemployment that may result if the claimant is unemployed for one week or longer. Transitional claims (filed by claimants as they start a new benefit year in a continuing spell of unemployment) are excluded; therefore, these data more closely represent instances of new unemployment and are widely followed as a leading indicator of job market conditions.

Insured unemployment and *persons claiming benefits* both describe the average number of persons receiving benefits in the indicated month or year.

The *insured unemployment rate* for state programs is the level of insured unemployment as a percentage of employment covered by state programs.

Monthly averages in this book are averages, calculated by the editor, of the weekly data published by the Employment and Training Administration. Annual data are averages of the monthly data.

Data availability

Data are published in weekly press releases from the Employment and Training Administration. These releases are available on their Web site at <http://www.doleta.gov> under "Labor Market Data," as are historical data, under the category "UI/Program Statistics."

TABLES 10-8, 10-9, 10-14, 15-1, AND 15-2

Nonfarm Payroll Employment

SOURCE: U.S. DEPARTMENT OF LABOR, BUREAU OF LABOR STATISTICS (BLS)

These nonfarm employment data, as well as the hours and earnings data in Tables 10-9 through 10-13, 10-15 through 10-18, and 15-3 through 15-6, are compiled from payroll records. Information is reported monthly on a voluntary basis to BLS and its cooperating state agencies by a large sample of establishments, representing all industries except farming. These data, formally known as the Current Employment Statistics (CES) survey, are often referred to as the "establishment data" or the "payroll data." They are also known as the BLS-790 survey.

The survey, originally based on a stratified quota sample, has been replaced on a phased-in basis by a stratified probability sample. The new sampling procedure went into effect for wholesale trade in June 2000; for mining, construction, and manufacturing in June 2001; and for retail trade, transportation and public utilities, and finance, insurance, and real estate in June 2002. The phase-in was completed in June 2003, upon its extension to the service industries. The phase-in schedule was slightly different for the state and area series.

The sample has always been very large. Currently, it includes approximately 142,000 businesses and government agencies covering about 689,000 individual worksites, which account for about one-third of total benchmark employment of payroll workers. The sample is drawn from a sampling frame of over 9.7 million unemployment insurance tax accounts.

Data are classified according to the North American Industry Classification System (NAICS). BLS has reconstructed historical time series to conform with NAICS, to ensure that all published series have a NAICS-based history extending back to at least January 1990. NAICS-based history extends back to January 1939 for total nonfarm and other high-level aggregates. For more detailed series, the starting date for NAICS data varies depending on the extent of the definitional changes between the old Standard Industrial Classification (SIC) and NAICS.

Definitions

An *establishment* is an economic unit, such as a factory, store, or professional office, that produces goods or services at a single location and is engaged in one type of economic activity.

Employment comprises all persons who received pay (including holiday and sick pay) for any part of the payroll period that includes the 12th day of the month. The definition of the payroll period for each reporting respondent is that used by the employer; it could be weekly, biweekly, monthly, or other. Included are all fulltime and parttime workers in nonfarm establishments, including salaried officers of corporations. Persons holding more than one job are counted in each establishment that reports them. Not covered are proprietors, the selfemployed, unpaid volunteer and family workers, farm workers, domestic workers in households, and military personnel. Employees of the Central Intelligence Agency, the Defense Intelligence Agency, the National Geospatial-Intelligence Agency, and the National Security Agency are not included.

Persons on an establishment payroll who are on paid sick leave (when pay is received directly from the employer), on paid holiday or vacation, or who work during a portion of the pay period despite being unemployed or on strike during the rest of the period, are counted as employed. Not counted as employed are persons who are laid off, on leave without pay, on strike for the entire period, or hired but not paid during the period.

Intermittent workers are counted if they performed any service during the month. BLS considers regular fulltime teachers (private and government) to be employed during the summer vacation period, regardless of whether they are specifically paid during those months.

The *government* division includes federal, state, and local activities such as legislative, executive, and judicial functions, as well as the U.S. Postal Service and all governmentowned and governmentoperated business enterprises, establishments, and institutions (arsenals, navy yards, hospitals, state-owned utilities, etc.), and government force account construction. However, as indicated earlier, members of the armed forces and employees of certain national-security-related agencies are not included.

The monthly *diffusion index of employment change*, currently based on 271 private nonfarm NAICS industries, represents the percentage of those industries in which the seasonally adjusted level of employment in that month was higher than six months earlier, plus one-half of the percentage of industries with unchanged employment. Therefore, the diffusion index reported for September represents the change from March to September. *Business Statistics* uses the September value to represent the year as a whole, since it spans the year's midpoint. Diffusion indexes measure the dispersion of economic gains and losses, with values below 50 percent associated with recessions. The current

NAICS-based series begins with January 1991. For October 1976 through December 1990, an earlier series is available based on 347 SIC industries (there are more industries using the older classification system because in SIC manufacturing industries were represented in greater detail). September values from this series are used here to represent the years 1977 through 1990.

Production and nonsupervisory workers include all *production and related workers* in mining and manufacturing; *construction workers* in construction; and *nonsupervisory workers* in transportation, communication, electric, gas, and sanitary services; wholesale and retail trade; finance, insurance, and real estate; and services. These groups account for about four-fifths of the total employment on private nonagricultural payrolls. Previously, this category was called "production or nonsupervisory workers." The definitions have not changed.

Production and related workers include working supervisors and all nonsupervisory workers (including group leaders and trainees) engaged in fabricating, processing, assembling, inspecting, receiving, storing, handling, packing, warehousing, shipping, trucking, hauling, maintenance, repair, janitorial, guard services, product development, auxiliary production for plant's own use (such as a power plant), record keeping, and other services closely associated with these production operations.

Construction workers include the following employees in the construction division of the NAICS: working supervisors, qualified craft workers, mechanics, apprentices, laborers, and the like, who are engaged in new work, alterations, demolition, repair, maintenance, and other tasks, whether working at the site of construction or working in shops or yards at jobs (such as precutting and preassembling) ordinarily performed by members of the construction trades.

Nonsupervisory employees include employees (not above the working supervisory level) such as office and clerical workers, repairers, salespersons, operators, drivers, physicians, lawyers, accountants, nurses, social workers, research aides, teachers, drafters, photographers, beauticians, musicians, restaurant workers, custodial workers, attendants, line installers and repairers, laborers, janitors, guards, and other employees at similar occupational levels whose services are closely associated with those of the employees listed.

Notes on the data

Benchmark adjustments. The establishment survey data are adjusted annually to comprehensive counts of employment, called "benchmarks." Benchmark information on employment by industry is compiled by state agencies from reports of establishments covered under state unemployment insurance laws; these form an annual compilation of administrative data known as the ES-202. These tabulations cover about 97 percent of all employees on nonfarm payrolls. Benchmark data for the residual are obtained from alternate sources, primarily from Railroad Retirement Board

records and the Census Bureau's *County Business Patterns.* The latest benchmark adjustment, which is incorporated into the data in this volume, increased the employment level in the benchmark month March 2018 by 135,000 jobs, which was 0.1 percent.

The estimates for the benchmark month are compared with new benchmark levels for each industry. If revisions are necessary, the monthly series of estimates between benchmark periods are adjusted by graduated amounts between the new benchmark and the preceding one ("wedged back"), and the new benchmark level for each industry is then carried forward month by month based on the sample.

More specifically, the month-to-month changes for each estimation cell are based on changes in a matched sample for that cell, plus an estimate of net business births and deaths. The matched sample for each pair of months consists of establishments that have reported data for both months (which automatically excludes establishments that have gone out of business by the second month). Since new businesses are not immediately incorporated into the sample, a model-based estimate of net business births and deaths in that estimating cell is added. The birth/death adjustment factors are re-estimated quarterly based on the Quarterly Census of Employment and Wages.

Not-seasonally-adjusted data for all months since the last benchmark date are subject to revision.

Beginning in 1959, the data include Alaska and Hawaii. This inclusion resulted in an increase of 212,000 (0.4 percent) in total nonfarm employment for the March 1959 benchmark month.

Seasonal adjustment. The seasonal movements that recur periodically—such as warm and cold weather, holidays, and vacations—are generally the largest single component of monthtomonth changes in employment. After adjusting the data to remove such seasonal variation, basic trends become more evident. BLS uses X-12-ARIMA software to produce seasonal factors and perform concurrent seasonal adjustment, using the most recent 10 years of data. New factors are developed each month adding the most current data.

For most series, a special procedure called REGARIMA (regression with autocorrelated errors) is used before calculating the seasonal factors; this adjusts for the length of the interval (which can be either four or five weeks) between the survey weeks. REGARIMA has also been used to isolate extreme weather effects that distort the measurement of seasonal patterns in the construction industry, and to identify variations in local government employment due to the presence or absence of election poll workers.

Seasonal adjustment factors are directly applied to the component levels. Seasonally adjusted totals for employment series are then obtained by aggregating the seasonally adjusted components directly, while hours and earnings series represent weighted averages of the seasonally adjusted component series. Seasonally adjusted data are not published for a small number of series characterized by small seasonal components relative to their trend and/or irregular components. However, these series are used in aggregating to broader seasonally adjusted levels.

Revisions of the seasonally adjusted data, usually for the most recent five-year period, are made once a year coincident with the benchmark revisions. This means that these revisions typically extend back farther than the benchmark revisions.

Data availability

Employment data by industry division are available beginning with 1919. Data for each month usually are released on the first Friday of the following month in a press release that also contains data from the household survey (Tables 10-1 through 10-5). (The release date actually is determined by the timing of the survey week.) The *Monthly Labor Review* frequently contains articles analyzing developments in the labor force, employment, and unemployment. The *Monthly Labor Review*, press releases, and complete historical data are available on the BLS Web site at <http://www.bls.gov>.

Benchmark revisions and revised seasonally adjusted data for recent years are made each year with the release of January data in early February. Before 2004, the benchmark revisions were not made until June; the acceleration is due to earlier availability of the benchmark UI (ES-202) data.

References

References can be found at <http://www.bls.gov/ces> under the headings "Special Notices," "Benchmark Information," and "Technical Notes." Extensive changes incorporated in June 2003 are described in "Recent Changes in the National Current Employment Statistics Survey," *Monthly Labor Review*, June 2003. The latest benchmark revision is discussed in an article available on the Website. See also Chapter 2, "Employment, Hours, and Earnings from the Establishment Survey," *BLS Handbook of Methods*, Bulletin 2490 (April 1997).

TABLES 10-9 THROUGH 10-11, 10-15, 10-16, 15-3 AND 15-6

Average Hours Per Week; Aggregate Employee Hours

SOURCE: U.S. DEPARTMENT OF LABOR, BUREAU OF LABOR STATISTICS (BLS)

See the notes and definitions to Tables 10-8 and related tables for an overall description of the "establishment" or "payroll" survey that is the source of these earnings data.

Hours and earnings have been reported for production and non-supervisory workers, as defined above, since the inception of the CES. Beginning with data for March 2006, such data have also been collected for all payroll employees. With sufficient history for calculation of seasonal adjustment factors, BLS began publishing all-employee hours and earnings in February 2010, and these new data are presented in *Business Statistics* Tables 10-10 through 10-13.

Definitions

Average weekly hours represents the average hours paid per worker during the pay period that includes the 12th of the month. Included are hours paid for holidays and vacations, as well as those paid for sick leave when pay is received directly from the firm.

Average weekly hours are different from standard or scheduled hours. Factors such as unpaid absenteeism, labor turnover, part-time work, and work stoppages can cause average weekly hours to be lower than scheduled hours of work for an establishment.

An important characteristic of these data is that average weekly hours pertain to jobs, not to persons; thus, a person with half-time jobs in two different establishments is represented in this series as two jobs that have 20-hour workweeks, not as one person with a 40-hour workweek.

Overtime hours represent the portion of average weekly hours worked in excess of regular hours, for which overtime premiums were paid. Weekend and holiday hours are included only if overtime premiums were paid. Hours for which only shift differential, hazard, incentive, or other similar types of premiums were paid are excluded.

Aggregate hours provide measures of changes over time in labor input to the industry, in index-number form. The indexes are obtained by multiplying seasonally adjusted employment by seasonally adjusted average weekly hours, dividing the resulting series by their monthly averages for a base period, and multiplying the results by 100, so that the annual average for the base period equals 100. For total private, goods-producing, service-providing, and major industry divisions, the indexes are obtained by summing the seasonally adjusted aggregate weekly employee hours for the component industries, dividing by the monthly average for the base period, and multiplying by 100. For the series covering production and nonsupervisory workers, the base period is 2002; for the all-employee series, the base period is 2007.

Notes on the data

Benchmark adjustments. Independent benchmarks are not available for the hours and earnings series. However, at the time of the annual adjustment of the employment series to new benchmarks, the levels of hours and earnings may be affected by the revised employment weights (which are used in computing the industry averages for hours and earnings), as well as by the changes in seasonal adjustment factors introduced with the benchmark revision.

Method of computing industry series. "Average weekly hours" for individual industries are computed by dividing worker hours (reported by establishments classified in each industry) by the number of workers reported for the same establishments. Estimates for divisions and major industry groups are averages (weighted by employment) of the figures for component industries.

Seasonal adjustment. Hours and earnings series are seasonally adjusted by applying factors directly to the corresponding unadjusted series. Data for some industries are not seasonally adjusted because the seasonal component is small relative to the trendcycle and/or irregular components. Consequently, they cannot be separated with sufficient precision.

Special adjustments are made to average weekly hours to account for the presence or absence of religious holidays in the April survey reference period and the occasional occurrence of Labor Day in the September reference period. In addition, REGARIMA modeling is used prior to seasonal adjustment to correct for reporting and processing errors associated with the number of weekdays in a month (rather than to correct for the 4- and 5-week effect, which is less significant for hours than it is for employment). This is of particular importance for average weekly hours in the service-providing industries other than retail trade. For this reason, BLS advises that calculations of over-the-year changes (for example, the change for the current month from a year earlier) should use seasonally adjusted data, since the actual not-seasonally-adjusted monthly data may be distorted.

Data availability

See data availability for Tables 10-8 and related, above.

References

See references for Tables 10-8 and related, above.

TABLES 10-9, 10-12, 10-13, 10-17, 10-18, 15-4, AND 15-5

Hourly and Weekly Earnings

SOURCE: U.S. DEPARTMENT OF LABOR, BUREAU OF LABOR STATISTICS (BLS)

See the notes and definitions to Tables 10-8 and related for an overall description of the "establishment" or "payroll" survey that is the source of these earnings data.

Hours and earnings have been reported for production and non-supervisory workers, as defined above, since the inception of the CES. Beginning with data for March 2006, such data have also been collected for all payroll employees. With sufficient history for calculation of seasonal adjustment factors, BLS began publishing all-employee hours and earnings in February 2010, and these data are presented in new *Business Statistics* Tables 10-10 through 10-13.

Definitions

Earnings are the payments that workers receive during the survey period (before deductions for taxes and other items), including premium pay for overtime or late-shift work but excluding irregular bonuses and other special payments. After being previously excluded, tips were asked to be reported beginning in September 2005. This made little difference in most industries, and BLS asserts that many respondents had already been including tips. In two industries, full-service restaurants and cafeterias, there was a substantial difference, and the historical earnings data for those industries have been reconstructed to reflect the new higher level of earnings.

Notes on the data

The hours and earnings series are based on reports of gross payroll and corresponding paid hours for full and parttime workers who received pay for any part of the pay period that included the 12th of the month.

Total payrolls are before deductions, such as for the employee share of oldage and unemployment insurance, group insurance, withholding taxes, bonds, and union dues. The payroll figures also include pay for overtime, holidays, vacations, and sick leave (paid directly by the employer for the period reported). Excluded from the payroll figures are fringe benefits (health and other types of insurance and contributions to retirement, paid by the employer, and the employer share of payroll taxes), bonuses (unless earned and paid regularly each pay period), other pay not earned in the pay period reported (retroactive pay), and the value of free rent, fuel, meals, or other payment-in-kind.

Average hourly earnings data reflect not only changes in basic hourly and incentive wage rates, but also such variable factors as premium pay for overtime and lateshift work and changes in output of workers paid on an incentive basis. Shifts in the volume of employment between relatively high-paid and low-paid work also affect the general average of hourly earnings.

Averages of hourly earnings should not be confused with wage rates, which represent the rates stipulated for a given unit of work or time, while earnings refer to the actual return to the worker for a stated period of time. The earnings series do not represent total labor cost to the employer because of the inclusion of tips and the exclusion of irregular bonuses, retroactive items, the cost of employer-provided benefits, and payroll taxes paid by employers.

Average weekly earnings are not the amounts available to workers for spending, since they do not reflect deductions such as income taxes and Social Security taxes. It is also important to understand that average weekly earnings represent earnings per job, not per worker (since a worker may have more than one job) and not per family (since a family may have more than one worker). A person with two half-time jobs will be reflected as two earners with low weekly earnings rather than as one person with the total earnings from his or her two jobs.

Method of computing industry series. Average hourly earnings are obtained by dividing the reported total worker payroll by total worker hours. Estimates for both hours and hourly earnings for nonfarm divisions and major industry groups are employment-weighted averages of the figures for component industries.

Average weekly earnings are computed by multiplying average hourly earnings by average weekly hours. In addition to the factors mentioned above, which exert varying influences upon average hourly earnings, average weekly earnings are affected by changes in the length of the workweek, part-time work, work stoppages, labor turnover, and absenteeism. Persistent longterm uptrends in the proportion of part-time workers in retail trade and many of the service industries have reduced average workweeks (as measured here), and have similarly affected the average weekly earnings series.

Benchmark adjustments. Independent benchmarks are not available for the hours and earnings series. At the time of the annual adjustment of the employment series to new benchmarks, the levels of hours and earnings may be affected by the revised employment weights (which are used in computing the industry averages for hours and earnings), as well as by the changes in seasonal adjustment factors that were also introduced with the benchmark revision.

Seasonal adjustment. Hours and earnings series are seasonally adjusted by applying factors directly to the corresponding unadjusted series; seasonally adjusted average weekly earnings are the product of seasonally adjusted hourly earnings and weekly hours.

REGARIMA modeling is used to correct for reporting and processing errors associated with variations in the number of weekdays in a month (rather than for the 4- and 5-week effect, which is less significant for earnings than for employment). This is of particular importance for average hourly earnings in wholesale trade, financial activities, professional and business services, and other services. For this reason, BLS advises that calculations of over-the-year changes, for example the change for the current month from a year earlier, should use seasonally adjusted data, since the actual not seasonally adjusted monthly data may be distorted.

Data availability

See data availability for Tables 10-8 and related, above.

References

See references for Tables 10-8 and related, above.

TABLE 10-19

Hours at Work, Current Population Survey

This table presents annual average measures of hours worked per week reported by workers in the Current Population Survey, in response to questions asking for the hours worked in the survey week by each worker in the household on all his or her jobs combined. These measures are not published in regular BLS reports but are available on request. *Business Statistics* obtained these numbers from BLS staff and presents them here in annual average form.

Do CPS respondents overestimate the number of hours they work? One study made that claim, but a careful re-examination of the data by Harley Frazis and Jay Stewart in the June 2014 *Monthly Labor Review* indicates that respondents slightly under-report the hours that they actually worked during the reference week. (On the other hand, the authors also note that the reference week overestimates the average for the entire month because of the absence of holidays.) (*Monthly Labor Review* is available online at www.bls.gov.) In any event, the CPS data give a more accurate picture of trends in the workweeks of typical American workers than does the establishment (CES) survey.

TABLE 10-20

Median Usual Weekly Earnings of Full-Time Wage and Salary Workers

SOURCE: U.S. DEPARTMENT OF LABOR, BUREAU OF LABOR STATISTICS

These data are from the Current Population Survey, which was described in the notes to Tables 10-1 through 10-6. Because they are earnings per worker, not per job, and are limited to full-time workers, the data are not distorted by the increasing proportion of part-time workers, as the CES earnings data are.

Definitions

Full-time wage and salary workers are those workers reported as "employed" in the CPS who receive wages, salaries, commissions, tips, pay in kind, or piece rates, and usually work 35 hours or more per week at their sole or principal job. Both private-sector and public-sector employees are included. All self-employed persons are excluded (even those whose businesses are incorporated).

Usual weekly earnings are earnings before taxes and other deductions and include any overtime pay, commissions, or tips usually received. In the case of multiple jobholders they refer to the main job. The wording of the question was changed in January 1994 to better deal with persons who found it easier to report earnings on other than a weekly basis. Such reports are then converted to the weekly equivalent. According to BLS, "the term 'usual' is as perceived by the respondent. If the respondent asks for a definition of usual, interviewers are instructed to define the term as more than half the weeks worked during the past 4 or 5 months."

The *median* is the amount that divides a given earnings distribution into two equal groups, one having earnings above the median and the other having earnings below the median.

Race and ethnicity. See the notes to Tables 10-1 through 10-6 for the definitions of these categories.

Data availability

These data become available about 3 weeks after the end of each quarter in the "Usual Weekly Earnings of Wage and Salary Workers" press release, available on the BLS Web site at <http://www.bls.gov/cps>. Recent data are available at that location.

Also available are greater detail by demographic and age groups, by occupation, by union status, and by education; distributional data, by deciles and quartiles; and earnings for part-time workers. Also available are earnings in 1982 dollars using the CPI-U. (In the opinion of the editor, these give a less accurate depiction of longer-term trends, which is why *Business Statistics* provides the CPI-U-RS data explained above.) Historical data are available upon request from BLS by telephone at (202) 691-6555.

CHAPTER 11: ENERGY

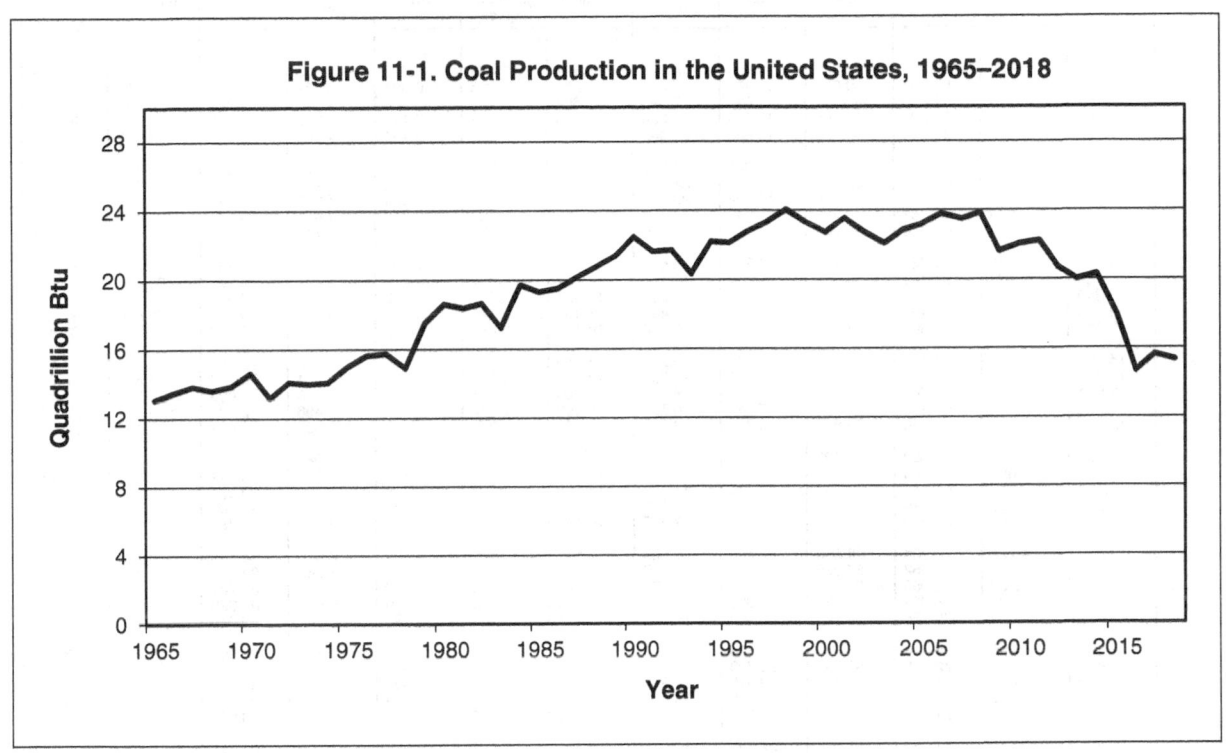

Figure 11-1. Coal Production in the United States, 1965–2018

- Coal production declined 1.9 percent in 2018 after increasing in 2017. Natural gas production attained its highest level in 2018, at 31.5 quadrillion Btu, while renewable energy production also continued to climb increasing from 6.221 Q Btus in 2005 to 11.722 quadrillion Btus in 2018. (Table 11-1)

- Energy consumption by end use is split with residential and commercial at 39.7 percent, industrial at 32.2 percent and transportation at 28.1 percent. (Table 11-1)

- Net imports of petroleum and products peaked in 2005 at 12.55 million barrels. By 2018, net imports were 2.340 million barrels, a decline of over 81 percent. As domestic production of energy increased (see Figure 11-1) imports fell. (Table 11-3)

- After reaching a high of $101.11 per barrel for crude oil in 2012, the price plummeted over 42 percent in 2018. However, between 2016 and 2018, the price increased 62 percent. Factors for this increase include strong demand and the agreement among members of the Organization of the Petroleum Exporting Countries (OPEC)—along with some key non-OPEC oil producers to reduce production by approximately 1.2 million barrels per day (b/d). (Table 11-3)

Table 11-1. Energy Supply and Consumption

(Quadrillion Btu.)

Year and month	Imports	Exports	Production, by source								Consumption, by end-use sector			
			Total	Fossil fuels					Nuclear electric power	Renewable energy, total	Total	Residential and commercial	Industrial	Transportation
				Total	Coal	Natural gas	Crude oil	Natural gas plant liquids						
1965	5.892	1.829	50.674	47.235	13.055	15.775	16.521	1.883	0.043	3.396	54.015	16.485	25.098	12.432
1966	6.146	1.829	53.532	50.035	13.468	17.011	17.561	1.996	0.064	3.432	57.014	17.493	26.422	13.100
1967	6.159	2.115	56.376	52.597	13.825	17.943	18.651	2.177	0.088	3.690	58.905	18.538	26.614	13.752
1968	6.905	1.998	58.220	54.306	13.609	19.068	19.308	2.321	0.142	3.773	62.415	19.665	27.883	14.866
1969	7.676	2.126	60.534	56.286	13.863	20.446	19.556	2.420	0.154	4.095	65.614	21.002	29.105	15.507
1970	8.342	2.632	63.495	59.186	14.607	21.666	20.401	2.512	0.239	4.070	67.838	22.112	29.628	16.098
1971	9.535	2.151	62.717	58.042	13.186	22.280	20.033	2.544	0.413	4.262	69.283	22.967	29.586	16.730
1972	11.387	2.118	63.904	58.938	14.092	22.208	20.041	2.598	0.584	4.382	72.688	24.040	30.930	17.717
1973	14.613	2.033	63.563	58.241	13.992	22.187	19.493	2.569	0.910	4.411	75.684	24.440	32.623	18.613
1974	14.304	2.203	62.345	56.331	14.074	21.210	18.575	2.471	1.272	4.742	73.962	24.048	31.787	18.120
1975	14.032	2.323	61.320	54.733	14.989	19.640	17.729	2.374	1.900	4.687	71.965	24.306	29.413	18.245
1976	16.760	2.172	61.561	54.723	15.654	19.480	17.262	2.327	2.111	4.727	75.975	25.474	31.393	19.101
1977	19.948	2.052	62.012	55.101	15.755	19.565	17.454	2.327	2.702	4.209	77.961	25.869	32.263	19.822
1978	19.106	1.920	63.104	55.074	14.910	19.485	18.434	2.245	3.024	5.005	79.950	26.644	32.688	20.617
1979	19.460	2.855	65.904	58.006	17.540	20.076	18.104	2.286	2.776	5.123	80.859	26.461	33.925	20.472
1980	15.796	3.695	67.175	59.008	18.598	19.908	18.249	2.254	2.739	5.428	78.067	26.332	32.039	19.697
1981	13.719	4.307	66.951	58.529	18.377	19.699	18.146	2.307	3.008	5.414	76.106	25.877	30.712	19.514
1982	11.861	4.608	66.569	57.458	18.639	18.319	18.309	2.191	3.131	5.980	73.099	26.391	27.614	19.089
1983	11.752	3.693	64.114	54.416	17.247	16.593	18.392	2.184	3.203	6.496	72.971	26.363	27.428	19.177
1984	12.471	3.786	68.840	58.849	19.719	18.008	18.848	2.274	3.553	6.438	76.632	27.403	29.570	19.656
1985	11.781	4.196	67.698	57.539	19.325	16.980	18.992	2.241	4.076	6.084	76.392	27.493	28.816	20.088
1986	14.151	4.021	67.066	56.575	19.509	16.541	18.376	2.149	4.380	6.111	76.647	27.581	28.274	20.789
1987	15.398	3.812	67.542	57.167	20.141	17.136	17.675	2.215	4.754	5.622	79.054	28.209	29.379	21.469
1988	17.296	4.366	68.919	57.875	20.738	17.599	17.279	2.260	5.587	5.457	82.709	29.711	30.677	22.318
1989	18.766	4.661	69.319	57.483	21.360	17.847	16.117	2.158	5.602	6.235	84.785	30.986	31.328	22.479
1990	18.817	4.752	70.704	58.560	22.488	18.326	15.571	2.175	6.104	6.040	84.485	30.257	31.802	22.419
1991	18.335	5.141	70.362	57.872	21.636	18.229	15.701	2.306	6.422	6.068	84.437	30.919	31.399	22.118
1992	19.372	4.937	69.955	57.655	21.694	18.375	15.223	2.363	6.479	5.821	85.782	30.796	32.571	22.415
1993	21.218	4.227	68.314	55.822	20.336	18.584	14.494	2.408	6.410	6.082	87.325	32.619	32.619	22.671
1994	22.307	4.035	70.725	58.044	22.202	19.348	14.103	2.391	6.694	5.987	89.040	32.208	33.520	23.319
1995	22.180	4.496	71.173	57.540	22.130	19.082	13.887	2.442	7.075	6.557	90.991	33.207	33.969	23.812
1996	23.633	4.613	72.485	58.387	22.790	19.344	13.723	2.530	7.087	7.011	94.000	34.674	34.903	24.419
1997	25.119	4.493	72.470	58.857	23.310	19.394	13.658	2.495	6.597	7.017	94.571	34.643	35.199	24.723
1998	26.473	4.237	72.875	59.314	24.045	19.613	13.235	2.420	7.068	6.493	94.982	34.919	34.841	25.225
1999	27.152	3.669	71.740	57.614	23.295	19.931	12.451	2.528	7.610	6.516	96.615	35.930	34.763	25.916
2000	28.865	3.962	71.330	57.366	22.735	19.662	12.358	2.611	7.862	6.102	98.776	37.596	34.662	26.516
2001	30.052	3.731	71.732	58.541	23.547	20.166	12.282	2.547	8.029	5.162	96.129	37.174	32.718	26.242
2002	29.331	3.608	70.710	56.834	22.732	19.382	12.160	2.559	8.145	5.731	97.605	38.131	32.660	26.808
2003	31.007	4.013	69.935	56.033	22.094	19.633	11.960	2.346	7.960	5.942	97.898	38.465	32.553	26.881
2004	33.492	4.351	70.228	55.942	22.852	19.074	11.550	2.466	8.223	6.063	100.073	38.737	33.515	27.827
2005	34.659	4.462	69.431	55.049	23.185	18.556	10.974	2.334	8.161	6.221	100.168	39.466	32.441	28.261
2006	34.649	4.727	70.735	55.934	23.790	19.022	10.767	2.356	8.215	6.586	99.464	38.378	32.390	28.697
2007	34.679	5.338	71.398	56.429	23.493	19.786	10.741	2.409	8.459	6.510	100.971	39.772	32.385	28.815
2008	32.970	6.949	73.205	57.587	23.851	20.703	10.613	2.419	8.426	7.192	98.825	40.070	31.333	27.422
2009	29.690	6.920	72.641	56.661	21.624	21.139	11.324	2.574	8.355	7.625	94.023	38.969	28.465	26.589
2010	29.866	8.176	74.970	58.222	22.038	21.806	11.596	2.781	8.434	8.314	97.608	39.953	30.670	26.978
2011	28.748	10.373	78.136	60.567	22.221	23.406	11.970	2.970	8.269	9.300	96.950	39.363	30.981	26.599
2012	27.068	11.267	79.282	62.334	20.677	24.610	13.801	3.246	8.062	8.886	94.480	37.291	31.061	26.126
2013	24.623	11.788	81.862	64.200	20.001	24.859	15.807	3.532	8.244	9.418	97.218	38.981	31.627	26.612
2014	23.241	12.270	87.746	69.642	20.286	26.718	18.542	4.096	8.338	9.766	98.382	39.709	31.798	26.869
2015	23.794	12.902	88.325	70.259	17.946	28.067	19.679	4.567	8.337	9.729	97.484	38.774	31.472	27.238
2016	25.378	14.119	84.362	65.507	14.667	27.576	18.494	4.770	8.427	10.428	97.445	38.211	31.453	27.786
2017	25.467	17.960	88.261	68.541	15.625	28.274	19.535	5.107	8.419	11.301	97.809	37.789	32.006	28.014
2018	24.838	21.202	95.705	75.541	15.333	31.536	22.826	5.846	8.441	11.722	101.239	40.229	32.618	28.391
2017														
January	2.315	1.382	7.311	5.620	1.382	2.281	1.568	0.389	0.765	0.926	8.982	4.021	2.733	2.226
February	1.959	1.387	6.741	5.209	1.300	2.078	1.456	0.376	0.665	0.867	7.623	3.141	2.438	2.047
March	2.195	1.467	7.401	5.698	1.299	2.354	1.622	0.423	0.681	1.023	8.430	3.332	2.721	2.379
April	2.112	1.429	7.023	5.433	1.184	2.281	1.560	0.409	0.593	0.997	7.452	2.632	2.558	2.265
May	2.264	1.459	7.340	5.663	1.252	2.353	1.626	0.432	0.641	1.035	7.800	2.710	2.668	2.423
June	2.117	1.430	7.302	5.610	1.335	2.295	1.558	0.422	0.701	0.991	7.964	2.918	2.643	2.401
July	2.129	1.459	7.425	5.747	1.271	2.400	1.638	0.438	0.746	0.932	8.433	3.272	2.723	2.432
August	2.153	1.392	7.526	5.895	1.424	2.400	1.640	0.432	0.757	0.874	8.298	3.098	2.705	2.491
September	1.993	1.481	7.234	5.670	1.269	2.357	1.630	0.414	0.712	0.852	7.630	2.755	2.573	2.301
October	2.067	1.686	7.603	5.988	1.336	2.470	1.721	0.461	0.690	0.924	7.838	2.762	2.674	2.403
November	2.027	1.671	7.559	5.941	1.296	2.455	1.735	0.456	0.697	0.921	8.130	3.111	2.734	2.286
December	2.136	1.718	7.795	6.066	1.277	2.552	1.781	0.455	0.771	0.959	9.229	4.037	2.833	2.360
2018														
January	2.231	1.602	7.748	5.965	1.259	2.502	1.768	0.437	0.781	1.002	9.655	4.567	2.830	2.255
February	1.864	1.516	7.203	5.576	1.223	2.302	1.637	0.415	0.678	0.949	8.077	3.510	2.515	2.053
March	2.119	1.691	7.964	6.230	1.329	2.574	1.850	0.476	0.701	1.033	8.684	3.519	2.757	2.411
April	2.126	1.782	7.593	5.940	1.177	2.500	1.793	0.471	0.618	1.034	7.884	2.984	2.587	2.318
May	2.145	1.750	7.964	6.197	1.241	2.611	1.851	0.493	0.704	1.063	8.019	2.820	2.739	2.462
June	2.176	1.760	7.875	6.095	1.249	2.541	1.827	0.478	0.729	1.051	8.144	3.041	2.670	2.430
July	2.161	1.832	8.098	6.394	1.278	2.678	1.934	0.504	0.758	0.946	8.607	3.345	2.756	2.501
August	2.194	1.720	8.380	6.666	1.406	2.735	2.003	0.522	0.756	0.958	8.694	3.286	2.859	2.544
September	2.001	1.753	7.976	6.425	1.266	2.684	1.963	0.512	0.677	0.874	7.860	2.914	2.632	2.312
October	1.976	1.897	8.246	6.713	1.345	2.800	2.044	0.523	0.621	0.912	8.095	2.911	2.775	2.411
November	1.887	1.948	8.168	6.572	1.273	2.752	2.041	0.505	0.669	0.928	8.476	3.422	2.731	2.324
December	1.959	1.951	8.488	6.768	1.286	2.856	2.116	0.511	0.749	0.970	9.043	3.907	2.767	2.371

Table 11-2. Energy Consumption Per Dollar of Real Gross Domestic Product

Year	Primary energy consumption (quadrillion Btu)			Gross domestic product (billions of chained [2012] dollars)	Energy intensity per real dollar of GDP (thousand Btu per chained [2012] dollar)		
	Total	Petroleum and natural gas	Other energy		Total	Petroleum and natural gas	Other energy
1950	34.616	19.284	15.332	2 289.5	15.12	8.42	1.01
1951	36.974	21.477	15.497	2 473.8	14.95	8.68	1.04
1952	36.748	22.505	14.243	2 574.9	14.27	8.74	1.00
1953	37.664	23.462	14.202	2 695.6	13.97	8.70	1.02
1954	36.639	24.169	12.470	2 680.0	13.67	9.02	0.91
1955	40.208	26.253	13.955	2 871.2	14.00	9.14	1.00
1956	41.754	27.551	14.203	2 932.4	14.24	9.40	1.00
1957	41.787	28.122	13.665	2 994.1	13.96	9.39	0.98
1958	41.645	29.190	12.455	2 972.0	14.01	9.82	0.89
1959	43.466	31.040	12.426	3 178.2	13.68	9.77	0.91
1960	45.086	32.305	12.782	3 260.0	13.83	9.91	0.92
1961	45.086	33.143	12.595	3 343.5	13.48	9.91	0.93
1962	47.826	34.780	13.047	3 548.4	13.48	9.80	0.97
1963	49.644	36.104	13.540	3 702.9	13.41	9.75	1.01
1964	51.815	37.589	14.226	3 916.3	13.23	9.60	1.08
1965	54.015	39.014	15.001	4 170.8	12.95	9.35	1.16
1966	57.014	41.396	15.618	4 445.9	12.82	9.31	1.22
1967	58.905	43.228	15.676	4 567.8	12.90	9.46	1.22
1968	62.415	46.189	16.225	4 792.3	13.02	9.64	1.25
1969	65.614	49.016	16.598	4 942.1	13.28	9.92	1.25
1970	67.838	51.315	16.523	4 951.3	13.70	10.36	1.21
1971	69.283	53.030	16.253	5 114.0	13.55	10.37	1.20
1972	72.688	55.645	17.043	5 383.3	13.50	10.34	1.26
1973	75.684	57.350	18.334	5 687.2	13.31	10.08	1.38
1974	73.962	55.186	18.776	5 656.5	13.08	9.76	1.44
1975	71.965	52.680	19.284	5 644.8	12.75	9.33	1.51
1976	75.975	55.523	20.452	5 949.0	12.77	9.33	1.60
1977	77.961	57.054	20.907	6 224.1	12.53	9.17	1.67
1978	79.950	57.963	21.987	6 568.6	12.17	8.82	1.81
1979	80.859	57.788	23.070	6 776.6	11.93	8.53	1.93
1980	78.067	54.440	23.627	6 759.2	11.55	8.05	2.05
1981	76.106	51.680	24.426	6 930.7	10.98	7.46	2.22
1982	73.099	48.588	24.511	6 805.8	10.74	7.14	2.28
1983	72.971	47.273	25.698	7 117.7	10.25	6.64	2.51
1984	76.632	49.447	27.185	7 632.8	10.04	6.48	2.71
1985	76.392	48.628	27.764	7 951.1	9.61	6.12	2.89
1986	76.647	48.790	27.857	8 226.4	9.32	5.93	2.99
1987	79.054	50.504	28.551	8 511.0	9.29	5.93	3.07
1988	82.709	52.671	30.038	8 866.5	9.33	5.94	3.22
1989	84.785	53.811	30.974	9 192.1	9.22	5.85	3.36
1990	84.484	53.156	31.330	9 365.5	9.02	5.68	3.47
1991	84.437	52.879	31.558	9 355.4	9.03	5.65	3.50
1992	85.782	54.239	31.544	9 684.9	8.86	5.60	3.56
1993	87.365	54.239	32.449	9 951.5	8.78	5.45	3.70
1994	89.087	56.286	32.802	10 352.4	8.61	5.44	3.81
1995	91.031	57.112	33.918	10 630.3	8.56	5.37	3.96
1996	94.021	58.760	35.261	11 031.4	8.52	5.33	4.14
1997	94.600	59.381	35.219	11 521.9	8.21	5.15	4.29
1998	95.018	59.648	35.370	12 038.3	7.89	4.95	4.48
1999	96.648	60.745	35.903	12 610.5	7.66	4.82	4.68
2000	98.817	62.090	36.727	13 131.0	7.53	4.73	4.88
2001	96.170	60.962	35.207	13 262.1	7.25	4.60	4.86
2002	97.643	61.736	35.908	13 493.1	7.24	4.58	4.96
2003	97.918	61.620	36.297	13 879.1	7.06	4.44	5.14
2004	100.090	63.150	36.940	14 406.4	6.95	4.38	5.32
2005	100.188	62.868	37.320	14 912.5	6.72	4.22	5.55
2006	99.485	62.062	37.422	15 338.3	6.49	4.05	5.77
2007	101.015	63.152	37.863	15 626.0	6.46	4.04	5.86
2008	98.891	60.750	38.141	15 604.7	6.34	3.89	6.02
2009	94.118	58.375	35.743	15 208.8	6.19	3.84	5.78
2010	97.445	60.064	37.381	15 598.8	6.25	3.85	5.98
2011	96.842	59.778	37.064	15 840.7	6.11	3.77	6.06
2012	94.416	60.105	34.312	16 197.0	5.83	3.71	5.89
2013	97.157	61.418	35.739	16 495.4	5.89	3.72	6.07
2014	98.329	62.264	36.065	16 912.0	5.82	3.68	6.20
2015	97.365	63.799	33.566	17 403.8	5.60	3.67	5.99
2016	97.397	64.366	33.030	17 688.9	5.52	3.64	5.99
2017	97.716	64.275	33.441	18 108.1	5.41	3.56	6.18
2018	100.128	68.756	31.372	18 638.2	5.39	3.70	5.82

Table 11-3. Petroleum and Petroleum Products—Prices, Imports, Domestic Production, and Stocks

(Not seasonally adjusted.)

Year and month	Crude oil futures price (dollars per barrel)		Imports				Supply (thousands of barrels per day)					Stocks (end of period, millions of barrels)		
			Total energy-related petroleum products (thousands of barrels)	Crude petroleum			Petroleum and products			Domestic production		Crude oil and petroleum products	Crude petroleum	
	Current dollars	2009 dollars		Thousands of barrels		Unit price (dollars per barrel)	Exports	Imports	Net imports	Crude oil	Natural gas plant liquids		Non-SPR	Strategic petroleum reserve
				Total	Average per day									
1983	30.66	69.52	...	1 293 819	3 545	29.51	739	5 051	4 312	8 688	1 559	1 454	344	379
1984	29.44	63.41	...	1 319 683	3 616	27.68	722	5 437	4 715	8 879	1 630	1 556	345	451
1985	27.89	57.07	...	1 260 856	3 454	26.20	781	5 067	4 286	8 971	1 609	1 519	321	493
1986	15.05	29.56	...	1 634 567	4 478	13.90	785	6 224	5 439	8 680	1 551	1 593	331	512
1987	19.15	36.38	...	1 744 977	4 781	16.80	764	6 678	5 914	8 349	1 595	1 607	349	541
1988	15.96	29.10	...	1 887 860	5 172	13.69	815	7 402	6 587	8 140	1 625	1 597	330	560
1989	19.58	34.69	...	2 146 552	5 881	16.49	859	8 061	7 202	7 613	1 546	1 581	341	580
1990	24.50	42.53	...	2 216 604	6 073	19.75	857	8 018	7 161	7 355	1 559	1 621	323	586
1991	21.50	37.24	2 828 953	2 146 064	5 880	17.46	1 001	7 627	6 626	7 417	1 659	1 617	325	569
1992	20.58	34.37	2 947 582	2 294 570	6 269	16.80	950	7 888	6 938	7 171	1 697	1 592	318	575
1993	18.48	29.82	3 257 008	2 543 374	6 968	15.13	1 003	8 620	7 618	6 847	1 736	1 647	335	587
1994	17.19	26.71	3 416 045	2 704 196	7 409	14.23	942	8 996	8 054	6 662	1 727	1 653	337	592
1995	18.40	27.76	3 361 882	2 767 312	7 582	15.81	949	8 835	7 886	6 560	1 762	1 563	303	592
1996	22.03	32.11	3 622 385	2 893 647	7 906	18.98	981	9 478	8 498	6 465	1 830	1 507	284	566
1997	20.61	28.95	3 802 574	3 069 430	8 409	17.67	1 003	10 162	9 158	6 452	1 817	1 560	305	563
1998	14.40	19.20	4 088 027	3 242 711	8 884	11.49	945	10 708	9 764	6 252	1 759	1 647	324	571
1999	19.30	24.44	4 081 181	3 228 092	8 844	15.76	940	10 852	9 912	5 881	1 850	1 493	284	567
2000	30.26	36.47	4 314 825	3 399 239	9 288	26.44	1 040	11 459	10 419	5 822	1 911	1 468	286	541
2001	25.95	30.48	4 475 026	3 471 067	9 510	21.40	971	11 871	10 900	5 801	1 868	1 586	312	550
2002	26.15	29.95	4 337 075	3 418 022	9 364	22.61	984	11 530	10 546	5 744	1 880	1 548	278	599
2003	30.99	34.41	4 654 638	3 676 005	10 071	26.98	1 027	12 264	11 238	5 649	1 719	1 568	269	638
2004	41.47	44.35	4 917 591	3 820 979	10 440	34.48	1 048	13 145	12 097	5 441	1 809	1 645	286	676
2005	56.70	58.57	5 004 339	3 754 671	10 287	46.81	1 165	13 714	12 549	5 181	1 717	1 682	308	685
2006	66.25	66.42	4 880 734	3 734 226	10 231	58.01	1 317	13 707	12 390	5 088	1 739	1 703	296	689
2007	72.41	71.01	4 807 811	3 690 568	10 111	64.28	1 433	13 468	12 036	5 077	1 783	1 648	268	697
2008	99.75	98.15	4 613 444	3 590 628	9 810	95.22	1 802	12 915	11 114	5 000	1 784	1 719	308	702
2009	62.09	62.09	4 266 007	3 314 787	9 082	56.93	2 024	11 691	9 667	5 350	1 910	1 758	307	727
2010	79.61	78.11	4 279 611	3 377 077	9 252	74.67	2 353	11 793	9 441	5 482	2 074	1 772	312	727
2011	95.11	91.25	4 164 117	3 321 918	9 101	99.82	2 986	11 436	8 450	5 645	2 216	1 725	308	696
2012	94.15	89.03	3 847 545	3 097 407	8 463	101.11	3 205	10 598	7 393	6 497	2 408	1 779	338	695
2013	98.05	91.17	3 549 945	2 813 770	7 709	96.95	3 621	9 859	6 237	7 465	2 606	1 728	327	696
2014	92.91	84.12	3 381 460	2 700 939	7 400	91.23	4 176	9 240	5 064	8 753	3 015	1 825	361	691
2015	48.79	44.54	3 384 807	2 661 968	7 293	47.26	4 738	9 449	4 650	9 408	3 342	1 982	449	695
2016	43.32	39.20	3 576 378	2 806 413	7 689	36.00	5 261	10 055	2 480	8 857	3 509	2 030	485	695
2017	50.80	45.09	3 639 778	2 883 753	7 901	45.97	6 343	10 075	3 723	9 355	3 736	1 895	421	662
2018	64.81	60.27	3 464 237	2 694 326	7 382	58.22	7 588	9 928	2 340	10 962	4 349	1 912	442	649
2017														
January	52.61	46.88	332 252	259 279	710	43.94	5 691	10 685	4 994	8 828	3 365	2 049	504	695
February	53.46	47.59	297 117	237 230	650	45.19	6 443	10 039	3 596	9 058	604	2 046	524	695
March	49.67	44.33	322 270	258 762	709	46.33	5 886	10 059	4 173	9 140	3 644	2 029	538	692
April	51.12	45.52	287 482	229 262	628	45.40	6 066	10 244	4 178	9 132	633	2 029	524	689
May	48.54	43.25	329 889	266 001	729	45.04	6 142	10 628	4 486	9 177	3 721	2 034	517	684
June	45.20	40.26	311 449	250 083	685	44.78	6 148	10 240	4 092	9 089	3 752	2 009	500	679
July	46.68	41.53	295 923	234 929	644	43.13	6 232	9 850	3 618	9 241	3 755	1 998	482	679
August	48.06	42.67	319 039	252 386	691	44.13	5 647	10 055	4 408	9 242	3 704	1 986	459	679
September	49.88	44.11	275 839	210 508	577	45.13	6 263	9 707	3 444	9 528	3 693	1 978	469	674
October	51.69	45.64	297 115	235 205	644	47.27	7 163	9 661	2 498	9 687	3 968	1 943	459	669
November	56.54	49.82	301 117	236 073	647	50.10	7 158	9 783	2 625	10 099	4 054	1 923	452	661
December	57.95	50.99	270 286	214 036	586	52.08	7 296	9 934	2 638	10 024	3 936	1 895	421	663
2018														
January	63.55	59.17	310 368	241 687	7 796	54.78	6 615	10 274	3 659	9 995	3 825	1 879	420	664
February	62.16	57.79	250 538	191 349	6 834	54.83	6 844	9 580	2 736	10 248	4 023	1 876	424	665
March	62.87	58.42	284 948	219 289	7 074	54.02	7 105	9 821	2 716	10 461	4 173	1 862	423	665
April	66.33	61.49	297 300	236 137	7 871	54.50	7 730	10 364	2 634	10 475	4 260	1 864	435	664
May	69.89	64.66	301 707	232 049	7 485	58.38	7 517	10 228	2 712	10 464	4 321	1 870	433	660
June	67.32	62.21	296 507	237 340	7 911	62.46	7 801	10 706	2 905	10 672	4 326	1 867	415	660
July	70.74	65.28	308 284	241 638	7 795	64.54	7 827	10 176	2 349	10 936	4 411	1 872	409	660
August	67.85	62.56	316 245	241 921	7 804	62.64	7 043	10 432	3 389	11 325	4 570	1 892	407	660
September	70.07	64.52	284 919	213 648	7 122	61.40	7 611	9 885	2 273	11 470	4 631	1 932	416	660
October	70.76	65.03	293 280	227 924	7 352	61.25	8 018	9 417	1 399	11 559	4 580	1 916	432	655
November	56.60	51.99	261 160	211 282	7 043	57.54	8 669	9 213	545	11 926	4 571	1 910	449	650
December	48.68	44.69	258 982	200 062	6 454	50.26	8 250	9 022	772	11 963	4 479	1 912	442	649

... = Not available.

NOTES AND DEFINITIONS, CHAPTER 11

TABLES 11-1 AND 11-2

Energy Supply and Consumption

SOURCES: U.S. DEPARTMENT OF ENERGY, ENERGY INFORMATION ADMINISTRATION; U.S. DEPARTMENT OF COMMERCE, BUREAU OF ECONOMIC ANALYSIS

This chapter presents energy data and statistics provided by the U.S. Energy Information Administration (EIA). Specific data presented include energy supply and consumption of nonrenewable and renewable sources plus imports and exports of these products; energy intensity as energy consumption per dollar and statistics of petroleum plus petroleum products such as prices, imports, domestic production and supply.

Sources of fossil fuel energy include coal, natural gas, crude oil and natural gas plant. Additional energy sources include nuclear electric power and renewable energy such as wind, solar and geothermal. Petroleum products include gasoline, distillates such as diesel fuel and heating oil, jet fuel, petrochemical feedstocks, waxes, lubricating oils, and asphalt.

The Department of Energy Organization Act of 1977 established EIA as the primary federal government authority on energy statistics and analysis, building upon systems and organizations first established in 1974 following the oil market disruption of 1973.

Definitions

The *British thermal unit (Btu)* is a measure used to combine data for different energy sources into a consistent aggregate. It is the amount of energy required to raise the temperature of 1 pound of water 1 degree Fahrenheit when the water is near a temperature of 39.2 degrees Fahrenheit. To illustrate one of the factors used to convert volumes to Btu, conventional motor gasoline has a heat content of 5.253 million Btu per barrel. For further information, see the Energy Information Administration's *Monthly Energy Review*, Appendix A.

Crude oil production is defined as the quantities of oil extracted from the ground after the removal of inert matter or impurities.

Renewable energy, total includes conventional hydroelectric power, geothermal, solar thermal and photovoltaic, wind, and biomass. Hydroelectric power includes conventional electrical utility and industrial generation. Biomass includes wood, waste, and alcohol fuels (ethanol blended into motor gasoline).

The sum of domestic energy *production* and net imports of energy (*imports* minus *exports*) does not exactly equal domestic energy *consumption*. The difference is attributed to inventory changes; losses and gains in conversion, transportation, and distribution; the addition of blending compounds; shipments of anthracite to U.S. armed forces in Europe; and adjustments to account for discrepancies between reporting systems.

Consumption by end-use sector is based on total, not net, consumption—that is, each sector's consumption includes its electricity purchases as well as its own energy production. End use sectors are residential, commercial, industrial and transportation. Electric utilities are not treated as a separate end-use sector. However, they are counted as primary producers in the production accounts, which measure only primary production.

A country's energy intensity is usually defined as energy consumption per unit of gross domestic product (GDP). Greater efficiency and structural changes in the economy have reduced energy intensity.

References and notes on the data

All of these data are published each month in Tables 1.1, 1.2, 1.7, and 2.1 in the *Monthly Energy Review*. Annual data before 1973 are published in the *Annual Energy Review*. These two publications are no longer published in printed form but are available, along with all current and historical data, on the EIA Web site at <http://www.eia.doe.gov>.

The real gross domestic product (GDP) data used to calculate energy consumption per dollar of real GDP are from the Bureau of Economic Analysis; see Table 1-2 and the applicable notes and definitions in this volume of *Business Statistics*. The GDP numbers shown in this table reflect the comprehensive revision of July 2013, which increased the overall level of GDP by including as investment research, development, and creation of intellectual property. This changed the average level of the energy-to-GDP ratio but had little effect on its downward trend. The GDP estimate shown here also reflects subsequent revisions and updates made through June 2016.

TABLE 11-3

Petroleum and Petroleum Products—Prices, Imports, Domestic Production, and Stocks

SOURCES: FUTURES PRICES—U.S. DEPARTMENT OF ENERGY, ENERGY INFORMATION ADMINISTRATION (EIA), AND U.S. DEPARTMENT OF COMMERCE, BUREAU OF ECONOMIC ANALYSIS; IMPORTS—U.S. DEPARTMENT OF COMMERCE, CENSUS BUREAU (SEE NOTES AND DEFINITIONS FOR TABLES 7-9 THROUGH 7-14); SUPPLY (NET IMPORTS AND DOMESTIC PRODUCTION) AND STOCKS—EIA.

Definitions and notes on the data

The *crude oil futures price* in *current dollars per barrel* is the price for next-month delivery in Cushing, Oklahoma (a pipeline hub), of light, sweet crude oil, as determined by trading on the New York Mercantile Exchange (NYMEX). Official daily closing prices are reported each day at 2:30 p.m., and are tabulated weekly in Table 13 of *EIA's Weekly Petroleum Status Report*. The monthly averages shown in this volume are the average prices for the nearest future from each trading day of the month. For example, for most days in January, the futures contract priced will be for February; for the last few days in January, the February contract will have expired and the March contract will be quoted. The annual averages are averages of the monthly averages.

The *crude oil futures price* in *2009 dollars* is calculated by the editors, by dividing each month's current-dollar price by that month's chain price index for total personal consumption expenditures (PCE), with the price index average for the year 2009 set at 1.0000. The PCE chain price index is compiled by the Bureau of Economic Analysis (BEA). It is described in the notes and definitions for Chapter 1 and is also presented in Chapter 8, Table 8-2, and discussed in its notes and definitions.

These *petroleum products* include gasoline, distillates such as diesel fuel and heating *oil*, jet fuel, petrochemical feedstocks, waxes, lubricating oils, and asphalt. A U.S. 42-gallon barrel of crude *oil* yields about 45 gallons of *petroleum products* in U.S. refineries because of refinery processing gain.

The import data in Columns 3 through 6 of this table are those published as Exhibit 17, "Imports of Energy-related Petroleum Products, including Crude Petroleum," in the monthly Census-BEA foreign trade press release, FT900. *Total energy-related petroleum products* includes the following Standard International Trade Classification (SITC) commodity groupings: crude oil, petroleum preparations, and liquefied propane and butane gas.

The data in Columns 7 through 11, on exports, imports, and net imports (imports minus exports) of petroleum and products and domestic production of crude oil and natural gas plant liquids (all expressed as thousands of barrels per day), and in Columns 12 through 14, depicting stocks of crude oil in millions of barrels, are derived from the Department of Energy's weekly petroleum supply reporting system. They are published in EIA's *Monthly Energy Review*, Tables 3.1 and 3.4, which can be reached by searching the Web site. Stock totals are as of the end of the period. Geographic coverage includes the 50 states and the District of Columbia.

Data availability

Data on futures prices, petroleum supply and stocks are available from the EIA Web site at <http://www.eia.doe.gov>, under the categories "Publications and Reports/Monthly Energy Review" and "Petroleum/Weekly Petroleum Status Report." The *Monthly Energy Review* is no longer published in printed form.

The import data are available in the FT900 report from the Census Bureau at www.census.gov. See the notes and definitions for Tables 7-9 through 7-14 for further information.

CHAPTER 12: MONEY, INTEREST, ASSETS, LIABILITIES, AND ASSET PRICES

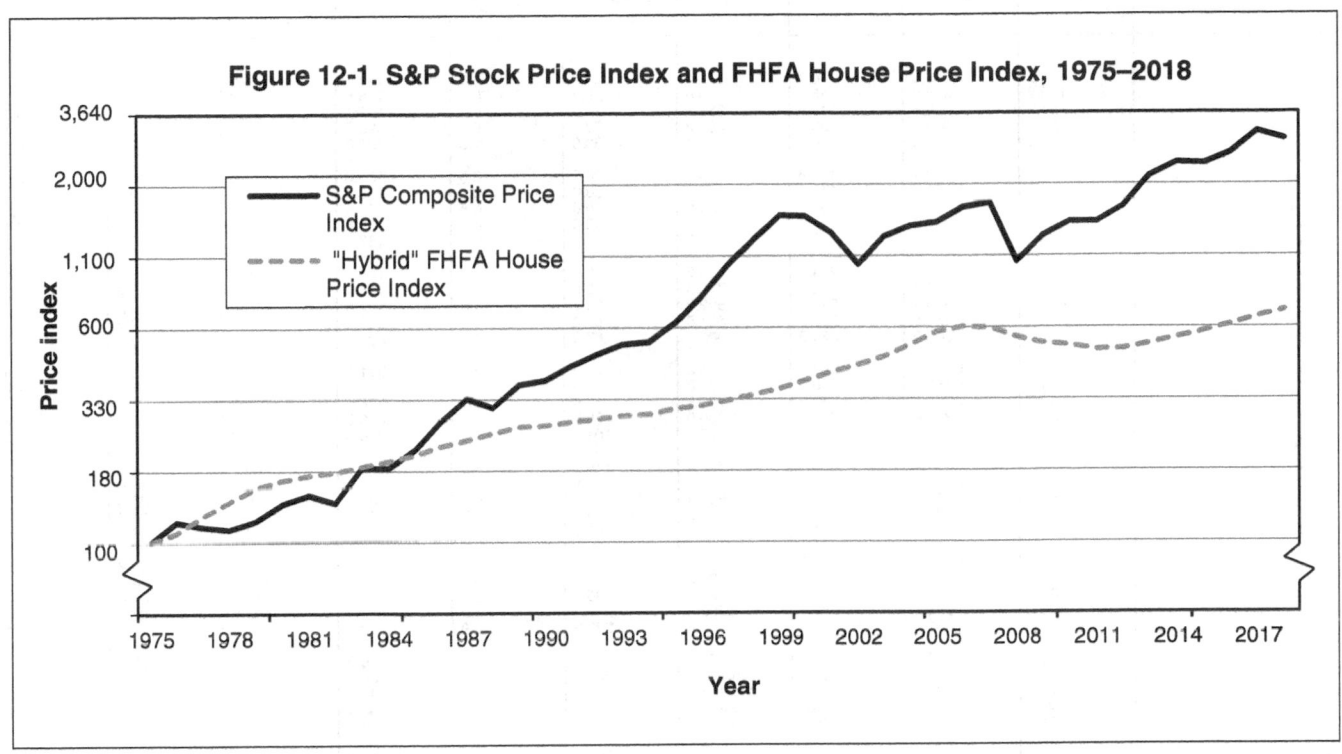

Figure 12-1. S&P Stock Price Index and FHFA House Price Index, 1975–2018

- The housing price index increased each year from 1975 to 2006 then decreased from 2007 to 2011 before increasing again each year from 2012 to 2018. The appreciation in 2017 and 2018 was the highest it has been since 2005. (Table 12-5)

- The money supply is commonly defined to be a group of safe assets that households and businesses can use to make payments or to hold as short-term investments. The monetary base is defined as the sum of currency in circulation and reserve balances (deposits held by banks and other depository institutions in their accounts at the Federal Reserve).

- The size of the monetary base grew rapidly between 2008 to 2014. But with financial institutions reluctant to lend and thereby create deposits, the money supply (both M1 and M2) has grown more slowly since then. From 2010 to 2011, the M1 money supply grew 18 percent. Meanwhile, from 2017 to 2018, the M1 money supply only grew 3.2 percent. (Table 12-1A)

- The stock price indexes all declined steeply during the Great Recession. From 2009 to 2018, the stock prices have typically increased or remained stable each year. However, 2018 was the first year that the Dow Jones, Standard and Poor's, and the Nasdaq composite all declined in the same year since 2008. (Table 12-5)

Table 12-1A. Money Stock Measures and Components of M1: Recent Data

(Billions of dollars, monthly data are averages of daily figures, annual data are for December.)

Year and month	Not seasonally adjusted		Seasonally adjusted					Other checkable deposits		
	M1	M2	M1	M2	Currency	Traveler's checks	Demand deposits	At commercial banks	At thrift institutions	Total
1960	144.5	315.3	140.7	312.4	28.7	0.3	111.6	0.0	0.0	0.0
1961	149.2	338.5	145.2	335.5	29.3	0.4	115.5	0.0	0.0	0.0
1962	151.9	365.8	147.8	362.7	30.3	0.4	117.1	0.0	0.0	0.0
1963	157.5	396.4	153.3	393.2	32.2	0.4	120.6	0.0	0.1	0.1
1964	164.9	428.3	160.3	424.7	33.9	0.5	125.8	0.0	0.1	0.1
1965	172.6	463.1	167.8	459.2	36.0	0.5	131.3	0.0	0.1	0.1
1966	176.9	483.7	172.0	480.2	38.0	0.6	133.4	0.0	0.1	0.1
1967	188.4	528.0	183.3	524.8	40.0	0.6	142.5	0.0	0.1	0.1
1968	202.8	569.7	197.4	566.8	43.0	0.7	153.6	0.0	0.1	0.1
1969	209.3	590.1	203.9	587.9	45.7	0.8	157.3	0.0	0.1	0.1
1970	220.1	627.8	214.4	626.5	48.6	0.9	164.7	0.0	0.1	0.1
1971	234.5	711.2	228.3	710.3	52.0	1.0	175.1	0.0	0.2	0.2
1972	256.1	803.1	249.2	802.3	56.2	1.2	191.6	0.0	0.2	0.2
1973	270.2	856.5	262.9	855.5	60.8	1.4	200.3	0.0	0.3	0.3
1974	281.8	903.5	274.2	902.1	67.0	1.7	205.1	0.2	0.4	0.6
1975	295.3	1 017.8	287.1	1 016.2	72.8	2.1	211.3	0.4	0.5	0.9
1976	314.5	1 153.5	306.2	1 152.0	79.5	2.6	221.5	1.3	1.4	2.7
1977	340.0	1 273.0	330.9	1 270.3	87.4	2.9	236.4	1.8	2.3	4.1
1978	367.9	1 370.8	357.3	1 366.0	96.0	3.3	249.5	5.3	3.1	8.4
1979	393.2	1 479.0	381.8	1 473.7	104.8	3.5	256.6	12.7	4.2	16.9
1980	419.5	1 604.8	408.5	1 599.8	115.3	3.9	261.2	20.8	7.3	28.1
1981	447.0	1 760.3	436.7	1 755.5	122.5	4.1	231.4	63.0	15.6	78.6
1982	485.8	1 914.3	474.8	1 906.4	132.5	4.1	234.1	80.5	23.6	104.1
1983	533.3	2 134.3	521.4	2 123.8	146.2	4.7	238.5	97.3	34.8	132.1
1984	564.6	2 318.8	551.6	2 306.8	156.1	5.0	243.4	104.7	42.4	147.1
1985	633.3	2 504.6	619.8	2 492.6	167.7	5.6	266.9	124.7	54.9	179.6
1986	739.8	2 741.9	724.7	2 729.2	180.4	6.1	302.9	161.0	74.2	235.2
1987	765.4	2 840.8	750.2	2 828.8	196.7	6.6	287.7	178.2	81.0	259.2
1988	803.1	3 003.0	786.7	2 990.6	212.0	7.0	287.1	192.5	88.1	280.6
1989	810.6	3 167.5	792.9	3 154.4	222.3	6.9	278.6	197.4	87.7	285.1
1990	842.7	3 286.0	824.7	3 272.7	246.5	7.7	276.8	208.7	85.0	293.7
1991	915.6	3 385.8	897.0	3 371.6	267.1	7.7	289.6	241.6	90.9	332.5
1992	1 045.6	3 439.7	1 024.9	3 423.1	292.1	8.2	340.0	280.8	103.8	384.6
1993	1 153.3	3 493.0	1 129.6	3 472.4	321.6	8.0	385.4	302.6	112.0	414.6
1994	1 174.5	3 506.4	1 150.7	3 485.0	354.5	8.6	383.6	297.4	106.6	404.0
1995	1 152.7	3 650.5	1 127.5	3 626.7	372.8	9.0	389.0	249.0	107.6	356.6
1996	1 105.8	3 826.1	1 081.3	3 804.9	394.6	8.8	402.1	172.1	103.6	275.7
1997	1 097.5	4 039.6	1 072.3	4 017.4	425.3	8.4	393.5	148.4	96.7	245.1
1998	1 121.2	4 381.0	1 095.0	4 356.6	460.4	8.5	376.3	143.9	105.9	249.8
1999	1 148.2	4 643.6	1 122.2	4 617.0	517.9	8.6	352.5	139.7	103.6	243.3
2000	1 111.7	4 931.3	1 087.8	4 903.7	531.3	8.3	309.6	133.2	105.5	238.7
2001	1 208.5	5 436.6	1 182.3	5 405.7	581.2	8.0	335.2	142.2	115.7	257.9
2002	1 245.5	5 772.5	1 219.2	5 740.4	626.2	7.8	305.9	154.3	125.0	279.3
2003	1 332.2	6 066.9	1 305.2	6 035.2	662.5	7.7	325.1	175.3	134.6	309.9
2004	1 401.2	6 417.5	1 375.5	6 388.4	697.8	7.6	342.3	187.0	140.8	327.8
2005	1 396.9	6 679.3	1 374.8	6 654.1	724.6	7.2	324.3	180.8	138.0	318.8
2006	1 387.3	7 070.1	1 368.3	7 046.0	750.2	6.7	305.6	176.8	128.9	305.7
2007	1 394.4	7 478.0	1 376.6	7 452.4	760.6	6.3	303.4	173.0	133.3	306.3
2008	1 631.9	8 207.5	1 607.1	8 177.0	816.2	5.5	473.3	178.8	133.3	312.1
2009	1 723.8	8 520.7	1 698.4	8 482.4	863.7	5.1	448.2	231.2	150.2	381.4
2010	1 870.8	8 831.4	1 841.8	8 782.9	918.8	4.7	519.3	235.8	163.2	399.0
2011	2 207.5	9 699.5	2 168.2	9 635.9	1 001.5	4.3	751.8	233.2	177.5	410.7
2012	2 505.4	10 499.5	2 457.7	10 423.6	1 090.5	3.8	920.5	244.6	198.3	442.9
2013	2 711.7	11 068.1	2 654.5	10 984.9	1 160.3	3.5	1 021.7	256.8	212.1	468.9
2014	2 976.7	11 719.9	2 910.4	11 630.5	1 252.2	2.9	1 165.9	267.2	222.0	489.2
2015	3 137.4	12 396.3	3 083.4	12 311.2	1 338.6	2.5	1 228.4	277.0	236.9	513.9
2016	3 378.4	13 273.0	3 326.7	13 180.8	1 419.7	2.2	1 353.3	293.8	257.7	551.5
2017	3 636.9	13 919.7	3 598.7	13 836.5	1 525.8	1.9	1 478.3	306.1	286.7	592.8
2018	3 752.8	14 369.4	3 795.8	14 455.0	1 626.7	1.7	1 495.3	333.0	296.1	629.1
2017										
January	3 387.7	13 274.9	3 397.7	13 277.0	1 431.9	2.1	1 406.6	289.9	267.1	557.1
February	3 346.5	13 288.2	3 393.0	13 320.1	1 434.8	2.1	1 400.3	288.4	267.3	555.8
March	3 467.5	13 476.0	3 443.8	13 393.1	1 446.7	2.1	1 430.3	293.2	271.5	564.7
April	3 473.2	13 542.9	3 443.5	13 448.9	1 457.6	2.1	1 421.0	287.1	275.7	562.8
May	3 503.6	13 454.2	3 502.9	13 508.7	1 467.6	2.1	1 457.4	301.4	274.5	575.9
June	3 508.5	13 496.5	3 503.9	13 543.6	1 477.0	2.0	1 453.9	294.9	276.1	570.9
July	3 539.3	13 569.2	3 530.6	13 615.3	1 485.4	2.0	1 469.6	294.6	279.1	573.6
August	3 574.6	13 632.9	3 563.7	13 665.5	1 493.8	2.0	1 487.2	301.1	279.6	580.7
September	3 533.3	13 688.4	3 564.0	13 708.6	1 503.2	2.0	1 476.4	300.7	281.8	582.5
October	3 588.6	13 728.9	3 591.6	13 755.9	1 512.3	1.9	1 490.5	305.0	281.9	586.9
November	3 584.6	13 791.8	3 606.5	13 785.7	1 518.1	1.9	1 492.1	307.5	286.9	594.4
December	3 636.9	13 919.7	3 598.7	13 836.5	1 525.8	1.9	1 478.3	306.1	286.7	592.8
2018										
January	3 653.1	13 868.1	3 652.7	13 855.5	1 537.1	1.9	1 512.3	314.2	287.7	601.9
February	3 622.5	13 890.6	3 566.5	13 841.4	1 542.0	1.9	1 479.1	312.8	286.7	599.5
March	3 656.3	13 941.2	3 688.7	14 023.0	1 550.6	1.9	1 499.9	314.0	290.0	604.0
April	3 660.2	13 974.2	3 698.5	14 064.6	1 559.7	1.8	1 488.8	317.5	292.4	609.9
May	3 654.6	14 035.4	3 655.9	13 984.8	1 569.8	1.8	1 470.3	322.1	290.6	612.7
June	3 655.0	14 107.7	3 654.6	14 079.4	1 580.2	1.8	1 460.7	318.7	293.8	612.4
July	3 676.9	14 148.7	3 678.0	14 114.0	1 589.2	1.8	1 469.0	323.6	293.3	616.8
August	3 679.8	14 190.6	3 686.3	14 170.5	1 597.9	1.8	1 458.1	328.1	293.9	622.0
September	3 704.4	14 225.9	3 671.7	14 205.7	1 608.7	1.7	1 469.4	328.8	295.7	624.5
October	3 722.4	14 248.3	3 718.7	14 209.0	1 615.0	1.7	1 468.7	339.2	297.9	637.1
November	3 701.6	14 266.1	3 676.4	14 262.4	1 620.9	1.7	1 450.0	334.1	294.9	629.0
December	3 752.8	14 369.4	3 795.8	14 455.0	1 626.7	1.7	1 495.3	333.0	296.1	629.1

Table 12-1B. Money Stock, Historical: 1892–1924

(Not seasonally adjusted, millions of dollars.)

Classification	1892	1893	1894	1895	1896	1897	1898	1899	1900	1901	1902
June 30											
Currency outside banks	1 015	1 081	972	971	974	1 013	1 150	1 181	1 331	1 395	1 431
Demand deposits adjusted	2 880	2 766	2 807	2 960	2 839	2 871	3 432	4 162	4 420	5 204	5 719
M1	3 895	3 847	3 779	3 931	3 813	3 884	4 582	5 343	5 751	6 599	7 150
Time deposits	1 929	2 007	1 994	2 088	2 220	2 305	2 397	2 617	3 015	3 315	3 565
M2	5 824	5 854	5 773	6 019	6 033	6 189	6 979	7 960	8 766	9 914	10 715
December 31											
Currency outside banks	. . .	. . .	. . .	. . .	. . .	. . .	. . .	. . .	. . .	. . .	. . .
Demand deposits adjusted	. . .	. . .	. . .	. . .	. . .	. . .	. . .	. . .	. . .	. . .	. . .
M1	. . .	. . .	. . .	. . .	. . .	. . .	. . .	. . .	. . .	. . .	. . .
Time deposits	. . .	. . .	. . .	. . .	. . .	. . .	. . .	. . .	. . .	. . .	. . .
M2	. . .	. . .	. . .	. . .	. . .	. . .	. . .	. . .	. . .	. . .	. . .

Classification	1903	1904	1905	1906	1907	1908	1909	1910	1911	1912	1913
June 30											
Currency outside banks	1 543	1 562	1 629	1 759	1 700	1 711	1 691	1 725	1 709	1 762	1 858
Demand deposits adjusted	5 962	6 256	7 069	7 504	7 872	7 384	7 768	8 254	8 668	9 156	9 140
M1	7 505	7 818	8 698	9 263	9 572	9 095	9 459	9 979	10 377	10 918	10 998
Time deposits	3 800	4 045	4 464	4 769	5 350	5 493	6 265	6 944	7 337	7 889	8 356
M2	11 305	11 863	13 162	14 032	14 922	14 588	15 724	16 923	17 714	18 807	19 354
December 31											
Currency outside banks	. . .	. . .	. . .	. . .	. . .	. . .	. . .	. . .	. . .	. . .	. . .
Demand deposits adjusted	. . .	. . .	. . .	. . .	. . .	. . .	. . .	. . .	. . .	. . .	. . .
M1	. . .	. . .	. . .	. . .	. . .	. . .	. . .	. . .	. . .	. . .	. . .
Time deposits	. . .	. . .	. . .	. . .	. . .	. . .	. . .	. . .	. . .	. . .	. . .
M2	. . .	. . .	. . .	. . .	. . .	. . .	. . .	. . .	. . .	. . .	. . .

Classification	1914	1915	1916	1917	1918	1919	1920	1921	1922	1923	1924
June 30											
Currency outside banks	1 533	1 575	1 876	2 276	3 298	3 593	4 105	3 677	3 346	3 739	3 650
Demand deposits adjusted	10 082	9 828	11 973	13 501	14 843	17 624	19 616	17 113	18 045	18 958	19 412
M1	11 615	11 403	13 849	15 777	18 141	21 217	23 721	20 790	21 391	22 697	23 062
Time deposits	8 350	9 231	10 313	11 543	11 717	13 423	15 834	16 583	17 437	19 722	21 259
M2	19 965	20 634	24 162	27 320	29 858	34 640	39 555	37 373	48 828	42 419	44 321
December 31											
Currency outside banks	. . .	. . .	. . .	. . .	. . .	. . .	. . .	. . .	. . .	3 726	3 696
Demand deposits adjusted	. . .	. . .	. . .	. . .	. . .	. . .	. . .	. . .	. . .	19 144	20 898
M1	. . .	. . .	. . .	. . .	. . .	. . .	. . .	. . .	. . .	22 870	24 594
Time deposits	. . .	. . .	. . .	. . .	. . .	. . .	. . .	. . .	. . .	20 379	22 232
M2	. . .	. . .	. . .	. . .	. . .	. . .	. . .	. . .	. . .	43 249	46 826

. . . = Not available.

Table 12-1C. Money Stock, Historical: January 1947–January 1959

(Averages of daily figures; seasonally adjusted, billions of dollars.)

Year and month	Money stock (M1)			Time deposits adjusted	M2 (M1 plus time deposits)
	Total	Currency component	Demand deposit component		
1947					
January	109.5	26.7	82.8	33.3	142.8
February	109.7	26.7	83.0	33.5	143.2
March	110.3	26.7	83.7	33.6	143.9
April	111.1	26.6	84.5	33.7	144.8
May	111.7	26.6	85.1	33.8	145.5
June	112.1	26.6	85.5	33.9	146.0
July	112.2	26.5	85.7	34.0	146.2
August	112.6	26.5	86.1	34.4	147.0
September	113.0	26.7	86.3	34.7	147.7
October	112.9	26.5	86.4	35.0	147.9
November	113.3	26.5	86.8	35.2	148.5
December	113.1	26.4	86.7	35.4	148.5
1948					
January	113.4	26.4	87.0	35.5	148.9
February	113.2	26.3	86.8	35.7	148.9
March	112.6	26.2	86.4	35.7	148.3
April	112.3	26.1	86.3	35.7	148.0
May	112.1	26.0	86.0	35.7	147.8
June	112.0	26.0	86.0	35.8	147.8
July	112.2	26.0	86.2	35.8	148.0
August	112.3	26.0	86.2	35.9	148.2
September	112.2	26.0	86.2	35.9	148.1
October	112.1	26.0	86.1	35.9	148.0
November	111.8	26.0	85.9	36.0	147.8
December	111.5	25.8	85.8	36.0	147.5
1949					
January	111.2	25.7	85.5	36.1	147.3
February	111.2	25.7	85.5	36.1	147.3
March	111.2	25.7	85.6	36.1	147.3
April	113.3	25.7	85.6	36.2	149.5
May	111.5	25.7	85.8	36.3	147.8
June	111.3	25.6	85.7	36.4	147.7
July	111.2	25.5	85.7	36.4	147.6
August	111.0	25.5	85.6	36.4	147.4
September	110.9	25.3	85.6	36.4	147.3
October	110.9	25.3	85.6	36.4	147.3
November	111.0	25.2	85.8	36.4	147.4
December	111.2	25.1	86.0	36.4	147.6
1950					
January	111.5	25.1	86.4	36.4	147.9
February	112.1	25.1	86.9	36.6	148.7
March	112.5	25.2	87.3	36.6	149.1
April	113.2	25.3	88.0	36.7	149.9
May	113.7	25.2	88.5	36.9	150.6
June	114.1	25.1	89.0	36.9	151.0
July	114.6	25.0	89.6	36.8	151.4
August	115.0	24.9	90.1	36.7	151.7
September	115.2	24.9	90.3	36.6	151.8
October	115.7	24.9	90.8	36.5	152.2
November	115.9	24.9	90.9	36.6	152.5
December	116.2	25.0	91.2	36.7	152.9
1951					
January	116.7	25.0	91.7	36.7	153.4
February	117.1	25.1	92.0	36.6	153.7
March	117.6	25.2	92.4	36.6	154.2
April	117.8	25.2	92.6	36.7	154.5
May	118.2	25.3	92.8	36.8	155.0
June	118.6	25.4	93.2	36.9	155.5
July	119.1	25.6	93.4	37.2	156.3
August	119.6	25.7	93.8	37.4	157.0
September	120.4	25.8	94.5	37.7	158.1
October	121.0	26.0	95.1	37.8	158.8
November	122.0	26.0	96.0	38.0	160.0
December	122.7	26.1	96.5	38.2	160.9
1952					
January	123.1	26.2	96.9	38.4	161.5
February	123.6	26.3	97.3	38.7	162.3
March	123.8	26.4	97.5	38.9	162.7
April	124.1	26.4	97.6	39.1	163.2
May	124.5	26.5	98.0	39.3	163.8
June	125.0	26.7	98.4	39.5	164.5
July	125.3	26.7	98.6	39.7	165.0
August	125.7	26.8	98.9	40.0	165.7
September	126.4	26.9	99.4	40.3	166.7
October	126.7	27.0	99.7	40.5	167.2
November	127.1	27.2	99.9	40.9	168.0
December	127.4	27.3	100.1	41.1	168.5

Table 12-1C. Money Stock, Historical: January 1947–January 1959—*Continued*

(Averages of daily figures; seasonally adjusted, billions of dollars.)

Year and month	Money stock (M1)			Time deposits adjusted	M2 (M1 plus time deposits)
	Total	Currency component	Demand deposit component		
1953					
January	127.3	27.4	99.9	41.4	168.7
February	127.4	27.5	99.9	41.6	169.0
March	128.0	27.6	100.4	41.9	169.9
April	128.3	27.7	100.7	42.1	170.4
May	128.5	27.7	100.7	42.4	170.9
June	128.5	27.7	100.7	42.6	171.1
July	128.6	27.8	100.8	42.9	171.5
August	128.7	27.8	100.9	43.2	171.9
September	128.6	27.8	100.8	43.5	172.1
October	128.7	27.8	100.9	43.9	172.6
November	128.7	27.8	100.9	44.2	172.9
December	128.8	27.7	101.1	44.5	173.3
1954					
January	129.0	27.7	101.3	44.8	173.8
February	129.1	27.7	101.5	45.2	174.3
March	129.2	27.6	101.6	45.6	174.8
April	128.6	27.6	101.0	46.1	174.7
May	129.7	27.6	102.1	46.5	176.2
June	129.9	27.5	102.3	46.8	176.7
July	130.3	27.5	102.8	47.3	177.6
August	130.7	27.5	103.2	47.8	178.5
September	130.9	27.4	103.5	47.9	178.8
October	101.5	27.4	104.1	48.1	179.6
November	132.1	27.4	104.7	48.2	180.3
December	132.3	27.4	104.9	48.3	180.6
1955					
January	133.0	27.4	105.6	48.5	181.5
February	133.9	27.5	106.4	48.7	182.6
March	133.6	27.5	106.0	48.8	182.4
April	133.9	27.5	106.3	49.0	182.9
May	134.6	27.6	107.0	49.0	183.6
June	134.4	27.6	106.8	49.2	183.6
July	134.8	27.7	107.2	49.3	184.1
August	134.8	27.7	107.0	49.3	184.1
September	135.0	27.7	107.3	49.6	184.6
October	135.2	27.8	107.4	49.7	184.9
November	134.9	27.8	107.1	49.9	184.8
December	135.2	27.8	107.4	50.0	185.2
1956					
January	135.5	27.9	107.7	49.9	185.4
February	135.5	27.9	107.7	49.9	185.4
March	135.7	27.9	107.8	50.1	185.8
April	136.0	27.9	108.1	50.3	186.3
May	135.8	27.9	107.9	50.4	186.2
June	136.0	27.9	108.1	50.7	186.7
July	136.0	28.0	108.0	50.9	186.9
August	135.7	28.0	107.8	51.2	186.9
September	136.2	28.0	108.2	51.5	187.7
October	136.3	28.0	108.2	51.6	187.9
November	136.6	28.1	108.4	51.8	188.4
December	136.9	28.2	108.7	51.9	188.8
1957					
January	136.9	28.2	108.6	52.6	189.5
February	136.8	28.2	108.6	53.1	189.9
March	136.9	28.2	108.7	53.7	190.6
April	136.9	28.2	108.7	54.0	190.9
May	137.0	28.2	108.8	54.5	191.5
June	136.9	28.3	108.6	54.8	191.7
July	137.0	28.3	108.7	55.3	192.3
August	137.1	28.3	108.8	55.7	192.8
September	136.8	28.3	108.4	56.1	192.9
October	136.5	28.3	108.2	56.6	193.1
November	136.3	28.3	108.0	57.0	193.3
December	135.9	28.3	107.6	57.4	193.3
1958					
January	135.5	28.3	107.2	57.6	193.1
February	136.2	28.2	107.9	59.2	195.4
March	136.5	28.2	108.3	60.5	197.0
April	137.0	28.2	108.7	61.5	198.5
May	137.5	28.3	109.2	62.3	199.8
June	138.4	28.3	110.1	63.2	201.6
July	138.4	28.4	110.0	64.0	202.4
August	139.1	28.4	110.7	64.6	203.7
September	139.5	28.5	111.1	64.8	204.3
October	140.1	28.5	111.6	64.9	205.0
November	140.9	28.5	112.4	65.2	206.1
December	141.1	28.6	112.6	65.4	206.5
1959					
January	142.2	28.7	113.5	66.3	208.5

Table 12-2. Components of Non-M1 M2

(Billions of dollars, seasonally adjusted; monthly data are averages of daily figures, annual data are for December.)

Year and month	Savings deposits			Small-denomination time deposits			Retail money funds	Total non-M1 M2
	At commercial banks	At thrift institutions	Total	At commercial banks	At thrift institutions	Total		
1960	58.3	100.8	159.1	9.7	2.8	12.5	. . .	171.7
1961	64.2	111.3	175.5	11.1	3.7	14.8	. . .	190.3
1962	71.3	123.4	194.8	15.5	4.6	20.1	. . .	214.9
1963	76.8	137.6	214.4	19.9	5.7	25.5	. . .	240.0
1964	82.9	152.4	235.2	22.4	6.8	29.2	. . .	264.4
1965	92.4	164.5	256.9	26.7	7.8	34.5	. . .	291.3
1966	89.9	163.3	253.1	38.7	16.3	55.0	. . .	308.1
1967	94.1	169.6	263.7	50.7	27.1	77.8	. . .	341.5
1968	96.1	172.8	268.9	63.5	37.1	100.5	. . .	369.4
1969	93.8	169.8	263.7	71.6	48.8	120.4	. . .	384.0
1970	98.6	162.3	261.0	79.3	71.9	151.2	. . .	412.1
1971	112.8	179.4	292.2	94.7	95.1	189.7	. . .	481.9
1972	124.8	196.6	321.4	108.2	123.5	231.6		553.0
1973	128.0	198.7	326.8	116.8	149.0	265.8	0.1	592.6
1974	136.8	201.8	338.6	123.1	164.8	287.9	1.4	627.9
1975	161.2	227.6	388.9	142.3	195.5	337.9	2.4	729.1
1976	201.8	251.4	453.2	155.5	235.2	390.7	1.8	845.8
1977	218.8	273.4	492.2	167.5	278.0	445.5	1.8	939.4
1978	216.5	265.4	481.9	185.1	335.8	521.0	5.8	1 008.7
1979	195.0	228.8	423.8	235.5	398.7	634.3	33.9	1 092.0
1980	185.7	214.5	400.3	286.2	442.3	728.5	62.5	1 191.3
1981	159.0	184.9	343.9	347.7	475.4	823.1	151.7	1 318.8
1982	190.1	210.0	400.1	379.9	471.0	850.9	180.5	1 431.5
1983	363.2	321.7	684.9	350.9	433.1	784.1	133.4	1 602.4
1984	389.3	315.4	704.7	387.9	500.9	888.8	161.7	1 755.2
1985	456.6	358.6	815.3	386.4	499.3	885.7	171.8	1 872.8
1986	533.5	407.4	940.9	369.4	489.0	858.4	205.3	2 004.5
1987	534.8	402.6	937.4	391.7	529.3	921.0	220.2	2 078.6
1988	542.4	383.9	926.4	451.2	585.9	1 037.1	240.4	2 203.9
1989	541.1	352.6	893.7	533.8	617.6	1 151.3	316.5	2 361.5
1990	581.3	341.6	922.9	610.7	562.6	1 173.3	351.8	2 448.0
1991	664.8	379.6	1 044.5	602.2	463.1	1 065.3	364.9	2 474.7
1992	754.2	433.1	1 187.2	508.1	359.7	867.7	343.2	2 398.2
1993	785.3	434.0	1 219.3	467.9	313.5	781.4	342.1	2 342.8
1994	752.8	398.5	1 151.3	503.6	313.9	817.5	365.6	2 334.3
1995	774.8	361.0	1 135.9	575.8	356.5	932.3	431.0	2 499.2
1996	906.0	368.7	1 274.8	594.2	353.7	947.9	500.9	2 723.6
1997	1 022.5	378.7	1 401.2	625.5	342.2	967.6	576.3	2 945.1
1998	1 187.4	416.1	1 603.6	626.4	324.9	951.3	706.7	3 261.6
1999	1 288.9	451.2	1 740.2	636.9	318.3	955.2	799.5	3 494.9
2000	1 426.1	454.5	1 880.6	700.8	345.3	1 046.0	889.2	3 815.9
2001	1 741.1	571.5	2 312.6	636.0	338.5	974.6	936.2	4 223.4
2002	2 058.5	713.3	2 771.9	591.3	303.5	894.7	854.6	4 521.2
2003	2 335.4	824.0	3 159.5	541.9	276.2	818.1	752.5	4 730.0
2004	2 631.4	875.7	3 507.1	551.9	276.5	828.4	677.3	5 012.9
2005	2 774.6	828.9	3 603.4	646.7	347.0	993.7	682.0	5 279.2
2006	2 911.8	783.7	3 695.5	780.7	425.4	1 206.0	776.2	5 677.7
2007	3 042.1	827.5	3 869.6	858.8	417.2	1 276.0	930.2	6 075.8
2008	3 322.1	769.3	4 091.4	1 078.8	378.8	1 457.6	1 020.9	6 569.9
2009	3 979.2	836.4	4 815.5	868.3	319.6	1 187.8	780.6	6 784.0
2010	4 410.0	923.6	5 333.7	661.7	270.8	932.5	675.0	6 941.2
2011	5 034.2	999.0	6 033.3	542.7	228.7	771.4	663.0	7 467.6
2012	5 727.0	959.0	6 686.0	460.4	176.7	637.2	642.7	7 965.9
2013	6 108.0	1 023.8	7 131.8	415.2	144.0	559.2	639.4	8 330.4
2014	6 502.6	1 081.8	7 584.4	378.4	131.5	509.9	625.9	8 720.2
2015	7 034.7	1 145.2	8 179.9	298.6	100.9	408.5	639.4	9 227.8
2016	7 567.2	1 260.0	8 827.1	246.7	100.2	346.5	680.0	9 854.1
2017	7 823.7	1 298.8	9 122.5	299.6	110.4	410.0	705.2	10 237.8
2018	7 940.5	1 336.9	9 277.4	419.6	106.3	525.9	813.4	10 616.6
2017								
January	7 578.8	1 266.4	8 845.2	249.0	99.6	348.6	685.4	9 879.2
February	7 614.0	1 278.2	8 892.2	252.0	98.1	350.1	684.8	9 927.1
March	7 613.2	1 295.3	8 908.6	255.4	96.8	352.2	688.5	9 949.3
April	7 655.0	1 298.8	8 953.8	261.0	96.4	357.4	694.2	10 005.4
May	7 636.5	1 312.7	8 949.2	264.7	98.1	362.8	693.8	10 005.7
June	7 660.8	1 317.6	8 978.3	270.2	99.6	369.8	691.6	10 039.7
July	7 696.8	1 318.3	9 015.1	276.4	100.8	377.2	692.4	10 084.7
August	7 700.4	1 318.2	9 018.6	281.9	103.4	385.2	698.0	10 101.8
September	7 736.2	1 312.2	9 048.4	288.3	107.9	396.2	700.0	10 144.6
October	7 758.9	1 301.8	9 060.7	292.4	108.9	401.3	702.3	10 164.3
November	7 774.6	1 295.7	9 070.3	295.8	109.4	405.2	703.7	10 179.2
December	7 823.7	1 298.8	9 122.5	299.6	110.4	410.0	705.2	10 237.8
2018								
January	7 785.8	1 309.2	9 095.0	307.5	109.6	417.2	702.8	10 214.9
February	7 813.9	1 320.2	9 134.1	315.9	106.9	422.8	711.2	10 268.1
March	7 796.6	1 340.4	9 137.0	327.1	105.2	432.2	715.7	10 284.9
April	7 836.9	1 309.1	9 146.0	343.4	100.0	443.4	724.6	10 314.0
May	7 859.3	1 325.1	9 184.4	353.6	101.5	455.1	741.3	10 380.8
June	7 907.4	1 325.4	9 232.8	364.8	102.7	467.5	752.4	10 452.7
July	7 904.4	1 328.3	9 232.7	374.7	103.8	478.5	760.6	10 471.8
August	7 921.7	1 331.1	9 252.8	384.7	104.8	489.5	768.5	10 510.8
September	7 911.6	1 329.2	9 240.8	396.6	105.0	501.6	779.2	10 521.6
October	7 893.0	1 332.4	9 225.4	403.0	105.1	508.2	792.3	10 525.8
November	7 916.0	1 327.9	9 243.9	410.5	105.9	516.4	804.2	10 564.5
December	7 940.5	1 336.9	9 277.4	419.6	106.3	525.9	813.4	10 616.6

. . . = Not available.

Table 12-3. Aggregate Reserves of Depository Institutions and the Monetary Base

(Not seasonally adjusted. Millions of dollars unless otherwise noted.)

Year and month	Reserves balances required			Reserves balances maintained			Reserves		Vault cash		
	Reserve balance	Top of penalty-free brand	Bottom of penalty-free band	Total	Balances maintaned to satisfy reserve balance requirements	Balances maintained that exceed the top of the penalty-free brand	Total	Required	Total	Used to satisfy required reserves	Surplus
1965	...	...	...	18 757	...	...	22 694	22 270	3 936	3 936	0
1966	...	...	...	19 550	...	...	23 785	23 446	4 235	4 235	0
1967	...	...	...	20 770	...	...	25 291	24 916	4 521	4 521	0
1968	...	...	...	22 455	...	...	27 192	26 766	4 737	4 737	0
1969	...	...	...	23 094	...	...	28 053	27 767	4 959	4 959	0
1970	...	...	...	23 907	...	...	29 246	28 998	5 340	5 340	0
1971	...	...	...	25 670	...	...	31 345	31 163	5 675	5 675	0
1972	...	...	...	25 317	...	...	31 415	31 131	6 098	6 098	0
1973	...	...	...	28 473	...	...	35 108	34 804	6 635	6 635	0
1974	...	...	...	29 682	...	...	36 861	36 603	7 179	7 179	0
1975	...	...	...	27 219	...	...	34 990	34 724	7 772	7 772	0
1976	...	...	...	26 689	...	...	35 237	34 964	8 548	8 548	0
1977	...	...	...	27 134	...	...	36 486	36 296	9 352	9 352	0
1978	...	...	...	31 347	...	...	41 678	41 446	10 331	10 331	0
1979	...	...	...	32 676	...	...	44 020	43 578	11 344	11 344	0
1980	...	...	...	27 354	...	...	40 660	40 146	18 149	13 306	4 843
1981	...	...	...	26 167	...	...	41 925	41 606	19 538	15 758	3 780
1982	...	...	...	24 806	...	...	41 855	41 354	20 392	17 049	3 343
1983	...	...	...	20 986	...	...	38 894	38 333	20 755	17 908	2 847
1984	20 898	...	...	21 734	...	...	40 693	39 858	22 193	18 960	3 233
1985	26 518	...	...	27 582	...	...	48 122	47 058	23 336	20 540	2 796
1986	35 991	...	...	37 165	...	...	59 369	58 196	24 538	22 204	2 334
1987	36 672	...	...	37 692	...	...	62 129	61 109	26 676	24 436	2 239
1988	36 703	...	...	37 765	...	...	63 678	62 616	28 204	25 913	2 291
1989	34 425	...	...	35 367	...	...	62 732	61 790	29 827	27 365	2 462
1990	28 562	...	...	30 227	...	...	59 122	57 457	31 789	28 895	2 894
1991	25 680	...	...	26 670	...	...	55 545	54 555	32 509	28 875	3 634
1992	24 246	...	...	25 401	...	...	56 578	55 423	34 541	31 177	3 364
1993	28 290	...	...	29 360	...	...	62 847	61 777	36 818	33 487	3 331
1994	23 500	...	...	24 671	...	...	61 359	60 188	40 382	36 688	3 694
1995	19 164	...	...	20 455	...	...	57 896	56 605	42 284	37 441	4 843
1996	11 913	...	...	13 332	...	...	51 176	49 757	44 527	37 844	6 683
1997	8 979	...	...	10 666	...	...	47 921	46 234	44 742	37 255	7 486
1998	7 512	...	...	9 025	...	...	45 208	43 695	44 295	36 183	8 112
1999	3 965	...	...	5 260	...	...	41 651	40 357	60 625	36 392	24 233
2000	5 594	...	...	6 920	...	...	38 371	37 045	45 246	31 451	13 795
2001	7 402	...	...	9 046	...	...	41 051	39 407	43 895	32 005	11 890
2002	7 916	...	...	9 925	...	...	40 271	38 263	43 363	30 347	13 016
2003	9 820	...	...	10 867	...	...	42 953	41 906	44 884	32 087	12 797
2004	10 136	...	...	12 045	...	...	46 847	44 938	48 555	34 802	13 754
2005	8 146	...	...	10 046	...	...	45 383	43 483	52 586	35 337	17 249
2006	6 616	...	...	8 479	...	...	43 282	41 419	53 054	34 803	18 251
2007	6 313	...	...	8 098	...	...	43 463	41 678	54 458	35 365	19 093
2008	16 312	...	...	783 631	...	...	820 876	53 558	56 029	37 245	18 784
2009	24 632	...	...	109 983	1	...	114 045	65 250	54 752	40 619	14 134
2010	28 438	...	...	103 507	4	...	107 800	71 365	55 708	42 927	12 781
2011	47 838	...	...	155 004	3	...	159 871	96 510	59 908	48 672	11 236
2012	58 675	...	...	151 742	5	...	157 038	111 634	63 846	52 959	10 887
2013	69 030	75 947	62 114	248 524	75 713	240 953	254 101	124 801	66 275	55 771	10 504
2014	82 770	91 061	74 481	260 670	90 852	251 584	266 593	142 007	69 116	59 236	9 880
2015	89 313	98 259	80 370	241 977	97 981	232 179	248 118	150 727	71 616	61 413	10 203
2016	105 944	116 551	95 338	203 100	116 285	191 472	209 528	170 224	74 751	64 280	10 471
2017	123 224	136 103	111 337	224 427	135 719	210 855	230 982	189 269	76 625	65 549	11 076
2018	123 703	136 085	111 322	169 139	135 698	155 569	175 985	192 165	80 794	68 462	12 333
2017											
January	110 483	121 546	99 424	209 277	121 169	197 160	215 854	176 249	75 793	65 766	10 027
February	108 252	119 092	97 415	223 798	118 775	211 920	230 293	173 208	76 277	64 956	11 320
March	108 174	119 005	97 345	232 625	118 757	220 749	238 777	169 699	72 019	61 525	10 493
April	113 224	124 561	101 890	228 142	124 219	215 720	234 481	176 616	74 055	63 392	10 663
May	115 464	127 023	103 906	222 576	126 547	209 922	228 893	178 630	72 954	63 167	9 787
June	118 187	130 018	106 357	220 654	129 695	207 685	226 968	181 324	72 881	63 137	9 745
July	120 636	132 712	108 561	223 332	132 296	210 102	229 692	184 242	73 571	63 606	9 965
August	120 091	132 113	108 071	234 401	131 641	221 237	240 717	183 249	74 398	63 158	11 241
September	122 109	134 332	109 887	229 573	133 950	216 178	235 787	184 248	73 039	62 140	10 900
October	116 873	128 574	105 176	224 570	128 204	211 749	231 090	182 074	76 477	65 200	11 277
November	123 141	135 467	110 816	231 445	134 981	217 947	237 828	186 968	74 567	63 827	10 741
December	123 720	136 103	111 337	224 427	135 719	210 855	230 982	189 269	76 625	65 549	11 076
2018											
January	126 792	139 484	114 101	221 460	138 807	207 579	228 101	193 209	77 203	66 417	10 786
February	124 006	136 420	111 594	223 877	135 786	210 298	230 525	190 487	78 784	66 481	12 304
March	120 285	132 326	108 245	216 700	131 778	203 523	223 087	184 151	75 293	63 866	11 427
April	126 415	139 070	113 762	208 619	138 481	194 771	215 104	191 270	75 666	64 855	10 811
May	128 144	140 970	115 318	202 245	140 316	188 213	208 657	192 267	73 914	64 124	9 790
June	125 503	138 065	112 942	198 822	137 568	185 065	205 274	190 028	74 641	64 525	10 117
July	125 736	138 323	113 152	194 982	137 861	181 196	201 461	190 529	75 437	64 792	10 645
August	123 620	135 994	111 246	191 116	135 544	177 562	197 609	188 549	76 483	64 929	11 554
September	126 756	139 444	114 069	187 386	138 965	173 490	193 801	190 901	74 942	64 145	10 797
October	123 821	136 215	111 428	183 001	135 731	169 428	189 579	189 602	75 865	65 781	10 085
November	126 537	139 202	113 872	177 529	138 700	163 659	184 082	192 074	76 554	65 537	11 017
December	123 703	136 085	111 322	169 139	135 698	155 569	175 985	192 165	80 794	68 462	12 333

See footnotes at end of table.

Table 12-3. Aggregate Reserves of Depository Institutions and the Monetary Base—Continued

(Not seasonally adjusted. Millions of dollars unless otherwise noted.)

Year and month	Monetary base			Borrowings from the federal reserve					
	Total	Total balances maintained	Currency in circulation	Total	Primary	Secondary	Seasonal	Other credit extensions	Nonborrowed reserves
1965	61 022	18 758	42 264	444	...	...	0	...	22 250
1966	64 185	19 551	44 634	532	...	...	0	...	23 252
1967	67 818	20 771	47 047	228	...	...	0	...	25 064
1968	73 139	22 456	50 683	746	...	...	0	...	26 446
1969	76 752	23 095	53 658	1 119	...	...	0	...	26 934
1970	80 987	23 907	57 079	332	...	...	0	...	28 914
1971	86 798	25 671	61 127	126	...	...	0	...	31 219
1972	91 381	25 318	66 063	1 050	...	...	0	...	30 365
1973	100 143	28 474	71 669	1 298	...	...	41	...	33 810
1974	108 640	29 682	78 957	727	...	...	32	...	36 134
1975	113 125	27 220	85 905	130	...	...	14	...	34 860
1976	120 337	26 690	93 647	53	...	...	13	...	35 184
1977	129 915	27 135	102 780	569	...	...	55	...	35 917
1978	144 721	31 348	113 373	868	...	...	135	...	40 810
1979	156 653	32 677	123 977	1 473	...	...	82	...	42 547
1980	163 376	27 355	136 021	1 690	...	...	116	...	38 970
1981	170 681	26 282	144 399	636	...	...	54	...	41 289
1982	180 717	25 241	155 476	634	...	...	33	...	41 221
1983	192 191	22 002	170 189	774	...	...	96	...	38 120
1984	204 812	23 060	181 752	3 186	...	...	113	...	37 507
1985	224 722	29 494	195 229	1 318	...	...	56	...	46 804
1986	248 625	39 492	209 133	827	...	...	38	...	58 543
1987	266 878	39 632	227 246	777	...	...	93	...	61 351
1988	283 789	39 344	244 445	1 716	...	...	130	...	61 962
1989	294 323	37 558	256 765	265	...	...	84	...	62 466
1990	315 270	32 343	282 927	326	...	...	76	...	58 796
1991	335 415	30 947	304 467	192	...	...	38	...	55 353
1992	361 903	31 418	330 486	124	...	...	18	...	56 454
1993	398 638	36 191	362 447	82	...	...	31	...	62 765
1994	428 489	29 483	399 006	209	...	...	100	...	61 150
1995	445 840	26 125	419 715	257	...	...	40	...	57 638
1996	464 955	20 615	444 340	155	...	...	68	...	51 021
1997	493 078	17 728	475 350	324	...	...	79	...	47 597
1998	526 435	15 969	510 467	117	...	...	15	...	45 091
1999	612 499	12 614	599 885	320	...	...	67	...	41 331
2000	598 305	14 053	584 252	210	...	...	111	...	38 161
2001	650 775	18 499	632 276	67	...	...	33	...	40 984
2002	699 216	20 906	678 310	80	...	...	45	...	40 191
2003	739 408	23 032	716 376	46	17	0	29	...	42 907
2004	776 279	22 785	753 494	63	11	0	52	...	46 784
2005	803 124	18 406	784 718	169	97	0	72	...	45 214
2006	826 731	15 586	811 145	191	111	0	80	...	43 091
2007	837 192	14 905	822 287	15 430	3 787	1	30	...	28 033
2008	166 636	788 042	878 323	653 565	88 245	52	3	0	167 311
2009	202 622	110 179	924 426	169 927	19 025	518	37	0	970 523
2010	201 700	103 733	979 663	45 488	41	3	26	0	103 251
2011	261 958	155 257	106 701	9 526	103	0	23	0	158 918
2012	267 594	151 742	115 852	795	12	0	23	0	156 958
2013	371 745	248 524	123 220	170	13	0	59	0	254 084
2014	393 445	260 670	132 775	102	22	0	80	0	266 583
2015	383 581	241 977	141 603	106	38	0	67	0	248 108
2016	353 156	203 100	150 055	39	13	0	25	0	209 524
2017	385 096	224 427	160 669	75	43	0	33	0	230 974
2018	340 082	169 139	170 943	76	18	0	58	0	175 978
2017									
January	359 545	209 277	150 267	16	11	0	5	0	215 852
February	374 640	223 798	150 842	21	17	0	4	0	230 291
March	385 628	232 625	153 003	12	7	0	5	0	238 776
April	382 169	228 142	154 027	45	21	0	24	0	234 476
May	377 441	222 576	154 864	56	11	0	44	0	228 888
June	376 278	220 654	155 623	95	15	0	80	0	226 958
July	379 543	223 332	156 210	166	12	0	154	0	229 676
August	391 002	234 401	156 600	220	7	0	213	0	240 695
September	387 449	229 573	157 875	224	10	0	214	0	235 765
October	382 991	224 570	158 420	151	4	0	146	0	231 075
November	390 773	231 445	159 328	65	11	0	54	0	237 821
December	385 096	224 427	160 669	75	43	0	33	0	230 974
2018									
January	382 479	221 460	161 019	58	51	0	7	0	228 096
February	385 505	223 877	161 628	20	7	0	14	0	230 523
March	380 060	216 700	163 359	16	4	0	12	0	223 085
April	372 711	208 619	164 092	51	16	0	35	0	215 099
May	367 475	202 245	165 230	94	25	0	69	0	208 648
June	365 048	198 822	166 226	143	17	0	126	0	205 260
July	361 827	194 982	166 845	224	20	0	203	0	201 438
August	358 447	191 116	167 331	261	18	0	243	0	197 583
September	355 989	187 386	168 602	290	38	0	252	0	193 772
October	352 095	183 001	169 093	209	21	0	189	0	189 558
November	347 640	177 529	170 111	97	11	0	86	0	184 073
December	340 082	169 139	170 943	76	18	0	58	0	175 978

. . . = Not available.

Table 12-4. Consumer Credit

(Outstanding at end of period, billions of dollars.)

Year and month	Seasonally adjusted			Not seasonally adjusted						
	Total	By major credit type		Total[1]	By major holder					
		Revolving	Non-revolving		Depository institutions	Finance companies[2]	Credit unions	Federal government[2]	Nonfinancial businesses	Securitized pools[3]
1965	96.0	...	96.0	97.5	49.1	23.9	6.5	0.0	18.1	0.0
1966	101.8	...	101.8	103.4	52.2	24.8	7.5	0.0	19.0	0.0
1967	106.8	...	106.8	108.6	55.8	24.6	8.3	0.0	19.9	0.0
1968	117.4	2.0	115.4	119.3	62.7	26.1	9.7	0.0	20.8	0.0
1969	127.2	3.6	123.6	129.2	67.8	27.8	11.7	0.0	21.9	0.0
1970	131.6	5.0	126.6	133.7	70.1	27.6	13.0	0.0	23.0	0.0
1971	146.9	8.2	138.7	149.2	79.0	29.2	14.8	0.0	26.2	0.0
1972	166.2	9.4	156.8	168.8	92.1	31.9	17.0	0.0	27.8	0.0
1973	190.1	11.3	178.7	193.0	108.1	35.4	19.6	0.0	29.8	0.0
1974	198.9	13.2	185.7	201.9	112.1	36.1	21.9	0.0	31.8	0.0
1975	204.0	14.5	189.5	207.0	116.2	32.6	25.7	0.0	32.6	0.0
1976	225.7	16.5	209.2	229.0	128.9	33.7	31.2	0.0	35.2	0.0
1977	260.6	37.4	223.1	264.9	152.1	37.3	37.6	0.5	37.4	0.0
1978	306.1	45.7	260.4	311.3	179.6	44.4	45.2	0.9	41.2	0.0
1979	348.6	53.6	295.0	354.6	205.7	55.4	47.4	1.5	44.6	0.0
1980	351.9	55.0	297.0	358.0	202.9	62.2	44.1	2.6	46.2	0.0
1981	371.3	60.9	310.4	377.9	208.2	70.1	46.7	4.8	48.1	0.0
1982	389.8	66.3	323.5	396.7	217.5	75.3	48.8	6.4	48.7	0.0
1983	437.1	79.0	358.0	444.9	245.2	83.3	56.1	4.6	55.7	0.0
1984	517.3	100.4	416.9	526.6	303.1	89.9	67.9	5.6	60.2	0.0
1985	599.7	124.5	475.2	610.6	354.8	111.7	74.0	6.8	63.3	0.0
1986	654.8	141.1	513.7	666.4	383.1	104.0	77.1	8.2	64.0	0.0
1987	686.3	160.9	525.5	698.6	399.4	140.0	81.0	10.0	68.1	0.0
1988	731.9	184.6	547.3	745.2	427.6	144.7	88.3	13.2	71.4	0.0
1989	794.6	211.2	583.4	809.3	445.8	138.9	91.7	16.0	69.6	47.3
1990	808.2	238.6	569.6	824.4	431.6	133.4	91.6	19.2	71.9	76.7
1991	798.0	263.8	534.3	815.6	412.3	121.6	90.3	21.1	67.3	103.0
1992	806.1	278.4	527.7	824.8	400.3	118.1	91.7	24.2	70.3	120.3
1993	865.7	309.9	555.7	886.2	433.6	116.1	101.6	27.2	77.2	130.5
1994	997.3	365.6	631.7	1 021.2	497.2	134.4	119.6	37.2	86.6	146.1
1995	1 140.7	443.9	696.8	1 168.2	542.4	152.1	131.9	43.5	85.1	213.1
1996	1 253.4	507.5	745.9	1 273.9	572.2	154.9	144.1	51.4	77.7	273.5
1997	1 324.8	540.0	784.8	1 344.2	562.2	167.5	152.4	57.2	84.4	320.5
1998	1 421.0	581.4	839.6	1 441.3	564.4	183.3	155.4	64.9	79.3	393.9
1999	1 531.1	610.7	920.4	1 553.6	569.5	201.6	167.9	81.8	76.1	456.7
2000	1 717.0	682.6	1 034.3	1 741.3	615.8	234.4	184.4	96.7	81.5	528.4
2001	1 867.9	714.8	1 153.0	1 891.8	639.5	280.0	189.6	111.9	73.1	597.8
2002	1 972.1	750.9	1 221.2	1 997.0	671.3	307.5	195.7	117.3	74.8	630.4
2003	2 077.4	768.3	1 309.1	2 102.9	747.3	393.0	205.9	102.9	59.1	594.8
2004	2 192.2	799.6	1 392.7	2 220.1	795.6	492.3	215.4	86.1	59.2	571.5
2005	2 290.9	829.5	1 461.4	2 320.6	816.1	516.5	228.6	89.8	59.6	609.9
2006	2 456.7	923.9	1 532.8	2 456.7	836.7	534.9	236.1	108.7	48.3	617.2
2007	2 609.5	1 001.6	1 607.9	2 609.5	894.9	577.9	236.6	115.7	49.4	652.5
2008	2 643.8	1 004.0	1 639.8	2 643.8	965.0	561.4	236.2	135.2	46.9	610.2
2009	2 555.0	916.1	1 638.9	2 555.0	906.3	480.8	237.1	232.6	43.2	572.5
2010	2 646.8	839.1	1 807.7	2 646.8	1 185.5	705.0	226.5	363.8	44.4	50.3
2011	2 756.6	840.4	1 916.2	2 756.6	1 192.6	687.6	223.0	494.8	45.4	46.2
2012	2 913.6	840.4	2 073.2	2 913.6	1 215.0	679.8	243.6	622.2	44.9	50.0
2013	3 091.4	854.7	2 236.7	3 091.4	1 271.2	679.1	265.6	735.5	39.1	49.1
2014	3 312.5	888.0	2 424.5	3 312.5	1 343.1	684.1	302.8	846.2	38.9	49.8
2015	3 411.0	906.7	2 504.3	3 411.0	1 428.3	561.3	342.3	949.7	38.5	46.0
2016	3 644.1	968.0	2 676.2	3 644.1	1 532.1	548.7	380.3	1 049.3	39.7	52.8
2017	3 828.3	1 022.1	2 806.1	3 828.3	1 611.9	541.0	418.4	1 145.6	38.6	37.6
2018	4 009.8	1 053.5	2 956.3	4 009.8	1 682.0	534.0	469.2	1 236.3	38.6	18.3
2017										
January	3 660.2	970.7	2 689.5	3 654.6	1 510.9	547.1	384.4	1 077.1	39.3	55.1
February	3 678.3	976.7	2 701.7	3 642.1	1 493.3	542.3	387.7	1 083.1	38.8	56.8
March	3 692.3	980.7	2 711.6	3 640.1	1 487.1	542.6	386.9	1 087.0	38.5	58.5
April	3 706.2	982.7	2 723.5	3 657.3	1 493.8	539.2	398.2	1 089.5	38.6	59.1
May	3 722.8	989.2	2 733.6	3 680.1	1 508.3	539.8	401.5	1 094.4	38.7	58.9
June	3 738.2	993.0	2 745.2	3 695.6	1 514.3	543.3	404.3	1 098.3	38.7	58.7
July	3 752.4	995.2	2 757.2	3 709.6	1 519.8	542.5	411.7	1 101.9	38.7	57.7
August	3 761.8	998.9	2 763.0	3 741.9	1 534.8	541.3	412.5	1 121.7	38.9	56.1
September	3 770.2	1 004.0	2 766.2	3 753.5	1 541.4	540.7	408.9	1 133.6	38.9	54.2
October	3 788.0	1 009.1	2 778.9	3 769.7	1 562.1	540.4	414.6	1 138.2	38.8	40.0
November	3 817.9	1 020.7	2 797.2	3 801.6	1 588.3	541.2	418.7	1 140.8	38.3	39.0
December	3 828.3	1 022.1	2 806.1	3 828.3	1 611.9	541.0	418.4	1 145.6	38.6	37.6
2018										
January	3 840.2	1 024.1	2 816.1	3 834.5	1 593.6	538.2	420.2	1 172.7	38.3	36.9
February	3 852.0	1 024.7	2 827.3	3 815.6	1 573.8	535.0	419.6	1 178.8	37.8	36.2
March	3 862.3	1 023.9	2 838.3	3 809.5	1 569.0	529.9	422.2	1 181.6	37.6	35.1
April	3 864.8	1 016.8	2 848.1	3 815.3	1 579.3	531.3	427.2	1 183.8	37.6	22.2
May	3 886.4	1 025.1	2 861.2	3 843.4	1 595.3	532.8	434.5	1 188.0	37.8	21.3
June	3 895.2	1 024.2	2 871.1	3 852.2	1 599.9	530.9	438.3	1 191.6	37.8	20.4
July	3 920.3	1 034.1	2 886.2	3 876.9	1 612.0	532.5	447.6	1 194.3	37.8	19.9
August	3 941.7	1 039.0	2 902.7	3 921.7	1 627.7	534.3	456.0	1 213.5	38.0	19.7
September	3 956.0	1 040.5	2 915.5	3 939.1	1 630.8	532.3	462.0	1 224.4	38.0	19.5
October	3 976.2	1 049.2	2 927.0	3 958.2	1 642.2	533.4	466.7	1 227.0	38.0	19.1
November	3 997.8	1 056.2	2 941.7	3 982.3	1 661.1	533.2	468.1	1 231.3	38.2	18.7
December	4 009.8	1 053.5	2 956.3	4 009.8	1 682.0	534.0	469.2	1 236.3	38.6	18.3

[1]Includes nonprofit and educational institutions, not shown.
[2]Student Loan Marketing Association (Sallie Mae) included in federal government sector until end of 2004, and in finance companies since then.
[3]Outstanding balances of pools upon which securities have been issued; these balances are no longer carried on the balance sheets of the loan originators.
... = Not available.

Table 12-5. Selected Stock and Housing Market Data

Year and month	Stock price indexes			FHFA House Price Indexes			
				House Price Index (purchases and refinance)		Purchase-Only Index	
	Dow Jones industrials (30 stocks)	Standard and Poor's composite (500 stocks) (1941–1943 = 10)	Nasdaq composite (Feb. 5, 1971 = 100)	Level at end of period (1980:I = 100)	Appreciation from same quarter one year earlier (percent)	Level at end of period (1991:I = 100)	Appreciation from same quarter one year earlier (percent)
1965	910.88	88.17	...	...	...	...	...
1966	873.60	85.26	...	...	...	...	...
1967	879.12	91.93	...	...	...	...	...
1968	906.00	98.70	...	...	...	...	...
1969	876.72	97.84	...	...	...	...	...
1970	753.19	83.22	...	...	...	...	...
1971	884.76	98.29	107.44	...	...	...	...
1972	950.71	109.20	128.52	...	...	...	...
1973	923.88	107.43	109.90	...	...	...	...
1974	759.37	82.85	76.29	...	...	...	...
1975	802.49	86.16	77.20	62.26	...	...	...
1976	974.92	102.01	89.90	67.26	8.03	...	...
1977	894.63	98.20	98.71	77.21	14.79	...	...
1978	820.23	96.02	117.53	87.43	13.24	...	...
1979	844.40	103.01	136.57	98.27	12.40	...	...
1980	891.41	118.78	168.61	104.77	6.61	...	...
1981	932.92	128.05	203.18	109.15	4.18	...	...
1982	884.36	119.71	188.97	112.22	2.81	...	...
1983	1 190.34	160.41	285.43	117.06	4.31	...	...
1984	1 178.48	160.46	248.88	122.47	4.62	...	...
1985	1 328.23	186.84	290.19	129.47	5.72	...	...
1986	1 792.76	236.34	366.96	138.87	7.26	...	...
1987	2 275.99	286.83	402.57	146.36	5.39	...	...
1988	2 060.82	265.79	374.43	154.60	5.63	...	...
1989	2 508.91	322.84	437.81	163.26	5.60	...	...
1990	2 678.94	334.59	409.17	165.19	1.18	...	...
1991	2 929.33	376.18	491.69	170.35	3.12	101.47	2.72
1992	3 284.29	415.74	599.26	174.47	2.42	104.23	2.72
1993	3 522.06	451.41	715.16	179.08	2.64	107.07	2.84
1994	3 793.77	460.42	751.65	181.90	1.57	110.11	2.66
1995	4 493.76	541.72	925.19	190.18	4.55	113.04	2.79
1996	5 742.89	670.50	1 164.96	195.03	2.55	116.19	3.30
1997	7 441.15	873.43	1 469.49	203.64	4.41	120.03	5.69
1998	8 625.52	1 085.50	1 794.91	213.95	5.06	126.86	6.18
1999	10 464.88	1 327.33	2 728.15	224.49	4.93	134.70	6.96
2000	10 786.85	1 320.28	2 470.52	240.38	7.08	144.08	6.75
2001	10 021.50	1 148.08	1 950.40	257.47	7.11	153.81	7.66
2002	8 341.63	879.82	1 335.51	274.72	6.70	165.59	7.83
2003	10 453.92	1 111.92	2 003.37	293.86	6.97	178.56	10.12
2004	10 783.01	1 211.92	2 175.44	324.14	10.30	196.63	10.18
2005	10 717.50	1 248.29	2 205.32	360.52	11.22	216.65	2.91
2006	12 463.15	1 418.30	2 415.29	376.85	4.53	222.95	-2.64
2007	13 264.82	1 468.36	2 652.28	372.57	-1.14	217.06	-10.10
2008	8 776.39	903.25	1 577.03	345.98	-7.14	195.13	-2.51
2009	10 428.05	1 115.10	2 269.15	327.87	-5.23	190.24	-4.01
2010	11 577.51	1 257.64	2 652.87	321.73	-1.87	182.61	-2.46
2011	12 217.56	1 257.60	2 605.15	311.06	-3.32	178.12	4.90
2012	13 104.14	1 426.19	3 019.51	312.97	0.61	186.85	6.94
2013	16 576.66	1 848.36	4 176.59	327.20	4.55	199.82	4.76
2014	17 823.07	2 058.90	4 736.05	344.05	5.15	209.33	5.61
2015	17 425.03	2 043.94	5 007.41	362.27	5.30	221.08	6.18
2016	19 762.60	2 238.83	5 383.12	382.96	5.71	234.75	6.76
2017	24 719.22	2 673.61	6 903.39	407.25	6.34	250.63	5.90
2018	23 327.46	2 506.85	6 635.28	431.95	6.07	265.41	...
2017							
January	19 864.09	2 278.87	5 614.79	...	...	...	...
February	20 812.24	2 363.64	5 825.44	...	...	...	...
March	20 663.22	2 362.72	5 911.74	386.47	5.70	238.42	6.21
April	20 940.51	2 384.20	6 047.61	...	...	...	...
May	21 008.65	2 411.80	6 198.52	...	...	...	...
June	21 349.63	2 423.41	6 140.42	396.29	6.31	242.60	6.59
July	21 891.12	2 470.30	6 348.12	...	...	...	...
August	21 948.10	2 471.65	6 428.66	...	...	...	...
September	22 405.09	2 519.36	6 495.96	403.07	6.22	246.54	6.64
October	23 377.24	2 575.26	6 727.67	...	...	...	...
November	24 272.35	2 647.58	6 873.97	...	...	...	...
December	24 719.22	2 673.61	6 903.39	407.25	6.34	250.91	6.76
2018							
January	26 149.39	2 823.81	7 411.48	...	...	...	...
February	25 029.20	2 713.83	7 273.01	...	...	...	...
March	24 103.11	2 640.87	7 063.45	413.14	6.90	255.83	7.30
April	24 163.15	2 648.05	7 066.27	...	...	...	...
May	24 415.84	2 705.27	7 442.12	...	...	...	...
June	24 271.41	2 718.37	7 510.30	423.24	6.80	259.17	6.83
July	25 415.19	2 816.29	7 671.79	...	...	...	...
August	25 964.82	2 901.52	8 109.54	...	...	...	...
September	26 458.31	2 913.98	8 046.35	430.15	6.72	262.58	6.51
October	25 115.76	2 711.74	7 305.90	...	...	...	...
November	25 538.46	2 760.17	7 330.54	...	...	...	...
December	23 327.46	2 506.85	6 635.28	431.95	6.07	265.70	5.89

. . . = Not available.

Table 12-6. Mortgage Debt Outstanding

(Billions of dollars, except as noted; end of period; not seasonally adjusted.)

| Year and quarter | Total | By type of property | | | | By type of holder | | | | | | |
| | | Home | Multi-family residences | Commercial | Farm | U.S.-chartered depository institutions | Life insurance companies | Federal and related agencies | Mortgage pools or trusts [1] | | | Other |
									Total	Federally related agencies	ABS issuers	
1955	130	88	13	19	9	69	29	5	0	0	0	26
1956	144	99	14	22	10	77	33	6	0	0	0	28
1957	157	107	15	24	10	83	35	7	0	0	0	30
1958	172	117	17	27	11	93	37	8	0	0	0	34
1959	191	130	19	30	12	105	39	10	0	0	0	37
1960	208	141	21	33	13	115	42	11	0	0	0	41
1961	229	154	24	37	14	127	44	12	0	0	0	46
1962	252	168	27	42	15	143	47	12	0	0	0	50
1963	279	185	30	47	17	164	51	11	1	1	0	53
1964	307	202	35	51	19	183	55	12	1	1	0	56
1965	335	219	38	56	21	202	60	13	1	1	0	59
1966	359	233	41	61	23	214	65	16	1	1	0	62
1967	382	246	45	66	25	228	68	19	2	2	0	65
1968	411	263	48	73	27	247	70	23	3	3	0	70
1969	440	279	53	79	29	264	72	28	3	3	0	73
1970	469	292	60	87	30	278	74	34	5	5	0	79
1971	518	318	70	97	32	313	75	37	10	10	0	83
1972	590	357	83	114	35	366	77	40	14	14	0	92
1973	667	400	93	134	40	418	81	47	18	18	0	102
1974	728	435	100	148	45	451	86	61	21	21	0	109
1975	786	474	101	161	50	485	89	73	29	29	0	110
1976	871	535	106	174	55	547	92	76	41	41	0	115
1977	999	628	114	193	64	637	97	84	57	57	0	125
1978	1 151	738	125	215	73	730	106	100	70	70	0	144
1979	1 317	856	135	239	87	808	118	121	95	95	0	175
1980	1 458	958	143	260	97	853	131	143	114	114	0	217
1981	1 580	1 030	142	300	107	890	138	160	129	129	0	262
1982	1 661	1 070	146	334	111	871	142	177	179	179	0	293
1983	1 851	1 186	161	389	114	950	151	188	245	245	0	316
1984	2 092	1 322	186	472	112	1 084	157	202	300	289	11	350
1985	2 369	1 527	206	542	94	1 189	172	213	392	368	25	402
1986	2 656	1 730	239	602	84	1 279	194	202	549	532	18	431
1987	2 954	1 928	258	692	76	1 403	212	189	701	669	31	450
1988	3 272	2 163	275	764	71	1 540	233	192	786	745	40	521
1989	3 524	2 370	287	798	69	1 609	254	198	922	870	53	540
1990	3 779	2 607	287	818	68	1 601	268	239	1 086	1 020	66	586
1991	3 931	2 775	284	804	67	1 527	260	266	1 270	1 156	113	608
1992	4 041	2 942	271	760	68	1 471	242	286	1 440	1 272	168	602
1993	4 172	3 101	268	734	68	1 494	224	326	1 561	1 357	204	567
1994	4 336	3 279	269	719	70	1 560	216	316	1 697	1 472	224	548
1995	4 522	3 446	274	730	72	1 642	213	308	1 812	1 571	241	547
1996	4 803	3 683	287	759	74	1 731	208	294	1 989	1 712	277	580
1997	5 116	3 918	299	821	79	1 840	207	285	2 167	1 826	340	617
1998	5 603	4 276	334	910	83	1 948	214	292	2 487	2 019	468	662
1999	6 210	4 701	375	1 046	87	2 133	231	320	2 832	2 294	539	693
2000	6 767	5 125	404	1 152	85	2 350	236	340	3 097	2 493	604	743
2001	7 450	5 678	446	1 237	89	2 511	243	372	3 532	2 832	701	792
2002	8 359	6 434	486	1 343	95	2 799	250	432	3 978	3 159	820	898
2003	9 367	7 261	561	1 462	83	3 084	261	694	4 330	3 343	987	997
2004	10 649	8 293	610	1 650	96	3 605	273	703	4 835	3 384	1 451	1 233
2005	12 117	9 450	675	1 887	105	4 055	286	665	5 712	3 548	2 163	1 399
2006	13 530	10 532	718	2 171	108	4 416	304	687	6 631	3 841	2 790	1 491
2007	14 613	11 253	808	2 440	113	4 659	327	725	7 436	4 464	2 972	1 466
2008	14 694	11 152	853	2 554	135	4 615	344	801	7 594	4 961	2 633	1 339
2009	14 450	10 963	866	2 475	146	4 372	327	816	7 651	5 377	2 275	1 283
2010	13 897	10 525	866	2 352	154	4 195	319	5 128	3 110	1 147	1 963	1 145
2011	13 572	10 283	866	2 255	167	4 050	335	5 034	3 036	1 312	1 724	1 118
2012	13 336	10 050	893	2 220	173	4 026	347	4 935	2 948	1 443	1 506	1 080
2013	13 345	9 959	933	2 267	185	3 983	366	4 993	2 774	1 574	1 201	1 228
2014	13 491	9 939	993	2 362	197	4 091	388	4 988	2 743	1 649	1 094	1 282
2015	13 860	10 064	1 095	2 493	209	4 297	431	5 037	2 792	1 775	1 016	1 304
2016	14 325	10 273	1 201	2 625	226	4 538	465	5 147	2 828	1 933	895	1 347
2017	14 891	10 584	1 309	2 760	238	4 698	507	5 315	2 973	2 127	846	1 398
2018	15 441	10 878	1 411	2 901	251	4 819	568	5 457	3 145	2 293	852	1 453
2016												
1st quarter	13 916	10 067	1 119	2 517	213	4 341	436	4 777	2 779	1 804	975	1 584
2nd quarter	14 038	10 131	1 143	2 547	217	4 432	445	4 877	2 788	1 844	944	1 495
3rd quarter	14 203	10 216	1 171	2 594	222	4 496	453	4 949	2 803	1 885	918	1 502
4th quarter	14 325	10 273	1 201	2 625	226	4 538	465	5 003	2 828	1 933	895	1 491
2017												
1st quarter	14 422	10 331	1 222	2 641	229	4 556	477	5 033	2 842	1 977	865	1 515
2nd quarter	14 566	10 408	1 242	2 683	232	4 622	490	5 112	2 876	2 023	853	1 465
3rd quarter	14 705	10 500	1 265	2 704	235	4 661	498	5 159	2 918	2 076	841	1 470
4th quarter	14 891	10 584	1 309	2 760	238	4 698	507	5 205	2 973	2 127	846	1 507
2018												
1st quarter	14 989	10 628	1 326	2 794	241	4 721	521	5 242	3 003	2 162	841	1 502
2nd quarter	15 160	10 716	1 351	2 849	244	4 768	536	5 304	3 068	2 200	867	1 485
3rd quarter	15 309	10 817	1 383	2 862	248	4 798	553	5 351	3 106	2 245	860	1 501
4th quarter	15 441	10 878	1 411	2 901	251	4 819	568	5 387	3 145	2 293	852	1 522

[1]Outstanding principal balances of mortgage-backed securities issued or guaranteed by the holder indicated.

Table 12-7. Derivation of U.S. Net Wealth

(Billions of dollars, amounts outstanding at the end of period.)

Year and month	Household net worth	Growth of domestic nonfinancial debt				
		Total	households	Business	State and local governments	Federal
2003	49 426	8.0	11.8	2.2	8.3	10.9
2004	55 951	9.0	11.1	5.6	11.4	9.0
2005	61 789	8.6	10.6	8.1	5.8	6.6
2006	66 073	8.4	10.5	9.8	3.9	3.9
2007	66 499	8.1	7.2	12.4	6.0	4.7
2008	56 292	5.8	0.0	5.7	1.4	21.4
2009	60 347	3.7	0.5	-3.9	4.7	20.4
2010	64 651	4.3	-0.6	-0.8	2.6	18.5
2011	66 403	3.6	0.0	2.6	-1.2	10.8
2012	72 364	4.8	1.0	5.0	0.0	10.1
2013	81 555	3.7	1.7	4.5	-1.7	6.7
2014	86 919	4.1	2.1	6.5	-1.2	5.4
2015	89 617	4.3	2.3	6.9	0.3	5.0
2016	95 086	4.5	3.3	5.4	1.1	5.6
2017	103 350	4.2	4.0	5.8	-0.1	3.7
2018	103 952	4.6	3.2	3.9	-1.7	7.6
2014						
1st quarter	80 940	4.1	1.9	6.1	-1.7	5.7
2nd quarter	82 627	4.2	5.2	5.1	0.1	3.5
3rd quarter	82 520	4.5	3.0	5.9	-1.7	6.0
4th quarter	84 189	4.7	2.2	7.0	1.5	5.9
2015						
1st quarter	85 798	2.7	2.0	7.4	1.6	-0.3
2nd quarter	86 434	4.4	3.8	8.0	0.5	2.7
3rd quarter	85 249	2.7	1.3	5.5	0.2	2.1
4th quarter	87 287	7.8	3.8	5.6	-1.2	15.4
2016						
1st quarter	86 811	5.5	2.5	9.2	0.7	6.2
2nd quarter	87 532	4.6	4.4	4.2	2.2	5.7
3rd quarter	89 750	5.0	3.5	6.1	0.8	6.3
4th quarter	91 583	3.1	3.7	2.4	0.4	3.6
2017						
1st quarter	97 147	3.3	3.8	6.0	-2.2	1.7
2nd quarter	98 742	4.8	4.2	6.6	-0.9	4.9
3rd quarter	100 804	4.9	2.7	6.2	-0.6	6.9
4th quarter	103 350	3.4	5.1	3.9	3.5	1.3
2018						
1st quarter	104 339	6.5	3.2	4.1	-2.9	13.4
2nd quarter	106 059	4.2	3.2	3.1	-0.3	6.9
3rd quarter	107 912	4.5	3.5	4.2	-1.4	6.8
4th quarter	103 952	2.7	2.8	3.9	-2.2	2.5

NOTES AND DEFINITIONS, CHAPTER 12

The Federal Reserve's mission is to promote the goals of maximum employment and stables prices. U.S. unemployment has declined significantly while inflation has been under the target rate of 2.0 percent. When the financial crisis of 2007–2009 hit, the Federal Reserve instituted three programs: 1) reducing interest rates to near-zero, 2) increasing the money supply with its Qualitative Easing policy and 3) Undergoing a major initiative called the Enhanced Financial Accounts. These actions are explained below.

The Z-1 *Financial Accounts* provide information on macrofinancial flows and aggregate balance sheets for major sectors of the economy, including households, financial institutions, nonfinancial businesses, governments, and the international sector. The *Financial Accounts* also make up the financial component of the *Integrated Macroeconomic Accounts*, which are published jointly with the Bureau of Economic Analysis and which relate production, income, and saving from the *National Income and Product Accounts* to changes in net worth and financial transactions across the major economic sectors of the U.S. economy. The basic framework of the FA allows understanding of the economic relationships underlying the major sections of the economy, as well as for monitoring the effects of macro-financial developments in the United States and globally.

As home-owners defaulted on their sub-prime mortgages in 2007 to 2009, the Federal Reserve played an important role in turning the economy around. One such program was Quantitative Easing where the Fed purchased financial assets in addition to conventional easing, which has consisted of reducing short-term interest rates to near zero and supporting that rate level with purchases and sales of short-term securities on the open market. In addition, the Federal Reserve began an initiative to provide a richer and more detailed picture of financial intermediation and interconnections.

Quantitative easing policy was discontinued in late 2014. By then the federal reserve had bought $4.5 trillion worth of bonds over five years. The goal of the program during the Great Recession was to lower interest rates and increase the money supply. In addition, the Fed held its benchmark rate to 0.25 which is effectively zero. This policy remained in place for seven years, until December 2015. The Fed raised interest rates to 0.5 percent. The feds fund rate controls short-term. These include bank's prime rate, adjustable-rate and interest-only loans and credit card rates.

The Federal Reserve Board announced on August 1, 2014 the Enhanced Financial Accounts (EFA) initiative which is an ambitious and long-term effort to enhance the "Financial Accounts" (FA) by providing additional detail and disaggregation, higher-frequency data, and additional documentation and analysis of financial data, in order to improve the picture of financial intermediation and activity in the United States. These large-scale and fundamental improvements will expand the detail, dimensionality, and scope of the FA in order to improve the overall picture of financial intermediation and activity in the United States.

Most of the data in this chapter are found on the Federal Reserve Board Web site, <http://www.federalreserve.gov>. Current releases and most historical data are found at that site by selecting Economic Research & Data/Statistical Releases and Historical Data and then selecting the appropriate report. This is the location for all data not otherwise specified.

Historical data not found online are taken from two printed volumes of statistical data that were published by the Board of Governors of the Federal Reserve System: *Banking and Monetary Statistics,* 1943, and *Banking and Monetary Statistics, 1941-1970,* 1976. These will be referred to as *B&MS* 1943 and *B&MS* 1976.

TABLES 12-1 AND 12-2

Money Stock Measures and Components

SOURCE: BOARD OF GOVERNORS OF THE FEDERAL RESERVE SYSTEM

Estimates of two monetary aggregates (M1 and M2) and the components of these measures are published weekly. The monthly data are averages of daily figures. In addition, these measures are revised in subsequent releases.

The Federal Reserve Board ceased publication of the M3 aggregate on March 23, 2006. Weekly publication was also discontinued for the following components of M3: large-denomination time deposits, repurchase agreements (RPs), and Eurodollars. The Board continues to publish institutional money market mutual funds as a memorandum item in this release. Measures of large-denomination time deposits continue to be published in the Financial Accounts (Z.1 release) and in the H.8 release weekly for commercial banks.

The Board stated that "M3 does not appear to convey any additional information about economic activity that is not already embodied in M2 and has not played a role in the monetary policy process for many years. Consequently, the Board judged that the costs of collecting the underlying data and publishing M3 outweigh the benefits." ("Discontinuance of M3," H.6, Money Stock Measures [November 10, 2005, revised March 9, 2006]. [Accessed November 6, 2006.]

Definitions

M1 consists of (1) currency, (2) traveler's checks of nonbank issuers, (3) demand deposits, and (4) other checkable deposits.

M2 consists of M1 plus savings deposits (including money market deposit accounts), smalldenomination time deposits, and balances in retail money market mutual funds.

Currency consists of paper notes and coins outside the U.S. Treasury, the Federal Reserve Banks, and the vaults of depository institutions.

Traveler's checks is the outstanding amount of U.S. dollar-denominated traveler's checks of nonbank issuers. Traveler's checks issued by depository institutions are included in demand deposits.

Demand deposits consists of funds at domestically chartered commercial banks, U.S. branches and agencies of foreign banks, and Edge Act corporations (excluding those amounts held by depository institutions, the U.S. government, and foreign banks and official institutions) less cash items in the process of collection and Federal Reserve float.

Other checkable deposits at commercial banks consists of negotiable order of withdrawal (NOW) and automatic transfer service (ATS) balances at domestically chartered commercial banks, U.S. branches and agencies of foreign banks, and Edge Act corporations.

Other checkable deposits at thrift institutions consists of NOW and ATS balances at thrift institutions, credit union share draft balances, and demand deposits at thrift institutions.

Savings deposits includes money market deposit accounts and other savings deposits at *commercial banks* and *thrift institutions*.

Small time deposits are deposits issued at *commercial banks* and *thrift institutions* in amounts less than $100,000. All Individual Retirement Account (IRA) and Keogh account balances at commercial banks and thrift institutions are subtracted from small time deposits.

Retail money funds are investment funds with capital invested by individual investors. Also IRA and Keogh account balances are excluded.

Institutional money funds are included in the money stock report for informational purposes. They are not part of M1 or M2.

Notes on the data

Seasonal adjustment. Seasonally adjusted M1 is calculated by summing currency, traveler's checks, demand deposits, and other checkable deposits (each seasonally adjusted separately). Seasonally adjusted M2 is computed by adjusting each of its non-M1 components and then adding this result to seasonally adjusted M1.

Revisions. Money stock measures are revised frequently and have a benchmark and seasonal factor review in the middle of the

year; this review typically extends back a number of years. The monetary aggregates were redefined in major revisions introduced in 1980.

Historical: January 1947–January 1959. These data are not currently maintained online and are found in *B&MS* 1976. They are not continuous with the current data series; for that reason the values for January 1959 are shown so that users can link to the current data. The *demand deposit component* includes demand deposits held in commercial banks by individuals, partnerships, and corporations both domestic and foreign, and demand deposits held by nonbank financial institutions and foreign banks. Note that this includes demand deposit liabilities to foreign governments, central banks, and international institutions—said to be "relatively small" in the source document. *Time deposits adjusted* is time and savings deposits at commercial banks, other than large negotiable certificates of deposit (CDs), and excluding all deposits due to the U.S. government and domestic commercial banks. The editor has added this to the *Money stock (M1)* to approximate *M2*, which is not shown as such in the source document, over this period.

Historical: 1892–1946. These data are from *B&MS* 1943 and *B&MS* 1976. They pertain only to the last day of June and, after 1922, the last day of December, the "call dates" on which banks reported to the federal government. For more data and analysis concerning monetary developments in this period, including monthly money supply estimates, see Milton Friedman and Anna Jacobson Schwartz, *A Monetary History of the United States, 1867–1960,* Princeton, Princeton University Press, 1963. In Table 12-1B, *Demand deposits adjusted* refers to the elimination of interbank and U.S. government deposits and cash items in process of collection, not to seasonal adjustment, which is not applicable to call report data because they are only reported once or twice a year. The editor has retitled as *M1* the "Total" money stock shown in the source document, and has added time deposits to it in order to approximate *M2*, which is not shown in the source document.

Data availability

Estimates are released weekly in Federal Reserve Statistical Release H.6, "Money Stock Measures." Current and historical data are available on the Federal Reserve Web site.

References

Board of Governors of the Federal Reserve System, *The Federal Reserve System: Purposes and Functions,* available online at <http://www.federalreserve.gov> in the category "About the Fed/Features," includes a chapter discussing monetary policy and the monetary aggregates and a glossary of terms as an appendix.

An explanation of the 1980 redefinition of the monetary aggregates is found in the *Federal Reserve Bulletin* for February 1980.

TABLES 12-3

Aggregate Reserves, Monetary Base, and FR Balance Sheet

SOURCE: BOARD OF GOVERNORS OF THE FEDERAL RESERVE SYSTEM

Background

Regulation D sets uniform requirements for all depository institutions to maintain reserve balances either with their Federal Reserve Bank or as cash. By amending Regulation D (Reserve Requirements of Depository Institutions), the simplification of reserves administration entered its final phase. This amendment permitted interest payments on certain balanced to be based on a daily rate than on a maintenance period average rate.

In July 2013 the Federal Reserve Board issued a substantially revised format for reporting reserves of depository institutions and the monetary base. The regulation D revisions were made in response to the substantial growth in reserves undertaken beginning in 2008 to mitigate the effects of the financial crisis and to recent changes in the administration of bank reserves.

With these revisions, the Federal Reserve adjusted the data back to 1965. As shown in Table 12-3, the penalty-free band and balance maintained to satisfy reserve balance requirements are the two new concepts are the result of amending Regulation D.

Seasonally adjusted data—no longer relevant in a situation of vastly expanded reserves—are no longer published. Several new aggregates and interest rates are published to "give the public insight into how depository institutions collectively manage their reserves within the current framework for the implementation of monetary policy."

The new format enables calculations of concepts similar to "excess reserves," and a new simplified monetary base is nearly identical with the historical time series of that name; both are without seasonal adjustment.

Definitions

Total Reserves consist of reserves with the Federal Reserve Banks plus vault cash used to satisfy reserve requirements.

Reserve Balances Requirement: The amount determined by applying the reserve ratios specified in Regulation D to an institution's reservable liabilities during the relevant computation period. The institution must satisfy its reserve requirement in the form of vault cash and/or balances maintained either directly with a Reserve Bank or in a pass-through arrangement.

Total Balance Requirements: That portion of the average end-of-day balance for a maintenance period that satisfies an institution's reserve balance requirement. The maximum value for an institution's balance maintained to satisfy reserve balance requirements is the top of the institution's penalty-free band.

Penalty-free band: A penalty-free band is a range on both sides of the reserve balance requirement within which an institution needs to maintain its average balance over the maintenance period in order to satisfy its reserve balance requirement. The top of the penalty-free band is equal to the reserve balance requirement plus a dollar amount prescribed by the Board. The bottom of the penalty-free band is equal to the reserve balance requirement minus a dollar amount prescribed by the Board.

Reserve Balances Maintained: A depository institution is required to satisfy its reserve requirement in the form of vault cash or if vault cash is insufficient to satisfy the requirement, in the form of a balance maintained either directly with a Reserve Bank or in a pass-through arrangement. The portion of the reserve requirement that is not satisfied by vault cash is called the reserve balance requirement. An institution is responsible for satisfying its reserve balance requirement by holding balances on average over a 14-day maintenance period in an account at the Federal Reserve.

Vault Cash is U.S. currency and coin owned by a depository institution. The average end-of-day holdings of vault cash over the computation period can be used to satisfy some or all of an institution's reserve requirement in the corresponding maintenance period.

Monetary Base: is the sum of currency (including coin) in circulation outside Federal Reserve Banks and the U.S. Treasury, plus deposits held by depository institutions at Federal Reserve Banks.

Primary credit is a lending program available to depository institutions that are in generally sound financial condition.

Secondary credit is available to depository institutions that are not eligible for primary credit.

Nonborrowed reserves equals total reserves less total borrowing from the Federal Reserve.

For further information, see "Aggregate Reserves of Depository Institutions and the Monetary Base - H.3" at <www.federalreserve.gov> under Economic Research & Data/Statistical Releases and Historical Data.

Data availability

Reserve and monetary base data are released weekly in Federal Reserve Release H.3, "Aggregate Reserves of Depository Institutions and the Monetary Base." Current and historical data are available on the Federal Reserve Web site.

The Federal Reserve balance sheet is published in the H.4.1 release, Factors Affecting Reserve Balances, released each Thursday, and available on the Federal Reserve Web site under Economic Research and Data/Statistical Releases and Historical Data. However, to obtain the full historical data for total assets, the user must go to Monetary policy/Credit and liquidity programs and the balance sheet/Recent balance sheet trends/Total assets/View as table/All/Copy data. Other explanatory material, including a monthly report on changes in these programs, is available at the Monetary policy location.

TABLE 12-4

Consumer Credit

SOURCE: BOARD OF GOVERNORS OF THE FEDERAL RESERVE SYSTEM

The consumer credit series cover most short and intermediateterm credit extended to individuals through regular business channels, excluding loans secured by real estate (such as first and second mortgages and home equity credit). In October 2003, the scope of this survey was expanded to incorporate student loans extended by the federal government and by SLM Holding Corporation (SLM), the parent company of Sallie Mae (Student Loan Marketing Association). The historical data have been revised back to 1977 to reflect this inclusion.

The failure to include home equity credit is an important limitation of this data set. The household debt series presented in Table 12-5 are more comprehensive, comprising both mortgage and consumer debt.

Consumer credit is categorized by major types of credit and by major holders.

Definitions and notes on the data

The major types of consumer credit are *revolving* and *nonrevolving*.

Revolving credit includes credit arising from purchases on credit card plans of retail stores and banks, cash advances and check credit plans of banks, and some overdraft credit arrangements.

Nonrevolving credit includes automobile loans, mobile home loans, and all other loans not included in revolving credit, such as loans for education, boats, trailers, or vacations. These loans may be secured or unsecured.

Debt secured by real estate (including first liens, junior liens, and home equity loans) is excluded. Credit extended to governmental agencies and nonprofit or charitable organizations, as well as credit extended to business or to individuals exclusively for business purposes, is excluded.

Categories of *holders* include *U.S.-chartered depository institutions* (comprising commercial banks and savings institutions), *finance companies, credit unions, federal government, nonfinancial businesses*, and *pools of securitized assets*. The Student Loan Marketing Association (Sallie Mae) is included in "Federal government" until the end of 2004, at which time it became fully privatized. Beginning with the end of 2004, Sallie Mae is included in "Finance companies." Retailers and gasoline companies are included in the nonfinancial businesses category.

Federal government includes student loans originated by the Department of Education under the Federal Direct Loan Program and the Perkins Loan Program, as well as Federal Family Education Loan Program loans that the government purchased under the Ensuring Continued Access to Student Loans Act.

Pools of securitized assets comprises the outstanding balances of pools upon which securities have been issued; these balances are no longer carried on the balance sheets of the loan originators.

The consumer credit series are benchmarked to comprehensive data that periodically become available. Current monthly estimates are brought forward from the latest benchmarks in accordance with weighted changes indicated by sample data. Classifications are made on a "holder" basis. Thus, installment paper sold by retail outlets is included in the figures for the banks and finance companies that purchased the paper.

The amount of outstanding credit represents the sum of the balances in the installment receivable accounts of financial institutions and retail outlets at the end of each month.

The estimates of the amount of credit outstanding include any finance and insurance charges included as part of the installment contract. Unearned income on loans is included in some cases when lenders cannot separate the components.

The seasonally-adjusted data are adjusted for differences in the number of trading days and for seasonal influences.

Data availability

Current data are available monthly in the Federal Reserve Statistical Release G.19, "Consumer Credit," available along with all current and historical data on the Federal Reserve Web site. In the autumn of each year there is a revision of several years of past data reflecting benchmarking and seasonal factor review.

TABLES 12-5

Stock Prices and Housing Market Data

SOURCES: BOARD OF GOVERNORS OF THE FEDERAL RESERVE SYSTEM; BUREAU OF ECONOMIC ANALYSIS; MOODY'S INVESTORS SERVICE; THE BOND BUYER; DOW JONES, INC.; STANDARD AND POOR'S CORPORATION; NEW YORK STOCK EXCHANGE;

Definitions and notes on the data

Interest rates and bond yields are percents per year and are averages of business day figures, except as noted. With a few exceptions, they are nominal rates or yields not adjusted for inflation.

The daily effective *federal funds rate*—a principal marker for Federal Reserve monetary policy—is a weighted average of rates on trades through New York brokers. Monthly figures include each calendar day in the month. Annualized figures use a 360-day year.

The *Federal Reserve discount rate* is the rate for discount window borrowing at the Federal Reserve Bank of New York. Monthly figures include each calendar day in the month. Annualized figures use a 360-day year. Before 1945, annual averages were calculated by the editor.

Beginning in January 2003, the rules governing the discount window programs were revised. "Adjustment credit," which had been extended at a below-market rate (as can be seen in the average discount rates from 1978 through 2002 shown in Table 12-9, which are below the federal funds rate), was replaced by a new type of credit called "primary credit." Primary credit is available for very short terms as a backup source of liquidity to depository institutions in generally sound financial condition, as judged by the lending Federal Reserve Bank. Primary credit is extended at a rate above the federal funds rate, eliminating the incentive for institutions to exploit the spread of money market rates over the discount rate.

The *U.S. Treasury bills, 3-month rate* and the *U.S. Treasury bills, 6-month rate* are the yields on these securities based on their prices as traded in the secondary market. The rates are quoted on a discount basis. Annualized figures use a 360-day year. Early data are taken from *B&MS* 1943 and 1976. From 1934 to 1944, the 3-month rate is based on dealers' quotations. From 1931 through 1933, it is the average rate on new issues. For 1920 through 1930, it is the rate on 3- to 6-month Treasury notes and certificate.

Commercial paper, 3-month rates are interpolated from data on certain commercial paper trades settled by the Depository Trust Company. This company is a clearinghouse and custodian for nearly all domestic commercial paper activity. The trades, which are on a discount basis, represent sales of commercial paper by dealers or direct issuers. Annualized figures use a 360-day year. From 1971 through September 1997, the series represented both nonfinancial and financial commercial paper. Since September 1997, rates have been reported separately for nonfinancial and financial companies; only rates for nonfinancial companies are shown here. This introduces a slight discontinuity in this series between August and September 1997. Before 1971, the prevailing rate in New York City for 4- to 6-month commercial paper; hence, this series is also discontinuous between December 1970 and January 1971, and is obtained from *B&MS* 1943 and 1976.

The *bank prime rate* is one of several base rates used by banks to price shortterm business loans. It is the rate posted by a majority of the top 25 (by amount of assets in domestic offices) insured U.S.-chartered commercial banks. Monthly figures include each calendar day in the month. Annualized figures use a 360-day year. Before 1949, the data are not on the Federal Reserve Web site but have been reproduced from *B&MS* 1976. In that volume, the prime rate is described as "the rate that banks charge their most creditworthy business customers on short-term loans", as posted by the largest banks. The same source goes on to write, "A nationally publicized and uniform prime rate did not emerge until the depression of the 1930s. The rate in that period—1 ½

percent—represented a floor below which banks were said to regard lending as unprofitable. The date shown [for the beginning of a changed rate] is that on which the new rate was put into effect by the first bank to make the change. The table shows a range of rates for 1929-1933 because no information is available to indicate when the rate changed in that period." For further information, the source document cites "The Prime Rate," *Monthly Review,* Federal Reserve Bank of New York, April and May 1962, pp. 54-59 and 70-73, respectively.

U.S. Treasury securities, constant maturities. The rates shown for 1-year, 5-year, 10-year, 20-year, and 30-year securities are yields on actively traded issues adjusted to constant maturities. Yields on Treasury securities at "constant maturity" are interpolated by the Treasury Department from the daily yield curve. This curve, which relates the yield on a security to its time to maturity, is based on the closing market bid yields on actively traded Treasury securities in the overthecounter market. These market yields are calculated from composites of quotations reported by U.S. Government securities dealers to the Federal Reserve Bank of New York. The constant maturity yield values are read from the yield curve at fixed maturities. For example, this method provides a yield for a 10-year maturity, even if no outstanding security has exactly 10 years remaining to maturity. The 30-year series was discontinued as of February 2002, because the Treasury Department was no longer issuing such bonds at that time. However, issuance of 30-year bonds was resumed in 2005 as an additional means of financing rising deficits, and the 30-year interest rate series resumes in 2006. The current 20-year series begins with 1993 and is not comparable with an earlier 20-year series. For further information, see the historical data series on the Federal Reserve Web site.

Before 1962, the *Treasury 10-year* is not available on the constant-maturity basis. The entries in that column are from *B&MS* 1943 and 1976 where they are labeled "United States government bonds (long-term)." Before 1941, they represent yields on bonds that were partly tax-exempt. In 1941, the yield for the partly tax-exempt series was 1.95 percent—comparable with the entries for previous years—while the entry comparable with subsequent years was 2.12 percent.

Domestic corporate bond yields, Aaa and Baa. The rates shown are for general obligation bonds based on Thursday figures, and are provided by Moody's Investors Service and republished by the Federal Reserve. The Aaa rates through December 6, 2001 are averages of Aaa utility and Aaa industrial bond rates. As of December 7, 2001, these rates are averages of Aaa industrial bonds only.

Stock price indexes and yields. The *Dow Jones industrial* average is an average price of 30 stocks compiled by Dow Jones, Inc. The *Standard and Poor's composite* is an index of the prices of 500 stocks that are weighted by the volume of shares outstanding, accounting for about 90 percent of New York Stock Exchange value, with a base of 1941-1943 = 10, compiled by Standard and

Poor's Corporation. (Before February 1957, these data are based on a conversion of an earlier 90-stock index, and are obtained from *B&MS* 1943 and 1976.) The *dividend-price ratio* is compiled by Standard and Poor's, covering the 500 stocks in the S&P index. It represents aggregate cash dividends (based on the latest known annual rate) divided by aggregate market value based on Wednesday closing prices. The *earnings/price ratio* measures earnings (after taxes) for four quarters, ending with the indicated quarter, as a ratio to stock prices for the last day of that quarter. Monthly data are averages of weekly figures; annual data are averages of monthly or quarterly figures. The *Nasdaq composite index* is an average price of over 5,000 stocks traded on the Nasdaq exchange.

The *Eurodollar deposits* rate shown is the bid rate at about 9:30 a.m. (EST) for 1-month Eurodollar deposits. Annualized figures use a 360-day year.

CDs (secondary market), 3-month rates are averages of dealer offering rates on nationally traded certificates of deposit. Annualized figures use a 360-day year.

Inflation-indexed yields. In recent years, the Treasury Department has issued Treasury Inflation-Protected Securities (TIPS), which are marketable long-term bonds whose redemption value is increased by the change in the CPI-U from the date of purchase. (See notes and definitions to Chapter 8.) The purchaser of these bonds, unlike with ordinary securities, is guaranteed that the real value of his or her principal will remain intact. He or she need not make his or her own estimate of future inflation in order to make a rational bid. Therefore, the observed purchase price represents a real rate of interest that purchasers and sellers are mutually willing to accept. Inflation-indexed yields are shown here for 5 and 20 years (adjusted to constant maturities by the Treasury) and for the long-term average, which is the unweighted average of the bid yields for all TIPS with remaining terms to maturity over 10 years.

The *fixed-rate first mortgage* rates are primary market contract interest rates on commitments for fixedrate conventional 30-year first mortgages. The rates are obtained by the Federal Reserve from the Federal Home Loan Mortgage Corporation (FHLMC, or Freddie Mac).

Data availability and references

Interest rates and bond yields are published weekly in the Federal Reserve's H.15 release, "Selected Interest Rates"; the release and current and historical data are available on the Federal Reserve Web site and in *B&MS* 1943 and 1976. The starting dates for individual interest rate series vary; some date back to 1911, and many begin in the 1950s and 1960s.

Stock market data are published monthly in *Economic Indicators,* and until 2014 annually in *Economic Report of the President,* both available online at <http://www.gpo.gov>. Some historical interest rate data that are not available on the Federal Reserve Web site were also taken from past *Economic Reports of the President.*

The FHFA house price indexes for the United States as a whole, regions, states, and metropolitan and sub-metropolitan groups are published every 3 months, approximately 2 months after the end of the previous quarter. The release and supporting data and explanatory material are available at <http://www.fhfa.gov>.

Table 12-6

Mortgage Debt Outstanding

SOURCE: BOARD OF GOVERNORS OF THE FEDERAL RESERVE SYSTEM

Definitions and notes on the data

By type of property

Home mortgages includes home equity loans; these are also shown separately elsewhere in this Federal Reserve statistical report.

Multifamily residences refers to mortgages on structures of five or more units.

By type of holder

Federal and related agencies show mortgages held directly by the federal government and government-sponsored enterprises.

Mortgage pools or trusts show mortgages that were refinanced by their holders through the issuance of mortgage-backed securities.

Other holders encompasses a variety of groups, including finance companies, individuals, state and local governments, credit unions, and others.

Data availability

These data are also published in the Federal Reserve's Statistical Release Z.1, "Financial Accounts of the United States," Table L.217.

Table 12-7

Household Net Worth and Growth of Domestic Nonfinancial Debt

SOURCE: BOARD OF GOVERNORS OF THE FEDERAL RESERVE SYSTEM.

Researchers and analysts have used the Financial Accounts to investigates the causes and consequences of the financial crisis and to create and analyze indicators of financial stability. However, they were not originally designed to be a comprehensive source of data for all the complex financial relationships and transactions whose importance to the macro-economy were made clear during the financial crisis of 2007–2009.

Data availability

The "Financial Accounts" are published online and in print four times per year, about 10 weeks following the end of each calendar quarter. The publication with series mnemonics and the guide are available online: www.federalreserve.gov/releases/Z1.

CHAPTER 13: INTERNATIONAL COMPARISONS

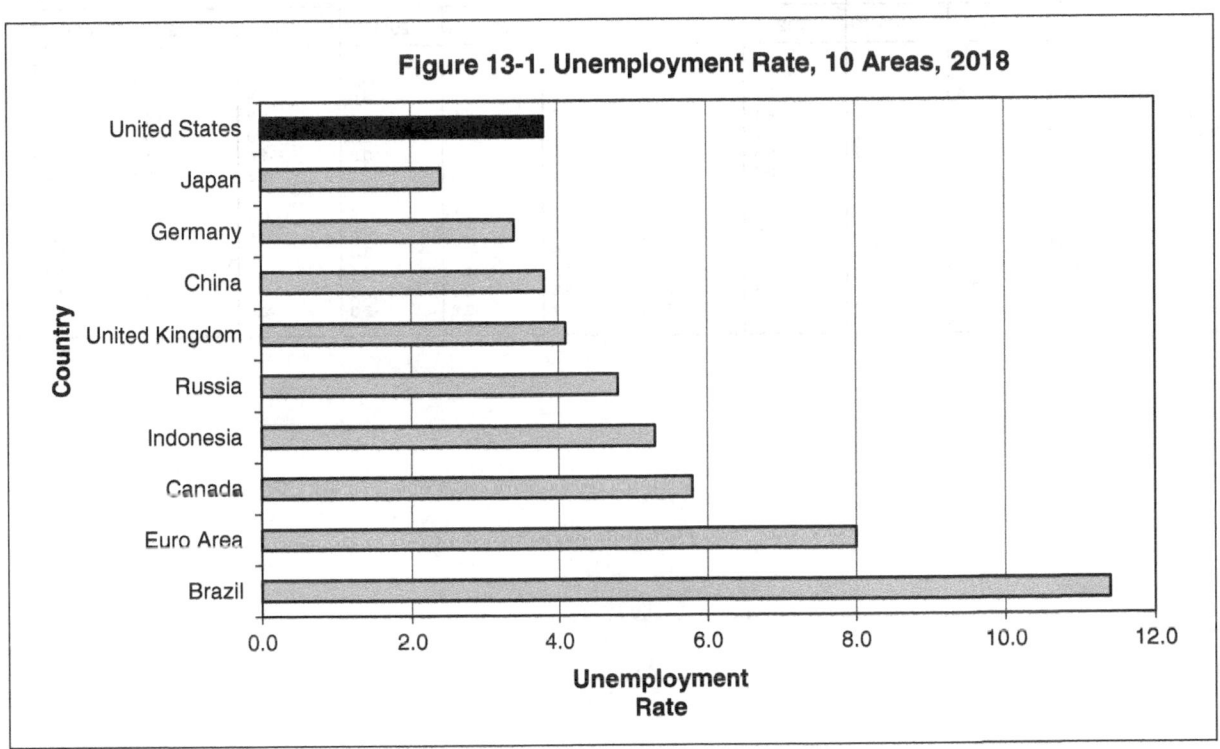

Figure 13-1. Unemployment Rate, 10 Areas, 2018

- The unemployment rate varies widely among the ten countries listed in Table 13-1. In 2018, Brazil had the highest unemployment rate at 11.4 percent followed by the Euro Area at 8.0 percent and Indonesia at 5.3 percent. The unemployment rate was 3.8 percent in the United States which was the same rate as in China and higher than only Japan and Germany. (Table 13-1)

- In 2018, inflation was the highest in Brazil at 3.7 percent followed by India at 3.5 percent and Indonesia at 3.2 percent. In 2012, inflation declined from 10.0 percent in India to 3.5 percent in 2018. Japan experienced the lowest inflation at 1.0 percent in 2018. Inflation was 2.4 percent in the United States. (Table 13-1)

- Listed in Table 13-2 are six 'major' currencies including: the Canadian dollar, the Euro, the Japanese yen, Swiss franc, and the British pound. These are all major industrial countries whose currencies are freely traded on world markets. All of these measures are defined as the foreign currency price of the U.S dollar. The other currency listed is the Chinese yuan, which is not freely traded. China is able to control the international values of their currencies (through, for example, direct capital controls) and keep their currencies from appreciating relative to the dollar to maintain their competitiveness in the U.S. market. (Table 13-2)

Table 13-1 International Comparisons: Percent Change in Real GDP, Consumer Prices, Current Account Balances, and Unemployment Rate, 2012, 2018, and 2020

(Annual percent change.)

Country	Gross domestic product			Consumer prices			Current account balance			Unemployment rate		
	2012	2018	2020	2012	2018	2020	2012	2018	2020	2012	2018	2020
Euro Area	-0.3	1.8	1.5	0.2	1.8	1.6	1.3	3.0	2.8	11.3	8.0	7.7
Germany	0.7	1.5	1.4	2.1	1.9	1.7	5.5	7.4	6.8	5.4	3.4	3.3
United Kingdom	1.5	1.4	1.4	2.7	2.5	2.0	-3.7	-3.9	-4.0	8.0	4.1	4.4
Brazil	1.9	1.1	2.5	5.4	3.7	4.1	-3.0	-0.8	-1.6	7.4	11.4	10.2
Canada	1.7	1.8	1.9	2.2	2.2	1.9	-3.6	-2.6	-2.8	7.3	5.8	6.0
China	7.9	6.6	6.1	2.6	2.1	2.5	2.5	0.4	0.3	4.1	3.8	3.8
India	5.5	7.1	7.5	10.0	3.5	4.2	-4.8	-2.5	-2.4	. . .	. . .	. . .
Indonesia	6.0	5.2	5.2	4.0	3.2	3.6	-2.7	-3.0	-2.6	6.1	5.3	5.0
Japan	1.5	0.8	0.5	-0.1	1.0	1.5	1.0	3.5	3.6	4.3	2.4	2.4
Russia	3.7	2.3	1.7	5.1	2.9	4.5	3.3	7.0	5.1	5.5	4.8	4.7
United States	2.2	2.7	1.9	2.1	2.4	2.7	-2.8	-2.3	-2.6	8.1	3.8	3.7

. . . = Not available.

Table 13-2 Foreign Exchange Rates

(Not seasonally adjusted.)

Year and month	Foreign currency per U.S. dollar					
	Canadian dollar	Chinese yuan	European currency unit	Japananese yen	Switzerland franc	British pound
1971	0.9993	. . .	. . .	320.07	3.9041	2.5266
1972	0.9968	. . .	. . .	301.24	3.7700	2.3449
1973	0.9994	. . .	. . .	280.18	3.2000	2.3174
1974	0.9882	. . .	. . .	300.41	2.6028	2.3294
1975	1.0139	. . .	. . .	305.67	2.6337	2.0221
1976	1.0183	. . .	. . .	294.70	2.4496	1.6784
1977	1.0973	. . .	. . .	241.02	2.0772	1.8546
1978	1.1798	. . .	. . .	195.96	1.6757	1.9861
1979	1.1700	. . .	. . .	240.37	1.5990	2.2007
1980	1.1968	. . .	. . .	209.49	1.7854	2.3459
1981	1.1851	1.7405	. . .	218.95	1.8152	1.9033
1982	1.2385	1.9445	. . .	241.94	2.0501	1.6160
1983	1.2469	1.9920	. . .	234.46	2.1983	1.4338
1984	1.3201	2.7953	. . .	247.96	2.5602	1.1861
1985	1.3955	3.2095	. . .	202.79	2.1042	1.4447
1986	1.3801	3.7314	. . .	162.05	1.6647	1.4393
1987	1.3075	3.7314	. . .	128.24	1.3304	1.8288
1988	1.1962	3.7314	. . .	123.61	1.4799	1.8258
1989	1.1613	4.1825	. . .	143.68	1.5686	1.5965
1990	1.1603	5.2352	. . .	133.89	1.2814	1.9219
1991	1.1467	5.4232	. . .	128.04	1.3855	1.8272
1992	1.2725	5.8106	. . .	124.04	1.4219	1.5510
1993	1.3308	5.8210	. . .	109.91	1.4634	1.4913
1994	1.3893	8.5033	. . .	100.18	1.3289	1.5587
1995	1.3693	8.3350	. . .	101.85	1.1631	1.5405
1996	1.3622	8.3290	. . .	113.98	1.3290	1.6639
1997	1.4271	8.3099	. . .	129.73	1.4090	1.6597
1998	1.5433	8.2780	. . .	117.07	1.3604	1.6708
1999	1.4722	8.2794	1.0110	102.58	1.5841	1.6132
2000	1.5219	8.2771	0.8983	112.21	1.6855	1.4629
2001	1.5788	8.2764	0.8912	127.59	1.6566	1.4413
2002	1.5592	8.2777	1.0194	121.89	1.4388	1.5863
2003	1.3128	8.2770	1.2298	107.74	1.2643	1.7516
2004	1.2189	8.2765	1.3406	103.81	1.1465	1.9286
2005	1.1615	8.0755	1.1861	118.46	1.3053	1.7458
2006	1.1532	7.8219	1.3205	117.32	1.2099	1.9629
2007	1.0021	7.3682	1.4559	112.45	1.1402	2.0161
2008	1.2337	6.8539	1.3511	91.28	1.1404	1.4854
2009	1.0537	6.8275	1.4579	89.95	1.0301	1.6226
2010	1.0081	6.6497	1.3221	83.34	0.9689	1.5595
2011	1.0235	6.3482	1.3155	77.80	0.9334	1.5587
2012	0.9898	6.2328	1.3119	83.79	0.9213	1.6145
2013	1.0639	6.0738	1.3708	103.46	0.8933	1.6383
2014	1.1532	6.1886	1.2329	119.32	0.9753	1.5644
2015	1.3713	6.4491	1.0889	121.64	0.9951	1.4981
2016	1.3339	6.9198	1.0545	116.00	1.0194	1.2483
2017	1.2769	6.5932	1.1836	112.94	0.9870	1.3404
2018	1.3436	6.8837	1.1380	112.20	0.9919	1.2664
2017						
January	1.3183	6.8907	1.0635	114.87	1.0075	1.2367
February	1.3109	6.8694	1.0650	112.91	1.0010	1.2495
March	1.3387	6.8940	1.0691	112.92	1.0015	1.2347
April	1.3437	6.8876	1.0714	110.09	1.0009	1.2639
May	1.3606	6.8843	1.1050	112.24	0.9867	1.2929
June	1.3295	6.8066	1.1233	110.91	0.9681	1.2810
July	1.2690	6.7694	1.1530	112.42	0.9604	1.2996
August	1.2608	6.6670	1.1813	109.83	0.9653	1.2952
September	1.2279	6.5690	1.1913	110.78	0.9625	1.3340
October	1.2607	6.6254	1.1755	112.91	0.9821	1.3202
November	1.2773	6.6200	1.1743	112.82	0.9915	1.3217
December	1.2769	6.5932	1.1836	112.94	0.9870	1.3404
2018						
January	1.2429	6.4233	1.2197	110.87	0.9604	1.3824
February	1.2588	6.3183	1.2340	107.97	0.9355	1.3961
March	1.2933	6.3174	1.2334	106.05	0.9480	1.3976
April	1.2732	6.2967	1.2270	107.66	0.9687	1.4079
May	1.2866	6.3701	1.1823	109.69	0.9969	1.3470
June	1.3125	6.4651	1.1679	110.06	0.9900	1.3294
July	1.3133	6.7164	1.1685	111.52	0.9948	1.3162
August	1.3042	6.8453	1.1547	111.00	0.9880	1.2878
September	1.3034	6.8551	1.1667	112.10	0.9683	1.3066
October	1.3004	6.9191	1.1488	112.72	0.9940	1.3012
November	1.3205	6.9367	1.1364	113.34	1.0011	1.2900
December	1.3436	6.8837	1.1380	112.20	0.9919	1.2664

. . . = Not available.

NOTES AND DEFINITIONS, CHAPTER 13

TABLE 13-1

International Comparisons: Gross Domestic Product Growth, Unemployment Rates, Inflation and Current Account Balances

SOURCE: THE APRIL 2017 WORLD ECONOMIC OUTLOOK (WEO), INTERNATIONAL MONETARY FUND AND THE FEDERAL RESERVE

The United States trades with many countries. Specific countries include neighbors Canada and Mexico; European countries like Britain and Germany; plus Asian countries such as China, Japan, and India and finally Russia. Table 13-1 provides international comparisons of these nations in terms of major economic indicators. These include real gross domestic product (GDP), employment rates, inflation, and current account balances as percent of GDP by country for the years 2012, 2018, and the International Monetary Fund projections to 2020.

Definitions and notes on the data

The basic measure of the overall health of an economy is the *gross domestic product* (GDP). It is the market value of all goods and services produced by labor and property located in the United States. *Real GDP* is an inflation-adjusted measure that reflects the value of all goods and services produced by an economy in a given year, expressed in base-year prices. See chapter 1 for in depth and historical data, notes and definitions on GDP.

The *Consumer Price Index* (CPI) measures the change in prices paid by consumers for goods and services. *Inflation/deflation* is the change over time of the CPI. See Chapter 8.

The *unemployment rate* represents the number unemployed civilians as a percent of the labor force. See Chapter 10.

Current account is all transactions other than those in financial and capital items. The major classifications are goods and services, income and current transfers. The focus of the BOP is on transactions (between an economy and the rest of the world) in goods, services, and income.

Table 13-1 presents the current account balance as a percent of GDP. This measure provides an indication on the level of international competitiveness of a country. Usually, countries recording a strong current account surplus have an economy heavily dependent on exports revenues, with high savings ratings but weak domestic demand. On the other hand, countries recording a current account deficit have strong imports, a low saving rates and high personal consumption rates as a percentage of disposable incomes. See Chapter 7.

References and notes on the data

This data is published by the International Monetary Fund (IMF) on an annual basis. Tables appear in Chapter 1's Annex.

TABLE 13-2

FOREIGN Exchange Rates

SOURCE: BOARD OF GOVERNORS OF THE FEDERAL RESERVE SYSTEM

Definitions and notes on the data

The price of a nation's currency in terms of another currency. An *exchange rate* thus has two components, the domestic currency and a foreign currency, and can be quoted either directly or indirectly. In a direct quotation, the price of a unit of foreign currency is expressed in terms of the domestic currency. In an indirect quotation, the price of a unit of domestic currency is expressed in terms of the foreign currency.

This table shows measures of the U.S. dollar relative to the currencies of some important individual countries and also relative to average values for major groups of countries. In *Business Statistics,* all of these measures are defined as the foreign currency price of the U.S. dollar. When the measure is relatively high, the dollar is relatively strong—but less competitive (in the sense of price competition)—and the other currency or group of currencies named in the measure is relatively weak and more competitive.

For consistency, this definition is used in *Business Statistics* even in the case of currencies that are commonly quoted in the financial press and elsewhere as dollars per foreign currency unit instead of foreign currency units per dollar. Notably, this is the case for the euro and for the British pound.

The definition of the dollar's value used in *Business Statistics* is the most useful for economic analysis from the U.S. point of view and for foreign tourists in the United States. For American tourists overseas, it is easier to use the inverse of this measure, the value of the other currency; for example, the traveler in Paris can more easily translate prices into dollars by multiplying by the dollar value of the euro than by dividing by the euro value of the dollar.

The foreign exchange rates shown are averages of the daily noon buying rates for cable transfers in New York City certified for customs purposes by the Federal Reserve Bank of New York.

The introduction of the euro was in January 1999 as the common currency for 11 European countries and marked a major change in the international currency system. Over time, the Eurozone has increased to 19 countries: Austria, Belgium, Cyprus, Estonia, Finland, France, Germany, Greece, Ireland, Italy, Latvia, Lithuania, Luxembourg, Malta, Netherlands, Portugal, Slovakia, Slovenia and Spain. In addition, the euro is also the national currency in Monaco, the Vatican City and San Marino, and is the de facto currency in Andorra, Kosovo, and Montenegro. The values of the currencies of these countries no longer fluctuate relative to each other, but the value of the euro still fluctuates relative to the dollar and to currencies for countries outside the EMU. The currency and coins of the individual countries continued to circulate from 1999 through the end of 2001, but in January 2002, new euro currency and coins were introduced, replacing the currency and coins of the individual countries. Once a country has entered the monetary union, its values relative to the <u>dollar</u> will continue to fluctuate—but only due to fluctuations in the value of the euro relative to the dollar.

There is no fully satisfactory historical equivalent to the euro. For comparisons over time, the Federal Reserve Board uses a "restated German mark," derived simply by dividing each historical value of the mark by the euro conversion factor, 1.95583. The G-10 dollar index described below includes five of the currencies that later merged into the euro, but also includes the currencies of Canada, Japan, the United Kingdom, Switzerland, and Sweden.

Data availability and references

Current press releases, historical data, and information on weights and methods for exchange rates and exchange rate indexes are available on the Federal Reserve Web site at <http://www.federalreserve.gov/>; go to Data, then "Foreign Exchange Rates H10/G5," and select DDP, the new Data Download Program. The dollar value indexes are described in the article "Indexes of the Foreign Exchange Value of the Dollar," *Federal Reserve Bulletin* (Winter 2005), also available on the Federal Reserve Web site.

Additional information on exchange rates can be found on the Federal Reserve Bank of St. Louis Web site at <http://www.stls.frb.org/fred/data/exchange.html>.

PART B: INDUSTRY PROFILES

CHAPTER 14: PRODUCT AND INCOME BY INDUSTRY

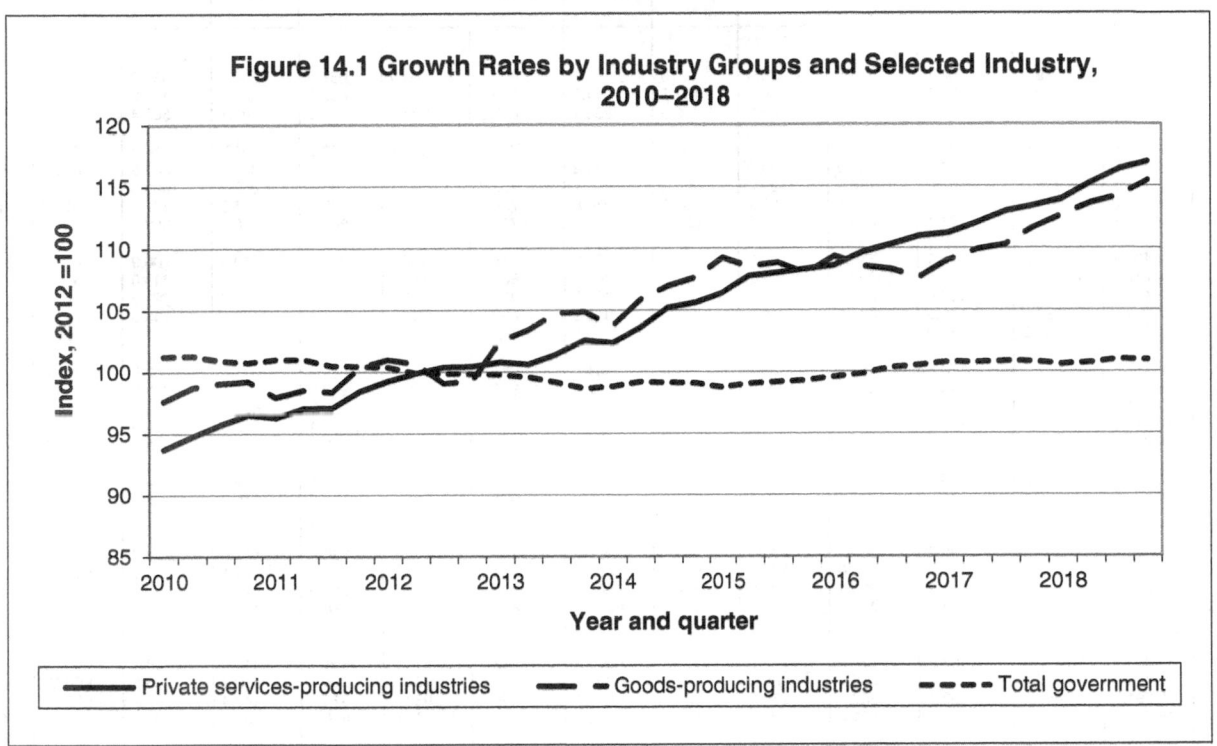

Figure 14.1 Growth Rates by Industry Groups and Selected Industry, 2010–2018

- Private services-producing industries, which accounted for 69.9 percent of total GDP in 2018, display less cyclical swing and more real growth than the goods sector, which was 17.9 percent of the total. Government production—the other 12.2 percent of total GDP—was comparatively stable, but this is an incomplete measurement of the counter-cyclical impact of government, as it does not include the effects of changes in tax receipts and payments. (Tables 14-1 and 14-2)

- Table 14-3 can be used to assess the share of each industry group in total gross domestic factor income and the shares paid to employees and accruing to land and capital in each industry. For these purposes factor income is a better measure than the BEA definition of value added because value added includes, but factor income excludes, taxes on production and imports.

- In private goods-producing industries in 2017, employee compensation was 52.3 percent of gross factor income, while in the less capital-intensive, private services-producing industries (excluding real estate and rental and leasing), compensation was 54.4 percent of total value. (Table 14-3)

- BEA publishes quarterly measures of GDP by industrial origin, in both current-dollar and real (quantity) terms. Industry output, sometimes referred to as "gross product originating in the industry," is here termed "value added".

Table 14-1. Gross Domestic Product (Value Added) by Industry

(Billions of dollars.)

Year	Total gross domestic product	Private industries Total	Agriculture, forestry, fishing, and hunting	Mining	Utilities	Construction	Manufacturing Durable goods	Manufacturing Nondurable goods	Wholesale trade	Retail trade	Transportation and warehousing	Information	Finance and insurance	Real estate and rental and leasing
1999	9 630.7	8 378.3	92.6	84.5	179.9	417.6	874.9	614.1	584.0	652.6	290.0	485.0	675.1	1 162.0
2000	10 252.3	8 929.3	98.3	110.6	180.1	461.3	924.8	625.4	622.6	685.5	307.8	471.3	743.1	1 231.7
2001	10 581.8	9 188.9	99.8	123.9	181.3	486.5	833.4	640.5	613.8	709.5	308.1	502.4	803.1	1 325.0
2002	10 936.4	9 462.0	95.6	112.4	177.6	493.6	832.8	635.7	613.1	732.6	305.7	550.6	816.4	1 400.5
2003	11 458.2	9 905.9	114.0	139.0	184.0	525.2	863.2	661.0	641.4	769.6	321.4	564.9	846.3	1 449.6
2004	12 213.7	10 582.5	142.9	166.5	199.2	584.6	905.1	703.0	697.1	795.6	352.1	620.4	872.4	1 516.7
2005	13 036.6	11 326.4	128.3	225.7	198.1	651.8	956.8	736.6	754.9	840.8	375.8	642.3	973.7	1 632.5
2006	13 814.6	12 022.6	125.1	273.3	226.8	697.1	1 004.4	789.4	811.5	869.9	410.4	652.0	1 034.2	1 709.7
2007	14 451.9	12 564.8	144.1	314.0	231.9	715.3	1 030.6	814.1	857.8	869.2	413.9	706.9	1 030.5	1 817.7
2008	14 712.8	12 731.2	147.2	392.2	241.7	648.9	999.7	801.1	884.3	848.7	426.8	743.0	873.2	1 889.5
2009	14 448.9	12 403.9	130.0	275.8	258.2	565.6	881.0	821.2	834.2	827.6	404.6	721.9	966.6	1 901.1
2010	14 992.1	12 884.1	146.3	305.8	278.8	525.1	964.3	832.7	888.9	851.5	433.0	753.3	1 003.6	1 939.4
2011	15 542.6	13 405.5	180.9	356.3	287.5	524.4	1 015.2	852.4	934.9	871.9	451.4	759.8	1 026.0	2 019.3
2012	16 197.0	14 037.5	179.6	358.8	279.7	553.4	1 061.7	865.3	997.4	908.4	472.0	759.0	1 162.8	2 098.2
2013	16 784.9	14 572.3	215.6	386.5	286.3	587.6	1 102.0	889.9	1 040.1	949.5	491.1	828.9	1 144.9	2 177.9
2014	17 521.7	15 250.0	200.8	413.0	298.2	636.4	1 130.7	916.6	1 088.2	973.8	521.9	840.6	1 282.0	2 271.0
2015	18 219.3	15 878.8	181.2	257.9	299.0	694.9	1 182.7	940.3	1 141.7	1 021.4	563.4	915.0	1 362.7	2 391.9
2016	18 707.2	16 319.4	164.9	216.2	302.7	745.5	1 182.0	903.1	1 136.6	1 052.0	577.4	998.1	1 432.7	2 497.2
2017	19 485.4	17 031.7	169.2	268.6	307.5	781.4	1 226.6	953.0	1 174.1	1 087.1	608.7	1 050.8	1 465.9	2 591.2
2018	20 494.1	17 989.4	164.2	321.1	319.4	840.0	1 300.6	1 034.0	1 234.0	1 132.5	648.0	1 125.9	1 514.2	2 725.1
2010														
1st quarter	14 721.4	12 634.5	135.8	307.3	275.0	519.4	911.0	843.0	862.5	838.4	414.5	734.2	1 015.2	1 908.2
2nd quarter	14 926.1	12 815.3	141.6	296.6	274.4	527.7	955.5	827.7	881.7	855.7	429.1	744.1	1 012.5	1 935.3
3rd quarter	15 079.9	12 966.1	149.5	300.2	281.8	527.7	988.3	825.8	905.9	853.9	441.8	766.9	984.6	1 942.5
4th quarter	15 240.8	13 120.4	158.3	319.3	284.2	525.7	1 002.5	834.2	905.5	858.0	446.7	767.9	1 002.1	1 971.4
2011														
1st quarter	15 285.8	13 156.5	183.5	334.3	275.0	510.9	1 002.7	833.0	911.7	861.8	446.2	758.1	1 013.8	1 962.8
2nd quarter	15 496.2	13 355.7	175.4	361.6	289.0	522.2	1 003.8	860.5	932.6	866.3	450.9	765.5	1 003.1	2 009.6
3rd quarter	15 591.9	13 449.4	183.1	357.7	289.5	529.1	1 011.1	858.2	933.8	873.2	449.4	755.8	1 034.5	2 028.6
4th quarter	15 796.5	13 660.5	181.8	371.6	296.4	535.5	1 043.1	857.7	961.6	886.4	459.2	760.0	1 052.7	2 076.1
2012														
1st quarter	16 019.8	13 871.1	182.6	369.7	272.4	548.3	1 050.8	864.4	979.9	901.5	465.5	755.4	1 126.1	2 069.4
2nd quarter	16 152.3	14 001.2	178.6	353.8	281.7	551.7	1 065.6	864.4	991.9	904.0	474.5	768.2	1 153.8	2 092.5
3rd quarter	16 257.2	14 096.1	175.8	349.6	283.2	552.0	1 063.7	874.1	1 003.1	910.6	474.2	760.1	1 184.5	2 115.5
4th quarter	16 358.9	14 181.7	181.3	362.1	281.3	561.6	1 066.8	858.5	1 014.5	917.6	473.6	752.1	1 186.8	2 115.2
2013														
1st quarter	16 569.6	14 375.2	218.3	369.9	287.4	571.5	1 090.0	887.9	1 027.9	943.1	483.4	812.1	1 109.3	2 146.7
2nd quarter	16 637.9	14 430.5	224.7	383.0	286.4	579.0	1 091.0	873.3	1 025.9	942.7	484.3	820.8	1 126.7	2 150.8
3rd quarter	16 848.7	14 633.7	222.2	401.3	283.2	594.3	1 107.2	882.3	1 045.0	953.8	491.6	830.2	1 143.9	2 196.1
4th quarter	17 083.1	14 850.1	197.2	391.8	288.3	605.4	1 119.6	916.3	1 061.5	958.2	505.1	852.6	1 199.8	2 217.8
2014														
1st quarter	17 102.9	14 855.6	194.5	409.8	300.6	616.5	1 098.0	892.0	1 049.7	951.0	497.2	835.2	1 235.8	2 215.2
2nd quarter	17 425.8	15 164.7	211.0	432.2	295.6	629.8	1 118.3	915.6	1 074.9	968.9	516.5	836.7	1 284.6	2 254.4
3rd quarter	17 719.8	15 440.6	197.2	428.0	298.1	643.4	1 151.9	931.6	1 108.4	981.2	532.3	841.0	1 295.5	2 296.6
4th quarter	17 838.5	15 539.1	200.6	382.0	298.3	656.1	1 154.7	927.2	1 119.6	993.8	541.7	849.7	1 312.0	2 317.7
2015														
1st quarter	17 970.4	15 652.1	177.9	284.1	299.7	672.6	1 175.0	934.1	1 131.5	1 003.5	554.5	875.1	1 349.5	2 348.7
2nd quarter	18 221.3	15 885.6	179.7	275.0	297.5	690.3	1 183.0	947.7	1 149.9	1 015.0	561.3	902.9	1 399.2	2 376.3
3rd quarter	18 331.1	15 981.6	187.1	248.1	302.4	702.8	1 189.8	957.2	1 143.6	1 031.5	568.5	928.4	1 352.4	2 409.3
4th quarter	18 354.4	15 996.0	180.2	224.3	296.5	714.1	1 183.1	922.2	1 141.8	1 035.5	569.2	953.8	1 349.7	2 433.4
2016														
1st quarter	18 409.1	16 043.6	168.7	195.1	296.0	731.3	1 177.3	898.0	1 132.8	1 042.4	569.4	974.9	1 356.1	2 461.5
2nd quarter	18 640.7	16 262.6	169.8	211.3	300.1	740.6	1 180.0	908.6	1 133.6	1 049.1	577.5	992.4	1 415.6	2 496.6
3rd quarter	18 799.6	16 403.8	164.0	221.7	308.4	748.5	1 183.6	905.0	1 137.6	1 053.9	576.5	1 009.8	1 465.9	2 506.8
4th quarter	18 979.2	16 567.4	157.1	236.8	306.4	761.7	1 187.3	901.0	1 142.3	1 062.4	586.5	1 015.1	1 493.2	2 523.8
2017														
1st quarter	19 162.6	16 729.0	172.9	255.1	300.5	770.9	1 200.5	923.7	1 151.8	1 071.5	594.4	1 018.4	1 466.6	2 549.7
2nd quarter	19 359.1	16 911.2	172.2	261.3	309.9	773.6	1 215.7	939.1	1 168.0	1 081.2	605.3	1 042.1	1 444.1	2 579.9
3rd quarter	19 588.1	17 126.0	165.9	265.6	306.6	784.0	1 235.3	959.2	1 180.4	1 092.7	613.7	1 062.5	1 472.4	2 603.4
4th quarter	19 831.8	17 360.6	165.9	292.5	313.0	797.1	1 255.0	990.0	1 196.3	1 103.1	621.5	1 080.0	1 480.6	2 632.0
2018														
1st quarter	20 041.0	17 561.6	166.6	301.8	315.3	816.1	1 268.4	1 003.5	1 196.4	1 113.9	633.6	1 085.9	1 489.1	2 668.0
2nd quarter	20 411.9	17 918.6	170.4	321.5	319.8	835.7	1 293.8	1 035.2	1 220.6	1 131.6	643.6	1 121.2	1 508.2	2 708.0
3rd quarter	20 658.2	18 142.6	157.6	333.9	317.4	848.5	1 311.3	1 041.8	1 243.1	1 141.9	650.6	1 141.5	1 539.6	2 745.9
4th quarter	20 865.1	18 334.6	162.1	327.3	325.2	859.7	1 329.0	1 055.5	1 276.1	1 142.5	664.4	1 154.9	1 519.8	2 778.5

Table 14-1. Gross Domestic Product (Value Added) by Industry—*Continued*

(Billions of dollars.)

Year	Professional, scientific, and technical services	Management of companies and enterprises	Administrative and waste management services	Educational services	Health care and social assistance	Arts, entertainment, and recreation	Accommodation and food services	Other services except government	Total government	Federal	State and local	Private goods-producing industries	Private services-producing industries	Information-communications-technology-producing industries[1]
1999	595.9	148.5	252.7	87.7	566.3	90.4	263.7	260.8	1 252.3	406.0	846.4	2 083.7	6 294.6	579.3
2000	651.8	171.1	282.1	95.2	600.2	99.0	287.5	279.7	1 323.0	423.5	899.5	2 220.4	6 708.9	632.9
2001	686.8	173.6	295.1	101.7	648.2	96.7	294.0	265.6	1 392.9	430.4	962.5	2 184.1	7 004.8	610.3
2002	714.7	175.0	300.2	106.5	700.5	103.8	309.7	284.9	1 474.4	462.5	1 011.9	2 170.1	7 291.9	620.9
2003	741.3	185.6	320.5	116.8	746.0	111.0	321.1	283.6	1 552.3	498.9	1 053.5	2 302.4	7 603.5	653.1
2004	792.4	203.1	345.5	129.1	798.3	117.9	343.3	297.3	1 631.3	525.5	1 105.8	2 502.2	8 080.3	706.2
2005	852.4	214.1	379.9	133.2	837.3	122.2	359.0	310.7	1 710.3	551.3	1 159.0	2 699.3	8 627.1	755.8
2006	917.1	230.5	399.1	142.9	892.5	130.5	381.1	325.0	1 792.0	575.5	1 216.4	2 889.4	9 133.2	795.2
2007	984.9	251.2	430.6	151.5	936.4	137.4	396.1	330.5	1 887.1	601.9	1 285.2	3 018.1	9 546.7	852.5
2008	1 082.1	256.4	438.6	167.8	1 017.0	143.2	399.5	330.3	1 981.6	632.2	1 349.4	2 989.1	9 742.1	904.2
2009	1 031.3	245.0	412.4	188.3	1 079.2	144.6	388.6	326.5	2 045.1	663.3	1 381.7	2 673.6	9 730.3	888.8
2010	1 063.9	265.4	437.6	198.9	1 111.8	152.2	403.5	328.0	2 108.0	697.4	1 410.5	2 774.3	10 109.8	937.4
2011	1 124.6	278.0	454.1	206.0	1 148.7	158.3	422.6	333.1	2 137.1	712.3	1 424.8	2 929.3	10 476.3	974.4
2012	1 188.5	302.2	473.9	213.9	1 193.5	171.2	450.1	348.0	2 159.5	713.9	1 445.6	3 018.8	11 018.7	985.4
2013	1 208.5	319.7	489.1	217.5	1 229.7	177.4	473.9	356.3	2 212.5	703.8	1 508.7	3 181.6	11 390.8	1 062.6
2014	1 267.0	333.2	517.8	226.2	1 266.4	188.8	501.3	376.3	2 271.7	716.1	1 555.7	3 297.6	11 952.4	1 094.2
2015	1 342.4	345.6	545.3	234.5	1 329.0	192.8	544.6	392.5	2 340.5	732.3	1 608.2	3 257.0	12 621.8	1 188.9
2016	1 381.1	348.4	569.6	244.1	1 395.3	203.6	567.2	401.8	2 387.8	744.8	1 643.0	3 211.8	13 107.6	1 283.9
2017	1 450.0	369.4	607.0	245.6	1 454.7	214.1	590.6	416.1	2 453.7	759.9	1 693.8	3 398.9	13 632.8	1 362.9
2018	1 548.3	387.1	640.8	254.2	1 526.1	223.5	615.6	434.7	2 504.7	763.6	1 741.1	3 659.9	14 329.5	...
2010														
1st quarter	1 035.3	257.7	425.7	195.0	1 091.5	148.2	393.9	322.8	2 086.8	687.2	1 399.6	2 716.5	9 918.0	...
2nd quarter	1 051.8	261.6	434.7	198.1	1 106.6	151.4	402.1	326.9	2 110.8	699.9	1 410.9	2 749.2	10 066.1	...
3rd quarter	1 076.0	266.8	442.0	200.5	1 120.6	153.1	407.4	330.8	2 113.8	699.7	1 414.1	2 791.5	10 174.6	...
4th quarter	1 092.5	275.4	447.8	201.9	1 128.5	156.3	410.8	331.6	2 120.4	703.0	1 417.5	2 839.9	10 280.5	...
2011														
1st quarter	1 101.3	278.1	448.4	204.0	1 131.3	155.5	413.9	329.8	2 129.3	708.8	1 420.5	2 864.5	10 292.0	...
2nd quarter	1 122.6	278.5	452.5	205.5	1 145.2	158.2	420.6	332.0	2 140.5	712.1	1 428.4	2 923.4	10 432.2	...
3rd quarter	1 132.0	276.5	456.5	206.8	1 154.4	161.0	424.4	333.8	2 142.4	713.6	1 428.8	2 939.4	10 510.0	...
4th quarter	1 142.3	278.7	459.1	207.8	1 163.7	158.5	431.5	336.7	2 135.9	714.5	1 421.5	2 989.7	10 670.8	...
2012														
1st quarter	1 169.8	291.9	471.4	212.5	1 184.2	167.6	444.4	343.3	2 148.7	715.1	1 433.6	3 015.8	10 855.3	...
2nd quarter	1 189.1	293.5	471.7	214.3	1 186.8	170.1	447.8	346.9	2 151.0	714.8	1 436.3	3 014.2	10 987.0	...
3rd quarter	1 191.4	302.8	474.9	214.5	1 194.1	172.0	450.4	349.6	2 161.0	714.2	1 446.8	3 015.1	11 081.0	...
4th quarter	1 203.8	320.5	477.7	214.2	1 208.8	175.3	458.0	352.0	2 177.2	711.5	1 465.7	3 030.3	11 151.4	...
2013														
1st quarter	1 201.8	313.9	479.9	214.7	1 222.3	174.5	467.5	353.1	2 194.4	706.5	1 487.9	3 137.6	11 237.6	...
2nd quarter	1 197.5	319.6	484.6	215.9	1 224.9	175.2	470.3	353.9	2 207.5	704.1	1 503.4	3 151.0	11 279.4	...
3rd quarter	1 208.6	321.1	492.9	217.8	1 231.7	177.7	476.4	356.4	2 215.1	698.4	1 516.7	3 207.3	11 426.4	...
4th quarter	1 226.3	324.2	498.9	221.4	1 240.0	182.2	481.5	361.8	2 233.1	706.3	1 526.8	3 230.4	11 619.6	...
2014														
1st quarter	1 229.7	331.6	503.4	223.2	1 239.3	185.0	483.2	364.5	2 247.3	710.7	1 536.6	3 210.9	11 644.7	...
2nd quarter	1 247.6	326.7	514.1	225.4	1 254.9	188.1	496.1	373.2	2 261.0	713.9	1 547.1	3 306.9	11 857.8	...
3rd quarter	1 290.6	333.4	523.1	227.4	1 279.1	191.3	508.5	382.0	2 279.2	718.0	1 561.2	3 352.1	12 088.5	...
4th quarter	1 299.9	341.1	530.6	228.7	1 292.1	190.7	517.4	385.3	2 299.3	721.6	1 577.7	3 320.6	12 218.5	...
2015														
1st quarter	1 320.3	344.3	533.9	230.8	1 306.7	190.6	531.2	388.2	2 318.4	727.1	1 591.3	3 243.6	12 408.5	...
2nd quarter	1 337.6	345.4	541.0	233.4	1 322.6	192.5	541.7	393.5	2 335.7	730.5	1 605.2	3 275.7	12 609.9	...
3rd quarter	1 353.6	345.9	549.8	235.8	1 338.7	193.3	549.8	393.4	2 349.5	734.7	1 614.8	3 284.9	12 696.6	...
4th quarter	1 358.0	346.7	556.5	237.7	1 348.0	194.6	555.6	395.0	2 358.4	737.1	1 621.3	3 223.8	12 772.1	...
2016														
1st quarter	1 365.7	345.8	558.8	241.1	1 372.8	198.6	559.5	397.7	2 365.5	738.8	1 626.7	3 170.5	12 873.1	...
2nd quarter	1 375.7	345.9	565.8	243.4	1 390.5	201.1	564.6	400.3	2 378.2	741.7	1 636.5	3 210.2	13 052.4	...
3rd quarter	1 381.9	348.3	571.6	244.1	1 397.2	206.4	569.8	403.0	2 395.8	746.8	1 649.0	3 222.7	13 181.1	...
4th quarter	1 400.9	353.5	582.0	247.6	1 420.7	208.5	574.6	406.1	2 411.8	752.0	1 659.9	3 243.8	13 323.6	...
2017														
1st quarter	1 421.2	359.5	595.9	243.0	1 435.4	208.4	580.9	408.7	2 433.5	757.8	1 675.7	3 323.2	13 405.8	...
2nd quarter	1 442.3	363.9	603.0	244.1	1 450.6	212.4	588.9	413.6	2 448.0	760.0	1 687.9	3 361.8	13 549.4	...
3rd quarter	1 461.1	372.6	611.5	246.6	1 460.5	218.7	594.0	419.4	2 462.1	761.5	1 700.6	3 410.0	13 716.0	...
4th quarter	1 475.4	381.5	617.5	248.5	1 472.3	217.1	598.6	422.7	2 471.2	760.3	1 711.0	3 500.5	13 860.1	...
2018														
1st quarter	1 503.8	382.3	623.2	251.4	1 495.1	216.6	604.4	426.3	2 479.5	758.8	1 720.6	3 556.4	14 005.2	...
2nd quarter	1 546.6	385.2	635.5	252.4	1 519.8	222.8	613.1	433.6	2 493.3	760.6	1 732.7	3 656.5	14 262.1	...
3rd quarter	1 561.4	387.9	646.1	256.2	1 534.9	226.6	619.8	436.5	2 515.6	765.4	1 750.2	3 693.0	14 449.6	...
4th quarter	1 581.3	393.1	658.5	256.7	1 554.5	228.0	625.2	442.4	2 530.5	769.5	1 761.1	3 733.6	14 601.0	...

[1]Consists of computer and electronic products manufacturing; publishing, including software; information and data processing services; and computer systems design and related services.
. . . = Not available.

Table 14-2. Chain-Type Quantity Indexes for Value Added by Industry

(New Series, 2012 = 100.)

Year	Total gross domestic product	Private industries												
		Total	Agriculture, forestry, fishing, and hunting	Mining	Utilities	Construction	Manufacturing		Wholesale trade	Retail trade	Transportation and warehousing	Information	Finance and insurance	Real estate and rental and leasing
							Durable goods	Nondurable goods						
1999	77.9	77.6	78.5	74.2	92.0	136.0	63.5	110.7	77.2	87.4	90.2	56.7	71.2	71.6
2000	81.1	81.1	90.1	65.8	93.2	141.5	70.9	111.7	81.1	90.3	90.0	55.6	77.8	73.7
2001	81.9	81.7	87.0	76.2	77.0	138.6	66.4	110.5	82.7	93.6	84.0	58.9	83.8	76.9
2002	83.3	83.1	90.0	78.2	79.7	134.1	67.8	109.7	83.5	97.7	80.9	64.6	83.2	78.2
2003	85.7	85.5	97.0	69.2	77.9	136.3	72.8	113.1	88.2	102.7	83.8	66.6	83.7	79.4
2004	88.9	89.0	104.7	69.6	82.7	141.2	78.0	120.9	91.9	104.5	90.8	74.3	83.6	81.4
2005	92.1	92.5	109.2	70.8	78.4	141.8	83.4	118.8	96.1	107.9	95.1	79.3	91.4	85.8
2006	94.7	95.5	111.0	81.7	83.3	138.8	89.8	122.5	98.7	108.7	100.7	82.1	95.2	87.4
2007	96.5	97.1	98.3	88.0	84.9	134.6	94.0	124.5	102.1	105.1	99.9	90.1	92.8	91.1
2008	96.3	96.5	100.4	85.2	89.5	121.4	94.5	118.1	102.0	101.3	99.0	95.9	79.1	93.2
2009	93.9	93.5	111.4	97.7	84.8	104.3	80.9	114.7	89.7	97.0	93.1	93.6	93.9	92.0
2010	96.3	95.9	108.0	86.2	95.0	98.9	91.1	112.4	95.0	99.1	97.6	98.9	92.7	94.7
2011	97.8	97.6	103.8	89.4	98.7	97.3	97.3	104.9	96.8	99.3	99.4	100.3	92.5	97.8
2012	100.0	100.0	100.0	100.0	100.0	100.0	100.0	100.0	100.0	100.0	100.0	100.0	100.0	100.0
2013	101.8	101.9	116.6	103.9	98.9	102.5	102.5	103.8	102.3	103.1	101.5	109.1	94.1	101.9
2014	104.3	104.6	117.2	114.3	94.8	103.9	103.3	105.1	106.4	104.7	104.2	111.4	98.9	103.8
2015	107.3	107.9	125.1	123.8	94.2	108.2	104.6	104.7	110.7	108.2	106.8	124.0	101.7	106.0
2016	109.0	109.6	130.4	117.9	98.9	111.8	104.1	102.6	109.4	112.3	108.1	137.0	101.7	107.8
2017	111.4	112.0	124.2	119.7	97.9	112.7	107.4	104.1	111.8	116.8	112.5	146.8	100.1	108.9
2018	114.6	115.3	119.1	121.5	100.0	116.0	113.2	107.6	115.4	121.3	115.2	159.3	97.0	111.6
2010														
1st quarter	95.2	94.6	105.9	84.1	95.1	98.4	85.5	118.4	91.2	97.8	94.3	96.6	95.0	93.0
2nd quarter	96.1	95.6	110.3	86.3	94.0	100.1	90.1	112.8	94.4	99.3	96.0	97.6	93.3	94.5
3rd quarter	96.8	96.4	109.5	86.3	95.0	99.0	93.4	110.2	97.4	99.5	99.0	100.7	90.6	95.0
4th quarter	97.2	97.1	106.1	88.1	96.0	98.2	95.6	108.1	97.2	99.7	101.0	100.5	91.8	96.3
2011														
1st quarter	97.0	96.7	103.9	85.7	93.3	96.1	95.9	106.1	96.7	99.6	99.1	99.8	92.9	95.6
2nd quarter	97.7	97.4	99.6	86.6	99.5	97.6	96.7	106.8	95.7	99.2	100.1	100.6	90.7	97.8
3rd quarter	97.7	97.4	102.2	89.8	97.8	97.8	97.1	103.5	95.2	98.4	98.7	100.2	92.4	97.9
4th quarter	98.8	98.9	109.5	95.4	104.2	97.9	99.5	103.2	99.6	99.9	99.6	100.5	93.9	99.8
2012														
1st quarter	99.6	99.6	105.4	96.9	98.7	100.1	99.8	103.8	98.1	100.5	100.5	100.0	97.6	99.3
2nd quarter	100.0	100.1	101.9	101.1	101.4	99.6	100.2	101.4	100.7	98.4	100.0	101.2	99.9	99.9
3rd quarter	100.1	100.1	97.1	100.1	100.8	99.1	99.6	98.3	101.0	100.7	99.8	99.7	101.7	100.7
4th quarter	100.3	100.2	95.6	101.9	99.1	101.2	100.3	96.5	100.2	100.4	99.7	99.1	100.8	100.1
2013														
1st quarter	101.1	101.1	109.1	101.4	101.7	102.0	102.3	102.1	101.5	103.6	101.2	106.8	93.5	101.0
2nd quarter	101.3	101.2	118.3	102.6	99.3	102.0	101.8	103.5	101.0	102.9	100.3	107.6	93.7	101.1
3rd quarter	102.1	102.1	121.3	104.0	98.5	103.3	102.7	105.1	102.6	102.7	101.1	109.4	93.2	102.6
4th quarter	102.9	103.1	117.7	107.8	96.2	102.6	103.0	104.6	104.1	103.3	103.2	112.6	95.9	103.0
2014														
1st quarter	102.6	102.6	113.6	105.1	91.8	102.5	100.9	105.2	102.7	103.7	102.0	109.8	97.0	102.7
2nd quarter	103.9	104.1	115.9	110.5	93.7	103.9	102.6	106.8	105.0	104.8	103.2	110.7	99.3	103.3
3rd quarter	105.2	105.6	117.9	116.7	97.3	103.9	105.1	104.8	108.6	105.4	105.9	111.6	99.1	104.5
4th quarter	105.7	106.1	121.5	124.9	96.2	105.2	104.6	103.6	109.2	104.8	105.6	113.4	100.3	104.8
2015														
1st quarter	106.5	107.0	122.7	130.3	90.3	106.0	104.8	106.6	109.9	106.5	104.4	118.0	101.9	105.3
2nd quarter	107.4	107.9	123.3	118.0	95.0	108.2	104.9	105.6	111.6	107.8	107.0	122.0	104.5	105.6
3rd quarter	107.7	108.2	127.3	122.7	96.1	109.0	104.9	104.5	110.8	109.2	108.4	125.5	100.3	106.4
4th quarter	107.8	108.3	126.9	124.3	95.4	109.6	103.9	102.2	110.4	109.1	107.6	130.4	100.0	106.7
2016														
1st quarter	108.2	108.8	129.6	131.8	97.8	111.7	103.7	103.7	108.0	111.1	105.7	133.3	99.2	107.1
2nd quarter	108.8	109.5	131.5	116.2	98.8	111.5	103.9	103.5	109.0	111.6	108.2	136.0	101.4	108.2
3rd quarter	109.3	109.9	131.2	112.7	99.8	111.9	104.5	102.3	110.5	112.6	108.5	138.8	102.9	108.0
4th quarter	109.8	110.3	129.5	111.0	99.2	112.3	104.3	100.7	110.2	113.9	110.0	140.1	103.6	107.8
2017														
1st quarter	110.3	110.8	128.0	112.2	96.6	112.7	105.2	104.0	110.7	114.3	111.1	141.2	101.9	108.3
2nd quarter	111.1	111.6	125.4	120.0	98.8	111.9	106.4	104.6	111.5	116.0	112.3	145.2	99.6	108.9
3rd quarter	111.9	112.4	122.4	124.0	97.2	112.2	108.0	103.1	111.9	118.3	113.3	149.1	100.5	108.9
4th quarter	112.5	113.1	120.9	122.4	99.2	114.0	109.8	104.8	113.0	118.6	113.2	151.7	98.5	109.4
2018														
1st quarter	113.1	113.7	118.3	116.5	97.3	115.2	111.1	107.8	113.2	120.8	115.0	153.3	97.1	110.2
2nd quarter	114.3	115.0	120.8	119.8	101.7	115.9	113.1	106.8	114.0	120.5	114.9	158.2	96.6	111.6
3rd quarter	115.2	115.9	118.1	119.8	100.4	116.7	113.6	107.8	116.0	122.3	115.6	161.1	97.9	112.0
4th quarter	115.9	116.7	119.0	129.8	100.6	116.1	114.8	108.1	118.6	121.6	115.1	164.6	96.4	112.8

Table 14-2. Chain-Type Quantity Indexes for Value Added by Industry—*Continued*

(New Series, 2012 = 100.)

Year	Private industries—Continued								Government			Private goods-producing industries	Private services-producing industries	Information-communications-technology-producing industries[1]
	Professional, scientific, and technical services	Management of companies and enterprises	Administrative and waste management services	Educational services	Health care and social assistance	Arts, entertainment, and recreation	Accommodation and food services	Other services except government	Total government	Federal	State and local			
1999	67.5	87.0	66.5	68.8	67.5	76.6	88.8	121.2	89.8	86.4	91.5	88.7	74.6	32.9
2000	71.1	91.8	71.3	71.4	70.0	80.5	94.5	124.0	91.6	87.4	93.7	94.0	77.6	38.9
2001	73.3	95.5	72.3	73.2	71.7	75.9	91.9	111.7	92.5	86.2	95.7	91.4	79.0	40.5
2002	75.1	94.6	72.0	73.6	74.9	79.3	93.8	114.8	94.2	87.9	97.3	91.6	80.8	42.8
2003	76.1	97.8	77.1	77.1	77.8	82.1	95.9	111.6	95.3	89.6	98.1	95.0	83.0	47.6
2004	78.7	92.2	81.1	81.8	81.3	84.9	100.6	113.0	96.2	91.2	98.6	100.5	85.9	54.5
2005	81.7	92.1	88.2	80.4	83.3	84.8	101.1	113.8	97.0	91.9	99.6	102.9	89.7	61.6
2006	84.9	91.8	90.2	82.4	86.9	87.5	103.7	114.4	97.6	92.3	100.2	107.4	92.3	68.1
2007	88.1	89.6	95.3	83.7	87.4	88.5	102.6	111.7	98.5	93.0	101.3	109.0	93.8	76.4
2008	94.9	90.2	95.4	89.1	93.0	89.6	99.1	107.6	100.4	95.5	102.9	104.9	94.2	84.8
2009	89.8	83.4	87.8	96.3	95.6	88.4	91.8	101.3	100.6	98.1	101.8	97.9	92.4	85.0
2010	92.0	89.2	93.8	99.1	96.3	92.9	94.9	99.4	101.1	100.3	101.4	98.7	95.2	92.7
2011	95.9	92.9	96.9	99.8	98.1	95.5	98.5	98.5	100.7	100.5	100.9	98.8	97.2	97.7
2012	100.0	100.0	100.0	100.0	100.0	100.0	100.0	100.0	100.0	100.0	100.0	100.0	100.0	100.0
2013	100.3	104.9	101.5	98.4	101.8	102.0	102.2	99.3	99.3	97.7	100.1	103.9	101.3	108.3
2014	104.5	111.8	105.2	99.6	103.8	106.3	105.1	102.0	99.1	97.1	100.0	106.1	104.2	110.0
2015	108.4	113.7	107.6	100.6	107.4	104.2	109.1	102.8	99.1	97.2	100.0	108.6	107.6	125.8
2016	110.9	114.8	108.8	101.9	110.3	106.4	109.1	102.1	100.1	98.0	101.1	108.5	109.9	139.0
2017	114.7	122.3	114.1	99.8	112.9	110.3	110.4	102.6	100.8	98.3	102.0	110.2	112.4	...
2018	121.0	128.1	118.3	101.2	116.5	112.6	111.1	104.2	100.8	97.9	102.2	114.0	115.6	...
2010														
1st quarter	90.0	87.1	91.9	98.1	95.3	90.8	92.8	98.4	101.3	99.9	101.9	97.6	93.7	...
2nd quarter	91.1	88.2	93.2	99.1	96.1	92.5	94.6	99.3	101.3	100.9	101.5	98.8	94.8	...
3rd quarter	92.9	89.6	94.3	99.4	96.8	92.9	96.0	100.1	100.9	100.4	101.2	99.1	95.7	...
4th quarter	93.9	92.1	95.8	99.9	97.0	95.4	96.2	99.8	100.8	100.1	101.1	99.2	96.5	...
2011														
1st quarter	94.2	93.2	95.9	100.1	97.1	94.1	97.4	98.7	101.0	100.7	101.2	98.0	96.3	...
2nd quarter	95.9	93.5	96.9	100.3	98.0	95.6	98.8	98.6	101.0	100.6	101.2	98.5	97.1	...
3rd quarter	96.3	92.2	97.4	99.5	98.5	97.0	99.1	98.3	100.5	100.3	100.6	98.4	97.1	...
4th quarter	97.0	92.7	97.5	99.2	98.8	95.1	98.7	98.4	100.5	100.3	100.5	100.4	98.5	...
2012														
1st quarter	99.0	97.3	100.0	100.3	99.9	99.2	100.2	99.6	100.4	100.4	100.4	101.0	99.3	...
2nd quarter	100.1	97.7	99.7	100.7	99.5	99.6	99.6	99.9	99.9	100.1	99.8	100.7	99.9	...
3rd quarter	100.0	99.8	100.0	100.0	99.7	99.7	99.4	100.2	99.9	99.9	99.8	99.1	100.4	...
4th quarter	100.8	105.2	100.4	99.0	100.8	101.5	100.9	100.3	99.8	99.5	100.0	99.3	100.4	...
2013														
1st quarter	100.0	101.2	100.2	98.1	101.6	101.3	102.5	99.4	99.8	99.0	100.2	102.5	100.8	...
2nd quarter	99.3	103.6	101.0	98.1	101.7	101.4	101.7	98.9	99.6	98.3	100.2	103.4	100.6	...
3rd quarter	100.3	106.5	102.1	98.3	101.8	101.4	101.8	99.0	99.2	97.3	100.1	104.8	101.4	...
4th quarter	101.5	108.4	102.8	99.3	102.2	103.9	102.7	99.7	98.7	96.1	99.9	104.9	102.6	...
2014														
1st quarter	101.6	111.3	103.5	99.3	102.2	105.1	103.2	99.9	98.8	97.4	99.5	103.8	102.3	...
2nd quarter	103.2	109.9	104.9	99.4	103.0	106.1	104.5	101.6	99.2	97.2	100.1	105.8	103.6	...
3rd quarter	106.4	112.8	105.9	99.9	104.5	107.3	106.1	103.2	99.1	97.0	100.2	107.0	105.2	...
4th quarter	106.6	113.2	106.6	99.8	105.3	106.8	106.6	103.2	99.1	96.7	100.2	107.7	105.6	...
2015														
1st quarter	107.2	113.5	106.6	100.0	106.5	104.5	107.9	102.9	98.7	97.1	99.5	109.2	106.4	...
2nd quarter	108.1	113.1	107.2	100.3	107.3	104.0	109.1	103.4	99.1	97.2	100.0	108.5	107.8	...
3rd quarter	109.1	114.0	108.2	100.8	107.9	104.2	109.8	102.6	99.2	97.2	100.1	108.8	108.0	...
4th quarter	109.3	114.0	108.3	101.2	108.2	104.1	109.5	102.2	99.3	97.3	100.3	108.0	108.3	...
2016														
1st quarter	110.1	113.7	107.7	102.1	109.6	104.8	108.8	102.0	99.6	97.6	100.6	109.4	108.6	...
2nd quarter	110.8	113.8	108.3	102.1	110.3	105.5	108.7	102.0	99.8	97.9	100.8	108.6	109.7	...
3rd quarter	110.7	114.9	109.0	101.5	110.0	106.5	109.1	102.1	100.4	98.2	101.4	108.3	110.3	...
4th quarter	112.1	116.6	110.3	102.0	111.2	109.0	109.7	102.1	100.5	98.3	101.6	107.6	111.0	...
2017														
1st quarter	112.6	118.7	112.3	99.4	112.0	106.9	110.1	101.6	100.8	98.4	101.9	109.0	111.2	...
2nd quarter	114.3	120.3	113.7	99.3	112.8	109.7	110.5	102.3	100.8	98.2	102.0	109.9	112.0	...
3rd quarter	115.4	123.5	114.7	100.1	113.1	112.8	110.4	103.1	100.8	98.3	102.1	110.3	112.9	...
4th quarter	116.4	126.6	115.5	100.4	113.6	111.8	110.4	103.4	100.8	98.1	102.1	111.7	113.4	...
2018														
1st quarter	118.1	126.1	115.8	101.1	114.9	110.3	109.7	103.4	100.6	97.7	102.0	112.7	113.9	...
2nd quarter	120.8	127.5	117.5	101.0	116.2	112.8	111.2	104.4	100.7	97.7	102.2	113.6	115.3	...
3rd quarter	122.0	129.0	119.3	101.5	117.1	113.7	111.7	104.2	101.0	98.1	102.4	114.2	116.4	...
4th quarter	123.1	129.9	120.9	101.0	117.8	113.4	111.9	104.6	100.9	97.9	102.4	115.4	117.0	...

[1]Consists of computer and electronic products manufacturing; publishing, including software; information and data processing services; and computer systems design and related services.

... = Not available.

Table 14-3. Gross Domestic Factor Income by Industry

(Billions of current dollars.)

NAICS industry	2005	2006	2007	2008	2009	2010	2011	2012	2013	2014	2015	2016	2017
Gross domestic factor income, total	12 155.1	12 869.1	13 469.6	13 715.7	13 480.5	13 984.8	14 498.8	15 119.0	15 655.9	16 338.9	17 006.6	17 465.3	18 199.5
Compensation of employees	7 077.7	7 491.3	7 889.4	8 068.7	7 767.2	7 933.0	8 234.0	8 575.4	8 843.6	9 258.6	9 707.9	9 968.9	10 420.6
Gross operating surplus	5 077.4	5 377.8	5 580.2	5 647.0	5 713.3	6 051.8	6 264.8	6 543.6	6 812.3	7 080.3	7 298.7	7 496.4	7 778.9
Private industries	10 427.9	11 059.6	11 563.9	11 713.7	11 412.9	11 856.1	12 339.1	12 937.0	13 418.8	14 042.0	14 641.8	15 051.6	15 720.6
Compensation of employees	5 694.4	6 047.1	6 369.8	6 475.5	6 126.3	6 241.3	6 532.0	6 871.6	7 095.3	7 467.0	7 859.4	8 079.9	8 478.3
Gross operating surplus	4 733.5	5 012.5	5 194.1	5 238.2	5 286.6	5 614.8	5 807.1	6 065.4	6 323.5	6 575.0	6 782.4	6 971.7	7 242.3
Agriculture, forestry, fishing, and hunting	142.8	131.6	146.6	148.8	131.8	147.9	180.2	178.5	214.3	197.3	179.5	166.1	168.1
Compensation of employees	34.6	38.0	41.8	42.1	42.3	41.4	41.2	48.1	48.9	51.0	50.9	53.8	55.3
Gross operating surplus	108.2	93.6	104.8	106.7	89.5	106.5	139.0	130.4	165.4	146.3	128.6	112.3	112.8
Mining	199.5	243.2	280.2	349.7	246.0	272.6	318.4	318.5	344.6	368.9	222.0	183.0	233.0
Compensation of employees	47.2	57.0	62.7	72.8	64.7	69.2	80.3	90.5	93.5	101.1	91.9	75.4	77.5
Gross operating surplus	152.3	186.2	217.5	276.9	181.3	203.4	238.1	228.0	251.1	267.8	130.1	107.6	155.5
Utilities	154.2	178.7	180.3	188.2	203.1	222.0	229.4	220.8	227.0	237.9	238.3	240.1	243.7
Compensation of employees	56.1	60.9	63.1	66.3	66.8	67.7	71.2	69.8	72.8	75.2	77.6	80.8	81.6
Gross operating surplus	98.1	117.8	117.2	121.9	136.3	154.3	158.2	151.0	154.2	162.7	160.7	159.3	162.1
Construction	643.8	688.8	707.0	641.0	558.3	517.9	517.2	545.9	579.6	628.1	686.1	736.3	771.7
Compensation of employees	387.5	421.3	440.0	433.9	368.4	343.8	347.3	365.9	388.0	422.4	458.5	486.9	517.1
Gross operating surplus	256.3	267.5	267.0	207.1	189.9	174.1	169.9	180.0	191.6	205.7	227.6	249.4	254.6
Durable goods manufacturing	933.7	980.6	1 005.4	974.0	854.9	939.1	989.9	1 036.3	1 073.2	1 101.8	1 153.4	1 153.6	1 196.9
Compensation of employees	590.0	614.8	627.4	615.5	539.2	546.2	578.7	604.4	612.0	638.6	662.8	666.7	696.3
Gross operating surplus	343.7	365.8	378.0	358.5	315.7	392.9	411.2	431.9	461.2	463.2	490.6	486.9	500.6
Nondurable goods manufacturing	700.1	751.7	775.0	761.4	772.0	781.7	798.9	811.6	834.3	861.1	884.2	846.5	895.2
Compensation of employees	305.4	309.2	316.2	316.4	296.0	300.9	303.7	312.6	319.8	334.5	343.8	347.3	359.9
Gross operating surplus	394.7	442.5	458.8	445.0	476.0	480.8	495.2	499.0	514.5	526.6	540.4	499.2	535.3
Wholesale trade	593.0	640.0	685.4	712.5	676.0	718.7	749.9	804.9	837.8	879.9	929.0	921.3	951.5
Compensation of employees	379.4	404.6	429.3	435.7	405.1	412.3	436.2	457.7	467.6	489.9	510.0	511.7	533.1
Gross operating surplus	213.6	235.4	256.1	276.8	270.9	306.4	313.7	347.2	370.2	390.0	419.0	409.6	418.4
Retail trade	672.5	690.9	688.2	670.1	661.7	673.1	684.8	713.8	744.1	759.8	800.6	825.7	852.0
Compensation of employees	477.7	492.2	506.4	500.9	474.3	478.9	495.7	510.1	526.7	545.5	573.8	588.2	606.9
Gross operating surplus	194.8	198.7	181.8	169.2	187.4	194.2	189.1	203.7	217.4	214.3	226.8	237.5	245.1
Transportation and warehousing	352.6	387.1	388.2	400.7	378.1	407.6	424.0	443.7	459.7	488.1	527.6	539.7	569.2
Compensation of employees	231.6	241.0	255.9	255.6	241.1	245.5	259.9	274.8	283.4	296.7	319.6	331.8	352.0
Gross operating surplus	121.0	146.1	132.3	145.1	137.0	162.1	164.1	168.9	176.3	191.4	208.0	207.9	217.2
Information	593.9	603.4	658.9	698.2	677.9	708.9	714.6	713.6	781.5	791.9	865.9	947.4	998.6
Compensation of employees	241.5	248.8	260.5	257.8	251.5	248.6	260.1	271.6	286.5	304.4	318.9	328.7	348.8
Gross operating surplus	352.4	354.6	398.4	440.4	426.4	460.3	454.5	442.0	495.0	487.5	547.0	618.7	649.8
Finance and insurance	935.0	994.0	988.2	829.5	922.9	959.5	980.9	1 117.5	1 097.8	1 224.9	1 300.0	1 367.2	1 408.9
Compensation of employees	534.6	579.2	617.4	612.1	548.0	573.3	606.5	630.1	641.3	681.1	715.7	734.5	788.3
Gross operating surplus	400.4	414.8	370.8	217.4	374.9	386.2	374.4	487.4	456.5	543.8	584.3	632.7	620.6
Real estate and rental and leasing	1 453.5	1 514.4	1 612.6	1 675.7	1 677.1	1 714.9	1 792.1	1 868.1	1 941.1	2 022.9	2 136.4	2 231.6	2 312.5
Compensation of employees	101.6	108.0	112.2	110.3	103.7	103.8	106.4	114.3	118.9	126.4	140.4	144.5	151.8
Gross operating surplus	1 351.9	1 406.4	1 500.4	1 565.4	1 573.4	1 611.1	1 685.7	1 753.8	1 822.2	1 896.5	1 996.0	2 087.1	2 160.7
Professional, scientific, and technical services	814.7	880.2	949.8	1 051.9	1 008.6	1 040.2	1 099.4	1 161.3	1 177.7	1 234.4	1 307.6	1 343.6	1 410.1
Compensation of employees	553.6	607.4	655.0	693.5	669.4	682.7	727.1	776.7	801.5	849.0	908.8	936.4	984.8
Gross operating surplus	261.1	272.8	294.8	358.4	339.2	357.5	372.3	384.6	376.2	385.4	398.8	407.2	425.3
Management of companies and enterprises	207.2	223.1	243.4	247.2	236.6	257.1	269.7	293.8	311.2	324.4	336.6	339.2	359.8
Compensation of employees	185.3	200.7	219.9	221.1	207.9	227.2	237.9	259.4	275.0	287.3	299.3	301.6	321.3
Gross operating surplus	21.9	22.4	23.5	26.1	28.7	29.9	31.8	34.4	36.2	37.1	37.3	37.6	38.5
Administrative and waste management services	371.7	390.4	421.3	429.1	403.1	427.7	443.7	463.1	477.6	505.8	532.8	556.5	593.4
Compensation of employees	274.8	292.4	310.1	314.1	289.2	306.5	321.9	341.3	353.7	376.9	398.3	409.9	438.8
Gross operating surplus	96.9	98.0	111.2	115.0	113.9	121.2	121.8	121.8	123.9	128.9	134.5	146.6	154.6
Educational services	127.8	137.0	145.2	160.9	180.7	191.0	198.1	205.9	209.2	217.7	225.6	234.7	235.9
Compensation of employees	105.9	113.1	120.9	129.4	139.9	145.5	152.8	161.4	165.2	174.8	182.3	190.1	190.3
Gross operating surplus	21.9	23.9	24.3	31.5	40.8	45.5	45.3	44.5	44.0	42.9	43.3	44.6	45.6
Health care and social assistance	817.1	871.3	913.5	993.6	1 054.6	1 087.0	1 127.8	1 169.6	1 205.4	1 240.9	1 301.1	1 365.5	1 422.9
Compensation of employees	687.5	733.2	775.1	824.6	862.4	888.5	922.0	963.6	997.3	1 028.4	1 084.5	1 135.0	1 184.6
Gross operating surplus	129.6	138.1	138.4	169.0	192.2	198.5	205.8	206.0	208.1	212.5	216.6	230.5	238.3
Arts, entertainment, and recreation	109.8	117.1	122.3	127.7	129.5	136.7	142.3	153.8	159.4	170.2	173.3	183.6	193.4
Compensation of employees	68.4	73.1	77.0	80.6	78.6	79.4	81.5	86.1	89.7	96.6	100.6	107.7	113.0
Gross operating surplus	41.4	44.0	45.3	47.1	50.9	57.3	60.8	67.7	69.7	73.6	72.7	75.9	80.4
Accommodation and food services	309.6	327.6	339.3	340.7	331.1	342.5	363.0	387.6	407.2	430.5	470.6	490.6	511.1
Compensation of employees	222.0	233.3	247.8	252.5	242.8	248.3	262.4	282.5	295.8	314.7	336.3	355.2	373.4
Gross operating surplus	87.6	94.3	91.5	88.2	88.3	94.2	100.6	105.1	111.4	115.8	134.3	135.4	137.7
Other services, except government	295.1	308.4	313.1	312.6	308.9	310.1	314.7	329.0	336.3	355.5	371.0	379.4	392.8
Compensation of employees	209.4	218.9	231.1	240.1	235.1	231.7	239.3	250.9	257.9	272.5	285.3	293.6	303.6
Gross operating surplus	85.7	89.5	82.0	72.5	73.8	78.4	75.4	78.1	78.4	83.0	85.7	85.8	89.2
Government	1 727.3	1 809.5	1 905.8	2 002.0	2 067.6	2 128.8	2 159.8	2 182.0	2 237.1	2 296.9	2 364.9	2 413.6	2 478.9
Compensation of employees	1 383.4	1 444.2	1 519.6	1 593.2	1 640.9	1 691.7	1 702.0	1 703.8	1 748.4	1 791.6	1 848.5	1 889.0	1 942.3
Gross operating surplus	343.9	365.3	386.2	408.8	426.7	437.1	457.8	478.2	488.7	505.3	516.4	524.6	536.6
Addenda:													
Private goods-producing industries	2 620.0	2 795.9	2 914.4	2 875.0	2 562.9	2 659.2	2 804.6	2 890.8	3 046.0	3 157.2	3 125.1	3 085.4	3 264.8
Compensation of employees	1 364.8	1 440.3	1 488.2	1 480.8	1 310.5	1 301.5	1 351.1	1 421.4	1 462.2	1 547.6	1 607.8	1 630.1	1 706.0
Gross operating surplus	1 255.2	1 355.6	1 426.2	1 394.2	1 252.4	1 357.7	1 453.5	1 469.4	1 583.8	1 609.6	1 517.3	1 455.3	1 558.8
Private services-producing industries	7 807.8	8 263.6	8 649.5	8 838.7	8 849.9	9 196.9	9 534.5	10 046.1	10 372.8	10 884.9	11 516.5	11 966.2	12 455.7
Compensation of employees	4 329.5	4 606.8	4 881.6	4 994.7	4 815.8	4 939.8	5 180.9	5 450.1	5 633.1	5 919.4	6 251.5	6 449.8	6 772.2
Gross operating surplus	3 478.3	3 656.8	3 767.9	3 844.0	4 034.1	4 257.1	4 353.6	4 596.0	4 739.7	4 965.5	5 265.0	5 516.4	5 683.5
Information-communications-technology-producing industries [1]	708.1	747.3	804.9	859.3	844.9	892.9	928.6	939.2	1 014.1	1 044.3	1 138.5	1 232.0	1 309.4
Compensation of employees	370.8	396.0	420.3	422.2	410.6	423.0	453.6	483.8	501.4	536.2	570.1	595.1	638.9
Gross operating surplus	337.3	351.3	384.6	437.1	434.3	469.9	475.0	455.4	512.7	508.1	568.4	636.9	670.5

[1]Consists of computer and electronic products manufacturing; publishing, including software; information and data processing services; and computer systems design and related services.

NOTES AND DEFINITIONS, CHAPTER 14

TABLES 14-1 THROUGH 14-3

Gross Domestic Product (VALUE ADDED) and Gross Factor Income by Industry

SOURCE: U.S. DEPARTMENT OF COMMERCE, BUREAU OF ECONOMIC ANALYSIS (BEA)

In the introduction to the notes and definitions for Chapter 1, it was observed that gross domestic product (GDP), while primarily measured as the sum of final demands for goods and services, is also the sum of the values created by each industry in the economy. In this chapter, selected data are presented from the industry accounts in the national income and product accounts (NIPAs). The industry accounts measure the contribution of each major industry to GDP.

New in the 24th edition of *Business Statistics* are quarterly estimates of current- and constant-dollar GDP by industry. These measures have been computed starting with the first quarter of 2005 and extending through a preliminary estimate for the first quarter of 2019; they are displayed for 2005–2018 in Tables 14-1A and 14-2A. They provide a complete accounting of the industrial origin of GDP, and measure the output of major industries on a more up-to-date basis than the annual measures previously available—within 120 days after the end of the reference quarter.

In recent years, estimates of GDP by industry have been prepared using a methodology integrated with annual input-output accounts in order to produce estimates of gross industry output, industry input, and the difference between the two—industry value added—with greater consistency and timeliness than was previously possible. The current integrated industry accounts also include a wealth of related information too extensive for inclusion here, such as quantity and price indexes for gross output and intermediate inputs, cost per unit of real value added allocated to the three components of value added, and components of domestic supply (domestic output, imports, exports, and inventory change).

The most precise estimates of GDP by NAICS industry begin with 1997, but the annual series were extended back to 1947, using approximations of current methods in earlier years when the source data were less comprehensive and were initially compiled on the older SIC classifications. Table 14-1B presents this "old series" current-dollar GDP for NAICS industry groups from 1947 through 2009; Table 14-2B presents "old series" indexes of real value added for each industry group.

The 2018 comprehensive benchmark revision incorporated three major types of revisions: 1) statistical changes to introduce new and improved technologies and newly available source data; 2) changes in definitions to more accurately portray the evolving U.S. economy to more accurately portray the evolving U.S. economy and to provide consistent comparisons with data for other national economies; and 3) changes in presentations to reflect the definitional and statistical changes, where necessary, or to provide additional data or perspective for users. Also output and price measures will be expressed with 2012 equal to 100.

In Table 14-3, the editor presents a measure derived from the components of current-dollar value added as published by BEA. This measure, "gross domestic factor income," enables users to obtain a clearer picture of the quantitative impact of each industry on the economy's factors of production and of the shares of capital and labor in each industry.

The 2004 revision incorporated a change in terminology. An industry's contribution to total GDP, formerly referred to as "gross product originating" (GPO) or "gross product by industry," is now called "value added." This is consistent with the use of the term "value added" in most economic writing. However, it should not be confused with a concept known as "Census value added," which is used in U.S. censuses and surveys of manufactures. Census value added is calculated at the individual establishment level and, for that reason, does not exclude purchased business services. This means that census value added, while still an important gauge of relative importance, is not a true measure of economic value added.

Definitions and notes on the data

An industry's *gross domestic product (value added)*, formerly *GPO*, is equal to the market value of its gross output (which consists of the value, including taxes, of sales or receipts and other operating income plus the value of inventory change) minus the value of its intermediate inputs (energy, raw materials, semi finished goods, and services that are purchased from domestic industries or from foreign sources).

In concept, this is also equal to the sum of *compensation of employees, taxes on production and imports less subsidies,* and *gross operating surplus.* (See Chapter 1 and its notes and definitions for more information.)

The GDP data from 1997 forward are not comparable with the "old series" data for earlier years presented in Tables 14-1B and 14-2B, the major reason being that they do not reflect the conceptual revisions introduced in the comprehensive 2018 revision of the NIPAs (see the Notes and Definitions to Chapter 1). In particular, the new data include, while the old data exclude, the capitalized values of research and development, intellectual property, and all real estate transfer costs. This does not change labor compensation values--the associated workers were always included; it was just that their costs were taken to be written off in the first year. But it does increase across the board the non-labor share of the total value of output, by the amount of capital consumption and net investment. This is seen in labor compensation shares derived from Table 14-3, compared with those that would have been calculated on the old basis last year.

Compensation of employees consists of wage and salary accruals and supplements to wages and salaries. This approximates the labor share of production, subject to the note below about proprietors' income.

Taxes on production and imports less subsidies. Although this is shown in BEA source data as a single net line item, it represents two separate components.

Taxes on production and imports are included in the market value of the goods and services sold to final consumers and therefore in the consumer valuation of those goods. Since they are not part of the payments to the labor and capital inputs in the producing industries, they must be underlined added to the sum of the returns to those inputs in order to account for the total value to consumers. Taxes that fall into this classification include property taxes, sales and excise taxes, and customs duties.

BEA allocates these taxes to the industry level at which they are assessed by law. Most sales taxes are considered by BEA to be part of the value added by retail trade. Some sales taxes, most fuel taxes, and all customs duties are allocated by BEA to wholesale trade. Residential real property taxes, including those on owner-occupied dwellings, are allocated by BEA to the real estate industry.

Subsidies to business by government are included in the labor and/or capital payments made by that industry. Since they are payments to the industry in addition to the market values paid by consumers, they are *subtracted* from the values of the labor and capital inputs to make them consistent with the market values as defined in value added. The role of subsidies is obvious in the source data for the agricultural sector, where the net "taxes on production and imports less subsidies" has a negative sign: farm subsidies more than offset this industry's taxes on production and imports, which mainly consist of property taxes, since sales, excise, and import taxes are not levied on farms.

For private sector businesses, *gross operating surplus* consists of business income (corporate profits before tax, proprietors' income, and rental income of persons), net interest and miscellaneous payments, business current transfer payments (net), and capital consumption allowances. For government, households, and institutions, it consists of consumption of fixed capital and (for government) government enterprises' current surplus. This approximates the share of the value of production ascribable to capital and land as measured in the NIPAs accounts; however, as BEA notes, "an unknown portion [of proprietors' income] reflects the labor contribution of proprietors." (*Survey of Current Business*, June 2004, p. 27, footnote 7.) Another aspect to be noted is that gross operating surplus includes the return to

owner-occupied housing in the real estate sector. Because there is in the NIPAs no employee compensation attributed to owner-occupied housing, the capital share in that industry as measured by gross operating surplus is very large.

Quantity indexes for value added. Measures of the constant-dollar change in each industry's gross output minus its intermediate input use are calculated by BEA, using a Fisher index-number formula which incorporates weights from two adjacent years. The changes for successive years are chained together in indexes, with the value for the year 2012 set at 100. The indexes are multiplied by 2012 current-dollar value added to provide estimates of value added in "chained 2012 dollars," but because the actual weights used change from year to year, components in chained 2012 dollars typically do not add up to total GDP in 2009 dollars—and do not contain any information not already summarized in the indexes. For that reason, only the indexes are published here.

Gross domestic factor income (not a category published as such in the NIPAs) is calculated by the editor as the sum of "compensation of employees" and "gross operating surplus," which the same as value added minus "taxes on production and imports less subsidies." The effect of this procedure is to take out the specified taxes, and to leave in the subsidies embedded in the employee compensation and gross operating surplus components. The editor believes that this provides a valuable alternative basis for assessing the importance of different industries in the economy, by measuring the value paid for by its labor and capital inputs, and for calculating the shares of labor and capital in each industry's output. The components shown in this table are found on the BEA website in a table entitled "Components of Value Added by Industry."

The editor's reasoning is based on the facts that more than half of these taxes are sales, excise, and import taxes, and the assignment of these taxes to industries is economically arbitrary. BEA assigns them to the industry with the legal liability to pay, not to the entity bearing the major incidence of the tax. Yet economists have demonstrated that most of the burdens of sales and excise taxes and import duties are not borne by the factors in the legally liable industry; instead, they are passed on to consumers. In addition, because the wholesale and retail trade industries are classified as "services-producing," BEA's allocation of those taxes has a very peculiar result: taxes on goods are represented as paid by "service" industries. The process adopted instead by the editor in Table 14-3, which excludes these taxes and focuses on "gross domestic factor income," has the effect (for example) of keeping the wholesale trade industry from appearing to be both larger and more heavily taxed than it really is.

Private goods-producing industries consists of agriculture, forestry, fishing, and hunting; mining; construction; and manufacturing.

Private services-producing industries consists of utilities; wholesale trade; retail trade; transportation and warehousing; information; finance, insurance, real estate, rental, and leasing; professional and business services; educational services, health care, and social assistance; arts, entertainment, recreation, accommodation, and food services; and other services, except government.

Information-communications-technology-producing industries is a category that cuts across the goods and services framework, consisting of computer and electronic products manufacturing; publishing industries (which includes software) from the information sector; information and data processing services; and computer systems design and related services.

Data availability and references

The comprehensive revision of the industry accounts from 1997 forward and the new quarterly statistics are described in "Industry Economic Accounts: Results of the Comprehensive Revision, Revised Statistics for 1997-2012" (*Survey of Current Business,* February 2014); "New Quarterly Statistics Detail Industries' Economic Performance," BEA news release, April 25, 2014; and "Prototype Quarterly Statistics on U.S. GDP by Industry, 2007-2011," June 2012, a "BEA Briefing." All are available on the BEA website, http://www.bea.gov. To access the historical data at <http://www.bea.gov>, click on "Annual Industry Accounts" and click on "Interactive Tables" under "Gross Domestic Product (GDP) by Industry."

"Old basis" annual GDP by industry data shown here were recalculated back to 1947 consistent with the 2009 comprehensive revision of the NIPAs. They were most recently presented and described in "Annual Industry Statistics: Revised Statistics for 2009-2011," *Survey of Current Business*, December 2012.

CHAPTER 15: EMPLOYMENT, HOURS, AND EARNINGS BY NAICS INDUSTRY

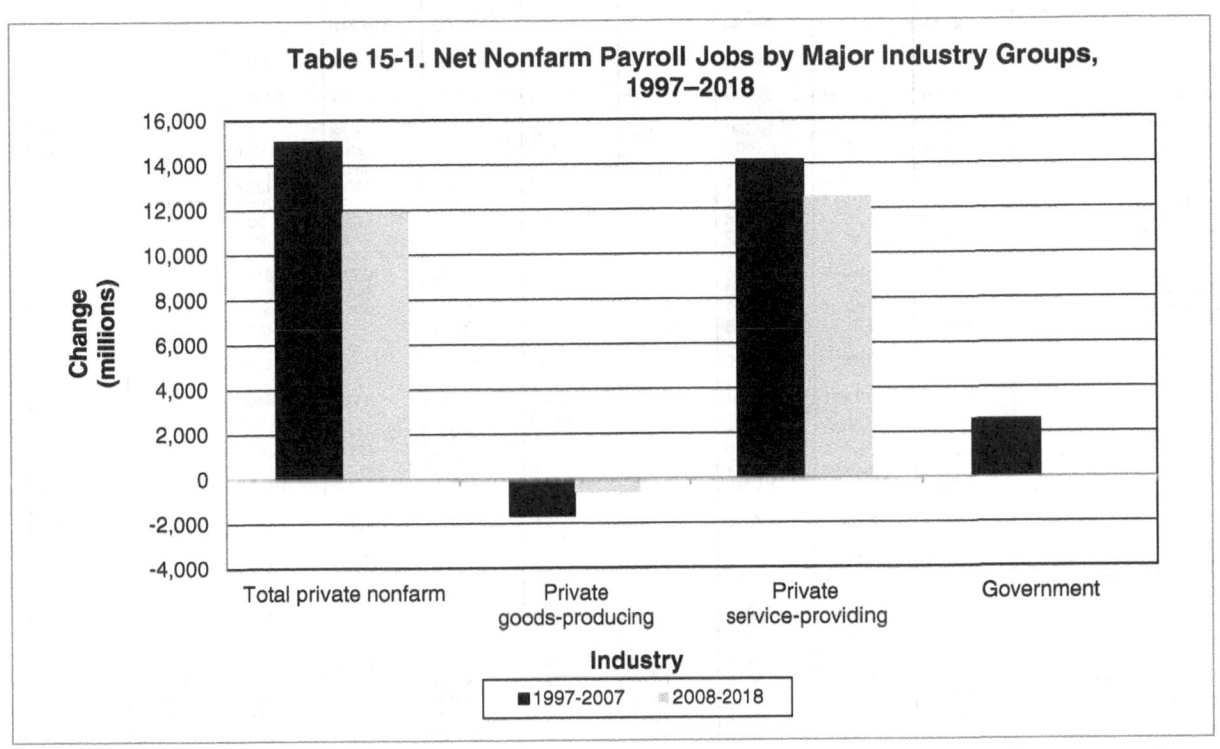

Table 15-1. Net Nonfarm Payroll Jobs by Major Industry Groups, 1997–2018

- Employment began to decline in February 2008 and didn't begin to increase until March 2010, nine months after the recession's end, which the National Bureau of Economic Research Business Cycle Dating Committee places at June 2009. (Preliminary monthly data indicate that the previous peak was finally reached in May 2014.) ("Business Cycle Perspectives" at the beginning of this volume, Table 10-9, Table 15-1, and recent data)

- Employment in 2014 finally exceeded 2007 employment. In 2007, there were 137.9 million nonfarm employees versus 138.9 in 2014, an increase of 0.7 percent. Nonfarm employment has increased each year since 2011. In 2018, nonfarm employment increased 1.7 percent from the previous year. (Table 15-1)

- Although the number of goods-producing jobs continued to increase in 2018, employment was still significantly lower than it was twenty years earlier in 1998. Meanwhile, the number of service-providing jobs increased over 26 percent from 1998 to 2018. (Table 15-1)

- The average workweek for production and nonsupervisory workers is up since 2009 in most industries and industry groups. In 2018, goods-producing workers averaged 41.5 hours per week while private servicing-providers averaged 32.4 hours. (Table 15-3)

- Average hourly earnings for production and nonsupervisory workers in 2018 ranged from a low of $13.87 in the leisure and hospitality sector to $40.32 in petroleum and coal products manufacturing. (Table 15-4)

Table 15-1. Nonfarm Employment by Sector and Industry

(Wage and salary workers on nonfarm payrolls, thousands.)

Industry	1997	1998	1999	2000	2001	2002	2003	2004	2005	2006	2007
TOTAL NONFARM	122 951	126 157	129 240	132 024	132 087	130 649	130 347	131 787	134 051	136 453	137 999
Total Private	103 287	106 248	108 933	111 235	110 969	109 136	108 764	110 166	112 247	114 479	115 781
Goods-Producing	23 886	24 354	24 465	24 649	23 873	22 557	21 816	21 882	22 190	22 530	22 233
Mining and logging	654	645	598	599	606	583	572	591	628	684	724
Logging	82.1	80.0	80.8	79.0	73.5	70.4	69.4	67.6	65.2	64.4	60.1
Mining	571.3	564.7	517.4	520.2	532.5	512.2	502.7	523.0	562.2	619.7	663.8
Oil and gas extraction	144.1	140.8	131.2	124.9	123.7	121.9	120.2	123.4	125.7	134.5	146.2
Mining, except oil and gas [1]	249.5	243.1	234.5	224.8	218.7	210.6	202.7	205.1	212.8	220.3	223.4
Coal mining	89.4	85.3	78.6	72.2	74.3	74.4	70.0	70.6	73.9	78.0	77.2
Support activities for mining	177.7	180.8	151.7	170.6	190.1	179.8	179.8	194.6	223.7	264.9	294.3
Construction	5 813	6 149	6 545	6 787	6 826	6 716	6 735	6 976	7 336	7 691	7 630
Construction of buildings	1 435.4	1 508.8	1 586.3	1 632.5	1 588.9	1 574.8	1 575.8	1 630.0	1 711.9	1 804.9	1 774.2
Heavy and civil engineering	824.9	865.3	908.7	937.0	953.0	930.6	903.1	907.4	951.2	985.1	1 005.4
Specialty trade contractors	3 552.6	3 775.1	4 049.6	4 217.0	4 283.9	4 210.4	4 255.7	4 438.6	4 673.1	4 901.1	4 850.2
Manufacturing	17 419	17 560	17 322	17 263	16 441	15 259	14 509	14 315	14 227	14 155	13 879
Durable goods	10 705	10 911	10 831	10 877	10 336	9 485	8 964	8 925	8 956	8 981	8 808
Wood products	597.8	611.6	622.7	615.4	576.3	557.0	539.5	551.6	561.0	560.6	517.1
Nonmetallic mineral products	525.7	535.3	540.8	554.2	544.5	516.0	494.2	505.5	505.3	509.6	500.5
Primary metals	638.8	641.5	625.0	621.8	570.9	509.4	477.4	466.8	466.0	464.0	455.8
Fabricated metal products	1 695.8	1 739.5	1 728.4	1 752.6	1 676.4	1 548.5	1 478.9	1 497.1	1 522.0	1 553.1	1 562.8
Machinery	1 495.9	1 514.1	1 468.3	1 457.0	1 370.6	1 231.8	1 151.6	1 145.2	1 165.5	1 183.2	1 187.1
Computer and electronic products [1]	1 803.3	1 830.9	1 780.5	1 820.0	1 748.8	1 507.2	1 355.2	1 322.8	1 316.4	1 307.5	1 272.5
Computer and peripheral equipment	316.7	322.1	310.1	301.9	286.2	250.0	224.0	210.0	205.1	196.2	186.2
Communications equipment	235.0	237.4	228.7	238.6	225.4	179.0	149.2	143.0	141.4	136.2	128.1
Semiconductors and electronic components	639.8	649.8	630.5	676.3	645.4	524.5	461.1	454.1	452.0	457.9	447.5
Electronic instruments	502.8	509.2	498.3	487.7	483.6	456.8	435.4	436.9	441.0	444.5	443.2
Electrical equipment and appliances	586.3	591.6	588.0	590.9	556.9	496.5	459.6	445.1	433.5	432.7	429.4
Transportation equipment [1]	2 027.6	2 078.4	2 088.6	2 057.1	1 939.1	1 830.0	1 775.1	1 766.7	1 772.3	1 768.9	1 711.9
Motor vehicles and parts	1 253.9	1 271.5	1 312.5	1 313.6	1 212.9	1 151.2	1 125.3	1 112.8	1 096.7	1 070.0	994.2
Furniture and related products	615.5	641.6	665.2	680.1	642.9	604.7	573.7	574.1	566.3	558.3	529.4
Miscellaneous manufacturing	718.2	726.8	724.0	728.0	709.5	683.3	658.3	650.6	647.2	643.7	641.7
Nondurable goods [1]	6 714	6 649	6 491	6 386	6 105	5 774	5 546	5 390	5 271	5 174	5 071
Food manufacturing	1 557.9	1 554.9	1 549.8	1 553.1	1 551.2	1 525.7	1 517.5	1 493.7	1 477.6	1 479.4	1 484.1
Textile mills	436.2	424.5	397.1	378.2	332.9	290.9	261.3	236.9	217.6	195.0	169.7
Textile product mills	236.4	234.7	232.4	229.6	217.0	204.2	187.7	183.2	176.4	166.7	157.7
Apparel	680.8	621.4	540.5	483.5	415.2	350.0	303.9	278.0	250.5	232.4	214.6
Paper and paper products	630.6	624.9	615.6	604.7	577.6	546.6	516.2	495.5	484.2	470.5	458.2
Printing and related support activities	821.1	827.9	814.6	806.8	768.3	706.6	680.4	662.6	646.3	634.4	622.0
Petroleum and coal products	136.0	134.5	127.8	123.2	121.1	118.1	114.3	111.7	112.1	113.2	114.5
Chemicals	986.8	992.6	982.5	980.4	959.0	927.5	906.1	887.0	872.1	865.9	860.9
Plastics and rubber products	932.7	941.4	947.0	950.9	896.2	846.8	814.3	804.7	802.3	785.5	757.2
Service-Providing	99 065	101 803	104 775	107 375	108 214	108 092	108 531	109 905	111 861	113 922	115 766
Private Service-Providing	79 401	81 894	84 468	86 585	87 096	86 579	86 948	88 284	90 057	91 949	93 548
Trade, transportation, and utilities	24 665	25 150	25 734	26 187	25 945	25 458	25 245	25 487	25 910	26 223	26 573
Wholesale trade	5 621.6	5 751.7	5 848.0	5 888.4	5 728.0	5 605.5	5 556.4	5 607.8	5 705.7	5 841.9	5 948.4
Durable goods	3 158.7	3 251.5	3 310.2	3 342.4	3 221.4	3 103.7	3 044.5	3 062.5	3 118.8	3 202.7	3 257.3
Nondurable goods	2 070.7	2 097.1	2 126.8	2 131.1	2 097.3	2 084.6	2 079.9	2 091.1	2 109.2	2 133.9	2 160.7
Electronic markets, agents, and brokers	392.2	403.2	411.1	414.9	409.3	417.3	432.1	454.2	477.8	505.3	530.4
Retail trade	14 392.5	14 613.1	14 973.9	15 283.6	15 242.4	15 029.0	14 921.7	15 062.7	15 284.5	15 358.5	15 525.5
Motor vehicle and parts dealers [1]	1 723.4	1 740.9	1 796.6	1 846.9	1 854.6	1 879.4	1 882.9	1 902.3	1 918.6	1 909.7	1 908.3
Automobile dealers	1 134.5	1 142.0	1 179.7	1 216.5	1 225.1	1 252.8	1 254.4	1 257.3	1 261.4	1 246.7	1 242.2
Furniture and home furnishings stores	484.7	499.1	524.4	543.5	541.2	538.7	547.3	563.4	576.1	586.9	574.6
Electronics and appliance stores	575.5	590.9	624.2	647.7	633.0	595.2	574.4	572.0	585.5	581.1	583.1
Building material and garden supply stores	1 043.6	1 062.9	1 101.6	1 142.7	1 152.3	1 177.1	1 185.7	1 227.8	1 276.8	1 324.8	1 310.1
Food and beverage stores	2 956.9	2 965.7	2 984.5	2 993.0	2 950.5	2 881.6	2 838.4	2 821.6	2 817.8	2 821.1	2 843.6
Health and personal care stores	853.3	876.0	898.2	927.6	951.5	938.8	938.1	941.1	953.7	961.1	993.1
Gasoline stations	956.2	961.3	943.5	935.7	925.3	895.9	882.0	875.6	871.1	864.1	861.5
Clothing and clothing accessories stores	1 235.9	1 268.6	1 306.6	1 321.6	1 321.1	1 312.5	1 304.5	1 364.3	1 414.6	1 450.9	1 500.0
Sporting goods, hobby, book, and music stores	545.1	555.1	582.7	602.7	601.1	591.8	584.7	585.9	597.9	606.0	623.3
General merchandise stores [1]	2 657.6	2 686.5	2 751.8	2 819.8	2 842.2	2 812.0	2 822.4	2 863.0	2 934.3	2 935.0	3 020.6
Department stores	1 638.9	1 664.4	1 694.2	1 739.5	1 752.5	1 668.5	1 605.2	1 589.8	1 579.5	1 541.5	1 575.0
Miscellaneous store retailers	913.2	950.3	985.5	1 007.1	993.3	959.5	930.7	913.5	899.9	881.0	865.4
Nonstore retailers	447.2	455.8	474.5	495.3	476.3	446.7	430.5	432.3	438.2	436.7	442.1

[1] Includes other industries, not shown separately.

Table 15-1. Nonfarm Employment by Sector and Industry—*Continued*

(Wage and salary workers on nonfarm payrolls, thousands.)

Industry	2008	2009	2010	2011	2012	2013	2014	2015	2016	2017	2018
TOTAL NONFARM	137 241	131 313	130 362	131 932	134 175	136 381	138 958	141 843	144 352	146 624	149 074
Total Private	114 732	108 758	107 871	109 845	112 255	114 529	117 076	119 814	122 128	124 275	126 625
Goods-Producing	21 335	18 558	17 751	18 047	18 420	18 738	19 226	19 610	19 750	20 084	20 710
Mining and logging	767	694	705	788	848	863	891	813	668	676	732
Logging	56.6	50.4	49.7	48.7	50.8	51.8	52.0	52.4	51.0	49.6	48.6
Mining	709.8	643.3	654.8	739.2	797.2	811.0	838.5	760.4	616.8	626.1	683.2
Oil and gas extraction	160.5	159.8	158.7	172.0	187.4	193.5	197.7	193.4	169.8	143.9	145.1
Mining, except oil and gas [1]	226.0	208.3	204.5	218.4	218.7	209.6	207.2	197.8	180.9	185.8	192.0
Coal mining	81.2	81.5	80.8	87.3	84.7	78.1	73.2	64.2	50.8	51.5	51.8
Support activities for mining	323.4	275.2	291.6	348.8	391.1	407.9	433.6	369.2	266.2	296.4	346.1
Construction	7 162	6 016	5 518	5 533	5 646	5 856	6 151	6 461	6 728	6 969	7 289
Construction of buildings	1 641.7	1 357.2	1 229.7	1 222.1	1 240.2	1 285.9	1 358.0	1 423.7	1 493.2	1 545.2	1 625.5
Heavy and civil engineering	964.5	851.3	825.1	836.8	868.3	885.0	912.3	937.6	951.6	996.4	1 052.0
Specialty trade contractors	4 555.8	3 807.9	3 463.4	3 474.4	3 537.1	3 684.4	3 880.6	4 099.7	4 283.0	4 427.2	4 611.6
Manufacturing	13 406	11 847	11 528	11 726	11 927	12 020	12 185	12 336	12 354	12 439	12 689
Durable goods	8 463	7 284	7 064	7 273	7 470	7 548	7 674	7 765	7 714	7 741	7 945
Wood products	457.7	360.2	342.1	337.1	339.1	353.2	371.6	382.5	392.7	397.0	406.4
Nonmetallic mineral products	465.0	394.3	370.9	366.6	365.3	373.4	384.1	398.0	406.1	409.8	415.0
Primary metals	442.0	362.1	362.3	388.3	402.0	395.3	398.7	394.0	374.5	371.3	378.0
Fabricated metal products	1 527.5	1 311.6	1 281.7	1 347.3	1 409.8	1 432.3	1 454.2	1 457.8	1 421.5	1 425.2	1 466.8
Machinery	1 187.6	1 028.6	996.1	1 055.8	1 098.5	1 104.5	1 127.3	1 120.0	1 070.5	1 079.1	1 120.0
Computer and electronic products [1]	1 244.2	1 136.9	1 094.6	1 103.5	1 089.0	1 065.6	1 048.7	1 052.8	1 048.4	1 039.2	1 055.4
Computer and peripheral equipment	183.2	166.4	157.6	157.4	157.4	157.5	159.8	160.3	162.7	155.6	156.4
Communications equipment	127.3	120.5	117.4	115.3	108.2	101.1	93.2	88.4	86.2	86.7	85.7
Semiconductors and electronic components	431.8	378.1	369.4	383.4	383.1	374.9	367.4	369.5	367.3	362.1	369.3
Electronic instruments	441.0	421.6	406.4	404.2	399.7	393.7	390.8	398.9	396.8	400.5	410.2
Electrical equipment and appliances	424.3	373.6	359.5	366.1	373.2	374.1	377.8	383.7	382.8	386.3	398.9
Transportation equipment [1]	1 608.0	1 347.9	1 333.1	1 381.5	1 461.1	1 508.8	1 559.0	1 604.9	1 630.3	1 643.2	1 702.4
Motor vehicles and parts	875.5	664.1	678.5	717.7	777.3	824.8	872.1	913.7	944.3	963.4	996.0
Furniture and related products	478.0	384.3	357.2	353.1	351.4	359.9	370.3	381.0	389.8	394.9	394.6
Miscellaneous manufacturing	628.9	584.4	566.8	573.7	580.1	580.5	581.9	589.9	590.9	594.5	607.8
Nondurable goods [1]	4 943	4 564	4 464	4 453	4 457	4 472	4 512	4 571	4 640	4 699	4 743
Food manufacturing	1 480.9	1 456.4	1 450.6	1 458.8	1 468.8	1 473.7	1 484.4	1 511.8	1 556.5	1 598.0	1 619.7
Textile mills	151.2	124.4	119.0	120.1	118.7	117.2	117.0	116.6	114.5	112.6	112.7
Textile product mills	147.2	125.7	119.0	117.6	116.3	114.4	115.0	116.1	116.2	116.2	116.2
Apparel	199.0	167.5	156.6	151.7	148.0	144.7	140.1	136.5	130.9	119.3	112.6
Paper and paper products	444.9	407.0	394.7	387.4	379.8	378.0	373.3	372.7	370.8	366.2	366.9
Printing and related support activities	594.1	521.9	487.6	471.8	461.8	452.0	453.7	450.3	447.6	440.2	430.9
Petroleum and coal products	117.4	115.3	113.9	111.8	112.1	110.4	111.6	112.9	112.5	114.6	115.6
Chemicals	847.1	804.1	786.5	783.6	783.3	792.7	803.0	807.3	811.9	823.8	837.8
Plastics and rubber products	729.4	624.9	624.8	635.2	645.1	659.2	674.3	689.4	702.4	716.9	730.9
Service-Providing	115 907	112 755	112 611	113 884	115 755	117 643	119 732	122 233	124 603	126 540	128 365
Private Service-Providing	93 398	90 201	90 121	91 798	93 835	95 791	97 850	100 204	102 379	104 191	105 916
Trade, transportation, and utilities	26 236	24 850	24 581	25 008	25 416	25 801	26 321	26 824	27 195	27 409	27 659
Wholesale trade	5 874.9	5 520.9	5 386.5	5 474.7	5 595.2	5 660.3	5 739.5	5 780.4	5 786.9	5 813.5	5 852.7
Durable goods	3 190.6	2 944.0	2 848.3	2 905.2	2 977.9	3 012.7	3 056.0	3 079.6	3 082.5	3 108.3	3 153.2
Nondurable goods	2 148.2	2 063.3	2 025.7	2 040.5	2 070.8	2 091.8	2 121.0	2 137.2	2 145.2	2 152.8	2 151.5
Electronic markets, agents, and brokers	536.1	513.7	512.5	528.9	546.6	555.9	562.5	563.6	559.2	552.5	547.9
Retail trade	15 288.8	14 527.9	14 446.0	14 673.7	14 846.8	15 084.7	15 363.4	15 610.9	15 831.6	15 845.7	15 833.1
Motor vehicle and parts dealers [1]	1 831.2	1 637.5	1 629.2	1 691.2	1 737.0	1 793.1	1 862.2	1 929.0	1 979.6	2 004.6	2 021.2
Automobile dealers	1 176.7	1 018.2	1 011.5	1 056.9	1 095.5	1 138.4	1 187.9	1 238.7	1 279.0	1 293.3	1 299.4
Furniture and home furnishings stores	531.1	449.2	437.9	438.9	439.4	445.8	455.9	466.6	470.9	476.2	481.4
Electronics and appliance stores	570.2	516.3	523.0	528.1	508.1	497.3	497.8	523.0	522.4	503.3	488.0
Building material and garden supply stores	1 248.8	1 156.4	1 132.6	1 146.6	1 175.1	1 208.5	1 229.0	1 235.3	1 267.8	1 276.7	1 307.0
Food and beverage stores	2 862.0	2 830.0	2 808.2	2 822.8	2 861.3	2 929.7	3 004.0	3 062.3	3 090.3	3 086.2	3 087.1
Health and personal care stores	1 002.8	986.0	980.5	980.9	997.9	1 015.8	1 022.5	1 033.7	1 053.3	1 066.7	1 061.2
Gasoline stations	842.4	825.5	819.3	831.0	843.5	866.3	881.3	905.3	923.1	929.8	934.0
Clothing and clothing accessories stores	1 468.0	1 363.9	1 352.5	1 360.9	1 391.4	1 390.9	1 370.4	1 353.7	1 359.3	1 374.4	1 365.6
Sporting goods, hobby, book, and music stores	621.9	589.2	579.1	577.9	582.2	602.5	618.8	623.2	620.4	606.4	575.8
General merchandise stores [1]	3 025.6	2 966.2	2 997.7	3 085.2	3 065.4	3 060.4	3 102.0	3 131.4	3 169.4	3 125.9	3 104.9
Department stores	1 524.1	1 456.9	1 485.2	1 521.7	1 439.0	1 334.0	1 335.3	1 312.5	1 267.2	1 173.6	1 150.8
Miscellaneous store retailers	842.5	782.4	761.5	772.4	794.0	802.5	818.3	828.0	831.8	828.4	833.5
Nonstore retailers	442.3	425.4	424.8	437.9	451.6	471.9	501.3	519.3	543.2	567.2	573.3

[1]Includes other industries, not shown separately.

Table 15-1. Nonfarm Employment by Sector and Industry—*Continued*

(Wage and salary workers on nonfarm payrolls, thousands.)

Industry	1997	1998	1999	2000	2001	2002	2003	2004	2005	2006	2007
Transportation and warehousing	4 029.7	4 171.3	4 303.7	4 413.7	4 375.4	4 227.1	4 189.3	4 252.8	4 365.4	4 474.3	4 546.0
Air transportation	542.0	562.7	586.3	614.4	615.3	563.5	528.3	514.5	500.8	487.0	491.8
Rail transportation	221.0	225.0	228.8	231.7	226.7	217.8	217.7	225.7	227.8	227.5	233.7
Water transportation	50.7	50.5	51.7	56.0	54.0	52.6	54.5	56.4	60.6	62.7	65.5
Truck transportation	1 308.5	1 354.7	1 391.8	1 406.1	1 387.1	1 339.6	1 325.9	1 352.1	1 397.9	1 436.2	1 439.6
Transit and ground passenger transportation	349.6	362.7	371.0	372.1	374.8	380.8	382.2	384.9	389.2	399.3	412.1
Pipeline transportation	49.7	48.1	46.9	46.0	45.4	41.7	40.2	38.4	37.8	38.7	39.9
Scenic and sightseeing transportation	24.5	25.4	26.1	27.5	29.1	25.6	26.6	27.2	28.8	27.5	28.6
Support activities for transportation	473.4	496.8	518.1	537.4	539.2	524.7	520.3	535.1	552.2	570.6	584.2
Couriers and messengers	546.0	568.2	585.9	605.0	587.0	560.9	561.7	556.6	571.4	582.4	580.7
Warehousing and storage	464.4	477.2	497.2	517.5	516.9	519.9	531.9	562.0	598.9	642.5	669.8
Utilities	621	613	608	601	599	596	577	564	554	548	553
Information	3 084	3 218	3 419	3 630	3 629	3 395	3 188	3 118	3 061	3 038	3 032
Publishing industries, except Internet	955.5	982.3	1 004.8	1 035.0	1 020.7	964.1	924.8	909.1	904.1	902.4	901.2
Motion picture and sound recording industries	353.0	369.5	384.4	382.6	376.8	387.9	376.2	385.0	377.5	375.7	380.6
Broadcasting, except Internet	313.0	321.2	329.4	343.5	344.6	334.1	324.3	325.0	327.7	328.3	325.2
Telecommunications	1 108.0	1 167.4	1 270.8	1 396.6	1 423.9	1 280.9	1 166.8	1 115.1	1 071.3	1 047.6	1 030.6
Data processing, hosting, and related services	268.4	282.8	307.1	315.7	316.8	303.9	280.0	267.1	262.5	263.2	267.8
Other information services [1]	85.5	95.3	121.9	157.1	146.5	123.6	115.9	116.9	117.7	120.8	126.3
Internet publishing and broadcasting	45.4	53.9	78.1	110.8	100.4	76.3	67.2	66.1	67.2	69.1	72.9
Financial activities	7 255	7 565	7 753	7 783	7 900	7 956	8 078	8 105	8 197	8 367	8 348
Finance and insurance	5 378.9	5 631.7	5 770.2	5 772.8	5 862.0	5 922.2	6 020.5	6 019.4	6 062.9	6 194.0	6 179.1
Monetary authorities–central bank	22.1	21.7	22.6	22.8	23.0	23.4	22.6	21.8	20.8	21.2	21.6
Credit intermediation and related activities [1]	2 433.6	2 531.9	2 591.0	2 547.8	2 597.7	2 686.0	2 792.4	2 817.0	2 869.0	2 924.9	2 866.3
Depository credit intermediation [1]	1 696.6	1 708.9	1 709.7	1 681.2	1 701.2	1 733.0	1 748.5	1 751.5	1 769.2	1 802.0	1 823.5
Commercial banking	1 277.9	1 286.0	1 281.2	1 250.5	1 258.4	1 278.1	1 280.1	1 280.8	1 296.0	1 322.9	1 351.4
Securities, commodity contracts, investments, and funds and trusts	674.5	734.5	782.2	851.1	879.0	836.4	803.9	813.1	834.2	868.9	899.6
Insurance carriers and related activities	2 248.8	2 343.7	2 374.5	2 351.1	2 362.3	2 376.4	2 401.5	2 367.5	2 338.9	2 379.1	2 391.6
Real estate and rental and leasing	1 875.9	1 933.7	1 982.5	2 010.6	2 038.4	2 033.3	2 057.5	2 085.5	2 133.5	2 172.5	2 169.1
Real estate	1 243.8	1 277.7	1 302.6	1 316.0	1 343.4	1 356.6	1 387.1	1 418.7	1 460.8	1 499.0	1 500.4
Rental and leasing services	609.5	630.8	653.1	666.8	666.3	649.1	643.1	641.1	645.8	645.5	640.3
Lessors of nonfinancial intangible assets	22.6	25.3	26.8	27.8	28.7	27.6	27.3	25.7	26.9	28.1	28.4
Professional and business services	14 371	15 183	15 994	16 704	16 514	16 016	16 029	16 440	17 003	17 619	17 998
Professional and technical services [1]	5 657.5	6 021.8	6 375.5	6 732.1	6 901.3	6 680.6	6 637.0	6 784.2	7 064.5	7 399.3	7 704.5
Legal services	987.5	1 021.1	1 051.4	1 065.7	1 091.3	1 115.3	1 142.1	1 163.1	1 168.0	1 173.2	1 175.4
Accounting and bookkeeping services	761.7	802.5	838.1	866.9	872.7	837.8	815.9	806.5	849.9	889.7	936.6
Architectural and engineering services	1 063.9	1 115.4	1 168.7	1 238.5	1 275.3	1 246.7	1 227.6	1 258.9	1 311.7	1 386.6	1 433.1
Computer systems design and related services	830.9	979.2	1 137.2	1 258.7	1 302.2	1 157.4	1 121.6	1 153.9	1 201.0	1 290.8	1 378.7
Management and technical consulting services	562.1	611.4	640.4	694.3	736.5	730.3	742.5	789.4	852.6	916.6	984.8
Management of companies and enterprises	1 733.3	1 759.8	1 777.5	1 799.7	1 782.7	1 709.3	1 691.5	1 729.0	1 763.8	1 816.1	1 872.0
Administrative and waste services	6 979.9	7 401.5	7 841.0	8 171.9	7 829.6	7 625.6	7 700.9	7 927.3	8 174.7	8 403.2	8 421.6
Administrative and support services [1]	6 689.4	7 102.1	7 530.5	7 859.0	7 512.3	7 307.3	7 378.8	7 598.7	7 837.0	8 055.1	8 066.6
Employment services [1]	2 954.6	3 246.5	3 582.3	3 850.0	3 468.9	3 274.0	3 327.3	3 456.4	3 607.8	3 681.9	3 547.1
Temporary help services	2 060.1	2 245.7	2 470.1	2 636.1	2 338.1	2 194.1	2 224.7	2 387.7	2 549.9	2 638.0	2 598.1
Business support services	735.2	773.6	781.8	788.1	781.2	758.0	751.3	759.5	768.2	794.8	819.5
Services to buildings and dwellings	1 424.1	1 460.0	1 534.7	1 570.5	1 606.2	1 606.1	1 636.1	1 693.7	1 737.5	1 801.4	1 849.5
Waste management and remediation services	290.5	299.3	310.5	312.9	317.3	318.3	322.1	328.6	337.4	348.1	355.0
Education and health services	14 185	14 570	14 939	15 252	15 814	16 398	16 835	17 230	17 676	18 154	18 676
Educational services	2 155.0	2 233.0	2 320.0	2 390.0	2 511.0	2 643.0	2 695.0	2 763.0	2 836.0	2 901.0	2 941.0
Health care and social assistance	12 030.0	12 336.9	12 618.3	12 861.1	13 302.9	13 755.2	14 139.6	14 467.8	14 840.4	15 253.2	15 734.5
Health care	10 358.0	10 540.9	10 690.9	10 857.8	11 188.1	11 536.0	11 817.1	12 055.3	12 313.9	12 601.8	12 946.8
Ambulatory health care services [1]	4 093.0	4 161.2	4 226.6	4 320.3	4 461.5	4 633.2	4 786.4	4 952.3	5 113.5	5 285.8	5 473.5
Offices of physicians	1 625.0	1 687.2	1 748.9	1 801.1	1 870.9	1 926.3	1 960.3	2 004.6	2 049.4	2 102.5	2 155.1
Outpatient care centers	387.0	399.6	413.1	425.2	440.0	454.5	469.0	493.7	517.3	537.9	558.4
Home health care services	703.0	659.5	629.6	633.3	638.6	679.8	732.6	776.6	821.0	865.6	913.8
Hospitals	3 822.0	3 892.4	3 935.5	3 954.3	4 050.9	4 159.6	4 244.6	4 284.7	4 345.4	4 423.4	4 515.0
Nursing and residential care facilities [1]	2 443.0	2 487.3	2 528.8	2 583.2	2 675.8	2 743.3	2 786.2	2 818.4	2 855.0	2 892.5	2 958.3
Nursing care facilities	1 475.0	1 489.3	1 501.0	1 513.6	1 546.8	1 573.2	1 579.8	1 576.9	1 577.4	1 581.4	1 602.6
Social assistance [1]	1 672.0	1 796.0	1 927.5	2 003.3	2 114.7	2 219.2	2 322.5	2 412.5	2 526.5	2 651.4	2 787.7
Child day care services	570.0	615.1	673.7	695.8	714.6	744.1	755.3	764.7	789.7	818.3	850.4

[1]Includes other industries, not shown separately.

Table 15-1. Nonfarm Employment by Sector and Industry—*Continued*

(Wage and salary workers on nonfarm payrolls, thousands.)

Industry	2008	2009	2010	2011	2012	2013	2014	2015	2016	2017	2018
Transportation and warehousing	4 513.4	4 241.4	4 195.7	4 306.8	4 421.3	4 503.6	4 666.6	4 877.0	5 020.2	5 194.7	5 418.9
Air transportation	490.7	462.8	458.3	456.9	459.2	444.3	444.2	458.6	477.5	491.5	501.3
Rail transportation	231.0	218.2	216.4	228.1	230.6	231.2	236.1	240.9	217.2	215.1	214.1
Water transportation	67.1	63.4	62.3	61.3	63.9	65.3	67.3	65.8	65.6	65.0	64.8
Truck transportation	1 389.4	1 268.6	1 250.8	1 301.0	1 349.8	1 382.6	1 417.7	1 452.7	1 448.1	1 457.0	1 491.9
Transit and ground passenger transportation	423.3	421.7	429.7	439.9	440.3	448.5	466.5	477.5	483.9	488.9	487.5
Pipeline transportation	41.7	42.6	42.3	42.9	43.6	44.5	47.0	49.7	49.8	48.8	48.5
Scenic and sightseeing transportation	28.0	27.6	27.3	27.5	28.0	29.1	30.6	32.9	34.5	35.2	34.2
Support activities for transportation	592.0	548.5	542.5	562.2	579.9	598.3	625.8	651.8	667.3	689.9	711.8
Couriers and messengers	573.4	546.3	528.1	529.2	534.1	543.9	576.5	612.8	644.8	676.1	725.4
Warehousing and storage	676.9	641.7	638.1	657.9	691.9	716.0	754.8	834.3	931.6	1 027.3	1 139.6
Utilities	559	560	553	553	553	552	552	556	556	555	555
Information	2 984	2 804	2 707	2 674	2 676	2 706	2 726	2 750	2 794	2 814	2 828
Publishing industries, except Internet	880.4	796.4	759.0	748.6	739.5	732.7	726.9	726.5	730.3	728.6	732.6
Motion picture and sound recording industries	371.3	357.6	370.2	362.1	362.3	370.5	379.3	397.9	426.1	432.6	436.3
Broadcasting, except Internet	318.7	300.5	290.3	283.2	285.1	283.7	282.8	276.7	270.5	268.0	270.4
Telecommunications	1 019.4	965.7	902.9	873.6	856.8	853.2	838.5	810.9	801.1	780.8	751.2
Data processing, hosting, and related services	260.3	248.5	243.0	245.8	254.9	269.6	279.5	296.2	303.9	318.0	329.8
Other information services [1]	133.5	135.0	141.7	160.0	177.2	190.2	219.0	241.4	202.3	205.0	207.0
Internet publishing and broadcasting	80.6	83.3	92.0	109.6	125.2	142.3	163.3	184.3	203.5	223.8	244.5
Financial activities	8 206	7 838	7 695	7 697	7 784	7 886	7 977	8 123	8 287	8 451	8 569
Finance and insurance	6 076.3	5 843.9	5 761.0	5 769.0	5 828.4	5 886.1	5 930.9	6 034.9	6 148.1	6 261.9	6 314.0
Monetary authorities–central bank	22.4	21.0	20.0	18.3	17.5	18.0	18.2	18.0	18.6	19.2	19.5
Credit intermediation and related activities [1]	2 732.7	2 590.2	2 550.0	2 554.1	2 583.3	2 614.1	2 564.0	2 570.7	2 609.7	2 644.9	2 647.2
Depository credit intermediation [1]	1 815.2	1 753.8	1 728.8	1 735.1	1 738.7	1 733.6	1 702.9	1 683.8	1 697.8	1 713.3	1 715.1
Commercial banking	1 357.5	1 316.9	1 305.9	1 314.5	1 320.6	1 313.9	1 292.8	1 284.7	1 309.5	1 323.6	1 320.4
Securities, commodity contracts, investments, and funds and trusts	916.2	862.1	850.4	860.1	859.3	865.0	882.9	907.8	927.1	938.3	956.9
Insurance carriers and related activities	2 405.1	2 370.6	2 340.6	2 336.4	2 368.3	2 388.9	2 465.8	2 538.3	2 592.7	2 659.6	2 690.4
Real estate and rental and leasing	2 129.6	1 994.0	1 933.8	1 927.4	1 955.2	2 000.2	2 045.7	2 088.4	2 138.8	2 189.2	2 254.9
Real estate	1 485.0	1 420.2	1 395.7	1 400.8	1 420.0	1 458.7	1 487.1	1 517.1	1 556.8	1 605.5	1 661.0
Rental and leasing services	616.9	547.3	513.5	502.2	511.0	517.7	535.0	547.5	558.3	560.1	570.6
Lessors of nonfinancial intangible assets	27.7	26.5	24.6	24.4	24.2	23.7	23.7	23.8	23.8	23.7	23.3
Professional and business services	17 792	16 634	16 783	17 389	17 992	18 575	19 124	19 695	20 114	20 508	20 999
Professional and technical services [1]	7 845.3	7 552.9	7 485.8	7 712.5	7 940.5	8 169.9	8 385.5	8 658.4	8 880.8	9 057.8	9 300.2
Legal services	1 161.5	1 124.9	1 114.2	1 115.7	1 124.0	1 128.5	1 119.0	1 118.6	1 123.7	1 136.8	1 140.8
Accounting and bookkeeping services	951.8	914.9	887.2	899.6	909.7	932.3	949.3	969.3	981.4	996.6	1 013.6
Architectural and engineering services	1 440.2	1 325.5	1 276.2	1 294.3	1 323.4	1 346.5	1 376.2	1 401.6	1 404.6	1 435.3	1 475.8
Computer systems design and related services	1 446.3	1 429.0	1 455.5	1 542.6	1 628.4	1 709.5	1 797.8	1 915.8	1 993.2	2 051.2	2 121.6
Management and technical consulting services	1 034.8	1 026.5	1 031.1	1 098.2	1 152.4	1 215.2	1 267.7	1 318.1	1 394.2	1 433.1	1 483.2
Management of companies and enterprises	1 910.1	1 872.4	1 877.8	1 939.3	2 029.3	2 109.0	2 174.6	2 213.6	2 251.6	2 308.1	2 371.8
Administrative and waste services	8 036.8	7 208.5	7 419.1	7 737.3	8 021.8	8 296.5	8 564.1	8 823.3	8 981.2	9 141.5	9 327.3
Administrative and support services [1]	7 680.0	6 856.8	7 061.8	7 372.0	7 649.9	7 919.1	8 178.0	8 426.8	8 577.1	8 726.3	8 889.8
Employment services [1]	3 134.1	2 481.8	2 723.5	2 943.2	3 135.2	3 269.7	3 408.1	3 528.3	3 537.9	3 595.1	3 678.8
Temporary help services	2 349.0	1 823.8	2 094.2	2 313.6	2 496.0	2 617.6	2 760.8	2 877.6	2 889.1	2 939.7	3 012.2
Business support services	834.4	822.1	810.7	816.7	830.7	858.3	879.7	895.9	912.0	904.4	898.9
Services to buildings and dwellings	1 839.8	1 753.3	1 745.0	1 788.6	1 830.0	1 884.7	1 943.9	2 006.6	2 062.8	2 111.2	2 153.3
Waste management and remediation services	356.8	351.7	357.3	365.3	371.9	377.5	386.1	396.5	404.1	415.2	437.6
Education and health services	19 228	19 630	19 975	20 318	20 769	21 086	21 439	22 029	22 639	23 188	23 667
Educational services	3 040.0	3 090.0	3 155.0	3 250.0	3 341.0	3 354.0	3 417.0	3 472.0	3 570.0	3 668.0	3 728.0
Health care and social assistance	16 188.2	16 539.9	16 820.0	17 068.6	17 428.1	17 731.1	18 022.2	18 557.4	19 068.8	19 519.6	19 939.2
Health care	13 289.9	13 543.0	13 776.9	14 025.9	14 281.6	14 491.5	14 676.5	15 042.3	15 413.5	15 716.6	16 006.1
Ambulatory health care services [1]	5 646.6	5 793.4	5 974.7	6 136.2	6 306.5	6 476.5	6 631.5	6 855.5	7 080.0	7 297.3	7 498.9
Offices of physicians	2 205.0	2 231.0	2 263.9	2 294.6	2 339.8	2 378.0	2 411.2	2 471.0	2 526.5	2 581.4	2 620.6
Outpatient care centers	580.8	605.6	648.7	670.2	699.0	731.1	763.7	805.7	853.0	897.2	934.3
Home health care services	961.4	1 027.1	1 084.6	1 140.3	1 185.0	1 230.3	1 262.4	1 314.7	1 365.3	1 419.6	1 472.7
Hospitals	4 627.3	4 667.4	4 678.5	4 721.7	4 779.0	4 785.8	4 786.8	4 895.8	5 015.2	5 071.8	5 145.1
Nursing and residential care facilities [1]	3 016.1	3 082.2	3 123.7	3 168.1	3 196.2	3 229.2	3 258.2	3 290.9	3 318.3	3 347.5	3 362.1
Nursing care facilities	1 618.7	1 644.9	1 657.1	1 669.6	1 662.8	1 653.8	1 650.3	1 648.3	1 641.5	1 627.1	1 609.0
Social assistance [1]	2 898.3	2 996.9	3 043.1	3 042.6	3 146.5	3 239.6	3 345.6	3 515.1	3 655.3	3 803.0	3 933.1
Child day care services	859.4	852.8	848.0	849.4	851.3	843.3	854.0	878.0	911.2	943.1	963.9

[1]Includes other industries, not shown separately.

Table 15-1. Nonfarm Employment by Sector and Industry—*Continued*

(Wage and salary workers on nonfarm payrolls, thousands.)

Industry	1997	1998	1999	2000	2001	2002	2003	2004	2005	2006	2007
Leisure and hospitality	11 018	11 232	11 543	11 862	12 036	11 986	12 173	12 493	12 816	13 110	13 427
Arts, entertainment, and recreation	1 599.9	1 645.2	1 709.1	1 787.9	1 824.4	1 782.6	1 812.9	1 849.6	1 892.3	1 928.5	1 969.2
Performing arts and spectator sports	349.6	350.0	361.1	381.8	382.3	363.7	371.7	367.5	376.3	398.5	405.0
Museums, historical sites, zoos, and parks	93.8	97.4	103.1	110.4	115.0	114.0	114.7	118.3	120.7	123.8	130.3
Amusements, gambling, and recreation	1 156.5	1 197.9	1 244.9	1 295.7	1 327.1	1 305.0	1 326.5	1 363.8	1 395.3	1 406.3	1 433.9
Accommodation and food services	9 417.9	9 586.2	9 833.7	10 073.5	10 211.3	10 203.2	10 359.8	10 643.2	10 923.0	11 181.1	11 457.4
Accommodation	1 729.5	1 773.5	1 831.7	1 884.4	1 852.2	1 778.6	1 775.4	1 789.5	1 818.6	1 832.1	1 866.9
Food services and drinking places	7 688.5	7 812.7	8 002.0	8 189.1	8 359.1	8 424.6	8 584.4	8 853.7	9 104.4	9 349.0	9 590.4
Other services	4 825	4 976	5 087	5 168	5 258	5 372	5 401	5 409	5 395	5 438	5 494
Repair and maintenance	1 169.3	1 189.2	1 222.0	1 241.5	1 256.5	1 246.9	1 233.6	1 228.8	1 236.0	1 248.5	1 253.4
Personal and laundry services	1 180.4	1 205.6	1 220.3	1 242.9	1 255.0	1 257.2	1 263.5	1 272.9	1 276.6	1 288.4	1 309.7
Membership associations and organizations	2 474.9	2 581.3	2 644.4	2 683.3	2 746.4	2 867.8	2 903.6	2 907.5	2 882.2	2 901.2	2 931.1
Government	19 664	19 909	20 307	20 790	21 118	21 513	21 583	21 621	21 804	21 974	22 218
Federal	2 806.0	2 772.0	2 769.0	2 865.0	2 764.0	2 766.0	2 761.0	2 730.0	2 732.0	2 732.0	2 734.0
Federal, except U.S. Postal Service	1 940.2	1 891.3	1 879.5	1 984.8	1 891.0	1 923.8	1 952.4	1 947.5	1 957.3	1 962.6	1 964.7
U.S. Postal Service	866.0	880.5	889.7	879.7	873.0	842.4	808.6	782.1	774.2	769.7	769.1
State government	4 582.0	4 612.0	4 709.0	4 786.0	4 905.0	5 029.0	5 002.0	4 982.0	5 032.0	5 075.0	5 122.0
State government education	1 904.0	1 922.2	1 983.2	2 030.6	2 112.9	2 242.8	2 254.7	2 238.1	2 259.9	2 292.5	2 317.5
State government, excluding education	2 677.9	2 690.2	2 725.6	2 755.9	2 791.8	2 786.3	2 747.6	2 743.9	2 771.6	2 782.0	2 804.3
Local government	12 276.0	12 525.0	12 829.0	13 139.0	13 449.0	13 718.0	13 820.0	13 909.0	14 041.0	14 167.0	14 362.0
Local government education	6 758.5	6 920.9	7 120.4	7 293.9	7 479.3	7 654.4	7 709.4	7 765.2	7 856.1	7 913.0	7 986.8
Local government, excluding education	5 516.9	5 603.9	5 708.6	5 844.6	5 970.0	6 063.2	6 110.2	6 144.1	6 184.6	6 253.8	6 375.5

1Includes other industries, not shown separately.

Table 15-1. Nonfarm Employment by Sector and Industry—*Continued*

(Wage and salary workers on nonfarm payrolls, thousands.)

Industry	2008	2009	2010	2011	2012	2013	2014	2015	2016	2017	2018
Leisure and hospitality	13 436	13 077	13 049	13 353	13 768	14 254	14 696	15 160	15 660	16 051	16 348
Arts, entertainment, and recreation	1 970.1	1 915.5	1 913.3	1 919.1	1 968.6	2 029.7	2 103.1	2 166.4	2 251.5	2 333.2	2 393.7
Performing arts and spectator sports	405.7	396.8	406.2	394.2	402.4	419.2	443.3	450.4	464.4	493.8	505.9
Museums, historical sites, zoos, and parks	131.6	129.4	127.7	132.7	136.1	140.3	146.9	153.0	159.6	165.6	170.3
Amusements, gambling, and recreation	1 432.8	1 389.2	1 379.4	1 392.2	1 430.1	1 470.2	1 512.9	1 563.1	1 627.5	1 673.8	1 717.4
Accommodation and food services	11 466.3	11 161.9	11 135.4	11 433.6	11 799.7	12 224.2	12 592.9	12 993.8	13 408.4	13 717.9	13 954.8
Accommodation	1 868.7	1 763.0	1 759.6	1 800.5	1 825.1	1 864.9	1 894.5	1 923.0	1 960.3	2 002.9	2 028.4
Food services and drinking places	9 597.5	9 398.9	9 375.8	9 633.1	9 974.6	10 359.2	10 698.3	11 070.8	11 448.1	11 715.0	11 926.4
Other services	5 515	5 367	5 331	5 360	5 430	5 483	5 567	5 622	5 691	5 770	5 845
Repair and maintenance	1 227.0	1 150.4	1 138.8	1 168.7	1 194.0	1 216.6	1 242.1	1 277.2	1 293.1	1 310.6	1 329.3
Personal and laundry services	1 322.6	1 280.6	1 265.3	1 288.6	1 313.6	1 341.7	1 371.2	1 405.0	1 444.3	1 478.4	1 509.6
Membership associations and organizations	2 965.7	2 936.0	2 926.4	2 903.0	2 922.4	2 924.6	2 953.6	2 939.5	2 953.3	2 981.0	3 005.8
Government	22 509	22 555	22 490	22 086	21 920	21 853	21 882	22 029	22 224	22 350	22 449
Federal	2 762.0	2 832.0	2 977.0	2 859.0	2 820.0	2 769.0	2 733.0	2 757.0	2 795.0	2 805.0	2 796.0
Federal, except U.S. Postal Service	2 014.4	2 128.5	2 318.1	2 227.6	2 209.2	2 174.5	2 140.4	2 160.0	2 185.8	2 189.4	2 187.8
U.S. Postal Service	747.4	703.4	658.5	630.9	611.2	594.9	593.0	596.9	608.8	615.4	608.6
State government	5 177.0	5 169.0	5 137.0	5 078.0	5 055.0	5 046.0	5 050.0	5 077.0	5 110.0	5 165.0	5 176.0
State government education	2 354.4	2 360.2	2 373.1	2 374.0	2 388.5	2 393.3	2 389.3	2 401.4	2 429.2	2 478.7	2 485.6
State government, excluding education	2 822.5	2 808.8	2 764.1	2 703.7	2 666.4	2 652.8	2 660.5	2 675.6	2 680.9	2 686.4	2 690.2
Local government	14 571.0	14 554.0	14 376.0	14 150.0	14 045.0	14 037.0	14 098.0	14 195.0	14 319.0	14 379.0	14 477.0
Local government education	8 083.9	8 078.8	8 013.4	7 872.5	7 777.9	7 776.7	7 814.9	7 870.9	7 905.1	7 921.5	7 963.3
Local government, excluding education	6 486.5	6 474.9	6 362.9	6 277.7	6 266.8	6 260.1	6 283.4	6 324.0	6 414.1	6 457.5	6 513.9

1Includes other industries, not shown separately.

Table 15-2. Production and Nonsupervisory Workers on Private Nonfarm Payrolls by Industry

(Wage and salary workers on nonfarm payrolls, thousands.)

Industry	1997	1998	1999	2000	2001	2002	2003	2004	2005	2006	2007
Total Private	84 313	86 516	88 647	90 547	90 214	88 664	87 964	89 247	91 443	93 776	95 260
Goods-Producing	17 698	18 008	18 067	18 169	17 466	16 400	15 732	15 821	16 145	16 559	16 405
Mining and logging	479	473	438	446	457	436	420	440	473	519	547
Construction	4 546	4 807	5 105	5 295	5 332	5 196	5 123	5 309	5 611	5 903	5 883
Manufacturing	12 673	12 729	12 524	12 428	11 677	10 768	10 189	10 072	10 060	10 137	9 975
Durable goods	7 599	7 721	7 651	7 659	7 164	6 530	6 152	6 140	6 220	6 355	6 250
Wood products	498	510	516	507	470	450	434	445	454	451	407
Nonmetallic mineral products	412	421	426	440	427	399	375	388	387	391	384
Primary metals	502	505	492	490	447	396	370	364	363	363	358
Fabricated metal products	1 285	1 320	1 305	1 326	1 254	1 147	1 092	1 109	1 129	1 162	1 171
Machinery	1 007	1 016	978	961	891	787	732	730	749	770	774
Computer and electronic products	951	965	933	949	876	744	673	656	700	756	744
Electrical equipment and appliances	428	432	433	433	402	352	320	307	300	303	305
Transportation equipment [1]	1 522	1 530	1 526	1 498	1 399	1 310	1 269	1 265	1 277	1 304	1 274
Motor vehicles and parts	1 062	1 050	1 076	1 073	987	931	906	903	894	873	804
Furniture and related products	490	512	532	544	509	475	444	444	436	433	409
Miscellaneous manufacturing	503	511	509	510	480	460	440	432	424	423	425
Nondurable goods [1]	5 075	5 008	4 872	4 769	4 513	4 238	4 037	3 932	3 841	3 782	3 725
Food manufacturing	1 228	1 228	1 229	1 228	1 221	1 202	1 192	1 178	1 170	1 172	1 184
Textile mills	367	357	334	315	276	242	217	194	174	158	137
Textile product mills	193	190	187	183	174	162	148	147	143	135	123
Apparel	594	534	458	404	341	286	242	219	193	182	173
Paper and paper products	489	484	474	468	446	421	393	374	365	357	350
Printing and related support activities	597	598	585	576	544	493	471	460	447	447	443
Petroleum and coal products	88	87	85	83	81	78	74	77	75	72	73
Chemicals	593	601	595	588	562	532	525	520	510	508	504
Plastics and rubber products	732	739	746	753	704	662	633	626	620	608	592
Private Service-Providing	66 615	68 508	70 580	72 378	72 748	72 264	72 232	73 426	75 298	77 217	78 855
Trade, transportation, and utilities	20 669	21 030	21 546	21 935	21 679	21 307	21 045	21 284	21 792	22 126	22 502
Wholesale trade	4 490	4 571	4 638	4 651	4 520	4 437	4 358	4 402	4 538	4 676	4 798
Retail trade	12 277	12 443	12 775	13 043	12 955	12 777	12 658	12 792	13 033	13 114	13 322
Transportation and warehousing	3 409	3 524	3 644	3 756	3 721	3 614	3 566	3 640	3 777	3 893	3 939
Utilities	494	492	489	485	483	478	464	450	443	443	444
Information	2 181	2 217	2 351	2 502	2 531	2 398	2 347	2 371	2 386	2 399	2 403
Financial activities	5 482	5 692	5 818	5 819	5 888	5 964	6 052	6 052	6 127	6 312	6 365
Professional and business services	11 924	12 594	13 214	13 819	13 617	13 080	12 942	13 321	13 892	14 487	14 828
Education and health services	12 566	12 903	13 217	13 491	13 998	14 491	14 753	15 018	15 401	15 832	16 318
Leisure and hospitality	9 780	9 947	10 216	10 516	10 662	10 576	10 666	10 955	11 263	11 568	11 861
Other services	4 013	4 124	4 219	4 296	4 373	4 449	4 426	4 425	4 438	4 494	4 578

[1]Includes other industries, not shown separately.

Table 15-2. Production and Nonsupervisory Workers on Private Nonfarm Payrolls by Industry —Continued

(Wage and salary workers on nonfarm payrolls, thousands.)

Industry	2008	2009	2010	2011	2012	2013	2014	2015	2016	2017	2018
Total Private	94 675	89 629	88 954	90 619	92 781	94 588	96 703	98 785	100 568	102 424	104 319
Goods-Producing	15 724	13 399	12 774	13 005	13 287	13 481	13 858	14 141	14 215	14 450	14 880
Mining and logging	574	510	525	594	641	636	653	592	471	490	544
Construction	5 521	4 567	4 172	4 184	4 246	4 423	4 640	4 866	5 074	5 230	5 438
Manufacturing	9 629	8 322	8 077	8 228	8 400	8 422	8 565	8 683	8 670	8 730	8 899
Durable goods	5 975	4 990	4 829	4 986	5 152	5 185	5 282	5 350	5 303	5 315	5 463
Wood products	358	278	269	269	272	283	298	306	309	311	319
Nonmetallic mineral products	363	303	284	278	273	275	280	297	306	305	310
Primary metals	348	272	275	301	317	306	310	307	293	292	294
Fabricated metal products	1 143	960	935	994	1 050	1 063	1 071	1 069	1 036	1 045	1 085
Machinery	772	641	616	662	700	699	716	711	686	691	718
Computer and electronic products	730	654	629	630	628	610	589	594	596	598	613
Electrical equipment and appliances	305	266	251	248	249	245	248	258	259	253	260
Transportation equipment [1]	1 177	948	937	972	1 024	1 053	1 103	1 139	1 150	1 149	1 188
Motor vehicles and parts	696	510	525	556	598	642	690	718	739	753	779
Furniture and related products	364	284	263	260	259	266	276	284	286	290	292
Miscellaneous manufacturing	416	382	370	373	380	386	390	384	382	381	384
Nondurable goods [1]	3 653	3 332	3 248	3 241	3 248	3 237	3 283	3 333	3 367	3 415	3 436
Food manufacturing	1 184	1 161	1 152	1 158	1 169	1 169	1 176	1 189	1 212	1 251	1 272
Textile mills	122	99	96	98	96	92	91	90	90	88	88
Textile product mills	115	98	92	89	85	83	86	88	89	89	86
Apparel	163	132	120	112	109	106	103	104	99	88	81
Paper and paper products	344	313	302	295	288	279	277	277	275	278	276
Printing and related support activities	424	369	342	327	316	310	312	310	312	305	295
Petroleum and coal products	77	70	70	70	72	70	72	74	76	80	77
Chemicals	512	479	474	480	491	490	497	507	516	525	548
Plastics and rubber products	572	476	472	482	487	497	518	532	536	543	548
Private Service-Providing	78 951	76 231	76 180	77 614	79 494	81 108	82 845	84 644	86 354	87 975	89 439
Trade, transportation, and utilities	22 292	21 072	20 829	21 188	21 569	21 826	22 232	22 581	22 855	23 107	23 393
Wholesale trade	4 768	4 453	4 324	4 388	4 505	4 563	4 638	4 644	4 634	4 660	4 698
Retail trade	13 139	12 476	12 430	12 652	12 798	12 928	13 112	13 269	13 433	13 486	13 529
Transportation and warehousing	3 935	3 692	3 631	3 708	3 824	3 890	4 036	4 222	4 342	4 513	4 721
Utilities	450	451	444	441	441	445	446	447	447	447	445
Information	2 388	2 240	2 170	2 148	2 164	2 194	2 209	2 226	2 252	2 268	2 278
Financial activities	6 320	6 066	5 942	5 900	5 986	6 068	6 155	6 278	6 430	6 571	6 637
Professional and business services	14 631	13 565	13 744	14 298	14 851	15 352	15 818	16 183	16 455	16 751	17 123
Education and health services	16 842	17 240	17 531	17 818	18 230	18 505	18 827	19 337	19 859	20 365	20 788
Leisure and hospitality	11 873	11 560	11 507	11 772	12 154	12 590	12 968	13 360	13 782	14 139	14 382
Other services	4 606	4 488	4 458	4 491	4 541	4 573	4 637	4 678	4 720	4 775	4 839

[1]Includes other industries, not shown separately.

Table 15-3. Average Weekly Hours of Production and Nonsupervisory Workers on Private Nonfarm Payrolls by Industry

(Hours.)

Industry	1997	1998	1999	2000	2001	2002	2003	2004	2005	2006	2007
Total Private	34.5	34.5	34.3	34.3	33.9	33.9	33.7	33.7	33.8	33.9	33.8
Goods-Producing	41.1	40.8	40.8	40.7	39.9	39.9	39.8	40.0	40.1	40.5	40.6
Mining and logging	46.2	44.9	44.2	44.4	44.6	43.2	43.6	44.5	45.6	45.6	45.9
Construction	38.9	38.8	39.0	39.2	38.7	38.4	38.4	38.3	38.6	39.0	39.0
Manufacturing	41.7	41.4	41.4	41.3	40.3	40.5	40.4	40.8	40.7	41.1	41.2
Overtime hours	5.1	4.9	4.9	4.7	4.0	4.2	4.2	4.6	4.6	4.4	4.2
Durable goods	42.6	42.1	41.9	41.8	40.6	40.8	40.8	41.3	41.1	41.4	41.5
Overtime hours	5.4	5.0	5.0	4.8	3.9	4.2	4.3	4.7	4.6	4.4	4.2
Wood products	41.4	41.4	41.3	41.0	40.2	39.9	40.4	40.7	40.0	39.8	39.4
Nonmetallic mineral products	41.9	42.2	42.1	41.6	41.6	42.0	42.2	42.4	42.2	43.0	42.3
Primary metals	44.3	43.5	43.8	44.2	42.4	42.4	42.3	43.1	43.1	43.6	42.9
Fabricated metal products	42.3	41.9	41.7	41.9	40.7	40.6	40.7	41.1	41.0	41.4	41.6
Machinery	44.0	43.1	42.3	42.3	40.9	40.5	40.8	42.0	42.1	42.4	42.6
Computer and electronic products	42.5	41.9	41.5	41.4	39.8	39.7	40.4	40.4	40.0	40.5	40.6
Electrical equipment and appliances	42.1	41.8	41.8	41.6	39.8	40.1	40.6	40.7	40.6	41.0	41.2
Transportation equipment [1]	44.2	43.3	43.6	43.3	41.9	42.5	41.9	42.5	42.4	42.7	42.8
Motor vehicles and parts	43.9	42.6	43.8	43.3	41.7	42.6	42.0	42.6	42.3	42.2	42.3
Furniture and related products	39.1	39.4	39.3	39.2	38.3	39.1	38.9	39.5	39.2	38.8	39.2
Miscellaneous manufacturing	39.7	39.2	39.3	39.0	38.8	38.7	38.4	38.5	38.7	38.7	38.9
Nondurable goods [1]	40.5	40.5	40.5	40.3	39.9	40.0	39.8	40.0	39.9	40.6	40.8
Overtime hours	4.6	4.6	4.6	4.5	4.1	4.2	4.1	4.4	4.4	4.4	4.2
Food manufacturing	39.8	40.1	40.2	40.1	39.6	39.6	39.3	39.3	39.0	40.1	40.7
Textile mills	41.6	41.0	41.0	41.4	40.0	40.6	39.1	40.1	40.3	40.6	40.3
Textile product mills	39.2	39.2	39.1	38.7	38.4	39.0	39.4	38.7	38.9	39.8	39.7
Apparel	35.6	35.5	35.4	35.7	36.0	36.7	35.6	36.1	35.8	36.5	37.2
Paper and paper products	43.9	43.6	43.6	42.8	42.1	41.8	41.5	42.1	42.5	42.9	43.1
Printing and related support activities	39.5	39.3	39.1	39.2	38.7	38.4	38.2	38.4	38.4	39.2	39.1
Petroleum and coal products	43.1	43.6	42.6	42.7	43.8	43.0	44.5	44.9	45.5	45.0	44.1
Chemicals	43.4	43.2	42.8	42.2	41.9	42.3	42.4	42.8	42.3	42.5	41.9
Plastics and rubber products	41.4	41.3	41.3	40.8	40.0	40.6	40.4	40.4	40.0	40.6	41.3
Private Service-Providing	32.8	32.8	32.7	32.7	32.5	32.5	32.4	32.3	32.4	32.4	32.4
Trade, transportation, and utilities	34.3	34.2	33.9	33.8	33.5	33.6	33.6	33.5	33.4	33.4	33.3
Wholesale trade	38.8	38.7	38.6	38.8	38.4	38.0	37.9	37.8	37.7	38.0	38.2
Retail trade	30.9	30.9	30.8	30.7	30.7	30.9	30.9	30.7	30.6	30.5	30.2
Transportation and warehousing	39.4	38.7	37.6	37.4	36.7	36.8	36.8	37.2	37.0	36.9	37.0
Utilities	42.0	42.0	42.0	42.0	41.4	40.9	41.1	40.9	41.1	41.4	42.4
Information	36.3	36.6	36.7	36.8	36.9	36.5	36.2	36.3	36.5	36.6	36.5
Financial activities	35.8	36.0	35.8	35.9	35.8	35.6	35.6	35.6	36.0	35.8	35.9
Professional and business services	34.3	34.3	34.4	34.5	34.2	34.2	34.1	34.2	34.2	34.6	34.8
Education and health services	32.2	32.2	32.1	32.2	32.3	32.4	32.3	32.4	32.6	32.5	32.5
Leisure and hospitality	26.1	26.2	26.1	26.1	25.8	25.8	25.6	25.7	25.7	25.7	25.5
Other services	32.7	32.6	32.5	32.5	32.3	32.1	31.4	31.0	30.9	30.9	30.9

[1]Includes other industries, not shown separately.

Table 15-3. Average Weekly Hours of Production and Nonsupervisory Workers on Private Nonfarm Payrolls by Industry—Continued

(Hours.)

Industry	2008	2009	2010	2011	2012	2013	2014	2015	2016	2017	2018
Total Private	33.6	33.1	33.4	33.6	33.7	33.7	33.7	33.7	33.6	33.7	33.8
Goods-Producing	40.2	39.2	40.4	40.9	41.1	41.3	41.5	41.2	41.2	41.3	41.5
Mining and logging	45.1	43.2	44.6	46.7	46.6	45.9	47.3	45.8	45.3	46.1	46.8
Construction	38.5	37.6	38.4	39.0	39.3	39.6	39.6	39.6	39.7	39.7	40.0
Manufacturing	40.8	39.8	41.1	41.4	41.7	41.8	42.0	41.8	41.9	41.9	42.2
Overtime hours	3.7	2.9	3.8	4.1	4.2	4.3	4.5	4.3	4.3	4.3	4.6
Durable goods	41.1	39.8	41.4	41.9	42.0	42.2	42.5	42.1	42.3	42.3	42.5
Overtime hours	3.7	2.7	3.8	4.2	4.3	4.4	4.6	4.3	4.4	4.4	4.7
Wood products	38.6	37.4	39.1	39.7	41.1	42.7	42.0	41.3	41.8	42.1	41.9
Nonmetallic mineral products	42.1	40.8	41.7	42.3	42.2	42.5	43.3	42.4	42.0	42.9	44.2
Primary metals	42.2	40.7	43.7	44.6	43.8	43.8	44.2	43.8	43.4	43.1	44.6
Fabricated metal products	41.3	39.4	41.4	42.0	42.1	42.2	42.6	42.3	42.1	42.2	42.2
Machinery	42.3	40.1	42.1	43.1	42.8	42.9	43.0	41.9	42.1	42.7	42.7
Computer and electronic products	41.0	40.4	40.9	40.5	40.4	40.5	40.8	40.9	41.2	41.2	40.9
Electrical equipment and appliances	40.9	39.3	41.1	40.8	41.6	41.8	41.7	42.2	42.9	43.1	42.6
Transportation equipment [1]	42.0	41.2	42.9	43.2	43.8	43.6	43.6	43.7	44.1	43.9	44.4
Motor vehicles and parts	41.4	40.1	43.4	43.4	44.2	43.7	43.8	44.1	44.7	44.0	44.7
Furniture and related products	38.1	37.7	38.5	39.9	40.0	40.3	40.9	39.8	40.0	39.3	39.3
Miscellaneous manufacturing	38.9	38.5	38.7	38.9	39.2	40.1	40.2	40.1	40.8	39.9	39.5
Nondurable goods [1]	40.4	39.8	40.8	40.8	41.1	41.2	41.3	41.4	41.2	41.3	41.6
Overtime hours	3.7	3.2	3.8	4.0	4.1	4.3	4.3	4.3	4.1	4.2	4.4
Food manufacturing	40.5	40.0	40.7	40.2	40.6	40.8	40.8	41.0	41.4	41.7	42.0
Textile mills	38.7	37.7	41.2	41.7	42.6	41.5	41.5	42.5	41.2	40.6	42.3
Textile product mills	38.6	37.9	39.0	39.1	39.7	38.4	38.0	37.0	37.7	38.4	39.3
Apparel	36.4	36.0	36.6	38.2	37.1	38.1	38.5	38.2	36.4	36.5	37.7
Paper and paper products	42.9	41.8	42.9	42.9	42.9	43.1	43.7	43.0	42.7	42.5	43.1
Printing and related support activities	38.3	38.0	38.2	38.0	38.5	38.6	39.0	39.8	39.2	39.5	39.3
Petroleum and coal products	44.6	43.4	43.0	43.8	47.1	46.4	46.1	45.2	44.2	45.0	45.6
Chemicals	41.5	41.4	42.2	42.5	42.4	42.9	42.7	42.6	41.8	41.7	42.1
Plastics and rubber products	41.0	40.2	41.9	42.0	41.8	41.8	42.2	42.4	42.4	42.6	42.1
Private Service-Providing	32.3	32.1	32.2	32.4	32.5	32.4	32.4	32.4	32.3	32.4	32.4
Trade, transportation, and utilities	33.2	32.9	33.3	33.7	33.8	33.7	33.6	33.7	33.5	33.8	33.9
Wholesale trade	38.3	37.7	37.9	38.5	38.7	38.7	38.6	38.6	38.6	39.0	38.9
Retail trade	30.0	29.9	30.2	30.5	30.6	30.2	30.0	30.1	29.7	30.2	30.4
Transportation and warehousing	36.4	36.0	37.1	37.8	38.0	38.5	38.5	38.8	38.8	38.4	38.4
Utilities	42.7	42.0	42.0	42.1	41.1	41.7	42.3	42.4	42.5	42.5	42.7
Information	36.7	36.6	36.3	36.2	36.0	35.9	35.9	35.7	35.5	35.8	35.6
Financial activities	35.9	36.1	36.2	36.4	36.8	36.7	36.7	37.1	36.9	37.0	37.0
Professional and business services	34.8	34.7	35.1	35.2	35.3	35.4	35.6	35.5	35.4	35.4	35.4
Education and health services	32.4	32.2	32.0	32.2	32.3	32.1	32.0	32.1	32.2	32.2	32.2
Leisure and hospitality	25.2	24.8	24.8	24.8	25.0	25.0	25.1	25.1	24.9	24.9	24.9
Other services	30.8	30.5	30.7	30.8	30.7	30.8	30.7	30.7	30.8	30.7	30.8

[1]Includes other industries, not shown separately.

Table 15-4. Average Hourly Earnings of Production and Nonsupervisory Workers on Private Nonfarm Payrolls by Industry

(Dollars.)

Industry	1997	1998	1999	2000	2001	2002	2003	2004	2005	2006	2007
Total Private	12.51	13.01	13.49	14.02	14.54	14.96	15.37	15.68	16.12	16.75	17.42
Goods-Producing	13.82	14.23	14.71	15.27	15.78	16.33	16.80	17.19	17.60	18.02	18.67
Mining and logging	15.57	16.20	16.33	16.55	17.00	17.19	17.56	18.07	18.72	19.90	20.97
Construction	15.67	16.23	16.80	17.48	18.00	18.52	18.95	19.23	19.46	20.02	20.95
Manufacturing	13.14	13.45	13.85	14.32	14.76	15.29	15.74	16.14	16.56	16.81	17.26
Excluding overtime [1]	12.38	12.70	13.08	13.55	14.06	14.54	14.96	15.29	15.68	15.96	16.43
Durable goods	13.83	14.07	14.46	14.93	15.38	16.02	16.45	16.82	17.33	17.68	18.20
Wood products	10.52	10.85	11.18	11.63	11.99	12.33	12.71	13.03	13.16	13.39	13.68
Nonmetallic mineral products	13.17	13.59	13.97	14.53	14.86	15.39	15.76	16.25	16.61	16.59	16.93
Primary metals	15.39	15.66	16.00	16.64	17.06	17.68	18.13	18.57	18.94	19.36	19.66
Fabricated metal products	12.64	12.97	13.34	13.77	14.19	14.68	15.01	15.31	15.80	16.17	16.53
Machinery	13.94	14.23	14.77	15.21	15.48	15.92	16.29	16.67	17.02	17.20	17.72
Computer and electronic products	13.24	13.85	14.37	14.73	15.42	16.20	16.68	17.27	18.39	18.94	19.94
Electrical equipment and appliances	12.24	12.51	12.90	13.23	13.78	13.98	14.36	14.90	15.24	15.53	15.93
Transportation equipment [3]	17.99	17.91	18.24	19.00	19.47	20.63	21.22	21.48	22.09	22.41	23.03
Motor vehicles and parts	18.43	18.21	18.49	19.11	19.66	21.09	21.68	21.71	22.26	22.13	22.00
Furniture and related products	10.50	10.89	11.28	11.73	12.14	12.62	12.99	13.16	13.45	13.80	14.32
Miscellaneous manufacturing	10.89	11.18	11.55	11.93	12.45	12.91	13.30	13.84	14.07	14.06	14.66
Nondurable goods [2]	12.04	12.45	12.85	13.31	13.75	14.15	14.63	15.05	15.26	15.33	15.67
Food manufacturing	10.77	11.09	11.40	11.77	12.18	12.55	12.80	12.98	13.04	13.13	13.55
Textile mills	10.22	10.58	10.90	11.23	11.40	11.73	11.99	12.13	12.38	12.55	13.00
Textile product mills	9.30	9.61	10.04	10.31	10.49	10.85	11.15	11.31	11.61	11.86	11.78
Apparel	7.76	8.05	8.35	8.61	8.83	9.11	9.58	9.77	10.26	10.65	11.05
Paper and paper products	14.76	15.20	15.58	15.91	16.38	16.85	17.33	17.91	17.99	18.01	18.44
Printing and related support activities	12.78	13.20	13.67	14.09	14.48	14.93	15.37	15.71	15.74	15.80	16.15
Petroleum and coal products	21.10	21.75	22.22	22.80	22.90	23.04	23.63	24.39	24.47	24.11	25.21
Chemicals	15.78	16.23	16.40	17.09	17.57	17.97	18.50	19.17	19.67	19.60	19.55
Plastics and rubber products	11.48	11.79	12.25	12.70	13.21	13.55	14.18	14.59	14.80	14.97	15.39
Private Service-Providing	12.07	12.61	13.09	13.62	14.18	14.58	14.98	15.28	15.73	16.40	17.00
Trade, transportation, and utilities	11.89	12.38	12.81	13.29	13.69	14.01	14.33	14.57	14.91	15.37	15.76
Wholesale trade	14.38	15.03	15.59	16.24	16.74	16.94	17.33	17.62	18.13	18.87	19.54
Retail trade	9.60	10.06	10.45	10.87	11.30	11.67	11.90	12.09	12.36	12.58	12.76
Transportation and warehousing	13.78	14.13	14.56	15.05	15.33	15.77	16.26	16.52	16.71	17.28	17.73
Utilities	20.59	21.48	22.03	22.75	23.58	23.96	24.77	25.61	26.68	27.40	27.88
Information	17.14	17.67	18.40	19.07	19.80	20.20	21.01	21.40	22.06	23.23	23.96
Financial activities	13.28	14.00	14.55	15.04	15.65	16.25	17.21	17.58	17.98	18.83	19.67
Professional and business services	13.58	14.28	14.86	15.53	16.34	16.82	17.22	17.49	18.09	19.14	20.16
Education and health services	12.53	12.96	13.40	13.91	14.58	15.15	15.56	16.07	16.62	17.28	17.99
Leisure and hospitality	7.32	7.67	7.96	8.32	8.57	8.81	9.00	9.15	9.38	9.75	10.41
Other services	11.29	11.79	12.26	12.73	13.27	13.72	13.84	13.98	14.34	14.77	15.42

[1]Derived by assuming that overtime hours are paid at the rate of time and one-half.
[2]Includes other industries, not shown separately.

Table 15-4. Average Hourly Earnings of Production and Nonsupervisory Workers on Private Nonfarm Payrolls by Industry—*Continued*

(Dollars.)

Industry	2008	2009	2010	2011	2012	2013	2014	2015	2016	2017	2018
Total Private	18.06	18.61	19.05	19.44	19.74	20.13	20.61	21.03	21.54	22.06	22.71
Goods-Producing	19.33	19.90	20.28	20.67	20.94	21.24	21.59	21.96	22.58	23.18	24.00
Mining and logging	22.50	23.29	23.82	24.50	25.79	26.80	26.84	26.48	26.97	27.44	28.30
Construction	21.87	22.66	23.22	23.65	23.97	24.22	24.67	25.20	25.97	26.74	27.74
Manufacturing	17.75	18.24	18.61	18.93	19.08	19.30	19.56	19.91	20.44	20.90	21.54
Excluding overtime [1]	16.97	17.59	17.78	18.03	18.16	18.34	18.57	18.93	19.43	19.87	20.43
Durable goods	18.70	19.36	19.81	20.11	20.18	20.35	20.66	20.97	21.48	21.89	22.51
Wood products	14.19	14.92	14.85	14.81	14.99	15.47	15.57	16.16	16.81	17.47	17.88
Nonmetallic mineral products	16.90	17.28	17.48	18.16	18.15	18.40	19.16	19.92	20.30	20.34	21.41
Primary metals	20.19	20.10	20.13	19.94	20.70	21.94	22.41	22.47	23.10	23.10	23.31
Fabricated metal products	16.99	17.48	17.94	18.13	18.26	18.35	18.68	19.02	19.66	20.15	20.55
Machinery	17.97	18.39	18.96	19.54	20.17	20.58	21.00	21.34	21.80	22.41	23.06
Computer and electronic products	21.04	21.87	22.78	23.32	23.34	23.41	23.36	23.30	24.27	24.58	25.09
Electrical equipment and appliances	15.78	16.27	16.87	17.96	18.03	18.04	18.28	18.85	19.29	19.72	20.59
Transportation equipment [2]	23.85	24.98	25.23	25.34	24.57	24.57	24.96	25.05	25.06	25.36	26.32
Motor vehicles and parts	22.21	21.86	22.02	21.93	21.27	21.08	21.38	21.49	21.59	21.72	22.77
Furniture and related products	14.54	15.04	15.06	15.24	15.46	15.59	15.67	16.09	16.82	17.52	17.86
Miscellaneous manufacturing	15.20	16.13	16.56	16.82	17.05	17.03	17.31	17.72	18.44	19.04	19.14
Nondurable goods [2]	16.15	16.56	16.80	17.06	17.29	17.56	17.75	18.16	18.74	19.32	19.96
Food manufacturing	14.00	14.39	14.41	14.63	15.02	15.44	15.55	15.90	16.52	16.92	17.49
Textile mills	13.58	13.71	13.56	13.79	13.51	13.90	14.15	14.72	15.71	15.97	16.46
Textile product mills	11.73	11.44	11.79	12.21	12.77	12.88	13.35	13.31	13.64	14.80	15.31
Apparel	11.40	11.37	11.43	11.96	12.89	13.21	13.51	13.65	13.73	14.35	15.32
Paper and paper products	18.89	19.29	20.04	20.28	20.42	20.32	20.35	21.46	21.71	21.75	21.88
Printing and related support activities	16.75	16.75	16.91	17.28	17.28	17.77	18.01	18.31	18.59	18.61	18.74
Petroleum and coal products	27.41	29.61	31.31	31.75	32.15	34.58	35.39	37.32	38.88	39.95	40.32
Chemicals	19.50	20.30	21.07	21.45	21.45	21.40	21.49	21.76	22.71	24.29	25.46
Plastics and rubber products	15.85	16.01	15.71	15.95	16.05	16.20	16.51	16.82	17.17	17.63	18.49
Private Service-Providing	17.75	18.33	18.78	19.18	19.48	19.90	20.40	20.84	21.32	21.82	22.44
Trade, transportation, and utilities	16.14	16.46	16.80	17.12	17.39	17.71	18.23	18.63	18.95	19.31	19.90
Wholesale trade	20.08	20.78	21.46	21.88	22.13	22.52	23.14	23.52	24.07	24.58	25.18
Retail trade	12.87	13.02	13.25	13.52	13.82	14.03	14.40	14.83	15.05	15.33	15.91
Transportation and warehousing	18.42	18.81	19.17	19.50	19.55	19.82	20.52	20.75	20.91	21.30	21.84
Utilities	28.83	29.48	30.04	30.82	31.61	32.27	32.86	34.02	35.33	36.22	36.77
Information	24.78	25.45	25.87	26.62	27.04	27.98	28.70	29.05	30.05	30.74	31.93
Financial activities	20.32	20.90	21.55	21.93	22.82	23.87	24.71	25.34	26.12	26.57	26.94
Professional and business services	21.19	22.36	22.80	23.14	23.31	23.74	24.31	24.81	25.43	26.05	26.81
Education and health services	18.73	19.34	19.95	20.60	20.91	21.29	21.64	22.09	22.52	23.03	23.65
Leisure and hospitality	10.84	11.12	11.31	11.45	11.62	11.78	12.09	12.41	12.85	13.38	13.87
Other services	16.09	16.59	17.06	17.32	17.59	18.00	18.51	19.01	19.36	20.10	20.78

[1]Derived by assuming that overtime hours are paid at the rate of time and one-half.
[2]Includes other industries, not shown separately.

Table 15-5. Average Weekly Earnings of Production and Nonsupervisory Workers on Private Nonfarm Payrolls by Industry

(Dollars.)

Industry	1997	1998	1999	2000	2001	2002	2003	2004	2005	2006	2007
Total Private	431.86	448.59	463.15	480.99	493.61	506.54	517.76	528.84	544.02	567.09	589.18
Goods-Producing	568.54	580.99	599.99	621.86	630.01	651.64	669.13	688.03	705.28	730.16	757.50
Mining and logging	720.07	727.19	721.77	734.88	757.96	741.97	765.94	804.01	853.87	907.95	962.63
Construction	609.48	629.75	655.11	685.78	695.86	711.82	727.00	735.55	750.37	781.59	816.23
Manufacturing	548.36	557.09	573.29	591.01	595.15	618.62	635.99	658.52	673.30	690.88	711.53
Durable goods	589.06	591.80	606.55	624.38	624.38	652.97	671.35	694.06	712.85	731.93	754.61
Wood products	435.66	449.77	461.46	476.97	481.33	492.07	514.06	530.15	526.65	533.12	539.41
Nonmetallic mineral products	551.65	573.04	587.42	604.76	618.91	647.00	664.96	688.33	700.63	712.67	716.78
Primary metals	681.64	681.47	700.80	734.83	723.86	749.17	767.15	799.69	815.90	843.79	843.42
Fabricated metal products	534.38	543.24	555.86	576.71	576.71	596.42	610.44	628.83	647.21	668.94	687.16
Machinery	613.19	613.73	625.07	643.94	632.68	645.38	664.48	699.45	716.27	728.84	754.34
Computer and electronic products	562.68	579.85	596.37	609.97	613.18	642.77	674.69	697.97	735.59	766.75	809.10
Electrical equipment and appliances	515.77	522.54	538.98	550.48	548.03	560.33	583.27	607.00	618.88	637.04	656.46
Transportation equipment [1]	795.75	774.77	795.57	817.56	816.74	877.50	889.42	912.56	937.72	957.47	986.75
Motor vehicles and parts	808.13	775.71	809.47	828.41	818.84	898.57	910.20	925.07	940.54	934.38	930.47
Furniture and related products	410.48	428.57	443.68	459.95	464.87	493.95	505.26	519.61	527.49	535.90	561.08
Miscellaneous manufacturing	431.72	437.95	454.14	464.80	483.07	499.01	510.54	533.27	545.01	555.87	570.12
Nondurable goods [1]	487.04	504.13	520.02	536.89	548.30	566.72	582.48	602.56	609.04	622.00	639.99
Food manufacturing	428.58	444.72	458.73	472.09	481.81	497.25	502.92	509.55	508.55	526.02	551.21
Textile mills	425.53	434.15	447.38	464.51	456.64	476.52	469.33	486.68	498.47	509.39	524.40
Textile product mills	364.12	376.37	392.46	398.93	402.53	423.05	438.67	437.89	451.36	472.10	467.80
Apparel	276.01	286.17	295.62	307.51	317.63	334.24	340.73	352.04	366.93	389.05	411.57
Paper and paper products	647.64	662.27	679.05	681.38	689.76	705.20	719.55	754.17	764.15	772.57	795.58
Printing and related support activities	504.46	518.32	534.15	552.15	560.89	573.05	587.58	603.97	604.73	618.92	632.02
Petroleum and coal products	908.50	949.28	947.60	973.53	1 003.34	990.88	1 052.32	1 095.00	1 114.51	1 085.50	1 112.73
Chemicals	685.43	700.53	701.06	721.41	735.57	759.56	784.26	819.93	831.79	833.84	819.51
Plastics and rubber products	475.04	487.04	505.45	517.56	528.62	550.14	572.53	589.99	591.59	608.37	635.63
Private Service-Providing	395.57	413.65	427.98	445.74	461.02	473.84	484.76	494.03	509.26	532.19	554.18
Trade, transportation, and utilities	407.19	422.93	433.92	449.56	459.14	470.76	480.66	487.79	497.82	513.90	525.41
Wholesale trade	558.06	581.48	601.78	630.09	642.08	643.40	656.36	665.96	683.99	717.42	747.08
Retail trade	296.04	310.43	321.74	333.48	346.31	360.92	367.28	371.23	377.58	383.25	385.18
Transportation and warehousing	542.68	546.95	548.03	562.59	562.63	580.06	598.53	615.08	618.70	637.10	655.29
Utilities	865.02	902.44	924.40	955.09	977.25	979.26	1 017.44	1 048.01	1 095.91	1 135.57	1 182.65
Information	622.37	646.52	675.47	700.92	731.18	737.94	760.84	776.72	805.11	850.64	874.45
Financial activities	475.03	503.94	520.74	540.39	560.46	578.94	611.82	625.53	646.48	673.63	706.52
Professional and business services	466.13	490.53	511.39	535.56	558.32	575.00	587.68	597.88	619.49	662.80	701.39
Education and health services	403.42	417.47	429.76	447.80	471.33	490.38	503.05	521.06	541.40	561.02	585.44
Leisure and hospitality	190.66	200.82	208.05	217.20	220.73	227.31	230.49	234.86	241.36	250.34	265.54
Other services	368.63	384.25	398.77	413.30	428.64	439.87	434.41	433.04	443.40	456.50	477.06

[1]Includes other industries, not shown separately.

Table 15-5. Average Weekly Earnings of Production and Nonsupervisory Workers on Private Nonfarm Payrolls by Industry—*Continued*

(Dollars.)

Industry	2008	2009	2010	2011	2012	2013	2014	2015	2016	2017	2018
Total Private	607.42	615.96	636.19	652.89	665.65	677.70	694.85	708.90	723.31	742.62	767.08
Goods-Producing	776.63	779.68	818.96	844.89	861.39	877.09	895.09	905.43	930.34	956.83	996.75
Mining and logging	1 014.69	1 006.67	1 063.11	1 144.64	1 201.69	1 229.70	1 270.91	1 211.91	1 221.69	1 265.92	1 323.22
Construction	842.61	851.76	891.83	921.84	942.14	958.72	977.11	998.02	1 031.88	1 061.98	1 108.49
Manufacturing	724.46	726.12	765.18	784.29	794.67	807.37	822.03	832.25	855.77	876.10	908.08
Durable goods	767.92	771.24	819.06	841.89	848.35	859.69	877.31	882.88	909.23	926.67	956.88
Wood products	547.53	557.65	580.74	587.77	615.76	661.10	653.87	666.82	703.47	735.06	749.60
Nonmetallic mineral products	711.11	705.54	728.22	768.35	766.16	782.69	829.79	844.77	852.66	873.01	946.72
Primary metals	851.12	817.70	880.46	889.34	907.23	960.61	991.13	984.77	1 003.42	995.41	1 038.34
Fabricated metal products	701.57	689.22	742.76	762.14	767.83	775.35	795.77	804.67	827.58	849.75	866.75
Machinery	760.09	737.97	797.62	842.96	864.16	883.06	903.42	894.65	918.23	956.97	985.58
Computer and electronic products	861.58	883.02	932.26	943.88	944.02	948.79	952.82	953.04	999.76	1 012.10	1 026.65
Electrical equipment and appliances	645.60	639.34	693.49	732.16	749.91	753.07	763.08	794.79	827.24	850.47	877.67
Transportation equipment [1]	1 000.87	1 028.51	1 081.60	1 094.46	1 075.00	1 070.88	1 088.27	1 094.01	1 105.96	1 113.99	1 168.48
Motor vehicles and parts	920.36	876.34	956.39	952.34	940.00	920.56	936.90	948.30	964.46	956.61	1 017.37
Furniture and related products	553.90	566.75	579.66	608.00	617.74	628.36	641.27	640.01	672.60	689.02	702.20
Miscellaneous manufacturing	591.95	620.74	640.85	654.90	668.95	682.97	695.67	709.56	751.91	759.06	756.74
Nondurable goods [1]	652.22	658.51	685.07	696.03	710.16	723.53	733.64	751.05	771.47	798.20	830.08
Food manufacturing	566.88	575.40	586.52	588.19	609.69	629.70	634.95	652.26	683.74	705.82	734.57
Textile mills	525.00	516.86	559.13	574.61	575.75	577.18	586.40	625.84	647.60	648.43	696.86
Textile product mills	453.06	433.09	459.46	477.49	507.00	494.76	507.19	493.03	513.89	568.47	602.48
Apparel	415.10	408.86	418.28	456.97	478.33	503.01	520.57	520.73	499.82	524.59	578.00
Paper and paper products	809.57	806.19	858.65	870.53	877.14	874.85	888.07	923.13	926.10	924.75	942.01
Printing and related support activities	642.50	635.68	646.11	655.81	665.45	685.33	701.55	729.57	728.41	734.43	736.05
Petroleum and coal products	1 222.07	1 284.44	1 345.72	1 390.80	1 513.36	1 605.02	1 630.84	1 685.27	1 719.89	1 796.54	1 837.44
Chemicals	809.29	841.18	888.25	910.88	910.02	918.69	917.96	927.83	950.00	1 012.35	1 071.90
Plastics and rubber products	649.02	643.91	658.55	669.54	671.28	677.72	697.67	713.74	728.16	750.22	779.04
Private Service-Providing	573.29	587.56	605.07	620.97	633.06	645.00	661.61	676.08	689.29	707.18	728.11
Trade, transportation, and utilities	535.18	541.12	558.75	576.56	587.68	596.13	612.45	627.45	634.80	652.76	675.21
Wholesale trade	768.14	782.62	814.04	842.06	857.06	871.89	894.06	907.72	929.09	957.95	980.54
Retail trade	386.44	388.74	400.38	412.29	422.35	423.44	431.97	446.01	447.69	463.10	482.90
Transportation and warehousing	670.46	677.80	711.22	737.38	742.71	762.57	789.61	804.95	810.89	817.13	838.89
Utilities	1 230.65	1 239.34	1 262.89	1 296.92	1 298.23	1 344.70	1 388.91	1 444.03	1 502.09	1 540.27	1 569.90
Information	908.78	931.08	939.85	964.85	973.52	1 003.65	1 030.17	1 038.10	1 068.23	1 100.03	1 137.84
Financial activities	729.64	754.90	780.19	798.68	840.04	875.04	908.19	939.68	962.95	982.50	997.43
Professional and business services	738.31	776.30	799.24	814.52	823.34	839.86	865.25	879.66	899.58	922.51	948.85
Education and health services	607.82	622.30	639.37	663.04	674.48	683.24	692.56	709.25	723.91	741.55	762.32
Leisure and hospitality	273.39	275.95	280.87	283.82	290.54	294.31	303.81	311.32	319.67	332.54	345.04
Other services	495.57	506.26	523.70	532.63	539.46	553.77	568.92	583.54	595.68	617.91	640.38

[1]Includes other industries, not shown separately.

Table 15-6. Indexes of Aggregate Weekly Hours of Production and Nonsupervisory Workers on Private Nonfarm Payrolls by Industry

(2002 = 100.)

Industry	1997	1998	1999	2000	2001	2002	2003	2004	2005	2006	2007
Total Private	97.0	99.4	101.4	103.5	102.0	100.0	98.7	100.3	102.8	105.8	107.4
Goods-Producing	111.2	112.3	112.6	113.1	106.6	100.0	95.7	96.8	98.9	102.5	101.7
Mining and logging	117.6	112.8	102.9	105.1	108.3	100.0	97.4	104.0	114.7	125.8	133.5
Construction	88.6	93.4	99.7	104.0	103.2	100.0	98.4	101.7	108.3	115.4	114.7
Manufacturing	121.4	121.0	119.0	117.8	108.1	100.0	94.5	94.3	93.9	95.6	94.4
Durable goods	121.6	122.0	120.5	120.4	109.3	100.0	94.3	95.2	96.1	98.9	97.4
Wood products	114.8	117.5	118.5	115.8	105.0	100.0	97.8	100.8	101.2	99.9	89.3
Nonmetallic mineral products	103.1	105.8	106.9	109.1	106.1	100.0	94.3	98.0	97.4	100.3	96.9
Primary metals	132.4	131.1	128.4	128.9	113.0	100.0	93.4	93.3	93.1	94.2	91.4
Fabricated metal products	116.6	118.6	116.7	119.1	109.3	100.0	95.3	97.7	99.2	103.1	104.5
Machinery	138.9	137.4	129.8	127.6	114.1	100.0	93.6	96.0	98.8	102.2	103.3
Computer and electronic products	136.8	136.7	131.1	133.1	117.9	100.0	92.1	89.8	94.8	103.6	102.2
Electrical equipment and appliances	127.7	127.8	128.3	127.7	113.4	100.0	92.0	88.7	86.4	88.0	89.1
Transportation equipment [1]	120.8	118.8	119.5	116.3	105.3	100.0	95.5	96.5	97.3	99.9	98.0
Motor vehicles and parts	117.5	112.8	118.8	117.3	103.6	100.0	95.9	97.0	95.2	92.9	85.8
Furniture and related products	102.9	108.4	112.6	114.7	104.8	100.0	93.0	94.3	91.9	90.4	86.1
Miscellaneous manufacturing	110.1	110.4	110.3	109.5	104.7	100.0	93.6	91.8	90.6	90.4	91.0
Nondurable goods [1]	120.9	119.4	116.1	113.3	106.0	100.0	94.7	92.7	90.3	90.4	89.6
Food manufacturing	102.6	103.3	103.7	103.4	101.4	100.0	98.4	97.1	95.8	98.6	101.1
Textile mills	155.2	148.9	139.1	132.4	112.2	100.0	86.3	79.0	71.3	65.1	56.3
Textile product mills	119.4	117.7	115.6	112.3	105.5	100.0	92.4	90.2	88.0	85.0	77.5
Apparel	201.3	180.9	154.6	137.5	117.1	100.0	81.9	75.1	65.7	63.4	61.5
Paper and paper products	121.6	119.6	117.2	113.5	106.6	100.0	92.5	89.2	88.0	86.9	85.8
Printing and related support activities	124.6	124.3	120.9	119.4	111.5	100.0	95.3	93.4	90.9	92.5	91.6
Petroleum and coal products	112.7	113.4	107.6	105.8	105.6	100.0	98.7	102.6	102.4	96.9	95.6
Chemicals	114.7	115.4	113.2	110.4	104.7	100.0	99.0	99.0	95.9	96.1	94.1
Plastics and rubber products	112.7	113.7	114.6	114.2	104.9	100.0	95.2	94.2	92.3	91.9	91.1
Private Service-Providing	93.0	95.7	98.3	100.8	100.7	100.0	99.5	101.1	103.8	106.7	108.9
Trade, transportation, and utilities	98.9	100.3	101.9	103.6	101.5	100.0	98.6	99.5	101.6	103.3	104.7
Wholesale trade	103.4	104.9	106.3	107.1	102.9	100.0	98.0	98.7	101.6	105.5	108.9
Retail trade	95.9	97.2	99.5	101.3	100.5	100.0	98.9	99.4	100.8	101.1	101.8
Transportation and warehousing	101.0	102.6	103.2	105.6	102.7	100.0	98.8	101.9	105.2	107.9	109.5
Utilities	106.1	105.7	105.0	104.1	102.3	100.0	97.4	94.2	93.1	93.8	96.2
Information	90.4	92.6	98.5	105.0	106.7	100.0	97.0	98.2	99.4	100.2	100.1
Financial activities	92.3	96.4	98.1	98.4	99.3	100.0	101.3	101.4	103.7	106.3	107.6
Professional and business services	91.5	96.7	101.6	106.5	104.0	100.0	98.7	101.8	106.4	112.2	115.4
Education and health services	86.2	88.6	90.4	92.6	96.5	100.0	101.7	103.8	107.0	109.6	113.2
Leisure and hospitality	93.4	95.5	97.8	100.5	100.6	100.0	100.0	103.0	106.2	108.8	110.8
Other services	91.9	94.3	96.2	97.8	99.0	100.0	97.4	96.1	96.2	97.4	99.3

[1] Includes other industries, not shown separately.

Table 15-6. Indexes of Aggregate Weekly Hours of Production and Nonsupervisory Workers on Private Nonfarm Payrolls by Industry—*Continued*

(2002 = 100.)

Industry	2008	2009	2010	2011	2012	2013	2014	2015	2016	2017	2018
Total Private	106.1	98.8	99.0	101.4	104.3	106.1	108.6	110.9	112.5	114.9	117.4
Goods-Producing	96.5	80.2	78.8	81.2	83.5	85.1	87.8	89.1	89.5	91.1	94.4
Mining and logging	137.6	117.2	124.5	147.4	158.7	155.0	164.3	144.1	113.4	120.2	135.2
Construction	106.5	85.9	80.2	81.7	83.6	87.7	92.0	96.5	100.9	104.0	108.8
Manufacturing	90.2	76.1	76.2	78.3	80.3	80.9	82.6	83.3	83.3	84.0	86.1
Durable goods	92.2	74.7	75.0	78.4	81.4	82.3	84.3	84.6	84.3	84.5	87.2
Wood products	76.7	57.8	58.5	59.3	62.1	67.3	69.6	70.2	72.0	72.9	74.4
Nonmetallic mineral products	91.2	73.8	70.5	70.1	68.9	69.7	72.4	75.0	76.7	78.1	81.7
Primary metals	87.4	66.1	71.7	80.1	82.7	79.8	81.7	80.1	75.8	74.9	77.9
Fabricated metal products	101.3	81.3	83.1	89.6	94.8	96.3	97.9	97.0	93.6	94.6	98.2
Machinery	102.3	80.6	81.2	89.6	94.0	94.0	96.6	93.4	90.6	92.5	96.2
Computer and electronic products	101.2	89.4	87.2	86.4	86.0	83.7	81.4	82.3	83.1	83.4	85.0
Electrical equipment and appliances	88.5	74.1	73.2	71.6	73.4	72.4	73.5	77.2	78.6	77.3	78.7
Transportation equipment [1]	88.6	70.1	72.1	75.3	80.4	82.4	86.3	89.3	91.1	90.6	94.7
Motor vehicles and parts	72.7	51.6	57.5	60.9	66.6	70.7	76.2	79.8	83.2	83.6	87.7
Furniture and related products	74.6	57.6	54.4	55.8	55.6	57.6	60.7	60.8	61.6	61.3	61.6
Miscellaneous manufacturing	89.2	81.1	79.0	80.0	82.2	85.3	86.3	84.9	85.8	83.8	83.7
Nondurable goods [1]	86.9	78.1	78.0	77.9	78.6	78.6	80.0	81.2	81.6	83.1	84.2
Food manufacturing	100.6	97.5	98.5	97.7	99.6	100.1	100.8	102.4	105.3	109.5	112.1
Textile mills	47.9	37.8	40.2	41.6	41.7	39.0	38.4	39.1	37.8	36.3	37.7
Textile product mills	70.5	58.5	56.5	55.2	53.3	50.5	51.5	51.5	53.3	54.1	53.3
Apparel	56.7	45.4	42.0	40.7	38.4	38.5	38.0	37.8	34.3	30.5	29.1
Paper and paper products	83.5	74.2	73.5	71.9	70.0	68.2	68.5	67.5	66.6	67.0	67.4
Printing and related support activities	86.1	74.1	69.0	65.6	64.3	63.2	64.2	65.2	64.6	63.7	61.3
Petroleum and coal products	102.4	90.2	89.1	91.2	101.6	96.8	98.5	100.1	100.7	107.0	105.2
Chemicals	94.6	88.3	88.9	90.6	92.6	93.6	94.4	96.1	96.1	97.4	102.6
Plastics and rubber products	87.1	71.3	73.6	75.3	75.8	77.3	81.4	84.0	84.6	86.0	85.9
Private Service-Providing	108.6	104.0	104.5	107.0	110.0	112.0	114.4	116.9	118.9	121.4	123.6
Trade, transportation, and utilities	103.2	96.7	96.7	99.6	101.8	102.6	104.3	106.2	106.9	109.1	110.8
Wholesale trade	108.3	99.5	97.4	100.2	103.5	104.8	106.4	106.4	106.2	107.8	108.6
Retail trade	99.8	94.3	95.0	97.7	98.9	98.8	99.5	101.0	101.1	103.1	103.9
Transportation and warehousing	107.7	100.1	101.3	105.5	109.3	112.6	116.8	123.2	126.6	130.2	136.4
Utilities	98.3	97.0	95.4	94.9	92.6	94.9	96.3	97.0	97.1	97.2	97.1
Information	100.0	93.5	90.0	88.9	88.9	89.8	90.5	90.8	91.4	92.6	92.6
Financial activities	106.8	103.1	101.3	101.2	103.7	104.7	106.5	109.6	111.6	114.4	115.7
Professional and business services	114.0	105.3	107.7	112.5	117.3	121.4	125.9	128.3	130.2	132.6	135.5
Education and health services	116.5	118.3	119.8	122.3	125.4	126.6	128.5	132.3	136.1	139.8	142.9
Leisure and hospitality	109.7	105.1	104.6	106.9	111.3	115.2	119.3	122.7	125.6	128.7	131.0
Other services	99.5	96.0	96.0	96.8	97.7	98.7	99.9	100.7	101.8	102.9	104.6

[1]Includes other industries, not shown separately.

NOTES AND DEFINITIONS, CHAPTER 15

TABLES 15-1 THROUGH 15-6

Employment, Hours, and Earnings by NAICS Industry

SOURCE: U.S. DEPARTMENT OF LABOR, BUREAU OF LABOR STATISTICS

See the notes and definitions for Tables 10-8 through 10-18 regarding definitions of *employment, production and nonsupervisory workers, average weekly hours, overtime hours, average hourly earnings, average weekly earnings*, and the *indexes of aggregate weekly hours*. Availability and reference information is also provided in those notes and definitions.

Definitions

The North American Industry Classification System (NAICS) is the standard used by federal statistical agencies in classifying business establishments for the purpose of collecting, analyzing, and publishing statistical data related to the U.S. business economy. NAICS defines industries according to a consistent principle: businesses that use similar production processes are grouped together.

NAICS is the product of a cooperative effort on the part of the statistical agencies of the United States, Canada, and Mexico. This system is the hierarchical, numerical system used by the federal government to classify businesses by industry, in order to collect, analyze, and publish statistical data related to the U.S. business economy.

There are 20 high-level industrial sections and are listed in the next column.

Sector	Description
11	Agriculture, Forestry, Fishing and Hunting
21	Mining, Quarrying, and Oil and Gas Extraction
22	Utilities
23	Construction
31-33	Manufacturing
42	Wholesale Trade
44-45	Retail Trade
48-49	Transportation and Warehousing
51	Information
52	Finance and Insurance
53	Real Estate and Rental and Leasing
54	Professional, Scientific, and Technical Services
55	Management of Companies and Enterprises
56	Administrative and Support and Waste Management and Remediation Services
61	Educational Services
62	Health Care and Social Assistance
71	Arts, Entertainment, and Recreation
72	Accommodation and Food Services
81	Other Services (except Public Administration)
92	Public Administration (*not covered in economic census*)

CHAPTER 16: KEY SECTOR STATISTICS

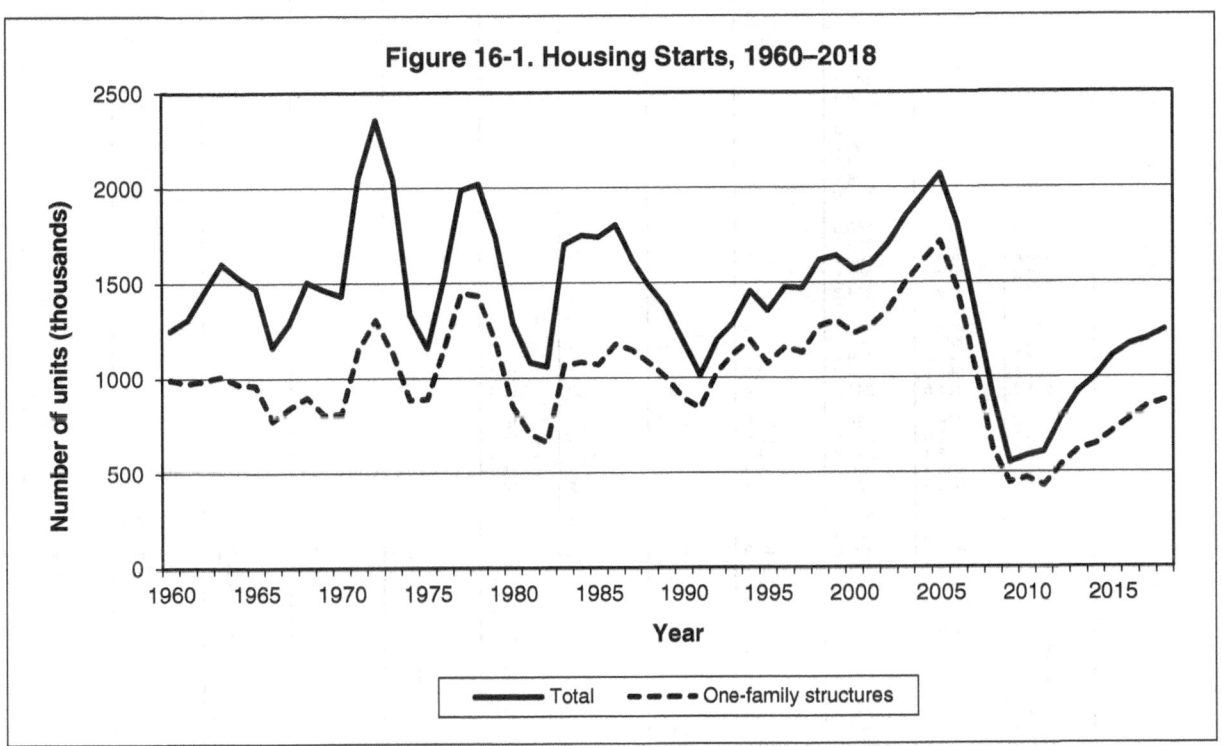

Figure 16-1. Housing Starts, 1960–2018

- The housing sector and associated financial excesses had the leading role in the boom and bust cycle of 2001–2009. In 2005, as shown in Figure 16-1, nearly 2.1 million housing units were started, the highest since 1972. Of those, 1.7 million were one-family homes, an all-time record. By 2009, both total and single-family starts had plunged to the lowest levels of their 50-year history. With very modest recovery in the years following 2009, starts remained at levels once viewed as recessionary. In 2018, the housing starts more than doubled the 2009 low to 1.25 million. (Table 16-2)

- New orders for nondefense capital goods excluding aircraft and parts at U.S. manufacturing firms is a widely followed advance indicator for production of investment goods. The value of such orders fell nearly 39 percent from 2007 to 2009 which was greater than their decline between 2000 and 2002. Their recovery in the four years following 2009 was faster than it was in the 4 years following 2002, and the value of new orders in 2013 surpassed the peak in 2007 but declined 4.3 percent between 2014 and 2018. However, new orders for nondefense capital goods increased 5.9 percent from 2017 to 2018. (Table 16-5)

- Sales of cars and light trucks fell to 10.4 million in 2009, down over 6.5 million from their recent high in 2005. By 2015, they exceeded their 2005 high. Since 2015, sales have increased by only 0.3 percent. (Table 16-7)

- E-commerce grew from 2.3 percent of total retail sales at the beginning of 2005 to 9.9 percent (seasonally adjusted) at the end of 2018. (Table 16-9)

Table 16-1. New Construction Put in Place

(Billions of dollars, monthly data are at seasonally adjusted annual rates.)

| Year and month | Total | Private | | | Commercial | | Health care | Educational | Amuse-ment and recreation | Transpor-tation | Commu-nication | Power | Manu-facturing |
		Total¹	Residential	Office	Total¹	Multi-retail							
1975	152.6	109.3	51.6	...	...	...	...	...	...	...	...	...	...
1976	172.1	128.2	68.3	...	...	...	...	...	...	...	...	...	...
1977	200.5	157.4	92.0	...	...	...	...	...	...	...	...	...	...
1978	239.9	189.7	109.8	...	...	...	...	...	...	...	...	...	...
1979	272.9	216.2	116.4	...	...	...	...	...	...	...	...	...	...
1980	273.9	210.3	100.4	...	...	...	...	...	...	...	...	...	...
1981	289.1	224.4	99.2	...	...	...	...	...	...	...	...	...	...
1982	279.3	216.3	84.7	...	...	...	...	...	...	...	...	...	...
1983	311.9	248.4	125.8	...	...	...	...	...	...	...	...	...	...
1984	370.2	300.0	155.0	...	...	...	...	...	...	...	...	...	...
1985	403.4	325.6	160.5	...	...	...	...	...	...	...	...	...	...
1986	433.5	348.9	190.7	...	...	...	...	...	...	...	...	...	...
1987	446.6	356.0	199.7	...	...	...	...	...	...	...	...	...	...
1988	462.0	367.3	204.5	...	...	...	...	...	...	...	...	...	...
1989	477.5	379.3	204.3	...	...	...	...	...	...	...	...	...	...
1990	476.8	369.3	191.1	...	...	...	...	...	...	...	...	...	...
1991	432.6	322.5	166.3	...	...	...	...	...	...	...	...	...	...
1992	463.7	347.8	199.4	...	...	...	...	...	...	...	...	...	...
1993	485.5	358.2	208.2	20.0	34.4	11.5	14.9	4.8	4.6	4.7	9.8	23.6	23.4
1994	531.9	401.5	241.0	20.4	39.6	12.2	15.4	5.0	5.1	4.7	10.1	21.0	28.8
1995	548.7	408.7	228.1	23.0	44.1	12.0	15.3	5.7	5.9	4.8	11.1	22.0	35.4
1996	599.7	453.0	257.5	26.5	49.4	13.3	15.4	7.0	7.0	5.8	11.8	17.4	38.1
1997	631.9	478.4	264.7	32.8	53.1	12.2	17.4	8.8	8.5	6.2	12.5	16.4	37.6
1998	688.5	533.7	296.3	40.4	55.7	13.3	17.7	9.8	8.6	7.3	12.5	21.7	40.5
1999	744.6	575.5	326.3	45.1	59.4	15.2	18.4	9.8	9.6	6.5	18.4	22.0	35.1
2000	802.8	621.4	346.1	52.4	64.1	14.9	19.5	11.7	8.8	6.9	18.8	29.3	37.6
2001	840.2	638.3	364.4	49.7	63.6	16.4	19.5	12.8	7.8	7.1	19.6	31.5	37.8
2002	847.9	634.4	402.0	44.3	62.5	15.6	27.1	73.9	17.3	25.8	18.5	36.8	22.9
2003	891.5	675.4	451.3	39.4	61.5	15.4	29.3	74.3	16.8	24.7	14.6	41.5	21.5
2004	991.4	771.2	538.4	42.4	67.1	18.8	32.2	74.3	16.7	25.1	15.5	35.6	23.4
2005	1 116.8	882.7	624.6	37.3	66.6	22.8	28.5	12.8	7.5	7.1	18.8	29.2	28.4
2006	1 161.3	905.9	607.8	45.7	73.4	29.2	32.0	13.8	9.3	8.7	22.2	33.7	32.3
2007	1 148.0	858.9	488.8	53.8	85.9	34.8	35.6	16.7	10.2	9.0	27.5	54.1	40.2
2008	1 077.4	768.6	359.2	55.5	82.7	32.0	38.4	18.6	10.5	9.9	26.3	69.2	53.6
2009	906.5	591.6	247.5	37.3	51.1	18.4	35.3	16.9	8.4	9.1	19.7	76.1	57.4
2010	809.3	505.3	242.0	24.4	37.2	12.5	29.6	13.4	6.5	9.9	17.7	66.1	40.6
2011	788.3	501.9	244.1	23.7	39.2	13.4	28.9	14.1	6.7	9.5	17.5	64.3	39.8
2012	850.5	571.1	269.8	27.4	44.3	14.9	31.4	16.6	6.2	10.9	16.0	86.4	46.8
2013	906.4	635.7	323.4	30.1	50.9	16.7	29.7	16.9	6.9	11.0	17.6	81.3	49.9
2014	1 005.6	729.5	369.8	38.9	60.9	19.5	28.9	16.6	7.7	12.2	17.1	98.2	58.1
2015	1 113.6	823.3	422.1	47.4	63.4	19.6	30.5	17.6	9.9	13.6	21.5	91.5	79.3
2016	1 191.8	899.3	467.2	59.6	75.0	22.1	31.5	19.8	12.6	13.1	22.0	91.7	75.7
2017	1 246.0	962.8	524.9	58.6	84.6	24.7	32.6	20.3	13.8	14.8	24.7	90.6	65.8
2018	1 294.0	992.5	539.8	63.4	85.9	22.2	32.9	20.8	14.6	17.0	24.5	94.2	64.7
2016													
January	1 147.5	846.1	446.6	50.6	70.5	20.5	30.1	18.4	10.4	12.9	23.6	80.3	75.0
February	1 157.7	856.5	450.6	54.6	71.3	21.2	30.7	19.0	10.9	12.6	21.3	81.9	76.1
March	1 183.7	885.2	468.1	55.2	71.8	20.4	32.4	19.0	11.4	12.6	20.4	86.0	77.9
April	1 171.6	877.3	457.1	57.3	72.4	20.9	31.0	18.8	12.4	13.0	20.4	87.2	76.5
May	1 178.6	886.3	458.7	56.7	74.0	21.9	31.0	18.9	12.5	13.4	21.1	90.1	78.5
June	1 195.1	896.8	460.3	59.9	73.2	21.8	31.0	20.0	13.1	16.1	21.1	92.6	77.3
July	1 192.2	904.0	461.8	61.8	75.1	22.1	31.7	20.6	12.7	13.9	21.5	95.4	77.8
August	1 194.1	908.4	464.0	62.5	76.8	22.3	31.4	19.7	12.9	12.8	22.0	98.4	76.6
September	1 204.4	914.9	468.0	63.6	76.6	22.3	32.8	20.7	13.1	12.6	22.2	98.6	73.8
October	1 209.6	919.9	480.3	63.3	76.8	23.6	31.5	20.6	13.6	11.9	22.6	96.0	72.0
November	1 235.5	942.9	491.5	64.3	78.9	23.6	31.5	21.0	14.1	12.1	23.4	99.3	74.0
December	1 234.6	947.4	494.5	63.8	82.0	24.1	32.1	21.0	13.3	12.8	24.0	99.2	72.4
2017													
January	1 231.1	952.1	503.2	62.6	80.7	23.9	32.3	21.2	12.8	13.3	25.3	98.6	69.2
February	1 241.8	955.4	512.1	59.6	80.0	22.9	32.1	20.9	13.3	13.9	25.5	99.0	66.9
March	1 256.9	968.9	523.4	59.4	84.1	23.7	32.4	20.4	13.1	14.0	24.5	96.4	68.7
April	1 242.3	963.1	522.2	58.4	86.0	25.5	31.8	20.3	14.0	13.6	23.9	94.1	67.0
May	1 253.6	963.1	519.6	57.5	87.0	25.3	32.5	20.3	14.1	14.1	24.7	93.1	68.5
June	1 241.3	957.7	522.1	57.7	86.8	25.2	32.2	20.2	13.7	14.3	24.6	90.2	63.4
July	1 242.8	961.6	524.9	58.9	86.3	25.0	32.4	19.3	14.0	14.6	24.5	89.6	64.9
August	1 237.5	959.5	527.5	56.4	85.2	25.3	32.9	19.1	13.8	15.0	24.4	89.1	64.1
September	1 240.4	962.0	529.4	56.7	84.0	24.8	33.4	20.6	14.1	15.1	24.6	87.5	64.9
October	1 247.5	961.6	529.5	57.1	84.2	24.5	33.1	20.3	13.9	16.3	24.9	85.6	64.8
November	1 257.3	971.0	538.2	59.1	84.9	25.0	33.3	19.9	13.7	16.4	25.0	84.7	64.3
December	1 272.6	983.1	543.7	60.2	86.1	24.8	33.4	21.3	14.4	16.9	25.1	85.5	64.0
2018													
January	1 276.3	985.9	541.8	61.4	85.8	24.1	33.8	20.6	13.9	16.5	25.6	90.1	63.8
February	1 305.5	1 013.3	560.2	62.8	89.6	24.9	33.9	20.8	14.3	16.4	26.0	91.0	64.5
March	1 293.1	997.9	546.6	62.3	89.3	24.4	33.0	20.5	14.1	16.2	27.3	91.2	63.3
April	1 314.7	1 015.3	563.4	62.1	88.3	23.9	32.9	20.4	14.2	16.5	25.1	96.2	61.3
May	1 324.3	1 016.8	561.9	63.1	89.7	23.8	32.6	21.3	14.2	16.6	24.6	97.8	60.8
June	1 314.8	1 009.4	551.7	62.7	89.4	23.9	32.4	20.9	14.7	17.5	24.4	97.1	64.2
July	1 317.7	1 012.6	555.4	63.3	87.8	23.0	32.1	20.7	15.2	17.5	24.2	96.8	65.1
August	1 311.8	999.2	540.9	63.4	84.9	22.0	32.9	20.7	15.2	17.5	23.9	97.6	67.1
September	1 287.9	983.2	530.0	64.1	84.2	20.2	32.4	20.8	15.2	17.6	23.6	95.3	65.0
October	1 290.6	984.1	526.1	67.0	84.3	20.0	33.6	22.1	14.8	17.2	23.8	93.1	65.6
November	1 273.1	971.1	522.7	64.6	79.8	18.9	32.0	21.3	14.9	18.0	23.5	92.4	66.2
December	1 276.0	978.7	527.6	65.0	80.2	18.0	33.0	20.2	14.7	17.2	24.5	92.3	67.8

¹Includes categories not shown separately.
... = Not available.

Table 16-1. New Construction Put in Place—*Continued*

(Billions of dollars, monthly data are at seasonally adjusted annual rates.)

| Year and month | Total | Public | | | | | | | | | | | | Federal |
| | | State and local | | | | | | | | | | | | |
		Total¹	Residential	Office	Health care	Educational	Public safety	Amusement and recreation	Transportation	Power	Highway and street	Sewage and waste disposal	Water supply	
1975	43.3	37.2	...	...	...	...	...	...	...	...	...	...	...	6.1
1976	44.0	37.2	...	...	...	...	...	...	...	...	...	...	...	6.8
1977	43.1	36.0	...	...	...	...	...	...	...	...	...	...	...	7.1
1978	50.1	42.0	...	...	...	...	...	...	...	...	...	...	...	8.1
1979	56.6	48.1	...	...	...	...	...	...	...	...	...	...	...	8.6
1980	63.6	54.0	...	...	...	...	...	...	...	...	...	...	...	9.6
1981	64.7	54.3	...	...	...	...	...	...	...	...	...	...	...	10.4
1982	63.1	53.1	...	...	...	...	...	...	...	...	...	...	...	10.0
1983	63.5	52.9	...	...	...	...	...	...	...	...	...	...	...	10.6
1984	70.2	59.0	...	...	...	...	...	...	...	...	...	...	...	11.2
1985	77.8	65.8	...	...	...	...	...	...	...	...	...	...	...	12.0
1986	84.6	72.2	...	...	...	...	...	...	...	...	...	...	...	12.4
1987	90.6	76.6	...	...	...	...	...	...	...	...	...	...	...	14.1
1988	94.7	82.5	...	...	...	...	...	...	...	...	...	...	...	12.3
1989	98.2	86.0	...	...	...	...	...	...	...	...	...	...	...	12.2
1990	107.5	95.4	...	...	...	...	...	...	...	...	...	...	...	12.1
1991	110.1	97.3	...	...	...	...	...	...	...	...	...	...	...	12.8
1992	115.8	101.5	...	...	...	...	...	...	...	...	...	...	...	14.4
1993	127.4	112.9	3.4	2.8	2.3	24.2	4.6	4.3	9.8	7.4	34.5	11.2	6.4	14.4
1994	130.4	116.0	4.2	3.0	2.4	25.3	4.5	4.7	9.2	5.1	37.3	12.0	6.4	14.4
1995	140.0	124.3	4.5	3.3	2.6	27.5	5.0	5.1	9.6	5.7	38.6	13.0	7.3	15.8
1996	146.7	131.4	3.7	3.6	2.8	31.0	5.5	5.0	10.4	4.8	40.6	13.6	7.7	15.3
1997	153.4	139.4	3.3	3.8	2.9	34.3	5.6	5.7	10.3	4.4	44.4	13.1	8.1	14.1
1998	154.8	140.5	3.2	3.7	2.3	35.1	6.1	6.2	10.5	2.8	45.4	13.2	8.9	14.3
1999	169.1	155.1	3.2	3.6	2.5	39.8	6.2	7.2	11.4	2.9	51.0	14.5	9.6	14.0
2000	181.3	167.2	3.0	4.5	2.8	46.8	5.9	7.6	13.0	5.5	51.6	14.0	9.5	14.2
2001	201.9	186.8	3.5	5.6	2.9	52.8	6.1	9.1	15.9	5.3	56.4	14.2	11.4	15.1
2002	213.4	196.9	3.8	6.3	3.5	59.5	6.0	9.2	17.3	3.8	56.7	15.3	11.7	16.6
2003	216.1	198.2	3.7	6.1	4.0	59.3	5.8	8.4	16.5	6.8	56.3	15.6	11.7	17.9
2004	220.2	201.8	4.1	6.0	5.0	59.7	5.5	7.8	16.4	7.0	57.4	17.1	12.0	18.3
2005	234.2	216.9	4.0	5.2	5.1	65.8	6.0	7.3	16.3	8.3	63.2	18.3	13.5	17.3
2006	255.4	237.8	4.3	5.6	5.6	69.8	6.6	9.4	17.7	7.8	71.0	21.5	14.3	17.6
2007	289.1	268.5	5.1	7.2	7.0	78.4	8.4	10.7	21.1	11.4	75.5	23.3	15.0	20.6
2008	308.7	285.0	4.9	8.5	7.0	84.5	9.7	10.9	23.2	11.0	80.4	24.1	16.0	23.7
2009	314.9	286.5	5.8	9.2	6.8	83.7	9.4	10.6	25.5	11.8	81.3	23.5	14.8	28.4
2010	304.0	272.8	7.6	8.3	6.2	71.9	7.6	9.7	26.5	10.8	81.3	24.6	14.4	31.1
2011	286.4	254.8	6.0	7.4	7.0	68.0	7.2	8.5	23.3	9.5	78.4	21.2	13.4	31.7
2012	279.3	252.4	4.7	6.1	7.0	65.6	7.6	8.9	24.8	9.8	79.7	20.9	12.7	26.9
2013	270.7	247.0	4.5	5.2	6.9	60.0	6.6	8.1	25.9	11.0	80.6	21.0	12.9	23.7
2014	276.1	253.7	4.1	5.4	6.2	61.2	6.4	8.7	27.8	10.7	84.0	21.9	12.7	22.4
2015	290.3	267.9	5.7	5.9	5.4	65.3	5.8	10.0	29.1	9.8	89.7	23.2	12.6	22.5
2016	292.5	270.3	6.1	5.6	5.6	68.9	5.7	10.2	27.8	8.4	91.7	22.2	12.7	22.2
2017	283.2	261.7	6.2	5.5	6.0	69.5	5.9	10.6	28.3	5.1	87.7	19.9	11.4	21.5
2018	301.5	279.8	5.9	6.6	6.3	71.8	7.0	11.5	32.4	5.4	91.4	21.4	13.3	21.6
2016														
January	301.0	279.0	5.7	5.3	5.4	67.2	5.8	9.9	27.2	9.7	100.0	24.9	12.6	22.4
February	301.1	278.3	5.9	5.4	5.3	68.6	5.9	10.0	27.8	10.8	94.1	25.6	13.1	22.8
March	298.5	277.1	5.9	5.6	5.6	69.8	5.4	10.1	28.2	8.2	94.6	24.4	13.0	21.4
April	294.3	272.7	5.9	5.6	5.9	68.9	5.7	10.6	26.9	8.7	91.4	24.1	12.7	21.6
May	292.3	269.9	5.8	5.5	5.4	68.3	5.7	10.5	27.4	10.0	90.2	22.0	12.9	22.4
June	298.4	275.5	5.8	5.6	5.6	72.1	5.7	10.7	27.9	10.3	91.1	21.6	13.5	22.9
July	288.2	266.5	5.9	5.5	5.6	68.1	5.5	10.2	28.8	7.9	90.2	21.1	12.7	21.7
August	285.7	264.2	6.4	5.8	5.5	69.5	5.5	10.0	26.4	7.5	88.4	21.8	12.2	21.6
September	289.5	267.7	6.1	5.6	5.7	68.4	5.8	9.8	28.2	7.1	91.5	21.1	12.9	21.8
October	289.7	268.3	6.6	5.3	5.7	68.0	5.7	10.4	28.4	7.4	92.6	21.1	12.0	21.4
November	292.6	270.5	6.3	6.0	5.7	68.4	5.7	10.3	28.0	6.6	94.3	20.8	13.2	22.1
December	287.2	263.3	6.3	5.6	5.5	68.8	5.6	10.2	29.2	6.5	88.7	19.6	12.3	24.0
2017														
January	279.1	258.2	6.3	5.6	5.2	70.8	5.3	10.3	26.2	6.2	87.0	19.5	11.1	20.8
February	286.4	265.4	6.5	5.7	5.8	70.8	5.8	11.4	27.3	5.7	90.6	19.9	11.2	21.0
March	287.9	265.9	6.8	5.6	6.0	70.1	5.9	11.7	28.1	5.6	89.9	19.6	11.9	22.0
April	279.1	258.2	5.6	4.8	5.9	68.2	5.8	10.5	28.9	3.9	87.9	19.1	12.3	20.9
May	290.5	267.2	6.3	5.4	6.1	72.5	5.9	11.1	28.9	4.9	88.4	20.2	12.1	23.3
June	283.6	262.6	6.2	5.2	6.1	71.4	5.5	11.3	28.3	4.2	87.3	19.7	11.7	21.0
July	281.2	260.0	6.0	5.0	6.1	68.8	5.8	11.1	27.0	4.6	89.6	19.7	11.0	21.1
August	277.9	257.4	6.0	5.5	6.3	68.4	6.3	10.3	27.3	4.4	86.1	19.9	11.3	20.5
September	278.4	257.6	6.1	5.5	5.9	66.3	6.0	10.1	29.0	5.2	86.4	19.9	11.0	20.8
October	285.9	262.7	6.0	5.8	6.3	68.7	6.2	9.3	29.3	4.9	88.6	20.3	11.1	23.2
November	286.3	264.0	6.4	6.1	6.1	69.1	6.1	10.3	29.1	6.9	87.2	20.4	11.0	22.3
December	289.5	267.6	6.5	6.3	6.0	70.4	6.5	10.3	31.2	5.4	86.6	20.7	11.7	21.9
2018														
January	290.3	268.2	6.2	6.1	6.2	70.3	6.6	10.4	31.1	5.1	87.3	21.5	11.7	22.1
February	292.2	271.3	6.2	6.1	6.3	70.8	6.4	10.4	31.5	4.5	89.2	21.4	12.5	20.9
March	295.2	274.2	6.2	6.1	6.3	70.7	6.6	10.6	30.9	5.5	91.1	21.8	12.3	21.0
April	299.4	278.1	6.1	6.3	6.2	71.4	6.7	10.7	31.6	5.6	93.1	21.2	12.6	21.3
May	307.5	285.8	6.4	7.0	6.8	73.2	7.0	10.8	32.1	6.0	94.2	21.5	14.0	21.7
June	305.4	284.1	5.9	7.1	6.4	68.8	7.1	11.5	33.2	5.6	94.7	21.9	14.9	21.3
July	305.1	283.8	6.0	6.9	6.6	68.4	7.2	12.3	33.5	5.7	94.8	21.1	14.2	21.4
August	312.6	290.0	6.0	7.2	6.3	72.0	7.3	12.2	33.8	5.3	95.7	21.6	14.2	22.6
September	304.7	283.0	5.7	6.6	6.6	73.5	7.0	12.4	32.9	4.9	90.4	21.3	13.7	21.7
October	306.4	284.1	5.6	6.8	6.3	74.8	7.5	12.5	33.0	5.6	89.6	21.2	13.7	22.3
November	302.1	279.7	5.3	6.5	6.1	74.9	7.1	11.8	32.8	5.5	88.7	21.5	12.5	22.4
December	297.3	275.9	5.0	6.6	5.8	74.6	7.1	11.8	31.7	5.3	86.7	21.3	12.7	21.4

¹Includes categories not shown separately.
... = Not available.

Table 16-2. Housing Starts and Building Permits; New House Sales and Prices

Year and month	Housing starts and building permits										New house sales and prices			
	New private housing units (thousands)									Shipments of manufactured homes (thousands, seasonally adjusted annual rate)	Seasonally adjusted			
	Started (not seasonally adjusted)			Seasonally adjusted annual rate							Sold (thousands, annual rate)	For sale, end-of-period (thousands)	Median sales price (dollars)	Price index (2005 = 100)
				Started			Authorized by building permits							
	Total¹	One-family structures	Five units or more	Total¹	One-family structures	Five units or more	Total¹	One-family structures	Five units or more					
1977	1 987	1 451	414	1 987	1 451	414	1 690	1 126	443	266	819	408	48 800	27.0
1978	2 020	1 433	462	2 020	1 433	462	1 801	1 183	487	276	817	419	55 700	30.9
1979	1 745	1 194	429	1 745	1 194	429	1 552	982	445	277	709	402	62 900	35.3
1980	1 292	852	331	1 292	852	331	1 191	710	366	222	545	342	64 600	38.9
1981	1 084	705	288	1 084	705	288	986	564	319	241	436	278	68 900	42.0
1982	1 062	663	320	1 062	663	320	1 000	546	366	240	412	255	69 300	43.0
1983	1 703	1 068	522	1 703	1 068	522	1 605	901	570	296	623	304	75 300	43.9
1984	1 750	1 084	544	1 750	1 084	544	1 682	922	617	295	639	358	79 900	45.7
1985	1 742	1 072	576	1 742	1 072	576	1 733	957	657	284	688	350	84 300	46.2
1986	1 805	1 179	542	1 805	1 179	542	1 769	1 078	583	244	750	361	92 000	48.0
1987	1 621	1 146	409	1 621	1 146	409	1 535	1 024	421	233	671	370	104 500	50.6
1988	1 488	1 081	348	1 488	1 081	348	1 456	994	386	218	676	371	112 500	52.5
1989	1 376	1 003	318	1 376	1 003	318	1 338	932	340	198	650	366	120 000	54.6
1990	1 193	895	260	1 193	895	260	1 111	794	263	188	534	321	122 900	55.7
1991	1 014	840	138	1 014	840	138	949	754	152	171	509	284	120 000	56.4
1992	1 200	1 030	139	1 200	1 030	139	1 095	911	138	211	610	267	121 500	57.2
1993	1 288	1 126	133	1 288	1 126	133	1 199	987	160	254	666	295	126 500	59.4
1994	1 457	1 198	224	1 457	1 198	224	1 372	1 068	241	304	670	340	130 000	62.9
1995	1 354	1 076	244	1 354	1 076	244	1 333	997	272	340	667	374	133 900	64.3
1996	1 477	1 161	271	1 477	1 161	271	1 426	1 069	290	363	757	326	140 000	66.0
1997	1 474	1 134	296	1 474	1 134	296	1 441	1 062	310	354	804	287	146 000	67.5
1998	1 617	1 271	303	1 617	1 271	303	1 612	1 188	355	373	886	300	152 500	69.2
1999	1 641	1 302	307	1 641	1 302	307	1 664	1 247	351	348	880	315	161 000	72.8
2000	1 569	1 231	299	1 569	1 231	299	1 592	1 198	329	250	877	301	169 000	75.6
2001	1 603	1 273	293	1 603	1 273	293	1 637	1 236	335	193	908	310	175 200	77.9
2002	1 705	1 359	308	1 705	1 359	308	1 748	1 333	341	169	973	344	187 600	81.4
2003	1 848	1 499	315	1 848	1 499	315	1 889	1 461	346	131	1 086	377	195 000	86.0
2004	1 956	1 611	303	1 956	1 611	303	2 070	1 613	366	131	1 203	431	221 000	92.8
2005	2 068	1 716	311	2 068	1 716	311	2 155	1 682	389	147	1 283	515	240 900	100.0
2006	1 801	1 465	293	1 801	1 465	293	1 839	1 378	384	117	1 051	537	246 500	104.7
2007	1 355	1 046	277	1 355	1 046	277	1 398	980	359	96	776	496	247 900	104.9
2008	906	622	266	906	622	266	905	576	295	82	485	352	232 100	99.5
2009	554	445	97	554	445	97	583	441	121	50	375	232	216 700	95.1
2010	587	471	104	587	471	104	605	447	135	50	323	188	221 800	95.0
2011	609	431	167	609	431	167	624	418	184	52	306	150	227 200	94.3
2012	781	535	234	781	535	234	830	519	285	55	368	148	245 200	97.6
2013	925	618	294	925	618	294	991	621	341	60	429	186	268 900	104.7
2014	1 003	648	342	1 003	648	342	1 052	640	382	64	437	212	288 500	110.2
2015	1 112	715	386	1 112	715	386	1 183	696	455	71	501	235	294 200	112.9
2016	1 174	782	381	1 174	782	381	1 207	751	421	81	561	257	307 800	120.4
2017	1 203	849	343	1 203	849	343	1 282	820	425	93	613	294	323 100	126.8
2018	1 250	876	360	1 250	876	360	1 329	855	434	97	617	348	326 400	132.4
2016														
January	74	50	23	1 114	765	331	1 179	726	418	89	514	235	288 400	. . .
February	84	58	25	1 208	839	357	1 195	732	432	93	518	239	305 800	. . .
March	91	62	28	1 115	756	350	1 106	728	343	108	526	243	303 200	119.0
April	106	73	33	1 160	770	377	1 162	747	382	101	564	243	318 300	. . .
May	105	70	34	1 131	737	389	1 192	741	419	102	556	241	295 200	. . .
June	112	75	35	1 191	765	408	1 194	745	418	111	563	244	311 200	119.0
July	115	73	42	1 232	775	449	1 185	723	430	88	631	237	297 400	. . .
August	103	67	35	1 159	721	420	1 195	745	414	113	586	241	298 900	. . .
September	95	67	26	1 063	780	269	1 276	752	485	106	569	244	314 800	122.4
October	115	73	40	1 325	871	444	1 276	778	465	104	580	248	302 800	. . .
November	88	61	27	1 150	825	322	1 249	777	431	103	571	248	315 000	. . .
December	87	53	33	1 287	822	453	1 263	817	407	84	560	253	327 000	123.7
2017														
January	82	53	29	1 221	801	417	1 329	798	501	93	591	258	315 200	. . .
February	88	59	28	1 292	881	392	1 248	825	378	92	603	261	298 000	. . .
March	97	70	27	1 179	825	345	1 279	825	418	109	627	266	321 700	124.7
April	105	77	27	1 152	826	310	1 255	796	423	96	586	267	311 100	. . .
May	106	77	28	1 124	793	317	1 205	784	386	107	604	273	323 600	. . .
June	116	84	32	1 232	855	371	1 312	813	462	109	624	274	315 200	126.4
July	112	79	32	1 196	843	342	1 258	817	399	85	559	278	322 900	. . .
August	103	78	24	1 167	874	285	1 300	803	461	114	565	283	314 200	. . .
September	104	73	30	1 163	836	310	1 254	831	387	101	642	282	331 500	128.4
October	110	76	32	1 261	884	359	1 343	854	454	109	631	286	319 500	. . .
November	98	69	28	1 299	951	340	1 323	864	418	104	715	288	343 400	. . .
December	81	55	26	1 219	854	361	1 320	877	405	80	656	292	343 300	127.8
2018														
January	92	60	31	1 335	883	439	1 366	870	451	98	628	293	329 600	. . .
February	90	62	26	1 295	906	371	1 323	886	391	92	644	298	327 200	. . .
March	107	73	34	1 332	889	429	1 377	851	486	102	654	296	335 400	129.8
April	118	85	31	1 267	892	354	1 364	863	460	99	629	299	314 400	. . .
May	124	89	34	1 332	937	383	1 301	843	424	109	650	302	316 700	. . .
June	112	84	28	1 180	854	316	1 292	853	403	103	618	309	310 500	131.0
July	112	82	30	1 184	860	318	1 303	873	402	89	609	314	327 500	. . .
August	114	81	32	1 279	889	373	1 249	827	387	116	604	318	321 400	. . .
September	110	75	34	1 236	880	347	1 270	854	376	98	607	324	328 300	132.5
October	106	75	29	1 211	865	327	1 265	847	382	116	557	333	328 300	. . .
November	91	59	32	1 202	804	387	1 322	848	435	103	615	334	308 500	. . .
December	76	53	22	1 142	814	307	1 326	829	460	79	564	346	329 700	135.1

¹Includes structures with 2 to 4 units, not shown separately.
. . . = Not available.

Table 16-3. Manufacturers' Shipments

(Millions of dollars, seasonally adjusted.)

Year and month	Total	Total durable goods [1]	Total nondurable goods [1]	Construction materials and supplies	Information technology industries	Capital goods Total	Nondefense Total	Nondefense Excluding aircraft and parts	Defense	Consumer goods Total	Consumer goods Durable	Consumer goods Nondurable
1993	3 020 497	1 604 544	1 415 953	303 391	241 680	580 859	493 875	463 753	86 984	1 127 629	274 813	852 816
1994	3 238 112	1 764 061	1 474 051	332 734	264 092	616 435	538 203	512 327	78 232	1 192 098	314 931	877 167
1995	3 479 677	1 902 815	1 576 862	353 198	291 885	666 167	590 578	565 729	75 589	1 256 611	324 036	932 575
1996	3 597 188	1 978 597	1 618 591	371 401	311 028	704 635	630 932	605 295	73 703	1 292 955	328 402	964 553
1997	3 834 699	2 147 384	1 687 315	399 880	349 846	779 232	702 971	665 074	76 261	1 358 516	360 193	998 323
1998	3 899 813	2 231 588	1 668 225	418 756	362 564	821 736	747 046	695 717	74 690	1 351 812	373 404	978 408
1999	4 031 887	2 326 736	1 705 151	434 138	374 384	839 754	768 799	713 042	70 955	1 424 828	412 646	1 012 182
2000	4 208 584	2 373 688	1 834 896	444 812	399 751	875 396	808 345	757 617	67 051	1 500 532	391 463	1 109 069
2001	3 970 499	2 174 406	1 796 093	424 517	353 237	801 999	728 495	678 288	73 504	1 480 495	367 522	1 112 973
2002	3 914 723	2 123 621	1 791 102	424 008	284 799	728 585	652 342	609 595	76 243	1 494 575	395 953	1 098 622
2003	4 015 388	2 142 589	1 872 799	429 183	274 829	719 602	634 273	600 616	85 329	1 584 329	418 821	1 165 508
2004	4 308 970	2 264 667	2 044 303	463 148	287 837	752 905	661 740	629 146	91 165	1 700 835	419 182	1 281 653
2005	4 742 077	2 424 844	2 317 233	509 865	295 447	821 906	730 368	686 939	91 538	1 895 119	422 555	1 472 564
2006	5 015 552	2 562 194	2 453 358	546 866	319 890	884 199	794 638	744 282	89 561	1 980 031	421 860	1 558 171
2007	5 319 457	2 687 028	2 632 429	560 208	322 846	936 404	836 426	768 662	99 978	2 106 081	421 822	1 684 259
2008	5 468 994	2 616 667	2 852 327	543 819	319 032	955 412	840 250	774 775	115 162	2 254 007	357 546	1 896 461
2009	4 423 779	2 056 829	2 366 950	426 663	271 678	820 521	697 469	642 555	123 052	1 849 113	268 036	1 581 077
2010	4 911 277	2 280 712	2 630 565	443 880	270 118	858 123	727 818	674 747	130 305	2 058 913	323 432	1 735 481
2011	5 491 894	2 479 086	3 012 808	471 598	274 110	923 876	803 248	740 848	120 628	2 373 512	344 299	2 029 213
2012	5 696 728	2 627 582	3 069 146	500 848	272 578	1 000 585	884 994	799 591	115 591	2 437 147	373 352	2 063 795
2013	5 809 745	2 695 807	3 113 938	523 707	271 444	1 007 645	891 684	799 052	115 961	2 496 085	402 805	2 093 280
2014	5 887 559	2 796 927	3 090 632	547 183	262 545	1 017 011	905 668	803 674	111 343	2 507 589	435 589	2 072 000
2015	5 519 020	2 772 027	2 746 993	554 070	258 110	988 095	877 630	773 609	110 465	2 256 499	461 623	1 794 876
2016	5 354 694	2 713 076	2 641 618	554 850	254 804	944 938	832 663	728 338	112 275	2 184 374	476 823	1 707 551
2017	5 604 916	2 813 550	2 791 366	581 813	269 572	994 587	870 046	766 771	124 541	2 331 113	475 759	1 855 354
2018	5 999 570	3 017 198	2 982 372	615 157	294 053	1 068 095	928 374	815 851	139 721	2 528 313	513 403	2 014 910
2015												
January	462 628	231 321	231 307	46 363	21 192	83 075	74 529	65 571	8 546	185 568	35 922	149 646
February	463 461	229 432	234 029	45 975	21 072	82 645	74 522	65 040	8 123	189 373	35 796	153 577
March	467 395	234 679	232 716	46 015	21 764	83 632	74 481	65 512	9 151	191 752	38 643	153 109
April	465 773	232 554	233 219	46 055	21 494	83 885	74 923	64 898	8 962	190 708	37 577	153 131
May	463 653	230 762	232 891	45 752	21 328	81 685	72 833	64 440	8 852	191 516	38 362	153 154
June	467 030	232 446	234 584	45 872	21 714	82 906	73 178	64 689	9 728	194 138	39 248	154 890
July	464 850	233 789	231 061	46 482	21 697	82 780	73 038	64 722	9 742	191 574	40 235	151 339
August	460 131	232 652	227 479	46 456	21 506	83 479	74 085	64 534	9 394	187 244	39 019	148 225
September	457 457	232 296	225 161	46 592	21 851	82 832	73 269	64 598	9 563	186 269	39 780	146 489
October	453 652	229 581	224 071	46 485	21 535	81 928	72 049	63 770	9 879	184 574	38 646	145 928
November	451 445	229 819	221 626	46 256	21 222	82 002	72 979	62 819	9 023	183 546	39 440	144 106
December	442 172	224 001	218 171	45 851	21 300	78 016	68 762	62 654	9 254	179 947	39 022	140 925
2016												
January	441 097	227 631	213 466	46 210	21 076	79 400	70 497	61 674	8 903	176 741	39 925	136 816
February	439 621	226 839	212 782	46 177	20 729	78 569	69 258	60 868	9 311	176 580	40 540	136 040
March	441 277	224 761	216 516	46 406	21 041	78 326	69 176	60 750	9 276	178 206	39 020	139 186
April	442 314	225 197	217 117	46 135	21 041	78 326	69 630	61 102	8 696	179 373	39 997	139 376
May	442 337	223 611	218 726	45 841	21 230	79 226	70 107	60 579	9 119	179 598	38 387	141 211
June	447 004	224 595	222 409	45 869	21 234	78 919	69 621	60 480	9 298	183 668	39 073	144 595
July	443 990	224 902	219 088	45 772	21 588	78 355	68 738	59 767	9 617	181 340	39 895	141 445
August	444 196	224 453	219 743	46 028	21 452	77 783	68 459	60 108	9 324	181 604	39 797	141 807
September	447 747	226 404	221 343	45 851	21 312	78 844	69 358	60 543	9 486	182 881	40 028	142 853
October	447 980	225 416	222 564	46 270	21 289	78 396	69 019	59 997	9 377	183 808	39 511	144 297
November	447 645	225 391	222 254	46 621	21 288	78 022	68 185	60 281	9 837	183 646	39 301	144 345
December	459 443	229 240	230 203	47 034	21 360	79 114	69 392	60 900	9 722	192 518	40 301	152 217
2017												
January	459 996	229 898	230 098	47 308	21 603	79 429	69 752	61 166	9 677	191 970	40 282	151 688
February	460 505	229 762	230 743	48 020	22 169	80 435	70 144	61 956	10 291	191 150	39 183	151 967
March	461 243	231 941	229 302	48 131	22 008	81 807	71 795	62 705	10 012	189 979	38 958	151 021
April	458 860	229 121	229 739	47 721	22 199	80 477	70 412	62 501	10 065	190 104	38 363	151 741
May	462 623	234 549	228 074	48 236	22 194	82 618	72 442	63 522	10 176	190 729	40 116	150 613
June	462 703	233 884	228 819	48 185	22 198	82 280	72 151	63 251	10 129	190 894	39 779	151 115
July	463 007	232 457	230 550	47 972	22 599	83 128	73 040	63 567	10 088	191 721	38 688	153 033
August	467 588	235 638	231 950	48 514	22 693	83 474	72 962	64 599	10 512	194 576	39 776	154 800
September	470 332	236 805	233 527	49 032	22 644	84 710	74 080	65 371	10 630	195 788	39 664	156 124
October	475 146	238 439	236 707	49 419	23 048	84 579	73 717	65 995	10 862	198 494	40 043	158 451
November	484 003	242 079	241 924	49 952	23 334	86 605	75 658	66 401	10 947	203 495	40 746	162 749
December	485 050	241 455	243 595	50 331	23 157	85 613	74 696	66 704	10 917	204 341	40 223	164 118
2018												
January	489 058	243 553	245 505	50 046	23 662	86 559	75 093	66 442	11 466	206 810	40 944	165 866
February	490 494	245 343	245 151	50 204	23 570	86 485	75 562	67 181	10 923	207 132	41 744	165 388
March	493 240	247 628	245 612	50 478	23 734	88 422	77 722	66 856	10 700	207 636	41 992	165 644
April	493 337	247 496	245 841	51 074	24 156	85 936	73 863	67 378	12 073	208 586	42 688	165 898
May	497 081	248 116	248 965	51 323	24 061	88 891	77 264	67 419	11 627	208 300	39 717	168 583
June	501 313	251 425	249 888	51 485	24 289	89 671	78 192	68 010	11 479	211 099	42 187	168 912
July	501 740	251 393	250 347	51 704	24 843	86 363	74 724	68 719	11 639	214 007	44 369	169 638
August	504 405	253 443	250 962	51 725	24 847	89 357	77 652	68 614	11 705	213 157	43 367	169 790
September	507 438	255 102	252 336	51 564	24 832	90 600	78 970	68 632	11 630	214 802	43 545	171 257
October	507 985	255 382	252 603	51 710	25 249	90 250	78 194	69 080	12 056	215 426	44 034	171 392
November	506 252	257 791	248 461	51 542	25 253	91 609	79 480	68 885	12 129	211 466	44 345	167 121
December	505 209	258 658	246 551	51 956	25 296	91 684	79 268	68 813	12 416	209 906	44 855	165 051

[1] Includes categories not shown separately.

Table 16-4. Manufacturers' Inventories

(Current cost basis, end of period; seasonally adjusted, millions of dollars.)

Year and month	Total	Total durable goods	Durables total by stage of fabrication			Total nondurable goods	Nondurables total by stage of fabrication		
			Materials and supplies	Work in process	Finished goods		Materials and supplies	Work in process	Finished goods
1993	379 806	238 719	72 633	101 948	64 138	141 087	54 233	23 394	63 460
1994	399 934	253 095	78 482	106 521	68 092	146 839	57 113	24 444	65 282
1995	424 802	267 437	85 545	106 634	75 258	157 365	60 778	25 776	70 811
1996	430 366	272 428	86 265	110 527	75 636	157 938	59 157	26 485	72 296
1997	443 227	280 730	92 253	109 714	78 763	162 497	60 175	28 523	73 799
1998	448 373	290 071	93 406	114 928	81 737	158 302	58 156	27 049	73 097
1999	463 004	296 009	97 722	113 805	84 482	166 995	60 975	28 791	77 229
2000	480 748	306 002	105 848	110 909	89 245	174 746	61 465	30 090	83 191
2001	427 353	267 409	91 168	93 768	82 473	159 944	55 671	27 157	77 116
2002	423 028	260 476	88 445	92 343	79 688	162 552	56 571	28 133	77 848
2003	408 302	246 922	82 330	88 608	75 984	161 380	56 866	27 251	77 263
2004	441 222	264 895	92 122	90 934	81 839	176 327	61 917	30 341	84 069
2005	474 639	283 712	98 652	98 400	86 660	190 927	67 224	33 226	90 477
2006	523 476	317 406	111 687	106 352	99 367	206 070	70 601	37 548	97 921
2007	562 714	334 556	116 560	117 386	100 610	228 158	75 606	45 273	107 279
2008	543 317	329 851	117 811	110 865	101 175	213 466	72 724	41 459	99 283
2009	505 452	294 854	100 697	106 260	87 897	210 598	72 074	42 399	96 125
2010	554 328	321 441	106 938	122 224	92 279	232 887	76 749	46 345	109 793
2011	606 839	352 710	119 493	133 100	100 117	254 129	82 097	50 855	121 177
2012	624 905	367 518	125 150	138 836	103 532	257 387	83 293	51 146	122 948
2013	630 267	370 693	125 470	141 812	103 411	259 574	84 566	52 781	122 227
2014	640 437	388 039	132 940	146 972	108 127	252 398	83 676	49 738	118 984
2015	635 783	391 485	129 210	152 493	109 782	244 298	84 072	46 491	113 735
2016	631 247	379 561	127 270	143 177	109 114	251 686	85 105	48 501	118 080
2017	659 418	396 648	133 210	149 242	114 196	262 770	90 279	53 762	118 729
2018	682 655	415 881	145 672	152 616	117 593	266 774	91 964	55 010	119 800
2015									
January	639 539	390 716	133 436	148 010	109 270	248 823	82 754	48 431	117 638
February	640 310	392 062	133 565	148 903	109 594	248 248	82 696	49 111	116 441
March	640 585	392 307	133 408	149 105	109 794	248 278	82 328	48 544	117 406
April	641 538	392 887	133 186	149 753	109 948	248 651	82 879	48 516	117 256
May	641 944	392 353	132 431	150 255	109 667	249 591	83 105	48 716	117 770
June	645 691	394 425	132 688	152 357	109 380	251 266	83 969	49 628	117 669
July	644 628	394 304	131 041	153 111	110 152	250 324	83 913	49 306	117 105
August	641 899	393 593	130 962	153 149	109 482	248 306	83 654	48 339	116 313
September	639 215	391 485	129 604	152 616	109 265	247 730	83 850	47 538	116 342
October	638 935	390 636	129 585	151 574	109 477	248 299	84 384	48 064	115 851
November	637 579	390 219	128 978	151 207	110 034	247 360	84 398	47 606	115 356
December	635 783	391 485	129 210	152 493	109 782	244 298	84 072	46 491	113 735
2016									
January	631 571	389 347	128 706	151 558	109 083	242 224	83 753	44 929	113 542
February	628 520	387 063	128 230	149 974	108 859	241 457	83 700	45 272	112 485
March	628 468	385 975	128 005	149 833	108 137	242 493	84 134	45 060	113 299
April	628 196	384 714	127 444	149 027	108 243	243 482	84 207	45 949	113 326
May	628 640	383 239	127 199	148 826	107 214	245 401	84 664	47 209	113 528
June	628 054	380 405	126 696	145 508	108 201	247 649	85 087	47 934	114 628
July	627 507	381 278	128 879	144 357	108 042	246 229	84 791	46 534	114 904
August	628 473	381 521	128 540	144 359	108 622	246 952	84 501	46 900	115 551
September	626 627	380 197	128 089	143 956	108 152	246 430	83 965	46 689	115 776
October	627 098	379 858	127 765	143 480	108 613	247 240	84 736	46 404	116 100
November	630 076	380 782	127 596	143 912	109 274	249 294	85 249	46 931	117 114
December	631 247	379 561	127 270	143 177	109 114	251 686	85 105	48 501	118 080
2017									
January	633 075	379 894	127 864	143 243	108 787	253 181	85 413	48 924	118 844
February	634 212	380 592	127 757	143 856	108 979	253 620	85 892	49 359	118 369
March	634 386	381 644	128 499	143 966	109 179	252 742	86 028	48 546	118 168
April	635 658	383 149	130 227	143 213	109 709	252 509	86 966	48 352	117 191
May	636 038	384 381	130 731	143 441	110 209	251 657	86 996	48 619	116 042
June	636 413	384 866	131 180	142 539	111 147	251 547	86 827	49 084	115 636
July	639 190	387 081	131 551	143 243	112 287	252 109	86 085	50 427	115 597
August	642 349	388 809	131 727	144 314	112 768	253 540	86 839	51 011	115 690
September	647 555	392 292	132 859	145 910	113 523	255 263	88 113	51 479	115 671
October	650 891	393 693	132 732	147 193	113 768	257 198	88 158	52 299	116 741
November	654 348	394 380	132 797	148 089	113 494	259 968	88 904	53 150	117 914
December	659 418	396 648	133 210	149 242	114 196	262 770	90 279	53 762	118 729
2018									
January	661 954	397 839	133 723	150 248	113 868	264 115	89 714	54 275	120 126
February	664 577	400 137	135 266	150 891	113 980	264 440	89 938	53 981	120 521
March	664 676	400 548	136 286	150 055	114 207	264 128	89 965	54 908	119 255
April	667 705	402 465	137 238	150 996	114 231	265 240	90 332	55 459	119 449
May	669 775	404 458	138 462	151 628	114 368	265 317	90 519	54 990	119 808
June	669 588	403 011	139 989	148 358	114 664	266 577	90 792	55 603	120 182
July	676 291	409 044	141 021	153 212	114 811	267 247	91 709	55 684	119 854
August	676 016	407 513	141 883	150 824	114 806	268 503	92 488	55 861	120 154
September	680 293	410 925	143 291	152 008	115 626	269 368	92 704	56 646	120 018
October	682 510	412 135	144 178	152 157	115 800	270 375	92 892	56 202	121 281
November	682 391	414 329	145 284	152 302	116 743	268 062	92 316	55 253	120 493
December	682 655	415 881	145 672	152 616	117 593	266 774	91 964	55 010	119 800

Table 16-4. Manufacturers' Inventories—Continued

(Current cost basis, end of period; seasonally adjusted, millions of dollars.)

Year and month	Construction materials and supplies	Information technology industries	Capital goods				Consumer goods		
			Total	Nondefense		Defense	Total	Durable	Nondurable
				Total	Excluding aircraft and parts				
1993	38 307	39 176	118 460	97 118	79 585	21 342	104 873	21 820	83 053
1994	40 879	41 538	123 368	103 320	85 717	20 048	109 536	23 948	85 588
1995	43 420	46 936	129 709	111 563	94 673	18 146	116 404	25 316	91 088
1996	44 124	43 556	132 746	115 211	93 441	17 535	116 351	24 752	91 599
1997	45 725	48 094	137 556	122 598	98 870	14 958	119 660	24 905	94 755
1998	46 500	45 333	145 048	127 108	97 362	17 940	117 343	24 895	92 448
1999	48 541	46 174	146 244	126 278	99 610	19 966	124 220	26 097	98 123
2000	50 518	52 719	148 992	131 655	110 122	17 337	130 823	27 295	103 528
2001	46 341	42 967	128 670	114 908	93 305	13 762	121 377	25 410	95 967
2002	46 698	39 083	122 538	108 147	88 529	14 391	124 904	25 687	99 217
2003	45 448	35 222	114 917	99 119	81 660	15 798	125 052	25 017	100 035
2004	51 244	33 132	117 611	98 900	83 687	18 711	134 390	26 240	108 150
2005	55 094	39 215	127 158	110 573	92 903	16 585	145 446	28 104	117 342
2006	60 864	39 914	139 499	122 442	101 760	17 057	154 185	27 686	126 499
2007	62 574	38 774	150 204	131 364	104 680	18 840	172 302	27 625	144 677
2008	61 472	38 141	150 764	134 359	107 786	16 405	155 218	24 086	131 132
2009	50 969	36 713	145 761	125 915	97 661	19 846	159 224	21 085	138 139
2010	53 341	37 625	160 199	139 598	103 786	20 601	175 239	21 922	153 317
2011	57 691	39 651	174 515	154 438	113 925	20 077	188 289	22 836	165 453
2012	60 149	39 406	184 454	164 114	118 371	20 340	191 309	24 997	166 312
2013	61 160	38 051	185 822	165 954	116 841	19 868	193 799	26 085	167 714
2014	65 034	36 893	192 467	172 847	120 816	19 620	186 663	27 721	158 942
2015	63 797	37 538	198 903	178 200	119 665	20 703	184 535	28 972	155 563
2016	64 513	36 675	189 907	169 218	115 957	20 689	190 593	28 697	161 896
2017	68 713	38 151	196 512	173 899	121 183	22 613	198 670	29 561	169 109
2018	74 175	39 845	205 031	182 333	127 534	22 698	199 256	30 578	168 678
2015									
January	65 133	36 996	193 541	173 870	121 669	19 671	184 050	27 980	156 070
February	65 316	37 319	194 725	174 465	121 500	20 260	184 091	27 974	156 117
March	65 845	37 358	194 644	174 601	121 393	20 043	185 189	28 494	156 695
April	65 680	37 707	195 949	175 199	121 754	20 750	185 689	28 659	157 030
May	65 062	37 724	196 179	175 240	121 641	20 939	187 144	28 650	158 494
June	65 150	38 006	198 157	176 174	121 494	21 983	189 000	28 749	160 251
July	64 693	37 976	199 215	177 188	121 712	22 027	188 207	28 645	159 562
August	64 214	37 840	199 575	176 973	121 022	22 602	186 548	28 643	157 905
September	64 241	37 526	198 475	176 528	120 333	21 947	186 208	28 476	157 732
October	63 993	37 503	197 701	176 537	119 741	21 164	188 104	29 000	159 104
November	63 801	37 701	197 848	176 962	119 809	20 886	187 284	28 842	158 442
December	63 797	37 538	198 903	178 200	119 665	20 703	184 535	28 972	155 563
2016									
January	63 631	37 079	198 262	176 622	118 508	21 640	182 704	29 085	153 619
February	63 484	36 836	196 659	175 542	117 881	21 117	181 560	28 939	152 621
March	63 435	36 877	195 811	175 114	117 441	20 697	182 882	28 785	154 097
April	63 314	36 823	195 470	174 540	116 856	20 930	183 983	28 655	155 328
May	63 439	36 837	194 466	173 599	116 344	20 867	184 844	28 191	156 653
June	63 469	36 661	190 985	170 561	116 010	20 424	188 132	29 199	158 933
July	63 670	36 877	191 734	170 406	116 211	21 328	186 211	28 855	157 356
August	64 064	36 727	191 185	170 340	116 263	20 845	186 980	28 888	158 092
September	64 262	36 751	190 238	169 419	115 732	20 819	186 153	28 737	157 416
October	64 434	36 566	190 083	169 011	115 481	21 072	186 246	28 646	157 600
November	64 475	36 698	190 272	169 347	115 836	20 925	188 232	28 686	159 546
December	64 513	36 675	189 907	169 218	115 957	20 689	190 593	28 697	161 896
2017									
January	64 906	36 803	190 236	169 228	116 344	21 008	191 557	28 350	163 207
February	65 161	36 832	190 492	169 172	116 524	21 320	191 455	28 136	163 319
March	65 035	37 170	191 292	170 247	117 156	21 045	190 494	28 416	162 078
April	65 942	37 049	191 114	169 348	117 189	21 766	190 053	28 549	161 504
May	66 207	37 164	191 482	169 728	117 712	21 754	188 955	28 523	160 432
June	66 574	37 387	191 453	169 699	118 542	21 754	188 976	28 839	160 137
July	66 947	37 515	192 360	169 848	119 090	22 512	189 463	29 040	160 423
August	67 466	37 638	193 408	170 674	119 427	22 734	190 652	29 033	161 619
September	67 706	37 586	194 425	171 843	120 026	22 582	192 673	29 311	163 362
October	67 968	37 805	195 389	172 668	120 407	22 721	193 667	29 237	164 430
November	68 419	37 806	195 308	172 905	120 443	22 403	196 136	29 156	166 980
December	68 713	38 151	196 512	173 899	121 183	22 613	198 670	29 561	169 109
2018									
January	69 122	37 987	196 996	174 553	121 329	22 443	200 255	29 689	170 566
February	69 587	38 175	198 581	175 975	122 013	22 606	200 365	29 693	170 672
March	70 185	38 222	197 997	175 439	122 338	22 558	199 985	29 818	170 167
April	70 723	38 400	198 908	176 663	122 675	22 245	200 787	29 835	170 952
May	71 147	38 670	199 866	177 410	123 214	22 456	200 951	30 111	170 840
June	71 826	38 839	197 499	174 589	123 749	22 910	201 317	30 155	171 162
July	72 407	38 815	201 991	179 032	123 931	22 959	201 539	30 225	171 314
August	72 682	38 895	199 929	177 399	124 600	22 530	202 430	30 488	171 942
September	73 422	39 299	202 430	179 818	125 780	22 612	202 705	30 285	172 420
October	73 571	39 274	203 307	180 489	126 097	22 818	203 324	30 330	172 994
November	73 846	39 448	204 061	181 178	126 728	22 883	201 082	30 696	170 386
December	74 175	39 845	205 031	182 333	127 534	22 698	199 256	30 578	168 678

Table 16-5. Manufacturers' New Orders

(Net, millions of dollars, seasonally adjusted.)

Year and month	Total [1]	Total durable goods [1]	Construction materials and supplies	Information technology industries	Capital goods Total	Capital goods Nondefense Total	Capital goods Nondefense Excluding aircraft and parts	Defense	Consumer goods Total	Consumer goods Durable	Consumer goods Nondurable
1994	3 199 686	1 725 635	335 962	265 010	616 252	542 094	523 461	74 158	1 192 584	315 417	877 167
1995	3 426 503	1 849 641	355 161	297 605	680 857	612 132	576 769	68 725	1 256 721	324 146	932 575
1996	3 567 384	1 948 793	373 536	310 074	737 268	648 797	607 174	88 471	1 293 537	328 984	964 553
1997	3 779 835	2 092 520	403 860	352 700	792 859	728 362	676 119	64 497	1 360 010	361 687	998 323
1998	3 808 143	2 139 918	419 330	365 723	809 727	745 600	698 279	64 127	1 352 708	374 300	978 408
1999	3 957 242	2 252 091	435 034	389 160	840 603	772 703	728 089	67 900	1 425 617	413 435	1 012 182
2000	4 161 472	2 326 576	446 792	409 500	910 933	831 335	767 754	79 598	1 501 810	392 741	1 109 069
2001	3 868 855	2 072 762	420 817	336 935	775 315	691 552	656 944	83 763	1 477 201	364 228	1 112 973
2002	3 823 145	2 032 043	423 561	273 788	698 865	622 403	589 148	76 462	1 494 106	395 484	1 098 622
2003	3 975 812	2 103 013	431 569	282 289	737 032	636 821	608 175	100 211	1 585 029	419 521	1 165 508
2004	4 289 001	2 244 698	468 746	293 836	787 748	688 813	643 997	98 935	1 701 565	419 912	1 281 653
2005	4 764 032	2 446 799	516 797	300 367	896 672	814 286	703 796	82 386	1 894 021	421 457	1 472 564
2006	5 089 646	2 636 288	549 467	332 649	992 499	889 370	775 156	103 129	1 979 165	420 994	1 558 171
2007	5 397 027	2 764 598	564 717	324 248	1 061 415	957 485	786 665	103 930	2 105 784	421 525	1 684 259
2008	5 452 124	2 599 797	543 218	318 571	1 006 627	879 429	781 810	127 198	2 252 724	356 263	1 896 461
2009	4 205 737	1 838 787	413 497	265 602	690 674	587 116	608 664	103 558	1 848 693	267 616	1 581 077
2010	4 895 899	2 265 334	450 921	272 817	885 135	747 708	685 492	137 427	2 058 565	323 084	1 735 481
2011	5 511 660	2 498 852	471 284	286 080	994 126	861 040	765 584	133 086	2 373 717	344 504	2 029 213
2012	5 709 710	2 640 564	500 617	274 254	1 051 261	932 086	803 477	119 175	2 437 479	373 684	2 063 795
2013	5 827 326	2 713 388	524 490	256 182	1 061 823	970 129	803 660	91 694	2 496 383	403 103	2 093 280
2014	5 925 991	2 835 359	546 885	257 049	1 085 147	976 917	804 526	108 230	2 507 676	435 676	2 072 000
2015	5 439 477	2 692 484	554 366	255 035	962 811	858 023	760 914	104 788	2 256 427	461 551	1 794 876
2016	5 292 735	2 651 117	555 512	257 496	929 830	810 116	727 259	119 714	2 184 557	477 006	1 707 551
2017	5 579 437	2 788 071	586 258	272 141	1 004 966	882 586	773 677	122 380	2 331 358	476 004	1 855 354
2018	5 994 457	3 012 085	620 391	297 479	1 086 199	934 982	820 706	151 217	2 527 988	513 078	2 014 910
2015											
January	459 368	228 061	46 335	20 778	83 603	75 177	65 854	8 426	185 471	35 825	149 646
February	453 300	219 271	45 802	20 709	80 119	72 287	64 254	7 832	189 270	35 693	153 577
March	460 761	228 045	46 084	22 301	82 520	74 633	64 321	7 887	191 849	38 740	153 109
April	459 461	226 242	45 945	21 489	82 976	74 739	63 970	8 237	190 837	37 706	153 131
May	450 741	217 850	45 532	20 977	75 237	67 284	62 649	7 953	191 522	38 368	153 154
June	464 064	229 480	46 422	21 292	84 820	74 858	63 225	9 962	194 100	39 210	154 890
July	459 887	228 826	46 546	21 401	83 770	73 757	63 979	10 013	191 435	40 096	151 339
August	453 145	225 666	46 401	21 602	81 918	74 104	63 465	7 814	187 262	39 037	148 225
September	445 211	220 050	46 776	21 272	71 919	66 251	63 285	5 668	186 173	39 684	146 489
October	450 242	226 171	46 165	20 899	83 516	74 967	63 011	8 549	184 675	38 747	145 928
November	445 703	224 077	46 017	20 766	81 416	69 636	62 254	11 780	183 644	39 538	144 106
December	439 237	221 066	46 509	21 140	72 218	62 321	59 922	9 897	179 943	39 018	140 925
2016											
January	439 864	226 398	46 355	21 482	81 834	72 770	61 849	9 064	176 785	39 969	136 816
February	430 664	217 882	46 333	21 205	73 419	66 038	60 788	7 381	176 584	40 544	136 040
March	433 442	216 926	46 186	20 837	74 326	62 474	60 517	11 852	178 244	39 058	139 186
April	442 914	225 797	46 160	21 517	83 165	71 371	60 503	11 794	179 331	39 955	139 376
May	439 577	220 851	46 028	21 629	79 502	70 887	59 218	8 615	179 641	38 430	141 211
June	431 399	208 990	46 083	21 225	66 725	61 066	59 724	5 659	183 639	39 044	144 595
July	435 826	216 738	45 701	21 631	73 436	67 124	60 456	6 312	181 399	39 954	141 445
August	440 397	220 654	45 870	21 547	77 791	66 523	61 821	11 268	181 639	39 832	141 807
September	441 079	219 736	45 594	21 321	76 845	65 468	60 102	11 377	182 928	40 075	142 853
October	453 867	231 303	46 892	21 189	88 601	79 022	60 055	9 579	183 879	39 582	144 297
November	441 828	219 574	46 923	21 512	75 076	62 121	60 578	12 955	183 560	39 215	144 345
December	454 311	224 108	46 814	22 144	79 630	65 616	60 470	14 014	192 554	40 337	152 217
2017											
January	453 147	223 049	47 709	22 156	77 172	68 170	62 355	9 002	191 999	40 311	151 688
February	452 766	222 023	47 481	22 533	75 478	71 656	62 687	3 822	191 341	39 374	151 967
March	457 555	228 253	49 142	22 029	80 822	71 706	63 108	9 116	189 904	38 883	151 021
April	457 825	228 086	48 049	22 127	81 629	70 534	63 572	11 095	190 124	38 383	151 741
May	455 275	227 201	48 432	22 169	79 235	70 415	63 719	8 820	190 694	40 081	150 613
June	472 648	243 829	48 418	22 227	96 823	85 220	63 456	11 603	190 927	39 812	151 115
July	455 539	224 989	48 467	22 641	79 296	68 770	64 123	10 526	191 772	38 739	153 033
August	461 856	229 906	48 934	22 411	80 980	71 175	64 555	9 805	194 696	39 896	154 800
September	476 345	242 818	49 717	23 267	93 647	78 029	68 024	15 618	195 741	39 617	156 124
October	472 563	235 856	49 726	23 466	85 146	74 898	67 140	10 248	198 495	40 044	158 451
November	480 900	238 976	50 589	23 767	86 380	76 207	65 101	10 173	203 581	40 832	162 749
December	489 350	245 755	50 779	23 441	88 434	77 463	66 707	10 971	204 269	40 151	164 118
2018											
January	481 873	236 368	50 335	23 531	83 382	74 194	66 237	9 188	206 880	41 014	165 866
February	493 481	248 330	50 866	23 640	92 492	79 194	67 522	13 298	207 110	41 722	165 388
March	497 308	251 696	51 279	23 973	93 073	82 773	66 828	10 300	207 755	42 111	165 644
April	494 350	248 509	51 861	24 340	88 728	77 301	67 882	11 427	208 541	42 643	165 898
May	499 354	250 389	51 920	24 473	92 222	78 388	68 034	13 834	208 224	39 641	168 583
June	499 951	250 063	51 810	24 853	88 908	77 217	68 784	11 691	211 195	42 283	168 912
July	498 246	247 899	51 976	25 185	85 649	74 788	69 973	10 861	214 076	44 438	169 638
August	509 334	258 372	52 116	25 109	96 595	81 181	69 477	15 414	212 976	43 186	169 790
September	514 541	262 205	51 613	25 003	93 948	78 730	69 184	15 218	214 736	43 479	171 257
October	503 155	250 552	52 112	25 804	88 538	76 295	69 904	12 243	215 360	43 968	171 392
November	500 392	251 931	51 898	25 759	89 725	75 415	69 052	14 310	211 369	44 248	167 121
December	501 352	254 801	52 433	25 454	91 320	77 551	67 964	13 769	209 769	44 718	165 051

[1] Includes categories not shown separately.

Table 16-6. Manufacturers' Unfilled Orders, Durable Goods Industries

(End of period, millions of dollars, seasonally adjusted.)

Year and month	Total [1]	Transportation equipment		Construction materials and supplies	Information technology industries	By topical categories				Consumer durable goods
		Nondefense aircraft and parts	Defense aircraft and parts			Capital goods				
						Total	Nondefense		Defense	
							Total	Excluding aircraft and parts		
1994	434 899	101 157	44 651	25 090	79 071	299 296	178 007	106 779	121 289	5 121
1995	447 510	109 506	42 360	27 155	84 919	313 963	199 613	118 006	114 350	5 282
1996	489 147	130 580	45 203	29 436	84 227	346 596	217 605	120 276	128 991	5 821
1997	513 317	143 183	40 842	33 584	87 704	360 636	243 413	132 076	117 223	7 330
1998	496 363	137 228	37 711	34 268	90 930	348 398	241 565	134 417	106 833	8 262
1999	505 667	126 256	35 616	35 246	105 883	349 497	245 603	149 679	103 894	9 035
2000	549 507	138 159	42 421	37 405	115 679	384 834	268 323	159 844	116 511	10 349
2001	509 910	124 656	52 002	33 580	98 908	357 539	230 560	137 637	126 979	6 988
2002	479 610	114 352	57 383	33 233	88 083	327 423	200 536	117 571	126 887	6 593
2003	506 423	109 736	63 441	35 701	95 736	344 824	203 047	125 283	141 777	7 296
2004	558 457	122 946	56 998	41 467	101 278	378 930	229 584	140 211	149 346	8 205
2005	653 074	187 614	54 421	48 493	106 404	451 740	311 502	157 208	140 238	7 245
2006	795 263	264 577	58 648	50 972	119 047	557 632	404 587	188 224	153 045	6 443
2007	943 780	375 665	61 854	55 438	120 264	681 230	523 730	206 329	157 500	6 139
2008	993 011	416 255	71 415	54 723	119 866	734 216	563 744	213 555	170 472	4 609
2009	825 121	342 290	61 003	41 208	113 702	606 615	455 468	179 027	151 147	4 162
2010	870 922	349 231	62 275	48 422	116 313	634 907	476 169	190 285	158 738	3 730
2011	954 483	389 224	65 547	48 173	128 490	705 905	533 929	215 509	171 976	3 904
2012	1 014 422	442 461	76 840	48 016	130 101	755 148	580 146	219 402	175 002	4 206
2013	1 075 850	526 019	70 844	48 902	115 160	808 236	657 723	224 201	150 513	4 505
2014	1 160 717	607 265	69 362	48 673	109 727	876 284	728 241	224 897	148 043	4 518
2015	1 129 060	602 069	65 700	49 053	106 678	851 451	709 607	211 839	141 844	4 490
2016	1 116 291	582 645	71 951	49 778	109 496	838 395	688 647	210 871	149 748	4 710
2017	1 138 054	588 969	72 430	54 400	111 884	848 282	702 041	217 680	146 241	5 012
2018	1 182 129	594 438	81 894	59 808	115 216	867 035	709 084	222 492	157 951	4 676
2015										
January	1 161 119	607 268	68 813	48 645	109 313	876 812	728 889	225 180	147 923	4 421
February	1 154 943	605 019	66 068	48 472	108 950	874 286	726 654	224 394	147 632	4 318
March	1 152 049	606 257	65 825	48 541	109 487	873 174	726 806	223 203	146 368	4 415
April	1 149 782	606 240	65 947	48 431	109 482	872 265	726 622	222 275	145 643	4 544
May	1 141 059	600 940	64 840	48 211	109 131	865 817	721 073	220 484	144 744	4 550
June	1 142 103	603 988	64 101	48 761	108 709	867 731	722 753	219 020	144 978	4 512
July	1 140 964	605 009	63 219	48 825	108 413	868 721	723 472	218 277	145 249	4 373
August	1 138 074	605 711	61 885	48 770	108 509	867 160	723 491	217 208	143 669	4 391
September	1 129 779	601 304	63 606	48 954	107 930	856 247	716 473	215 895	139 774	4 295
October	1 130 136	602 960	64 615	48 634	107 294	857 835	719 391	215 136	138 444	4 396
November	1 128 314	598 932	67 090	48 395	106 838	857 249	716 048	214 571	141 201	4 494
December	1 129 060	602 069	65 700	49 053	106 678	851 451	709 607	211 839	141 844	4 490
2016										
January	1 131 625	604 624	66 276	49 198	107 084	853 885	711 880	212 014	142 005	4 534
February	1 126 361	601 765	65 850	49 354	107 560	848 735	708 660	211 934	140 075	4 538
March	1 122 441	595 367	66 983	49 134	107 575	844 609	701 958	211 701	142 651	4 576
April	1 126 727	597 981	69 567	49 159	108 051	849 448	703 699	211 102	145 749	4 534
May	1 127 577	600 585	68 739	49 346	108 450	849 724	704 479	209 741	145 245	4 577
June	1 115 560	592 183	68 844	49 560	108 441	837 530	695 924	208 985	141 606	4 548
July	1 111 248	589 736	68 472	49 489	108 484	832 611	694 310	209 674	138 301	4 607
August	1 111 441	586 283	69 301	49 331	108 579	832 619	692 374	211 387	140 245	4 642
September	1 108 582	582 976	68 618	49 074	108 588	830 620	688 484	210 946	142 136	4 689
October	1 118 578	592 368	68 889	49 696	108 488	840 825	698 487	211 004	142 338	4 760
November	1 117 038	586 882	71 968	49 998	108 712	837 879	692 423	211 301	145 456	4 674
December	1 116 291	582 645	71 951	49 778	109 496	838 395	688 647	210 871	149 748	4 710
2017										
January	1 114 093	578 788	72 084	50 179	110 049	836 138	687 065	212 060	149 073	4 739
February	1 109 962	579 258	72 412	49 640	110 413	831 181	688 577	212 791	142 604	4 930
March	1 110 276	578 356	73 058	50 651	110 434	830 196	688 488	213 194	141 708	4 855
April	1 113 012	577 245	74 588	50 979	110 362	831 348	688 610	214 265	142 738	4 875
May	1 109 539	574 873	74 325	51 175	110 337	827 965	686 583	214 462	141 382	4 840
June	1 123 529	587 443	73 125	51 408	110 366	842 508	699 652	214 667	142 856	4 873
July	1 119 909	581 938	73 774	51 903	110 408	838 676	695 382	215 223	143 294	4 924
August	1 118 010	579 946	73 421	52 323	110 126	836 182	693 595	215 179	142 587	5 044
September	1 128 009	580 995	73 337	53 008	110 749	845 119	697 544	217 832	147 575	4 997
October	1 129 263	581 303	72 375	53 315	111 167	845 686	698 725	218 977	146 961	4 998
November	1 129 987	583 200	71 147	53 952	111 600	845 461	699 274	217 677	146 187	5 084
December	1 138 054	588 969	72 430	54 400	111 884	848 282	702 041	217 680	146 241	5 012
2018										
January	1 134 870	587 738	71 196	54 689	111 753	845 105	701 142	217 475	143 963	5 082
February	1 141 748	590 496	70 789	55 351	111 823	851 112	704 774	217 816	146 338	5 060
March	1 149 702	596 948	70 052	56 152	112 062	855 763	709 825	217 788	145 938	5 179
April	1 154 634	600 705	69 534	56 939	112 246	858 555	713 263	218 292	145 292	5 134
May	1 160 971	602 193	70 038	57 536	112 658	861 886	714 387	218 907	147 499	5 058
June	1 163 755	601 121	72 536	57 861	113 222	861 123	713 412	219 681	147 711	5 154
July	1 164 255	600 190	71 938	58 133	113 564	860 409	713 476	220 935	146 933	5 223
August	1 173 195	603 640	72 546	58 524	113 826	867 647	717 005	221 798	150 642	5 042
September	1 184 472	602 995	79 741	58 573	113 997	870 995	716 765	222 350	154 230	4 976
October	1 183 745	600 530	80 013	58 975	114 552	869 283	714 866	223 174	154 417	4 910
November	1 181 915	596 146	81 980	59 331	115 058	867 399	710 801	223 341	156 598	4 813
December	1 182 129	594 438	81 894	59 808	115 216	867 035	709 084	222 492	157 951	4 676

[1]Includes categories not shown separately.

Table 16-7. Motor Vehicle Sales and Inventories

(Units.)

Year and month	Retail sales of new passenger cars						Retail inventories of new domestic passenger cars (thousands of units, end of period)		
	Thousands of units, not seasonally adjusted			Millions of units, seasonally adjusted annual rate			Not seasonally adjusted	Seasonally adjusted	Inventory to sales ratio
	Total	Domestic	Foreign	Total	Domestic	Foreign			
1976	9 994.0	8 492.0	1 502.0	9.994	8.492	1.502	1 465.0	1 494.0	1.900
1977	11 046.0	8 971.2	2 074.8	11.046	8.971	2.075	1 731.0	1 743.0	2.300
1978	11 164.0	9 163.9	2 000.1	11.164	9.164	2.000	1 729.0	1 731.0	2.300
1979	10 558.8	8 230.1	2 328.7	10.559	8.230	2.329	1 691.0	1 667.0	2.400
1980	8 981.8	6 581.4	2 400.4	8.982	6.581	2.401	1 448.0	1 440.0	2.600
1981	8 534.3	6 208.8	2 325.5	8.535	6.209	2.326	1 471.0	1 495.0	3.600
1982	7 979.4	5 758.2	2 221.2	7.979	5.758	2.221	1 126.0	1 127.0	2.200
1983	9 178.6	6 793.0	2 385.6	9.179	6.793	2.386	1 352.0	1 350.0	2.000
1984	10 390.2	7 951.7	2 438.5	10.391	7.952	2.439	1 415.0	1 411.0	2.100
1985	10 978.4	8 204.7	2 773.7	10.979	8.205	2.774	1 630.0	1 619.0	2.500
1986	11 405.7	8 215.0	3 190.7	11.406	8.215	3.191	1 499.0	1 515.0	2.000
1987	10 170.9	7 080.9	3 090.0	10.171	7.081	3.090	1 680.0	1 716.0	2.800
1988	10 545.6	7 539.4	3 006.2	10.545	7.539	3.006	1 601.0	1 601.0	2.300
1989	9 776.8	7 078.1	2 698.7	9.777	7.078	2.699	1 669.0	1 687.0	3.100
1990	9 300.2	6 896.9	2 403.3	9.300	6.897	2.403	1 408.0	1 418.0	2.600
1991	8 175.0	6 136.9	2 038.1	8.175	6.137	2.038	1 283.0	1 296.0	2.600
1992	8 214.4	6 276.6	1 937.8	8.215	6.277	1.938	1 276.0	1 288.0	2.300
1993	8 517.7	6 734.0	1 783.7	8.518	6.734	1.784	1 364.9	1 377.3	2.359
1994	8 990.4	7 255.2	1 735.2	8.990	7.255	1.735	1 436.6	1 445.3	2.319
1995	8 620.4	7 114.1	1 506.3	8.620	7.114	1.506	1 618.5	1 629.8	2.565
1996	8 478.6	7 206.3	1 272.3	8.478	7.206	1.272	1 363.4	1 422.5	2.478
1997	8 217.7	6 862.3	1 355.4	8.218	6.862	1.356	1 329.9	1 399.7	2.368
1998	8 084.9	6 705.1	1 379.8	8.085	6.705	1.380	1 324.4	1 369.7	2.222
1999	8 637.7	6 918.8	1 718.9	8.638	6.919	1.719	1 367.6	1 406.6	2.369
2000	8 777.7	6 761.6	2 016.1	8.778	6.762	2.016	1 377.0	1 397.2	2.810
2001	8 352.1	6 254.4	2 097.7	8.352	6.254	2.098	955.7	976.9	2.267
2002	8 042.1	5 816.5	2 225.6	8.043	5.817	2.226	1 132.0	1 229.6	2.533
2003	7 555.6	5 472.6	2 083.0	7.556	5.473	2.083	1 115.8	1 226.9	2.713
2004	7 482.5	5 333.6	2 148.9	7.482	5.333	2.149	1 034.2	1 144.6	2.405
2005	7 660.4	5 473.6	2 186.8	7.661	5.474	2.187	930.6	1 012.0	2.289
2006	7 761.3	5 416.7	2 344.6	7.762	5.417	2.345	1 031.7	1 105.3	2.497
2007	7 562.1	5 197.1	2 365.0	7.562	5.197	2.365	906.8	985.1	2.271
2008	6 769.2	4 491.0	2 278.2	6.769	4.491	2.278	1 144.2	1 215.3	4.115
2009	5 401.5	3 558.4	1 843.1	5.401	3.558	1.843	686.4	733.0	2.162
2010	5 635.7	3 791.5	1 844.2	5.636	3.792	1.844	759.1	796.5	2.402
2011	6 092.9	4 146.0	1 946.9	6.093	4.146	1.947	813.0	806.5	2.254
2012	7 245.2	5 119.8	2 125.3	7.246	5.120	2.125	1 069.5	1 067.6	2.374
2013	7 586.3	5 433.2	2 153.2	7.586	5.433	2.153	1 259.3	1 231.0	2.812
2014	7 708.0	5 609.9	2 098.1	7.708	5.610	2.098	1 249.4	1 217.3	2.520
2015	7 516.8	5 595.1	1 921.7	7.252	5.595	1.922	1 178.6	1 157.4	2.594
2016	6 872.7	5 145.6	1 727.2	6.873	5.146	1.727	1 209.8	1 147.2	2.689
2017	6 080.9	4 593.0	1 488.0	6.081	4.593	1.488	967.6	874.7	2.456
2018	5 303.6	4 086.9	1 216.7	5.302	4.085	1.217	802.6	745.0	2.213
2016									
January	472.6	361.4	111.2	7.234	5.499	1.735	1 193.3	1 136.9	2.481
February	555.2	427.3	127.9	7.114	5.356	1.758	1 196.0	1 129.7	2.531
March	665.9	502.1	163.8	6.882	5.122	1.759	1 172.2	1 128.3	2.643
April	608.4	447.8	160.7	6.837	5.033	1.805	1 149.7	1 112.9	2.654
May	625.3	468.3	157.0	6.894	5.105	1.789	1 093.7	1 103.4	2.594
June	604.5	452.3	152.2	6.850	5.079	1.771	1 119.3	1 119.2	2.644
July	590.7	437.2	153.5	6.844	5.136	1.708	1 049.5	1 158.6	2.707
August	579.9	433.4	146.5	6.706	5.039	1.667	1 066.7	1 176.5	2.802
September	555.5	419.6	135.9	6.964	5.261	1.703	1 116.5	1 183.1	2.699
October	506.4	377.3	129.2	6.674	5.005	1.670	1 172.8	1 175.0	2.817
November	508.0	374.9	133.1	6.675	4.994	1.681	1 251.9	1 184.1	2.845
December	600.2	444.2	156.0	6.798	5.119	1.679	1 209.8	1 147.2	2.689
2017									
January	412.4	307.6	104.8	6.321	4.693	1.628	1 240.9	1 172.6	2.998
February	483.5	368.8	114.7	6.209	4.637	1.572	1 245.0	1 168.4	3.024
March	589.9	446.6	143.3	6.125	4.609	1.516	1 251.5	1 172.1	3.052
April	537.7	411.0	126.7	6.215	4.753	1.462	1 222.1	1 176.6	2.971
May	561.8	433.9	127.9	6.041	4.612	1.429	1 188.0	1 179.6	3.069
June	521.4	399.6	121.8	5.874	4.478	1.396	1 191.7	1 179.5	3.161
July	501.3	376.2	125.1	5.926	4.501	1.425	1 055.9	1 125.3	3.000
August	522.8	394.2	128.6	5.871	4.434	1.437	1 025.4	1 100.0	2.977
September	534.8	406.6	128.2	6.412	4.870	1.542	982.4	1 026.5	2.530
October	458.5	345.3	113.2	6.239	4.736	1.503	986.7	968.9	2.455
November	461.4	341.0	120.4	6.026	4.518	1.508	1 012.0	927.0	2.462
December	495.5	362.1	133.4	5.712	4.274	1.438	967.6	874.7	2.456
2018									
January	365.7	280.5	85.3	5.514	4.205	1.310	988.2	899.3	2.567
February	421.8	329.0	92.9	5.436	4.156	1.280	995.9	898.5	2.594
March	537.0	416.1	120.9	5.367	4.122	1.246	971.1	887.4	2.583
April	428.5	326.9	101.6	5.318	4.081	1.236	973.1	906.9	2.666
May	513.1	394.9	118.2	5.389	4.080	1.309	917.9	882.4	2.595
June	482.9	372.4	110.5	5.240	4.013	1.228	891.1	856.4	2.561
July	416.7	321.3	95.4	5.126	3.994	1.132	815.6	854.7	2.568
August	440.8	338.1	102.7	4.982	3.844	1.138	775.7	822.4	2.567
September	429.8	329.6	100.3	5.276	4.034	1.241	752.1	775.7	2.307
October	419.8	329.2	90.7	5.583	4.399	1.185	790.5	749.4	2.044
November	400.8	306.4	94.4	5.226	4.058	1.168	821.5	729.5	2.157
December	446.5	342.6	103.9	5.171	4.040	1.131	802.6	745.0	2.213

Table 16-7. Motor Vehicle Sales and Inventories—Continued

(Units.)

Year and month	Retail sales of new trucks and buses							Unit sales of cars and light trucks (millions of units, seasonally adjusted annual rate)		
	Thousands of units, not seasonally adjusted				Millions of units, seasonally adjusted annual rate					
	Total	0–14,000 pounds		14,001 pounds and over	Total	0–14,000 pounds		14,001 pounds and over	Total	Domestic	Foreign
		Domestic	Foreign			Domestic	Foreign				
1976	3 300.5	2 738.3	237.5	324.7	3.296	2.733	0.239	0.324	12.966	11.225	1.741
1977	3 813.0	3 112.8	323.1	377.1	3.818	3.116	0.324	0.378	14.486	12.087	2.399
1978	4 256.8	3 481.1	335.9	439.8	4.249	3.469	0.340	0.440	14.973	12.633	2.340
1979	3 589.7	2 730.2	469.4	390.1	3.599	2.740	0.469	0.390	13.768	10.970	2.798
1980	2 487.4	1 731.1	484.6	271.7	2.482	1.731	0.480	0.271	11.193	8.312	2.881
1981	2 255.6	1 581.7	447.6	226.3	2.255	1.585	0.444	0.226	10.564	7.794	2.770
1982	2 562.8	1 967.5	410.4	184.9	2.569	1.971	0.413	0.185	10.363	7.729	2.634
1983	3 117.3	2 465.2	463.3	188.8	3.130	2.480	0.461	0.189	12.120	9.273	2.847
1984	4 093.1	3 207.2	607.7	278.2	4.086	3.199	0.609	0.278	14.199	11.151	3.048
1985	4 741.7	3 618.4	828.3	295.0	4.760	3.634	0.831	0.295	15.444	11.839	3.605
1986	4 912.1	3 671.4	967.2	273.5	4.918	3.676	0.969	0.273	16.051	11.891	4.160
1987	4 991.5	3 792.0	912.2	287.3	4.978	3.783	0.907	0.288	14.861	10.864	3.997
1988	5 231.9	4 199.7	697.9	334.3	5.225	4.194	0.697	0.334	15.436	11.733	3.703
1989	5 055.9	4 113.6	630.3	312.0	5.065	4.123	0.629	0.313	14.529	11.201	3.328
1990	4 837.0	3 956.8	602.7	277.5	4.840	3.960	0.602	0.278	13.862	10.857	3.005
1991	4 355.4	3 605.6	528.8	221.0	4.361	3.612	0.528	0.221	12.315	9.749	2.566
1992	4 892.2	4 247.0	395.9	249.3	4.893	4.247	0.398	0.248	12.860	10.524	2.336
1993	5 667.8	5 000.5	364.5	302.8	5.658	4.991	0.365	0.302	13.874	11.725	2.149
1994	6 407.3	5 658.2	396.3	352.8	6.408	5.659	0.395	0.354	15.044	12.914	2.130
1995	6 496.4	5 705.9	402.1	388.4	6.498	5.706	0.402	0.390	14.728	12.820	1.908
1996	6 977.6	6 179.8	438.7	359.1	6.976	6.180	0.439	0.357	15.097	13.386	1.711
1997	7 280.4	6 324.7	579.5	376.2	7.280	6.325	0.579	0.376	15.122	13.187	1.935
1998	7 882.4	6 802.0	656.1	424.3	7.883	6.802	0.656	0.425	15.543	13.507	2.036
1999	8 777.3	7 480.7	775.3	521.3	8.777	7.481	0.775	0.521	16.894	14.400	2.494
2000	9 033.9	7 719.7	852.3	461.9	9.033	7.720	0.852	0.461	17.350	14.482	2.868
2001	9 120.4	7 789.0	981.3	350.1	9.120	7.789	0.981	0.350	17.122	14.043	3.079
2002	9 096.5	7 707.8	1 066.3	322.4	9.096	7.708	1.066	0.322	16.817	13.525	3.292
2003	9 411.9	7 856.3	1 227.2	328.4	9.411	7.856	1.227	0.328	16.639	13.329	3.310
2004	9 815.8	8 138.0	1 246.2	431.6	9.813	8.138	1.246	0.429	16.866	13.471	3.395
2005	9 784.4	8 072.4	1 215.5	496.5	9.784	8.072	1.215	0.497	16.948	13.546	3.402
2006	9 287.0	7 396.0	1 346.6	544.4	9.288	7.396	1.347	0.545	16.505	12.813	3.692
2007	8 897.9	7 138.7	1 388.1	371.1	8.900	7.139	1.388	0.373	16.089	12.336	3.753
2008	15 408.6	5 329.2	9 780.9	298.5	6.724	5.329	1.097	0.298	13.195	9.820	3.375
2009	5 200.5	4 116.5	884.2	199.8	5.200	4.117	0.884	0.199	10.402	7.675	2.727
2010	6 136.6	5 020.3	898.7	217.6	6.138	5.021	0.899	0.218	11.556	8.813	2.743
2011	6 955.5	5 666.5	982.4	306.6	5.668	5.668	0.982	0.306	12.743	9.814	2.929
2012	7 534.3	6 127.3	1 060.7	346.3	7.535	6.128	1.061	0.346	14.435	11.249	3.186
2013	8 296.4	6 704.5	1 239.3	352.6	8.296	6.704	1.239	0.352	15.530	12.137	3.392
2014	9 151.8	7 384.3	1 359.9	407.7	9.150	7.384	1.360	0.406	16.452	12.994	3.458
2015	10 328.8	8 097.4	1 782.1	449.3	10.329	8.097	1.782	0.449	17.131	13.693	3.439
2016	10 993.0	8 436.2	2 155.8	401.0	10.994	8.436	2.156	0.402	17.465	13.582	3.883
2017	11 470.5	8 651.8	2 403.6	415.0	11.470	8.652	2.404	0.415	17.136	13.245	3.892
2018	12 397.8	9 159.0	2 750.9	487.9	12.370	9.138	2.745	0.487	17.185	13.223	3.962
2016										
January	716.2	551.9	132.9	31.4	10.882	8.454	1.965	0.463	17.653	13.953	3.700
February	819.1	637.9	148.0	33.1	10.808	8.391	1.978	0.439	17.483	13.747	3.736
March	948.1	729.9	177.9	40.4	10.455	8.056	1.937	0.462	16.875	13.179	3.697
April	915.7	711.2	170.0	34.4	10.853	8.378	2.065	0.410	17.280	13.411	3.870
May	927.4	702.8	190.6	34.0	10.820	8.134	2.283	0.403	17.310	13.238	4.072
June	944.4	724.0	184.6	35.7	10.876	8.264	2.221	0.392	17.335	13.343	3.992
July	956.1	728.0	197.9	30.1	11.189	8.574	2.237	0.379	17.654	13.709	3.945
August	959.9	726.6	199.8	33.6	11.081	8.501	2.202	0.378	17.409	13.540	3.869
September	906.6	700.6	173.7	32.4	11.069	8.488	2.198	0.382	17.650	13.749	3.901
October	890.6	692.0	168.8	29.9	11.306	8.811	2.129	0.366	17.614	13.815	3.799
November	891.4	680.8	180.8	29.8	11.144	8.523	2.247	0.374	17.445	13.517	3.928
December	1 117.7	850.5	230.8	36.4	11.445	8.662	2.407	0.376	17.868	13.781	4.086
2017										
January	751.9	567.3	158.9	25.7	11.357	8.645	2.343	0.368	17.310	13.338	3.971
February	868.6	667.4	173.2	28.0	11.442	8.750	2.307	0.385	17.265	13.387	3.879
March	992.8	752.0	205.6	35.2	11.005	8.386	2.233	0.386	16.744	12.995	3.749
April	912.0	694.9	185.1	32.0	11.014	8.311	2.295	0.408	16.821	13.064	3.757
May	982.0	741.2	206.2	34.6	11.159	8.364	2.397	0.398	16.803	12.976	3.827
June	981.4	742.8	200.9	37.7	11.321	8.519	2.390	0.412	16.783	12.997	3.786
July	939.7	704.4	203.0	32.4	11.168	8.414	2.345	0.409	16.685	12.915	3.770
August	989.3	738.1	213.0	38.2	10.996	8.286	2.281	0.429	16.438	12.719	3.718
September	1 018.3	766.4	215.9	36.1	12.103	9.060	2.603	0.439	18.075	13.930	4.145
October	927.1	699.9	190.8	36.5	12.063	9.129	2.498	0.436	17.866	13.865	4.001
November	963.2	721.3	206.5	35.3	11.927	8.929	2.554	0.444	17.508	13.446	4.062
December	1 144.1	856.0	244.5	43.5	12.090	9.027	2.599	0.464	17.338	13.302	4.036
2018										
January	816.0	600.8	184.5	30.7	12.027	8.934	2.667	0.426	17.115	13.138	3.977
February	906.3	665.0	206.9	34.4	11.957	8.726	2.759	0.472	16.922	12.882	4.039
March	1 150.6	844.5	265.5	40.5	12.341	9.101	2.764	0.476	17.233	13.223	4.010
April	962.7	714.6	210.5	37.7	12.330	9.090	2.793	0.447	17.201	13.172	4.029
May	1 113.4	824.5	248.9	40.0	12.266	8.987	2.819	0.461	17.195	13.067	4.128
June	1 103.7	811.3	249.4	42.9	12.469	9.144	2.841	0.485	17.225	13.156	4.069
July	986.4	728.7	217.5	40.2	12.084	8.967	2.623	0.493	16.716	12.961	3.755
August	1 086.6	803.6	237.8	45.2	12.214	9.158	2.548	0.509	16.688	13.002	3.686
September	1 045.2	774.8	227.5	42.9	12.690	9.334	2.817	0.539	17.426	13.368	4.058
October	986.1	729.4	211.1	45.6	12.519	9.283	2.706	0.529	17.573	13.682	3.891
November	1 021.4	756.1	225.6	39.7	12.714	9.435	2.780	0.499	17.441	13.492	3.948
December	1 219.4	905.7	265.6	48.1	12.833	9.493	2.828	0.511	17.492	13.533	3.958

Table 16-8. Retail and Food Services Sales

(All retail establishments and food services; millions of dollars; not seasonally adjusted.)

Year and month	Retail and food services, total [1]	Retail (NAICS industry categories)											Food services and drinking places
		GAFO (department store type goods), total [2]	Motor vehicles and parts	Furniture and home furnishings	Electronics and appliances	Building materials and garden	Food and beverages	Health and personal care	Gasoline	Clothing and accessories	General merchandise	Nonstore retailers	
1994	2 330 235	616 347	541 141	60 416	57 266	157 228	384 340	96 359	171 222	129 083	285 190	96 280	225 000
1995	2 450 628	650 040	579 715	63 470	64 770	164 561	390 386	101 632	181 113	131 333	300 498	103 516	233 012
1996	2 603 794	682 613	627 507	67 707	68 363	176 683	401 073	109 554	194 425	136 581	315 305	117 761	242 245
1997	2 726 131	713 387	653 817	72 715	70 061	191 063	409 373	118 670	199 700	140 293	331 363	126 190	257 364
1998	2 852 956	757 936	688 415	77 412	74 527	202 423	416 525	129 582	191 727	149 151	351 081	133 904	271 194
1999	3 086 990	815 665	764 204	84 294	78 977	218 290	433 699	142 697	212 524	159 751	380 179	151 797	283 900
2000	3 287 537	862 739	796 210	91 170	82 206	228 994	444 764	155 233	249 816	167 674	404 228	180 453	304 261
2001	3 378 906	882 700	815 579	91 484	80 240	239 379	462 429	166 532	251 383	167 287	427 468	180 563	316 638
2002	3 459 077	912 707	818 811	94 438	83 740	248 539	464 856	179 983	250 619	172 304	446 520	189 279	330 525
2003	3 612 457	946 114	841 588	96 736	86 442	263 463	474 385	275 187	178 694	178 694	468 771	206 359	349 726
2004	3 846 605	1 003 891	866 372	103 757	93 896	295 274	490 380	324 006	324 006	190 253	497 382	228 977	373 557
2005	4 085 746	1 059 599	888 307	109 120	100 461	320 802	508 484	378 923	378 923	200 969	528 385	255 579	396 463
2006	4 294 359	1 110 155	899 997	112 795	105 477	334 130	525 232	421 976	421 976	213 189	554 256	284 343	422 786
2007	4 439 733	1 143 426	910 139	111 144	106 599	320 854	547 837	451 822	451 822	221 205	578 582	308 767	444 551
2008	4 391 580	1 136 376	785 865	98 720	105 317	301 833	569 276	246 573	503 639	215 583	595 041	319 152	456 265
2009	4 064 476	1 088 197	671 772	84 749	95 364	261 637	568 418	252 794	391 234	204 475	588 918	311 152	452 005
2010	4 284 968	1 114 374	742 913	85 205	97 396	260 566	580 530	260 435	448 349	213 286	603 757	340 957	466 920
2011	4 598 302	1 155 666	812 938	87 586	99 928	269 480	609 137	271 612	533 457	228 606	624 766	376 344	495 350
2012	4 826 390	1 191 843	886 494	91 542	102 060	281 533	628 205	274 000	555 419	239 493	642 313	408 171	524 161
2013	5 001 763	1 212 493	959 294	95 349	102 998	301 797	640 847	281 840	549 613	244 722	651 874	433 126	543 313
2014	5 215 656	1 238 694	1 020 851	99 718	103 518	318 352	669 165	299 263	538 790	250 409	667 163	470 867	576 216
2015	5 349 487	1 258 154	1 094 112	106 570	103 658	331 611	685 381	315 244	444 027	255 798	674 889	509 652	623 494
2016	5 509 323	1 262 110	1 142 261	110 695	99 297	349 372	699 362	327 031	418 684	259 840	675 374	561 556	657 549
2017	5 740 610	1 269 077	1 174 417	113 783	99 401	365 651	725 915	333 219	452 856	258 472	683 854	629 562	693 716
2018	6 021 090	1 301 988	1 205 167	116 611	101 217	377 514	754 656	344 065	511 398	270 409	704 030	688 251	737 815
2016													
January	401 982	89 351	80 151	8 181	7 599	21 980	56 837	26 088	30 018	15 694	49 182	41 289	50 092
February	415 189	94 359	89 395	8 384	7 840	23 177	54 016	26 742	27 991	18 939	51 119	41 455	51 502
March	461 198	103 795	100 382	9 264	7 941	29 942	57 929	28 299	32 406	21 492	55 822	45 098	55 888
April	451 468	98 242	96 991	8 610	7 074	32 435	56 104	26 905	34 003	20 428	53 280	42 896	56 349
May	470 430	103 495	98 252	9 002	7 581	34 702	59 047	27 432	37 363	21 656	56 080	44 592	57 025
June	465 023	101 667	97 880	9 113	7 760	34 219	58 312	27 264	38 521	20 160	55 083	44 345	54 626
July	462 162	102 392	99 151	9 177	7 827	30 122	59 621	26 391	38 169	20 667	55 350	42 591	56 023
August	471 996	106 892	103 502	9 567	8 313	29 936	58 370	27 699	37 150	22 388	55 913	45 619	55 201
September	448 749	98 042	94 921	9 529	7 854	28 844	57 174	26 911	36 294	19 790	51 554	44 509	54 307
October	453 088	101 302	92 173	9 048	7 490	28 561	58 430	26 372	37 243	20 500	55 157	45 411	55 734
November	467 765	114 303	89 469	9 950	9 796	28 553	58 850	26 488	34 151	23 644	60 949	54 828	53 307
December	540 273	148 270	99 994	10 870	12 222	26 901	64 672	30 440	35 375	34 482	75 885	68 923	57 495
2017													
January	421 560	89 152	84 240	8 307	7 523	23 383	57 414	26 492	34 136	15 663	49 248	47 188	52 902
February	417 983	89 752	88 429	8 277	7 192	23 538	54 131	25 479	32 457	17 745	48 937	44 468	52 747
March	483 059	102 699	105 194	9 698	8 024	30 610	59 855	28 352	37 070	21 028	55 043	51 259	60 143
April	466 058	100 102	96 381	8 630	7 151	32 748	59 241	26 627	37 393	20 852	54 942	47 223	58 601
May	495 264	103 623	104 634	9 470	7 675	37 341	61 774	28 233	39 128	21 606	55 898	51 313	60 211
June	482 456	102 587	101 671	9 411	7 761	35 043	60 255	27 674	38 384	20 322	56 026	49 678	58 751
July	475 984	102 026	101 248	9 276	7 601	31 480	61 357	26 749	38 154	20 535	55 684	48 275	58 613
August	491 090	108 136	103 770	9 897	8 042	31 759	61 192	28 417	39 554	22 536	57 167	51 406	58 686
September	470 406	100 178	98 997	9 530	7 853	30 354	60 117	27 293	39 671	19 720	53 874	48 499	57 313
October	476 925	102 530	96 494	9 494	7 719	31 081	60 649	28 509	40 035	20 307	55 815	51 147	58 408
November	499 446	118 765	93 956	10 536	10 487	30 477	62 032	28 087	38 512	24 438	63 143	63 108	55 933
December	560 379	149 527	99 403	11 257	12 373	27 837	67 898	31 307	38 362	33 720	78 077	75 998	61 408
2018													
January	445 484	90 719	87 735	8 699	7 851	24 501	60 510	28 292	37 558	15 881	50 235	53 839	54 795
February	437 005	92 717	89 300	8 505	7 501	24 437	56 682	26 155	35 713	18 585	50 488	49 564	55 073
March	510 380	108 455	109 339	10 013	8 157	30 843	64 077	28 975	40 958	22 404	58 712	55 611	63 868
April	482 412	99 037	99 132	9 175	7 343	33 127	59 907	27 617	41 476	20 616	53 570	53 291	60 734
May	530 082	110 704	108 914	9 910	7 942	39 115	64 958	29 460	46 292	23 764	60 018	56 082	64 705
June	510 029	106 233	104 752	9 713	8 052	36 045	63 240	28 016	45 838	21 589	57 871	52 749	64 236
July	508 010	105 477	103 894	9 700	7 914	34 099	63 862	27 934	46 327	21 919	57 145	54 326	64 033
August	523 933	111 994	108 537	10 170	8 439	33 022	64 109	29 418	47 187	23 381	59 769	56 161	65 152
September	481 094	100 886	96 316	9 546	7 809	30 142	61 664	27 392	43 933	20 260	54 765	52 466	60 492
October	506 360	105 621	99 487	9 693	7 860	33 344	63 038	29 981	46 609	21 473	57 558	57 717	62 089
November	522 804	122 433	95 799	10 550	10 388	30 658	64 068	29 389	41 673	25 831	66 046	70 567	59 401
December	563 497	147 712	101 962	10 937	11 961	28 181	68 541	31 436	37 834	34 706	77 853	75 878	63 237

[1] Includes store categories not shown separately.
[2] Includes furniture, home furnishings, electronics, appliances, clothing, sporting goods, hobby, book, music, general merchandise, office supplies, stationery, and gifts.

Table 16-8. Retail and Food Services Sales—*Continued*

(All retail establishments and food services; millions of dollars; seasonally adjusted.)

Year and month	Total	Retail and food services											
		Retail (NAICS industry categories)											
		Total	GAFO (department store type goods) [2]	Motor vehicles and parts	Furniture and home furnishings	Electronics and appliances	Building materials and garden	Food and beverages			Health and personal care	Gasoline
								Total [1]	Groceries	Beer, wine, and liquor		
1994	2 330 235	2 105 231	616 347	541 141	60 416	57 266	157 228	384 340	350 523	22 101	96 359	171 222
1995	2 450 628	2 217 613	650 040	579 715	63 470	64 770	164 561	390 386	356 409	22 007	101 632	181 113
1996	2 603 794	2 361 546	682 613	627 507	67 707	68 363	176 683	401 073	365 547	23 157	109 554	194 425
1997	2 726 131	2 468 765	713 387	653 817	72 715	70 061	191 063	409 373	372 570	24 081	118 670	199 700
1998	2 852 956	2 581 761	757 936	688 415	77 412	74 527	202 423	416 525	378 188	25 382	129 582	191 727
1999	3 086 990	2 803 088	815 665	764 204	84 294	78 977	218 290	433 699	394 250	26 476	142 697	212 524
2000	3 287 537	2 983 275	862 739	796 210	91 170	82 206	228 994	444 764	402 515	28 507	155 233	249 816
2001	3 378 906	3 062 267	882 700	815 579	91 484	80 240	239 379	462 429	418 127	29 621	166 532	251 383
2002	3 459 077	3 128 552	912 707	818 811	94 438	83 740	248 539	464 856	419 813	29 894	179 983	250 619
2003	3 612 457	3 262 731	946 114	841 588	96 736	86 442	263 463	474 385	427 987	30 469	275 187	178 694
2004	3 846 605	3 473 048	1 003 891	866 372	103 757	93 896	295 274	490 380	441 136	32 189	324 006	324 006
2005	4 085 746	3 689 283	1 059 599	888 307	109 120	100 461	320 802	508 484	457 667	33 567	378 923	378 923
2006	4 294 359	3 871 573	1 110 155	899 997	112 795	105 477	334 130	525 232	471 699	36 016	421 976	421 976
2007	4 439 733	3 995 182	1 143 426	910 139	111 144	106 599	320 854	547 837	491 360	38 128	451 822	451 822
2008	4 391 580	3 935 315	1 136 376	785 865	98 720	105 317	301 833	569 276	511 222	39 504	246 573	503 639
2009	4 064 476	3 612 471	1 088 197	671 772	84 749	95 364	261 637	568 418	510 033	40 245	252 794	391 234
2010	4 291 968	3 818 048	1 114 374	742 913	85 205	97 396	260 566	580 530	520 750	41 401	260 435	448 349
2011	4 598 302	4 102 952	1 155 666	812 938	87 586	99 928	269 480	609 137	547 476	42 392	271 012	503 457
2012	4 826 390	4 302 229	1 191 843	886 494	91 542	102 060	281 533	628 205	563 645	44 365	274 000	555 419
2013	5 001 763	4 458 450	1 212 493	959 294	95 349	102 998	301 797	640 847	574 547	46 076	281 840	549 613
2014	5 215 656	4 639 440	1 238 694	1 020 851	99 718	103 518	318 352	669 165	599 603	48 286	299 263	538 790
2015	5 349 487	4 725 993	1 258 154	1 094 112	106 570	103 658	331 611	685 381	613 159	50 550	315 244	444 027
2016	5 509 323	4 851 774	1 262 110	1 142 261	110 695	99 297	349 372	699 362	624 097	53 135	327 031	418 684
2017	5 740 610	5 046 894	1 269 077	1 174 417	113 783	99 401	365 651	725 915	648 504	55 183	333 219	452 856
2018	6 021 090	5 283 275	1 301 988	1 205 167	116 611	101 217	377 514	754 656	674 201	57 998	344 065	511 398
2016												
January	448 546	395 370	104 355	92 466	9 110	8 332	29 042	57 530	51 498	4 225	26 840	33 428
February	452 413	398 143	105 971	94 522	9 153	8 448	29 080	57 559	51 340	4 343	27 344	31 844
March	450 750	396 856	105 070	92 472	9 082	8 289	29 809	57 058	50 918	4 299	27 211	33 034
April	453 171	398 989	104 928	93 273	9 101	8 293	28 848	57 699	51 540	4 337	27 426	33 935
May	454 440	400 234	104 840	93 363	9 130	8 349	28 366	57 921	51 690	4 381	27 350	34 692
June	459 030	404 730	105 780	93 812	9 280	8 335	29 150	58 301	52 029	4 426	27 567	35 668
July	458 833	404 494	104 941	95 319	9 214	8 398	28 914	57 979	51 702	4 435	27 519	35 309
August	459 341	404 578	105 296	95 667	9 199	8 517	28 860	58 236	51 961	4 449	27 371	34 430
September	461 721	406 362	104 511	96 275	9 444	8 207	29 026	58 474	52 124	4 522	27 404	35 271
October	463 213	407 921	104 564	96 440	9 261	8 097	29 134	58 745	52 404	4 505	26 558	36 158
November	462 889	406 599	103 971	96 066	9 222	7 919	29 339	58 819	52 442	4 550	26 756	35 911
December	467 773	412 436	104 313	98 489	9 119	8 110	29 384	58 943	52 688	4 395	26 890	37 553
2017												
January	473 676	416 299	105 985	98 355	9 386	8 222	30 124	59 130	52 755	4 515	26 978	38 098
February	472 456	415 060	104 483	97 421	9 395	8 145	30 305	59 402	53 045	4 496	27 221	38 230
March	471 493	414 214	105 264	95 311	9 443	8 358	30 107	59 841	53 477	4 513	27 446	37 865
April	474 643	417 247	105 642	96 510	9 360	8 423	30 470	60 057	53 621	4 584	27 593	37 581
May	472 453	415 000	104 252	96 852	9 404	8 306	30 023	60 257	53 810	4 584	27 652	36 163
June	474 446	417 072	105 725	97 717	9 458	8 248	29 997	60 104	53 678	4 589	27 897	35 442
July	475 122	417 658	105 245	97 695	9 417	8 164	30 188	60 232	53 811	4 590	27 864	35 459
August	475 980	418 332	105 671	95 872	9 471	8 173	30 286	60 592	54 098	4 627	28 108	36 556
September	485 679	427 552	106 352	100 247	9 473	8 275	31 302	61 207	54 747	4 625	28 166	38 703
October	487 071	428 663	106 549	101 069	9 658	8 327	30 835	61 465	54 897	4 706	28 339	38 907
November	490 781	431 718	107 143	100 071	9 747	8 437	31 160	61 705	55 129	4 704	28 257	40 242
December	492 980	433 706	107 171	99 812	9 688	8 417	31 254	62 339	55 809	4 681	28 205	41 161
2018												
January	494 208	434 453	107 587	99 973	9 698	8 469	30 677	62 028	55 396	4 764	28 320	41 731
February	495 028	434 970	108 529	98 816	9 698	8 563	31 275	62 203	55 544	4 765	28 003	42 065
March	494 681	434 200	107 514	99 306	9 740	8 488	30 920	62 413	55 716	4 813	28 213	41 624
April	496 768	436 873	108 286	99 932	9 951	8 608	30 937	62 760	56 082	4 822	28 413	41 559
May	502 983	441 651	109 975	100 488	9 764	8 521	31 452	62 898	56 205	4 839	28 911	42 626
June	503 209	441 025	108 877	100 552	9 781	8 548	31 842	62 868	56 184	4 826	28 588	42 640
July	506 494	443 469	109 622	100 680	9 788	8 491	31 866	63 279	56 559	4 851	28 798	43 135
August	506 017	442 824	108 725	100 042	9 686	8 516	31 645	63 142	56 471	4 816	28 983	43 490
September	504 647	442 540	108 336	99 856	9 741	8 361	31 655	63 294	56 537	4 860	28 925	43 156
October	510 549	448 211	109 106	101 219	9 674	8 397	32 067	63 565	56 769	4 864	29 221	45 208
November	510 233	448 292	109 395	102 032	9 652	8 317	31 414	63 601	56 808	4 902	29 389	43 364
December	499 879	438 003	107 739	102 467	9 569	8 209	31 552	63 431	56 701	4 898	28 295	40 726

[1] Includes store categories not shown separately.
[2] Includes furniture, home furnishings, electronics, appliances, clothing, sporting goods, hobby, book, music, general merchandise, office supplies, stationery, and gifts.

Table 16-9. Quarterly U.S. Retail Sales: Total and E-Commerce

Year and quarter	Retail sales (millions of dollars)		E-commerce as a percent of total sales	Percent change from prior quarter		Percent change from same quarter a year ago	
	Total	E-commerce		Total sales	E-commerce sales	Total sales	E-commerce sales
NOT SEASONALLY ADJUSTED							
2017							
1st quarter	1 161 416	97 511	8.4	-10.5	-19.9	3.6	14.2
2nd quarter	1 273 412	104 407	8.2	9.6	7.1	4.4	15.8
3rd quarter	1 270 345	106 241	8.4	-0.2	1.8	4.3	15.6
4rd quarter	1 368 553	141 719	10.4	7.7	33.4	5.5	16.5
2018							
1st quarter	1 225 399	113 244	9.2	-10.5	-20.1	5.5	16.1
2nd quarter	1 341 878	120 479	9.0	9.5	6.4	5.4	15.4
3rd quarter	1 330 972	121 411	9.1	-0.8	0.8	4.8	14.3
4rd quarter	1 417 243	158 548	11.2	6.5	30.6	3.6	11.9
SEASONALLY ADJUSTED							
2005							
1st quarter	898 561	20 801	2.3	0.8	6.0	6.2	24.6
2nd quarter	916 525	22 233	2.4	2.0	6.9	7.2	26.9
3rd quarter	934 239	23 653	2.5	1.9	6.4	7.6	27.8
4th quarter	937 255	24 364	2.6	0.3	3.0	5.2	24.1
2006							
1st quarter	963 388	26 417	2.7	2.8	8.4	7.2	27.0
2nd quarter	967 539	27 367	2.8	0.4	3.6	5.6	23.1
3rd quarter	972 764	28 842	3.0	0.5	5.4	4.1	21.9
4th quarter	972 870	30 138	3.1	0.0	4.5	3.8	23.7
2007							
1st quarter	985 713	31 728	3.2	1.3	5.3	2.3	20.1
2nd quarter	994 838	33 524	3.4	0.9	5.7	2.8	22.5
3rd quarter	1 001 953	34 841	3.5	0.7	3.9	3.0	20.8
4th quarter	1 014 852	35 784	3.5	1.3	2.7	4.3	18.7
2008							
1st quarter	1 007 189	36 017	3.6	-0.8	0.7	2.2	13.5
2nd quarter	1 011 101	36 544	3.6	0.4	1.5	1.6	9.0
3rd quarter	999 159	36 292	3.6	-1.2	-0.7	-0.3	4.2
4th quarter	910 187	33 042	3.6	-8.9	-9.0	-10.3	-7.7
2009							
1st quarter	888 604	34 132	3.8	-2.4	3.3	-11.8	-5.2
2nd quarter	892 284	35 282	4.0	0.4	3.4	-11.8	-3.5
3rd quarter	912 858	37 402	4.1	2.3	6.0	-8.6	3.1
4th quarter	919 242	38 084	4.1	0.7	1.8	1.0	15.3
2010							
1st quarter	933 310	39 289	4.2	1.5	3.2	5.0	15.1
2nd quarter	949 219	41 303	4.4	1.7	5.1	6.4	17.1
3rd quarter	952 783	43 472	4.6	0.4	5.3	4.4	16.2
4rd quarter	981 429	45 046	4.6	3.0	3.6	6.8	18.3
2011							
1st quarter	1 004 675	46 936	4.7	2.4	4.2	7.6	19.5
2nd quarter	1 021 076	48 680	4.8	1.6	3.7	7.6	17.9
3rd quarter	1 029 188	49 994	4.9	0.8	2.7	8.0	15.0
4rd quarter	1 047 074	52 946	5.1	1.7	5.9	6.7	17.5
2012							
1st quarter	1 068 644	54 788	5.1	2.1	3.5	6.4	16.7
2nd quarter	1 065 153	56 009	5.3	-0.3	2.2	4.3	15.1
3rd quarter	1 073 489	58 103	5.4	0.8	3.7	4.3	16.2
4th quarter	1 089 483	60 395	5.5	1.5	3.9	4.1	14.1
2013							
1st quarter	1 108 399	61 985	5.6	1.7	2.6	3.7	13.1
2nd quarter	1 108 026	63 866	5.8	0.0	3.0	4.0	14.0
3rd quarter	1 117 927	65 744	5.9	0.9	2.9	4.1	13.2
4th quarter	1 124 171	68 222	6.1	0.6	3.8	3.2	13.0
2014							
1st quarter	1 133 004	70 425	6.2	0.8	3.2	2.2	13.6
2nd quarter	1 161 330	73 381	6.3	2.5	4.2	4.8	14.9
3rd quarter	1 169 320	75 807	6.5	0.7	3.3	4.6	15.3
4th quarter	1 170 377	77 793	6.6	0.1	2.6	4.1	14.0
2015							
1st quarter	1 160 120	80 500	6.9	-0.9	3.5	2.4	14.3
2nd quarter	1 180 338	83 314	7.1	1.7	3.5	1.6	13.5
3rd quarter	1 191 816	86 335	7.2	1.0	3.6	1.9	13.9
4th quarter	1 189 054	88 913	7.5	-0.2	3.0	1.6	14.3
2016							
1st quarter	1 188 935	92 145	7.8	0.0	3.6	2.5	14.5
2nd quarter	1 205 521	95 644	7.9	1.4	3.8	2.1	14.8
3rd quarter	1 216 704	98 858	8.1	0.9	3.4	2.1	14.5
4th quarter	1 230 389	101 402	8.2	1.1	2.6	3.5	14.0
2017							
1st quarter	1 248 299	105 387	8.4	1.5	3.9	5.0	14.4
2nd quarter	1 257 077	110 478	8.8	0.7	4.8	4.3	15.5
3rd quarter	1 270 977	114 186	9.0	1.1	3.4	4.5	15.5
4th quarter	1 303 222	118 216	9.1	2.5	3.5	5.9	16.6
2018							
1st quarter	1 309 686	122 534	9.4	0.4	3.3	4.9	16.3
2nd quarter	1 328 203	126 985	9.6	1.4	3.6	5.7	14.9
3rd quarter	1 339 375	130 068	9.7	0.8	2.4	5.3	13.8
4rd quarter	1 345 363	132 992	9.9	0.4	2.2	3.1	12.1

Table 16-10. Retail Inventories

(All retail stores; end of period, millions of dollars.)

Year and month	Not seasonally adjusted			Seasonally adjusted								
	Total	Excluding motor vehicles and parts	Motor vehicles and parts	Total	Excluding motor vehicles and parts	Motor vehicles and parts	Furniture, home furnishings, electronics, and appliances	Building materials and garden	Food and beverages	Clothing and accessories	General merchandise	
											Total	Department stores [1]
1994	299 812	211 223	88 589	304 519	218 757	85 762	20 397	24 857	28 112	29 518	56 581	41 908
1995	317 304	220 921	96 383	322 216	228 842	93 374	21 817	26 337	28 718	29 298	59 530	43 455
1996	328 160	227 888	100 272	333 255	236 015	97 240	22 253	27 447	29 658	29 778	60 591	44 124
1997	338 790	234 270	104 520	343 825	242 518	101 307	22 109	28 897	29 891	31 079	60 715	44 309
1998	351 255	245 400	105 855	356 518	253 888	102 630	22 665	30 921	30 837	32 308	61 542	43 438
1999	378 813	259 879	118 934	384 078	268 567	115 511	24 096	33 005	32 635	33 735	64 297	43 835
2000	400 394	268 982	131 412	405 931	277 829	128 102	25 622	34 269	32 108	36 753	64 929	42 667
2001	387 859	266 370	121 489	393 638	274 796	118 842	24 497	34 235	33 165	35 627	64 790	40 464
2002	409 314	271 757	137 557	415 174	280 277	134 897	25 853	36 123	32 907	37 401	66 006	38 750
2003	425 748	277 471	148 277	431 199	285 668	145 531	27 152	37 398	32 504	38 503	66 655	36 823
2004	454 848	297 561	157 287	460 335	305 941	154 394	30 043	41 593	33 433	41 417	70 969	37 230
2005	465 791	310 227	155 564	471 616	318 849	152 767	30 550	45 072	33 784	43 003	74 179	38 030
2006	480 338	324 005	156 333	486 403	332 972	153 431	30 807	47 010	34 700	47 272	75 592	37 133
2007	494 372	334 566	159 806	500 489	343 476	157 013	31 356	49 074	36 484	47 381	75 820	36 428
2008	471 330	323 360	147 970	477 148	331 531	145 617	27 772	46 496	37 273	45 958	72 777	33 090
2009	423 380	114 772	114 772	429 159	315 870	113 289	25 710	42 707	37 254	40 811	69 575	30 817
2010	449 059	320 671	128 388	454 548	327 909	126 639	28 006	40 737	38 613	41 931	73 290	30 933
2011	464 953	331 178	133 775	470 759	338 844	338 844	26 707	43 277	40 144	44 225	76 565	31 165
2012	499 533	340 900	150 025	505 316	349 021	349 021	27 524	44 492	41 029	46 618	77 926	30 035
2013	537 199	357 271	179 928	543 376	366 013	366 013	27 691	47 328	42 545	49 561	80 718	29 827
2014	554 006	366 715	187 291	560 416	375 515	375 515	27 680	48 471	43 982	50 690	81 094	28 231
2015	580 662	380 986	199 676	587 438	389 823	389 823	27 584	50 175	45 383	51 970	82 914	28 324
2016	603 836	389 192	214 644	610 966	397 803	397 803	27 509	52 082	46 830	52 190	82 213	27 253
2017	617 426	397 477	219 949	624 988	405 743	405 743	28 154	53 946	48 278	51 284	79 365	25 502
2018	643 776	404 243	239 533	651 540	412 544	412 544	27 827	57 757	48 659	51 595	80 099	23 906
2015												
January	552 117	365 637	186 480	559 477	375 433	375 433	27 667	47 995	44 183	50 754	81 222	28 394
February	556 113	369 370	186 743	561 434	377 366	377 366	27 580	48 596	44 143	50 552	81 663	28 448
March	560 941	373 061	187 880	562 637	377 285	377 285	27 635	48 555	44 849	50 896	80 582	28 024
April	567 626	375 901	191 725	568 238	380 963	380 963	27 496	48 996	44 352	51 288	82 813	28 591
May	559 254	372 235	187 019	565 838	379 546	379 546	27 296	49 000	44 403	50 832	81 988	28 540
June	562 363	373 718	188 645	570 055	381 786	381 786	27 487	49 569	44 672	51 050	82 545	28 607
July	561 091	375 517	185 574	574 677	383 675	383 675	27 515	49 465	44 922	51 043	82 552	28 432
August	564 416	381 358	183 058	576 136	385 093	385 093	27 520	49 731	45 028	51 669	82 555	28 494
September	585 665	396 977	188 688	582 187	387 722	387 722	27 711	49 928	45 325	51 910	83 080	28 999
October	611 264	415 672	195 592	584 582	389 258	389 258	27 798	50 269	45 576	51 898	83 141	28 806
November	616 738	418 504	198 234	585 103	389 823	389 823	28 338	50 418	45 535	51 930	82 955	28 558
December	580 662	380 986	199 676	587 438	389 823	389 823	27 584	50 175	45 383	51 970	82 914	28 324
2016												
January	580 862	380 559	200 303	588 122	390 279	390 279	27 450	50 260	45 522	52 148	82 838	27 980
February	585 793	382 997	202 796	590 729	391 019	391 019	27 234	50 425	45 872	52 158	82 905	28 012
March	597 261	388 342	208 919	598 576	392 511	392 511	27 538	50 693	45 757	52 022	82 878	27 804
April	597 891	387 320	210 571	598 302	392 657	392 657	27 582	50 655	45 856	52 287	82 727	27 715
May	593 459	386 319	207 140	600 449	394 116	394 116	27 691	51 007	45 975	52 406	82 961	27 612
June	595 377	386 716	208 661	602 547	394 681	394 681	27 738	50 667	46 359	52 414	82 861	27 566
July	587 332	385 920	201 412	601 838	394 690	394 690	27 749	50 782	46 387	52 443	82 693	27 904
August	592 510	391 261	201 249	604 479	394 973	394 973	27 781	51 184	46 374	51 648	82 622	27 401
September	611 229	405 581	205 648	608 563	396 745	396 745	27 668	51 298	46 657	52 164	82 682	27 288
October	631 606	421 171	210 435	605 794	395 048	395 048	27 577	51 373	46 582	51 862	82 073	27 188
November	642 265	425 319	216 946	611 169	397 102	397 102	27 679	52 027	46 585	52 263	82 061	27 151
December	603 836	389 192	214 644	610 966	397 803	397 803	27 509	52 082	46 830	52 190	82 213	27 253
2017												
January	606 387	387 524	218 863	613 158	397 002	397 002	27 592	52 170	46 755	52 005	81 551	27 190
February	610 411	388 618	221 793	614 390	396 169	396 169	27 695	52 106	46 699	51 994	80 907	26 827
March	619 436	393 914	225 522	620 043	397 814	397 814	27 675	52 274	46 792	52 097	81 325	26 923
April	616 066	390 895	225 171	615 894	396 028	396 028	27 620	52 218	46 716	51 666	80 972	26 917
May	612 160	389 405	222 755	618 771	397 105	397 105	27 917	52 290	47 200	51 701	79 811	26 341
June	616 219	391 813	224 406	622 932	399 586	399 586	27 910	52 338	47 127	51 663	80 811	26 200
July	606 633	390 424	216 209	621 925	399 343	399 343	27 938	52 792	47 207	51 462	80 213	26 115
August	614 428	397 461	216 967	626 414	401 155	401 155	27 883	52 760	47 558	51 392	79 952	25 869
September	623 608	409 338	214 270	621 499	401 000	401 000	28 036	53 076	47 378	51 378	79 892	25 736
October	647 167	428 533	218 634	622 068	402 690	402 690	27 973	53 405	47 597	51 277	79 694	25 619
November	654 186	432 239	221 947	623 781	404 380	404 380	27 897	53 665	47 970	51 261	79 629	25 477
December	617 426	397 477	219 949	624 988	405 743	405 743	28 154	53 946	48 278	51 284	79 365	25 502
2018												
January	621 188	396 942	224 246	627 547	406 055	406 055	27 916	54 133	48 471	51 270	79 572	25 622
February	627 611	399 890	227 721	630 656	407 004	407 004	27 909	54 695	48 693	51 738	79 922	25 663
March	629 334	401 795	227 539	629 418	405 417	405 417	27 949	54 853	48 111	51 413	79 269	25 115
April	631 802	402 011	229 791	631 324	406 941	406 941	27 823	55 101	47 548	51 814	79 917	25 514
May	628 049	399 957	228 092	634 341	407 571	407 571	27 639	55 570	47 890	51 460	80 289	25 580
June	627 657	399 998	227 659	634 222	407 657	407 657	27 324	56 366	47 848	51 255	80 137	25 445
July	621 536	399 132	222 404	637 581	408 412	408 412	27 626	56 711	47 952	51 178	80 454	25 327
August	629 712	405 018	224 694	641 918	408 882	408 882	27 840	57 031	48 232	51 155	80 719	25 242
September	642 722	415 649	227 073	641 781	407 848	407 848	27 857	57 098	48 048	50 809	80 432	25 274
October	672 876	437 430	235 446	647 878	411 623	411 623	28 876	57 281	48 224	51 423	81 135	25 334
November	674 759	434 365	240 394	645 145	407 524	407 524	27 447	57 252	48 226	51 067	79 380	23 989
December	643 776	404 243	239 533	651 540	412 544	412 544	27 827	57 757	48 659	51 595	80 099	23 906

[1]Excluding leased departments.

Table 16-11. Merchant Wholesalers—Sales and Inventories

(Millions of dollars.)

Year and month	Not seasonally adjusted						Seasonally adjusted					
	Sales			Inventories (current cost, end of period)			Sales			Inventories (current cost, end of period)		
	Total	Durable goods establish-ments	Nondurable goods establish-ments	Total	Durable goods establish-ments	Nondurable goods establish-ments	Total	Durable goods establish-ments	Nondurable goods establish-ments	Total	Durable goods establish-ments	Nondurable goods establish-ments
1994	1 974 899	1 037 638	937 261	222 826	139 941	82 885	1 974 899	1 037 638	937 261	221 978	141 975	80 003
1995	2 158 980	1 141 701	1 017 279	239 275	151 709	87 566	2 158 980	1 141 701	1 017 279	238 392	154 089	84 303
1996	2 284 343	1 190 342	1 094 001	241 396	154 207	87 189	2 284 343	1 190 342	1 094 001	241 058	156 683	84 375
1997	2 377 845	1 256 384	1 121 461	258 900	165 371	93 529	2 377 845	1 256 384	1 121 461	258 454	168 089	90 365
1998	2 427 120	1 306 545	1 120 575	272 575	175 994	96 581	2 427 120	1 306 545	1 120 575	272 297	178 918	93 379
1999	2 599 159	1 406 371	1 192 788	290 382	187 738	102 644	2 599 159	1 406 371	1 192 788	290 182	190 899	99 283
2000	2 814 554	1 486 673	1 327 881	309 710	198 525	111 185	2 814 554	1 486 673	1 327 881	309 191	201 797	107 394
2001	2 785 152	1 422 195	1 362 957	298 577	182 521	116 056	2 785 152	1 422 195	1 362 957	297 536	185 517	112 019
2002	2 835 528	1 421 503	1 414 025	302 715	182 150	120 565	2 835 528	1 421 503	1 414 025	301 310	185 048	116 262
2003	2 978 282	1 466 158	1 512 124	310 115	186 435	123 680	2 978 282	1 466 158	1 512 124	308 274	189 325	118 949
2004	3 330 011	1 689 588	1 640 423	341 642	213 590	128 052	3 330 011	1 689 588	1 640 423	340 128	216 704	123 424
2005	3 638 496	1 830 347	1 808 149	369 380	233 199	136 181	3 638 496	1 830 347	1 808 149	367 978	236 438	131 540
2006	3 941 257	2 005 139	1 936 118	400 297	256 731	143 566	3 941 257	2 005 139	1 936 118	398 924	260 115	138 809
2007	4 223 473	2 101 969	2 121 504	426 254	263 195	163 059	4 223 473	2 101 969	2 121 504	424 344	266 628	157 716
2008	4 524 357	2 126 715	2 397 642	446 109	280 403	165 706	4 524 357	2 126 715	2 397 642	445 529	284 176	161 353
2009	3 829 378	1 745 557	2 083 821	399 322	233 720	165 602	3 829 378	1 745 557	2 083 821	397 699	237 136	160 563
2010	4 337 359	1 999 247	2 338 112	444 639	256 969	187 670	4 337 359	1 999 247	2 338 112	442 154	260 923	181 231
2011	4 885 076	2 229 683	2 655 393	489 804	286 171	203 633	4 885 076	2 229 683	2 655 393	488 061	290 810	197 251
2012	5 208 023	2 389 576	2 818 447	525 855	311 262	214 593	5 208 023	2 389 576	2 818 447	524 005	316 609	207 396
2013	5 370 550	2 461 025	2 909 525	546 197	326 427	219 770	5 370 550	2 461 025	2 909 525	545 175	332 288	212 887
2014	5 564 180	2 556 641	3 007 539	578 644	348 845	229 799	5 564 180	2 556 641	3 007 539	577 344	355 111	222 233
2015	5 292 436	2 520 142	2 772 294	586 037	349 537	236 500	5 292 436	2 520 142	2 772 294	585 167	355 811	229 356
2016	5 228 485	2 506 047	2 722 438	597 647	349 858	247 789	5 228 485	2 506 047	2 722 438	596 302	355 957	240 345
2017	5 557 898	2 677 535	2 880 363	617 230	365 808	251 422	5 557 898	2 677 535	2 880 363	615 722	371 854	243 868
2018	5 936 960	2 868 963	3 067 997	661 301	405 489	255 812	5 936 960	2 868 963	3 067 997	659 673	411 736	247 937
2015												
January	413 973	197 571	216 402	583 261	355 016	228 245	444 727	215 158	229 569	577 906	356 831	221 075
February	391 355	182 035	209 320	584 499	358 106	226 393	438 613	207 491	231 122	581 033	358 532	222 501
March	451 779	217 700	234 079	587 258	358 739	228 519	438 588	210 152	228 436	582 156	360 029	222 127
April	452 140	212 790	239 350	585 573	360 193	225 380	444 065	211 203	232 862	583 023	359 542	223 481
May	442 123	203 494	238 629	582 325	361 393	220 932	444 606	210 793	233 813	586 401	360 869	225 532
June	468 527	222 633	245 894	582 839	361 184	221 655	445 881	209 513	236 368	589 112	361 165	227 947
July	451 454	212 515	238 939	583 227	363 243	219 984	445 188	210 766	234 422	588 202	360 251	227 951
August	434 373	207 160	227 213	579 948	361 853	218 095	440 141	208 516	231 625	588 970	360 514	228 456
September	451 506	220 133	231 373	587 519	361 018	226 501	440 299	210 012	230 287	590 664	359 138	231 526
October	461 893	220 815	241 078	595 394	361 203	234 191	440 257	208 658	231 599	590 318	359 381	230 937
November	424 472	203 846	220 626	591 200	357 829	233 371	434 345	207 957	226 388	586 797	357 428	229 369
December	448 841	219 450	229 391	586 037	349 537	236 500	429 277	206 839	222 438	585 167	355 811	229 356
2016												
January	382 121	181 104	201 017	590 368	352 214	238 154	425 216	204 356	220 860	584 933	354 077	230 856
February	391 226	188 002	203 224	583 401	352 000	231 401	421 745	206 074	215 671	580 291	352 295	227 996
March	451 567	220 011	231 556	586 918	350 560	236 358	427 311	207 620	219 691	582 101	351 413	230 688
April	424 345	201 513	222 832	588 556	352 902	235 654	429 169	205 811	223 358	586 495	352 448	234 047
May	438 996	205 558	233 438	582 771	353 420	229 351	429 343	206 568	222 775	587 612	353 082	234 530
June	460 282	223 021	237 261	581 912	351 990	229 922	437 388	209 696	227 692	587 808	352 049	235 759
July	416 995	199 033	217 962	582 305	355 110	227 195	435 777	209 750	226 027	587 430	352 021	235 409
August	457 959	219 319	238 640	577 120	353 826	223 294	438 035	207 873	230 162	585 805	352 404	233 401
September	448 571	218 491	230 080	584 302	352 624	231 678	437 045	207 589	229 456	586 870	350 804	236 066
October	450 006	214 751	235 255	591 677	352 413	239 264	441 182	209 404	231 778	587 732	350 818	236 914
November	446 566	212 606	233 960	597 026	354 452	242 574	442 646	210 081	232 565	592 152	353 993	238 159
December	459 851	222 638	237 213	597 647	349 858	247 789	452 395	215 947	236 448	596 302	355 957	240 345
2017												
January	422 331	199 089	223 242	600 798	353 481	247 317	456 773	218 302	238 471	594 672	355 266	239 406
February	406 908	190 749	216 159	599 862	355 451	244 411	458 056	218 193	239 863	595 588	355 618	239 970
March	481 453	231 614	249 839	602 444	357 305	245 139	458 530	218 719	239 811	597 272	358 173	239 099
April	438 347	208 945	229 402	597 405	359 292	238 113	457 328	221 064	236 264	595 409	359 092	236 317
May	475 246	225 154	250 092	593 612	361 545	232 067	452 124	220 021	232 103	598 344	361 402	236 942
June	480 099	234 192	245 907	596 562	362 507	234 055	456 084	219 956	236 128	602 064	362 678	239 386
July	437 699	209 145	228 554	599 722	368 683	231 039	456 796	220 478	236 318	604 987	365 448	239 539
August	486 731	237 239	249 492	600 895	370 295	230 600	463 017	223 966	239 051	609 993	368 799	241 194
September	467 501	230 272	237 229	610 217	371 863	238 354	468 025	225 833	242 192	613 171	370 149	243 022
October	495 938	242 070	253 868	612 909	371 027	241 882	472 279	228 562	243 717	610 076	369 388	240 688
November	484 580	234 968	249 612	618 376	371 458	246 918	480 951	232 170	248 781	614 141	371 085	243 056
December	481 065	234 098	246 967	617 230	365 808	251 422	488 119	234 232	253 887	615 722	371 854	243 868
2018												
January	457 074	216 771	240 303	628 135	370 173	257 962	481 495	231 509	249 986	621 149	371 979	249 170
February	431 140	205 793	225 347	631 354	375 722	255 632	485 732	235 123	250 609	625 490	375 747	249 743
March	500 911	244 350	256 561	633 911	378 242	255 669	488 298	236 874	251 424	627 707	379 170	248 537
April	480 831	229 368	251 463	630 164	379 988	250 176	489 732	236 665	253 067	627 672	379 751	247 921
May	529 591	246 561	283 030	624 627	381 169	243 458	501 154	239 680	261 474	630 056	381 068	248 988
June	510 173	246 567	263 606	623 849	383 942	239 907	498 754	239 441	259 313	630 059	384 202	245 857
July	491 711	234 581	257 130	627 681	391 031	236 650	498 589	239 556	259 033	633 725	387 432	246 293
August	525 757	256 583	269 174	630 379	392 947	237 432	501 363	242 415	258 948	640 809	391 534	249 275
September	487 449	239 978	247 471	640 563	396 887	243 676	500 571	242 270	258 301	644 299	395 234	249 065
October	537 910	262 839	275 071	652 102	403 343	248 759	499 205	241 537	257 668	649 922	401 681	248 241
November	498 433	243 500	254 933	656 840	404 968	251 872	495 399	240 649	254 750	652 442	404 650	247 792
December	485 980	242 072	243 908	661 301	405 489	255 812	491 945	242 024	249 921	659 673	411 736	247 937

. . . = Not available.

Table 16-12. Manufacturing and Trade Sales and Inventories

Year and month	Sales, billions of dollars					Inventories, billions of dollars, end of period, seasonally adjusted				Ratios, inventories to sales, seasonally adjusted [1]			
	Not seasonally adjusted, total	Seasonally adjusted				Total	Manufac-turing	Retail trade	Merchant wholesalers	Total	Manufac-turing	Retail trade	Merchant wholesalers
		Total	Manufac-turing	Retail trade	Merchant wholesalers								
1998	8 908.7	8 906.4	3 897.4	2 582.7	2 426.3	1 077.2	448.4	356.5	272.3	1.44	1.39	1.62	1.32
1999	9 434.1	9 431.7	4 033.3	2 801.4	2 597.0	1 137.3	463.0	384.1	290.2	1.40	1.35	1.59	1.30
2000	10 006.4	9 998.9	4 202.4	2 979.4	2 817.1	1 195.9	480.7	406.0	309.2	1.41	1.35	1.59	1.29
2001	9 817.9	9 819.3	3 971.2	3 062.2	2 785.9	1 118.6	427.4	393.7	297.5	1.42	1.38	1.58	1.32
2002	9 878.8	9 881.5	3 917.4	3 129.6	2 834.5	1 139.5	423.0	415.2	301.3	1.36	1.29	1.55	1.26
2003	10 256.4	10 255.7	4 016.6	3 261.6	2 977.4	1 147.8	408.3	431.2	308.3	1.34	1.25	1.56	1.22
2004	11 112.0	11 071.7	4 295.3	3 460.9	3 315.6	1 241.7	441.2	460.4	340.1	1.30	1.19	1.56	1.17
2005	12 069.9	12 073.8	4 743.7	3 686.6	3 643.6	1 314.3	474.6	471.7	368.0	1.27	1.17	1.51	1.17
2006	12 828.4	12 839.3	5 016.8	3 876.6	3 946.0	1 408.8	523.5	486.4	398.9	1.28	1.20	1.49	1.17
2007	13 538.1	13 538.6	5 320.1	3 997.4	4 221.1	1 487.6	562.7	500.6	424.3	1.28	1.22	1.49	1.17
2008	13 928.7	13 882.4	5 448.2	3 927.6	4 506.6	1 466.0	543.3	477.2	445.5	1.31	1.26	1.52	1.20
2009	11 865.6	11 869.5	4 427.8	3 613.0	3 828.6	1 332.3	505.5	429.2	397.7	1.38	1.39	1.47	1.29
2010	13 066.7	13 064.2	4 914.4	3 816.7	4 333.0	1 451.0	554.3	454.5	442.2	1.27	1.28	1.39	1.15
2011	14 479.9	14 489.7	5 496.3	4 102.0	4 891.3	1 565.7	606.8	470.8	488.1	1.26	1.29	1.35	1.15
2012	15 207.0	15 187.6	5 691.8	4 296.8	5 199.0	1 654.6	624.9	505.7	524.0	1.28	1.30	1.38	1.17
2013	15 638.7	15 636.0	5 811.8	4 458.5	5 365.7	1 719.1	630.3	543.6	545.2	1.29	1.29	1.41	1.19
2014	16 091.2	16 078.2	5 887.4	4 634.0	5 556.7	1 778.5	640.4	560.7	577.3	1.31	1.31	1.43	1.22
2015	15 537.4	15 527.0	5 519.6	4 721.3	5 286.0	1 808.8	635.8	587.9	585.2	1.39	1.39	1.46	1.33
2016	15 439.5	15 403.5	5 344.7	4 841.5	5 217.3	1 839.6	631.2	612.0	596.3	1.42	1.41	1.49	1.35
2017	16 236.5	16 261.0	5 611.1	5 081.9	5 568.1	1 901.6	659.4	626.5	615.7	1.38	1.37	1.47	1.30
2018	17 252.0	17 251.5	5 997.6	5 321.7	5 932.2	1 995.2	682.7	652.9	659.7	1.36	1.35	1.44	1.29
2014													
January	1 225.6	1 311.2	486.2	373.0	452.0	1 726.6	632.4	546.1	548.0	1.32	1.30	1.46	1.21
February	1 201.7	1 326.3	494.0	378.0	454.3	1 734.6	637.3	545.6	551.6	1.31	1.29	1.44	1.21
March	1 362.9	1 337.7	492.6	382.0	463.1	1 740.1	638.2	545.7	556.2	1.30	1.30	1.43	1.20
April	1 356.4	1 345.6	402.0	386.5	466.2	1 740.9	640.1	548.1	560.8	1.30	1.30	1.42	1.20
May	1 395.5	1 347.4	492.7	386.9	467.8	1 756.1	644.4	549.6	562.1	1.30	1.31	1.42	1.20
June	1 376.8	1 348.9	494.4	387.9	466.6	1 759.3	645.1	551.2	563.0	1.30	1.30	1.42	1.21
July	1 361.5	1 355.4	497.6	388.2	469.5	1 765.8	645.2	556.7	563.9	1.30	1.30	1.43	1.20
August	1 371.3	1 355.5	496.4	391.3	467.8	1 769.7	646.0	555.9	567.8	1.31	1.30	1.42	1.21
September	1 362.9	1 350.1	493.4	389.8	466.9	1 770.9	646.4	556.3	568.2	1.31	1.31	1.43	1.22
October	1 390.4	1 342.6	488.1	390.6	463.9	1 776.8	645.4	559.5	571.9	1.32	1.32	1.43	1.23
November	1 289.5	1 335.9	482.0	391.5	462.3	1 779.5	645.4	557.3	576.7	1.33	1.34	1.42	1.25
December	1 396.7	1 321.7	477.1	388.2	456.3	1 778.5	640.4	560.7	577.3	1.35	1.34	1.44	1.27
2015													
January	1 188.1	1 292.8	462.6	385.5	444.7	1 777.4	639.5	559.9	577.9	1.37	1.38	1.45	1.30
February	1 163.0	1 285.9	463.5	383.9	438.6	1 783.7	640.3	562.4	581.0	1.39	1.38	1.47	1.32
March	1 337.0	1 296.8	467.4	390.8	438.6	1 785.4	640.6	562.6	582.2	1.38	1.37	1.44	1.33
April	1 308.4	1 300.3	465.8	390.5	444.1	1 793.7	641.5	569.2	583.0	1.38	1.38	1.46	1.31
May	1 319.9	1 302.7	463.7	394.5	444.6	1 795.2	641.9	566.9	586.4	1.38	1.38	1.44	1.32
June	1 361.3	1 308.3	467.0	395.4	445.9	1 805.8	645.7	571.0	589.1	1.38	1.38	1.44	1.32
July	1 311.7	1 308.0	464.9	398.0	445.2	1 808.1	644.6	575.2	588.2	1.38	1.39	1.45	1.32
August	1 308.3	1 297.9	460.1	397.6	440.1	1 807.8	641.9	576.9	589.0	1.39	1.40	1.45	1.34
September	1 310.9	1 293.9	457.5	396.2	440.3	1 812.8	639.2	582.9	590.7	1.40	1.40	1.47	1.34
October	1 319.5	1 289.1	453.7	395.2	440.3	1 814.7	638.9	585.4	590.3	1.41	1.41	1.48	1.34
November	1 256.4	1 282.3	451.4	396.5	434.3	1 810.0	637.6	585.6	586.8	1.41	1.41	1.48	1.35
December	1 352.9	1 268.8	442.2	397.4	429.3	1 808.8	635.8	587.9	585.2	1.43	1.44	1.48	1.36
2016													
January	1 132.2	1 261.4	441.1	395.1	425.2	1 805.5	631.6	589.0	584.9	1.43	1.43	1.49	1.38
February	1 178.3	1 258.2	439.6	396.9	421.7	1 800.8	628.5	592.0	580.3	1.43	1.43	1.49	1.38
March	1 328.4	1 265.6	441.3	397.0	427.3	1 808.8	628.5	598.3	582.1	1.43	1.42	1.51	1.36
April	1 259.1	1 270.6	442.3	399.1	429.2	1 814.0	628.2	599.3	586.5	1.43	1.42	1.50	1.37
May	1 303.8	1 272.9	442.3	401.2	429.3	1 818.1	628.9	601.9	587.6	1.43	1.42	1.50	1.37
June	1 345.5	1 289.6	447.0	405.2	437.4	1 819.7	628.1	603.8	587.8	1.41	1.41	1.49	1.34
July	1 245.0	1 284.8	444.0	405.1	435.8	1 817.5	627.5	602.5	587.4	1.41	1.41	1.49	1.35
August	1 339.9	1 287.1	444.2	404.9	438.0	1 820.1	628.5	605.9	585.8	1.41	1.41	1.50	1.34
September	1 311.5	1 291.5	447.7	406.7	437.0	1 822.9	626.6	609.4	586.9	1.41	1.40	1.50	1.34
October	1 300.8	1 298.1	448.0	408.9	441.2	1 821.7	627.1	606.9	587.7	1.40	1.40	1.48	1.33
November	1 299.7	1 298.4	447.6	408.1	442.6	1 834.4	630.1	612.2	592.2	1.41	1.41	1.50	1.34
December	1 395.4	1 325.2	459.4	413.3	452.4	1 839.6	631.2	612.0	596.3	1.39	1.37	1.48	1.32
2017													
January	1 213.3	1 334.0	460.0	417.2	456.8	1 843.0	633.1	615.3	594.7	1.38	1.38	1.47	1.30
February	1 200.9	1 333.6	460.5	415.1	458.1	1 846.7	634.2	616.9	595.6	1.38	1.38	1.49	1.30
March	1 397.7	1 335.8	461.2	416.0	458.5	1 851.7	634.4	620.0	597.3	1.39	1.38	1.49	1.30
April	1 297.4	1 335.3	458.9	419.1	457.3	1 848.7	635.7	617.7	595.4	1.38	1.39	1.47	1.30
May	1 391.4	1 332.4	462.6	417.6	452.1	1 855.5	636.0	621.1	598.3	1.39	1.37	1.49	1.32
June	1 397.6	1 338.7	462.7	419.9	456.1	1 863.3	636.4	624.8	602.1	1.39	1.38	1.49	1.32
July	1 298.0	1 340.4	463.0	420.6	456.8	1 867.8	639.2	623.6	605.0	1.39	1.38	1.48	1.32
August	1 411.3	1 351.8	467.6	421.2	463.0	1 881.1	642.3	628.7	610.0	1.39	1.37	1.49	1.32
September	1 368.5	1 368.9	470.3	430.6	468.0	1 883.7	647.6	623.0	613.2	1.38	1.38	1.45	1.31
October	1 403.1	1 379.6	475.1	432.2	472.3	1 884.3	650.9	623.3	610.1	1.37	1.37	1.44	1.29
November	1 404.2	1 399.5	484.0	434.6	481.0	1 893.4	654.3	624.9	614.1	1.35	1.35	1.44	1.28
December	1 453.2	1 411.0	485.1	437.8	488.1	1 901.6	659.4	626.5	615.7	1.35	1.36	1.43	1.26
2018													
January	1 302.4	1 405.9	489.1	435.4	481.5	1 912.6	662.0	629.5	621.1	1.36	1.35	1.45	1.29
February	1 270.3	1 413.4	490.5	437.2	485.7	1 923.6	664.6	633.6	625.5	1.36	1.35	1.45	1.29
March	1 468.9	1 418.6	493.2	437.1	488.3	1 922.0	664.7	629.6	627.7	1.35	1.35	1.44	1.29
April	1 395.4	1 422.5	493.3	439.5	489.7	1 927.6	667.7	632.2	627.7	1.36	1.35	1.44	1.28
May	1 512.7	1 441.7	497.1	443.5	501.2	1 934.8	669.8	635.0	630.1	1.34	1.35	1.43	1.26
June	1 484.7	1 444.4	501.3	444.4	498.8	1 934.3	669.6	634.6	630.1	1.34	1.34	1.43	1.26
July	1 421.0	1 447.0	501.7	446.7	498.6	1 948.2	676.3	638.1	633.7	1.35	1.35	1.43	1.27
August	1 516.3	1 452.2	504.4	446.4	501.4	1 958.9	676.0	642.1	640.8	1.35	1.34	1.44	1.28
September	1 427.6	1 454.3	507.4	446.3	500.6	1 967.6	680.3	643.0	644.3	1.35	1.34	1.44	1.29
October	1 511.6	1 458.2	508.0	451.0	499.2	1 980.5	682.5	648.1	649.9	1.36	1.34	1.44	1.30
November	1 461.2	1 452.9	506.3	451.2	495.4	1 980.4	682.4	645.6	652.4	1.36	1.35	1.43	1.32
December	1 479.9	1 440.2	505.2	443.1	491.9	1 995.2	682.7	652.9	659.7	1.39	1.35	1.47	1.34

[1] Annual data are averages of monthly ratios.

Table 16-13. Real Manufacturing and Trade Sales and Inventories

(Billions of chained [2012] dollars, ratios; seasonally adjusted; annual sales figures are averages of seasonally adjusted monthly data.)

NIPA Tables 1BU, 2BU, 3BU

Year and month	Sales, monthly average				Inventories, end of period				Ratios, end-of-period inventories to monthly average sales			
	Total	Manufac- turing	Retail trade	Merchant wholesalers	Total	Manufac- turing	Retail trade	Merchant wholesalers	Total	Manufac- turing	Retail trade	Merchant wholesalers
2000	1 127.1	501.2	304.7	325.9	1 580.9	677.4	479.2	429.8	1.40	1.35	1.57	1.32
2001	1 111.0	473.4	311.7	328.0	1 525.2	644.5	466.5	418.6	1.37	1.36	1.50	1.28
2002	1 128.4	470.9	320.6	338.1	1 558.3	642.3	496.8	419.7	1.38	1.36	1.55	1.24
2003	1 150.6	471.9	332.8	345.9	1 570.5	631.6	515.8	421.3	1.37	1.34	1.55	1.22
2004	1 195.0	481.7	346.0	367.0	1 631.0	638.3	543.3	446.3	1.37	1.33	1.57	1.22
2005	1 248.2	503.5	358.9	385.9	1 689.2	665.4	549.3	472.6	1.35	1.32	1.53	1.23
2006	1 283.1	509.8	370.2	402.9	1 742.7	687.6	560.5	493.2	1.36	1.35	1.51	1.22
2007	1 311.3	522.0	375.4	414.0	1 781.6	709.1	568.2	502.9	1.36	1.36	1.51	1.22
2008	1 261.8	494.7	358.7	408.8	1 741.2	692.1	535.3	513.1	1.38	1.40	1.49	1.26
2009	1 144.5	428.0	338.5	377.6	1 604.0	662.5	478.7	462.6	1.40	1.55	1.41	1.23
2010	1 198.7	449.8	349.0	399.9	1 663.9	687.1	493.1	483.7	1.39	1.53	1.41	1.21
2011	1 238.6	465.5	359.8	413.3	1 709.4	711.7	494.6	503.1	1.38	1.53	1.38	1.22
2012	1 279.7	473.9	371.7	434.1	1 784.0	730.8	522.1	531.0	1.39	1.54	1.41	1.22
2013	1 319.9	483.2	387.1	449.6	1 878.7	758.4	558.9	561.2	1.42	1.57	1.44	1.25
2014	1 358.6	486.2	404.7	467.8	1 947.2	776.9	574.5	595.7	1.43	1.60	1.42	1.27
2015	1 389.0	481.4	428.5	479.4	2 039.9	815.3	602.2	622.2	1.47	1.69	1.41	1.30
2016	1 411.5	477.1	448.6	486.1	2 068.3	814.6	627.4	624.6	1.47	1.71	1.40	1.29
2017	1 450.9	483.5	465.8	502.0	2 108.0	824.8	640.8	640.4	1.45	1.71	1.38	1.28
2018	1 484.0	494.4	482.1	508.1	2 161.8	826.8	663.2	669.0	1.46	1.67	1.38	1.32
2014												
January	1 328.4	485.2	388.2	455.0	1 884.5	759.7	561.1	563.5	1.42	1.57	1.45	1.24
February	1 340.0	490.5	394.2	455.3	1 890.8	763.0	561.7	566.0	1.41	1.56	1.43	1.24
March	1 350.7	486.8	398.7	465.3	1 893.9	762.9	561.8	569.1	1.40	1.57	1.41	1.22
April	1 352.6	484.1	402.6	466.0	1 901.1	765.2	563.0	572.8	1.41	1.58	1.40	1.23
May	1 356.3	484.7	403.5	468.3	1 908.1	770.0	564.2	573.8	1.41	1.59	1.40	1.23
June	1 355.8	486.4	404.1	465.5	1 912.5	772.1	565.0	575.3	1.41	1.59	1.40	1.24
July	1 364.4	489.0	404.9	470.7	1 921.5	773.3	570.7	577.3	1.41	1.58	1.41	1.23
August	1 367.0	487.6	409.2	470.4	1 928.1	775.7	569.8	582.5	1.41	1.59	1.39	1.24
September	1 366.8	485.9	408.5	472.6	1 930.6	777.6	569.2	583.7	1.41	1.60	1.39	1.24
October	1 368.6	484.5	410.6	473.7	1 938.6	777.9	572.5	588.1	1.42	1.61	1.39	1.24
November	1 372.2	483.3	415.3	473.9	1 944.8	779.5	571.7	593.5	1.42	1.61	1.38	1.25
December	1 380.2	486.9	416.3	477.2	1 947.2	776.9	574.5	595.7	1.41	1.60	1.38	1.25
2015												
January	1 383.6	482.4	421.7	479.8	1 954.1	779.9	575.2	598.8	1.41	1.62	1.36	1.25
February	1 376.6	483.7	419.0	474.2	1 967.7	786.4	577.2	604.1	1.43	1.63	1.38	1.27
March	1 384.0	486.2	423.4	474.7	1 975.2	790.7	578.6	605.9	1.43	1.63	1.37	1.28
April	1 393.4	486.4	425.0	482.3	1 986.7	795.7	583.6	607.3	1.43	1.64	1.37	1.26
May	1 382.0	478.9	426.0	477.3	1 991.5	798.7	582.2	610.9	1.44	1.67	1.37	1.28
June	1 383.8	481.8	426.0	476.2	2 004.3	805.7	585.2	613.9	1.45	1.67	1.37	1.29
July	1 386.9	480.7	429.1	477.4	2 009.9	807.1	589.3	613.9	1.45	1.68	1.37	1.29
August	1 388.3	479.5	430.6	478.5	2 016.1	807.8	591.6	616.9	1.45	1.69	1.37	1.29
September	1 402.9	482.8	433.5	487.0	2 027.0	809.3	597.0	620.5	1.45	1.68	1.38	1.27
October	1 398.4	479.9	432.8	486.2	2 034.6	812.7	599.6	622.2	1.46	1.69	1.39	1.28
November	1 394.9	479.7	435.4	480.0	2 035.0	814.3	599.8	620.9	1.46	1.70	1.38	1.29
December	1 393.4	474.9	439.3	479.4	2 039.9	815.3	602.2	622.2	1.46	1.72	1.37	1.30
2016												
January	1 395.6	477.4	439.1	479.3	2 041.8	814.3	603.9	623.2	1.46	1.71	1.38	1.30
February	1 404.6	479.0	446.6	479.1	2 041.1	814.3	606.1	620.3	1.45	1.70	1.36	1.30
March	1 407.6	479.4	443.9	484.6	2 049.8	816.6	610.9	621.7	1.46	1.70	1.38	1.28
April	1 404.3	477.6	443.9	483.1	2 053.9	817.0	612.0	624.1	1.46	1.71	1.38	1.29
May	1 398.6	474.2	444.8	479.8	2 054.6	816.2	614.6	622.8	1.47	1.72	1.38	1.30
June	1 406.2	475.0	448.5	482.8	2 053.0	814.8	616.6	620.5	1.46	1.72	1.38	1.29
July	1 406.3	472.3	450.4	483.9	2 048.9	812.7	616.2	618.9	1.46	1.72	1.37	1.28
August	1 415.6	474.6	450.9	490.6	2 051.5	813.4	618.6	618.2	1.45	1.71	1.37	1.26
September	1 419.8	477.2	452.6	490.4	2 055.6	811.9	623.8	618.2	1.45	1.70	1.38	1.26
October	1 419.9	475.4	452.7	492.1	2 053.3	812.5	621.8	617.5	1.45	1.71	1.37	1.26
November	1 420.7	476.4	452.6	492.0	2 063.5	815.0	625.7	621.2	1.45	1.71	1.38	1.26
December	1 438.6	486.7	457.1	495.1	2 068.3	814.6	627.4	624.6	1.44	1.67	1.37	1.26
2017												
January	1 435.0	483.2	457.7	494.4	2 070.2	814.0	632.1	622.1	1.44	1.69	1.38	1.26
February	1 432.3	481.7	457.8	493.2	2 070.3	812.2	632.8	623.2	1.45	1.69	1.38	1.26
March	1 433.2	480.8	458.1	494.6	2 072.7	810.3	635.8	624.4	1.45	1.69	1.39	1.26
April	1 428.9	476.2	461.7	491.4	2 067.9	810.1	633.3	622.3	1.45	1.70	1.37	1.27
May	1 435.6	480.7	461.7	493.6	2 073.1	809.3	636.1	625.3	1.44	1.68	1.38	1.27
June	1 441.4	480.0	464.6	497.3	2 082.1	809.9	639.7	629.9	1.44	1.69	1.38	1.27
July	1 447.5	480.9	465.5	501.5	2 088.9	813.2	638.9	634.3	1.44	1.69	1.37	1.27
August	1 454.5	483.3	464.3	507.4	2 099.1	816.4	642.0	638.3	1.44	1.69	1.38	1.26
September	1 463.1	483.0	471.1	509.6	2 099.9	820.4	638.0	639.4	1.44	1.70	1.35	1.26
October	1 468.5	486.4	473.6	509.0	2 098.1	821.4	639.0	635.5	1.43	1.69	1.35	1.25
November	1 481.1	492.7	475.8	513.0	2 103.9	822.4	640.7	638.7	1.42	1.67	1.35	1.25
December	1 489.4	493.3	477.5	519.1	2 108.0	824.8	640.8	640.4	1.42	1.67	1.34	1.23
2018												
January	1 473.7	494.6	475.8	503.7	2 115.0	825.2	643.5	644.2	1.44	1.67	1.35	1.28
February	1 476.2	494.0	476.3	506.4	2 122.3	825.0	647.4	647.7	1.44	1.67	1.36	1.28
March	1 478.6	495.2	476.3	507.5	2 119.3	823.1	645.6	648.3	1.43	1.66	1.36	1.28
April	1 475.5	492.2	478.1	505.7	2 119.8	824.1	647.5	645.9	1.44	1.68	1.35	1.28
May	1 482.0	489.9	482.1	510.6	2 120.6	823.9	648.7	645.6	1.43	1.68	1.35	1.26
June	1 480.7	492.2	481.0	508.0	2 117.2	821.7	649.1	643.9	1.43	1.67	1.35	1.27
July	1 482.5	491.8	483.2	508.2	2 128.1	826.7	652.7	646.3	1.44	1.68	1.35	1.27
August	1 489.2	494.4	483.5	511.9	2 134.2	825.0	654.6	652.1	1.43	1.67	1.35	1.27
September	1 491.4	496.8	484.3	510.8	2 139.1	828.2	653.9	654.5	1.43	1.67	1.35	1.28
October	1 488.1	493.7	489.1	505.9	2 149.2	828.8	658.8	659.0	1.44	1.68	1.35	1.30
November	1 494.4	496.0	491.9	507.3	2 147.8	827.0	656.8	661.4	1.44	1.67	1.34	1.30
December	1 496.0	502.0	483.3	511.0	2 161.8	826.8	663.2	669.0	1.45	1.65	1.37	1.31

Table 16-14. Selected Services—Quarterly Estimated Revenue for Employer Firms

(Millions of dollars.)

2012 NAICS code	Kind of business	4th quarter 2017	1st quarter 2018	2rd quarter 2018	3rd quarter 2018	4th quarter 2018	1st quarter 2019
SEASONALLY ADJUSTED							
51	**Information**	390 301	398 416	408 804	416 653	421 540	425 679
5112	Software publishers	56 285	61 729	64 338	65 717	66 040	71 246
512	Motion picture and sound recording industries	25 911	26 886	28 664	28 300	27 593	27 150
54	**Professional, Scientific, and Technical Services**	469 116	476 608	486 307	491 518	494 957	496 247
5411	Legal services	79 521	78 801	82 683	82 972	82 901	84 437
5412	Accounting, tax preparation, bookkeeping, and payroll services	43 634	44 539	43 644	44 974	45 778	45 751
56 pt	**Administrative and Support and Waste Management and Remediation Services**	236 955	240 661	245 965	252 540	257 255	256 184
5613	Employment services	95 189	96 957	101 502	106 042	107 431	107 635
5615	Travel arrangement and reservation services	12 019	12 062	12 482	12 625	12 161	12 460
562	Waste management and remediation services	24 442	24 859	24 847	25 456	26 052	26 413
622	**Hospitals**	285 162	283 693	288 550	293 394	290 646	300 633
NOT SEASONALLY ADJUSTED							
51	**Information**	407 865	390 846	405 125	409 987	440 088	417 165
511	Publishing industries (except Internet)	86 073	81 887	86 869	86 034	95 375	90 440
51111	Newspaper publishers	6 923	5 961	6 157	6 022	6 596	5 836
51112	Periodical publishers	7 413	6 559	6 765	6 692	7 120	6 211
5111 pt	Book, directory and mailing list, and other publishers	10 724	8 811	10 252	11 217	10 072	8 501
5112	Software publishers	61 013	60 556	63 695	62 103	71 587	69 892
512	Motion picture and sound recording industries	27 362	26 752	28 320	27 225	29 111	26 933
515	Broadcasting (except Internet)	43 569	41 391	41 582	41 755	46 659	42 007
5151	Radio and television broadcasting	21 639	20 031	19 913	20 506	24 240	20 794
5152	Cable and other subscription programming	21 930	21 360	21 669	21 249	22 419	21 213
517	Telecommunications [1]	157 772	152 542	154 001	157 664	161 043	156 309
5171	Wired telecommunications carriers	78 559	77 139	77 446	78 127	79 005	77 491
5172	Wireless telecommunications carriers (except satellite)	67 848	63 812	64 740	67 127	69 563	66 454
54	**Professional, Scientific, and Technical Services** [1]	478 029	467 552	490 684	487 094	504 361	486 818
5411	Legal services	88 984	70 369	81 939	82 557	92 932	75 402
5412	Accounting, tax preparation, bookkeeping, and payroll services	37 831	52 734	46 350	39 757	39 735	54 307
5413	Architectural, engineering, and related services	81 796	84 573	85 419	86 365	83 430	83 108
5415	Computer systems design and related services	103 518	102 324	107 836	110 624	114 630	110 205
5416	Management, scientific, and technical consulting services	68 201	65 305	67 305	68 824	70 581	67 111
5417	Scientific research and development services	42 293	40 010	43 999	43 658	46 236	44 377
5418	Advertising and related services	27 423	24 667	26 670	26 149	27 845	24 019
56	**Administrative and Support and Waste Management and Remediation Services** [1]	239 738	233 724	246 950	255 436	260 866	248 628
561	Administrative and support services	215 247	210 083	221 705	229 242	234 710	223 483
5613	Employment services	99 472	96 181	98 558	105 300	112 265	106 451
5615	Travel arrangement and reservation services	11 418	11 604	12 906	13 357	11 504	11 962
562	Waste management and remediation services	24 491	23 641	25 245	26 194	26 156	25 145
61	**Educational Services**	16 914	16 992	18 330	17 950	18 061	18 236
62	**Health Care and Social Asistance**	653 530	647 088	661 614	661 383	677 469	678 592
71	**Arts, Entertainment, and Recreation**	67 699	60 622	69 412	74 735	71 822	65 904
81	**Other services (Except Public Administration)**	151 955	131 013	133 283	139 133	146 850	142 439

[1]Includes components not shown separately.

NOTES AND DEFINITIONS, CHAPTER 16

TABLE 16-1

Construction Put in Place

SOURCE: U.S. DEPARTMENT OF COMMERCE, CENSUS BUREAU

The Census Bureau's estimates of the value of new construction put in place are intended to provide monthly estimates of the total dollar value of construction work done in the United States.

The United States Code, Title 13, authorizes the Survey of Construction and provides for voluntary responses. Construction statistics are collected from building permits, housing starts and housing completions.

Definitions and notes on the data

The estimates cover all construction work done each month on new private residential and nonresidential buildings and structures, public construction, and improvements to existing buildings and structures. Included are the cost of labor, materials, and equipment rental; cost of architectural and engineering work; overhead costs assigned to the project; interest and taxes paid during construction; and contractor's profits.

The total value put in place for a given period is the sum of the value of work done on all projects underway during this period, regardless of when work on each individual project was started or when payment was made to the contractors. For some categories, estimates are derived by distributing the total construction cost of the project by means of historic construction progress patterns. Published estimates represent payments made during a period for some categories.

The statistics on the value of construction put in place result from direct measurement and indirect estimation. A series results from direct measurement when it is based on reports of the actual value of construction progress or construction expenditures obtained in a complete census or a sample survey. All other series are developed by indirect estimation using related construction statistics. On an annual basis, estimates for series directly measured monthly, quarterly, or annually accounted for about 71 percent of total construction in 1998 (private multifamily residential, private residential improvements, private nonresidential buildings, farm nonresidential construction, public utility construction, all other private construction, and virtually all of public construction). On a monthly basis, directly measured data are available for about 55 percent of the value in place estimates.

Beginning in 1993, the Construction Expenditures Branch of the Census Bureau's Manufacturing and Construction Division began collecting these data using a new classification system, which bases project types on their end usage instead of on building/nonbuilding types. Data collection on this system for federal construction began in January 2002.

With the changes in project classifications, data presented in these tables for 1993 to date are not directly comparable with data for previous years, except at aggregate levels. For that reason, *Business Statistics* shows earlier historical data only at these aggregate levels. Although some categories, such as lodging, office, education, and religion, have the same names as categories in previously published data, there have been changes within the classifications that make these values non-comparable. For example, private medical office buildings were classified as "office" buildings previously, but are categorized as "health care" under the new classification.

The seasonally adjusted data are obtained by removing normal seasonal movement from the unadjusted data to bring out underlying trends and business cycles, which is accomplished by using the Census X-12-ARIMA method. Seasonal adjustment accounts for month-to-month variations resulting from normal or average changes in any phenomena affecting the data, such as weather conditions, the differing lengths of months, and the varying number of holidays, weekdays, and weekends within each month. It does not adjust for abnormal conditions within each month or for year-to-year variations in weather. The seasonally adjusted annual rate is the seasonally adjusted monthly rate multiplied by 12.

Residential consists of new houses, town houses, apartments, and condominiums for sale or rent; these dwellings are built by the owner or for the owner on contract. It includes improvements inside and outside residential structures, such as remodeling, additions, major replacements, and additions of swimming pools and garages. Manufactured housing, houseboats, and maintenance and repair work are not included.

Office includes general office buildings, administration buildings, professional buildings, and financial institution buildings. Office buildings at manufacturing sites are classified as *manufacturing,* but office buildings owned by manufacturing companies but not at such a site are included in the *office* category. In the state and local government category, *office* includes capitols, city halls, courthouses, and similar buildings.

Commercial includes buildings and structures used by the retail, wholesale, farm, and selected service industries. One of the subgroups of this category is *multi-retail,* which consists of department and variety stores, shopping centers and malls, and warehouse-type retail stores.

Health care includes hospitals, medical buildings, nursing homes, adult day-care centers, and similar institutions.

Educational includes schools at all levels, higher education facilities, trade schools, libraries, museums, and similar institutions.

Amusement and recreation includes theme and amusement parks, sports structures not located at educational institutions, fitness centers and health clubs, neighborhood centers, camps, movie theaters, and similar establishments.

Transportation includes airport facilities; rail facilities, track, and bridges; bus, rail, maritime, and air terminals; and docks, marinas, and similar structures.

Communication includes telephone, television, and radio distribution and maintenance structures.

Power includes electricity production and distribution and gas and crude oil transmission, storage, and distribution.

Manufacturing includes all buildings and structures at manufacturing sites but not the installation of production machinery or special-purpose equipment.

Included in *total private construction*, but not shown separately in these pages, are lodging facilities (hotels and motels), religious structures, and private public safety, sewage and waste disposal, water supply, highway and street, and conservation and development spending.

Included in *total state and local construction,* but not shown separately in these pages, are state and local construction of commercial buildings, conservation and development (dams, levees, jetties, and dredging), lodging, religious facilities, and communication structures.

Public safety includes correctional facilities, police and sheriffs' stations, fire stations, and similar establishments.

Highway and street includes pavement, lighting, retaining walls, bridges, tunnels, toll facilities, and maintenance and rest facilities.

Sewage and waste disposal includes sewage systems, solid waste disposal, and recycling.

Water supply includes water supply, transmission, and storage facilities.

Among the data sources for construction expenditures are the Census Bureau's Survey of Construction, Building Permits Survey, Consumer Expenditure Survey (conducted for the Bureau of Labor Statistics), Annual Capital Expenditures Survey, and Construction Progress Reporting Survey; also included are data from the F.W. Dodge Division of the McGraw-Hill Information Systems Company, the U.S. Department of Agriculture, and utility regulatory agencies.

Data availability

Each month's "Construction Spending" press release is released on the last workday of the following month. The release, more

detailed data, and a discussion of methodologies can be found on the Census Bureau's Web site at <http://www.census.gov/constructionspending>. Current data at this site may reflect benchmarking since Table 16-1 was compiled.

TABLE 16-2

Housing Starts and Building Permits; New House Sales and Prices

SOURCES: U.S. DEPARTMENT OF COMMERCE, CENSUS BUREAU

These data are mainly found in two major Census Bureau reports, "New Residential Construction" and "New Residential Sales." They cover new housing units intended for occupancy and maintained by the occupants, excluding hotels, motels, and group residential structures. Manufactured home units are reported in a separate survey.

Definitions

A *housing unit* is a house, an apartment, or a group of rooms or single room intended for occupancy as separate living quarters. Occupants must live separately from other individuals in the building and have direct access to the housing unit from the outside of the building or through a common hall. Each apartment unit in an apartment building is counted as one housing unit. As of January 2000, a previous requirement for residents to have the capability to eat separately has been eliminated. (Based on the old definition, some senior housing projects were excluded from the multifamily housing statistics because individual units did not have their own eating facilities.) Housing starts exclude group quarters such as dormitories or rooming houses, transient accommodations such as motels, and manufactured homes. Publicly owned housing units are excluded, but units in structures built by private developers with subsidies or for sale to local public housing authorities are both classified as private housing.

The *start* of construction of a privately owned housing unit is when excavation begins for the footings or foundation of a building primarily intended as a housekeeping residential structure and designed for nontransient occupancy. All housing units in a multifamily building are defined as being started when excavation for the building begins.

One-family structures includes fully detached, semi-detached, row houses, and townhouses. In the case of attached units, each must be separated from the adjacent unit by a ground-to-roof wall to be classified as a one-unit structure and must not share facilities such as heating or water supply. Units built one on top of another and those built side-by-side without a ground-to-roof wall and/or with common facilities are classified by the number of units in the structure.

Apartment buildings are defined as buildings containing *five units or more*. The type of ownership is not the criterion—a condominium apartment building is not classified as one-family structures but as a multifamily structure.

A *manufactured* home is a moveable dwelling, 8 feet or more wide and 40 feet or more long, designed to be towed on its own chassis with transportation gear integral to the unit when it leaves the factory, and without need of a permanent foundation. Multiwides and expandable manufactured homes are included. Excluded are travel trailers, motor homes, and modular housing. The shipments figures are based on reports submitted by manufacturers on the number of homes actually shipped during the survey month. Shipments to dealers may not necessarily be placed for residential use in the same month as they are shipped. The number of manufactured "homes" used for nonresidential purposes (for example, those used for offices) is not known.

Units authorized by building permits represents the approximately 97 percent of housing in permit-requiring areas.

The *start* occurs when excavation begins for the footing or foundation. Starts are estimated for all areas, regardless of whether permits are required.

New house *sales* are reported only for new single-family residential structures. The sales transaction must intend to include both house and land. Excluded are houses built for rent, houses built by the owner, and houses built by a contractor on the owner's land. A sale is reported when a deposit is taken or a sales agreement is signed; this can occur prior to a permit being issued.

> Once the sale is reported, the sold housing unit drops out of the survey. Consequently, the Census Bureau does not find out if the sales contract is cancelled or if the house is ever resold. As a result, if conditions are worsening and cancellations are high, sales are temporarily overestimated. When conditions improve and the cancelled sales materialize as actual sales, the Census sales estimates are then underestimated because the case did not re-enter the survey. In the long run, cancellations do not cause the survey to overestimate or underestimate sales; but in the short run, cancellations and ultimate resales are not reflected in this survey, and fluctuations can appear less severe than in reality.

A house is *for sale* when a permit to build has been issued (or work begun in non-permit areas) and a sales contract has not been signed nor a deposit accepted.

The *sales price* used in this survey is the price agreed upon between the purchaser and the seller at the time the first sales contract is signed or deposit made. It includes the price of the improved lot. The *median sales price* is the sales price of the house that falls on the middle point of a distribution by price of the total number of houses sold. Half of the houses sold have a price lower than the median; half have a price higher than the median. Changes in the *sales price* data reflect changes in the distribution of houses by region, size, and the like, as well as changes in the prices of houses with identical characteristics.

The *price index* measures the change in price of a new single-family house of constant physical characteristics, using the characteristics of houses built in 2005. Characteristics held constant include floor area, whether inside or outside a metropolitan area, number of bedrooms, number of bathrooms, number of fireplaces, type of parking facility, type of foundation, presence of a deck, construction method, exterior wall material, type of heating, and presence of air-conditioning. The indexes are calculated separately for attached and detached houses and combined with base period weights. The price measured includes the value of the lot.

Notes on the data

Monthly permit authorizations are based on data collected by a mail survey from a sample of about 9,000 permit-issuing places, selected from and representing a universe of 20,000 such places in the United States. The remaining places are surveyed annually. Data for 2014 to the present includes approximately 20,100 places while data for 2004 to 2013 included 19,300 places.

Data for 1994 through 2003 represented 19,000 places; data for 1984 through 1993 represented 17,000 places; data for 1978 through 1983 represented 16,000 places; data for 1972 through 1977 represented 14,000 places; data for 1967 through 1971 represented 13,000 places; data for 1963 through 1966 represented 12,000 places; and data for 1959 through 1962 represented 10,000 places.

Housing starts and sales data are obtained from the Survey of Construction, for which Census Bureau field representatives sample both permit-issuing and non-permit-issuing places.

Effective with the January 2005 data release, the Survey of Construction implemented a new sample of building permit offices, replacing a previous sample selected in 1985. As a result, writes the Census Bureau, "Data users should use caution when analyzing year over year changes in housing prices and characteristics between 2004 and 2005." In the newer sample, land may be more abundant, lot sizes larger, and sales prices lower.

For 2004, the permit data were compiled for both the new 20,000 place universe and the old 19,000 place universe. Ratios of the new estimates to the old were calculated by state for total housing units, structures by number of units, and valuation. For the United States as a whole, the new estimate was 100.9 percent of the old estimate.

Effective with the data for April 2001, the Census Bureau made changes to the methodology used for new house sales, including discontinuing an adjustment for construction in areas in which building permits are required without a permit being issued. It was believed that such unauthorized construction has virtually ceased. The upward adjustment was not phased out but dropped completely in revised estimates as of January 1999. The total effect of these changes was to lower the number of sales by about 2.9 percent relative to those published for earlier years.

The data used in the price index are collected in the Survey of Construction, through monthly interviews with the builders or owners. The size of the sample is currently about 20,000 observations per year.

Data availability and references

Housing starts and building permit data have been collected monthly by the Bureau of the Census since 1959.

The monthly report for "New Residential Construction" (permits, starts, and completions) is issued in the middle of the following month. The monthly report and associated descriptions and historical data can be found at <http://www.census.gov/construction/nrc>.

The monthly report for "New Residential Sales" (sales, houses for sale, and prices) is issued toward the end of the following month. The monthly report and associated descriptions and historical data can be found at <http://www.census.gov/construction/nrs>.

The manufactured housing data (not seasonally adjusted) and background information can be found at <http://www.manufacturedhousing.org/statistics>. Data with and without seasonal adjustment can be found at <http://www.census.gov/construction/mhs>.

Data and background on the price index for new one-family houses can be found at <http://www.census/gov>, in the alphabetical index under the category "Construction price indexes."

TABLES 16-3 THROUGH 16-6

Manufacturers' Shipments, Inventories, and Orders

SOURCE: U.S. DEPARTMENT OF COMMERCE, CENSUS BUREAU

These data are from the Census Bureau's monthly M3 survey, a sample-based survey that provides measures of changes in the value of domestic manufacturing activity and indications of future production commitments. The sample is not a probability sample. It includes approximately 4,300 reporting units, including companies with $500 million or more in annual shipments and a selection of smaller companies. Currently, reported monthly data represent approximately 60 percent of shipments at the total manufacturing level.

One important technology industry, semiconductors, is represented in the shipments and inventories data in this report but not in new or unfilled orders. This affects the new and unfilled orders totals for computers and electronic products, durable goods industries, and total manufacturing. Based on shipments data, semiconductors accounted for about 15 percent of computers and electronic products, 3 percent of durable goods industries,

and 1.5 percent of total manufacturing. Since semiconductors are intermediate materials and components rather than finished final products, the absence of these data does not distort new and unfilled orders data for important final demand categories, such as capital goods and information technology.

Definitions and notes on the data

Shipments. The value of shipments data represent net selling values, f.o.b. (free on board) plant, after discounts and allowances and excluding freight charges and excise taxes. For multi-establishment companies, the M3 reports are typically company- or division-level reports that encompass groups of plants or products. The data reported are usually net sales and receipts from customers and do not include the value of interplant transfers. The reported sales are used to calculate month-to-month changes that bring forward the estimates for the entire industry (that is, estimates of the statistical "universe") that have been developed from the Annual Survey of Manufactures (ASM). The value of products made elsewhere under contract from materials owned by the plant is also included in shipments, along with receipts for contract work performed for others, resales, miscellaneous activities such as the sale of scrap and refuse, and installation and repair work performed by employees of the plant.

Inventories. Inventories in the M3 survey are collected on a current cost or pre-LIFO (last in, first out) basis. As different inventory valuation methods are reflected in the reported data, the estimates differ slightly from replacement cost estimates. Companies using the LIFO method for valuing inventories report their pre-LIFO value; the adjustment to their base-period prices is excluded. In the ASM, inventories are collected according to this same definition.

Inventory data are requested from respondents by three stages of fabrication: finished goods, work in process, and raw materials and supplies. Response to the stage of fabrication inquiries is lower than for total inventories; not all companies keep their monthly data at this level of detail. It should be noted that a product considered to be a finished good in one industry, such as steel mill shapes, may be reported as a raw material in another industry, such as stamping plants. For some purposes, this difference in definitions is an advantage. When a factory accumulates inventory that it considers to be raw materials, it can be expected that that accumulation is intentional. But when a factory—whether a materials-making or a final-product producer—has a buildup of finished goods inventories, it may indicate involuntary accumulation as a result of sales falling short of expectations. Hence, the two types of accumulation can have different economic interpretations, even if they represent identical types of goods.

Like total inventories, stage of fabrication inventories are benchmarked to the ASM data. Stage of fabrication data are benchmarked at the major group level, as opposed to the level of total inventories, which is benchmarked at the individual industry level.

New orders (durable goods), as reported in the monthly survey, are net of order cancellations and include orders received and filled during the month as well as orders received for future delivery. They also include the value of contract changes that increase or decrease the value of the unfilled orders to which they relate. Orders are defined to include those supported by binding legal documents such as signed contracts, letters of award, or letters of intent, although this definition may not be strictly applicable in some industries.

New orders (nondurable goods) are equal to shipments, as order backlogs are not reported for these industries.

Unfilled orders (durable goods) includes new orders (as defined above) that have not been reflected as shipments. Generally, unfilled orders at the end of the reporting period are equal to unfilled orders at the beginning of the period plus net new orders received less net shipments.

Series are adjusted for seasonal variation and variation in the number of trading days in the month using the X-12-ARIMA version of the Census Bureau's seasonal adjustment program.

Benchmarking and revisions

The data shown in the volume have been benchmarked to the 2007 Economic Census and to ASMs through 2011. In each benchmark revision, new and unfilled orders are adjusted to be consistent with the benchmarked shipments and inventory data, seasonal adjustment factors are revised and updated, and other corrections are made.

Data availability and references

Data have been collected monthly since 1958.

The "Advance Report on Durable Goods Manufacturers' Shipments, Inventories and Orders" is available as a press release about 18 working days after the end of each month. It includes seasonally adjusted and not seasonally adjusted estimates of shipments, new orders, unfilled orders, and inventories for durable goods industries.

The monthly "Manufacturers' Shipments, Inventories, and Orders" report is released on the 23rd working day after the end of the month. Content includes revisions to the advance durable goods data, estimates for nondurable goods industries, tabulations by market category, and ratios of shipments to inventories and to unfilled orders. Revisions may affect selected data for the two previous months.

Press releases, historical data, descriptions of the survey, and documentation are available on the Census Bureau Web site at <http://www.census.gov>, under the category "Manufacturing" in the alphabetic listing there.

TABLE 16-7

Motor Vehicle Sales and Inventories

SOURCE: U.S. DEPARTMENT OF COMMERCE, BUREAU OF ECONOMIC ANALYSIS

The Bureau of Economic Analysis (BEA) collects data on retail sales and inventories of cars, trucks, and buses from data from the American Automobile Manufacturers Association, Ward's Automotive Reports, and other sources. Seasonal adjustments are recalculated annually. Data are available on the BEA Web site at <http://www.bea.gov> as a part of the national income and product accounts data set; they are found under the "Supplemental Estimates" heading. They are also available on the STAT-USA subscription Web site at <http://www.stat-usa.gov>.

Definitions

Product	Description
Autos	All passenger cars, including station wagons.
Light trucks	Sport utility vehicles. Prior to the 2003 Benchmark Revision light trucks were up to 10,000 pounds.
Heavy trucks	Trucks more than 14,000 pounds gross vehicle weight. Prior to the 2003 Benchmark Revision heavy trucks were more than 10,000 pounds
Domestic sales	United States (U.S.) sales of vehicles assembled in the U.S., Canada, and Mexico
Foreign sales	U.S. sales of vehicles produced elsewhere
Domestic auto production	Autos assembled in the U.S
Domestic auto inventories	U.S. inventories of vehicles assembled in the U.S., Canada, and Mexico.

Data Availability

Auto and truck sales are updated by the end of the third business day after the end of the month. Auto production and inventories are available by the end of the second week after the end of month.

TABLES 16-8 AND 16-10

Retail and Food Services Sales; Retail Inventories

SOURCE: U.S. DEPARTMENT OF COMMERCE, CENSUS BUREAU

Every month, the Census Bureau prepares estimates of retail sales and inventories by kind of business, based on a mail-out/mail-back survey of about 5,500 retail businesses with paid employees.

In June 2019, the monthly retail sales estimate is based on the results of the 2017 Annual Trade Survey and the service annual service. The annual retail trade survey (ARTS) produces national estimates of total annual sales, e-commerce sales, sales taxes, end-of-year inventories, purchases, total operating expenses, gross margins, and end-of-year accounts receivable for retail businesses located in the united states.

Retail sales and inventories are now compiled using the new NAICS classification system, which replaced the old SIC system. Historical data have been restated on the NAICS basis back to January 1992. In NAICS, Eating and drinking places and Mobile food services have been reclassified out of retail trade and into sector 72, Accommodation and food services, which also includes Hotels. The retail sales survey still collects and publishes sales data for Food services and drinking places. It no longer includes them in the Retail total, but they are included in a new Retail and food services total.

Subtotals of durable and nondurable goods are no longer published. They were always imprecise for retail sales, since general merchandise stores (including department stores) were included in nondurable goods, yet obviously sold substantial quantities of durable goods.

Definitions

Sales is the value of merchandise sold for cash or credit at retail or wholesale. Services that are incidental to the sale of merchandise, and excise taxes that are paid by the manufacturer or wholesaler and passed along to the retailer, are also included. Sales are net, after deductions for refunds and merchandise returns. They exclude sales taxes collected directly from customers and paid directly to a local, state, or federal tax agency. The sales estimates include only sales by establishments primarily engaged in retail trade, and are not intended to measure the total sales for a given commodity or merchandise line.

Inventories is the value of stocks of goods held for sale through retail stores, valued at cost, as of the last day of the report period. Stocks may be held either at the store or at warehouses that maintain supplies primarily intended for distribution to retail stores within the organization.

Leased departments consists of the operations of one company conducted within the establishment of another company, such as jewelry counters or optical centers within department stores. The values for sales and inventories at department stores in Tables 16-9 and 16-10 exclude sales of leased departments.

GAFO (department store type goods) is a special aggregate grouping of sales at general merchandise stores and at other stores that sell merchandise normally sold in department stores—clothing and accessories, furniture and home furnishings, electronics, appliances, sporting goods, hobby, book, music, office supplies, stationery, and gifts.

Notes on the data

The data published here have been benchmarked to the 2012, 2007, 2002, 1997, and 1992 Economic Censuses and the Annual Retail Trade Surveys for 2012 and previous years. Each year, the monthly series are benchmarked to the latest annual survey and new factors are incorporated to adjust for seasonal, trading-day, and holiday variations, using the Census Bureau's X-12-ARIMA program.

The survey sample is stratified by kind of business and estimated sales. All firms with sales above applicable size cutoffs are included. Firms are selected randomly from the remaining strata. The sample used for the end-of-month inventory estimates is a sub-sample of the monthly sales sample, about one-third of the size of the whole sample.

New samples, designed to produce NAICS estimates, were introduced with the 1999 Annual Retail Trade Survey and the March 2001 Monthly Retail Trade Survey. On November 30, 2006, another new sample was introduced, affecting the data for September 2006 and the following months. The sample is updated quarterly to take account of business births and deaths.

Data availability and references

An "Advance Monthly Retail Sales" report is released about nine working days after the close of the reference month, based on responses from a sub-sample of the complete retail sample.

The revised and more complete monthly "Retail Trade, Sales, and Inventories" reports are released six weeks after the close of the reference month. They contain preliminary figures for the current month and final figures for the prior 12 months. Statistics include retail sales, inventories, and ratios of inventories to sales. Data are both seasonally adjusted and unadjusted.

The "Annual Benchmark Report for Retail Trade" is released each spring. It includes updated seasonal adjustment factors; revised and benchmarked monthly estimates of sales and inventories; monthly data for the most recent 10 or more years; detailed annual estimates and ratios for the United States by kind of business; and comparable prior-year statistics and year-to-year changes.

All data are available on the Census Bureau Web site at <http://www.census.gov/retail>.

TABLE 16-9

Quarterly Retail Sales: Total and E-Commerce

SOURCE: U.S. DEPARTMENT OF COMMERCE, CENSUS BUREAU

Beginning with the fourth quarter of 1999, the Census Bureau has conducted a quarterly survey of retail e-commerce sales from the Monthly Retail Trade Survey sample. (The monthly survey does not report electronic shopping separately; it is combined with mail order.)

E-commerce sales are the sales of goods and services in which an order is placed by the buyer or the price and terms of sale are negotiated over the Internet or an extranet, Electronic Data Interchange (EDI) network, electronic mail, or other online system. Payment may or may not be made online. The quarterly release is issued around the 20th of February, May, August, and November, and is available along with full historical data on the Census Bureau Web site at <http://www.census.gov>. It can be located under the heading "Retail" in the alphabetical Web site index, under "E" for "Economic data and information."

These estimates reflect the NAICS definition of retail sales, which excludes food service. Online travel services, financial brokers and dealers, and ticket sales agencies are not classified as retail and are not included in these estimates; they are, however, included in the annual survey of selected services.

TABLE 16-11

Merchant Wholesalers–Sales and Inventories

SOURCE: U.S. DEPARTMENT OF COMMERCE, CENSUS BUREAU

The Annual Wholesale Trade Survey (AWTS) produces national estimates of total annual sales, e-commerce sales, end-of-year inventories, inventories held outside of the United States, purchases, total operating expenses, gross margins, and commissions (for electronic markets, agents, and brokers) for wholesale businesses located in the United States. The sample consists of about 8,400 establishments, with a response rate of about 62 percent; missing reports are imputed based on reports of similar reporters.

These data are now based on the new NAICS classification system, which replaced the old SIC system. Historical data have been restated on the NAICS basis back to January 1992.

Classification changes in NAICS

NAICS shifts a significant number of businesses from the Wholesale to the Retail sector. An important new criterion for classification concerns whether or not the establishment is intended to solicit walk-in traffic. If it is, and if it uses mass-media advertising, it is now classified as Retail, even if it also serves business and institutional clients.

Definitions

Merchant wholesalers includes merchant wholesalers that take title of the goods they sell, as well as jobbers, industrial distributors, exporters, and importers. The survey does not cover marketing sales offices and branches of manufacturing, refining, and mining firms, nor does it include NAICS 4251: Wholesale Electronic Markets and Agents and Brokers.

Notes on the data

Inventories are valued using methods other than LIFO (last in, first out) in order to better reflect the current costs of goods held as inventory.

A survey has been conducted monthly since 1946. New samples are drawn every 5 years, most recently in 2009. The samples are updated every quarter to add new businesses and to drop companies that are no longer active.

Data availability and references

"Monthly Wholesale Trade, Sales and Inventories" reports are released six weeks after the close of the reference month. They contain preliminary current-month figures and final figures for the previous month. Statistics include sales, inventories, and stock/sale ratios, along with standard errors. Data are both seasonally adjusted and unadjusted.

The "Annual Benchmark Report for Wholesale Trade" is released each spring. It contains estimated annual sales, monthly and yearend inventories, inventory/sales ratios, purchases, gross margins, and gross margin/sales ratios by kind of business. Annual estimates are benchmarked to annual surveys and the most recent census of wholesale trade. Monthly sales and inventories estimates are revised consistent with the annual data, seasonal adjustment factors are updated, and revised data for both seasonally adjusted and unadjusted values are published.

Data and documentation are available on the Census Bureau Web site at <http://www.census.gov>, under "Economic Indicators" and in the alphabetic index under "E" for economic data.

TABLES 16-12 AND 16-13

Manufacturing and Trade Sales and Inventories

SOURCES: U.S. DEPARTMENT OF COMMERCE, CENSUS BUREAU (CURRENT-DOLLAR SERIES) AND U.S. DEPARTMENT OF COMMERCE, BUREAU OF ECONOMIC ANALYSIS (BEA; CONSTANT-DOLLAR SERIES)

The current-dollar data on which these tables are based bring together summary data from the separate series on manufacturers' shipments, inventories, and orders; merchant wholesalers' sales and inventories; and retail sales and inventories, all of which are included in this chapter. Generally, current-dollar inventories are collected on a current cost (or pre-LIFO [last in, first out]) basis. See the notes and definitions for Tables 16-3, 16-4, 16-8 16-10, and 16-11 for further information about these data.

Based on these current-dollar values and relevant price data, BEA makes estimates of real sales, inventories, and inventory-sales ratios. Note, however, that annual figures for sales are shown as annual <u>totals</u> in Table 16-12 but as <u>averages</u> of the monthly data in Table 16-13, reflecting the practices of the respective source agencies. Also note that constant-dollar detail may not add to constant-dollar totals because of the chain-weighting formula; see the discussion of chain-weighted measures in the notes and definitions for Chapter 1. The constant-dollar estimates shown in this volume of *Business Statistics* are stated in 2009 dollars.

Inventory values are as of the end of the month or year. In Table 16-13, annual values for monthly current-dollar inventory-sales ratios are averages of seasonally adjusted monthly ratios. However, for the real ratios in Table 16-14, annual figures for inventory-sales ratios are calculated by BEA as year-end (December) inventories divided by the monthly average of sales for the entire year; this means that the annual ratios will not be equivalent to averages of month ratios. In all cases, the ratios in these two tables (like those in Table 1-8) represent the number of months' sales on hand as inventory at the end of the reporting period.

Data availability

Sales, inventories, and inventory-sales ratios for manufacturers, merchant wholesalers, and retailers are published monthly by the Census Bureau in a press release entitled "Manufacturing and Trade Inventories and Sales." Recent and historical data are available on the Census Bureau Web site at <http://www.census.gov/mtis/www/mtis.html>. They can also be found by going to the general Census website, <http://www.census.gov>, going to the alphabetical index, finding "Economic Data and Information" under "E" and then finding "Manufacturing and trade" under "Manufacturing" or "Sales."

Sales and inventories in constant dollars are available on the BEA Web site at <http://www.bea.gov>. To locate these data on that site, click on "National Economic Accounts." Scroll down to "Supplemental Estimates," click "Underlying Detail Tables," and then click on "List of Underlying Detail Tables." For the most recent data, if there is more than one table with the same title, select the last table listed.

TABLE 16-14

Selected Services, Quarterly: Estimated Revenue for Employer Firms

SOURCE: U.S. DEPARTMENT OF COMMERCE, CENSUS BUREAU

Census data on quarterly revenue for selected service industries are based on information collected from a probability sample that has been expanded over recent years to approximately 19,000 employer firms (firms with employees), chosen from the sample for the larger Service Annual Survey and expanded to represent totals—for employer firms only—for the selected industries. The sample is updated quarterly to account for business births,

deaths, and other changes. Industries are defined according to the 2012 NAICS.

The scope of the survey and the size of the sample have increased several times since the inception of this survey in 2004. More industries are now available than are shown here in Table 16-15, which focuses on industries with a longer statistical history.

Definitions and notes on the data

These data are collected in current dollars only. See the Producer Price Indexes in Chapter 8 for indicators of price trends in various sectors.

Both taxable and tax-exempt firms are covered unless otherwise specified.

Generally, government enterprises are not within the scope (that is, the industries it is designed to cover) of the survey. *Utilities* excludes government-owned utilities and *Transportation and warehousing* excludes the U.S. Postal Service.

Private industries that are not within the scope of the survey are NAICS 482, rail transportation; 525, funds, trusts, and other financial vehicles; 51112, offices of notaries; 6111, 6112, and 6113, elementary, secondary, and post-secondary schools; 8131, religious organizations; 81393 (labor unions and similar), 81394 (political organizations, and 814 (private households).

Totals shown for sectors and subsectors may include data for kinds of business that are in the scope of the survey but are not shown separately in the detail beneath.

Data for selected industry groups are adjusted for seasonal variation using the Census Bureau's X-13 ARIMA-SEATS program.

Data availability and references

The quarterly release "U.S. Government Estimates of Quarterly Revenue for Selected Services" is available on the Census Web site at <http://www.census.gov/>, as "Quarterly Services Survey" listed under Q in the alphabetic index; historical, technical, and background information, including sampling errors, is also available there.

INDEX